D1123801

THE CHEMISTRY OF OUR ENVIRONMENT

The Chemistry of Our Environment

R. A. HORNE

A WILEY-INTERSCIENCE PUBLICATION

JOHN WILEY & SONS, New York · Chichester · Brisbane · Toronto

Library of Congress Cataloging in Publication Data:

Horne, Ralph Albert, 1929-
The chemistry of our environment.

"A Wiley-Interscience publication."
Includes bibliographies and index.
1. Environmental chemistry. I. Title.

QD31.2.H64 301.31 77-1156
ISBN 0-471-40944-8

Printed in the United States of America

10 9 8 7 6 5 4 3 2 1

Dedicated

to

HENRY DAVID THOREAU

Who thought he lived
just in the nick of time
. . . and who was right

On earth creatures shall be seen who are constantly killing one another. Their wickedness shall be limitless; their violence shall destroy the world's vast forests; and even after they have been sated, they shall in no wise suspend their desire to spread carnage, tribulations, and banishment among all living beings. Their overreaching pride shall impel them to lift themselves toward heaven. Nothing shall remain on the earth, or under the earth, or in the waters, that shall not be hunted down and slain, and what is in one country, dragged away into another; and their bodies shall become the tomb and the thoroughfare for all the living things they have ruined. . . .

The fertile earth, following the law of growth, will eventually lose the water hidden in her breast, and this water, passing through the cold and rarefied air, will be forced to end in the element of fire. Then the surface of the earth will be burned, and that will be the end of all terrestrial nature.

LEONARDO Da VINCI

PREFACE

If you bemoan technological compartmentalization, take hope! Consider environmental chemistry. Man is a chemical mote aswim in a chemical soup. All the changes about him, all the changes within him are chemical. Everything he sees and touches, everything he does, and the effect of what he does are chemical. If there is anything in the real world that does not fit under the rubric of environmental chemistry, then I am at a loss to find it.

In the grand old days when textbooks were written in verse (this one, I apologize, is in prose) and men of learning were generalists, what we call science was known by the broader, finer-sounding term "natural philosophy." Perhaps because of its all-embracive nature and because of its supreme importance in human affairs, environmental chemistry may encourage a more expansive view of ourselves and the world about us—in a word, natural philosophy.

The world took little notice at my earlier field of endeavor—the kinetics and mechanism of electron-exchanger reactions—perhaps rightly so. But environmental chemistry is different. There is something new and alien to scientists. Environmental chemistry is necessarily and properly inescapably controversial. The public is interested. The public should be interested. Environmental chemistry is so important to us all that it is frightening. What the environmental chemist does and finds and says can be of terrible importance. If he errs in one direction, we may be poisoned; if he miscalculates in another, we may lose our jobs. Our baby may be born with no arms, we may starve and freeze in the dark, we may start dropping atomic bombs on one another, all because some environmental scientists made a mistake.

Scientists should not run from responsibility; they should not cower in their academic retreats. But neither should they spotlight their naivety on public platforms nor try to ban the research of their colleagues because it is out of step with their own political tastes.

How can scientists best be responsible to the society that their work has helped to shape? One way is to expose their students, future scientists, to facts *and* issues, to break down the insulating wall that has been erected between the work of scientists and its consequences.

This is what I have tried to do in this book. I have cast aside the traditional pose of objectivity—it was always phoney and now, it appears, it can be pernicious. I do not pretend that I do not care about the subject matter of this text. I would like to start a modest revolution. I would like to see scientific texts written as if the authors cared about the subject matter, as if they thought it important enough to venture an opinion. Opinions as well as facts *belong* in scientific papers, monographs, and texts. *The Chemistry of Our Environment* is full of them. The author does owe the reader a duty, however, to give him notice and to try and carefully distinguish which statements are fact and which opinions.

This book was written to instruct. It was also written to provoke. I hope it is successful on both counts.

R. A. HORNE

Deerfield, New Hampshire
January 1977

CONTENTS

THE CHEMISTRY
OF OUR ENVIRONMENT

I

INTRODUCTION

Man is a paradoxical creature prone to extremes. On the one hand, he busies himself with practical expediences, whereas on the other, he dreams about the remotest worlds in time and space. This peculiar tendency is exemplified even in his earliest attempts at scientific speculation. The natural philosophers of ancient Greece directed their attention, not upon the more immediate problems of chemistry, mechanics, and geology, but upon those most abstruse areas—cosmogony, cosmology, and the fundamental nature of substance. This tendency is evident even in Aristotle, that most level headed and practical of philosophers.

Until the romantic movement in the nineteenth century, man was interested in the Earth largely in terms of the exploitation of its mineral riches, in the oceans as routes of commerce, and in the skies, although ever a victim of weather, hardly at all. Even his interest in his fellow creatures appears to have been motivated by a sort of curiosity for the bizarre. He did not discover his environment until the unregulated growth of his numbers threatened to exhaust its resources. He did not appreciate the delicate ecological processes of which he is a part until he had grossly damaged many of them. He has been like a spoiled child, pampered and nurtured in a comfortable home called Earth. He did not care, he did not even realize that the rent was not free, that bills must be paid, the roof kept in repair, the drains not clogged. But now at last he has grown up. The house has become overcrowded. The wear and tear and neglect of its fabric has become intolerable. He realizes now that repairs must be made, care taken, that he must become a husband in the older and more responsible sense of that term.

At long last, perhaps too late, man is beginning to discover that the Earth is home and that he must treat his home with care and respect if he and his progeny are to enjoy a pleasant life therein.

With respect to their attitude toward the natural order, the Indian and the occidental mind are profoundly different. The goal of Eastern philosophy has been to consent to Nature and so become a part of it, whereas Western philosophy has sought to understand Nature in order to conquer it. The Greek was unique in setting man apart from Nature and often opposed to it. He was the inventor of individualism in a more universal sense, as well as in the social and political senses. But he was also aware of a certain Necessity beyond the alteration of men or even gods. Thus his struggle against the Natural Order was always rational, inasmuch as he was cognizant of the unavoidable constraints that it imposes. Man was the center of man's universe, but man's universe only. But the constraints were forgotten and the estrangement from Nature became complete with the ascendency of Christian dogma, which not only placed man at the center of *the* universe but even had the presumption to imagine that there was a superhuman being who has a particular interest in man's fate. The discoveries of Copernicus rendered the myth of homocentricity untenable, yet many people, even the educated, have persisted in believing and acting according to its corollaries. They continue to look upon the Earth as man's bauble to use and abuse as he will. They continue to presume that they have the right, even the mandate, to reproduce indiscriminately even though their species has already become a pest that has exterminated many forms of life from this planet forever and threatens to destroy many species more.

Despite man's depredations, our world is still so rich that it is fun to write about. However, I cannot escape a certain feeling of futility in writing these pages, for science and technology can have no solutions to the problem. They can at best only temporarily relieve some of the symptoms. The central problem is not that man pollutes or even that man pollutes too much, but rather that too many men pollute. The fate of the biosphere, the quality of future human life, and the viability of our most precious political ideals will be determined not in the laboratory but in the bedroom.

All living creatures alter their environment. Environmental modification is intrinsic to the life processes and may be taken as one of its characteristics. Any living organism, whether one celled or many celled, whether a virus or a man, removes chemical substances from its surroundings and releases others. In the long course of biological evolution, Nature has developed an amazingly complex and delicate system of checks and balances. In many instances, some life factories actually use the wastes of others as their raw materials, sometimes even in a reciprocal manner. The best known example of this is the utilization of oxygen, a waste from photosynthesis in plants, by animals and the utilization of carbon dioxide, a waste from respiration in animals, by plants. But because the system is delicate and complex it is subject to disruption. It can easily go awry. An organism can become a plague; its environmental context can become seriously disrupted. Such a catastrophic imbalance is called *pollution.*

Francis Bacon observed that knowledge is power. Science has enabled man to escape *apparently* some of Nature's checks and balances. This ability to hold the balance of Nature in temporary abeyance has enabled him along among all living things to pollute on a truly massive and catastrophic scale.

Man's knowledge has made him foolish. Science can only teach us how to *use* the natural order. It cannot conquer it. On this planet, man has become a criminal. He is "breaking" natural law, and he thinks that he is getting away with it. But he is not. He is guilty, and his sentence is being written everywhere, every day. And sooner or later he is going to have to pay a terrible penalty.

Why? Why can't man escape indefinitely? Simply because he is violating one of the most profoundly fundamental laws of physics and chemistry, a law so obvious and all evasive that it is often fortotten. The Principle of Le Chatelier states that when a stress is applied to a system, the system reponds in such a manner as to relieve that stress. In physics, this principle governs all mechanical processes. Newton's celebrated laws of motion are little more than particularized statements of the principle. In chemistry, it is the fundamental principle governing all chemical equilibria and disequilibria. It also governs all biological systems from the simplest tactile response of microorganisms to the most complex psychological states of the human mind. In environmental science, we can best describe it as the first rule of ecology.

When an organism becomes a plague, the Principle of Le Chatelier catches up to it—the organism exhausts its supply of nutrients and/or perishes in its own toxic products. The one difficulty with the Principle of Le Chatelier is that its operance can exhibit widely variable time lags. When a mechanical system is stressed (such as a nail struck by a hammer) the response is nearly instantaneous. In chemical systems, the response time can be very short (such as the rapid exhaustion of reactants in an explosion) or very long (slow chemical reactions such as the fading of wallpaper), and in biological systems with their exceedingly complex chemical relationships, again the response time can be very short (death from carbon monoxide poisoning) or very long (death from cigarette smoking). In the case of geological processes, the response time can be hundreds, thousands, and even millions of years. But nothing can ever escape the inexorable operation of the Principle of Le Chatelier. Man has not escaped this principle. He is only wasting stupidly a precious time-lag.

This book touches on matters philosophical, social, moral, political, and even religious. It is full of opinions; it may even contain some prejudices. I expect I may be severely scolded by some of my older colleagues. But I submit here that the "objectivity" of science is not only a myth, it is a pernicious myth. When science began, it was called natural philosophy (and frequently it was even written in verse!). It was the passion of great, curious, and committed minds. Alas! it has become the business of oftentimes very small and narrow and timid minds. The province of the scientist is the universe and all and everything in it. Nothing should be barred or hidden from his scrutiny because it is "beneath" science (modern theory of probability derived from a consideration of games of chance) or because it is "sacred" (efforts are currently being made to suppress scientific study of possible racial differences). The epigone have shown a disinclination approaching cowardice to think about the meaning and consequences of their scientific researches. This is not surprising. To look is easy; to think is difficult and may even be hazardous. But, as Bobby Dylan sang, "the times they are a changin'." The world with all its nasty problems has intruded upon the ivied tower. There is no longer any room for the isolated savant. We live in times so tumultuous, so grim, that the scientist can not longer insist on the luxury of "objectivity." Every man today, and most especially the scientist, has a clear and unavoidable moral responsibility to press his training, his knowledge, his understanding, his opinions, and his emotions into the service of attacking the horrendous environmental and social problems that man's ignorance has created.

II

THE EXOSPHERE

1

THE COSMIC ORIGIN AND DISTRIBUTION OF THE ELEMENTS

[The atoms] move in the void and catching each other up jostle together, and some recoil in any direction that may chance, and others become entangled with one another in various degrees according to the symmetry of their shapes and sizes and positions and order, and they remain together and thus the coming into being of composite things is effected.

SIMPLICIUS

1.1 INTRODUCTION

Our total environment may be divided into five major zones (Figure 1.1): lithosphere, hydrosphere, biosphere, atmosphere, and exosphere. Environmental chemistry is the study of the chemical composition of these zones and of the chemical processes occurring in them and, even more important, in the interfacial regions that mark their boundaries. Clearly then, there is very little going on in our world that is not a part of environmental chemistry so defined. With respect to our more immediate terrestrial environment, most of this chemistry occurs at the solid/liquid, gas/liquid, and, to a lesser extent, gas/solid and liquid/liquid interfaces, and most of this chemistry involves, or at least occurs in, the presence of water. The chemistry of our more immediate environment (and ourselves) is thus largely the surface chemistry of aqueous solutions.

To return to our classification of our total environment, we could quite legitimately take the biosphere of which we are a part as the center of the cosmic onion (Figure 1.2), but we might easily fall prey to the fallacy of anthropocentricity that we condemned in our introduction. We also would be taking as our point of departure chemical systems that are the most complex and highly organized of any in the universe. Scientists are guilty of a lesser parochialism and can better maintain the pretense of objectivity if they take the solid Earth upon which they stand as the core of their environment. The Earth or LITHOSPHERE is covered by two interpenetrating thin films—the HYDROSPHERE and the BIOSPHERE. These in turn are enveloped by the Earth's gaseous mantle, the ATMOSPHERE, which gradually tails off as we leave Earth into everything else—the EXOSPHERE.

Chemical reactions are associated with energy transformations. In our immediate environment, the major source of energy that keeps the Earth's surface chemistry running is our Sun. Solar energy as it beams down upon Earth creates two chemical

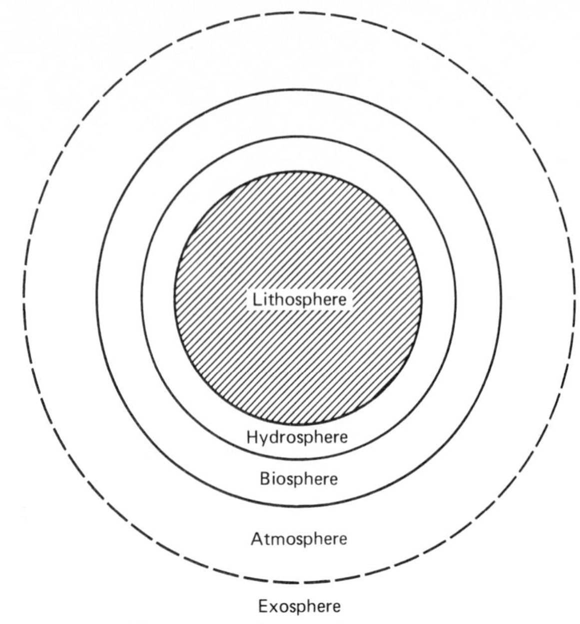

Figure 1.1. Our total environment.

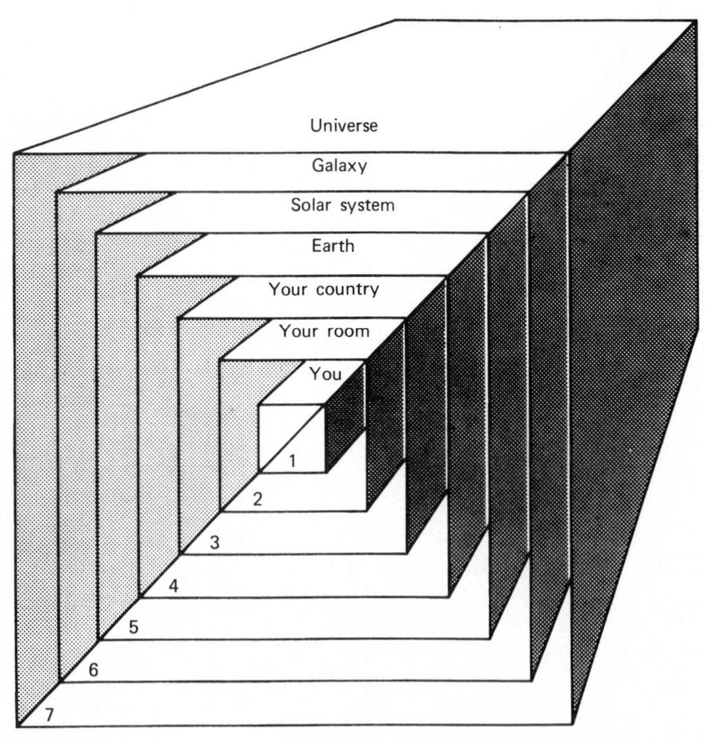

Figure 1.2. Our cosmic onion.

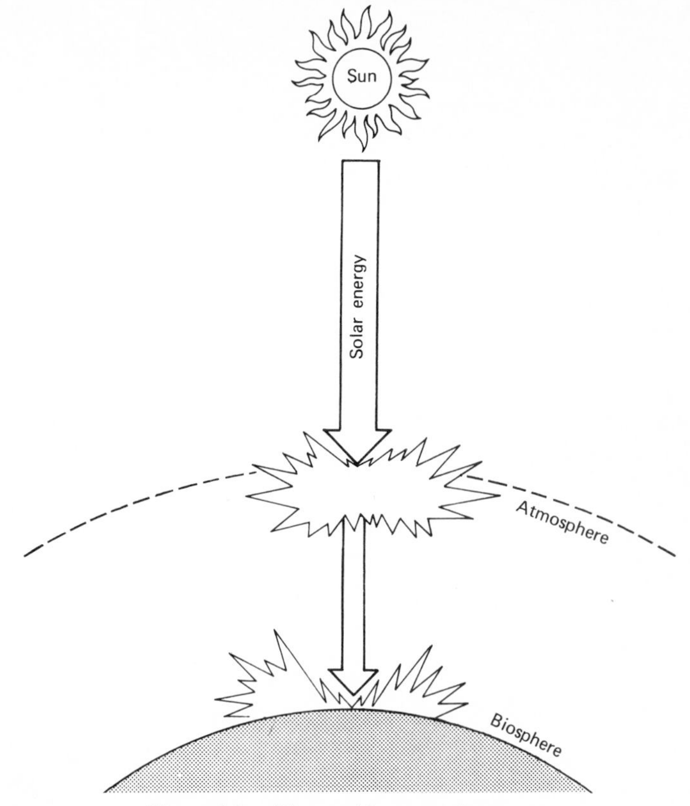

Figure 1.3. The earthly energy hotspots.

"hot spots" (Figure 1.3): the first as this energy hits the chemical species of the upper atmosphere and the second when it impinges on the biosphere and is utilized by photosynthesis. It is curious to note that the impingement of this energy directly upon the lithosphere, while playing a secondary role in weathering processes, produces very few chemical reactions because of the great stability of crustal materials, whereas the only major transformation that it effects upon the hydrosphere proper is evaporation

$$H_2O(l) \rightarrow H_2O(g) \tag{1.1}$$

Although life probably would be impossible without the radiation filtering action of the first of these chemical "hot spots," the second is far more important to us, not only because we are a part of it, but also because the major chemical features of this planet have been determined by it. The oygen content of the atmosphere, the highly oxidized composition of crustal rocks, even the chemical composition of the dissolved species in seawater are all the consequence of the biosphere. Ours is a planet with an oxidized surface, and photosynthesis is the basic process that has released and mobilized oxygen in this environment. It is most crucial to bear this point in mind for there are those who try to minimize the potential danger of man's polluting activities. His effect on the massive environment, they argue, is negligible. Remember that the biosphere, that organisms far more minute and infinitely less destructive than man, has already profoundly altered the chemistry of the Earthly environment.

We can best judge where we are in time and from whence we came; we can best form an image of what we are, what sort of chemical anomaly we and our world represent by examining the chemical milieu from which we emerged. So having criticized the ancients, we will begin just where they began—with cosmology, with the exosphere.

1.2 INTERSTELLAR MATTER

This is one of the most difficult sections in this book to write, and the reason is a very happy one. The different sciences advance at different rates in different times, and the most exciting times are those when one sience suddenly bursts forth in a splendid explosion of new discoveries. Right now (1971–1972) particle physics is in a doldrums, the earth sciences are struggling to pull themselves together and suffer from a lack of talent, molecular biology has somehow fallen short of expectation. Yet the pages of the vanguard journal *Nature* every week are full of marvelous findings, of mysterious radiation sources in distant space, of great quantities of intergalactic gas and dust, of

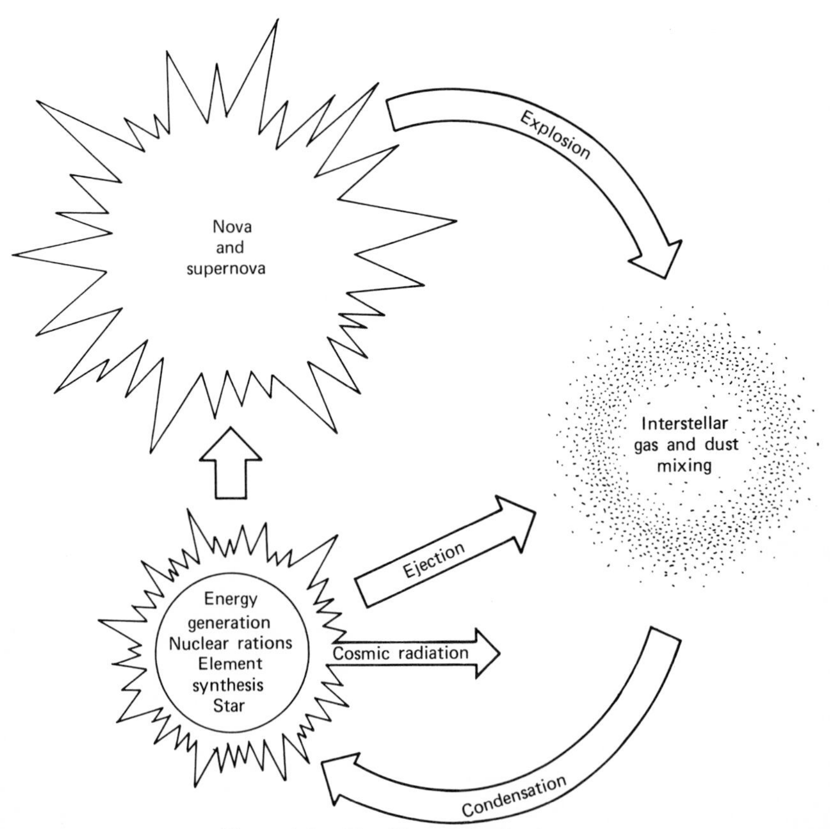

Figure 1.4. The life and death of stars.

quite complex chemical aggregates in interstellar space. In the long run, these discoveries may extend our knowledge of the physical–chemical nature of the universe even more widely than will our recent spectacular conquest of the Moon and our probes of our neighbors in our solar system. I hardly know how to begin to digest and summarize these findings. In terms of our knowledge, astronomy has become a nova!

Space is not empty; it is filled with diffuse matter, both gaseous and particulate. This interstellar material is the stuff from which stars are made (Figure 1.4), the stuff that their nuclear fires eject, the debris they scatter when they explode. In our part of our galaxy (the outer half), the total stellar and interstellar masses of material are comparable (Donn, 1970). The interstellar medium is largely neutral hydrogen with an average density over our whole galaxy of about 1 atom/cm^3. The distribution of this matter is irregular with gas and dust grains being concentrated in dense regions or "clouds" that are in turn surrounded by regions of less dense, hotter gas. These clouds occupy between 5 and 10% of the volume of our galaxy. They vary in size from 0.1 to 50 light yr, in gas density from 10 to $>10^5$/cm^3, in kinetic temperature from 20 to 200°K, and in turbulent velocity from 0.2 to 20 km/sec. In the more dense clouds, the ratio of molecular to atomic hydrogen is greater, that is to say, the tendency to form molecules of many types is enhanced by greater cloud density. The dust grains are presumed to be small, about 0.2 μ, and typically represent about 1% of the cloud mass or about one dust grain for about every 10^{12} hydrogen molecules. Between the clouds the gas density is roughly 0.1/cm^3 and the temperature 100 to greater than 1000°K (Rank, Townes, and Welch, 1971). A number of ideas have been advanced as to the chemical nature of the dust grains (Table 1.1). Spectral absorption at 10 μm in the Crab Nebula shows a slight excess flux that has been attributed to thermal emission from graphite dust grains formed in supernova remnants (Aitken and Polden, 1971; Hoyle and Wickramasinghe, 1970). Ultraviolet observations have also yielded evidence of interstellar graphite (Stecher, 1969). The diffuse absorption lines in supernova spectra have been attributed to solid state effects in grains of Fe^{2+} and Fe^{3+} containing silicates (Huffman, 1970; Manning, 1970), to X- and cosmic-ray-irradiated quartz or cristobalite particles (Wickramasinghe, 1971), and even to Na, Ca, Ti, and Mn impurities in solid hydrocarbons such as benzene and toluene (Duley, 1968; Graham and Duley, 1971). It is argued that these aromatic hydrocarbons because of their considerable chemical and thermal

Table 1.1 Some Proposed Chemical Compositions of Interstellar Dust Grains

Trace Material	Base Material	
Fe ions or other metallic impurities in X- or cosmic-ray-irradiated	Silica or silicates Quartz	SiO_2
H_2(s) layer on	Silicates Graphite	
	Graphite	C
Na, Ca, Ti, or Mn in	Solid aromatic hydrocarbons	HC
	Iron particles	Fe

stability are not unlikely constituents of the interstellar medium (Donn, 1968) despite the intense fluxes of destructive ultraviolet radiation. The search for ice in the interstellar spectral bands has been unsuccessful (Knacke et al., 1969). Small iron particles could account for the observed reddening curves (Schalen, 1965) but would involve quantities far greater than expected. Still another hypothesis postulates a layer of solid hydrogen on graphite (Hoyle and Wickramasinghe, 1967; Wickramasinghe and Reddish, 1968) or silicate (Feldman, Rees, and Werner, 1969) particles. This could account nicely for the variability in small particle extinction from star to star, but there is some doubt as to whether the particles are sufficiently cold to maintain such layers (Greenberg and de Jong, 1969). Stein and Gillett (1971) have calculated that less than one-third of the interstellar grains can be metallic silicates; thus the question of the chemical composition of this material remains largely unresolved.

Even more surprising than the interstellar dust grains is the growing list of radicals and molecules, some of them quite complex, that have been detected in deep space (Donn, 1970; Kaplan and Pikelner, 1970; Wick, 1970; Buhl, 1971; Buhl and Snyder, 1971; Rank, Townes, and Welch, 1971). With the exception of one report of OH (Weliachew, 1971), to date all of the radicals and molecules found have been restricted to within our galaxy, and the incidence of these materials seems to be concentrated in the galactic plane. Table 1.2 presents this list and the chronology of discovery of these chemicals, and the list is still growing. In 1970, Wick mentioned as expected but then not yet detected the species SiO, SO, H_2CS (thioformaldehyde), H_2C_2O (ketane), H_2COO (formic acid), and NO, and we can see from the Table 1.2 that many of these have since been discovered. The microwave spectrum of the S-analog of formaldehyde, thioformaldehyde ($\mathrm{H{-}\overset{\overset{\textstyle S}{\|}}{C}{-}H}$ was described by Johnson and Powell (1970). The cosmic S/O ratio is 1/40, yet initial attempts to detect this compound were not successful (Evans et al., 1970). Silicates, although proposed as a constituent of the dust grains, were similarly elusive (Stein and Gillett, 1971). Formaldehyde has now been detected in a great number of radio sources in the direction of our galactic center and appears to be dispersed in a series of spiraling concentric bands with the inner ones contracting and the outer ones expanding (Snyder et al., 1969; Wick, 1970). The most complicated molecule for which some evidence has been found is a magnesium porphyrin (Johnson, 1970).

Isotopic ratios have also been examined in interstellar material (Table 1.3), and the results hold some surprises. The $^{12}C/^{13}C$ ratio in the formaldehyde, for example, is close, not to the stellar value of about 4, but to the terrestrial value of 89. The $^{16}O/^{18}O$ ratio is also close to its terrestrial value. Other molecules, however, do have $^{12}C/^{13}C$ ratios in the range of stellar values: 3.5 for HCN and 2 for CO (Wick, 1970). Formaldehyde in dark clouds is further strange inasmuch as it appears to be at a temperature well below the expected 2.7°K (Palmer et al., 1969), and the cause of this refrigeration is unknown, although collision with molecular hydrogen has been suggested as a mechanism (Townes and Cheung, 1969). Another unexpected finding is that the properties of interstellar OH are peculiar, so strange that it was at first called "mysterium," and it has been hypothesized that some sort of cosmic-scale maser-type amplification is responsible (Wick, 1970; Rank et al., 1971). Similar effects have been observed in the cases of H_2O and NH_3. All H_2O maser action so far

Table 1.2 Molecules Found in the Interstellar Medium (From Buhl, 1971)

Year	Molecule	Symbol	Wavelength	Telescope	Initial Discovery
1937		CH	4300 Å	Mt Wilson 100 inch	Dunham (Mt Wilson)
1940	Cyanogen	CN	3875 Å	Mt Wilson 100 inch	Adams (Mt Wilson)
1941		CH^+	3745–4233 Å	Mt Wilson 100 inch	Adams (Mt Wilson)
1963	Hydroxyl	OH	18, 6.3, 5.0, and 2.2 cm	Lincoln Lab 84 foot	MIT/Lincoln Lab
1968	Ammonia	NH_3	1.3 cm	Hat Creek 20 foot	Berkeley
1968	Water	H_2O	1.4 cm	Hat Creek 20 foot	Berkeley
1969	Formaldehyde	H_2CO	6.2, 2.1, and 1 cm; 2.1 and 2.0 mm	NRAO 140 foot NRAO 36 foot	University of Virginia, NRAO, University of Maryland and University of Chicago
1970	Carbon monoxide	CO	2.6 mm	NRAO 36 foot	Bell Labs
1970	Cyanogen	CN	2.6 mm	NRAO 36 foot	Bell Labs
1970	Hydrogen	H_2	1100 Å	UV rocket camera	NRL
1970	Hydrogen cyanide	HCN	3.4 mm	NRAO 36 foot	University of Virginia and NRAO
1970	X-ogen	?	3.4 mm	NRAO 36 foot	NRAO and University of Virginia
1970	Cyano-acetylene	HC_3N	3.3 cm	NRAO 140 foot	NRAO
1970	Methyl alcohol	CH_3OH	36 and 1 cm; 3 mm	NRAO 140 foot	Harvard
1970	Formic acid	$CHOOH$	18 cm	NRAO 140 foot	University of Maryland and Harvard University
1971	Carbon monosulphide	CS	2.0 mm	NRAO 36 foot	Bell Labs and Columbia University
1971	Formamide	NH_2CHO	6.5 cm	NRAO 140 foot	University of Illinois
1971	Silicon oxide	SiO	2.3 mm	NRAO 36 foot	Bell Labs and Columbia University
1971	Carbonyl sulphide	OCS	2.7 mm	NRAO 36 foot	Bell Labs and Columbia University
1971	Acetonitrile	CH_3CN	2.7 mm	NRAO 36 foot	Bell Labs and Columbia University
1971	Isocyanic acid	$HNCO$	3.4 mm; 1.4 cm	NRAO 36 foot	University of Virginia and NRAO
1971	Hydrogen isocyanide	HNC	3.3 mm	NRAO 36 foot	University of Virginia and NRAO
1971	Methyl-acetylene	CH_3C_2H	3.5 mm	NRAO 36 foot	University of Virginia and NRAO
1971	Acetaldehyde	CH_3CHO	28 cm	NRAO 140 foot	Harvard University
1971	Thioformaldehyde	H_2CS	9.5 cm	Parkes 210 foot	CSIRO, Australia

Table 1.3 Observed Microwave Resonances from Interstellar Molecules[a] (From Rank et al., © 1971 by the American Association for the Advancement of Science with permission of the AAAS.)

Discovery	Molecule	Rotational Quantum Numbers	Transition Type	ν (Ghz)	Hfs	Spectrum
1963	$^{16}OH(^2\pi_{3/2})$	$J = 3/2$	Λ	1.665	+	E, A
1969		$= 5/2$	Λ	6.035	+	E
1970		$= 7/2$	Λ	13.441		E
1968	$^{16}OH(^2\pi_{1/2})$	$J = 1/2$	Λ	4.766		E
1969		$= 5/2$	Λ	8.136		E
1966	$^{18}OH(^2\pi_{3/2})$	$J = 3/2$	Λ	1.637	+	A
1968	$^{14}NH_3$(para)	$(JK) = 1.1$	ID	23.694		E
	(para)	$= 2.2$	ID	23.723		E
	(ortho)	$= 3.3$	ID	23.870		E
	(para)	$= 4.4$	ID	24.139		E
	(ortho)	$= 6.6$	ID	25.056		E
1969	$H_2{}^{16}O$(ortho)	$J_{K_{-1}K_1} = 5_{23} - 6_{16}$	R	22.235		E
1969	$H_2{}^{12}C^{16}O$(ortho)	$J_{K_{-1}K_1} = 1_{11} - 1_{10}$	R	4.830	+	A, E
		$= 2_{12} - 2_{11}$	R	14.488		A
		$= 3_{13} - 3_{12}$	R	28.974		A
1971	$H_2{}^{12}C^{16}O$(ortho)	$= 1_{11} - 2_{12}$	R	140.839		E
	$H_2{}^{12}C^{16}O$(para)	$= 1_{01} - 2_{02}$	R	145.603		E
	$H_2{}^{12}C^{16}O$(ortho)	$= 1_{10} - 2_{11}$	R	150.498		E
1969	$H_2{}^{13}C^{16}O$(ortho)	$= 1_{10} - 1_{10}$	R	4.593		A
1970	$^{12}C^{16}O$	$J = 0 - 1$	R	115.271	None	E
1971	$^{13}C^{16}O$	$J = 0 - 1$	R	110.201		E
1971	$^{12}C^{18}O$	$J = 0 - 1$	R	109.782	None	E
1970	$^{12}C^{14}N$	$J = 0 - 1$	R	113.492	+	E
1970	$H^{12}C^{14}N$	$J = 0 - 1$	R	88.632	+	E
1970	$H^{13}C^{14}N$	$J = 0 - 1$	R	86.339	+	E

(Continued)

detected is from sources that show OH maser effects as well, yet many OH masers show no H_2O emission (Rank, Townes, and Welch, 1971). Intense maser OH and H_2O emission corresponds to small sources of great energy ($>10^{30}$ ergs/sec) and high density ($10^8/cm^3$), indicating an object in the throes of gravitational collapse. Through collision with hydrogen molecules the H_2O and OH could be cooling the hydrogen by maser emission as these objects collapse into protostars (Buhl, 1971).

The stability of interstellar molecules in such intense ultraviolet and cosmic radiation fields as encountered in space also defies explanation. Interstellar dust and/or greater gas density could provide some shielding from destructive radiation. The estimated lifetime of interstellar H_2O, NH_3, CH_4, and H_2CO exposed to the ultraviolet field is less than 100 yr (Cheung et al., 1968). In the case of formaldehyde, the lifetime estimates may be far too long (Gentieu and Mentall, 1970). In any event, as Table 1.4 shows, the concentrations of these molecular species are impressive, and consideration of hydrogen content leads to the conclusion that a substantial fraction of the elements C, O, and N in the clouds is in molecular form. In the cases of C and O, the molecular forms may even exceed the total amounts of those elements

Table 1.3 (*continued*).

Discovery	Molecule	Rotational Quantum Numbers	Transition Type	ν (Ghz)	Hfs	Spectrum
1970	X-ogen (unknown, possibly HCO^+)			89.190		E
1970	$H^{12}C_3{}^{14}N$	$J = 0 - 1$	R	9.098	+	E
1970	$^{12}CH_3{}^{16}OH$	$J_{K_{-1}K_1} = 1_{11} - 1_{10}$	R	834		E
1971	$^{12}CH_3{}^{16}OH$	$(JK) = 4.1 - 4.2$	R	24.933		E
		$= 5.1 - 5.2$	R	24.959		E
		$= 6.1 - 6.2$	R	25.018		E
		$= 7.1 - 7.2$	R	25.125		E
		$= 8.1 - 8.2$	R	25.294		E
1970	$H^{12}C^{16}O^{16}OH$	$J_{K_{-1}K_1} = 1_{11} - 1_{10}$	R	1.639		E
1971	$^{12}C^{32}S$	$J = 2 - 3$	R	146.969	None	E
1971	$^{28}Si^{16}O$	$J = 2 - 3$	R	130.268	None	E
1971	$^{12}CH_3{}^{12}C_2H$(ortho)	$(JK) = 4.0 - 5.0$	R	85.457		E
1971	$H^{14}N^{12}C^{16}O$	$J_{K_{-1}K_1} = 3_{03} - 4_{04}$	R	87.925		E
		$= 0_{00} - 1_{01}$	R	21.982		E
1971	$^{16}O^{12}C^{32}S$	$J = 8,9$	R	109.463	None	E
1971	$^{12}CH_3{}^{12}C^{14}N$(ortho)	$(JK) = 5.0 - 6.0$	R	110.384		E
	(para)	$= 5.1 - 6.1$	R	110.381		E
	(para)	$= 5.2 - 6.2$	R	110.375		E
	(ortho)	$= 5.3 - 6.3$	R	110.364		E
	(para)	$= 5.4 - 6.4$	R	110.349		E
	(para)	$= 5.5 - 6.5$	R	110.330		E
1971	X_2(unknown, possibly HNC)			90.665		E
1971	$^{14}NH_2H^{12}C^{16}O$	$J_{K_{-1}K_1} = 2_{12} - 2_{11}$	R	4.619	+	E
1971	CH_3HCO	$J_{K_{-1}K_1} = 1_{10} - 1_{11}$	R	1.065		E

[a] Transition types are indicated by Λ for Λ doublet, ID for inversion doublet, and R for rotational. A plus sign indicates that the hyperfine structure (Hfs) has been detected. Emission and absorption are denoted by E and A, respectively.

previously thought to be available (Penzias, Jefferts, and Wilson, 1971). The free radicals OH and CN are rare compared to the more stable diatomic molecule CO (interstellar lifetime 1000 yr) (Table 1.4).

Rank et al. (1971) list four, among many, possible mechanisms for the formation of interstellar molecules:

1. Building up of molecules by binary collisions in the gaseous medium of interstellar space.
2. Formation of molecules on surfaces of dust grains from atoms or simpler molecules impinging on the surfaces.
3. Formation of molecules in the dense atmospheres of stars, perhaps largely by many-body collisions, and their subsequent expulsion into the interstellar medium.
4. Evaporation or decomposition of dust grains by bombardment, shock waves, or other heating processes. This suggestion reduces to the older problem of explaining the origin of dust grains; perhaps these originate in stellar atmospheres.

Table 1.4 Present Estimates of Column Density (number of molecules per square centimeter) Along the Line of Sight for Molecules Found in the Directions of Sagittarius B2 and the Orion Nebula.[a] (From Rank et al., © 1971 by the American Association for the Advancement of Science with permission of the AAAS.)

Molecule	Column Density in Sgr B2 (cm^{-2})	Column Density in Orion (cm^{-2})	Comment
H_2	$\geq 10^{22}$	$\sim 2 \times 10^{23}$	Indirect determination
OH	$>5 \times 10^{16}$	?	T assumed 29°K for Sgr B2; primarily maser radiation from Orion
CO	$\sim 10^{19}$	$\sim 10^{18}$	Optically dense clouds, hence column density uncertain
CN	Not detected	$\sim 10^{15}$	T assumed 50°K
CS	$\sim 10^{14}$	$2 \times 10^{13} - 5 \times 10^{14}$	
SiO	$\sim 4 \times 10^{13}$	Not detected	T assumed 30°K
H_2O	?	?	Maser radiation
HCN (hydrogen cyanide)	Not determined	$\sim 10^{15}$	T assumed 20°K for Orion
OCS (carboxyl sulfide)	$\geq 3 \times 10^{15}$	Not detected	
NH_3 (ammonia)	$\geq 10^{17}$	Not detected	T assumed 35°K
H_2CO (formaldehyde)	$\sim 2 \times 10^{15}$	$\sim 3 \times 10^{14}$	T assumed 3°K
HNCO (isocyanic acid)	Not determined	Not detected	
HC_3N (cyanoacetylene)	$\sim 2 \times 10^{16}$	Not detected	T assumed 50°K
HCOOH (formic acid)	$10^{13} - 3 \times 10^{15}$	Not detected	Line interfered with by ^{18}OH transitions
CH_3OH (methyl alcohol)	?	$\sim 5 \times 10^{16}$	Cloud sizes unknown; possibly maser radiation from Sgr B2; very small cloud in Orion
CH_3CN (methyl cyanide)	$\sim 2 \times 10^{14}$	Not detected	
CH_3C_2H (methyl acetylene)	Not determined	Not detected	
X-ogen (unknown)	Not detected	$\sim 10^{15}$	Frequency 89 and 190 Mhz; rough estimate of abundance
X_2 (unknown)	Not detected	Not detected	Frequency 90 and 665 Mhz; found in sources W51 and DR21
NH_2HCO (formamide)	Not determined	Not detected	

[a] The density of H_2 is inferred from excitation processes. Other column densities come primarily from integrated line intensities.

Whichever mechanism (or mechanisms) is responsible, it must be capable of producing molecular species as fast as they disappear. While the dust grains may protect molecules from radiation and catalyze molecule formation, the grains may also remove molecules from the gas phase by "freezing-out" or surface absorption. If grains of about 0.2-μ diameter comprise about 1% of the mass of a cloud, then the lifetime of a molecule before striking and sticking to a grain is about 10^5 yr (Rank et al., 1971).

Here we have been concerned with the chemistry of the interstellar matter. The importance of these discoveries to astronomy—to unraveling movements, temperatures, ages, radiation and particle fields, and spectral emission and absorption phenomena in our galaxy—is enormous. These discoveries have also aroused the enthusiasm of biologists, not so much from the idea that life on Earth was a consequence of a rain or injection of interstellar matter, but rather and simply because they have underscored the ubiquity of the molecules that form the first stepping stones to the life substance. Their presence throughout space indicates that life is probably the rule rather than the exception in the cosmos, that life on other worlds is not only possible, it is very highly probable. Already amino acids have been synthesized in the laboràtory from mixtures of molecular interstellar constituents (Fox and Windsor, 1970; Wollin and Ericson, 1971), but, like Horatio, we need not look to the stars; as will become clear in Chapter 11, extraterrestrial inoculation is a redundant hypothesis for biogenesis on Earth.

1.3 STELLAR CHEMISTRY

Before I begin the subject of the life and composition of the stars, including our own Sun, I would like to describe our physical place in the universe in more sicentific terms than Figure 1.2 and assign some dimensions to creation, or at least to our corner of it.

The visible universe is composed mostly of hydrogen with a little helium and trace amounts of the other elements (Table 1.5 and Table A.16). Far from being distributed evenly, this material is aggregated in clusters of an enormous range of configurations and densities. The largest material organizations within our comprehension are the galaxies and supergalaxies. The galaxies are of the order of 1,000,000,000,000,000,000 km in diameter, and they are spaced about 10,000,000,-000,000,000,000 km apart. Our galaxy, like Andromeda its closest neighbor in space some 8,000,000,000 light yr away, is a spiraling disklike structure some 100,000 light yr (1 light yr = 10^{13} km) in diameter, 15,000 light yr thick at its center, and from 3000 to 6000 light yr thick elsewhere. Our Sun, one of its 200,000,000,000 stars is located at a distance of some 32,600 light yr or two-thirds from the galactic center and rotating about that center at a velocity of 275 km/sec to give a period of 220,000,000 yrs.

In addition to being strewn with clots and diffuse clouds of matter, our galaxy is permeated with various kinds of radiation and high speed particles—X-rays, radiowaves, ultraviolet and infrared radiation, and cosmic rays. The last are largely protons and heavy nuclei, and the energy density of all such particles is about 10^{-12} erg/cm^3. Most of the particles are in the lower energy range of the spectrum, but some have energies up to 100,000,000,000,000,000,000 eV/nucleon. The rush of these enormous quantities of corpuscular matter creates and is perturbed by massive electrical, magnetic, and gravitational fields.

How were the galaxies formed? When and where did the universe come from? Since antiquity, cosmogenic opinion has fallen into two camps: the universe had a beginning in time ("Genesis," or creation *ex nihilo*, or its modern ramification the "big bang" hypothesis) or it did not have a beginning in time (Democritus—the "continuous creation" hypothesis). When I was younger, the big bang hypothesis held the field, but since then, thanks to the vigorous advocacy of Hoyle and others, continuous creation for a while enjoyed the greater popularity. Most recently continuous creation has fallen out of favor. The immensity of these questions boggles my imagination. I am not even sure that they are meaningful, and I am even less sure that these two hypotheses are either/or alternatives. The universe as a whole may be without beginning or end in space and time, but perhaps there are continuously occurring within it localized big bangs. We can, however, bring our speculations back within the pale of scientific verification when we treat the smaller scale of the birth, evolution, and death of stars such as our Sun, and while there may be much that must lie forever beyond the bounds of our comprehension, what little evidence we do have indicates that our corner of the universe is not too different chemically from the rest.

Stars can be classified on the basis of their color or spectral type:

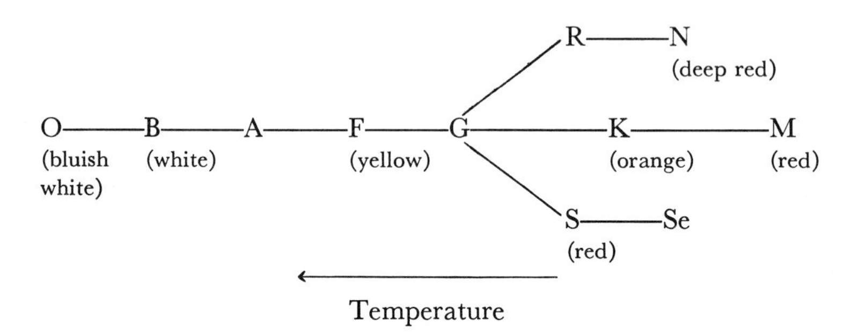

and, inasmuch as the color is related to their temperatures, and temperature in turn is an indication of their masses and where they are along the path of stellar evolution, the classification can be a highly useful one. If the absolute stellar magnitudes are now plotted versus spectral type, the so-called "Hertzsprung-Russell diagram" (see Hack, 1966), the majority of stars fall within a narrow zone or "main sequence" with their eccentric brothers, the giants and the dwarfs, clustering in areas above and below, respectively (Figure 1.5). Our Sun is a fairly small and undistinguished member of the main sequence. Frequently, stars may also be classified with respect to their apparent ages with younger ones, such as our Sun, falling into Population I, and older ones, into Population II. The latter are of special interest for if the material exchange between their cores and atmospheres is poor, then their atmospheres should reflect the chemical composition of the galactic medium when they were young. Indeed, they appear to do so, being deficient in metals (Wallerstein et al., 1963). Now let us trace the life cycle of a typical star (Figure 1.6). We must remember that time scales are mass dependent so that large stars hasten through their life cycle much faster than smaller ones. A star's life cycle begins with the condensation and aggregation of interstellar gas and dust (1) by gravitational attraction over a period of roughly 10,000,000 yr to form the star and its attendant planets (2). With further compaction, the star enters the main sequence where it remains an equilibrium for about 8,000,000,000 yr gradually consuming hydrogen (3). During the next 100,000,000 yr, the star departs from the main sequence and becomes a red giant (4), consuming its planets as it does so. After a few thousand years pulsating as a variable star (5), it at last explodes into a supernova (6) throwing much of its matter out into space and completing the material cycle (Figure 1.4) and dies by collapsing into a

Table 1.5 Relative Abundance of Elements (From Calvin, © Oxford University Press, 1969, with permission of the publisher. For a more detailed compilation of cosmic abundances see Table A.16)

		Earth		Life	
Element	Cosmos	Atmosphere Hydrosphere	Crust	Plant (%)	Animal (%)
Hydrogen	1000.0	2.0	0.03	10.0	10.0
Helium	140.0				
Oxygen	0.680	9.978	0.623	79.0	65.0
Carbon	0.300	0.0001	0.0005	3.0	18.0
Neon	0.280				
Nitrogen	0.091	0.003		0.28	3.0
Magnesium	0.029		0.018	0.08	0.05
Silicon	0.017		0.211	0.12	
Iron	0.008		0.019	0.02	0.004
Argon	0.004				
Sulphur	0.003	0.0005		0.01	0.25
Aluminum	0.0019		0.064		
Calcium	0.0017		0.019	0.12	2.0
Sodium	0.0017	0.0008	0.026	0.03	0.15
Nickel	0.0005				
Phosphorus	0.0003			0.05	1.0
Potassium	0.00008			0.32	0.35
Others	0.00015	0.011	0.020	0.04	0.156

white dwarf (7). In our galaxy, supernovae occur about once every 250 yr, and they may irradiate the surface of Earth with sufficient radiation to induce discontinuities in biological evolution (Terry and Tucker, 1968). The Crab Nebula is the remnants of such a stellar explosion that was actually observed by Chinese astronomers in 1054 A.D.

The several stages of stellar evolution are associated with different nuclear-chemical processes, and these reactions in turn have given rise to the observed distribution of chemical elements in the cosmos (Tables 1.5 and A.16) (Alpher and Herman, 1953; Fowler, 1957; Craig, Miller, and Wasserburg, 1964; Ahrens, 1968). Two features of Table 1.5 are noteworthy; hydrogen is, as we have said, by far the most common element in the cosmos, and, with the exception of the "noble" gases, the elements most common on a cosmic scale—H, C, N, O, and P—are the very ones most common in the biomaterial. Further important features become apparent if we plot cosmic abundances versus mass number (Figure 1.7):

1. An overall rapid exponential decrease in abundance with increasing mass number;
2. Difficult to see in Figure 1.7, but a tendency for even atomic numbers to be more abundant than odd ones, the so-called "Oddo-Harkins rule";

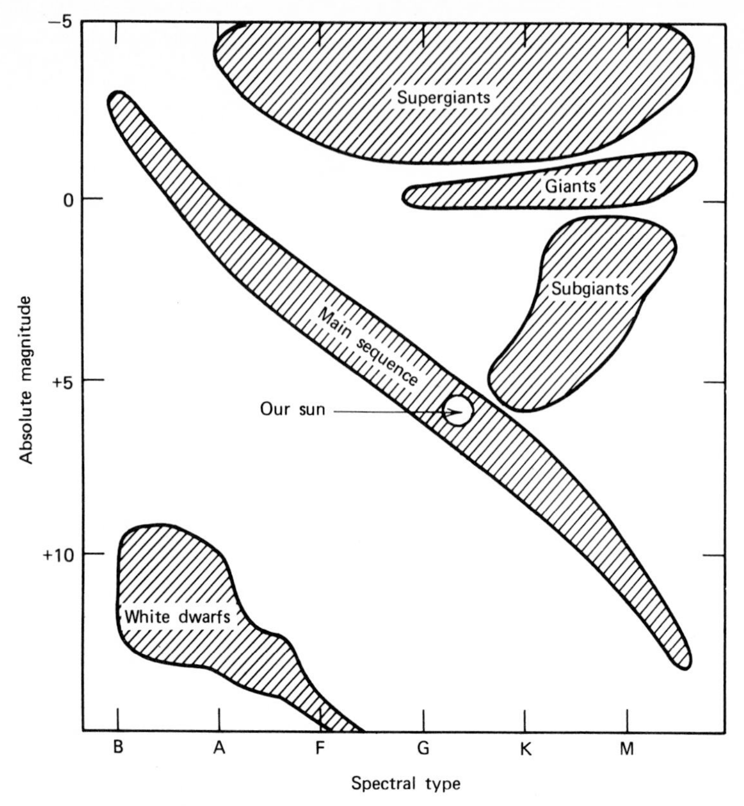

Figure 1.5. Hertzsprung–Russell diagram.

3. The very low abundances of Li, Be, and B; and
4. The marked abundance peaks in the iron group.

But before we examine the chemical processes and the synthesis of elements in stars, let us address ourselves to the possibility of prestellar nucleosyntheiss in an expanding universe prior to the formation of galactic structures. According to the big bang hypothesis, there was first a gigantic explosion of the highly dense, undifferentiated primordial substance or "ylem" (North, 1965; Zel'dovich, 1965), and the matter of the universe is still hurtling outward. Alpher, Bethe, and Gamow (1948) early recognized (and the idea in a poor pun has been called the α-β-γ theory in their honor) that the cosmic abundances of the elements (Figure 1.7) bore some relation to the information then rapidly accumulating on neutron-capture cross sections. In the primordial fireball, neutrons could decay to protons, then successive neutron captures could account for the synthesis of all the elements, or at least one could bridge the mass gap at 5 and 8 (Figure 1.7) and synthesize elements up to ^{20}Ne (Alpher and Herman, 1950; Alpher, Follin, and Herman, 1953; Hayashi and Nishida, 1956). The big bang cosmogenic model found supporting evidence in the observations that the helium content of the universe is greater than one would expect from stellar synthesis alone (Hoyle and Taylor, 1964) and that the cosmic background temperature is 3°K (Penzias and Wilson, 1965) very close to the value of 5°K

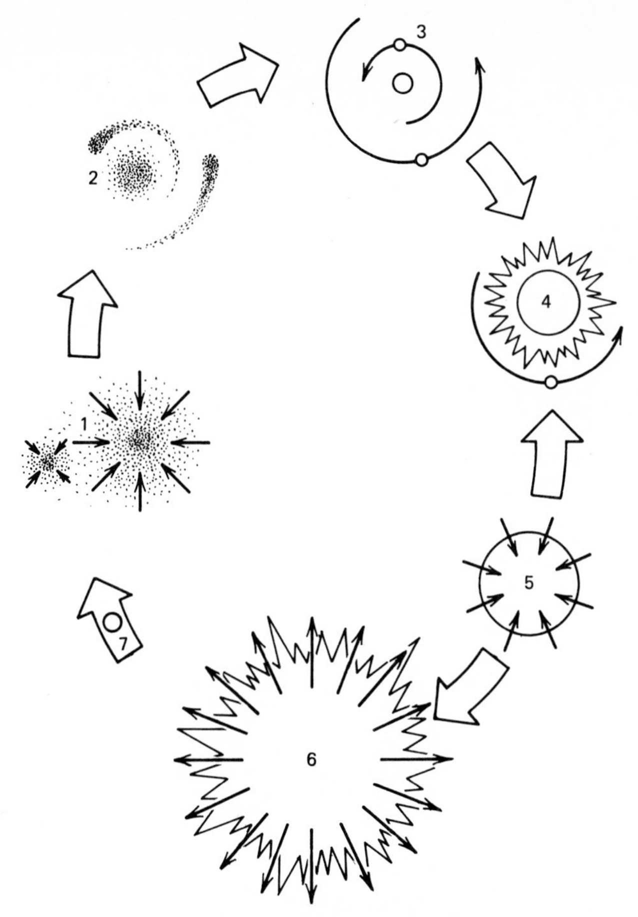

Figure 1.6. Life cycle of a star.

predicted by Alpher and Herman (1948, 1949) on the basis of fireball expansion cooling (see also Dicke et al., 1965). However, more detailed analysis revealed that only D, ^3He, ^4He, and ^7Li could have been produced in the initial explosion and that even the possible helium synthesis was inadequate (Wagoner, Fowler, and Hoyle, 1967). These and other inadequacies of the big bang hypothesis of prestellar nucleosynthesis have led to the formulation of the hypothesis of continuous creation and to the elaboration of the processes involved in stellar element synthesis (Hoyle, 1954; Hoyle, et al., 1956). In recent years the theory of nucleosynthesis has continued to change and evolve.

What is the source of the enormous energy output of the stars? Clearly, it cannot be chemical combustion for the temperatures are much too high and the stars are composed mostly of hydrogen with chemically inert helium and very little oxidant to "burn" the hydrogen and the cosmic abundances of the expected products (such as H_2O) of any such combustion processes are far too low. In the case of the nearest star, our Sun, early hypotheses for the origins of its heat energy included conversion of the kinetic energy of meteors falling into it and the conversion of potential into thermal energy as a consequence of the Sun's gravitational contraction. However,

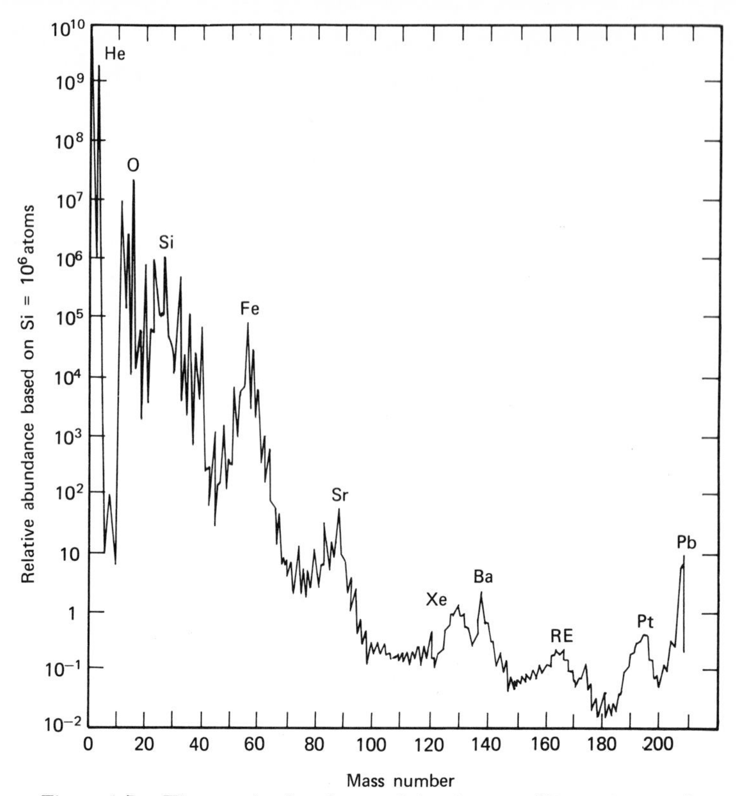

Figure 1.7. The cosmic abundance of the elements (Si = 10^6 atoms).

the Sun's maximum life calculated from this shrinkage is only 5×10^7 yr. Although this is longer than the Sun's age estimated on the basis of chemical energy (10^5 yr), it is still far too short, shorter even than the age of the Earth (about 10^9 yr). In 1920, Sir Arthur Eddington developed the suggestion of Perrin that proton fusion (or "hydrogen burning") to form helium is a major source of stellar energy.

$$4\,{}^{1}\mathrm{H} \rightarrow {}^{4}\mathrm{He} + \text{energy} \tag{1.2}$$

The mass loss in the overall reaction is converted to energy in accordance with the Einstein equation

$$E = mc^2 \tag{1.3}$$

where c is the velocity of light. Taking the mass of our Sun to be 1.985×10^{33} g and its radiation energy to be 3.78×10^{33} erg/sec, and assuming that it was initially composed of hydrogen, we calculate that our Sun can burn at its present rate for about 10^{11} yr. This view is somewhat oversimplified, since not all the H can be converted to He.

In greater detail, the conversion of H to He at the high temperatures (2×10^{7}°K for the Sun) within the stars of the principal series is believed to occur by the so-called C–N chain reaction

$$ {}^{1}\mathrm{H} + {}^{12}\mathrm{C} \rightarrow {}^{13}\mathrm{N} + \gamma \tag{1.4}$$

$$^{13}N \rightarrow {}^{13}C + e^+$$

$$e^+(\text{positron}) + e^-(\text{electron}) \rightarrow \gamma \quad (1.5)$$

$$^{1}H + {}^{13}C \rightarrow {}^{14}N + \gamma \quad (1.6)$$

$$^{1}H + {}^{14}N \rightarrow {}^{15}O + \gamma$$

$$^{15}O \rightarrow {}^{15}N + e^+$$

$$e^+ + e^- \rightarrow \gamma \quad (1.7)$$

$$^{1}H + {}^{15}N \rightarrow {}^{12}C + {}^{4}He \quad (1.8)$$

in which ^{12}C is regenerated and thus acts as a catalyst. In some stars (G and cooler with central temperatures below 1.5×10^{7}°K), the H–He reaction sequence can give rise to the same overall conversion of protons to helium

$$^{1}H + {}^{1}H \rightarrow {}^{2}D + e^+ + \gamma$$

$$e^+ + e^- \rightarrow \gamma \quad (1.9)$$

$$^{1}H + {}^{2}D \rightarrow {}^{3}He + \gamma \quad (1.10)$$

$$^{3}He + {}^{4}He \rightarrow {}^{7}Be + \gamma \quad (1.11)$$

$$^{7}Be \xrightarrow{\text{fast}} {}^{7}Li + e^+ + \gamma$$

$$e^+ + e^- \rightarrow \gamma \quad (1.12)$$

$$^{1}H + {}^{7}Li \xrightarrow{\text{fast}} 2\,{}^{4}He \quad (1.13)$$

Helium mixing between envelope and core is poor, and as the hydrogen of a star is consumed the star will cease to be homogeneous. Helium ash will accumulate in its core while hydrogen continues to burn in its exceedingly hot shell (30,000,000°K). With these compositional changes, the further evolution of the star will carry it out of the main sequence. But the extinction of the hydrogen furnace does not result in a temperatue drop, for the subsequent contraction of the core converts enormous quantities of potential gravitational energy into heat. The sudden rise in core temperature causes the envelope to expand and the increased surface area of the envelope enables it to radiate energy at a lower surface temperature—the star has become a red giant (Figures 1.5 and 1.6).

In due course, the temperatures and densities of the core of a red giant become so great (100,000,000°K and 100,000 g/cm^3) that coulombic repulsion between two helium nuclei is no longer adequate to prevent their interaction to form ^{8}Be. But ^{8}Be is unstable and interacts with another helium (α-particle) to give an overall reaction corresponding to the production of ^{12}C from the collision of three α-particles

$$3\,{}^{4}He \rightarrow {}^{12}C \quad (1.14)$$

The ^{12}C can capture an additional helium nucleus to give ^{16}O in the overall reaction

$$4\,{}^{4}He \rightarrow {}^{16}O \quad (1.15)$$

Thus we have gone from mass 4 to masses 12 and 16 skipping masses 5 and 8 (Salpeter, 1952), and, as we saw in Figure 1.7, these are exactly the elements of very low cosmic abundance. This bridging of the 5 and 8 mass gap was a major step in our understanding of stellar nucleosynthesis.

Because the mass converted into energy by helium burning is not large (0.07%), a red giant will not be stable for an extended period. At this point, the star can either continue along the sequence of stellar evolution, or it can become unstable and eject unburned hydrogen and helium along with synthesized ^{12}C and ^{16}O. These nuclides will mix with the interstellar matter (Figure 1.4) and may eventually condense and be taken up in the formation of a second or latter generation star. In such a star, an additional path, the so-called CNO bicycle, is available for the conversion of H to He (Caughlan and Fowler, 1962). As the temperature increases, C and O can themselves burn to produce nuclei such as ^{20}Ne, ^{24}Mg, ^{28}Si, and ^{32}S (Reeves and Salpeter, 1959).

The coulombic repulsion between nuclei of charge numbers 10 to 16 is sufficiently strong to discourage further burning by simple fusion. Instead, the so-called α-process becomes important—the intermediate nuclei photodisintegrate with the emission of high energy α-particles, which in turn can be captured by the surviving nuclei to build higher elements, for example.

$$^{28}Si \rightarrow 7\,^{4}He \text{ (at 3,000,000,000°K)} \tag{1.16}$$

then

$$^{28}Si + 7\,^{4}He \rightarrow \,^{56}Ni \tag{1.17}$$

to give an overall

$$2\,^{28}Si \rightarrow \,^{56}Ni \tag{1.18}$$

^{56}Ni is unstable and decays through ^{56}Co to give ^{56}Fe (or ^{54}Fe, Truran, Cameron, and Gilbert, 1966). Other nuclei near ^{56}Fe, such as ^{46}Ti and ^{62}Ni, may be produced in this general way. The α-process, together with an e-process resulting from β-decay, account for the large observed abundances of the iron group elements (Figure 1.7). The process probably occurs during the catastrophic collapse of the core of a red giant in its terminal throes. With core collapse, unspent nuclear fuel from the envelope ignites, and the whole star may explode as a supernova (Figure 1.6) hurling its matter out into space (Minkowski, 1964). From the relative abundances of iron isotopes, it is possible to estimate that the presupernova e-process build-up time may be about 9 hr.

Beyond the iron group nuclei, heavy element synthesis is accomplished primarily by neutron capture processes (Cameron, 1955; Clayton et al., 1961; Seeger, Fowler, and Clayton, 1965) thereby avoiding strong coulombic repulsion, building up the heavy elements gradually unit by unit following the thin path of isotopic stability that cuts its way diagonally across the isotope chart until in the case of the heaviest nuclei neutron-induced fission terminates the synthetic path. The abundance curve (Figure 1.7) evidences two neutron-capture processes at work—a slow process (s-process) compared to β-decay and a fast process (r-process). The former takes place in the red giant stage of stellar evolution, whereas the latter is believed to occur in supernovae explosions, which last 1 to 100 sec (Hoyle and Fowler, 1960). The observation of ^{99}Tc in certain stellar spectra (Merrill, 1952) indicates that the s-process is still or has recently occurred in these bodies, for with its short half-life of 200,000 yr, it would have long since disappeared if formed only in primordial creation. This observation has been cited as supporting the theory of continuous creation in the universe in contrast to the big bang hypothesis. Similarly, the cosmic $^{232}Th/^{238}U$ and $^{235}U/^{238}U$ ratios also indicate continuous galactic synthesis over the past 12,000,000,000 yr (plus 3,000,000 yr for stars to reach the supernova r-process stage).

In summary, let us return to our abundance curve (Figure 1.7). Element synthesis begins with H-burning. After boron, this is followed for the more abundant elements by He- and C, O-burning, and the α -process, or by continued H-burning and the s-process to form the iron group of elements. Elements immediately beyond Fe are created by the e-process, and further beyond, by r- and s-processes until fusion terminates element synthesis at around a nuclear mass of 200 (Burbidge et al., 1957; Greenstein, 1961). Recently, the possibility of superheavy element (proton numbers 110 to 110) synthesis by an explosive r-process has been raised by Schramm and Fowler (1971).

In the foregoing duscussion of stellar evolution and cosmochemistry, I have tried to sketch the major events and seqeuences. Many objects in the heavens have "anomalous" chemical compositions: to mention only a few, ^{13}C enrichments with $^{13}C/^{12}C$ = 1/2 rather than the terrestrial value of 1/90 (Wyller, 1965); F and G dwarf stars with $^{6}Li/^{7}Li$ = 1/2 rather than the terrestrial value of 1/12 possibly due to ^{6}Li produced by spallation processes (Herbig, 1964) and young G stars with Li enriched about 100 times over our Sun (Bonsack and Greenstein, 1960), strange A stars with intense magnetic fields and anomalous compositions as a result of surface nuclear reactions (Sargent, 1964); abundant promethium from the fission of transuranic elements in the surface of a peculiar A star (Kuchowicz, 1971); old giants enriched in elements heavier than strontium (Burbidge and Burdidge, 1957), and Li depletion in K dwarfs (compared to G dwarfs), which may be related to the object's mass and rotational breaking (van den Heuvel and Conti, 1971).

One extremely important subject closely related to cosmochemistry and currently causing great excitement in astronomy and physics, but which we cannot treat in detail here, is the enormous sources and fluxes of particles and electromagnetic energy both within and outside our galaxy. These fields can take the form of enormous accelerators that make the largest of our earthly cosmotrons, masers, and X-ray apparatus look like the most trivial of toys. Another marvel of our unverse that I cannot pass without at least mentioning is the fantastic state of matter in some stellar objects—temperatures, pressures, and densities beyond our comprehension. White dwarfs have densities around 10^8 g/cm^3, whereas in burned out and collapsed neutron stars or pulsars, the compression is so great that densities as large as 10^{14} g/cm^3 are obtained. Such a superdense neutron fluid is virtually incompressible (10^{22} times less compressible than steel). At still greater densities negative muons replace electrons and "strange" particles make their appearance. The magnetic field of a neutron star can be 10^{12} times stronger than that of Earth (Ruderman, 1971).

The knowledge that man has fashioned concerning the nature of the universe in which he finds himself is as marvelous as the enormity and complexity of that universe itself. Nevertheless, for all the questions he has answered, countless more remain, and overriding them all, that persistent question of questions—did it all begin at some finite point in time with a big bang, as the expanding universe and the remarkable consensus of cosmic ages (Table 1.6) suggest, or has the universe existed without beginning, without end, with galaxies, stars, and planets, in the words of the ancient cosmologists, coming into being and passing away, with elements being synthesized and destroyed, as indicated by the apparent different states of stellar evolution represented by the heavenly bodies all around us? How can these two patently divergent views be reconciled? By different rates? By localized big bands? Perhaps we shall never know.

Table 1.6 Some Age Estimates

Location	Method	Age (yr)
Sun	H/He ratio	10.7×10^9 (max)
		4.3×10^9
Pleiades	H/He ratio	3.0×10^9
Nebulae	Red shift	$3.5–7.8 \times 10^9$
Galaxy	^{232}Th, ^{235}U, and ^{238}U abundances	12×10^9 (min)
Galaxy	Red shift (Hubble constant)	$10–13 \times 10^9$
Galaxy	Various evolutionary models	$6–20 \times 10^9$
Galaxy	Steady-state continuous creation	∞
Stellar Clusters (in our galaxy)	Fusion rates	15×10^9 (max)
Earth	^{232}Th, ^{235}U, and ^{238}U abundances	3×10^9
Earth	Pb isotope ratios	3.4×10^9
Earth's crust	He and Pb contents	$2–3 \times 10^9$
Crystallization of Earth's crust	Isotope ratios	3.5×10^9
Meteorites	^{232}Th, ^{235}U, and ^{238}U abundances	4.6×10^9
Meteorites	$^{206}Pb/^{204}Pb$ and $^{207}Pb/^{204}Pb$ ratios	4.5×10^9
Meteorites	K/A ratio	4.5×10^9

BIBLIOGRAPHY

L. H. Ahrens (ed.), *Origin and Distribution of the Elements*, Pergamon, Oxford, 1968.

D. K. Aitken and P. G. Polden, *Nature Phys. Sci.*, *234*, 72 (1971).

R. A. Alpher, H. A. Bethe, and G. Gamow, *Phys. Rev.*, *73*, 803 (1948).

R. A. Alpher, J. W. Follin, and R. C. Herman, *Phys. Rev.*, *92*, 1347 (1953).

R. A. Alpher and R. C. Herman, *Nature*, *162*, 774 (1948).

R. A. Alpher and R. C. Herman, *Phys. Rev.*, *75*, 1089 (1949).

R. A. Alpher and R. C. Herman, *Rev. Mod. Phys.*, *22*, 153 (1950).

R. A. Alpher and R. C. Herman, *Ann. Rev. Nucl. Sci.*, *2*, 1 (1953).

W. K. Bonsack and J. L. Greenstein, *Astrophys. J.*, *131*, 83 (1960).

D. Buhl, *Nature*, *234*, 332 (1971).

D. Buhl and L. D. Snyder, *Tech. Rev.* (*M.I.T.*), *73*(6), 55 (April 1971).

E. M. Burbidge and G. R. Burbidge, *Astrophys. J.*, *126*, 357 (1957).

E. M. Burbidge, G. R. Burbidge, W. A. Fowler, and F. Hoyle, *Rev. Mod. Phys.*, *29*, 547 (1957).

M. Calvin, *Chemical Evolution*, Oxford Univ. Press, New York, 1969.

A. G. W. Cameron, *Astrophys. J.*, *121*, 144 (1955).

G. R. Caughlan and W. A. Fowler, *Astrophys. J.*, *136*, 453 (1962).

A. C. Cheung, D. M. Rank, C. H. Townes, D. C. Thornton, and W. J. Welch, *Phys. Rev. Lett.*, *21*, 1701 (1968).

D. D. Clayton, W. A. Fowler, T. E. Hull, and B. A. Zimmerman, *Ann. Phys.*, *12*, 331 (1961).

H. Craig, S. L. Miller, and G. J. Wasserburg, *Isotope and Cosmic Chemistry*, North-Holland, Amsterdam, 1964.

R. H. Dicke, P. J. E. Peebles, P. G. Roll, and D. T. Wilkinson, *Astrophys. J.*, *142*, 414 (1965).

B. Donn, *Astrophys. J. Lett.*, *152*, L129 (1968).
B. Donn, *Science*, *170*, 149 (1970).
W. W. Duley, *Nature*, *218*, 153 (1968).
N. J. Evans, C. H. Townes, H. F. Weaver, and D. R. W. Williams, *Science*, *169*, 680 (1970).
P. A. Feldman, M. J. Rees, and M. W. Werner, *Nature*, *224*, 752 (1969).
W. A. Fowler, *Sci. Monthly*, *84*, 84 (1957).
S. W. Fox and C. R. Windsor, *Science*, *170*, 984 (1970).
E. P. Gentieu and J. E. Mentall, *Science*, *169*, 681 (1970).
W. R. M. Graham and W. W. Duley, *Nature Phys. Sci.*, *232*, 43 (1971).
J. M. Greenberg and T. de Jong, *Nature*, *223*, 251 (1969).
J. L. Greenstein, *Amer. Sci.*, *49*, 449 (1961).
M. Hack, *Sky and Telescope*, 260 (May 1966); 332 (June 1966).
C. Hayashi and M. Nishida, *Prog. Theor. Phys.* (*Kyoto*), *16*, 613 (1956).
G. H. Herbig, *Astrophys. J.*, *140*, 702 (1964).
E. P. van den Heuvel and P. S. Conti, *Science*, *171*, 895 (1971).
F. Hoyle, *Astrophys. J. Suppl.*, *1*(5), 121 (1954).
F. Hoyle and W. A. Fowler, *Astrophys. J.*, *132*, 565 (1960).
F. Hoyle, W. A. Fowler, G. R. Burbidge, and E. M. Burbidge, *Science*, *124*, 611 (1956).
F. Hoyle and R. J. Taylor, *Nature*, *206*, 1108 (1964).
F. Hoyle and N. C. Wickramasinghe, *Nature*, *214*, 969 (1967).
F. Hoyle and N. C. Wickramasinghe, *Nature*, *226*, 62 (1970).
D. R. Huffman, *Astrophys. J.*, *161*, 1157 (1970); *Nature*, *225*, 833 (1970).
D. R. Johnson and F. K. Powell, *Science*, *169*, 679 (1970).
F. M. Johnson, *Bull. Amer. Astron. Soc.*, *2*, 323 (1970).
S. A. Kaplan and S. B. Pikelner, *The Interstellar Medium*, Harvard Univ. Press, Cambridge, 1970.
R. F. Knacke et al., *Astrophys. J.* *158*, 151 (1969).
B. Kuchowicz, *Nature*, *232*, 551 (1971).
P. G. Manning, *Nature*, *225*, 833 (1970); *226*, 829 (1970).
P. W. Merrill, *Astrophys. J.*, *116*, 21 (1952).
R. Minkowski, *Ann. Rev. Astron. Astrophys.*, *2*, 247 (1964).
Y. Miyake, *Elements of Geochemistry*, Maruzen, Tokyo, 1965.
J. D. North, *The Measure of the Universe*, Clarendon, Oxford, 1965.
P. Palmer, B. Zuckerman, D. Buhl, and L. E. Snyder, *Astrophys. J.*, *156*, L147 (1969).
A. A. Penzias, K. B. Jefferts, and R. W. Wilson, *Astrophys. J.*, *165*, 229 (1971).
A. A. Penzias and R. W. Wilson, *Astrophys. J.*, *142*, 419 (1965).
D. M. Rank, C. H. Townes, and W. J. Welch, *Science*, *174*, 1083 (1971).
H. Reeves and E. E. Salpeter, *Phys. Rev.*, *116*, 1505 (1959).
M. A. Ruderman, *Sci. Amer.*, *224*, No. 2, 24 (February 1971). (1929).
E. E. Salpeter, *Astrophys. J.*, *115*, 326 (1952).
W. L. W. Sargent, *Ann. Rev. Astron. Astrophys.*, *2*, 297 (1964).
C. Schalen, *Ark. Astron.*, *4*(1), (1965).
D. N. Schramm and W. A. Fowler, *Nature*, *231*, 103 (1971).
P. A. Seeger, W. A. Fowler, and D. D. Clayton, *Astrophys. J. Suppl.*, Ser. 11 (97), 121 (1965).

L. E. Snyder, D. Buhl, B. Zuckerman, and P. Palmer, *Phys. Rev. Lett.*, *22*, 679 (1969).

T. P. Stecher, *Astrophys. J. Lett.*, *157*, L125 (1969).

W. A. Stein and F. C. Gillett, *Nature Phys. Sci.*, *233*, 72 (1971).

K. D. Terry and W. H. Tucker, *Science*, *159*, 421 (1968).

C. H. Townes and A. C. Cheung, *Astrophys. J.* *157*, L103 (1969).

J. W. Truran, A. G. W. Cameron, and A. Gilbert, *Can. J. Phys.*, *44*, 563 (1966).

R. V. Wagoner, W. A. Fowler, and F. Hoyle, *Astrophys. J.*, *148*, 3 (1967).

G. Wallerstein, J. L. Greenstein, R. Parker, H. L. Hefler, and L. H. Aller, *Astrophys. J.*, *137*, 280 (1963).

L. Weliachew, *Astrophys. J.*, *167*, L47 (1971).

G. L. Wick, *Science*, *170*, 149 (1970).

N. C. Wickramasinghe, *Nature Phys. Sci.*, *234*, 7 (1971).

N. C. Wickramasinghe and V. C. Reddish, *Nature*, *218*, 661 (1968).

G. Wollin and D. B. Ericson, *Nature*, *233*, 615 (1971).

A. A. Wyller, *Astrophys. J.*, *143*, 829 (1965).

Y. B. Zel'dovich, *Adv. Astron. Astrophys.*, *3*, 241 (1965).

ADDITIONAL READING

L. H. Aller, *Abundance of the Elements*, Interscience, New York, 1961.

L. H. Aller and D. B. McLaughlin, *Stellar Structure*, Univ. Chicago Press, Chicago, 1965.

H. Bondi, *Cosmology*, 2nd ed., Cambridge Univ. Press, Cambridge (England), 1961.

R. Cayrel and G. Cayrel de Strobel, *Ann. Rev. Astron. Astrophys.*, *4*, 1 (1966).

W. A. Fowler, *Chem. Engr. News*, 90 (March 16, 1964).

O. Gingerich (ed.), *Theory and Observation of Normal Stellar Atmospheres*, M.I.T. Press, Cambridge, 1969.

M. Hack (ed.), *Mass Loss from Stars*, Springer-Verlag, New York, 1969.

C. Hayashi, R. Hoshi, and D. Sugimoto, *Prog. Theor. Phys. (Kyoto) Suppl.*, *22*, 1 (1962).

S. A. Kaplan and S. B. Pikelner, *The Interstellar Medium*, Harvard Univ. Press, Cambridge, 1970.

B. Mason, *Principles of Geochemistry*, 2nd ed., Chapter 2, Wiley, New York, 1958.

E. Novotny, *An Introduction to Stellar Atmospheres and Interiors*, Oxford Univ. Press, New York, 1971.

M. Schwarzchild, *Structure and Evolution of the Stars*, Princeton Univ. Press, Princeton, 1958.

H. E. Suess and H. C. Urey, *Rev. Mod. Phys.*, *28*, 53 (1956).

2

THE SOLAR SYSTEM

2.1 SOLAR CHEMISTRY AND THE ORIGIN OF THE PLANETS

Let us focus our attention now on our little corner of the cosmos. The Sun is one of some 200,000,000,000 stars in our galaxy and is located about 32,600 light yr from the galactic center. It is a dwarf G_0 type star with a surface temperature of about 60,000 K and an equatorial radius of 695,000 km (Table 2.1). Its mass is more than 330,000 times greater than that of Earth and represents 99.8% of the total mass of the solar system. Its specific gravity (1.4) is greater than that of water (1.0) but less than that of aluminum (2.7). About 27 days are required for the Sun to rotate completely. With respect to chemical composition, except for its somewhat higher oxygen content, the Sun appears to resemble the rest of cosmic matter (compare Tables 1.5, 2.2 and A.16). The more recent findings of Goldberg, Muller, and Aller (1960) suggest that the O and Fe contents are higher and the Mg content lower than given in Table 2.2.

The spectrum of the Sun's atmosphere has been intensively studied for many years (Zirin, 1966), and despite interference and other difficulties, more than two-thirds of the known naturally occurring elements have been detected there. In fact, one, helium, was discovered in the solar spectrum before it was detected on Earth. The chemistry of the Sun, and the sources of its energy have already been discussed in Chapter 1. However until very recently, it is interesting to note, there was no *direct* experimental evidence for energy production by nuclear reactions in the Sun's interior. Hitherto, our direct knowledge of the chemistry of the Sun and other stars was based entirely on their envelopes for the particles produced by the central nuclear furnace are all altered before they reach the Sun's surface. But not quite all, neutrinos alone by virtue of their massless and chargeless nature can travel through matter with very little attenuation, but this same property makes them exceedingly difficult to detect. Neutrinos from nuclear transformations comprise about 3% of the energy released by the Sun, and they are formed both by pp chain and CN chain reactions (see Chapter 1.1). A former colleague of mine, Dr. Raymond Davis, Jr., of Brookhaven National Laboratory, has been patiently working for about 20 yr trying to "see" a solar neutrino. Using 378,000 liter of CCl_4 (to form ^{37}A from a neutrino impact with ^{37}Cl) 1.48 km down in a South Dakota lead mine shaft as a detector, he isolates 25 (yes, 25!) atoms of ^{37}A from neutrino influx from 10^{30} atoms of fluid, corresponding to a rate of about one solar neutrino every 2 days (Wick, 1971). The persistence and skill required for this experimental quest can hardly be imagined. I once heard a visitor to Brookhaven remark that if Ray Davis worked as hard looking for the Holy Grail as he does for solar neutrinos he would find it. The neutrinos detected by Davis are fewer than predicted by theory, and this has far-reaching astrophysical implications. Already the predominance of relatively low-energy

Table 2.1 Properties of the Solar System

	Sun	Mercury	Venus	Earth	Moon	Mars	Jupiter	Saturn	Uranus	Neptune	Pluto
Distance from Sun (Mean)	0.000	0.87	0.723	1.000[a]	1.000	1.524	5.202	9.539	19.182	30.058	39.439
Mass (Earth = 1)	[c]	0.056	0.817	1.000[b]	0.012	0.108	318.0	95.2	14.6	17.3	0.9(?)
Radius (Earth = 1)	[d]	0.39	0.97	1.00	0.27	0.53	11.19	9.47	3.69	3.50	1.1
Volume (Earth = 1)	1.3 × 10	0.06	0.92	1.00		0.15	1312	734	64	43	0.1
Density (g/cm^3)	1.41	5.42	5.25	5.51	3.34	3.96	1.33	0.68	1.60	1.65	3(?)
Temperature (approx. °K)	6000	450	235,600	240		220	100	75	50	40	40
Observed Gases in Atmosphere	H, He		CO_2, H_2O	N_2, O_2, CO_2, H_2O		CO_2, H_2O	H_2, CH_4, NH_3	H_2, CH_4	H_2, CH_4	H_2, CH_4	
Sidereal period (tropical years)		0.241	0.615	1.000		1.881	11.862	29.458	84.013	164.794	247.686
Surface gravity (Earth = 1)		0.36	0.87	1.000	0.16	0.38	2.64	1.13	1.07	1.41	?
Rotation period (Earth days)		88	?	1.000	27.3	1.03	0.41	0.43	0.45	6.39(?)	27.3

[a] Mean Earth, Sun distance = 1.495×10^8 km.
[b] Mass of Earth = 5.98×10^{24} kg.
[c] Mass of Sun = 1.99×10^{30} kg.
[d] Radius of Sun = 6.96×10^5 km.

Table 2.2 Abundances of the Elements in the Solar Atmosphere (Based on Unsöld, 1950)

Element	Atomic number	Abundance (atoms/10^4 atoms Si)
H	1	5.1×10^8
He	2	1×10^8
C	6	10,000
N	7	21,000
O	8	2.8×10^6
Na	11	1000
Mg	12	17,000
Al	13	1100
Si	14	10,000
S	16	4300
K	19	81
Ca	20	870
Sc	21	1.1
Ti	22	47
V	23	5.9
Cr	24	200
Mn	25	150
Fe	26	27,000
Co	27	55
Ni	28	470
Cu	29	8.7
Zn	30	31

compared to high-energy neutrinos has shown that the pp chain process is probably far more important in our Sun than the CN chain.

Mason (1958) lists five major features of our solar system for which any satisfactory hypothesis of its origin must account:

1. Over 99.8% of the mass of the system resides in the Sun (discussed previously), yet
2. Most of the angular momentum of the system resides in the planets.
3. All the planets rotate about the Sun in the same direction and very nearly in the same plane, and
4. Except for Uranus, the planets rotate about their axes in the same direction.
5. The planetary distances from their central Sun (Table 2.1) are regular (Bode's law), and with respect to their physical and chemical compositions, the planets fall into two distinct and quite different groups: the small, dense "terrestrial planets"—Mercury, Venus, Earth, and Mars— and the large, low specific gravity "major planets"—Jupiter, Saturn, Uranus, and Neptune.

Pluto is an oddity. It does not belong among the major planets and may very well be a strayed satellite of Neptune or an object that has joined the solar system from the outside. There appears to be a planet missing between Mars and Jupiter. The belt of asteroids and meteors in this region according to one hypothesis are fragments from its disintegration.

Older hypotheses by Buffon (1749), Laplace (1796), and the philosopher Kant (1755) visualized the planets as being formed by the cooling and condensation of incandescent gaseous material thrown out by or pulled out from the Sun. More modern thinking on the subject has tended to return to the view of the ancient Greeks and visualize the formation of the planets more or less concurrently with the formation of the Sun by the condensation, accretion, and aggregation of interstellar material with subsequent heating up from gravitational pressure and radioactive decay. Inasmuch as double star systems, like our solar system but unlike single stars, have high angular momenta, the idea has been advanced recently that double stars might become solar systems if one component of the pair disintegrates. Another fresh approach to the problem of cosmogony has been sketched by Alfven (1971). Now that we have found that space, far from being void and structureless, is "filled with plasmas, intersected by sheathlike discontinuities, and permeated by a complicated pattern of electric currents and electric and magnetic fields" perhaps we are equipped to go back to the initial steps of solar system formation (Figure 2.1) and construct a general theory of *hetegonic processes* or the formation of companion systems from the primordial plasma.

2.2 THE COMPOSITION OF THE PLANETS

The chemistry of Earth is the subject of this book, so we will skip that planet here and examine the chemistry of the other planets in order.

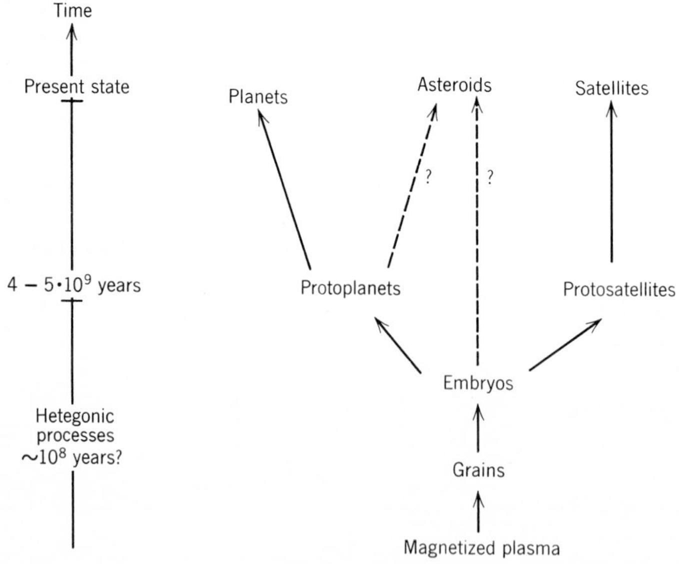

Figure 2.1. Events leading to the formation of the solar system (from Alfven © 1971 by the American Association for the Advancement of Science with permission of the AAAS).

Mercury

This planet appears to have little atmosphere (Table 2.1). Because of its small mass and high surface temperatures ((Table 2.1), even its dark side has an average temperature of 111 K (Murdock and Ney, 1970)), any products of degassing should rapidly escape into space. Its density, and thus presumably its gross chemical composition, is similar to that of Earth.

Venus

In contrast to naked Mercury, Venus is shrouded by clouds so heavy that its surface is hidden from visual observation (it can, however, be examined by radar (Goldstein and Rumsey, 1970)). But again the planet's density indicates that it probably does not differ greatly compositionally from the Earth. The cloud cover consists mostly of carbon dioxide with some particulate material. The nature of these particulates is presently the subject of some speculation. Since CO_2^+ is believed to be the major ionic constituent of the Venusian atmosphere (McElroy, 1969; Herman, Hartle, and Bauer, 1971), Aikin (1972) has proposed that micrometer-sized aggregates could be formed by the clustering of CO_2^+ with CO_2 and other molecules including N_2 and O_2 (Whitten, Poppoff, and Sims, 1971) and even H_2O. Kuiper (1968) has attributed the haze to particles of NH_4Cl, but Hansen and Arking (1971) eliminate this possibility along with SiO_2, NaCl, and $FeCl_2$. They also point out that the refractive indices of pure water droplets and ice also eliminate them. This leaves as candidates polymers of carbon suboxide, $(C_3O_2)_n$, and aqueous solutions of HCl, but these authors conclude that "the likelihood of either is not high. A new look at the question of the Venus cloud composition seems in order."

Mars

More than any other object in the exosphere, even the Sun and the Moon, Mars has incited the imagination of man, for of all the worlds within his ken, it remains the one most likely to be inhabited by life. As for the internal structure of Mars, after reviewing earlier theories (Jeffreys, 1937; Ramsay, 1948; Bullen, 1949; Urey, 1952; Kovach and Anderson, 1965; Reynolds and Summers, 1969; Binder, 1969) and in the light of recent findings Ringwood and Clark (1971) have proposed the Martian model diagrammed in Figure 2.2. The elemental composition is similar to that of Earth, as in the case of Earth there has been differentiation into a mantle and a core in contrast to some earlier models, but the core is not liquid (hence the absence of a magnetic field). The major dissimilarity between the two planets accounting for the observed density and moments of inertia differences is that overall Mars is in a more highly oxidized condition than the Earth. This nicely accounts for its reddish color (ferric iron in minerals such as limonite, olivine, and clinopyroxenes) and the presence of CO_2 and H_2O in its atmosphere. The atmosphere of Mars is quite attenuated—about one-tenth that of Earth—and consists mostly of CO_2 with a trace of water vapor (Owen and Mason, 1965; Anon., 1971) with the water vapor tending to be concentrated over the polar regions (Hanel et al., 1972), but no nitrogen (Delgarno and McElroy, 1970), this gas having escaped (Brinkman, 1971). Carbon monoxide and monatomic hydrogen and oxygen have been observed by ultraviolet measurements on the upper layers of the atmosphere (Barth et al., 1972). In addition to

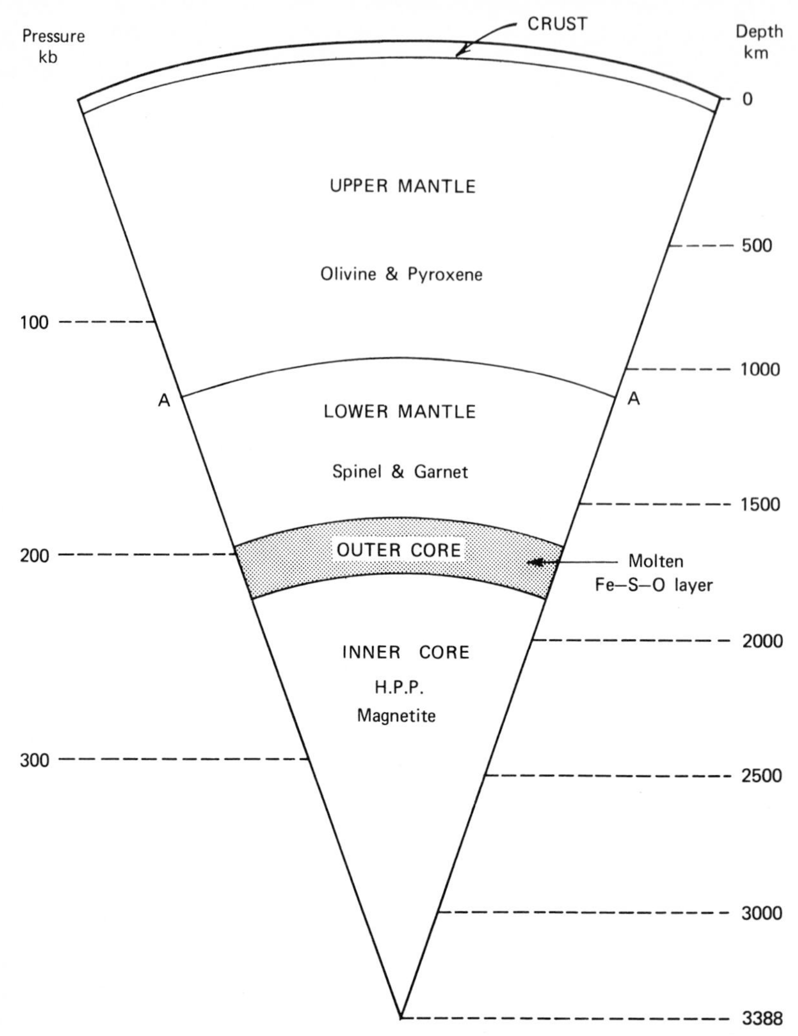

Figure 2.2. Hypothetical section through Mars (from Ringwood and Clark, 1971, with permission of the publisher and authors).

implying the absence of extensive photosynthetic activity, the predominance of CO_2 in the Martian atmosphere is also commensurate with the paucity of available water on the planet, since on Earth CO_2 reacts with crustal silicates in the presence of water. There can be a great deal of particulate material, mostly SiO_2 (55–65%, Hanel et al., 1972; Steinbacher et al., 1972) in the Martian atmosphere (Chase et al., 1972), and recent rocket studies encountered an enormous planet-wide dust storm (Masursky et al., 1972; Parkinson and Hunten, 1972).

The seasonal advance and retreat of the white polar caps of Mars appear to be accompanied by color changes in other parts of the planet. In particular, the so-called "maria" darken as the polar caps diminish. It has been theorized that moisture from the melting caps produces blooms of low plant forms such as fungi and lichens, but it is also possible to explain these color changes in purely inorganic terms. The polar caps are believed to be rather thin, more a frost than an ice cap, and there has been a great deal of speculation as to their chemical composition with suggestions including, in addition to water and dry ice, clathrates of carbon dioxide (Miller and Smythe, 1970) and ozone trapped in solid CO_2 (Broida et al., 1970; Barth and Hord, 1971).

But, the big question is, of course, Is there liquid water on Mars? For liquid water is necessary for life (Horne, 1971—I am what Professor Sagan would call a "water chauvinist") and where there is water there may be life. Ingersoll (1970) has argued that because the evaporation rate is so high there is little likelihood of pure liquid water formation from the melting of ice, unless liquid water exists in the form of concentrated aqueous solutions of strongly deliquescent salts. Schorn et al. (1969) have found fairly abundant water vapor on Mars—the humidity can exceed 50%—and have suggested that there may be transitory appearances of liquid water possibly as some form of precipitation (recent photographs have shown sinuous markings that look very much like water-eroded river courses). Aerosol precipitation could be responsible for the abrupt clearings of the Martian blue haze (Wells and Hale, 1971). For all its scarcity of liquid water, there may still be life on our nearest neighboring planet. Such a possibility is well worth exploring (Radmer and Kok, 1971), but in the light of recent discoveries, compared to Earth, Mars now appears to be a geologically young and active planet (Hammond, 1972), so the first men on Mars can expect to find, not the ruins of a dead civilization, but rather rudimentary forms of life. They will find that they have not gone forward but rather backward in the biological evolutionary time scale.

Jupiter

The first of the major planets, Jupiter, like its brothers presents an entirely different chemical aspect from the inner terrestrial planets. It has a small, solid(?) core enveloped by a gigantic atmosphere thousands of miles thick of reduced gases, hence its low density (Table 2.1). This atmosphere consists mostly of H_2 and, by inference from the observed mean molecular weight of 3.3, He with about 0.3 mole % CH_4 and 0.015 mole % NH_3. Vividly colored changing striations indicate current belts flowing parallel to the equator of the planet. Many hypotheses have been advanced to account for these colorations including solutions of organic molecules or Na in liquid NH_3 droplets; purines and pyrimidines; ammonium sulfide, ammonium hydrosulfide (NH_4SH), and ammonium polysulfides ($(NH_4)_2S_x$) (Lewis, 1969; Lewis and Prinn, 1970); and polymeric compounds formed by the reaction of HCN with NH_3 (Woeller and Ponnamperuma, 1969; Münch and Neugebauer, 1971). In any event, it is becoming increasingly clear that there may be a great deal of very complex organic chemistry going on in the upper atmosphere of Jupiter, including the possible synthesis of biologically significant substances (Sagan et al., 1967). The most famous and mysterious feature of the planet is its Red Spot, an enormous discontinuity some 30,000 miles long and 7000 to 8000 miles wide that appears to be floating in the Jovian atmosphere and whose color varies from a dull gray to an intense brick red. Could it be a raft of solid molecular hydrogen colored by trace impurities and drifting on a sea of helium-rich liquid hydrogen (Streett, 1969; Smoluchowski, 1970)? The nature of Jupiter's core is also unsettled, and suggestions have ranged all the way from a dense metallic core similar to Earth's and covered by a mantle of ice to a liquid (and hence the observed magnetic field) metallic hydrogen–helium core encased in a double mantle of solid metallic hydrogen and solid H_2–He (Smoluchowski, 1967).

Lewis (1971) has considered the satellites of Jupiter. The model he proposes is a dense core of hydrous silicates and iron oxides and a deep liquid mantle covered by a thick crust of ice. Temperatures above the ice–ammonia eutectic are maintained in their interiors, he believes, by the decay of radioisotopes of K, U, and Th.

Saturn, Uranus, and Neptune

These planets are, generally speaking, chemically similar to Jupiter. The most spectacular feature of Saturn is, of course, its magnificent rings. These are believed to be composed of irregularly shaped particulate grains, possibly ice crystals. Not listed in Table 2.1 as constituents of these planetary atmospheres are He and NH_3. The He is probably there, but the NH_3 is frozen out. The large difference in density between Uranus and Neptune (Table 2.1) is noteworthy: Water may undergo a transition to a metallic state at pressures between the central pressures of the two planets.

Pluto

As we noted, Pluto is a stranger. Its characteristics are not well known (Table 2.1). It may be a chunk of basalt-like material.

2.3 METEORITES, ASTROIDS, AND COMETS

On an average night, a sky watcher may see some 5 to 10 "shooting stars" per hour. These firey trails signal the death of meteors as they enter the Earth's atmosphere some 70 to 110 km in altitude with velocities ranging from 10 to 70 km/sec (Hawkins, 1964) and are consumed. At times, spectacular showers of meteors are observed, and these phenomena appear to be associated with the appearance of comets. Other meteors are not so linked; in general, the smaller meteors may be frequently associated with comets, whereas the larger meteors are not and may be associated with the fragmentation of asteroids. The range of meteor sizes is wide—extremely fine dust (Buddhue, 1950), on the one hand, to chunks miles in diameter on the other. The former rain is detectable by special means only while collisons with large objects (Krinov, 1966) have gouged enormous craters in the surface of the Earth (and much more visibly on the face of the Moon) and, so speculation runs, may even have produced catastrophes so great as to alter the poles of this planet, its climate, and the course of biological evolution (Gallant, 1964). The total quantity of material swept out of space by the Earth as it races around the Sun is very large—estimates place the material increment of the Earth by meteoritic infall between 1000 and 10,000 tons daily (Mason, 1958).

Prior to our conquest of the Moon, our knowledge of the chemical nature of the exosphere was based largely on remote techniques such as spectroscopy. Meteorites were the sole exception, the only samples of matter from the "outside" that we had in hand for laboratory examination. It is not surprising then to find that their chemical composition has been intensively studied. I might mention in passing that spectral examination of the tracks of shooting stars has also contributed mightily to our knowledge of the chemistry of the upper atmosphere of Earth (Whipple, 1952). We must also bear in mind that passage through the atmosphere alters the morphology and surface chemistry even of those space visitors who survive the experience and that those falls not recovered immediately are further altered by terrestrial weathering (Buddhue, 1957). Chemically, meteorites fall into two main (and a third intermediate) groups—iron meteorites and stony meteorites (Table 2.3)—that are believed according to one hypothesis to correspond to the core and crust, respectively, of a terrestrial type planet probably once occupying the position of the astroid belt between Mars and Jupiter. In recent times the notion of the asteroids as the debris of a shattered planet has lost adherents and the two types of meteorite are traced now to the core and crust of asteroids rather than to a planet. Of special recent interest are carbonaceous chondrites. Some scientists suspect they are very primitive matter having condensed out of the original nebula at the birth of the solar system. The ratio of iron to stony meteorites found is about 5 to 4, but this is misleading. Iron meteorites are much easier to spot because their appearance differs from that of common crustal rocks. Actually, about 20 times more stony than iron meteorites fall. The chemical compositions of stony and iron meteorites are summarized in Tables 2.4 and 2.5 (see also Urey, 1964). In Table 2.4, note the similarity to terrestrial igneous rocks. Etching of the cut surface of iron meteorites reveals handsome Widmanstätten patterns that are typical for the slow crystallization of iron–nickel alloys at high temperature (Axon, 1968). A great deal of effort has also been devoted to age determination of meteorites and some of these findings are listed in Tables 2.6 and 2.7 (see also Patterson, 1956; Anders, 1962; Fireman and Goebel, 1970).

Reflection spectra of Vesta, Pallas, and other asteroids have yielded evidence of Fe(II) in a magnesium pyroxene. The composition of these asteroids is similar to certain basaltic achondrites, but their mineralogy is "distinctly different from that of other meteorite types and from samples of the lunar surface" (McCord, Adams, and Johnson, 1970).

As objects from the "outside," a great deal of fascination has focused on meteorites as possibly containing traces, if not of life, at least of the synthesis of organic molecules (Duchesne, 1969). Carbonaceous chondrites contain hydrocarbons (Hayes, 1967). Careful examination of the recently fallen and relatively uncontaminated Murchison meteorite (Kvenvolden et al., 1970) has revealed the presence of pyrimidines, but no purines or triazines (Folsome et al., 1971)—all heterocyclic compounds of great importance in pre-biological chemical evolution. Pering and Ponnamperuma (1971), using gas chromatography and mass spectrometry, have detected 14 polynuclear aromatic hydrocarbons in this meteorite including naphthalene and a number of its methylated compounds, phenanthrenes, anthracene, and pyrene. However, it appears to be fairly certain that these materials are the products of nonbiological high-temperature synthesis. Carbon also occurs in carbonaceous chondrites as carbonates (DuFresne and Anders, 1962) and, curiously enough, the isotopic carbon ratio is not only greater for the carbonates than for the hydrocarbons ($\delta^{13}C = +40$ to $+70$ compared with -15 to -17/mil) (Deines, 1968; Smith and Kaplan, 1970), but the difference between the two types of carbon forms is much greater than for terrestrial organic matter and carbonates. A number of possible mechanisms for this isotopic fractionation have been proposed (Clayton, 1963), in particular Studier, Hayatsu,

Table 2.3 Types of Meteorites (Reprinted from *Meteorites* by Heide (1964) by permission of The University of Chicago Press.)

I. Stones[a]

Silicates Predominate Over Metallic Constituents

	Principal Minerals	Examples
Chondrites		
Enstatite chondrites	Enstatite with some nickel-iron	Hvittis, Finland
Bronzite-olivine-chondrites	Bronzite, olivine, and some nickel-iron	Forest City, Iowa Pultusk, Poland
Hypersthene-olivine-chondrites	Hypersthene, olivine, and some nickel-iron	Holbrook, Arizona Mocs, Rumania
Achondrites		
Calcium-poor		
Aubrites	Enstatite	Aubres, France
Ureilites	Clinobronzite, olivine	Novo-Urei, U.S.S.R.
Diogenites[b]	Hypersthene	Johnstown, Colorado
Amphoterites[c] & Rodites	Hypersthene, olivine	Rhoda, Spain
Chassignites	Olivine	Chassigny, France
Calcium-rich		
Angrites	Augite	Angra dos Reis, Brazil
Nakhlites	Diopside, olivine	Nakhla, Egypt
Eucrites[d] and Shergottites	Clinohypersthene, anorthite; the same with maskelynite instead of plagioclase	Stannern, Czechoslovakia Shergotty, India
Howardites[e]	Hypersthene, olivine, clinohypersthene, anorthite	Pasamonte, New Mexico
Siderolites[f]		
Transitional between stones and irons: silicates predominant		
Lodranites	Bronzite, olivine, nickel-iron	Lodran, India
Mesosiderites	Bronzite, olivine, nickel-iron	Estherville, Iowa
Grahamites[g]	Bronzite, olivine, nickel-iron, with plagioclase	Vaca Muerta, Chile

(*Continued*)

Table 2.3 (*continued*).

II. Irons Metallic Constituents Predominant or Exclusive		
	Principal Minerals	Examples
Lithosiderites[f]		
Transitional between irons and stones; metal predominant		
Siderophyres[h]	Bronzite, tridymite, troilite, nickel-iron	Steinbach, Germany
Pallasites[i]	Olivine, nickel-iron troilite	Krasnojarsk, U.S.S.R.
Hexahedrites	Kamacite, troilite, schreibersite	Braunau, Czechoslovakia
Octahedrites (with coarsest, coarse, medium, fine, and finest lamellae)	Kamacite, taenite, troilite, schreibersite	Sikhote-Alin, U.S.S.R. Ogg, Canyon Diablo, Arizona, Og. Toluca, Mexico, Om. Gibeon, S. W. Africa, Of. Bristol, Tennessee, Off.
Ataxites		
Nickel-poor ataxites	Nickel-iron	Chesterville, S. Carolina
Nickel-rich ataxites	Nickel-iron	Babbs Mill, Tennessee Hoba, S. W. Africa

[a] May be subdivided further according to color (white, intermediate, gray, black), structure (crystalline, brecciated, veined), and peculiarities of composition (carbonaceous).

[b] After Diogenes of Apollonia.

[c] After Greek amphoteroi = containing both (olivine and hypersthene)

[d] After Greek eucritos = plain (mineral content easily determinable because of large grain size).

[e] After the chemist Howard

[f] After Greek sideros = iron, lithos = stone

[g] After the chemist Graham

[h] After Greek sideros = iron, phyrao = knead

[i] After the explorer Pallas

and Anders (1968) have shown that nearly all of the organic compounds that have been found in meteorites can be produced by a Fischer–Tropsch type synthesis

$$nCO + (2n + 1)H_2 \rightarrow C_nH_{2n+2} + nH_2O \quad (2.1)$$

$$2nCO + (n + 1)H_2 \rightarrow C_nH_{2n+2} + nCO_2 \quad (2.2)$$

between CO, H_2, and NH_3 in the presence of an iron–nickel catalyst, and more recently Lancet and Anders (1970) have demonstrated in the laboratory that such a

Table 2.4 Chemical Composition of Stony Meteorites (Reprinted from *Meteorites* by Heide (1964) by permission of the University of Chicago Press.)

Element	Chondrite (%)	Igneous Rocks (%)
Oxygen (O)	34.84	46.60
Iron (Fe)	25.07	5.00
Silicon (Si)	17.78	27.72
Magnesium (Mg)	14.38	2.09
Sulfur (S)	2.09	0.05
Calcium (Ca)	1.39	3.63
Nickel (Ni)	1.34	0.08
Aluminum (Al)	1.32	8.13
Sodium (Na)	0.68	2.83
Chromium (Cr)	0.25	0.02
Carbon (C)	0.1	0.03
Potassium (K)	0.084	2.59
Cobalt (Co)	0.08	0.002
Titanium (Ti)	0.066	0.44
Phosphorus (P)	0.05	0.12

Table 2.5 Chemical Composition of Iron Meteorites (Reprinted from *Meteorites* by Heide (1964) by permission of the University of Chicago Press.)

	Percent		Percent
Iron (Fe)	89.7	Phosphorus (P)	0.18
Nickel (Ni)	9.10	Carbon (C)	0.12
Cobalt (Co)	0.62	Sulfur (S)	0.08
Copper (Cu)	0.04		

[These values were determined in selected samples of virtually pure metal phase, as nearly free from troilite, schreibersite, and cohenite inclusions as possible. However, such inclusions, though very unevenly distributed, make up an appreciable part of the meteorite; for example, the average troilite content of irons has been variously estimated as 1.4 to 5%, corresponding to 0.5 to 2.2% sulfur.]

synthesis at 400°K does indeed give rise to a kinetic isotopic fractionation similar to that observed in meteorites.

While least obvious to the untrained observer, micrometeorites representing the smallest range of meteoritic material sizes may have the largest effect on the chemistry of Earth. The nickel content of deep sea sediments is far greater than one would expect from the usual terrestrial sources and may be of such an extraterrestrial origin (Petterson and Rotschi, 1952); however, this hypothesis has been the subject of considerable criticism, and other sources have been proposed (Laevastu and Mellis,

Table 2.6 Helium Ratio Ages of Meteorites (Reprinted from *Meteorites* by Heide (1964) by permission of The University of Chicago Press.)

No.	Meteorite	Class[a]	^4He 10^{-6} cm^3/g at STP	$^3He/^4He$	Cosmic Ray Exposure Age (million yr)
1	Aroos, U.S.S.R.	C. Oct.	25.4	0.258	530
2	Braunau, Czechoslovakia	Hex.	0.34	0.103	8
3	Grant, New Mexico	F. Oct.	19.9	0.276	590
4	Sikhote-Alin, U.S.S.R.	C. Oct.	1.65	0.231	220
5	Treysa, Germany	M. Oct.	21.0	0.326	310
6	Kunashak, U.S.S.R.	Ch.	1.5	0.033	2.8
7	Beardsley, Kansas	Ch.	13.6	0.0067	6
8	Bjurböle, Finland	Ch.	16.7	0.010	12
9	Holbrook, Arizona	Ch.	18.3	0.015	18
10	Richardton, North Dakota	Ch.	15.1	0.022	22
11	St. Michel, Finland	Ch.	6.25	0.050	21
12	Ensisheim, Alsace	Ch.	13.4	0.022	20
13	Farmington, Kansas	Ch.	1.7	0.003	0.2
14	L'Aigle, France	Ch.	9.4	0.015	9
15	McKinney, Texas	Ch.	1.3	0.041	4
16	Pultusk, Poland	Ch.	13.0	0.0077	7
17	St. Marks, South Africa	Enst. Ch.	6.0	0.0022	1
18	Sioux Co., Nebraska	Ho	54.0	0.0044	16
19	Stannern, Czechoslovakia	Eu	79.2	0.0045	24

[a] C. Oct., course octahedrite; Hex, hexahedrite; F, fine; M, medium; Ch, chondrite; Enst. Ch., Enstatite chondrite; Ho, howardite; Eu, eucrite.

1955; Öpik, 1955; Smales and Wiseman, 1955; Smales, Mapper, and Wood, 1957; Yamakoshi and Tazawa, 1971). In any event, moving up the size scale a bit, magnetic Fe–Ni–Co spherules of probable extraterrestrial origin have been found widely distributed over the entire Earth, in fresh water and marine sediments and on the land masses (Castaing and Fredriksson, 1958; Petterson and Fredriksson, 1958; Nriagu and Bowser, 1969; and Millard and Finkelman, 1970).

Equally controversial is the origin of glasslike spherules or tektites (and even smaller, <1 mm, particles called microtektites) with O'Keefe (1963, 1970a, b, c, 1971) proposing that they may represent material spallated from the Moon by meteoritic impact or spewed up by lunar volcanoes, and Urey (1971a, b) insisting with some vigor that they are terrestrial in origin. Tektites and microtektites have been found in many places including deep sea sediment cores (Glass, 1968; Cassidy, Glass, and Heezen, 1969); however, the facts that their distribution on Earth, unlike the iron spherules, appears to be spotty and that their chemical composition differs from the lunar samples but does resemble that of terrestrial material (Zähringer, 1963; Frey, Spooner, and Baedecker, 1970) (Figure 2.3) argues strongly in support of the view of Barnes (1969) and Urey (1971a, b) that they originate here on Earth. Compositional differences among tektites indicate that they may have more than one possible origin (Glass, 1970).

Table 2.7 Potassium–Argon Ages of Meteorites (Reprinted from *Meteorites* by Heide (1964) by permission of The University of Chicago Press.)

No.	Meteorite and Class[a]	Potassium (Parts per million)	Argon 10^{-6} cm^3/g at STP	Potassium–Argon Age AE	Uranium–Helium Age AE
1	Kunashak, U.S.S.R., Ch., gray part	650	2.4	0.72	0.55
2	Kunashak, U.S.S.R., Ch., black part	900	6.5	1.2	—
3	Pervomaisky, U.S.S.R., Ch., gray part	800	2.5	0.65	0.63
4	Pervomaisky, U.S.S.R., Ch., black part	1,250	15.2	1.8	0.94
5	Beardsley, Kansas, Ch.	1,010	71.3	4.3	3.6
6	Forest City, Iowa	831	53.4	4.15	4.1
7	Bjurböle, Finland, Ch.	840	60.5	4.32	4.2
8	Holbrook, Arizona, Ch.	880	66.5	4.4	4.4
9	Marion, Iowa, Ch.	870	54.0	4.08	2.35
10	Mocs, Rumania, Ch.	870	62.0	4.30	2.4
11	Modoc, Kansas, Ch.	830	55.0	4.18	3.0
12	Richardton, North Dakota, Ch.	830	54.0	4.15	3.8
13	St. Michel, Finland, Ch.	910	54.0	4.00	1.9
14	Chateau Renard, France, Ch.	835	1.93	0.51	0.3
15	Ensisheim, Alsace, Ch.	199	8.0	3.48	3.75
16	Farmington, Kansas, Ch.	850	3.55	0.83	0.71
17	L'Aigle, France, Ch.	903	50.8	4.03	3.0
18	McKinney, Texas, Ch.	826	15.0	2.32	0.45
19	Pultusk, Poland, Ch.	770	40.8	3.93	3.42
20	St. Marks, South Africa, Ch.	757	36.6	3.78	2.45
21	Sioux Co., Nebraska, Ho.	326	11.4	3.26	2.95
22	Stannern, Czechoslovakia, Eu.	690	30.0	3.79	3.6

[a] Ch, chondrite; Ho, howardite; Eu, eucrite.

I neglected to mention that there is more He in meteorites than one expects. This excess He results from cosmic radiation of the meteoritic material to produce α-particles (Bauer, 1947, 1948) and considerably complicates determining the age of these materials by the helium ratio technique.

A fascinating feature concerning the distribution of meterorites is that no meteorite or meteoritic remains have been found in geological deposits older than the Pliocene, and despite the great quantities of coal mined in the last century, no meteorites have been found therein. Furthermore, collections of strange objects assembled by Stone Age men include tektites but no iron meteorites despite their unusual appearance. Thus the advent of meteorites may be relatively recent in geological time, and as the Earth sweeps its orbit clear of such material, meteoritic falls may cease entirely.

Another of the heaven's spectaculars are comets (Porter, 1952; Lyttleton, 1953; Richter, 1963; Ley, 1969). Morphologically, a typical comet (Figure 2.4) consists of a nucleus surrounded by a coma, with a gas tail always directed away from the Sun

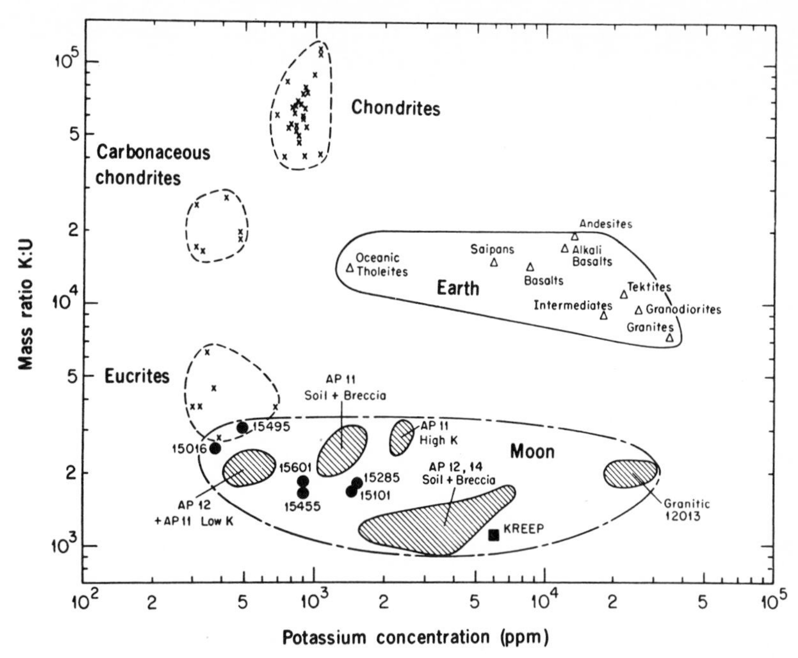

Figure 2.3. Values of the mass ratio K:U as a Function of the Concentration of K (from O'Keefe et al., © 1972 by the American Association for the Advancement of Science, with permission of the AAAS).

and a dust tail. Multiple nuclei and tails are not unusual. The nucleus may be a condensed mass, say 1 km in diameter. The tail of the 1910 Halley's Comet was 3×10^7-km long. Solar energy vaporizes the material of the nucleus to form the coma and tails, and spectral analysis has revealed a variety of free radicals and ions in the coma, including CN, NH, NH_2, H, OH, OH^+, CH, CH_2, C_2, and C_3, where the material has a sufficiently low density to prevent the reaction of the free radicals, and in the gas tail CO^+, N_2^+, CO_2^+ and molecules such as CO, CO_2, and N_2 (Hawkins, 1964). As for the chemical composition of the nucleus, Whipple (1950) has proposed that it might consist of a conglomeration of various ices and hydrates. Vaporization

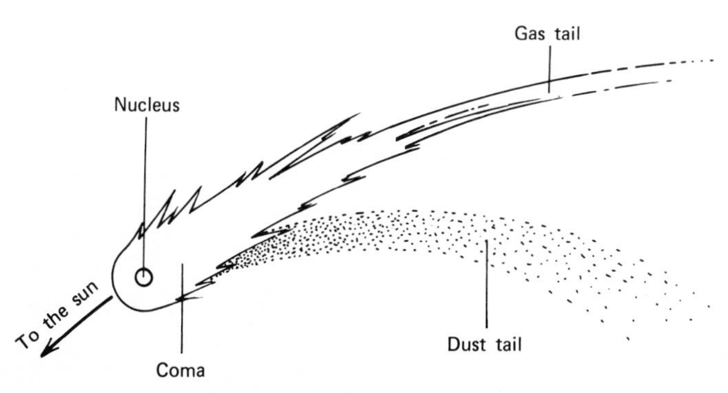

Figure 2.4. Structure of a comet.

of these ices followed by photodissociation of H_2O and photoexcitation of H and OH can then account for the observed hydrogen and hydroxyl halo (Delsemme, 1971). A second model pictures a comet as a sort of flying gravel bank or swarm of small particles of composition similar to that of the interstellar medium.

2.4 LUNAR CHEMISTRY

As the largest object in the night sky and as a heavenly light of great romantic and peculiar beauty, man has wondered about the topography and chemistry of the Moon ever since the ancient Greeks persuaded him that it was not a goddess. Children once were told that it is made of cheese, and in the fantasy of H. G. Wells the first men on the Moon were captured by its antlike inhabitants and bound with fetters of pure gold, a common metal there. In recent years, our planet's companion in space has been the subject of a great deal of more scientific speculation. But we will ignore this material; we have no need of it now. For on July 26, 1969, the American space exploration craft Apollo 11 brought back to Earth samples collected from the Moon (for a highly readable account of the recovery, analyses, and early discussion of these lunar rocks see Cooper, 1970). The unanchored speculation was over, the hard work begun. Except for meteorites, these samples were the first extraterrestrial objects that we have been able to get our hands on. In this section on lunar chemistry, we will restrict ourselves entirely to the results of the first examinations of these and subsequent samples.

The enormity of the human achievement represented by those first unprepossessing-looking, dust-smeared rocks staggers the mind. Verbal communication, social organization, mastery of fire, the invention of the wheel, the alphabet, science, the mechanical engine, the computer, mastery of the atom, and now, and in many ways the most momentous of all, the first short but ever so marvelous step in the exploration of the exosphere. I am filled with the most enormous gratitude and pride because my own species, my own countrymen even, in my own times, have so advanced our understanding of the infinitesimally small atom and the infinitely vast universe. If only Lucretius could be summoned back from the majority for his tongue alone could begin to celebrate these conquests adequately!

Unfortunately our success in making this step, in leaving our planet for the first time and reaching into space, while illustrating the glory and the power of human intelligence, has also illumined new depths of human stupidity.* Not only has the significance of the achievement entirely escaped some critics, they have had the temerity to suggest publically that the effort might have been better spent on certain political goals. We do not know what was or was not in the stomach of Periclean Athenians, but we are the heirs of what was in their minds. Rome's bread lines are of no consequence now, but her fallen ruins enrich our lives. The mission of our species is not anyone's comfort; it is the mastery, that is to say the exploration and understanding, of the natural order. As for price tags, no cost is too great for this end.

* The criticism and fear excited by human achievement is even reflected in mythology. Prometheus, the first to master fire, is punished by the gods. I wonder what did happen to the first man to use fire, to the man who invented the wheel? If they were murdered by their ignorant fellows, I would not be in the least surprised.

Far from resolving the mysteries of the Moon, the samples brought back by Apollo 11, 12, 14, and 15 missions have raised as many questions as they have answered. Lunar material is distinctly different from terrestrial material (Figure 2.3), and it is quite clear that the geological history of the Earth's satellite is quite different from that of our planet, but whether the "hot moon" (Baldwin, 1970) or the "cold moon" (Urey, 1952, 1971a, b) hypothesis of the Moon's history is the more accurate remains unsettled. It now appears that although the Moon may once have been hot, if has in any event been cold and inactive for a long period of time. Similarly, we cannot say whether the Moon was torn from the Earth (largely from the lighter crust with relatively little of the heavier core material, thus accounting for the Moon's lesser specific gravity), whether the Earth acquired its companion from space, or whether both bodies accreted from space material more or less contemporaneously although the majority opinion presently seems to incline toward the last hypothesis.

Residual magnetism evidences that the Moon's center may have once been molten, yet massive unmixed subsurface lumps or "mascons" (gravity anomalies) would indicate that the melting might not have been general. There are several lines of very convincing evidence pointing to the conclusion that extensive chemical differentiation has taken place on the Moon, and seismic experiments hint that the crust is layered. The upper 1 to 2 km of crust appears to be composed of broken rocks and rock fragments. Below this and to a depth of about 25 km, the sound velocity is comparable to that of the basaltic material that characterizes the lunar maria, and the maria may have been filled by past outflows of this material. A third layer extending down to about 65 km has a higher sound velocity. There is an electrical conductivity discontinuity at 250 km (Sonnett et al., 1971) possibly marking an Fe–S layer between core and mantle (Murthy, Evenson, and Hall, 1971). Strangely enough, radioactive materials in the Moon's surface appear to be highly concentrated in a few areas such as the *Mare Imbrium*, and the heat flux from the Moon's interior is much higher than expected on the basis of radioactive heating. The thermal activity of the Moon appears to have been mostly confined to the early periods of its history: (1) widespread melting at about the time of the Moon's formation (4.6×10^9 yr), (2) partial melting (4.1×10^9 yr) possibly in a series of incidents to form KREEP basalts enriched in potassium, rare earths, and phosphorus, and (3) latter* lava flooding of preexisting lunar basins, possibly caused by large meteoritic impact, to form the maria, finally followed by a period, continuing to the present time, of thermal inactivity and slow cooling. The iron-rich basalts of the maria have low, the lunar highland plagioclase-rich materials high Al/Si ratios. The latter are relatively poor in Fe and rare earths. It should be noted that, because of the high meteorite impact rate, very little of the original crustal rocks of the Moon may have survived intact.

Turning now in greater detail to the chemical and mineralogical composition of the lunar materials, especially the samples from Tranquillity Base brought back by Apollo 11, these samples were composed of basaltic, igneous rocks, microbreccias (a mixture of soil particles and rock fragments compacted into a coherent rock), and soil. The soil in addition to crystalline fragments contained glassy fragments and small fragments of iron meteorites (LSAPT, 1970). The average lunar meteoritic influx

* [3.1×10^9 to 3.8×10^9 yr on the basis of K/Ar, Rb/Sr, and Pb isotopic ratios (Schaeffer et al., 1970; Albee et al., 1970; Tatsumoto, 1970; Silver, 1970). For a series of papers on age determinations on Apollo 11 samples, see *Science, 167* (3918) (Jan. 30, 1970) and for a review paper see Wetherill, 1971.]

rate has been estimated to be about 4×10^{-9} g/cm^2 yr (Ganapathy, Keays, and Anders, 1970). There was a controversy whether or not tektite glass occurs in Apollo 12 samples (O'Keefe, 1970a, b, c; King, Martin, and Nance, 1970). The Moon's surface is very dusty. Much of this dust is certainly the product of the extensive non-hydrothermal erosion processes to which the Moon's surface materials are continuously subjected, but the high density of nuclear tracks in an appreciable fraction of the dust particles may indicate that they were irradiated in space and are thus extralunar in origin (Barber, Hutcheon, and Price, 1971). Pits and other surface erosion by high velocity particulate impacts are much in evidence on the surface of lunar rocks. The surface of lunar material is subjected to a great deal of abuse from bombardment by solar wind (Brandt, 1970), meteorites, and cosmic rays (Marti, Lugmair, and Urey, 1970; Stoenner, Lyman, and Davis, 1970; Eberhardt et al., 1970; Crozas et al., 1970; Shedlovsky et al., 1970; Chao et al., 1970; Arrhenius et al., 1970; von Engelhardt et al., 1970; Quaide, Bunch, and Wrigley, 1970; Short, 1970; Sclar, 1970; Bibring et al., 1972; James, 1972). Many rocks are glazed with a glassy coating whose elemental composition may be significantly different from their interiors (Morgan et al., 1971), and, in addition to the impacts just noted, other hypotheses have been advanced for the production of these glazes ranging from a giant solar outburst (Gold, 1969, 1970; Mueller and Hinsch, 1970) to volcanic action (Green, 1970). As noted earlier, detailed chemical analysis has revealed dissimilarities between lunar materials and tektites found on Earth (Showalter et al., 1972). The bombardment and radiation of lunar surface materials produces a number of isotopes within them, including noble gases (Eberhardt et al., 1970; Funkhouser et al., 1970; Heymann et al., 1970; Hintenberger et al., 1970; Kirsten et al., 1970; Pepin et al., 1970; Reynolds et al., 1970; Stoenner, Lyman, and Davis, 1970; Marti and Lightner, 1972; Podosek, Huneke, and Wasserburg, 1972). These results have opened up a whole new realm of the chemistry of the interaction of objects in space with the surrounding exosphere, a chemistry from which the Earth's atmosphere spares it and of which, because of their ablation as they enter the Earth's atmosphere, previous studies of meteorites could at best provide only a hint.

Table 2.8 lists the percentages of the main lithic fragments found in the lunar soil, Table 2.9, their bulk compositions, and Table 2.10, some of the minerals found in the samples. The most common minerals are pyroxene (often highly zoned with iron-rich rims), plagioclase, ilmenite, olivine, and cristobalite. Apollo 12 samples contained more orthopyroxene-calcic plagioclase fragments than did the Apollo 11 soil (Fuchs, 1970). For a description of the Apollo 14 samples see LSPET (1971) and for Apollo 15, AFPET (1972). Free metallic iron and troilite, both very rare on Earth, are common acessory minerals in the igneous lunar rocks; in addition, three new minerals were discovered—pyromanganite (a triclinic pyroxene-like mineral), ferropseudobrookite, and a chromium-titanium spinel (LSAPT, 1970*a*). In the absence of hydrothermal weathering such as occurs on Earth, the lunar silicate minerals tend to be very handsome and clear. These silicate minerals crystallized under very dry highly reducing (10^{-13} atm partial pressure of oxygen) conditions (LSAPT, 1970*b*). Figure 2.5 compares elemental abundances in the lunar materials with cosmic abundances. Volatile elements to the right of the periodic table are conspicuously depleted, whereas elements to the left are enriched. Pursuing a suggestion of Anders (1970), Singer and Bandermann (1970) interpret these depletions as resulting from the slow accretion of volatile materials from a solar nebula rather than from a heating and

Table 2.8 Proportions of Rock Types Among 1676 Lithic Fragments in the Apollo 11 Soil Sample (From J. A. Wood, *J. Geophys. Res.*, *75*, 6497 (1970), © American Geophysical Union with permission of the AGU and the author.)

Rock Type	Percent	Totals
Crystalline basalts	37.4	
Basaltic glasses	4.3	
Basaltic breccias	52.4	94.1
Crystalline anorthosites	2.0	
Anorthositic glasses	1.5	
Anorthositic breccias	1.5	5.0
Other (including meteorite fragments)		0.9
		100.0

evaporation process (Walter and Carron, 1969; Brown and Peckett, 1971), and they argue that this accretion occurred not while the Moon was orbiting about the Earth but rather prior to its capture by our planet. On the other hand, despite certain dissimilarities (Figure 2.3), Fanale and Nash (1971) and Fisher (1971) hold that the potassium–uranium systematics of lunar and terrestrial rocks are compatible with Earth and Moon accreting from the same portion of the presolar cloud. Mercury is of particular interest, since it volatilizes at temperatures realized during the lunar day (Reed, Goleb, and Jovanovic, 1971). Some elements depleted in lunar soils may be enriched in particular locations, possibly due to meteroritic influx; Bi and Cd are observed examples (Ganapathy, Keays, and Anders, 1970). The unexpected enrichment of Ti caused a great deal of consternation to Professor Urey and other proponents of a cold moon and glee among hot moon advocates (Cooper, 1970). However we must bear in mind that when we speak of chemical differentiation and *igneous* lunar rocks that it has not been possible to distinguish unambiguously between once widespread molten moon material and localized material that has been melted by particular lunar events such as massive meteoretic impacts. Table 2.11 gives elemental abundances in several types of Apollo 11 materials.

The Lunar Sample Preliminary Examination Team (1970*a*, *b*) not unexpectedly found no evidence of water in the Apollo 11 and 12 samples, no traces of aqueous alteration, not even any hydrated silicates. The escape time of a water molecule from the Moon is estimated to be less than 1 yr (Opik, 1962), but some scientists have been reluctant to abandon the hope that there might be water (and thus life?) on the Moon, if not at the present time, at least in the satellite's geological past. Sinuous rills, maria fillings, and other lunar surface features have been attributed to a now vanished hydrosphere (Gilvarry, 1960; Lingenfelter, Peale, and Schubert, 1968; Urey, 1968, 1969), and it has even been proposed that some of this ancient water may have survived in shadow areas (Watson, Murray, and Brown, 1961), permafrost layers (Gold, 1964), or ice-filled pingos. More recently, Anders (1970) has reexamined the question, "Did the Moon ever contain large amounts of water?" On the basis of

Table 2.9 Bulk Compositions of Crystalline and Vitreous Anorthositic Fragments in Apollo 11 Soil (From J. A. Wood, *J. Geophys. Res.*, *75*, 6497 (1970), © American Geophysical Union with permission of the AGU and the author.)

	Anorthosite (37–9)[a]	Anorthosite (19–1)[a]	Anorthosite (37–7)[a]	Gabbroic Anorthosite (37–7)[a]	Anorthositic Gabbro (19–57)[a]	Anorthositic Gabbro (9–8)[a]	Colorless Glasses[b]
SiO_2	44.2	46.6	45.4	43.9	47.8	46.0	46.1
TiO_2	0.0	0.1	0.0	0.3	0.4	0.3	0.9
Al_2O_3	34.6	32.5	33.8	32.9	25.5	27.3	24.9
Cr_2O_3	0.0	0.0	0.0	0.1	0.1	0.2	0.1
FeO	1.1	1.6	2.8	4.0	6.2	6.2	5.9
MnO	0.0	0.0	0.1	0.0	0.1	0.1	0.1
MgO	1.3	2.3	1.7	3.8	7.2	7.9	7.3
CaO	18.3	17.7	17.5	17.0	14.6	14.1	14.2
Na_2O	0.7	0.6	0.4	0.5	0.4	0.3	0.6
K_2O	0.1	0.0	0.0	0.1	0.1	0.0	0.1
NiO	0.0	0.0	0.0	0.1	0.0	0.0	0.0
S	0.0	0.1	0.0	0.1	0.1	0.0	0.0
Sum	100.3	101.5	101.7	102.8	102.5	102.4	100.2

[a] Average of 7 to 10 defocused beam (50 μm) electron microprobe analyses, randomly placed on sectioned fragment.

[b] Average of defocused beam analyses of 10 colorless glasses.

Table 2.10 List of Minerals in Apollo 11 Samples (From G. M. Brown, *J. Geophys. Res.*, *75*, 6480 (1970). © American Geophysical Union with permission of the AGU and the author.)

Mineral Name	General Formula	Comments
Clinopyroxene	$Ca(Mg, Fe)Si_2O_6$	Most abundant mineral. Zoned from calcic augite to subcalcic ferraugite. Pigeonite and hedenbergite also present.
Pyroxferroite	$(Fe, Ca)_2Si_2O_6$	Triclinic pyroxenoid. New mineral.
Olivine	$(Mg, Fe)_2SiO_4$	Fo_{80} to Fo_{50} range. Low Ni, high Cr. Fayalite (Fo_0) as rare, interstitial grains.
Plagioclase feldspar	$CaAl_2Si_2O_8$–$NaAlSi_3O_8$	An_{96} to An_{70} range. Generally zoned.
Alkalic feldspar	$(K, Na)AlSi_3O_8$	Rare interstitial grains, potassic variety.
Cristobalite	SiO_2	Common. Inverted to low-temperature form.
Tridymite	SiO_2	Rare needles, in same rocks as cristobalite. Usually inverted to quartz.
Quartz	SiO_2	Very rare.
Fluorapatite	$Ca_5(PO_4)_3F$	Present in most rocks in small amounts. Chlorapatite also recorded.
Whitlockite	$Ca_3(PO_4)_2$	Rare.
Ilmenite	$FeTiO_3$	Abundant. Some magnesian (geikielo-ilmenite) varieties.
Armalcolite	$(Fe, Mg)Ti_2O_5$	New mineral.
Ulvöspinel	Fe_2TiO_4	Al and Cr also significant.
Titanochrome spinel	$Fe(Cr, Ti)_2O_4$	Mg, Al also significant. No terrestrial equivalent. Charge balance may require Ti^{3+}.
Rutile	TiO_2	Lamellae in ilmenite and residua needles.
Spinel	$MgAl_2O_4$	Very rare.
Troilite	FeS	Hexagonal.
Kamacite	Fe	Low Ni in igneous rocks.
Taenite	Fe, Ni	High Ni(15 wt %) in soil, breccias, and spherules.
Copper	Cu	Tentative identification of traces in several samples.
Baddelyite	ZrO_2	Rich in Hf.
Zircon	$ZrSiO_4$	Rare.
Dysanalyte	$(Ca, Na, Fe, Ce)(Nb, Ti)O_3$	May be another member of the perovskite group.
Graphite	C	One grain.
Schreibersite	$(Fe, Ni)_3P$	Very rare intergrowths with taenite.
Cohenite	Fe_3C	Very rare intergrowths with taenite.
Amphibole	$Na_3(Fe, Mg)_5Si_8O_{22}(OH, F)$	Sodic richterite. One grain in vug.
Aragonite	$CaCO_3$	One fragment, from coarse basalt.
Magnetite	Fe_3O_4	Titaniferous variety. Mössbauer spectra suggests doubtful trace.
Mica	$K_2(Mg, Fe)_5(Al, Ti)(Si, Al)_8O_{20}(OH)_4$	Formula assumes Ti^{3+} and takes analysis as showing Si:Al = 5:3. One grain, tentative analysis.

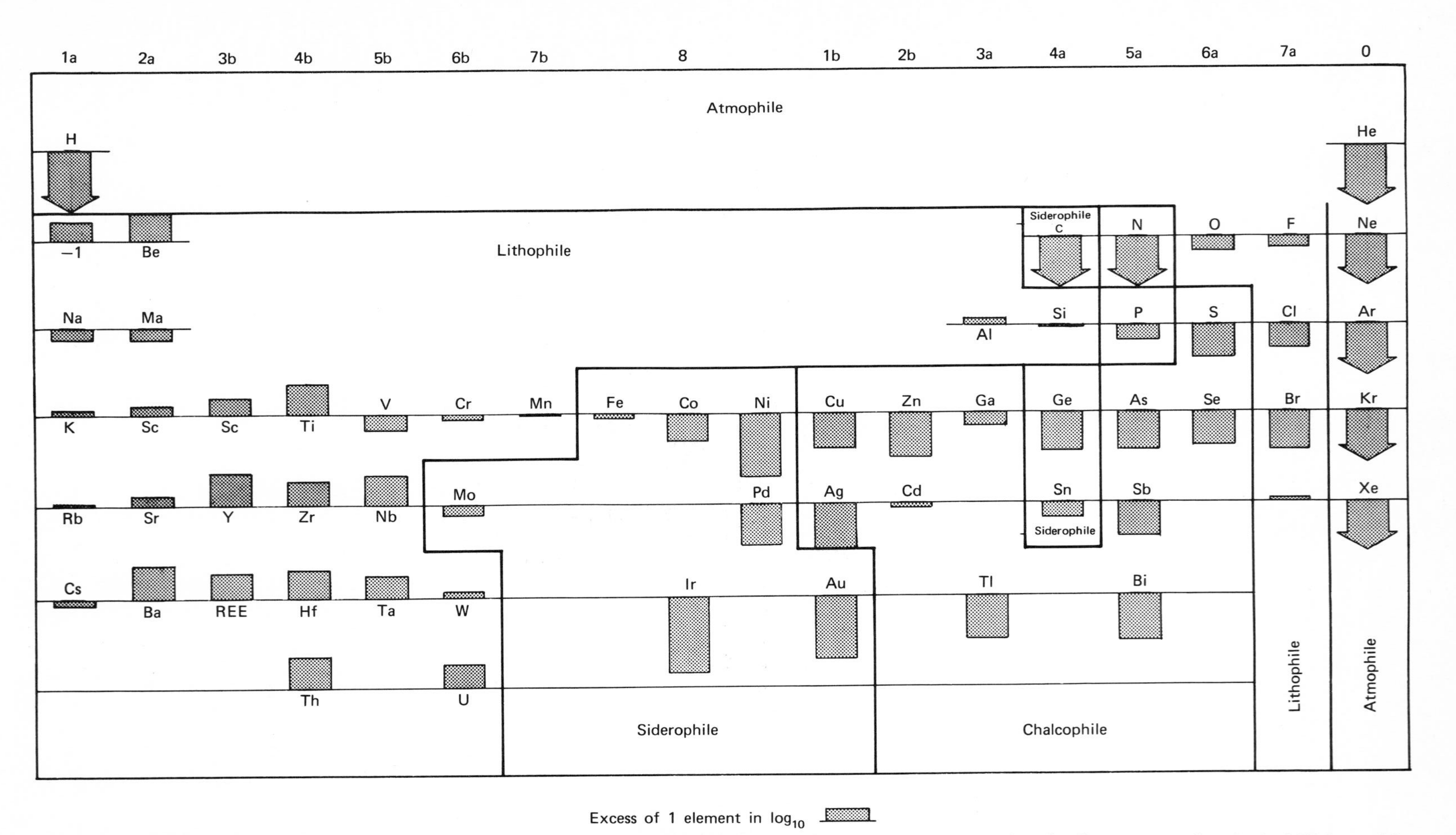

Figure 2.5. Excess of lunar (Apollo 11) over cosmic abundances (from O'Keefe, J. A., *J. Geophys. Res.*, *75*, 6565 (1970), © American Geophysical Union, with permission of the AGU and the authors).

Table 2.11 Some Elemental Abundances in the First Lunar Rock Samples (Weight %)

	Fine-grained	Coarse-grained	Breccia	Soil
Si	19–22	18–21	19–20	20
Al	4–6	5–6	6	7
Ti	6–7	5–6	5–6	4
Fe	15–16	13–15	14–15	13
Mg	4–5	3–6	3–5	5
Ca	10	10–11	10–11	10
Na	0.3–0.4	0.3–0.4	0.3	0.3
K	0.05–0.3	0.09–0.2	0.09–0.2	0.1
Mn	0.2	0.1–0.2	0.2	0.2
Cr	0.2	0.2	0.1–0.2	0.2
Zr	0.04–0.06	0.04–0.06	0.04–0.06	0.04
Ni	0.04–0.06	0.007–0.008	0.005–0.007	0.02
V	0.004–0.006	0.004–0.007	0.006	0.008
P	0.04–0.07	0.02–0.1	0.03–0.06	0.1
		Selected Trace Elements (ppm)		
F	70–100	50–220	30–80	66
Cl	50–150	50–520	16–150	350
Sc	84–86	64–87	70–97	60
Sr	130–170	170–180	160–180	200
Y	130–250	150–190	180–210	150
Ba	96–300	140–280	100–250	220
Nd	43–120	70–72	82–96	46

an accretion model of lunar origin and the observed low abundances of relatively volatile metals such as Pb, Bi, In, and Tl in lunar rocks, he sets a maximum initial hydrosphere layer depth of 3.7 m, which is the equivalent to a carbonaceous chondritic layer 9-m thick, and he adds that "in reality, the amount [of water] would be even less." But the question is by no means settled—on March 7, 1971, suprathermal detectors left at the Apollo 12 and 14 landing sites observed what may have been a small cloud of water released from the Moon's surface, possibly by some sort of seismic event.

As for that other crucial ingredient of life, carbon and its compounds, Apollo 11 fines contained 64 to 230 ppm of carbon (Moore et al., 1970). Apollo 12 fines contained about 110 $\mu g/g$ of carbon of which 7 to 21 $\mu g/g$ was in the form of carbides and 2 $\mu g/g$ as methane (Chang et al., 1971; Nagy et al., 1971), and the ^{13}C studies indicated that extensive isotopic fractionation had occurred. Possibly this great enrichment of ^{13}C is a consequence of "hydrogen-stripping" by solar wind (Kaplan and Smith, 1970; Berger, 1970). Besides methane, the samples also contained CO and C_2H_6. Burlingame et al. (1971) found the former to be the major carbon species present in the Apollo 11 fines. The observed carbon levels are approximately what one expects from the action of the solar wind (LSAPT, 1970*b*). This also accounts for the much larger carbon content of the fine lunar surface material than of lunar rocks (10 to 50 ppm) (Holland et al., 1972; Pillinger et al., 1972).

The summary of the Apollo 11 findings (LSAPT, 1970b) notes that "the search for important protobiological compounds . . . was carried on with some of the most sophisticated and sensitive analytic techniques ever devised. Nevertheless, no unambiguous identification of indigenous compounds was made at extremely low levels of detection (usually < 10 ppb)." Rho et al. (1972) found no porphyrins, yet Ponnamperuma et al. (1970) reported minute traces (approx. 10^{-4} $\mu g/g$), possibly contamination from rocket exhaust (Simoneit et al., 1969). In my own opinion, the abiotic high temperature synthesis of these complex and biologically crucial substances is in itself quite as significant to exobiology and terrestrial biogenesis as would have been their discovery on the Moon. Subsequent investigations (Nagy et al., 1970; Harada et al., 1970) however revealed the presence of amino acids, and substances hydrolyzable to amino acids in Apollo 11 and 12 materials at levels in the range 20 to 70 ppb. The principle amino acids detected in these experiments were glycine and alanine. The results are comparable to those obtained for the Murchison meteorite (discussed previously), and the amino acid profile of the lunar specimen is distinctly different from that of a human handprint, thus rendering unlikely that source of contamination.

Even while I have been writing this chapter (May 1972), a new large load of lunar materials has been returned to Earth. These samples may resolve some of the mysteries concerning the origin and evolution of the Moon, but, like their predecessors. they will almost certainly raise as many questions as they settle. Of particular interest is a sample taken from a permanently "shaded" location—perhaps it will not have lost all its volatiles, perhaps it may even contain traces of more complex organic molecules.

BIBLIOGRAPHY

AFPET, *Science*, *175*, 363 (1972).

A. C. Aikin, *Nature Phys. Sci.*, *235*, 10 (1972).

A. L. Albee et al., *Science*, *167*, 463 (1970).

H. Alfven, *Science*, *172*, 991 (1971).

E. Anders, *Rev. Mod. Phys.*, *34*, 287 (1962).

E. Anders, *Science*, *169*, 1309 (1970).

Anon., *Nature*, *234*, 500 (1971).

G. Arrhenius et al., *Science*, *167*, 659 (1970).

H. J. Axon, *Metallurgy of Meteorites*, Pergamon, New York. 1968.

R. B. Baldwin, *Science*, *170* 1297 (1970).

D. J. Barber, I. Hutcheon, and P. B. Price, *Science*, *171*, 372 (1971).

V. E. Barnes, *Geochim. Cosmochim. Acta*, *39*, 1121 (1969).

C. A. Barth and C. W. Hord, *Science*, *173*, 197 (1971).

C. A. Barth et al., *Science*, *175*, 309 (1972).

C. A. Bauer, *Phys. Rev.*, *72*, 354 (1947); *72*, 501 (1948).

R. Berger, *Nature*, *226*, 738 (1970).

J. P. Bibring et al., *Science*, *175*, 753 (1972).

A. B. Binder, *J. Geophys. Res.*, *74*, 3110 (1964).

J. C. Brandt, *Introduction to the Solar Wind*, Freeman, San Francisco, 1970.

R. T. Brinkman, *Science*, *174*, 944 (1971).
H. P. Broida et al., *Science*, *170*, 1402 (1970).
G. M. Brown, *J. Geophys. Res.*, *75*, 6480 (1970).
G. M. Brown and A. Peckett, *Nature*, *234*, 262 (1971).
J. D. Buddhue, *Meteoritic Dust*, Univ. New Mexico Press, Albuquerque, 1950.
J. D. Buddhue, *The Oxidation and Weathering of Meteorites*, Univ. New Mexico Press, Albuquerque, 1957.
K. E. Bullen, *Mon. Not. Roy. Astron. Soc.*, *109*, 457, 688 (1949).
A. L. Burlingame et al., *Science*, *167*, 751 (1972).
W. A. Cassidy, B. P. Glass, and B. C. Heezen, *J. Geophys. Res.*, *74*, 382 (1969).
R. Castaing and K. Fredriksson, *Geochim. Cosmochim. Acta*, *14*, 114 (1958).
S. Chang, K. Kvenvolden, J. Lawless, and C. Ponnamperuma, *Science*, *171*, 474 (1971).
E. C. T. Chao et al., *Science*, *167*, 644 (1970).
S. C. Chase et al., *Science*, *175*, 308 (1972).
R. N. Clayton, *Science*, *140*, 192 (1963).
H. S. F. Cooper, Jr., *Moon Rocks*, Dial, New York, 1970.
G. Crozas et al., *Science*, *167*, 563 (1970).
A. Dalgarno and M. B. McElroy, *Science*, *170*, 167 (1970).
P. Deines, *Geochim. Cosmochim. Acta*, *32*, 613 (1968).
A. H. Delsemme, *Science*, *172*, 1126 (1971).
J. Duchesne, *Sci. J.*, *5*, 33 (1969).
E. R. DuFresne and E. Anders, *Geochim. Cosmochim. Acta*, *26*, 1085 (1962).
P. Eberhardt et al., *Science*, *167*, 558 (1970).
W. von Engelhardt et al., *Science*, *167*, 669 (1970).
F. P. Fanale and D. B. Nash, *Science*, *171*, 282 (1970); *172*, 1167 (1971).
E. L. Fireman and R. Goebel, *J. Geophys. Res.*, *75*, 2115 (1970).
D. E. Fisher, *Science*, *172*, 1166 (1971).
C. E. Folsome et al., *Nature*, *232*, 108 (1971).
F. A. Frey, C. M. Spooner, and P. A. Baedecker, *Science*, *170*, 845 (1970).
L. H. Fuchs, *Science*, *169*, 866 (1970).
J. G. Funkhouser et al., *Science*, *167*, 561 (1970).
R. L. Gallant, *Bombarded Earth*, Baker, London, 1964.
R. Ganapathy, R. R. Keays, and E. Anders, *Science*, *170*, 533 (1970).
J. J. Gilvarry, *Nature*, *188*, 886 (1960).
B. P. Glass, *Science*, *161*, 891 (1968).
B. P. Glass, *Science*, *169*, 766 (1970).
T. Gold, in P. J. Brancazio and A. G. W. Cameron (eds.), *The Origin and Evolution of Atmospheres and Oceans*, Wiley, New York, 1964.
T. Gold, *Science*, *165*, 1345 (1969); *168*, 611 (1970).
L. Goldberg, E. A. Muller, and L. H. Aller, *Astrophys. J. Suppl. Ser.* *5*, *45*, 1 (1960).
R. M. Goldstein and H. Rumsey, Jr., *Science*, *169*, 975 (1970).
J. Green, *Science*, *168* 608 (1970).
A. L. Hammond, *Science*, *175*, 286 (1972).
R. A. Hanel et al., *Science*, *175*, 305 (1972).
J. E. Hansen and A. Arking, *Science*, *171*, 669 (1971).

K. Harada et al., *Science*, *167*, 433 (1970).
G. S. Hawkins, *The Physics and Astronomy of Meteors, Comets, and Meteorites*, McGraw-Hill, New York, 1964.
J. M. Hayes, *Geochim. Cosmochim. Acta*, *31*, 1395 (1967).
F. Heide, *Meteorites*, Univ. Chicago Press, Chicago, 1964.
J. R. Herman, R. E. Hartle, and S. J. Bauer, *Planet. Space Sci.*, *19*, 443 (1971).
D. Heymann et al., *Science*, *167*, 555 (1970).
H. Hintenberger et al., *Science*, *167*, 543 (1970).
P. T. Holland et al., *Nature Phys. Sci.*, *235*, 106 (1972).
R. A. Horne, *Space Life Sci.*, *3*, 34 (1971).
A. P. Ingersoll, *Science*, *168*, 972 (1970).
O. B. James, *Science*, *175*, 432 (1972).
H. Jeffreys, *Mon. Not. Roy. Astron. Soc.*, *Geophys. Suppl.*, *4*, 62 (1937).
I. R. Kaplan and J. W. Smith, *Science*, *167*, 541 (1970).
E. A. King, Jr., R. Martin, and W. B. Nance, *Science*, *170*, 199 (1970).
T. Kirsten et al., *Science*, *167*, 571 (1970).
R. L. Kovach and D. L. Anderson, *J. Geophys. Res.*, *70*, 2873 (1965).
E. L. Krinov, *Giant Meteorites*, Pergamon, Oxford, 1966.
G. Kuiper, *Comm. Lunar Planet. Lab.*, Nos. 100–104, *6*, 229 (Univ. Arizona), (1968–69).
K. Kvenvolden et al., *Nature*, *228*, 5273 (1970).
T. Laevastu and O. Mellis, *Trans. Amer. Geophys. Union*, *36*, 385 (1955).
M. S. Lancet and E. Anders, *Science*, *170*, 980 (1970).
J. S. Lewis, *Icarus*, *10*, 365 (1969).
J. S. Lewis, *Science*, *172*, 1127 (1971).
J. S. Lewis and R. G. Prinn, *Science*, *169*, 472 (1970).
W. Ley, *Visitors from Afar: The Comets*, McGraw-Hill, New York, 1969.
R. E. Lingenfelter, S. J. Peale, and G. Schubert, *Science*, *161*, 266 (1968).
LSAPT, *Science*, *167*, 1325 (1970*a*).
LSAPT, *Science*, *167*, 449 (1970*b*).
LSPET, *Science*, *165*, 1211 (1969); *167*, 1325 (1970).
LSPET, *Science*, *173*, 681 (1971).
R. A. Lyttleton, *The Comets and Their Origin*, Cambridge Univ. Press, Cambridge (England), 1953.
K. Marti and B. D. Lightner, *Science*, *175*, 421 (1972).
K. Marti, G. W. Lugmair, and H. C. Urey, *Science*, *167*, 548 (1970).
B. Mason, *Principles of Geochemistry*, 2nd ed., Wiley, New York, 1958.
H. Masursky et al., *Science*, *175*, 294 (1972).
T. B. McCord, J. B. Adams, and T. V. Johnson, *Science*, *168*, 1445 (1970).
M. B. McElroy, *J. Geophys. Res.*, *74*, 29 (1969).
H. T. Millard, Jr., and R. B. Finkelman, *J. Geophys. Res.*, *75*, 2125 (1970).
S. L. Miller and W. D. Smythe, *Science*, *170*, 531 (1970).
C. B. Moore et al., *Science*, *167*, 495 (1970).
J. W. Morgan et al., *Science*, *172*, 556 (1971).
G. Mueller and G. W. Hinsch, *Nature*, *228*, 254 (1970).
G. Münch and G. Neugebauer, *Science*, *174*, 940 (1971).

T. L. Murdock and E. P. Ney, *Science, 170*, 535 (1970).
V. R. Murthy, N. M. Evenson, and H. T. Hall, *Nature, 234*, 267 (1971).
B. Nagy et al., *Science, 167*, 770 (1970).
B. Nagy et al., *Nature, 232*, 94 (1971).
J. O. Nriagu and C. J. Bowser, *Water Res., 3*, 833 (1969).
J. A. O'Keefe (ed.), *Tektites*, Univ. Chicago Press, Chicago, 1963.
J. A. O'Keefe, *Science, 168*, 1209 (1970*a*).
J. A. O'Keefe, *Science, 170*, 200 (1970*b*).
J. A. O'Keefe, *J. Geophys. Res., 75*, 6565 (1970*c*).
J. A. O'Keefe, *Science, 171*, 313 (1971).
G. D. O'Kelley et al., *Science, 175*, 440 (1972).
E. J. Opik, *Nature, 176*, 926 (1955).
E. J. Opik, *Planet. Space Sci., 9*, 211 (1962).
T. Owen and M. P. Mason, *Science, 165*, 3895 (1965).
T. D. Parkinson and D. M. Hunten, *Science, 175*, 323 (1972).
C. Patterson, *Geochim. Cosmochim. Acta, 10*, 230 (1956).
R. O. Pepin et al., *Science, 167*, 550 (1970).
K. L. Pering and C. Ponnamperuma, *Science, 173*, 237 (1971).
H. Petterson and K. Fredriksson, *Pacific Sci., 12*, 71 (1958).
H. Petterson and H. Rotschi, *Geochim. Cosmochim. Acta, 2*, 81 (1952).
C. T. Pillinger et al., *Nature Phys. Sci., 235*, 108 (1972).
F. A. Podosek, J. C. Huneke, and G. J. Wasserburg, *Science, 175*, 423 (1972).
C. Ponnamperuma et al., *Science, 167*, 760 (1970).
J. G. Porter, *Comets and Meteor Showers*, Wiley, New York, 1952.
W. Quaide, T. Bunch, and R. Wrigley, *Science, 167*, 671 (1970).
R. Radmer and B. Kok, *Science, 174*, 233 (1971).
W. H. Ramsay, *Mon. Not. Roy. Astron. Soc., 108*, 406 (1948).
G. W. Reed, J. A. Goleb, and S. Jovanovic, *Science, 172*, 258 (1971).
J. H. Reynolds et al., *Science, 167*, 545 (1970).
R. T. Reynolds and A. L. Summers, *J. Geophys. Res., 74*, 2494 (1969).
J. H. Rho et al., *Science, 167*, 754 (1972).
N. B. Richter, *The Nature of Comets*, Methuen, London, 1963.
A. E. Ringwood and S. P. Clark, *Nature, 234*, 89 (1971).
C. E. Sagan et al., *Nature, 213*, 273 (1967).
O. A. Schaeffer, J. G. Funkhouser, and J. Zähringer, *Science, 170*, 161 (1970).
R. A. Schorn et al., *Icarus, 11*, 286 (1969).
C. B. Sclar, *Science, 167*, 675 (1970).
J. P. Shedlovsky et al., *Science, 167*, 574 (1970).
N. M. Short, *Science, 167*, 673 (1970).
D. L. Showalter et al., *Science, 175*, 170 (1972).
L. Silver, *Geochim. Cosmochim. Acta, Suppl. 2*, 1533 (1970).
B. R. Simoneit et al., *Science, 166*, 733 (1969).
S. F. Singer and L. W. Bandermann, *Science, 170*, 438 (1970).
A. A. Smales, D. Mapper, and A. J. Wood, *Analyst, 82*, 75 (1957).

A. A. Smales and J. D. H. Wiseman, *Nature*, *175*, 464 (1955).
J. W. Smith and I. R. Kaplan, *Science*, *167*, 1376 (1970).
R. Smoluchowski, *Nature*, *215*, 691 (1967).
R. Smoluchowski, *Science*, *168*, 1340 (1970).
C. P. Sonnett et al., *Nature*, *230*, 359 (1971).
R. H. Steinbacher et al., *Science*, *175*, 293 (1972).
R. W. Stoenner, W. J. Lyman, and R. Davis, Jr., *Science*, *167*, 553 (1970).
W. B. Streett, *J. Atmos. Sci.*, *26*, 924 (1969).
M. H. Studier, R. Hayatsu, and E. Anders, *Geochim. Cosmochim. Acta*, *32*, 151 (1968).
M. Tatsumoto, *Geochim. Cosmochim. Acta*, *Suppl. 2*, 463 (1970).
A. Unsöld, *Trans. Internat. Astron. Union*, *7*, 460 (1950).
H. C. Urey, *The Planets*, Yale Univ. Press, New Haven, 1952.
H. C. Urey, *Rev. Geol. Chem.*, *2*, 1 (1964).
H. C. Urey, *Naturwissenschafe*, *2*, 49 (1968).
H. C. Urey, *Science*, *164*, 1088 (1969).
H. C. Urey, *Science*, *171*, 312 (1971*a*).
H. C. Urey, *Science*, *172*, 403 (1971*b*).
L. S. Walter and M. K. Carron, *Geochim. Cosmochim. Acta*, *28*, 937 (1964).
K. Watson, B. Murray, and H. Brown, *J. Geophys. Res.*, *66*, 1598 (1961).
E. H. Wells and D. P. Hale, *Nature*, *232*, 324 (1971).
G. W. Wetherill, *Science*, *173*, 383 (1971).
F. Whipple, *Astrophys. J.*, *111*, 375 (1950).
F. Whipple, *Bull. Amer. Meteor. Soc.*, *33*, 13 (1952).
R. C. Whitten, I. G. Poppoff, and J. S. Sims, *Planet. Space Sci.*, *19*, 243 (1971).
G. L. Wick, *Science*, *173*, 1011 (1971).
F. Woeller and C. Ponnamperuma, *Icarus*, *10*, 386 (1969).
J. A. Wood, *J. Geophys. Res.*, *75*, 6497 (1970).
K. Yamakoshi and Y. Tazawa, *Nature*, *233*, 542 (1971).
J. Zähringer, in *Radioactive Dating*, p. 289, International Atomic Energy Agency, Vienna, 1963.
H. Zirin, *The Solar Atmosphere*, Blaisdell, Waltham, Mass., 1966.

ADDITIONAL READING

I. Adler and J. I. Tromka, *Geological Exploration of the Moon and Planets*, Springer-Verlag, New York, 1970.
H. Brown and C. C. Patterson, *J. Geol.*, *56*, 85 (1948).
D. E. Evans, D. E. Pitts, and G. L. Kraus, *Venus and Mars Nominal Natural Environment for Advanced Manned Planetary Mission Program*, Sci. Tech. Info. Div., NASA, U.S. Gov. Print. Off., Washington, D.C., 1965.
G. Fielder, *Geology and Physics of the Moon*, Univ. Lancaster Press, Lancaster (England), 1971.
V. A. Firsoff, *Strange World of the Moon*, Basic Books, New York, 1960.
S. Glasstone, *The Book of Mars*, Sci. Tech. Info. Div., NASA, U.S. Gov. Print. Off., Washington, D.C., 1968.
F. L. Jackson and P. Moore, *Life on Mars*, Norton, New York, 1956.

R. Jastrow and S. I. Rasoo, *The Venus Atmosphere*, Gordon and Breach, New York, 1969.

H. Jeffreys, *Proc. Roy. Soc., London Ser. A, 214*, 281 (1952).

H. Jeffreys, *The Earth*, Cambridge Univ. Press, New York, 1970.

W. W. Kellogg and C. Sagan, *The Atmospheres of Mars and Venus*, Nat. Acad. Sci. U.S.A.–Nat. Res. Council Pub. 944 (1961).

L. R. Koenig (ed.), *Handbook of the Physical Properties of the Planet Venus*, Sci. Tech. Info. Div., NASA, U.S. Gov. Print. Off., Washington, D.C., 1967.

Z. Kopal, *An Introduction to the Study of the Moon*, Reidel, Dordrecht, Holland, 1966.

G. P. Kuiper (ed.), *The Atmospheres of the Earth and Planets*, Univ. Chicago Press, Chicago, 1952.

A. A. Levinson (ed.), *Lunar Science Conferences, Houston, Texas, 1970 and 1971.* (1) Pergamon, New York, 1970; (2) M.I.T. Press, Cambridge, 1971.

A. A. Levinson and S. R. Taylor, *Moon Rocks and Minerals*, Pergamon, New York, 1971.

B. Mason, *Meteorites*, Wiley, New York, 1962.

B. Mason and W. G. Melson, *The Lunar Rocks*, Wiley-Interscience, New York, 1970.

B. M. Middlehurst and G. P. Kuiper (eds.) *The Moon, Meteorites, and Comets*, Univ. Chicago Press, Chicago, 1963.

P. Moore, *The Planet Venus*, Faber & Faber, London, 1956.

C. B. Moore (ed.), *Research on Meteorites*, Wiley, New York, 1962.

T. A. Mutch, *Geology of the Moon*, Princeton Univ. Press, Princeton, 1970.

T. Page and L. W. Page (eds.), *Neighbors of Earth: Planets, Comets, and the Debris of Space*, Macmillan, New York, 1965.

B. M. Peek, *The Planet Jupiter*, Faber & Faber, London, 1958.

A. E. Ringwood, *J. Geophys. Res., 75*, 6453 (1970).

J. W. Salisbury and P. E. Glaser (eds.), *Lunar Surface Materials Conference, Boston, 1963*, Academic, New York, 1964.

W. Sanders, *The Planet Mercury*, Macmillan, New York, 1963.

H. C. Urey, *Phys. Rev. 80*, 295 (1950).

H. C. Urey, *Geochim. Cosmochim. Acta, 1*, 209 (1951).

H. C. Urey, *Geochim. Cosmochim. Acta, 2*, 269 (1952).

H. C. Urey, *J. Geophys. Res., 61*, 394 (1956).

H. C. Urey and H. Craig, *Geochim. Cosmochim. Acta, 4*, 36 (1953).

H. C. Urey and G. J. F. MacDonald, in Z. Kopal (ed.), *Physics and Astronomy of the Moon*, 2nd ed., Academic, New York, 1971.

K. H. Wedepohl, *Geochemistry*, Holt, Rinehart, and Winston, New York, 1971.

F. L. Whipple, *Earth, Moon, and Planets*, 3rd ed., Harvard Univ. Press, Cambridge, 1968.

J. A. Wood, *Meteorites and the Origin of Planets*, McGraw-Hill, New York, 1968.

III

THE LITHOSPHERE

3

THE CHEMICAL COMPOSITION OF THE EARTH

3.1 INTRODUCTION

The subject matter of this book is the chemistry of our own planet, Earth. Even though we have just examined the most remote part of the environment, the exosphere, in some, perhaps too much, detail, the central emphasis of this book is upon *our* environment, the environment most immediate to the origin, well-being, and fate of our species. Those aspects of the lithosphere closest to our interests are dealt with elsewhere—sedimentation, ion-exchange equilibria, and other surface geological processes in Chapter 13 on the lithosphere–hydrosphere interactions and the chemical relationships between the biosphere and lithosphere, including the effects of human agriculture, in Chapter 14. Consequently, the present chapter may seem somewhat truncated to many readers. I will be sketchy just where some authors have been most detailed, and I can afford to do this because of the numerous excellent and very highly detailed monographs available in the field of geochemistry (a few of the best known are listed as references and in the additional reading bibliography at the end of this chapter).

3.2 THE STRUCTURE AND COMPOSITION OF INNER EARTH

Man has now set foot on the Moon, nearly 400,000 km away, but he has penetrated the Earth under our very feet only a few kilometers. He has sent probes to Venus, Mars, and the more distant planets, yet he has barely scratched the surface of his home planet. Our knowledge of the structure and chemical composition of the interior of Earth is inferred almost entirely from seismic measurements, with a little help from our chemical analyses of meteorites, heat flow, and gravitational data, and the precession of the equinoxes (Birch, 1952). Seismic waves from earthquakes are of three types: (*a*) primary, longitudinal, small amplitude, compressional *P*-waves; (*b*) secondary, transverse, larger amplitude, shearing *S*-waves; and (*c*) large amplitude surface waves. The refraction and absorption of these waves has revealed that the Earth is a complex onion consisting of a solid innermost core surrounded by a liquid core (Figure 3.1). This core material is then covered by a thick mantle, the superficial skin of which is called the crust, or together with the upper mantle in more exact terminology—the lithosphere proper. Table 3.1 compares the thicknesses and some physical properties of these regions. The core in the table is the outer liquid core that has a density ranging from 9.4 to 14.2 g/cm^3.

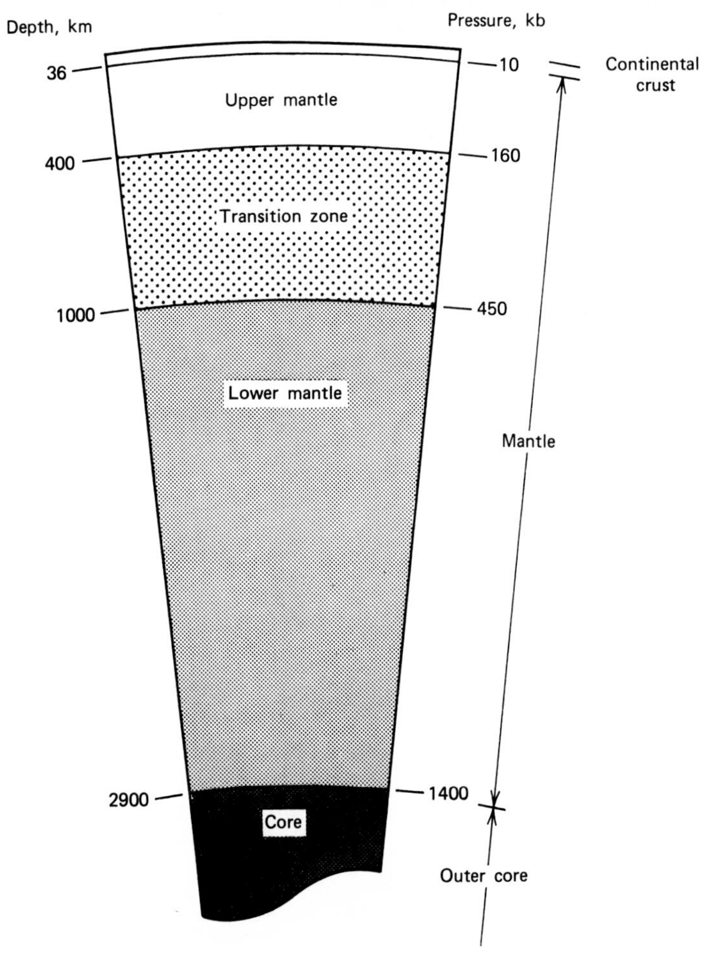

Figure 3.1. The internal structure of the Earth (from Mason, 1966).

Measurements in mines and boreholes have shown that the temperature increases with depth about 30°C/km. The source of this heat is the radioactive decay of isotopes of U and Th and ^{40}K, and the heat flux to the surface has been estimated to be 50 cal/cm^2 yr, a value very small compared with the surface influx of solar energy yet large compared with the thermal energy released by global volcanic activity. The Earth's crust beneath the oceans is thinner and less radioactive than beneath the continents. Yet, surprisingly enough, the average oceanic heat flow is not appreciably less than the continental value. We infer from this finding that whereas the continental heat flow originates from the crust, the oceanic source is the mantle and that the radionuclide composition of the mantle must differ beneath the oceans and continents, being less in the latter case. Alternatively, some other means of heat transfer, such as convection may be operating. The temperatures at the bottom of the mantle cannot exceed the melting point of the silicates of which it is largely composed, roughly 10,000°C or possibly, considering the pressure effects on melting points (Boyd, 1964), considerably less. A value of 7500 ± 2000°C is

Table 3.1 Comparison of Various Parts of Earth (From Holland, 1964)

	Thickness (km)	Volume ($\times 10^{27}$ cm^3)	Mean Density (g/cm^3)	Mass ($\times 10^{27}$ g)	Mass (%)
Atmosphere	—	—	—	0.000005	0.00009
Hydrosphere	3.80 (mean)	0.00137	1.03	0.00141	0.024
Crust	30	0.015	2.8	0.043	0.7
Mantle	2870	0.892	4.5	4.056	67.8
Core	3471	0.175	10.7	1.876	31.5
Whole earth	6371	1.083	5.52	5.976	100.00

consistent with the molten iron core. The melting point of iron at core boundary pressures is 5500°C (Mason, 1966).

Of course, no one has ever obtained a fresh sample of the deeper inner Earth for analysis and probably never will; however, it is fairly certain that it must be some sort of iron–nickel alloy. We noted in Chapter 2 that the meteoritic fragments from the core of the disintegrated terrestrial-type planet (or the asteroids) between Mars and Jupiter have this composition. Venus and Mercury are also believed to possess extensive iron cores. Furthermore, recall the spectacularly high cosmic abundance of iron (Figure 1.7). Finally, if we take iron to be the substance of inner Earth, we can readily construct multilayer models of the Earth that can account for its observed density structure and moment of inertia. Well almost, pure iron or Fe–Ni alloys yield a density for the outer core that is 8 to 15% too dense (Birch, 1961; Knopoff and MacDonald, 1960; Birch, 1964; McQueen and Marsh, 1966) and a seismic velocity that is too low (Anderson, Sammis, and Jordan, 1971). In order to decrease the density and increase the sound speed, the addition of a lighter alloying element appears to be necessary, and the two most mentioned candidates are sulfur (Murthy and Hall, 1970) and silicon (Ringwood, 1966*a*, *b*). The choice of the former entails the presumption of a rather cool Earth in order to avoid the escape of volatiles such as sulfur, whereas the latter presumes a hot origin. Density measurements have been made on pressure-shocked Fe–Si alloys (Balchan and Cowan, 1966) but not on Fe–S mixtures. Some of the objections to a Fe–S or chondritic Earth may have been resolved recently (Hall and Murthy, 1971; Lewis, 1971). Although molten, the outer core may have some chemical zoning (Anderson, Sammis, and Jordan, 1971). The temperature-pressure conditions in the Earth's interior may closely approximate the melting curve for iron, and Gardiner and Stacey (1971) and Jain and Evans (1972) have considered the pressure dependence of the electrical conductivity of iron to arrive at estimates of the resistivity of the Earth's core, 100 to 200 μohm-cm, estimates essential to theories of our planet's geomagnetic field. The nature of the inner core is even more of a mystery than that of the outer core. Its boundary appears to be quite sharp (Bolt and Quamar, 1970) at a radius of 1220 km, but more recent seismic overtone work (Julian, Davies, and Sheppard, 1972) has shown that although not molten, the iron may be "soft," comparable to gold or lead with a ratio of shear to bulk modulus of 0.08. Although some models of the inner Earth set its density greater than 17,000 kg/m^3 (see Spar, 1965), Bolt and Quamar (1970) have shown that the density just inside the inner core does not exceed 13,500 kg/m^3.

3.3 THE MANTLE

The mantle of the Earth accounts for 67% of its total mass and 90% of its volume. Although we now have a good idea of its elemental composition, the exact nature of the compounds and phases involved remains a subject of considerable controversy. Seismic measurements have shown that the structure of the mantle is exceedingly complex with possible lateral heterogeneities as well as a number of well-defined layers. Bullen (1967) has identified four of these layers: the upper mantle (33 to 410 km) with normal S and P seismic velocity gradients, a transition region (410 to 1000 km) with greater than normal gradients, another normal gradient zone between 1000 and 2700 km forming the first shell of the lower mantle, and the second shell of the lower mantle (2700 to 2900 km) with gradients near zero.

Our knowledge of the chemical composition of the mantle, like our speculations concerning the Earth's core, is derived from extraterrestrial abundances; density, sound velocity, and other geophysical considerations; and experimental studies of phase transitions under high pressure and temperature. But to this information, another most important aid can be added to resolve the composition of the upper mantle, namely the petrology of ultramafic and basaltic igneous rocks derived from the mantle and brought to the surface by processes such as tectonic movements and the volcanic extrusion of molten magma. The surface specimens taken to be representative of the composition of the mantle must be carefully selected; for example, they must contain the same level of radioisotopes believed to be responsible for the heating and local melting (to form magma) of the mantle. Table 3.2 compares the estimated compositions of the mantle with the core and whole Earth. Table 3.3 lists estimates derived from ultramafic rocks. In summarizing these findings, Wyllie

Table 3.2 Estimated Compositions of the Whole Earth, the Core, and Mantle Using Extraterrestrial Information (weight %) (From Wyllie, 1971*a*)

	Earth			Core		Mantle			
	Washington, 1925	Mason, 1966	Ringwood, 1966*a*, *b*	Mason, 1966	Ringwood, 1966*a*, *b*	Mason, 1966	Oxide	Mason, 1966*a*, *b*	Ringwood, 1966
O	28	30	30			44	SiO_2	48	43
Fe	40	35	31	86	84	9.9	MgO	31	38
Si	15	15	18	—	11	23	FeO	13	9.3
Mg	8.7	13	16	6.0		19	Fe_2O_3	—	—
S	0.64	1.9	—	7.4	—	—	Al_2O_3	3.0	3.9
Ni	3.2	2.4	1.7		5.3	—	CaO	2.3	3.7
Ca	2.5	1.1	1.8			1.7	Na_2O	1.1	1.8
Al	1.8	1.1	1.4			1.6	Cr_2O_3	0.55	—
Na	0.39	0.57	0.9			0.84	MnO	0.43	
Cr	0.20	0.26	—			0.38	P_2O_5	0.34	
Mn	0.07	0.22	—			0.33	K_2O	0.13	
P	0.11	0.10	—			0.14	TiO_2	0.13	
Co	0.23	0.13	—	0.40	—				
K	0.14	0.07	—			0.11			
Ti	0.02	0.05	—			0.08			

Table 3.3 Compositions of Mantle-Derived Ultramafic Rocks (From Wyllie, 1971*a*)

Weight Percent	Continents		Oceans		Nodules		Hypothetical	
	(1)	(2)	(3)	(4)	(5)	(6)	(7)	(8)
SiO_2	44.77	44.65	39.82	43.56	44.18	41.10	42.71	40.3
MgO	39.22	41.66	48.60	41.53	40.95	46.33	41.41	32.7
FeO	8.21	6.81	7.86[a]	7.77[a]	7.34	9.31	6.51	7.1
Fe_2O_3	—	—	1.00[a]	1.00[a]	1.16	1.24	1.57	1.8
Al_2O_3	4.16	3.50	0.87	2.36	2.81	0.56	3.30	3.7
CaO	2.42	2.02	0.37	2.51	2.49	0.17	2.11	2.1
Na_2O	0.22	0.23	0.37	0.32	0.22	0.23	0.49	0.5
K_2O	0.05	0.04	[b]	[b]	0.04	0.03	0.18	0.0(2)
Cr_2O_3	0.40	0.59	0.46	0.40	0.3	0.35	0.45	0.3
NiO	0.24	0.29	0.46	0.34	0.27	0.44	0.42	0.2
CoO	—	—	—	—	—	—	0.02	—
MnO	0.11	0.14	0.10	0.10	0.14	0.15	0.13	0.1
P_2O_5	—	—	0.08	0.07	—	—	0.06	0.1
TiO_2	0.19	0.08	0.01	0.04	0.09	0.08	0.47	0.4
H_2O^+	—	—	—	—	—	—	0.17	9.7
CO_2	—	—	—	—	—	—	—	0.8
Cl	—	—	—	—	—	—	—	0.2
Total	99.99	100.01	100.00	100.00	99.99	99.99	100.00	100.00

[a] Ferrous ferric ratio adjusted so that Fe_2O_3 is 1%.
[b] means less than 0.005.

(1) Green (1964). Average composition for the Lizard peridotite.
(2) Carswell (1968). Mean of three garnet peridotites from Ugelvik, Norway.
(3) Hess and Otalora (1964). Average (D + E)-type serpentinite, recalculated water free, residual type.
(4) Hess and Otalora (1964). Average C-type serpentinite, recalculated water-free.
(5) Harris et al. (1967). Mean of five high calcium, high aluminum olivine nodules.
(6) Harris et al. (1967). Average of three olivine nodules with CaO and Al_2O_3 contents less than 1%, residual nodules.
(7) Green and Ringwood (1963). Pyrolite with 4:1 of, respectively, average anhydrous dunite and the mean of average normal tholeiite and normal alkali basalt.
(8) Nicholls (1967). Composition of volatile-rich parts of the upper mantle, such as may occur beneath the midoceanic ridges.

(1971*a*) notes that (*a*) "more than 90% by weight of the mantle is represented by the system FeO–MgO–SiO_2, and no other oxide exceeds 4%"; (*b*) if the components Na_2O–CaO–Al_2O_3 are added, then this accounts for more than 98%; and (*c*) "no other oxide reaches a concentration of 0.6% in the mantle." He adds that the mineralogy of the mantle can be represented by the system CaO–MgO–Al_2O_3–SiO_2 with FeO substituting for MgO in most magnesium minerals and Na_2O for CaO in plagioclase (see the Mineral Glossary in Appendix A.17) at low pressures and in jadeitic pyroxene at high pressures.

Table 3.4 shows the various minerals and transitions based on the peridotite mantle model of Clark and Ringwood (1964) and Ringwood (1969*a*, *b*, 1970), while Figure 3.2 is a schematic representation of the layered structure of the upper peridotite or eclogite mantles in two different tectonic environments (*L* indicates liquid and *V* water vapor). Phase diagrams for the peridotite system superimposed on the geotherm (dotted line, and which we must remember is different under the continents than under the oceans) are shown in Figure 3.3. Notice that the presence of water appreciably alters the picture. Incipient melting of peridotite or eclogite due to traces of water can nicely account for the mantle's low sound seismic velocity zone (Kushiro, Syono, and Akimoto, 1968; Lambert and Wyllie, 1968, 1970;

Table 3.4 Mineralogy of the Mantle Peridotite as a Function of Depth: 1969 Model of A. E. Ringwood[a] (From Wyllie, 1971*a*)

Depth (km)	Mineral Assemblages and Transformations	Weight Percent Mineral	Coordination of Elements	Zero Pressure Density
To 80	*Plagioclase peridotite*	Table		
	Spinel peridotite	6-7		
80–350	*Garnet peridotite*		Si-4	3.38
	olivine	57	Mg, Fe, Ca,	
	orthopyroxene	17	-6, 8	
	clinopyroxene	12		
	garnet	14		
350–450	Pyroxene → garnet			
	Olivine → spinel[b]			
450–600	Spinel[b]	57	Si-4, 6	3.66
	Garnet	39		
	Jadeite	4		
600–700	Spinel[b] → Sr_2PbO_4 structure or MgO + ilmenite structure			
	Garnet → ilmenite + perovskite structures			
	Jadeite → calcium ferrite structure			
700–1050	Ilmenite solid solution	36	Si, Mg, Fe	3.99
	Strontium plumbate structure	55	-6	-4.03
	Perovskite, $CaSiO_3$	6.5		
	Calcium ferrite, $NaAlSiO_4$	2.5		
1050–1150	Transformations into phases denser than isochemical oxide mixture			
1150–2900	Speculative. Depends on germanate analog systems and shock wave studies		Si, -6 Mg, Fe 6	7% higher than mixed oxides

[a] See also Ringwood, 1970.

[b] Includes beta-phase.

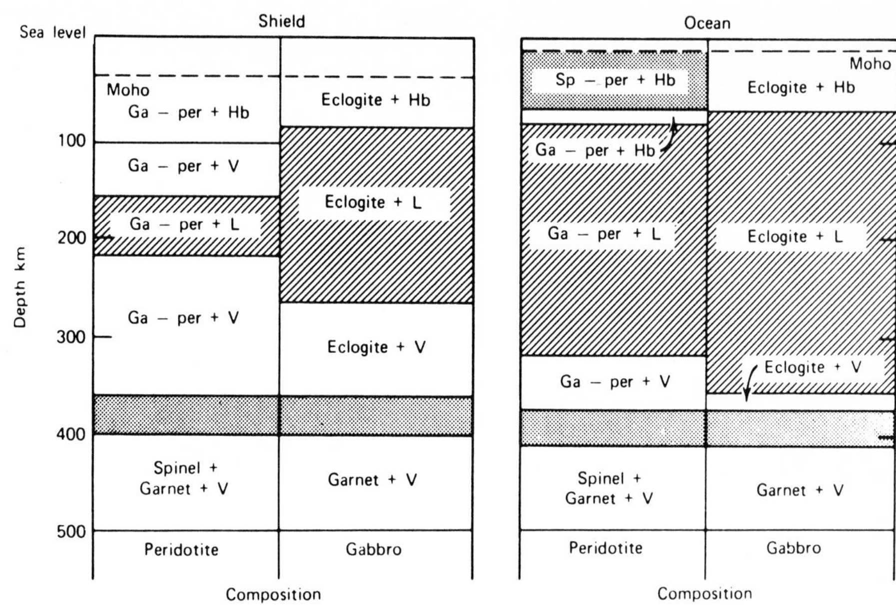

Figure 3.2. Schematic sections through the upper mantle in two different tectonic environments, for mantle material composed of either peridotite or eclogite, in the presence of traces of water (from Wyllie, 1971*a*).

Anderson and Sammis, 1969; Ringwood, 1969*a*, *b*). The content of volatiles, such as H_2O and CO_2 in the mantle, is probably highly variable both vertically and laterally. Wyllie (1971*a*) places the value at less than 0.1%, but in special locations, it may be much higher—10% H_2O and 0.8% CO_2 in mantle peridotite beneath mid-ocean ridges (Nicholls, 1967). I find the point of view that water is present in the mantle attractive inasmuch as the oceans are believed to be the accumulation of water released from the mantle derived magma (see Chapter 13).

How well do the peridotite or eclogite models fit the observed geophysical data? In an interesting exercise Press (1968, 1969), using a Monte Carlo statistical procedure, fed values of seismic parameters into a computer, and out of some 5,000,000 density models, he obtained only six models that satisfied the restrictions of known geophysical data. Plausibility tests further reduced the survivors to three (Figure 3.4). The ecologite model alone is consistent with the factual constraints between 80 and 150 km, and the pyrolite (a hypothetical peridotite) model is favored in the region near 300 km.

The closest approximation to "fresh" samples of inner Earth that we can obtain are the materials spewed up by volcanic action, but we are never certain exactly where these samples come from, how representative they are, and what the effects of cooling, depressurization, fractional crystallization, and other physical and chemical processes are upon them (Gorshkov, 1970; Thompson, 1972). Ash and blocks as well as molten lava may be ejected by volcanoes. Mafic, subsilicic lava flows are extruded at temperatures exceeding 1100°C and are relatively thin in consistency, whereas the more silicic, felsic flows because of their lower temperature, greater silica content, and more extensive polymerization of the silicic material are more

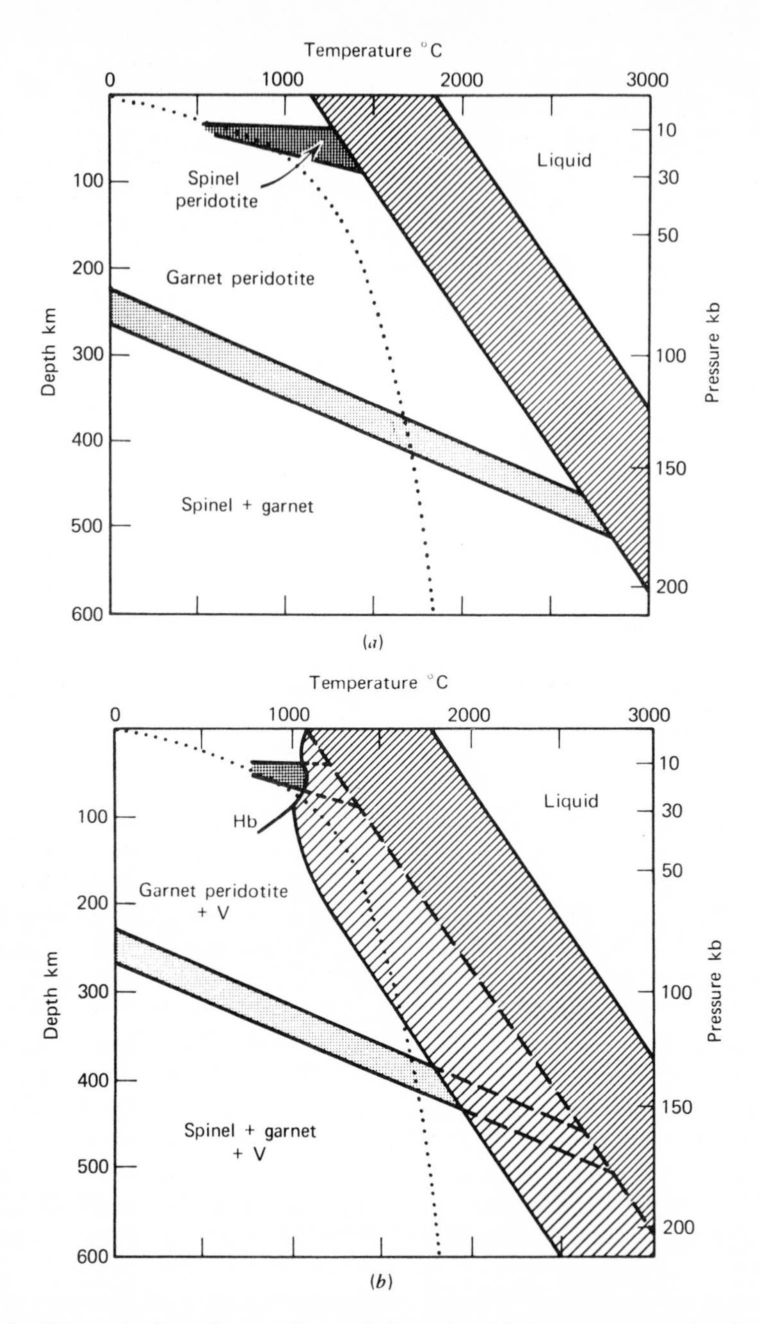

Figure 3.3. Schematic phase diagrams for peridotite and peridotite–water extrapolated to pressures corresponding to 600-km depth in the mantle (from Wyllie, 1971*a*): *a*) dry peridotite and *b*) isopleth for peridotite with 0.1% water. The melting interval consists of two parts: The light-shaded band represents incipient melting and the dark-shaded band is almost equivalent to dry melting. The depth scale and the geotherm (dotted) relate the phase diagram to upper mantle of specific compositions.

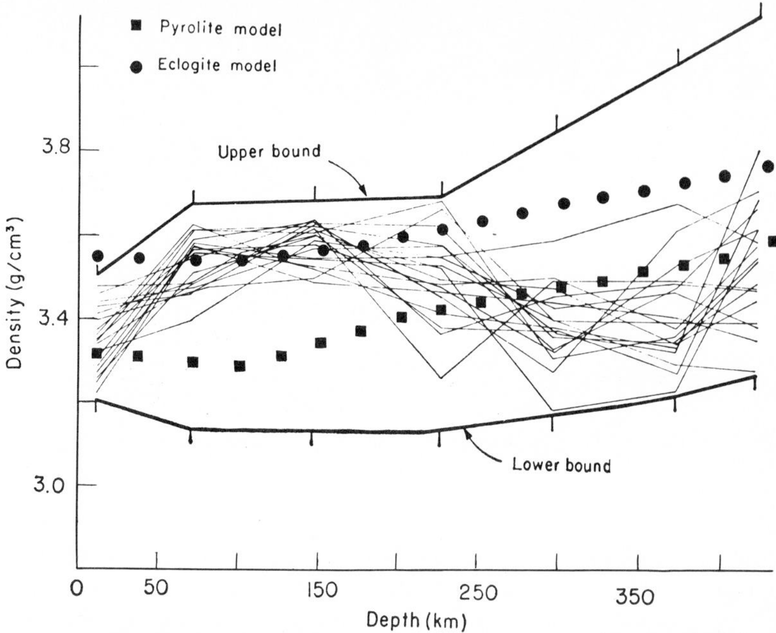

Figure 3.4. Successful density models for the suboceanic upper mantle using the Monte Carlo method (from Press, © 1969 by the American Association for the Advancement of Science with permission of the AAAS).

viscous (Ernst, 1969). The escape of water and other volatiles from hot lava upon depressurization (discussed further in Chapters 13 and 15) can result in a very great porosity as in the case of pumice. The chemical compositions of volcanic and plutonic rocks are very similar, although the relative amount of the higher valence state of iron appears to be consistently larger in the former (Miyake, 1965). Analyses of alkali olivine basalts from an eruption of Beerenberg volcano on Jan Mayen gave 47.4% SiO_2, 15.8% Al_2O_3, 9.76% CaO, 7.97% FeO, 5.60% MgO, 3.26% Fe_2O_3, 3.25% TiO_2, 3.25% Na_2O, 2.60% K_2O, 0.50% P_2O_5, and 0.20% MnO (Weigand et al., 1972), whereas the rare earth composition fell well within the range of that for other basalts (Gast, 1967). Although Jan Mayen basalts are unusual among oceanic volcanic rocks, these findings are of special interest because Jan Mayen is one of the few exposed volcanic islands lying directly on a midocean ridge, and its material may be relatively primordial, being raised from the asthenosphere by a deep mantle convective plume (Morgan, 1971). (See Figure 3.11.)

3.4 THE MOHOROVICIC DISCONTINUITY AND THE CRUST

The layered structure of the Earth is marked by several sharp discontinuities in seismic properties. The sharpest and most discussed of these boundaries is the Mohorovicic discontinuity (discovered in 1909) or "Moho," which separates the brittle, highly heterogeneous crust of the Earth from its presumably more homogeneous and plastic upper mantle. The depth of the Moho under the continental masses is considerably greater (35 km) than under the oceans (5 km, Figure 3.5).

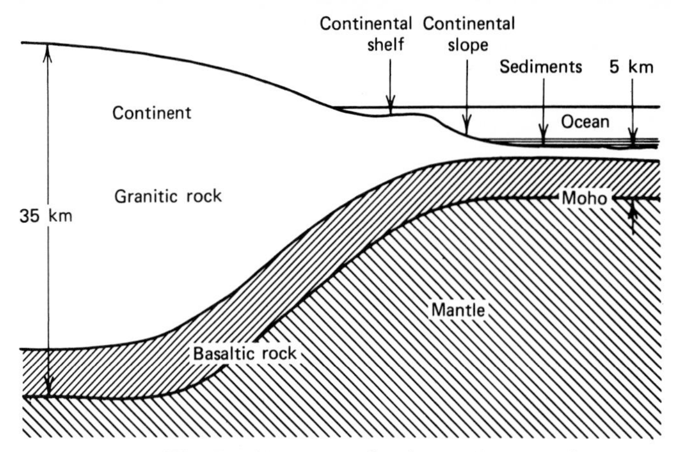

Figure 3.5. The Earth's crust under the continents and oceans.

Beneath the great mountain ranges, the Moho falls even deeper in order to maintain isostatic equilibrium ("floating," McConnell, 1965) and gives rise to mountain roots (Figure 3.6). Because of the crust–mantle boundary's chemical importance and role in major tectonic movements, geoscientists would dearly like to penetrate to the Moho and resolve the many arguments, which follow, about it. It makes sense to drill where the crust is the thinnest—namely beneath the oceans. Unfortunately, "Project Mohole" proved to be a disastrous epic of politicing (both governmental and scientific) and mismanagement. However, the situation has been redeemed somewhat by the many fine scientific findings revealed by the cruises of the *Glomar Challenger* (The Deep Sea Drilling Project—Tracey et al., 1971, see also van Andel, 1968), findings that, while hardly approaching the Moho, have shed new light upon such major and magnificent geological processes as the drift of the continents and the spreading of the ocean floor (Wilson, 1963; Emilia and Heinrichs, 1969; Maxwell, 1970).

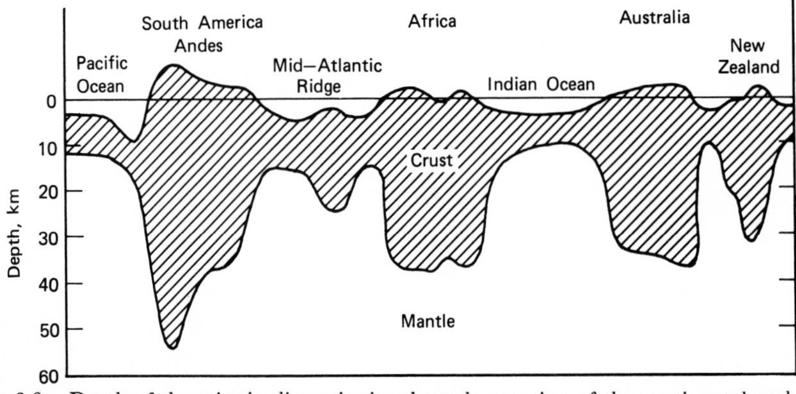

Figure 3.6. Depth of the seismic discontinuity through a section of the continental and oceanic crusts showing the foundations of the continental blocks and the "roots" of the Andes Mountains.

But to return to the Moho, much of the controversy about it hinges on the question of whether it represents simply a phase transition or a chemical boundary. Wyllie (1971*a*) has presented an excellent review of the matter, and while fairly noncommittal, gives greater emphasis in his discussion to the gabbro-eclogite phase transition. Figure 3.7 shows some of the Moho models he examines while Figure 3.8 represents some details of the proposed crustal structure and upper mantle of a central United States section. Let us hope that someday scientists and administrators will get themselves together, and a successful Mohole will be drilled. If we can go up 400,000 km to get samples, then we should be able to go down 5 km.

We infest the Earth's crust, so, of course, in many ways, the crust is that part of the lithosphere of most immediate concern to us. The crust is heterogeneous and exceedingly complex. Upon our first visit to the mineralogical halls of a museum, there appears to be an enormous host of different minerals. While the number of exotic minerals is great, about a dozen minerals, composed of only eight of the chemical elements, account for more than 99% of the volume of the crust (Tables 3.5 and 3.6). Table 3.6 is particularly instructive: Oxygen alone accounts for 47% of the Earth's crust by weight, 63 atom %, and over 93% by volume. This clearly points to the most important chemical generalization that can be made about the Earth as a whole and its crust in particular, "the crust of the Earth is essentially a packing of oxygen anions, bonded by silicon and the ions of the common metals" (Mason, 1966). Goldschmidt (1954) has characterized the chemistry of the crust in

Table 3.5 Abundances of Main Rock Types and Minerals in the Crust (From Wyllie, 1971*a*, after Ronov and Yaroshevsky, 1969)

Rocks	% Volume of Crust	Minerals	% Volume of Crust
Sedimentary		Quartz	12
Sands	1.7	Alkali feldspar	12
Clays and shales	4.2	Plagioclase	39
Carbonates (including salt-bearing deposits)	2.0	Micas	5
		Amphiboles	5
Igneous		Pyroxenes	11
Granites	10.4	Olivines	3
Granodiorites, diorites	11.2	Clay minerals (+ chlorites)	4.6
Syenites	0.4	Calcite (+ aragonite)	1.5
Basalts, gabbros, amphibolites,		Dolomite	0.5
eclogites	42.5	Magnetite (+ titanomagnetite)	1.5
Dunites, peridotites	0.2	Others (garnets, kyanite, andalusite, sillimanite, apatite, etc.)	4.9
Metamorphic			
Gneisses	21.4		
Schists	5.1		
Marbles	0.9		
Totals		Totals	
Sedimentary	7.9	Quartz + feldspar	63
Igneous	64.7	Pyroxene + olivine	14
Metamorphic	27.4	Hydrated silicates	14.6
		Carbonates	2.0
		Others	6.4

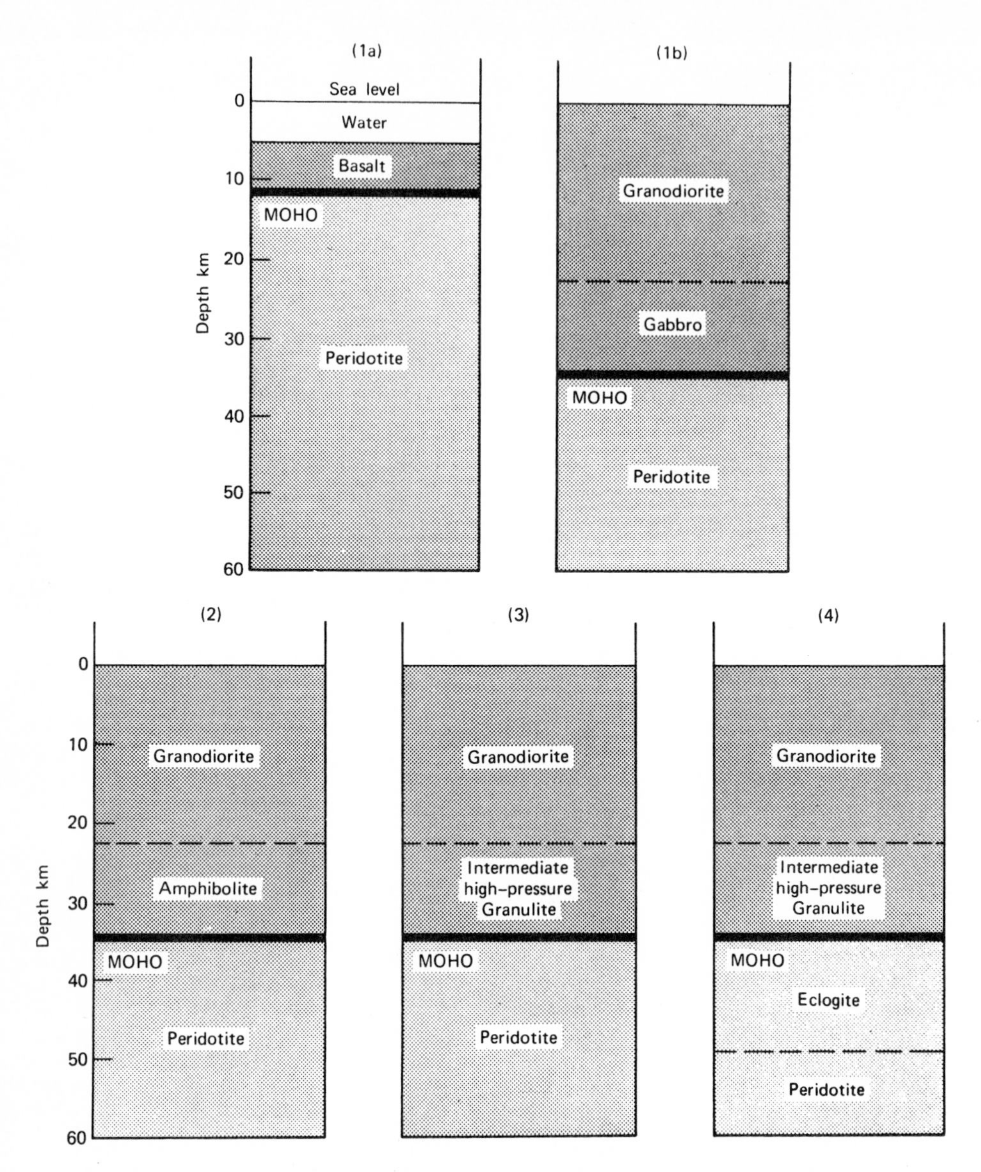

Figure 3.7. Models of the Mohorovicic discontinuity (from Wyllie, 1971*a*).

a nutshell by calling it the "oxysphere." But before saying more about the chemistry of the crust, let me insert a word or two here about its major structural features and their movements. Far from being a rigid structure, overwhelming evidence has accumulated in recent years showing that the Earth's crust is a dynamic system. Thin slabs or plates of the rigid lithosphere (and now we are using the term in its narrowest technical sense as the crust and the uppermost part of the mantle—a sheet about 100-km thick) move about on the low strength asthenosphere, which in turn rests on the bulk of the remainder of the mantle or mesosphere (Runcorn, 1962; Wilson, 1972). In particular, one gigantic continental raft, called "Pangaea" by Wegener (1966) broke up into two primordial land masses—Laurasia in the northern

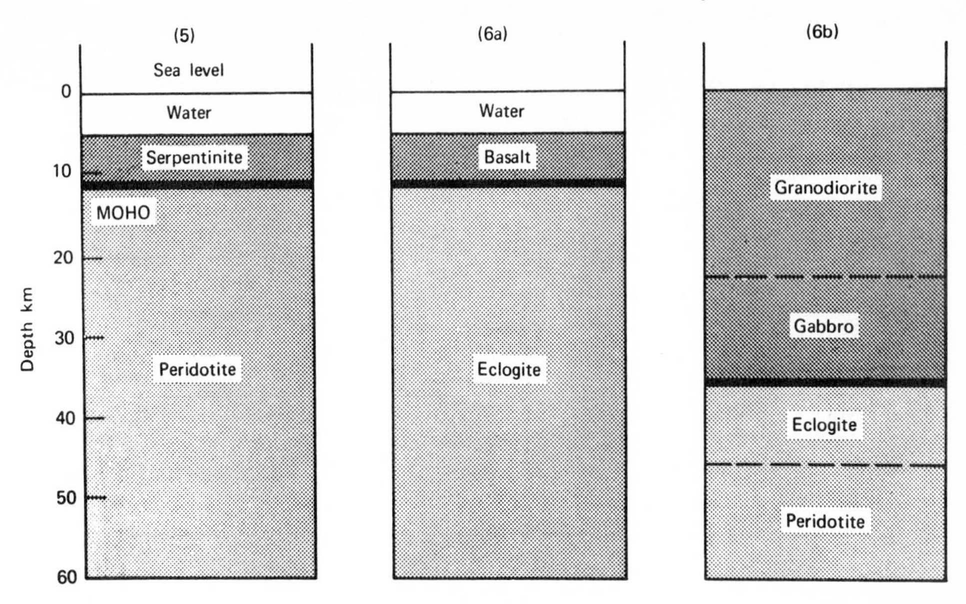

Figure 3.7. (*Continued*).

hemisphere and Gondwanaland in the south, which further began to split up and drift apart in the Jurassic Period (see Table A.19 for the geological periods) to form the present continental masses and open up the Atlantic Ocean. Evidence for such a movement comes from many sources; among the most convincing arguments are the good fit of the continental coasts (Figure 3.9) and the excellent age match between rocks of South America and rocks from the corresponding coast of Africa (Figure 3.10). The process is still continuing, and recent magnetic studies and deep ocean cores provide still further and very convincing evidence of the spreading of the sea

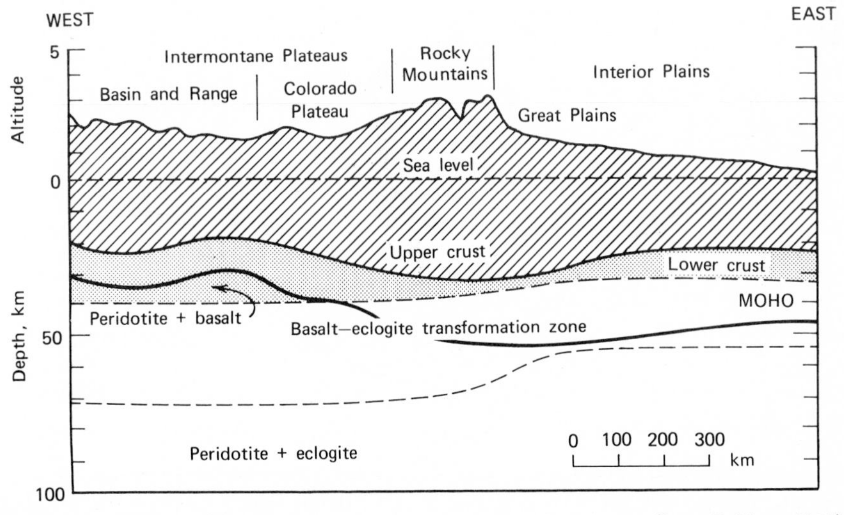

Figure 3.8. Profile of the Earth's crust in the central United States (from Pakiser, 1965).

Table 3.6 The Commoner Chemical Elements in the Earth's Crust (From Mason, 1966)

	Weight Percent	Atom Percent	Radius (Å)	Volume Percent
O	46.60	62.55	1.40	93.77
Si	27.72	21.22	0.42	0.86
Al	8.13	6.47	0.51	0.47
Fe	5.00	1.92	0.74	0.43
Mg	2.09	1.84	0.66	0.29
Ca	3.63	1.94	0.99	1.03
Na	2.83	2.64	0.97	1.32
K	2.59	1.42	1.33	1.83

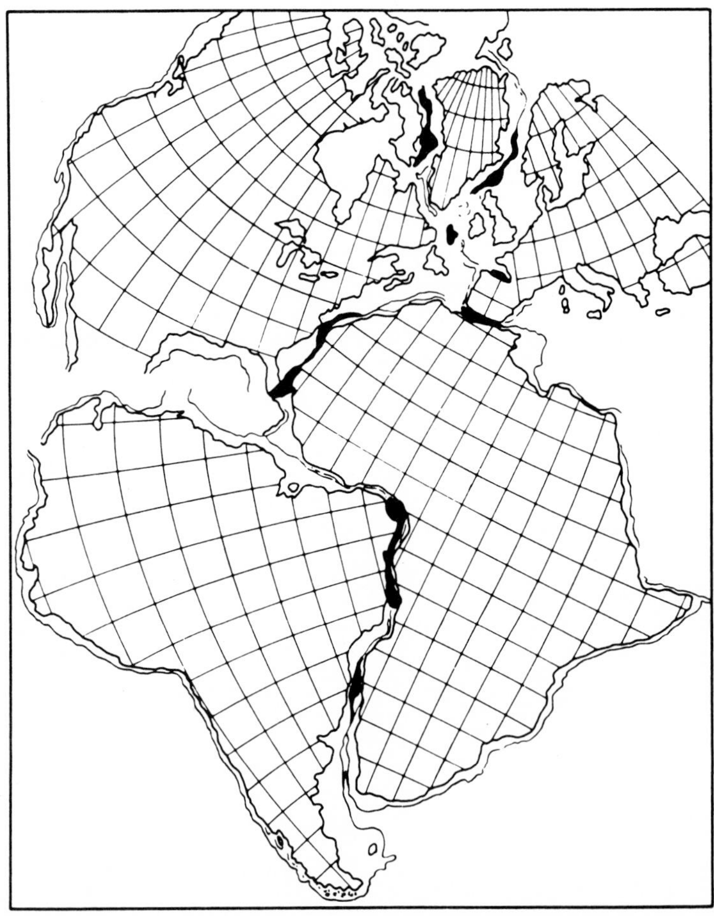

Figure 3.9. Computer fit of the continents around the Atlantic Ocean. Black areas show the overlap of continental shelves (from Wyllie, 1971*a*, after Bullard et al., 1965).

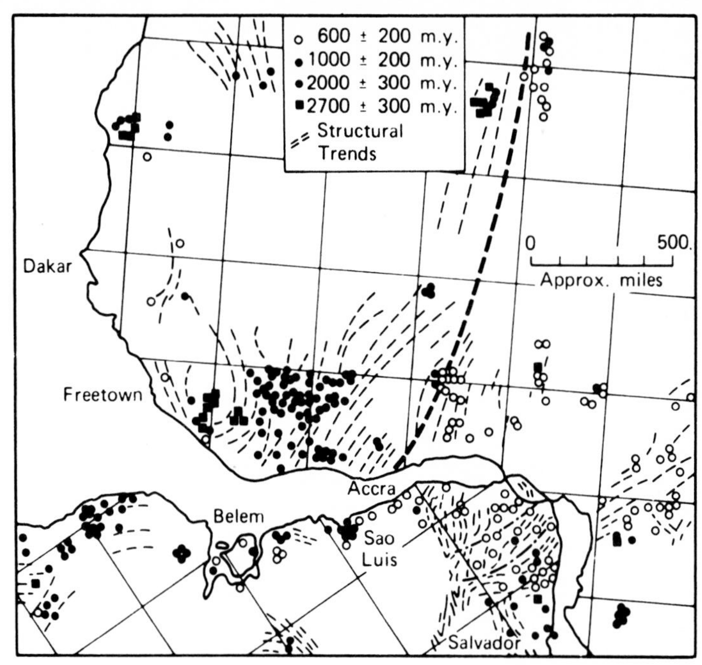

Figure 3.10. West Africa and South America shown fitted together. The age measurements for Brazil appear to show the same age provinces as those in West Africa with the boundary at the predicted location (from Wyllie, 1971*a*).

floor. These movements are of interest to us here primarily because they provide a mechanism, albeit a very slow one, for the movement both vertical and lateral of chemical substances in the Earth's upper regions. Material rises to the surface along the midocean ridges (Figure 3.11) and is carried away from the ridges with the spreading ocean floor until finally with its burden of sedimentary material it slides under the continental plates. Here it is forced down to greater and greater depths and finally metamorphized and melted into the magma. The possibility that exposure of the fresh uprising crustal material at the midocean ridges may contribute to the salt content of seawater has been discussed in a fine semipopular article by MacIntyre (1970). But this exposure at the ridge area is not the only site in the cycle for significant geochemical processes. As the lithosphere is forced downward beneath the continental plates, increased temperature and pressure, especially in the presence of water, will result in chemical transformations and phase changes (Oxburgh and Turcotte, 1968; Hamilton, 1969; Hatherton and Dickinson, 1969). Elsewhere in the Earth's crust, except for the minor effect of vulcanism and alterations of sedimentary rocks (see the discussion of Figure 3.12), there is relatively little chemistry going on, which, if one pauses to think about it for a moment, is just what one would expect in view of the high state of oxidation and resulting chemical stability of crustal materials. The chemistry that is occurring, such as rock weathering, is very slow by human standards.

A number of different approaches have been taken for estimating the composition of the crust including an average of more than 5000 "superior" continental igneous

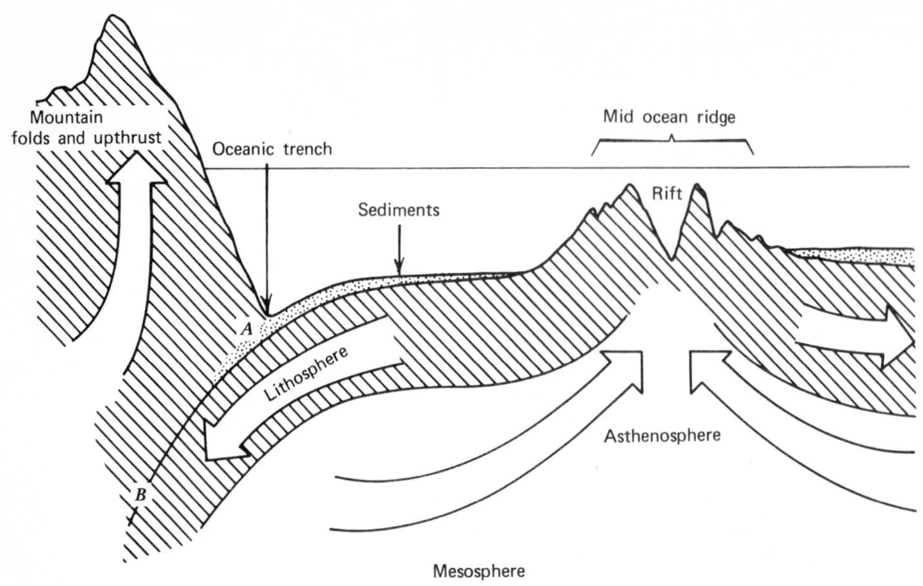

Figure 3.11. Ocean floor spreading and the movement of crustal material.

rocks (Clark and Washington, 1924; Row A in Table 3.7), analyses of very fine rock "flour" deposited as glacial clay (Goldschmidt, 1954; Row B in Table 3.7) and averages of analyses of samples carefully chosen to be representative of the four major geological divisions (Poldervaart, 1955; Row C in Table 3.7). Also included in Table 3.7 (Row D) are the more recent estimates of Ronov and Yaroshevsky (1969) based on the major crustal layers for both continental and oceanic environments and of McBirney (1969) based on island arc calc-alkaline lavas (Row E). Apart from some minor differences largely arising from the relative importance assigned to the oceanic samples, the values in Table 3.7 are in remarkable agreement. Table A.16 summarizes the crustal levels of all of the chemical elements and holds some surprises—elements such as Cu, Pb, and Hg long known and used by man are actually quite rare, whereas many unfamiliar elements such as Rb, V, Sc, and Hf are more abundant. In addition to comparative ease of recovery, part of the explanation for these differing patterns of human utilization lies in distinguishing between elemental abundance and elemental availability. Some of the more abundant elements are highly dispersed, that is to say they occur at low levels throughout the

Table 3.7 Average Chemical Composition of the Earth's Crust

Author (see text)	SiO_2	Al_2O_3	$FeO + Fe_2O_3$	MgO	CaO	Na_2O	K_2O	TiO_2	P_2O_5	H_2O
A	60.18	15.61	7.02	3.56	5.17	3.91	3.19	1.06	0.30	
B	59.12	15.82	6.99	3.30	3.07	2.05	3.93	0.79	0.22	3.02
C	55.2	15.3	8.6	5.2	8.8	2.9	1.9	1.6	0.3	
D	59.3	15.9	7.0	4.0	7.2	3.0	2.4	0.9	0.2	
E	58.7	17.3	7.0	3.1	7.1	3.2	1.3	0.8	0.2	

majority of common crustal minerals. Mason (1966) cites Rb in K-minerals and Ga in Al-minerals as examples. Otherwise, as in the cases of Ti and Zr, they form specific minerals that are again widely dispersed in small amounts in common crustal rocks. The availability of an element, on the other hand, depends on its tendency to form individual minerals (ores) in which it is a major constituent. Ease of extraction, as we have mentioned, is a further very important consideration—clay minerals are a widely distributed potential source of aluminum, but it is much easier to extract this metal from the much rarer ore bauxite.

The rocks of the Earth's crust may be divided into three main types: igneous rocks, sedimentary rocks, and metamorphic rocks. We have already seen (Table 3.5) that of these, the igneous rocks are the more important, comprising more than 60% of the Earth's crust, with sedimentary rocks accounting for less than 8%. Table 3.8 is intended to give an idea of which of these rock types form the large structural units of the crust.

Far from being separate, enduring species, these three rock types on the geological time scale are slowly transformed from one to another by a definite scheme of processes known as the geochemical cycle (Figure 3.12). Some of these major processes are discussed elsewhere in this book, notably sedimentation and weathering in Chapter 13. In Figure 3.11, I have tried to indicate two of the crustal regions where these processes are occurring—A, the transformation of sediments into metamorphic rock by compaction, cementation, substitution, and recrystallization; and *B*, the transformation and melting of this material under great pressure and temperature

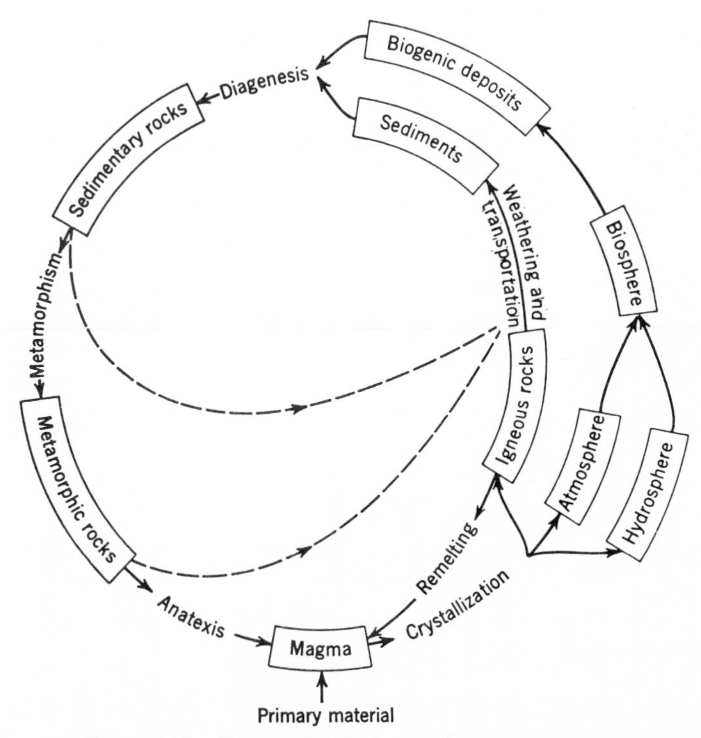

Figure 3.12. The geochemical cycle (from Mason, 1966).

Table 3.8 Distribution of Rock Types in Large Structural Units of the Crust and the Crustal Layers (Data from Ronov and Yaroshevsky, 1969) (From Wyllie, 1971*a*)

Crustal Layers in Different Crustal Units	Average Thickness (km)	Volume (km^3)	Mass (10^{24} g)	Types of Rocks and Abundances (percent volume of layer, except percent area for oceanic Layer 1)
Continental Platform				
Sedimentary	1.8	135	0.35	Sands, 23.6; clays, 49.5; carbonates, 21.0; evaporites, 2.0; basalts, 3.9
Continental Geosynclinal Folded Belts				
Sedimentary	10.0	365	0.94	Sands, 18.7; clays and shales, 39.4; carbonates, 16.3; evaporites, 0.3; basalts, 12.6; andesites, 10.2; rhyolites, 2.5
Subcontinental, Shelf, and Slope				
Sedimentary	2.9	190	0.48	Similar to above groups
Continental and Subcontinental				
Granitic		3590	9.81	Granites, 18.1; granodiorites, 19.9
Continental	20.1			Syenites, 0.3; gabbro, 3.7
Subcontinental	9.1			Peridotites, 0.1; gneisses, 37.6; schists, 9.0; marbles, 1.5; amphibolites, 9.8
Basaltic		3760	10.91	Acid igneous and metamorphic rocks, 50.0; basic igneous and metamorphic rocks, 50.0
Continental	20.1			
Subcontinental	11.7			
Oceanic				
Layer 1 Sedimentary	0.4	120	0.19	Terrigenous, 7.3; calcareous, 41.5; siliceous, 17.0; red clays, 31.2
Layer 2	0.6	175	0.44	Sediments, 50.0
	0.6	175	0.52	Basalts, 50.0
Layer 3	5.7	1700	4.92	Oceanic tholeiitic basalts, 99.0; alkaline differentiates, 1.0

(10 kb, 1000°C) to form magma from which, in turn, igneous rocks can crystallize. Diagenesis, a sort of prelude to metamorphosis, may also extend into region *A*, but this preliminary step of the conversion of sedimentary deposits into rock begins to occur on the ocean floor before the material begins to sink beneath the continental margin. The asthenosphere is a low-sound velocity, partially molten layer about 100-km thick underlying the Earth's crust. While often not shown as such, the geochemical or rock cycle and the hydrologic cycle (Chapter 9) are linked, and water is also involved in the former. Sediments containing hydrous minerals and pore fluids can be carried deep into the crust by tectonic movements. This water may escape to the surface by uplift and subsequent erosive exposure. Alternatively, it may migrate to the surface following its release during progressive rock metamorphosis. Other water will remain in the rock, and magma formed by anatexis or partial fusion of metamorphic rock probably will be more "wet" than mantle-derived magma. Even so, juvenile water is still introduced from mantle magma, so the effusion of volcanoes, hot springs, and other such sources will inject a highly variable mixture of once meteoretic and juvenile water into the surface hydrologic cycle. The rock cycle summarizes the important chemistry of the Earth's crust, and it is in the course of this cycle that most of the observed chemical differentiation and isotopic fractionation occurs.

A most fascinating problem has been pointed out in connection with the sequence of events represented in Figure 3.11. The figure implies that the convection cells responsible for sea floor spreading are conservative cycles of material. Ringwood (1969*a*, *b*) raised the possibility that the cycle might "leak," losing the more refractory material of the lithosphere to the mesosphere instead of returning it all via the asthenosphere. Dickinson and Luth (1971) have explored the question further and have come up with a surprising possibility—the Earth's crust and mantle may be evolving *irreversibly* with a gradual loss of material into the deeper mantle. When the asthenosphere, which may represent the last of the original material of the primordial Earth, is exhausted, global plate tectonic processes will cease, perhaps within the next 10^9 yr.

For the greater part, crustal geochemistry consists of phase changes and cation substitutions in the pervasive oxygen–lattice matrix, and the overall chemical composition of the major constituents is not highly variable (Figure 3.13). Sedimentary rocks are a little richer in Ca and Si and poorer in Al, Mg, and Fe. In Figure 3.13, we see, not surprisingly, that metamorphic rocks are intermediate in composition between sedimentary and igneous rocks. Dennen and Moore (1971) have pointed out that the Al–Fe ratio is remarkably constant (= ca. 2) in mature detrital sedimentary rocks despite wide compositional variations in their predecessors but that the Si content changes in a regular way and can be used to identify the state of maturation along the path of protometamorphosis. Rather than become involved in the details of rock chemistry, I feel that it will probably be more useful to study carefully the somewhat complicated diagrams presented by Wyllie (1971*a*), summarizing the physical and chemical processes that igneous, metamorphic, and sedimentary rocks undergo (Figures 3.14, 3.15, and 3.16).

We will return to the topic of diagenesis again in Chapter 13. In the course of metamorphosis, although the crystalline lattice can be altered, the bulk chemical composition of the rock may remain constant. This is called isochemical metamorphism. While there may be some short range transport of material (<1 mm) during

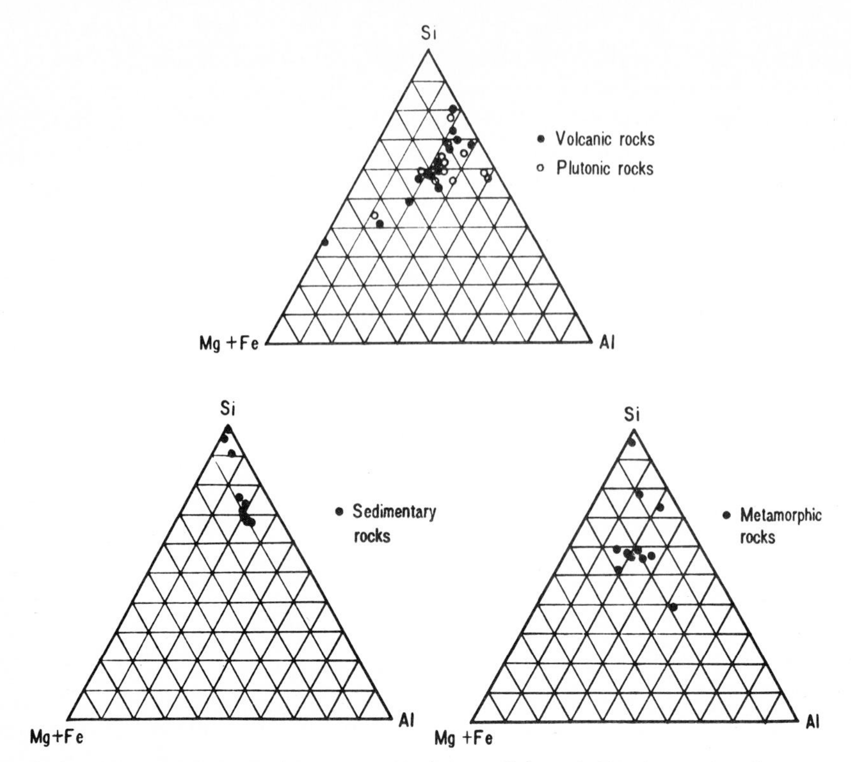

Figure 3.13. Relation of atomic fractions of Si, (Mg + Fe), and Al in igneous, sedimentary, and metamorphic rocks (from Miyake, 1965, with permission of the author).

these processes, such transport is to be distinguished from the long-range transport resulting in bulk compositional changes during allochemical metamorphism (or "metasomatism"). Just how chemical substances migrate these great distances in the solid phase remains unclear. Diffusion, even along crystal boundaries, would appear to be too slow. Again the small amount of water in crustal and mantle materials may play an important role (Durney, 1972).

Let us take as an example of metamorphism the formation of wollastonite from calcium carbonate and silica

$$CaCO_3 + SiO_2 \rightleftharpoons CaSiO_3 + CO_2 \tag{3.1}$$

a reaction that has been studied in some detail in the laboratory. In a closed system, the reaction goes as written in the *PT* region above curve *AB* in Figure 3.17 and reverses below this curve. CO_2 is a volatile component and in a real situation may escape favoring the right-hand side of equilibrium 3.1 and bringing the reaction curve down to much lower temperatures (*AC* in Figure 3.17).

It is useful to classify mineral substances in terms of the *PT* conditions under which they are formed. A mineral *facies* is defined as the family of rocks originating under *PT* conditions so similar that a definite chemical crystallization has resulted regardless of the mode (from magma, aqueous solution, or direct) of crystallization. Figure

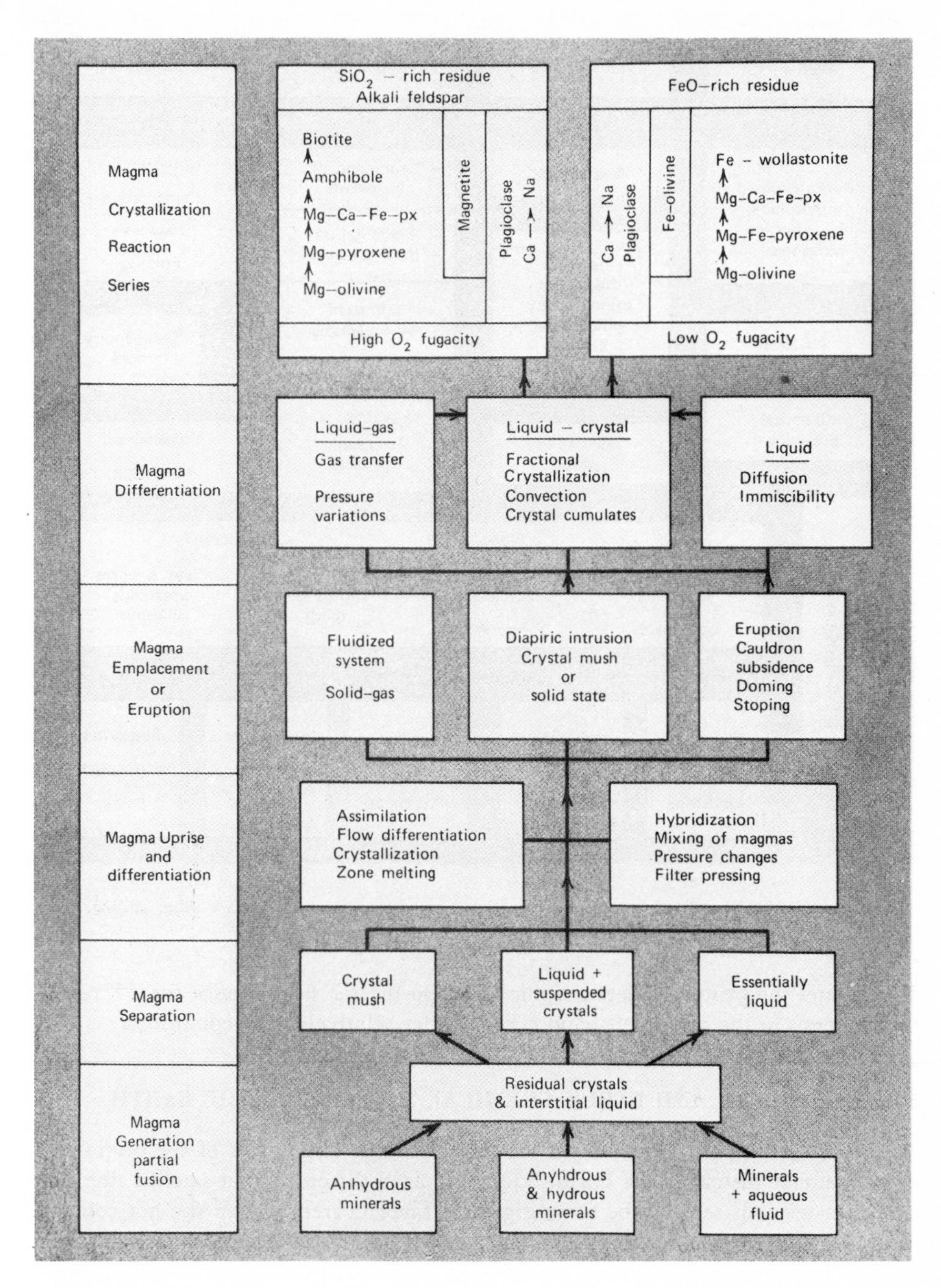

Figure 3.14. Schematic representation of igneous processes (from Wyllie, 1971*a*).

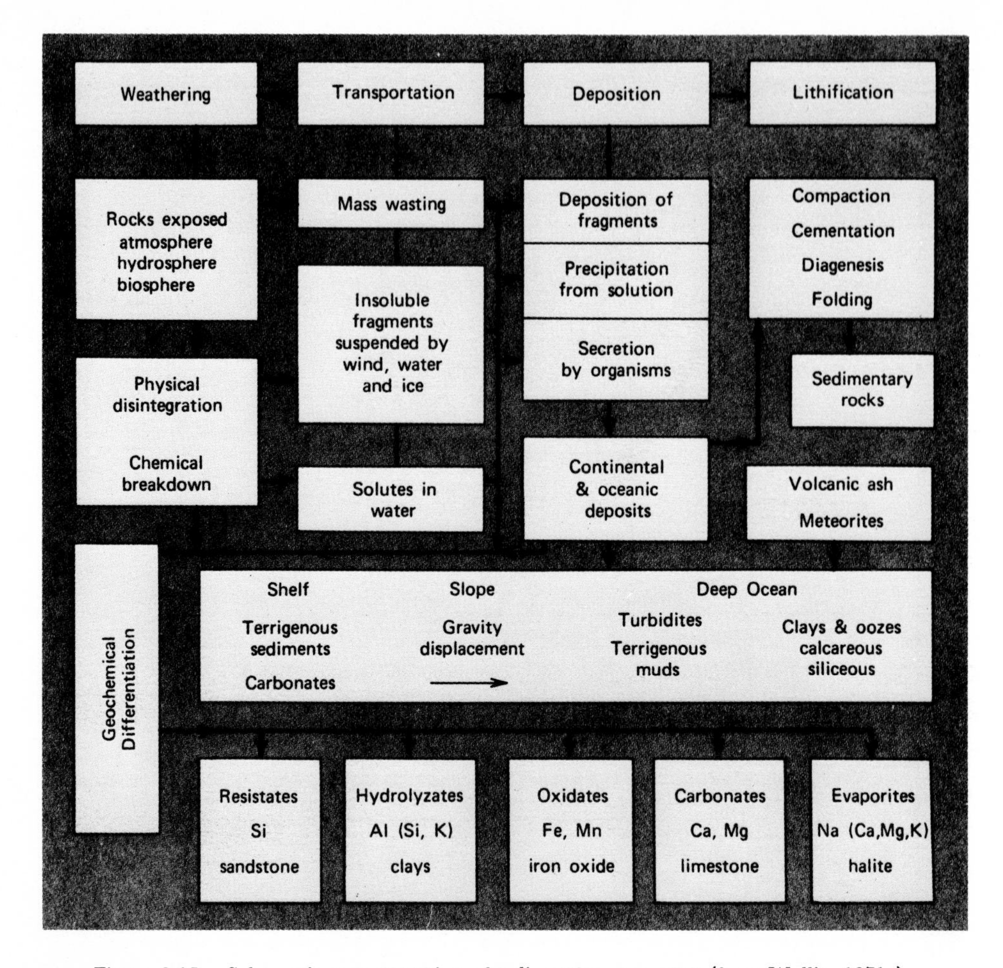

Figure 3.15. Schematic representation of sedimentary processes (from Wyllie, 1971*a*).

3.18 locates the principle metamorphic facies on the *PT* field. Notice the *PT* region of diagenesis in the upper left-hand corner under relatively mild conditions.

3.5 THE ORIGIN AND EARLY CHEMICAL HISTORY OF THE EARTH

As in the case of the Moon, there are several viable hypotheses of the origin and early chemical history of the Earth. But again, as in the case of our satellite, the cold accretion versions seem to be finding greater favor currently than the hot condensation view. They can more readily account for the low abundances of light gases such as hydrogen and helium, so common in the Sun, as well as krypton and xenon and the Earthly ubiquity of so volatile a substance as water. In any event, the layered structure of the Earth makes it clear that at least at one point in its early history it was entirely, or almost entirely, molten. Goldschmidt (1954) was the first to examine in great detail the processes responsible for chemical differentiation in the primitive Earth. He took as his point of departure from the condensation of our planet from hot gases (nevertheless many of his conclusions are equally appropriate to a cold Earth—subsequent melting hypothesis). In the resulting gigantic liquid droplet, the

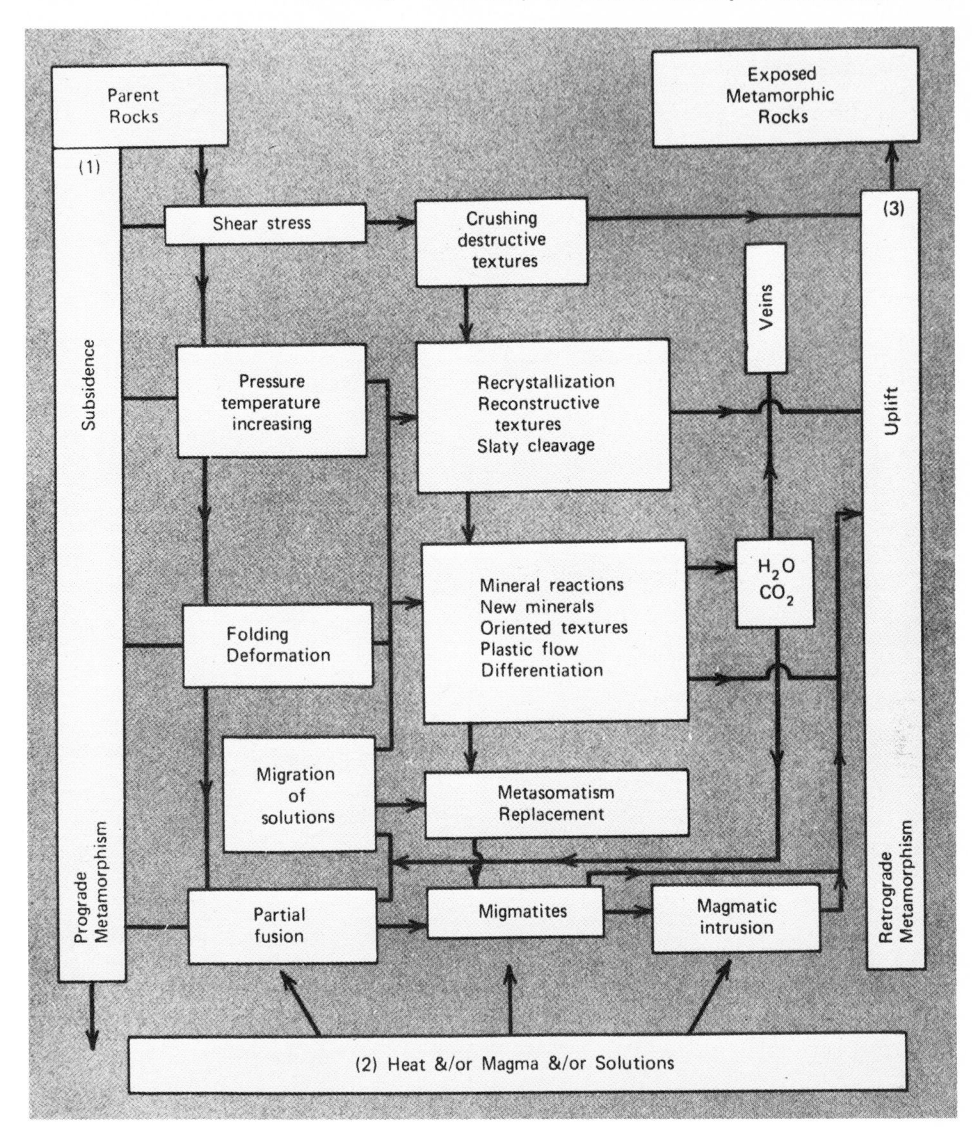

Figure 3.16. Schematic representation of metamorphic processes (from Wyllie, 1971*a*).

various chemical phases separated out with the heavy iron at the center and a crust of the lighter silicates much like the formation of slag in a blast furnace.

Iron–Magnesium Silicate
Iron Sulfide
Free Iron

While present opinion may disagree with his point of departure, Goldschmidt here again displays his remarkable insight for sensing the most important chemical prin-

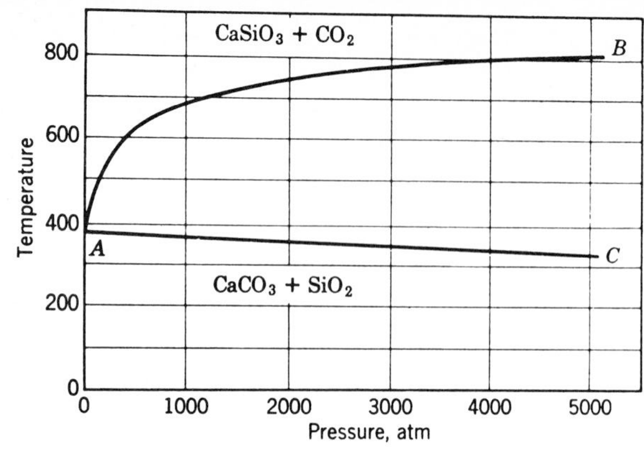

Figure 3.17. Pressure–temperature curves for the reaction: $CaCO_3 + SiO_2 \rightleftharpoons CaSiO_3 + CO_2$ (from Mason, 1966).

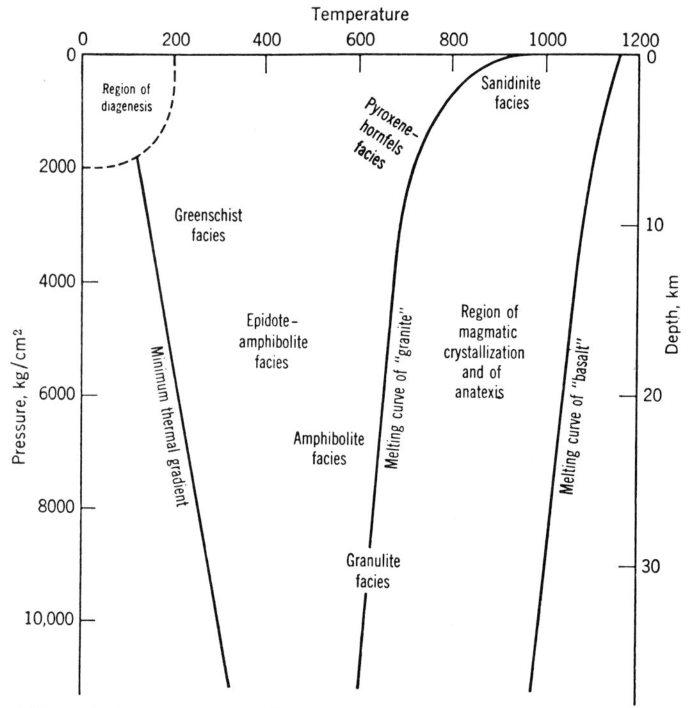

Figure 3.18. The principal metamorphic facies in relation to temperature and pressure (from Mason, 1966).

ciples at work. It is not, he points out, simply a matter of the heavier metals sinking to the center of the Earth and the lighter elements rising to the surface. On the contrary, while the major phases stratify according to their densities, the distribution of the overwhelming majority of elements is determined, not by their densities, but by their solubilities in these phases. Accordingly, largely from their distribution in meteorites, the chemical elements can be classified on the basis of in which phase—

Table 3.9 Geochemical Classification of the Elements (based on distribution in meteorites) (From Mason, 1966)

Siderophile	Chalcophile	Lithophile	Atmophile
Fe,[a] Co,[a] Ni[a]	(Cu), Ag	Li, Na, K, Rh, Cs	(H), N, (O)
Ru, Rh, Pd	Zn, Cd, Hg	Be, Mg, Ca, Sr, Ba	He, Ne, Ar, Kr, Xe
Os, Ir, Pt	Ga, In, Tl	B, Al, Sc, Y, La, Lu	
Au, Re,[b] Mo[b]	(Ge), (Sn), Pb	Si, Ti, Zr, Hf, Th	
Ge,[b] Sn,[a] W[b]	(As), (Sb), Bi	P, V, Nb, Ta	
C,[c] Cu,[a] Ga[a]	S, Se, Te	O, Cr, U	
Ge,[a] As,[b] Sb[b]	(Fe), Mo, (Os), (Ru, (Rh), (Pd)	H, F, Cl, Br, I, (Fe), Mn, (Zn), (Ga)	

[a] Chalcophile and lithophile in the Earth's crust.
[b] Chalcophile in the Earth's crust.
[c] Lithophile in the Earth's crust.

iron (siderophile), sulfide (chalcophile), silicate (lithophile), or gaseous (atmophile)—they are normally found as components of compounds or dissolved as alloy constituents (Table 3.9). A number of rules have been advanced correlating the geochemical nature of the elements and properties, such as atomic volumes and the heats of formation of the oxides. The behavior is far from simple, and a given element may behave differently under different situations. In addition to liquid–liquid equilibrium solubility criteria, a second very important factor in determining geochemical behavior is the ease of crystalline lattice substitution. Thus the *size*, as well as the outer electronic configuration of an element's ions, is critical.

The succession of condensation of chemical substances from a cooling gas of cosmic composition has been examined in some detail (Larimer, 1967; Larimer and Anders, 1967). Silicates condense first and metals such as iron, late in the process; hence, a subsequent melting and reversal of phase stratification is necessary to give the Earth its present structure. Alternatively, in order to get the iron to the core the Earth would have to remain molten throughout the accretion period (Hanks and Anderson, 1969). Views currently popular envision heating, largely by radioactivity, and partial or total melting following cold accretion of planetesimals and other particulate material. About 5×10^9 yr ago, radiogenic heat production may have been six times greater than at present, mostly due to ^{40}K and ^{235}U, relatively short-lived radionuclides whose abundances are now much reduced (Figure 3.19). ^{40}K, however, is still largely responsible for the radioactivity of seawater.) The hot accretion theories visualize the Earth's core as forming either during or immediately after the slow accretion process (Ringwood, 1960, 1966*a*, *b*; Hanks and Anderson, 1969; Cameron, 1970), whereas in the cold accretion models (Kuiper, 1952; Urey, 1952, 1957; Runcorn, 1962; Wood, 1962; Elsasser, 1963; Birch, 1965; Larimer and Anders, 1967) the core was formed by subsequent radioactive heating and melting. On the basis of the distribution of lead in the metallic and silicate phases, Overby and Ringwood (1972) have recently argued that the time interval between the accretion of the Earth and core formation was relatively brief ($<5 \times 10^8$ yr) and that the

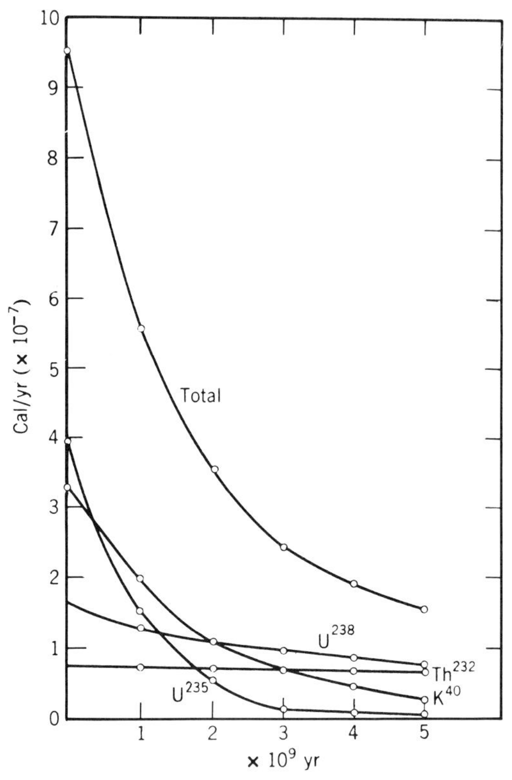

Figure 3.19. Radiogenic heat formation in the Earth by U, Th, and ^{40}K (from Mason, 1966).

cold accretion models are therefore untenable. To return to the cold Earth model, Elsasser (1963) has sketched out an imaginative hypothesis of melting and differentiation. At a zone several hundred kilometers deep over a period of 100,000 yr, the iron of the formerly cold and homogeneous Earth melted to form a large droplet that then slowly sank to the center to form the core and forced the silicates upward (the formation and fall of such droplets can be readily observed in those popular two immiscible fluid television lamps). A feature of this hypothesis that I find particularly satisfying is that such an unsymmetrical process nicely accounts for the otherwise baffling unequal distribution of the continental land masses, that is to say, the location of the original protocontinent, in the northern hemisphere in the crustal region above the great fallen drop. It might also explain the "bottomless" Pacific Ocean, a feature so strange that some have speculated that the Moon was torn from the Earth and that the Pacific is the wound. The sinking of the droplet and the fractionation and rise of the silicate material could have easily disrupted and swallowed up any original surface "crust," thus no crustal rocks older than 3.5×10^9 yr have been found. Eventually, as the radioactive heating diminished, the churning Earth quieted down, the mantle film solidified, and the great stable crustal blocks began to take their form. On the other hand, other geologists imagine the crust initially solidifying as a more or less uniform basaltic cover with little hint of the shapes of future continental and ocean basin differentiation.

BIBLIOGRAPHY

T. H. van Andel, *Science*, *160*, 1419 (1968).

D. L. Anderson and C. Sammis, *Geofis. Internac.*, *9*, 3 (1969).

D. L. Anderson, C. Sammis, and T. Jordan, *Science*, *171*, 1103 (1971).

A. S. Balchan and G. R. Cowan, *J. Geophys. Res.*, *71*, 3577 (1966).

F. Birch, *J. Geophys. Res.*, *57*, 227 (1952).

F. Birch, *Geophys. J. Roy. Astron. Soc.*, *4*, 295 (1961).

F. Birch, *J. Geophys. Res.*, *69*, 4377 (1964).

F. Birch, *Bull. Geol. Soc. Amer.*, *76*, 133 (1965).

B. A. Bolt and A. Quamar, *Nature*, *228*, 148 (1970).

F. R. Boyd, *Science*, *145*, 13 (1964).

E. C. Bullard, J. E. Everett, and A. G. Smith, *Phil. Trans. Roy. Soc. London, Ser. A*, *258*, 41 (1965).

K. E. Bullen, in T. F. Gaskell (ed.), *The Earth's Mantle*, Academic, New York, 1967.

A. G. W. Cameron, *Trans. Amer. Geophys. Union*, *51*, 628 (1970).

D. A. Carswell, *Contrib. Mineral. Petrol.*, *19*, 97 (1968).

F. W. Clark and H. S. Washington, U.S. Geol. Surv. Prof. Paper No. 127 (1924).

S. P. Clark and A. E. Ringwood, *Rev. Geophys.*, *2*, 35 (1964).

W. H. Dennen and B. R. Moore, *Nature Phys. Sci.*, *234*, 127 (1971).

W. R. Dickinson and W. C. Luth, *Science*, *174*, 400 (1971).

D. W. Durney, *Nature*, *235*, 315 (1972).

W. M. Elsasser, in J. Geiss and E. Goldberg (eds.), *Earth Science and Meteoritics*, Wiley, New York, 1963.

D. A. Emilia and D. F. Heinrichs, *Science*, *166*, 1267 (1969).

W. G. Ernst, *Earth Materials*, Prentice-Hall, Englewood Cliffs, 1969.

B. B. Gardiner and F. D. Stacey, *Phys. Earth Planet. Interiors*, *4*, 406 (1971).

P. W. Gast, *Geochim. Cosmochim. Acta*, *32*, 1057 (1968).

V. M. Goldschmidt, *Geochemistry*, Clarendon, Oxford, 1954.

G. S. Gorshkov, *Volcanism and the Upper Mantle*, Plenum, New York, 1970.

D. H. Green, *J. Petrol.*, *5*, 134 (1964).

D. H. Green and A. E. Ringwood, *J. Geophys. Res.*, *68*, 937 (1963).

H. T. Hall and V. R. Murthy, *Earth Planet. Sci. Lett.*, *11*, 239 (1971).

W. Hamilton, *Geol. Soc. Amer. Bull.*, *80*, 2409 (1969).

T. Hanks and D. L. Anderson, *Phys. Earth Planet. Interiors*, *2*, 19 (1969).

P. G. Harris et al., *J. Geophys. Res.*, *72*, 6359 (1967).

T. Hatherton and W. R. Dickinson, *J. Geophys. Res.*, *74*, 5301 (1969).

H. H. Hess and G. Otalora, in C. A. Burk (ed.), *A Study of Serpentinite*, Nat. Acad. Sci. U.S.A.–Nat. Res. Council Pub. No. 1188 (1964).

H. H. Holland, in P. J. Brancazio and A. G. W. Cameron (eds.), *The Origin and Evolution of Atmospheres and Oceans*, Wiley, New York, 1964.

A. Jain and R. Evans, *Nature Phys. Sci.*, *235*, 165 (1972).

B. R. Julian, D. Davies, and R. M. Sheppard, *Nature*, *235*, 317 (1972).

L. Knopoff and G. J. F. MacDonald, *Geophys. J. Roy. Astron. Soc.*, *3*, 68 (1960).

G. P. Kuiper, *The Atmospheres of the Earth and Planets*, Univ. Chicago Press, Chicago, 1952.

I. Kushiro, Y. Syono, and S. Akimoto, *J. Geophys. Res.*, *73*, 6023 (1968).

I. B. Lambert and P. J. Wyllie, *Nature*, *219*, 1240 (1968).

I. B. Lambert and P. J. Wyllie, *Science*, *169*, 764 (1970).

J. W. Larimer, *Geochim. Cosmochim. Acta*, *31*, 1215 (1967).

J. W. Larimer and E. Anders, *Geochim. Cosmochim. Acta*, *31*, 1239 (1967).

J. F. Lewis, *Earth Planet. Sci. Lett.*, *11*, 130 (1971).

F. MacIntyre, *Sci. Amer.*, *223*(5), 104 (November 1970).

B. Mason, *Principles of Geochemistry*, 3rd ed., New York, 1966.

A. E. Maxwell (ed.), *The Sea*, Vol. 4, Wiley-Interscience, New York, 1970.

A. R. McBirney, *Oregon Dept. Geol. Mineral Ind. Bull.*, *65*, 185 (1969).

R. K. McConnell, Jr., *J. Geophys. Res.*, *70*, 5171 (1965).

R. G. McQueen and S. P. Marsh, *J. Geophys. Res.*, *71*, 1751 (1966).

Y. Miyake, *Elements of Geochemistry*, Maruzen, Tokyo, 1965.

W. J. Morgan, *Nature*, *230*, 42 (1971).

V. R. Murthy and H. T. Hall, *Phys. Earth Planet. Interiors*, *2*, 276 (1970).

G. D. Nicholls, in S. K. Runcorn (ed.), *Mantles of the Earth and Terrestrial Planets*, Wiley-Interscience, New York, 1967.

V. M. Overby and A. E. Ringwood, *Nature*, *234*, 463 (1972).

E. R. Oxburgh and D. L. Turcotte, *J. Geophys. Res.*, *73*, 2643 (1968).

L. C. Pakiser, "The Basalt-Eclogite Transition and the Crustal Structure in the Western U.S.," *U.S. Geol. Surv. Prof. Paper 525-B*, 1 (1965).

A. Poldervaart (ed.), *Crust of the Earth*, Geol. Soc. Amer. Special Paper 62 (1955).

F. Press, *J. Geophys. Res.*, *73*, 5223 (1968).

F. Press, *Science*, *160*, 1218 (1968); *165*, 174 (1969).

A. E. Ringwood, *Geochim. Cosmochim. Acta*, *20*, 241 (1960).

A. E. Ringwood, *Geochim. Cosmochim. Acta*, *30*, 41 (1966a).

A. E. Ringwood, in P. M. Hurley (ed.), *Advances in Earth Sciences*, M.I.T. Press, Cambridge, 1966b.

A. E. Ringwood, in P. J. Hart (ed.), *The Earth's Crust and Upper Mantle*, Geophys. Monogr. No. 13, Amer. Geophys. Union, Washington, D.C., 1969a.

A. E. Ringwood, *Earth Planet. Sci. Lett.*, *5*, 401 (1969b).

A. E. Ringwood, *Phys. Earth Planet. Interiors*, *3*, 109 (1970)

S. K. Runcorn (ed.), *Continental Drift*, Academic, New York, 1962.

A. B. Ronov and A. A. Yaroshevsky, in P. J. Hart (ed.), *The Earth's Crust and Upper Mantle*, Geophys. Monogr. No. 13, Amer. Geophys. Union, Washington, D.C., 1969.

J. K. J. Spar, *Earth, Sea, and Air*, Addison-Wesley, Reading, Mass., 1965.

R. N. Thompson, *Nature*, *236*, 106 (1972).

J. I. Tracey, Jr., et al., *Initial Reports of the Deep Sea Drilling Project* (JOIDES), NSF-Scripps Inst. Ocean., Superintendent Doc., Washington, D.C., 1971.

H. C. Urey, *The Planets*, Yale Univ. Press, New Haven, 1952.

H. C. Urey, in L. Ahrens, F. Press, K. Rankama, and S. K. Runcorn (eds.), *Physics and Chemistry of the Earth*, Vol. 2, Pergamon, New York, 1957.

H. S. Washington, *Amer. J. Sci.*, *9*, 351 (1925).

A. Wegener, *The Origin of Continents and Oceans*, Dover, New York, 1966.

P. W. Weigand et al., *Nature Phys. Sci.*, *235*, 31 (1972).

J. T. Wilson, *Sci. Amer.*, *208*(4), 86 (April, 1963).
J. T. Wilson (ed.), *Continents Adrift*, Freeman, San Francisco, 1972.
J. A. Wood, *Nature*, *194*, 127 (1962).
P. J. Wyllie, *The Dynamic Earth*, New York, 1971.

ADDITIONAL READING

W. S. Broecker and V. M. Overby, *Chemical Equilibria in the Earth*, McGraw-Hill, New York, 1971.
K. E. Bullen, *An Introduction to the Theory of Seismology*, 3rd ed., Chapters 12 and 13, Cambridge Univ. Press, Cambridge (England), 1963.
S. P. Clark and A. E. Ringwood, *Rev. Geophys.*, *2*, 35 (1964).
R. H. Dott, Jr., and R. L. Batten, *Evolution of the Earth*, McGraw-Hill, New York, 1971.
W. G. Ernst, *Earth Materials*, Prentice-Hall, Englewood Cliffs, 1969.
M. Fleischer (gen. ed.), *The Data of Geochemistry*, U.S. Geol. Survey Prof. Paper No. 440, U.S. Gov. Print. Off., Washington, D.C., 1962.
J. G. Heacock (ed.), *The Structure and Physical Properties of the Earth's Crust*, Amer. Geophys. Union, Washington, D.C., 1971 (Geophys. Monogr. No. 14).
H. H. Hess and A. Poldervaart (eds.), *Basalts*, Interscience, New York, 1967.
P. M. Hurley (ed.), *Advances in Earth Sciences*, M.I.T. Press, Cambridge, 1966.
H. Jeffreys, *The Earth*, 4th ed., Cambridge Univ. Press, Cambridge (England), 1959.
L. Knopoff, C. L. Drake, and P. J. Hart (eds.), *The Crust and Upper Mantle of the Pacific Area*, Geophys. Monogr. No. 12, Amer. Geophys. Union, Washington, D.C., 1968.
K. B. Krauskopf, *Introduction to Geochemistry*, McGraw-Hill, New York, 1967.
G. P. Kuiper (ed.), *The Earth as a Planet*, Univ. Chicago Press, Chicago, 1954.
R. A. Phinney (ed.), *The History of the Earth's Crust*, Princeton Univ. Press, Princeton, 1968.
K. Rankama and T. G. Sahama, *Geochemistry*, Univ. Chicago Press, Chicago, 1950.
A. E. Ringwood and D. H. Green (eds.), "Phase Transformations and the Earth's Interior" (Symp. Proc.), *Phys. Earth Planet Interiors*, Vol. 3 (1970).
A. Rittmann, *Volcanoes and Their Activity*, Wiley, New York, 1962.
J. S. Steinhart and T. J. Smith (eds.), *The Earth Beneath the Continents*, Geophys. Monogr. No. 10, Amer. Geophys. Union, Washington, D.C., 1966.
S. R. Taylor, *Geochim. Cosmochim. Acta*, *28*, 1273 (1964).
K. H. Wedepohl (ed.), *Handbook of Geochemistry*, Springer-Verlag, Heidelberg, 1969.
H. Williams, F. J. Turner, and C. M. Gilbert, *Petrography*, Freeman, San Francisco, 1954.
P. J. Wyllie (ed.), *Ultramafic and Related Rocks*, Wiley, New York, 1967.
A. Volborth, *Elemental Analysis in Geochemistry*, Elsevier, New York, 1969.

4

THE EXPLOITATION OF MINERAL RESOURCES AND THE ABUSE OF THE EARTH

4.1 INTRODUCTION

The trouble with books is that they contain so much information that the most important statements tend to get lost in the crowd. In order to avoid this difficulty, let me list right here what I feel are the five most crucial summary statements concerning our environment:

1. Until space travel becomes commonplace, the Earth is a closed system with respect to materials.[1]
2. Human perturbation of the environment has now become comparable in magnitude to natural processes.
3. The Principle of Le Chatelier is inescapable and governs all processes in the environment.
4. Water is the central actor in the environmental drama, and the chemistry of our environment is largely the chemistry of aqueous solution and of gases, colloids, and solids in the presence of water.
5. Human activity affects the environment in the forms of resource depletion and pollution.

In this chapter, I wish to dwell on the importance of the fifth statement. We must constantly bear in mind the twofold nature of the environmental problem—pollution *and* resource depletion. Man, like nearly all other living things, moves chemicals about in the environment; he takes them from their natural location (resource depletion) and redeposits them somewhere they do not belong (pollution). Presently, pollution is the object of a great deal of public concern, but in the larger and longer view, resource depletion is probably an even greater threat to the continued existence of human civilization. In the United States, for example, population growth and the resulting pollution have noticeably degraded the quality of life over the past 20 yr or more that I can remember. But for all the ugliness and unpleasantness, it has created pollution is still far from representing a serious menace to the survival of

[1] But this is not true with respect to energy. This statement also ignores the relatively minor material transport from gas escape and meteroritic accretion.

our species. I rather doubt that, with the exception of radioactive materials, man is likely to pollute his environment to the verge of his own extinction. It is a useful exaggeration to describe man as a "threatened species." What is threatened is not man's survival, but something every bit as important—the survival of civilization. The material demands of our technological culture are enormous and growing. Any thought of some sort of a return to pretechnological culture is stupid, even a slowing of the pace of technological advance will be exceedingly difficult, if possible. At the present time, we are in sight of resource depletion that will profoundly alter our way of living. To be more specific, at the present time, how many people die of pollution compared to the many that die of hunger and malnutrition as a consequence of improper resource utilization? How many years does air pollution subtract from the GWLE (gross world life expectancy) compared to malnutrition and famine? Millions of the world's hungry would welcome contaminated grain. As for the strains upon our technological civilization, although the era of the great shortages has not begun yet (we did get a foretaste of it during World War II), as our society gets worn and torn apart, the gloomy prophecies of Spengler appear closer.

The implications of resource depletion are far more grievous than simply electricity rationing and no more automobiles. The competition for remaining resources will become more intense. In order to satisfy their consumption demands, nations will become more stern, more rapacious, both internally and externally. The nature of society will change; the benevolence of human morality will wither away as the pinch becomes stronger. Not only will garages be empty, but privacy will be extinguished, natural aesthetic values will be abandoned, state control will become total, and human rights will be indefinitely suspended. Resource-rich underdeveloped nations will fall victims of genocide. These are the sort of prices that the relentless operation of the Principle of Le Chatelier will exact. Far from being distant apocalyptic visions, a perceptive observer can detect the omens of these future events in the attrition of altruism in the foreign policy of our own country and, within, the growth of federal regulation. The choice between clean air and freedom is a difficult one to make, and we have our own fertility to blame for forcing us to make it.

With respect to pollution the spoilation of the atmosphere and hydrosphere have been most obvious, but the full brunt of resource depletion has been borne by the biosphere (species depletion and extinction) and especially by the lithosphere. The lithosphere is exploited for virtually all of the inorganic material needs of the human race. In this chapter on the use and abuse of the Earth, we will deal first with the exploitation of mineral resources, then with the physical alteration of the face of the land, and finally with waste disposal in and on the lithosphere.

4.2 EARTH RESOURCES

The chemical resources of the lithosphere fall into several broad categories:

1. Elements for metal production and technology,
2. Building materials,
3. Minerals for the chemical industry,
4. Minerals for agriculture,
5. Fossil fuels,

6. Nuclear fuels, and
7. Water.

Table A.20 lists the principle sources, estimated annual consumption, and value of the chemical elements. Conventionally, resources have been classified as renewable or nonrenewable; however, the distinction can be unclear and is rapidly becoming more so. Food and plant materials are seasonally regenerated, and we therefore speak of their supply as being renewable, although, of course, since the influx of solar energy is constant and the world's potentially productive land area is steadily decreasing, the supply is by no means infinite. On the other hand, the geological fossilation processes are so slow that the supply of fossil fuels such as coal, natural gas, and petroleum are for all practical purposes fixed and not renewable, and if these resources continue to be exploited in an unmindful manner they will be exhausted. Most mineral resources are regarded as nonrenewable, and once removed from the Earth, they are returned in the form of useless wastes—iron as junk cars and dumps full of cans, building material as building rubble. In principle, a nonrenewable resources can be made a renewable one by recycling, but presently recycling is not always economically feasible, especially for those very materials consumed in greatest amounts. Also, it must be remembered that even the best recycling systems are leaky, so some material will always be lost. Again in principle, water is a renewable resource thanks to the hydrologic cycle, but in fact, unwise use and pollution commonly transform the water into a nonrenewable resource.

Analysis and evaluation of our potential reserves of nonrenewable resources is at best an exceedingly difficult and uncertain business involving consideration of a great number of factors (see Lovering, 1969). A given chunk of crustal rock, just like the oft-cited cubic mile of seawater, contains just about all of the naturally occurring chemical elements, but neither the rock nor the seawater can be a panacea for our material needs for the concentraton of most elements in both is so small that enormous quantities of the source material would have to be processed with the expenditure of ridiculously large amounts of energy. In the real world, man gets his metals, for example, from crustal materials in which much of the work of concentration has already been done by nature, that is to say, from enriched materials or ores. As a rich ore is exhausted, man is forced to turn to poorer sources: As we exhaust our crude oil supplies, we will have to turn more and more to tar sands and oil shales in addition to developing alternative energy sources. Rising labor costs can make the further exploitation of an ore uneconomic. The ghost towns of the American west are testimony to the impact of extraction costs. There is still "gold in them thar hills," but following unionization, mining and processing costs became too expensive. Similarly, unionization has brought the subsurface anthracite coal mining industry in the United States to the verge of extinction, encouraged environmentally disasterous strip mining, and accelerated the growth of use of alternative fuels such as oil and gas. On the other hand, a new need can create new economic ores and, even more important, technological advance by developing new extraction techniques can convert a previously useless ore into a profitable one.

In addition to the unavoidable arbitrariness in distinguishing the fluctuating boundary that divides a possible source from an economic ore, the difficulty of estimating potential reserves is compounded quite simply by our lack of geochemical knowledge. The longest-worked deposits tend to be located in the better explored regions of the Earth, but, although spectacular discoveries of vast mineral riches continue to be made, Skinner (1969) warns that "unfortunately, there is no evidence that discovery rates of such rich, easily found deposits will long continue; on the contrary, in the more carefully prospected areas such as the United States and Europe, the discovery rate has already decreased drastically." Except for Al and Cr (metals that the Romans did not use), all of the mines now operated in that portion of Europe contained in the Roman Empire were known to the Romans. "Despite close settlement and keen observation by members of a society aware of the importance of metals, no major discoveries were made during a period of almost 2,000 years."

Those deposits which have been discovered are largely the more obvious surface deposits. There are also subsurface or "blind" ore deposits. But there is not a great deal of optimism concerning our ability to find such deposits. Prospecting clearly is going to become a much more sophisticated and difficult activity making greater and greater demands on geological science and improved technology.

Metals

Skinner (1969) divides metallic mineral resources into abundant metals, including Fe, Al, Cr, Mn, Ti, and Mg, and scarce metals such as Cu, Pb, Zn, Sn, W, Au, Ag, Pt, U, Hg, Mo, and so on. Of the first class, we will discuss here as our examples iron and aluminum.

The processing and use of iron may be said to characterize technological civilization, and scholars have very properly divided human prehistory into the Ages of Stone, Bronze, and Iron. Prior to the discovery of iron ore smelting by the Sumerians some 5000 yr ago, iron was known only in its meteoritic form, in fact the ancient Egyptian name for iron, "Ba-en-pet," means "metal of the sky." We are still living in the Iron Age. Iron still accounts for more than 95% of all metals consumed, and of the remainder many such as Ni, Cr, W, V, Co, and Mn are used largely to alloy with iron.

In the period 1930 to 1960, United States' production of iron ore increased only slightly, remaining around 100,000,000 ton/yr, but world production climbed from 200,000,000 to more than 500,000,000 ton/yr. About 40% of obsolete steel in the United States gets recycled. The mean lifetime of steel products varies widely from a few weeks for cans to more than 50 yr for girders and concrete reinforcement with a mean lifetime of 25 to 30 yr (Brown, 1970). In 1966, the United States imported about 34% of its iron ore with much of it coming from Canada (16%), Venezuela (9%), Liberia (3%), Brazil (2%), Chile (2%), and Peru (1%) (Skinner, 1969). Fast growing Japan is totally dependent on imported iron. Because of improved transportations systems, it is no longer economically necessary to manufacture iron near its raw materials. In addition to the iron ore, these materials are coke and limestone. Air injected into the blast furnace (Figure 4.1) partially burns the coke to form carbon monoxide

$$2C + O_2 \rightarrow 2CO \tag{4.1}$$

and any CO_2 formed is reduced back to CO by the hot coke higher in the furnace. About 6 to 8 ft below the top of the furnace at temperatures of about 500°C, the iron ore, say hematite, Fe_2O_3, undergoes its first reductive step

$$3Fe_2O_3 + CO \rightarrow 2Fe_3O_4 + CO_2 \tag{4.2}$$

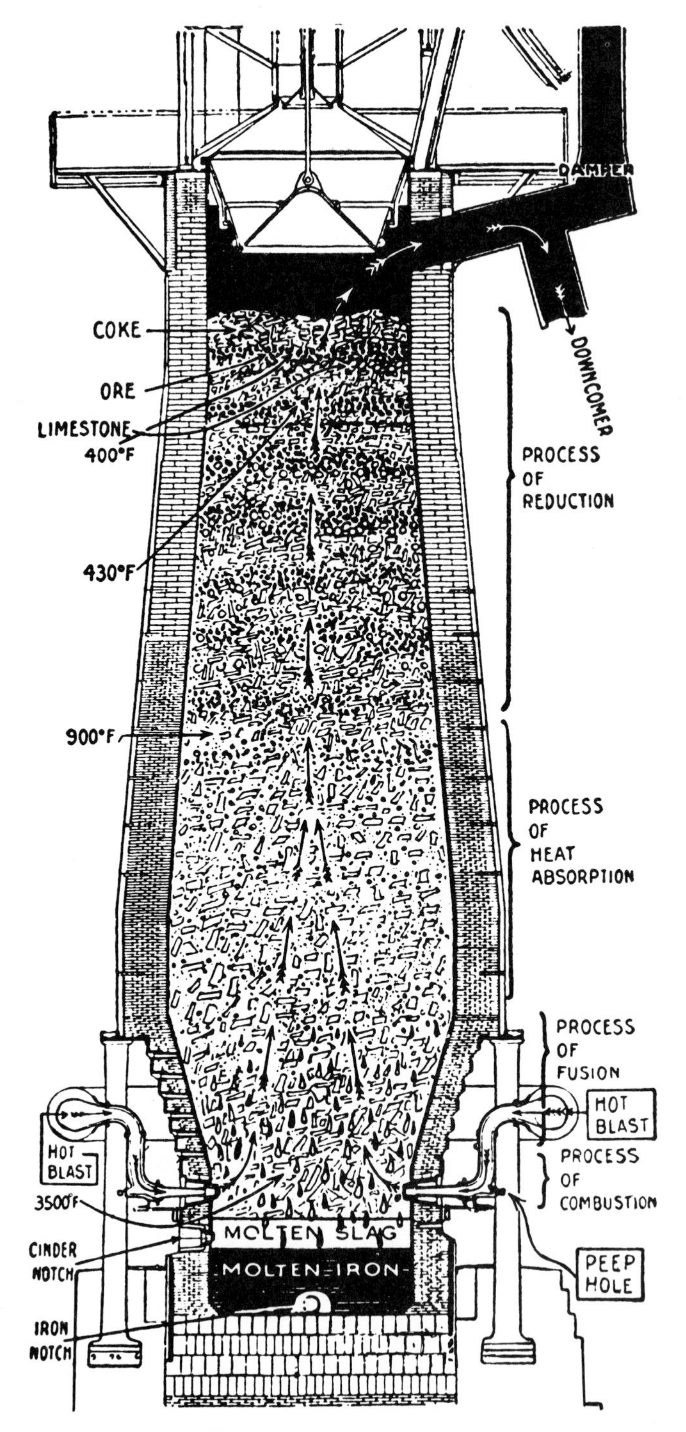

Figure 4.1. The blast furnace.

At 20 ft below the top at temperatures of about 850°C, the second reduction step occurs

$$Fe_3O_4 + CO \rightarrow 3FeO + CO_2 \quad (4.3)$$

and at temperatures of about 1000°C, the final reduction step takes place

$$FeO + CO \rightarrow Fe + CO_2 \quad (4.4)$$

About halfway down the furnace, the limestone is converted to lime

$$CaCO_3 \rightarrow CaO + CO_2 \quad (4.5)$$

which in turn converts impurities such as silica and alumina into molten slag

$$CaO + SiO_2 \rightarrow CaSiO_3 \quad (4.6)$$

$$CaO + Al_2O_3 \rightarrow Ca(AlO_2)_2 \quad (4.7)$$

Pig iron, the crude product of the blast furnace, contains 92 to 94% iron along with impurities such as S, P, Mn, Si, and C. S, P, Si, and varying amounts of C can be removed from the pig iron by a number of burning-off processes (open hearth, Bessemer, duplex, crucible, or electric furnace) to make steels. Finally, alloying elements such as Mn, Cr, V, and Ni can be added to the steel to give it desireable special properties (such as stainless steel).

As we have seen, the cosmic abundance of iron is high, and the element is widely distributed in the Earth's crust in a great many mineral forms. However only four of these—magnetite (Fe_3O_4), hematite (Fe_2O_3), goethite or limonite ($HFeO_2$), and siderite ($FeCO_3$)—are important ores. Other common iron-bearing minerals such as pyrite (FeS_2) and chamosite ($Fe_2Al_2SiO_5(OH)_4$) probably will not become significant iron resources because of the great difficulty of extraction of the metal from these materials. With respect to their mode of formation, iron ore deposits tend to fall into three categories: (a) deposits associated with igneous rocks, (b) residual deposits, and (*c*) sedimentary deposits. Three mechanisms may be further distinguished in the first category—magmatic segregation, contact metamorphasis, and hydrothermal deposits. The contact metamorphic deposit of magnetite near Cornwall, Pennsylvania, was formed when diabase, a mafic igneous rock, intruded during the Triassic Period into Cambrian sedimentary rocks. The great Kiruna deposit in Swedish Lapland is an extraordinary example of magmatic segregation. Residual iron deposits are formed when the Fe(II) in rock is oxidized up to the relatively insoluble compounds of Fe(III), and, while widespread, the individual deposits of such brown ores (color due to geothite) tend to be small. Thus although they were the first ores to be worked by man, their relative importance is diminishing.

Most of the iron today comes from sedimentary deposits. As we will see later, with the advent of photosynthesis the Earth's atmosphere and hydrosphere were transformed from reducing to oxidizing environments. The oxygen dissolved in seawater and oxidized the dissolved Fe(II) up to more insoluble forms of Fe(III) which then, some 3.2 to 1.7 billion yr ago, were laid down on the ocean floor in finely banded sediments. In addition to these enormous banded iron formations or so-called Lake Superior ores, deposition continued to occur in the post-Cambrian period. But whereas the former appears to have occurred over large shallow seas, the latter appears to have been restricted to certain basins. The original sediment may have been enriched by natural leaching processes, as in the case of the Cerro Bolivar, Venezuela, deposits, but the enriched layer is commonly underlaid with an unconcentrated iron formation. Only recently has improved technology and the exhaustion of enriched ore deposits made the mining and benefication of these relatively low grade "taconites" commercially feasible on a large scale. In 1968, 40% of the iron in the United States came from taconite ores, and by 1978, it has been estimated that the percentage will increase to 75%. Happily enough, inasmuch as there are numerous taconite reserves throughout the world with potentials in excess of 10^{11} to 10^{12} tons of iron each, for all the iron he consumes and wastes it seems most unlikely that man will exhaust the supply of this vital metal. Nevertheless, we must remember that while 18 rich nations totaling 680,000,000 people have an annual per capita steel consumption of 300 to 700 kg, 13 poor nations representing 1,400,000,000 people have an annual per capita consumption of only 10 to 25 kg. In addition, there are 400,000,000 people even poorer and another about 440,000,000 in the intermediate range. While not exhausting the iron in the Earth, continued population growth and continued national development could easily exceed the human capability of extracting sufficient metal. "Such a demand would clearly place enormous strains on the Earth's resources and would greatly accentuate rivalries between nations for the Earth's remaining deposits of relatively high grade ores" (Brown, 1970).

We have dubbed our own times as the Atomic Age or the Space Age, but with his broader perspective, some future archaeologist (possibly from another planet) digging in our dumps and cemeteries (for it is from them that the future will know us) may very well style us as the beginning of the Age of Aluminum. In contrast to the rather even iron consumption in the period 1920 to 1960, the per capita consumption of Al in the United States leaped from less than 2 lb/yr to more than 22 lb/yr. Aluminum consumption is characteristic of the most highly developed countries: In 1964, the United States with only 299 of the world's 3356 million people had a per capita Al consumption of 33 lb/yr compared with the world average of only 3.6 lb/yr. The enormous and fast growing consumption of this metal is even more spectacular in the light of the fact that prior to this century, although discovered by Oersted in 1825, Al was a very rare and costly (about $160/lb) metal. I have seen jewelry settings and elaborate dinner plate in Al that, while no doubt status symbols of the very rich in their day, now, because of the cheapness of the metal have a rather ridiculous air. Once material for the smallest jewelry, now the world's two tallest buildings, the World Trade Center in New York, are made of aluminum. The great advance in the technology of this metal occurred in 1886 when Charles Hall, a young student at Oberlin College, showed that Al can be recovered by the electrolysis of its oxide dissolved in cryolite. The bauxite ore (see following discussion) is purified by digestion under pressure with NaOH and hydrolyzed to precipitate alumina (Bayer process)

$$Al0_2^- + 2H_2O \rightarrow Al(OH)_3\downarrow + OH^- \tag{4.8}$$

The hydroxide is converted to the oxide by ignition. The oxide is then dissolved in molten cryolite (Na_3AlF_6) or a mixture of fluorides ($6NaF: 3CaF_2: 2AlF_3$) and the solution electrolyzed to produce the metal

$$2Al_2O_3 \rightarrow 4Al + 3O_2 \tag{4.9}$$

If even greater purity is required, the Al may be further refined electrolytically in a Hoopes cell. Aluminum production uses prodigious quantities of electrical power. In 1965, the aluminum industry consumed about 3% of all the power generated in the United States.

Among current and potentially important Al ore minerals are boehmite and diaspore ($HAlO_2$ dimorphs); gibbsite (H_3AlO_3); andalusite, kyanite, and sillimanite (Al_2SiO_5 trimorphs); kaolinite ($Al_2Si_5O_5(OH)_4$, the most aluminous clay); anorthite ($CaAl_2Si_2O_8$, the most aluminous feldspar); and nepheline. While Al is very plentiful in crustal materials, Al-rich ores are formed under very special conditions. Minerals formed by igneous and metamorphic processes deep in the Earth's crust usually contain little or no water. When brought to the surface these highly anhydrous materials are no longer stable and are slowly hydrolyzed by chemical weathering. The relatively soluble compounds of elements such as Na, K, Ca, and Mg are leached away leaving a laterite capping (Figure 4.2). Most laterites are Fe-rich, but those rich in Al are called Bauxites (Valeton, 1972). Many low silica content rocks, including clay-containing limestones, can be the parents of bauxite; however, because of the sensitivity of their formation conditions (if the leaching water is too acid, all the kaolinite will dissolve instead of just the silica component) and the vulnerability of such surface deposits to erosion (glaciation, for example, will readily scrape away deposits), extensive bauxite deposits tend to be restricted to relatively young geological formations in tropical or once tropical regions of the world. After World

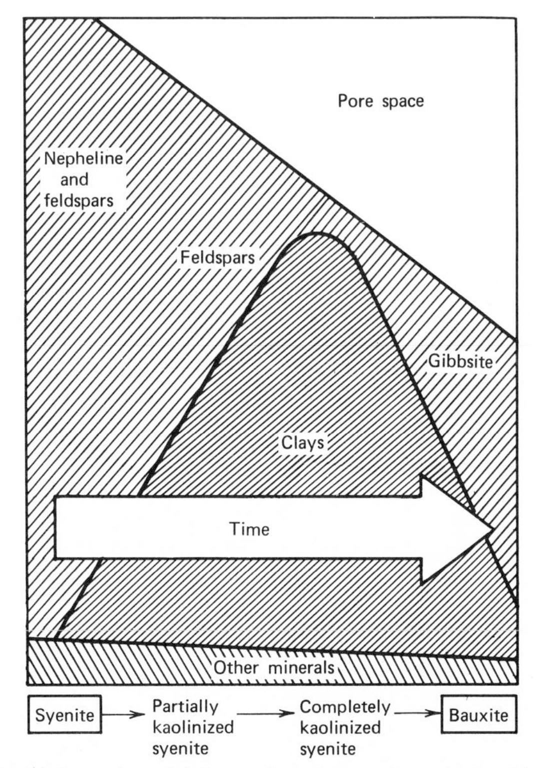

Figure 4.2. Steps in the formation of Arkansas bauxite ores from the leaching of nepheline syenite.

War II, extensive exploration revealed large reserves of rich bauxites in the tropics, and the measured and indicated world reserve is now estimated to be some 5.8×10^9 tons (Patterson, 1967). While very large, if the aluminum consumption of the developed nations continues to escalate and the number of developed countries continues to increase, even these enormous reserves will become depleted in the next century. In the meantime, we can confidently expect that improved technology will make the utilization of poorer ores economic, thus insuring an adequate source of this metal for the foreseeable future.

Skinner (1969) further classifies the scarce metals on the chemical basis of their major ore deposits: (*a*) metals that commonly form sulfide minerals (Cu, Pb, Zn, Ni, Mo, Ag, As, Sb, Bi, Cd, Co, and Hg), (*b*) metals that commonly occur in the native state (Pt, Pd, Rh, Ir, Ru, Os, and Au), and (*c*) metals that commonly form oxide or silicate minerals (W, Ta, V, Nb, Sn, and Be). Before discussing two scarce metals, copper and silver, as our examples, I would like to make a couple of points regarding the geological distribution of scarce metals. The scarce metals tend to be widely distributed in common rocks and rarely form separate minerals. Often an ion of the scarce metal substitutes for a similarly sized and charged cation of a more abundant metal in the mineral crystal lattice; thus Ni substitutes for Mg in olivine (Mg_2SiO_4), In for Zn in sphalerite (ZnS), Ag for Cu in tetrahedrite ($Cu_{12}Sb_4S_3$) and chalcocite (Cu_2S), and for Pb in galena (PbS). If the amount of scarce element substitution is small, the resulting rock will be highly stable, and the recovery of the metal correspondingly difficult. As the amount of scarce element substitution into the lattice becomes larger, strains will be induced, the crystal structure will become

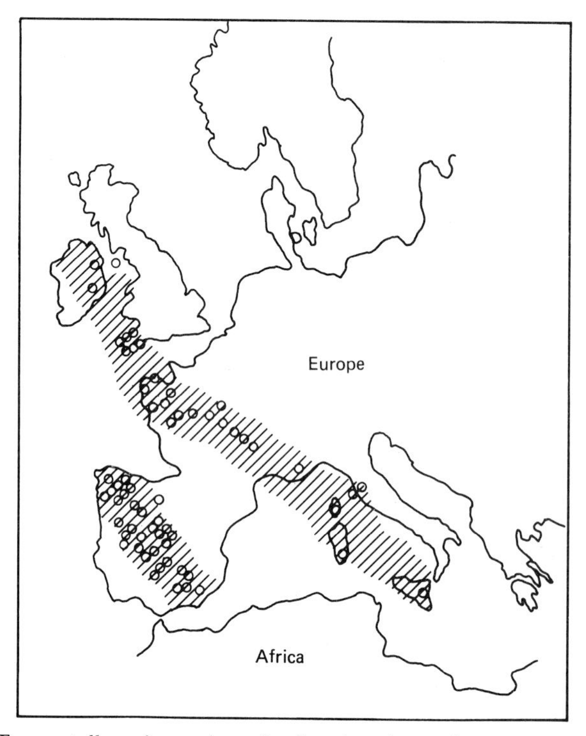

Figure 4.3. Two metallogenic provinces for tin minerals running across southern Europe.

less stable—the substituting element may even form a separate mineral—and beneficiation both by natural and technological processes can be facilitated. Another practical consequence of substitution, the demand for many scarce metals, can be met largely as a by-product from the production of other metals. For example, in 1966, of the ten largest Ag producers in the United States, five were Pb or Zn producers, three Cu producers, and only two were primarily silver producers. Perhaps surprising the deposits of scarce metals are not randomly distributed; on the contrary, one or more metals often tend to be grouped in distinct geographic belts called *metallogenic provinces*, and two tin metallogenic provinces in Europe are shown in Figure 4.3. In many instances, the geologic processes responsible for these formations is still inadequately understood.

Now to return to our two examples of scarce metals, although aluminum can perform many of the uses of copper, notably electrical, thereby stretching our copper reserves, the United States production of copper has climbed erratically and slowly over the past 40 yr (Figure 4.4), whereas the quality of the ore to which we have had to resort has diminished, not so much because we have exhausted the rich vein ores but rather because the high costs of subterranean mining has made the exploitation of lower grade surface porphyry copper deposits by open pit techniques attractive. Porphyry copper, commonly as the mineral chalcopyrite ($CuFeS_2$), appears to be hydrothermal deposits, and about 10 yr ago we gained some insight into the possible mechanism for the formation of such deposits when metal-rich hot brines were discovered in deep drilled wells near the Salton Sea and in pools at the bottom of the Red Sea (Degens and Ross, 1969). Copper ores are also found in stratiform deposits, the most famous of which is the Kupferschiefer shale laid down as highly organic mud rich in the sulfides of Cu, Pb, and Zn in Eastern Europe from a shallow sea in Permian times. About 50% of the world's copper production comes from porphyry copper, largely from the United States, and 25% from stratiform deposits, principally in Zambia (Table 4.1). Brown (1970) has estimated that world copper reserves will last a little beyond the year 2000 (Table 4.2), and Brooks (1972) has jointed his

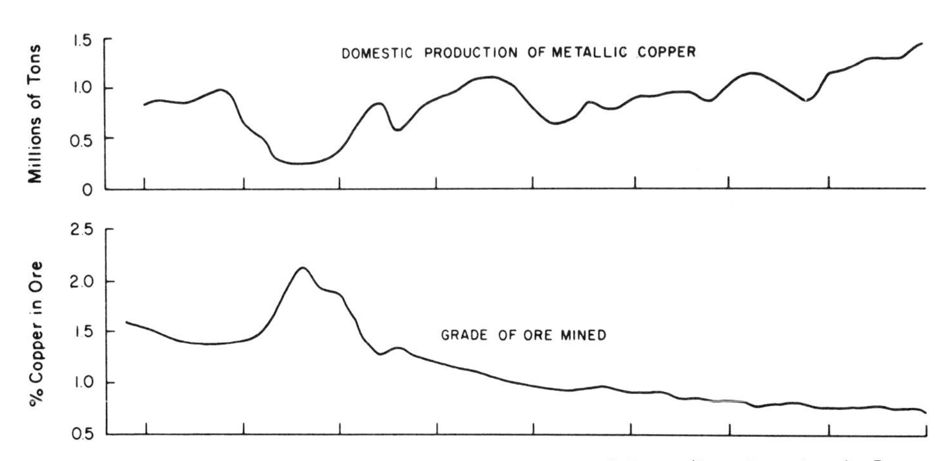

Figure 4.4. Copper production and ore quality in the United States (from Lovering, in *Resources and Man: A Study and Recommendations by the Committee on Resources and Man of the Division of Earth Sciences*, National Academy of Sciences–National Research Council, with cooperation of the Division of Biology and Agriculture. Freeman, © 1969, with permission of the publisher).

Table 4.1 Copper Production and Reserves

Country	Production (1967) (in thousands of tons)	Reserves (1965) (in thousands of tons)
U.S.A.	954	32,500
U.S.S.R.	880	35,000
Chile	732	46,000
Zambia	730	25,000
Canada	603	8,400
Congo	353	20,000
Peru	200	12,500
Others	884	33,000
	5436	212,400

Table 4.2 Estimated Lifetime of Metal Reserves (based on assumptions of continued population growth, rising per capita demand, and new reserve discovery)

	World Reserves to the Year	United States Reserves to the Year
Iron and Iron-Alloy Metals		
Fe	2500	2150
Mn	2100	1970
Cr	2500	1975
Ni	2100	1975
Mo	2100	2150
W	2000	1980
Co	2100	1990
Nonferrous Metals		
Cu	2000	1990
Pb	1990	1975
Zn	1990	1985
Sn	1990	1970
Al	2150	1975
Precious Metals		
Au	1990	1975
Ag	1990	1975
Pt	1990	1970

voice to the many others warning us that we simply *cannot* continue to consume materials, including metals, at present rates.

Our second example, silver, is a case in point. Silver minerals commonly occur in hydrothermal vein deposits associated with Pb, Zn, and Cu or as substituted ions in Pb and Cu minerals. Very few deposits are rich enough to be worked for the silver alone, hence most of the present demands are being met, as we have mentioned, by by-product Ag. The photographic and electrical industries are making ever increasing demands on Ag production, and as the cost of Ag soars higher it may become economic to reactivate old mines and to process poorer ores. However Skinner (1969) concludes that "there seems little hope that silver production will grow to meet all demands; rather it is likely that future uses will have to be curtailed to meet the limited supply."

Building Materials

Our steel, aluminum, and glass cities certainly do not imply that traditional building materials such as stone have fallen into disuse. On the contrary, enormous quantities of these and related materials are still used (Table 4.3). Some natural stones are among the most beautiful materials that have delighted civilized man for centuries, and now that the austere influence of the Bauhaus, has begun to wane the appreciation of the merits of stone is being expressed in an increasing degree in important

Table 4.3 Production and Consumption of Construction and Related Materials (From data in Skinner (1969) taken from U. S. Bureau of Mines statistics)

		tons (in 1967 for U.S. unless otherwise noted)
Building Stone	Granite	630,000
	Limestone	560,000
	Sandstone	350,000
	Slate	150,000
	Marble	70,000
	Other	190,000
Crushed Rock		811,000,000 (in 1966)
		91,456,000 (largely limestone for concrete)
Sand and Gravel		905,162,000
Portland Cement Production		93,500,000 (U.S.S.R.
		72,500,000 (U.S.A.)
		47,700,000 (Japan)
		529,000,000 (World)
Plaster		49,629,000 (World)
Clay		54,664,000
Asbestos Production		1,479,000 (Canada) in 1966
		925,000 (U.S.S.R.) in 1966
		3,350,000 (World) in 1966

architectural structures in the United States and elsewhere. But concrete too, apart from its economic and engineering advantages, is not without its aesthetic virtues, especially when used on its own textural terms in the style which has been sometimes called "the new brutalism." In the 50-yr period from 1910 to 1960, cement consumption in the United States has grown fivefold (Figure 4.5) and has proved to be an accurate barometer of the economic condition of the nation generally. In the production of 1 ton of Portland cement (so-called because it resembled English Portland stone), 2530 lb of dolomitic limestone (75% $CaCO_3$, 4% $MgCO_3$) are heated in a kiln with 670 lb of shale or clay (14% SiO_2, 5% Al_2O_3, 1% Fe_2O_3). The architectural possibilities of concrete are great; it enabled the Romans (who built in concrete, rubble, and brick with a marble veneer) to erect much larger, if more poorly proportioned structures, than the Greeks (who raised their perfect temples in solid marble blocks—the great columns of the Temple of Apollo at Corinth, for example, are even monoliths). One can hardly conceive of what an imaginative artist like Antonio Gaudi could have done with modern concrete technique.

Crushed rock and sand and gravel are used primarily in concrete aggregates and in roadbeds, clays are used in brickmaking and for structural and other ceramics. Glass is made by melting together silica, CaO from limestone, Na_2O from sodium carbonate, and borax. Asbestos, a fibrous form of minerals such as chrysotile ($Mg_3Si_2O(_5OH)_4$), finds many uses, not only in shingle and tile manufacture but also for electrical and thermal insulating materials.

Rocks, cements, and stones not only produce pollution problems—air pollution from cement kilns, dust from stone crushers, respiratory illness from asbestos fiber inhalation, but they are also the victims of pollution. In the good old days, as Bernard Berenson somewhere remarks, the worst enemies of art were prosperity and war. Today we must add air pollution. In the few years that it has stood in New York's Central Park, the Cleopatra's Needle obelisk has deteriorated far more than it did during the many centturies that it lay in Egypt. The Elgin marbles in the British Museum are now in much better condition than the frieze fragments left on the Parthenon exposed to Athens' polluted aïr. The great cathedrals of Europe have turned black and are being eaten away by the noxious air. The havoc is particularly

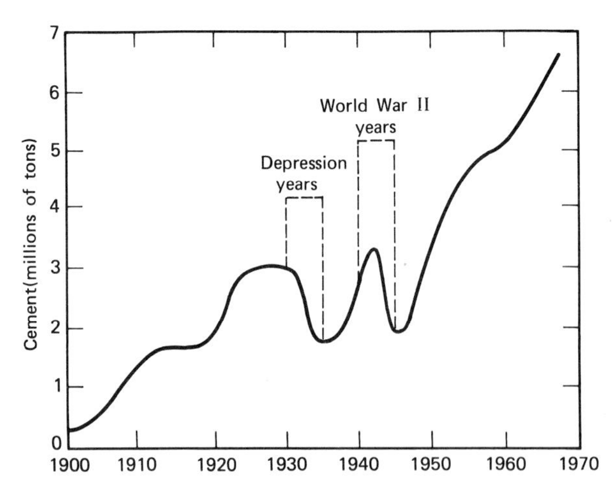

Figure 4.5. Portland cement consumption in the United States.

mournful in Venice where sulfur dioxide and other contaminants of the industrially polluted air have reduced much of the city's once beautiful staturary (carved in a soft marble particularly susceptible to air pollution) to shapeless lumps. No people love art more than the Florentines, but as I have watched the traffic swarm about the darkening walls of the once magnificent Duomo, it became clear to me that they love their Fiats and scooters more. Stone dressing chemicals have been developed to preserve stone, and although fairly effective, we can hardly paint all the great buildings, ruins, and statuary of the world with them.

Minerals for the Chemical Industry and Agriculture

With the exception of plantstuffs and a small but growing list of materials from the atmosphere and hydrosphere, the chemicals consumed by world industry and agriculture come from the lithosphere. We will return to a more detailed examination of the impact of agriculture on our environment in Chapter 14, but here I would like to mention the extraction of mineral fertilizers from the Earth. These necessary elements are depleted both in place in the soil by growing plants and by mining for agricultural purposes. The three most important fertilizing elements are K, N, and P, and their consumption is increasing rapidly as the world's population and food demands grow. The world consumption of K salts, for example, increases about 10%/yr. In 1967, the world production of potassium salts (expressed as the oxide K_2O) was 16,861,000 tons, with the United States leading with a production of 3,299,000 tons. Potassium, as we have seen, is an abundant and widely distributed element, and there appears to be reserves ample for centuries yet to come. Most of our K comes from deposits of mineral evaporites, laid down by the evaporation of and/or precipitation from ancient isolated ocean basins. As seawater evaporates, its dissolved salts separate out stepwise in a definite sequence (Figure 4.6). The K salts do not crystallize out until the very last phases—polyhalite (K_2SO_4, $MgSO_4$, $2CaSO_4$, $2H_2O$) when the volume of the remaining solution is only 4 to 9.5% of its initial value, and finally minerals such as sylvite (KCl) and carnallite (KCl, $MgCl_2$, $6H_2O$) come down out of the bitterns when the residue solution volume falls below 4%. Thus K and Mg minerals are precipitated out during the final phases of the evaporative process and are consequently, relatively speaking, the rarest evaporite minerals (Borchert, 1965; Braitsch, 1971). Evaporate deposits are still forming, but these formations are trivial compared to the enormous deposits laid down in the past, especially during the Permian, Pennsylvanian, and Devonian Periods.

The exploitation of Chilean mineral nitrates has now been largely superceded by the fixation of atmospheric nitrogen. Phosphorus minerals were also deposited from the oceans by precipitation, notably during Permian times, and areas of phosphate nodules are found on the ocean floor (Pratt and McFarlin, 1966). The most important mineral in phosphate rock is apatite ($Ca_5(PO_4)_3OH$), but because of the relative insolubility of this material for agricultural purposes, it is necessary to treat it with dilute sulfuric acid to form a soluble "superphosphate" material such as $Ca(H_2PO_4)_2$ whose P is accessible to growing plants. According to the U.S. Bureau of Mines in 1966 the United States production of phosphate rock was 39,050,000 tons with the U.S.S.R. second in production (32,190,000 tons) and Morocco third (10,405,000 tons). Morocco, however, has the largest estimated reserves (21,000,000,000 tons compared with the United States' 14,500,000,000 tons). While the world's reserves

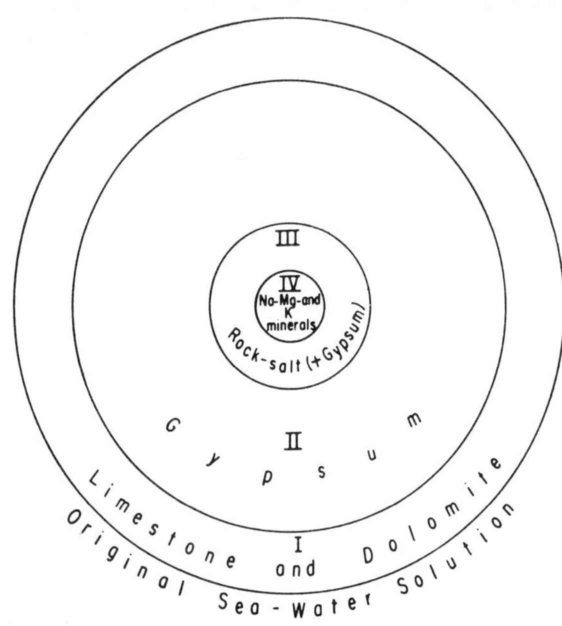

Figure 4.6. Principal stages in the evaporation of seawater (from Borchert, 1965, with permission of Academic Press, Inc., and the author).

of phosphate rock appear to be ample, because of their spotty distribution, some countries, such as Australia, are experiencing difficulties marshalling adequate supplies to insure future agricultural growth.

We have noted the use of sulfuric acid to make superphosphate. In fact, about 40% of the world's sulfur production (27,011,000 tons in 1967) is used in the manufacture of superphosphate and ammonium sulfate (another fertilizer), whereas another 20% is used by the chemical industry that in turn produces insecticides and fungicides to protect crops. Sulfuric acid forms the basis of a whole web of industrial processes; it is one of those chemicals whose production and consumption can be used as an index to monitor the technological level of a nation. Sulfur is also one of the worst pollutants—in the atmosphere from combustion processes and in our rivers and streams from paper manufacture. There are very large reserves of evaporite sulfate ($CaSO_4$) deposits, but it is much cheaper to exploit native sulfur. In addition to being deposited in some volcanic regions, anaerobic bacteria can produce S from $CaSO_4$ under geological conditions the same as those required for petroleum formation

$$CaSO_4 + 2C \text{ (from biomaterial)} + H_2O \xrightarrow{\text{bacteria}} CaCO_3 + H_2S + CO_2 \quad (4.10)$$

followed by the oxidation of the H_2S

$$H_2S + \tfrac{1}{2}O_2 \rightarrow H_2O + S \quad (4.11)$$

Native sulfur in the Gulf of Mexico area accounts for about 25% of the world's production. The free sulfur occurs sometimes on the top of salt domes and is brought to the surface by the Frasch process in which superheated water (170°C, 100 lb/in²) is forced down through an outer pipe to melt the sulfur, which is then forced up an inner concentric pipe by compressed air (Figure 4.7). While the reserves of sulfate are large, those of free sulfur are not, and some countries have had to actively develop new S-sources, such as recovery from the "sour gas" (H_2S) content of natural gas and recovery as a by-product from the processing of sulfide metal ores.

Two other familiar substances exploited from the Earth in such large quantities that we tend to overlook them are limestone (its use in iron production has already been noted) and, of course, salt. Salt (NaCl), both mined and from the evaporation

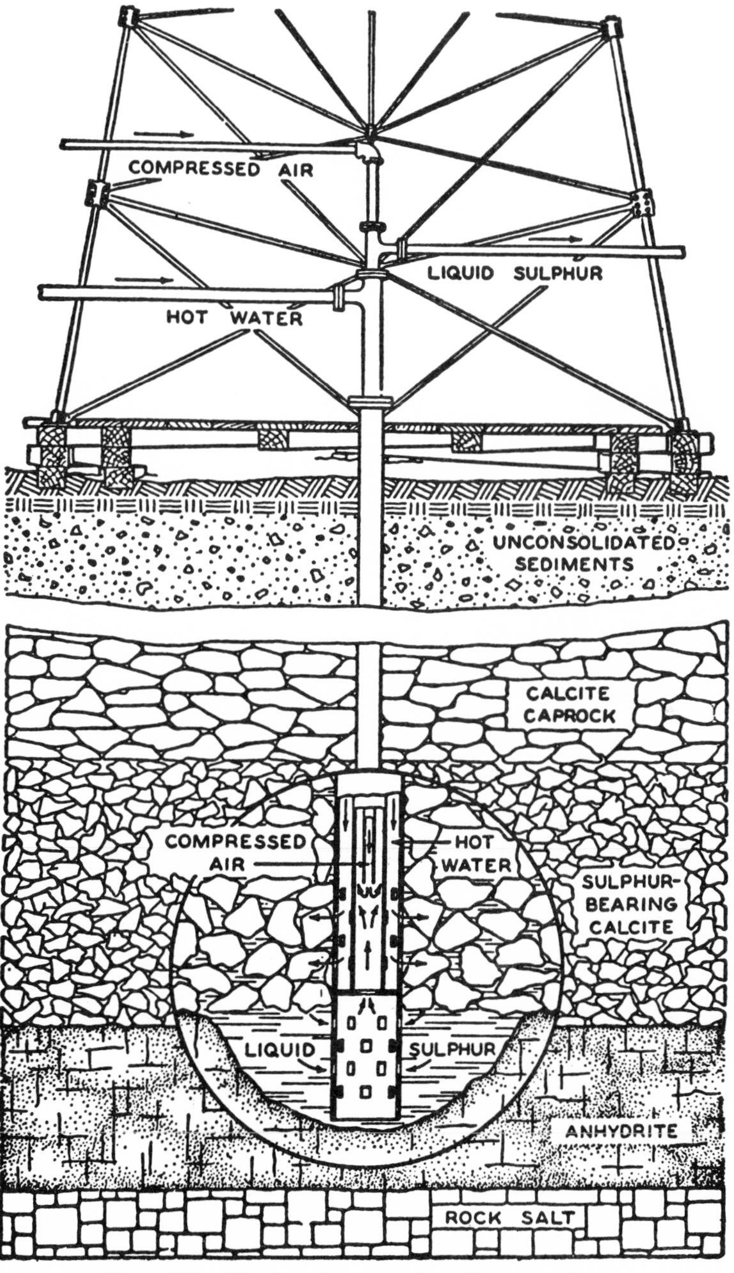

Figure 4.7. The Frasch process for sulfur extraction.

of seawater, is consumed in enormous quantities (123,000,000 tons in 1966) and, like sulfuric acid, forms the basis of a web of crucial chemical industries (Figure 4.8).

Here we have mentioned only a selected few of the Earth-materials utilized by man, and there are many, many, many more. Table 4.4 lists a few more, giving production figures for France, the United States, and the world for the years 1959 and 1964. While world production increased significantly over the 5-yr period, the national productions actually decreased in a number of cases. Far from representing decreased consumption, these decerases represent a greater dependency upon foreign imports. This dangerous tendency is made clearer in Table 4.5 and especially in Figures 4.9 and 4.10. Our nation is hogging and wasting the world's mineral resources at an unacceptable rate, and as the awareness of the Earth's exploited peoples grows, we are going to find ourselves entrapped in difficulties, technological, economic, and political, of ever increasing severity.

The lithosphere gives us not only the materials of our civilization but also the fossil and nuclear fuels on which technology runs. Here we concern ourselves only with the distribution of fossil fuels. The processes responsible for their formation is discussed later, and the horrendous pollution problems caused by their escalating consumption is a theme that appears throughout this book. Of the fossil fuels, we focus our attention on the three most heavily used—coal, oil, and natural gas. As Figure 4.11 shows, the relative importance of coal as an energy source in the United States has declined sharply over the last 50 yr, whereas oil has increased steadily and natural gas, strongly. The major uses of fossil fuels are (*a*) conversion into electrical energy, (*b*) conversion into mechanical energy (transportation, etc.) and conversion to thermal energy (heating), and (*c*), a nonfuel use, in the production of numerous, especially organic, chemicals.

Coal is formed by the slow carbonization of organic material, and as the noncarbon material is expelled, the calorific value or "rank" as a fuel increases (Table 4.6 and

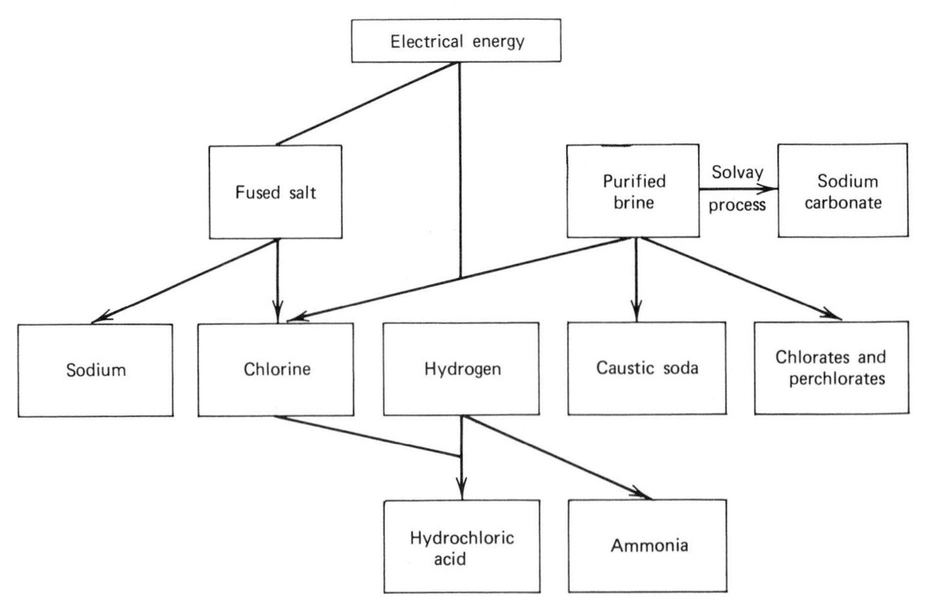

Figure 4.8. Salt-based industries.

Table 4.4 Mineral Production Statistics (From *Statistical Summary of the Mineral Industry, 1959–1964,* H.M. Stationery Office, London, 1966)

	Production (in long tons)					
	France		United States		World	
Mineral	1959	1965	1959	1964	1959	1964
Pumice	1,842	902	2,032,250	2,478,993		
Garnet			13,007	14,400		
Bauxite	1,729,283	2,395,000	1,700,000	1,061,000	14,600,000	20,100,000
Asbestos	20,857	22,000	40,588	90,261	2,000,000	3,100,000
Borates			553,523	693,106		
Portland Cement	8,847,996		56,918,800	61,877,700	290,000,000	407,000,000
China Clay	142,000	123,000	2,263,816	2,974,412		
Copper Ore	529	263	736,470	1,113,196	3,600,000	4,600,000
Diatomaceous Earth	100,734	130,981	401,589	504,000		
Feldspar	79,200		548,390	629,894	1,400,000	1,900,000
Fuller's Earth			365,734	492,755		
Gold (f.t.oz.)	42,159	54,140	1,602,931	1,456,308	32,000,000	39,700,000
Gypsum	3,684,162	3,716,979	9,700,000	9,539,000	41,000,000	45,000,000
Dolomite	686,241	1,102,505	1,774,792	1,935,288		
Mica (lb)	670,000	646,000	203,082,000	229,458,000	282,000,000	295,000,000
Phosphate Rock	75,642	42,428	15,869,000	22,960,000	37,000,000	58,000,000
Potash Minerals	9,382,550		3,601,000			
Pyrites	289,738	118,319	1,056,617	847,495	17,300,000	18,000,000
Salt (brine)	2,086,440	2,357,000	13,413,000	16,399,000	78,000,000 (total salt)	97,000,000 (total salt)
Kyanite			29,000	46,000		
Sulfur	428,024	1,487,000	4,553,634	5,278,207		
Talc	172,793	197,000	571,771	673,072	2,270,000	3,120,000
Vermiculite		184,446	202,053		233,000	306,000

Table 4.5 Production and Consumption Statistics (From Frasche, 1962, with permission of the National Academy of Sciences and the author.)

Summary Statistics, Metallic Elements, 1950–1960

				Production				Ratios for 1960			
				1950		1960		U. S. prod.	U. S. cons.		
		U. S. Consumption		United		United		U. S. cons.	World prod.	Reserves	
Commodity	Unit	1950	1960	States	World	States	World	(%)	(%)	United States	World
Iron ore	Thous. l.t.	106,610	108,050	98,045	241,000	88,777	507,089	82	21	5,500,000	80,000,000
Copper (excl. secondary)	Thous. s.t.	1,447	1,193	909	2,770	1,140	4,503	96	26	32,500	210,000
Lead (excl. secondary)	Thous. s.t.	885	582	431	1,840	244	2,374	42	25	7,700	49,000
Zinc	Thous. s.t.	1,101	950	623	2,359	432	3,428	45	28	25,000	84,500
Tin (excl. secondary)	Thous. l.t.	71	52	—	166	—	180	0	29	5	5,000
Mercury	Thous. flasks (76 #)	49	49	4.5	143	32	230	65	21	300	6,000
Antimony (excl. secondary)	Thous. s.t.	15	13	2.5	48	0.6	61	5	21	50	2,000
Bismuth	Thous. s.t.	1	0.8	0.35	1.5	0.25	2.6	31	29	15	25
Manganese ore (+35 % Mn)	Thous. s.t.	1,650	1,950	134	6,200	80	13,600	4	14	1,000	1,000,000
Chromite	Thous. s.t.	980	1,220	0.4	2,600	107	4,920	9	25	500	3,000,000
Nickel	Thous. s.t.	100	108	0.9	160	13	354	12	31	400	45,000
Cobalt	Thous. s.t.	4.1	4.8	0.4	8	1	18	20	27	50	2,200
Vanadium	Thous. s.t.	1.3	2.0	2.3	3.1	5	7	226	32	680	1,000
Molybdenum	Thous. s.t.	13	17.5	14.2	15.9	34.1	37.5	195	47	1,500	2,450
Tungsten	Thous. s.t.	3.3	5.8	2.0	15.8	3.3	33.1	57	18	70	1,275
Columbium	Short tons	490	789	1	480	0	2,100 ?	0	38	50,000	6,000,000
Tantalum	Short tons	30 ?	289	1	50	0	400 ?	0	72		100,000
Rhenium	Thous. lb.		?			2		100	?	1,000 ?	1,500
Aluminum (bauxite)	Thous. l.t.	3,600	8,100	1,335	8,041	2,096	23,710	26	29	50,000	5,000,000
Magnesium	Thous. s.t.	18	37	16	45	40	104	108	36	Unlimited	
Titanium ores	Thous. s.t.	691	1,113	475	740	795	2,340	71	48	64,000	2,000,000
Beryllium (beryl)	Thous. s.t.	3	9.3	0.5	6.7	0.5	11.1	5	81	10	200 ?
Zirconium Hafnium (zircon + baddelyite)	Thous. s.t.	30 ?	50 ?	25	75	25 ?	100 ?	50	50	12,000	35,000
Cadmium	Thous. s.t.	4.8	5.1	4.6	6.6	5.1	10.9	100	47	50	400
Germanium	Sh. tons		40			27	50	68	80		
Indium	Thous. oz (troy)	1		126							
Gold	Thous. oz (troy)	2,795	2,500	2,394	32,700	1,658	44,000	66	6	50,000	1,000,000
Silver	Million oz (troy)	110	100	42.5	199	31.2	224.3	31	44	750	5,000
Platinum	Thous. oz (troy)	309	325	18	600	10	1,101	3	70	150	25,000
Other Pt Group Metals	Thous. oz (troy)	187	451	13		10					
Yttrium and Rare Earths	Thous. s.t.	1.4	1.8	0.46	2	.8	?	45	?	5,700	Large

Arsenic (As_2O_3)	Thous. s.t.	32.1	28.1	13.3	51.8	11.0	62.0	40	47	2,500	Large
Boron (borates)	Thous. s.t.	505.2	640.6	647.7	650 ?	950 ?	1,000 ?	150	65	121,000	135,000 ?
Lithium (minerals)	Thous. s.t.		51	9.3	?	?	?			Very large	
Sodium and Na Salts	Thous. s.t.		28,000 ?			28,000		100	?	Very large	
Potassium (K_2O)	Thous. s.t.	1,410	4,000	1,288	4,400	2,638	10,000	66	40	400,000	5,000,000
Cesium and Niobium	Sh. tons		?			2	10				75,000
Calcium and Compounds	Thous. s.t.		1,000 ?					100			
Strontium ($SrSO_4$)	Thous. s.t.	8.6	3.3	.25	10	0	12 ?	0			
Fluorine (fluorspar)	Thous. s.t.	426	644	301	834	230	2,160	36	30	15,000	
Chlorine	Thous. s.t.	500						100		Unlimited	
Bromine	Thous. lbs.		165	98		175		100		Unlimited	
Iodine	Thous. lbs.	1,392	1,944	6,673		?	?	10			
Sulfur	Thous. l.t.	4,988	5,860	5,986	10,600	6,661	17,995	113	33	125,000	700,000
Selenium	Thous. lb.	775	650	559		620	1,777	95	35		
Tellurium	Thous. lb.		320	60		260	390	81	82		
Phosphate Rock	Thous. l.t.	8,581	13,673	11,114	21,250	17,516	40,100	128	33	13,500,000	46,800,000
Asbestos	Thous. s.t.	729	709	42	1,200	45	2,400	6	30	1,000	60,000 ?
Barite	Thous. s.t.	786	1,190	695	1,200	771	3,100	69	36	100,000	Very large
Diamond (total)	Million carats			0	15.3	0	27.3	0		None	Large
Gem	Million carats			0	2.5	0	6.6	0			
Industrial	Million carats	10.8	12.0	0	12.8	0	20.7	0	58		
Graphite	Thous. s.t.	20.9	37.3	5.1	148	5 ?	465	13	8		
Gypsum	Thous. s.t.	11,382	17,000	8,200	20,700	10,900	41,930	64	41	50,000,000	
Magnesite	Thous. s.t.	367	511	429	2,150	499	7,100	98	7	Very large	
Mica (total)	Thous. s.t.		125	70	116	120	205				
Sheet	Thous. s.t.	13.9	4.6	0.28		0.28		6		Small	Large
Ground	Thous. s.t.	72.2	120.4	70		120.3		100	59	Large	
Talc and Pyrophyllite	Thous. s.t.	621	722	621	1,470	734	2,450	100	30	50,000	Very large
Chromite (refractory)	Thous. s.t.	354	391			—	600 ?	0	65	Very small	10,000
Quartz Crystals	Thous. lb.		230	—		—	500	0	?	None	Moderate ?
Battery Grade Mn Ore	Thous. s.t.	41.3	27.2	11.5	56	9.1	?	37	?	Small	Moderate

Source: P. W. Guild, U. S. Geological Survey.

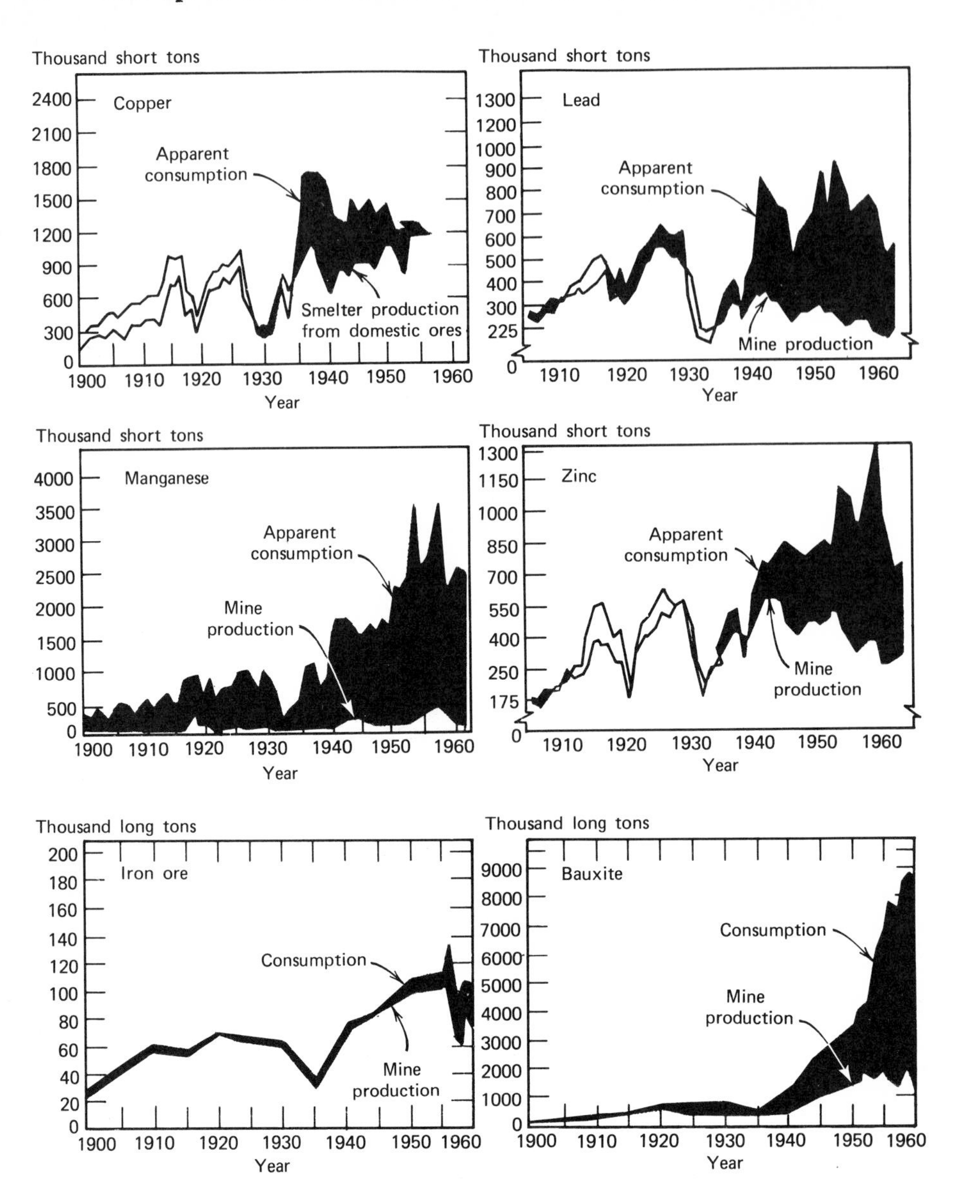

Figure 4.9. Increasing United States' dependence on mineral imports (from Frasche, 1962, with permission of the National Academy of Sciences and the author.

Figure 4.12), so the most desirable coals (bituminous and anthracite) are the oldest ones. Coal deposits are mostly found in the remains of ancient fresh water sedimentary basins, and they are scattered throughout the world. Geologists believe that all of the world's major coal basins have been discovered, and Averitt (1967) has estimated an inferred world reserve of recoverable coal of 8,415,000,000,000 tons. In the last hundred years, world coal production has climbed sharply (Figure 4.13), but in the United States, coal production has fluctuated erratically since about 1920. Unlike its position with respect to other mineral resources, the United States is in a

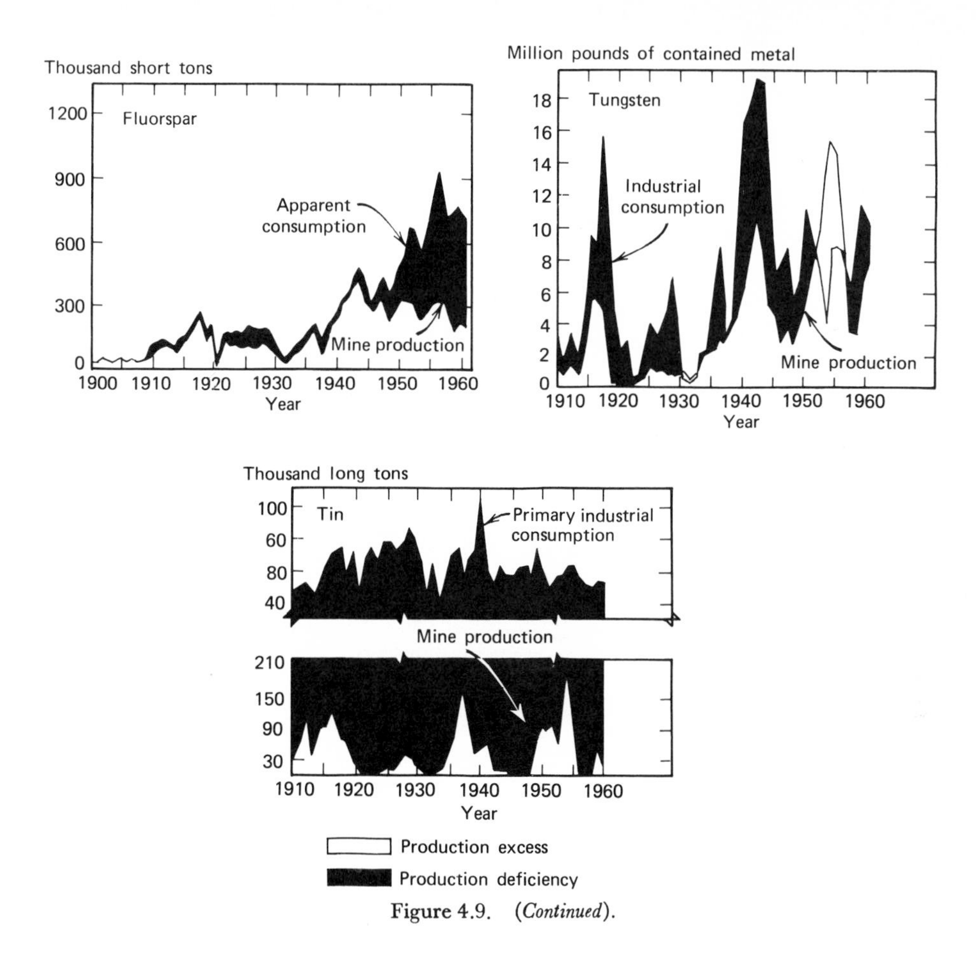

Figure 4.9. (*Continued*).

most peculiar position—much of the world's recoverable coal reserves are in North America (nearly 2,500,000,000,000 tons; Asia has the largest with 5,500,000,000,000 tons). Yet with the increasing use of oil and gas, the United States consumes relatively little coal. In 1964, only about 25% of United States energy came from coal compared to more than 85% for Poland, 80% for India, and more than 60% for the United Kingdom and West Germany. Skinner (1969) gives some interesting notes on the history of coal usage; although used on a small scale by the Chinese as long

Table 4.6 Rank and Energy Content

	Rank (% C)	Calorific Value (Btu/lb)
Peat	60	10,000
Lignite	70	12,000
Bituminous Coal	80	14,000
Anthracite	90	15,000

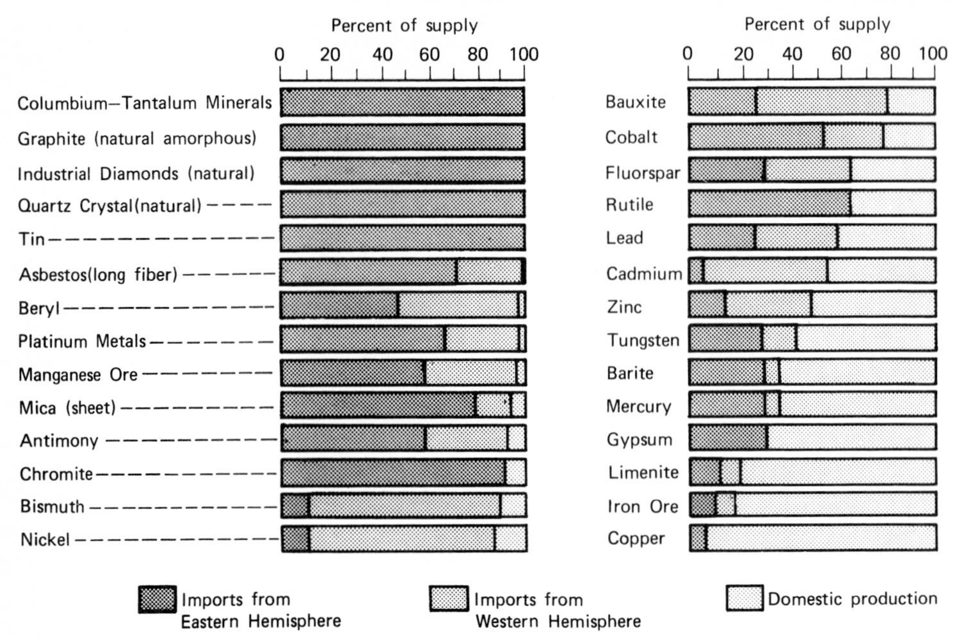

Figure 4.10. Mineral imports vital to United States' industry (from Frasche, 1962, with permission of the National Academy of Sciences and the author).

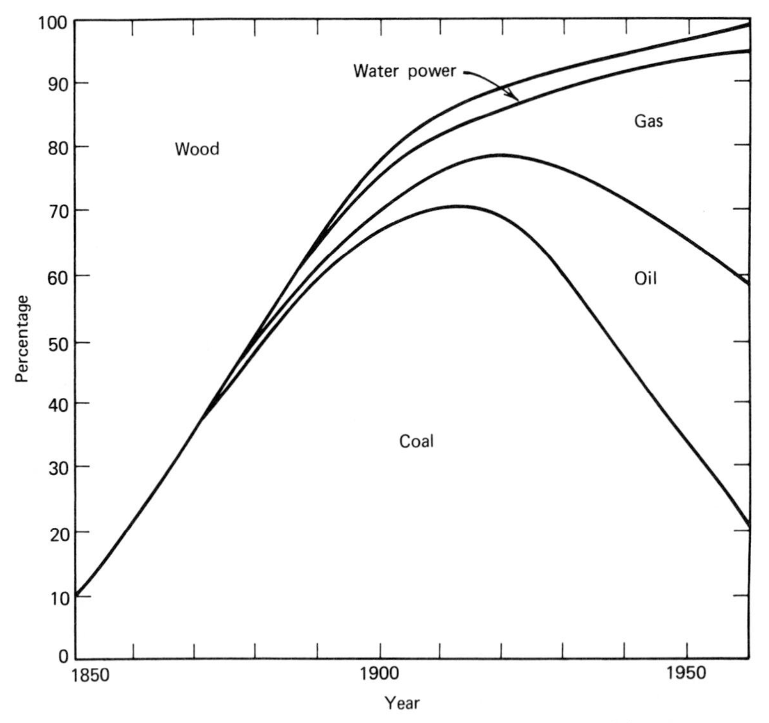

Figure 4.11. Fuels for energy production in the United States.

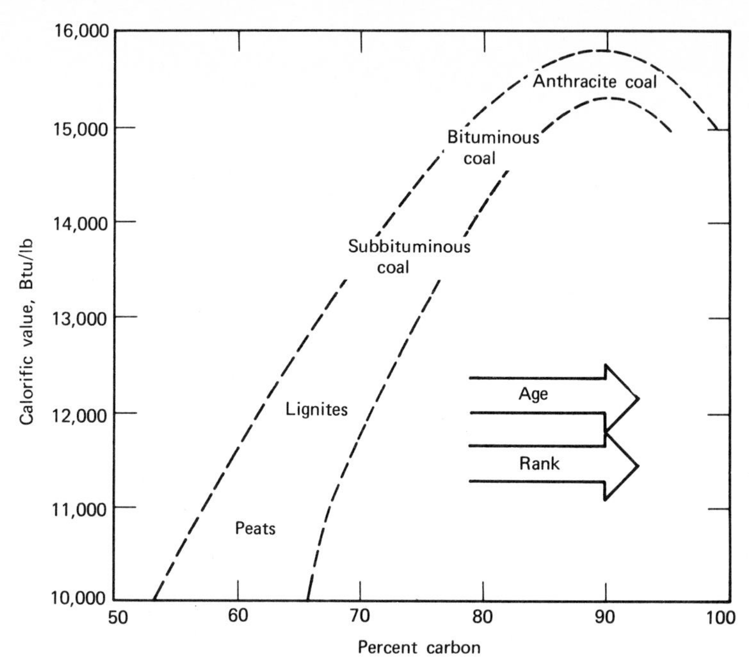

Figure 4.12. Calorific value, composition, age, and rank of solid carbonaceous fuels.

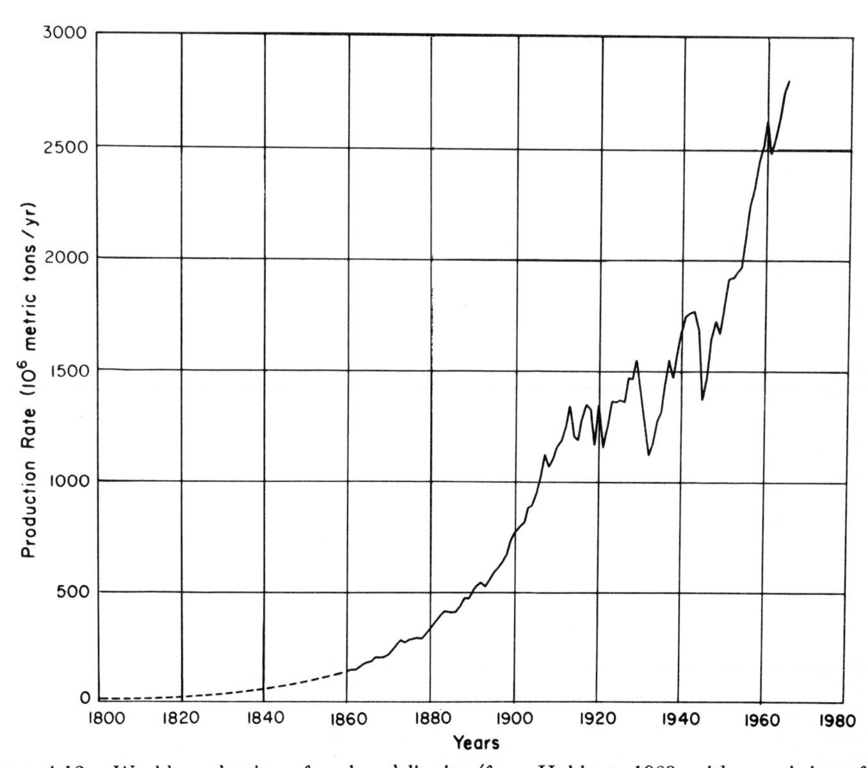

Figure 4.13. World production of coal and lignite (from Hubbert, 1968, with permission of the National Academy of Sciences and the author).

ago as 2000 yr, coal did not become an important fuel in Europe until the twelfth century when the English, faced with rapidly diminishing forests, began to burn the black rocks or "sea coals" that had been exposed by the weathering of coastal cliffs. By 1273, Londoners were already complaining of the smell and the air pollution produced by coal burning.

While much of the coal was formed in fresh water sedimentary basins, most of the world's crude oil (see Chapter 14 for its composition and the chemistry of its formation) appears to have been formed in marine basins. While coal was the prime fuel of technology's recent past, oil probably will become the fuel of its future—but only for a while. Although the world's estimated recoverable petroleum (oil and gas) reserves are 1500×10^9 to 3500×10^9 barrels (1 bbl is about 310 lb and is equivalent to about 5,400,000 to 6,000,000 Btu), there is good reason to believe, despite increasingly frantic prospecting activity, that the greatest period of oil discovery has passed (Zapp, 1962). The only encouraging feature of Figure 4.14 is the exclusion of

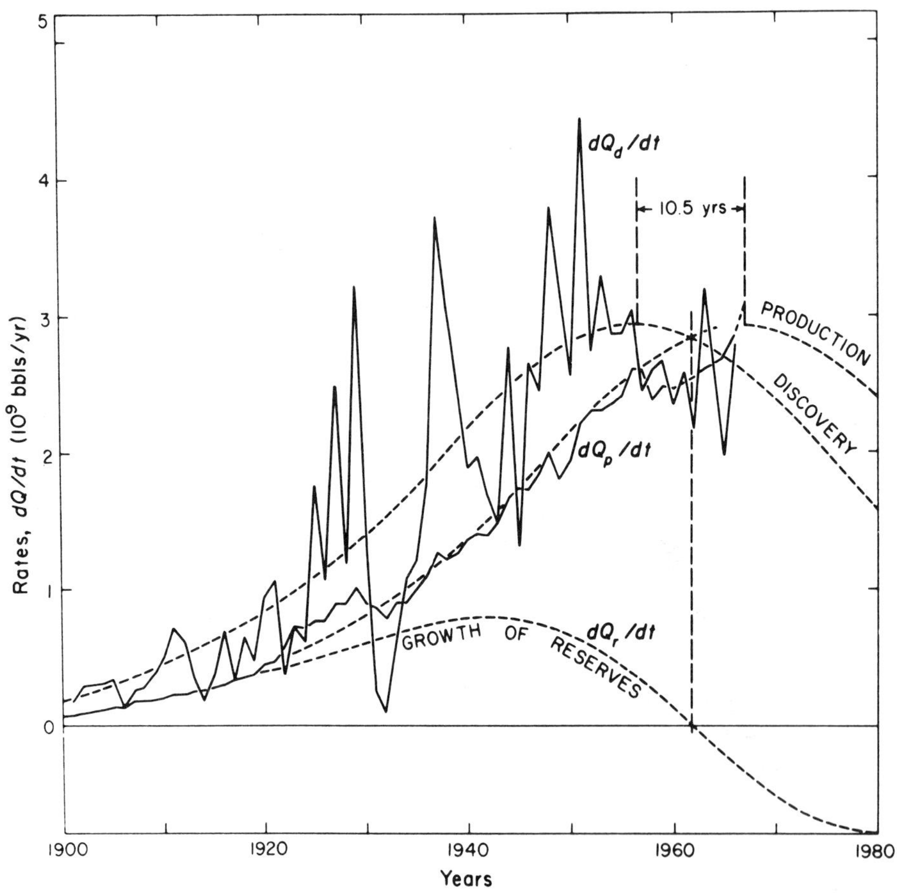

Figure 4.14. Rates of proved discovery (dQ_d/dT), production (dQ_p/dT), and increase of reserves (dQ_r/dT) of crude oil in the United States, exclusive of Alaska. Dashed curves are analytical derivatives; solid curves are actual yearly data (from Hubbert, 1968, with permission of the National Academy of Sciences and the author).

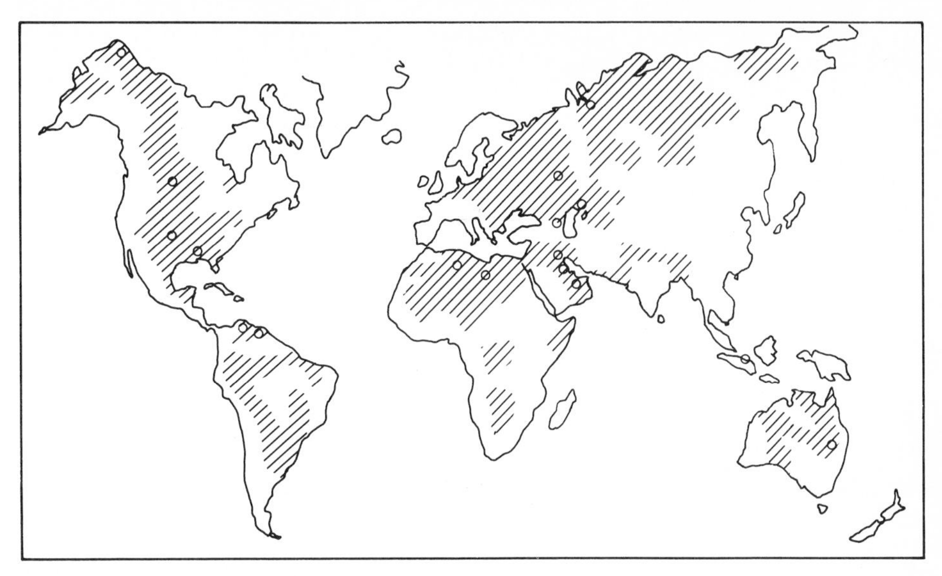

Figure 4.15. Major sedimentary basins of the world and the more important petroleum deposits located (as of 1968) within them.

Alaska. Oil fields are scattered throughout the world (Figure 4.15), and the even greater expanses of sedimentary basin and continental shelf areas hold some hope that further reserves will be discovered. Recent rich discoveries have included the North Slope of Alaska and the North Sea. Exploratory drilling is even going on off the coast of New England, but the public hue and cry that has been raised over this activity and the trans-Alaska pipeline presage the political as well as technological difficulties in exploiting these reserves. Our technological society *must* have oil and on this issue the environmentalists will surely lose—Americans are not and never will be prepared to walk to work to save a few caribou, and mass transportation alternatives in the United States seem to be diminishing rather than increasing. As the machinery of our culture begins to thirst for oil, we shall be forced to reappraise agonizingly some of our international political loyalties, we shall become increasingly at the mercy of oil-rich developing countries, and we may be forced to become a predator people, taking political and military action of unacceptable severity. Mankind has not been able to control his fertility, and as a consequence, we are being thrust into a terrible phase of our history where the struggle for materials, for food and fuel will reach a hitherto unequalled and perhaps terminal intensity.

Oil reserves in forms less convenient than crude oil and gas, such as tar sands and oil shale, are large (Table 4.7) and may give us a temporary stay. Our best hope to escape from these problems that overpopulation and technology have created is that technology will find improved ways of reducing birth rates and new sources of energy—that the exploitation of tars and shales will be made economic, that oil from coal and even garbage, both technically feasible, will be made economically feasible, that alternative energy sources (wind, tidal power, hydroelectric power, and solar energy) will be more fully utilized, and that nuclear fuels will replace fossil fuels for many uses. Another lithospheric energy source is geothermal energy (Bowen and Groh, 1971; Hammond, 1972). The exploitation of this potential resource has been

Table 4.7 Potential World Resources of Fossil Fuels (Expressed as Energy Equivalent in Btu $\times 10^{15}$)

		Will last[a]:
Coal	197,000	1160 yr
Oil and Gas	14,250	85 yr
Oil Shale	5,700	33 yr
Tar Sands	3,400	20 yr

[a] At the 1966 energy consumption rate of 170×10^{15} Btu/yr.

comparatively slight, which, considering the mess man usually makes out of everything he touches, may be just as well.

Our best immediate hope to keep our civilization running lies in the utilization of nuclear fuels. Unfortunately, a great deal of irrational hysteria is aroused by mention of anything radioactive, and while this ignorance cannot stop the development of nuclear power because that development is necessary, it can and is slowing it down. A particularly ironic turn of this tragedy is that nuclear power has a far less deleterious effect on the environment than the combustion of fossil fuels. Far more people are dying right now as the consequence of air pollution than by radioactivity, and the health record in the handling of radioactive substances by the users (if not the miners, see following discussion) is excellent (Sagan, 1972). Nevertheless, there are real problems associated with nuclear energy, and the atom certainly is no panacea to all our problems. Although much publicized, thermal pollution from nuclear power plants appears to be a relatively minor problem, and, as mentioned later, neither are we likely to exhaust our nuclear resources. It has become increasingly clear that the limiting factor in the utilization of nuclear energy is the problem created by the disposal of nuclear waste—a topic to which we return in this chapter.

There are two important nuclear energy processes—*fission* of a heavy radionuclide to produce two or more lighter fragments and *fusion* of two or more light elements to produce a heavier one. In both processes, the mass differences yield enormous quantities of energy; the fissioning of one pound of uranium yields a heat equivalent to nearly 6000 bbl of crude oil. Power is presently obtained by fission while work to devise a method of harnessing the even greater power potential of fusion is being carried on at a rather feverish pitch. ^{235}U, which accounts for only 0.72% of all naturally occurring uranium, is the most important fissionable atom in nature from a power standpoint. It decays naturally in the so-called actinouranium series, and the Th/U ratios resulting from this decay is one of the major tools for establishing geochemical chronologies. ^{235}U is scarce and the expense of concentrating it from its ores high; however, in a "breeder reactor" the more abundant isotopes ^{238}U and ^{232}Th can be converted to fissionable ^{239}Pu and ^{233}U, respectively.

U(IV) is readily oxidized up to U(VI). Compounds of the latter tend to be more soluble, but in a reducing environment, such as in the presence of organic matter, they are reduced back down to U(IV) and precipitate out in minerals such as uraninite (or pitchblend). Many uranium deposits appear to have been formed in this manner. U^{4+} ions can also be precipitated as a substitutional ion in certain minerals, for example, substituting for Ca^{2+} in apatite $(Ca_5(PO_4)_3(OH,F)$. The U

Table 4.8 U. S. Uranium Resources and Reserve Prices (Based on U. S. A.E.C., 1963 information)

Price/lb. of oxide (U_3O_8)	Potential Resources (tons)	Potential Power Equivalent (barrels of crude oil)
<$10	650,000	$8,100 \times 10^9$
$10–30	600,000	$7,500 \times 10^9$
$30–100	20,000,000	$250,000 \times 10^9$
$100–500	1,700,000,000	$21,000,000 \times 10^9$

reserves in the United States are large—145,000 tons, second only to Canada's 180,000 tons (reliable data are not available for communist nations, and because of the strategic nature of the element numbers released by non-Communist governments are not above suspicion.) The useable reserves are highly dependent on the economics of ^{235}U fuel (Table 4.8). It is noteworthy that the 20,000,000-ton value in this table corresponds to a power equivalent about seven times greater than the world's total fossil fuel supply. The mounting of a successful breeder reaction program will enable lower grade and thus more abundant U resources to be exploited.

Uranium mining has produced some rather bizarre pollution hazards. Despite their radioactivity, uranium mine tailings have been used by contractors as a cheap source of landfill and concrete mix. In summer months, when its classrooms are closed, the radiation level in the Pomona (Colorado) Elementary School Annex reaches values 18 times greater than that stipulated in guidelines set by the U.S. Surgeon General. Whole uranium producing towns have become radioactive. In the Grand Junction area, the incidence of cleft lip and palate is almost twice as great as in the rest of Colorado, birth rates are lower, and the death rate from congenital deformities is 50% higher.

As for water, the most essential resource, the people of the United States use an estimated 1.1×10^{14} gal/yr or 8% of the total rainfall or 27% of the total stream flow. The human race uses far more water than any other material in our environment. I have not made the calculation, but I would be very surprised if we do not use more water than all other materials combined (I would guess that atmospheric oxygen occupies second place). Were not water a recoverable resource civilization would die of thirst in a couple of years. Because of its prime importance, water occurs again and again in this book, and I do not want to extend this already over long chapter by treating lithospheric water in detail here.

4.3 CHANGING THE FACE OF THE LAND

In nature, physical and chemical processes are hardly ever separable, so at this point I would like to insert a very brief discussion of some of the physical changes that man impresses on the face of Earth and that, in some instances, have very grave chemical as well as physical consequences.

Until a few years ago, a widespread price of unquestioned "progress" was the "improvement" of wetlands by filling them up or draining them. In the period from

1850 to 1968, 239 mi^2, or about 31%, of the water–marshland area of magnificent San Francisco Bay disappeared (Anon., 1971a). This is particularly tragic because the west coast of the United States had relatively little wetlands to begin with. For the hard-pressed coastal municipalities filling in marshland is a cheap way of disposing of solid waste, while to develop it is an easy way to make a quick dollar. But the price the world pays is high—a spacious open vista is replaced by an eyesore, and the delicate balance of the local marine ecosystem is devastated. Other incursions into the realm of the ocean in the guise of coastal "improvements" are dikes, breakwaters, and sea walls. Whenever I see an enormous Atlantic breaker hurl itself against the shore, crush a scruffy beach house to kindling wood, and suck the churning fragments back out to sea it never fails to fill me with the most sublime feeling of justice and satisfaction. Some of man's damage is not so deliberately malevolent. The dams we build trap sedimentary material behind them—the sand and gravel never reach the coast and the beaches disappear. Dams also damage the characteristics of the hydrosphere. The waters trapped behind dams can become anoxic, the oxygen content of ground water can also be decreased, and the temperature of the dammed river is generally increased (Trueb, 1971). The weight of water behind a large dam is enormous; is it enormous enough to cause an earthquake (Anon., 1971c)? The extensive damming and redistribution of the waters of Russia's northflowing rivers for irrigation purposes has changed ground water distribution, evaporative patterns, and even the climate. In fact, the specter of an advancing icecap has been raised, and some scientists are worried that extensive tinkering with the world's rivers could perturb the rotation of the Earth (Goldman, 1970). Similarly, extensive harnessing of tidal power (Gray and Gashus, 1972) could alter this planet's movement and by increasing drag effects eventually cause the Earth to slow down and spiral into the Sun. As a consequence of dams and the removal of sand and gravel by contractors (as much as 120,000 m^3/yr), the Black Sea coast in the Soviet Republic of Georgia is vanishing. In some places, the sea has moved 40 m inland, and near the resort area of Adler, hospitals, hotels, and a sanitorium have collapsed as the shoreline gave way (Goldman, 1970). Plundering the environment, it should be noted, is not a vice restricted to laissez faire capitalism, and it is perversely reassuring to see that communist nations are having their problems too. Even man's best intentioned efforts seem to have a wayward will to harm: his attempts to stabilize barrier dune systems along the coasts, for example, appear to have accelerated rather than retarded destructive geological and ecological changes (Dolan, 1972).

Modern man has a deep-seated adversion for natural beauty. The will to destroy is a part of the psyche of the bureaucrat. Since 1954, more than 8000 mi of the United States' beautiful meandering streams have been channelized and converted into ugly, barren ditches. If present plans are carried out, over 300,000 acres of forested wildlife habitat will be wiped out in the southeastern United States alone (Gillette, 1972). One more perceptive government officer, Mr. M. P. Reed, has been quoted as declaring, "stream channel alteration under the banner of 'improvement' is undoubtedly one of the most destructive water management practices . . . the aquatic version of the dust-bowl disaster." Which brings us to another havoc wrecked by man on the face of the land—bad agricultural practice that has metamorphosed fertile farmland into deserts and accelerated erosion. A combination of injurious farming techniques and climatic changes is responsible for the drying and spread of the Sahara Desert over the past 7000 yr (Cloudsley-Thompson, 1971; Flohn and

Ketata, 1971). For example, since 1881 the number of date palms grown on the border of the Tunisian Sahara has increased 51%, and this increase, together with artesian wells, has resulted in a drop in the subterranean water level of 5 cm/yr. Agriculture-produced erosion is now largely averted in the United States by such simple means as contour plowing. But the lesson of contour plowing was not readily recognized by the road builders, and only recently for aesthetic and conservation reasons have attempts been made to let highways follow the natural undulations of the landscape rather than cutting through in gigantic scars.

While ubiquitous, road scars are only the tiniest of scratches compared to the monstrous wounds opened up in the Earth's surface by strip and pit mining. Somewhere Professor Goldberg has called attention to the fact that man has become a significant geological agency, since his mining activities now remove as much mineral material each year as does the natural weathering of crustal rocks. Strip mining is cheaper and safer than subsurface mining. In the United States at the present time, 37% of the annual coal production is strip mined, and already more than 1,800,000 acres have been ripped up and left as poisoned, sterile, unsightly wastelands. Some stripping shovels can scoop up 200 tons of earth in a single bit! As power demands grow, stripping will worsen. In one western United States area alone, 42,000 mi^2 are threatened. In some European countries, the law requires that the strippers restore the land to the condition in which they found it, but in the United States, such restoration is estimated to cost more than $2000/acre.

Still a far more threatening disfigurement of the Earth's face is the rapid sprawl of our cities. Despite irrigation projects and other advances in agricultural technology, every year the *potentially* cultivatable land area of the world actually diminishes as it is swallowed up by population growth propelled urban sprawl. In the light of this terrible fact, talk about increasing world food production to feed the added numbers resulting from uncontrolled population growth becomes a grotesque lie. Every day scenic, food- and oxygen-producing, water-conserving, fertile green space disappears and is replaced by ugly, food- and oxygen-consuming, pollution-producing, water-wasting, barren gray space. In many regions, such as parts of Califronia, the best farming land is being consumed at an even greater rate than poorer land. At the same time in tropical regions of the world, the great green belt whose contribution to the world's supply of oxygen is second only to marine plant life, is also disappearing as a consequence of slash and burn farming (Gomez-Pompa, Vazquez-Yanes, and Guevara, 1972). Far from representing an increment to the world's cultivateable areas, because of the inability of tropical jungle soil to withstand such abuse, in a few years these areas become deserts. To return to the problem of the disappearance of precious land beneath urban sprawl, The Netherlands offer a dismal example. The Dutch are a resourceful people who for centuries have brought a characteristic mixture of ingenuity and good sense to bear on their problems. To accommodate their growing population, the Dutch have an ambitious program of land reclaimation from the Ysselmeer. By the year 2000, they will have thus added some 550,000 acres to their nation from beneath the sea. But at the present time, The Netherlands is losing 7500 acres/yr of agricultural land to roads, airports, and urban sprawl (Butler, 1972). At this rate, simple arithmetic shows that unless the Dutch learn to eat asphalt, by the year 2045 the reclaimed land will have been lost, and they are going to be in rather desperate straights.

4.4 THE EARTH AS A DUMP

Every man, woman, and child in the United States generates over 100 lb of solid waste per day, and by 1980, the number is expected to increase to 150 lb/day. Of this gigantic national urban solid waste output, about half is incinerated and about half is dumped on land in dumps and landfills (Kenehan, 1971). Coastal cities try to hide some of their trash in the sea (see Chapter 8), but the lithosphere bears almost the entire brunt of this pollution, of these "resources out of place." Urban solid waste falls into three categories and annual United States production in each is summarized in Table 4.9. The ubiquity of solid waste in its most aesthetically offensive form, litter, is astonishing. Oceanographers have found even the deep ocean floor littered, and climbers found Alaska's remote and forbidding Mt. McKinley strewn with litter—they collected and carried down nearly 400 lb of the stuff from the 18,200 ft level (Anon., 1971b). As of this year (1972) there are some 2600 pieces of astrojunk orbiting the Earth! While Americans make a fettish of personal cleanliness, with respect to their environment their habits are unspeakably dirty. They are the world's worst nest foulers, and their cities are among the filthiest on Earth.

Figure 4.16 shows the sort of materials comprising urban waste (see also Floyd and Lefke, 1971), and it is interesting to note that about half of the stuff is paper and paper products (just think of the forests consumed!). Solid waste technology generally falls into three steps: first collection and compaction, by far the most expensive step, $8 to $20/ton and often representing more than half of the total waste management expense; then further concentration and possibly some reclaimation of materials; and finally disposal (Hershaft, 1972). Incineration (Corey, 1969) is an expensive alternative, land dumping the cheapest, and ocean dumping, fortunately, is more than twice as expensive as land dumping. Nevertheless because of the unavailability of suitable land dumping areas (New York City will run out of landfill sites in 1975), 20 to 25% of the stuff in the United States is incinerated, and by the year 2000 the percentage is expected to increase to 30%. This increase may represent short-range

Table 4.9 United States Solid Waste Output (Data From Kenehan, 1971)

Category	Output
URBAN REFUSE (domestic, commercial, municipal, and industrial)	400,000,000 ton/yr
	including 60,000,000,000 cans; 36,000,000,000 bottles; 58,000,000 tons of paper and paper products; 4,000,000 tons of plastics; 1,000,000 abandoned automobiles; 180,000,000 tires; and so on. Collection and disposal of urban wastes alone costs $6,000,000,000/yr.
MINERAL WASTE	1,700,000,000 ton/yr
	which is added to a past accumulation of 23,000,000,000 tons. For example, the production of one ton of copper produces 500 tons of waste earth and rock.
AGRICULTURAL WASTE	2,000,000,000 ton/yr
	including farming, slaughterhouse, and animal waste. An average steer, for example, generates 10 ton/yr of solid waste

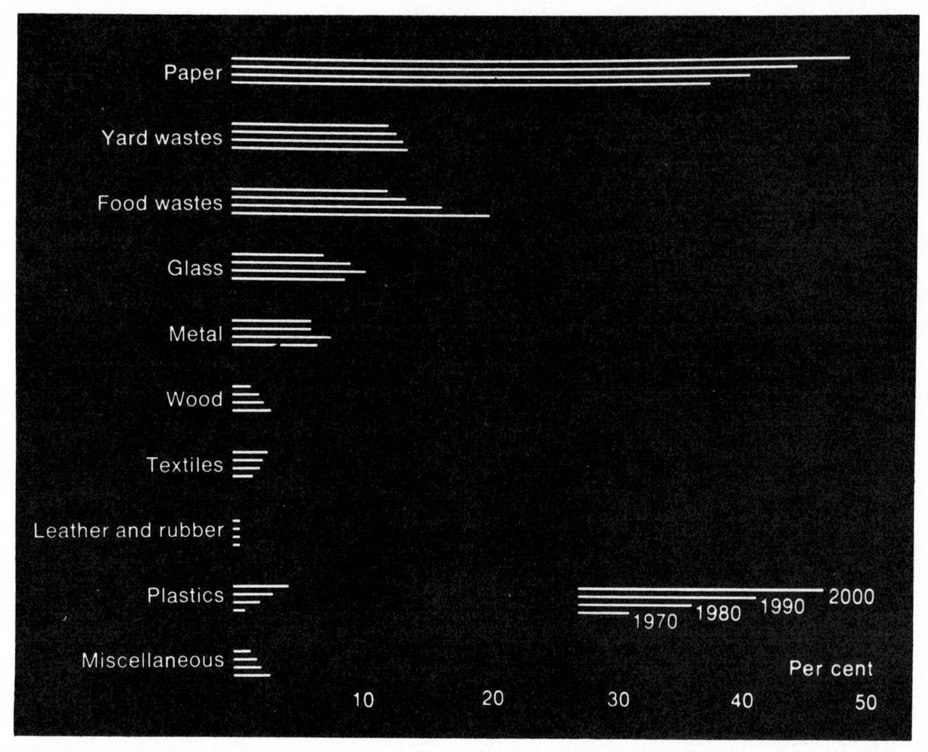

Figure 4.16. Composition of urban waste (from Niessen, *Technology Review*, edited at the Massachusetts Institute of Technology, © 1972 by the Alumni Association of M.I.T., with permission).

thinking rather than improvement. Incineration exchanges a lithosphere pollution problem for an air pollution one, and if we must pollute, land pollution is preferable.

A pollutant can be diluted and dispersed or it can be concentrated and confined. Presently, the preferred approach is dispersal, since this readily gets the pollutant out of sight, but this is certainly unwise. For in dispersing a pollutant, two things happen: (*a*) control is lost over the pollutant and the capability of surveilance over the pollutant diminishes, and (*b*) the pollutant is really "thrown away," and any possibility of future utilization of the pollutant is lost. Remember, a pollutant is a resource out of place, and as discussed in the next section, someday we may have to mine our dumps. The three potential disposal sinks—atmosphere, hydrosphere, and lithosphere—differ greatly with respect to their viscosities and the velocity of the transport of chemical substances within them. Thus dispersal in the atmosphere is immediate, in the hydrosphere rapid, and in the lithosphere exceedingly slow. On the basis of these principles, we can conclude that in terms of the long-range view air disposal is always the least and land disposal always the most desireable alternative. In modern industrial societies, there have been far more deaths from air pollution than even from water pollution (excluding epidemic disease).

Volume even more than weight is the important parameter in solid waste treatment, and the treatment process can be looked upon as a series of volume-reducing steps. In this respect, overlooking the air pollution it produces, incineration is highly

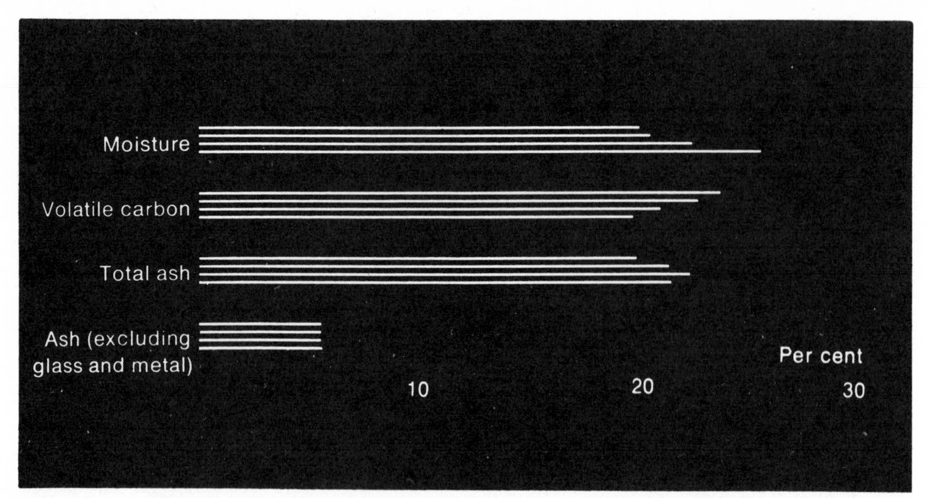

Figure 4.17. Weight percentage composition of urban waste as incinerated for the years 1970, 1980, 1990, and 2000 (from Niessen, *Technology Review*, edited at the Massachusetts Institute of Technology, © 1972 by the Alumni Association of M.I.T., with permission).

effective, especially if glass and metals are recovered (Figure 4.17), and it has the added advantage of sterilizing the waste.

Quite apart from the lack of suitable sites, the so-called "sanitary landfill" (Sorg and Hickman, 1970; Anon., 1972) in which the layers of waste are carefully covered with layers of sand is not without its chemical problems. The organic material decomposes, and provision must be made for venting the resulting gas (Fungaroli and Steiner, 1971). Decomposition and compaction results in instability and settling. Many other problems, often very difficult to solve, are caused by the leaching of the waste material by rain percolation and ground waters (Qasim and Burchinal, 1970; Salvato, Wilkie, and Mead, 1971). and in the past, landfills, both "sanitary" and otherwise, have been sources of serious water pollution (Hopkins and Popalisky, 1970).

As illustrated in Table 4.9, mineral solid wastes are discarded in amounts more than four times greater than urban wastes. Fortunately, much of this marerial, such as mine tailings, is inert and innocuous chemically. However, this material is often eminently weatherable and leachable, and in instances where radioactive elements or poisonous heavy metals, such as mercury are involved, dangerous contamination hazards have been created. One such special case, acid mine drainage, will be examined in detail in a subsequent chapter.

In the cases of very concentrated and dangerous solid and liquid wastes, the importance of the principle of confinement discussed previously and generally ignored for less threatening wastes is recognized. While some of this potent stuff is containerized and sneaked into the oceans (see Chapter 8), the preferred method of disposal now appears to be injection into the lithosphere or storage in abandoned mines and other deep shafts (Caswell, 1970). Intensely radioactive nuclear wastes are a problem of particular severity and may, as I have suggested, impose limits on the development of atomic energy. Underground disposal at first seemed our answer for the permanent (^{239}Pu requires 250,000 yr to decay to safe levels) storage of radiowastes, but now serious doubts have arisen (Hambleton, 1972). The "solid" Earth is far from solid. For example, the Lyons, Kansas, salt mine site surveyed by the Atomic Energy Commission turned out to be "like a piece of Swiss cheese" with considerable circulation of ground water (Holden, 1971; Hambleton, 1972). Also any kind of settling or seismic event that could open up fissures could be disastrous. Disasters have already occurred in the United States from nerve gas escape from underground storage. Deep well injection is being used for highly corrosive and toxic liquid industrial wastes (Tofflemire and Brezner, 1971). But again this technology is no panacea and is fraught with hazards. Meddling with the subterranean hydraulic situation can be a very dangerous business. Intensive water withdrawal by deep-drill wells can cause salt water invasion. Venice's sinking is aggravated by ground fluid withdrawal in the Adriatic area. But perhaps most frightening of all, high pressure fluid injection designed to facilitate oil recovery into the previously faulted and stressed Baldwin Hills (California) produced an earthquake that ruptured a reservoir dam and damaged or destroyed nearly 300 homes. In the conclusion to their discussion of this catastrophe, Hamilton and Meehan (1971) warn that "although fluid injection operations may be carried out for beneficial purposes, the effect of such injection on the geological fabric can be serious and far reaching." Reservoir loading and even mining can also trigger earthquakes (Dudley and Riecker, 1972).

4.5 RECYCLING

Perhaps the most compelling argument for land disposal is that the waste is concentrated so that we can keep an eye on it, and, even more important, when the need arises, as it surely will if present trends continue, the stuff will be there and accessible for recovery. It costs New York City $100,000,000/yr to collect solid waste and another $25,000,000/yr to dump it, but although thrown away, at current dealers' prices, the recoverable scrap in this refuse is worth an estimated $34,000,000. We can afford to throw many materials away today; in the future, this will not be the case. Mining of our dumps can be looked upon as a sort of delayed cycling. The importance of cycling cannot be overemphasized. In the absence of checks on population growth, cycling could extend the life expectancy of modern civilization, and with careful control of population, cycling could permit our way of life to be continued and probably improved.

Table 4.10 lists various types of recyclable waste materials. The amounts actually recycled (last column) varies from 0 to 100%. The value of the material is the most important parameter in determining how much is recycled: Copper is expensive thus over 60% of the total available copper is recycled. Paper, on the other hand, the major waste material (Figure 4.16), is less than 20% recycled (Table 4.11). Recycling reduces waste; in practice, it cannot eliminate it. Recycling buys time. Recycling does not prevent pollution or the exhaustion of natural resources, but as the three computer scenarios developed for copper show (Figure 4.18b) recycling can alleviate the severity of the problem.

Recycling means far more than simply recovering materials back out of solid waste. It means making new materials out of waste and putting waste to new uses.

Table 4.10 Types of Recyclable Materials (Reprinted with permission from H. Ness, *Envir. Sci. Tech.*, *6*, 700 (1972), © The American Chemical Society.)

Material Type	Examples	Condition of Scrap	Recycle Rate (%)
Manufacturing Residues	Drosses, slags, skimmings	25–75% recoverable	Over 75
Manufacturing Trimmings	Machining wastes, blanking and stamping trimmings, casting wastes	90% recoverable	Nearly 100
Manufacturing Overruns	Obsolete new parts, extra parts	Variable compositions	Nearly 100
Manufacturing Composite Wastes	Galvanized trimmings, blended textile trimmings, coated paper wastes	Often not all constituents are recovered	0–100
Flue Dusts	Brass mill dust, steel furnace dust	Often not economical to recover	Under 25
Chemical Wastes	Spent plating solutions; processing plant sludges, residues, and sewage	Often recoverable	Under 10
Old "Pure" Scrap	Cotton rags, copper tubing	Over 90% recoverable material	Over 75
Old Composite Scrap	Iron die castings, auto radiators, paper-base laminates	Often not economical to recover valuable materials	0–100
Old Mixed Scrap	Auto hulks, appliances, storage batteries	Not all materials are recovered	Under 50
Solid Wastes	Municipal refuse, industrial trash, demolition debris	Very low recovery rates now	Under 1

Table 4.11 Less Than One Fourth of Paper Consumed Is Recycled (million tons) 1969 (Reprinted with permission from H. Ness, *Envir. Sci. Tech.*, *6*, 700 (1972), © The American Chemical Society.)

	Newspapers	Containerboard	Pulp Substitutes and Mixed Papers	Permanent End Use	Total United States' Consumption of Paper and Paperboard
Total Paper Consumed	9800	15,000	22,500	10,000	58,200
Paper Recycled	2400	3,900	5,100	—	11,400
% Recycled	24	24	23	0	19
Paper Not Recycled	7400	12,000	17,400	10,000	46,800
% Not Recycled	76	76	77	100	81

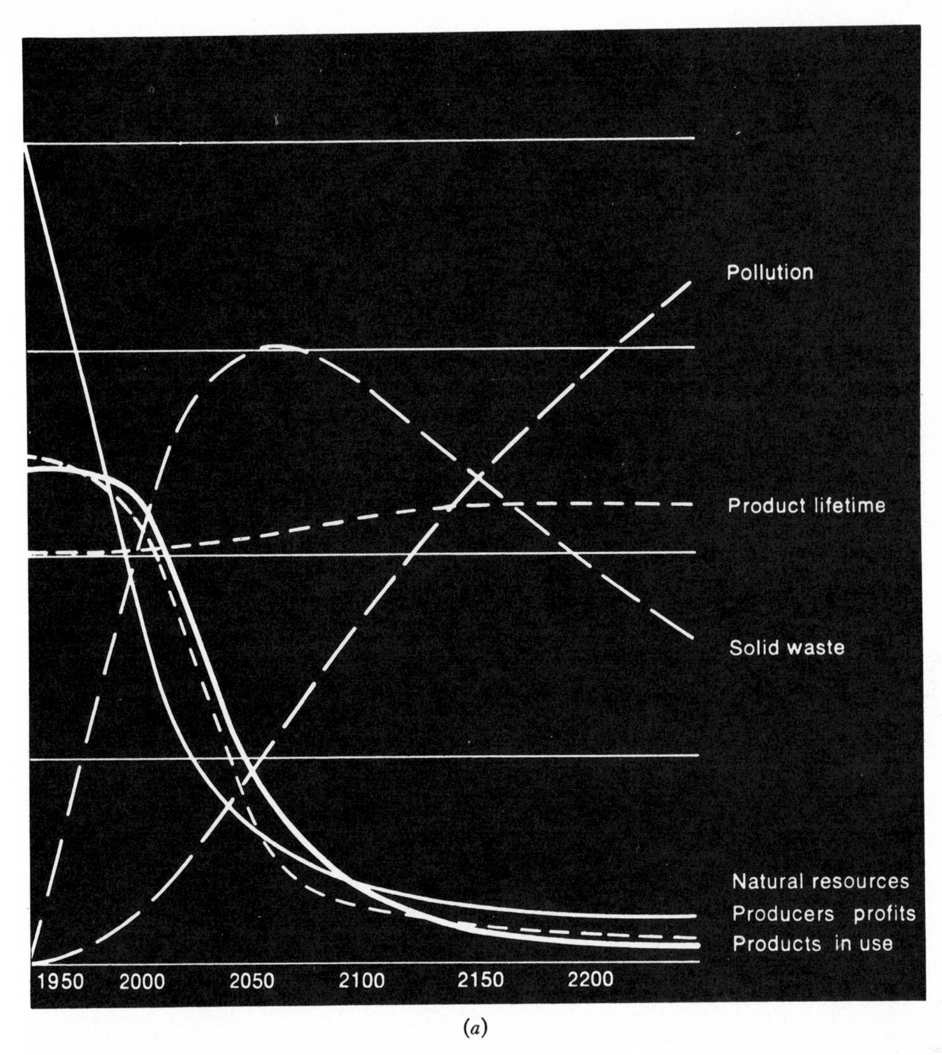

(*a*)

Figure 4.18. Projected resource scenarios (from Randers and Meadows, *Technology Review*, edited at the Massachusetts Institute of Technology, © 1972 by the Alumni Association of M.I.T., with permission).

Modern technology has been applied to this goal, and it is making significant progress. Bricks and mineral wool have been made from glass recovered from waste. Garbage and waste paper have been converted into crude oil. Domestic refuse and animal manures thus transformed could represent 2,000,000,000 bbl/yr of oil (Kenehan, 1971). Refuse containing organic matter and metals can be used to reduce non-magnetic taconite iron ore (an iron mining waste) into high-quality magnetic iron ore. Progress has also been made in rehabilitating areas buried under mine tailings so that they once again will support vegetation; in some cases, using flyash from incinerators or the wastes from another industry to do the job. Fungi can produce high-quality protein from cellulose (paper) waste (Rogers et al., 1972).

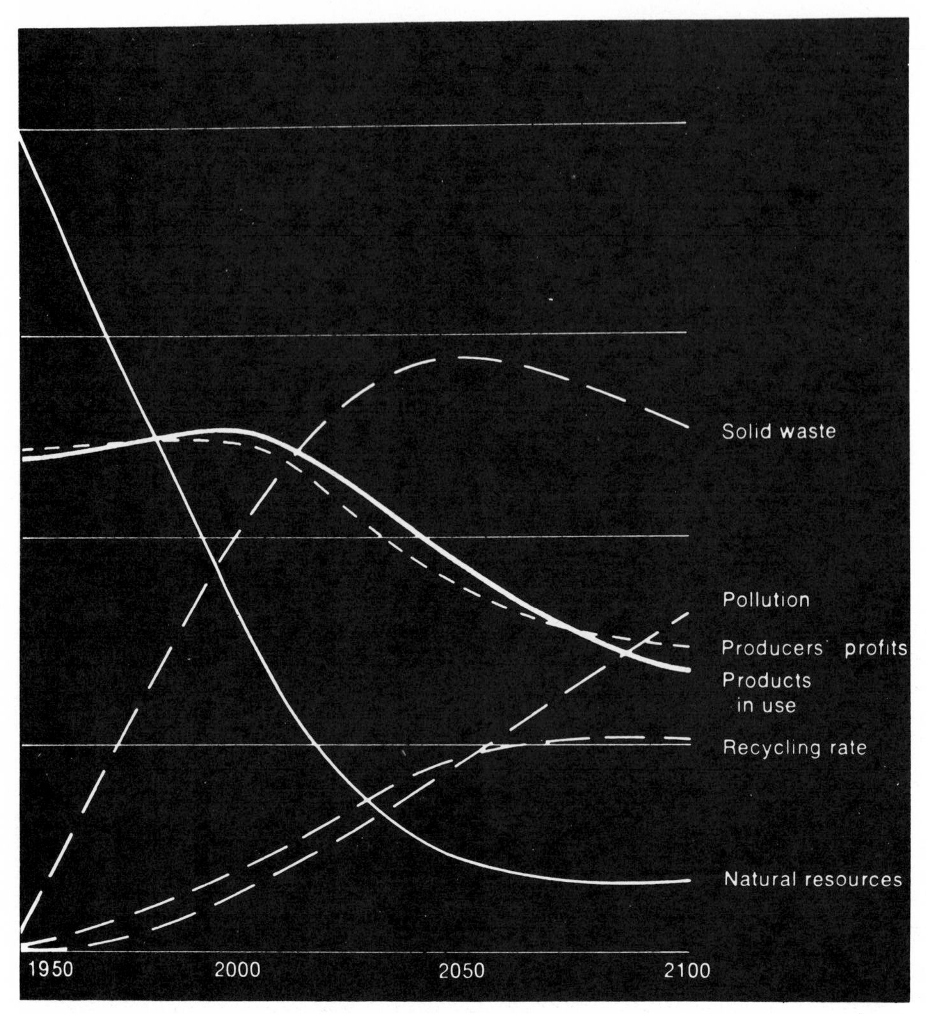

Figure 4.18. (*Continued*).

4.6 WHAT CAN AND MUST BE DONE

In order to conserve the Earth's fast diminishing supply of essential resources and insure the continuation of modern civilization and the political idealism upon which our culture is founded, certain steps have become necessary. Some of these steps are technological but others are sociological. In the past, technology has always come to man's rescue. It cannot continue to do so indefinitely, and difficult social decisions and adjustments are going to have to be made. While some of my scientific colleagues might raise criticisms, I am convinced that it is futile to write a book about our environment without at least listing some of the sociological measures that are necessary to preserve it. In the past, the negligence of scientists and technocrats to

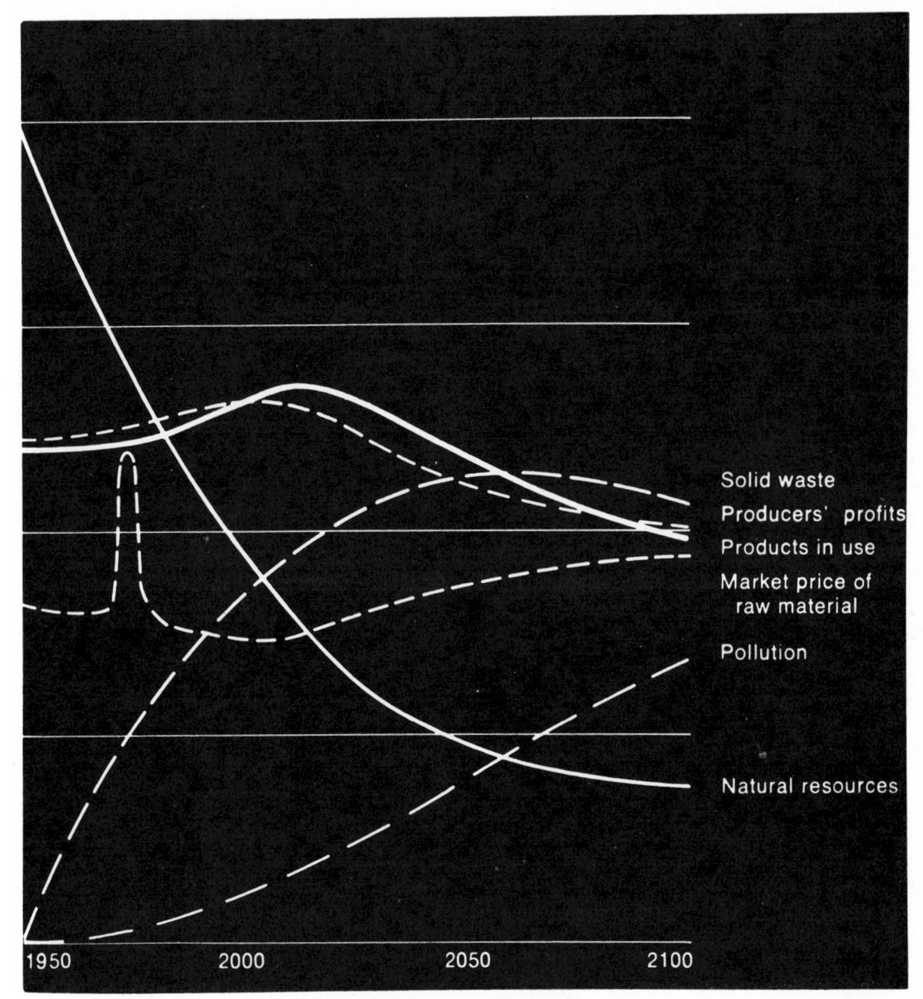

Figure 4.18. (*Continued*).

confront the social implications of their work has aggravated the world's problems. Some of the more important Earth resource conservation steps are:

1. Regulation of population;
2. Exploitation of nonfossil fuel energy sources such as nuclear and solar power;
3. Cycling and reuse of materials;
4. Discouragement of the development of undeveloped nations;
5. Birth control and environmental assistance to less developed nations;
6. Peace (war is a great waste of materials);
7. Government regulation of the use of materials (a national materials inventory and rationing);

8. Elimination of nonnecessities and of wasteful useages (like air-conditioning in northern cities);
9. Development of mass transportation systems and the restriction of private transportation;
10. Development of mass housing and the elimination of new one-family units;
11. Substitution of renewable resource materials for nonrenewable ones, that is, plastics for metals;
12. Strictly enforced conservation and antipollution laws with deterring penalties;
13. Strict government regulations of all land use; and
14. Much better long-range planning.

If we do not take acceptable measures (2, 3, 5, 6, 11, 12, and 14) now, we will be forced to take less acceptable measures (1, 4, 7, 10, and 13) later. Even the acceptable alternatives may not be easy. For example, while measure 11 may sound like a good idea, it has two serious drawbacks: plastics tend to create particularly durable wastes, and if the raw materials for the plastics is fossil plantstuffs, this cuts into energy reserves. On the other hand, if they are living plantstuffs, this diminishes the available acreage for food production. If a plastics production technology could be developed based on the utilization of atmospheric CO_2 this might appear to be a panacea, but again it would compete with photosynthesis for a material and because of the chemical stability of CO_2 would make large energy demands. If something seems to be conspiring against us in the resolution of our difficulties, let me say that there certainly is—a conspiracy of the laws of thermodynamics, the Principle of Le Chatelier and sex.

BIBLIOGRAPHY

Anon., *World Dredging Mar. Constr.*, *7*, 33 (December 1971*a*).

Anon., *Time* (June 7, 1971*b*).

Anon., *Nature*, *232*, 27 (1971*c*).

Anon., *Environ. Sci. Tech.*, *6*, 408 (1972).

P. Averitt, *Coal Resources of the United States*, U.S. Geol. Survey Bull. No. 1275 (1967).

H. Borchet, in J. P. Riley and G. Skirrow (eds.), *Chemical Oceanography*, Vol. 2, Chapter 19, Academic, London, 1965.

R. G. Bowen and E. A. Groh, *Tech. Rev.* (*M.I.T.*), *74*, 42 (October/November 1971).

O. Braitsch, *Salt Deposits: Their Origin and Composition*, Springer-Verlag, Berlin, 1971.

H. Brooks, *Met. Trans.*, *3*, 759 (1972).

H. Brown, *Sci. Amer.*, *223*(3), 195 (September 1970).

M. Butler, *Science*, *176*, 1002 (1972).

C. A. Caswell, *Environ. Sci. Tech.*, *4*, 642 (1970).

J. L. Cloudsley-Thompson, *Internat. J. Environ. Stud.*, *2*, 35 (1971).

R. C. Corey, *Principles and Practices of Incineration*, Wiley-Interscience, New York, 1969.

E. T. Degens and D. A. Ross (eds.), *Hot Brines and Recent Heavy Metal Deposits in the Red Sea*, Springer-Verlag, New York, 1969.

R. Dolan, *Science*, *176*, 286 (1972).

P. P. Dudley and R. E. Riecker, *Science*, *177*, 87 (1972).
H. Flohn and M. Ketata, World Meteorology, Organ. Tech. Note 116, Geneva, 1971.
E. P. Floyd and L. W. Lefke, *J. Water Poll. Control Fed.*, *43*, 1043 (1971).
D. F. Frasche, *Mineral Resources*, Nat. Acad. Sci. U.S.A.–Nat. Res. Council, Washington, D.C., 1962.
W. R. Fungaroli and R. L. Steiner, *J. Water Poll. Control Fed.*, *43*, 252 (1971).
R. Gillette. *Science*, *176*, 890 (1972).
M. I. Goldman, *Science*, *170*, 37 (1970).
A. Gomez-Pompa, C. Vazquez-Yanes, and S. Guevara, *Science*, *177*, 762 (1972).
T. J. Gray and O. K. Gashus (eds.), *Tidal Power*, Plenum, New York, 1972.
W. W. Hambleton, *Tech. Rev.* (M.I.T.), *74*(5), 15 (March/April 1972).
A. L. Hammond, *Science*, *177*, 978 (1972).
D. H. Hamilton and R. L. Meehan, *Science*, *172*, 333 (1971).
A. Hershaft, *Environ. Sci. Tech.*, *6*, 143 (1972).
C. Holden, *Science*, *172*, 249 (1971).
G. J. Hopkins and J. R. Popalisky, *J. Water Poll. Control Fed.*, *42*, 431 (1970).
M. K. Hubbert, in Committee on Resources and Man, *Resources and Man*, Nat. Acad. Sci. U.S.A.–Nat. Res. Council, Freeman, San Francisco, 1969.
C. B. Kenehan, *Environ. Sci. Tech.*, *5*, 594 (1971).
T. S. Lovering, in *Resources and Man, Committee on Resources and Man of the Division of Earth Sciences*, Nat. Acad. Sci. U.S.A.–Nat. Res. Council, Freeman, San Francisco, 1969.
H. Ness, *Environ. Sci. Tech.*, *6*, 700 (1972).
W. R. Niessen, *Tech. Rev.* (*M.I.T.*), *74*(5), 10 (March/April 1972).
S. Patterson, *U.S. Geol. Survey Bull.* No. 1228 (1967).
R. M. Pratt and P. F. McFarlin, *Science*, *151*, 1080 (1966).
S. R. Qasim and J. C. Burchinal, *J. Water Poll. Control Fed.*, *42*, 371 (1970).
J. Randers and D. L. Meadows, *Tech. Rev.* (*M.I.T.*), *74*(5), 20 (March/April 1972).
C. J. Rogers et al., *Environ. Sci. Tech.*, *6*, 715 (1972).
L. A. Sagan, *Science*, *177*, 487 (1972).
J. A. Salvato, W. G. Wilkie, and B. E. Mead, *J. Water Poll. Control Fed.*, *43*, 2084 (1971).
B. J. Skinner, *Earth Resources*, Prentice-Hall, Englewood Cliffs, 1969.
T. J. Sorg and H. L. Hickman, Jr., *Sanitary Landfill Facts*, 2nd ed., U.S. Gov. Print. Off., Washington, D.C., 1970.
T. J. Tofflemire and G. P. Brezner, *J. Water Poll. Control Fed.*, *43*, 1468 (1971).
E. Trueb, *Gas Wasser, Abwasser*, *51*, 317 (1971).
T. Valeton, *Bauxites*, Elsevier, Amsterdam, 1972.
A. D. Zapp, *Future Petroleum Producing Capacity of the United States*, U.S. Geol. Survey Bull. No. 1142-H (1962).

ADDITIONAL READING

Anon., *Proc. Second Mineral Waste Utilization Symposium*, ITT Res. Inst., Chicago, 1970.
Anon., *Solid Wastes*, Amer. Chem. Soc., Washington, D.C., 1971.
Anon., *Reuse and Recycle of Wastes*, Technomic, Stamford, Conn., 1971.

Anon., *Effective Technology for Recycling Metal*, Nat. Assoc. Secondary Materials Ind., New York, 1971*c*.

Anon., *Industry Survey on Disposal of Industrial Wastes by Combustion*, ASME, New York, 1971*d*.

Anon., *Minerals Handbook*, published annually by U.S. Bur. Mines.

A. M. Bateman, *The Formation of Mineral Deposits*, Wiley, New York, 1951.

R. L. Bates, *Geology of Industrial Rocks and Minerals*, Harper, New York, 1960.

H. J. Barnett and C. Morse, *Scarcity and Growth*, Johns Hopkins Press, Baltimore, 1963.

D. B. Brooks, *Supply and Competition in Minor Metals*, Johns Hopkins Press, Baltimore, 1965.

H. Brown, J. Bonner, and J. Weir, *The Next Hundred Years*, Viking, New York, 1957.

E. N. Davies and G. A. Northedge, *Mining and Minerals*, Oxford Univ. Press, New York, 1967.

N. L. Drobny, H. E. Hull, and R. F. Testin, *Recovery and Utilization of Municipal Solid Waste*, Environ. Prot. Agency, U.S. Gov. Print. Off., Washington, D.C., 1971.

P. T. Flawn, *Mineral Resources*, Rand-McNally, Chicago, 1966.

P. T. Flawn, *Environmental Geology*, Harper & Row, New York, 1970.

A. I. Levorsen, *Geology of Petroleum*, 2nd ed., Freeman, San Francisco, 1967.

H. H. Landsberg, *Natural Resources for U.S. Growth*, Johns Hopkins Press, Baltimore, 1964.

T. S. Lovering, *Minerals in World Affairs*, Prentice-Hall, Englewood Cliffs, 1943.

S. L. McDonald, *Petroleum Conservation in the United States*, Johns Hopkins Press, Baltimore, 1971.

C. F. Park, Jr., *Affluence in Jeopardy*, Freeman, San Francisco, 1968.

C. F. Park, Jr., and R. A. MacDiarmid, *Ore Deposits*, Freeman, San Francisco, 1964.

W. E. Small, *Third Pollution[a] The National Problem of Solid Waste*, Praeger, New York, 1971.

T. J. Sorg, *J. Water Poll. Control Fed.*, *43*, 1039 (1971).

R. L. Stanton, *Ore Petrology*, McGraw-Hill, New York, 1972.

W. H. Voskuil, *Minerals in World Industry*, McGraw-Hill, New York, 1955.

I. A. Williamson, *Coal Mining Geology*, Oxford Univ. Press, New York, 1967.

IV

THE ATMOSPHERE

5

ATMOSPHERIC CHEMISTRY

5.1 INTRODUCTION

We are benthic organisms crawling around at the bottom of a great sea of air. The philosopher–scientist Empedocles (fl. ca. 440 B.C.) demonstrated experimentally that air is a substance, and the ancients made air, along with water, earth, and fire, one of their four elements. Modern experimental chemistry began with the analysis of air, the resolution of its components, the elucidation of the nature of combustion by Lavoisier (born 1743, beheaded in 1794), and the subsequent overthrow of the phlogiston theory. Except for the discovery of the chemical elements (Weeks, 1945), the analysis of lithospheric constituents had little comparable influence on the chemical sciences, and analysis of the hydrosphere has had virtually none.

Earlier we mentioned that the trails of meteors as they enter the Earth's atmosphere have given us clues as to its composition. However since World War II, our knowledge of atmospheric chemistry, especially the upper atmosphere, has been enormously enlarged thanks to balloon, rocket, and instrument technology. In fact, it is probably not an exaggeration to say that our rocket and space programs have yielded just as much valuable information about the atmosphere of our own planet as about the Moon and other planets. Figure 5.1 shows the altitude ranges of some of the techniques for exploring the atmosphere.

5.2 THE STRUCTURE OF THE EARTH'S ATMOSPHERE

Far from being simply a homogeneous gaseous mantle whose density trails off as it imperceptibly passes into the exosphere the Earth's atmosphere is a region of the cosmic onion (Figures 1.1 and 1.2) of splendid complexity (Figure 5.2—notice that the scales in this figure are not linear). The multilayered lithosphere is hardly more complex, and the marine hydrosphere is by comparison a paragon of simplicity. As the least viscous of the Earth's zones, the atmosphere is the most changeable, and its changes are the most rapid, depending not only on terrestrial position and influences but upon the position of the Sun and upon the details of that star's activities. To deal with the complexity of the atmosphere and the rapid accummulation of data about it, a number of different systems of nomenclature have been proposed over the past several years—the classification that we adopt here (Figure 5.2) is something of a conglomerate.

With increasing altitude gas pressure (Figure 5.3) and density (Figure 5.4) gradually attenuate although not in a simple linear manner. The portion of the atmosphere

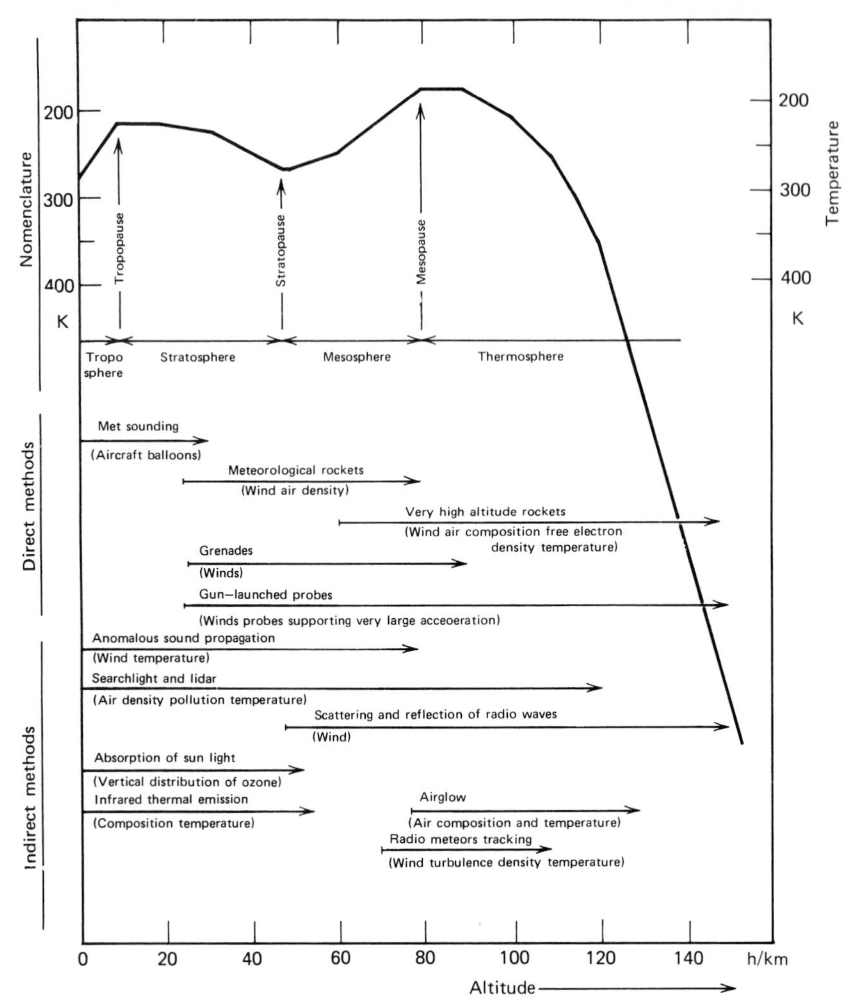

Figure 5.1. Experimental methods for exploring the atmosphere (from Grobecker, 1972).

most familiar to us is, of course, the lowest layer or troposphere, which extends 8 to 12 km up above the Earth's surface (Figure 5.2). Air temperature in the troposphere decreases with height at an average rate of about 0.5°C/km (Miyake, 1965). Convection mixes the troposphere vigorously (Figure 5.5) so that its chemical composition with respect to major constituents is homogeneous. As we ascend, the strong winds cease abruptly (Figure 5.5), and the temperature stays fairly constant at about −55°C; we have passed the tropopause (Figure 5.2). As every air traveler knows, there is little turbulence in the lower stratosphere; however, there can be a strong constant wind, the "jet stream," Between 35 and 50 km, there is a temperature inversion with a "warm" layer at about 0°C around 50 km called the stratopause. Another inversion begins above 80 km (Figure 5.2), and the temperature begins to climb steadily until at 400 to 500 km the temperature may be as high as 2000 to 3000°K. The region between 80 and 400 to 500 km is variously known as the thermosphere, the heterosphere, and the ionosphere (reflections from the layers of the ionosphere make long distance radio communication possible). All three of these

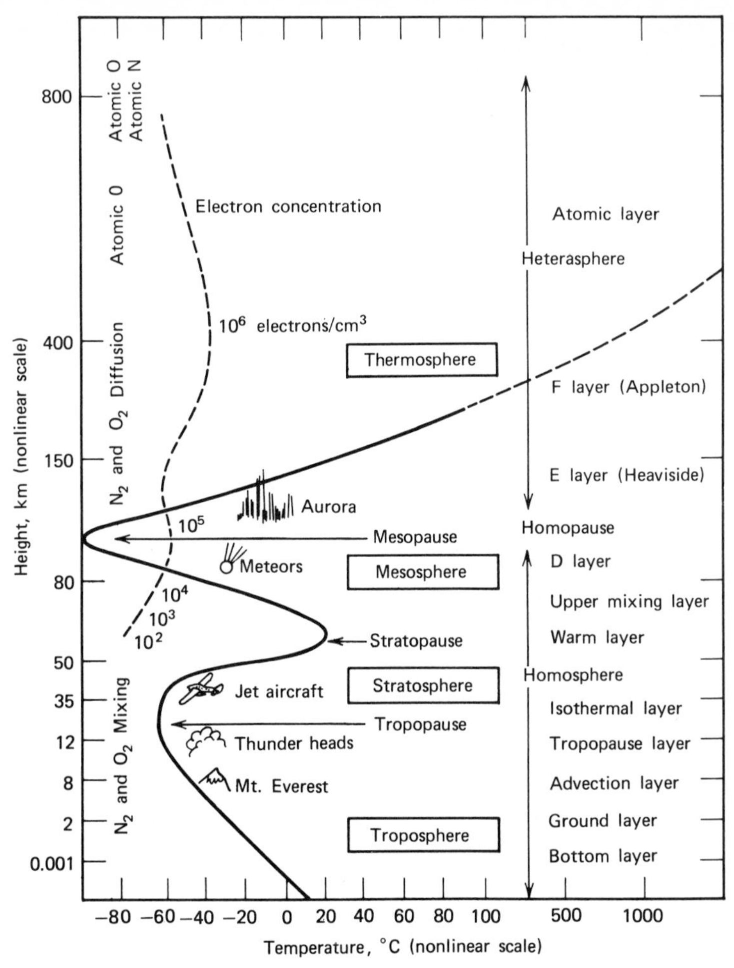

Figure 5.2. Structure of the Earth's atmosphere.

names are appropriate, for this region is characterized by high and increasing temperature; chemical heterogeneity because of photochemical reactions, slow mixing, and low density; and the presence of charged species. I would like to emphasize that the temperature profile in Figure 5.2 is a generalized one; not only are some of the segments of the curve very highly variable (bottom layer temperatures range from −50 to +80°C, upper mixing layer from −80 to +70°C, F-layer from +60 to +1000°C), but the overall shape of the curve is strongly dependent on latitude and on terrestrial and solar conditions. The warming of air near the Earth's surface is due to the absorption of solar energy by the more dense atmosphere and radioactive and conductive heating from the ground, which in turn is heated by the absorption of solar energy from without and by the heat from radioactive decay within. The temperature maximum in the stratosphere is attributed to the absorption of solar ultraviolet rays by ozone and molecular oxygen, while more drastic chemical reactions produced by ionizing solar radiation are the cause of the warming of the thermosphere. The temperature of a gas, it should be noted, is defined in terms of corpuscular collisions, and thus at the greater altitudes where the atmosphere is

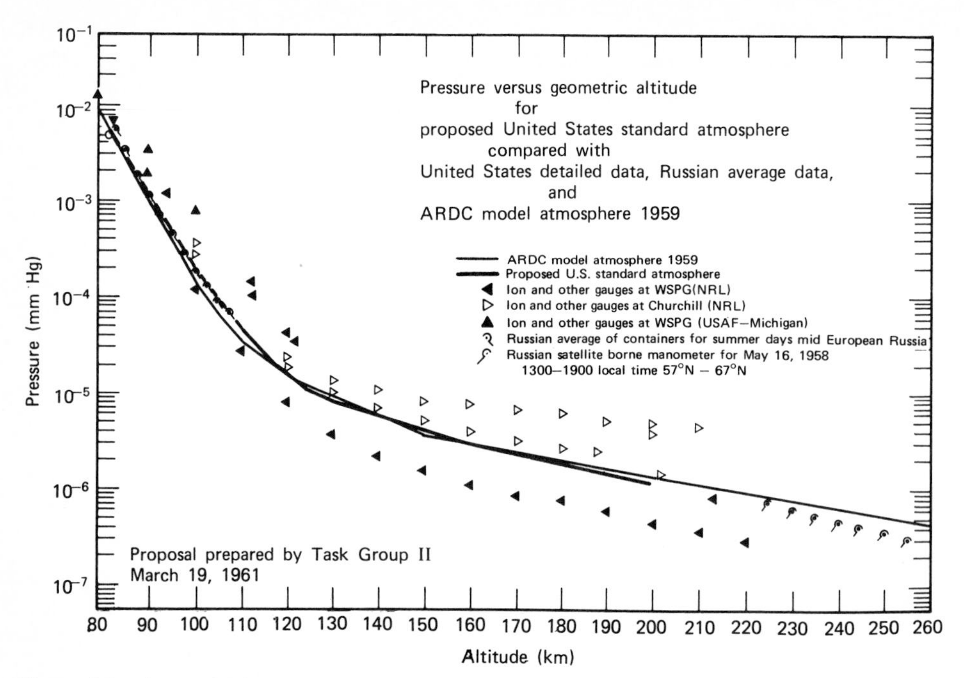

Figure 5.3. Atmospheric pressure as a function of altitude (from Nawrocki and Papa, 1963, with permission of Prentice-Hall, Inc.).

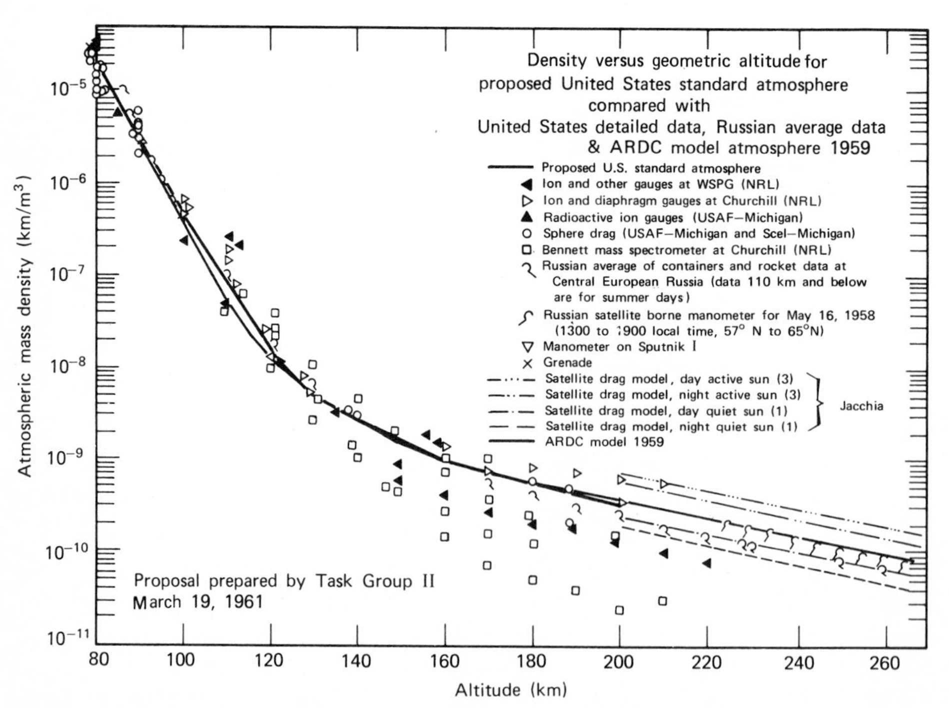

Figure 5.4. Atmospheric density as a function of altitude (from Nawrocki and Papa, 1963, with permission of Prentice-Hall, Inc.).

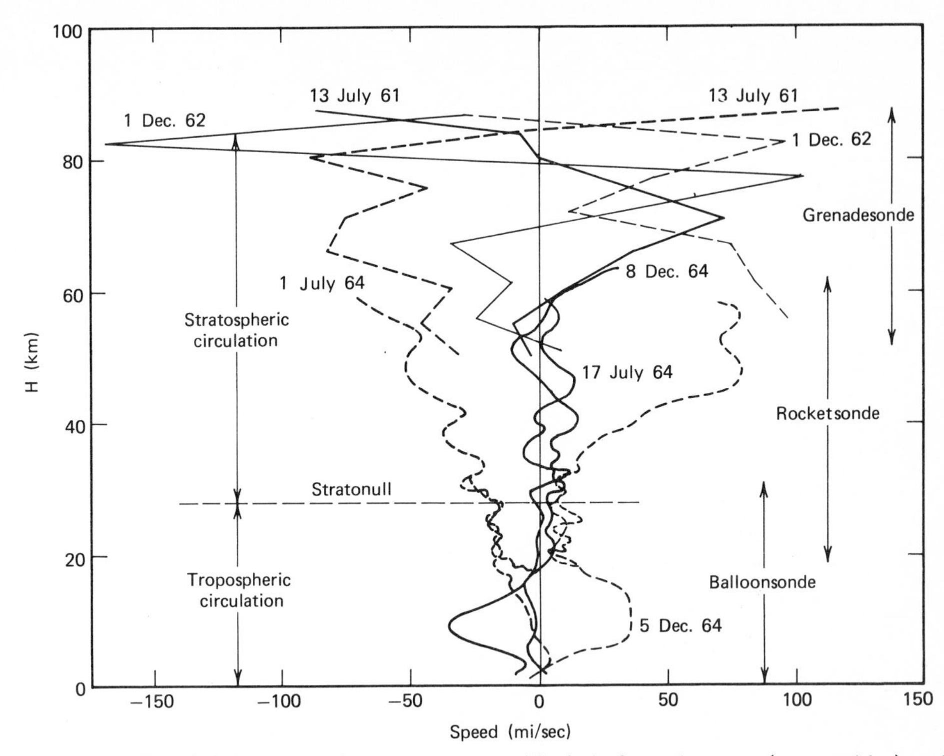

Figure 5.5. Wind data presented in components with dashed zonal curves (west positive) and solid meridianal curves indicating wind direction (from Webb, 1966, with permission of Academic Press, Inc. and the author).

very dilute and collisions infrequent, the concept of temperature becomes a bit unclear.

The space surrounding the Earth in addition to being a complex, stratified chemical reactor is also, thanks to the planet's magnetic field and the radiative flux from the Sun and other astronomical features, a remarkable particle accelerator. In recent years, rocket research has revealed the existence of two concentric tori of intense radiation girdling the Earth—the Van Allen belts (Figure 5.6)—high above the

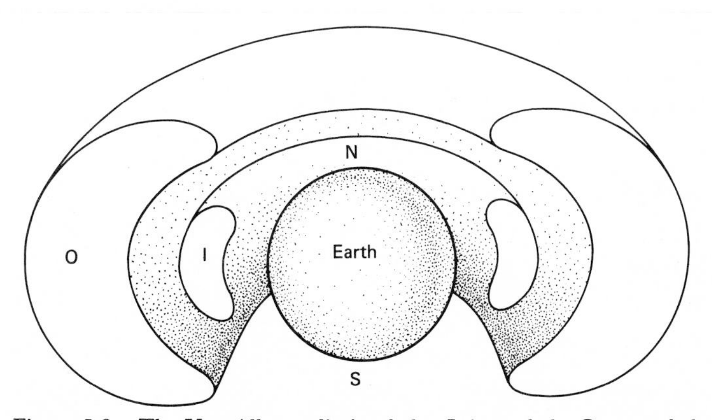

Figure 5.6. The Van Allen radiation belts. I, inner belt; O, outer belt.

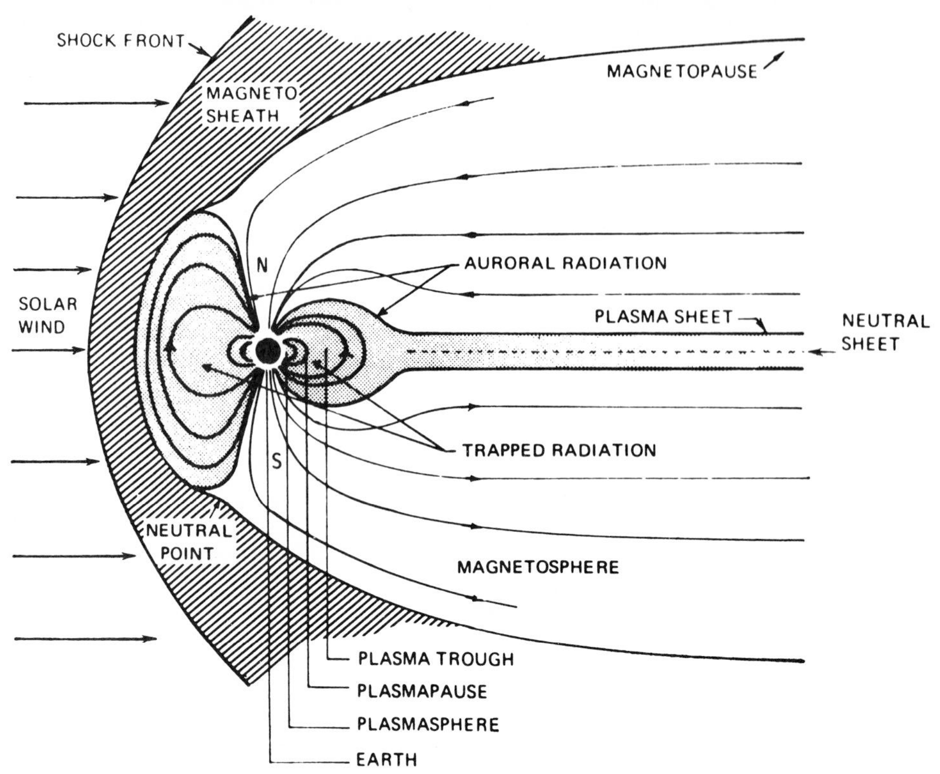

Figure 5.7. The Earth's wake in the solar flux (from Johnson, Young, and Holmes, © 1971 by the American Association for the Advancement of Science with permission of the AAAS).

F-layer, and a new name, the magnetosphere, has been coined for this spectacular region. The Earth's purturbation of the solar flux on an even grander scale is shown in Figure 5.7. The Van Allen belts are composed of high-energy electrons and protons. The inner belt (at about 3000 km) is believed to be produced by cosmic rays, while the corpuscles in the outer belt (16,000 to 100,000 km) are theorized to be plasma clouds thrown into space by solar eruptions and trapped in the belt by the Earth's magnetic field.

5.3 CHEMISTRY OF THE LOWER ATMOSPHERE

Because solar radiation capable of inducing chemical reactions has been previously absorbed out, there is relatively little chemistry involving the major constituents of the lower atmosphere (troposphere). An exception are reactions between oxygen and nitrogen produced by electrical discharges during thunderstorms. This is not to say that there is no chemistry in the lower atmosphere; on the contrary, there are at least two types of chemical processes occurring involving minor constituents that are of most profound importance to human affairs—these are the chemistry of air pollution (to which we devote Chapter 6) and the phase changes of water, changes that are, of course, the most obvious features of our "weather."

Table 5.1 The Average Composition of the Atmosphere (From Mason, 1966)

Gas	Composition by Volume (ppm)	Composition by Weight (ppm)	Total Mass ($\times 10^{20}$ g)
N_2	780,900	755,100	38.648
O_2	209,500	231,500	11.841
Ar	9,300	12,800	0.655
CO_2	300	460	0.0233
Ne	18	12.5	0.000636
He	5.2	0.72	0.000037
CH_4	1.5	0.94	0.000043
Kr	1	2.9	0.000146
N_2O	0.5	0.8	0.000040
H_2	0.5	0.035	0.000002
O_3[a]	0.4	0.7	0.000035
Xe	0.08	0.36	0.000018

[a] Variable, increases with height.

The average composition of the lower atmosphere is given in Table 5.1, and in the well-mixed region below 60 km, the deviation from these averages is small. Above 100 km, gravitational fractionation according to molecular weight may become significant. As for isotopic fractionation, despite earlier reports of the fractionation of ^{14}N and ^{15}N in the atmosphere, subsequent measurements (Dole et al., 1954) failed to detect any ^{16}O–^{18}O fractionation. Figure 5.8 is an impressive one. It shows the concentration profiles of a great number of species in the atmosphere over a wide altitude range, yet Grobecker (1972) who reproduces this figure remarks that it shows only about two-thirds of the number of stratospheric constituents that should be considered in assessing the possible impact of aircraft such as the SST on global climate. There is a detailed but now somewhat old monograph on the chemistry of the lower atmosphere by Junge (1963).

The chemical composition of the Earth's atmosphere has (and is) changed in the course of geological time. Ancient changes are discussed in Chapter 15, while Chapter 10 treats current changes, largely brought about by human activity.

As discussed in Table 5.1, the air we breathe is about 78% N_2 and 21% O_2. The oxygen we need for respiration is diluted with nitrogen, which is a relatively inert gas chemically. Patients suffering breathing difficulties are often treated with an oxygen-rich environment—a procedure not without its hazards, premature babies were frequently put in oxygen tents until a high rate of permanent blindness was traced to the destruction of the retina's capillaries by this "precaution."

Of the minor constituents of the lower atmosphere, by far the most important are carbon dioxide and water vapor. The concentrations of the minor constituents, unlike the major ones, especially when they are pollutants such as CO_2, can be highly variable in time and space. Table 5.1 is for the dry atmosphere. H_2O is so highly variable it is usually deleted from such tables. We discuss CO_2 in chapter 6 on air pollution and again in Chapter 15 when we discuss the carbon dioxide–carbonate

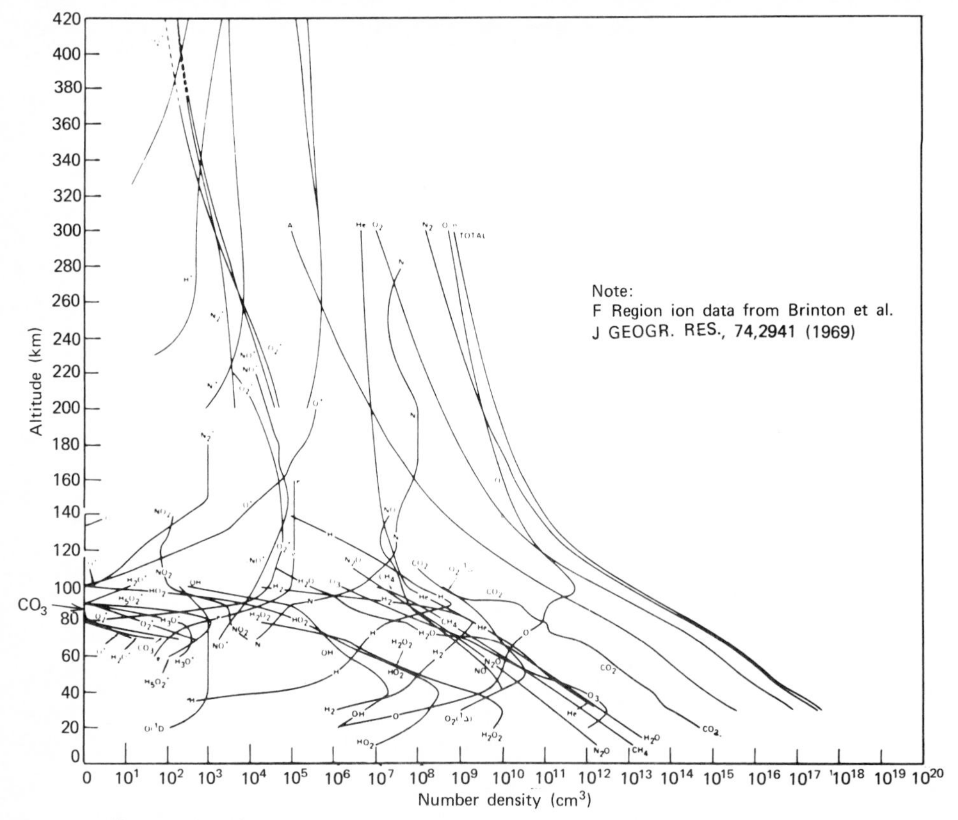

Figure 5.8. Composition of the daylight atmosphere (from Grobecker, 1972).

cycle. Turning now to water, the air can be exceedingly dry as in desert regions, or it can be supersaturated to the point where the water eventually precipitates out as rain, fog, dew, snow, hail, or sleet. There is some highly interesting but as yet still imperfectly understood physical chemistry involved in atmospheric precipitation that we briefly touch on. Local and global atmospheric circulation while certainly relevant lies outside the main theme of this book, however we return to the subject of atmospheric water when we discuss the total hydrologic cycle in Chapter 9.

Table 5.2 shows the equilibrium vapor pressure of water as a function of temperature. The relative humidity is the ratio of the actual to the equilibrium vapor pressure, corresponding to saturation, expressed as a percentage and at a given temperature. When the equilibrium vapor pressure is exceeded (the dew point), precipitation should occur; however, supersaturation is common. Thus as an air mass is cooled, we expect rain or snow. Air, like most other gases and gaseous mixtures, to a very good approximation obeys the perfect gas law

$$PV = (\text{constant})\ T \tag{5.1}$$

under environmental conditions. In the lower atmosphere, the important variables determining air temperature are pressure, cloud cover, the influx of solar energy, and, because of the relationship described in Figure 5.2, altitude. As warm air rises

Table 5.2 Water Vapor Pressure

Temperature (°C)	Pressure (mm Hg)
−30	0.3
−20	0.8
−10	2.0
0 (F.P.)	4.6
10	9.2
20	17.5
30	31.8
40	55.3
50	92.5
60	149.4
70	233.7
80	355.1
90	525.8
100 (B.P.)	760.0

because of its lower density, the pressure decreases, the air mass tends to expand, and consequently cool (Equation 5.1) and precipitation often ensues. Mountain masses can also force air to rise so while lush forests often clothe the windward side of mountain chains, parched deserts are found beyond their leeward flanks (Figure 5.9). Another familiar mechanism of precipitation induction is the formation of great thunderheads as the warm air over sunny plains rises (Figure 5.10). Thunderheads, which can extend up to and beyond 15 km (Lee and McPherson, 1971), can inject enormous quantities of water into the upper troposphere and lower stratosphere. Kuhn, Lojko, and Petersen (1971) have estimated that thunderstorms over Oklahoma

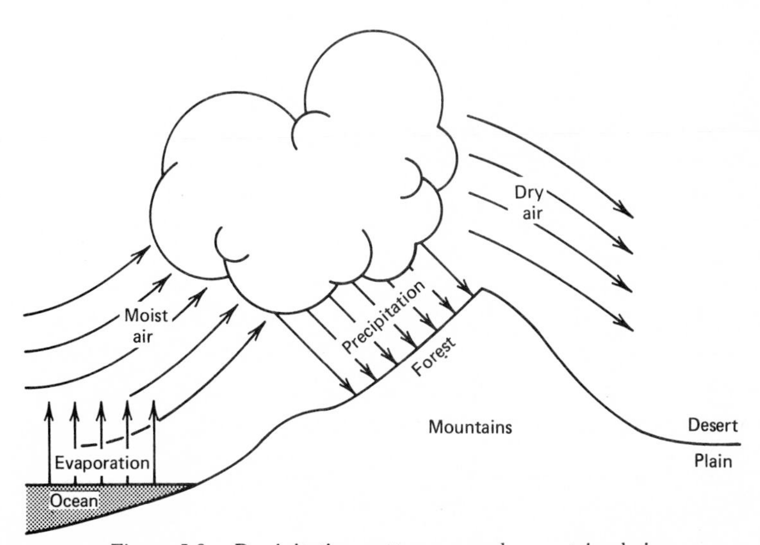

Figure 5.9. Precipitation pattern around mountain chains.

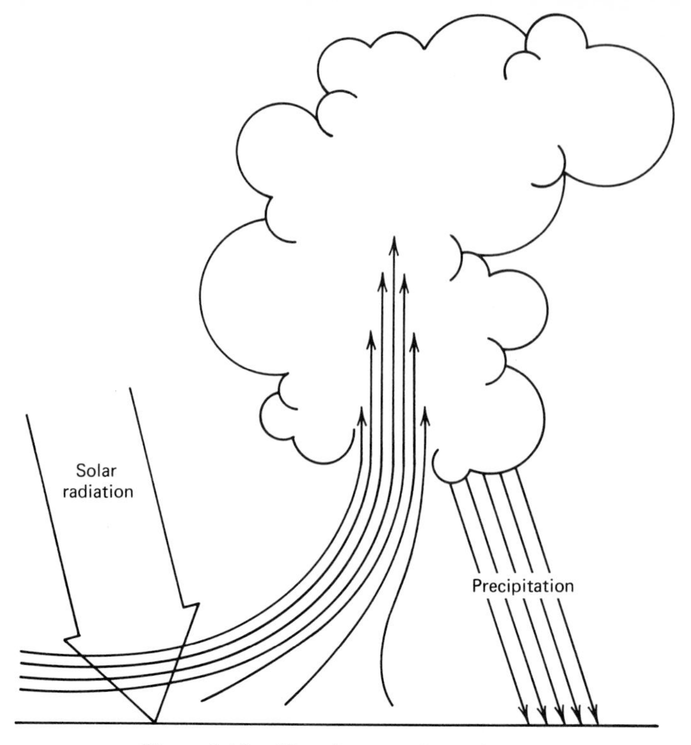

Figure 5.10. Thunderstorm formation.

vertically transport some 1.6×10^9 g/day of water vapor. The chemical content of rain is treated in Chapter 6.

The vertical distribution of the average water content of the tropopause and lower stratosphere is shown in Figure 5.11 and, as we would expect from Table 5.2, it parallels the thermal profile of the lower atmosphere (Figure 5.2). Water dominates the chemistry of our environment, and this is no less true of the atmosphere than elsewhere. Not only is it the substance of "weather," but in the words of Junge (1963) "condensation of water is of paramount importance for the cleansing of the atmosphere and thus for the cycle of numerous trace substances." Water vapor also plays a very significant role in determining the heat balance of the lower atmosphere as well as of the surface of Earth (Brunt, 1962; Kondratyev, 1969).

Many of the other minor gaseous constituents of the lower atmosphere, such as the oxides of sulfur and nitrogen, are pollutants, and we deal with them in Chapter 6. I must insert a word here about the atmosphere as a resource. Nearly all of our combustion processes, including the respiration of our own bodies, run on a chemical from the air, oxygen, a fact whose enormous significance is easy to forget. We have mentioned once and we mention again the fixation of atmospheric nitrogen for agricultural purposes, but air itself forms the source of an indespensible family of industries—these are the air liquification and fractionation industries (Simpson, 1969). liquid air is a bluish fluid that boils at about −190°C. Oxygen (B.P. −182.96°C), nitrogen (B.P. −195.8°C), argon, and the other noble gases are prepared commercially by the careful fractionation of liquid air. Liquid air, nitrogen,

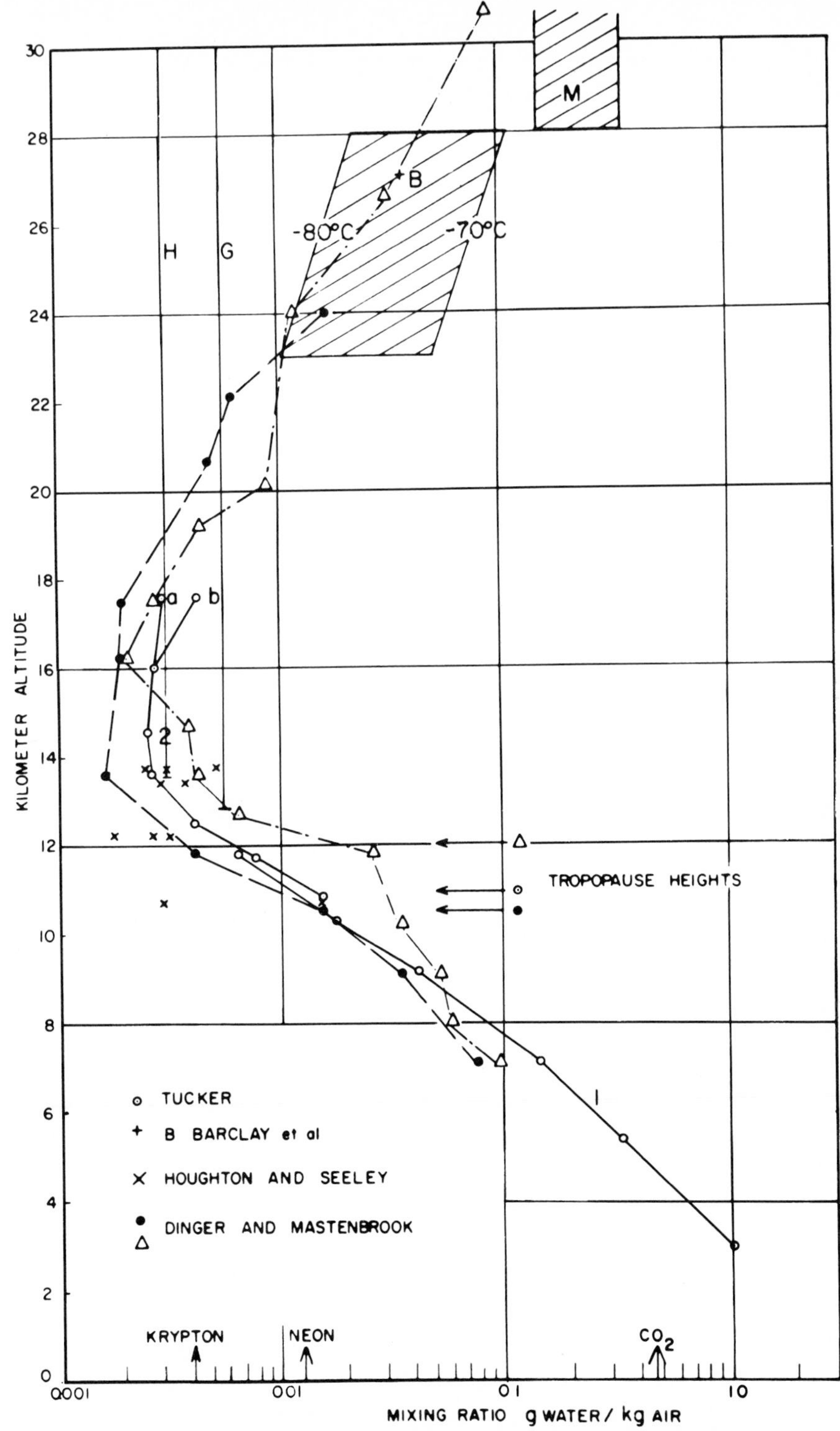

Figure 5.11. Water vapor mixing ratio as a function of altitude. Hatched area between 23 and 28 km indicates range of frost points between −70 and −80°C for mother-of-pearl clouds (from Junge, 1963, with permission of Academic Press, Ltd. and the author).

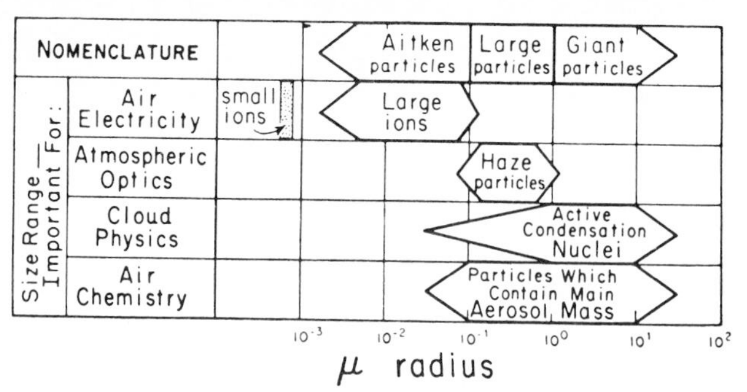

Figure 5.12. Nomenclature of natural aerosols and the importance of particle sizes for various fields of meteorology (from Junge, 1963, with permission of Academic Press, Ltd. and the author).

and even liquid He and H_2 are essential to low-temperature research, and fuming Dewar flasks are a familiar feature of many chemistry and physics laboratories. Familiar also to television audiences are the great plumes of evaporating liquid oxygen (LOX) from our rocket launches.

In addition to its gaseous constituents, the lower reaches of the atmosphere contain quantities of suspended material both liquid and solid, and although their concentrations are relatively small, these aerosols play a disproportionately important role in the chemistry of the troposphere and stratosphere. We treat the origin and chemical composition of atmospheric aerosols in Chapter 6, confining ourselves here to physical properties and distribution. Figure 5.12 summarizes a nomenclature that has been worked out for variously sized atmospheric aerosols, while Figure 5.13 shows some size distribution curves for the lower atmosphere over Germany. These curves can be highly variable, but not uncommonly there is a maximum around a radius of

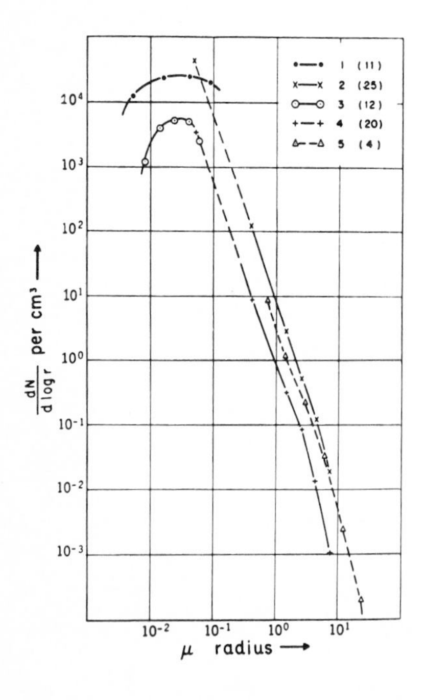

Figure 5.13. Complete size distributions of natural aerosols over Germany (from Junge, 1963, with permission of Academic Press, Ltd. and the author).

Table 5.3 Particle Concentrations per Cubic Centimeter for the Two Model Size Distributions and for the Stratosphere (From Junge, 1963, with permission of Academic Press, Ltd., and the author.)

	Particle Concentration of Radius (μm)							
	<0.01	0.01–0.032	0.032–0.10	0.10–0.32	0.32–1.0	1.0–3.2	>3.2	Total Number
Surface Air								
Continent	1600	6800	5800	940	29	0.94	0.029	15,169
Ocean	3[a]	83[a]	105[a]	14[a]	2	0.47	0.029	207
Stratosphere								
1 km[b]	2[a]	24[a]	9.1[a]	0.19	0.019	0.001	—	35
4 km[b]	0.08[a]	3.4	1.7[a]	0.06	0.006	0.0003	—	5.4

[a] These values are estimated from the total number of Aitken particles and the assumptions about the size distribution.

[b] Above tropopause.

0.1 μm. Perhaps more instructive is Table 5.3 from which we see the differences in sizes of continental and oceanic particles and getween the upper and lower atmosphere. Generally speaking, the particle concentration decreases rapidly with increasing altitude (Figure 5.14), but there does seem to be some sort of a maximum in the concentrations of Aitken and large particles associated with the tropopause at above 10 km (see also Figures 46 and 47 in Junge, 1963).

The aerosol particles can be charged or uncharged. The smaller particles affect the electrical and optical properties of the atmosphere, but only the large and giant particles (Figure 5.12) serve as nuclei for the condensation of atmospheric water. As I have said, there is some exceedingly complex and interesting physical chemistry involved in the formation of atmospheric precipitation, and although we know a great deal about these processes, there is still clearly a great deal that we have still to learn (Pruppacher, 1963; Kassner, 1968; Kuhns and Mason, 1968; Pena and de Pena, 1970). Of one thing I am convinced, however, the key to these processes is the structure of liquid water and the effect of solutes, ionic, nonpolar, and gaseous, and of surfaces on that structure, and the relationship of these modifications to the structure of Ice I. Man has recently projected himself into this act, and his experiments: Seeding clouds with silver iodide and dry ice have demonstrated a capability of modifying the weather on a major scale. This capability fills some of us with trepidation. Seeding experiments may have been responsible for the 10 in. of rain that fell in 4 hr on Rapid City, South Dakota, in 1972, killing more than 200 persons and causing extensive damage. The specter of weather modification as a military weapon has raised its ugly head (Shapley, 1972); certainly against a heavily overpopulated, undeveloped country it could be a weapon of catastrophic effectiveness, although the certainties with respect to confinement and control of the phenomenon

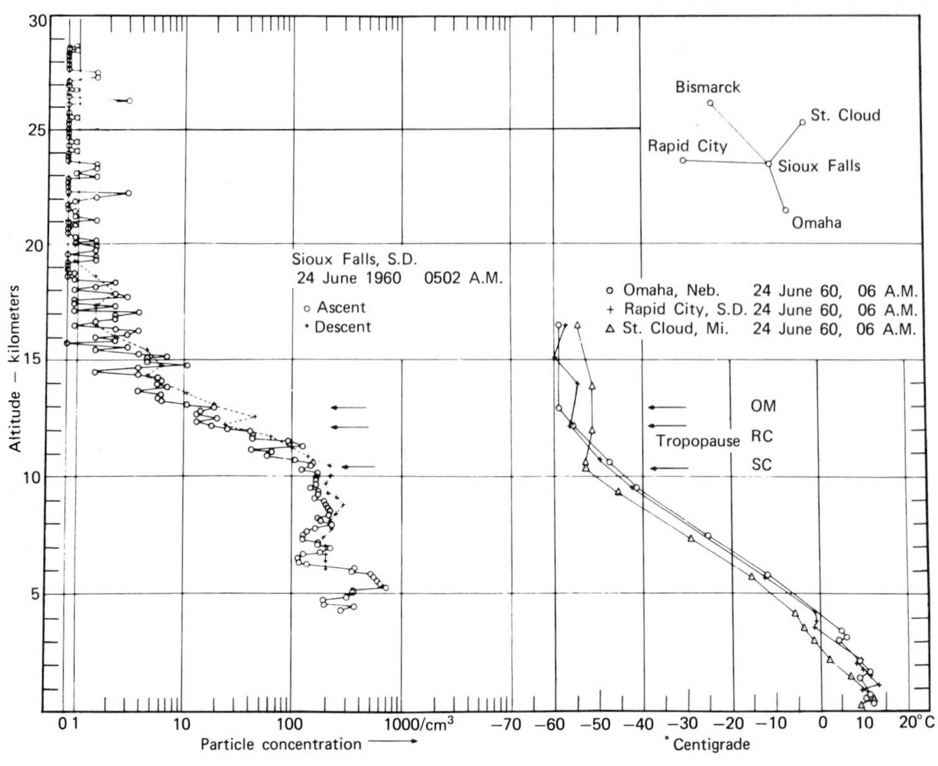

Figure 5.14. Profile of Aitken particles in Sioux Falls, North Dakota, and temperature profiles at the nearest radiosonde station (from Junge, 1963, with permission of Academic Press, Ltd. and the author).

are still poor. Seeding of their centers over the ocean before they reach populated coastal areas might reduce the havoc caused by hurricanes, and the risks involved have been discussed by Howard, Matheson, and North (1972). Remember the first environmental principle that we set down in the introduction to Chapter 4; we cannot give a resource (rain) to the farms, homes, and factories of one region without taking it away from people downwind. The potential legal entanglements are horrendous. While I cannot condemn carefully controlled cloud seeding as an experimental tool to further our understanding of atmospheric processes, large scale weather modification is a Pandora's box best left securely shut.

5.4 CHEMISTRY OF THE UPPER ATMOSPHERE

As we ascend into the atmosphere from the Earth just above the tropopause at an altitude of about 20 km, we find ourselves entering one of those hot spots (Figure 1.3) where the influx of solar radiation produces relatively intense chemical activity. The concentration of the highly reactive and very poisonous gas ozone, O_3, increases markedly (Figure 5.15). The chemical reactions of atmospheric ozone are many, but fortunately only four account for the majority of ozone production and destruction.

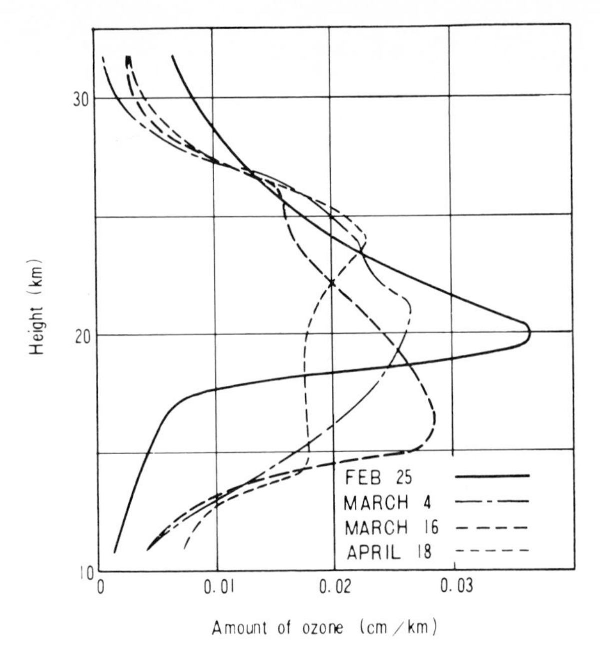

Figure 5.15. Vertical distribution of atmospheric ozone in New Mexico (from Miyake, 1965, with permission of the author).

Molecular oxygen absorbs solar energy in the 1760- and 2030 Å ranges to form atomic oxygen (Goody, 1964).

$$O_2 + h\nu \rightarrow 2O \tag{5.2}$$

then in the presence of a third body, M, which may be O_2, N_2, or NO_x and which helps to dissipate excess energy, the atomic oxygen reacts with molecular oxygen to produce ozone (Cadle and Allen, 1970)

$$O_2 + O + M \rightarrow O_3 + M \tag{5.3}$$

The ozone can be destroyed by further absorption of energy either between 2000 and 3200 Å in the ultraviolet, between 4500 and 7000 Å in the visible, and even to a lesser extent in the infrared

$$O_3 + h\nu \rightarrow O_2 + O \tag{5.4}$$

or it can encounter an atomic oxygen to produce an excited oxygen molecule

$$O_3 + O \rightarrow 2O_2^* \tag{5.5}$$

Ozone is a very powerful oxidant and is thus very unstable in the lower atmosphere where the concentration of reduced species (such as CO, H_2S, CH_4, etc.) is relatively high. O_3 is also destroyed in the stratosphere by reaction with hydrogen from a

$$O_3 + H \rightarrow OH + O_2 \tag{5.6}$$

variety of sources (Clyne and Thrush, 1963; Phillips and Schiff, 1962).* The residence time of tropospheric ozone is between 1 and 2 months; in the stratosphere, 1 to 2 yr (Junge, 1963).

* Recently the possible destruction of ozone by fluorohydrocarbons and other man-made pollutants has been the subject of much controversy.

The absorption of ultraviolet light in the 2000- to 3200-Å range by ozone is so great that the atmospheric ozone layer effectively prevents this radiation from reaching the Earth's surface. While solar ultraviolet radiation may have played a role in prebiological chemical synthesis on this planet, it is highly lethal to complexly organized biomolecules. Thus ancient life forms could not leave the protective seas until photosynthesis had produced the oxygen from which in turn the ozone shield was formed. Absorption reaction 5.4 produces a great deal of thermal energy, and contributes to the warming of the stratosphere, although the temperature maximum (Figure 5.2) occurs at a considerable distance above the ozone maximum (Figure 5.15).

As one might expect of a chemical whose formation and destruction is dependent on solar as well as terrestrial conditions, the atmospheric ozone profile is very variable in space and time. Seasonal fluctuations are evident in Figure 5.15, while a great deal of shorter time span variation is shown in Figure 5.16. The ozone concentration also depends on location (Figure 5.17). The concentration of ozone in the lower atmosphere is very low, very variable, and because of the presence of oxidizeable aerosols very transitory (Figure 5.18). The chemistry of ozone removal from the lower atmosphere is still unclear. Earth-produced methane and H_2S can account only for a minute fraction of the total quantity disappearing (Dillemuth, Skidmore, and Schubert, 1960; Cadle, 1960). Although mass transfer through the tropopause is slight, suitable cyclonic conditions can produce sufficient vertical convection to bring ozone down. The morning minimum and afternoon maximum in the ozone level of the lower air mass tend to correspond to daily maxima and minima in vertical mixing in the lower troposphere (Atkins, Cox, and Eggleton, 1972). The "clean smell" of the air following thunderstorms has been attributed to very low levels of ozone formation by lightning. In larger concentrations, ozone has an offensive, astringent, and unmistakable odor and is sometimes noticeable around electrical discharge machinery. Ozone can be a pollutant, and it can be produced in the troposphere by photochemical reactions of smog (Leighton, 1961; Wayne, 1962, Junge, 1963).

As we ascend higher through the mesosphere and mesopause into the thermosphere, we enter a region of even more intense chemical activity caused by the impact of bombardment, mostly ultraviolet and X-radiation (Goody, 1964), from the Sun sufficiently energetic to not only excite but to ionize atoms and molecules by literally knocking their superficial electrons off (notice the free electron concentration profile in Figure 5.2). Thus we notice in Figure 5.8 that while the major species below about

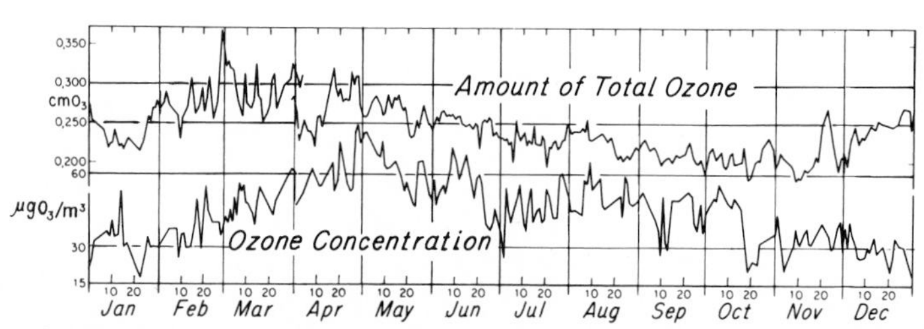

Figure 5.16. Ozone data for Arosa, April 1950 to March 1951, 1860 m above sea level. Upper curve: total ozone; lower curve: daily maximum ozone concentration at ground level in γ/m^3 = $\mu g/m^3$ (from Junge, 1963, with permission of Academic Press, Ltd. and the author).

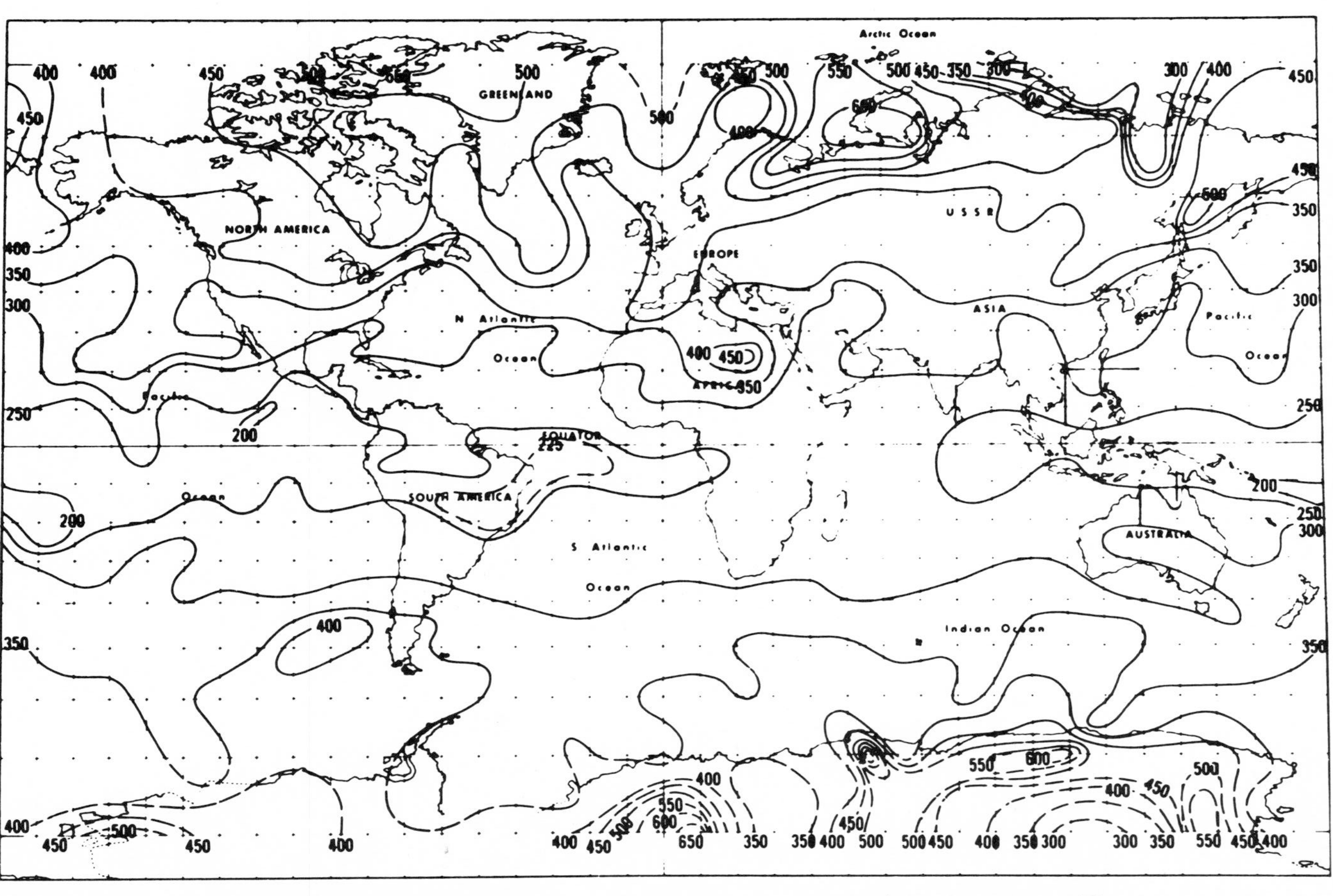

Figure 5.17. Total ozone content (10^{-3} cm, STP), April 28, 1969 (from Grobecker, 1972).

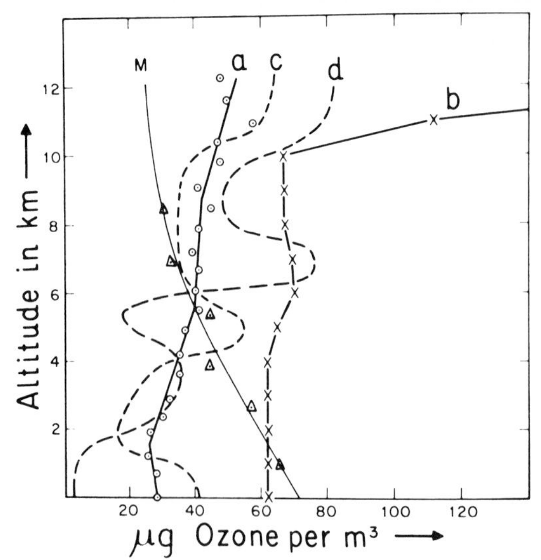

Figure 5.18. Vertical distributions of ozone concentrations in the troposphere. Curve a: Average value, England, October 1952 to November 1953. Curve b: Average value, Norway, June to July 1955. Curves c and d: individual flights over England. Curve M: constant mixing ratio. The triangles represent average values of tropospheric ozone in Boston during Winter and Spring, 1960–1961 (from Junge, 1963, with permission of Academic Press, Ltd. and the author).

80 km are neutral, there is a rich predominance of charged species above 80 km. Notice too that most of these ions are positively charged. This might be an artifact. The passage of a rocket through the atmosphere generates a negative charge that repells negative ions from its samplers. Consequently, very little is known about negatively charged species in the upper atmosphere (Donahue, 1968). With such a myriad of highly activated charged particles dashing around in such an intense radiation field, the number and variety of possible chemical permutations and interactions is bewilderingly great. Crutzen (1972) has simply listed what he feels may be some of the more important reactions (Table 5.4), while Nicolet (1972) (Figure 5.19) and Hudson (1972) (Figure 5.20) have tried to summarize this information with schematic reaction diagrams. This long list and these complicated diagrams, it should be noticed, are given in discussions of the stratosphere, not even the ionosphere! Donahue (1968) has been able to sift some simplified conclusions out of the chemistry of the thermosphere (ionosphere), and I follow the lead of his discussion here. To repeat, solar radiation frees electrons from atmospheric atoms and molecules. Some of these photoelectrons have sufficient energy to cause further ionization before they reach thermal equilibrium. A major process for free electron disappearance is recombination with a diatomic positive ion to form two fast atoms. As for the gaseous composition of the upper atmosphere, it is appreciably different from that of the lower atmosphere with its turbulent mixing. The gases diffuse freely in the Earth's gravitational field, and as a consequence of the resulting density fractionation, the concentrations of the lighter gases trail off more slowly than the heavier ones. At 120 km, there is more atomic oxygen than O_2; at 200 km, it is even more abundant than N_2

Table 5.4 Reactions and Reaction Coefficients (From Crutzen, 1972) (in cm-molecule-sec. units)

Reactants		Products	Coefficients
$O_2 + h\nu$	→	$O + O$	$\lambda < 2420$ Å, $\mathcal{J}_1 < 10^{-9}$
$O + O_2 + M$	→	$O_3 + M$	$k_2 = 2.04 \times 10^{-35} \exp(1050/T)$
$O_3 + h\nu$	→	$O(^1D) + O_2(^1\Delta_g)$	$\lambda \leq 3100$ Å, $\mathcal{J}_{3a} < 5 \times 10^{-3}$
$O_3 + h\nu$	→	$O + O_2$	3100 Å $\leq \lambda \leq 10{,}400$ Å
$O(^1D) + M$	→	$O + M$	$k_4 = 5 \times 10^{-11}$
$O + O_3$	→	$2O_2$	$k_5\ 1.33 \times 10^{-11} \exp(-2100/T)$
$O + OH$	→	$H + O_2$	$k_6\ 5 \times 10^{-11}$
$H + O_3$	→	$OH + O_2$	$k_7 = 2.6 \times 10^{-11}$
$H + O_2 + M$	→	$HO_2 + M$	$k_8 = 4 \times 10^{-32}$
$HO_2 + O$	→	$OH + O_2$	$k_9 = 2 \times 10^{-11}$
$OH + O_3$	→	$HO_2 + O_2$	$k_{10} \leq 10^{-16}$
$OH + CO$	→	$H + CO_2$	$k_{11} = 10^{-13}$
$HO_2 + NO$	→	$OH + NO_2$	
$HO_2 + HO_2$	→	$H_2O_2 + O_2$	$k_{13} = 8 \times 10^{-11} \exp(-1000/T)$
$H_2O_2 + h\nu$	→	$2OH$	$\lambda < 5650$ Å; $J_{14} \geq 5 \times 10^{-6}$
$H_2O + h\nu$	→	$H + OH$	$\lambda < 2240$ Å; $J_{15} < 10^{-8}$
$H_2O + O(^1D)$	→	$2OH$	$k_{14} = 3 \times 10^{-10}$
$H_2O + h\nu$	→	H_2O	$\lambda \approx 1.4\ \mu m$
$H_2O + O$	→	$2OH$	
$OH + OH$	→	$H_2O + O$	$k_{19} \approx k_{13}$
$OH + HO_2$	→	$H_2O + O_2$	$k_{20} \geq 10^{-11}$
$OH + H_2O_2$	→	$HO_2 + H_2O$	$k_{21} = 6 \times 10^{-12} \exp(-600/T)$
$NO + O_3$	→	$NO_2 + O_2$	$k_{22} = 1.33 \times 10^{-12} \exp(-1250/T)$
$NO_2 + O$	→	$NO + O_2$	$k_{23} = 1.67 \times 10^{-11} \exp(-300/T)$
$NO_2 + O_3$	→	$NO_3 + O_2$	$k_{24} = 10^{-11} \exp(-3500/T)$
$NO_3 + h\nu$	→	$NO + O_2$	$J_{25a} \approx 10^{-2}$ (?)
$NO_3 + h\nu$	→	$NO_2 + O$	$J_{25b} \approx 10^{-2}$, $\lambda < 5710$ Å
$NO_2 + h\nu$	→	$NO + O$	$J_{26} \approx 5 \times 10^{-3}$, $\lambda < 4000$ Å
$NO_3 + NO$	→	$2NO_2$	$k_{27} \approx 10^{-11}$
$NO + h\nu$	→	$N + O$	$J_{28} < 5 \times 10^{-6}$
$N + NO$	→	$N_2 + O$	$k_{29} = 2 \times 10^{-11}$
$N + O_2$	→	$NO + O$	$k_{30} = 1.2 \times 10^{-11} \exp(-3525/T)$
$N + O_3$	→	$NO + O_2$	$k_{31} = 3 \times 10^{-11} \exp(-1200/T)$
$N + OH$	→	$NO + H$	$k_{32} = 7 \times 10^{-11}$
$N_2O + O(^1D)$	→	$2NO$	$k_{33a} = 1 \times 10^{-10}$
$N_2O + O(^1D)$	→	$N_2 + O_2$	$k_{33b} = 1 \times 10^{-10}$
$N_2O + h\nu$	→	$N_2 + O$	$J_{34} < 5 \times 10^{-7}$, $\lambda < 3370$ Å

(which is not appreciably dissociated by ultraviolet radiation, although it is ionized by X-rays from the Sun's corona); and above 1000 km, atomic hydrogen and helium, both very minor constituents in the lower atmosphere, become major ingredients (Figure 5.21). With altitude, the electron density (Figure 5.2) shows several features and these fluctuations are designated as the D, E, and F regions. These features are shown in somewhat greater detail in Figure 5.22, which also, unlike Figure 5.8, shows how the constituent concentrations trail off into the exosphere between 500 and 1000 km. A most conspicuous feature of Figure 5.22 is the enormous O^+ peak. The NO^+

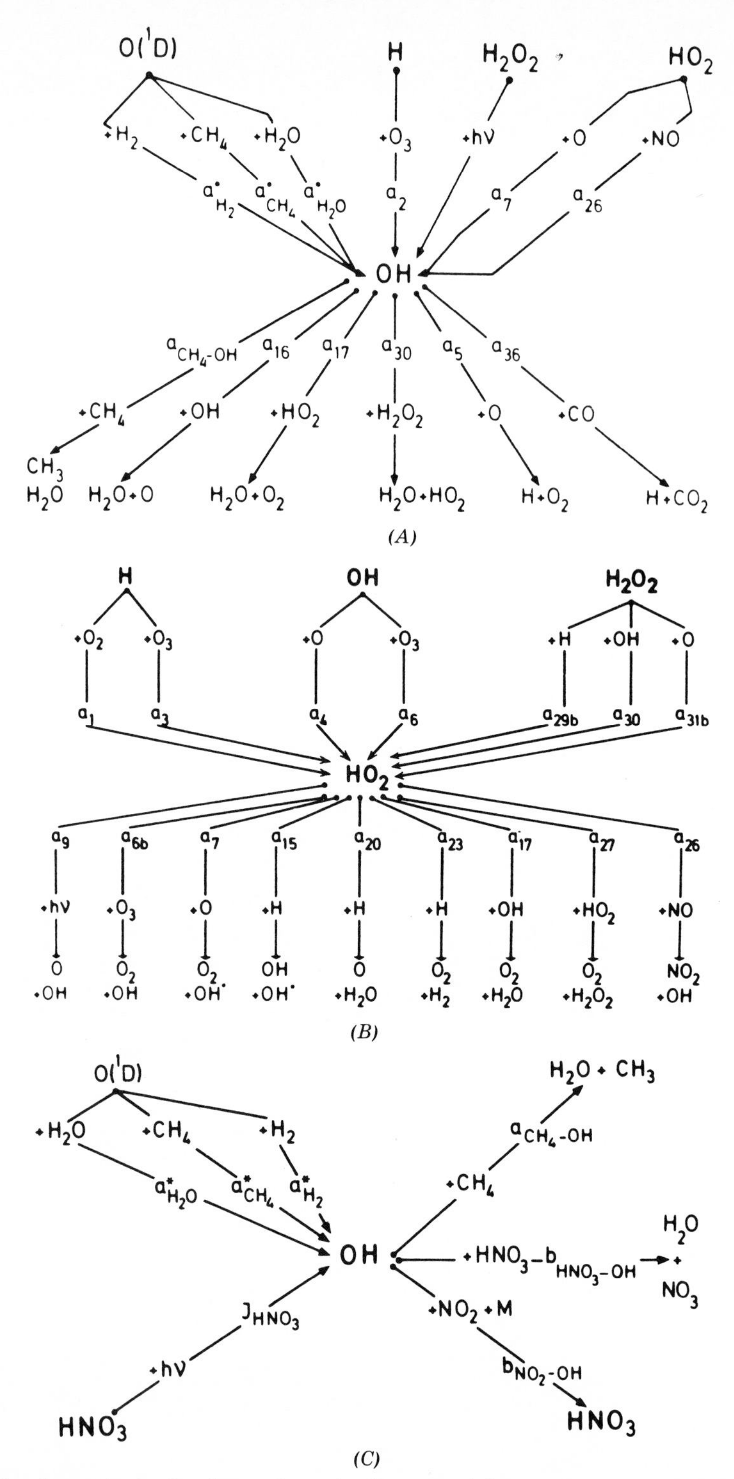

Figure 5.19. Atmospheric chemistry schematic diagrams. (*A*) Reaction scheme of the hydroxyl radical in a hydrogen–oxygen atmosphere. (*B*) Reaction scheme of the hydroperoxyl radical in a hydrogen–oxygen atmosphere. (*C*) Scheme of the principal reactions of hydroxyl radicals above the tropopause in the lower stratosphere. (*D*) Reaction scheme showing the water vapor cycle in the stratosphere. (From Nicolet, 1972.)

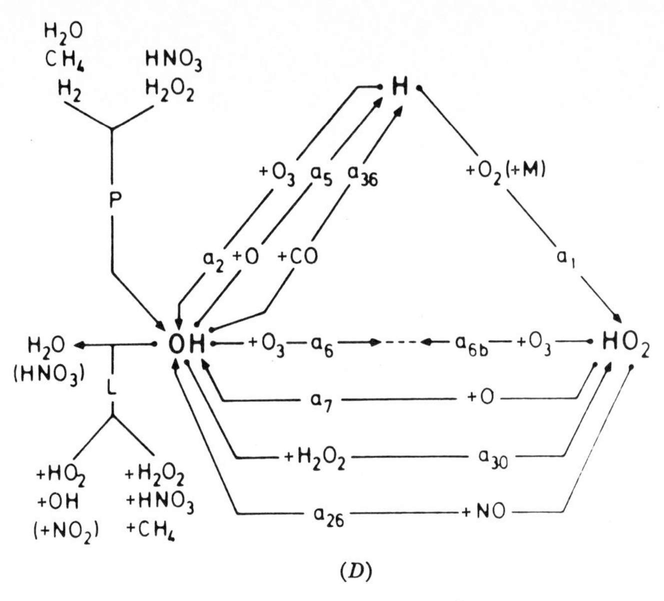

(*D*)

Figure 5.19. (*Continued*).

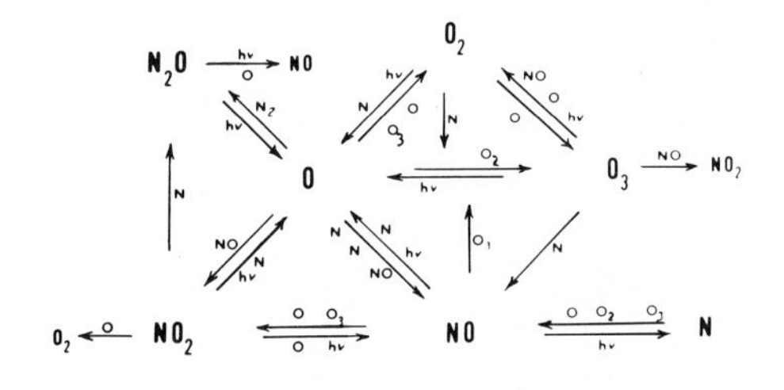

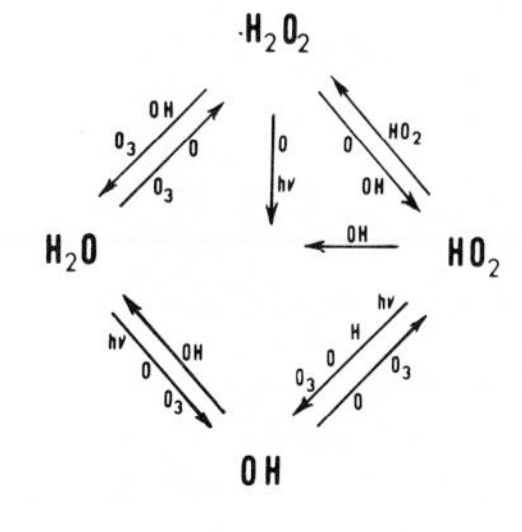

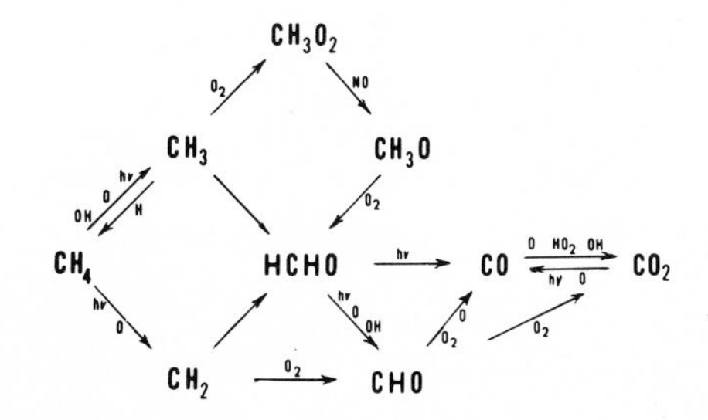

Figure 5.20. Important reactions among atmospheric species (from Hudson, 1972).

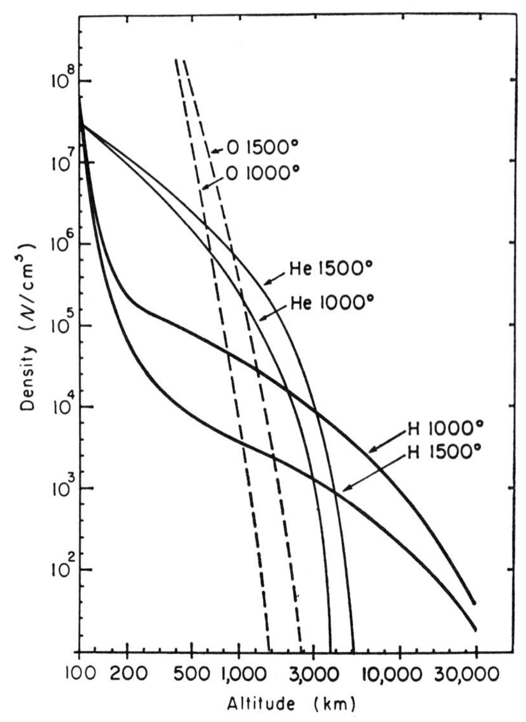

Figure 5.21. Oxygen, hydrogen, and helium profiles in the upper atmosphere (from Donahue, © 1968 by the American Association for the Advancement of Science with permission of the AAAS).

peak is also noteworthy inasmuch as nitric oxide itself is only a minor constituent of the atmosphere. On the basis of four fast reactions,

$$N_2^+ + O_2 \rightarrow O_2^+ + N_2 \tag{5.7a}$$

$$O^+ + O_2 \rightarrow O_2^+ + O \tag{5.7b}$$

$$O^+ + N_2 \rightarrow NO^+ + N \tag{5.7c}$$

$$N_2^+ + O \rightarrow NO^+ + N \tag{5.7d}$$

and the model represented in Figure 5.23, where the circles are ion reservoirs (with ion densities shown), the arrows in are rates of production by solar flux and reactions 5.7*a*,*b*,*c*, and *d* and the arrows out are losses by reactions and recombination. Donahue (1968) has been able to calculate a fair fit of the observed daylight density curve of the oxygen species: However, a second model shows other advantages, while the difficulties with both models underscore the incompleteness of our understanding of the chemistry of the upper atmosphere. The F2 maximum in O^+ in Figure 5.22 marks the point where diffusion becomes more important in O^+ loss than does chemical loss.

Only in the last few years have hydrogen and helium data been obtained from the upper atmosphere (McDonald, 1963; Reber and Nicolet, 1965; Hedin and Nier, 1966; Hoffman, 1967), and the results are rather mystifying. The He content of the atmosphere is much lower than one expects from production from terrestrial radio-

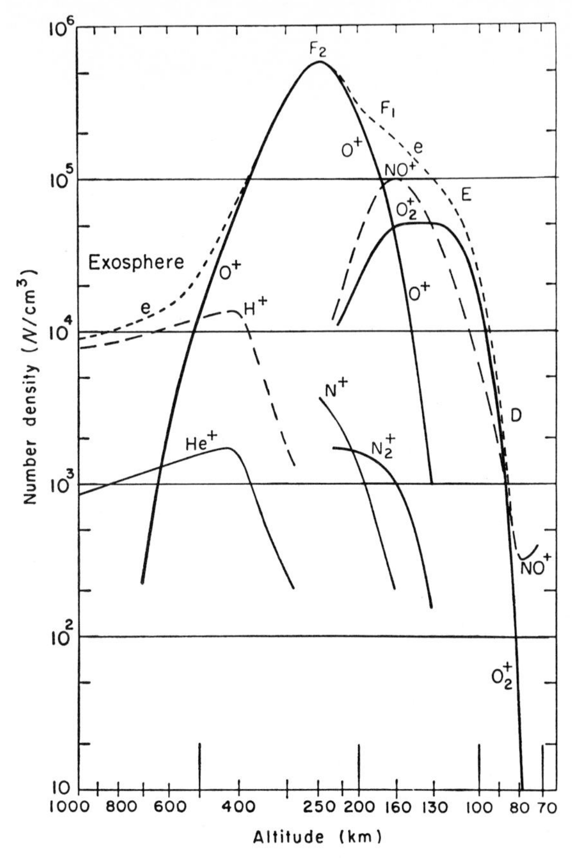

Figure 5.22. Density of ionosphere ions at altitudes between 70 and 1000 km for atmospheric model 1 (in Figure 5.23) in the Daytime (from Donahue, © 1968 by the American Association for the Advancement of Science with permission of the AAAS).

active decay, and the calculated escape rate into the exosphere, while the H^+ density indicates higher hydrogen abundances than expected.

While our knowledge of the upper regions is in fair shape, the chemistry of the lower E and D regions, although closer to us, is complicated by negative ion formation and is very poorly understood. For example, at night the free electron density in the D layer is much greater than it ought to be (Donahue, 1968). The number of potential reactions becomes vastly enlarged (Table 5.4 and Figures 5.19 and 5.20) with even water entering into the chemical melee. These lower regions mark the lower range of the aurora polaris (McCormac and Omholt, 1969). The main spectral lines associated with these magnificent luminous phenomena are summarized in Table 5.5, in particular the very characteristic wierd green glow is due to a forbidden transition in oxygen. Auroral displays may be enhanced by particle acceleration in fields arising from electrostatic double layers (for a discussion of the electrical double layer in aqueous solutions see Chapter 13) in the ionosphere (Alfven, 1958; Albert and Lindstrom, 1970). Far less spectacular is the permanent aurora or night sky "airglow," and its spectrum reveals some chemical surprises. Not only is the sodium D-line very much in evidence but in the 90- to 95-km region there are sharp spectral peaks

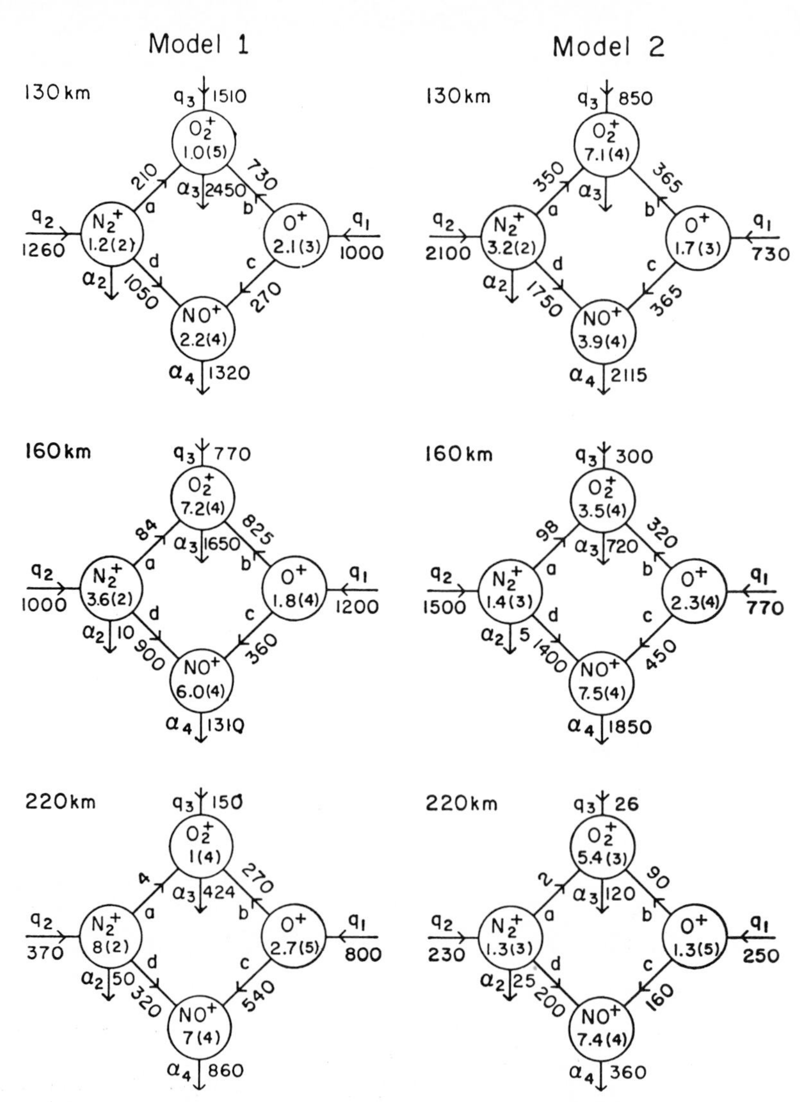

Figure 5.23. Flow diagrams showing rates of creation, transfer, and removal of species N_2^+, O_2^+, O^+, and NO^+, together with their densities at various altitudes (from Donahue, © 1968 by the American Association for the Advancement of Science with permission of the AAAS).

corresponding to Si^+, Ca^+, and Mg^+. The source of these materials is speculated to be the evaporation of meteoric dust (Donahue, 1968). There is also a hydrogen geocorona, and recent studies (Johnson, Young, and Holmes, 1971) have detected evidence that the Earth is surrounded by a large volume of glowing helium ions.

In the past few years, our knowledge of the upper atmosphere has exploded, and upon the construction of the joint Canadian–United States ground-based radar, upper atmosphere observatory (Evans, 1972), this knowledge should continue to make great strides forward; in fact, the atmosphere's chemistry soon may become much better understood than the chemistry of the lithosphere or of the hydrosphere.

Table 5.5 Main Spectral Lines of Aurora (From Miyake, 1965, with permission of the author.)

Luminous Substances	Transition	Explanation
O	$^1S \rightarrow {}^1D$ (5577 Å)	Green line of aurora, most intense, forbidden transition
	$^1D \rightarrow {}^3P$ (6300, 6363, 6392 Å)	Very intense Forbidden transition
N	$^2P^0 \rightarrow {}^4S^0$ (3466 Å)	Forbidden transition
N_2^+	$A'^2\Sigma \rightarrow X'^2\Sigma$ (Bluish violet to ultraviolet) λλ 5864–2987 Å	First negative band, most intense among luminescences of nitrogen
N_2	$B^3\Pi \rightarrow A^3\Sigma$ (Red to infrared) λλ 14,700–5030 Å	First positive band, most numerous
	$C^3\Pi \rightarrow B^3\Pi$ (Ultraviolet) λλ 5440–2690 Å	Second positive band, weak
	$A^3\Sigma \rightarrow X^1\Sigma$ (Bluish violet to ultraviolet) λλ 3400–2300 Å	Vegard–Kaplan band, very weak Forbidden transition

BIBLIOGRAPHY

R. D. Albert and P. J. Lindstrom, *Science*, *170*, 1398 (1970).

H. Alfven, *Tellus*, *10*, 104 (1958).

D. H. F. Atkins, R. A. Cox, and A. E. J. Eggleton, *Nature*, *235*, 372 (1972).

F. R. Barclay et al., *Quart. J. Roy. Meteorol. Soc.*, *86*, 358 (1960).

D. Brunt, *Physical and Dynamical Meteorology*, 2nd ed., Cambridge Univ. Press, Cambridge, (England), 1952.

R. D. Cadle, personal communication (1960), cited by Junge (1963).

R. D. Cadle and E. R. Allen, *Science*, *167*, 243 (1970).

M. A. A. Clyne and B. A. Thrush, *Proc. Roy. Soc. London Ser. A*, *275*, 559 (1963).

P. J. Crutzen, in A. E. Barrington (ed.), *Proc. Surv. Conf. Climatic Impact Assessment Program*, U.S. Dept. Trans. Rept. No. DOT-TSC-OST-72-13 (1972).

F. J. Dillemuth, D. R. Skidmore, and C. C. Schubert, *J. Phys. Chem.*, *64*, 1496 (1960).

M. Dole, G. A. Lane, D. P. Rudd and D. A. Zaukelies, *Geochim. Cosmochim. Acta*, *6*, 65 (1954).

T. M. Donahue, *Science*, *159*, 489 (1968).

J. V. Evans, *Science*, *176*, 463 (1972).

R. M. Goody, *Atmospheric Radiation*, Clarendon, Oxford, 1964.

A. J. Grobecker, in A. E. Barrington (ed.), *Proc. Surv. Conf. Climatic Impact Assessment Program*, U.S. Dept. Trans. Rept. No. DOT-TSC-OST-72-13

A. E. Hedin and A. O. Nier, *J. Geophys. Res.*, *71*, 412 (1966).

J. H. Hoffman, *J. Geophys. Res.*, *72*, 1883 (1967).

J. T. Houghton and J. S. Seeley, *Quart. J. Roy. Meteorol. Soc.*, *86*, 358 (1960).

R. A. Howard, J. E. Matheson, and D. W. North, *Science*, *176*, 1191 (1972).

F. P. Hudson, in A. E. Barrington (ed.), *Proc. Surv. Conf. Climatic Impact Assessment Program*, U.S. Dept. Trans. Rept. No. DOT-TSC-OST-72-13 (1972).

C. Y. Johnson, J. M. Young, and J. C. Holmes, *Science*, *171*, 379 (1971).

C. E. Junge, *Air Chemistry and Radioactivity*, Academic, New York, 1963.

J. L. Kassner, Jr., "Experimental and Theoretical Studies of Nucleation Phenomena," Final Rept. to Nat. Sci. Found., 1968.

K. Y. Kondratyev, *Radiation in the Atmosphere*, Academic, New York, 1969.

P. M. Kuhn, M. S. Lojko, and E. V. Petersen, *Science*, *174*, 1319 (1971).

I. E. Kuhns and B. J. Mason, *Proc. Roy. Soc. London, Ser. A*, *302*, 437 (1968).

J. T. Lee and A. McPherson, in *Proc. Internat. Conf. Atm. Turbulence*, Roy. Aeronaut. Soc., London, 1971.

P. A. Leighton, *Photochemical Air Pollution*, Academic, New York, 1961.

B. Mason, *Principles of Geochemistry*, 3rd ed., Wiley, New York, 1966.

H. J. Mastenbrook and J. H. Dinger, "The Measurement of Water-Vapor Distribution in the Stratosphere," U. S. Nav. Res. Lab., Rept. No. 5551,1–36 (1960).

B. M. McCormac and A. Omholt (eds.), *Atmospheric Emissions*, Van Nostrand-Reinhold, New York, 1969.

G. J. F. McDonald, *Rev. Geophys.*, *1*, 305 (1963).

Y. Miyake, *Elements of Geochemistry*, Maruzen, Tokyo, 1965.

P. J. Nawrocki and R. Papa, *Atmospheric Processes*, Prentice-Hall, Englewood Cliffs, 1963.

M. Nicolet, in A. E. Barrington (ed.), *Proc. Surv. Conf. Climatic Impact Assessment Program*, U.S. Dept. Trans. Rept. No. DOT-TSC-OST-72-13 (1972).

J. Pena and R. G. de Pena, *J. Geophys. Res.*, *75*, 2831 (1970).

L. F. Phillips and H. I. Schiff, *J. Chem. Phys.*, *37*, 1233 (1962).

H. R. Pruppacher, *J. Geophys. Res.*, *68*, 4463 (1963).

C. A. Reber and M. Nicolet, *Planetary Space Sci.*, *13*, 617 (1965).

D. Shapley, *Science*, *176*, 1216 (1972).

C. H. Simpson, *Chemicals from the Atmosphere*, Doubleday, Garden City, 1969.

G. B. Tucker, *Meteorol. Res. Papers*, Air Ministry, London, No. 1052, 1–31 (1957).

L. G. Wayne, *The Chemistry of Urban Atmosphere*, L.A. County Air Poll. Dist., Los Angeles, 1962.

W. L. Webb, *Structure of the Stratosphere and Mesopause*, Academic, New York, 1966.

M. E. Weeks, *Discovery of the Elements*, 5th ed., J. Chem. Educ., Easton, Pa., 1945.

ADDITIONAL READING

Anon., *Inadvertent Climate Modification*, M.I.T. Press, Cambridge, 1971.

R. Claiborne, *Climate, Man, and History*, New York, 1970.

R. A. Craig, *The Upper Atmosphere*, Academic, New York, 1965.

R. M. Goody, *Physics of the Stratosphere*, Cambridge Univ. Press, Cambridge (England), 1954.

G. P. Kuiper, *The Atmospheres of the Earth and Planets*, 2nd ed., Univ. Chicago Press, Chicago, 1952.

H. S. W. Massey and R. L. F. Boyd, *The Upper Atmosphere*, Hutchinson, London, 1958.

H. Riehl, *Introduction to the Atmosphere*, McGraw-Hill, New York, 1965.

6

THE CHEMISTRY OF AIR POLLUTION

Tomorrow morning when you get up take a nice deep breath. It'll make you feel rotten.

CITIZENS FOR CLEAN AIR, INC. (New York)
Advertisement, 1967.

6.1 INTRODUCTION

In December 1952, a particularly severe smog aggravated by an extended temperature inversion shrouded the city of London, and during it and in the following two months it is believed to have killed 4000 persons, while 8000 more persons died of respiratory ailments believed to be caused and/or worsened by the smog (for a recent analysis of this catastrophe see Naden and Leeds, 1971). This was the worst killer smog in modern times, but it is not an isolated example. Death from air pollution curses most of the world's larger cities. A 1956 London smog killed 1000 and one in 1962, 300. In New York City, a 1953 smog killed 200; a 1963 smog, 400; and a 1968 smog, 80. Big cities are not the only victims; in October 1948, a smog settled on the industrial town of Donora, Pennsylvania, that killed 20 people and an assortment of dogs, cats, and canaries and made 5910 of the town's 14,000 inhabitants sick before a long rain washed it away. The Minamata disaster (see Chapter 7), the worst incident of hydrosphere pollution in modern times, resulted in 43 deaths; thus we see that, if we exclude chemicals deliberately ingested such as ethanol (Chapter 11), far more deaths have resulted from specific air pollution situations than from any other type of chemical pollution. But in addition to outright deaths, the ailments and less definable attrition of the GWLE (gross world life expectancy) caused by air pollution are believed to be enormous (Lee, 1972). Lave and Seskin (1970) have given a long and dismal review of the increased morbidity and mortality from air pollution. They conclude that "air pollution accounts for a doubling of the bronchitis mortality rate for urban, as compared to rural, areas." Among the evidence they cite is the study of Winkelstein et al. (1967) of 50- to 69-yr-old white males in the Buffalo, New York area, which revealed that the mortality rate for asthma, bronchitis, and emphysema increased by more than 100% as the air pollution level increased from one to four units. Death statistics analyses in Great Britain and the United States have shown that the death rate from lung cancer is 1.56 (Haenzel et al., 1962),to as much as ten times (Stocks and Campbell, 1955) greater in polluted urban areas than in less polluted rural areas. Urban air pollution also doubles the death rates from stomach cancer and heart disease, and infant mortality correlates with such pollution indices as the sulfate and particulate levels. Another disease caused by dirty air but not usually associated with air pollution is rickets. In smoky cities, lack of sunlight interferes with the production of calciferol (vitamin D) in the body. This substance is

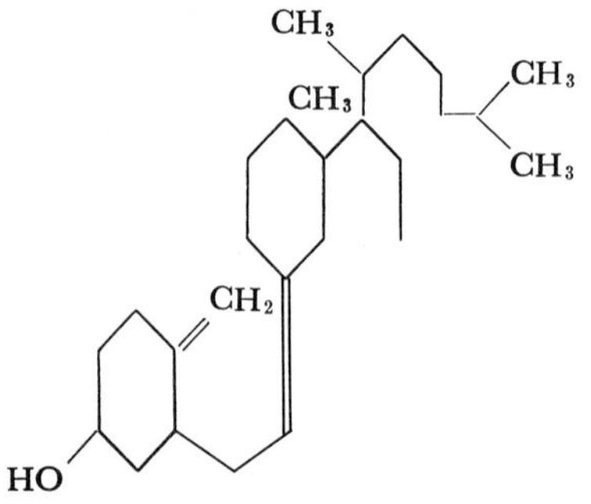

Calciferol

crucial in the hormonal control of the level of calcium in the blood. The evolution of characteristic racial skin coloration, it is interesting to note, was also determined by the body's calciferol requirements—in sunny tropical areas dark skin pigmentation prevents the overproduction of calciferol. Farms, orchards, and forests also suffer from air pollution (Anon., 1969*a*; Anon., 1970*a*; Jacobson and Hill, 1970), and park departments are hard pressed to find species of trees that can survive in noxious city air. Songbirds and zoo animals are also victims, and we have mentioned that even buildings and art treasures succumb to air pollution—Venice looses an estimated 6% of its marble works per year (Behrman, 1972)! In Chicago and Los Angeles, smog disintegrates nylon stockings, and air pollution from an oil refinery stripped the paint from houses in Whiting, Indiana. Steel corrodes two to four times faster in polluted urban air than in rural areas. Air pollution is responsible for $11,000,000,000/yr property damage in the United States and adds a cleaning bill of $600/yr to the average New York City family of four.

But quite apart from its nuisance and lethality, air pollution is particularly insidious for two further reasons. It is the form of pollution most difficult for the individual to protect himself against. You can exercise some control over what you ingest but very little over the air you breathe. Japanese school children, policemen, and other city dwellers often wear face masks, but this clearly is *not* an acceptable way of life. The second reason is the ease of spreading of the contamination. Industries in one country or state pollute the air of their neighbors. Abatement efforts in one area are futile if its neighbors are dirty. In the United States and other developed countries, top priority has been assigned to attacking the problem of air pollution, rather than even to water pollution, and in the light of the above observations properly so. Within the United States, it is important that air quality standards be imposed and enforced at the federal, not the state, level, while on the larger scale international cooperation is most crucial. This is particularly true because air pollution has altered local urban climates (Landsberg, 1970) and has the potential of adversely changing the climate on a global scale (Singer, 1970; Matthews, Kellogg, and Robinson, 1971).

Although I have titled this chapter "The Chemistry of Air Pollution," a more appropriate title might be "The Chemistry of the Atmosphere's Minor Constituents," for many of the substances that we examine may derive from natural sources as well as from human activity. I might note in passing that whether he likes it or not, man is a part of Nature. Nevertheless, I will continue to make a distinction between "human" and "natural" forces for, while in a sense artificial, man's excesses have become so great, he has become so dangerous, that he should be segregated out of Nature just as the criminal used to be segregated out of society.

6.2 OXIDES OF CARBON

In addition to natural sources, such as animal respiration, fermentation processes, and slow oxidation of carbonaceous material, man's activities pour enormous quantities of pollutants into our atmosphere. Among the major sources are the combustion of fossil fuels for power, heat, and transportation (Figure 6.1). The chemical nature of

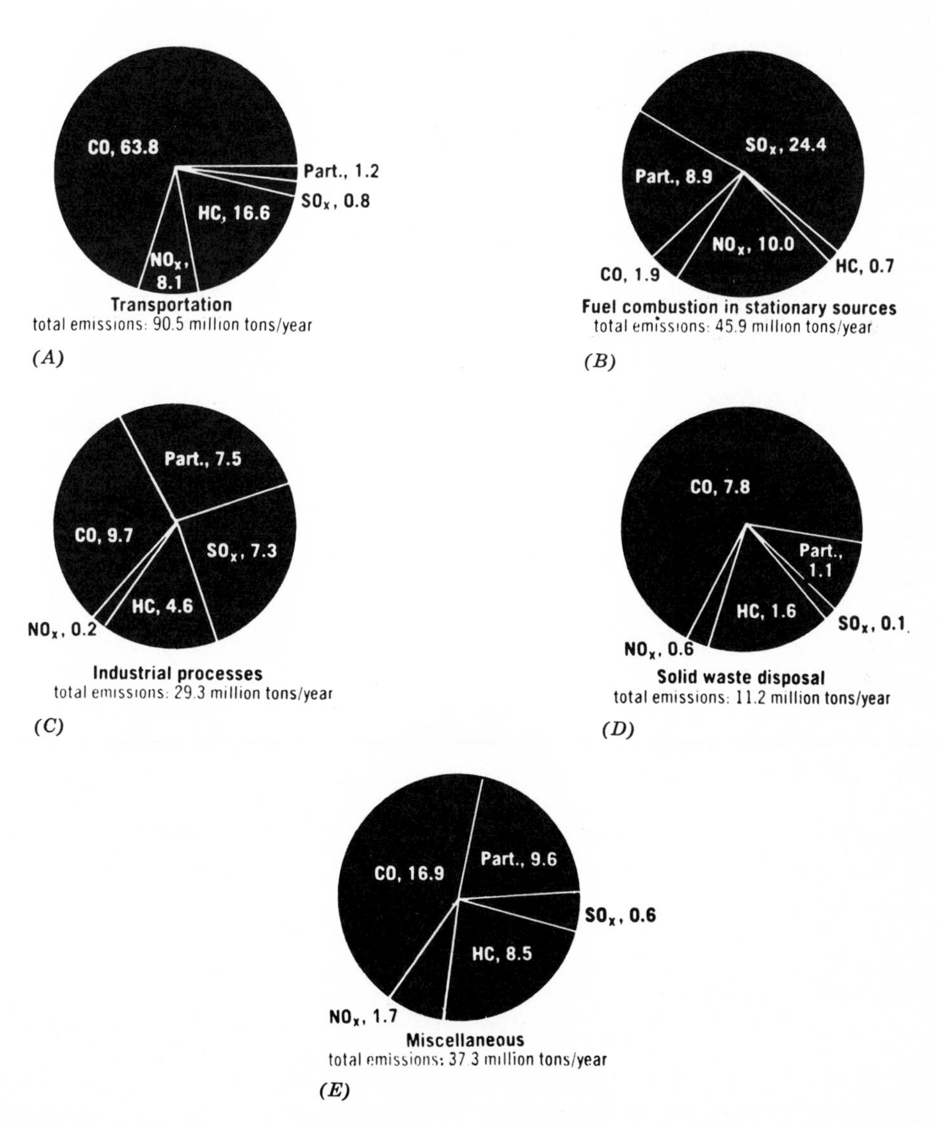

Figure 6.1. Sources of air pollution emissions for the United States in 1968 (Reprinted with permission from G. A. Mills et al., *Environ. Sci. Tech.*, *5*, 30 (1971), © The American Chemical Society).

these emissions depends on the conditions of combustion and on the chemical composition of the fossil fuels, especially their impurities. The combustion of a ton of coal produces about 3.7 tons of carbon dioxide. Coal, particularly cheap low-grade coal, is commonly heavily contaminated with sulfur so its burning produces quantities of sulfur dioxide (Figures 6.1 and 6.2), a lot of particulates, and a little carbon monoxide from incomplete combustion. Petroleum products, on the other hand, especially gasoline, are highly refined, and therefore their combustion produces little SO_2 (Figure 6.2). The combustion of a ton of representative hydrocarbon produces about 3 tons of CO_2 and 1.4 tons of water vapor. Because the residence times of the reactions in the combustion chamber of an internal combustion engine is often much shorter than, say, in a home furnace, there is less opportunity for the reactions to go to completion and the quantity of CO escaping from transportation sources will be greater than from stationary sources (Compare Figures 6.1*A* and *B*). We have more to say about emissions from the internal combustion engine when we discuss oxides of nitrogen. It is important to bear these differences in mind, for as the pattern of fossil fuel utilization changes, the nature of the air pollution problems produced thereby will correspondingly alter. Thus in the United States as oil continues to displace coal as the major fuel, the number one air pollution hazard may cease being particulates (and SO_2) and become partially oxidized hydrocarbons and oxides of nitrogen, largely from transportation.

The production of water from hydrocarbon combustion is hardly an immediate environmental hazard; in effect, all it does is remove water from the fossil biosphere, and this water is replaced by photosynthesis in the current biosphere. Of course, there is a long time lag between the fossil and current biospheres. Some scientists have raised fears that water vapor from stratospheric flights of SSTs could contribute to the upsetting of global climate (Matthews, Kellogg, and Robinson, 1971). In my

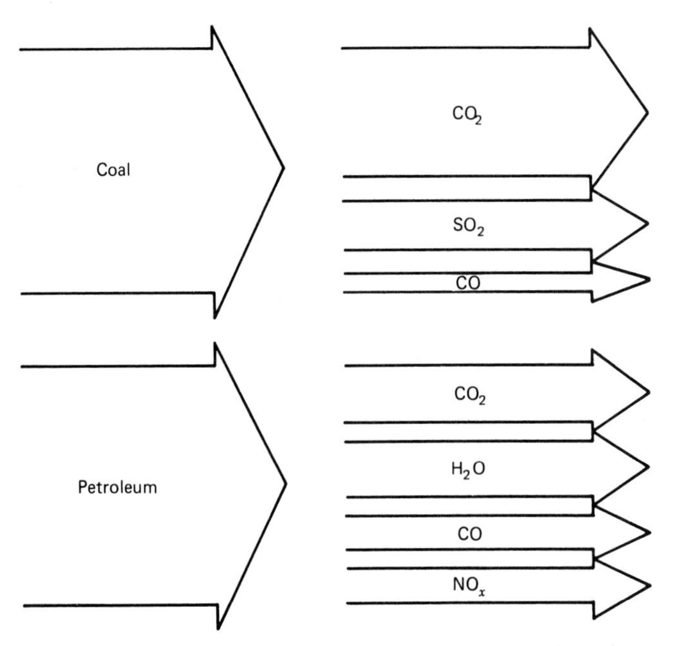

Figure 6.2. A comparison of coal and petroleum combustion products (not to scale).

opinion, this is a most remote possibility, and I suspect that some scientists have raised this specter to reinforce their weak political reasons for opposing the development of these aircraft. Such an abuse has seriously damaged the credulity of scientists as a whole in the United States. Singer (1970) has estimated that water vapor injection into the atmosphere by the oxidation of methane from natural sources is about twice as great as that from the envisioned SST fleet (Daniels, 1970). One interesting fact: Fossil fuels are depleted in the isotope ^{14}C, thus enabling one to distinguish between "old" and "new" carbon in the environment. Human activity by releasing quantities of old carbon diminishes the relative atmospheric level of ^{14}C (Suess effect), and the recent history of this pollution can be neatly traced in the ^{14}C content of the growth rings of trees (Suess, 1955; Fergusson, 1958). This same technique can be applied to determining the age and residence time of water masses and informulating oceanic mixing models (Broecker et al., 1961). In recent years, nuclear devices have injected ^{14}C into the atmosphere, but a correction is applied for this increment.

The concentration and residence time of CO_2 and other minor atmospheric constituents are shown in Table 6.1. The carbon dioxide cycle is treated in Chapter 15. Except possibly at very high concentrations, CO_2 is not harmful to us, and because the oceans are so well buffered (Chapter 8) there is little likelihood that the dissolution of excess CO_2 from the atmosphere will alter the pH of seawater. On the basis of a model that makes provision for both deep oceans and shallow seas, Cramer and Myers (1972) have esitmated that the level of atmospheric CO_2 will reach 380 ppm by the year 2000, and by the year 2100, the level will have increased to about 495 ppm. The nimbus weather satellite now provides a good means for keeping watch on atmospheric CO_2 (Wick, 1971). In addition to not being harmful to animals in concentration encountered in the environment, CO_2 as, along with water, a major reactant of photosynthesis, is beneficial to plant life. CO_2 enrichment has increased crop yields in greenhouses (Kretchman and Howlett, 1970) and computer simulation has indicated the application of this agricultural technique to field conditions (Allen et al., 1971).

Carbon monoxide (Seiler and Junge, 1970; Anon., 1970*b*), on the other hand, is one of the most toxic gases known to man. It is particularly insidious because, unlike the poisonous gases H_2S and HCN, it has no warning odor. CO combines very strongly with hemoglobin

$$Hb\cdot H_2O + CO \rightleftharpoons Hb\cdot CO + H_2O \tag{6.1}$$

interfering with the essential oxygen-bearing function of the respiratory pigments. An hour's exposure to 1500 ppm of CO can endanger a man's life, while an hour's exposure to only 120 ppm can impair his ability to operate a motor vehicle, yet concentrations as high as 100 ppm are not rare in tunnels, garages, expressways, and even the streets of New York, Chicago, Detroit, and London. In Tokyo, where smog warnings were issued on 154 days in 1966 alone, traffic policemen in the most heavily polluted districts return to the station house every half hour to breathe pure oxygen in an attempt to counteract the effects of CO poisoning

$$Hb\cdot CO + O_2 \rightleftharpoons Hb\cdot O_2 + CO \tag{6.2}$$

In urban air, the CO content is closely related to traffic volume (Figure 6.3) yet over the past 50 yr, the CO level in the streets of Paris and New York appears to

Table 6.1 Atmospheric Gases Other Than Oxygen and Nitrogen (From Junge, 1969, with permission of Academic Press, Ltd., and the author.)

Gas		Conversion of Units			Atmospheric Values		
					Ground Level		
Name	Formula	1 ppm = $x \times \mu g/m^3$ STP	1 $\mu g/m^3$ STP = $x \times 10^{-4}$ ppm	1 $\mu g/m^3$ STP = $x \times 10^{-7}$ mm Hg Partial Pressure	ppm	$\mu g/m^3$ STP	Residence Time
Argon	Ar	1,784	5.61	4.23	9300	1.6×10^7	—
Neon	Ne	900	11.11	8.37	18	1.6×10^4	—
Helium	He	178	56.20	42.40	5.2	920	$\sim 2 \times 10^6$ yr
Krypton	Kr	3,708	2.70	2.04	1.1	4100	—
Xenon	Xe	5,851	1.71	1.29	0.086	500	—
Water Vapor	H_2O	800	12.50	9.40	$(0.4\text{–}400) \times 10^2$	$(3\text{–}3000) \times 10^4$	10 days
Ozone	O_3	2,140	4.67	3.55	$(0\text{–}5) \times 10^{-2}$	0–100	~ 2 yr
Hydrogen	H_2	89	112.00	84.50	0.4–1.0	36–90	—
Carbon Dioxide	CO_2	1,960	5.10	3.87	$(2\text{–}4) \times 10^2$	$(4\text{–}8) + 10^5$	4 yr
Carbon Monoxide	CO	1,259	8.10	6.10	$(1\text{–}20) \times 10^{-2}$	$(1\text{–}20) \times 10^1$	~0.3 yr
Methane	CH_4	712	14.05	10.70	1.2–1.5	$(8.5\text{–}11) \times 10^2$	~100 yr
Formaldehyde	CH_2O	1,340	7.46	5.67	$(0\text{–}1) \times 10^{-2}$	0–16	—
Nitrous Oxide	N_2O	1,960	5.10	3.87	$(2.5\text{–}6.0) \times 10^{-1}$	$(5\text{–}12) \times 10^2$	~4 yr
Nitrogen Dioxide	NO_2	2,050	4.88	3.71	$(0\text{–}3) \times 10^{-3}$	0–6	—
Ammonia	NH_3	760	13.15	10.00	$(0\text{–}2) \times 10^{-2}$	0–15	—
Sulfur Dioxide	SO_2	2,850	3.51	2.61	$(0\text{–}20) \times 10^{-3}$	0–50	~5 days
Hydrogen Sulfide	H_2S	1,520	6.58	5.00	$(2\text{–}20) \times 10^{-3}$	3–30	~40 days
Chlorine[a]	Cl_2	3,165	3.16	2.41	$(3\text{–}15) \times 10^{-4}$	1–5	—
Iodine[b]	I_2	11,300	0.88	0.67	$(0.4\text{–}4) \times 10^{-5}$	0.05–0.5	—

[a] Gaseous Cl compound; not proved to be Cl_2.

[b] Fraction of I_2 likely to be adsorbed on aerosols.

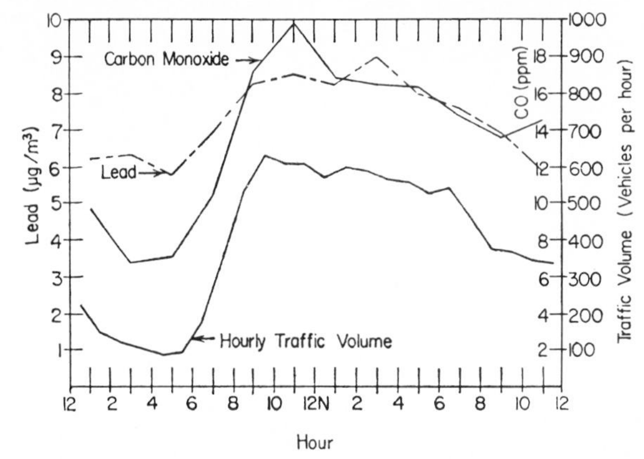

Figure 6.3. Hourly-averaged lead (g/m³) and carbon monoxide (ppm) concentrations and hourly traffic (vehicles/hr) at East 45th St., New York, N.Y. (from Bove and Siedenberg, © 1970 by the American Association for the Advancement of Science with permission of the AAAS).

have declined (Eisenbud and Ehrlich, 1972)! This happy trend is attributed to the improved combustion efficiency of heating units and engines, to the replacement of coal furnaces with oil and gas furnaces, and to the disappearance of highly inefficient coal and wood-burning stoves. The discontinuance of manufacturer's gas in New York City in 1956 also resulted in a very marked decline in suicide and accidental deaths from CO. Among primitive peoples, improper heating devices may still expose them to very high CO levels. Thanks to improved combustion technology these authors conclude, perhaps over-optimistically, that "it is very possible that the concentration of this gas is now lower than at any time since man first learned to use fire" (Eisenbud and Ehrlich, 1972). Still acute CO poisoning claims more than 1000 lives each year in the United States, and regression analysis has shown a significant association between mortality and CO levels such as encountered in Los Angeles air (Hexter and Goldsmith, 1971).

But while CO is still clearly a most dangerous *local* pollutant, recent investigations into this gas as a global atmospheric pollutant have been rather reassuring. Maugh (1972*a*) has concluded that natural sources of carbon monoxide "dwarf" man's output. In the populous northern hemisphere, natural production of CO is more than ten times greater than man-made pollutant (Table 6.2), while Weinstock and Niki (1972) have estimated the tropospheric production of CO to be 5×10^{11} g/yr or 25 times the rate from combustion processes. The tropospheric burden of CO is normally about 530,000,000 tons. Where does all this gas come from and where does it go? Inasmuch as the $^{18}O/^{16}O$ and $^{13}C/^{12}C$ ratios in CO depend on the source, mass spectroscopic isotopic studies have enabled several different sources to be distinguished (Table 6.2). The average residence time of CO is only between 0.1 and 0.3 yr (McConnell et al., 1971; Weinstock, 1972). Much of the CO comes from the incomplete oxidation of methane (mean residence time 1.5 yr) by hydroxyl radicals in the troposphere (McConnell et al., 1971; Weinstock and Niki, 1972), and the methane in turn comes from the subaqueous decay of organic matter in swamps, rice paddies, estuarine zones, and other wetlands. The estuarine zones alone represent an enormous area (Table 6.3) and a huge potential source of CO.

Table 6.2 Identifiable Species of Atmospheric CO in Rural Illinois (From Maugh, © 1972 by the American Association for the Advancement of Science with permission of the AAAS.)

Variety	^{18}O Enrichment[a] (%)	^{13}C Depletion[b] (%)	Principal Occurrence	Source	Production Rate in Northern Hemisphere
AGA[c]	2.46	2.74			
1	0.5	3.0	Principal species everywhere. Increased abundance in summer.	Methane	$>3 \times 10^9$ ton/yr (varieties 1 and 2)
2	0.5	2.4	In varying amounts with 1. Increased concentration in winter and spring. Also in marine air of low northern latitudes.	Probably methane	
3	1.6–1.8	2.8	Lesser abundant heavy oxygen species during summer.	Unknown	$\sim 5 \times 10^7$ ton/month during summer
4	2.6–3.3	2.2–2.6	Major species during autumn.	Degradation of chlorophyll	2–5 $\times 10^8$ tons during autumn
5	2.0–2.5	2.7	Major species during winter and early spring.	Primarily anthropogenic	3–6 $\times 10^7$ ton/month during winter

[a] With respect to the accepted oxygen isotopic standard, standard mean ocean water.
[b] With respect to the accepted carbon isotopic standard, Peedee belemnite.
[c] AGA, average global automobile.

[Source: Argonne National Laboratory]

Table 6.3 Areas of Estuary and Estuarine Marsh (From Woodwell et al., 1972)

	Coast (km)	Estuary (km²)[a] Water	Marsh	Total
World Coastlines[b]				
United States	40,600	107,200	31,700	138,900
Canada	14,500	38,300	11,300	49,600
Mexico	23,500	62,100	18,300	80,400
South America	39,300	103,700	30,700	134,400
Africa	56,300	148,600	43,900	192,500
Europe[c]	42,300	111,700	33,000	144,700
Asia	76,400	201,700	59,600	261,300
Australia; New Zealand	34,600	91,300	27,000	118,300
Other[d]	29,800	78,700	23,200	101,900
Total	357,300	943,300	278,700	1,222,000
Bays, Deltas, Seas				
United States (Chesapeake Bay)				13,300
Canada (St. Lawrence Gulf)				33,200
South America				52,400
Africa (Nile, Niger)				19,300
Europe (Baltic Sea)				382,000
Asia		420,000	105,000	24,800
Total		420,000	105,000	525,000
Grand Total		1,363,300	383,700	1,747,000

[a] Ratios are same as those applied to U. S. in National Environment Pollution Study (1970).
[b] Except where indicated, measurements do not include coastlines north of 60°N latitude.
[c] Includes Baltic sea coast, Norway west coast 60 to 70°N latitude, U.K.
[d] Crude coastline estimates of major portions of Philippines, Indonesia, and Madagascar.

An average swamp can produce 3000 lbs CH_4/yr acre (equivalent to 5000 lb CO/yr acre). Unlike methane, whose concentration in rainwater is near equilibrium, rainwater may be as much as 200-fold supersaturated with CO, suggesting a very large but as yet unknown natural source of CO (Swinnerton et al., 1971; cf., Weinstock and Niki, 1972; who argue that CO formation *and* removal by OH can account for the tropospheric CO balance). Carbon monoxide may also be produced by plant growth as well as decomposition (flatulence of domesticated ruminants accounts for an estimated 85,000,000 tons CH_4/yr—How's that for an interesting observation), and the oceans are another source of great quantities of CO, about 150,000,000 ton/yr (Swinnerton et al., 1970; Wilson and Levy, 1970). And where does the CO go? Remember, CO is a relatively reactive gas. Hydroxyl radicals that produce the CO from CH_4 remove it by further oxidation

$$CO + OH \rightarrow CO_2 + H \tag{6.3}$$

thereby, according to Weinstock and Niki (1972), maintaining the tropospheric CO balance. The hydroperoxyl radical may also remove CO

$$CO + HO_2 \rightarrow CO_2 + OH \qquad (6.4)$$

and this reaction may be even faster than reaction 6.3 (Westenberg, 1972). In the lower atmosphere, 0.05%/hr of the CO is oxidized giving a lifetime of 0.3 yr (Table 6.1), and Dimitriades and Whisman (1971) point out that this rate of removal is too great to be attributed to oxidation by oxygen atoms or ozone. In addition, soil appears to be an effective sink for CO that is bacterially converted there to CO_2 (Inman et al., 1971). In temperate regions, soil is capable of removing an average of 210 ton CO/yr mi^2. Fungi can convert CO to CO_2 (Kummler, 1971).

The "average" man engaged in "light work" inhales about 13 to 23 m^3 air/day, and this corresponds to a consumption of 800 to 1000 liter O_2/day (Nat. Acad. Sci., 1971; Kaufman et al., 1970). Each ton of coal or hydrocarbon burned consumes, respectively, about 2.6 and 3.6 ton of atmospheric oxygen. Does this together with animal respiration, chemical weathering, slow oxidation, biomaterial decay, and other processes represent a greater drain upon the atmosphere's oxygen resources than photosynthesis can restore? The answer happily is no. There is *no* evidence that the level of atmospheric oxygen has changed over the period 1910 to 1970 (Machta and Hughes, 1970). Broecker (1970) reassuringly points out that if we were to burn *all* known fossil fuel reserves, we would consume thereby less than 3% of the available oxygen, while if all photosynthesis should cease and animals and bacteria destroy *all* the organic debris, including the humus stored in soil, complete oxidation of this carbon would take only a fraction of a percent of the atmospheric oxygen. He concludes, "We are faced with so many real environmental crises that there is no need to increase the public concern by bringing out bogeymen." On the other hand, can we expect the oxygen content of the atmosphere to increase as it clearly did at some time in the Earth's geological past? Again the answer appears to be no. Net photosynthesis inhibition, passage of hydrogen through the oxidizing part of the atmosphere, burial of reduced carbon in anaerobic waters, and possibly other major processes still undiscovered appear to have stabilized the oxygen content of the atmosphere (van Valen, 1971; see also Gregor, 1971).

Transportation, as we have seen (Figure 6.1*A*), accounts for a lion's share of atmospheric pollution by CO, and as discussed in Sections 6.3 and 6.7, the motor vehicle is the central villain responsible for smog formation. Another environmental hazard for which much of the blame has been placed on the internal combustion engine is lead poisoning and I conclude this section with some remarks on the lead problem, reserving the discussion of other heavy metal atmospheric pollutants until Section 6.6.

The efficiency of high-compression internal combustion engines is seriously degraded by rapid detonation or "knock" at the end of the combustion cycle. The more highly branched the hydrocarbon fuel, the less the knocking. An octane scale has been established ranging from 0 for *n*-heptane to 100 for highly branched "isooctane" or 2,2,4-trimetylpentane. Treatment of petroleum with high temperature and pressure ("cracking") fragments the molecules and allows them to rearrange into more highly branched hydrocarbons. The addition of lead antiknock compounds can increase the octane rating by as much as 15 points.

The major antiknock lead alkyls added to gasoline since 1923 are tetraethyl lead, $Pb(C_2H_5)_4$ (2 to 4 g/gal), and tetramethyl lead, $Pb(CH_3)_4$. The antiknock additives also contain scavengers to prevent deposit formation, such as ethylene dichloride or dibromide, hence the lead material largely exhausts as volatile inorganic salts, chiefly PbBrCl (Hirschler et al., 1957; Habibi, 1970; Ter Haar and Bayard, 1971). In the atmosphere these compounds can react to form lead carbonates, such as $2PbCO_3 \cdot Pb(OH)_2$ and oxide particulates (Ter Haar and Bayard, 1971). The range of lead aerosol particle sizes appears to be 0.2 to 0. 9μ (Lundgren, 1970; Lee et al., 1968; Robinson and Ludwig, 1967), and these particles are even found embedded in the bark of trees near highways (Heichel and Hankin, 1972). Br appears to be depleted with respect to Cl; gaseous bromine formation may account for this greater loss rate (Axelrod et al., 1971; Moyers et al., 1972; Robbins and Snitz, 1972). The amount of lead exhausted into the atmosphere varies between 25 and 75% of the lead content of the burned fuel depending on driving conditions. At low speeds much of the Pb is retained in the vehicle's exhaust system, but this is subsequently exhausted when the engine is run at higher speeds (Hall, 1972). Vehicles traveling through Boston's Sumner Tunnel emit 0.36 g particulates/mi of which 0.031 g/mi or 8.6% are lead particulates (Larsen and Konopinski, 1962). In the United States, the petroleum industry consumes about 20% of the total lead production, second only to the electrical storage battery industry (40%). Whereas the lead in batteries is largely recycled and is a major portion of the 45% of the total consumed Pb recovered (battery makers, however, can pollute the atmosphere with lead—John, et al., 1972; McIntire and Angle, 1972), the gasoline lead additive goes directly into the atmosphere (Figure 6.3), more than 300,000 ton/yr. Soils, particularly those rich in clay and organic matter, bind lead very strongly (Goldschmidt, 1937; Swaine and Mitchell, 1960; Dedolph et al., 1970), and the lead content of soils as one approaches a heavily traveled highway increases steeply (Danes et al., 1970; John et al., 1972). But other heavy metals, including Cd, Ni, and Zn, not added to gasoline, also increase so there must be other sources from vehicular traffic, such as tire wear. Plants, including animal grazing plants and human food plants, can take up lead from the soil (Cannon and Bowles, 1961; Kloke and Riebartsch, 1964; Patterson, 1965; Ter Haar, 1970; Motto et al., 1970, Rains, 1971), but commonly the Pb does not concentrate in the edible fruited part of the plant (Motto et al., 1970). Urban trees also concentrate Pb (Smith, 1972) as do microbes such as found in lake waters and soils (Tornabene and Edwards, 1972). Lead in rain water has little effect on crops (Ter Haar et al., 1969). Although much (50% in 7 days, 90% in 22 days) of the atmospheric Pb is washed out by precipitation (Ter Haar et al., 1967; Lazrus et al., 1970) and the mean residence time of Pb in the atmosphere is only 7 to 30 days (Burton and Steward, 1960; Francis et al., 1970), and although some Pb may be solubilized by chelating agents in natural waters (Gregor, 1972), there is little lead in river waters and drinking waters (see Table 7.7) (Durfor and Becker, 1964) because lead, as we have said, like mercury and some other heavy metals, is so tightly bound by the organic components of silts and sediments. Some plants, such as Spanish moss, have been suggested as possible lead sensors (Martinez et al., 1971). Much of the atmospheric Pb pollution is particulate and probably settles to Earth not very far from its source; however, finer particles and possibly hydrated heavy metal ions remain in the atmosphere longer and can be carried further. The lead content of the global atmosphere appears to have increased significantly on the basis of the pollution

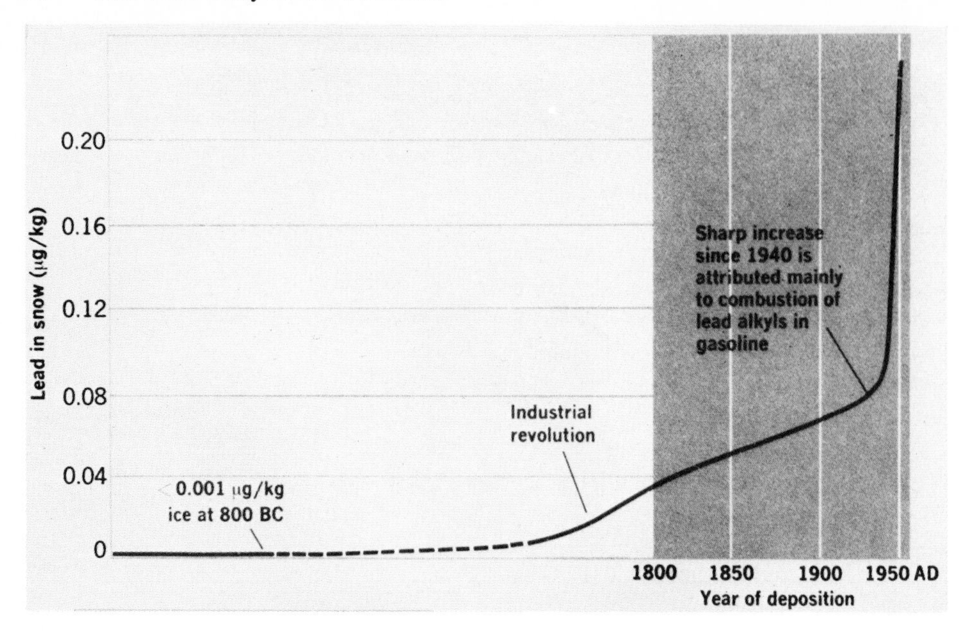

Figure 6.4. The lead content of Greenland snow (Reprinted with permission from S. K. Hall, *Environ. Sci. Tech.*, *6*, 31 (1972), © The American Chemical Society).

record preserved in Greenland snow and ice deposits (Figure 6.4) (Mauozumi et al., 1969). Yet Dickson (1972) has raised some serious objections to the interpretation of mercury and lead in Greenland ice. The pristine atmosphere Pb level based on geochemical evidence is estimated to have been about 0.0006 $\mu g/m^3$ (Patterson, 1965), while Chow et al. (1972) propose 0.008 $\mu g/m^3$ as representing the present average baseline concentration for the United States. Young ocean surface waters contain 0.07 to 0.4 μg Pb/liter, while old deep waters contain about only 0.02 μg Pb/liter (Tatsumoto and Patterson, 1963; Chow and Patterson, 1966; Patterson, 1971). Between what is dumped in the rivers and what is carried by the atmosphere, man's emissions appear to have raised the Pb level of the mixed layer of the northern hemisphere oceans by about a factor of 5. Lake sediments also show increased Pb in their upper layers. To return to the case of the Greenland ice, the Pb level began to climb with the beginning of the Industrial Revolution, and in the first half of the twentieth century, with the advent of leaded gasoline, climbed even more steeply. As is the case with scientific issues that become public issues (Bazell, 1971; Gillette, 1971), some of the information circulated on Pb seems to be contradictory. The Panel on Lead of the U.S. National Research Council Committee on the Biologic Effects of Atmospheric Pollutants found, on the one hand, that the average lead content in the air over major American cities has *not* increased greatly over the last 15 yr, while at the same time noting that the Pb content of the air over these cities (except Los Angeles, which is much worse with an average of 2.5 μg Pb/m^3 compared with 1.4 and 1.6 μg Pb/m^3 for Philadelphia and Cincinnati, respectively) is 20 times greater than the air over rural areas and 2000 times greater than over the mid-Pacific Ocean (Nat. Acad. Sci. U.S.A., 1971). The explanation may be that urban lead levels have been high since the beginning of The Industrial Revolution, hence no marked increase in recent years. Despite greater consumption of leaded gasoline the atmospheric Pb content of Cincinnati air actually decreased over the period 1941 to 1962 (Cholak, 1964). Clearly, there are other important sources of atmospheric lead, such as incinerators and smelters, in addition to motor vehicles. Differences in isotopic compositions should make it possible to distinguish fossil fuel lead from other sources (Chow and Earl, 1972; Rabinowitz and Wetherill, 1972). Nevertheless on a small scale, the Pb content of the air can be directly related to high traffic density (Colluci et al., 1969), and samples taken near a given site can reveal appreciable increases, such as the 5%/yr increase found by Chow and Earl (1970) in San Diego. Strong seasonal variations of the Pb level also occur (Chow et al., 1972). Even in the face of all these difficulties, certain important conclusions do seem to be emerging. Apart from specialized occupational hazards, such as traffic policemen, parking garage attendants, and smelter workers, pollution of the atmosphere by lead does not appear to be a serious menace to human health, even of city dwellers. In fact, in the period 1871 to 1971, human exposure to lead, as indicated by analyses of hair, actually decreased significantly (164 to 16 $\mu g/g$) (Weiss et al., 1972). This at first unexpected finding makes sense when we remember that most of the Pb that gets into our bodies does so by ingestion, largely from food (Kehoe, 1961; Schroeder and Balassa, 1961) and prepared drinks, since the Pb level in drinking water is usually low, rather than from inhalation (Goldwater, 1967; Kehoe, 1968). Still in some communities, Hall (1972) has argued that respiratory intake of Pb is comparable to and "even surpasses the contribution made by dietary intake." The disuse of lead plumbing, leaded paints, leaded ceramic glazes, and other sources of exposure thus has more than compensated for the increased atmospheric burden of Pb. The "safe" blood level for Pb is highly controversial but is often taken to be 0.4 $\mu g/g$ (clinical symptoms appear at about 0.8 $\mu g/g$). The average urbanite in the United States has about 0.15 μg Pb/g blood. There is remarkably little variation in Pb levels in human blood around the world (Goldwater and Hoover, 1967). New Guinea aborigines, far from any autos and industry, and who use wooden utensils, have blood lead levels similar to residents of congested cities. Street dust and dirt in the city may contain more than 2000 $\mu g/g$ of Pb. Reportedly, this stuff has been eaten by small children and zoo animals (Bazell, 1971), but then, lead pencils are not for eating either (Pichirallo, 1971). Banning leaded gasoline because improperly supervised children eat dirt off the streets and get sick makes about as much sense as banning the automobile because impropely supervised children play in the streets and get killed.

There remains a serious problem with respect to leaded gasoline—the Pb poisons the catalytic systems designed to remove noxious automobile exhaust emissions (Sorensen and Nobe, 1972). If we remove the antiknock lead, increased valve seat wear and increased exhaust emissions result, as well as increasing vehicle operating costs and decreasing vehicle life expectancy. Certain aromatic antiknock compounds, such as benzene and toluene, can be substituted for lead, but aromatics and the products of their partial oxidation, such as phenols, are known carcinogens. Lead removal will increase the cost of gasoline by an estimated 10%. Unleaded gasoline also results in greater reduction of atmospheric clarity and more soiling (Pierrard, 1972). Clearly then, the problem of leaded gasoline is not a simple one.

Atmospheric lead can be used as a research tool to study air pollution, for example, radioactive ^{212}Pb from the decay of atmospheric radon has been used to estimate air transport across the upper boundary layer of the troposphere (Assaf and Biscaye, 1972). It has been proposed that water molecules cluster around the lead and other heavy metal ions formed from the decay of radon and thoron (Castleman and Tang, 1971). I might note in passing that in the case of the proton clusters as large as $H(H_2O)_5^+$ and $H(H_2O)_6^+$ have been observed in the summer mesopause (Johannessen et al., 1972).

The physiological impact of lead is the subject of a great deal of controversy. It is not known whether or not lead serves any essential biological function; however, Pb is ubiquitous in the environment, and this is reason enough for supposing that it follows the usual sort of biosystem impact curve (Figure 7.31). On the basis of regression analysis, Goldsmith and Hexter (1947) claimed that the blood level of Pb increases with increasing lead level in the respired air. This oft cited study has been criticized on many grounds. The more considered and reasonable viewpoint now seems to be that the body is equipped with a mechanism perfectly adequate for maintaining Pb balance, excreting any excess, unless it is upset by an overwhelming exposure (Kehoe, 1961; Barry, 1970; Thompson, 1971; Goldwater, 1972). The body has three lead-storage pools: (*a*) a relatively large but immobile skeletal pool, (*b*) a smaller mobile skeletal pool, and (*c*) a still smaller and less mobile soft tissue pool (Barry and Mossman, 1970; Hammond, 1971). Biologically inert Pb also accumulates in teeth, hair, and nails. Inner city children have nearly five times more Pb in their teeth than their suburban contemporaries (Needleman et al., 1972).

Massive exposure to Pb can result in lead poisoning or plumbism. Lead (and other heavy metals) can interact with sulfhydryl groups (–SH) in biomaterials, thus rendering them inaccessable to perform their biochemical function and interferring with cellular metabolism. In particular, hemoglobin synthesis can be impaired (White and Harvey, 1972), since lead depresses the activity of delta-aminolevulinic acid dehydrase (ALAD). Zn, Fe, Mo, and inorganic sulfates also disturb the functioning of this "copper enzyme." Anemia is a symptom of plumbism. Exposure to excessive Pb levels over an extended period of time can result in chronic nephritis or scarring and shrinking of kidney tissues. Chronic overexposure to Pb can also damage the myelin sheath of the nerve fibers.

The treatment of lead poisoning involves some interesting chemistry. Powerful chelating agents such as EDTA and BAL are administered to compete for the Pb and tie it up in forms inaccessible to the bioprocesses. Before the availability of these chelators, 66% of all children with lead encephalopathy died. The mortality rate has now been reduced to less than 5%, but the incidence of brain damage in the

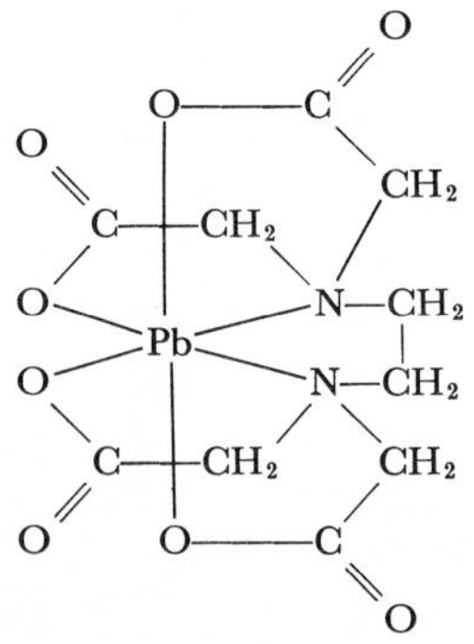

Pb-EDTA

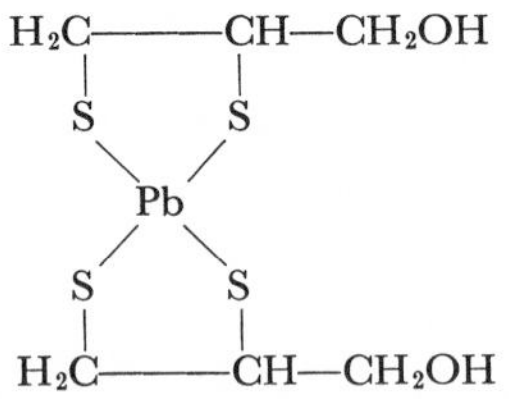

Pb-BAL

survivors has not been substantially reduced. Of one group of children with encephalopathy, 82% were left with permanent mental retardation, convulsive disorders, cerebral palsy, or blindness.

It has been argued with some cogency that lead poisoning contributed to the decline of the Roman Empire (Gilfillan, 1965). The ruling aristocracy had lead plumbing in their homes and drank exotic wines from lead-lined casks. As a consequence, this class may have suffered from impotency, while the inferior classes, without the advantage of plumbing and rare wines, continued to propagate with undiminished fertility. The resulting social imbalance, so the argument runs, set the empire tottering.

In more recent times, consumers of moonshine whiskey are sometimes the victims of lead poisoning, whose symptoms by the way, are similar to acute alcoholism, since home stills are often fabricated out of lead-containing automobile parts. Fanciers of arty pottery also court plumbism for some earthenware is improperly glazed with lead. Chisolm (1971) tells of a physician who poisoned himself by drinking a cola beverage (and 3.2 mg of Pb) every evening for 2 yr out of a mug made for him by his son.

While the danger of lead to public health seems to be diminishing, there is one important exception and this is plumbism among children, especially ghetto children. It is ironic that a disease that cursed the very rich in ancient Rome in our day is a curse of the very poor. Large scale screening programs in Chicago and New York City have revealed that 5 to 10% of the children tested evidenced asymptomatic, increased lead absorption and indicated that between 1 and 2% had unsuspected plumbism (Chisolm, 1971). The incidence of the disease is even higher in slum neighborhoods where children living in old decaying buildings may actually nibble (pica behavior) at peeling paint, plaster, and lead-containing putty. Since 1940, TiO_2 has replaced lead pigments in paint, so the problem is at its worst in older housing. Some of the newest lead-free, latex-based paints contain a mercury fungicide that can cause high Hg levels in rooms painted with them (Foote, 1972). Is the choice then between living in an old house and being poisoned by lead or in a new one and being poisoned by mercury?

The problems of lead and of automobile emissions provide several illustrations of the tendency of Americans to place the blame for our environmental problems where it is politically expedient to do so rather than where it belongs. While no one can deny the efficacy of the private car in devastating our environment and the quality of life, the blame should not rest solely on the Ethyl Corporation and General Motors. Diesel buses and trucks contribute mightily to the general stench, and if you drive up the Southeast Expressway into Boston as the city stretches out before you in view, it will be obvious that the brown cloud lying over the city does not come from street traffic but from about a dozen smoke stacks, including those of utilities and even municipal incinerators. The lead content of Boston air is not particularly high for a city of its size, but the vanadium content is. Now vanadium is a relatively rare element in nature but not in the Venezuelean oil burned by the local utility companies. Although the control that can be exercised over small children has its limits, the blame for infant plumbism lies not with the paint manufacturers or even the landlords, but rather with ignorant and irresponsible parents. Pica behavior is related to improper child care, to improper diet, to inadequate attention and affection, and even to child abuse. A mother picketing city hall protesting leaded paint might much better devote her time watching her children to see that they do not chew paint off the walls and eat dirt off the streets. Although intended as health measures, codes banning leaded paints have become another device enabling militant tenant organizations to harass landlords. The city is quick to enforce the law on small property owners while "overlooking" the big slumlords and failing to clean its own house—the schools.

While I am on the subject of misplaced blame, let me digress to mention one of the most blatant examples. Although very few fatalities are attributable to mechanical failure, automobile manufacturers are being pressured to make "safer" cars, yet little is being done to strengthen and enforce the law against the major cause of automobile fatalities—the drunken driver. The automobile certainly deserves a lot of blame, but it should not be made the scapegoat of *all* of our environmental problems.

6.3 OXIDES OF NITROGEN

Most of the pollutants belched into the atmosphere by combustion processes come from the fuels and their contaminants, but one very important family of atmospheric pollutant comes from the oxidant, comes from the atmosphere itself, and this is the oxides of nitrogen, often represented as NO_x. At high temperatures, the nitrogen and oxygen of the air react to form nitric oxide, NO, which can then be further oxidized in the atmosphere to nitrogen dioxide, NO_2. NO_2 is a highly poisonous gas that dissolves in water to form nitric acid, HNO_3. In air at room temperature, about 5% of the NO_2 is hydrolyzed to HNO_3 (Cadle and Allen, 1970). Atmospheric HNO_3 is also formed by the rapid reaction of water vapor with N_2O_5 from the oxidation of NO_2 by ozone. Unfortunately the formation of NO_x is often the worst under just those operating conditions where the combustion of fuel is optimized (Figure 6.5). But vehicle engines are by no means the only man-made sources of oxides of nitrogen as Table 6.4 shows. Washed down by rain, the nitrates thus formed in the atmosphere can build up significant concentrations in the dust and soil of high smog areas (Figure 6.6). Nitrates and nitrites in the soil are valuable fertilizers and even nitrogen dioxide in the atmosphere does not seem to have a deleterious effect on plants such as citrus groves (Thompson et al., 1971). In Los Angeles, however, the rain of nitrates is heavy enough even to damage telephone equipment (Hermance et al., 1971)!

Dinitrogen trioxide, N_2O_3; dinitrogen pentoxide, N_2O_5; nitrogen trioxide, NO_3; and dinitrogen hexoxide, N_2O_6, are all unstable and presumably cannot exist in the atmosphere for an appreciable time under normal conditions (Junge, 1963). Nitrous oxide, N_2O, is a stable gas in the troposphere (estimated lifetime of 4000 days at 10 km); however, it may be photodissociated at higher levels (estimated lifetime 20 days at 40 km) (Bates and Witherspoon, 1952). Schütz et al. (1970) report a lifetime of 70 yr. Near the ground, its residence time is about 4 yr, or comparable to CO_2 (Table 6.2). Much of this gas appears to originate on the ground from a variety of biological processes, such as the decomposition of nitrogen compounds by soil bacteria. The rate of production from soil increases with decreasing aeration and is thus especially high for moist and/or submerged soils. The photochemical decomposition of N_2O_3 in the stratosphere may make a minor contribution (Bates and Witherspoon, 1952), yet the overwhelming trend seems to be that N_2O is released from the

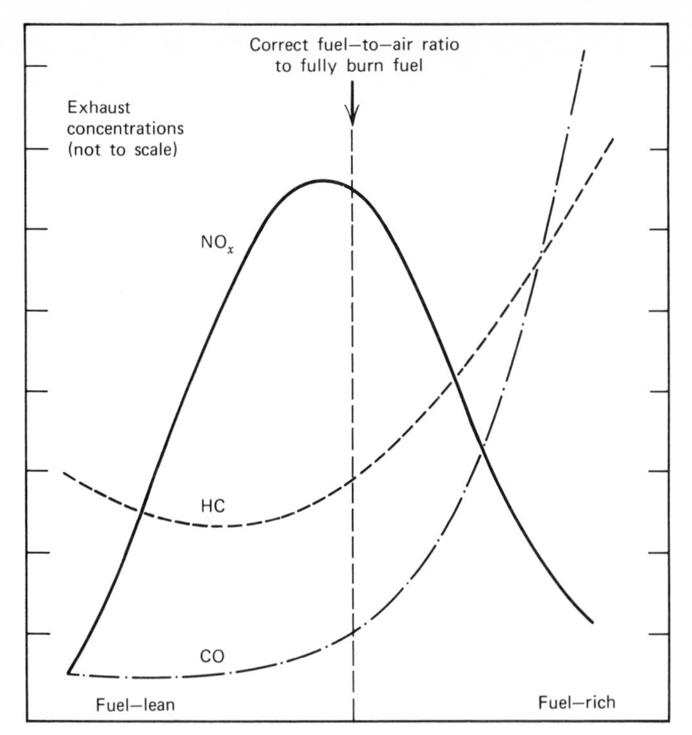

Figure 6.5. Emissions from an automobile engine as a function of fuel–air ratio in the combustion mixture. No single operating point reduces all three pollutants (from Heywood, *Technology Review*, edited at the Massachusetts Institute of Technology, © 1971 by the Alumni Association of M.I.T., with permission).

ground and destroyed in the atmosphere (Bates and Hayes, 1967). The question has been raised whether oxides of nitrogen from SST exhaust might threaten the ozone shield (Johnston, 1971).

Many analytic procedures fail to distinguish between NO and NO_2, and at the time Junge (1963) reviewed the chemistry of the troposphere, the question of the presence of appreciable amounts of the former in the lower atmosphere was unsettled. Perhaps the recently developed Zeeman modulation laser detector (Kaldor et al., 1972) will prove helpful in establishing atmospheric profiles of NO. Cadle and Allen (1970) estimate that the NO level in the troposphere is about 10^{-2} ppm. NO is oxidized slowly by O_2 but rapidly by O_3 (Ford, 1957)

$$O_3 + NO \rightarrow NO_2 + O_2 \tag{6.5}$$

NO_2, as we might expect, is an important constituent of polluted urban air, and as Figures 6.6 and 6.7 show, its level drops as we move away from urban centers. The average lifetime of NO_2 in the atmosphere is only about 2 months, probably because it is so readily washed down as nitrate by rain. Again, anaerobic processes in soil are an important natural source of NO_2. Abeles et al. (1971) have suggested that the soil may also be a sink for NO_2 along with SO_2 and ethylene. NO_2 gas is also formed by electrical discharges in air, but the atmospheric contribution from lightning storms

Table 6.4 Emission Factors for Nitrogen Oxides During the Combustion of Fuels and Other Materials (From Faith and Atkisson, 1972)

Source	Average Emission Factor
Fuels	
Coal	
Household and commercial	8 lb/ton
Industry	20 lb/ton
Utility	20 lb/ton
Fuel oil	
Household and commercial	12–72 lb/10^3 gal
Industry	72 lb/10^3 gal
Utility	104 lb/10^3 gal
Natural gas	
Household and commercial	116 lb/10^6 ft^3
Industry	214 lb/10^6 ft^3
Utility	390 lb/10^6 ft^3
Wood	11 lb/ton
Combustion Sources	
Gas engines	
Oil and gas production	770 lb/10^6 ft^3
Gas plant	4300 lb/10^6 ft^3
Pipeline	7300 lb/10^6 ft^3
Refinery	4400 lb/10^6 ft^3
Gas turbines	
Gas plant	200 lb/10^6 ft^3
Pipeline	200 lb/10^6 ft^3
Refinery	200 lb/10^6 ft^3
Waste disposal	
Open burning	11 lb/ton
Conical incinerator	0.65 lb/ton
Municipal incinerator	2 lb/ton
On-site incinerator	2.5 lb/ton
Other combustion	
Coal refuse banks	8 lb/ton
Forest burning	11 lb/ton
Agricultural burning	2 lb/ton
Structural fires	11 lb/ton
Chemical industries	
Nitric acid manufacture	57 lb/ton HNO_3 product
Adipic acid	12 lb/ton product
Terephthalic acid	13 lb/ton product
Nitrations	
large operations	0.2–14 lb/ton HNO_3 used
small batches	2–260 lb/ton HNO_3 used

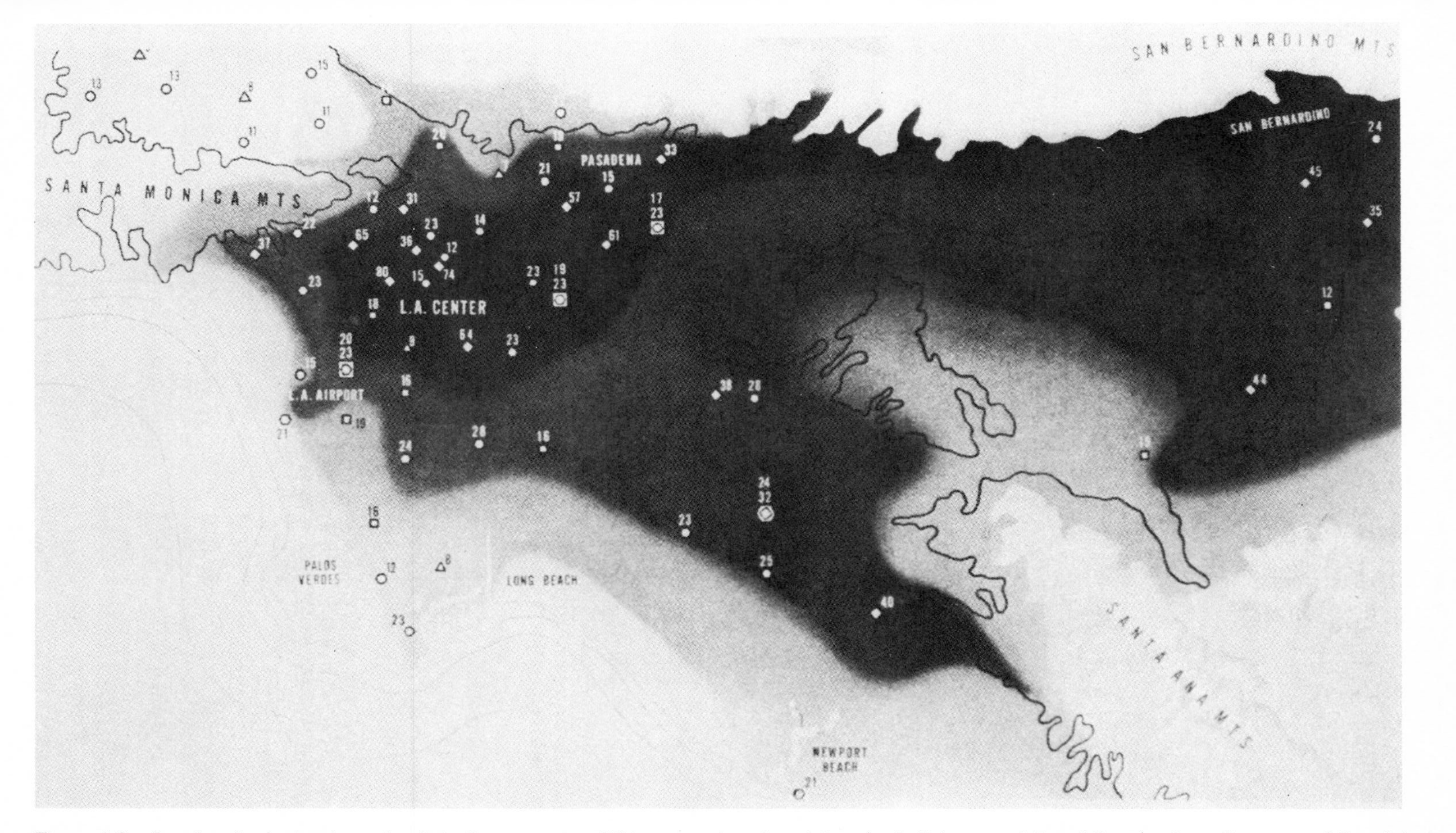

Figure 6.6. Los Angeles basin nitrate levels in 8-yr exposure. White areas, less than 1.5 $\mu g/cm^2$; light grey, 1.5 to 2.3 $\mu g/cm^2$; medium grey, 2.3 to 3.1 $\mu g/cm^2$; dark grey, 3.1 to 4.6 $\mu g/cm^2$; black, 4.6 $\mu g/cm^2$. (Reprinted with permission from H. W. Hermance et al., *Environ. Sci. Tech.*, *5*, 781 (1971), © The American Chemical Society.)

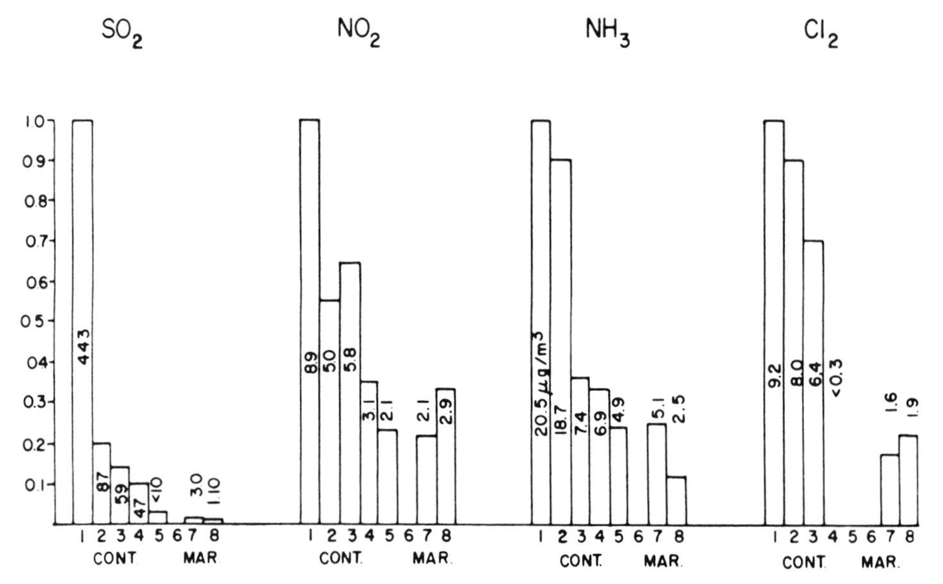

Figure 6.7. Concentrations of various trace gases: (1) Frankfurt am Main (November through March); (2) Frankfurt am Main (April through October); (3) Kleiner Feldberg/Taunus (November through January); (4) Zugspitz (August); (5) St. Moritz (August); (6) Round Hill; (7) Florida; (8) Hawaii. The bar heights are normalized for Frankfurt (November through March) = 1.0. The absolute concentrations are given in the bars ($\mu g/m^3$). The stations are arranged in order from industrial-polluted to unpolluted maritime areas (from Junge, 1963, with permission of Academic Press, Ltd. and the author).

is believed to be relatively small (Viemeister, 1960) as is also the contribution from the oxidation of ammonia.

Figure 6.7 also contains data on NH_3 levels. It is not clear whether the oceans are a source or a sink for ammonia, but soil may be either a source or a sink depending on the pH. NH_3 is released from alkaline soils. Notice in Figure 6.7 that in the Frankfort area the air contains almost as much NH_3 in the winter as in the summer. This seasonal independence suggests the importance of anthropogenic production. Although readily washed out by precipitation, ammonia is otherwise fairly stable in the troposphere. It may react with atomic oxygen in the upper atmosphere (Wong and Potter, 1963).

6.4 HYDROGEN SULFIDE AND OXIDES OF SULFUR

Sulfur dioxide is perhaps the most dangerous atmospheric pollutant. It was probably the or at least one of the killer constituents in the great fatal smogs cited at the beginning of this chapter. Glasser and Greenburg (1971) found a statistically significant rise in deaths in New York City with increased levels of SO_2 and a corresponding decrease following washout of this pollutant by rainfall. In contrast to NO_2 (see previous discussion) SO_2 is very deleterious to plant life (Tanaka et al., 1972). For example, it kills the tips of evergreen trees, causes their needles to drop, and deforms their growth. In 1966, two large coal-burning power plants were put into operation in Mount Storm, West Virginia, and in the next 3 yr the harvest of Christmas trees in the area fell by 90% (Anon., 1969*b*). Because SO_2 is such a villain in polluted air the sulfur cycle has been diligently studied, but this is not to say that all of its mysteries have been solved. Figure 6.8 shows the world sulfur budget. Rodhe's (1972) study is noteworthy because it illustrates the necessity of *international* monitoring and control of air quality. About half of the sulfur deposited in Sweden originates from anthropogenic emissions from surrounding industrial countries, while a large fraction of the sulfur emitted by Sweden falls on its neighbors. In the air over northern Europe, anthropogenic sulfur sources outweigh "natural" sources. Similarly, sulfur isotope studies (Grey and Jensen, 1972) of the air near Salt Lake City, Utah, have shown that while bacteriogenic sources are large, the industrial contribution is larger. However, in the air over Japan, the situation seems to be reversed (Koyama et al., 1965).

Power plant emissions are frequently blamed for SO_2 pollution, and estimates have claimed that 30 to 80% of the SO_2 emissions from United States industrial cities comes from this source (see Figure 6.9). In the period 1967 to 1968, electrical power generation poured an estimated 66×10^6 ton of SO_2 into the atmosphere compared with 27×10^6 ton contributed by other major industrial sources (Kellogg et al., 1972). Projections for the year 2000 are 275×10^6 ton SO_2/yr into the atmosphere. On the other hand, a modeling study (Golden and Mongan, 1971) of air quality in highly polluted Cook County, Illinois, indicated that, thanks to the improved design of their stacks, power plants contribute a constant low-level SO_2 background distributed over a wide area, whereas excessive ground level concentrations of SO_2 are probably attributable to space heating. This conclusion is supported by the fact that SO_2 ground levels are commonly lower in the summer than in the winter, even though power output is the same or even higher in summer weather. Table 6.5 shows that world sulfur emissions are large and increasing with smelters adding a generous fraction, while Table 6.6 gives further details on the emissions from smelters in the western United States. Figure 6.9 summarizes the major sulfur emission sources. Although only a small fraction of solid and other wastes are presently disposed by incineration (Figure 6.10), incinerators are another important source of sulfur

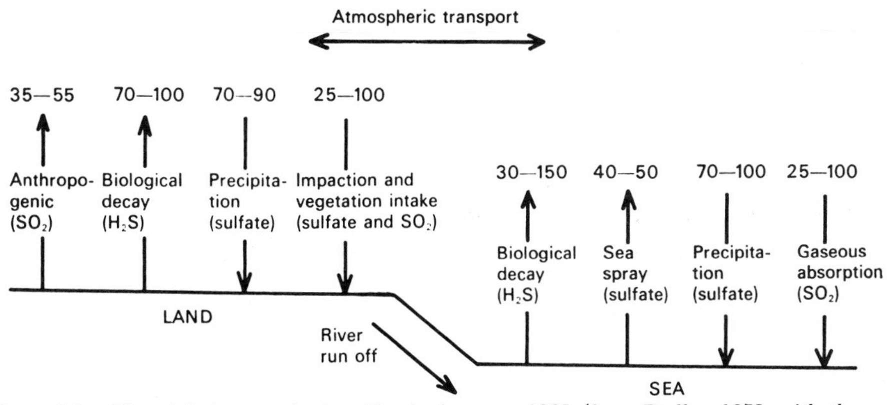

Figure 6.8. The global atmospheric sulfur budget, ca. 1965 (from Rodhe, 1972, with the publisher's permission).

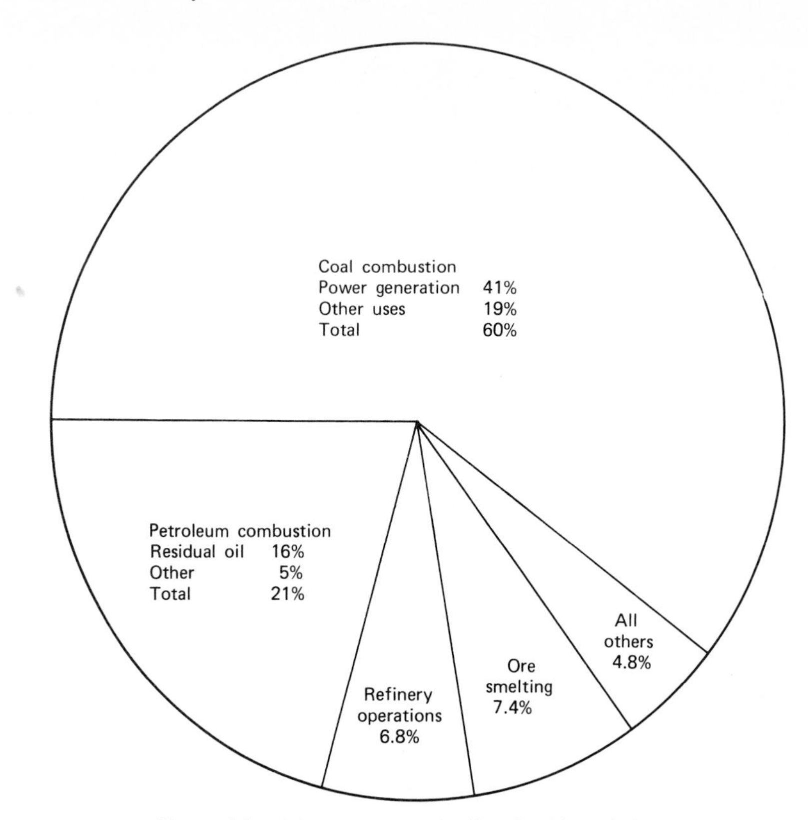

Figure 6.9. Major sources of sulfur dioxide emissions.

emissions, and the amount of pollutant being dispersed into the air is expected to increase fivefold by the year 2000 (Table 6.7). With proper design incinerators can be made relatively nonpolluting, but proper design is expensive, and a proper incinerator is just about as common as a proper sanitary landfill. Table 6.8 is an interesting one; it shows how much SO_2 is produced per weight of material processed for a number of industries, including sulfuric acid manufacture and paper mills.

Some controversy exists concerning the total amount of sulfur mobilized: Bertine and Goldberg (1971) estimated that weathering contributes 40 times more sulfur than human sources, while Friend (1972) has argued that anthropogenic sulfur emissions are about twice as great as natural ones. The discrepancy may lie in the

Table 6.5 World Sulfur Emission to the Atmosphere, Per Year, Given as SO_4 (From *Air Pollution Handbook*, P. L. Magill *et al.* (eds.), © 1956 by McGraw-Hill. Used with permission of McGraw-Hill Book Co.)

Year	Smelters (tons)	Crude Oil (tons)	Coal (tons)	Total (tons)
1937	21.1×10^6	12.2×10^6	69.8×10^6	103.1×10^6
1943	25.4×10^6	13.5×10^6	77.2×10^6	116.1×10^6

Table 6.6 Sulfur Oxide Generation and Recovery in Western Smelters[a] (Reprinted with permission from L. P. Argenbright and B. Preble, *Envir. Sci. Tech.*, *4*, 554 (1970). © The American Chemical Society.)

	Generated (long tons per year)	Recovered (long tons per year)	Percent Recovered
Copper smelters			
Roasters	307,000	99,000	
Reverberatory furnaces	344,000	0	
Converters	914,000	185,000	
Total	1,565,000	284,000	18.1
Zinc smelters			
Roasters	370,000	165,000	
Sintering machines	15,000	0	
Sinter-roast machines	54,000	0	
Cokers and retorts	1,000	0	
Total	440,000	165,000	37.5
Lead smelters			
Sintering machines	152,000	42,000	
Blast furnaces	3,000	0	
Other	5,000	0	
Total	160,000	42,000	26.3
All smelters	2,165,000	491,000	22.7

[a] Approximate annual rates (as sulfur equivalent) in the first half of 1969.

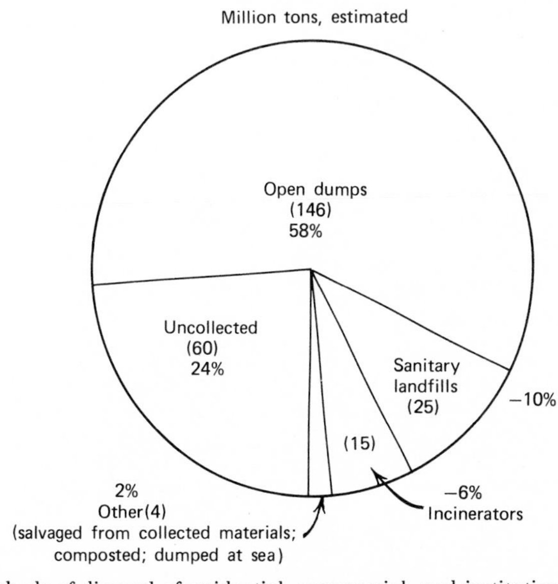

Figure 6.10. Methods of disposal of residential, commercial, and institutional solid wastes in the United States in 1969. Total, 250 million tons (Source: Bureau of Solid Waste Management, HEW).

Table 6.7 Incinerator Furnace Emissions (thousands of tons) (Reprinted with permission from Anon., *Envir. Sci. Tech.*, *4*, 78 (1970). © The American Chemical Society.)

	1968	2000
Mineral Particulates	90	708
Combustible Particulates	38	131
Carbon Monoxide	280	829
Hydrocarbons	22	64
Sulfur Dioxide	32	161
Nitrogen Oxides	26	147
Hydrogen Chloride	8	219
Volatile Metals (lead)	0.3	0.055
Polynuclear Hydrocarbons	0.01	0.03
Est. Total	496	2259

Table 6.8 Generalized Emission Factors for Sulfur Dioxide from Selected Sources[a] (From Faith and Atkisson, 1972)

Source	Emission factor
Combustion of coal	(38 times % S by wt) lb SO_2/ton of coal
Combustion of fuel oil	(158.8 times % S by wt) lb SO_2/1000 gal of oil
Open-burning dumps and municipal incinerators	1.2–2.0 lb SO_2/ton of refuse
Sulfuric acid manufacture	20–70 lb SO_2/ton of 100% acid
Copper smelting (primary)	1400 lb SO_2/ton of concentrated ore
Lead smelting (primary)	660 lb SO_2/ton of concentrated ore
Lead smelting (secondary cupola)	64 lb SO_2/ton of metal charged
Zinc smelting (primary)	1090 lb SO_2/ton of concentrated ore
Kraft mill recovery furnace	2.4–13.4 lb SO_2/ton of air-dried pulp
Sulfite mill recovery furnace (assuming 90% recovery)	40 lb SO_2/ton of air-dried pulp

assumptions made concerning what fraction of the sulfur remains in the ash and what fraction is volatilized. Does atmospheric SO_2 in turn effect weathering? Possibly not. In contrast to the stones of Venice discussed previously, washout of acidic atmospheric sulfur compounds apparently has relatively little effect on chemical weathering, at least in New England where there appears to have been no excessive removal of cationic lithospheric constituents (Johnson et al., 1972).

Wet muds can produce H_2S by decay processes analogous to those that produce methane, thus mobilizing the sulfur in biogenic organic material. An even more important source of H_2S, especially in marine wetlands (see Table 6.3 for their areas) where the supply of sulfate ions is ample, is the action of sulfate reducing bacteria. Nature is resourceful: If oxygen is available, she will use that powerful oxidant for her metabollic processes, but in the absence of oxygen, she will resort to nitrate (denitrification), and when nitrate (and nitrite) is exhausted, she will as a last resort use sulfate ion as an oxidant. The H_2S thereby produced can escape, be oxidized both in the waters (Chen and Morris, 1972) and in the atmosphere (Sidebottom et al., 1972), and be added to the atmospheric SO_2 burden. The process is aggravated by water pollution and, as we just saw, Koyama et al. (1965) have suggested that most of the SO_2 in the air over Japan comes from polluted marshlands. This raises the possibility that the barene or mudflats surrounding Venice may contribute as much or perhaps more of the lung-attacking, marble-wasting SO_2 to that city's air as does the much blamed mainland industrial complex. While marshlands and tidal flats are an important source of H_2S, little is believed to escape from the open ocean because of its oxidation in the water column (Östlund and Alexander, 1963). Generally speaking, the oceans are a sink for SO_2, but they may be a source under special conditions (Kellogg et al., 1972). Salt spray from the oceans (discussed in a subsequent chapter) is an important source of sulfate in the marine aerosol. Lovelock et al. (1972) have shown that soils also release dimethyl sulfide, $(CH_3)_2S$, and they suggest that this compound may represent an even more important contribution to the atmospheric sulfur burden than H_2S.

Rural areas release most of their sulfur in the summer in the form of H_2S, whereas urban areas add more of their sulfur to the atmosphere in the winter heating season as SO_2. The soil is an important sink of sulfur as well as a source (Abeles et al., 1971); if the sulfur content of the soil is low, there will be a net uptake of sulfur, while if it is high, there will be a net loss of sulfur to the atmosphere (Ericksson, 1959).

Table 6.9 shows the SO_2 and H_2S concentrations in the air over Bedford, Massachusetts. In general, there was 2 to 3.5 times as much SO_2 as H_2S. The northwest wind value is probably representative of unpolluted air over the northeast United States, but the southeast wind is polluted by the city of Boston, 15 mi in that direction. Monthly average values over Sweden range from 2.3 to 20.1 $\mu g\ SO_2/m^3$ (Egner and Eriksson, 1955). The ratio of gaseous to particulate sulfur ranges from about 3 to 15 and in the winter air over industrial cities may be as high as 25.

The relative concentration of sulfate particles is particularly high in polar regions (Cadle et al., 1968). It comes presumably from SO_3

$$SO_3 + H_2O \rightleftharpoons H_2SO_4 \tag{6.6}$$

which rapidly combines with atmospheric ammonia

$$H_2SO_4 + 2NH_3 \rightleftharpoons (NH_4)_2SO_4 \tag{6.7}$$

Table 6.9 SO_2 and H_2S Concentrations at Bedford, Massachusetts (From Junge, 1963, with permission of Academic Press, Ltd., and the author.)

Winds	Number of Measurements	Concentration ($\mu g/m^3$)		Ratio
		SO_2	H_2S	SO_2/H_2S
Northwest Sector	12	17.5	8.3	2.12
Southeast Sector	13	24.8	9.4	2.64
Indifferent	7	31.1	8.9	3.50
Total Average	32	23.5	8.9	2.65

so the level of free SO_3 in the atmosphere remains low. SO_2 also dissolves in water to form sulfurous acid, H_2SO_3, which is rapidly oxidized by dissolved O_2, especially in the presence of certain salts commonly present in aerosol droplets, and this is probably an important mechanism for the oxidation of further SO_2 in the lower atmosphere (Kellogg et al., 1972). Persulfate particles are found in the stratosphere, and these and sulfate particles are the only sulfur species that have been detected in that region. There appears to be a world-wide layer of sulfate particles just above the tropopause whose concentration increases following the injection of sulfur into the atmosphere by volcanic eruptions (Cadle and Allen, 1970). Volcanic emissions contain mostly SO_2 and smaller amounts of H_2S, SO_3, and sulfates (Shepherd, 1938; Krauskopf, 1967; Cadle et al., 1971; Berner, 1971), but Kellogg et al. (1972) estimate that sulfur from this source amounts to only 1.5×10^6 ton/yr or about two orders of magnitude less than from human pollution. Yet just how the H_2S gets oxidized to SO_2 and SO_2 to SO_3 is still not entirely clear (Urone and Schroeder, 1969; Cadle and Allen, 1970) but appears, at least in part in the case of the troposphere, to involve the complex photochemistry of smog formation (see Section 6.7). The direct oxidation of H_2S by ozone in the gas phase is slow, but in the presence of water, such as in fog and cloud droplets, the reaction may be fast (Kellogg et al., 1972). The small tropospheric concentration of atomic oxygen from the photolysis of O_3 and NO_2 may oxidize H_2S

$$H_2S + O \rightarrow OH + HS \tag{6.8}$$

and in the presence of a third body, especially in polluted air and the stratosphere, SO_2 (Cadle and Powers, 1966)

$$SO_2 + O + M \rightarrow SO_3 + M \tag{6.9}$$

Unexcited SO_2 is not oxidized by O_2 in the gas phase nor by O_3 in the absence or presence of water, however excited SO_2 is. SO_2 can be oxidized nonphotochemically by O_2 in the presence of certain metal oxides common in airborne dust (Urone et al., 1968).

Earlier we mentioned ion clustering: like water, SO_2 can cluster on oxonium and nitric oxide ions to form species such as $NO(SO_2)^+$, $NO(SO_2)_2^+$, $NO(SO_2)_3^+$, $NO(H_2O)(SO_2)^+$, $NO(H_2O)(SO_2)_2^+$, $NO_2(SO_2)^+$, $NO_2(H_2O)(SO_2)^+$, $O_2(SO_2)^+$, $O_2(SO_2)_2^+$, $SO_2(SO_2)_2^+$, $NO_2(H_2O)(SO_2)_2^+$, and $H_3O(H_2O)_nSO_2^+$ (Castleman et al., 1971). Figure 6.11 summarizes the major types of chemical mechanisms involved in the atmospheric sulfur cycle. The major mechanism for the removal of sulfur from the atmosphere, which is should be noted, is washout by rain.

An element chemically related to sulfur but very poisonous is selenium. The Se/S ratio in the Greenland ice sheet, however, actually seems to be decreasing (Weiss et al., 1971*a*, *b*), possibly because Se released from increased fossil fuel combustion is less mobile in the atmosphere than S, travels less far, and comes down more readily in solid form, whereas the volatility of biogenic organo-Se and -S compounds are comparable.

6.5 AEROSOLS

In addition to gases, the atmosphere is heavily laden with a rich variety of condensed phases or aerosols, both liquid and solid, including aqueous solutions, smokes, spores (Laseter and Valle, 1971), pollen, asbestos fibers (Richards and Badami, 1971),

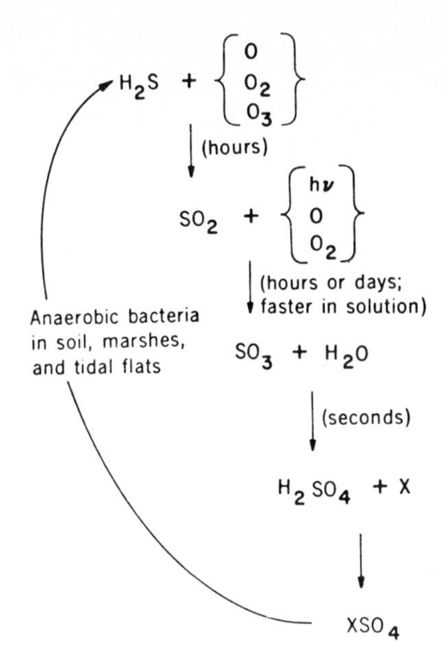

Figure 6.11. Schematic representation of the chemical processes involving environmental sulfur, with indications of the mean lifetime of each compound in the lower atmosphere (from Kellogg et al., © 1972 by the American Association for the Advancement of Science with permission of the AAAS).

continental dust, talc (Windom et al., 1967), volcanic emissions, and so on. Since the response of most Americans to their environment is an aesthetic rather than a hygenic one, and this suspended material, unlike the polluting gases, is often visible, in the past as much effort has been spent on reducing smoke as on reducing the atmospheric level of dangerous poisons such as sulfur dioxide. I certainly do not wish to imply, however, that particulate material does not contribute, and mightily, to respiratory diseases. One look at the shriveled, blackened lungs from the autopsy of life-long city dwellers is almost enough to make one pull up stakes and move to the woods. The carbon particles that coat the lungs of urbanites may sorb dangerous gases such as SO_2 on their surfaces, thus enabling these gases to penetrate deep into the respiratory system. Particulate material has long been recognized as a most serious occupational hazard in certain mining and industrial operations.

Nearly all measurements of particulate levels have been made in the open atmosphere, yet we spend much of our lives in buildings. In passing, I would like to call your attention to a paper by Schaefer et al. (1972) who made the interesting experimental observation that the particulate fallout in American homes with windows closed is correlated with and roughly one-tenth as great as the particulate fallout in the outside air, with the kitchen, not surprisingly, being the dirtiest room.

The chemical composition of suspended matter in the atmosphere, its concentrations, and its particle size distributions are highly variable, and in the atmosphere, as in other sectors of our environment, interfacial processes appear to be of prime importance with aerosols acting as catalysts for much of the chemistry that occurs in the lower atmosphere. The size ranges and distributions of aerosols were treated in Chapter 4 (see Figures 5.12, 5.13, and 5.14 and Table 5.3, and for a recent review of particulate sizes see Lee, 1972). Here we will concentrate on their chemical composition and their origin. The role of aerosols in atmospheric nucleation and condensation and precipitation processes is discussed in Chapter 9. We focus here on aerosols of continental origin, deferring the treatment of the marine aerosol to Chapter 15.

Aitken particles have received the most attention and Junge (1963) divides them into three types based on their probable origin:

1. Condensation of low vapor pressure materials released from heating and combustion processes, partly anthropogenic but including such natural sources as volcanoes and forest fires;
2. Products of the reactions of atmospheric trace gases, especially in the presence of water, for example, the formation of NH_4Cl from NH_3 and HCl and the oxidation of SO_2 to H_2SO_4; and
3. Dispersal of material from the Earth's surface, both salt spray from the oceans and mineral dust from the continental land masses.

Generally speaking, the particles from type 3 sources are larger than from types 1 and 2. Because they have similar growth rates and because they exchange matter, it is believed that the Aitken, large, and giant particles have similar compositions. Particles differ with respect to their relative amounts of soluble and insoluble material, with continental aerosols (source 3) having a relatively large amount of the latter. Under type 2, we might include the possibility of the formation of aerosols from the photolysis of water *vapor* in the upper atmosphere (Clark and Noxon, 1971) as well as the formation of particulate sulfur compounds from gaseous SO_2 (Bufalini, 1971). Anthropogenic sources 1 are largely from energy production rather than from transportation with gasoline powered vehicles, contributing less than 1.8% of the total man-made particulates entering the atmosphere (Public Health Service, 1968).

The major ionic constituents of the large nuclei are NH_4^+ and SO_4^{2-} with the level of Na^+ and Cl^- depending upon how much marine aerosol is mixed in. The large nuclei themselves rarely contain Na^+, Cl^-, or NO_3^-, but the giant particles are rich in these ions (Junge, 1963). These differences are summarized in Figure 6.12. Tables 6.10 and 6.11 give particulate analyses for some other aerosol constituents in urban and rural air (for more recent studies see Brar et al., 1970; Morrow and Brief, 1971; and Rhodes et al., 1972). The high levels of acetone-soluble organic material is also of note, and we return to this subject in Section 6.6. Junge (1963) summarizes the findings reported in these tables by saying that in polluted areas most of the water insoluble aerosol material is organic and ash dust, whereas in unpolluted air, the ashes are replaced by mineral dust (we return to the subject of aeolian dust and the marine aerosol in Chapter 15). Dust from the Sahara Desert, which is found in central Europe and as far away as Florida, contains 37 to 75% SiO_2, 0 to 20% Al_2O_3, 1 to 16% $CaCO_3$, 6 to 22% Fe_2O_3, 2 to 4% Mn_3O_4, and 0.4 to 3% MgO with small amounts of K, Na, Cu, H_2SO_4, and HCl. Sixty-eight percent of the particles are $\leq 0.5\ \mu$, about 24% are between 0.5 to 1.0 μ and 4% $\geq 1\ \mu$ radius. Most aerosols are readily washed out of the lower atmosphere by precipitation thereby giving them a residence time, about 10 days, similar to that of gas pollutants that suffer the same fate.

Again in stratospheric aerosols (recall Figure 5.14 and Table 5.3 for particle distributions), sulfate ion is much in evidence, but the level of NH^+ is 70% less. Whether the other cation is Na^+ and/or H^+ is not clear. Table 6.12 gives some information on the chemical composition of stratospheric aerosols.

We have already stressed the importance of aerosols in atmospheric chemistry; in fact, at least for the lower atmosphere, atmospheric chemistry, like marine chemistry (Horne, 1969), may be largely surface chemistry (Carabine, 1972). If aerosols are

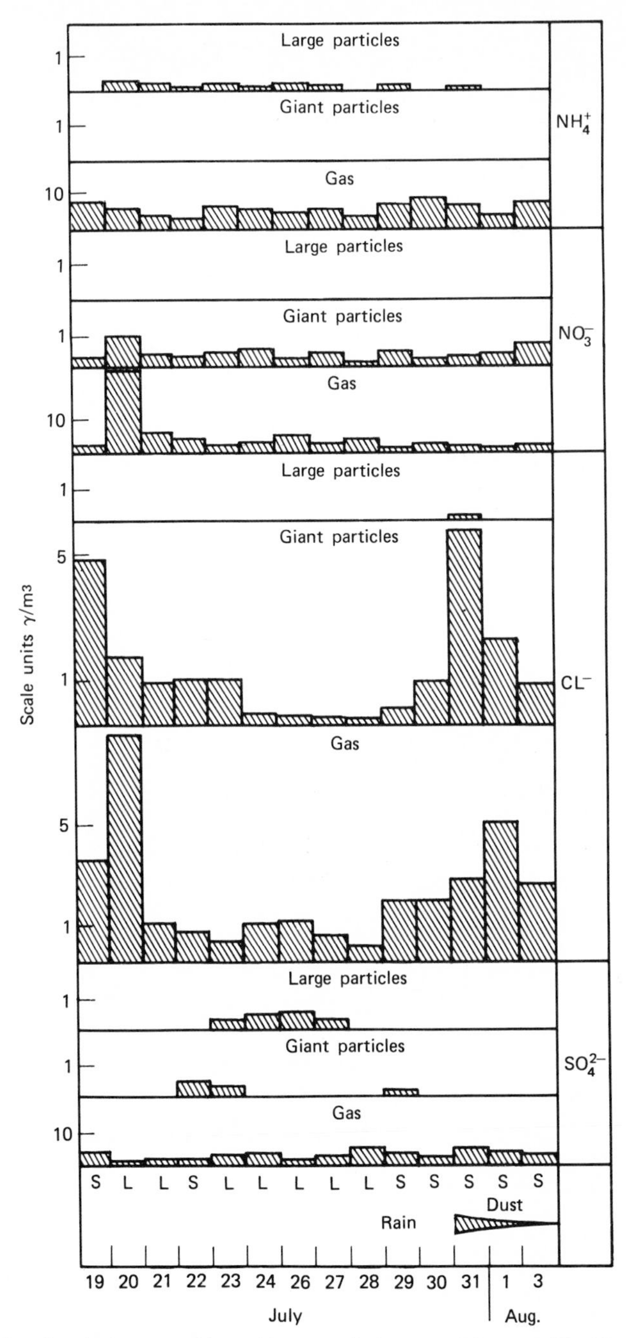

Figure 6.12. The chemical composition of large and giant particles over Florida, and the concentrations of the gases NH_3, NO_2, Cl_2(?), and SO_2 measured simultaneously and plotted at NH_4^+, NO_3^-, Cl^-, and SO_4^{2-}. S and L indicate sea and land breeze, respectively. Note the different scales for gases and aerosols (from Junge, 1963, with permission of Academic Press, Ltd. and the author).

Table 6.10 Particulate Analyses in $\mu g/m^3$ from Cities Having Populations Between 500,000 and 2,000,000 (From Junge, 1963, with permission of Academic Press, Inc., and the author.)

	Cincinnati	Kansas City	Portland (Oregon)	Atlanta	Houston	San Francisco	Minneapolis
Total Load	176	146	143	137	129	104	120
Acetone Soluble	31.4	18.4	32.1	24.2	18.5	19.4	15.8
Fe	4.5	4.1	5.1	3.3	4.0	2.4	4.4
Pb	1.6	1.0	1.2	1.8	1.0	2.4	0.5
F^-	0.21	0.01	Nil	0.05	Nil	0.37	0.06
Mn	0.24	0.08	0.23	0.12	0.23	0.11	0.08
Cu	0.18	0.04	0.05	0.01	0.02	0.07	0.60
V	0.09	0.002	0.009	0.024	0.001	0.002	0.002
Ti	0.06	0.21	0.24	0.12	0.29	0.04	0.11
Sn	0.03	0.03	0.01	0.03	0.02	0.02	0.01
As	.02	0.02	0.02	<0.01	0.01	0.01	0.01
Be	0.0002	0.0003	0.0003	0.0002	0.0002	0.0001	0.0002
SO_4^{2-}	5.6	1.5	0.8	1.0	2.4	1.8	0.8
NO_3^-	1.0	0.6	0.2	0.8	1.0	3.4	1.3

Table 6.11 Particulate Analyses in $\mu g/m^3$ from Nonurban Areas (From Junge, 1963, with permission of Academic Press, Inc., and the author.)

	Boonsboro	Salt Lake City	Atlanta	Cincinnati	Portland (Oregon)
Total Load	68	55	71	45	86
Acetone Soluble	8.7	6.2	9.3	9.0	12.6
Fe	3.7	4.1	2.7	2.4	3.6
Pb	0.1	0.1	0.9	0.4	0.3
F^-	—	—	Nil	0.26	—
Mn	0.00 ·	0.04	0.11	0.07	0.04
Cu	Nil	0.28	0.01	0.19	<0.01
V	0.003	Nil	0.004	<0.001	0.002
Ti	0.26	Nil	0.13	0.01	Nil
Sn	<0.01	<0.01	<0.01	0.01	<0.01
As	0.01	0.03	0.01	<0.01	0.04
Be	0.0001	<0.0001	0.0002	0.0001	0.0001
SO_4^{2-}	0.3	<0.01	0.5	1.9	0.4
NO_3^-	—	—	—	0.7	—

Table 6.12 Absolute and Relative Composition of Stratospheric Aerosols for Elements of Atomic Number 12 to 30[a] (From Junge, 1963, with permission of Academic Press, Inc., and the author.)

Component	*A*	*B*	*C*	*D*
Mg	4	0.000	0.000	0.006
Al^+	4	0.43	0.056	0.006
Si	12	0.11	0.014	0.003
P	4	0.000	0.000	0.01
SO_4	11	6.82	0.890	0.001
Cl	4	0.01	0.0013	0.0013
K	7	0.13	0.017	0.01
Ca	8	0.10	0.013	0.0005
Ti	3	0.000	0.000	0.05
V	4	0.000	0.000	0.01
Cr	4	0.000	0.000	0.01
Mn	4	0.000	0.000	0.005
Fe	12	0.071	0.0093	0.004
Co	4	0.000	0.000	0.006
Ni	7	0.000	0.000	0.01
Cu	4	0.000	0.000	0.01
Zn	4	0.000	0.000	0.02
Total	12	7.68	—	—

[a] *A*, number of samples; *B*, absolute concentration (10^{-15} g/cm C^6); relative composition; *D*, limit of dectability relative to *C*.

indeed the catalysts of tropospheric chemistry, then their surface area becomes an important parameter. Corn et al. (1971) have found that the specific surface areas of suspended particulates in Pittsburgh air varies from a low of 1.9 m^3/g in the spring to a high of 3.1 m^3/g in the winter. Cambell (1972) has theorized that atmospheric aerosols may represent a "leak" in the protective ozone shield, transferring downward to the biosphere via highly reactive chemical species such as free radicals and peroxides the effects of high energy radiation.

While an increase in the global level of atmospheric CO_2 could possibly warm up the surface air temperature of our planet appreciably (the "greenhouse" effect); on the other hand, an increase of atmospheric turbidity by pollution could reduce the intensity of incident solar radiation and thus reduce the temperature of the lower atmosphere. Shall we then boil or freeze to death? Neither, we shall be pressed to death by our numbers first. Solar radiation measurements made at the Mauna Loa Observatory (Hawaii) over the last 13 yr gave "no evidence that human activities affect atmospheric turbidity on a global scale" (Ellis and Pueschell, 1971). Natural sources of atmospheric turbidity are more significant than anthropogenic ones, especially volcanic activity, which, while isolated and sporadic, nevertheless appears to eject enormous quantities of gas and dust 15 to 20 km into the stratosphere (Cronin,

1971). The 1883 eruption of Krakatoa threw enough dust up into the atmosphere to darken daylight skies for hundreds of miles and to produce brilliant red sunsets the world over for months afterwards. Note the increase in atmospheric turbidity in Figure 6.13 following the eruption of Mt. Agung, Bali, on March 17, 1963. It is such sources, rather than pollution, which account for the present slow increase in the atmospheric aerosol background level (Fischer, 1971). Similar conclusions are reached by Laudsberg (1970) in his review of man-made climatic changes. He finds that the evidence, although somewhat obscured by natural short-range climatic fluctuations, points in the direction that man has succeeded only in modifying climate locally, such as in urban areas. Nevertheless, he allows the possibility that in the future human activity might modify climate on a global scale, and he expresses the opinion that man-made aerosols are probably a greater threat than man-made CO_2. Rasool and Schneider (1971) agree that aerosols represent a greater potential threat than CO_2, and they estimate that if man's ability to pollute increases six- to eightfold in the next 50 yr as has been projected, then the global surface temperature may decrease by as much as 3.5°C—enough to trigger an ice age. But they end on the hopeful note that "by that time, nuclear power may have largely replaced fossil fuels as a means of energy production," thus averting disaster. Studies of the polar ice sheets indicate that, indeed, there appears to be a correlation between atmospheric turbidity and climatic changes (Hamilton and Seliga, 1972). Over the past 10^5 yr cold periods are associated with volcanic dust.

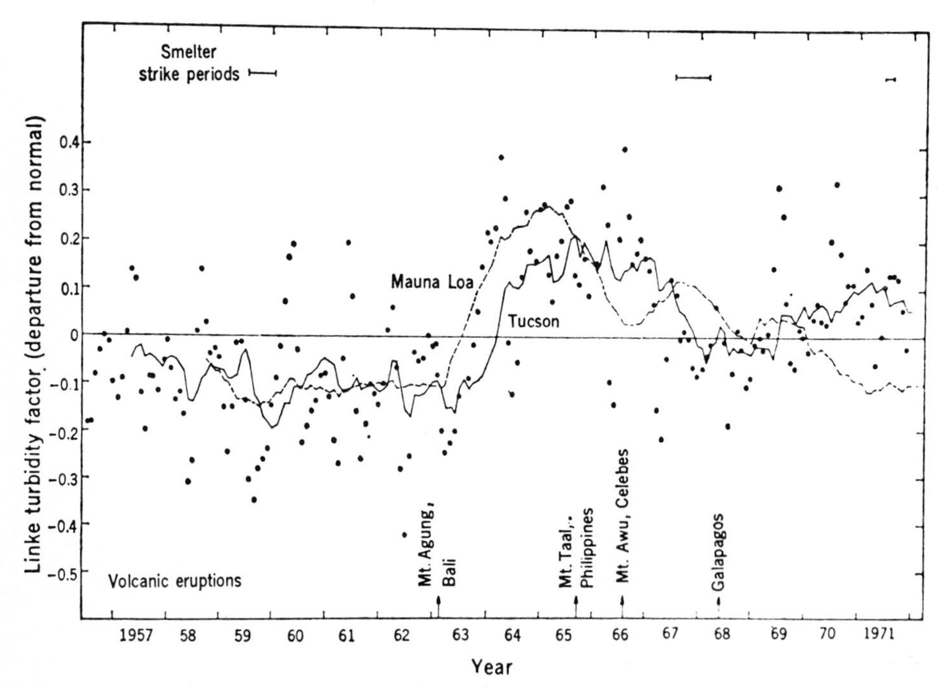

Figure 6.13. Atmospheric turbidity variations at Tucson, Arizona, and Mauna Loa, Hawaii. The curves indicate average turbidity for the preceding 12 months, the points indicate monthly turbidity values for Tucson (from Heidel, © 1972 by the American Association for the Advancement of Science with permission of the AAAS).

Radon scattering has evidenced a large influx of dust into the upper atmosphere (>30 km), some 10^4 ton (Clemesha and Nakamura, 1972). In the previous chapter, we noted the presence of metal ions in the upper atmosphere, possibly from the vaporization of meteors, and, similarly, particulate material might come from exospheric sources. There appears to be at least several "dust" layers in the upper atmosphere. Hemenway et al. (1972) have proposed that the heavy metal particles, including La, Os, Y, and Ta, in high noctilucent clouds may be of solar origin.

6.6 OTHER ATMOSPHERIC CHEMICALS

The high organic content of atmospheric particulates has already been noticed (Tables 6.10 and 6.11). There is an enormous variety of organic chemicals entering the atmosphere. Figure 6.14 gives gas chromatograms for an automobile exhaust sample and there are more than 60 peaks! But man is not alone in enriching the organic content of the lower atmosphere. An estimated 10^8 ton annually of terpene-like hydrocarbons and partially oxygenated hydrocarbons are released into the atmosphere from the biosphere (Went, 1960). The highly aromatic sagebush of the southwestern United States alone releases some 10^6 ton/yr of volatile organic matter (Table 6.13).

In recent years, man has been injecting a new poison into the atmosphere—pesticides. In the United States' air, the level of pesticides, largely particulate, ranges from 0.1 to 2520 ng/m^3 with DDT being the most ubiquitous (Stanley et al., 1971), indicating that the aeolian path in making its contribution in distributing these harmful chemicals throughout our environment (Woodwell et al., 1971). However, our lack of knowledge concerning what happens to DDT in the atmosphere, specifically its residence time, leaves the relative importance of this contribution unclear (Stewart, 1972). Another large slug of organic material in the atmosphere comes from the evaporation of gasoline and the many organic solvents used by industries such as paint and varnish makers, cleaning establishments, and so on.

Aromatic as well as aliphatic hydrocarbons are present in polluted air in both vapor and particulate forms. *n*-Butane, *n*-pentane, and isopentane contribute to tropospheric ozone-formation (see Section 6.7 on smog-formation), whereas sub-

Table 6.13 Estimated Release of Terpenelike and Other Hydrocarbons

Type of Vegetation	Percent of Earth's Surface	Estimated Release (ton/yr)
Coniferous Forest	14	5×10^7
Hardwood Forest	6	
Cultivated Land	6	5×10^7
Steppes	6	
Carotene-Decomposition of Organic Material		7×10^7
Total		17×10^7

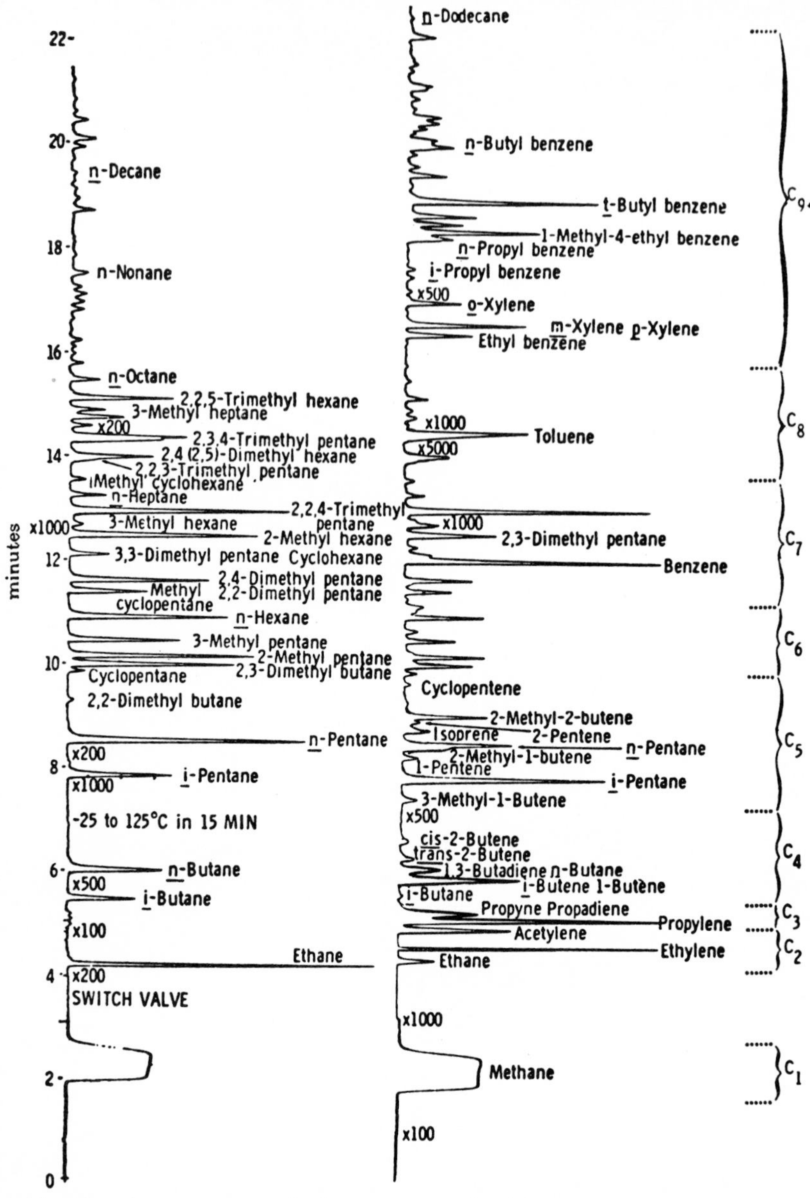

Figure 6.14. Gas chromatograms of exhaust gas sample: (*a*) unsaturates removed by mercuric perchlorate subtractor column and (*b*) complete sample (from Berry and Lehman, 1971, reproduced with permission from *Annual Review of Physical Chemistry*, Vol. 22, © 1971 by Annual Reviews, Inc. All rights reserved).

stances formed from the partial oxidation of ethylene, toluene, and xylene are possible eye irritants (Altshuller et al., 1971). Of special concern are chemicals like benzo(a)-pyrene (BaP) known to be carcinogenic

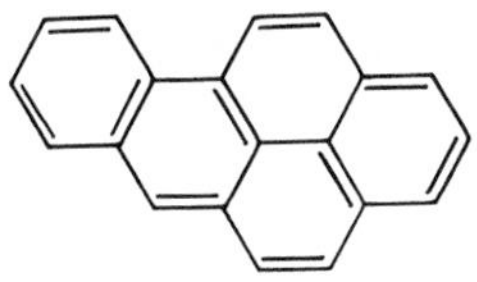

Benzo (a)-pyrene 2 (BaP)

Figure 6.15 compares the day and night levels of BaP and BaA (benzo(a)–anthracene) and other pollutants at various street locations in New York City. It may be some compensation for New Yorkers to learn that the level of BaP in their air is lower than for many other cities (Colucci and Begeman, 1971).

For some time, it was not clear whether ozone in the lower atmosphere had to be transported by mixing processes from the upper atmosphere or whether it could be formed in the troposphere. Went's (1960) suggestion that ozone might be formed by natural processes in the troposphere has been confirmed by ozone synthesis by the photochemical reaction of NO_2 or formaldehyde with "terpenoids" such as α-pinene (Ripperton et al., 1971). Rasmussen and Went (1965) have proposed that the

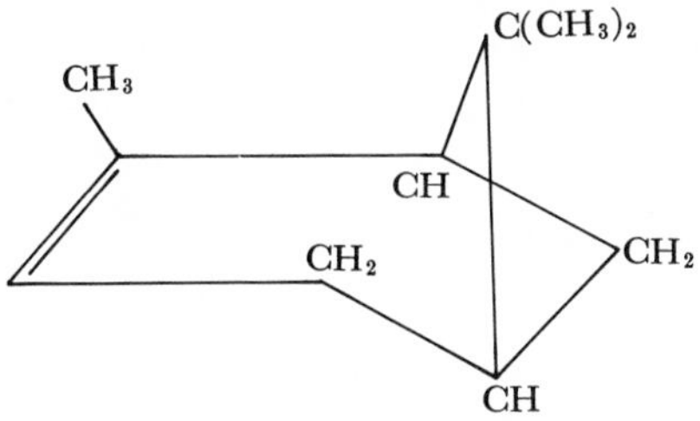

α-pinene

characteristic "blue haze" of heavily forested mountain areas might be due to aerosol terpenes, and Went (1960) even went so far as to theorize that the eventual settling and accummulation of this organic matter might be a possible source of petroleum deposits. Perhaps even more fascinating, Fish (1972) has attributed the blue haze to airborne wax particles ejected by the tips of pine needles when exposed to electrical potential gradients. The oceans may be an additional natural source of much simpler atmospheric hydrocarbons such as ethylene and propylene (Wilson et al., 1972).

The presence of these organic substances greatly complicates the chemistry of the lower atmosphere, not only participating in ozone and smog formation, but also in a medley of other as yet still very imperfectly understood reactions. Levy (1971) has summarized some of the more important photochemical reactions in the *unpolluted* atmosphere at ground level in Figure 6.16, and on the basis of this model, he has estimated daytime concentrations of formaldehyde as high as 5×10^{10} molecules/cm^3. Such reactions, he argues, rapidly remove CO from the lower atmosphere giving it a lifetime as short as 0.2 yr. The photolsyis of formaldehyde, an important pollutant resulting from partial oxidation, may be a significant source of hydrogen atoms and

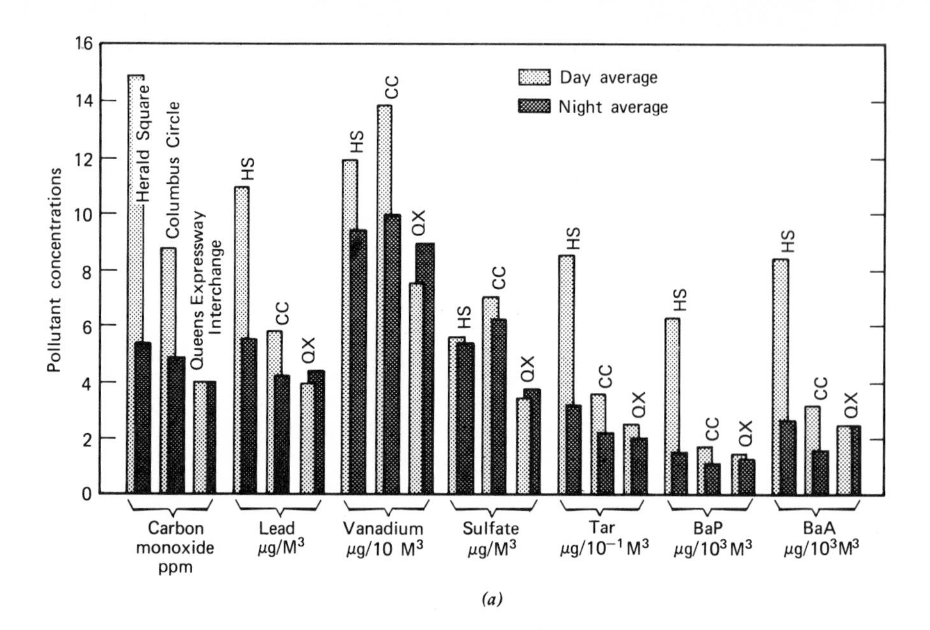

(a)

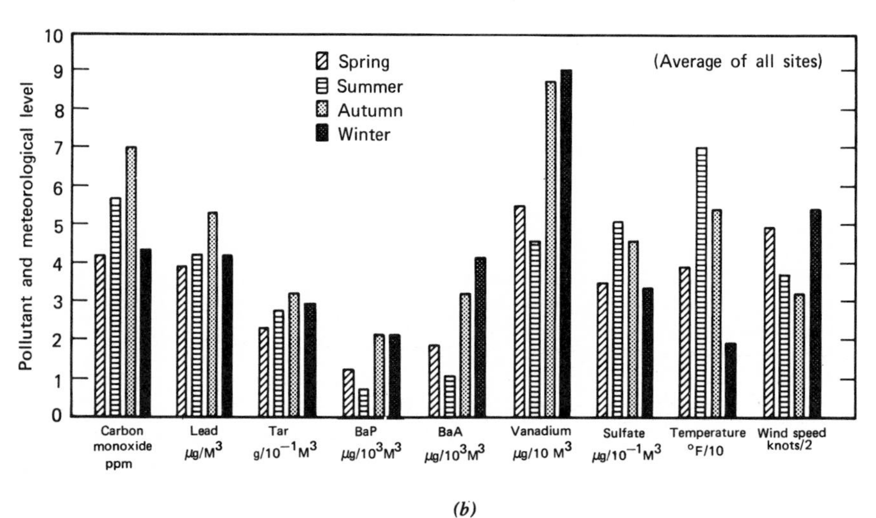

(b)

Figure 6.15. Pollutant levels in New York City: (*A*) annual average day and night pollutant concentrations and (*B*) seasonal pollutant concentrations and meteorological measurements (reprinted with permission from J. M. Colucci and C. R. Begeman, *Environ. Sci. Tech.*, *5*, 145 (1971), © The American Chemical Society).

additional CO (Calvert et al., 1972). Finally, Jones and Adelman (1972) have proposed that studies of the photosulfoxidation of hydrocarbons in the liquid phase should have similarities with atmospheric aerosol chemistry.

Fluorine appears to be another sinister atmospheric pollutant (Anon., 1971*a*), and again it can originate both from natural sources and human activity. Near Tampa, Florida, the rain of fluorides from phosphate plants has damaged citrus groves and other crops. Even when not visibly damaged, crop productivity can be damaged by

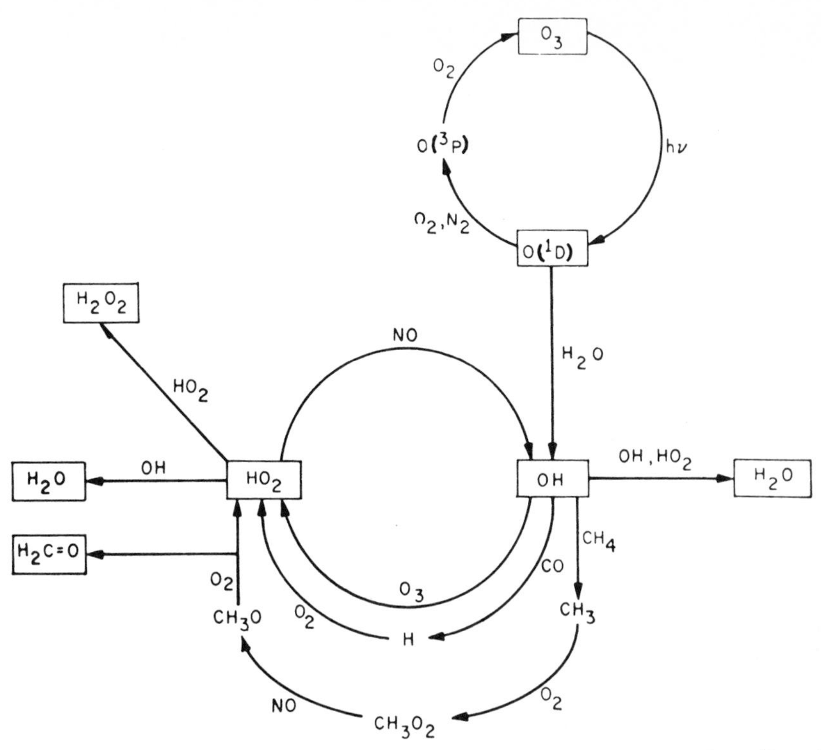

Figure 6.16. Simplified photochemical reaction model for the normal atmosphere (from Levy, © 1971 by the American Association for the Advancement of Science with permission of the AAAS).

this pollutant (Pack, 1971). Cattle grazing on fluoride-contaminated grass have suffered such deformities of bone joints that they cannot chew or even stand. Following the eruption of the Volcano Hekla in 1970, great numbers of sheep and cattle perished in Iceland from eating fluoride-poisoned grass. Another cause for alarm are freons and other fluorintated hydrocarbons. These highly stable and very persistent gases and high vapor pressure liquids are manufactured in large quantities for refrigeration and other purposes, and most of this production eventually escapes to the atmosphere and ends up somewhere in our environment possibly endangering the ozone shield.

We come now to the question of pollution of the atmosphere by heavy metals. Heavy metals are highly toxic, and, if we except ethanol, as pollution causes of human death and disease (Lee, 1972*a*) as a group, they are possibly second only to sulfur dioxide. While Hg and Pb have attracted the most attention in this regard, they certainly are not alone. Bertine and Goldberg (1971) have compared the quantities of elements, including heavy metals, mobilized by the combustion of fossil fuels with the quantities mobilized by natural weathering processes (Table 6.14). While smaller, the quantities are nevertheless very considerable, especially when one realizes that fossil fuel combustion is not the only way that anthropogenic heavy metals can find their way into our atmosphere. Any sort of heating, combustion, smelting, or calcining of mineral materials injects heavy metals into the atmosphere. For example, cement manufacture releases an estimated 10^8 g/yr of mercury into the atmosphere (Weiss et al., 1971). Other mining, ore dressing, and industrial

Table 6.14 Amounts of Elements Mobilized into the Atmosphere as a Result of Weathering Processes and the Combustion of Fossil Fuels (ppm) (From Bertine and Goldberg, © 1971 by the American Association for the Advancement of Science with permission of the AAAS.)

Element	Fossil Fuel Concentration (ppm)		Fossil Fuel Mobilization (× 10^9 g/yr)			Weathering Mobilization (× 10^9 g/yr)	
	Coal	Oil	Coal	Oil	Total	River flow	Sediment
Li	65		9			110	12
Be	3	0.0004	0.41	0.00006	0.41		5.6
B	75	0.002	10.5	0.0003	10.5	360	
Na	2,000	2	280	0.33	280	230,000	57,000
Mg	2,000	0.1	280	0.02	280	148,000	42,000
Al	10,000	0.5	1400	0.08	1400	14,000	140,000
P	500		70			720	
S	20,000	3400	2800	550	3400	140,000	
Cl	1,000		140			280,000	
K	1,000		140			83,000	48,000
Ca	10,000	5	1400	0.82	1400	540,000	70,000
Sc	5	0.001	0.7	0.0002	0.7	0.14	10
Ti	500	0.1	70	0.02	70	108	9,000
V	25	50	3.5	8.2	12	32	280
Cr	10	0.3	1.4	0.05	1.5	36	200
Mn	50	0.1	7	0.02	7	250	2,000
Fe	10,000	2.5	1400	0.41	1400	24,000	100,000
Co	5	0.2	0.7	0.03	0.7	7.2	8
Ni	15	10	2.1	1.6	3.7	11	160
Cu	15	0.14	2.1	0.023	2.1	250	80
Zn	50	0.25	7	0.04	7	720	80
Ga	7	0.01	1	0.002	1	3	30
Ge	5	0.001	0.7	0.0002	0.7		12
As	5	0.01	0.7	0.002	0.7	72	
Se	3	0.17	0.42	0.03	0.45	7.2	
Rb	100		14			36	600
Sr	500	0.1	70	0.02	70	1,800	600
Y	10	0.001	1.4	0.0002	1.4	25	60
Mo	5	10	0.7	1.6	2.3	36	28
Ag	0.5	0.0001	0.07	0.00002	0.07	11	0.6
Cd		0.01		0.002			0.5
Sn	2	0.01	0.28	0.002	0.28		11
Ba	500	0.1	70	0.02	70	360	500
La	10	0.005	1.4	0.0008	1.4	7.2	40
Ce	11.5	0.01	1.6	0.002	1.6	2.2	90
Pr	2.2		0.31			1.1	11
Nd	4.7		0.65			7.2	50
Sm	1.6		0.22			1.1	13
Eu	0.7		0.1			0.25	21
Gd	1.6		0.22			1.4	13
Tb	0.3		0.042			0.29	
Ho	0.3		0.042			0.36	23
Er	0.6	0.001	0.085	0.0002	0.085	1.8	5.0
Tm	0.1		0.014			0.32	0.4
Yb	0.5		0.07			1.8	5.3
Lu	0.07		0.01			0.29	1.5
Re	0.05		0.007				0.001
Hg	0.012	10	0.0017	1.6	1.6	2.5	1.0
Pb	25	0.3	3.5	0.05	3.6	110	21
Bi	5.5		0.75				0.6
U	1.0	0.001	0.14	0.001	0.14	11	8

processing can create heavy metal containing dust. Particularly serious offenders are smelters and metal processing installations. Heavy concentrations of Pb, Cd, Ni, and Zn are commonly found in the soil near smelters (Burkitt et al., 1972). John et al. (1972) found Cd levels as high as 95 ppm in soil samples taken near a Canadian battery smelter, and they noted that oats grown on the contaminated soil had very high amounts of Cd in their roots but less in their above ground portions. Ashton (1972) found 0.17 to 0.32% Ni in dry grass and 0.05 to 0.06% Ni in the soil near a large nickel–copper works in South Wales. Airborne pollution from this source caused trees to wither and deterioration of livestock within a radius of about a mile. Mosses, by the way, are good indicator organisms for heavy metal contamination. Mosses near smelters and slag heaps in Swansea, Wales, contained as much as 1200 ppm Zn, 800 ppm Pb, and 40 ppm Ni, as well as Cd and Cu (Anon., 1971*b*). As for our old enemy lead, it finds its way into the air, not only from the burning of leaded gasoline, but also from factories where antiknock compounds are made (Lee, 1972*b*). While particular industries are particular offenders, heavy metal contamination (Cd, Cr, Co, Cu, Fe, Pb, Mn, Hg, Ni, Ag, and Zn) seems to be generally concomitant with industrial activity, and as pollutants, they often occur en suite (Lazrus et al., 1970; Klein, 1972). As for another old enemy, mercury, Joensuu (1971) has correctly pointed out that the value in Table 6.14 is conservative, and he places the value at 3.0×10^9 g Hg/yr mobilized by fossil fuel combustion. About 90% of the Hg in pulverized coal goes up the stack upon combustion, while 10% remains in the ash (Billings and Matson, 1972). An Associated Press news release in 1971 reported that human organs taken from autopsies between 1913 and 1970 preserved at the University of Michigan upon examination showed a decline in mercury levels, possibly as a consequence of the replacement of coal by oil as a fuel. On the other hand, Weiss et al., (1971*a*, *b*) found evidence in Greenland ice of an increase in the Hg level as a consequence of human activity (see Figure 6.4). The Hg level in the unpolluted atmosphere is 1 to 10 ng/m^3, thus making the total atmospheric burden of Hg some 4×10^9 g. Hg is one of the pollutants readily washed out of the atmosphere, so its residence time is only about 10 days. The flux of Hg from the continents to the atmosphere is 1.5×10^{11} g/yr, but the authors do not make it clear whether this includes the release of methyl mercury from coastal zones. The human increment is only about 15% of the total. Most of the Hg in our atmosphere comes from natural sources. The metal occurs in its elemental state in nature, and its vapor pressure is high. Similarly, some of its naturally occurring compounds are also relatively volatile and/or readily decomposable. Most of the atmospheric mercury then comes from the gradual degassing of the Earth's crust. The Hg content of vapors from Hawaiian fumarole vents is as high as 27 μg/m^3, 90% of it being either gaseous or as particles smaller than 0.3 μm (diam.) (Eshleman et al., 1971). The mercury "scare" a few years ago was followed by a smaller cadmium scare. Cadmium produces a disease known accurately in Japan as "Ouch-ouch." Another metal that has received much attention because of its very great toxicity is beryllium (Anon., 1971*c*, *d*).

6.7 SMOG

Photochemical smog, the most notorious form of air pollution and long associated with the city of Los Angeles, is not simply a chemical soup; it is a complex meteorological phenomenon. It has to be experienced to be believed, and unfortunately (or perhaps fortunately), more and more people in more and more cities throughout the world are doing exactly that.

In areas where there is inadequate movement of the warm daytime air mass photochemical, smog formation is a cyclic process, reforming each day and disappearing in the evening. Berry and Lehman (1971) have described this cycle in terms of five steps:

1. Emission of hydrocarbons (about 65% in the form of exhaust gases (Figure 6.14) and 35% from evaporation and "blow-up" with a composition similar to that of gasoline but enriched in the lighter components), the fuel of smog formation, and oxides of nitrogen, which provide chemical energy to drive the reactions:
2. Absorption of solar energy and the photodissociation of NO_2;
3. Consumption of the oxides of nitrogen and the buildup of oxidants, including ozone, oxygen atoms, excited molecular oxygen, and possibly peroxides;
4. Oxidation of the hydrocarbons by these oxidants; and
5. Termination of the oxidation process and dispersal of the pollutants.

These steps and their accompanying formation and consumption of chemical substances can be clearly seen both in smog produced in the laboratory (Figure 6.17) and in the real situation (Figure 6.18). Table 6.15 gives some typical photochemical smog constituent concentrations. While a great deal is known about smog reactions (see Altshuller and Bufalini, 1971, for a detailed review), several important questions have defied resolution including the mechanism of the rapid conversion of NO to NO_2

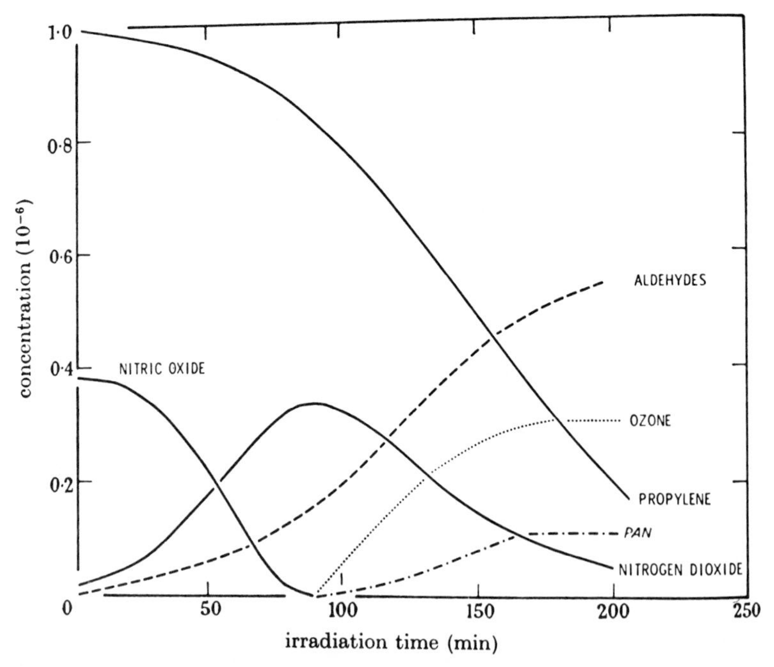

Figure 6.17. Typical concentration changes in a photochemical smog reaction. PAN, Peroxyacetylnitrate. (From Berry and Lehman, 1971, reproduced with permission from *Annual Review of Physical Chemistry*, Vol. 22, © 1971 by Annual Reviews, Inc. All rights reserved.)

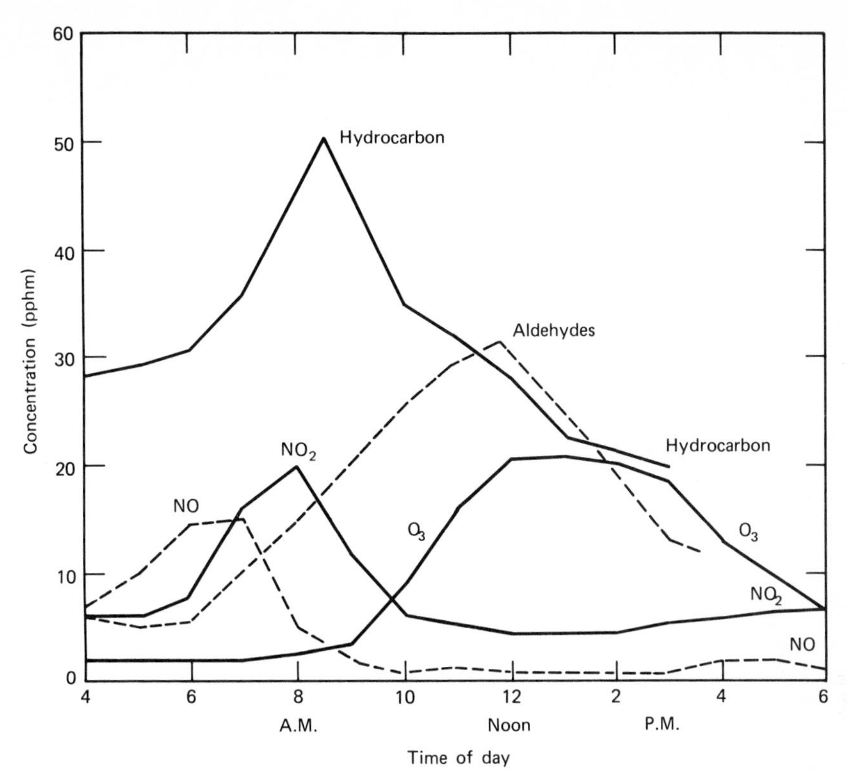

Figure 6.18. Typical variation of components in photochemical smog for a day of intense smog (ppm), Parts of Air by Volume (from Cadle and Allen, © 1970 by the American Association for the Advancement of Science with permission of the AAAS).

Table 6.15 Typical Concentrations of Trace Constituents in Photochemical Smog (From Cadle and Allen, 1970)

Constituent	Concentration (pphm)
Oxides of Nitrogen	20
NH_3	2
H_2	50
H_2O	2×10^6
CO	4×10^3
CO_2	4×10^4
O_3	50
CH_4	250
Higher paraffins	25
C_2H_4	50
Higher olefins	25
C_2H_2	90
C_6H_6	10
Aldehydes	60
SO_2	20

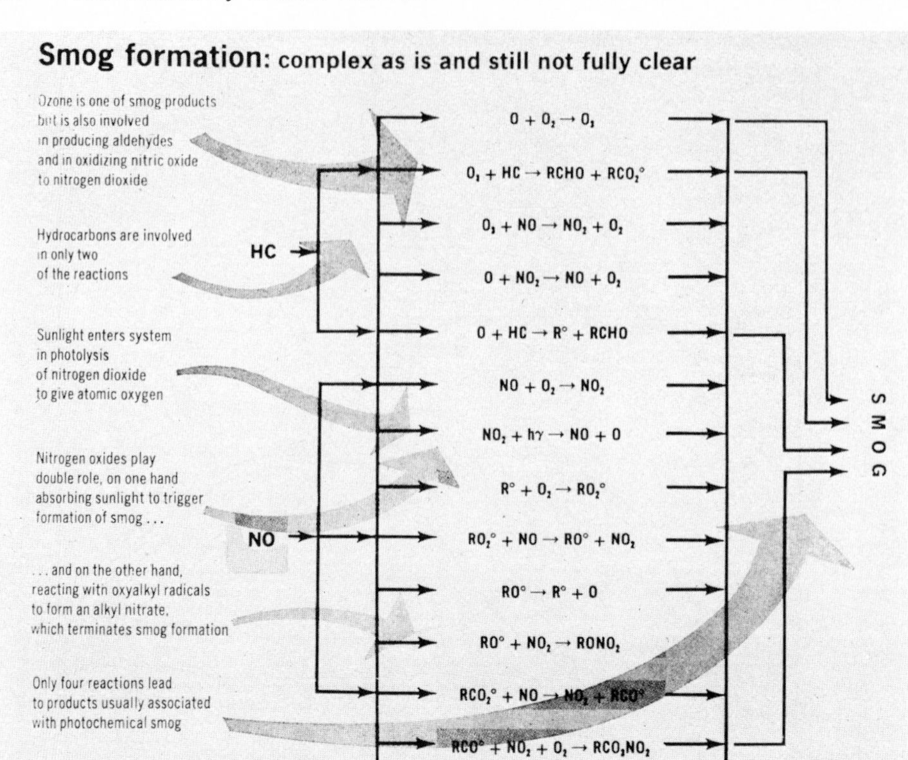

Figure 6.19. Some of the major chemical processes in smog formation (reprinted with permission from Anon., *Chem. Engr. News* (May 16, 1966), p. 57, © The American Chemical Society).

NO —Nitric oxide	O Atomic Oxygen	RCO° —Acyl radical
HC Hydrocarbon	O_3 —Ozone	RCO°_2 —Oxyacyl radical
NO_2—Nitrogen dioxide	R° —Alkyl radical	$RONO_2$ —Alkyl nitrate
O_2 Oxygen	RO° —Oxyalkyl radical	RCO_3NO—Cpd X, acyl peroxynitrate
$h\gamma$ —Sunlight energy	RO°_2—Peroxyalkyl radical	RCHO —Aldehyde

(the total nitrogen balance in smog has been recently resolved, Gay and Bufalini, 1971) and the "excess rate" of the oxidation of the hydrocarbons. Figure 6.19 is an attempt to summarize the more important smog reactions. Among the effects of smog are eye irritation, vegetation damage, reduction of visibility, and deterioration of materials such as rubber (Faith and Atkisson, 1972). Ozone, which can also reduce the germination of corn pollen (Mumford et al., 1972), is the guilty ingredient for the last, while peroxyacetylnitrate, or PAN, is one of the ingredients responsible for both eye irritation and plant damage—tobacco plants are particularly sensitive

$$CH_3\overset{\overset{\displaystyle O}{\|}}{C}OONO_2$$

PAN

(Ordin et al., 1971). A more powerful eye irritant but, fortunately, present in much lower concentrations than PAN is peroxybenzoylnitrate or PBzN

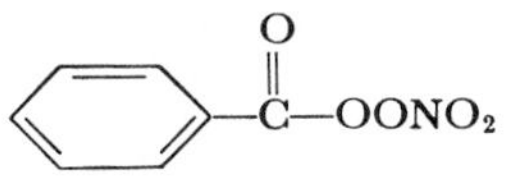

$$C_6H_5\text{—}\overset{\overset{\displaystyle O}{\|}}{C}\text{—}OONO_2$$

PBzN

The products of hydrocarbon oxidation are commonly less volatile than the original hydrocarbons and condense into small aerosol droplets, thus producing the reduced visibility associated with smog. These aerosols in turn affect the smog reactions; for example, they alter the rate of hydrocarbon oxidation. Carbon monoxide may aggravate smog formation (Westberg et al., 1971), but the presence of SO_2, it is interesting to note, especially in dry atmospheres, has an inhibitory effect on smog formation (Wilson and Levy, 1970) and may actually decrease eye irritation (Wilson et al., 1972). Nevertheless, the ozone formed in the troposphere in connection with smog formation apparently does oxidize atmospheric sulfur up to sulfate so that sulfuric acid is an important constituent of the smog aerosol (Stephens and Price, 1970; Atkins et al., 1972). This also raises the question as to whether the ozone level of the lower atmosphere might be increasing on a world wide basis (Komhyr et al., 1971). Nitric oxide can also inhibit the photooxidation of hydrocarbons, and this and other considerations lead to the practical conclusion that it is more important to control hydrocarbon than oxides of nitrogen emissions from automobiles (Altshuller et al., 1970; Glasson and Tuesday, 1970).

6.8 RADIOACTIVITY AND FALLOUT

Far from being an unmitigated disaster, man's contamination of the Earth's atmosphere with radioactive substances has been in at least two important respects an unexpected blessing. Studies of radioisotopes have enormously increased our knowledge of atmospheric mixing processes and of the distribution and movement of chemical elements in the environment generally. Also the handling of radioactive materials has demonstrated that man has an entirely adequate competence for dealing with environmental contaminants, even cooperating on an international level, *if* he believes that the potential threat is great enough.

The radionuclides in the Earth's atmosphere come from three sources:

1. Natural, very long half-lived nuclides that have persisted since the formation of the planet and their shorter-lived daughter nuclides that are continually renewed by decay,
2. Natural, relatively short-lived nuclides, continually formed by processes such as cosmic radiation in the atmosphere, and
3. Artificial nuclides resulting from human activity, especially atomic weapon testing.

The three important decay schemes responsible for the first of these sources are:

The Uranium Series

$$^{238}\text{U} \xrightarrow[\alpha]{4.51 \times 10^{9}\text{ yr}} {}^{234}\text{Th} \xrightarrow[\beta^-]{24.10\text{ d}} {}^{239}\text{Pa} \xrightarrow[\beta^-]{1.18\text{ mo}} {}^{234}\text{U} \xrightarrow[\alpha]{2.48 \times 10^{5}\text{ yr}}$$

(99.274% of U) (0.056% of U)

$$^{230}\text{Th} \xrightarrow[\alpha]{8.0 \times 10^{4}\text{ yr}} {}^{226}\text{Ra} \xrightarrow[\alpha]{1622\text{ yr}} {}^{222}\text{Rn} \xrightarrow[\alpha]{3.825\text{ d}} {}^{218}\text{Po} \rightarrow \rightarrow {}^{210}\text{Pb} \rightarrow$$

The Thorium Series

$$^{232}\text{Th} \xrightarrow[\alpha]{1.39 \times 10^{10}\text{ yr}} {}^{228}\text{Ra} \xrightarrow[\beta]{6.7\text{ yr}} {}^{228}\text{Ac} \xrightarrow[\beta^-]{6.13\text{ hr}} {}^{228}\text{Th} \longrightarrow$$

(100% of Th)

The Actinouranium Series

$$^{235}\text{U} \xrightarrow[\alpha]{7.1 \times 10^{8}\text{ yr}} {}^{231}\text{Th} \xrightarrow[\beta^-]{95.64\text{ hr}} {}^{231}\text{Pa} \xrightarrow[\alpha]{3.43 \times 10^{4}\text{ yr}} {}^{227}\text{Ac} \xrightarrow[\beta^-]{21.89\text{ yr}} {}^{227}\text{Th} \longrightarrow$$

(0.720% of U)

Table 6.16 Estimated Values for Radon and Thoron Exhalation from Soil (From Junge, 1963, with permission of Academic Press, Inc., and the author.)

	Unit	Uranium-238 (radium)	Thorium-232
Decay rate of mother substance	sec^{-1}	4.8×10^{-18}	1.6×10^{-18}
Range of average content of mother substance in igneous and sedimentary rocks	g/g	1.4×10^{-6}	4.15×10^{-6}
Accepted value of average content for model calculations, soil density = 2 g/cm^3	g/cm^3	4×10^{-6}	20×10^{-6}
	atoms/cm^3	1×10^{16}	5×10^{16}
	ci/cm^3[a]	1.3×10^{-12}	2.2×10^{-12}
		Radon	Thoron
Decay rate λ of emanations	sec^{-1}	2.1×10^{-6}	1.27×10^{-2}
Concentration $c_{so} = a/\lambda$ of emanation in undisturbed soil air (= 0.1 × equilibrium concentration)	atoms/cm^3	2.3×10^{3}	6.5×10^{-1}
	ci/cm^3	1.3×10^{-13}	2.2×10^{-13}
Production rate a of emanation in soil	atoms/cm^3 sec	4.9×10^{-3}	8.3×10^{-3}
	ci/cm^3 sec	2.7×10^{-19}	2.8×10^{-15}
Exhalation rate $E = a(d/\lambda)^{1/2}$[a]	atoms/cm^2 sec	7.4×10^{-1}	1.7×10^{-2}
	ci/cm^2 sec	4.0×10^{-17}	5.6×10^{-15}
Concentration of emanation in air near ground, $c_{so} = c_{so}(d/D)^{1/2}$[a]	atoms/cm^3	2.3×10^{0}	6.5×10^{-4}
	ci/cm^3	1.3×10^{-16}	2.2×10^{-16}
Depth where $c_s = c_{so}/2$, $z = \ln 2(d/\lambda)^{1/2}$[a] below ground	cm	100	1.5
Height where $c_a = c_{ao}/2$, $h = \ln 2(D/\lambda)^{1/2}$[a] above ground	m	1000	14

[a] Constants used for calculation: $d = 0.05$ cm^2/sec; $D = 5 \times 10^{4}$ cm^2/sec. 1 ci = 3.7×10^{10} d/sec.

Of particular interest to us here are the isotopes of radon, ^{222}Rn and ^{220}Rn (sometimes called "Thoron"), which both decay by α-emission with half-lives of 3.825 d and 51.5 sec, respectively, for this element is gaseous and escapes into the atmosphere from the lithosphere where it is formed. Table 6.16 gives estimated radon exhalation rates from soils. The values listed are not too different from the average observed values of 3×10^{-13} Ci/cm^3 for the ^{222}Rn content of soil, 4×10^{-17} Ci/cm^2 sec for the exhalation rate, and 70×10^{-18} to 330×10^{-18} Ci/cm^3 for the concentration of ^{222}Rn in near ground air when one considers that the soil exhalation rate depends strongly on soil condition, noteably moisture content, and that the concentration in near ground air depends in a complex manner, not only on the exhalation rate but on turbulent transport into the upper layers of the atmosphere. The radon concentrations over land are much higher than over the oceans, since virtually all of the tropospheric Rn comes from the continental land masses, in fact, the seas seem to act as a sink rather than a source for radon. Because the relative surface area of exposed land masses is less in the Southern Hemisphere and atmospheric mixing is slow across the equator, the Rn content in southern air is much lower than in northern air. Radon is gaseous, but its decay products are solid, commonly charged solids, and quickly form molecular clusters or become attached to aerosol particles and share the fate of aerosols, being precipitated and washed down with them and showing relatively short residence times, 1 to 40 days.

Table 6.17 gives some production rates of cosmic ray produced radioisotopes, while Table 6.18 summarizes their inventories in the stratosphere, troposphere, ocean mixed layer, deep ocean, and marine sediments. The more important of these nuclides include ^{14}C produced by neutron capture by nitrogen, ^{3}H ("tritium") produced by the spallation of air molecules and by the interaction of secondary neutrons with nitrogen, ^{32}Si probably produced from the spallation of argon, and ^{10}Be from atmospheric nitrogen and oxygen. ^{32}P, ^{33}P, and ^{35}S come from interactions involving

Table 6.17 Cosmic-Ray-Produced Radioisotopes (From Junge, 1963, with permission of Academic Press, Inc., and the author.)

Radio-isotope	Calculated Average Total Production Rate (atoms/cm^2 year)	Half-lifetime	Estimated Steady-State activity in the Troposphere (disintegrations/cm^2 and sec)	Detected and Measured in:
^{10}Be	2.6×10^6	2.7×10^6 yr	1.1×10^{-9}	Deep sea sediments
^{14}C	6.3×10^7	5.7×10^3 yr	—	$^{14}CO_2$, all organic materials
^{32}Si	6.3×10^3	710 yr	8×10^{-9}	Marine sponges
^{3}H(T)	7.9×10^6	12.5 yr	—	HTO, HT
^{22}Na	$\sim 5 \times 10^2$	2.6 yr	1.7×10^{-7}	Rain
^{35}S	4.2×10^4	87 days	1.6×10^{-4}	Rain, air
^{7}Be	2.4×10^6	53 days	1.5×10^{-2}	Rain, air
^{33}P	2.0×10^4	25 days	2.6×10^{-5}	Rain
^{32}P	2.4×10^4	14.3 days	5.6×10^{-5}	Rain

Table 6.18 Steady-State Fractional Inventories and Decay Rates of Cosmic-Ray-Produced Radioisotopes in the Exchange Reservoirs (From Lal, 1967, with permission of the International Atomic Energy Agency.)

Exchange Reservoir	Radioisotope												
	^{10}Be	^{26}Al	^{36}Cl	^{14}C	^{32}Si	^{39}Ar	^{3}H	^{22}Na	^{35}S	^{7}Be	^{37}Ar	^{33}P	^{32}P
Stratosphere	3.7×10^{-7}	1.3×10^{-6}	10^{-6}	3×10^{-3}	1.9×10^{-3}	0.16	6.8×10^{-2}	0.25	0.57	0.60	0.63	0.64	0.60
Troposphere	2.3×10^{-8}	7.7×10^{-8}	6×10^{-8}	6×10^{-2} [a]	1.1×10^{-4}	0.83	4×10^{-3}	1.7×10^{-2}	8×10^{-2}	0.11	0.37	0.16	0.24
Mixed Oceanic Layer	8×10^{-6}	2×10^{-5}	2×10^{-2}	2.3×10^{-2}	5×10^{-3}	5×10^{-4}	0.50 [b]	0.62	0.34	0.28	0	0.19	0.16
Deep Oceanic Layer	1.4×10^{-4}	10^{-4}	0.98	0.91	0.96	3×10^{-3}	0.43	0.11	6×10^{-3}	3×10^{-3}	0	10^{-3}	1.5×10^{-4}
Oceanic Sediments	0.999	0.999	0	10^{-2}	4×10^{-2}	0	0	0	0	0	0	0	0
(yr^{-1})	2.8×10^{-7}	9.4×10^{-7}	2.2×10^{-6}	1.24×10^{-4}	1.4×10^{-3}	2.3×10^{-3}	5.6×10^{-2}	0.27	2.9	4.8	7.2	10	17.7
Normalizing Factor for obtaining disintegrations/min in cm^2 column in different reservoirs	2.7	1.4×10^{-2}	0.21	108	9.6×10^{-3}	0.34	15	3.4×10^{-3}	8.4×10^{-2}	4.9	0.05	4.1×10^{-2}	4.9×10^{-2}

[a] Includes amount present in biosphere and humus.
[b] Includes amount present in the continental hydrosphere.

atmospheric argon. Most of the radioisotopes are formed by secondary, low-energy neutrons, and their production rate is proportionally higher in the polar atmosphere where the neutron flux is about four times greater than at the equator (Figure 6.20). Figure 6.20 also shows that production is at a maximum at altitudes of about 12 to 15 km, where the product of decreasing air density and increasing neutron flux is at a maximum. The nongaseous radionuclides, like the Rn decay products, quickly identify themselves with aerosols, while those such as ^{3}H, ^{14}C, and possibly ^{35}S that form gases behave in a different manner. Carbon-14 is of particular interest. About 90% of it gets oxidized to CO (Pandow et al., 1960) and subsequently to CO_2 whereupon it enters the Earth's carbon cycle and provides the basis of carbon dating—

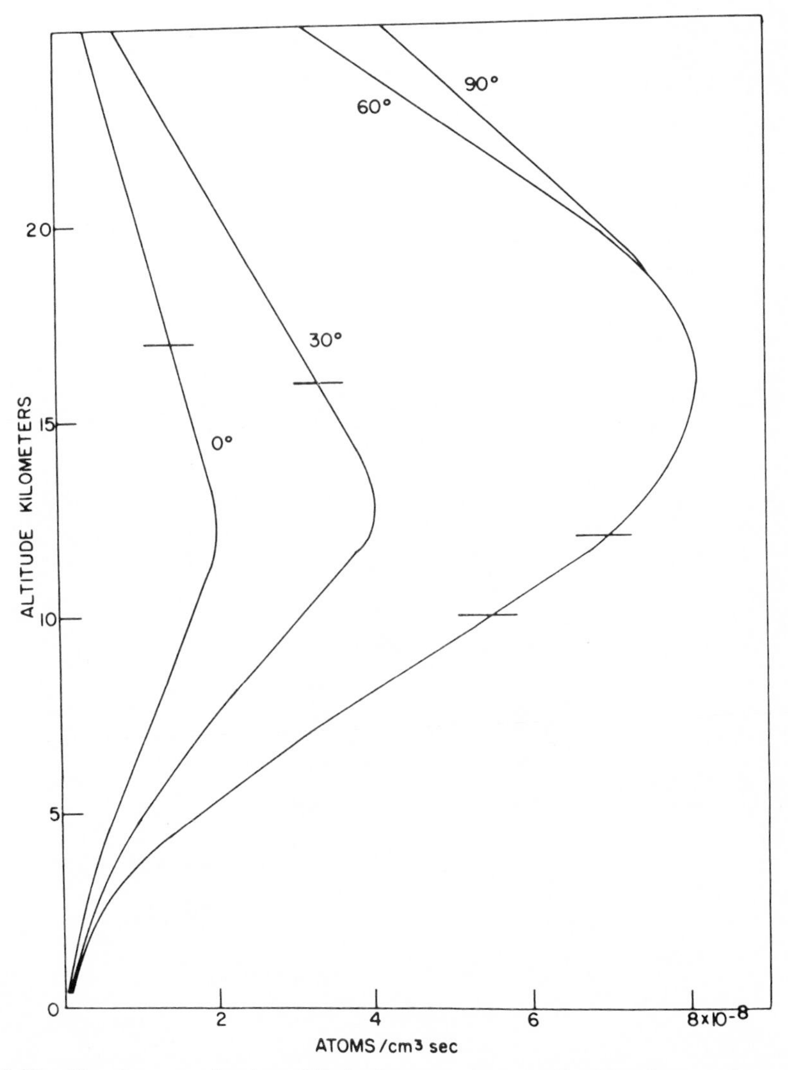

Figure 6.20. Cosmic ray production of ^{7}Be in the atmosphere for 0°, 30°, 60°, and 90° geomagnetic latitude as a function of altitude. The horizontal lines mark the tropopause heights (from Junge, 1963, with permission of Academic Press, Ltd. and the author).

a most powerful scientific tool. Carbon-14 from nuclear weapons testing follows the same route, enters the carbon cycle, and in due course appears in our bodies (Figure 6.21) where it may be a potential genetic hazard (Pauling, 1958). Figure 6.21 is a rather scary curve, especially when you remember that the half-life of ^{14}C is nearly 60 centuries. This is perhaps an appropriate place to mention that the effects of low-level radiation are a most controversial topic (see, for example, the recent paper of Rossi and Kellerer, 1972). The popular notion, shared by some scientists who really should know better, is that the effects of radiation are incremental and accumulative so that even a little of a bad thing is bad. Yet people who live at high altitudes appear to be none the worse for their exposure to a higher background level of natural radiation. Mutations induced by natural radiation may even provide the mechanism for past and continued biological evolution. My own prejudice is that curves such as that presented in Chapter 7 for chemicals (Figure 7.33) see also Horne, 1972) are probably equally applicable for various types of radiation.

Like ^{14}C, tritium too provides a powerful tool for studying geochemical and geophysical processes (see Suess, 1969), but not as powerful as might have been. Man's

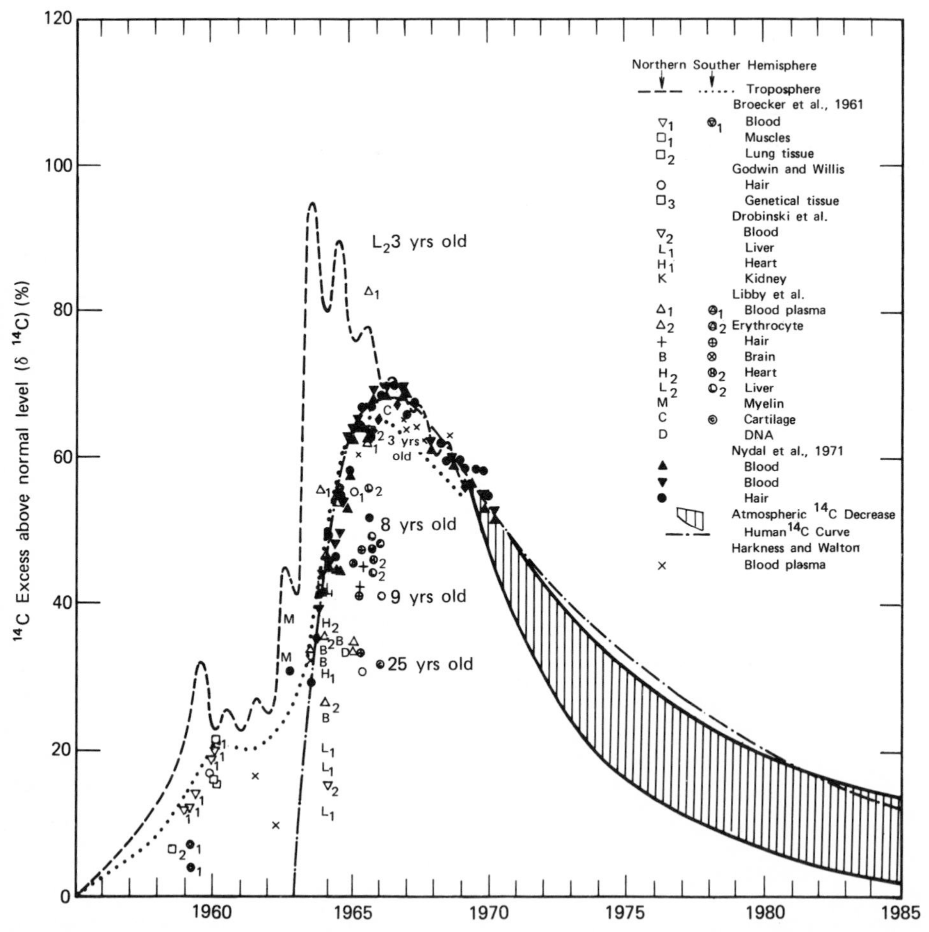

Figure 6.21. Radiocarbon in the troposphere and the human body (from Nydal et al., 1971, with permission of the publisher and the authors).

contamination of our environment with artificial tritium has forever precluded the possibility of certain invaluable low-level counting experiments. Anthropogenic sources of radioactive contamination in our environment are largely weapon testing; nuclear reactors including power generators; and the production, processing, and various applications of radioactive materials. Except for the first, the escape of radioactivity into our environment is now very carefully scrutinized and controlled. The most severe problem of the ultimate disposal of highly radioactive wastes resulting from these usages has already been mentioned as have also been a few of the hazards that have resulted from negligent mining and processing in the past (see also Shapley, 1971). Figures 6.22 and 6.23 diagram the human exposure routes of radioactive gaseous and liquid emissions and effluents. Radiopollution of natural fresh waters is touched on briefly in Chapter 7, while the radiochemistry of the oceans is dealt with in Chapter 10 in my book *Marine Chemistry* (1969) and was reviewed earlier by Picciotto (1961), Miyake (1963), and Burton (1965). The major culprits contributing to the radioactive legacy from the slow neutron fission of ^{235}U after

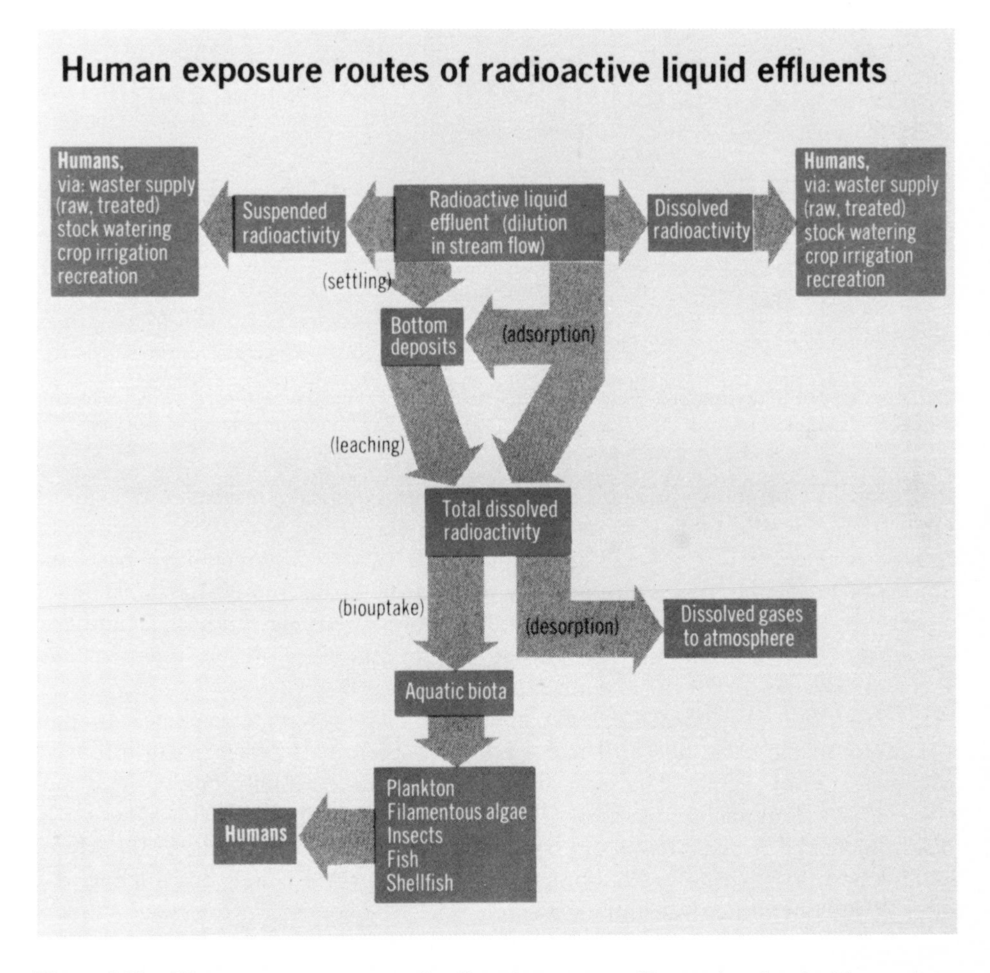

Figure 6.22. Human exposure routes of radioactive gaseous effluents (reprinted with permission from E. C. Tsivoglou, *Environ. Sci. Tech.*, *5*, 404 (1971), © The American Chemical Society).

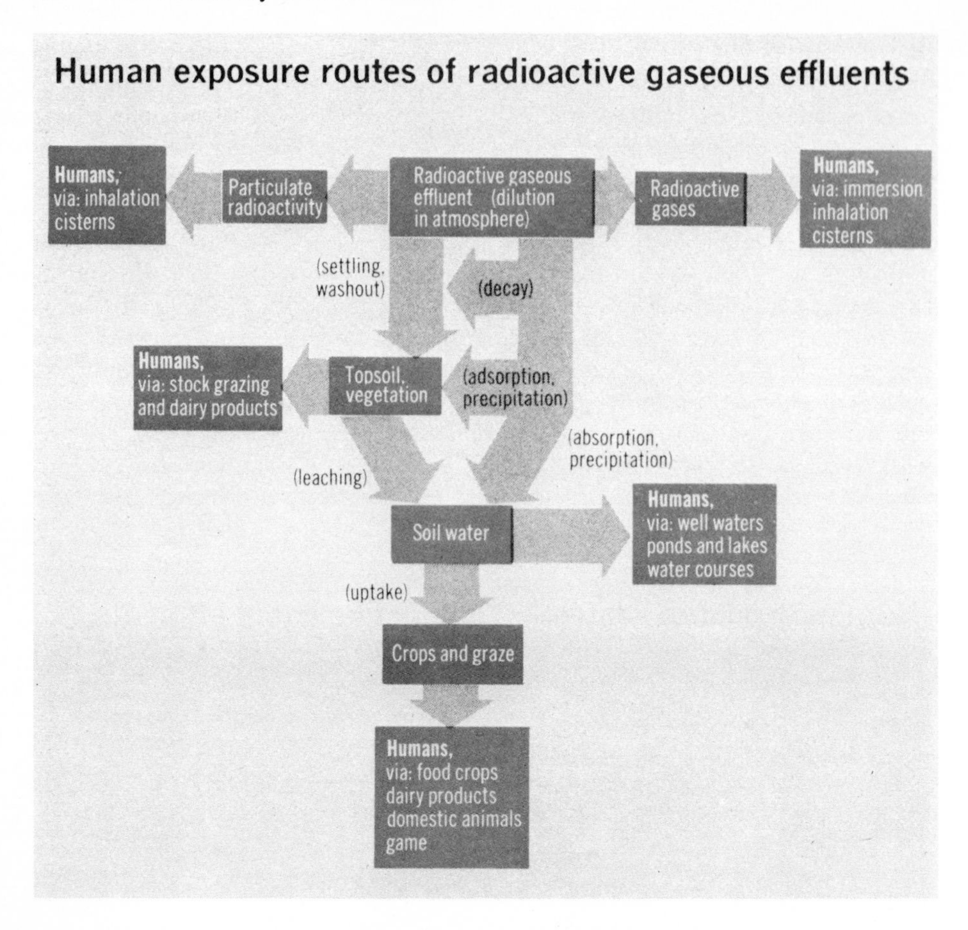

Figure 6.23. Human exposure routes of radioactive liquid effluents (reprinted with permission from E. C. Tsivoglou, *Environ. Sci. Tech.*, *5*, 404 (1971), © The American Chemical Society).

2 days and 1 yr are shown in Table 6.19. After 20 yr most of the residual radioactivity comes from ^{90}Sr–^{90}Y (48%), ^{137}Cs–^{137}Ba (45%), and ^{147}Pm and ^{157}Sm. Because of their persistence ^{90}Sr ($t_{1/2}$ = 28.8 yr) and ^{137}Cs ($t_{1/2}$ = 27.7 yr) have been the subjects of particular concern, especially the former. Strontium-90 has a tendency to concentrate and remain in bone material, notably of growing children. In addition to the fission products themselves, neutron capture and other nuclear reactions add other radioisotopes to the list of contaminants such as ^{3}H, ^{14}C, ^{35}S, ^{57}Co, ^{58}Co, ^{60}Co, ^{65}Zn, ^{54}Mn, ^{59}Fe, ^{55}Fe, ^{89}Sr, ^{95}Zr, ^{140}Ba, and ^{144}Ce. There is even a very small amount of uranium in the atmosphere (0.1 to 1.5 ng/m^3) presumably from natural sources (McEachern et al., 1971). The most abundant man-made radioisotope in the troposphere is ^{85}Kr from nuclear reactors. In the period 1960 to 1970, the tropospheric ^{85}K content increased from 200 to about 600 dpm/mmole (Schröder et al., 1971). Many chemists have a "favorite" radioisotope. Mine is ^{55}Fe, since I used this nuclide as a tracer in my doctorate dissertation research. The ^{55}Fe content in the blood of Norwegians, including Lapps, has been investigated and some interesting differences between the sexes have appeared that are attributed to the higher metabolic rate of women (Lekven, 1972).

Table 6.19 The Percentage of Activity in Curies of the Principal Radioisotopes from Slow Neutron Fission of ^{235}U (1 kg) (From Miyake, 1963, with permission of John Wiley & Son)

Twenty Days	9.8×10^6 Curies		One Year	3.1×10^4 Curies	
Nuclide	Activity (ci) (10^5)	Percent	Nuclide	Activity (ci) (10^3)	Percent
^{140}La	13.6	13.9	^{144}Ce–^{144}Pr	164	52.8
^{140}Ba	11.8	12.0	^{95}Nb	45.6	14.7
^{143}Pr	11.8	12.0	^{95}Zr	22.4	7.2
^{141}Ce	9.5	9.7	^{147}Pm	17.7	5.7
^{133}Xe	6.2	6.3	^{91}Y	11.8	3.8
^{95}Zr	5.8	5.9	^{89}Sr	8.4	2.7
^{91}Y	5.5	5.6	^{106}Ru–^{106}Rh	15.2	4.9
^{131}I	5.5	5.6	^{90}Sr–^{90}Y	11.5	3.7
^{89}Sr	4.9	5.0	^{137}Cs–^{137}Ba	9.0	2.9
^{147}Nd	4.9	5.0	^{103}Ru	2.5	0.8
^{103}Ru	4.3	4.4	^{103}Rh	2.5	0.8
^{95}Nb	4.1	4.2			
^{144}Ce	2.3	2.3			
^{144}Pr	2.6	2.6			
^{90}Mo	1.3	1.3			
^{131}I	1.0	1.05			

Fallout carried down by precipitation over the land, both inert and radioactive, is strongly taken up by soil or, if it drains eventually into the sea, by organisms and/or sediments in the coastal zone. That which rains down upon the open ocean, because of the slow vertical mixing in the oceans, tends to linger in the upper mixed layer. Table 6.18 summarized marine levels of radioisotopes produced in the atmosphere and Table 6.20 summarizes the levels of natural radionuclides in seawater. Potassium-40 is responsible for more than 90% of the natural radioactivity of seawater. As for man-made radioisotopes in surface ocean waters the concentrations of ^{147}Pm, ^{144}Ce, ^{137}Cs, ^{90}Sr, ^{36}Cl, ^{14}C, and ^{3}H are 0.2×10^{-17} to 3×10^{-17}, 0.1×10^{-17} to 2.5×10^{-17}, 0.5×10^{-15} to 1.2×10^{-15}, $<5 \times 10^{-18}$, —3×10^{-16}, and 0.1×10^{-15} to 1×10^{-15} g/liter, respectively.

A small amount of radiation is allowed to escape in the flue gases from nuclear installations, but of primary concern to us here are the great quantities of radioactivity spewed into the atmosphere from weapons testing. The more responsible nuclear powers no longer test in the atmosphere, but some of the more backward nations still persist in thinking that they have something to "prove" to the world by testing devices above ground, despite international fear and indignation.

If the device is detonated in a tower on the surface, quantities of the lithosphere will be vaporized and swept up into the fireball. For a land nuclear shot, 20 to 50% of the fission products will remain in the stratosphere and for a water surface shot 30 to 80% (Junge, 1963). As the fireball cools, a variety of materials, gaseous and

Table 6.20 Natural Radionuclides Present in the Ocean (From Picciotto, 1961)

Nuclide	Half-life (yr)	Concentration (g/ml)	Isotopic Abundance (%)	Disintegrations (per sec and per ml)
^{3}H	1.2×10^{1}	3.2×10^{-21}	1.0×10^{-16}	$1.1 \times 10^{-6}\ \beta$
^{14}C	5.5×10^{3}	3.1×10^{-17}	1.3×10^{-10}	$5.2 \times 10^{-6}\ \beta$
^{10}Be	2.7×10^{6}	1×10^{-16}		$7 \times 10^{-8}\ \beta$
^{40}K	1.3×10^{9}	4.5×10^{-8}	1.2×10^{-2}	$1.1 \times 10^{-2}\ \beta + \gamma$
^{87}Rb	5.0×10^{10}	3.4×10^{-8}	27.8	$1.0 \times 10^{-4}\ \beta$
^{238}U	4.5×10^{9}	2×10^{-9}	99.3	$2.5 \times 10^{-6}\ \alpha$
^{230}Th	8.0×10^{4}	6×10^{-16}	$>3 \times 10^{-3}$	$4 \times 10^{-7}\ \alpha$
^{226}Ra	1.6×10^{3}	8×10^{-17}	~100	$2.9 \times 10^{-6}\ \alpha$
^{235}U	7.1×10^{8}	1.4×10^{-11}	0.7	$1.1 \times 10^{-6}\ \alpha$
^{231}Pa	3.4×10^{4}	5×10^{-17}	~100	$8 \times 10^{-8}\ \alpha$
^{227}Th (RdAc)	—	7×10^{-23}	—	$8 \times 10^{-8}\ \alpha$
^{232}Th	1.4×10^{10}	2×10^{-11}	~100	$8 \times 10^{-8}\ \alpha$
^{228}Th (RdTh)	1.9	4.0×10^{-21}	—	$1.2 \times 10^{-7}\ \alpha$
^{228}Ra (MsTh)	6.7	1.4×10^{-20}	$\sim 1 \times 10^{-2}$	$1.2 \times 10^{-7}\ \beta$

particulate, are formed and condense out. The larger particles (1 to 10 μm) can be carried great distances and come down in the following weeks as tropospheric fallout. As particle size decreases, settling becomes less and washout by precipitation increasingly more, important (U. S. Atomic Energy Commission, 1970). The smaller particles once injected into the stratosphere can remain there for a very long time, being slowly removed by mixing processes between stratosphere and troposphere and subsequent removal from the troposphere by precipitation. Figure 6.24 is composed of a series of three interesting diagrams showing the change in the distribution of ^{90}Sr in the stratosphere following weapons testing. Again it illustrates the poor mixing across the equator and a sort of "leakage" into the troposphere in the polar regions. The exchange of radioisotopes between stratosphere and troposphere has many similarities with the peregrinations of ozone in the atmosphere, and from O_3, ^{90}Sr, ^{185}W, and ^{102}Rh data, Junge (1963) has evolved a general picture of lower atmosphere exchange processes (Figure 6.25). The seasonal variations of fallout are greater in polar than in tropical regions. Residence times for the low polar stratosphere are of the order of 0.5 yr, of the low tropical stratosphere 1 yr, and of the higher (30 km) tropical stratosphere perhaps as long as several years (Junge, 1963).

The greater part of the ^{90}Sr in rain is in soluble form, and it is mostly retained (60 to 80%) in the upper 5 cm of undisturbed soil. In temperate latitudes, it might be noted, precipitation accounts for 80 to 90% of the total fallout deposition. For the year 1958, the following ^{90}Sr budget was estimated:

2.3 MCi total surface burden,

1.0 MCi total stratospheric burden,

3.6 MCi total stratospheric injection, that is, total production minus local and tropospheric fallout. The latter is only a few percent.

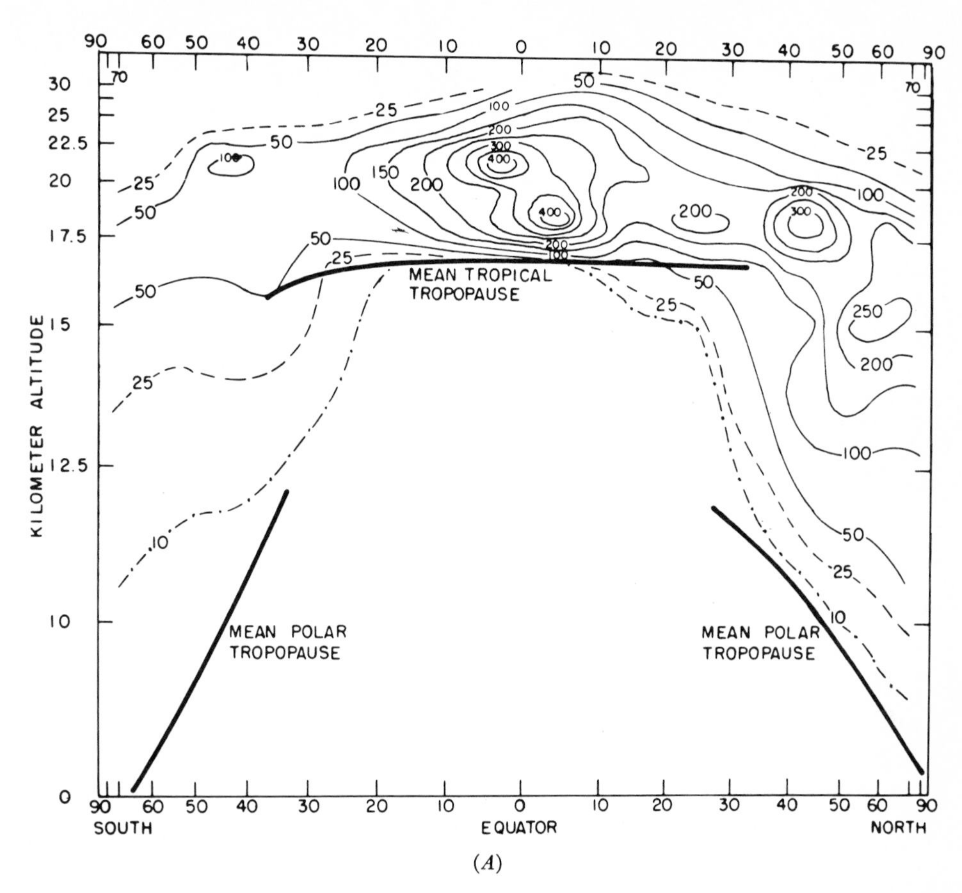

(*A*)

Figure 6.24. Mean atmospheric strontium-90 distributions in disintegrations per minute per 1000 ft^3 STP. (*A*) November 1957 through December 1958; (*B*) January through August 1959; (*C*) January through March 1960. (From Junge, 1963, with permission of Academic Press, Ltd. and the author).

The cumulative world deposition of ^{90}Sr is shown in Figure 6.26 while Figure 6.27 shows the level of ^{90}Sr activity in ground level air over the period 1964 to 1969. The latter shows pronounced peaks corresponding to atmospheric bomb tests, but it is some consolation that at least the general trend of the curve is downward. Carbon-14 and tritium, as we have noted, form gases, so their fate in the atmosphere is somewhat different; they are more slowly removed by precipitation and their residence times, therefore, tend to be longer.

There has been a feeling in some quarters that the scientific community should feel some remorse for unleashing the might of the atom. This is utter nonsense. (*a*) It would be the most extreme cowardice and a denigration of man's highly specialized gifts to fail to pursue any course of exploration and discovery. (*b*) The behavioral, social, and moral sciences have conspicuously failed to produce any deterrent to war comparable to the demonstrated effectiveness of the nuclear deterrent. Even with the accelerated time scale of modern history and despite an endless succession of minor skirmishes for the past 20 yr or more, we have enjoyed a *pax atomica*. (*c*) Nuclear

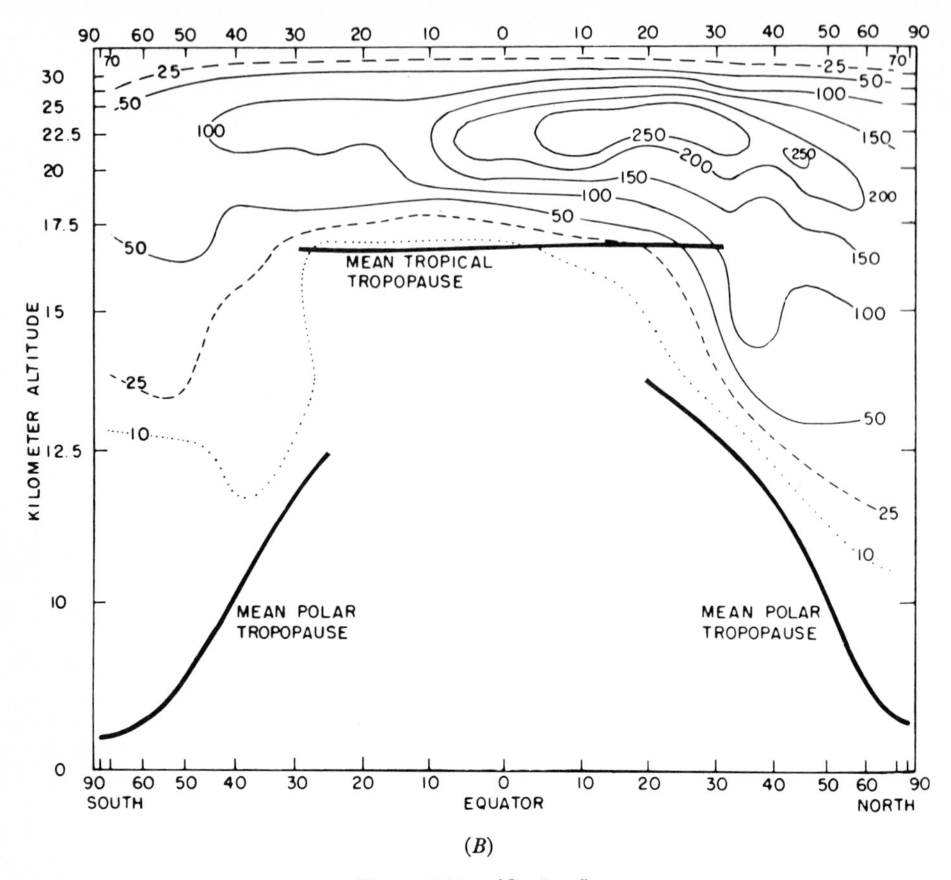

(*B*)

Figure 6.24. (*Continued*).

power provides the only feasible means of providing the long-range energy requirements of our technological civilization with a minimum of environmental havoc. (*d*) And if human fertility is not controlled, nuclear holocaust will provide a solution rather than pose a threat. I prefer to have the history of my species end in a brilliant flash of light created by our scientific knowledge than in a frenzied pack of starving beasts tearing at one another's flesh with tooth and nail and killing and eating the old, the young, and the weak.

6.9 AIR POLLUTION ABATEMENT

The lower atmosphere is cleansed primarily by precipitation washout, but because this chapter is already over long, we defer discussion of this most important topic until Chapter 9 and conclude here with a brief examination of methods of air pollution control, remembering here as throughout this book that air pollution is just like any other form of pollution and that no technology, however elaborate and expensive, can control air pollution if population growth continues to explode. The majority of this planet's people, safe to say, still see air pollution as a symbol of

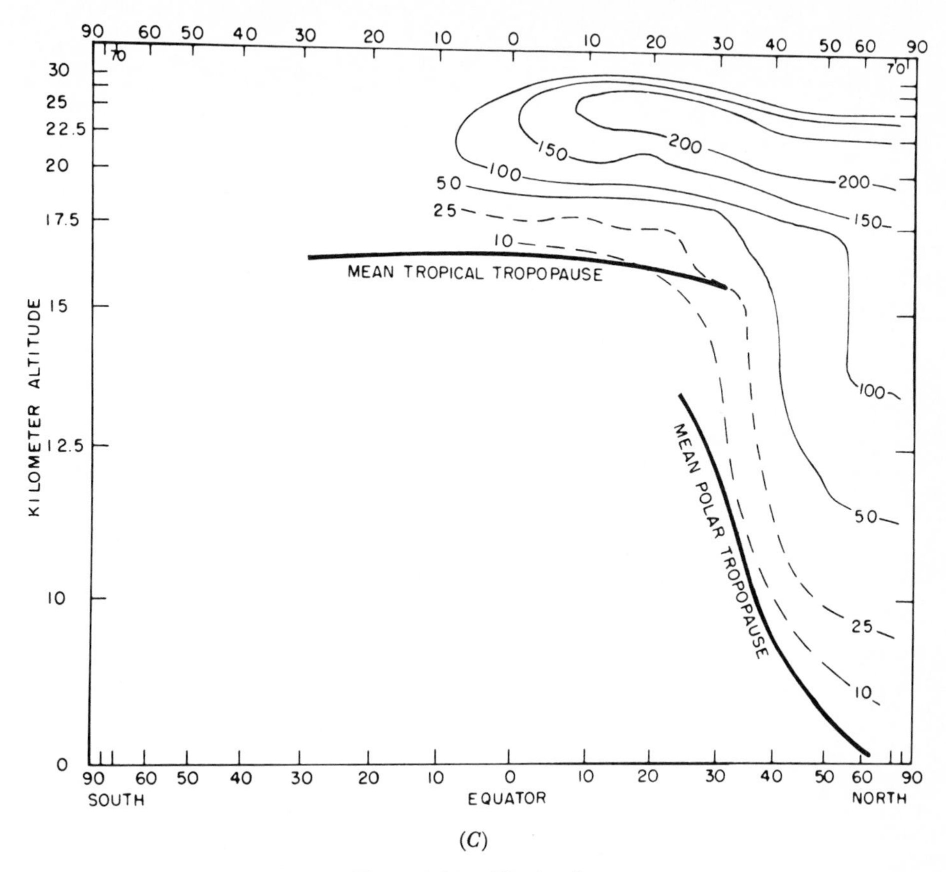

(*C*)

Figure 6.24. (*Continued*).

progress. While the few most developed nations are finally beginning to worry about the messes they create, if compelling action is not taken, their poorer neighbors will increasingly dirty everybody's air.

As a rough rule of thumb, if you can see it and/or smell it, complain (an exception, steam vapor plumes), and if you cannot, do not worry (an exception, carbon monoxide). However, on a less subjective level, there is a great variety of instrumentation now available, not only for detecting and measuring quantitatively air pollution, but equally important, for monitoring air pollution on a continuous basis (see Faith and Atkinson, 1972; Maugh, 1972*a*). Some of these devices are highly sophisticated using such products of the most advanced technology as masers (Kreuzer et al., 1972) and lasers (Hinkley and Kelley, 1971; Maugh, 1972*b*). Clean air is frequently expensive. The cost of air pollution control equipment can exceed the cost of the basic processing equipment (see Table 6.21). How should the cost of clean air be carried? In the case of automobile emissions, the cost will be immediately borne by the public in terms of increased gasoline and new car costs and decreased engine life expectancy, whereas in the instance of SO_2 emissions from stationary sources in terms of the use of low S fuels and/or the removal of SO_2 from smoke stacks. President Nixon has proposed a tax on pollutors based on their SO_2 outpourings. Wilson (1972) has suggested that

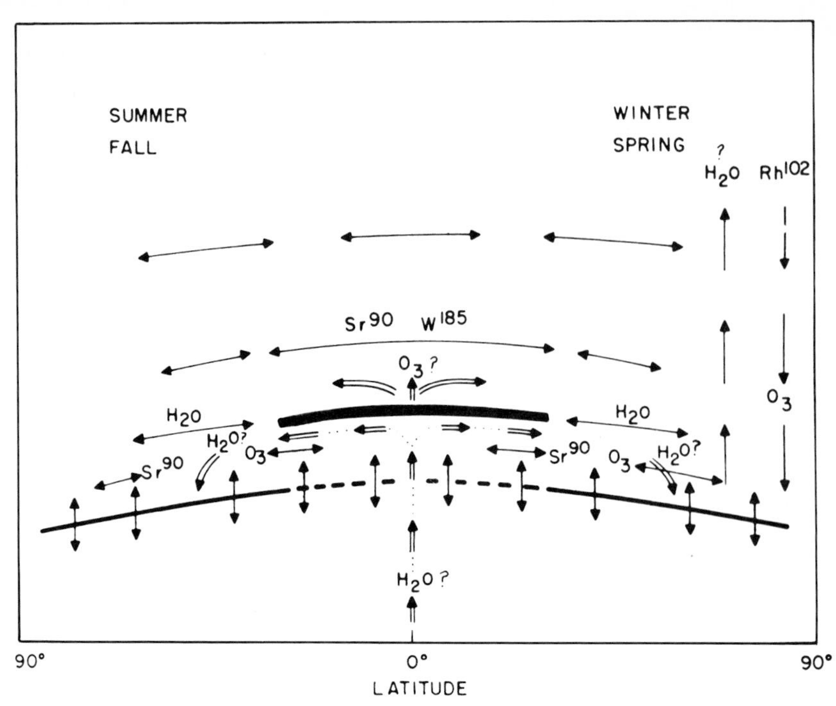

Figure 6.25. Schematic diagram of the stratospheric exchange processes. Double-headed arrows indicate mixing (from Junge, 1963, with permission of Academic Press, Ltd. and the author).

Table 6.21 Relative Costs of Basic and Control Equipment for Selected Operations (From Faith and Atkisson, 1972)

Source	Size of Equipment	Cost of Basic Equipment ($)	Type of Control Equipment	Cost of Control Equipment ($)
Airblown Asphalt System	550 bbl/batch	10,000	Afterburner	3,000
Asphalt Saturator	6 × 65 × 8 ft	40,000	Scrubber and electrostatic precipitator	50,000
Bulk Gasoline Loading Rack	667,000 gal/day	88,000	Vapor control system	50,000
Carbon Black Plant	2000 gal/day	5,000	Baghouse	5,000
Gray Iron Cupola	48 in. ID	40,000	Baghouse and quench tank	67,000
	27 in. ID	25,000	Baghouse and quench tank	32,000
Steel Electric-Arc Furnace	18 ton/heat	75,000	Baghouse	45,000
Open-Hearth Steel Furnace	60 ton/heat	200,000	Electrostatic precipitator	150,000
Oil–Water Separator	300,000 bbl/day	170,000	Floating roof	80,000

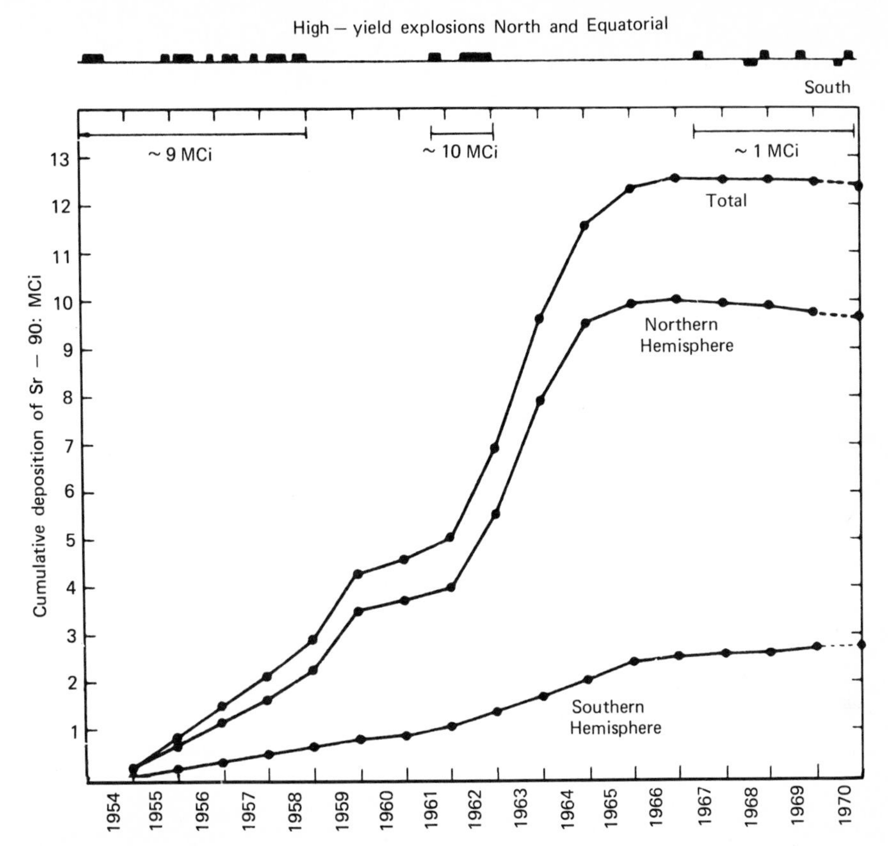

Figure 6.26. Cumulative deposition of strontium-90 (from Pierson, 1971, with permission of the publisher and author).

instead the tax should be based on the product of the pollutant concentration and the population at risk (but would this result in the removal of dirty activities to unspoiled areas?). Although the life expectancy of people is more important than the life expectancy of engines, it is far from clear at this time whether the American people are willing to pay the price necessary for a clean environment. For a stable population, present technology can keep the air clean, but only if strict laws are passed and rigorously enforced. Thanks to strict air pollution laws, London now has 50% more sunshine than 10 yr ago, songbirds are returning to the city's parks, and even fish are being caught in the Thames again. In Pittsburgh, Pennsylvania, once known as the "Smoky City," smoke was reduced 70% between 1945 and 1953, while in St. Louis, where streetlights were not infrequently needed at midday, strict law enforcement reduced smoke by 75% (Faith and Atkisson, 1972).

Turning now to that villain the automobile, cogent attacks on that problem lie in the direction of, again, population control, restriction of private vehicle use, and the substitution of nonpolluting means (i.e., not deisel busses) of public transportation for private cars. Short-range nonsolutions are presently primarily directed to regulating and reducing emissions (Shinnar, 1972). Figure 6.28 compares the national expectations for hydrocarbon, carbon monoxide, and oxides of nitrogen emissions on the

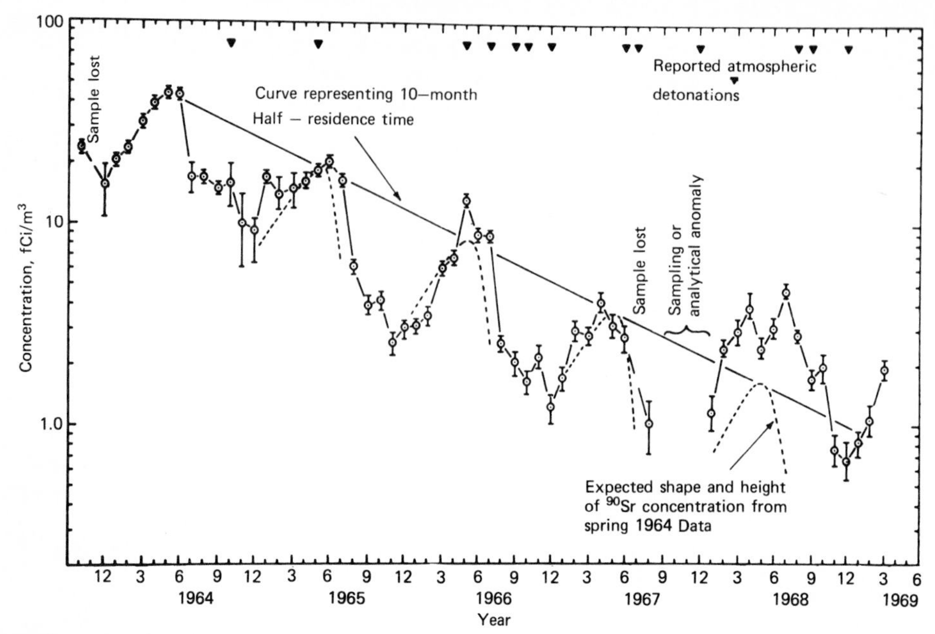

Figure 6.27. Strontium-90 in ground level air (reprinted with permission from B. Shleien et al., *Environ. Sci. Tech.*, *4*, 598 (1970), © The American Chemical Society).

bases of no controls, present controls, and emission standards set down by the Clean Air Act Amendments of 1970. This all looks fine, but if car population continues to grow, the curves will resume their upward climb. In the United States, leaded gasoline, more for political than environmental reasons, appears due to be banned. The problems that this will aggravate have already been mentioned. Absorbents have been proposed for the evaporative losses from fuel tanks and carburetors and various blow-by schemes for crankcase emissions (Faith and Atkisson, 1972). As for exhaust emissions, the brunt of the attack has depended on various catalytic systems (see Crouse, 1971). The problem is not a simple one, for the chemistry of cleaning up the hydrocarbons is exactly opposite to that of getting rid of the oxides of nitrogen. The former is a much more simple chore and can be accomplished by oxidizing catalytic systems; the latter is much more difficult and presumably will involve exhaust gas recirculation or catalytic reduction (Jones et al., 1971; Klimisch and Barnes, 1972). Lead, as we have noted, tends to poison the catalysts, but if Pb is banned, that aspect of the problem will be removed. The more promising catalysts are noble metals, and there has been talk of "the platinum car" and a certain amount of heady speculation on the Pt market on the strength of such talk. Other catalysts proposed are copper oxide alumina convertors and a mixture of rare earth and Co and Mn oxides (Libby, 1971; Pederson and Libby, 1972; Voorhoeve, 1972). The name of Dr. Libby should be noted. It is reassuring to find such a distinguished scientist genuinely concerned with practical environmental problems. I hope that future federal support of science will be directed in such a way as to encourage academics to become more involved with the nitty-gritty of the real world. Just how emissions are going to be monitored, just how the continued effective operation of the emission control systems is going to

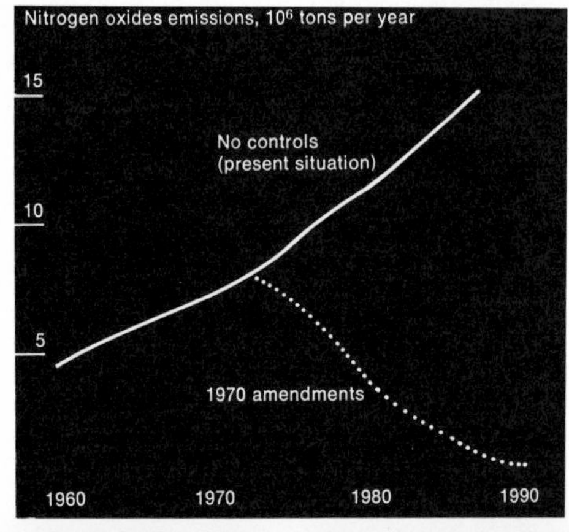

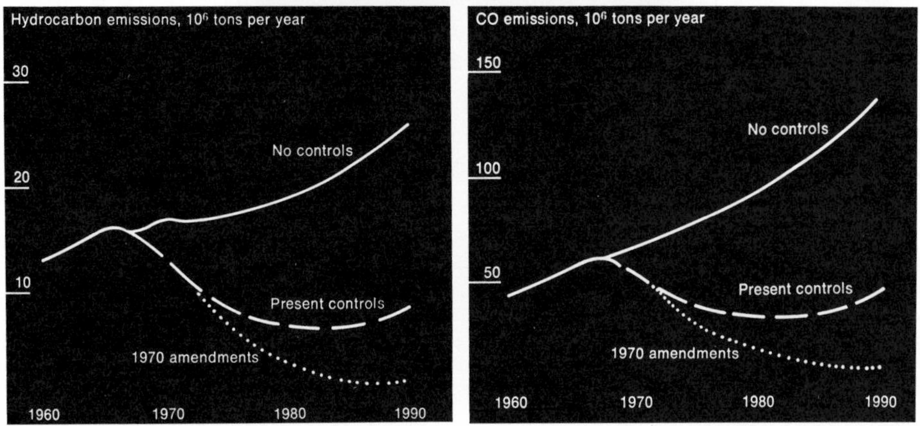

Figure 6.28. Projected nationwide emissions of hydrocarbons, CO, and NO_x showing expected reductions from the Clean Air Act Amendments of 1970 (from Heywood, *Technology Review*, edited at the Massachusetts Institute of Technology, © 1971 by the Alumni Association of M.I.T. with permission).

be insured as the car gets old, just how the public as a whole is going to like picking up the bill for cleaner air, and just how poor people in particular are going to react to having to install expensive systems on their old model cars are all horrendous problems into which we are going to have to plunge before we all choke to death.

Fossil fuels burned in stationary sources, largely for power and heat, are another enormous can of worms, and here the emphasis has been for the most part on getting the sulfur out, either before or after burning, either by using low-sulfur fuels or by removing SO_2 from stack gases. The S-content of coal is highly variable (Table 6.22), and, in the absence of controls at the present rate of increase by 1990, about 50×10^6 metric tons of SO_2 will be dumped into the atmosphere from coal combustion in the United States (Health, Education, and Welfare, 1969). The sulfur in coal is in the form of organic compounds and iron pyrites. Because of their great insolubility in the usual solvents, removal of the iron pyrites by chemical pretreatment is difficult. Meyers et al. (1972) claim that treatment with an aqueous ferric solution can convert 40 to 75% of the pyrite sulfur up to soluble sulfate. Substitution by lignitic coal

Table 6.22 Typical Analysis of Coals (From Faith and Atkisson, 1972)

	Proximate Analysis (dry)				Ultimate Analysis (dry)					
	Moisture as Received (%)	Volume Matter (%)	Fixed C (%)	Ash (%)	C (%)	H_2 (%)	O_2 (%)	N_2 (%)	S (%)	Gross Heating Value (Btu/lb) (dry)
Lignite	33.7	44.1	44.9	11.0	64.1	4.6	18.3	1.2	0.8	11,084
Subbituminous C	16.3	42.8	39.0	18.2	60.7	4.6	14.5	0.9	1.1	10,582
Subbituminous B	17.0	42.8	54.4	2.8	75.0	4.9	15.5	1.3	0.5	13,248
Subbituminous A	10.6	38.9	56.4	4.7	75.1	5.0	12.8	1.4	1.0	13,595
High-Volatile Bituminous, C	7.8	36.4	54.5	9.1	73.1	4.8	8.9	1.5	2.6	13,469
High-Volatile Bituminous, B	3.6	39.2	55.4	5.4	78.3	5.2	8.2	1.5	1.4	14,108
High-Volatile Bituminous, A	1.4	34.3	59.2	6.5	79.5	5.2	6.1	1.4	1.3	14,396
Bituminous, Medium Volume	3.4	22.2	74.9	2.9	86.4	4.9	3.6	1.6	0.6	15,178
Bituminous, Low Volume	3.6	16.0	79.1	4.9	85.4	4.8	2.6	1.5	0.8	15,000
Semianthracite	2.4	13.0	74.6	12.4	78.3	3.6	2.3	1.4	2.0	13,580
Anthracite	3.3	3.4	87.2	9.4	84.2	2.8	2.2	0.8	0.6	13,810
Meta-anthracite	2.8	1.2	90.7	8.1	86.8	1.6	2.0	0.6	0.9	13,682

could improve the situation, but even its S-content is still too high (Table 6.22) to meet emission standards. Oil and gas can and are being substituted for coal, but their supplies are limited. Those limits together with increased use has already produced serious shortages in the United States. The sulfur content of oil, like that of coal, is highly variable. 0.5 to 3%, but unlike coal, oil is readily desulfurized at a cost (Mills et al., 1971). Sulfur dioxide can be removed from stack gases by a number of processes discussed later, but all too commonly, this approach merely substitutes land and/or water pollution for air pollution. More recently, a great deal of hope has been placed in the production of synthetic fuels from coal (Mills et al., 1971). Sulfur-free liquid fuels can be produced from coal at costs presently approaching those for the refinery production of gasoline from petroleum. The carbon in coal and/or naphtha (coal's petroleum fraction) can be gasified and subsequently desulfurized (Figure 6.29). Oil shale and coal can even be gasified underground to provide clean fuels (Mills et al., 1971).

Any well-stocked engineering library will contain a wealth of books, journals, pamphlets, and monographs on the sampling, continuous-monitoring, and control of industrial particulate and gaseous emissions (see, for example, Strauss, 1971; Faith and Atkisson, 1972). Dusts and smokes are produced in a variety of particles sizes (Figure 6.30) by a multitude of human activities (Tables 6.23 and 6.24), and because they are highly visible and thus highly objectionable, a number of techniques are widely used to cope with the problem of their removal, including settling chambers, cyclone separators, filters and baghouses, wet collectors, and various types of scrubbers electrostatic precipitators, and even ultrasonic agglomerators (Table 6.25). Volatile solvents are a problem in a number of industries, and, of course, our old enemy sulfur dioxide recurs again and again, especially in the smelting industry (Tables 6.5, 6.6, and 6.8). Often it is economically feasible to recover the sulfur removed as free sulfur (Table 6.6) or as sulfuric acid, especially in smelter gas where the SO_2 may be as concentrated as 5 to 10%. As for power plants, the emissions after removal of 99% of the flyash by an electrostatic precipitator can be passed over a V_2O_5 catalyst to oxidize as much as 90% of the SO_2 up to SO_3, which then can be converted to H_2SO_4. A H_2SO_4 mist can also form in the electrostatic precipitator itself and seriously

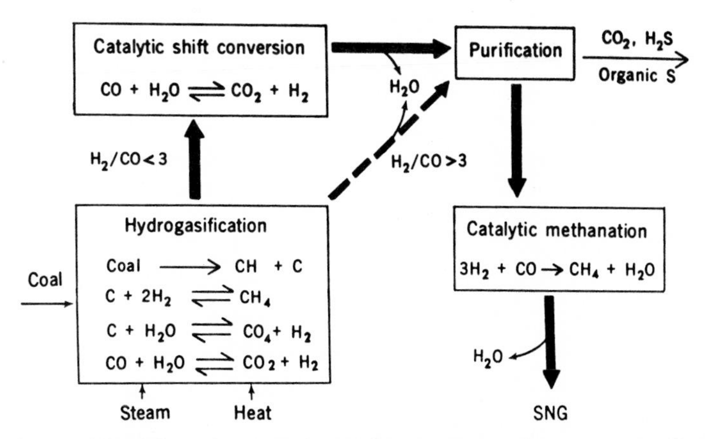

Figure 6.29. A generalized flow chart for the gasification of coal (from Maugh, © 1972 by the American Association for the Advancement of Science with permission of the AAAS).

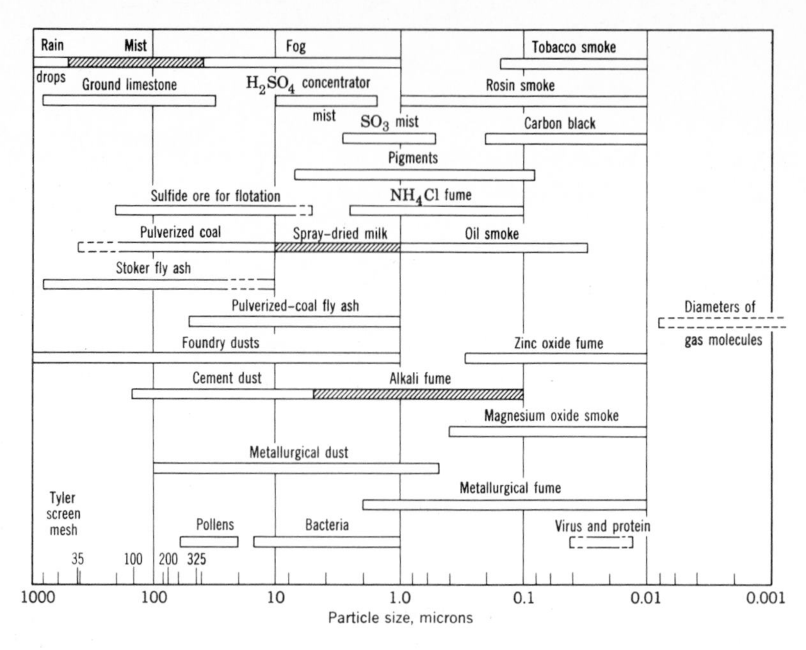

Figure 6.30. Particle size range for aerosols, dusts, and fumes (from Faith and Atkisson, 1972).

Table 6.23 Sources of Atmospheric Dust (From Faith and Atkisson, 1972)

Combustion	Materials Handling and Processing	Earth Moving	Miscellaneous
Fuel burning Incineration Open fires Burning dumps Forest fires	Loading and unloading (sand, gravel, ores, coal, lime, bulk chemicals) Mixing and packaging (fertilizers, chemicals, feed) Crushing and grinding (ores, gravel, chemicals, cement) Food processing (milling, e.g., flour, cornstarch; drying; handling grain) Cutting and forming (sawmills, wallboard, plastics, etc.) Manufacturing and processing solids (cement, chemicals, carbon black) Metallurgical (smelters, blast furnaces, foundries)	Construction (roads, dams, buildings, site clearance) Mining (blasting, sorting, refuse disposal) Agricultural operations (land preparation, soil tilling) Natural (winds)	House cleaning Sand blasting Crop spraying Poultry feeding Rubber-tire abrasion Engine exhaust

Table 6.24 Particulate Emission Factors for Selected Uncontrolled Sources (From Faith and Atkisson, 1972)

Source	Rate
Fuel combustion	
General (pulverized)	(16 × % ash in coal)/ton of coal burned
Solid waste disposal	
Open burning dump	16 lb/ton of refuse burned
Municipal incinerator	17 lb/ton of refuse burned
Single-chamber incinerator	10 lb/ton of refuse burned
Multiple-chamber incinerator	3 lb/ton of refuse burned
Flue-fed incinerator	28 lb/ton of refuse burned
Chemical industry	
Sulfuric acid manufacturers	0.3–7.5 lb/ton of acid produced
Food and agricultural industries	
Direct-fired coffee roaster	7.6 lb/ton of green beans
Cotton-ginning	11.7 lb/500-lb bale of cotton
Starch flash drier	8 lb/ton of starch
Primary metal industry	
Iron and steel manufacture	
Sinter plant gases	20 lb/ton of sinter
Open-hearth furnace	
Oxygen lance	22 lb/ton of steel
No oxygen lance	14 lb/ton of steel
Blast furnace	
Ore charging	110 lb/ton of iron
Agglomerate charging	40 lb/ton of iron
Coking	2 lb/ton of coal
Secondary metal industry	
Brass and bronze smelting	
Reverberatory furnace	26.3 lb/ton of metal charged
Gray iron foundry	
Cupola	17.4 lb/ton of metal charged
Lead smelting	
Cupola	300 lb/ton of metal charged
Mineral product industry	
Asphalt batch plant drier	5.0 lb/ton of mix
Cement manufacture	
Dry process kiln	46 lb/barrel of cement
Wet process kiln	38 lb/barrel of cement
Concrete batch plant	0.2 lb/yard of concrete
Lime production	
Rotary kiln	200 lb/ton of lime
Vertical kiln	20 lb/ton of lime
Rock, gravel, and sand production	
Crushing	20 lb/ton of product
Conveying, screening, shaking	1.7 lb/ton of product
Petroleum industry	
Fluid catalytic cracker	0.1–0.2 lb/ton of catalyst circulated
Kraft pulp industry	
Smelt tank (uncontrolled)	20 lb/ton of dry pulp
Lime kiln	94 lb/ton of dry pulp
Recovery furnace with primary	
Stack gas scrubber	150 lb/ton of dry pulp

Table 6.25 Collection Equipment—Solid and Liquid Aerosols (From Faith and Atkisson, 1972)

Type	Dust Characteristics Type	Particle size (μm)	Specific Gravity	Pressure drop (in. H_2O)	Advantages	Disadvantages	Percent Efficiency (wt. basis)
CENTRIFUGAL COLLECTORS							
Simple Cyclone	Wood dust	50–1000	0.4–0.7	0.5–2.0	Simple in construction	Low efficiency	70–90
	Grain dust	10–200	0.9–1.1				60–80
	Mineral dust	10–500	2.0–3.0				70–90
	Pulverized chemicals	10–500	1.5–3.0				70–90
High-Efficiency Cyclone	Catalyst dust	2–80	1.5–3.5	2.0–6.0	Relatively high efficiency	Subject to abrasion damage	65–80
	Flyash	0.1–100	0.4–1.5				50–70
	Other fine dust	5–200	1.0–3.0				85–98
Impeller	Foundry dust	10–300	2.5–4.0	Acts as own fan	Low-space requirement	Impeller abrasion, causing unbalance	70–90
ELECTROSTATIC PRECIPITATORS							
Single-Stage	Gray iron	0.5–50	3–6	0.25–0.5	High efficiency under severe conditions	High initial cost, operating difficulties	90–97
	Cupola fume	0.1–20	5–7				
	Electric steel Furnace fume	0.1–20	5–7				90–97
	Open hearth steel Furnace fume	0.1–3	5–7				96–99
	Catalyst dust	2–80	1.5–3.5				85–98
Two-Stage	Oil mist	10–400	~1	0.25–0.5	High-efficiency for low-dust loading, safe in operation	Limited in use	85–99
	Air conditioning	0.2–10	—	—			95–99

			CLOTH FILTERS				
Tubular	Metallurgical fume			0.5–6.0	High efficiency over wide-particle size range	Caking from moisture	
	Nonferrous	0.03–1.0	~5				98–99.5
	Ferrous	0.10–50.0	3–6				97–99.5
Screen or Frame	Ceramic dusts	1–50	1–3	0.5–4.0	Somewhat self-cleaning	Higher stresses on filter mediums	95–99.0
	Metallurgical fume	0.1–50	3–6				94–97
Reverse Flow (standard cloths)	Same as screen type shown above			0.5–3.0	Higher dust loadings possible	—	—
						—	—
Reverse Jet (felt mediums)	Carbon black	0.1–10	1.5	1.0–6.0	High-filter ratios possible	Bag wear	99.5
	Flour dust	5–100	0.9				99.9
			WET COLLECTORS				
Spray Chamber	Rock dust	40–500	2–3	0.5–1.0	Low-pressure drop	High-nozzle pressure required for good collection	60–75
	Asphalt mist	10–400	~1				70–80
	Acid mist	20–500	1.1–1.3				70–90
Inertial	Al and Mg Grinding dust	50–1000	1.7–2.8	2.0–4.0	No nozzle maintenance	Higher-pressure drop	80–95
	Foundry dust	10–300	2.5–4.0				70–90
Centrifugal Spray	Rock and sand Dust	20–500	2–3	1.0–4.0	Combined scrubbing and centrifugal action	Abrasion	75–95
	Acid and caustic mist	20–500	1.1–1.3				75–95
Venturi Scrubber	Sulfuric acid Mist from concentrator	2–10	~1.5	10.0–15.0	High-efficiency, low-water rate	High-power consumption	85–95
	Chemical fume	0.1–50	1.5–3.5				60–85

reduce precipitator efficiency (Matteson et al., 1972). More commonly the SO_2 from power plants is removed by some inexpensive basic substance such as chalk, limestone, dolomites, or oxides of calcium or magnesium, either as a slurry (Figure 6.31) or by direct dry injection (Murthi et al., 1971). The sulfur ends up as a calcium or magnesium sulfite or sulfate, which then has to be dumped—either on land or in the closest river.

Novel solutions to our air pollution problems are being proposed ranging from zinc-chlorine-powered automobiles to home fuel cells. While imagination is certainly welcome, and determination even more so, at the risk of sounding cynical, we probably shall be stuck basically with the partial solutions at hand for the foreseeable future. In addition to fusion power, another ray of hope on the horizon is the use of hydrogen as a primary fuel. This fills me with a certain uneasiness for I have not forgotten the Principle of LeChatlier, I have not forgotten the continued growth of this planet's population, and I have not forgotten the enormous potential danger

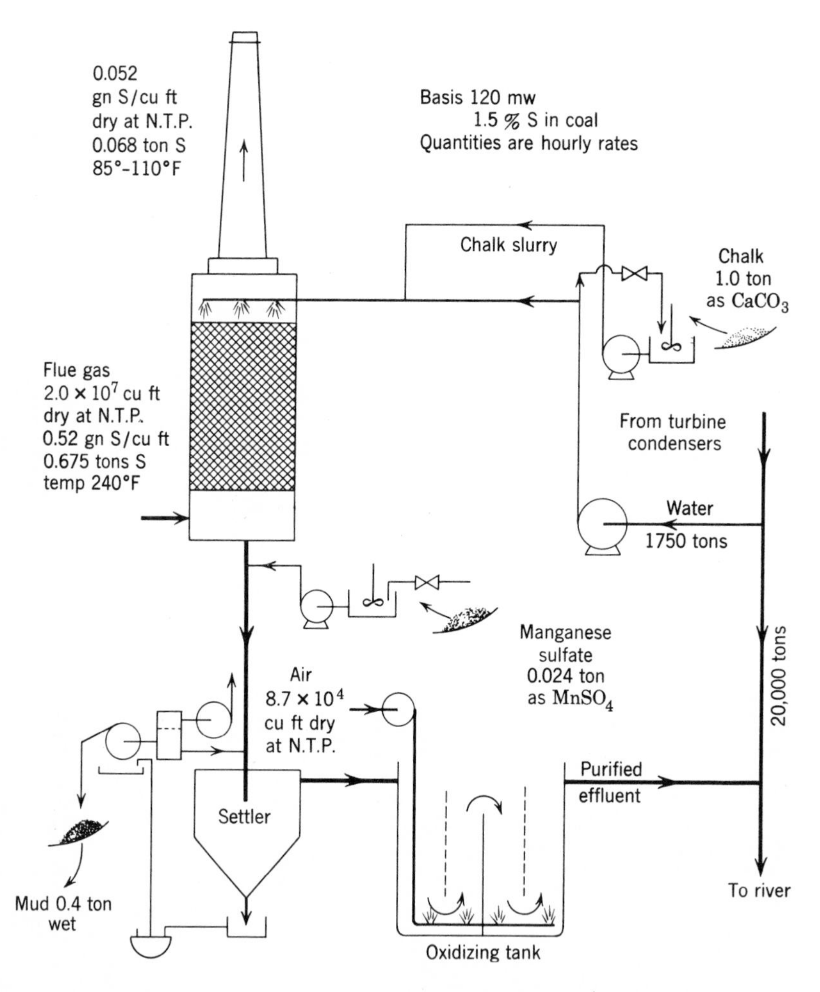

Figure 6.31. Typical arrangement of modified Battersea effluent process (from Faith and Atkisson, 1972).

implied in hydrogen from water in man's tampering with the most important chemical transport system in our environment—the hydrologic cycle.

In the United States, strict air pollution control laws have been inacted and strict goals set. There is at present and there will be increasing pressures to ignore and/or weaken these laws. Resist! And be willing to pick up the bill.

The anti-intellectuals have suggested a "posttechnological culture" as a panacea, but historians have given such catastrophic setbacks of civilizations a far more appropriate name—The Dark Ages. No society needs a lobectomy. The problems created by technology are insignificant compared to those created by its absence.

BIBLIOGRAPHY

F. B. Abeles, et al., *Science*, *173*, 914 (1971).

L. H. Allen, Jr., et al., *Science*, *173*, 256 (1971).

A. P. Altshuller, et al., *Environ. Sci. Tech.*, *4*, 44 (1970).

A. P. Altshuller et al., *Environ. Sci. Tech.*, *5*, 1009 (1971).

A. P. Altshuller and J. J. Bufalini, *Environ. Sci. Tech.*, *5*, 39 (1971).

Anon., *Chem. Engr. News*, 57 (May 16, 1966).

Anon., *First Europ. Congr. Influence Pollution Plants and Animals*, Centre Agricul. Pub. Doc., Wageningen, The Netherlands, 1969*a*.

Anon., *Time*, 51 (Dec. 19, 1969*b*).

Anon., *Environ. Sci. Tech.*, *4*, 635 (1970*a*).

Anon., *Carbon Monoxide*, Nat. Acad. Sci. U.S.A., Washington, D.C., 1970*b*.

Anon., *Environ. Sci. Tech.*, *4*, 718 (1970*c*).

Anon., *Fluorides*, Nat. Res. Council, Div. Med. Sci., Nat. Acad. Sci. U.S.A., Washington, D.C., 1971*a*.

Anon., *Nature*, *233*, 44 (1971*b*).

Anon., *Environ. Sci. Tech.*, *5*, 584 (1971*c*).

Anon., *Beryllium and Air Pollution: An Annotated Bibliography*, Supt. Doc., U.S. Gov. Print. Off., Washington, D.C., 1971*d*.

L. P. Argenbright and B. Preble, *Environ. Sci. Tech.*, *4*, 554 (1970).

W. M. Ashton, *Nature*, *237*, 46 (1972).

G. Assaf and P. E. Biscaye, *Science*, *175*, 890 (1972).

D. H. F. Atkins et al., *Nature*, *235*, 372 (1972).

Atomic Energy Commission U.S., *Precipitation Scavenging*, Nat. Tech. Inform. Serv., Springfield, Va., 1970.

H. D. Axelrod et al., *Environ. Sci. Tech.*, *5*, 420 (1971).

P. Barry, *Brit. J. Ind. Med.*, *27*, 339 (1970).

P. Barry and D. B. Mossman, *Brit. J. Ind. Med.*, *27*, 339 (1970).

D. R. Bates and P. B. Hayes, *Planet. Space Sci.*, *15*, 189 (1967).

D. R. Bates and A. E. Witherspoon, *Mo. Not. Roy. Astron. Soc.*, *112*, 101 (1952).

R. J. Bazell, *Science*, *173*, 130 (1971).

R. J. Bazell, *Science*, *174*, 574 (1971).

D. Behrman, *Realites*, *42* (June 1972).

R. A. Berner, *J. Geophys. Res.*, *76*, 6597 (1971).
R. S. Berry and P. A. Lehman, *Ann. Rev. Phys. Chem.*, *22*, 47 (1971).
K. K. Bertine and E. D. Goldberg, *Science*, *173*, 233 (1971).
C. E. Billings and W. R. Matson, *Science*, *176*, 1232 (1972).
J. L. Bove and S. Siebenberg, *Science*, *167*, 986 (1970).
J. R. Brar et al., *Environ. Sci. Tech.*, *4*, 50 (1970).
W. S. Broecker, *Science* *168* 1537 (1970).
W. S. Broecker et al. in M. Sears (ed.) *Oceanography*, Amer. Assoc. Adv. Sci., Pub. No. 67, Washington, D.C., 1961.
M. Bufalini, *Environ. Sci. Tech.*, *5*, 685 (1971).
A. Burkitt et al., *Nature*, *238*, 327 (1972).
J. D. Burton, in J. P. Riley and G. Skirrow (eds.), *Chemical Oceanography*, Academic, London, 1965.
W. M. Burton and N. G. Steward, *Nature*, *186*, 584 (1960).
R. D. Cadle and E. R. Allen, *Science*, *167*, 243 (1970).
R. D. Cadle and J. W. Powers, *Tellus*, *18*, 176 (1966).
R. D. Cadle et al., *J. Atmos. Sci.*, *25*, 100 (1968).
R. D. Cadle et al., *Geochim. Cosmochim. Acta*, *35*, 503 (1971).
J. G. Calvert et al., *Science*, *175*, 751 (1972).
H. L. Cannon and F. M. Bowles, *Science*, *137*, 765 (1962).
M. J. Cambell, *Nature*, *236*, 26 (1972).
M. D. Carabine, *Chem. Soc. Rev.*, *1*, 411 (1972).
A. W. Castleman and I. N. Tang, *Nature Phys. Sci.*, *234*, 129 (1971).
A. W. Castleman et al., *Science*, *173*, 1025 (1971).
K. Y. Chen and J. C. Morris, *Environ. Sci. Tech.*, *6*, 529 (1972).
J. J. Chisolm, Jr., *Sci. Amer.*, *224*(2), 15 (February 1971).
J. Cholak, *Arch. Environ. Health*, *8*, 314 (1964).
T. J. Chow and J. L. Earl, *Science*, *169*, 577 (1970).
T. J. Chow and J. L. Earl, *Science*, *176*, 510 (1972).
T. J. Chow and C. C. Patterson, *Earth Planet. Sci. Lett.*, *1*, 397 (1966).
T. J. Chow et al., *Science*, *178*, 401 (1972).
I. D. Clark and J. F. Noxon, *Science*, *174*, 941 (1971).
B. R. Clemesha and Y. Nakamura, *Nature*, *237*, 328 (1972).
J. M. Colucci and C. R. Begeman, *Environ. Sci. Tech.*, *5*, 145 (1971).
J. M. Colucci et al., *J. Air Pollut. Control. Assoc.*, *19*, 255 (1969).
M. Corn et al., *Environ. Sci. Tech.*, *5*, 155 (1971).
J. Cramer and A. L. Myers, *Atmos. Environ.*, *6*, 563 (1972).
J. F. Cronin, *Science*, *172*, 847 (1971).
W. H. Crouse, *Automotive Emission Control*, McGraw-Hill, New York, 1971.
R. H. Danes et al., *Environ. Sci. Tech.*, *4*, 318 (1970).
G. M. Daniels, *Astronaut. Aeronaut.*, *8*, 22 (1970).
R. Dedolph et al., *Environ. Sci. Tech.*, *4*, 217 (1970).
E. M. Dickson, *Science*, *177*, 536 (1972).
D. Dimitriades and M. Whisman, *Environ. Sci. Tech.*, *5*, 219 (1971).
C. Durfor and E. Becker, *J. Amer. Water Works Assoc.*, *56*, 237 (1964).

H. Egner and E. Eriksson, *Tellus*, *7*, 134 (1955).
M. Eisenbud and L. R. Ehrlich, *Science*, *176*, 193 (1972).
H. T. Ellis and R. F. Pueschell, *Science*, *172*, 845 (1971).
E. Eriksson, *Tellus*, *11*, 375; *12*, 63 (1959).
A. Eshleman et al., *Nature*, *233*, 471 (1971).
W. L. Faith and A. A. Atkisson, Jr., *Air Pollution*, Wiley-Interscience, New York, 1972.
G. J. Fergusson, *Proc. Roy. Soc.* (*London*) *Ser. A*, *243*, 561 (1958).
W. H. Fischer, *Science*, *171*, 828 (1971).
B. R. Fish, *Science*, *175*, 1239 (1972).
H. W. Ford et al., *J. Chem. Phys.*, *26*, 1337 (1957).
R. S. Foote, *Science*, *177*, 513 (1972).
C. W. Francis et al., *Environ. Sci. Tech.*, *4*, 586 (1970).
J. P. Friend, *Science*, *175*, 1278 (1972).
B. W. Gay, Jr. and J. J. Bufalini, *Environ. Sci. Tech.*, *5*, 422 (1971).
S. C. Gilfillan, *J. Occup. Health*, *7*, 53 (1965).
R. Gillette, *Science*, *174*, 800 (1971).
M. Glasser and L. Greenburg, *Arch. Envir. Health*, *22*. 334 (1971).
W. A. Glasson and C. S. Tuesday, *Environ. Sci. Tech.*, *4*, 37 (1970).
J. Golden and T. R. Mongan, *Science*, *171*, 381 (1971).
V. M. Goldschmidt, *J. Chem. Soc.*, 655 (1937).
J. R. Goldsmith and A. C. Hexter, *Science*, *158*, 132 (1967).
L. J. Goldwater, *Arch. Environ. Health*, *15*, 60 (1967).
L. J. Goldwater, *Ind. Med.*, *41*, 13 (1972).
L. J. Goldwater and A. W. Hoover, *Arch. Environ. Health*, *15*, 60 (1967).
B. Gregor, *Science*, *174*, 316 (1971).
C. D. Gregor, *Environ. Sci. Tech.*, *6*, 278 (1972).
D. C. Grey and M. L. Jensen, *Science*, *177*, 1099 (1972).
K. Habibi, *Environ. Sci. Tech.*, *4*, 239 (1970).
W. Haenzel et al., *J. Nat. Cancer Inst.*, *28*, 947 (1962).
S. K. Hall, *Environ. Sci. Tech.*, *6*, 31 (1972).
W. L. Hamilton and T. A. Seliga, *Nature*, *235*, 320 (1972).
P. B. Hammond, *Toxic. Appl. Pharm.*, *18*, 296 (1971).
Health, Education, and Welfare, *Air Quality Criteria for Sulfur Oxides*, Nat. Air Pollut. Control Adm. Pub. No. AP-50 (January 1969).
G. H. Heichel and L. Hankin, *Environ. Sci. Tech.*, *6*, 1121 (1972).
K. Heidel, *Science*, *177*, 882 (1972).
C. L. Hemenway et al., *Nature*, *238*, 256 (1972).
H. W. Hermance et al., *Environ. Sci. Tech.*, *5*, 781 (1971).
J. B. Heywood, *Tech. Rev.* (*M.I.T.*), *73*(8), 21 (June 1971).
A. C. Hexter and J. R. Goldsmith, *Science*, *172*, 265 (1971).
E. D. Hinkley and P. L. Kelley, *Science*, *171*, 635 (1971).
D. A. Hirschler et al., *Ind. Engr. Chem.*, *49*, 1131 (1957).
R. A. Horne, *Marine Chemistry*, Wiley-Interscience, New York, 1969.
R. A. Horne, *Science*, *177*, 1152 (1972).
R. E. Inman et al., *Science*, *172*, 1229 (1971).

J. S. Jacobson and A. C. Hill (eds.), *Recognition of Air Pollution Injury to Vegetation*, Air Pollut. Control Assoc., Pittsburgh, 1970.

O. I. Joensuu, *Science, 172*, 1027 (1971).

A. Johannessen et al., *Nature, 235*, 215 (1972).

M. K. John, *Environ. Sci. Tech., 5*, 1199 (1971).

M. K. John et al., *Environ. Sci. Tech., 6*, 555 (1972).

N. M. Johnson et al., *Science, 177*, 514 (1972).

H. Johnston, *Science, 173*, 517 (1971).

J. H. Jones et al., *Environ. Sci. Tech., 5*, 790 (1971).

P. W. Jones and A. H. Adelman, *Environ. Sci. Tech., 6*, 933 (1972).

C. E. Junge, *Air Chemistry and Radioactivity*, Academic, New York, 1963.

A. Kaldor et al., *Science, 176*, 508 (1972).

W. C. Kaufman et al., *Aerospace Med., 41*, 591 (1970).

R. A. Kehoe, *J. Roy. Inst. Pub. Health Hygen., 24*, 1, 101, 129, 177 (1961).

R. A. Kehoe, *Science, 159*, 1000 (1968).

W. W. Kellogg et al., *Science, 175*, 587 (1972).

D. H. Klein, *Environ. Sci. Tech., 6*, 560 (1972).

R. L. Klimisch and G. J. Barnes, *Environ. Sci. Tech., 6*, 543 (1972).

A. Kloke and K. Riebartsch, *Naturwissenschaft, 51*, 367 (1964).

W. D. Komhyr et al., *Nature, 232*, 390 (1971).

T. Koyama et al., *J. Earth Sci., 13*, 1 (1965).

K. B. Krauskopf, *Introduction to Geochemistry*, McGraw-Hill, New York, 1967.

D. W. Kretchman and F. S. Howlett, *Trans. Amer. Soc. Agric. Engr., 13*, 252 (1970).

L. B. Kreuzer et al., *Science, 177*, 347 (1972).

R. H. Kummler, *Environ. Sci. Tech., 5*, 1140 (1971).

D. Lal, in *Radioactive Dating*, Internat. Atomic Energy Agency, Vienna, 1967.

H. E. Laudsberg, *Science, 170*, 1265 (1970).

R. I. Larsen and V. J. Konopinski, *Arch. Environ. Health, 5*, 597 (1962).

J. L. Laseter and R. Valle, *Environ. Sci. Tech., 5*, 631 (1971).

L. B. Lave and E. P. Seskin, *Science, 169*, 723 (1970).

A. L. Lazrus et al., *Environ. Sci. Tech., 4*, 55 (1970).

D. H. K. Lee (ed.), *Environmental Factors in Respiratory Disease*, Academic Press, New York, 1970.

D. H. K. Lee, *Metallic Contaminants and Human Health*, Academic, New York, 1972*a*.

J. A. Lee, *Nature, 238*, 165 (1972).

R. E. Lee, Jr., *Science, 178*, 567 (1972).

R. F. Lee et al., *Environ. Sci. Tech., 2*, 288 (1968).

J. Lekven, *Nature, 235*, 284 (1972).

H. Levy, *Science, 173*, 141 (1971).

W. F. Libby, *Science, 171*, 499 (1971).

J. E. Lovelock et al., *Nature, 237*, 452 (1972).

D. A. Lundgren, *J. Air Poll. Control Assoc., 20*, 603 (1970).

L. Machta and E. Hughes, *Science, 168*, 9582 (1970).

P. L. Magill, F. R. Holden, and C. Ackley (eds.), *Air Pollution Handbook*, McGraw-Hill, New York, 1956.

J. D. Martinez et al., *Nature*, *233*, 210 (1971).

M. J. Matteson, *Environ. Sci. Tech.*, *6*, 895 (1972).

W. H. Matthews, W. W. Kellogg, and G. D. Robinson (eds.), *Man's Impact on Climate*, M.I.T. Press, Cambridge, 1971.

T. H. Maugh, *Science*, *177*, 338 (1972*a*).

T. H. Maugh, *Science*, *177*, 685 (1972*b*).

T. H. Maugh, *Science*, *177*, 1091 (1972*c*).

T. H. Maugh, *Science*, *178*, 44 (1972*d*).

M. Mauozumi et al., *Geochim. Cosmochim. Acta*, *33*, 1247 (1969).

J. C. McConnell et al., *Nature*, *233*, 187 (1971).

P. McEachern et al., *Environ. Sci. Tech.*, *5*, 700 (1971).

M. S. McIntire and C. R. Angle, *Science*, *177*, 520 (1972).

R. A. Meyers et al., *Science*, *177*, 1187 (1972).

G. A. Mills et al., *Environ. Sci. Tech.*, *5*, 30 (1971).

Y. Miyake, in M. N. Hill (ed.), *The Sea*, Vol. 2, Interscience, New York, 1963.

N. L. Morrow and R. S. Brief, *Environ. Sci. Tech.*, *5*, 786 (1971).

H. L. Motto et al., *Environ. Sci. Tech.*, *4*, 231 (1970).

J. L. Moyers et al., *Environ. Sci. Tech.*, *6*, 68 (1972).

R. A. Mumford et al., *Environ. Sci. Tech.*, *6*, 427 (1972).

K. S. Murthi et al., *Environ. Sci. Tech.*, *5*, 776 (1971).

R. A. Naden and J. V. Leeds, *Environ. Sci. Tech.*, *5*, 522 (1971).

Nat. Acad. Sci. U.S.A., *Airborne Lead in Perspective*, U.S. Gov. Print. Off., Washington, D.C., 1971.

R. Nydal et al., *Nature*, *232*, 418 (1971).

L. Ordin et al., *Environ. Sci. Tech.*, *5*, 621 (1971).

H. G. Ostlund and J. Alexander, *J. Geophys. Res.*, *68*, 3995 (1963).

M. R. Pack, *Environ. Sci. Tech.*, *5*, 1128 (1971).

M. Pandow et al., *J. Inorg. Nucl. Chem.*, *14*, 153 (1960).

C. C. Patterson, *Arch. Environ. Health*, *11*, 344 (1965).

C. C. Patterson, in D. Hood (ed.), *Impingement of Man on the Oceans*, Wiley, New York, 1971.

L. Pauling, *Science*, *128*, 3333 (1958).

L. A. Pedersen and W. F. Libby, *Science*, *176*, 1355 (1972).

D. H. Pierson, *Nature*, *234*, 79 (1971).

E. E. Picciotto, in M. Sears (ed.), *Oceanography*, Amer. Assoc. Adv. Sci., Pub. No. 67, Washington, D.C., 1961.

J. Pichirallo, *Science*, *173*, 509 (1971).

J. M. Pierrard, *Science*, *175*, 615 (1972).

Public Health Service, *Nationwide Inventory of Air Pollution Emissions*, U.S. Dept. HEW, Raleigh, N.C., 1968.

M. B. Rabinowitz and G. W. Wetherill, *Environ. Sci. Tech.*, *6*, 705 (1972).

R. A. Rasmussen and F. W. Went, *Proc. Nat. Acad. Sci. U.S.A.*, *53*, 215 (1965).

S. I. Rasool and S. H. Schneider, *Science*, *173*, 138 (1971).

J. R. Rhodes et al., *Environ. Sci. Tech.*, *6*, 922 (1972).

A. L. Richards and D. V. Badami, *Nature*, *234*, 43 (1971).

L. A. Ripperton et al., *Environ. Sci. Tech.*, *5*, 246 (1971).

J. A. Robbins and F. L. Snitz, *Environ. Sci. Tech.*, *6*, 164 (1972).
E. Robinson and F. L. Ludwig, *J. Air Pollut. Control Assoc.*, *17*, 664 (1967).
H. Rodhe, *Tellus*, *24*, 128 (1972).
H. H. Rossi and A. M. Kellerer, *Science*, *175*, 200 (1972).
V. J. Schaefer et al., *Science*, *175*, 173 (1972).
J. Schroder et al., *Nature*, *233*, 614 (1971).
H. A. Schroeder and J. J. Balassa, *J. Chronic Dis.*, *14*, 408 (1961).
K. Schütz et al., *J. Geophys. Res.*, *75*, 2230 (1970).
W. Seiler and C. Junge, *J. Geophys. Res.*, *75*, 2217 (1970).
D. Shapley, *Science*, *174*, 569 (1971).
E. S. Shepherd, *Amer. J. Sci.*, *235A*, 311 (1938).
R. Shinnar, *Science*, *175*, 1357 (1972).
B. Shleien et al., *Environ. Sci. Tech.*, *4*, 598 (1970).
H. W. Sidebottom et al., *Environ. Sci. Tech.*, *6*, 72 (1972).
S. F. Singer (ed.), *Global Effects of Environmental Pollution*, Springer-Verlag, New York, 1970.
S. F. Singer, *Nature*, *233*, 543 (1971).
W. H. Smith, *Science*, *176*, 1237 (1972).
L. L. Sorensen and K. Nobe, *Environ. Sci. Tech.*, *6*, 239 (1972).
C. W. Stanley et al., *Environ. Sci. Tech. 5*, 430 (1971).
E. R. Stephens and M. A. Price, *Science*, *168*, 1584 (1970).
C. A. Stewart, Jr., *Science*,, *177*, 724 (1972).
P. Stocks and J. M. Campbell, *Brit. Med. J.*, *2*, 923 (1955).
W. Strauss (ed.), *Air Pollution Control*, New York, 1971.
H. E. Suess, *Science*, *122*, 415 (1955).
H. E. Suess, *Science*, *163*, 1405 (1969).
D. J. Swain and R. L. Mitchell, *J. Soil Sci.*, *11*, 347 (1960).
J. W. Swinnerton et al., *Science*, *167*, 984 (1970).
J. W. Swinnerton et al., *Science*, *172*, 943 (1971).
H. Tanaka et al., *Water, Air, Soil Pollut.*, *1*(2), (1972).
M. Tatsumoto and C. C. Patterson, *Earth Sciences and Meteorites*, North Holland, Amsterdam, 1963.
G. Ter Haar, *Environ. Sci. Tech.*, *4*, 226 (1970).
G. Ter Haar and M. A. Bayard, *Nature*, *232*, 553 (1971).
G. Ter Haar et al., *Nature*, *216*, 353 (1967).
G. Ter Haar et al., *Environ. Res.*, *2*, 267 (1969).
C. R. Thompson et al., *Environ. Sci. Tech.*, *5*, 1017 (1971).
J. A. Thompson, *Brit. J. Ind. Med.*, *28*, 189 (1971).
T. G. Tornabene and H. W. Edwards, *Science*, *176*, 1334 (1972).
E. C. Tsivoglou, *Environ. Sci. Tech.*, *5*, 404 (1971).
P. Urone and W. H. Schroeder, *Environ. Sci. Tech.*, *3*, 436 (1969).
P. Urone et al., *Environ. Sci. Tech.*, *2*, 611 (1968).
L. van Valen, *Science*, *171*, 439 (1971).
P. E. Viemeister, *J. Meteorol.*, *17*, 681 (1960).
R. J. H. Voorhoeve et al., *Science*, *177*, 353 (1972).

B. Weinstock, *Science*, *176*, 290 (1972).

B. Weinstock and H. Niki, *Science*, *176*, 290 (1972).

D. Weiss et al., *Science*, *178*, 69 (1972).

H. V. Weiss et al., *Science*, *172*, 261 (1971*a*).

H. V. Weiss et al., *Science*, *174*, 692 (1971*b*).

F. W. Went, *Proc. Nat. Acad. Sci. U.S.A.*, *46*, 212 (1960).

K. Westberg et al., *Science*, *171*, 1013 (1971).

A. A. Westenberg, *Science*, *177*, 255 (1972).

J. M. White and D. R. Harvey, *Nature*, *236*, 71 (1972).

G. L. Wick, *Science*, *172*, 1222 (1971).

D. F. Wilson et al., *Science*, *168*, 1577 (1970).

R. Wilson, *Science*, *178*, 182 (1972).

W. E. Wilson, Jr. and A. Levy, *J. Air Pollut. Control Assoc.*, *20*, 385 (1970).

W. E. Wilson, Jr. et al., *Environ. Sci. Tech.*, *6*, 423 (1972).

W. Winkelstein, Jr. et al., *Arch. Envir. Health*, *4*, 162 (1967).

H. Windom et al., *Environ. Sci. Tech.*, *1*, 923 (1967).

G. M. Woodwell et al., *Science*, *174*, 1101 (1971).

G. M. Woodwell et al., *Symp. Carbon Biosphere*, Brookhaven Nat. Lab., Upton, N.Y., 1972.

E. L. Wong and A. E. Potter, Jr., *J. Chem. Phys.*, *39*, 2211 (1963).

ADDITIONAL READING

Anon., *Air Quality and Lead*, Amer. Petrol. Inst., New York, 1972.

Anon., *Sources, Abundance, and Fate of Gaseous Atmospheric Pollutants*, Amer. Petrol. Inst., New York, 1969.

Anon., *Hydrocarbons and Air Pollution: An Annotated Bibliography*, Superint. Doc., U.S. Gov. Print. Off., Washington, D.C., 1970.

Anon., *Particulate, Polycyclic Organic Matter*, Nat. Res. Council–Nat. Acad. Sci. U.S.A., Washington, D.C., 1972.

R. U. Ayres and R. P. McKenna, *Alternatives to the Internal Combustion Engine*, Johns Hopkins Univ. Press, Baltimore, 1972.

R. S. Cambray, E. M. R. Fisher, W. L. Brooks, and D. H. Peirson, *Radioactive Fallout in Air and Rain*, H.M. Stationary Office, London, 1971.

L. V. Cralley, L. J. Cralley, G. D. Clayton, and J. A. Jurgiel (eds.), *Industrial Environmental Health*, Academic, New York, 1972.

R. H. Daines et al., in N. C. Brady (ed.), *Agriculture and the Quality of Our Environment*, Amer. Assoc. Adv. Sci. Pub. No. 85, Washington, D.C., 1967.

P. B. Downing (ed.), *Air Pollution and the Social Sciences*, Praeger, New York, 1971.

H. M. Englund and W. T. Berry (eds.), *Proc. 2nd Internat. Clean Air Congress*, Academic, New York, 1971.

G. M. Grosvenor (ed.), *As We Live and Breathe: The Challenge of Our Environment*, Nat. Geograph. Soc., Washington, D.C., 1971.

G. H. Hagevik, *Decision-Making in Air Pollution Control*, Praeger, New York, 1970.

G. M. Hidy (ed.), *Aerosols and Atmospheric Chemistry*, Academic, New York, 1972.

P. A. Leighton, *The Photochemistry of Air Pollution*, Academic, New York, 1961.

P. L. Magill, F. R. Holden, and C. Ackley (eds.), *Air Pollution Handbook*, McGraw-Hill, New York, 1956.

G. Mamantov and W. D. Shults (eds.), *Determination of Air Quality*, Plenum, New York, 1972.

National Air Pollution Control Association, *Nitrogen Oxides: An Annotated Bibliography*, U.S. Gov. Print. Off., Washington, D.C., 1970.

R. D. Ross (ed.), *Air Pollution and Industry*, Van Nostrand, New York, 1972.

A. R. Smith, *Air Pollution*, Pergamon, New York, 1966.

E. S. Starkman (ed.), *Combustion-Generated Air Pollution*, Plenum, New York, 1971.

A. C. Stern (ed.), *Air Pollution*, Academic, New York, 1968.

C. S. Tuesday (ed.), *Chemical Reactions in Urban Atmospheres*, American Elsevier, New York, 1971.

V

THE HYDROSPHERE

7

FRESH WATER CHEMISTRY

7.1 INTRODUCTION

Easily the most striking feature of our planet and one conspicuously lacking from its neighbors in space such as Venus, Mars, and the Moon is its extensive hydrosphere. Not only does this enormous quantity of water cover most of the surface of the globe and riddle the continents with lakes and networks of streams and rivers, not only are the highest mountain ranges and the polar regions buried beneath trackless sheets of snow and ice, but moisture also extends up into the atmosphere and water penetrates deeply into the lithosphere (Table 7.1). In fact, as we shall see, the Earth's hydrosphere probably originated from the lithosphere and is in all likelihood still accreting from that source. In this and Chapter 8, we confine our attention to what we have designated in Table 7.1 as the "hydrosphere proper." The role of water in atmosphere, biosphere, lithosphere, and even exosphere is discussed elsewhere in this book.

The importance of the Earth's hydrosphere is so enormous and overwhelming that it is easy to fail to grasp its full significance. The hydrosphere is huge in bulk and extent. It is the medium, catalyst, and/or participant in nearly all of the chemical reactions occurring in our environment, including those of the life processes. Indeed, water is a necessary condition for life. The biosphere, lithosphere, and atmosphere commonly interact with one another via the action of water. Even the peculiar fluid and solvent properties of water endow it with a unique environmental significance for if there is a common coinage which extends throughout the Earthly environment that coin is water. Water is the carrier in many of the great natural cycles that transform energy and carry chemical materials from place to place. And now we have begun to appreciate still another type of role played by water in our environment; it is the principle carrier and repository of man-made pollution.

In the past, the hydrosphere has determined the course of human history. The first great civilizations sprang up along the banks of rivers. Later cultures spread their influence by the seas and oceans. It is no accident of history that the ruins of the famous cities of Africa and Asia are often buried beneath the desert sands or that the fallen metropolises of central America are pock-marked with seasonally dry wells. The hydrosphere has also played more subtle roles in the human experience; it has been theorized that the geographical regions of greatest intellectual stimulation and thus of the most advanced culture move with the regions of cyclonic storms (or atmospheric water exchange). As the press of uncontrolled human fertility makes water more precious, the role of water in determining the fate of men, nations, and civilizations will inevitably become enlarged. Perhaps some future historians, if there are any historians in the future, will write (if anybody can read or write) a history of the human race in terms of water use and abuse.

Table 7.1 The Total Hydrosphere

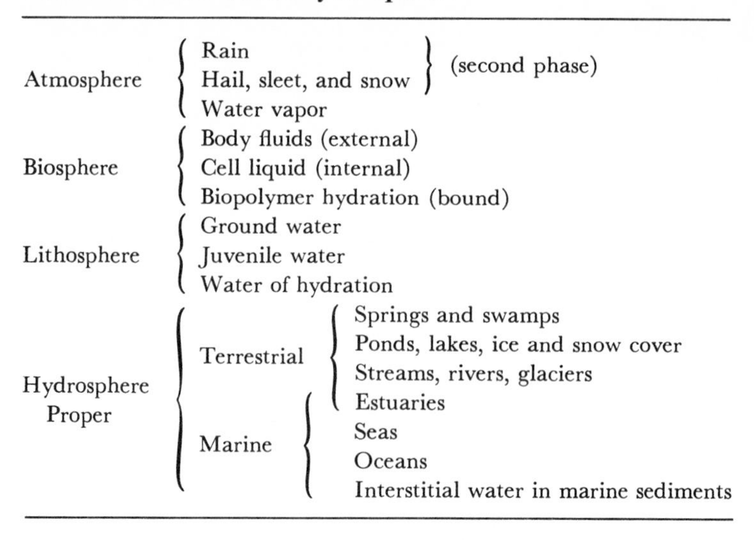

Atmosphere	Rain		(second phase)
	Hail, sleet, and snow		(second phase)
	Water vapor		
Biosphere	Body fluids (external)		
	Cell liquid (internal)		
	Biopolymer hydration (bound)		
Lithosphere	Ground water		
	Juvenile water		
	Water of hydration		
Hydrosphere Proper	Terrestrial	Springs and swamps	
		Ponds, lakes, ice and snow cover	
		Streams, rivers, glaciers	
	Marine	Estuaries	
		Seas	
		Oceans	
		Interstitial water in marine sediments	

7.2 THE STRUCTURE AND PROPERTIES OF LIQUID WATER

The substance of the hydrosphere is water. Water is so familiar a chemical that we tend to take it for granted, and the overwhelming majority of chemists concentrate their attention on more exotic, and far less important, substances. In so doing, they avoid a lot of difficulties—and miss a lot of fun—for water without question is the most extraordinary, the most complex substance known to man. Liquid helium is dull and well-behaved compared to liquid water. Liquid water is anomalous in every single one of its physical–chemical properties. Now water, of course, is not the only

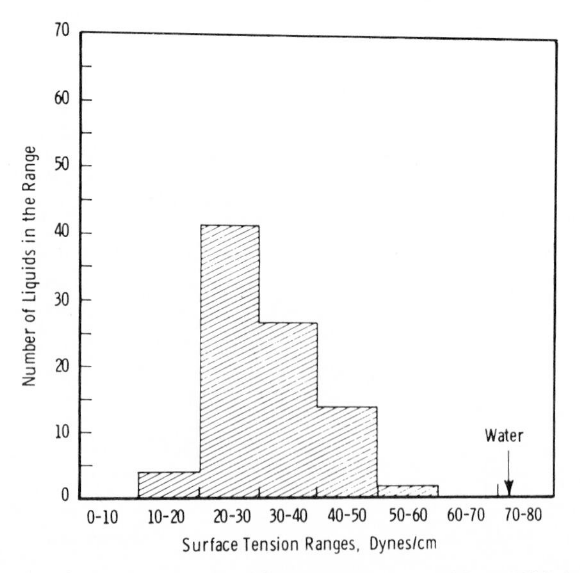

Figure 7.1. Range of surface tension of some 90 representative liquids at 20°C (from Horne, 1969).

strange acting liquid; other liquids share one or a few of its pecularities, but no other liquid has so many eccentricities and, in many cases, to such a degree. Table 7.2, based in part on the classic book *The Oceans* by Sverdrup, Johnson, and Fleming (1942), lists a few of these anomalities together with their environmental significance, while Table 7.3 compares some properties of water with those of *n*-heptane, a "normal" liquid of comparable boiling point. Liquid water is strange in its surface as well as its bulk properties (Table 7.2 and Figure 7.1). Further tables of water properties are listed in the appendix of this book. In molecular terms, the majority of these anomalies all reflect the same water property—namely the ability of water molecules to cling tenaciously to one another and to other substances. The latter, or the ability of water to hydrate other materials and readily form solutions of them, is discussed in Chapter 8. The tendency of water molecules to stick together is nicely illustrated by

Table 7.2 Some of the Anomalous Physical-Chemical Properties of Water and Their Environmental and Biological Significance

Property	Comparison with Normal Liquids	Significance
State	Liquid rather than gas like H_2S, H_2Se, and H_2Te	Provides life media
Heat Capacity	Very high	Moderates environmental temperatures, good heat transport medium
Latent Heat of Fusion	Very high	Moderating effect, tends to stabilize liquid state
Latent Heat of Vaporitation	Very high	Moderation effect, important in atmospheric physics and in precipitation-evaporation balance
Density	Anomalous maximum of 4°C (for pure water)	Freezing from the surface and controls temperature distribution and circulation in bodies of water
Surface Tension	Very high	Important in surface phenomena, droplet formation in the atmosphere, and many physiological processes including transport through biomembranes
Dielectric Constant	Very high	Thus good solvent
Hydration	Very extensive	Thus good solvent and mobilizer of environmental pollutants, alters the biochemistry of solutes
Dissociation	Very small	Provides a neutral medium but with some availability of both H^+ and OH^- ions
Transparency	High	Thickens biologically productive euphatic zone
Heat Conduction	Very high	Can provide an important heat transfer mechanism in stagnant systems such as cells

Table 7.3 Comparison of Physical Properties of Water and *n*-Heptane (From Dietrich, 1963, with permission of John Wiley & Sons)

	Water $(H_2O)_n$	Normal Heptane (C_7H_{16})
Molecular Weight	$(18)_n$	100
Dipole Moment, e.s.n.	1.84×10^{-18}	$>0.2 \times 10^{-18}$
Dielectric Constant	80	1.97
Density (g/cm³)	1.0	0.73
Boiling Point (°C)	100	98.4
Melting Point (°C)	0	−97
Specific Heat (cal/g/°C)	1.0	0.5
Heat of Evaporation (cal/g)	540	76
Melting Heat (cal/g)	79	34
Surface Tension at 20°C (dyne/cm)	73	25
Viscosity at 20°C (poise)	0.01	0.005

the large amount of energy required to melt and evaporate the substance and the high temperatures at which these molecular separation processes occur (Figure 7.2). In Figure 7.2, the transition temperatures of the family of Group VIA (in the periodic table in the Appendix) hydrides, of which water or hydrogen oxide is a member, are plotted versus molecular weight. In going from hydrogen telluride to hydrogen sulfide, the melting and boiling points decrease in an orderly manner with decreasing molecular weight. Then abruptly the curve leaps up to the very high melting and

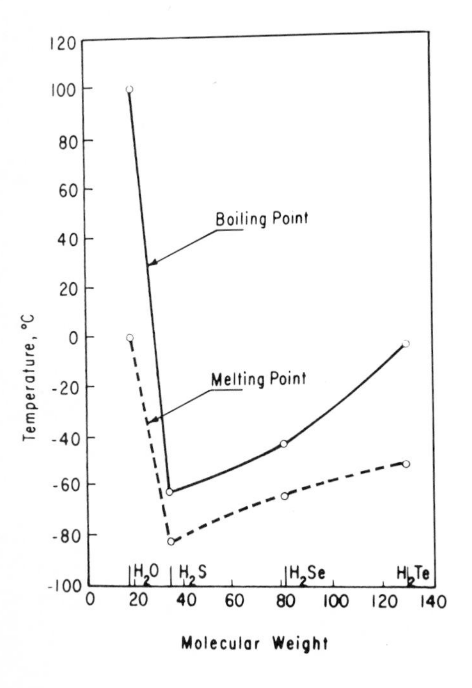

Figure 7.2. Transition temperatures of the group VIA hydrides (from Horne, 1968, with permission of Academic Press, Inc.).

boiling points of water. Clearly some very considerable force is holding the water molecules together and preventing their separation, and now we must turn our attention to the nature of this force. But before we do, reflect a moment. According to Figure 7.2, water should be a gas at all terrestrial temperatures. Just imagine what this planet would be like if such were the case. There would be no oceans, no rivers, no clouds, no oxygen to breathe, no rain, no grass, no trees, no life of any kind, no you, no gently rolling hills but only a desolate, barren, jagged, lunarlike landscape. That will give you some ikling of the importance of water and its peculiar properties.

The "glue" that holds the water molecules together is the so-called hydrogen bond. Water is far from being the only substance capable of forming intermolecular hydrogen bonds; its H bond is not even the strongest (Table 7.4), and many other liquids in addition to water, contrary to what we may have been taught in elementary chemistry, exhibit some structuring. However the *unique* structural property of water appears to be its ability to form extended, three-dimensional, polymorphic, H-bonded structures in the liquid. Figure 7.3 summarizes a few properties and geometrical features of the water molecule and the H-bond in water.

The question of why water H-bonds is best answered by examining the electronic configuration of the water molecule. The water molecule (Figure 7.4) can be represented as a sort of truncated jack contained in a distorted cube. Two arms of the jack contain protons and are thus regions of positive electrification, while the other two arms formed by *p*-electronic orbitals represent regions of negative electrification. The strength of the H-bond in water arises from the interaction of these unlike charged polar regions and is further fortified by several possible resonance configurations:

I: $\underset{a}{H{-}\!:\!\overset{H}{\ddot{\underset{\cdot\cdot}{O}}}\!:}\quad \underset{b}{H{-}\!:\!\overset{H}{\ddot{\underset{\cdot\cdot}{O}}}\!:}$

II: $\underset{a}{H{-}\!:\!\overset{H}{\ddot{\underset{\cdot\cdot}{O}}}\!:}\quad \underset{b}{H^{\oplus}\ :\!\overset{H}{\ddot{\underset{\cdot\cdot}{O}}}\!:^{\ominus}}$

III: $\underset{c}{H{-}\!:\!\overset{H}{\ddot{\underset{\cdot\cdot}{O}}}\!:}\quad \underset{a}{H{-}\!:\!\overset{H}{\ddot{\underset{\cdot\cdot}{O}}}\!:^{\oplus}{-}H}\quad \underset{b}{:\!\overset{H}{\ddot{\underset{\cdot\cdot}{O}}}\!:^{\ominus}}\quad \underset{d}{H{-}\!:\!\overset{H}{\ddot{\underset{\cdot\cdot}{O}}}\!:}$

Table 7.4 Hydrogen-Bond Properties (From Horne, 1969, based on Pauling, 1948)

Bond	Substance	Bond Energy (kcal/mole)	Bond Length (Å)
F—H—F	H_6F_6	6.7	2.26
O—H---O	H_2O (ice)	4.5	2.76
	H_2O_2	4.5	
	Alcohols, (ROH)	6.2	2.70
	$(HCOOH)_2$	7.1	2.67
	$(CH_3COOH)_2$	8.2	—
C—H---N	$(HCN)_2$	3.2	—
	$(HCN)_3$	4.4	—
N—H---N	NH_3	1.3	3.38
N—H---F	NH_4F	5	2.63
O—H---Cl	*o*-C_6H_5OHCl	3.9	—

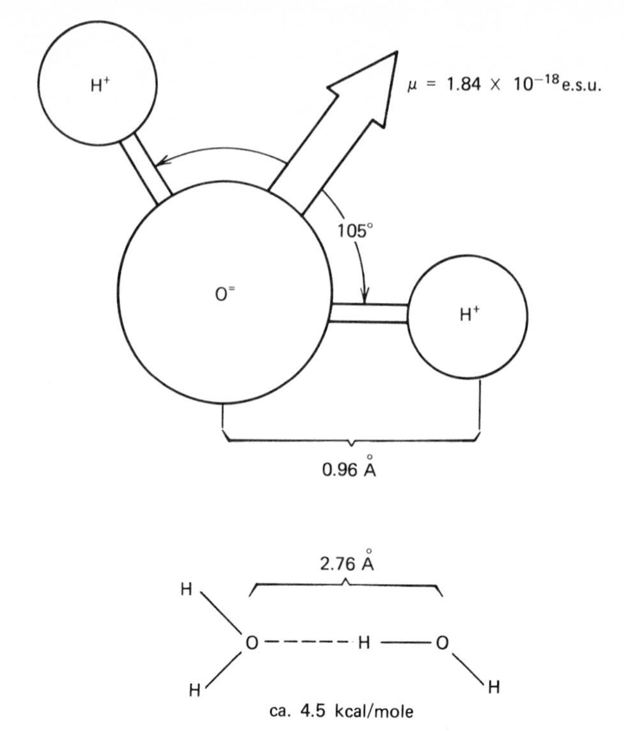

Figure 7.3. Schematic summary of the structural features of the water molecule and the hydrogen bond (from Horne, 1969).

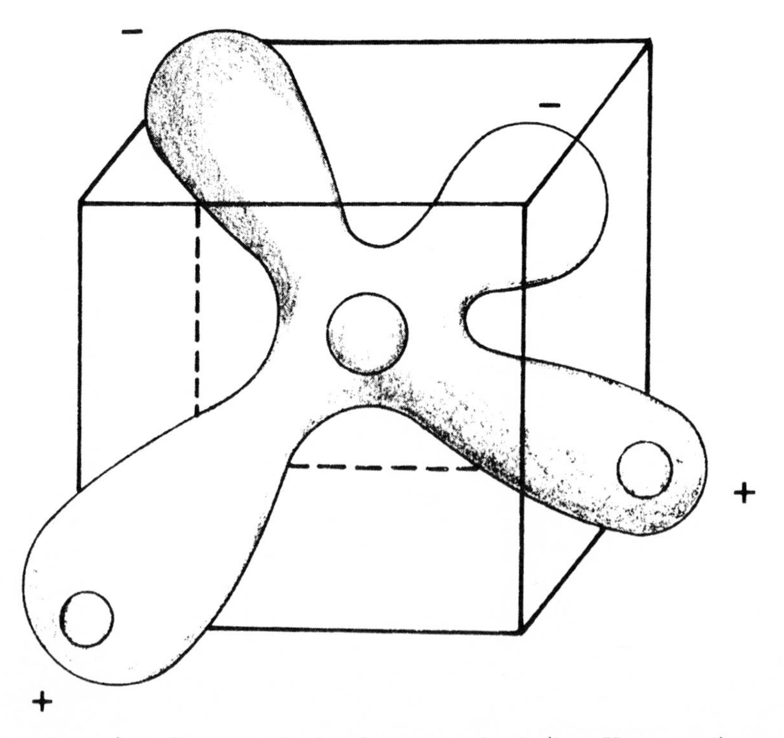

Figure 7.4. Electronic cloud of the water molecule (from Horne, 1969).

On our planet, water is very unusual in that it appears in all three states—gaseous, liquid, and solid—at normally occurring terrestrial temperatures. This important observation is probably one of the principle reasons why ancient Greek scientists such as Thales of Miletus (ca. 600 B.C.) made it their primal element. There are, if you pause to think about it a bit, very, very few naturally occurring substances, except for water, mercury, and petroleum, which are liquid at terrestrial temperatures. In the gaseous state, the water molecules are so excited and widely spaced that their mutual interactions are slight. The water in water vapor is largely monomeric with perhaps an occasional dimer or even rarer trimer. In the solid state, depending on the pressure, there are a number of well-characterized crystalline structural forms that ice can assume. But when we come to the liquid, the situation becomes much more complicated and much less understood.

Now the classical method of investigating the crystal structure of substances is X-ray diffraction. If we apply this powerful technique to liquid water we find, not the featureless curve (dashed line in Figure 7.5) expected of unstructured liquids, but rather a series of peaks that appear to be a sort of smeared-out version of the pattern for ordinary ice. That is to say, there is still a certain amount of short-range, icelike structural order in the liquid.

We will now turn our attention to the structure of Ice-I (the form obtained under ordinary circumstances at 1 atm pressure), not only because it is the material of the solid portion of the hydrosphere, but also because many of the modern theories of the structure of liquid water have taken Ice-I as their point of departure.

In Ice-I, each water molecule is surrounded by four closest neighbors in a tetrahedral arrangement. In the H-bond between a given water molecule and an immediate neighbor, the proton occupies two equivalent positions between the oxygen

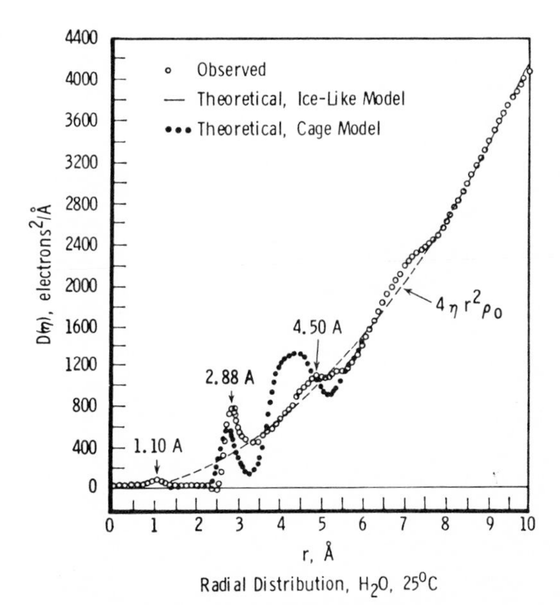

Figure 7.5. X-Ray radial distribution curve for liquid water (reprinted with permission from M. D. Danford and H. A. Levy, *J. Amer. Chem. Soc.*, *84*, 3965 (1962), © the American Chemical Society).

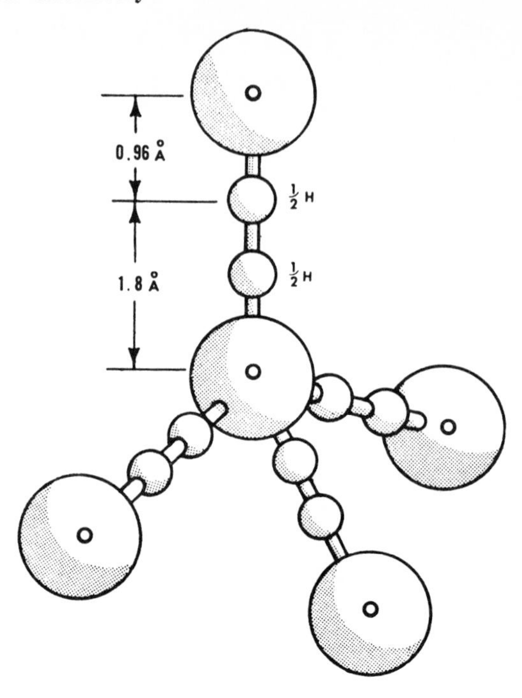

Figure 7.6. Position of the hydrogen atoms in the ice lattice (from Horne, 1969).

centers as shown in Figure 7.6. The extended Ice-I lattice can be represented schematically as a series of parallel and interconnected sheets of joined hexagonal rings (Figure 7.7). Kamb (1972) has reviewed the structure of the various ices, Jaccard (1972) transport processes in Ice-I, and Jellinck (1972) the surface properties of Ice-I. The high pressure ices exist only at very elevated pressures (greater than 1000 atm) and are thus unlikely to occur naturally on Earth or on any of the other planets. However their foremost student has advanced the hypothesis that one or more of them might be a component of the liquid water "mix" (Kamb, 1968). On the other hand, the transport properties, especially mass and heat transport, and surface properties of Ice-I have great environmental significance for they determine such phenomena as avalanche formation, the movement of glaciers, and the characteristics of permafrost and other frozen soils.

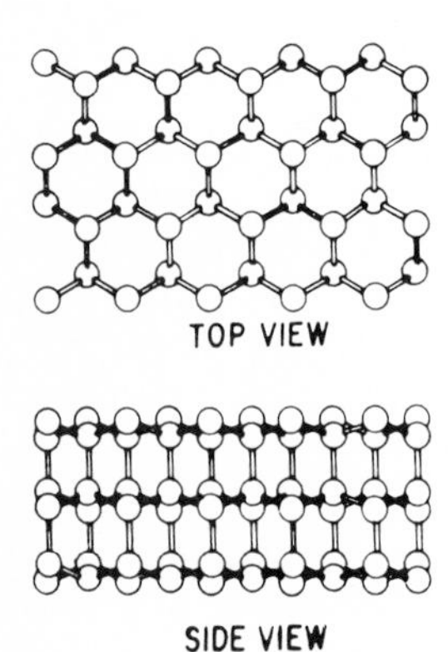

Figure 7.7. Schematic representation of the ice-I_h structure (from Davis and Litovitz, 1965, with permission of the American Institute of Physics and the authors).

Whereas the structure of Ice-I has been well resolved, the structure of liquid water remains unknown despite very massive experimental and theoretical efforts brought to bear on the question. The subject continues to be a very controversial one even today, nearly 40 yr after Bernal and Fowler (1933) proposed their model in the pages of the first volume of *The Journal of Chemical Physics*, and a steady procession of new theories of liquid water structure continue to appear. These theories have been reviewed numerous times elsewhere (Nemethy and Scheraga, 1962; Luck, 1963; Kavanau, 1964; Samoilov, 1965; Wicke, 1966; Drost-Hansen, 1967; Franks, 1968; Horne, 1968, 1969, 1970; Eisenberg and Kauzmann, 1969). Of these many models, one of the most popular has been that proposed by Frank and Wen (1957). These authors picture liquid water as consisting of a mixture of hydrogen bonded clusters and "free" or monomeric water. The polymers or clusters are said to be "flickering," that is to say they are constantly forming and disintegrating with thermal fluctuations in the microregions of the liquid so that there estimated half-life is about 10^{-1}- to 10^{-12} sec. The clusters have been described as Icelike, and, while the liquid clearly exhibits many of the structural features of the solid, Frank and Wen (1957) are careful to avoid committing themselves to any specific lattice. They are equally cautious in their description of the nonclustered water as "free." Nemethy and Scheraga (1962) have developed quantitatively the Frank–Wen model and Figure 7.8 is an accurate, scale, two-dimensional diagram based on their work. Recently, the analysis of Nemethy and Scheraga (1962) has been subjected to considerable criticism, the gist of which is that they overestimated the percentage of free water. At the present time, the thinking about liquid water seems to be moving away from the coherent clusters to a more loose, extended, random network sort of picture. In terms of Figure 7.8, this new direction can be readily accommodated simply by moving a very small fraction of the total H-bonds about. Such minor alterations can also bring Figure 7.8 into line with the broken-down-ice type structure proposed by Samoilov (1965). The readiness with which Figure 7.8 can be made to conform with any of a number of the current theories leaves one wondering if these models are really as different as they once appeared to be, and the distinction between "mixture" and "continuum" theories of water structure becomes very blurred indeed.

We defer our discussion of the effect of solutes on the structure of water to Chapter 8 on the marine hydrosphere and of the structure of water near interfaces to Chapter 13. But I would like to conclude this section with a brief mention of the effects of temperature and pressure on water structure.

Viscosity is a good indicator of the extent of H-bonding or structure in liquid water—the more and the tighter the molecules are bound together, the more difficult their movement and the higher the viscosity. As temperature increases the viscosity decreases (Table 7.5 and Figure 7.9) indicating that the polymeric clusters are melting. There is great disagreement among the experts as to just how many H-bonds there are at any given temperature (see Falk and Ford, 1966), but they do all concur that the number of H-bonds decreases with increasing temperature and that the decrease in going from the freezing point to the boiling point is surprisingly slight. Nemethy and Scheraga (1962) estimate that there are about 65 water molecules in the average cluster at 0°C but only 12 at 100°C. Since the clusters fragment with increasing temperature as well as "melt" the number of clusters increases, the decrease in the fraction of unbroken H-bonds is significantly less than the decrease in average cluster size.

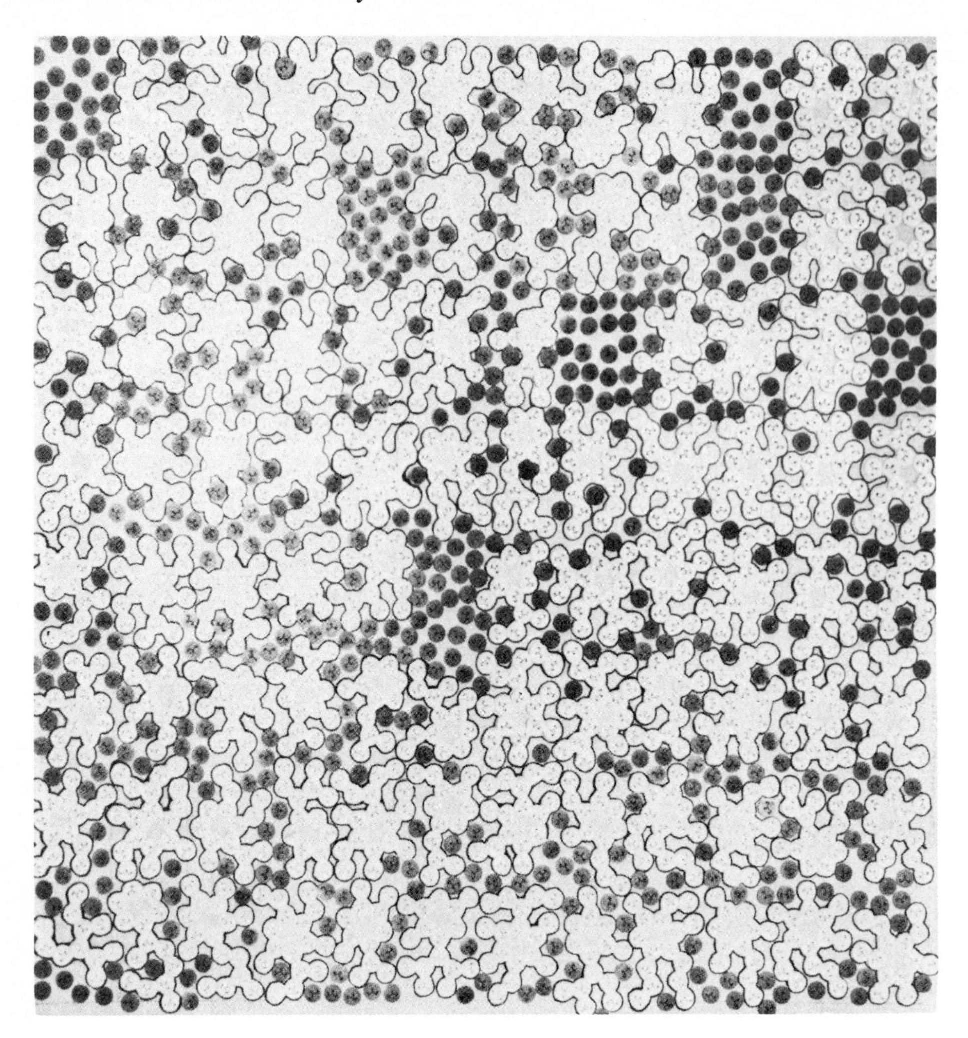

Figure 7.8. The structure of liquid water at 20°C and 1 atm. The water clusters are outlined and the "free" waters shaded (from Courant et al., 1972).

If hydrostatic pressure is applied to liquid water, the viscosity first decreases and then at higher pressures begins to increase (Figure 7.10). It is the only substance known that exhibits such an anomalous minimum in the relative viscosity versus pressure curve (Figure 7.10). The initial decrease in this curve evidently is due to the break-up of the clusters. This is not unexpected for the specific volume of the clusters is greater than that of the free water (just as the specific volume of Ice-I is greater than that of liquid water at 0°C), thus the process

$$\underset{\text{clustered}}{(H_2O)_n} \rightleftharpoons \underset{\text{free}}{nH_2O} \tag{7.1}$$

as written is accompanied by a volume decrease, and the equilibrium is therefore shifted to the right by the application of pressure. After the more open polymers have

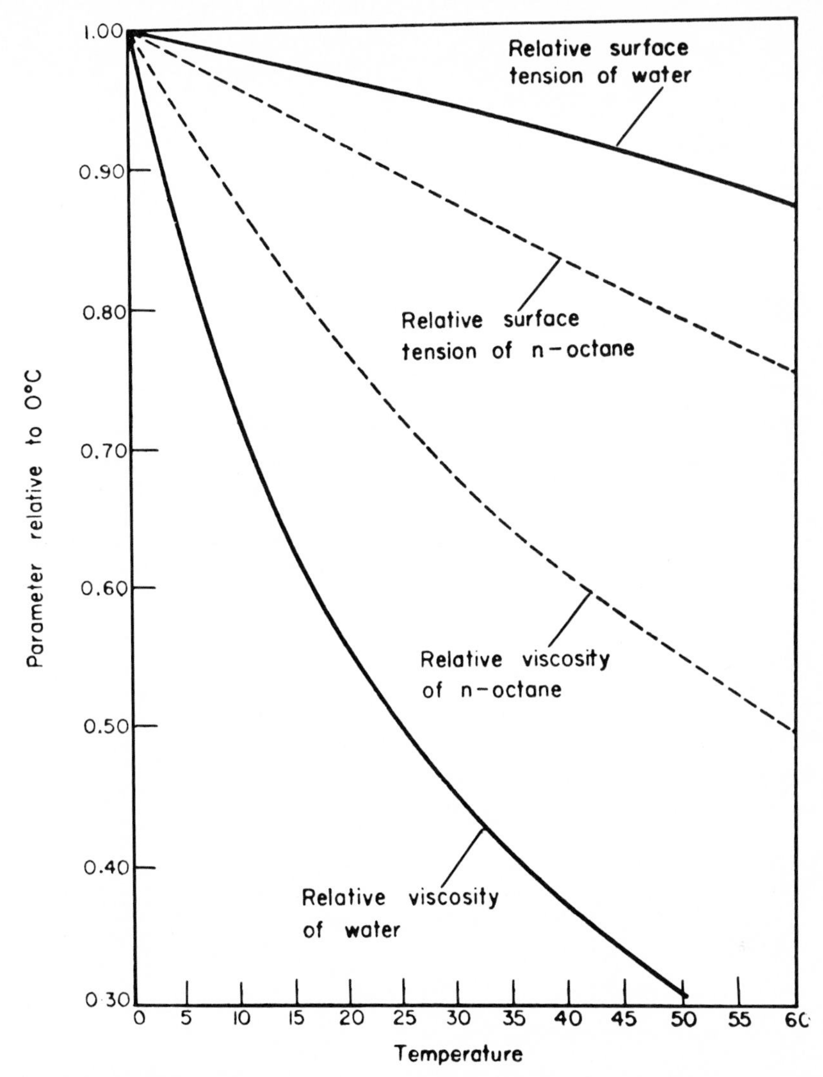

Figure 7.9. Comparison of the temperature dependence of a bulk (viscosity) with a vicinal (surface tension) water structure parameter and with a "normal" liquid (*n*-octane) (from Horne et al., 1968, with permission of Pergamon Press, Ltd.).

been destroyed the application of further pressure results in an increase in viscosity, just as one expects for a "normal" liquid, as the molecules are crowded closer and closer together and are less free to move about.

Adding something to water also alters its structure, but we defer our discussion of the effects of solutes on water structure until the chapter on seawater and the marine hydrosphere. Finally, an interface or second phase alters water structure. Since hydrosphere, lithosphere, biosphere, and atmosphere all interact with one another only through such interfaces, the importance of interfacial water structure (or vicinal water) cannot be overemphasized, and we deal with this topic in some detail in Part VII.

Table 7.5 Selected Physical Properties of Pure Water (From Robinson and Stokes, 1959)

Temperature (°C)	Density (g/ml)	Specific Volume (ml/g)	Vapor Pressure (mm Hg)	Dielectric Constant	Viscosity (cP)
0	0.99987	1.00113	4.580	87.740	1.787
5	0.99999	1.00001	6.538	85.763	1.516
10	0.99973	1.00027	9.203	83.832	1.306
15	0.99913	1.00087	12.782	81.945	1.138
20	0.99823	1.00177	17.529	80.103	1.002
25	0.99707	1.00293	23.753	78.303	0.8903
30	0.99568	1.00434	31.824	76.546	0.7975
40	0.99224	1.00782	55.338	73.151	0.6531
50	0.98807	1.01207	92.56	69.910	0.5467
60	0.98324	1.01705	149.57	66.813	0.4666
70	0.97781	1.02270	233.81	63.855	0.4049
80	0.97183	1.02899	355.31	61.027	0.3554
90	0.96534	1.03590	525.92	58.317	0.3156
100	0.95838	1.04343	760.00	55.720	0.2829

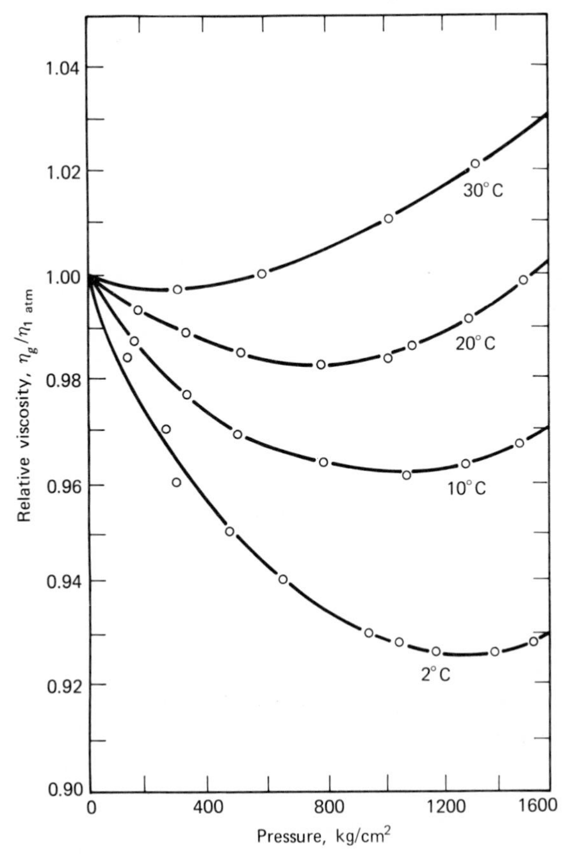

Figure 7.10. Pressure dependence of the relative viscosity of water (based on data from Stanley and Batten, 1969).

7.3 LAKES, RIVERS, PONDS, AND STREAMS

Unlike the oceans of Earth, the lakes, rivers, ponds, and streams of the continents are not immediately connected to one another. They remain for the most part unmixed with one another and tend to retain their separate chemical identities. Consequently, it is less easy to generalize about fresh water systems than about seawater. Nevertheless, although the terrestrial hydrosphere is more variable than its marine counterpart, including in its compass waters ranging from great purity to lakes far saltier than the sea and with enormous variations in some of the minor chemical constituents, the prospect it presents is not one of complete disorder. On the contrary, the major constituents find their way from the atmosphere and lithosphere into the terrestrial hydrosphere by more or less the same geochemical processes throughout the world, in particular by leaching of roughly the same crustal material, and thus the deviations of their "average" values for different continental regions from the world mean are not great (Table 7.6). In the terrestrial hydrosphere, the major anionic constituents are carbonate, sulfate, and silicate, whereas in seawater, chloride is easily the only major anionic species (sulfate is present in concentrations about 100 times smaller). In fresh water systems, the important cation is calcium, whereas in the marine hydrosphere it is of course sodium. Finally, there is roughly 2000 times less dissolved salts in fresh water than in seawater (always excepting, of course, highly saline lakes). Table 7.7 lists some selected concentrations of minor elements in fresh waters, and, as expected, the values are highly variable. These values are for the most part "natural" values. In particular locations, the concentrations of some of the toxic heavy metals such as mercury have been escalated by industrial activity far beyond the point where they represent a serious pollution hazard.

In order to give you some idea of the great chemical variety of lakes, Table 7.8 lists a few selected extraordinary lakes.

Before continuing our discussion of the chemistry of river and lake waters, it is necessary to mention the classification of such terrestrial hydrosphere systems.

Table 7.6 The Chemical Composition of Salt in River and Lake Waters (wt. %) (From Miyake, 1965, with permission of the author.)

	Japan	North America	South America	Europe	Asia	Africa	World Mean
CO_3^{2-}	33.9	33.4	32.5	40.0	36.6	32.8	35.2
SO_4^{2-}	14.9	15.3	8.0	12.0	13.0	8.7	12.1
Cl^-	8.8	7.4	5.8	3.4	5.3	5.7	5.7
SiO_2	12.4	8.6	18.9	8.7	9.5	17.9	11.7
NO_3^-	1.6	1.2	0.6	0.9	1.0	0.6	0.9
Ca^{2+}	12.9	19.4	18.9	23.2	21.2	19.0	20.4
Mg^{2+}	4.5	4.9	2.6	2.4	3.4	2.7	3.4
Na^+	8.2	7.5	5.0	4.3	6.0	4.9	5.8
K^+	2.5	1.8	2.0	2.8	2.0	2.4	2.1
$(Fe, Al)_2O_3$	0.3	0.7	5.7	2.4	2.0	5.5	2.7
Salinity (mg/liter)	83	341	—	205	—	151	—

Table 7.7 Minor Constituents in Lake and River Waters (Selected data from Livingstone, 1963)

Element	Locality	Concentration
Bromine	Dead Sea	4600 ppm
	U.S.S.R. rivers	0.019 ppm
	Australian lakes	5 to 272 ppm
	Russian lakes	0.002 to 0.01 ppm
Iodine	U.S.S.R. rivers	0.007 ppm
	Finnish lakes	0.00001 ppm
Boron	River Tone, Japan	0.345 ppm
	Great Salt Lake, Utah	43.5 ppm
	Florida streams	0.019 ppm
	U.S.S.R. rivers	0.013 ppm
Lithium	Major North American rivers	0.0033 ppm
	Japanese rivers	0.0002 to 0.005 ppm
	Lake Tanganyika, Africa	0.4 ppm
	Salton Sea, California	1.9 ppm
Rubidium	Japanese rivers	0.0003 to 0.002 ppm
Cesium	Japanese rivers	0.00005 to 0.0002 ppm
Beryllium	Atchafalaya River, Louisiana	0.0 to 1 ppb
Strontium	Housatonic River, Connecticut	45.6 ppb
	Hudson River, New York	107.9 ppb
	Trout Lake, Wisconsin	44.4 ppb
	Lake Erie	137 ppb
	Sebago Lake, Maine	12.1 ppb
	Great Salt Lake, Utah	2100 ppb
	Major North American rivers	90 ppb
Barium	Linsley Pond, Connecticut	10 ppb
	Major North American rivers	54 ppb
Radium	Mississippi River, Missouri	1.2×10^{-10} to 2.9×10^{-10} ppm
	River Thames, England	0.1×10^{-10} ppm
	Nashua River, Massachusetts	0.14×10^{-10} ppm
	Normal surface water (U.S.)	3.6×10^{-10} to 34.1×10^{-10} ppm
	Hudson River, New York	
Selenium	Se-rich ponds, South Dakota	21.4 to 85.5 ppb
Arsenic	California waters (<2000 ppm total dissolved solids)	0.4 ppb
	California waters (>2000 ppm total dissolved solids)	225 ppb
Argon	Japanese pond	0.29 cc/liter (96% sat.)
Gallium	Linsley Pond, Connecticut	0.1 to 1 ppb
Gold	Waters near gold deposits	0.06 ppb
Mercury	Saale River	0.066 ppb
Cadmium	Urov River	9.66 to 80.5 ppb (high ?)
Copper	Linsley Pond, Connecticut	11 to 383 ppb (total)
	Maine lakes	0.07 to 140 ppb

(Continued)

Table 7.7 (*Continued*)

Element	Locality	Concentration
Copper (*Continued*)	Saale River	15 ppb (total)
	U.S.S.R. rivers	10.5 ppb
Cobalt	Linsley Pond, Connecticut	0.05 to 0.105 ppb (total)
	Lake Baikal, U.S.S.R.	2.3 ppb
	Major North American rivers	0.89 ppb
Nickel	Lake Michigan	2 ppb
	Maine lakes	0.01 to 7 ppb
	Lake Baikal, U.S.S.R.	5 ppb
	Major North American rivers	11.7 ppb
Silver	United States water supplies	28 ppb
Zinc	Lake Michigan	200 to 300 ppb
	Maine lakes	0.25 to 34 ppb
	Japanese rivers	36 ppb
	U.S.S.R. rivers	45 ppb
Titanium	Lake Michigan	70 ppb
	Linsley Pond, Connecticut	50 ppb
	Maine lakes	1.60 ppb
	Major North American rivers	13.2 ppb
Zirconium	Maine lakes	0.05 to 22.5 ppb
Tin	Lake Michigan	40 ppb
	Maine lakes	0.038 ppb
Lead	Lake Michigan	2 ppb
	Maine lakes	0.03 to 115.0 ppb
	Major North American rivers	6.6 ppb
Vanadium	Lake Michigan	20 ppb
	Argentina water supplies	320 ppb
	Maine lakes	0.112 ppb
Chromium	Lake Michigan	2 ppb
	Maine lakes	0.177 ppb
	Major North American rivers	10.8 ppb
Molybdenum	Maine lakes	0.023 ppb
	Major North American rivers	0.84 ppb
Manganese	Maine lakes	0.02 to 87.5 ppb
	North Germany lakes	25 ppb
	Wisconsin lakes	3 to 23 ppb (as high as 1200 ppb in deep waters)
	Mississippi River, Iowa	80 to 120 ppb
	U.S.S.R. rivers	11.9 ppb
Uranium	Ohio River, Pennsylvania	<2.5 ppb
	Great Salt Lake, Utah	5 ppb
	Hudson River, New York	0.022 ppb
	Major North American rivers	0.016 to 0.040 ppb
	Danube River, Vienna	47 ppb
	Lake Mendota, Wisconsin	0.4 ppb

Table 7.8 Some Extraordinary Lakes

Lake Baikal, Siberia	Deepest lake in the world (1741 m). Surface area of 31,500 km^2. One of the world's purest bodies of water, now threatened with pollution largely by pulp and paper industry.
Lake Tanganyika, Africa	World's second deepest lake (1435 m). Surface area of 31,900 km^2.
Lake Titicaca, Peru	World's highest (4000 m) large lake. 281-m deep, 7600 km^2 in area.
Caspian Sea, U.S.S.R.	World's largest lake, 135,000 km^2.
Lake Superior, U.S.–Canada	World's second largest lake, 82,000 km^2.
Ranu Klindungan, Java	Very high Mn maximum (15,000 mg/m^3) at 9-m depth, possibly due to manganiferous spring (Ruttner, 1931).
Bear Lake, Idaho	Zn-rich waters (650 mg/m^3) (Kemmerer, Bovard, and Boorman, 1923).
Goodenough Lake, British Columbia	Saline lake of very high phosphorus content (208 g/m^3)
Son-sakesar-kahar, Pakistan	Closed, oxygen-less lake containing a high concentration of H_2S (79.9 mg/liter near bottom) whose waters turn pink due to blooms of the purple sulfur bacterium *Lamprocystis roseopersina*.
Lake Erie, United States–Canada	One of the world's largest and most polluted lakes.
Great Salt Lake, Utah	Saline lake with a salt content of 150 to 280 g/liter.
Dead Sea, Jordan	Saline lake with a salt content of 200 to 260 g/liter.
Red Lake, Wyoming	Saline lake with a salt content of 300 g/liter. In 1888 when this usually dry lake contained water the sulfate concentration was the highest ever reported (60 g/liter).
Kata-numa, Japan	Volcanic lake with pH 1.7. The waters contain suspended elemental sulfur and 474 mg SO_4/liter (Yoshimura, 1934).
Lake Nakuru, Kenya	Effluentless alkaline lake of pH 12.0 (Jenkin, 1932).
Awoe, Indonesia	Crater lake with 0.005 N HCl and H_2SO_4.
Kawah Idjen, Indonesia	Crater lake with pH possibly under 0.7.
Mud Lake, Michigan	Edged with sphagnum mat with highly acid (pH 3.3) pools (Jewell and Brown, 1929).

Hydrological regions may be classified into three types:

Exorheic Regions from which rivers reach to the sea,

Endorheic Regions within which rivers arise but disappear into dry water courses or closed lake basins before they reach the sea, and

Arheic Regions in which no rivers arise.

Very large lakes may receive most of their water from precipitation, but lakes in endorheic regions tend to receive most of their waters from influents. Lakes and ponds are perhaps best classified on the basis of their thermal characteristics, and the following scheme has been proposed (Hutchinson, 1957):

Polar Zone
- Amictic, rare ice-sealed lakes found in Antarctica and ocasionally in high mountains.
- Cold Monomictic, temperature never above 4°C at all depths, freely circulating in summer, ice covered with inverse thermal stratification in the winter.

Temperate Zone
- Dimitic, freely circulating twice a year in spring and fall, inversely stratified in winter, directly in summer. The common type of lake in the cooler temperate zones.
- Warm Monomictic, water never below 4°C, freely circulating in winter, direct stratification in summer.

Tropical Zone
- Oligomictic, water well above 4°C, rare and irregular circulation.
- Polymictic, continually circulating near 4°C. Found in high mountains in equatorial latitudes.

Notice in passing that the thermal and circulation characteristics of lakes and ponds that form the basis of this system of classification are a consequence of one of liquid water's best-known anomalies, namely the maximum in its density at 4°C and the resulting smaller density of the solid than of the liquid.

Generally speaking (and of course there are many exceptions), the thermal structure of lakes and ponds is much more pronounced, more highly variable, and more important in their aquatic chemistry than is the thermal structure of the water column in the oceans. Figure 7.11 shows the seasonal variations in much studied Linsley Pond, Connecticut, while Figure 7.12 shows the temperature distribution in Moosehead Lake, Maine, one of New England's largest bodies of fresh water.

The ionic composition of Moosehead Lake and some other New England waters is shown in Table 7.9, Table 7.10 is of interest for it is intended to show the dependence of the chemical composition (expressed as $pC \equiv -\log C$) on the type of

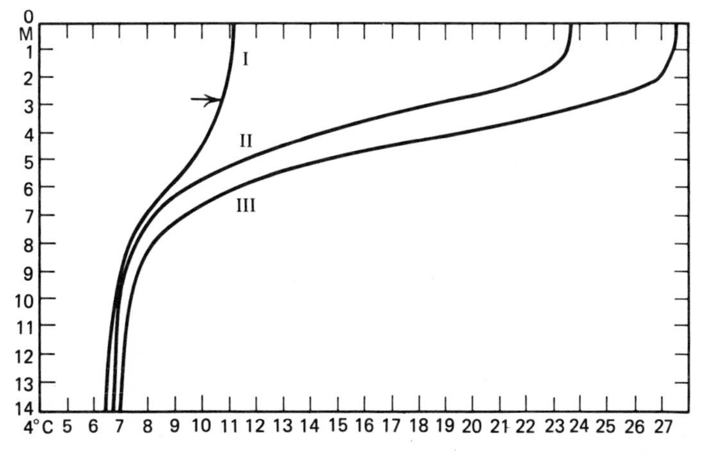

Figure 7.11. Temperature curves for Linsley Pond, Connecticut: I. April 30, 1936; II. June 1–15, 1936; III. August 3–17, 1936 (from Hutchinson, 1957).

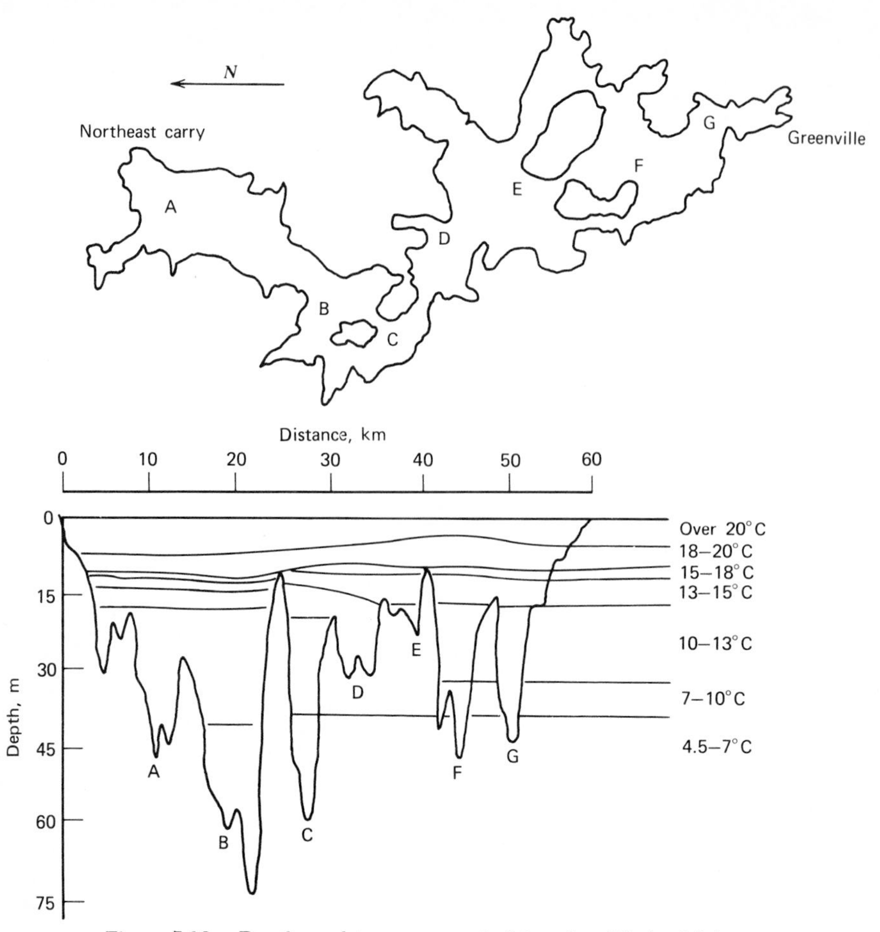

Figure 7.12. Depths and temperatures in Moosehead Lake, Maine.

Table 7.9 Inorganic Composition of Some New England Fresh Waters (Selected values from Brooks and Deevey, 1963)

	Milliequivalents per Liter							Total
	Na^+	K^+	Ca^{++}	Mg^{++}	HCO_3^-	$SO_4^=$	Cl^-	(mg/liter)
Moosehead Lake (Maine)	0.078	0.015	0.110	0.025	0.80	0.043	0.056	16.0
Lake Winnipesaukee (New Hampshire)	0.087	0.005	0.195	0.041	0.131	0.108	0.065	21
Upper Connecticut River (New Hampshire)	0.035	0.010	0.165	0.082	0.125	0.121	0.003	17.9
Lower Connecticut River (Connecticut)	0.304	0.056	0.847	0.162	0.690	0.457	0.211	102

Table 7.10 Examples of Natural Waters (From Stumm and Morgan, 1970

	1	2	3	4	5	6	7	8	9	10	11
Type	Stream	Stream	Stream	Lake Erie	Ground Water	Ground Water	Ground Water	Ground Water	Ground Water	Ground Water	Closed Basin Lake
Type of Rocks Being Drained	Granite	Quartzite	Sandstone		Granite	Gabbro-plagioclase	Sandstone	Shale	Limestone	Dolomite	Soda Lake
pH	7.0	6.6	8.0	7.7	7.0	6.8	8.0	7.3	7.0	7.9	9.6
pNa	4.0	4.6	4.3	3.4	3.4	3.0	3.3	2.6	3.0	3.5	0.0
pK	4.7	5.1	4.8	4.3	4.0	4.5	4.0	4.2	3.7	—	1.7
pCa	4.0	4.3	3.1	3.0	3.5	3.1	3.0	2.5	2.7	2.8	4.5
pMg	4.6	5.1	4.0	3.4	3.8	3.2	3.5	2.5	3.4	2.8	4.6
pH_4SiO_4	3.8	4.2	4.1	4.7	3.2	3.0	3.9	3.5	3.7	3.4	2.8
$pHCO_3$	3.6	4.0	2.9	2.7	2.9	2.5	2.6	2.1	2.3	2.2	0.4
pCl	5.3	5.8	5.3	3.6	4.0	3.5	3.7	4.0	3.2	3.3	0.3
pSO_4	4.5	4.7	3.7	3.6	4.2	4.0	3.2	2.2	3.4	4.7	2.0
−log (ionic strength)	3.5	3.8	2.7	2.5	2.8	2.4	2.4	1.7	2.2	2.2	0.0

lithospheric material drained by the waters. Depending upon the mineralogical nature of the drainage region, the waters can be high in one or more of the inorganic constituents thus giving rise to chloride, sulfate, or carbonate waters (Table 7.11). If the concentrations are excessively high, the water can be rendered unfit for drinking purposes, agriculture, or even industrial use. Perhaps the most familiar constituent excess is that of the cations magnesium and calcium in high carbonate waters. While not rendering the waters unfit, the resulting *hardness* is a definite common and expensive nuisance. Hard waters may be classified into two types: temporary hardness and permanent hardness. The insoluble carbonates of magnesium, calcium, and iron can be dissolved by the action of CO_2 in natural waters to form the more soluble bicarbonates

$$H_2O + CO_2 \rightleftharpoons H_2CO_3 \tag{7.2}$$

$$H_2CO_3 + CaCO_3 \rightleftharpoons Ca^{2+} + 2HCO_3^- \tag{7.3}$$

Boiling reverses these equilibria and with the escape of CO_2 the carbonates are reformed and precipitated to form a deposit or scale in teakettles and boilers. The scale is a poor thermal conductor, and its formation can seriously reduce the efficiency of steam boiler systems in addition to worsening corrosion problems. Hard water is also a problem in bathroom and laundry, since conventional soaps composed of the sodium salts of fatty acids such as stearic acid instead of lathering form an insoluble scum

$$2C_{17}H_{35}COO^- + Ca^{2+} \rightleftharpoons Ca(C_{17}H_{35}COO)_2\downarrow \tag{7.4}$$

Modern detergents based on alkane sulfonic and phosphoric acids react in a similar fashion. The hardness of the water depends on local mineralological conditions. New England, for example, is fortunate in having ample supplies of soft waters, but there are hard water areas particularly as one moves down into southwest Connecticut (Figure 7.13).

Temporary hardness can be removed by boiling the water, although this is rarely feasible. Permanent hardness, however, is caused by the presence of sulfates rather than carbonates of magnesium and calcium and cannot be removed by boiling.

Hard water, either temporary or permanent, can be softened by a number of chemical processes. On a large scale, stoichiometric amounts of slaked lime can be added to temporary hard water and the resulting carbonate precipitates filtered off.

$$Ca^{2+} + 20H + Ca^{2+} + 2HCO_3^- \rightleftharpoons 2CaCO_3\downarrow + 2H_2O \tag{7.5}$$

$$\underbrace{Ca^{2+} + 20H}_{\text{slaked lime}} + Fe^{2+} + 2HCO_3^- \rightleftharpoons FeCO_3\downarrow + CaCO_3\downarrow + 2H_2O \tag{7.6}$$

The addition of the proper amount of Na_2CO_3 will remove permanent hardness by replacing the troublesome Ca with Na in solution

$$2Na^+ + CO_3^{2-} + Ca^{2+} + SO_4^{2-} \rightleftharpoons CaCO_3\downarrow + 2Na^+ + SO_4^{2-} \tag{7.7}$$

Both permanent and temporary hardness can be removed by the addition of crude NaOH

$$Ca^{2+} + 2HCO_3^- + 2OH^- \rightleftharpoons CaCO_3\downarrow + CO_3^{2-} + 2H_2O \tag{7.8}$$

Table 7.11 Composition of Mineral Salts of Chloride, Sulfate, and Carbonate Lakes and Their Influents (mg/liter) (From Hutchinson, 1957)

	Na	K	Mg	Ca	CO_3	SO_4	Cl	SiO_2 A	$(AlFe)_2O_3$	Salinity
Chloride Waters										
Bear River										
Upper, Wyoming	4.49		6.86	23.69	52.68	5.76	2.68	3.84	—	185
Lower, Utah	20.54		4.76	10.12	21.53	8.16	32.36	—	2.53	637
Great Salt Lake, Utah	33.17	1.66	2.76	0.17	0.09	6.68	55.48	—	—	203.490
Jordan at Jericho	18.11	1.14	4.88	10.67	13.11	7.22	41.47	1.95	1.45	7.700
Dead Sea	11.14	2.42	13.62	4.37	Trace	0.28	66.37[a]	Trace	—	226.000
Sulfate Waters										
Montreal Lake, Saskatchewan	4.9	2.3	10.8	16.8	56.5	1.8	2.5	3.9	0.5	150.5
Redberry Lake, Saskatchewan	12.0	0.85	12.3	0.56	2.58	70.5	1.1	0.03	0.07	12.898
Little Manitou Lake	16.8	1.0	10.9	0.48	0.47	48.4	21.8	0.019	0.21	106.851
Carbonate Waters										
Silvies River, Oregon	10.42	2.45	3.13	12.88	34.76	7.35	2.88	25.13	0.08	163
Malheur Lake, Oregon	24.17	5.58	4.13	5.58	44.63	7.64	4.55	2.89	Trace	484
Pelican Lake, Oregon	29.25	3.58	2.62	2.27	30.87	22.09	7.97	1.21	0.02	1,983
Bluejoint Lake, Oregon	37.70	2.62	0.63	0.53	38.68	5.67	13.85	0.55	0.02	3,640
Moses Lake, Washington	19.86		7.25	8.41	51.56	2.87	3.88	5.06	1.11	2,966

[a] And 1.78% Br.

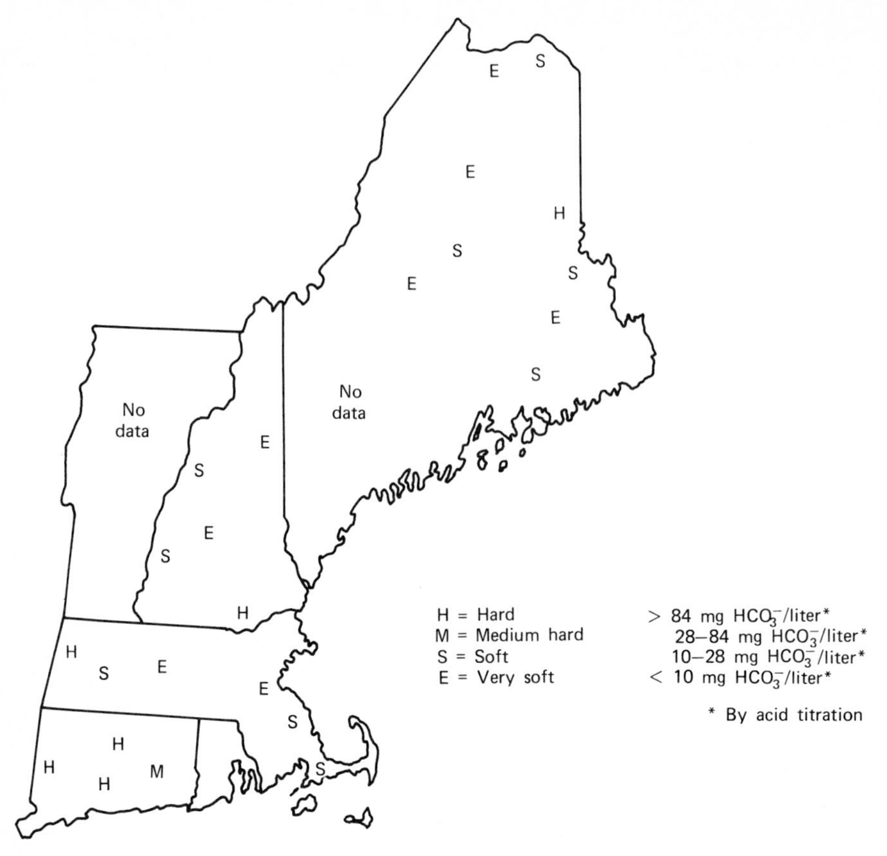

Figure 7.13. Hardness of New England lake waters.

since the Na_2CO_3 formed then reacts with $CaSO_4$ as indicated previously (Equation 7.7). In industrial practice, a varying mixture of soda ash and quicklime is used to control hardness.

The foregoing processes are not very helpful to the housewife, since they involve the addition of the correct amount of chemical reagents and the removal of precipitate products. Household ammonia can be used to soften temporary hard water in the home

$$Ca^{2+} + 2HCO_3^- + 2NH_3 \rightleftharpoons CaCO_3\downarrow + 2NH_4^+ + CO_3^{2-} \qquad (7.9)$$

as can also borax ($Na_2B_4O_7 \cdot 10H_2O$), trisodium phosphate ($Na_3PO_4 \cdot 12H_2O$), and complex hexametaphosophates ($(NaPO_3)_6$). Many commercial wash powders have a water softening agent added.

The problem of exact stoichiometry and product removal in water softening can be avoided completely by ion exchange. Aristotle had noted that seawater loses some of its salt when percolated through certain sands, and the phenomenon was rediscovered in 1850 by Thompson who observed cation exchange in soils. An ion exchanger is an insoluble substance with exchangeable ions. Passing hard water through a sodium aluminosilicate (the natural material is zeolite, the synthetic material permutite) results in a replacement of troublesome Ca^{2+} ions by Na^+ ions

$$2NaAlSI_2O_6 + Ca^{2+} \rightleftharpoons Ca(AlSi_2O_6)_2 + 2Na^+ \quad (7.10)$$

The process is reversible, and the material can be regenerated by treatment with a concentrated NaCl solution. By the use of a mixture of an anion-exchanger and a cation-exchanger water can be completely desalinated. A great number of synthetic ion exchange resins are available with a wide selection of properties. The exchange site of the cation exchangers is commonly a sulfonate group, $—R—SO_3^-$, while the anion exchangers are often based on quarternary amino groups, $—R'—N(R)_3^+$. The physical chemistry of these materials has been intensively examined since World War II (when they found use in connection with chemical separation procedures involved in the Manhattan Project), and their technology is highly exploited (Helfferich, 1962). Ion exchange, largely in clay minerals, is one of the major chemical processes occurring in our natural environment, and we return to this topic later in our discussion of the interaction of the hydrosphere with the lithosphere.

There is one advantage to hard water that I might mention—activated sludge sewage treatment removes greater quantities of phosphate from hard than soft waters thus lessening the dangers of eutrofication,* due to precipitation of calcium phosphate, which becomes entrapped and removed in the activated sludge floc matrix (Menar and Jenkins, 1970).

To return to the chemistry of lakes and rivers, let us consider some of the major chemical species and processes in fresh waters. If there is any single key to fresh water (and marine) chemistry that key is oxygen. Aquatic chemistry and life are probably more sensitive to this variable than to any other, even temperature and pH. Like other gases that do not react with water and unlike most ionic solutes, the solubility of oxygen in water decreases with increasing temperature (Table A.2), consequently oxygen stratification tends to be established concurrently with thermal stratification, and the oxygen and temperature profiles tend to be the inverse of one another (Figure 7.14). The water equilibrates with the oxygen of the atmosphere at the air–water interface, the rate determining step being the diffusion of the gaseous solute molecules through the boundary layer of vicinal water. Because gas solubility increases with increasing pressure, the saturation concentration will increase with depth. Also it should be noted that the saturation value at the lake's surface will depend on the height of the body of water above sea level, since the atmospheric pressure and thus the oxygen partial pressure decreases with altitude. In waters of high biological activity, either supersaturation or undersaturation may occur depending on whether oxygen production by photosynthesis or oxygen consumption by respiration and the oxidation of organic matter is the dominant process. Photosynthetic activity can give rise to marked diurnal fluctuations in the oxygen content of the water (Figure 7.15). In summer, the formation of a warm layer or epilimnion above the thermocline will result in oxygen loss. In deep unproductive lakes, the oxygen content may continue to increase with depth giving rise to an orthograde oxygen distribution, but more commonly small productive lakes exhibit a summer clinograde profile with the oxygen decreasing with increasing depth due to oxygen consumption by biological activity in the hypolimnion or subthermocline zone. The intermediate

* See Chapter 16 for further material on sewage treatment and eutrofication.

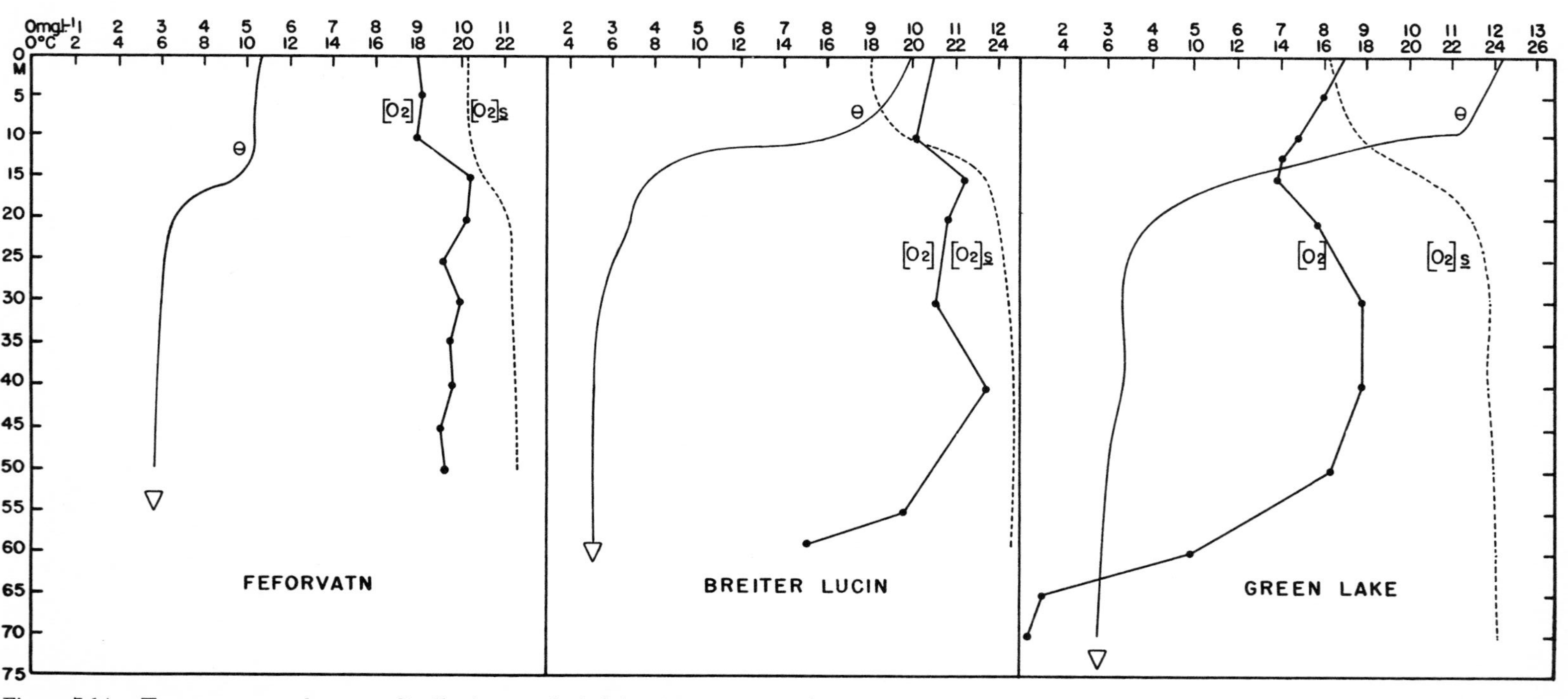

Figure 7.14. Temperature and oxygen distributions at the height of summer stratification in three dimictic lakes, ranging from biologically sterile to highly productice, 50–75-m deep: Feforvatn, a Norwegian Mountain Lake; Breiter Lucin, Northern Germany; Green Lake, Wisconsin, U.S.A. (from Hutchinson, 1957).

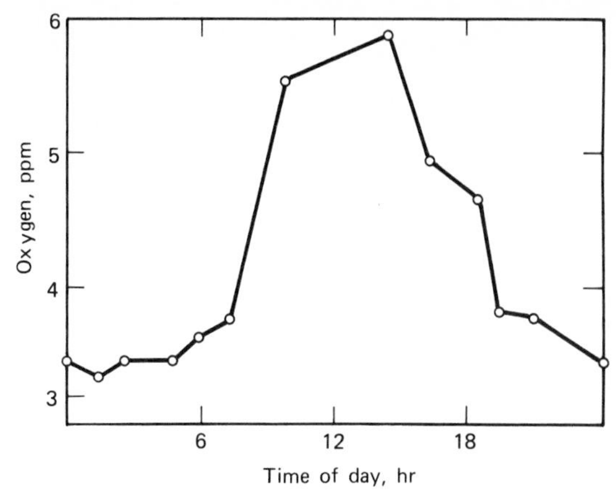

Figure 7.15. Diurnal oxygen change in Silver Springs, Florida (from Livingstone, 1963).

region, the metalimnion, may develop a summer maximum if there is deep photosynthetic activity or a minimum if there is a net demand of oxygen by the rain of organic particles from the epilimnion above. If the oxygen in the near-bottom waters becomes completely depleted, anoxic conditions with the accompanying production of reduced chemical species, such as hydrogen sulfide, methane, hydrogen, ammonia, and even hydroxylamine, will obtain (Tanaka, 1954).

Perhaps equally as important as oxygen in the chemistry of both the terrestrial and marine hydrosphere is carbon dioxide. Inasmuch as it reacts with water in a series of equilibrium steps, CO_2 is about 200 times more "soluble" in fresh water than is oxygen (see Table A.3).

$$CO_2\text{ (g)} + H_2O \rightleftharpoons CO_2\text{ (dissolved)} + H_2O \quad (7.11)$$

$$CO_2\text{ (dissolved)} + H_2O \rightleftharpoons H_2CO_3\text{ (carbonic acid)} \quad (7.12)$$

$$H_2CO_3 \rightleftharpoons H^+ + HCO_3^-\text{ (bicarbonate ion)} \quad (7.13)$$

$$HCO_3^- \rightleftharpoons H^+ + CO_3^{2-}\text{ (carbonate ion)} \quad (7.14)$$

These equilibria are but a part of the complex carbon dioxide–carbonate system (Figure 7.16), which plays an important role in each of the elements of the total environment—atmosphere, lithosphere, hydrosphere, and biosphere. We return to a more detailed discussion of CO_2 in our treatment of the chemistry of seawater in Chapter 8. Suffice it here to say that the dominant form of carbonate in all natural waters is controlled by the pH (Figure 7.17), while by the same token the pH of both fresh waters and the oceans is largely controlled by equilibria with atmospheric CO_2 and carbonate rocks (with perhaps a lesser contribution from silica mineral substances). The carbonate equilibria tend to buffer natural water systems thereby restraining them from becoming either highly acid or highly alkaline. For the majority of open lakes, the pH range is from 6.0 to 9.0 with seepage lakes and lakes in regions of acid igneous rocks having values below 7.0 and very calcareous waters having values in excess of 8.0 (for some exceptionally high and low pH values see Table 7.8). Biological processes can give rise to cyclic fluctuations in the pH (Figure 7.18).

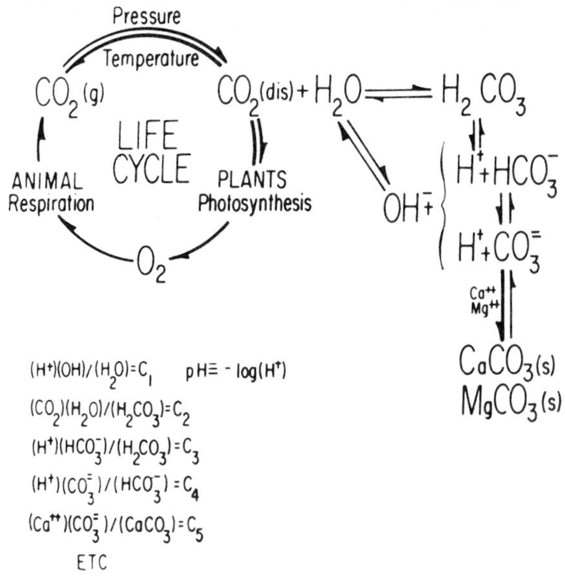

Figure 7.16. The carbon dioxide–carbonate cycle (from Horne, 1969).

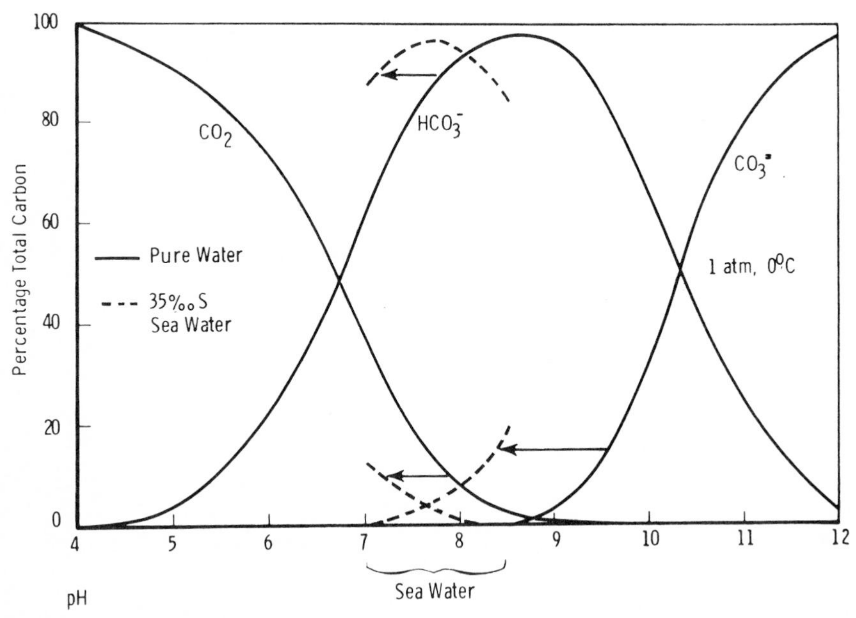

Figure 7.17. Speciation of the CO_2–HCO_3^-–CO_3^{2-} system in pure water and seawater as a function of pH at 1 atm (from Horne, 1969).

Often the calcium and bicarbonate concentrations in lake waters may be in excess of the level corresponding to saturation with respect to the atmosphere suggesting the presence of appreciable amounts of suspended colloidal $CaCO_3$. Lee (1970) has proposed that the absence of suitable nucleation centers in very pure lake waters may be responsible for the failure of mineral species such as hydroxyapatite and $CaCO_3$ to precipitate. During summer stagnation, the vertical distribution of CO_2

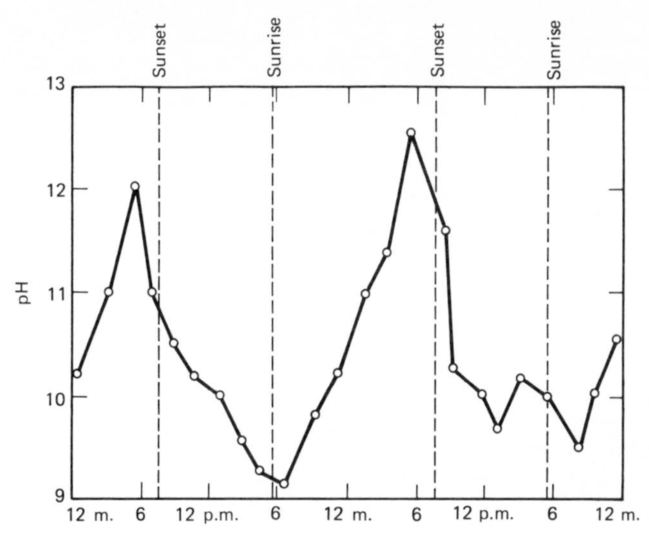

Figure 7.18. Diurnal pH changes in a small fresh water lake near Cape Town (from Livingstone, 1963).

and HCO_3^- often varies as the inverse of oxygen profile. This imbalance with respect to photosynthetic and respiratory cycles hints that additional bicarbonate may be finding its way into the waters from anaerobic processes in the underlying bottom muds, possibly as ammonium bicarbonate and as bicarbonates of iron, manganese, calcium, and magnesium.

We noted previously that silica, in addition to carbonate, equilibria determine the pH of natural waters. At ordinary pH's, silica is primarily in the form of undissociated orthosilicate, with perhaps some colloidal SiO_2 and complex aluminosilicate ions such as $Al(OH)(HSiO_3)^+$ or $Al(HSiO_3)^{2+}$. The silica levels listed in Tables 7.6, 7.9, 7.10, and 7.11 show a considerable range of values, and levels as high as 77.5 mg/liter have been reported in open lake water. Tropical fresh waters tend to exhibit somewhat higher silica levels than do waters in the temperate zone. In addition to silica added to lakes by rivers from the solution of crustal rocks, silica is also added to waters from bottom sediments. A major mechanism, over and above inorganic precipitation, for the removal of silica from lake waters is provided by diatom blooms. The siliceous skeletal remains of these organisms may settle to the bottom before they can redissolve.

Of the nitrogen and phosphorus cycles in fresh water systems, especially in lakes and ponds, we have more to say in our discussion of eutrofication in Chapter 16. Hutchinson (1957) observes that "phosphorus is in many ways the element most important to the ecologist, since it is more likely to be deficient, and therefore to limit the biological productivity of any region of the earth's surface, than are the other major biological elements."* Reduced forms of phosphorus, such as phosphine (P_2H_4), occur only rarely in nature and under extreme reducing conditions. The eerie will-of-the-wisp (*ignis fatuus*) may be due to the combustion of bacterially produced phosphine in stagnant swamp waters.

* But see what we shall have to say about the trace element iron.

It is useful to classify the total phosphate into several forms:

Soluble phosphate P
Organic soluble (and colloidal) P
Sestonic P { organic
acid soluble (mainly ferric and perhaps calcium phosphate)

The soluble phosphate is commonly a highly variable but small fraction, about 10% of the total (Table 7.12). At any given time a sizeable portion of the total P will be contained in living organisms. Additional total P values are given in Table 7.13. Brown waters rich in humic materials tend to have higher P levels than clear waters. Local geology also influences the phosphate content, the P-level tending to be higher for waters exposed to lowland sedimentary rocks and lower for waters draining the crystalline rocks of mountain ranges.

All phytoplanktonic organisms remove phosphorus from the water column, and the element is returned to the waters upon their death and decomposition. The stoichiometry of phosphate (and nitrate) utilization can be represented by

$$106CO_2 + 16NO_3^- + HPO_4^{2-} + 122H_2O + 18H^+ \underset{\text{respiration}}{\overset{\text{photosynthesis}}{\rightleftharpoons}} C_{106}H_{263}O_{110}N_{16}P(\text{algae}) + 138O_2 \quad (7.15)$$

Table 7.12 Fractionation of Total Phosphorus in Wisconsin Lakes and in Linsley Pond (From Hutchinson, 1957)

	Mean, N. Wisconsin (mg/m^3)	Mean, Linsley Pond (mg/m^3)
Soluble Phosphate P	3	2
Organic Soluble P	14	6
Sestonic P	6	13
Total P	23	21

Table 7.13 Regional Variation in Total Phosphorus in Surface Waters (From Hutchinson, 1957)

Region	Mean (mg/m^3)	Extremes (mg/m^3)
Northeastern Wisconsin	23.0	8–140
Connecticut, E. highland	10.8	4–21
Connecticut, W. highland	13.0	7–31
Connecticut, C. lowland	20.0	10–31
Japan	14.8	4.4–43.5
Austrian Alps	20.0	0–46
Prov. Uppland, S. Sweden	38.3	2–162
Prov. Dalarna, S. Sweden	26.0	4–92
N. Sweden	21.1	7–64

and as a consequence of the rain of this material down through the waters "the plankton acts as a conveyor of P into the deep water layers" (Stumm and Morgan, 1970). Figure 7.19 summarizes some of the major processes in the complex P-cycle. This figure implies extensive exchange of phosphorus between water and bottom sediments, and it has been suggested that bottom muds may act as a reservoir for P. We return to this topic again in our discussion of lithosphere–hydrosphere interaction (Chapter 13) and of eutrophication (Chapter 16). During periods of stagnation the oxygen-deficient, near-bottom waters may become enriched in phosphate by the decomposition of sedimented plankton and the reduction of phosphate-containing precipitates of ferric iron. An oxidized mud layer at the sediment–water boundary will tend to hold phosphate as sell as block the diffusion of phosphate and iron(II) from deeper mud layers.

The residence times for P in its various forms in lake waters range from 5 min (for exchange between dissolved phosphate and phytoplankton) through 8 hr (for exchange between inorganic phosphate and dissolved organic P) to 3 days (for bacterially accelerated water–sediment exchange) and 15 days (for abiotic water–sediment exchange) (Hayes and Phillips, 1958). Hutchinson (1957) tabulates even longer water–sediment turnover times ranging from 39 to 176 days.

Nitrogen, like oxygen, dissolves from the atmosphere in equilibrium with water at the surface (for nitrogen solubilities see Table A.4). But unlike oxygen, nitrogen gas itself is quite inert chemically. Nevertheless, nitrogen may exist in a number of

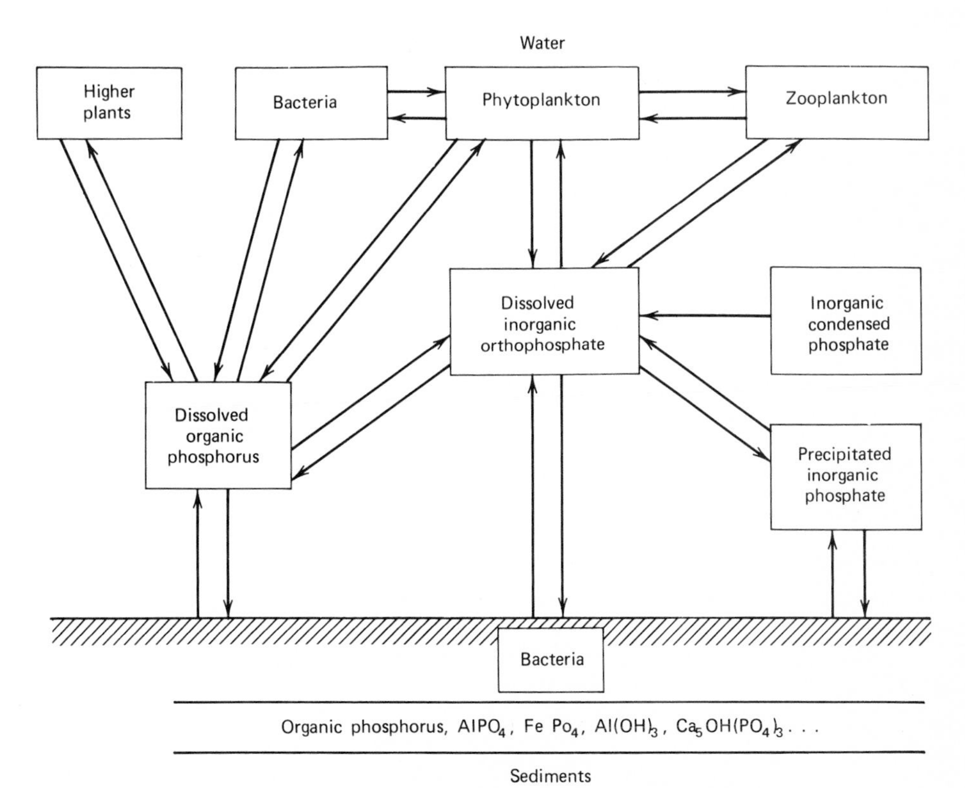

Figure 7.19. The phosphorus cycle (from Phillips, 1964).

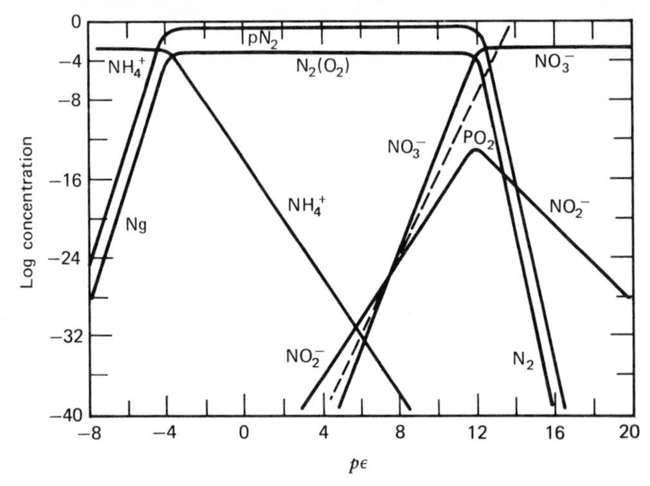

Figure 7.20. Equilibrium forms of nitrogen at pH 7 as a function of redox conditions, *pε* (from Stumm and Morgan, 1970).

different valence states in the natural environment depending on the prevailing redox conditions (Figure 7.20). Except in lakes where large blooms of the nitrogen-fixing algae *Anabaena* occur, most of the nitrogen in lake waters has inflowing water as its source, even though nitrogen-fixing bacteria are found in the waters as well as the sediments. Table 7.14 compares the N balance of some lakes of differing productivity. It is not surprising, then, to find that the disturbance of the ecology of the watershed feeding the waters can profoundly alter their nitrogen content. When a

Table 7.14 Forms of Nitrogen in the Influents and Effluents of Lakes in the English Lake District (From Hutchinson, 1957)

Lake	Total N in Mud (% dry wt.)	Influent			Effluent		
		$N \cdot NO_3$	Org. N	Tot. N	$N \cdot NO_3$	Org. N	Tot. N
		LESS PRODUCTIVE					
Wastwater	0.45	0.10	0.13	0.23	0.11	0.10	0.21
Ennerdale	0.37	0.06	0.12	0.18	0.05	0.16	0.21
Buttermere	0.34	0.07	0.17	0.24	0.06	0.20	0.26
Crummock	0.40	0.06	0.20	0.26	0.08	0.22	0.30
Hawes Water	—	0.08	0.13	0.21	0.08	0.19	0.27
Thirlmere	—	0.07	0.44	0.51	0.15	0.38	0.53
Mean	0.39	0.07	0.20	0.27	0.09	0.21	0.30
		MORE PRODUCTIVE					
Loweswater	0.88	0.36	0.26	0.62	0.09	0.26	0.35
Esthwaite	0.85	0.29	0.87	1.16	0.19	0.55	0.74
Blelham Tarn	1.03	0.35	0.40	0.75	0.16	0.30	0.46
Mean	0.92	0.33	0.51	0.84	0.15	0.37	0.52

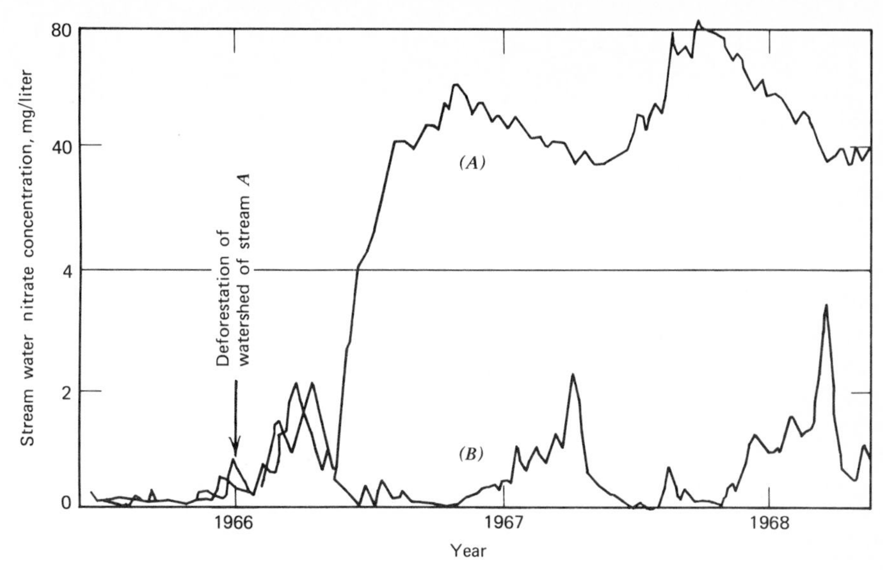

Figure 7.21. Effect of deforestration (*A*) on the normal cycle (*B*) of nitrate content of stream water (adapted from Bormann and Likens, 1970).

watershed is deforested, summer vegetation fails to take up this nutrient, and the highly leachable forms of nitrogen are carried away by the waters and lost to the ecosystem (Figure 7.21). Other ionic constituents, such as potassium and calcium, are similarly lost.

In the course of the decomposition of organic matter, heterotrophic bacteria produce ammonia. Surface absorbed ammonia can then be oxidized up to nitrate. The reverse process, or denitrification, can also occur, and, although bacteria are known that can liberate molecular nitrogen, most nitrate reduction produces ammonia. These oxidation–reduction processes involving nitrogen are highly dependent on pH, the concentrations of electrons or $p\epsilon$, and biological activity and as a consequence show very pronounced seasonal fluctuations (Figure 7.22). Nitrite and hydroxylamine intermediate species in these redox processes are sometimes detectable in lake waters.

In addition to silicate, nitrate, and phosphate, still another important anionic nutrient in lake waters is sulfate. The sulfate content of fresh waters, unlike seawater, often exceeds the chloride content (Tables 7.6, 7.9, 7.10, and 7.11) reflecting the high SO_4^{2-}/Cl^- ratio of rain water and may be very high (see Red Lake in Table 7.8)—sometimes with tragic results as in the case of certain shallows of African lakes where thousands of young flamingos perish from heavy shackles of gypsum ($CaSO_4 \cdot 2H_2O$) that grow about their legs and prevents their flight. We sometimes forget that Nature can be as indifferent as man and wreck chemical havoc in the environment by natural processes. In the absence of the more accessible oxygen forms such as dissolved molecular oxygen and nitrate, certain bacteria can utilize sulfate as a hydrogen acceptor for the metabolic oxidation of organic material. This process, together with the release of sulfur combined in reduced form in proteinaceous organic material, can produce H_2S during summer stagnation in hypolimnetic lake muds, but unless the waters are quite acid, FeS precipitates and the H_2S does not

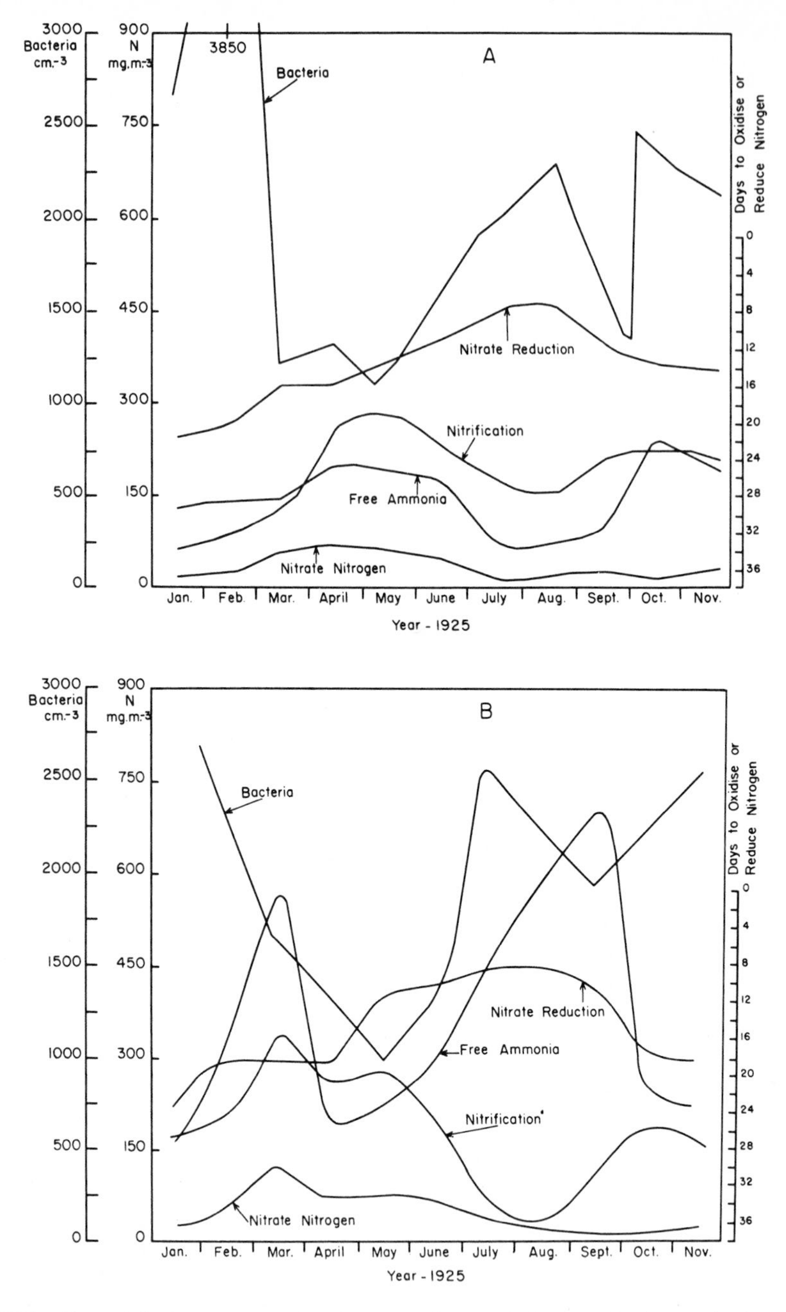

Figure 7.22. Bacterial counts, ammonia, nitrate, and rate of nitrate reduction and of surface (*A*) and bottom (*B*) water samples from Lake Mendota, 1925 (from Hutchinson, 1957).

appear in the waters of the deep hypolimnion. If the sulfate influx into anoxic waters is unusually great as the consequence of either natural (such as in Big Soda Lake, Nevada) or polluted (such as paper mill waste residues) conditions, considerable quantities of H_2S can be generated (786 mg/liter at the bottom of Big Soda Lake).

We have mentioned iron both in connection with the phosphorus and sulfur cycles. Iron "is probably the key element that controls phosphorus concentrations in many natural waters" (Lee, 1970), yet ferric phosphate has never been isolated from sedimentary material. What we may very well be dealing with here is the sorption of phosphate on hydrous ferric oxide. The chemistry of iron in the environment is as complex as it is important, and Figure 7.23 attempts to summarize some of the more significant of these processes together with some manganese chemistry—another strategic element in the environment. With respect to the environmental importance of iron, Sillen (1954) has hypothesized that the reaction

$$12FeOOH(s) \rightleftharpoons 4Fe_3O_4(s) + 6H_2O + O_2(g) \tag{7.16}$$

may control the partial pressure of oxygen in this planet's atmosphere or at least act as a "brake" to slow the increase of atmospheric oxygen from global photosynthetic

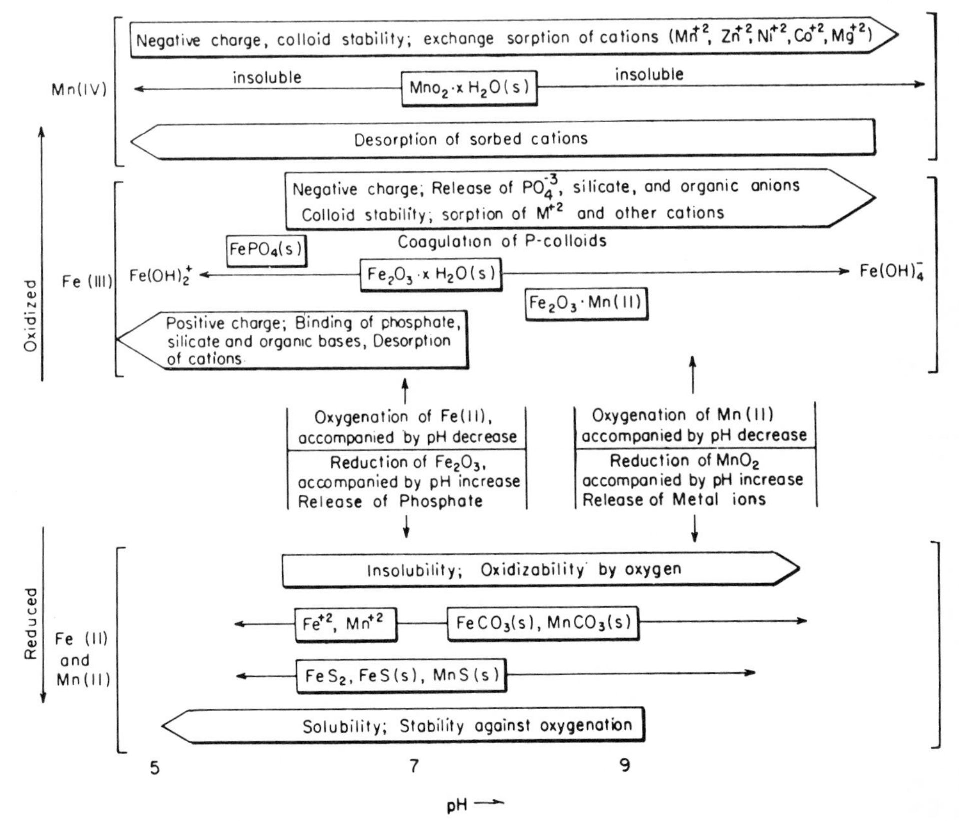

Figure 7.23. Summary of the major features of the chemistry of iron, manganese, and phosphorus (from Stumm and Morgan, 1970).

activity. A numer of years ago, Cooper (1948) proposed that Fe(III) phosphate and organic complexes may be important in marine biochemistry. If anything, this insight may be a modest understatement for more recent evidence seems to indicate that iron, and in the proper organically complexed form to make it biologically accessible, rather than the familiar nutrients N and P, may control biological productivity in the oceans (Menzel and Ryther, 1961; Barber and Ryther, 1969; Ryther and Guillard, 1959). Finally Baas Becking, Kaplan, and Moore (1960) have listed the redox reactions of iron and sulfur, along with photosynthesis and respiration, as being chiefly responsible for determining the all-important E_H–pH characteristics of natural environmental systems. The oxidation–reduction characteristics of the environment are closely related to and every bit as important as the acid-base characteristics (Stumm, 1967; Morris and Stumm, 1967), and in order to emphasize their similarity, in addition to the redox potential, E_H, a second parameter, $p\epsilon$, has been defined as an expression of the concentration of electrons just as pH is a quantitative measure of the concentration of protons. The essential acidity and redox relationships are defined and compared in Table 7.15.

The range of redox conditions in natural waters is just as broad and variable as the range of acidity conditions and include on the one extreme waters supersaturated with molecular oxygen to anoxic, anaerobic strongly reducing waters. Also, as we shall see again when we examine the history of the chemical evolution of the Earth's environment, the principle event in that history was the advent of photosynthesis and, as a consequence of the oxygen thus produced, the transition from a reducing to an oxidative environment. To dwell a bit on the importance of that event, life could probably not have begun under oxidative conditions, yet life almost certainly could not have evolved to the highest (and respiratory) forms without later oxidative conditions. The major milestones in the essential energy utilization mechanisms of biological evolution—fermentation, photosynthesis, and respiration—are all chemical.

When corrected to pH 7, the potential of most epilimnetic lake waters is between 0.4 and 0.5 V, a range slightly below 0.52 V, the value corresponding to the equilibration of air with water at 1 atm and 25°C. As long as there is some O_2 in the waters, the potential is relatively insensitive to the amount; thus some lakes exhibit a fairly constant potential with depth in their hypolimnion. On the other hand, in many other lakes, especially smaller ones, hypolimnion potentials are below those of the epilimnion, and there is a very rapid fall-off in potential near the bottom (Figure 7.24) due to reducing substances, notably Fe(II), in the mud. The presence of reducing organic matter in the water column itself also reduces the potential. Deep water sediments usually represent reducing environments; however, if the muds are exposed to oxygenated waters, an oxidized surface layer may be formed. Perhaps you have seen such brown layers (colored by hydrated ferric oxide) in ponds and even in home aquaria.

To return to the fresh water chemistry of iron, Figure 7.24 shows the verticle distribution of Fe(II) in Linsley Pond along with temperature, E_H, and dissolved oxygen profiles, while Figure 7.25 shows a much more complex total iron profile obtained in Oyster Pond on Cape Cod (Horne and Woernle, 1972). The bottom waters of this small coastal pond are anoxic (Emery, 1969), and the iron profile (and manganese also) exhibits a sharp maximum in the zone between oxygenated and anoxic waters. We were fortunately making measurements when the pond happened to be mixed by a violent coastal storm so that we were able to witness not only the

Table 7.15 pH and $p\epsilon$ Relationships[a] (From Stumm and Morgan, 1970)

$\mathrm{pH} = -\log \{H^+\}$		$\mathrm{p}\epsilon = -\log \{e\}$	
Acid-base reaction: $HA + H_2O = H_3O^+ + A^-$; K_1	(1*a*)	Redox reaction: $Fe^{3+} + \frac{1}{2}H_2(g) = Fe^{2+} + H^+$; K_1	(1*b*)
Reaction (1*a*) is composed of two steps:		Reaction (1*b*) is composed of two steps:	
$HA = H^+ + A^-$; K_2	(2*a*)	$Fe^{3+} + e = Fe^{2+}$; K_2	(2*b*)
$H_2O + H^+ = H_3O^+$; K_3	(3*a*)	$\frac{1}{2}H_2(g) = H^+ + e_6$ K_3	(3*b*)
According to thermodynamic convention: $K_3 = 1$		According to thermodynamic convention: $K_3 = 1$	
Thus: $K_1 = K_2 = K_2K_3 = \{H^+\}\{A^-\}/\{HA\}$	(4*a*)	Thus: $K_1 = K_2 = K_2K_3 = \{Fe^{2+}\}/\{Fe^{3+}\}\{e\}$	(4*b*)
or $\mathrm{pH} = \mathrm{p}K + \log [\{A^-\}/\{HA\}]$	(5*a*)	or $p\epsilon = p\epsilon^\circ + \log [\{Fe^{3+}\}/\{Fe^{2+}\}]$	(5*b*)
Since $\mathrm{p}K = -\log K = \Delta G^\circ/2.3\,RT$		Since $p\epsilon^\circ = \log K = -\Delta G^\circ/2.3\,RT$	
$\mathrm{pH} = \Delta G^\circ/2.3\,RT + \log [\{A^-\}/\{HA\}]$	(6*a*)	$p\epsilon = -\Delta G^\circ/2.3\,RT + \log [\{Fe^{3+}\}/\{Fe^{2+}\}]$	(6*b*)
or for the transfer of 1 mole of H^+ from acid to H_2O:		or for the transfer of 1 mole of *e* from oxidant to H_2:	
$\Delta G/2.3\,RT = \Delta G^\circ/2.3\,RT + \log [\{A^-\}/\{HA\}]$	(7*a*)	$-\Delta G/2.3\,RT = -\Delta G^\circ/2.3\,RT + \log [\{Fe^{3+}\}/\{Fe^{2+}\}]$	(7*b*)
For the general case where *n* protons are transferred:		For the general case where *n* electrons are transferred:	
$H_nB + nH_2O = nH_3O^+ + B^{-n}$; β^*	(8*a*)	$\mathrm{ox} + \{n/2\}H_2 = \mathrm{red} + nH^+$; $\mathrm{ox} + ne = \mathrm{red}$; K^*	(8*b*)
$\mathrm{pH} = \{1/n\}\mathrm{p}\beta^* + \{1/n\} \log [\{B^{-n}\}/\{H_nB\}]$	(9*a*)	$\mathrm{p}\epsilon = \{1/n\} \log K^* + \{1/n\} \log [\{\mathrm{ox}\}/\{\mathrm{red}\}]$; $p\epsilon = p\epsilon^\circ + \{1/n\} \log [\{\mathrm{ox}\}/\{\mathrm{red}\}]$	(9*b*)
$\mathrm{pH} = \Delta G/n2.3\,RT$			
$= \Delta G^\circ/n2.3\,RT + \{1/n\} \log [\{B^{-n}\}/\{H_nB\}]$	(10*a*)	$\mathrm{p}\epsilon = -\Delta G^\circ/n2.3\,RT + \{1/n\} \log [\{\mathrm{ox}\}/\{\mathrm{red}\}]$	(10*b*)
$\Delta G = -nFE$ (E = acidity potential)	(11*a*)	$\Delta G = -nFE_H$ (E_H = redox potential)	(11*b*)
$\mathrm{pH} = E/\{2.3\,RTF^{-1}\}$		$\mathrm{p}\epsilon = E_H/\{2.3\,RTF^{-1}\}$	
$= -E^\circ/\{2.3\,RTF^{-1}\} + \{1/n\} \log [\{B^{-n}\}/\{H_nB\}]$	(12*a*)[a]	$= E^\circ/2.3\,RTF^{-1} + \{1/n\} \log [\{\mathrm{ox}\}/\{\mathrm{red}\}]$	(12*b*)[a]
Acidity potential:		Redox potential (Peters–Nernst equation):	
$E = E^\circ + \{2.3\,RT/nF\} \log [\{H_nB\}/\{B^{-n}\}]$	(13*a*)	$E_H = E_H + \{2.3\,RT/nF\} \log [\{\mathrm{ox}\}/\{\mathrm{red}\}]$	(13*b*)

[a] At 25°C, $2.3\,RTF^{-1} = 0.059$ (V/eq). From (10) and (12): 25°C, $p\epsilon = E/0.059$, $p\epsilon^\circ = E_H/0.059$.

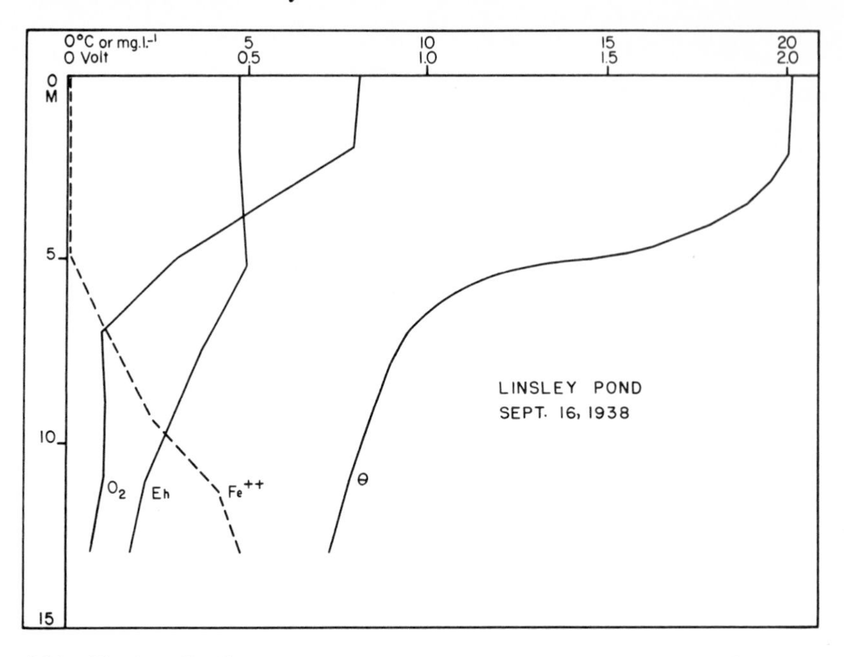

Figure 7.24. Vertical distributions of temperature, oxygen, redox potential, and ferrous iron in Linsley Pond, Connecticut, a lake with a clinograde oxygen curve (from Hutchinson, 1957).

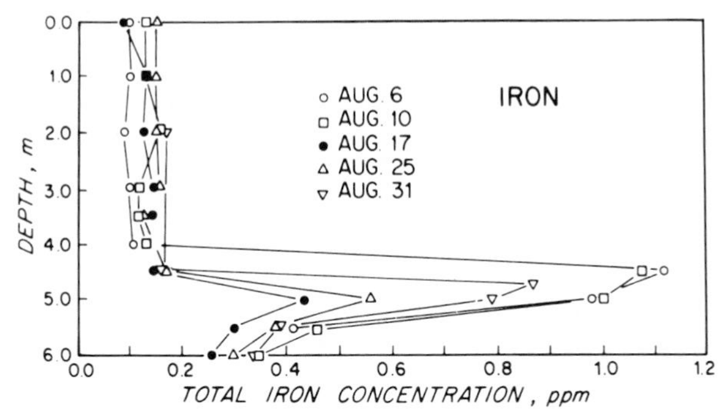

Figure 7.25. Total iron profile in Oyster Pond, Cape Cod (from Horne and Woernle, 1972, with permission of the publisher).

destruction but also thd reestablishment of the chemical profiles of the water column. We hypothesized that suspended pyrite, FeS_2, or hydrotroilite, FeS (Berner, 1964), is carried upward by diffusion or other mixing processes (Figure 7.26) and is oxidized to the Fe(III) state when it reaches the oxygen-containing waters with the precipitation of free sulfur

$$12H^+ + 4FeS_2 + 3O_2 \rightarrow 4Fe^{3+} + S_8 + 6H_2O \tag{7.17}$$

The amorphous FeOOH formed by the hydrolysis of Fe^{3+} may be colloidal and remain suspended at the boundary layer giving rise to the observed maxima. Upon further coagulation, the larger particles would settle, be converted to sulfides, and return to the bottom sediments.

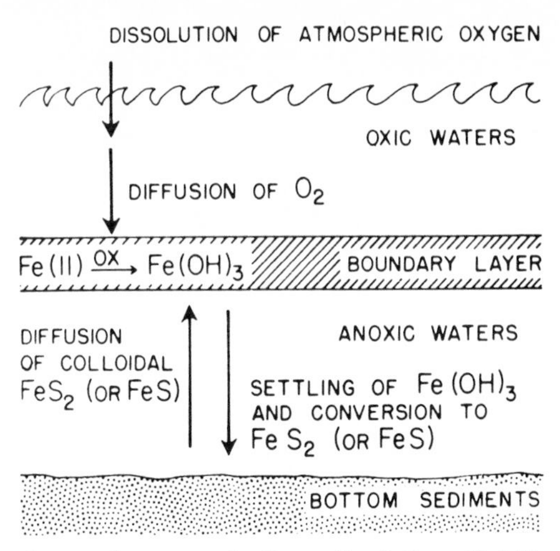

Figure 7.26. Iron chemistry and transport in Oyster Pond, Cape Cod (from Horne and Woernle, 1972, with permission of the publisher).

The iron in fresh waters can be present in several forms and valence states (Table 7.16). Phosphate complexes have already been mentioned. The suggestion recurs constantly that the iron may be tied up with humic acid (Lamar, 1968; Ishiwatari, 1969; Schnitzer, 1969) and other organic material from soils and swamps (Shapiro, 1957, 1964; Christman, 1967; Bruckert and Jackquin, 1969; Kahn, 1969; Takkar, 1969). The possibility has been raised that the iron-humic acid content of inflowing streams may be related to the deadly Florida red tide (*Gymnodinium breve*) (Ingle and Martin, 1971). As for the so-called and poorly understood humic acids, they complex organic material as well as inorganic ions, and Ogner and Schnitzer (1970) have proposed that one of them, fulvic acid, play a role in the transport and eventual immobilization of toxic pollutants in aquatic systems.

McMahon (1969) found a complex annual and diurnal variation of the acid-soluble Fe(II) profile in a small dimictic lake but was unable to resolve whether the

Table 7.16 Distribution of Various Fractions of Total Iron in the Surface Water of Linsley Pond (From Hutchinson, 1957)

	Range (mg/m^3)	Mean (mg/m^3)
Total Iron	70–300	170
Ferric Iron in Suspension	30–50	40
Ferric Iron in Solution	Not detectable	
Ferrous Iron in Suspension	10–40	20
Ferrous Iron in Solution	0–5(?)	<5
Nonreducible Iron in Suspension	4–180	75
Nonreducible Iron in Solution	20–50	30

observed cycles were due to photochemical and/or metabolic processes. The presence of Fe(II) in oxygen-containing waters is surprising, but it appears that in the absence of suitable surfaces the oxidation of this lower valence state can be very slow.

This brings us to the matter of the organic content of lake waters, and in this connection it is sometimes useful to distinguish between *autochthonous* organic matter actually produced IN the waters by photosynthesis and other biological processes and *allochthonous* organic material originating OUTSIDE the lake's basin and carried to the waters by streams and rivers. Among important sources of the latter are marginal and headwaters swamps and bogs and leaves and twigs fallen from trees in forested regions.

Table 7.17 gives a breakdown of the physical forms of the organic material. In Furesø, most of the nonsestonic organic matter is not colloidal, and most of the colloidal material is inorganic in nature. A recent and more microscopic examination of aquatic humus by ultrafiltration revealed that 10% of the organic carbon and 1% of the colored material have molecular weights below 1000, and 50 and 90%, respectively, have molecular weights in excess of 20,000 (Gjessing, 1970). Insight into the chemical nature of the organic material can be obtained from studying Tables 7.18 and 7.19 also taken from Hutchinson's (1957) classical treatise. As we might expect, much of the nitrogenous organic matter is composed of amino acids and their degradation products. In a series of Wisconsin waters investigated, the C:N ratio tended to increase from about 12 to 30 as the level of organic C increased from 1 to about 30 mg/liter. Inasmuch as allochthonous matter contains little N (C:N $\approx$ 47:1) and tends to be proportional to the amount of seston, the C:N ratio enables us to make an estimate of the relative importances of the antochthonous and allochthonous components of the organic content. Among specific organic compounds that have been found in fresh waters, vitamins might be mentioned; values as high as 0.89 mg/liter have been recorded for niacin, and the level seems to reach a maximum value under winter ice cover (Hutchinson, 1957).

7.4 RIVER CHEMISTRY, POLLUTION, AND AERATION

Most of the foregoing discussion has been devoted to ponds and lakes rather than to streams and rivers. From a chemical standpoint, the latter differ from the former in three very significant respects: (*a*) their water volume can be subject to much more drastic changes, (*b*) their surface to volume ratio is generally much larger, and (*c*) their mixing characteristics are much different being more subject to flow and

Table 7.17 Proximate Composition of Seston, Colloidal, and Soluble Organic Matter of Furesø (From Hutchinson, 1957)

Constituent	Fraction (mg %)	Protein (%)	Fat (%)	Carbohydrate (%)
Sestonic organic matter	1.56	53.2	9.9	36.9
Colloidal organic matter	0.67	41.5	11.7	46.8
Soluble organic matter	8.8	37.5	—	62.5

Table 7.18 Fractionation of Organic Nitrogen (mg/m³) in Various Lakes in Wisconsin (From Hutchinson, 1957)

Lake	Plankton N	Total Sol. N	Phospho-tung. Ppt. N	Free Amino N	Peptide N	Non-amino N
Michigan	20.5	383.4	58.2	40.5	56.9	45.9
February 28, 1924						
Devils						
October 27, 1922	38.3	337.7	136.1	21.8	103.6	176.4
July 11, 1922	15.2	281.4	132.4	74.9	72.4	71.3
October 5, 1923	14.3	314.3	110.4	69.5	67.7	39.3
Turtle						
January 18, 1923						
Bass	33.6	600.6	176.6	63.3	177.7	185.3
December 20, 1923						
Green						
July 18, 1923	42.3	424.0	130.3	93.3	96.0	120.5
October 17, 1923	48.7	460.8	136.8	70.9	122.7	102.8
Geneva	50.9	457.7	135.5	74.5	135.9	126.8
July 16, 1923						
Rock	73.1	683.0	143.0	120.5	176.7	252.7
July 6, 1923						
Madeline	95.6	616.5	150.5	82.2	151.1	221.3
December 12, 1923						
Waubesa	299.0	842.6	367.2	115.5	294.0	334.4
July 20, 1923						
Monona	388.8	885.8	326.6	49.0	315.8	347.3
July 20, 1923						
Kegonsa	696.6	839.4	379.0	99.7	299.3	331.3
July 20, 1923						
Wingra	882.0	896.0	298.0	78.9	299.5	343.6
July 6, 1923						

Table 7.19 Proximate Composition of Organic Matter from Northeastern Wisconsin Waters Containing Varying Amounts of Total Organic Carbon (From Hutchinson, 1957)

Carbon Content (mg/liter)	Organic Seston (mg/liter)	Organic Matter in Solution (mg/liter)	Crude Protein (%)	Ether Extract (%)	Carbo-hydrate (%)	C–N Ratio
1.0–1.9	0.62	3.09	24.3	2.3	73.6	12.2
5.0–5.9	1.27	10.33	19.4	1.3	79.0	15.1
10.0–10.9	1.89	20.48	14.4	0.4	85.2	20.1
15.0–15.9	2.32	31.30	12.9	0.2	86.9	22.4
20.0–25.9	2.22	48.12	9.9	0.2	89.9	29.0

less subject to stratification. In those parts of the United States with an ample supply of both types of water systems, lakes and rivers have been subjected to very different patterns of human abuse. In New England, for example, the industrial utility of the rivers was recognized in the eighteenth century, and the aesthetic and recreational virtues of lakes, in the nineteenth. As a consequence, while their shores are ringed with ugly cottages crowded shoulder to shoulder, the waters of our lakes remain surprisingly clean. The waters of our rivers, on the other hand, have been unspeakably befouled. Perversely enough, this has proved to be an asset for the filth of their waters has discouraged the development of the rivers' banks. You must travel many miles into the country (further every year) to find patches of unspoiled nature comparable to those between river and railroad tracks, between dumps and junkyards, within sight of industrial smoke stacks. In such by-passed places, I have watched heron and red-winged black birds nest, muskrats work, rabbits play, skunk cabbage unfold, and a host of other natural wonders within the Boston–Cambridge city limits.

Mr. Thoreau was enamoured of a pond, and Mr. Melville understood oceans, but rivers are my particular favorites, so I hope you will bear with me a bit longer as I digress further about their neglected virtues.

We tend to have only the most tenuous grasp of the influences of our fresh waters on our country's history and present. Our water resources have affected even our mores and social attitudes. For example, seaboard New Englanders look upon residents of the western part of the region as New Yorkers and objects of suspicion and distaste. The reason for this deep-seated attitude, in part, is the fact that the rivers of New England tend to run north and south, thus, prior to the advent of the railroads east–west communication, unlike coastal commerce, was relatively difficult.

It may even be said without too much exaggeration that New England's rivers shaped the course of the history of the United States. Water power and transportation promoted the rapid industrialization of New England to take place making these colonies self-sufficient to a high degree, and this self-sufficiency in turn made them self-confident and put them in a position where it was economically feasible to challenge the authority of the mother country. This same water power, coupled with the ruggedness of the land, in contrast to the southern colonies, encouraged manufacturing, discouraged large-scale farming, and gave rise to an extremely powerful and urbane nonagricultural mercantile elite. Although northern mills still devoured southern cotton, there ceased to be an immediate need for slave field labor in the north. Even more important, the flourishing New England manufacturing and shipping economy encouraged, albeit sometimes grudgingly, an intellectual elite conscious of European moral and political philosophy and at the same time free to face national moral problems squarely, to whom the southern way of life based on slavery was an abomination. It is no coincidence that the shores of the Merrimack River produced such Abolitionists as Whittier and Garrison. Were it not for the rivers of New England, the American Revolution and the War of 1812 would have been long delayed, and, quite possibly, The Civil War would never have been fought. It is quite possible that, were it not for the fast flowing waters of the Merrimack, the black man in America would still be in a more profound bondage than he presently finds himself.

But now to get back to business and the aforementioned differences between rivers and lakes. While lake levels may fluctuate somewhat, the volume of water in rivers often undergoes violent variations ranging from dryness in periods of drought to flood stage (Figure 7.27). Such traumatic experiences in the life of a river produce equally striking dislocations of its chemical composition (Table 7.20). The amount of pollutant relative to the volume of receiving water is the crucial factor in rational waste disposal so the planner and engineer must first come to grips with the hydrography of the river. This can be done with only a limited measure of success for the river's behavior for a given year may not correspond to the expected mean behavior (Figure 7.28), and its volume change can even fall outside the range of monthly means that can be expected for 80% of the years (Figure 7.28).

Where does a municipality throw its sewage in a period of extended drought when there is simply no or not enough water to carry the pollution away? Floods can be equally disasterous with enormous quantities of raw sewage being washed into the waterway when the volume capacity of the city's treatment facilities and storm drains is exceeded. In the great flood in Florence, extensive damage was done to building facades and art treasures by fuel oil flooded out of domestic storage tanks. Flood

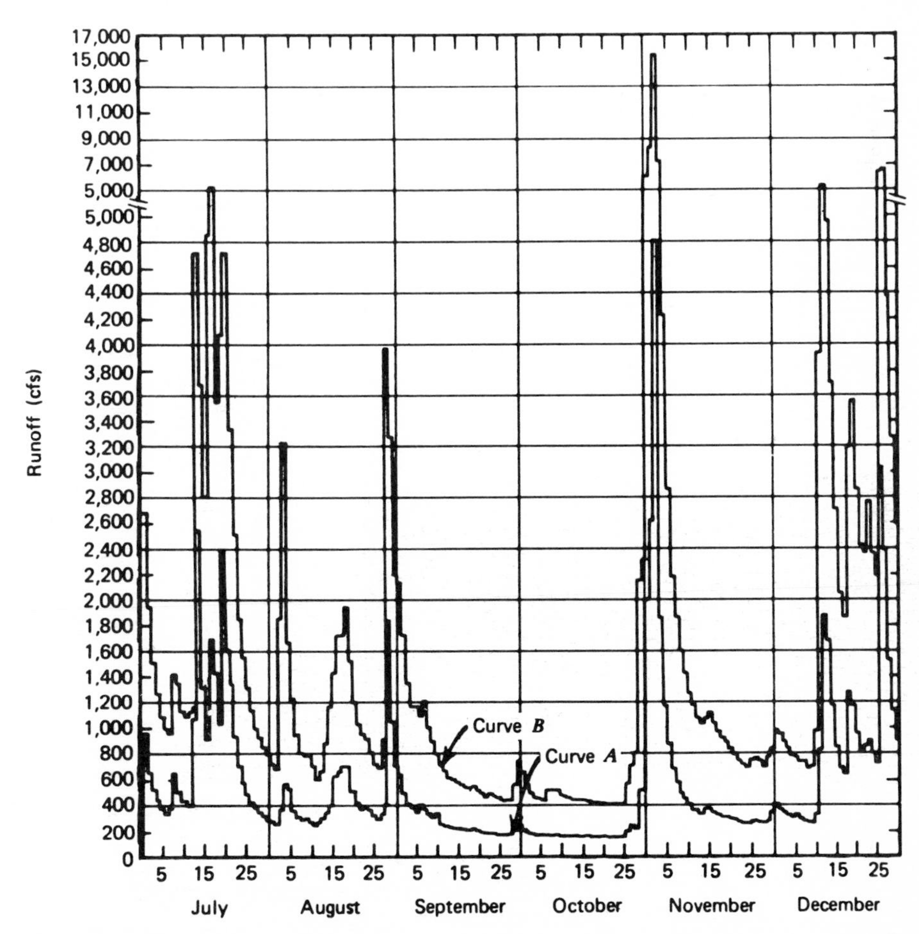

Figure 7.27. Daily hydrograph of discharge for the James River and a tributary, the Jackson River, for 1949: (*A*) Jackson River (409-mi² drainage area) at the Falling Springs Gage; (*B*) James River (1369-mi² drainage area) at the Lick Run Gage (from Velz, 1970).

Table 7.20 Moreau River at Bixby, South Dakota, Showing Changes in Chemical Composition of a Stream in a Semiarid Region[a] (From Livingston, 1963)

Date of Sample	Mean discharge (ft^3/sec)	pH	Percent													Total ions (ppm)
			SiO_2	Fe	Ca	Mg	Na	K	CO_3	HCO_3	SO_4	Cl	F	NO_3	B	
1949																
October 1–3	3.2	8.9	0.2	0.002	0.5	1.0	29	0.3	2.6	22	43	0.9	0.02	0.05	0.02	3400
October 4–12	16	8.9	0.3	0.002	0.6	0.6	29	0.2	3.2	34	31	0.7	0.02	0.08	0.02	2430
October 13–26	7.0	8.5	1.3	0.010	1.7	0.5	26	0.5	1.7	38	30	0.6	0.03	0.22	0.02	1200
October 27–31	8.0	8.3	1.0	0.005	1.9	0.2	28	0.4	1.7	39	28	0.6	0.02	0.09	0.02	1450
November 1–30	4.1	8.4	0.8	0.005	2.3	0.5	27	0.3	1.2	37	31	0.6	0.02	0.06	0.02	1740
December 1–21	3.2	8.4	0.4	0.002	1.3	0.9	27	0.3	1.2	37	32	0.6	0.01	0.02	0.02	3810
1950																
March 6–9	123	8.4	3.0	0.003	3.0	0.8	25	0.9	0.0	31	35	0.7	0.07	0.39	0.02	540
March 10	100	7.3	1.5	0.004	4.4	1.4	23	0.8	0.0	20	47	1.9	0.04	0.40	0.01	890
March 14–April 1	126	7.2	3.2	0.009	4.9	1.4	21	1.1	0.0	32	35	0.7	0.09	0.54	0.02	410

April 3	2900	7.2	8.1	—	7.5	1.4	15	3.2	0.0	35	25	0.3	0.25	2.73	0.06	160
April 4–6	2800	7.1	6.4	0.015	9.2	2.8	16	1.5	0.0	33	38	1.2	0.08	0.80	0.04	250
April 7	6590	7.4	5.4	0.027	10.1	2.6	11	1.3	0.0	49	19	0.7	0.07	0.64	0.03	300
April 11–14	1260	7.2	5.0	0.011	9.1	2.5	15	1.2	0.0	31	35	0.7	0.06	0.61	0.03	340
April 15–17	7600	7.4	5.6	0.014	8.2	2.1	15	1.2	0.0	37	29	0.9	0.07	0.37	0.04	270
April 18–20	1350	7.5	3.1	0.005	9.1	2.5	15	1.1	0.0	25	43	0.6	0.05	0.66	0.03	390
April 21	363	7.1	2.1	0.007	8.6	3.2	16	0.9	0.0	22	48	0.6	0.04	0.26	0.04	540
April 22–26	280	7.3	2.1	0.003	6.7	2.5	19	0.8	0.0	24	45	0.7	0.03	0.28	0.02	610
April 27	160	7.5	1.5	0.004	7.3	2.5	18	0.6	0.0	21	48	0.6	0.02	0.14	0.02	810
April 28–May 15	470	7.5	2.0	0.002	7.2	2.8	18	0.7	0.0	18	50	0.6	0.03	0.18	0.01	800
May 16–31	45	8.0	1.5	0.001	5.6	2.6	20	0.6	0.0	23	46	0.6	0.02	0.14	0.01	1170
June 1–30	30	7.9	0.5	0.002	3.8	4.5	23	0.5	0.0	19	50	0.6	0.03	0.07	0.01	2000
July 1–31	14	7.9	0.5	0.002	4.6	2.6	22	0.5	0.0	18	52	0.6	0.03	0.06	0.01	1300
August 1–5	4.1	7.9	0.6	0.002	2.1	1.6	25	0.5	0.0	24	45	0.9	0.04	0.07	0.02	1800
August 6–8	75	7.9	2.1	0.024	3.1	0.7	24	0.9	0.0	33	35	1.1	0.08	0.53	0.66	660
August 9–31	6.7	8.3	0.7	0.004	1.8	0.8	26	0.5	0.6	32	37	0.7	0.03	0.15	0.03	1430
September 1–19	4.5	8.3	0.6	0.003	1.4	1.1	27	0.4	0.8	29	39	0.8	0.03	0.08	0.03	1760
September 20–24	34	7.8	1.9	0.054	2.1	0.4	26	0.7	0.0	33	35	0.9	0.07	0.26	0.04	920
September 25–30	5.2	8.3	0.7	0.004	1.8	0.8	27	0.5	0.8	34	34	0.7	0.04	0.08	0.00	1440

[a] The drainage area above the sampling station is 1570 mi^2 and the data, which cover the water year October 1949–September 1950, have been recalculated from U.S. Geological Survey (1955b).

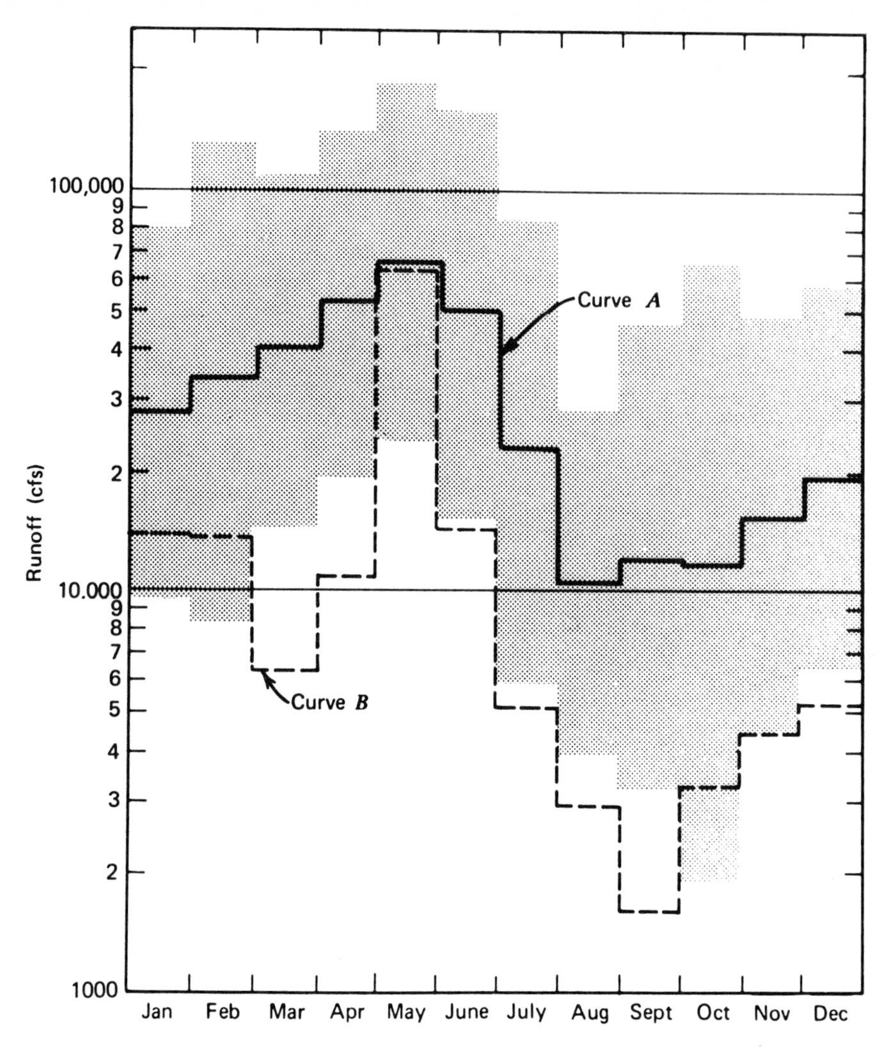

Figure 7.28. Season pattern (1934–1955) of run-off for the Arkansas River at Little Rock: (*A*) most probable monthly mean and (*B*) recorded monthly means for 1954. The shaded area is the range within which monthly means can be expected for 80% of years (from Velz, 1970).

control attempts often create more pollution problems than they solve, such as the stagnation of water behind dams and other impoundments.

Water volume is not the only factor strongly permuting the chemical content of river waters. Park et al. (1970) have studied the chemistry of the Columbia River in some detail. Simply, the quantities of chemicals involved in this large river are impressive. In 1966, 1.8×10^{14} liter of water, 8.3×10^{7} moles of phosphate, 2.1×10^{9} moles of nitrate, 2.7×10^{10} moles of reactive silicate, and 1.9×10^{11} moles of total CO_2 (largely in the form of HCO_3^-, since the pH range of the river's waters is 6.63 to 8.68, see Figure 7.17) passed a point 87 km above the river's mouth. The nutrients phosphate, nitrate, and silicate are strongly influenced by in-stream primary production, and their seasonal changes show a winter maximum and a summer minimum. Not only their total amount, but their relative amounts vary seasonally: The summer nitrate to phosphate ratio is 3:1, whereas for the rest of the year it is in excess of 19:1. In addition to Table 7.7, further data on the trace metal content of United States rivers can be found in Kopp and Kroner (1967).

A plentiful oxygen supply is essential to maintaining the vitality of natural waters. Inasmuch as the oxygen comes largely from atmospheric exposure (with a lesser contribution from photosynthesis), their large surface-to-volume ratio (along with their more violent agitation) is clearly an advantage that rivers have over ponds and lakes.

With respect to pollution, the simple and perfectly obvious fact that rivers flow gives them a tremendous chemical advantage for they possess therein an effective means of self-purification. Three types of polluting waste material may be distinguished: chemical, bacterial, and organic. Stream self-purification differs for each type. In the case of stable chemical wastes that suffer no or little change as they are carried along by the river, the primary ameliorating factor is dilution. In the case of bacterial contamination, as from sewage, in addition to dilution, organism mortality is improtant for self-purification. The diminuation of the biopopulations depends on time and temperature. In terms of the quantities involved, unstable organic matter is usually the most important of the three waste types, and here self-purification depends, still upon dilution, and upon the processes of biochemical decay. Removal by decay depends on the availability of dissolved oxygen, time, and temperature. The decay rate increases with increasing temperature, but unfortunately this advangage tends to be offset by the decrease in the solubility of oxygen.

A great deal of attention has been focused on organic waste material and organic self-purification or the reoxygenation of stream and river waters. This subject forms the heart of river sanitary engineering, and elaborate mathematical formalisms have been developed for the accurate analysis of such situations. The reader is referred to such standard texts on the subject as Phelps (1944) and Velz (1970). Suffice it here to remark that these authors both describe self-purification in accounting terms with the dissolved oxygen being the streamflow's assets and the waste material, expressed as biological oxygen demand (BOD), its liabilities. The rate of BOD satisfaction, or the amortization of the organic debt, is controlled by microorganisms. Under "normal" conditions at 20°C about 20% of the outstanding debt remaining from the previous day is amortized. If the oxygen demand exceeds the supply, then the river waters are in danger of bankruptcy. The bookkeeping is complicated by many factors; for example, in addition to the BOD of carbonaceous material, nitrogenous matter can draw upon the oxygen assets to be oxidized to nitrate. Such nitrification is the work of organisms that, unlike the hardy and ubiquitous carbon-consuming bacteria, are highly specialized, relatively delicate, and rare. Consequently, under natural conditions, nitrification is seldom important compared to the oxidation of carbonaceous material; nevertheless, it is often encountered in the determination of BOD rates in the laboratory. We return to this subject in Chapter 16 in our discussion of the chemistry of sewage and water pollution.

Inasmuch as oxygen is the coin of river solvency, the addition of oxygen artificially to the waters suggests itself as an obvious approach to decreasing the BOD and improving water quality. Speece (1971) has described a system for the aeration of a lake's hypolimnion without disturbing stratification, and Whipple et al. (1969) have studied in-stream aeration of the Passaic River in New Jersey in detail. We see one unexpected finding of the latter investigation: Oxygen aeration greatly accelerates the rate of deoxygenation below the point of injection, possibly by triggering nitrogenous biochemical oxygen demand. Reed (1971) has asked "why must our water be barren?" and has argued that "the problem with most polluted bodies of water is not the concentration of nutrients in the water, but lack of oxygen." He has proposed relatively inexpensive means such as building special weirs and cascades and modifying existing dams for increasing the aeration of our streams and rivers.

Let us pause at this point to consider a particular river. The Merrimack River is one of the few blue rivers in the world. It is a waterway of great natural beauty, and it has been justly celebrated in American literature by writers such as Whittier and Thoreau. In addition to Whittier, its native sons include Daniel Webster, Whistler, Mary Baker Eddy, and more recently Jack Kerouac. But it flows past many villages, towns, factories, and industrial cities as well as meadows and mountains. It is a grossly polluted river, and the loss in value from this spoilation has been estimated to be about $37,000,000/yr (Pahren et al., 1966). The drainage basin of the Merrimack is shown in relation to other New England rivers in Figure 7.29, while more detailed Figure 7.30 shows the large number of streams, smaller rivers, ponds, and lakes in

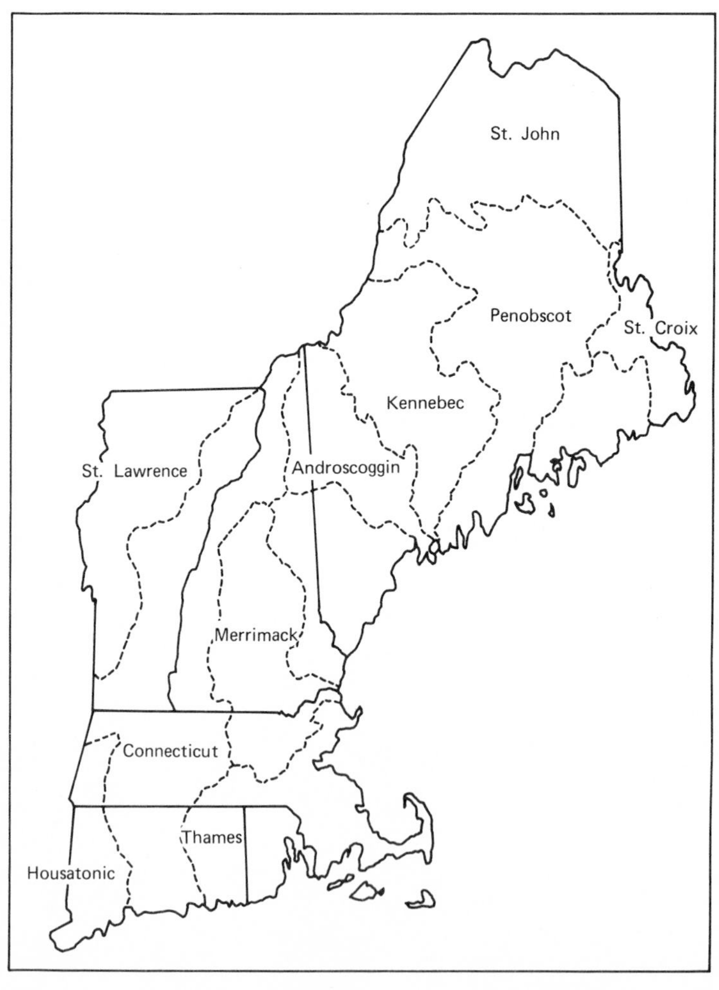

Figure 7.29. New England River drainage basins.

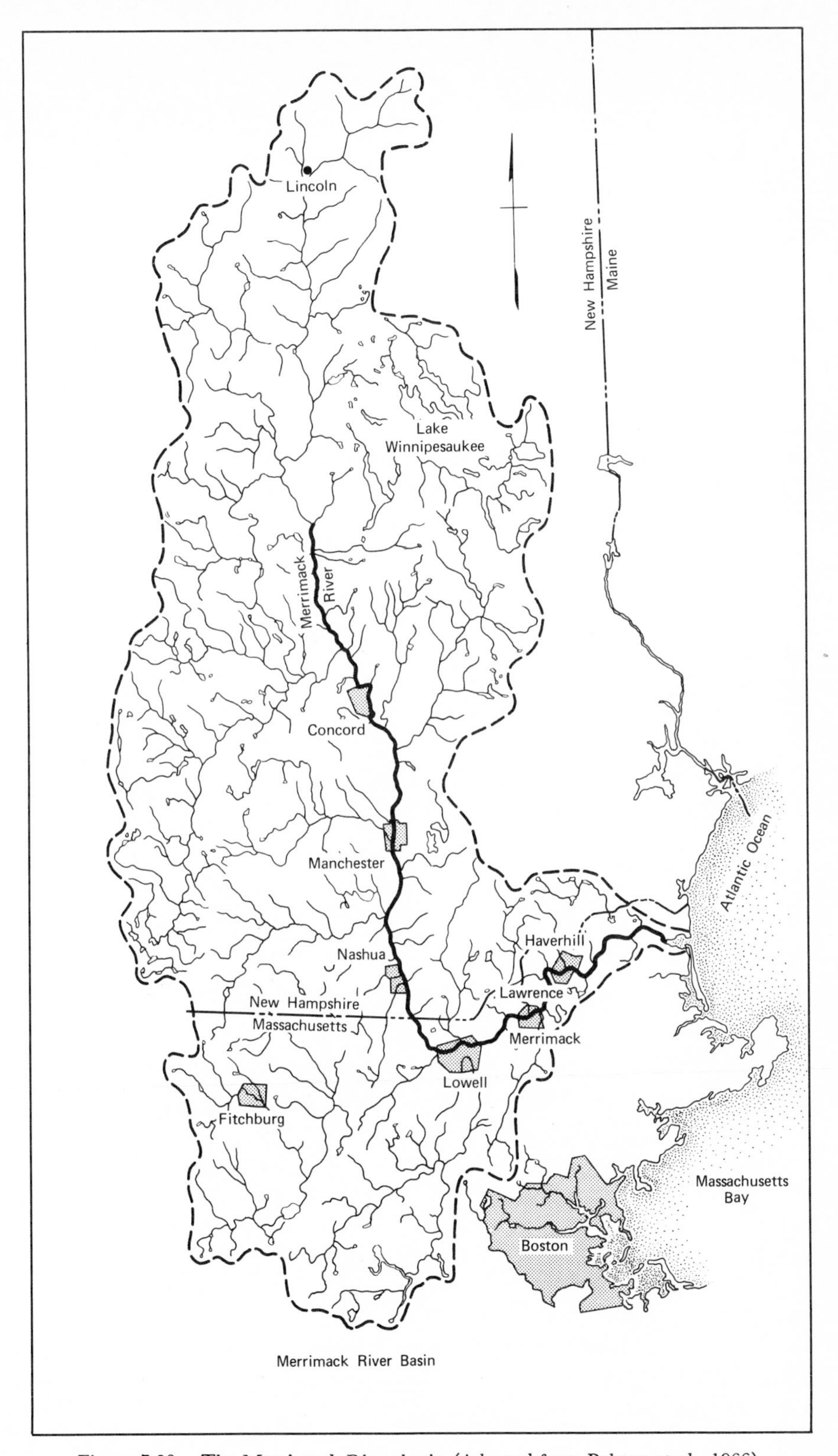

Figure 7.30. The Merrimack River basin (Adapted from Pahren et al., 1966).

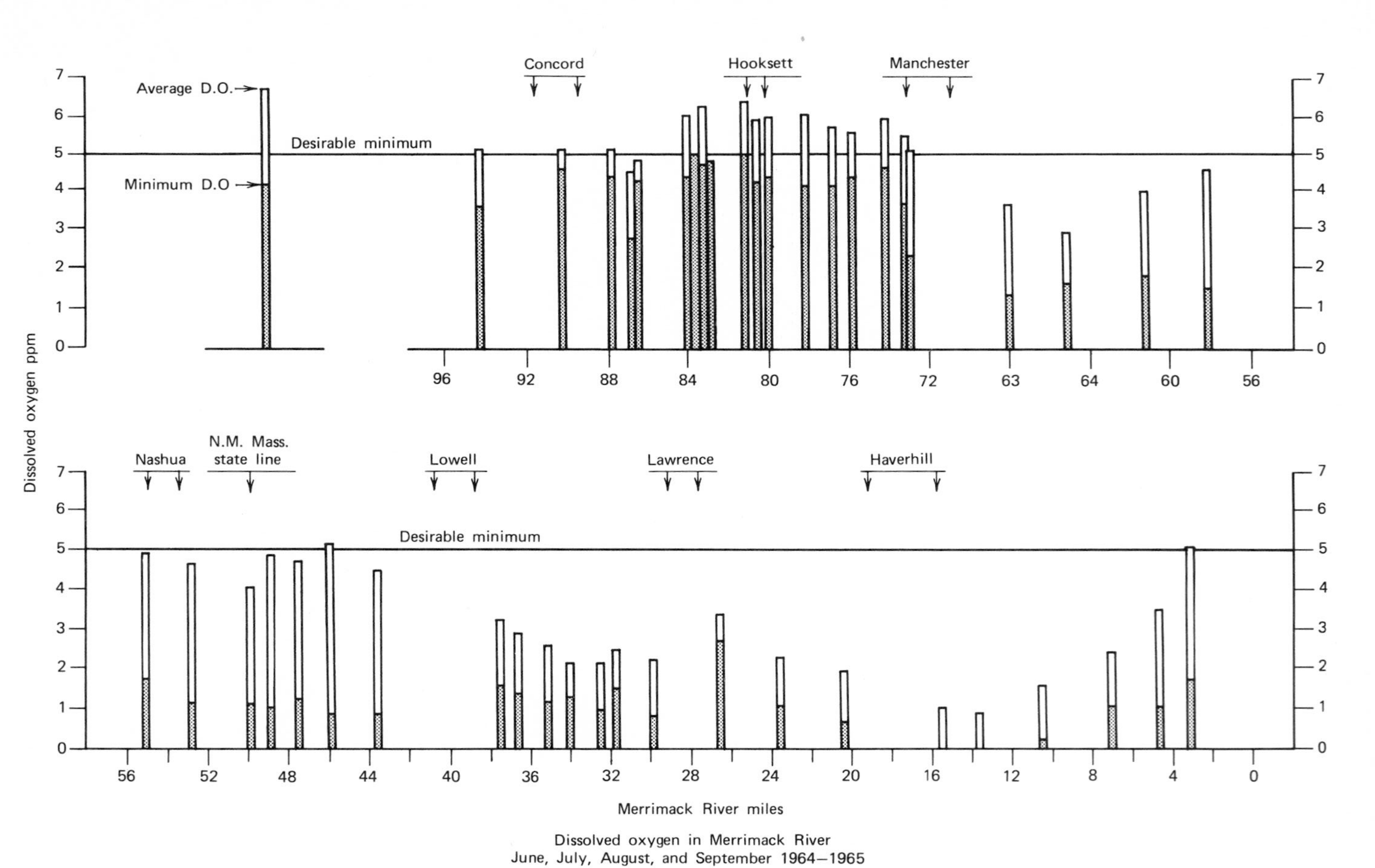

Figure 7.31. Dissolved oxygen in the Merrimack River (from Pahren et al., 1966).

the Merrimack basin. Table 7.21 lists the major polluters of the Merrimack in terms of population equivalents of the wastes dumped. Many of the smaller towns do not appear since their contributions are less than 1000 units each. The chief villains are the larger towns and cities (encircled) and industries, especially paper, tanning, and textiles. Particularly unfortunate is the enormous dose of pulp and paper filth that the river receives in Lincoln, New Hampshire, in a sense getting the waters off to a very bad start. The jobs to filth ratio of the pulp and paper industry is very low, but also included in the table is an example of a relatively clean industry, the large Western Electric operation in North Andover, Massachusetts, with a large jobs to filth ratio. The abuse of the river is reflected in its dissolved oxygen content, and it is clear from Figure 7.31 that by the time the Merrimack flows out of Manchester, New Hampshire, it is a very sick river. Coliform populations (Figure 7.32) confirm our picture that the lower waters of this once magnificent river are now little more than an open sewer. I am happy to say, however, that various state and federal agencies are presently hard at work trying to clean up the Merrimack, and the results of their efforts are beginning to show. In recent years, trade unionism has forced the textile industry out of the Merrimack valley to low labor cost regions in the south and overseas; so today the Merrimack is much cleaner than when I was a boy. Still the picture is not entirely hopeful. Due to low taxation, southern New Hampshire is one of the fastest growing areas in the nation. A feature of this rapid growth has been the proliferation of rural slums of house trailers that have so marred the natural beauty of the region as to endanger the state's important tourist industry and that commonly result in the uncontrolled passage of sewage directly into the

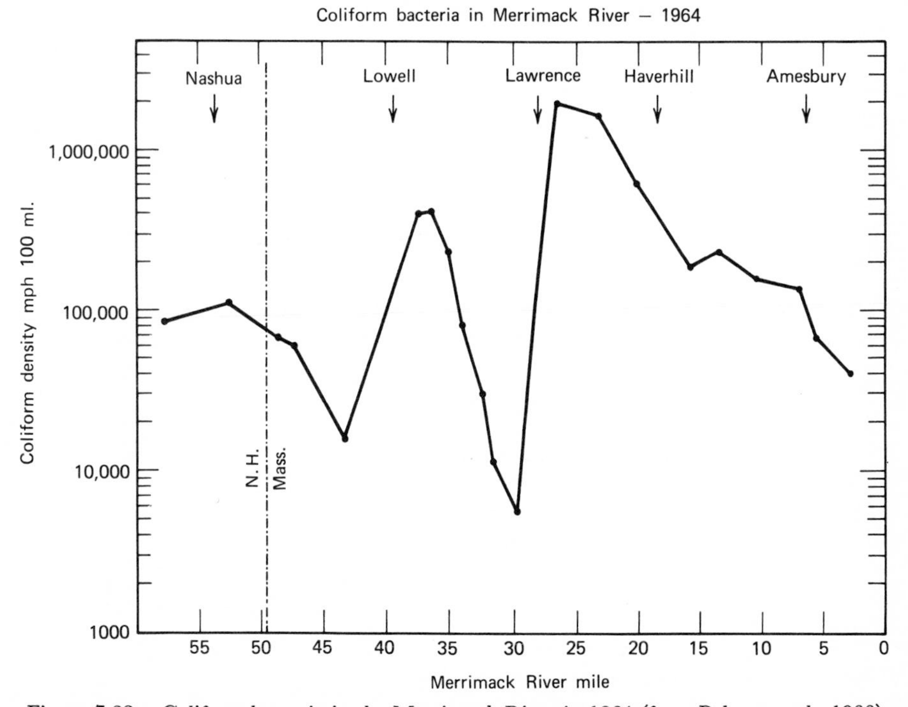

Figure 7.32. Coliform bacteria in the Merrimack River in 1964 (from Pahren et al., 1966).

Table 7.21 Estimated Characteristics of Sewage and Industrial Wastes Discharged to Merrimack River and Tributaries (Abstracted from Pahren et al., 1966)

Source	River Discharged to	Population Equivalents Discharged		
		Bacterial	Suspended Solids	Oxygen Demand
New Hampshire				
Franconia Paper Corp., Lincoln	Pemigewasset East Branch	—	200,000	400,000
Franklin	Winnipesaukee	4,500	4,500	4,500
Brezner Tanning Corp., Boscawen	Contoocook	—	2,500	1,500
Concord (Penacook Village)	Merrimack	2,000	50,000	32,000
Concord	Merrimack	24,000	24,000	24,000
Pembroke	Merrimack	1,800	1,800	1,800
Allenstown	Merrimack	1,250	1,250	1,250
Hooksett	Merrimack	1,000	1,000	1,000
French Bros. Beef Co., Hooksett	Merrimack	—	380	1,080
Manchester	Merrimack	72,500	72,500	72,500
M. Schwer Realty Co., Manchester	Merrimack	—	650	6,500
Granite State Packing Co., Manchester	Merrimack	—	19,000	46,000
MKM Knitting Mills Inc., Manchester	Merrimack	—	400	4,000
Seal Tanning Co., Manchester	Merrimack	—	8,000	5,000
Stephens Spinning Co., Manchester	Merrimack	—	400	4,000
Waumbec Mills Inc., Manchester	Merrimack	—	700	7,200
Foster Grant Co., Manchester	Merrimack	—	110	15,000
Merrimack Leather Co., Merrimack	Souhegan	—	12,000	7,500
New England Pole and Wood Treating Corp., Merrimack	Merrimack	—	—	—
Wilton	Souhegan	1,000	1,000	1,000
Hillsborough Mills, Wilton	Souhegan	—	7,000	3,500
Milford	Souhegan	3,000	3,000	3,000
Granite State Tanning Co., Nashua	Nashua	—	12,000	16,500
Sanders Associates, Nashua	Nashua	—	850	1,200
Nashua	Merrimack	28,500	28,200	30,300
Total		141,300	454,280	693,000
Massachusetts				
Weyerhaeuser Paper Co., Fitchburg	North Nashua	—	184,600	39,650
Fitchburg Paper Co., Fitchburg	North Nashua	—	108,200	37,060
Simonds Saw and Steel Co., Fitchburg	North Nashua	—	—	5,800
Falulah Paper Co., Fitchburg	North Nashua	—	115,400	27,940
Fitchburg	North Nashua	18,900	20,700	19,500
Mead Corp., Leominster	North Nashua	—	30,300	5,700
Foster Grant Co., Leominster	North Nashua	—	16,600	2,500
Leominster	North Nashua	3,000	5,200	12,140
Clinton	South Nashua	1,300	1,560	1,040
Hollingsworth and Vose Co., Groton	Nashua	—	1,470	6,650

(Continued)

Table 7.21 (*Continued*)

Source	River Discharged to	Population Equivalents Discharged		
		Bacterial	Suspended Solids	Oxygen Demand
Groton Leather Board Co., Groton	Nashua	—	5,880	2,120
St. Regis Paper Co., Pepperell	Nashua	—	64,700	16,200
Southwell Combing Co., Chelmsford	Merrimack	—	30,800	22,100
H. E. Fletcher Co., Chelmsford	Merrimack	—	2,940	150
Gilet Wool Scouring Corp., Chelmsford	Stony Brook	—	13,600	19,700
Dracut	Beaver Brook	1,000	1,000	1,000
Westborough	Assabet	300	1,760	2,900
Hudson Combing Co., Hudson	Assabet	—	1,000	950
Hudson	Assabet	70	1,080	720
Maynard	Assabet	510	1,020	680
No. Billerica Co., Billerica	Concord	—	1,410	5,530
Lowell Rendering Co., Billerica	Concord	—	5,300	11,000
Ames Textile, Lowell	Merrimack	—	18	1,850
Vertipile Inc., Lowell	Merrimack	—	210	2,220
Jean-Alan Products Co., Lowell	Merrimack	—	2,040	940
Robinson Top & Yarn Dye Works, Lowell	Merrimack	—	8	1,100
Middlesex Worsted Spinning Co., Lowell	Merrimack	—	18	1,550
Suffolk Knitting Co., Lowell	Merrimack	—	1,270	5,700
Commodore Foods Inc., Lowell	Merrimack	—	4,300	4,400
Lowell	Merrimack	90,000	95,000	112,000
Andover	Shawsheen	8,400	12,600	8,400
Mead Corp., Lawrence	Merrimack	—	22,500	9,300
Oxford Paper Co., Lawrence	Merrimack	—	51,100	32,100
Merrimack Paper Co., Lawrence	Merrimack	—	5,100	4,400
Lawrence Wool Scouring Co., Lawrence	Merrimack	—	13,500	9,180
Loom Weave Corp., Lawrence	Merrimack	—	440	1,760
Lawrence	Merrimack	70,000	149,000	120,000
Western Electric Co., North Andover	Merrimack	—	400	135
North Andover	Merrimack	9,000	18,800	13,600
Methuen	Merrimack	17,000	18,000	23,800
Continental Can Co., Haverhill	Merrimack	—	77,000	47,000
Hoyt & Worthen Tanning Corp., Haverhill	Merrimack	—	7,000	4,400
Haverhill	Merrimack	44,000	71,000	50,000
Groveland	Merrimack	1,000	1,000	1,000
Amesbury Fibre Corp., Amesbury	Merrimack	—	6,820	3,530
Merrimack Hat Co., Amesbury	Merrimack	—	235	1,120
Amesbury Metal Products Co., Amesbury	Merrimack	—	—	—
Amesbury	Powwow	7,200	14,000	11,000
Newburyport	Merrimack	140	7,700	10,000
Salisbury	Merrimack	1,250	1,100	1,620
Total		274,897	1,198,465	729,490
Grand Total		416,197	1,652,745	1,422,490

river and its tributaries. Pollution of the environment is inextricably linked to social, political, and economic conditions, and I think that these last two instances serve as examples to illustrate these sometimes not so devious links—how, on the one hand, high labor costs have helped to clean a river, while on the other, low taxes and political permissiveness have dirtied it.

7.5 WATER ADDITIVES

In addition to the unwanted materials that man simply dumps into the most convenient waters, our lakes and rivers are also contaminated with chemical substances that man adds to them deliberately. In some cases, this human intervention into aquatic ecology is entirely warranted on the basis of compelling reasons of public health, but in other instances, the validity of the apology for adding these chemicals is far from clear. Throughout the civilized world, chlorine is used to disinfect drinking water supplies. Such chlorination is very effective at reducing the coliform population and this precaution has undoubtedly saved inestimateable thousands of people from the ravages of epidemic typhoid, cholera, and dysentery. Chlorination is also used to prevent the growth of fouling organisms in power plant cooling water (Hamilton et al., 1970).*

Chemicals such as lime and ferrous or aluminum sulfate that form flocculent precipitates may be added to waters to accelerate the removal of excessive turbidity, while copper sulfate is sometimes added to reservoirs to control the algae which impart a "fishy" taste to drinking waters. In these instances, the dividing line between the protection of public health and aesthetics is not always sharply defined. Even less clear have been the scientific facts hidden somewhere in the current controversy over the fluoridation of drinking water (but for a careful presentation of these facts see McClure, 1970). Affronted that anyone should question their opinion the medical–dental profession has reacted to criticism with a vengeance and has been quite successful in getting their opponents dismissed as crackpots.

As we have already mentioned in the case of iron, many chemical substances in minute concentrations are essential to life, but they become deleterious and even fatal in larger concentrations. This is particularly true of certain trace elements, such as Fe, Co, Ni, Cu, Zn,† As, Sn, and Pb, but applies to ALL chemicals, both natural and artificial, even including such common substances as water (the symptoms of water intoxification include headache, tremor, ataxia, convulsions, stupor, coma, and death (Wakim, 1963), while a massive overdose is called drowning) and air (blindness in premature babies was traced to the indiscriminate use of oxygen tents). The effects of a chemical on organisms might be represented by a hypothetical diagram such as Figure 7.33. Health cannot be assured if its concentration falls below a certain point *A*, and there is a level, always unknown, of optimum concen-

* Recently (1976) disturbing evidence has come to light that carcinogens in lower Mississippi River drinking water may be formed from anthropogenic organic pollutants upon chlorination.

† Powerful metal complexers such as the chelating agent EDTA ethylene-diamine tetraacetic acid) thus can be toxic, and their use in detergents (as has been proposed in the case of nitriloacetic acid) is cause for concern. As an illustration, EDTA injected into pregnant rats results in young with gross congenital malformations unless their diet is supplemented with zinc (Swenerton and Hurley, 1971).

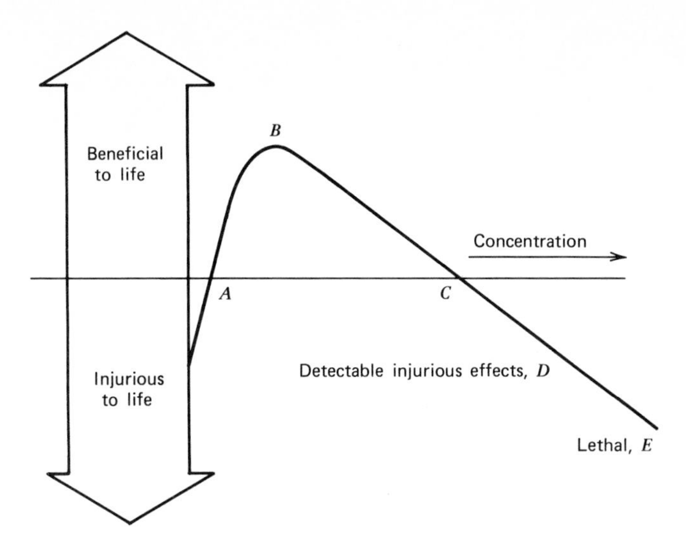

Figure 7.33. The dose–response curve. The influence of a chemical substance on a biosystem.

tration (point *B*). The difficulty in treating these materials in the environment and in the rational evaluation of their potential hazard is that commonly we know only the approximate lethal concentration (*E*) or at best the concentration level (*D*) capable of producing deleterious effects serious enough to be obvious. Ideally, our permissible level standards should be set at point *C*. The maximum *safe* dosage (*C*) is probably not known for *any* chemical substance, and such values will be most difficult to establish. So we are left with a wide concentration region between *C* and *D* in which small but possibly incremental "hidden" damage can be done. Perhaps the only rational rule to establish in setting concentration levels of pollutants is that the pollutant level should never be allowed to exceed the maximum value of that pollutant found naturally in unpolluted and life rich environments. In some cases where even naturally occurring levels produce disease, as in the case of teeth mottling in certain fluoride-rich districts, the permissible level of the pollutant would have to be set even lower than defined by the "natural" limit.

Fluoride ion is a dangerous chemical. Calcium fluoride is the active ingredient in some rat poisons, and overexposure to fluoride can produce a host of unpleasant diseases. Yet it is virtually impossible to establish any long-range deleterious effects of fluoride in the CD range of Figure 7.33. The massive administration of chemical substances to drinking water is warranted only when there exist compelling reasons of public health to do so. Many people are not convinced that tooth decay is a serious enough malady to warrant the risk.

In addition to chemicals added specifically for public health purposes, man sometimes injects poisons into natural waters in a deliberate attempt to alter the ecology and make imagined "improvements." Among the chemicals so employed are sodium arsenite, used to kill aquatic plants, and toxaphene (a mixture of chlorinated terpenes used to control populations of undesireable fish. The detoxification of lakes following the application of the latter has been studied (Fukano and Hooper, 1958; Mayhew, 1959; Johnson, Lee, and Spyridakis, 1966). The poison rapidly disappears out of the water and into the sediments where it is very persistent (Veith and Lee, 1971).

Toxic substances, including mercury preparations, have also been added to waters to prevent fouling in power plants and other installations.

Another chemical that man adds indiscriminately in enormous and ever increasing amounts to the environment and which soon finds its way into the hydrosphere is salt used for de-icing roads. Over the years from 1940 to 1970 the amounts of $CaCl_2$* and NaCl spread on United States roads and highways have climbed from 100,000 to nearly 10,000,000 metric tons per year (Bubeck et al., 1971). In places the salt has created a roadside corridor of dead trees, it has poisoned wells and reservoirs (Hanes, Zelanzy, and Blaser, 1970). In some communities, the salt level of the drinking water has exceeded the level hazardous to victims of heart and other diseases who must adhere to low salt diets. No one appears to know how many motor vehicle accidents have been caused by salt spray-obscured windows, although the rationale for dumping this material on the roads is accident prevention. Cases have even been documented where de-icing salt invasion was sufficiently severe to alter the mixing characteristics of a small Michigan lake (Judd, 1970) and the stratification of a bay of Lake Ontario (Bubeck et al., 1971). Finally, we might mention the organic materials added to reservoirs to form a thin surface film barrier to prevent evaporative water loss.

7.6 ISOTOPES

Both hydrogen and oxygen have three naturally occurring istopic forms, hence ordinary water is a mixture of eighteen different water molecules

$HO^{16}H$	$DO^{16}D$	$DO^{16}T$
$HO^{18}H$	$DO^{18}D$	$DO^{18}T$
$HO^{17}H$	$DO^{17}D$	$DO^{17}T$
$HO^{16}D$	$HO^{16}T$	$TO^{16}T$
$HO^{18}D$	$HO^{18}T$	$TO^{18}T$
$HO^{17}D$	$HO^{17}T$	$TO^{17}T$

Table 7.22 summarizes the relative isotopic content of natural waters, Table 7.23 compares some physical–chemical properties of ordinary (H_2O) and heavy (D_2O) water, and Table 7.14 lists the isotopic content of selected river waters. The molecular weight of water is small, therefore the differences in the masses of the different isotopic waters will be relatively large, and, as a consequence of these differences in mass, there will be a significant "isotope effect" or fractionation of these isotopic forms in chemical and biological processes and in the course of phase changes (Weston, 1955; Roth, 1963; Van Hook, 1968). These fractionations will give rise to differences in isotopic ratios in the natural environment. Other significant differences reflect the length of atmospheric exposure: Continental rain and river water that has had a longer time to equilibrate with atmospheric tritium, for example, has a greater tritium content than seawater or maritime rain (5.2×10^{-18} to 5.5×10^{-18} compared with 0.54×10^{-18} tritium atoms/protium atom, see also Table 7.25).

* Calcium chloride is interesting inasmuch as it has completed the cycle from nuisance to useful chemical and back again to nuisance. A waste product of the Solvay process for the production of sodium bicarbonate and carbonate, no one knew how to get rid of the stuff until it found use in road de-icing. As is so often the case, in solving one problem we have created another.

Table 7.22 Approximate Distribution of Various Kinds of Molecules in Natural Water (From Hutchinson, 1957)

Molecule	Mole Percent
$H_2{}^{16}O$	99.745
$H_2{}^{18}O$	0.198
$H_2{}^{17}O$	0.042
$HD^{16}O$	0.015
$HD^{18}O$	0.000029
$HD^{17}O$	0.000006
$D_2{}^{16}O$	0.000002
$D_2{}^{18}O$	0.000000004
$D_2{}^{17}O$	0.0000000008

Table 7.23 Physical and Chemical Properties of H_2O and D_2O (From Hutchinson, 1957)

Property	H_2O	HDO	D_2O
Freezing Point (°C)	0		3.82
Boiling Point (°C)	100	101.76	101.42
Specific Gravity at 0°C			
Vapor Pressure at 20°C	17.535	16.27	15.2
Temperature of Maximum Density (°C)	4		11.6
Critical Temperature (°C)	374.1		371.5
Critical Pressure (atm)	217.7		218.6
Ionization Product, k_w, at 25°C	1×10^{-14}		0.3×10^{-14}
Dielectric Constant at 20°C	82		80.5
Viscosity at 20°C (ml)	10.09		12.6
Surface Tension at 19°C (dynes/cm)	73.66		72.83
Refractive Index, $n_D{}^{20}$	1.33300		1.32844
Representative Solubilities at 25°C (g/g water)			
NaCl	0.359		0.305
$BaCl_2$	0.357		0.289

Contamination of the natural environment by radionuclides continues to be a matter of grave concern, and various aspects of the problem are discussed elsewhere in this book. Within the continental hydrosphere of the United States, attention has focused on two rivers—the Columbia and the Savannah (Georgia). An index of the rate of delivery of radionuclides into the Pacific Ocean from the Columbia River is the radionuclide flux past Vancouver (Table 7.26). Inasmuch as the atomic reactors involved are directly cooled by river water, the situation represented is the worse possible and fortunately a quite exceptional one. Table 7.27 shows the radionuclide concentration in Columbia River waters near the point of discharge at Richland and

Table 7.24 Isotopic Content of River Waters (Selected data from Livingstone, 1963)

Locality and Date	Concentration
Oxygen-18	
Me-Yome River, Siam, 1953	19.72 ^{18}O atoms/10^4 atoms ^{16}O
Ravi River, Pakistan, 1953	19.77 ^{18}O atoms/10^4 atoms ^{16}O
Mulunguzi Stream, Nyasaland, 1953	19.97 ^{18}O atoms/10^4 atoms ^{16}O
Perana River, Argentina, 1953	19.80 ^{18}O atoms/10^4 atoms ^{16}O
San Juan River, New Mexico, 1953	19.92 ^{18}O atoms/10^4 atoms ^{16}O
Deuterium	
Columbia River, British Columbia, 1943	13.3 ^2H atoms/10^5 atoms ^1H
Mississippi River, Louisiana, 1948	14.9 ^2H atoms/10^5 atoms ^1H
Connecticut River, Connecticut, 1948	14.5 ^2H atoms/10^5 atoms ^1H
Rio Grande River, Texas, 1948	15.3 ^2H atoms/10^5 atoms ^1H
Colorado River, Arizona, 1948	13.8 ^2H atoms/10^5 atoms ^1H
Tritium	
Mississippi River, Missouri, 1953	4.5 to 7.3 ^3H atoms/10^{18} atoms ^1H
Rhone River, France, 1953	2.6 ^3H atoms/10^{18} atoms ^1H
Rhine River, Germany, 1953	2.1 ^3H atoms/10^{18} atoms ^1H
Rio Grande River, New Mexico, 1954	6.6 ^3H atoms/10^{18} atoms ^1H

Table 7.25 Tritium in Natural Waters (Selected values from Libby, 1961)

Source	Date	$T/H \cdot 10^{18}$
Rain, Bedford, Mass.	April 1, 1959	292 ± 15
	December 28–29, 1959	42 ± 1.7
	February 4–8, 1960	15 ± 0.8
	April 1–6, 1960	11 ± 2.0
Rain, Recife, Brazil	May 2, 1959	22 ± 1
Rain, Buenos Aires, Argentina	April 1959	17 ± 1.3
Rain, Fiji Islands	August 6, 1959	1.5 ± 0.4
Mississippi River at St. Louis	October 16, 1960	119 ± 2
Colorado River	February 1960	52 ± 1.4
Lake Michigan	February 1, 1960	28 ± 3
Pacific Ocean at Santa Monica, California	March 9, 1960	14 ± 2.3
	April 4, 1961	2.7 ± 1
Atlantic Ocean at Atlantic City Pier	April 1960	23 ± 1.3
Dead Sea	February 1, 1960	8.5 ± 0.3
Mediterranean Sea near Sicily and South of Sardinia	February 1, 1960	5.5 ± 0.6

Table 7.26 Annual Average Rate of Transport of Selected Radionuclides Past Vancouver, 1961–1963 (From Seymour and Lewis, 1964)

	Ci/day		
Year	1961	1962	1963
^{51}Cr	840	650	860
^{65}Zn	44	29	28
^{32}P	29	13	12
^{239}Np	67	31	[a]
Total	980	723	900

[a] Incomplete data.

Table 7.27 Annual Average Concentration of Several Radionuclides in Columbia River Water—1966 (From Joseph et al., 1971)

Radionuclides	Richland	Bonneville Dam
R.E. + Y.[a]	270	Insufficient Data (I.D.)
^{24}Na	2600	I.D.
^{32}P	140	23
^{51}Cr	3600	1300
^{64}Cu	1400	I.D.
^{65}Zn	200	43
^{76}As	420	I.D.
^{90}Sr	1	I.D.
^{131}I	18	3
^{239}Np	770	I.D.

[a] R.E. + Y = Rare Earth + Yttrium = ^{140}La, ^{152}&En, ^{153}Sm, ^{165}Dy, ^{90}Y, ^{91}Y, ^{93}Y, ^{141}Ce, ^{143}Ce, ^{144}Cc, ^{142}Pr, ^{143}Pr, ^{147}Nd, ^{147}Pm, ^{149}Pm, ^{151}Pm, ^{152}Eu, ^{156}Eu, ^{153}Gd, ^{159}Gd, ^{160}Tb, ^{161}Tb, ^{166}Ho, ^{169}Er, ^{171}Er.

at Bonneville Dam about 20-mi downriver. The radioisotopes originate from the activation of elements present in the entering cooling waters and from corrosion products. The decrease in concentration downstream is due largely to dilution by inflowing streams, radioactive decay of the isotopes, and absorption of the radioactive material on bottom sediments. Nevertheless, a great deal of radioactivity is carried downstream and into the Pacific where it can be incorporated into and concentrated by marine organisms. For example, commercially valuable Wallapa Bay oysters concentrate the ^{65}Zn by a factor of 11,300. It takes about 200 days to get rid of 50% of the ^{65}Zn taken up by the shellfish (Seymour and Lewis, 1964).

7.7 MERCURY POLLUTION

We conclude our examination of fresh water chemistry with a discussion of a particular pollutant, albeit one that is not confined to the hydrosphere, let alone to fresh waters.

In 1953, a mysterious disease broke out among the inhabitants, especially fishermen and their families, of Minamata City, Japan. Three years later, the malady had reached epidemic proportions. Between 1953 and 1960, over 111 cases of the sickness were reported, and the mortality rate was very high—43 deaths. These cases included 19 infants born with congenital defects. A second outbreak occurred in 1961 in Niigata, Japan, in which six out of the 30 cases reported resulted in death. Even cats and rats contracted the disease. The symptoms of "Minamata disease" included numbness of lips and limbs; impairment of vision, hearing, and speech; and difficulty in walking and performing simple manipulations.

These symptoms were reminiscent of toxic heavy metal poisoning, and a number of elements were suspect, including selenium, until the disease was traced to the consumption of fish and shellfish taken from bay waters contaminated with mercury (Kurland, Faro, and Siedler, 1960; Tokuomi, 1969). The mercury was discharged into the waters as an organomercury compound from a large plastics plant that used mercuric chloride as a catalyst for the synthesis of vinyl chloride and also as a concentrated acidic solution of mercuric sulfate which was the spent catalyst for the production of acetaldehyde, an intermediate in the manufacture of octanol and butanol (Ui, 1969).

Once mercury became suspect and people began to be on the lookout for this toxic metal, it soon became evident that human negligence once again had managed to seriously contaminate the hydrosphere. In 1970, the U.S. Federal Drug Administration ordered 1,000,000 cans of tuna fish (of an estimated value of $84,000,000) off the United States market (Anon., 1970*a*; Goldwater, 1971; Hammond, 1971; Keckes and Miettinen, 1970). The average mercury level in the food samples tested was 0.37 ppm.* Nor was the problem confined to the marine environment; in the same year, pike, perch, and pickerel taken from heavily fished Lake St. Clair (connecting Lakes Huron and Erie) contained 1.3 ppm Hg or more than 0.8 ppm above the legal limit. The metal had been concentrated by as much as 4000 times in the food chain. As a consequence of these findings, a 50,000,000 lb/yr fishing industry was endangered and Ontario and Michigan officials closed and/or threatened to close the paper and chemical plants responsible (Anon., 1970*b*, *c*; Turney, 1971).

There is a trace (about 0.02 $\mu g/m^3$) of mercury in the atmosphere, probably either as the vapor of the metal or as dimethyl mercury (Saukov, 1953), and the content under smoggy conditions may be as high as 0.05 $\mu g/m^3$ (Williston, 1968). The Hg content of rain water is 0.2 to 2 ppb, and the residence time in the atmosphere is estimated to be about 2 yr (Saukov, 1953). The Hg content of igneous rocks is low, 10 to 100 ppb and of sedimentary rocks somewhat higher (Pecora, 1970). Dried sediments from the California coast have 0.02 to 1.0 ppm Hg (Klein and Goldberg, 1970). Boström and Fisher (1969) give 1 to 400 ppb Hg as the range for Pacific marine sediments. Ground waters normally contain 0.01 to 0.07 ppb Hg, and lake and river

* I note with relief a recent report that the mercury level in fish protein concentrate (a foodstuff that has often been mentioned to feed the exploding world population) does not appear to be excessive (Beasley, 1971).

waters, between 0.08 and 0.12 ppb (Pecora, 1970). Recent measurements indicate that the mercury level in surface seawater is in the 0.013 to 0.018 μg/liter range, but there is some scant evidence that the values in deep water samples may be markedly lower (Leatherhead et al., 1971; Burton and Leatherhead, 1971). Klein and Goldberg (1970) found "hotspots" of mercury near a sewer outfall, and comparable findings have been made by Smith, Nicholson, and Moore (1971) near outfalls in the tidal Thames. With respect to the water column, the latter authors distinguish between dissolved and particulate mercury, and they found that 82 to 97% of the element is in the particulate form.

Goldberg (1970) has claimed that as much mercury may be entering the environment from human activity as from natural sources. World mercury consumption over the past decade has been in excess of 100,000,000 lb (Holmes, 1971), and it is a sobering thought that much of this toxic metal may still be circulating in the environment. In 1966, the world production of Hg was 9200 metric tons, and about half this amount per year is being dispersed into the environment. Major industries using mercury are (Bidstrup, 1964):

1. Chlorine and alkaline plants for electrolytic production of chlorine and caustic soda;
2. Cellulose industries for preserving the wet pulp from bacterial and fungal biodeterioration;
3. Plastic industries for catalytic reactions;
4. Electric industries for production of relays, switches, batteries, rectifiers, lamps, and so on;
5. Pharmaceutical industries for production of diuretics, antiseptics, cathartics, some contraceptive, and drugs for treatment of congestive heart failure;
6. Paint industries, mainly for production of anticorrosive paints;
7. Metal refinement by amalgamation;
8. Power plants in special heat engines instead of steam;
9. Plants treating nuclear wastes for electrolytic purification of wastes, and
10. Industries producing industrial and control instruments such as thermometers, barometers, mercury pumps, and so on.

and consumption statistics for the United States alone are given in Table 7.28. The worst offender is the chlorine-caustic soda industry. This electrolytic process discharges 0.25 to 0.5 lb Hg into the environment for every ton of caustic soda produced, and it is this industry that was largely guilty for the contamination of Lake St. Clair. The tactics of the guilty parties in such cases is interesting: First, the specter of job losses is trundled out—BASF Wyandotte announced it was closing its 22-yr old Michigan mercury-cell chlor-alkali plant because of the great expense of permanent pollution control (Anon., 1971*a*), then the public gesture—BASF Wyandotte announced that it was investing $500,00 in antipollution equipment to bring the Hg discharge of three of its plants below the 0.5 lb Hg/day level set by the Federal Water Quality Administration (FWQA), and then finally and almost parenthetically the truth of the matter—the chlorine market is "slow . . . in fact a number of economically marginal plants have been shut down" (Anon., 1971*b*).

Table 7.28 United States Consumption of Mercury (From Anon., 1970*d*) (Reprinted with permission from *Envir. Sci. Tech.*, *4*, 891 (1970). © The American Chemical Society.)

Source	Thousand Pounds
Electrolytic chlorine	1572
Electrical apparatus	1382
Paint	739
Instruments	391
Catalysts	221
Dental preparations	209
Agriculture	204
General laboratory use	126
Pharmaceuticals	52
Pulp and paper making	42
Amalgamation	15
Other	1082
Total	6035

Although not so large quantity-wise (Table 7.18), the agricultural use of mercury compounds as herbicides, pesticides, slimicides, and fungicides, often in the form of seed dressing, is particularly worrisome, since it represents a direct addition to the environment. The United States alone is responsible for about 25% of the total world application in this form (Smart, 1968). However due to increasing public concern, there has been an increase in vigilence on the part of the responsible government agencies and an increase in fastidiousness on the part of industry with respect to throwing mercury about. For example, the use of mercury fungicides and slimicides in the United States paper and pulp industry declined from 182 ton/yr in 1959 to 21 ton/yr in 1966, and in Scandinavian countries, this use of mercury has been discontinued.

The fields of stability for inorganic mercury under environmental conditions are shown in Figure 7.34. Because of the great affinity of Hg for —SH groups and the common occurrance of proteinous material in natural waters, not to mention ion exchange and absorption on minerals and other sedimentary materials, only a small fraction of the mercury is in free solution or inorganic form except near the point of pollutant discharge. Of the methylmercury, $(CH_3)_2Hg$, in food, 98% is absorbed by tissue, whereas only 1% of the inorganic mercury is absorbed (Anon., 1970*d*). Of the environmental forms of mercury, methylmercury easily represents the greatest hazard. In both fresh and marine waters, mercury once released soon disappears out of the water and into the bottom muds and sediments. Mercury uptake depends on the type of sediment (Bayev, 1968) with peaty sediments with a large organic content holding the element much more strongly than a sandy sedimentary deposit. How persistent mercury is in the solid phase and how it gets back into the food chain are not clear. Because of the strength with which it is held in the sediments, one

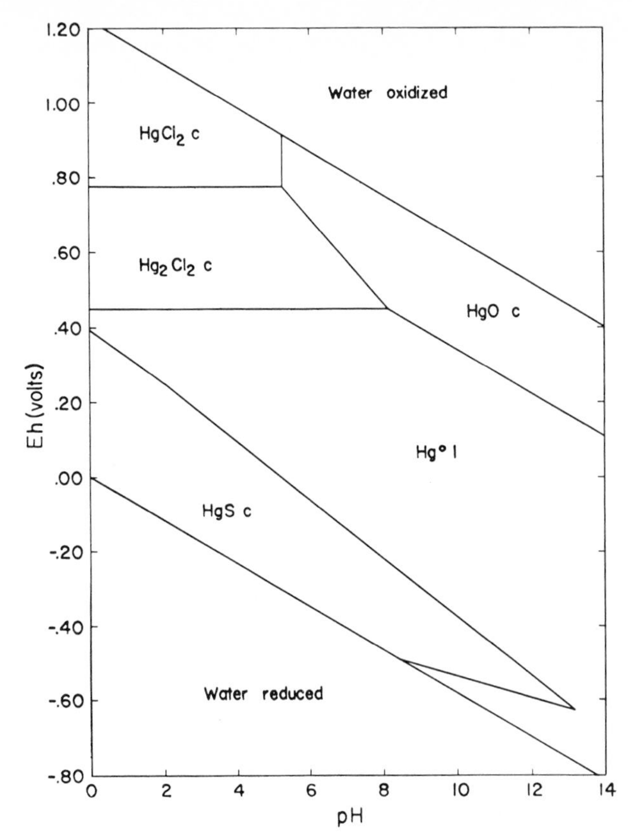

Figure 7.34. Fields of stability for solid (c) and liquid (l) mercury species at 25°C and 1 atm. System includes water containing 36 ppm Cl^- and total sulfur 96 ppm as SO_4^{2-} (from J. D. Hem, 1970).

would expect the exchange of mercury from the bottom material back into the water column to be slight. Yet the equilibria can be shifted and the exchange rates accelerated by a number of processes. For example, we have demonstrated in our laboratories that some of the sediment-bound mercury can be released by the addition of NaCl or $CaCl_2$ in concentrations corresponding to what one might expect from the run-off of road de-icing salt mentioned previously. Also the microbial methylation of mercury by bottom microorganisms almost certainly accelerates the leakage of mercury from the bottom sediments back into the water column and food webs.

Mercury contamination in fish is nearly always in the form of methylmercury (Westoo, 1966). The details of the methylation process are not entirely clear, and there is some difference of opinion. Jensen and Jernelöv (1969) and Kitamura (1969) report that microorganism-containing muds convert phenylmercury or inorganic mercury to methylmercury. Methylation of mercury ion by anaerobiosis has also been reported by Wood, Kennedy, and Rosen (1968). Yet Rissanen, Erkama, and Miettinen (1970) found no (i.e., less than 0.1% of the total mercury added) methylation of inorganic Hg(II) by anaerobic muds in 34 days.

Figure 7.35 summarizes the chemical forms and routes of mercury in the environment, while Figure 7.36 shows in somewhat greater detail the cycling of this element

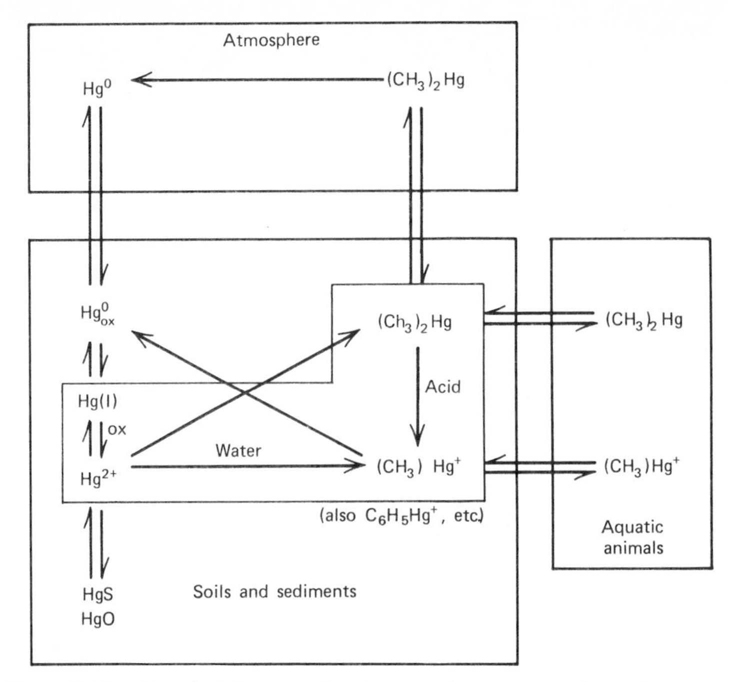

Figure 7.35. Chemical forms and pathways of mercury in the environment.

in the total environment. It is very important to notice that mercury (and lead) differ from the majority of the other heavy metals in the relative significance of the airborne path of the contamination. Not only does the combustion of fossil fuels inject mercury directly into the atmosphere (Joensuu, 1971), along with very appreciable amounts of other contaminants (Bertine and Goldberg, 1971), but highly volatile methylated mercury escapes from sediments up through the water column and into the atmosphere where it may be photoreduced to the free metal and washed back into the soils and the hydrosphere in rain. Due to this atmospheric cycling of mercury, lakes and ponds remote from any specific source of pollutant can be contaminated. There may also be an appreciable input to atmospheric mercury levels from vulcanism (Eshleman, Siegel, and Siegel, 1971).

There is little unambiguous evidence that the mercury level in the open seas or even in ocean fish has increased as the result of human activities (Goldwater, 1971), and the cases of clear contamination that have been uncovered thus are presumably the result of localized pollution. And just as well too! For inasmuch as mercury can reduce planktonic photosynthesis (Harriss, White, and Macfarlane, 1970), marine contamination by this toxin on a global scale could be disasterous. The situation is further complicated in that the "permissible" level of the metal of 0.5 ppm, established by the FDA in the absence of necessary information of the type exemplified by Figure 15.33, may not only be very near to the natural background, but may leave very little margin of safety (Holmes, 1971). Until Figure 7.33 is know for the more important pollutants, and we have a good sense of their complete inventory in the environment, we shall continue to be subject to one pollution scare after another until a weary public no longer heeds the cry of "Wolf!" Just when people are be-

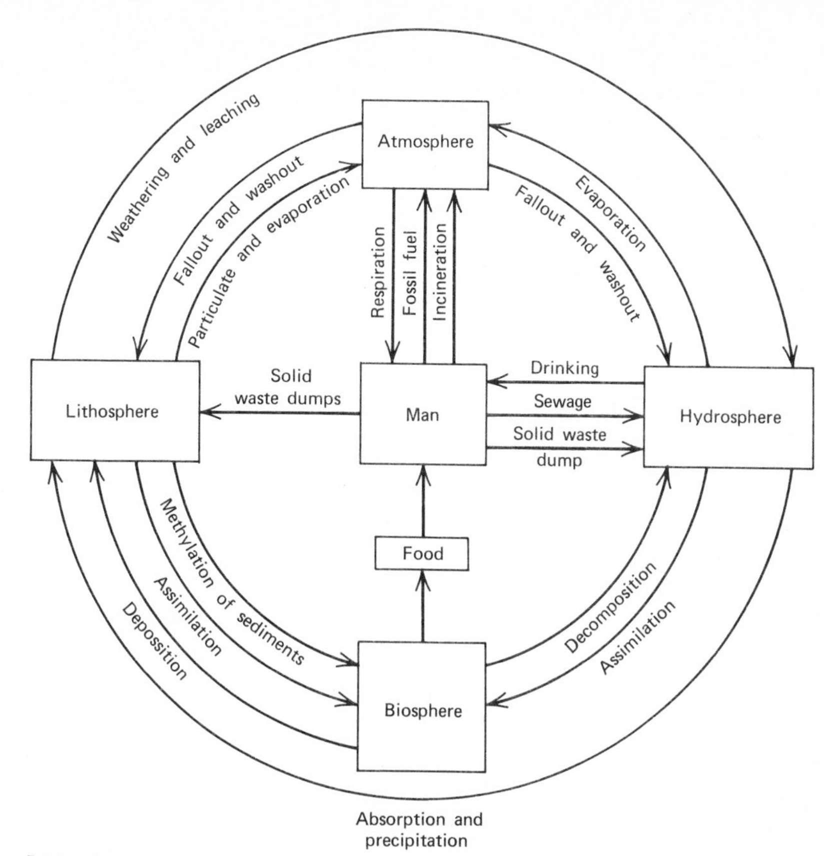

Figure 7.36. Mercury cycling in the total environment. The boxes represent levels, and the arrows, rates of exchange and transport.

ginning to feel a little less uneasy about mercury the specter of cadmium contamination of Hudson River fish is raised (Anon., 1971*c*).

BIBLIOGRAPHY

Anon., *New York Times* (December 16, 1970*a*).

Anon., *Time* (May 4, 1970*b*).

Anon., *Science*, *168*, 232 (1970*c*).

Anon., *Environ. Sci. Tech.*, *4*, 891 (1970*d*).

Anon., *Chem. Week* (January 27, 1971*a*).

Anon., *Chem. Week* (February 24, 1971*b*).

Anon., *Boston Herald Traveler* (June 13, 1971*c*).

L. G. M. Baas Becking, I. R. Kaplan, and D. Moore, *J. Geol.*, *68*, 243 (1960).

R. T. Barber and J. H. Ryther, *Deep-Sea Res.*, *7*, 276 (1969).

Y. G. Bayev, *Dokl. Acad. Nauk SSSR*, *181*, 211 (1968).

T. M. Beasley, *Environ. Sci. Tech.*, *5*, 634 (1971).

J. D. Bernal and R. H. Fowler, *J. Chem. Phys.*, *1*, 515 (1933).

R. A. Berner, *J. Geol.*, *72*, 293 (1964).

K. K. Bertine and E. D. Goldberg, *Science*, *173*, 233 (1971).

P. L. Bidstrup, *The Toxicity of Mercury and Its Compounds*, Elsevier, New York, 1964.

F. H. Bormann and G. E. Likens, *Sci. Amer.*, *223*(4), 92 (1970).

U. Boström and D. H. Fisher, *Geochim. Cosmochim. Acta*, *33*, 743 (1969).

J. L. Brooks and E. S. Deevey, Jr., in D. G. Frey (ed.), *Limnology in North America*, Univ. Wisconsin Press, Madison, 1963.

S. Bruckert and F. Jackquin, *C.R. Acad. Sci.* (*D*), *269*, 17 (1969).

R. C. Bubeck, W. H. Diment, B. L. Deck, A. L. Baldwin, and S. D. Lipton, *Science*, *271*, 1128 (1971).

J. D. Burton and T. M. Leatherhead, *Nature*, *231*, 440 (1971).

R. F. Christman, *Environ. Sci. Tech.*, *1*, 302 (1967).

L. H. N. Cooper, *J. Mar. Biol. Assoc. U.K.*, *27*, 314 (1948).

R. A. Courant, B. J. Ray, and R. A. Horne, *Zhur. Struk. Ihim. 4*, 581 (1972).

M. D. Danford and H. A. Levy, *J. Amer. Chem. Soc.*, *84*, 3965 (1962).

C. M. Davis, Jr., and T. A. Litovitz, *J. Chem. Phys.*, *42*, 2563 (1965).

G. Dietrich, *General Oceanography*, Interscience New York, 1963.

W. Drost-Hansen, in W. Stumm (ed.) *Equilibrium, Concepts in Natural Water Systems*, Adv. Chem. Ser. No. 67, Amer. Chem. Soc., Washington D.C., 1967.

D. Eisenberg and W. Kauzmann, *The Structure and Properties of Water*, Oxford Univ. Press, Cambridge (England), 1969.

K. O. Emery, *A Coastal Pond*, American Elsevier, New York, 1969.

A. Eshleman, S. M. Siegel, and B. Z. Siegel, *Nature*, *233*, 471 (1971).

M. Falk and T. A. Ford, *Canad. J. Chem.*, *44*, 1699 (1966).

H. S. Frank and W. Y. Wen, *Disc. Faraday Soc.*, *24* 133 (1957).

F. Franks, *Chem. Ind.*, 560 (May 1968).

K. G. Fukano and F. F. Hooper, *Prog. Fish Cult.*, *20* 189 (1958).

E. T. Gjessing, *Environ. Sci. Tech.*, *4*, 437 (1970).

E. D. Goldberg, paper presented at FAO(UN) Conf. Mar. Pollut., Rome, 1970.

L. J. Goldwater, *Sci. Amer.*, *224*(5), 15 (1971).

D. H. Hamilton, Jr., D. A. Flemer, C. W. Keefe, and J. A. Mihursky, *Science*, *169*, 197 (1970).

A. L. Hammond, *Science*, *171*, 788 (1971).

R. E. Hanes, L. W. Zelanzy, and R. E. Blaser, *Effects of Deicing Salts on Water Quality and Biota*, Nat. Coop. Highw. Res. Program, Rept. No. 91, Nat. Acad. Sci. U.S.A., Washington, D.C., 1970.

R. C. Harriss, D. B. White, and R. B. Macfarlane, *Science*, *170*, 736 (1970).

F. R. Hayes and J. E. Phillips, *Limnol. Oceanogr.*, *3*, 459 (1958).

F. Helfferich, *Ion Exchange*, McGraw-Hill, New York, 1962.

J. D. Hem, in W. T. Pecora (ed.), *Mercury in the Environment*, U.S. Geol. Surv. Prof. Paper No. 713, U.S. Gov. Print. Off., Washington, D.C., 1970.

J. Holmes, *Esquire*, *75*(5), 135 (1971).

R. A. Horne, *Surv. Prog. Chem.*, *4*, 1 (1968).

R. A. Horne, *Marine Chemistry*, Wiley-Interscience, New York, 1969.

R. A. Horne, in A. Standen (ed.), *Kirk-Othmer Encyclopedia of Chemical Technology*, Wiley-Interscience, New York, 1970.

R. A. Horne, A. F. Day, R. P. Young, and N-T Yu, *Electrochim. Acta*, *13*, 397 (1968).

R. A. Horne and C. H. Woernle, *Chem. Geol.*, *9*, 299 (1972).

G. E. Hutchinson, *A Treatise on Limnology*, Vol. I Wiley, New York, 1957.

R. M. Ingle and D. F. Martin, *Environ. Lett.*, *1* 69 (1971).

R. Ishiwatari, *Soil Sci.*, *107*, 53 (1969).

C. Jaccard, in R. A. Horne (ed.), *Water and Aqueous Solutions*, Wiley-Interscience, New York, 1972.

H. H. G. Jellinek, in R. A. Horne (ed.), *Water and Aqueous Solutions*, Wiley-Interscience, New York, 1972.

P. M. Jenkin, *Ann. Mag. Nat. Hist.*, Ser. 10, *18*, 133 (1932).

S. Jensen and A. Jernelöv, *Nature*, *223*, 753 (1969).

M. E. Jewell and H. W. Brown, *Ecology*, *10*, 427 (1929).

O. I. Joensuu, *Science*, *172*, 1027 (1971).

W. D. Johnson, G. F. Lee, and D. Spyridakis, *J. Air Water Pollut.*, *10*, 555 (1966).

A. Joseph et al., in A. H. Seymour (ed.), *Radioactivity in the Marine Environment*, Nat. Acad. Sci. U.S.–Nat. Res. Council, Washington, D.C. in press, 1971.

J. H. Judd *Water Res.*, *4*, 521 (1970).

S. V. Kahn, *Proc. Amer. Soc. Soil Sci.*, *33*, 6 (1969).

B. Kamb, in R. A. Horne (ed.), *Water and Aqueous Solutions*, Wiley-Interscience, New York 1972.

B. Kamb, in A. Rich and N. Davidson (eds.), *Structural Chemistry and Molecular Biology*, Freeman, San Francisco 1968.

J. L. Kavanau, *Water and Solute-Water Interactions*, Holden-Day, San Francisco, 1964.

S. Keckes and J. K. Miettinen, paper presented FAO(UN), Conf. Mar. Poll., Rome, 1970.

G. Kemmerer, J. F. Bovard, and W. R. Boorman, *Bull. U.S. Bur. Fish.*, *39*, 51 (1923).

S. Kitamura, *Jap. J. Hygen.*, *24*, 132 (1969).

D. H. Klein and E. D. Goldberg, *Environ. Sci. Tech.*, *4*, 765 (1970).

J. F. Kopp and R. C. Kroner, *Trace Metals in Waters of the United States*, FWPCA, U.S. Dept. Interior, Div. Poll. Surv., Cincinnati, Ohio, 1967.

L. T. Kurland, S. N. Faro, and H. S. Siedler, *World Neurol.*, *1*, 320 (1960).

W. L. Lamar, *U.S. Geol. Surv. Prof. Paper*, 600-D (1968).

T. M. Leatherhead, J. D. Burton, M. J. McCartney, and F. Culkin, *Nature*, *232*, 112 (1971).

G. F. Lee, "Factors Affecting the Transfer of Materials between Water and Sediments," *Univ. Wisconsin Water Resources Center Lit. Rev.*, No. 1 (July 1970).

W. F. Libby, *J. Geophys. Res.*, *66*, 3767 (1961).

D. A. Livingstone, in M. Fleischer (ed.), *Data of Geochemistry -G*, Geol. Survey Prof. Paper No. 440-G, U.S. Gov. Print. Off., Washington, D.C., 1963.

W. Luck, *Fortschr. Chem. Forsch.*, *4*, 653 (1963).

J. Mayhew, *Proc. Iowa Acad. Sci.*, *66*, 513 (1959).

F. J. McClure, *Water Fluoridation*, Nat. Inst. Dental Health, Bethesda, Md., 1970.

J. W. McMahon, *Limnol. Oceanogr.*, *14*, 357 (1969).

A. B. Menar and D. Jenkins, *Environ. Sci. Tech.*, *4*, 1115 (1970).

D. W. Menzel and J. H. Ryther, *Deep-Sea Res.*, *7*, 276 (1961).

Y. Miyake, *Elements of Geochemistry*, Maruzen, Tokyo, 1965.

J. E. Morris and W. Stumm, in R. F. Gould (ed.), *Equilibrium Concepts in Natural Water Systems*, Amer. Chem. Soc., Washington, D.C., 1967.

G. Nemethy and H. A. Scheraga, *J. Chem. Phys.*, *36*, 3382 (1962).

G. Ogner and M. Schnitzer, *Science*, *170*, 317 (1970).

H. R. Pahren, D. R. Smith, M. O. Knudson, C. D. Larson, and H. S. Davis, *Report on Pollution of the Merrimack River and Certain Tributaries* (Part II), FWPCA, Dept. Interior, Merrimack River Project (August 1966).

P. K. Park, M. Calalfomo, G. R. Webster, and B. H. Reid, *Limnol. Oceanogr.*, *15*, 70 (1970).

L. Pauling, *The Nature of the Chemical Bond*, 2nd ed., Cornell Univ. Press, Ithaca, N.Y., 1948.

W. T. Pecora, *Mercury in the Environment*, U.S. Geol. Surv. Prof. Paper No. 713, U.S. Gov. Print. Off., Washington, D.C., 1970.

E. B. Phelps, *Stream Sanitation*, Wiley, New York, 1944.

J. E. Phillips, in H. Heukelekian and N. C. Dondero (eds.), *Principles and Applications in Aquatic Microbiology*, Wiley, New York, 1964.

F. E. Reed, *Environ. Sci. Tech.*, *5*, 385 (1971).

K. Rissanen, J. Erkama, and J. K. Miettinen, Paper presented FAO(UN), Conf. Mar. Pollut., Rome, 1970.

R. A. Robinson and R. H. Stokes, *Electrolyte Solutions*, 2nd ed., Butterworths, London, 1959.

E. Roth, *J. Chem. Phys.*, *60*, 339 (1963).

F. Ruttner, *Arch. Hydrobiol. Suppl.*, *8*, 197 (1931).

J. H. Ryther and R. R. L. Guillard, *Deep-Sea Res.*, *6*, 65 (1959).

O. Y. Samoilov, *The Structure of Aqueous Electrolyte Solutions and the Hydration of Ions*, Consultants Bur., New York, 1965.

A. A. Saukov, *Geochemie*, VEB-Verlag Technik, Berlin, 1953.

M. Schnitzer, *Proc. Amer. Soil Sci. Soc.*, *33*, 75 (1969).

A. H. Seymour and G. B. Lewis, *Radionuclides of Columbia River Origin*, U.S. Atomic Energy Commission, Div. Tech. Inform. UWFL-86 (December, 1964).

J. Shapiro. *Limnol. Oceanogr.*, *2*, 161 (1957).

J. Shapiro, *J. Amer. Water Works Assoc.*, *56*, 1062 (1964).

L. G. Sillen, *Acta Chem. Scand.*, *18*, 1016 (1954).

N. A. Smart, *Residue Rev.*, *23*, 1 (1968).

J. D. Smith, R. A. Nicholson, and P. J. Moore, *Nature*, *232*, 393 (1971).

R. E. Speece, *J. Amer. Water Works Assoc.*, *63*, 6 (1971).

E. M. Stanley and R. I. Batten, *J. Phys. Chem.*, *73*, 1187 (1969).

W. Stumm, *Adv. Water Pollut. Res.*, *1*, 283 (1967).

W. Stumm and J. J. Morgan, *Aquatic Chemistry*, Wiley-Interscience, New York, 1970.

H. U. Sverdrup, M. W. Johnson, and R. H. Fleming, *The Oceans*, Prentice-Hall, Englewood Cliffs, 1942.

H. Swenerton and L. S. Hurley, *Science*, *176*, 62 (1971).

P. N. Takkar, *Soil Sci.*, *108*, 2 (1969).

M. Tanaka, *Bull. Chem. Soc. Jap.*, *27*, 200 (1954).

H. S. Thompson, *J. Roy. Agri. Soc. Engl.*, *11*, 68 (1850).

J. H. Todd, *Sci. Amer.*, in press, 1971.

H. Tokuomi, *Rev. Internat. Ocean. Med.*, *13–14*, 37 (1969).

W. G. Turney, *J. Water Pollut. Control Fed.*, *43*, 1427 (1971).

J. Ui, *Rev. Internat. Ocean. Med.*, *13–14*, 37 (1969).

W. A. Van Hook, *J. Phys. Chem.*, *72*, 1234 (1968).

G. D. Veith and G. F. Lee, *Environ. Sci. Tech.*, *5*, 230 (1971).

C. J. Velz, *Applied Stream Sanitation*, Wiley-Interscience, New York, 1970.

K. G. Wakim, in S. Licht (ed.), *Medical Hydrology*, Elizabeth Licht, New Haven, 1963.

R. Weston, *Geochim. Cosmochim. Acta*, *8*, 281 (1955).

G. Westoo, *Acta. Chem. Scand.*, *20*, 2131 (1966).

W. Whipple, Jr., J. V. Hunter, R. Davidson, F. Dittman, and S. Yu, *Instream Aeration of Polluted Rivers*, Water Res. Inst., Rutgers Univ., August 1969.

E. Wicke, *Angew. Chem.*, *5*, 106 (1966).

S. H. Williston, *J. Geophys. Res.*, *73*, 7051 (1968).

J. M. Wood, F. S. Kennedy, and C. G. Rosen, *Nature*, *220*, 173 (1968).

S. Yoshimura, *Arch. Hydrobiol.*, *26*, 197 (1934).

8

MARINE CHEMISTRY

8.1 INTRODUCTION

Over 97% of the Earth's hydrosphere is in its oceans (Table 8.1), excluding water buried more deeply in the Earth's crust (Table 8.2). This vast sheet of liquid water covers more than 70% of our planet's surface or some 36,000,000 km^2. The mean depth of the oceans of Earth is nearly 3800 m; thus the quantity of water involved is enormous, more than 1,300,000,000 km^3 or 1,400,000,000,000,000,000 kg of seawater (Sverdrup, Johnson, and Fleming, 1942). (See Table A-12.) Quite apart from being the most spectacular surface feature of our planet (Horne, 1970*a*), the oceans of Earth represent one of the most severe and certainly the most extraordinary environment in our (and perhaps many other) solar systems (Table 8.3), being characterized by moderately low temperatures, high hydrostatic pressures, and a moderately concentrated aqueous electrolytic solution.

To describe seawater chemically in as few words as possible, we can say that it is a roughly 0.5 *M* NaCl solution, approximately 0.05 *M* in $MgSO_4$ and containing a trace of just about everything else imaginable. Figure 8.1 attempts to classify the substances in seawater, for the most part on the basis of the size and character of their hydration envelopes. Before explaining the later, I would like to describe briefly the physical aspects of the marine environment. The important environmental parameters in the oceans are temperature, salinity (to be defined presently), and pressure. The water column at any point in the oceans at any given time will have a complex thermal and salinity structure (Figure 8.2). The nature of these profiles is highly variable in space and time, being highly dependent on currents and other mixing processes and local meteorological conditions, but they do tend to have certain shapes characteristic of given regions at given seasons of the year. Both profiles tend to be highly variable near the surface with even a diurnal cycle. At greater depths, there is a long region of slowly decreasing temperature and salinity, called the main thermocline and halocline, respectively, which may fluctuate slowly with the seasons of the year. Finally, at greater depths, both temperature and salinity become very stable. The deep oceans, it should be noticed in passing, are an excellent thermostat.

The hydrostatic pressure in the water column increases steadily with increasing depth at a rate of about 0.1 atm/m (Figure 8.3). The deepest trenches in the oceans correspond to pressures as great as 1000 atm (or bars).

8.2 THE HYDRATION OF SOLUTES

Unlike the majority of the terrestrial hydrosphere, the water of the marine hydrosphere contains appreciable concentrations of dissolved and suspended material of

Table 8.1 Distribution of the Hydrosphere of Earth and the World Water Budget (From Todd, 1970, with permission of the publisher.)

Water Item	Volume (thousands) Cubic Miles	Volume (thousands) Cubic Kilometers	Percent of Total Water
Water in land areas			
Fresh water lakes	30	125	0.009
Saline lakes and inland seas	25	104	0.008
Rivers (average instantaneous volume)	0.3	1.25	0.0001
Soil moisture and vadose water	16	67	0.005
Ground water to depth of 4000 m (about 13,100 ft)	2,000	8,350	0.61
icecaps and glaciers	7,000	29,200	2.14
Total in land area (rounded)	9,100	37,800	2.8
Atmosphere	3.1	13	0.001
World ocean	317,000	1,320,000	97.3
Total, all items (rounded)	326,000	1,360,000	100
Annual evaporation[a]			
From world ocean	85	350	0.026
From land areas	17	70	.005
Total	102	420	0.031
Annual precipitation			
On world ocean	78	320	0.024
On land areas	24	100	.007
Total	102	420	0.031
Annual runoff to oceans from rivers and icecaps	9	38	0.003
Ground water outflow to oceans[b]	.4	1.6	.0001
Total	9.4	39.6	0.0031

[a] Evaporation (420,000 km^3) is a measure of total water participating annually in the hydrological cycle.

[b] Arbitrarily set equal to about 5% of surface runoff.

Table 8.2 Comparison of the Water Content of the Atmosphere, Hydrosphere, and Earth's Crust (After Kulp, 1951)

Atmosphere		$<0.01 \times 10^{24}$ g
Hydrosphere		1.4×10^{24} g
Lithosphere	Sedimentary rocks	0.09×10^{24} g
	Granite and metamorphic rocks	0.02×10^{24} g
	Basalt	0.73×10^{24} g

Table 8.3 Solar System Environments (From Horne, 1969)

Environment	Medium	Temperature (°C)	Pressure (atm)	Gravitation Field
Earth, sea	aq 0.5 *M* NaCl	0 to +25	1 to 1000	1
Earth, air	O_2, N_2, CO_2	−607 to +57	0 to 1	1
Earth, center	Fe	+4000	3,700,000	0
Space	Near-vacuum	—	0	Variable
Moon	Near-vacuum	−153 to +134	0	0.165
Sun	H_2, He	+4700 to +5700	~1	—
Mercury	Near-vacuum	−253 to +340	0.003	0.37
Venus	CO_2, H_2O, N_2O_4	+427	0.1	0.89
Mars	CO_2, H_2O, N_2	−100 to +10	0.1	0.38
Jupiter	H_2, He, NH_2, CH_4	−138	—	27
Saturn	CH_4, NH_3	−153	—	11
Uranus	CH_4	−170	—	0.96
Neptune	CH_4	−170	—	15

every type imaginable (Figure 8.1). In chapter 7, we examined the structure of pure liquid water, and now we must turn our attention to the effect of solutes on that structure. We noted then that viscosity is a good indication of the extent of structure in liquid water. The effect of electrolyte addition on the viscosity of water is complex (Figure 8.4). Some electrolytes, such as NaCl and especially $MgSO_4$, increase the viscosity of water and thus appear to be increasing or enforcing the water structure, whereas other electrolytes such as KCl appear to be water structure breakers, since their addition reduces the viscosity.

The thermodynamics of solution also can provide useful clews as to structural changes. If we add an electrolyte, that is, a species that forms charged ions in solution, the thermodynamics of the process is dominated by the energetics of transferring a charge from a low (air) to a high (water) dielectric constnt medium, and any structural effects get swamped out. However if we add a nonpolar solute to water (such as H_2, O_2, CO, N_2, or the noble gases), the entropy changes that result clearly indicate that order or structure is being induced in the system, presumably by the formation of a structured "iceberg" surrounding the solute molecules (Frank and Evans, 1945). Hydrocarbons and other nonpolar organic solutes also structure the surrounding water, and in many instances, it has been possible to isolate a clathrate type hydrate, or cage of water molecules inclosing the "guest" solute molecule (Figure 8.5). Even elemental mercury is believed to be surrounded by such a structure in aqueous solution (Spencer and Voight, 1968). We refer to the water structure, possibly of a clathrate nature, surrounding nonpolar solutes as "hydrophobic hydration" to distinguish it from the "coulombic hydration" surrounding charged species such as ions in solution.

I digress here a moment and mention some of the roles that clathrate hydrates might play in the environment and within ourselves. With respect to the latter, Pauling (1961), and Miller (1961, 1963) have proposed that the ordering of water in the cerebrospinal fluid into clathrate-like structures could form the basis of the anaesthetic properties of certain nonpolar solutes by impeding the transmission of

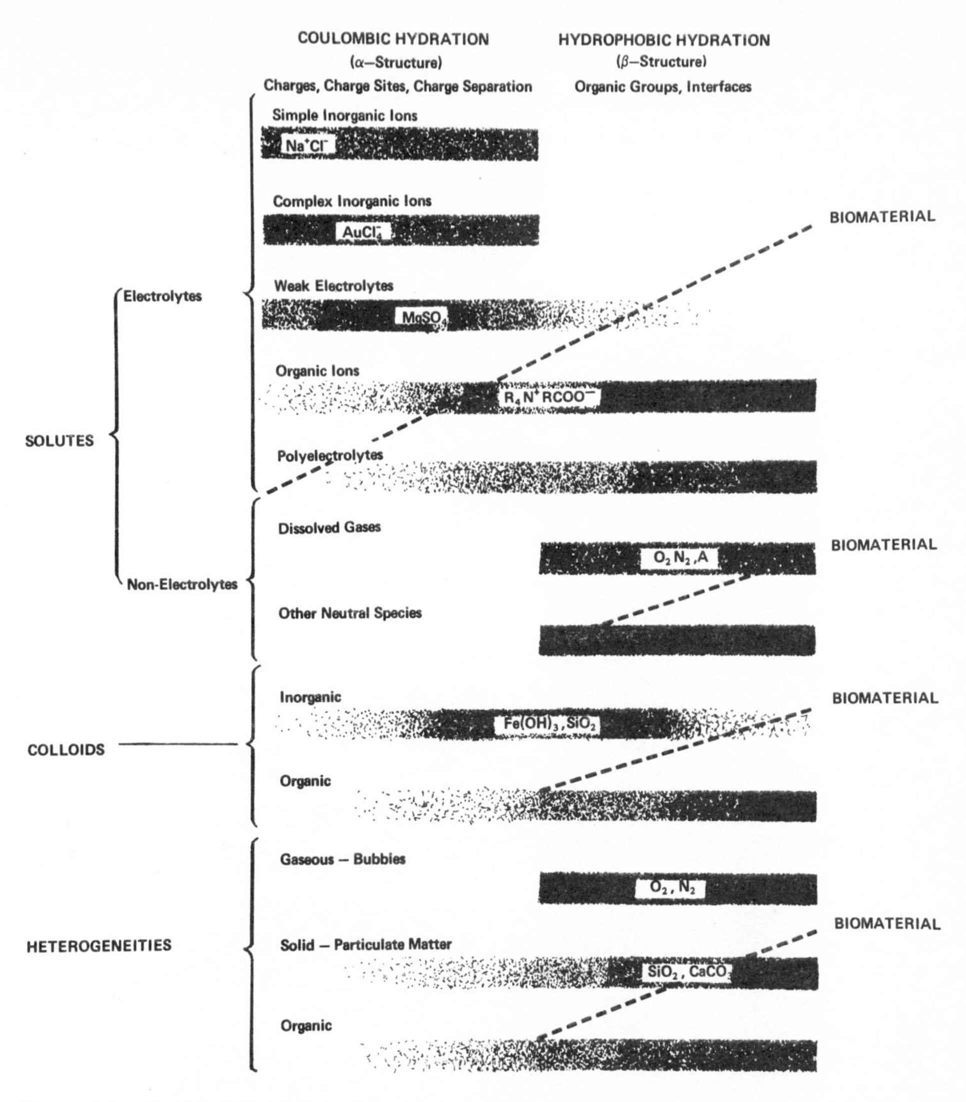

Figure 8.1 Classification of the substances in seawater (reprinted from Horne, 1971, by courtesy of Marcel Dekker, Inc.).

neural signals. More recently, Miller (1969) has attributed the disappearance of air bubbles in Antarctic ice at depths greater than 800 to 1200 m (Gow, Ueda, and Garfield, 1968) to the crystallization of the constituents of air as clathrate hydrates such as $(N_2, O_2) \cdot 6H_2O$. Miller has also discussed the occurrence of hydrates on other planets (Miller, 1961), and he has proposed that the Martian polar cap may contain the clathrate hydrate of carbon dioxide (Miller and Smythe, 1970).

Returning now to ionic solutes, the complexity of the effects of ionic solution on the viscosity and apparent structure of water (Figure 8.4) led Frank and Wen (1957) to propose a two zone model for the structure of the hydration atmospheres of ions in solution (Figure 8.6) consisting of an inner zone (A) of strongly bound, electro-

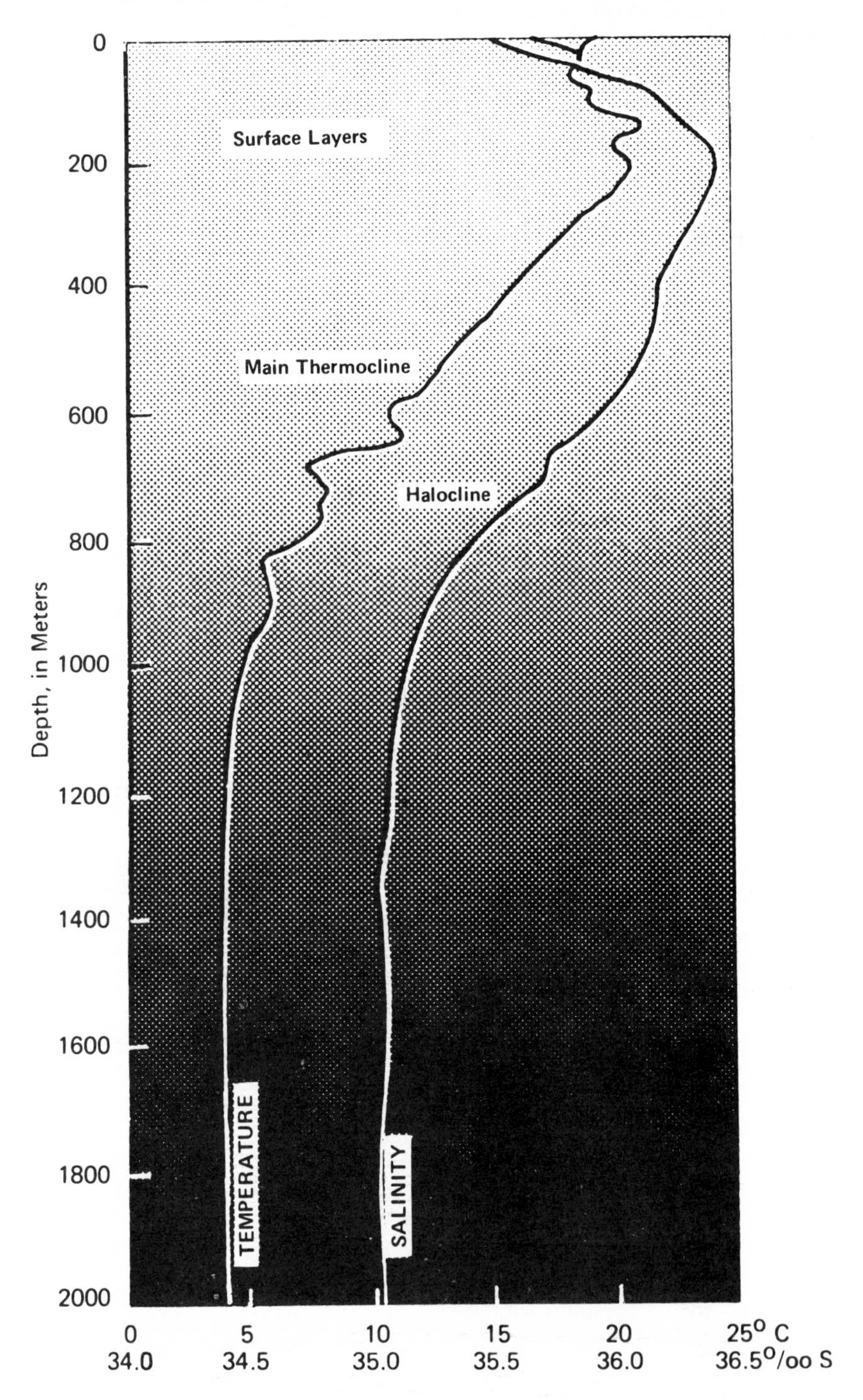

Figure 8.2 Typical temperature and salinity profiles in the oceans (from Horne, 1970*a*, with permission of Academic Press, Inc.).

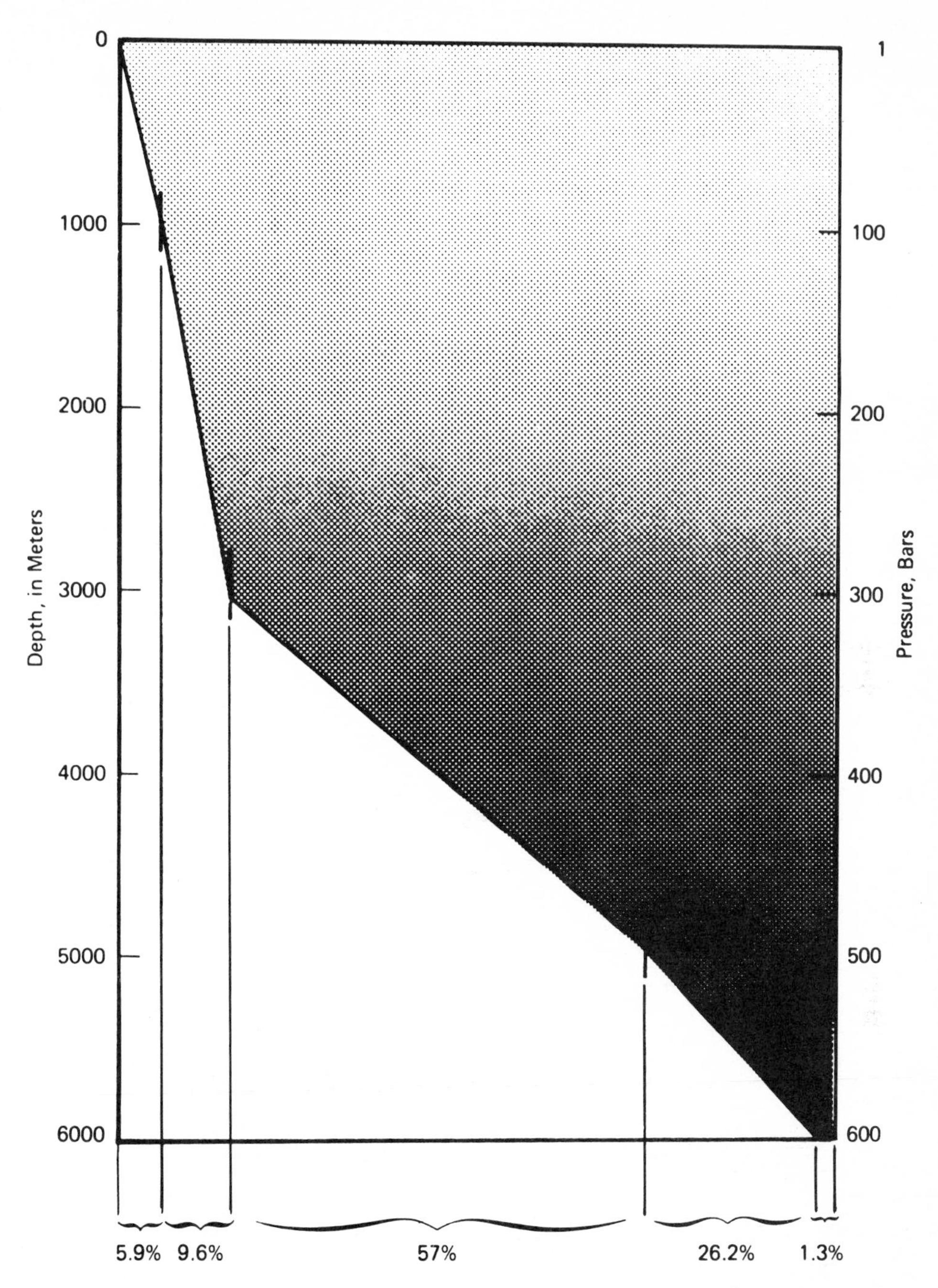

Figure 8.3 Percentage of ocean areas in a given depth and pressure range (from Horne, 1970*a*, with permission of Academic Press, Inc.).

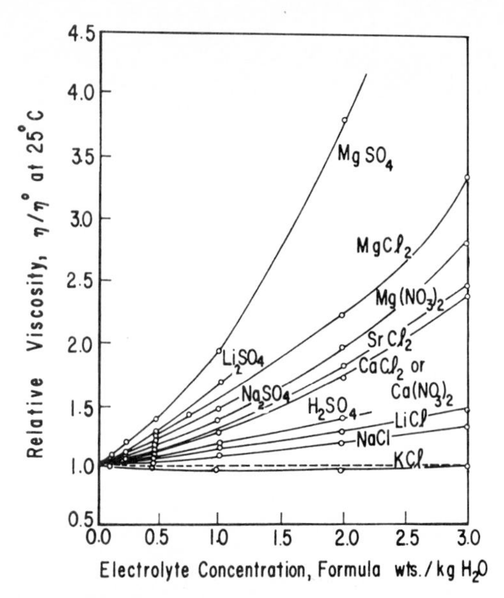

Figure 8.4 The effect of electrolytes on the viscosity of water (from Horne, *Water Resources Res.*, *1*, 263 (1965), © American Geophysical Union, with permission of the AGU).

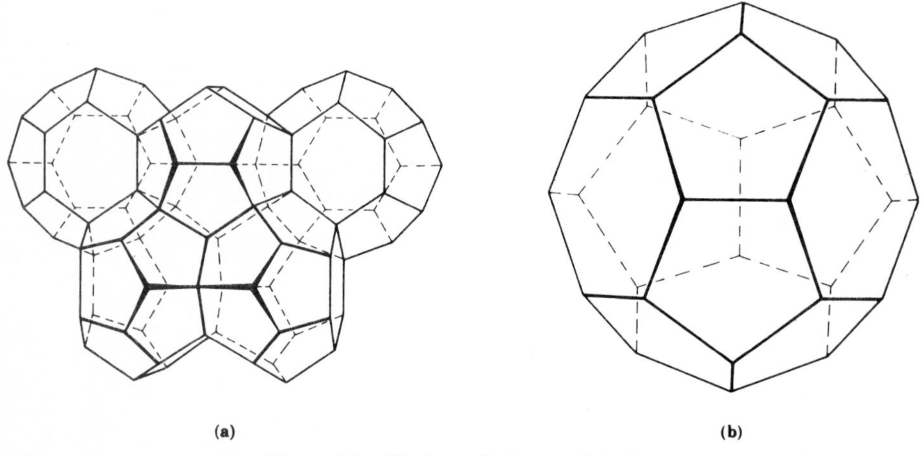

Figure 8.5 Clathrate hydrate polyhedra.

stricted* water molecules surrounded by a region (B) in which the coulombic field of the ion is sufficiently strong to disrupt the structure of the bulk water (C) but is too weak to reorient the water molecules into a new pattern. In the case of structure makers, zone A is the more important; for structure breakers, zone B. We have further analyzed zone A into two subregions (Horne and Birkett, 1967); an innermost shell

* When dissolved in water, an ion may interact so strongly with the polar water molecules, crowding them in about itself, that an overall volume decrease results. This phenomenon is known as electrostriction.

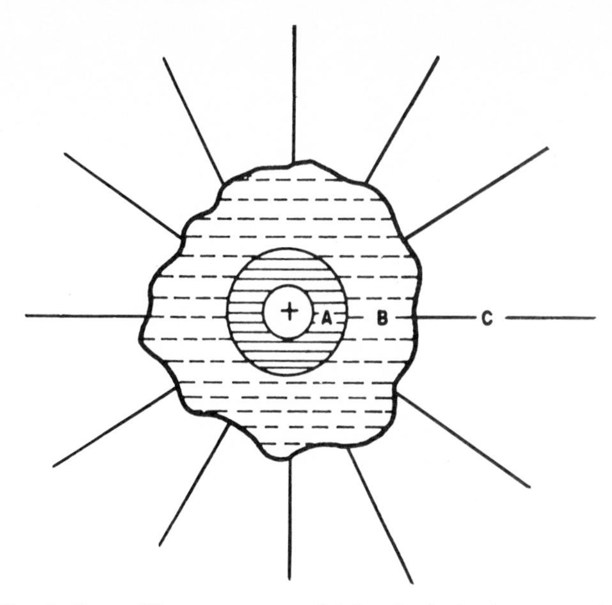

Figure 8.6 The Frank–Evans–Wen two-zone model for the hydration atmosphere of an ion in aqueous solution (from Frank and Wen, 1957, with permission of The Faraday Society and the authors).

of so-called "primary hydration" (*A* in Figure 8.7), which may be inseparable from the ion even upon evaporation of the solution (water of hydration or crystallization), surrounded by what appears to be a Frank–Wen cluster (*B* in Figure 8.7). In addition to this *coulombic hydration* surrounding ions and polar solutes, a quite different type of water structure—*hydrophobic hydration*—appears to envelope nonpolar solutes and hydrophobic segments of macromolecules (see Figure 8.1).

The nature of the hydration envelope is not an either/or situation. We can describe a solute as having so much hydrophobic and so much coulombic hydration just as we can describe a chemical bond as having so much ionic and so much covalent character. In this respect, although unimportant in natural waters, the tetraalkylammonium cations form an interesting family for as n in $(C_nH_{2n+1})_4N^+$ increases, the ion charge density decreases, and there is a gradual progression from coulombic to hydrophobic character (Horne and Young, 1968). A complex molecule, such as a biomacromolecule, will have a hybrid hydration sheath with coulombic hydration surrounding polar sites (open circles in Figure 8.8) and hydrophobic hydration (shaded areas in Figure 8.8) encasing the hydrophobic backbone of the molecule.

The constancy of the Walden product (electrical conductivity times viscosity), even for solutions as concentrated as seawater (Horne and Courant, 1964) would appear to indicate that temperature has relatively little effect on the hydration atmospheres of ions in solution, or at least on that part of its hydration envelope that a moving solute species carries along with it. However, in the case of ionic solutes and hydrostatic pressure, Walden's rule fails, and the situation is more complex (Horne, 1963). Pressure, unlike temperature, appears capable of reducing the effective hydrodynamic radii of ions in solution, presumably by breaking up their hydration atmospheres (Zisman, 1932; Horne, 1969), although there is an alternative,

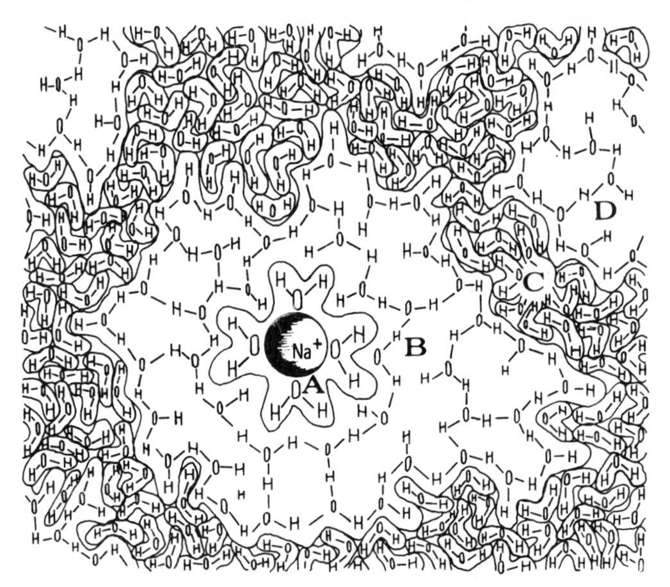

Figure 8.7 The local water structure near a cation in aqueous solution: *A*, Innermost, dense, tightly bound, electrostricted region (primary hydration); *B*, rarified, bound, Fran–Wen clusterlike region; *C*, broken water structure region; *D*, bulk water—Frank–Wen clusters and "monomeric" water, $D_A > D_D > D_B$. (Reprinted from Horne, 1971, by courtesy of Marcel Dekker, Inc.).

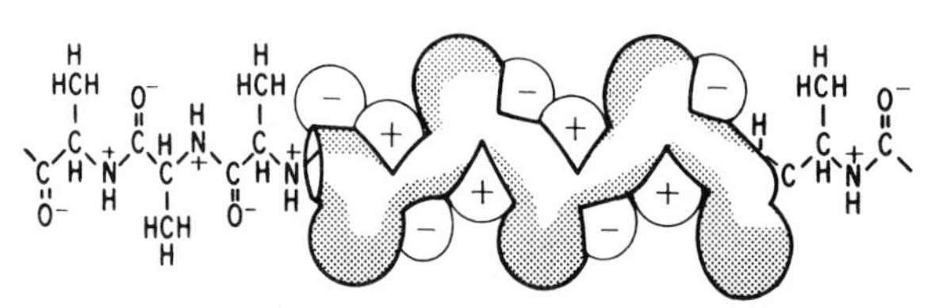

Figure 8.8 Hybrid hydration sheath of a biomacromolecule. Coulombic hydration (open circles) surrounds polar sites and hydrophobic hydration (shaded circles) encases nonpolar regions (from Horne, 1970*b*).

and I feel a less likely explanation for the observed conductance phenomena under high pressure (Adams and Hall, 1931; Gancy and Brummer, 1969). In contast to coulombic hydration, there is some evidence that the application of pressure may have little effect on hydrophobic hydration (Horne and Young, 1968). If such is indeed the case, then in order to achieve the necessary high density to insure relative pressure stability the spaces in the clathrate cage must be occupied by additional water molecules.

In the water column in the ocean, the mixing of chemical species is accomplished almost entirely by physical processes, such as currents, turbulence, and avection Organisms may also transport substances both horizontally and vertically in the seas. A "diffusion coefficient" appears in the expressions that have been developed for ocean mixing models, but closer scrutiny soon shows that it is a catch-all term that includes all sorts of mixing mechanisms and is, thus, not a token of our knowledge but of our ignorance of the detailed mechanisms, both microscopic and macroscopic, of material transport in the oceans. The contribution of real or molecular diffusion to mixing in the oceans is relatively insignificant: The self-diffusion coefficient of water at 25°C, for example, is only 2.57×10^{-5} cm²/sec (Wang, 1965), but the Gulf Stream off Chesapeake Bay is capable of transporting water with velocities as great as 300 cm/sec. Nevertheless, *within* a given moving water mass, the rates of chemical processes still can be diffusion controlled. Diffusional processes can also be rate controlling in the surface boundary layer of the oceans and in the interstitial water in marine sediments. As we have emphasized elsewhere, the bulk of interesting chemistry of the oceans occurs in just those regions that mark the boundaries of the marine hydrosphere and describe the thin zone of chemical interaction with the atmosphere, lithosphere, and biosphere.

With the exception of protons, hydroxyl ions, and very large solutes, all solutes have similar diffusion coefficients and equivalent electrical conductivities in aqueous solutions. Furthermore, the activation energies of viscous flow, diffusion, electrical conductivity, and dielectric relaxation are all similar, strongly suggesting that the same fundamental molecular mechanism forms the basis of all of these solute and solvent movement processes (Horne, 1969, Chapter 3). The classical theory of electrical conductivity in aqueous solution is based on the model of a charged sphere moving through a viscous continuum, and, although such a picture "works" remarkably well, it is clearly grossly oversimplified. Modern theories tend to the point of view that the fundamental rate-determining step of transport processes is the breaking of H-bonds and the formation of a "hole" or vacancy in the solvent. A nearby solute species then jumps in to occupy this hole (Glasstone, Laidler, and Eyring, 1941).

Electrolyte addition decreases the dielectric constant of water in a nearly linear fashion

$$\epsilon = \epsilon_0 + \frac{C}{2} \sum n_i \delta_i \tag{8.1}$$

where ϵ_0 is the dielectric constant of the pure solvent, C is the concentration of electrolyte, n_i is the number of the ion i formed on dissociation, and δ_i is a constant characteristic of the ion i. At 25°C ϵ_0, is 79.3, thus taking seawater to be simply an 0.5 M NaCl solution we calculate a dielectric constant of 76.6 for seawater at this temperature.

Some diffusion coefficients of seawater electrolytes and gas diffusion in pure water are given in Tables A.8 and A.9. Inasmuch as NaCl and $MgSO_4$ both bend to increase the structure of liquid water, we might expect diffusion coefficients, equivalent conductances, and fluidity (reciprocal of viscosity) of water all to decrease with increasing salt content. Table 8.4 gives the relative viscosity of seawater as a function of temperature and salinity (or total salt concentration, to be defined in Section 8.4), and Table 8.5, the specific conductivity. The electrical conductivity of seawater (and fresh waters as well) is one of its most important physical–chemical properties for the ease with which this property can be measured with relatively great accuracy, even under adverse conditions in situ, ideally suits it as being the single most useful parameter for chemically describing seawater, and a number of instruments, or salinometers, have been developed for this purpose (Paquette, 1959; Brown and Hamon, 1961; Carritt, 1963).

Increasing the temperature facilitates transport processes, increasing the conductivity (Table 8.5 and Figure 8.9) and decreasing the viscosity (Table 8.4). The

Table 8.4 Viscosity of Seawater at 1 atm (in centipoise)

Temperature (°C)	5‰ *S*	10‰ *S*	20‰ *S*	30‰ *S*	40‰ *S*
0	1.803	1.817	1.844	1.887	1.884
5	1.528	1.542	1.567	1.592	1.617
10	1.319	1.331	1.403	1.380	1.403
15	1.149	1.160	1.183	1.206	1.230
20	1.015	1.026	1.047	1.070	1.092
25	0.901	0.911	0.931	0.953	0.974
30	0.811	0.822	0.840	0.860	0.877

Table 8.5 Specific Conductivity of Seawater[a] (Values calculated from the equation of P. Weyl, *Limnol. Oceanog.*, *9*, 75 (1964), based on data of B. D. Thomas, T. G. Thompson, and C. L. Utterback, *J. Conseil Perm. Intern. Exploration Mer.*, *9*, 28 (1934).)

	Temperature (°C)						
S (‰)	30	25	20	15	10	5	0
10	(19.127)	17.345	15.628	13.967	12.361	10.816	9.341
20	(35.458)	32.188	29.027	25.967	23.010	20.166	17.456
30	(50.856)	46.213	41.713	37.351	33.137	29.090	25.238
31	(52.360)	47.584	42.954	38.467	34.131	29.968	26.005
32	(53.859)	48.951	44.192	39.579	35.122	30.843	26.771
33	(55.352)	50.314	45.426	40.688	36.110	31.716	27.535
34	(56.840)	51.671	46.656	41.794	37.096	32.588	28.298
35	(58.323)	53.025	47.882	42.896	38.080	33.457	29.060
36	(59.801)	54.374	(49.105)	(43.996)	(39.061)	(34.325)	(29.820)
37	(61.274)	55.719	(50.325)	(45.093)	(40.039)	(35.190)	(30.579)
38	(62.743)	57.061	(51.541)	(46.187)	(41.016)	(36.055)	(31.337)
39	(64.207)	58.398	(52.754)	(47.278)	(41.990)	(36.917)	(32.094)
40	(65.667)	(59.732)	(53.963)	(48.367)	(42.962)	(37.778)	(32.851)

[a] Conductivity in ohm^{-1} cm^{-1} $\times$ 1000.

effect of hydrostatic pressure is not so straightforward; as the pressure is increased, the relative conductance of aqueous solutions of strong one-one electrolytes, such as NaCl, first increase, go through a maximum, and then decrease (Figure 8.10). Notice in the figure that increasing the temperature tends to wipe out this pressure feature. Analysis of the initial increase reveals unexpectedly that the increase in conductance is greater than one expects from the pressure-induced viscosity and volume changes (Horne, 1969*b*), that is to say, Walden's rule is breaking down (Horne, 1963; Horne and Courant, 1964). Two explanations of this anomalous conductive increment have been offered: (*a*) further dissociation of the strong electrolytes under pressure (Adams and Hall, 1931; Gancy and Brummer, 1969), and (*b*) pressure-induced dehydration of the ions resulting in a decrease in their effective hydro-

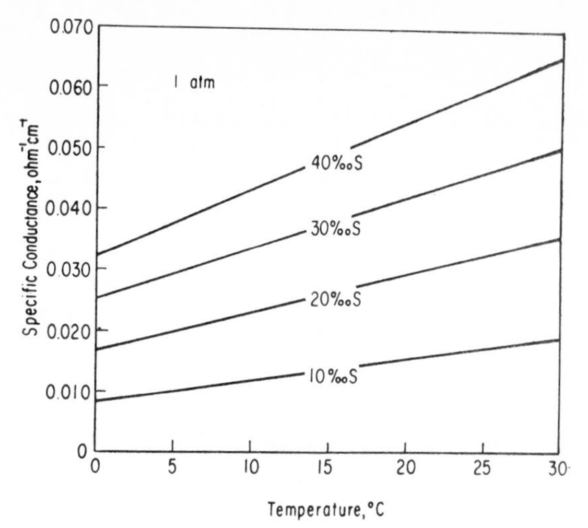

Figure 8.9 The effect of temperature on the specific electrical conductivity of seawater of various salinities at 1 atm (from Horne, 1969).

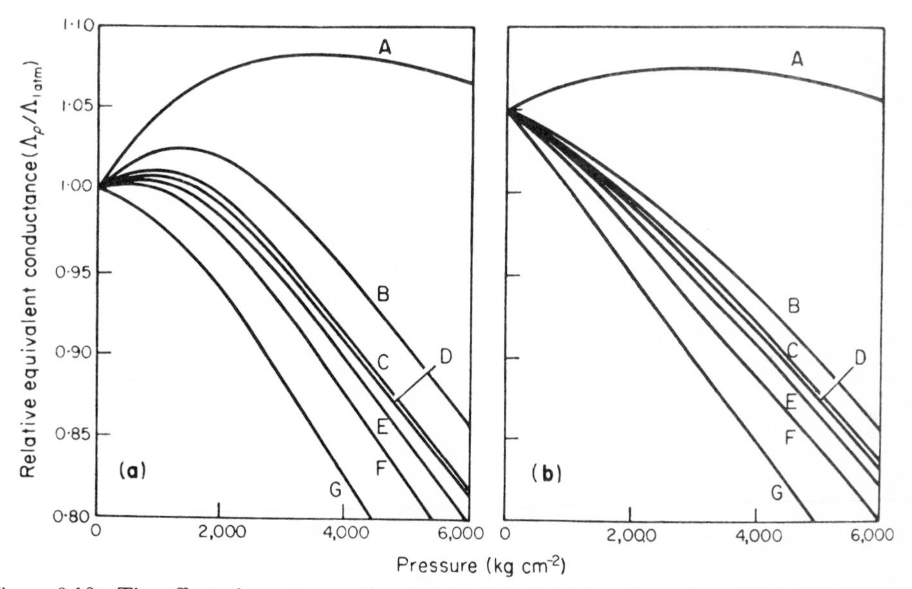

Figure 8.10 The effect of pressure on the electrical conductivity of aqueous electrolytic solutions at (*a*) 30°C and (*b*) 75°C. *A*, HCl; *B*, LiCl; *C*, KCl; *D*, NaCl; *E*, RbCl; *F*, CsCl; *G*, NaI. (From Horne, 1969*b*, with permission of Academic Press, Inc. Based on data of Zisman, 1932.)

dynamic radii (Zisman, 1932; Horne, 1969*b*). The effect of pressure on the electrical conductivity of seawater has been examined experimentally by Horne and Frysinger (1963) and Bradshaw and Schleicher (1965), and the results of the latter are shown in Table 8.6. In the case of seawater, a further conductive increment is contributed by the increased dissociation of the weak electrolyte $MgSO_4$ under pressure. This equilibrium is discussed in more detail subsequently. It should be noted that the greatest hydrostatic pressures encountered in the ocean deeps, about 1000 atm, are

Table 8.6 Effect of Pressure on the Conductivity of Seawater[a] (From A. Bradshaw and K. E. Schleicher, *Deep-Sea Res.*, *12*, 151 (1965), with permission of Pergamon Press, Ltd.)

Pressure (dbar)	Temperature (°C)	S (‰) 31	35	39	Temperature (°C)	S (‰) 31	35	39
1,000	0	1.599	1.556	1.512	15	1.032	1.008	0.985
2,000		3.089	3.006	2.922		1.996	1.951	1.906
3,000		4.475	4.354	4.233		2.895	2.830	2.764
4,000		5.759	5.603	5.448		3.731	3.646	3.562
5,000		6.944	6.757	6.569		4.506	4.403	4.301
6,000		8.034	7.817	7.599		5.221	5.102	4.984
7,000		9.031	8.787	8.543		5.879	5.745	5.612
8,000		9.939	9.670	9.401		6.481	6.334	6.187
9,000		10.761	10.469	10.178		7.031	6.871	6.711
10,000		11.499	11.188	10.877		7.529	7.358	7.187
1,000	5	1.368	1.333	1.298	20	0.907	0.888	0.868
2,000		2.646	2.578	2.510		1.755	1.718	1.680
3,000		3.835	3.737	3.639		2.546	2.492	2.438

[a] Percentage increase compared with the conductivity at 1 atm.

still very slight when compared with the scale in Figure 8.10, thus, except in the case of $MgSO_4$, the effect of ambient pressure on condensed phase equilibria and other physical–chemical processes in the oceans (but not in the lithospheric crust) is slight: Temperature is a far more significant variable in marine chemistry.

The relative viscosity versus pressure curves for water and aqueous solutions of strong one-to-one electrolytes exhibit minima (Figure 7.10). Water is the only substance known that exhibits such an anomaly, and the feature is smeared out by increasing the temperature or adding electrolyte (Horne and Johnson, 1967). The Frank–Wen clusters are less dense than the "free" water; hence, the application of hydrostatic pressure tends to favor the more dense and destroy the larger volume form

$$\underset{\text{cluster}}{(H_2O)_n} \overset{P}{\rightleftharpoons} \underset{\text{free}}{n(H_2O)} \tag{8.2}$$

and this results in a decrease in the relative viscosity until all the clusters are destroyed. Then the relative viscosity increases with increasing pressure, just as expected of a "normal" liquid (Horne and Johnson, 1966*a*). The maximum in the conductivity versus pressure curve does *not* coincide with the minimum in the viscosity versus pressure curve, indicating that the situation is not as simple as one might hope (Horne, 1970*a*, *b*). Horne and Johnson (1966*b*) and Stanley and Batten (1969) have measured the viscosity of seawater under pressure and the results of the latter are summarized in Figure 7.10 and Table 8.7. The effect of pressure on diffusional processes in the sea and in marine sediments is believed to be slight (Horne, Young, and Day, 1969). Table A.10 compares some transport processes in pure water and seawater.

Table 8.7 Relative Viscosity, η_1/η_P atm of 19.374‰ *Cl* Seawater (Values selected from Stanley and Batten, 1969)

P (kg/cm²)	−0.02°C	2.22°C	6.00°C	10.01°C	15.01°C	20.01°C
176	0.98	0.99	0.99	0.99	0.99	1.00
352	0.97	0.97	0.98	0.99	0.99	1.00
527	0.96	0.97	0.98	0.98	0.99	1.00
703	0.96	0.96	0.97	0.98	0.99	1.00
1055	0.95	0.96	0.99	0.99	1.00	1.01
1406	0.95	0.96	0.99	0.99	1.01	1.02

8.3 COLLIGATIVE AND PVT PROPERTIES OF SEAWATER

The colligative properties of solutions are determined by the total concentration of dissolved solutes. Figure 8.11 (see also Table A.11) shows the dependence of some of these properties on the salt content of seawater. Robinson (1954) gives a value of 25.5 atm for the osmotic pressure of 35‰ *S* seawater; Miyake (1939), a freezing point depression of 1.98°C; and Fabuss and Korosi (1966) and Stoughton and Lietzke (1967), a value of 0.53°C for the boiling point elevation.

An important thermodynamic question is how ideal a solution is seawater? As we shall see, seawater is a roughly 0.5 *M* NaCl solution. While not a concentrated solution, seawater is not sufficiently dilute so that we might expect the formalisms that have been developed for ideal, that is, very dilute, solutions to be applicable. If we

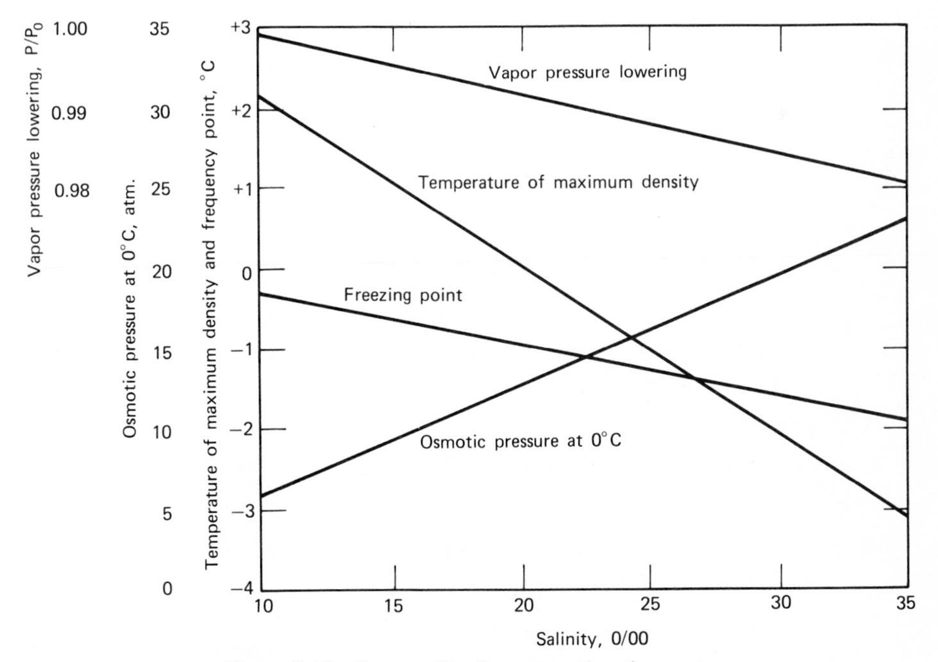

Figure 8.11 Some colligative properties of seawater.

consider a solvent property such as the vapor pressure, the deviation from ideality of seawater is only 0.4%; however, a solute property such as the equivalent electrical conductivity is 26% lower for seawater than for an infinitely dilute NaCl solution (Horne, 1969, p.55). Nevertheless, despite such difficulties, for practical purposes, it is often possible to approximate satisfactorily seawater properties with those of a 0.5 *M* NaCl solution (Table 8.8).

While some scientists might be reluctant to admit it, our store of scientific knowledge has profited greatly from the demands of military necessity. The best known historical example is perhaps the formulation of the principle of the mechanical equivalent of heat as a consequence of cannon boring. But a case more to the point here is the many very accurate measurements of the PVT properties of seawater undertaken in view of the importance of underwater sound transmission in antisubmarine warfare. In the marine hydrosphere, pressure as well as temperature is an important variable. In the *S* and *T* range of oceanographic interest the pressure field in the oceans (Figure 8.12) increases so regularly with depth that, as a quick yet accurate rule of thumb, one can say that the pressure in decibars is numerically equal to the geometric depth in meters. Thus the mean depth of the world's oceans (3795 m, see Table A.12) corresponds to a pressure of about 4000 dbar. Sound velocity, specific volume, and temperature of maximum density data for seawater can be found in Tables A.13, A.14, and A.15, while further specific volume, com-

Table 8.8 Comparison of Pure Water and Seawater Properties (From Horne, 1969)

Property	Seawater (35‰ *S*)	Aqueous NaCl Solution (0.50 *M*)	Aqueous NaCl Solution (0.6 *M*)	Pure Water
Density, 25°C (g/cm^3)	1.02412	1.01752	1.02172	1.0029
Equivalent Conductivity, 25°C ($cm^2\ ohm^{-1}\ equiv^{-1}$)	—	93.62	91.58	—
Specific Conductivity, 25°C ($ohm^{-1}\ cm^{-1}$)	0.0532	0.0468	0.0458	—
Viscosity, 25°C (mP)	9.02	9.32	9.41	8.90
Vapor Pressure (mm Hg at 20°C)	17.4	17.27	17.18	17.52
Isothermal Compressibility, 0°C (unit vol/atm)	46.4×10^{-6}	46.6×10^{-6}	45.9×10^{-6}	50.3×10^{-6}
Temperature of Maximum Density (°C)	−3.52	—	—	+3.98
Freezing Point (°C)	−1.91	−1.72	−2.04	0.00
Surface Tension, 25°C (dyne/cm)	72.74	72.79	72.95	71.97
Velocity of Sound, 0°C (m/sec)	1450	—	—	1407
Specific Heat, 17.5°C (joule/g°C)	3.898	4.019	3.998	4.182

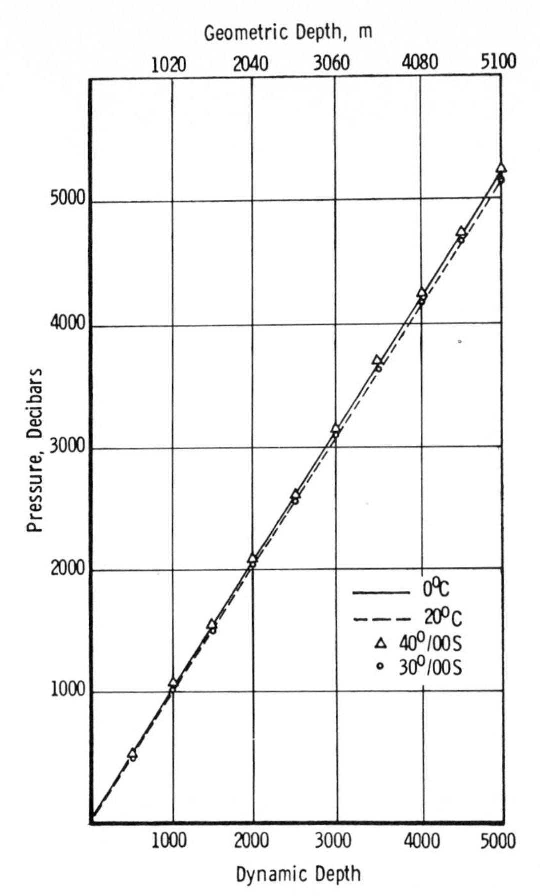

Figure 8.12 Hydrostatic pressure as a function of dynamic and geometric depth in the seas (from Horne, 1969).

pressibility, and thermal expansion data as a function of salinity, pressure, and temperature can be found in Tables A.7, A.8, and A.9 of Horne (1969).

As for an equation-of-state of seawater, Li (1967) recommends the Tait–Gibson equation

$$V_P = V_{1\ \mathrm{atm}} - (1 - S \times 10^{-3})C \log \frac{B^* + P}{B^* + 1} \tag{8.3}$$

where $C = 0.315\ V_{0°\mathrm{C},\ 1\ \mathrm{atm}}$ and B^* is given in Figure 8.13.

8.4 THE CHEMICAL CONSTITUENTS OF SEAWATER

There is at least just a little bit of just about everything imaginable in seawater. The oceans are the final sink for many of the substances involved in the myriad natural geochemical processes, and they are all too often the ultimate cesspool of human activity. Not only do the oceans receive the materials leached from the continents and washed from the skies, but, inasmuch as they are also the habitat of a large

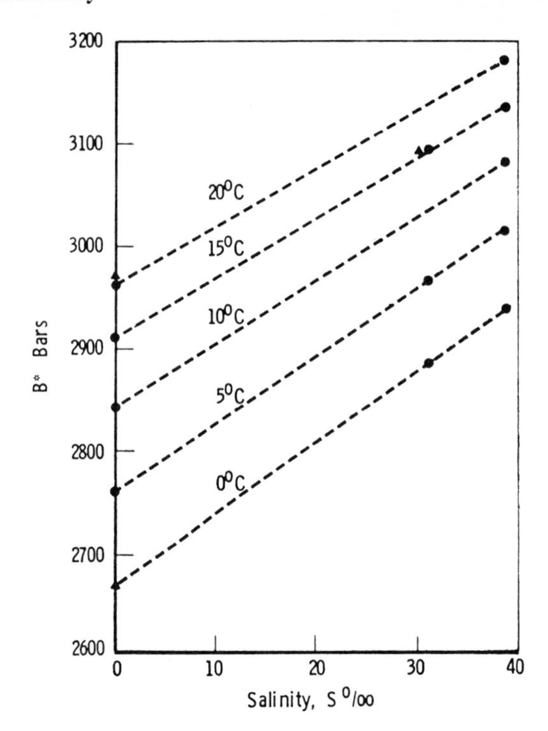

Figure 8.13 B^* for seawater as a function of salinity (adapted from Li, 1967).

portion of the Earth's biosphere, they contain, as well all the nutrients, all the intermediate products, all the excreted and eliminated substances, and all the decay products characteristic of the life processes. We can try to order and classify the constituents of such a soup in many ways, and we have already seen one such classification scheme in Figure 8.1. That figure was arranged on the basis of the size of the constituents, ranging as we read down from solutes, both polar and nonpolar, up the size spectrum through colloids, to substances present as a definite second phase. The shading was intended to convey the nature of the hydration envelope of the species, and we discover that we have gained thereby some insight into the nature of the chemistry of life since biomaterial tends to fall in the lower right of each of the major size categories; that is to say, biosubstances are large, which is no surprise and their hydration sheaths have a great deal of hydrophobic character. On an abundance basis, our classification is very distorted since (excepting H and O) only three chemical elements—Na, Cl, and Mg—account for almost 90% of the stuff in seawater, and if we add K, Ca, and S (largely as SO_4^{2-}), we leave only about 7% for all the other elements in seawater. All the major constituents are crowded into the first category of our classification; they are present as simple inorganic ions, Na^+, K^+, Mg^{2+}, Ca^{2+}, Cl^-, and SO_4^{2-}, and their combinations. Nature has been kind to the marine chemist for the oceans are well mixed. As a consequence, in contrast to the wide variability of the composition of fresh waters, the *ratios* of the major constituents of seawater are *nearly* the same throughout all the world's oceans, even though the total salt content or "salinity" (defined subsequently) is highly variable in both space and time. Table 8.9 compares the ratios of the major con-

Table 8.9 Major Constituent Concentration-to-Chlorinity Ratios for Various Oceans and Seas (From Horne, 1969)

Ocean or Sea	Na / ‰ *Cl*	Mg / ‰ *Cl*	K / ‰ *Cl*	Ca / ‰ *Cl*	Sr / ‰ *Cl*	SO_4 / ‰ *Cl*	Br / ‰ *Cl*
North Atlantic	—	—	0.02026	—	—	—	0.0337–0.00341
Atlantic	0.5544–0.5567	0.0667	0.01953–0.0263	0.02122–0.02126	0.000420	0.1393	0.00325–0.0038
North Pacific	0.5553	0.06632–0.06695	0.02096	0.02154	—	0.1396–0.1397	0.00348
West Pacific	0.5497–0.5561	0.06627–0.0676	0.02125	0.02058–0.02128	0.000413–0.000420	0.1399	0.0033
Indian	—	—	—	0.02099	0.000445	0.1399	0.0038
Mediterranean	0.5310–0.5528	0.06785	0.02008	—	—	0.1396	0.0034–0.0038
Baltic	0.5536	0.06693	—	0.02156	—	0.1414	0.00316–0.00344
Black	0.55184	—	0.0210	—	—	—	—
Irish	0.5573	—	—	—	—	0.1397	0.0033
Puget Sound	0.5495–0.5562	—	0.0191	—	—	—	—
Siberian	0.5484	—	0.0211	—	—	—	—
Antarctic	—	—	—	0.02120	0.000467	—	0.00347
Tokyo Bay	—	0.0676	—	0.02130	—	0.1394	—
Barents	—	0.06742	—	0.02085	—	—	—
Arctic	—	—	—	—	0.000424	—	—
Red	—	—	—	—	—	0.1395	0.0043
Japan	—	—	—	—	—	—	0.00327–0.00347
Bering	—	—	—	—	—	—	0.00341
Adriatic	—	—	—	—	—	—	0.00341

stituents and Sr and Br to the chlorinity for various ocean areas. This very important generalization was first observed by Marcet in 1819 and was confirmed by chemical analyses of the water samples collected on the famous oceanographic cruise of the *H.M.S. Challenger* (1873–1876). Marcet's principle has remained one of the most useful generalizations in the earth sciences.

Table 8.10, which in turn summarizes two tables published by Goldberg (1963, 1965) lists the abundances of the chemical elements in 35‰ S seawater, the major form that they are believed to be in, and an estimate of their residence times in the oceans. While probably representing the best information available at the time, the reader should be cautioned that the principal species and residence times are the subject of controversy and will continue to be so. Worse still, an unpublished interlaboratory check reported at a Gordon Conference in 1971 suggested strongly that the concentration of *no* trace metal in seawater is known with any degree of confidence. Because seawater is such a complex soup, because of the difficulties of sample collection and storage, and because of the very low concentrations involved, the analytical problems are most formidable and in some instances may be insurmountable. The analytical chemistry of seawater has been reviewed by Barnes (1959), Culkin (1965), and Riley (1965). The mess in the analytical chemistry of seawater trace elements contrasts sharply with measurements of its physical–chemical properties where, in most instances, recent work has tended to underscore the truely remarkable accuracy of older measurements.

The operancy of Marcet's principle together with the small quantities of the less than major constituents in seawater enables us to define a useful term characteristic of a given seawater sample: the *salinity* (S) for the total grams of solid per kilogram of seawater (‰). Open ocean water commonly has a salinity of about 34 to 36‰. Seawater used to be analyzed by the classical wet Volhard method for chloride and other halides, and we have inherited a second concentration term, the *chlorinity* (Cl in ‰), which is related to salinity by the expression

$$S = 0.03 + 1.805\ Cl \tag{8.4}$$

but inasmuch as S does not go to zero as Cl goes to zero, this convention is clearly an unsatisfactory one and is not applicable to dilute seawater. Yet the term is still so widely used that a textbook such as this is not an appropriate place to scrap it.

The salinity of seawater, as I have said, is highly variable in space and time. The salinity structure of the oceanic water column commonly resembles the temperature profile (Figure 8.2), and they tend to alter with one another. While the temperature and salinity of deep waters are highly constant, T and S in the surface waters can fluctuate widely either due to particular meteorological conditions and/or a diurnal cycle, while the intermediate thermocline and halocline tend to waiver more gently on a seasonal cycle. Surface values of T and S depend strongly no the balance between evaporation and precipitation and may take on characteristic values in certain latitudes. Physical oceanographers often use T-S diagrams to identify water masses and their movements. In addition to rainfall, river influx and ice melting can reduce the salinity of seawater. In some localities, geological circumstances can produce anomalous salinities as in the case of the very hot (45°C) very saline (>43‰ S) waters discovered near the bottom of the Red Sea (Miller, 1964; Swallow and Crease, 1965; Degens, 1969).

Still within our first category in Figure 8.1, a second group of very important simple inorganic ions in seawater are the so-called nutrients, such as ammonium ion, nitrate, phosphate, and silicate. In strong contrast to the major constituents discussed previously, these species are highly dependent upon biological activity and thus vary widely in the Earth's oceans, in fact the major oceans tend to have, not the same, but characteristic nutrient ratios (Table 8.11). These ratios are a sort of biological clock. Bioactivity takes up and removes phosphate from the water and incorporates it into complex organic forms while (upon death) releasing silica, hence the "young" waters of the well-circulated Atlantic are relatively rich in the former and poor in the latter, whereas the tired "old" waters of the more sluggishly circulating Pacific are rich in the products but poor in the reactants of biological activity.

Another type of solute tabulated in Figure 8.1 are gases such as oxygen, nitrogen, and carbon dioxide. These gases and their related species, such as bicarbonate, ammonium ion, nitrate, and nitrite, are also involved in the chemistry of the marine life cycle, and we defer discussion of them to our treatment of the interactions between atmosphere, hydrosphere, and biosphere in Chapter 16.

The second and third solute types listed in Figure 8.1 are weak electrolytes or ion-pairs and complex ions. Sodium chloride is a strong electrolyte, that is to say, it is very nearly completely dissociated in aqueous solutions, but magnesium sulphate, for example, is not. Therefore it becomes necessary to examine the *forms* in which the major chemical elemental constituents of seawater are present. We must formulate a chemical model for seawater. On the basis of available equilibria data, Garrels and Thompson (1962) have calculated the species partitioning given in Table 8.12. Because the marine temperature range is narrow, the variations in these values as a consequence of temperature fluctuations are slight, and the pressure effects are insignificant (however, see the subsequent discussion). Ion association reduces the ionic strength* of 34.8‰ *S* seawater down from the expected value of 0.72 to 0.67, a more than 7% decrease (Courant, Horne, and Kester, 1971). The successful calculation of the colligative and other properties of sea water from the Garrels–Thompson model demonstrates that it is a highly satisfactory chemical description of seawater. However, with the possible exception of some of the carbonate equilibria, these ionic reactions are all fast, and it is dangerous in the extreme to attempt to extend the equilibrium model approach to the less than major species in seawater for seawater is almost certainly *not* in an equilibrium state with respect to most of its constituents. We are not even certain, for example, of the relative importance of the equilibria that buffer seawater to keep its pH so nearly constant (pH 7.5 to 8.4): Are they carbonate equilibria involving the dissolution of atmospheric CO_2 or are they equilibria such as Sillen (1967) proposed in his multiphase chemical model involving mineral silicate substances? Professor Sillen's equilibrium chemical model of the oceans was invaluable in stimulating marine chemists to try and think more exactly about the air/ocean/sediments system; it opened the way for the development of probably more realistic steady state models of the oceans (Siever, 1968; Broecker, 1971).

* Ionic strength, μ, is defined by $\mu = \sum \frac{1}{2} m_i z_i^2$ for all ionic species, i, where m_i is the molal concentration of i and z_i the charge. Empirically, the ionic strength of seawater is given by $\mu = 0.0054 + 0.01840\, S‰ + 1.78 \times 10^{-5}\, (S‰)^2 - 3.0 \times 10^{-4}\, (25 - T°C) + 7.6 \times 10^{-6}\, (P\ \text{atm}^{-1})$.

Table 8.10 The Elemental Composition of Seawater (Compiled from tables of Goldberg, 1963, 1965)

Element	Abundance (mg/liter)	Principal Species	Residence Time (yr)
H	108,000	H_2O	—
He	0.000005	He (g)	—
Li	0.17	Li^+	2.0×10^7
Be	0.0000006	—	1.5×10^2
B	4.6	$B(OH)_3$; $B(OH)_2O^-$	—
C	28	HCO_3^-; H_2CO_3; CO_3^{2-}; organic compounds	—
N	0.5	NO_3^-; NO_2^-; NH_4^+; N_2 (g) organic compounds	—
O	857,000	H_2O; O_2 (g); SO_4^{2-} and other anions	—
F	1.3	F^-	—
Ne	0.0001	Ne (g)	—
Na	10,500	Na^+	2.6×10^8
Mg	1,350	Mg^{2+}; $MgSO_4$	4.5×10^7
Al	0.01	—	1.0×10^2
Si	3	$Si(OH)_4$; $Si(OH)_3O^-$	8.0×10^3
P	0.07	HPO_4^{2-}; $H_2PO_4^-$; PO_4^{3-}; H_3PO_4	—
S	885	SO_4^{2-}	—
Cl	19,000	Cl^-	—
A	0.6	A (g)	—
K	380	K^+	1.1×10^7
Ca	400	Ca^{2+}; $CaSO_4$	8.0×10^6
Sc	0.00004	—	5.6×10^3
Ti	0.001	—	1.6×10^2
V	0.002	$VO_2(OH)_3^{2-}$	1.0×10^4
Cr	0.00005	—	3.5×10^2
Mn	0.002	Mn^{2+}; $MnSO_4$	1.4×10^3
Fe	0.01	$Fe(OH)_3(s)$	1.4×10^2
Co	0.0005	Co^{2+}; $CoSO_4$	1.8×10^4
Ni	0.002	Ni^{2+}; $NiSO_4$	1.8×10^4
Cu	0.003	Cu^{2+}; $CuSO_4$	5.0×10^4
Zn	0.01	Zn^{2}; $ZnSO_4$	1.8×10^5
Ga	0.00003	—	1.4×10^3
Ge	0.00007	$Ge(OH)_4$; $Ge(OH)_3O^-$	7.0×10^3
As	0.003	$HAsO_4^{2-}$; $H_2AsO_4^-$; H_3AsO_4; H_3AsO_3	—
Se	0.004	SeO_4^{2-}	—
Br	65	Br^-	—
Kr	0.0003	Kr (g)	—
Rb	0.12	Rb^+	2.7×10^5
Sr	8	Sr^{2+}; $SrSO_4$	1.9×10^7
Y	0.0003	—	7.5×10^3
Zr	—	—	—
Nb	0.00001	—	3.0×10^2
Mo	0.01	MoO_4^{2-}	5.0×10^5
Tc			
Ru			

(Continued)

Table 8.10 (*Continued*)

Element	Abundance (mg/liter)	Principal Species	Residence Time (yr)
Rh	—	—	—
Pd	—	—	—
Ag	0.00004	$AgCl_2^-$; $AgCl_3^{2-}$	2.1×10^6
Cd	0.00011	Cd^{2+}; $CdSO_4$	5.0×10^5
In	<0.02	—	—
Sn	0.0008	—	1.0×10^5
Sb	0.0005	—	3.5×10^5
Te	—	—	—
I	0.06	IO_3^-; I^-	—
Xe	0.0001	Xe (g)	—
Cs	0.0005	Cs^+	4.0×10^4
Ba	0.03	Ba^{2+}; $BaSO_4$	8.4×10^4
La	1.2×10^{-5}	—	4.4×10^2
Ce	5.2×10^{-6}	—	8.0×10^1
Pr	2.6×10^{-6}	—	3.2×10^2
Nd	9.2×10^{-6}	—	2.7×10^2
Pm	—	—	—
Sm	1.7×10^{-6}	—	1.8×10^2
Eu	4.6×10^{-7}	—	3.0×10^2
Gd	2.4×10^{-6}	—	2.6×10^2
Tb	—	—	—
Dy	2.9×10^{-6}	—	4.6×10^2
Ho	8.8×10^{-7}	—	5.3×10^2
Er	2.4×10^{-6}	—	6.9×10^2
Tm	5.2×10^{-7}	—	1.8×10^3
Yb	2.0×10^{-6}	—	5.3×10^2
Lu	4.8×10^{-7}	—	4.5×10^3
Hf	—	—	—
Ta	—	—	—
W	0.0001	WO_4^{2-}	1.0×10^3
Re	—	—	—
Os	—	—	—
Ir	—	—	—
Pt	—	—	—
Au	0.000004	$AuCl_4^-$	5.6×10^5
Hg	0.00003	$HgCl_3^-$; $HgCl_4^{2-}$	4.2×10^4
Tl	<0.00001	Tl^+	—
Pb	0.00003	Pb^{2+}; $PbSO_4$	2.0×10^3
Bi	0.00002	—	4.5×10^5
Po	—	—	—
At	—	—	—
Rn	0.6×10^{-15}	Rn (g)	—
Fr	—	—	—
Ra	1.0×10^{-10}	Ra^{2+}; $RaSO_4$	—
Ac	—	—	—
Th	0.00005	—	3.5×10^2
Pa	2.0×10^{-9}	—	—
U	0.003	$UO_2(CO_3)_3^{4-}$	5.0×10^5

Table 8.11 Concentration Ratios of Inorganic Nutrient Anions in the Deep Waters of the Major Oceans (From Chow and Mantyla, 1965, with permission of the publisher and authors.)

Ocean	Silicate/Phosphate	Silicate/Nitrate	Nitrate/Phosphate
Southeast Pacific	55–65	3–5	13–14
Equatorial Indian	40–50	3	15
North Atlantic	20–40	1–2	12–16

As a consequence of electrostriction hydrated magnesium and sulfate ions are less voluminous than hydrated magnesium sulfate ion-pair, and the application of hydrostatic pressure therefore tends to enhance the dissociation of $MgSO_4$

$$MgSO_4 \overset{P}{\rightleftharpoons} Mg_2^+ + SO_4^{2-} \tag{8.5}$$

When an acoustic signal or pressure pulse is transmitted through the sea, in addition to spreading loss and scattering and absorption by heterogenities, it looses energy and gets absorbed by at least two pressure-sensitive chemical equilibria (Equations 7.1 and 8.5). Fisher (1962, 1965) has studied the effect of pressure on reaction 8.5 in detail. The phenomenon, it should be noted, is not restricted to $MgSO_4$ but is well marked for aqueous solutions of all two-two electrolytes (Verma, 1959). Fisher has also theorized that a very low-frequency acoustic absorption more recently observed might be due to pressure-dependent carbonate equilibria (Thorp, 1965).

Many of the trace elements in seawater are undoubtedly in the form of species in which they are complexed to both inorganic (such as Cl^- and SO_4^{2-}) and organic ligands (second category in Figure 8.1, see also Table 8.10). As I have indicated, our knowledge of the distribution of the trace elements in the marine water column is very unsatisfactory, but since they are known to be taken up by organisms and by particulate material, there is every reason to suspect that their concentrations in seawater will be highly variable. I have already mentioned the complexity of the iron and manganese profiles in a coastal pond, Dobrzanskaya and Kovalevsky (1959) have discussed the vertical distribution of iron in the Black Sea in terms of particle sizes, and Morris (1971) has observed fluctuations of dissolved and particulate Zn, Cu, and Mn in the waters of the Menai Straights that may be associated with planktonic blooms.

Apart from fast simple ionic metathesis equilibria among the major constituents, very little interesting chemistry occurs in seawater itself. Most of the important chemical processes in chemical oceanography occur at the interfaces that are the surfaces of demarcation of the oceans and their atmospheric, lithospheric, and biospheric boundaries. As we proceed down our classification scheme, it becomes increasingly difficult to resist the temptation to digress and discuss the interfacial interactions.

8.5 ORGANIC MATTER AND SUSPENDED MATERIAL

Moving down now to the lower half of Figure 8.1, a sample of deep Atlantic water, in addition to 30 to 40 g/liter of dissolved salts, contained about 0.003 g/liter of solid

Table 8.12 Principal Species Present in 19‰ *Cl* Seawater at 25°C and 1 atm[a] (From Horne (1969) according to the chemical model of Garrels and Thompson, 1962)

Element					
Sodium	0.4752 m	99% Na^+	1.2% $NaSO_4^-$	0.01% $NaHCO_3$	—
Magnesium	0.0540 m	87% Mg^{2+}	11% $MgSO_4$	1% $MgHCO_3^+$	0.3% $MgCO_3$
Calcium	0.0104 m	91% Ca^{2+}	8% $CaSO_4$	1% $CaHCO_3^+$	0.2% $CaCO_3$
Potassium	0.0100 m	99% K^+	1% KSO_4^-	—	—
Anionic Species					
Sulfate	0.0284 m	54% SO_4^{2-}	3% $CaSO_4$	21.5% $MgSO_4$	21% $NaSO_4^-$
Bicarbonate	0.0024 m	69% HCO_3^-	4% $CaHCO_3^+$	19% $MgHCO_3^+$	8% $NaHCO_3$
Carbonate	0.0003 m	9% CO_3^{2-}	7% $CaCO_3$	67% $MgCO_3$	17% $NaCO_3^-$

Table 8.13 Distribution of Organic and Inorganic Fractions in Suspended Particulate Matter (From Horne, 1969)

Area	Total Suspended Matter (mg/liter)	Organic Matter (%)	Reference
Offshore, Pacific	10.5	62	Fox, Oppenheimer, and Kittredge (1953)
Offshore, Pacific	3.8	29	Fox, Oppenheimer, and Kittredge (1953)
North Sea	6.0	27	Postma (1954)
Wadden Sea	18.0	14	Postma (1954)
Oceanic average	0.8–2.5	20–60[a]	Lisitzen (1959)
Bering Sea	2–4	—	Lisitzen (1959)
Indian Ocean	—	6–36[a]	Lisitzen (1959)
Long Island Sound	2–7	20–45	Riley (1959)

material (Groot and Ewing, 1963), and it is not uncommon for the level of suspended solid material to fall below this value by a factor of 10 (Jacobs and Ewing, 1965). While the quantity of particulate suspended material is very small compared to the major dissolved salts, it can very well be disproportionately important in the chemistry of the oceans, as I have suggested, for the majority of the more significant environmental chemical processes are heterogeneous reactions, and in the instance of the marine environment, these processes occur not only at the air–sea and sea-sediment boundaries of the oceans but also on the surfaces of suspended solids and gel-like material, both inorganic and organic, both inanimate and biological. The suspended material ranges from about 6 to 62% organic (Table 8.13). The inorganic fraction tends to reflect the composition of the underlying bottom sediments and, in addition to SiO_2 and $CaCO_3$ of biotic origin, is composed of minerals such as illite,

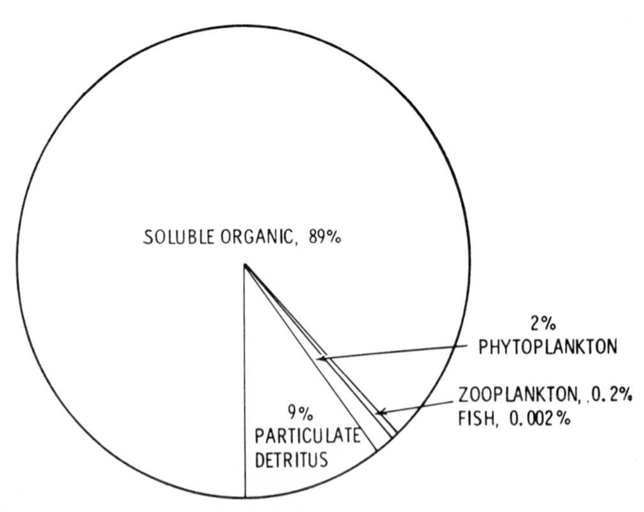

Figure 8.14 Distribution of organic material in the euphotic zone of the marine hydrosphere (from Horne, 1969).

kaolinite, chlorite, talc, quartz, feldspar, and so on, while the organic fraction is largely detritus (Figure 8.14), that is to say, a highly variable biological debris, including cell fragments, exoskeletons, fecal pellets, pollen grains, and the products of human pollution. The detritus content of the seas is very changeable in both space and time, especially in coastal areas. It is usually high in the euphotic zone and low below (Riley, van Hemert, and Wangersky, 1965).

The chemical forms of silicon in the sea, both dissolved and particulate, have been the subject of some interest. Because both geological and biological processes are involved, its marine chemistry is quite complex. Armstrong (1965) has described it as the most variable element in the sea, and the exact chemical nature of its dissolved and particulate forms remains something of a mystery. Largely because of its role in biological cycles, the silicate profile of the water column is complex and displays strong seasonal dependencies. Figure 8.15 is an attempt to represent the silicon cycle in the oceans—its uptake by growing organisms and its release upon their death, its dissolution, and its slow rain upon the ocean floor where it adds to the sedimentary accumulation. As mentioned previously, siliceous (and carbonate) materials buffer seawater and keep its pH within a narrow range (Garrels, 1965; Sillen, 1967).

The majority of the organic content of seawater is dissolved (Figure 8.14), and this material includes just about every naturally occurring organic chemical imaginable, which, inasmuch as the oceans are a media for all the major life processes, is not in the least surprising. The biomaterial may be divided into three categories (Figure

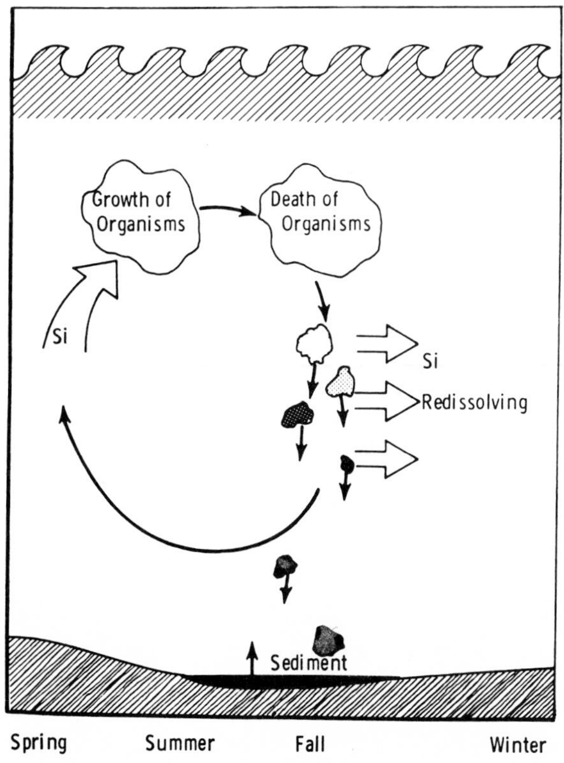

Figure 8.15 Seasonal changes in the silicon content of surface waters of the English Channel.

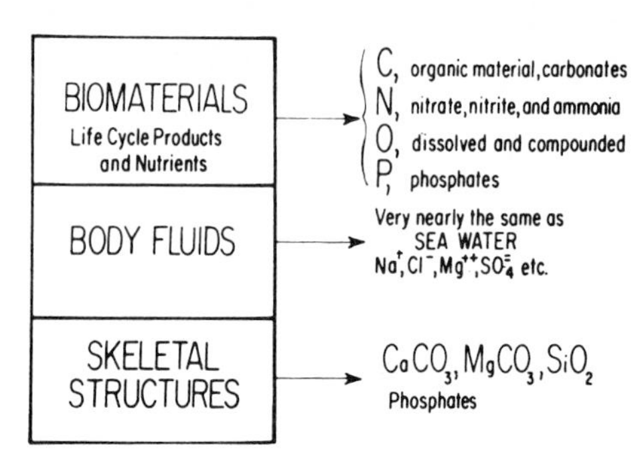

Figure 8.16 The three types of materials constituting marine organisms (from Horne, 1969).

8.16): (*a*) the biosubstance proper, (*b*) the body fluids, and (*c*) the supporting and/or containing skeletal structures. We have already discussed one of the last, silica, and another, calcium carbonate, is treated elsewhere. The formation from or the release into seawater of body fluids has very little effect on the chemistry of seawater, since the composition of the two solutions are similar (Table 8.14). The slight differences that are found in some cases, however, are most fascinating for they can give us insights, not only into the nature of certain crucial biological functions, but into the evolutionary history of certain species and even into the chemical composition of ancient seas (see Horne, 1969*b*). The biosubstance proper can in turn be further subdivided into the major structural elements, such as C, H, O, P, N, and S, and trace catalytic elements, such as Cu, Zn, Fe, V, and Br. The former group appear in seawater as carbohydrates, proteins, lipids, and the full spectrum of their decomposition products. Lipids are found in somewhat greater proportion since they are more resistant to decomposition than carbohydrates or proteins (Duursma, 1965). The amino acid content of surface waters has been examined by Tatsumoto and coworkers (1958, 1961) and of deep waters by Park and co-workers (1962); among the more abundant compounds found are glutamic acid, lysine, aspartic acid, serine, alanine, leucine, valine, and threonine. Both saturated and to a lesser extend unsaturated fatty acids are found in seawater, the greater concentrations being as high as 20 μg/liter for the C_{16} and C_{18} compounds (Williams, 1961), while concentrations of simple carboxylic acids—acetic, formic, and lactic—as great as 1.4 mg/liter have been observed in near-shore waters (Koyama and Thompson, 1959). Walsh (1965) found the dissolved carbohydrate content of waters off Cape Cod to range from 0.4 to 1.0 mg/liter. Generally speaking, the dissolved organic substances all follow more or less the same behavior; larger and very variable concentrations in shallow, especially coastal waters and smaller more constant concentrations in deep waters. The former seasonal fluctuations depend strongly on biological activity, notably, in the case of carbohydrates, on phytoplankton growth (Provasoli, 1963; Strickland, 1965). The distributions of more complicated biosubstances dissolved in seawater, such as vitamins (Belser, 1963; Duursma, 1965; Natarajan and Dugdale, 1966), chlorophyll (Parsons, 1963), and ATP (Holm-Hansen and Booth, 1966), have also received attention. The concentration of trace elements by organisms in some

Table 8.14 A. Comparison of the Composition of Seawater and Human Blood[a] (From Dietrich, 1963, with permission of John Wiley and Sons)

	Major Constituents			
	Seawater		Human Serum	
Species	g/kg	%	g/kg	%
Na^+	10.75	30.7	3.00	34.9
K^+	0.39	1.1	0.20	2.3
Ca^{2+}	0.416	1.2	0.10	1.2
Mg^{2+}	1.295	3.7	0.025	0.3
Cl^-	19.345	55.2	3.55	41.3
SO_4^{2-}	2.701	7.7	0.02	0.2
HPO_4^-	0.000185	0.0005	0.10	1.2
HCO_3^-	0.145	0.4	1.60	18.6

	Minor Constituents		
Element	Seawater (mg/m)	Human Serum (mg/kg)	Serum/Seawater
Fe	50	1	20
Zn	5	3.3	660
Cu	5	1.7	340
		In Total Blood	Blood/Seawater
As	15	0.6	40
Mn	5	0.3–1.5	60–300
Al	120	Up to 2	17
Pb	5	4–7	800–1400
I	50	0.03–0.1	0.6–2
F	1400	0.5–1	0.36–0.71

[a] For a very detailed survey of the composition of body fluids see Altman and Dittmer (1961).

instances (such as iron) is capable of affecting the vertical distribution in the water column and in still other instances (Cd, Zn, Cu, and Pb) is so enormous that it has alarmed those concerned with heavy metal and radioactive pollution.

The nutrient element phosphorus (silicon has already been mentioned and nitrogen is discussed in Chapter 15) is equally important in the marine environment as in fresh water (examined in Chapter 7). As one of the primary nutrients for phytoplankton, which in turn provide food for zooplankton (Figure 8.17), phosphorus levels are closely associated with photosynthetic activity, and particulate phosphate profiles tend to parallel the chlorophyll profiles, as one might expect (Ketchum and Corwin, 1965). Most of the phosphorus in seawater is dissolved, a large fraction as HPO_4^{2-}, and with appreciable amounts of acid-soluble organic phosphorus and

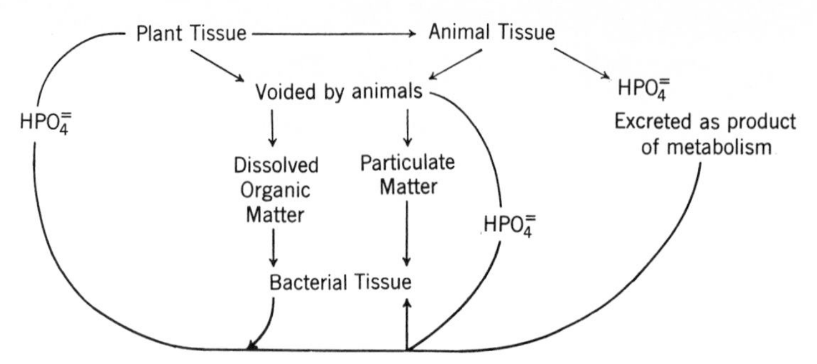

Figure 8.17 The phosphorus cycle in the oceans (from Horne, 1969).

phospholipids (Correll, 1965). Some years ago Harvey (1926), whose little book (1966) on the chemistry of seawater is highly recommended, pointed out that phytoplankton simultaneously utilize phosphate and nitrate and that as a consequence the N/P ratio in the seas should remain relatively constant. In fact, except for restricted areas and surface waters, it does, the ratio being about 15 atoms of N to one of P (Figure 8.18).

In the oceans, as in lakes and ponds, in the near-surface photosynthetic zone, oxygen is produced, but in deeper waters there is a net consumption of oxygen by animal respiration and the decomposition and biochemical oxidation of organic material. If circulation is poor, such as in the Black Sea, certain Norwegian fjords, British Columbian and Chilean inlets, and the Gulf of Cariaco in the Caribbean Sea, the rain of organic material from surface waters may totally deplete the oxygen content of deeper waters (Richards, 1965). The resulting stagnant anoxic regions are literally marine deserts, devoid of the familiar fish and microscopic forms that populate oxygenated waters. But there is life in these deserts nevertheless for life is

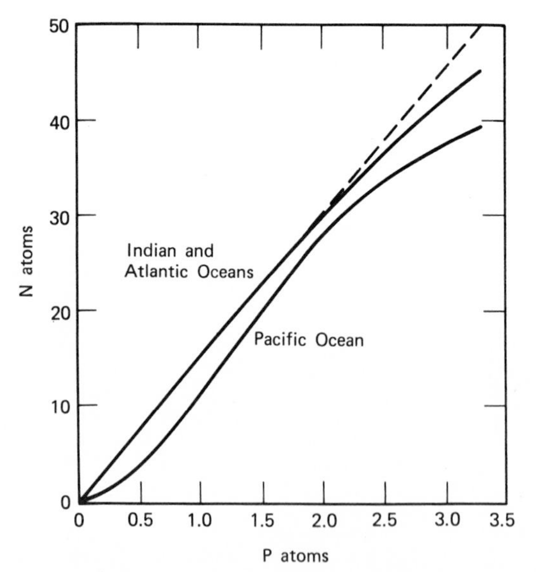

Figure 8.18 The nitrogen/phosphorus ratio in the oceans.

very persistent, and in the absence of oxygen as a free energy source, it turns to other chemicals. Instead of the normal oxidation of the biomaterial in seawater, which can be represented by

$$(CH_2O)_{106}(NH_3)_{16}H_3PO_4 + 138O_2 \rightarrow 106CO_2 + 122H_2O + 16HNO_3 + H_3PO_4 \quad (8.6)$$

first nitrogen is utilized ("denitrification")

$$(CH_2O)_{106}(NH_3)_{16}H_3PO_4 + 84.8HNO_3 \rightarrow 106CO_2 + 42.4N_2 + 148.4H_2O + 16NH_3 + H_3PO_4 \quad (8.7)$$

(in addition to N_2 and NH_3 nitrite ion, NO_2^-, may also be produced as an intermediate), and, in the absence of nitrate or nitrite, sulfate bacterial reduction can occur

$$(CH_2O)_{106}(NH_3)_{16}H_3PO_4 + 53SO_4^{2-} \rightarrow 106CO_2 + 53S^{2-} + 16NH_3 + 106H_2O + H_3PO_4 \quad (8.8)$$

with the production of H_2S that characterizes so many stagnant waters. Methane has also been detected in anoxic waters along with H_2S, formed presumably by processes analogous to the reduction of CO_2 to form "marsh gas" (Richards et al., 1965*b*).

At this point, I must mention the radioisotopes present in seawater. With respect to their sources they fall into three categories:

1. Natural, long-life nuclides such as ^{238}U, ^{230}Th, ^{226}Ra, and ^{231}Pa from the uranium and thorium series, which have been decaying since the birth of this planet, and their shorter half-life decay products;
2. Natural shorter half-life nuclides, such as ^{14}C, ^{3}H, ^{32}Si, and ^{10}Be continually formed by processes such as cosmic radiation in the atmosphere of Earth; and
3. Artificial nuclides, such as ^{90}Sr, ^{137}Cs, ^{147}Pm, ^{157}Sm, ^{60}Co, ^{54}Mn, and ^{55}Fe, from weapons testing and other human pollution (Burton, 1965; Miyake, 1963).

Of the natural radioactivity of seawater the radioisotope ^{40}K (which decays by β-emission and K-electron capture) accounts for more than 90%, with ^{87}Rb being a poor second with a less than 1% contribution. Just as the nutrient ratios (discussed previously) can characterize a water mass in the oceans and be used as a biological clock to give us an indication of the age of the waters, radionuclides, especially ^{14}C, provide a nuclear tag and clock that has proved most useful in resolving the movement of water masses and mixing processes in the oceans (Broecker, 1963).

8 6 OCEAN DUMPING

Having reviewed briefly the chemistry of seawater, we now turn our attention to the impact of human activity on that chemistry. The attitude of the human race to the oceans has been determined largely by the immensity of the Earth's marine hydrosphere. The oceans, until very recently, were universally accepted as an inexhaustable cupboard to feed the world's run-away population and an incorruptable sewer in which to dump all our wastes. We are presently in that painful awakening stage where we realize that nothing could be further from the truth. Far from being a near-infinite soup, the biopopulation density of the oceans is just as variable and just as restricted in distribution as upon the continental land masses. Vast expanses of the open oceans are, relatively speaking, deserts, while the highly fertile zones of the sea tend to be the very coastal regions most vulnerable to human damage. Man has been just as reckless and relentless at destroying marine species as he has been in the case of land plants and animals. His depredations have brought even the magnificent whale to the verge of extinction, and in his rapacity, aided by his technological advances, he has literally fished-out areas that were once among the richest fishing gounds. In addition to this pillage of the sea, he has further managed to endanger this planet's limited marine resources by his misuse of the oceans as the ultimate dump. Not only has he managed to pollute the oceans on a global scale—no mean "achievement"—but in at least one instance, which we examine, he has succeeded in completely killing a sizeable area of the ocean floor.

The problems of solid waste disposal were examined in a previous chapter. Even although most of the materials are incinerated or disposed on land, an enormous quantity, about 48,000,000 tons in 1968 (see Table 8.15) is dumped into the sea, largely on the continental shelf.

In the late nineteenth century, The Congress of the United States charged the Corps of Engineers of the Department of the Army with the responsibility of keeping navigable waters open and free for shipping. Among the tasks entailed in this responsibility is the dredging of silted-in channels and harbors and the disposal of the resulting dredge spoil. Over the years, a great number of dump grounds were established in the coastal waters of the United States, not only for dredge spoil but also for other solid wastes as well, including sewer sludge, rubble and cellar dirt, ammunition and explosives, and liquids, such as acid waste and toxic chemicals. One of the largest of these dumping areas is that serving the metropolitan New York area located in the waters of the New York bight outside the harbor of the city, and the locations of the various dumps within this site is shown in Figure 8.19. The waste burden of the New York bight alone is 9,600,000 ton/yr (Table 8.16), which corresponds to 1 ton/person-yr or 6 lb/person-day, and, if we except the Gulf of Mexico, this material represents the largest source of sediment discharging directly into the north Atlantic Ocean from the North American Continent—much larger than

Table 8.15 Ocean Dumping: Types and Amounts, 1968 (in tons) (From Train, Cahn, and MacDonald, 1970)

	Atlantic	Gulf	Pacific	Total	
Dredge Spoils	15,808,000	15,300,000	7,320,000	38,428,000	80
Industrial Wastes	3,013,200	696,000	981,300	4,690,500	10
Sewage Sludge	4,477,000	0	0	4,477,000	9
Construction and Demolition Debris	574,000	0	0	574,000	<1
Solid Waste	0	0	26,000	26,000	<1
Explosives	15,200	0	0	15,200	<1
Total	23,887,400	15,966,000	8,327,300	48,210,700	100

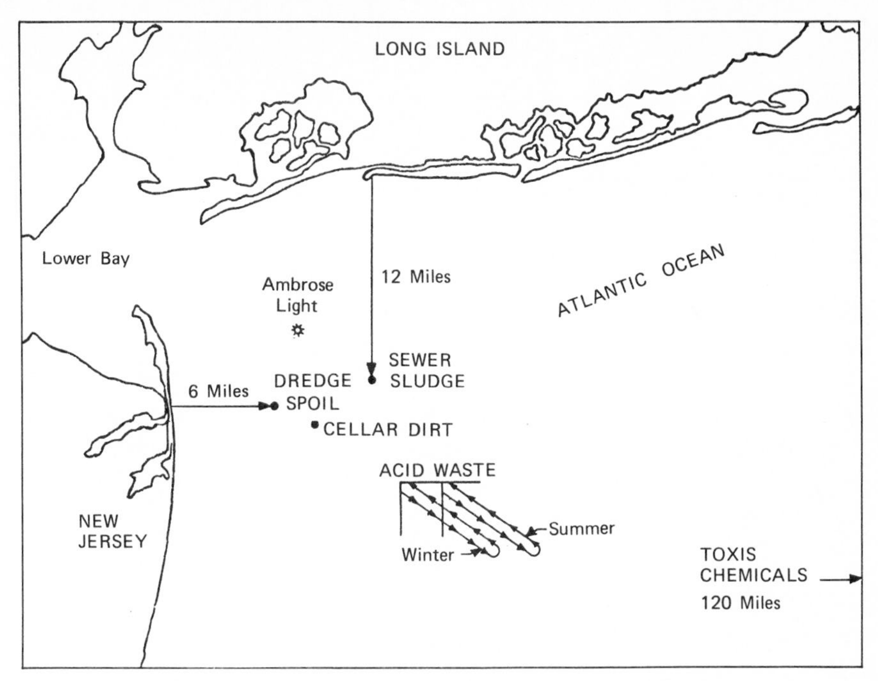

Figure 8.19 Waste dumps in the waters of the New York bight.

Table 8.16 Average Amount of Waste Solids from New York Metropolitan Region Dumped in Western Long Island Sound and on Continental Shelf (New York Bight) (Data from Gross, 1970)

	Millions of Tons Per Year	
	1960–1963	1964–1968
Western Long Island Sound	0.74[a]	1.9[a]
New York Bight		
Mud Disposal Area		
Private Contractors	2.2[a]	3.4[a]
Federal Dredges	4.2[b]	3.4[b]
Cellar Dirt Disposal Area	0.67[a]	0.59[a]
Sewage Sludge Disposal Area	0.11[c]	0.15[c]
Waste Chemical Disposal Area	0.10[d]	0.18[d]
Total	8.02	9.62

[a] Estimated bulk density 1.1 g/cm^3.

[b] Estimated bulk density 1.3 g/cm^3.

[c] Discharged as liquids containing 4.5% solids by weight, assumed density of solids 1 g/cm^3.

[d] Discharged as liquid, estimated solid content 5% by weight.

natural sedimentation from rivers (Gross, 1970). Worst still, the amount of solid wastes dumped at sea is steadily increasing, in the case of the New York bight at a rate of about 4%/yr (Gross, 1970). With the increasing population of the coastal regions, the quantity of solid wastes dumped in United States' coastal waters is expected to exceed 100,000,000 ton/yr by the year 2000 (Train, Cahn, and MacDonald, 1970). In the New York bight, the situation is exacerbated by the fact that the wastes are being dumped into waters that are already grossly polluted. The Hudson River is burdened with ten times more pollution than it can accommodate by natural biological decomposition (Ketchum, 1970); over 350,000,000 gal/day of raw sewage pour into the Hudson and East Rivers, which eventually finds its way to the bight (Peter, 1970). The choice of the dumping sites in the bight is particularly unfortunate for the direction of the bottom currents in the area is onshore much of the year (Horne, Mahler, and Rossello, 1971). While the bight situation is bad and, despite considerable concern on the part of responsible government agencies, will probably continue to get worse (Anon., 1971*a*) the situation on the west coast of the United States is considerably brighter for there some type of dumping operations (excluding outfalls) have actually declined or been discontinued over the period 1968 to 1971 (Table 8.17). However the relative cheapness of marine disposal together with the rapid disappearance of land fill sites and the tightening of air pollution measures could alter the picture and encourage an increase in ocean dumping (Smith and Brown, 1971). The dredge spoil disposal problem is not confined to the marine environment: for example, a total of 8.5×10^6 m^3/yr of spoil are dredged from over 100 Great Lake harbors. Most of this material is open-lake dumped and results in the expected destruction of benthic biota (Gannon and Beeton, 1971).

What has the enormous injection of wastes done to the marine environment of the New York bight? Because of its dumping license function, the Corps of Engineers realized that it could not avoid environmental responsibility, although not specifically charged by law with environmental protection. Research sponsored by the Corps indicated that the dumping activity had resulted in the creation of a "dead sea"—a spreading area covering nearly 20 mi^2 in which the bottom waters are deficient in oxygen and at the center of this artificial anoxic area is a zone totally devoid of benthic life (Peter, 1970; ad hoc Committee, 1970; Horne, Mahler, and Rossello, 1971).

The dredge spoil consists for the most part of mineral material, much of it native to the marine environment, and thus should in itself not constitute a pollution problem. But the spoil is contaminated with oxygen-consuming organic material and with petrochemicals. As in most ports, the oil pollution in New York harbor is very severe. The petroleum coats solid particulate material and is thereby carried to the bottom where it accumulates in a water–oil–sediment emulsion–slurry with the consistency of mayonnaise (Pearce, 1969). The hexane-extractable chemicals alone in this mess are in the lethal concentration range, and to make matters worse, petroleum contamination appears to solvent extract and concentrate pesticides and other organic compounds and toxic heavy metals as organometallic complexes (Pearce, 1969; Hartung and Klingler, 1970). As legal action stops heavy metal pollution at its source and limits the indiscriminate application of pesticides, the pollution problem posed by the disposal of dredge spoil may decline.

A far more serious long-range problem is the disposal of sewage sludge, which also carries a burden of toxic metal contamination (Figure 8.20) and whose large organic

Table 8.17 Summary of the Type, Amount, and Number of Individual Ocean Dumping Operations for the Pacific Coast—1968 and 1971 (From Brown and Shenton, 1971, with permission of the Marine Technology Society.)

Type of Waste (Industrial[c])	1968[a] Annual Tonnage	Number[b] of Individual Dumping Operations 1968	1971 Annual Tonnage	Number of Individual Dumping Operations 1971
Spent Steel Pickling Acid (sulfuric and hydrochloric)	41,700	1	—	Discontinued February 1971
Refinery Wastes	164,160	2	2,160	Discontinued December 1971
Toxic Chemicals[d]	506	3	500	2
Paper Mill Wastes	116,534	1	—	Discontinued 1970
Oil Drilling Muds	653,100	1	—	Discontinued December 1970
Waste Oil	5,300	1	—	Discontinued 1970
Cannery Wastes	20,000	1	20,000	1
Vessel Refuse and Garbage	6,200	3	1,200	1[e]
Filter Cake	—	—	—	Discontinued 1970[f]
Total All Wastes	1,007,500	13	23,860	4

[a] Based on 1968 BSWM data (Smith and Brown, in press) excluding dredge spoils, explosives and radioactive wastes.

[b] Based on private survey conducted in May 1971.

[c] Includes bulk and containerized wastes.

[d] Includes cadmium, copper and chromium cyanide, laboratory wastes, and other unidentified industrial wastes.

[e] U.S. Naval dumping operations were discontinued in 1968 and 1970 for San Diego and Long Beach, respectively.

[f] 346,480 tons of filter cake were dumped in 1969–1970.

content depletes of oxygen content of the bottom waters even more considerably than dredge spoil (Figure 8.21). The centers of the dump areas are totally devoid of macroscopic benthic biota; in fact, living specimens could not be found on which to make heavy metal and bacteriological analyses. Nevertheless, the influence of Soviet fish-out, market demands, weather conditions, and other factors is so great that analysis of New York landing statistics failed to reveal any unambivalent evidence that the massive pollution of the waters of the bight is adversely effecting fish yields (Horne, Mahler, and Rossello, 1971). It is clear, however, that shell fisheries have suffered damage (Smith and Brown, 1971). If infection and chemical contamination is finding its way into food webs, it is probably so doing from the peripheral rather than the central areas of the dumps.

With the construction of more water treatment facilities throughout the country, the sewer sludge disposal problem will be enlarged. Because of its greater concentra-

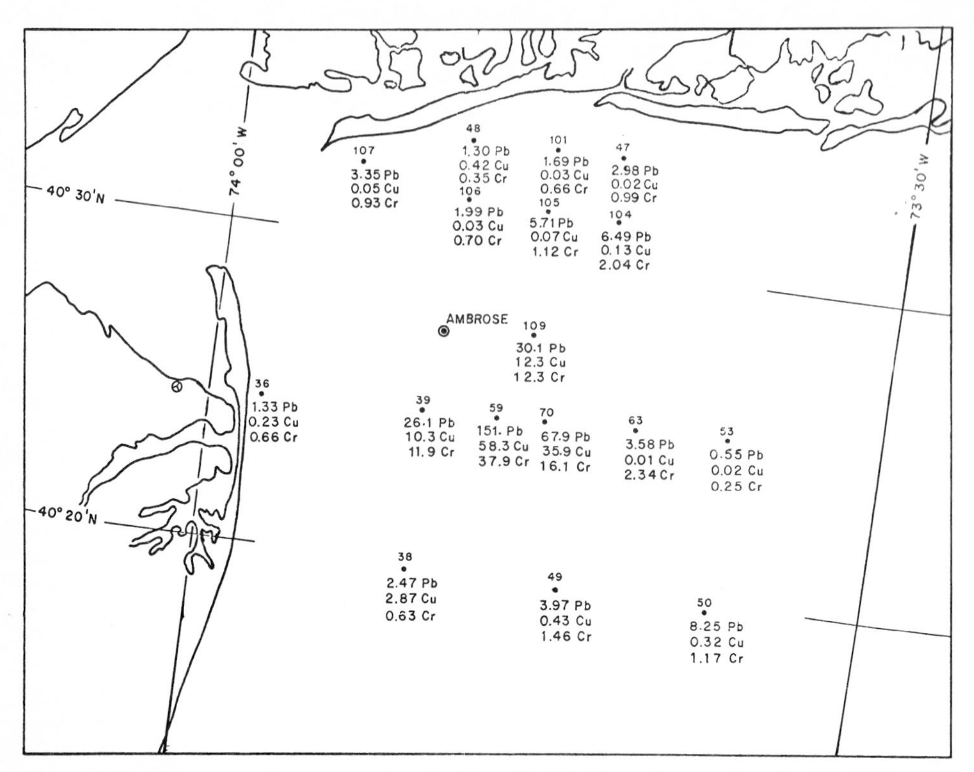

Figure 8.20 Heavy metal concentrations in New York bight sediments (ppm), 2% HNO_3 Extraction (from Pearce, 1969).

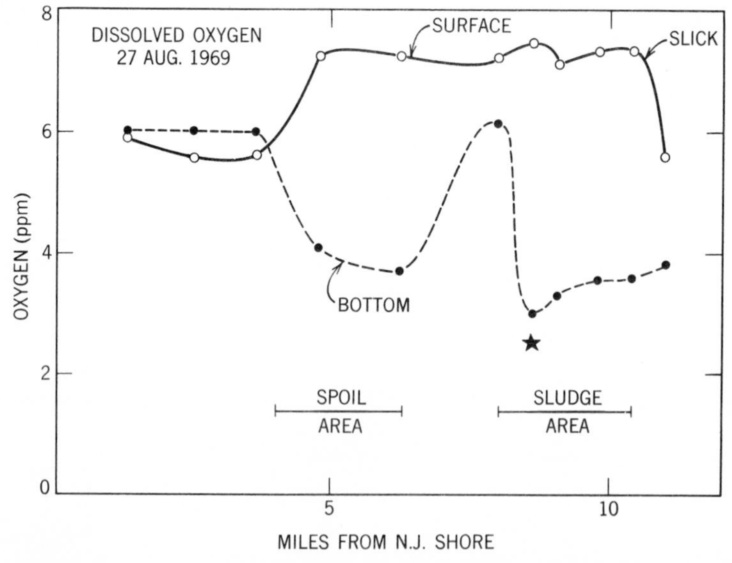

Figure 8.21 Dissolved oxygen content of surface and bottom New York bight waters. The latter is now a permanent feature, but the former could reflect a recent disposal (from Ketchum, 1970, with permission of the Water Pollution Control Federation).

tion, barged sludge appears to have a more adverse effect on the marine environment than does raw sewage or partially treated sewer effluent. The enormous Hyperion Plant (Los Angeles) sewer sludge outfall where the sludge is carried and diluted with ample water appears to have done surprisingly little environmental damage in its 20 yr of operation. Alternative sludge disposal techniques have not been very promising: for example, the fertilizing capacity of the sludge is poor compared with commercial fertilizers (Burd, 1968; Dean, 1969; Train, Cahn, and MacDonald, 1970). The possibility of relocating the dumps on the edge or off the continental shelf has been raised. Ketchum (1970) has pointed out that such an alternative would not only endanger the delicate biota of "the last unpolluted space available on Earth," but, in view of the lower temperature of the deeper waters, might amount to "merely refrigerating waste materials and preserving them [and the problem] for posterity," a suggestion substantiated by the slow rate of microbial degradation of organic matter in the deep sea (Jannasch et al., 1971).

In the noise, politicking, confusion, and sometimes near-hysteria that surrounds ocean dumping (and other environmental issues), several important points have tended to get lost. The waters of the New York bight have died, not because waste materials have been dumped there, but because materials have been dumped at such a rate and in such a concentration that the natural capacity of the waters to receive them has been grossly exceeded. As the result of our own studies of the bight problem (Horne, Mahler, and Rossello, 1971), we concluded (but these conclusions were censored at Woods Hole Oceanographic Institution and did not appear in the report as issued) that rational, *controlled* dumping might be an acceptable short-range expedient. The dumping would have to be carefully supervised, perhaps alternating two or more sites in much the same way agricultural crops are rotated so that the oxidative capacity of the waters is not exceeded. The possible positive use or beneficial effects of sewage waste on the marine environment is open to question. Dr. Ryther has suggested using secondary sewage as a nutrient source for oyster beds and claims some promising experimental results in this direction. Biostimulation or overfertilization is a threat at certain pollution levels. Pearce (1969) found no evidence for any fertilizing effect of the large amounts of organic matter in sludge on benthic communities. Some interesting success has been achieved, however, in the implacement of inert solid wastes (such as junk autos) to form artificial reef habitats for aquatic life (Smith and Brown, 1971). Another ingenious but probably impractical scheme for waste disposal utilizes the slow folding under of the Earth's crust into the continental trenches.

But in the larger view, the dumping of sewage into the seas is undesirable not because it pollutes the sea, but because it impoverishes the continents. The overall result of human activity is to move chemicals about in the environment; to take them from where they belong to where they should not be. The present case affords an excellent example. The end effect of our population distribution and sewage disposal practice is an enormous pump (Figure 8.22) that drains nitrogen, phosphorus, and other nutrients from the center of the continental land mass and dumps them into the ocean. The application of artificial fertilizers is now extensive, but the return of precious nutrients back to the food-producing areas by natural processes is, relatively speaking, a slow trickle. If man does not desist in his foolishness he will find himself on a crowded barren land surrounded by a putrid sea.

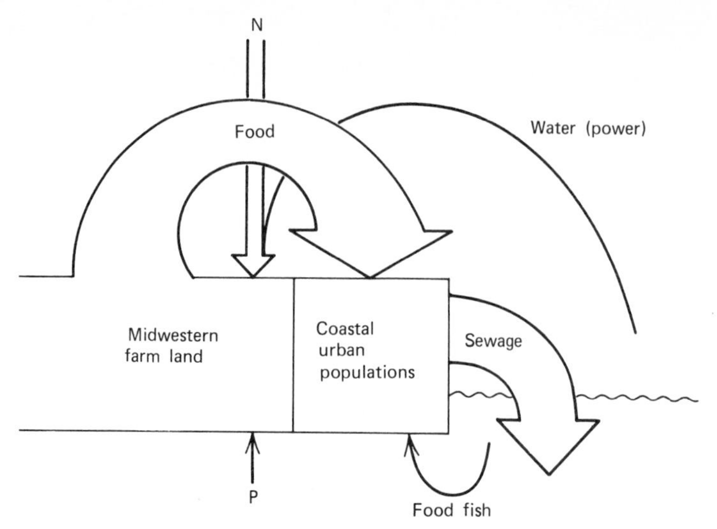

Figure 8.22 The nutrient pump.

8.7 THE ESTUARINE AND COASTAL ZONE

Most of the marine-disposed solid waste discussed previoulsy is dumped on the continental shelf in waters relatively close to shore. This is particularly unfortunate. Just as the land masses have their stretches of arid deserts and fertile valleys much of the ocean is relatively barren, while the oceans' "fertile valleys" are largely confined to coastal waters. Many of the world's fisheries are located in this coastal zone, and the coastal zone includes the estuarine environment, which is not only rich in life itself but also provides a crucial link in the whole ecology of coastal waters (Lauff, 1968; Nat. Acad. Sci. U.S.A., 1970) and/or a coastal zone pollution bibliography, see Sinha, 1970). Here the balance of intricate relationships among living things is most delicate, and it is in this sparse precious realm that man treads the most heavily, burying essential wetlands with his "improvements," dumping his sewage, destroying birds sanctuaries and marine life hatcheries, and polluting shellfish beds. Consider the case of Galveston Bay (Carter, 1970). The largest estuarine environment on the Texas coast, it covers some 500 mi^2. In the past 30 yr, the population of the area has tripled and the bay has borne the full brunt of this population growth and industrialization. Until recently, millions of cubic yards of shellfish reef were dredged up each year for highway construction and cement manufacture. Water diversion has upset the ecology of the bay. One diversion dam being built alone will destroy 20,000 acres of prime shrimp and finfish nursery wetlands. The bottom of the Houston Ship Channel is covered with a blanket of putrid slude 2-ft thick, and passing ships constantly stir up this oil. The pollution burden of the channel corresponds to the raw sewage of a city of 2 to 3 million people. Little wonder that nearly half of the bay is closed to oyster harvesting because of pollution.

Like coastal wetlands, fresh water marshlands are constantly being diminished by the hands of man, by land-fill operations and drainage. The area of the Earth's marshlands is not inconsiderable; one estimates it at 3,500,000 km^2 or 2.3% of the total land area. In some parts of the Soviet Union as much as 20 to 22% of the land is marshy (Bulavko, 1971).

We confine our attention here to the principle chemical features of the estuarine zone. As fresh water flows into the sea two major changes occur in the chemical environment:

1. The ionic strength increases by many orders of magnitude, and
2. The concentration of sodium and chloride ions increases to as much as 0.5 *M*.

Increasing the ionic strength by adding electrolyte is a well-known technique for precipitating flocculent and other suspended material. Colloids are stabilized in suspension by the repulsion of their like charges. Facilitation of charge neutralization by the electrolyte permits particles to aggregate until they grow to sufficient size to settle out (water treatment by synthetic flocculants involves a bridging mechanism as well as charge neutralization, Reis and Meyers, 1968). Chloride ion is a powerful complexing ligand. We might, therefore, expect solubilization of some substances, such as metallic cations, by complex formation. It should be noted that these two processes are in opposite directions, the one tending to precipitate material out of the water, the other to solubilize it in. In both laboratory simulation experiments (Kuziemska, 1970) and the real world, the former process evidently dominates, and the overall effect is for material to be laid down on the bottom of the estuarine environment. As is so often the case in the chemistry of the hydrosphere, iron is the crucial element (silicon is also important, Burton, 1970) because of the ability of its hydrous oxide to co-precipitate just about everything imaginable. The Fraser River in British Columbia discharges about 1.9×10^{-5} kg/day of particulate and soluble iron, and Williams and Chan (1966) found that nearly all of this iron is immediately laid down in the river's delta. This does not necessarily mean that there is a large accumulation of trace elements in estuarine and very little in deep sea sediments for another phenomenon is at work. The absorption of trace elements on particulate matter increases as particle size decreases; yet the smaller the particle, the further out to sea it will be carried before it finally settles to the bottom. Thus fine-grained deep sea sediments are often richer in trace elements than more coarse-graned near-shore material (Krauskopf, 1956; Chester, 1965; Turekian, 1965; Horne, 1969*b*). Nelson (1959) noted that the colloidal mineral content of waters decreases from about 0.2 to 0.005 mg/liter upon flowing into Chesapeake Bay, but here again we must remember that the observed decrease in concentration is almost entirely due to dilution by seawater rather than precipitation processes. Finally, the increasing Na^+ concentration as fresh water meets seawater will alter cation-exchange equilibria involving mineral ion exchangers.

In unconscious acknowledgement of their importance to him, man has built many of his largest cities near estuarine waters, and as a consequence the chemistry of estuarine waters is often dominated by the chemistry of sewage. This topic is treated more fully in Chapter 16, but we spell out the four distinct progressive phases of the pollution of estuarine waters by sewage here:

1. Low pollution levels can act as nutrients for algal growth and cause eutrophication in coastal waters.

Then as the level increases, Torpey (1967) found in a comparative study of New York harbor and the Thames (England) estuary a three step decline of the oxygen content of the waters:

2. When pollution loading increases to a rate requiring 20 lb O_2/day/acre, instability develops, O_2 level drops sharply, fish migrate.

3. At a pollution loading level requiring a rate of 20 to 132 lb O_2/day/acre, the dissolved O_2 remaining substantially constant at between 25 to 50% of saturation. This plateau is homeostatic because symbiotic algae and bacteria are able to maintain this O_2 level.
4. At loading rates exceeding 132 lb O_2/day/acre, the O_2 is exhausted and anaerobic conditions obtain.

In both areas of studies, the catastrophic fourth stage has been reached. Two of these steps deserve further mention because of their clear applied implications, not so much in the direction of a ban on ocean dumping, which would be expensive and impractical since no alternative disposal means is presently available, as in the direction of modifications of our dumping habits to make the practice more acceptable to the marine environment. Ryther and Dunstan (1971) have found that in New York coastal waters, because of the low N/P ratio in terrigenous contributions including human waste and the more rapid regeneration of P than NH_3 from decomposing organic material, nitrogen rather than phosphorus is the crucial nutrient limiting algal growth and eutrophication. In the light of this finding, they warn that the "removal of phosphate from detergents is therefore not likely to slow the eutrophication of coastal marine waters, and its replacement by N-containing nitriloacetic acid detergents may worsen the situation." Of even more far reaching significance is step 3. Torpey's work and later Gould's (1968) points to the existence of a broad plateau in which the dissolved oxygen content remains relatively stable even as the loading rate is increased over a broad range (Figure 8.23). Such being the case, "large expenditures of funds for treatment works might be reflected in only modest improvements in the quality of the waters as measured by dissolved oxygen" (Wolman

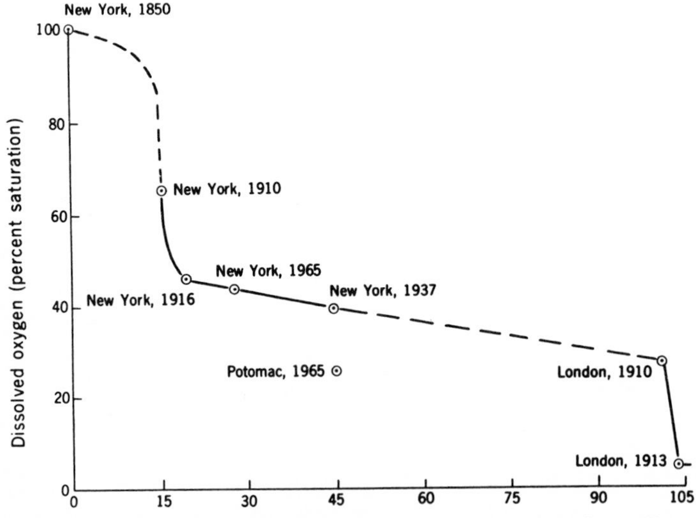

Figure 8.23 Sequence of changes in dissolved oxygen in the Hudson (N.Y.) and Thames (England) Rivers relating dissolved oxygen to BOD loading (from Wolman, © 1971 by The American Association for the Advancement of Science with permission of the AAAS).

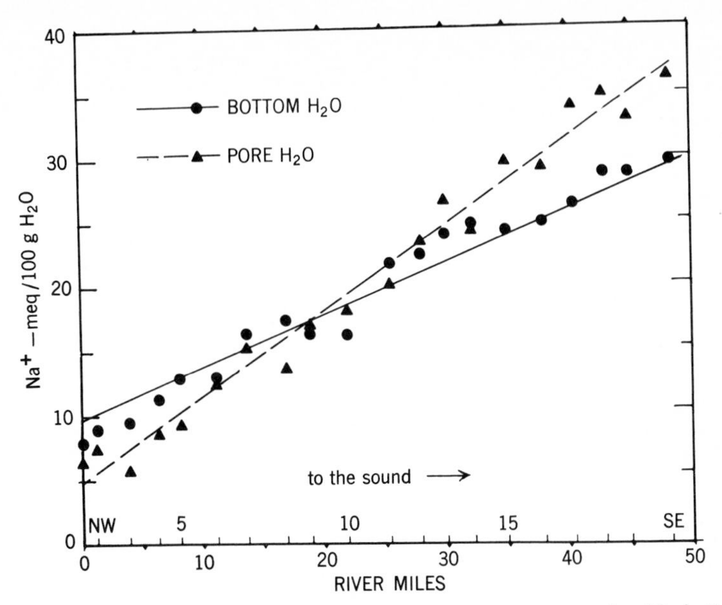

Figure 8.24 The Na^+ concentration of estuarine pore waters compared with bottom waters (reprinted with permission from D. A. Dobbins et al., *Environ. Sci. Tech.*, *4*, 743 (1970), © The American Chemical Society).

1971). Thus far we have been lucky. Even in the Hudson River "serious fouling and deoxygenation have so far been avoided for most of the river" (Howells, Kneipe, and Eisenbud, 1970). Perhaps because the plateau (Figure 8.24) has bought us time, but the ecological relationships involved are complex and delicate, and if we do not mend our ways we may suddenly push our estuarine waters over the edge.

Amino acids in the estuarine environment have been studied by Hobbie, Crawford, and Webb (1968) and Hall, Weimer, and Lee (1970). Fatty acids are also found in seawater and marine sediments (Williams, 1965), and the conversion of oleic acid to saturated fatty acids has been observed in the sediments of the estuary of the River Severn (Rhead et al., 1971). Dobbins, Ragland, and Johnson (1970) have examined cation–clay interactions in estuarine sediments and among their findings is an increase in the ratio of Na^+ concentrations in pore water to bottom water as one proceeds downstream (Figure 8.24) probably due to the entrapment of relatively saline water as the suspended sediment load settles after storm agitation. Chemical changes in interstitial waters in sediments from deltas, estuaries, and marshes have also been investigated by Friedman and Gavish (1970). Jitts (1959) examined phosphate absorption by estuarine sediments and Martin and Bella (1971) have found that oxygen uptake by estuarine bottom deposits is largely controlled by the release of biodegradable stuff from the bottom to the overlying water by physical mixing. Because of the importance of oxygen content and stagnation conditions in the chemistry of estuarine sediments, the use of the redox potential or E_h has been proposed as an operational parameter for chemically characterizing the estuarine environment (Whitfield, 1969).

8.8 OIL POLLUTION

Man's indifference to the oceans has even exceeded his disrespect for the land. Finally, it is becoming clear that, large though the oceans may be, man has the capability of exhausting their resources and degrading them to a ruinous condition. Many marine species are now dangerously close to extinction or greatly reduced in numbers thanks to man's plundering, and in some instances the productivity of the ocean and the edibility of its food resources is being threatened by man-made pollution. A long, long list could be compiled of the marine pollution problems created by man ranging from fragile-shelled sea bird eggs due to pesticides to the disappearance of our beaches due to the obstruction by dams of the natural transport of particulate material from the rivers into the sea. One of the most highly publicized ocean pollutants has been oil (Hepple, 1969; Hoult, 1969).

Contamination of the marine environment by petrochemicals has been a constant and growing threat ever since oil started to be used as a vessel fuel. One of the earliest recorded oil spills was in December 1907, when the seven-masted *Thomas W. Lawson*, the largest schooner ever built, while on her maiden voyage spilled her cargo of 2,000,000 gal of crude oil into the sea off the Isles of Scilly (Murphy, 1970). A nearby colony of puffins, numbering about 100,000 birds was severely affected. Sixty years later when the *Torrey Canyon* went aground on Seven Stones Reef near the same location, spilling some 117,000 tons of Kuwait crude oil into the waters, only about 100 birds remained. Oil pollution stands accused as a major cause of wildfowl population decline. As early as 1925, shellfish authority F. W. Lane (Lane et al., 1925) declared that "oil is, gallon for gallon as thrown out, the most destructive to aquatic life of all the foreign substances now entering our coastal waters."

The *Torrey Canyon* disaster (Zuckerman et al., 1967) roused world concern with the problem of the pollution of the marine environment by petrochemicals and caused particular anguish in the scientific community for the detergents and other chemicals dumped into the sea in a frantic effort to keep the beaches aesthetic appear to have wrecked greater ecological havoc than the oil itself (Smith, 1968). Another spill of a completely different kind, which also was important in galvanizing public indignation, especially in the United States, was an oil well leak near Santa Barbara, California. One day in January 1969, mud began to flow up the drill pipe of Union Oil Well No. 21, located 5.5 mi south of Santa Barbara on federally leased ocean bottom. The well was capped, but gas and oil began to boil up from the ocean floor through the waters near the platform. For many days despite desperate efforts to stop the leakage as much as 21,000 gal/day of oil leaked into the Santa Barbara Channel doing widespread damage. It is estimated that in all some 3,250,000 gal of oil were spilled (Holmes, 1969).

Petroleum is, of course, a constituent of the natural environment, and in places such as the Santa Barbara Channel, there is a certain amount of natural seepage (Allen, Schlueter, and Mikolaj, 1970). Nature has developed means of coping with this natural leakage in the form of microorganisms capable of metabolizing the constituents of oil (ZoBell, 1963; Pilpel, 1968), thus the sediments on the bottom of Lake Maracaibo, Venezuela, despite very dirty drilling operations, are relatively clean, presumably because this area of high natural leakage has a healthy population of oil eating microorganisms. However, it is the same old story: Man is dumping petrochemicals into the sea at a *rate* far exceeding the natural capacity of the waters to accommodate them. As a consequence, the oceans of Earth are being polluted with oil on a massive and global scale. While coastal waters suffer the worse, even the open ocean is befouled. In the Sargasso Sea some 600 mi or more from shore, the R/V *Chain* had to discontinue towing for surface organisms because the research tow nets became fouled with oil, sometimes in lumps up to 3 in. in diameter (Blumer, 1969; see also Horn, Teal, and Backus, 1970). B. F. Morris (1971) has estimated that there are about 27,000 and 50,000 tons of pelagic tar in the north Atlantic Ocean and Mediterranean Sea, respectively.

World demand for petroleum is enormous and escalating (Table 8.18) thereby, not only threatening an invaluable natural resource with depletion, but since a large and increasing fraction of the oil is produced by off-shore wells and/or transported by sea (Table 8.18) also constituting a most alarming menace to the marine environment. Table 8.19 lists estimates of the contributions to marine oil pollution from various sources. There are some 7500 oil pollution incidents annually in United States waters alone, and port losses from collisions and handling operations amount to 1,000,000 ton/yr (Blumer, 1970). While more carefully controlled handling can reduce the present port influx substantially, it must be remembered that a certain

Table 8.18 World Petroleum Production and Transportation (From Goldberg, 1970)

	In Million Metric Tons		
	1967	1970	1980
Production, Total	1850	2,200	4000
Production, Offshore		440	1300
Tanker Cargoes, Crude Oil	650	950	
Crude Oil and Products		1,500	2900
Petroleum Reserves		75,000	?[a]

[a] Ultimate world offshore potential to 350-m waterdepth estimated at 220,000 × 10^6 metric tons.

Table 8.19 Annual Sources of Oil Pollution in the Sea (From Hunt and Blumer, 1970)

	Million Metric Tons
Washing of Oil Cargo Tanks	1.5–3
Bilge Pumping (45,000 ships)	0.5–0.5
Tanker Spills (15,000/year)	1 –2
Oil Drilling, Pipeline, and Offshore Storage Tank Spills (>20,000 wells)	1 –3
Sewage and Industrial Outfalls	1 –1.5
Total	5 10
Natural Hydrocarbons Produced	1

fraction of the total oil handled will still find its way into the marine environment despite all due precautions. In the relatively new English oil port Milford Haven, considerable care has been given to minimize oil spills since the port is adjacent to a national park, yet in 1966, 0.01% of the total oil handled found its way into the waters (10 to 20% of this from a single accident.)

Coastal areas near the major shipping routes (Figure 8.25) are the most frequent victims of oil pollution, and the construction of enormous supertankers (the *Europoort* is 253,000 tons and 1,000,000 ton tankers are envisioned) holds threats of future spills far more catastrophic than any of those that have already been experienced.

In addition to being the consumer of much of the sea-transported oil, the internal combustion engine is a further villain in marine pollution in many other ways. Nearly 2,000,000 tons of used lubrication oil goes unaccounted for each year in the United States alone (Blumer, 1970), and an appreciable fraction of this must end up in coastal waters. The products of incomplete combustion are carried to the sea via the atmosphere, as is also an enormous amount of hydrocarbons loss by evaporation from storage, not to mention countless spills at countless gasoline stations throughout the world.

The composition of crude oil is discussed in greater detail in Chapter 4. Suffice it here to say that it is a complex mixture of saturated and unsaturated, aliaphatic and aromatic hydrocarbons (Figure 8.26) with a wide range of boiling points.

Sometimes it is possible, if suitable equipment happens to be available fast enough, to contain or scoop up an oil spill by mechanical means. Often the oil slick is too thin or too fragmented to be burned off; however, methods have been devised using

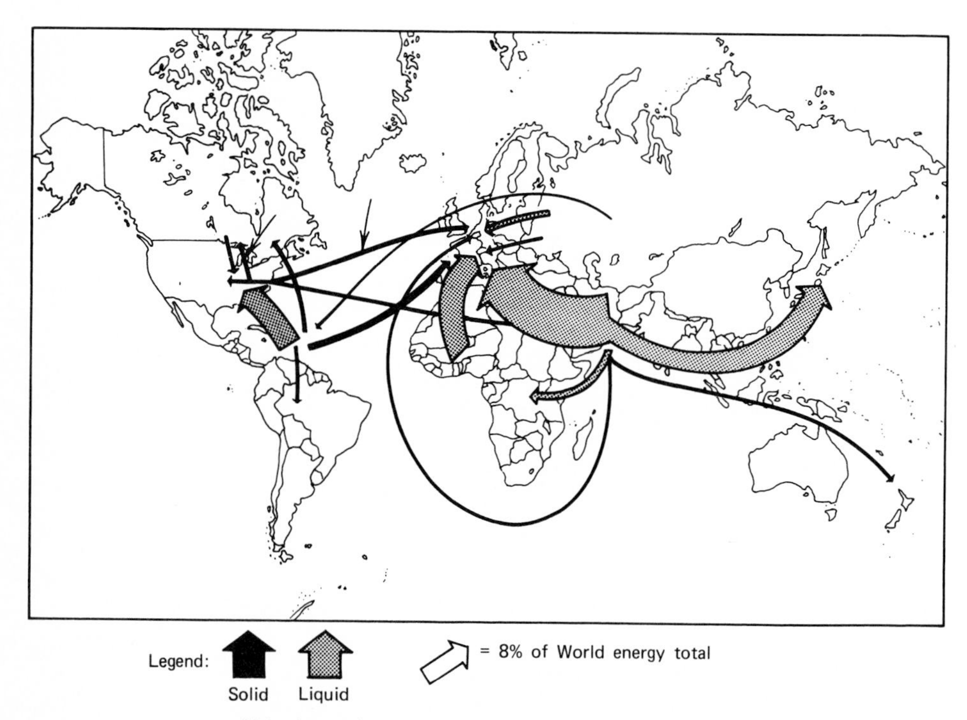

Figure 8.25 Major world routes of liquid fuel transportation, 1965.

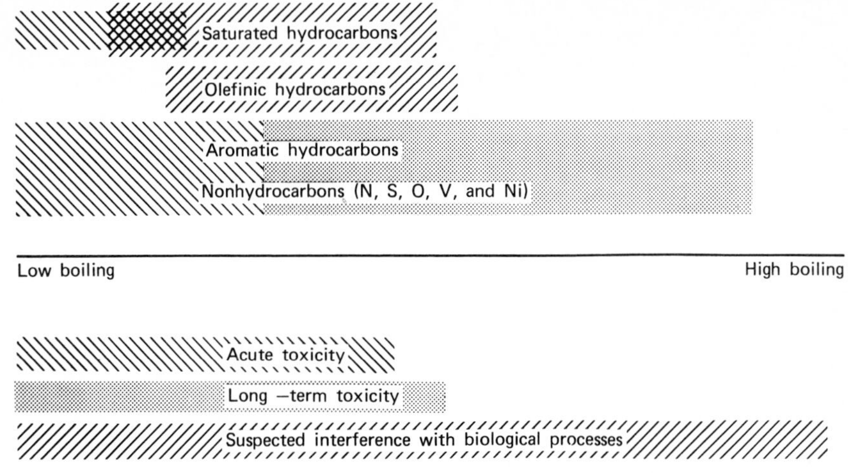

Figure 8.26 Toxicity of petroleum constituents.

bouyant porous materials to absorb the oil for concentrating the oil sufficiently for combustion. Detergents have also been used to disperse oil spills, as in the case of the *Torrey Canyon* disaster, but this is only a cosmetic approach, and while it pacifies resort hotel owners, far from containing the contamination, it mixes and spreads it into the marine environment. Nature's technique is microbial decomposition of the petroleum, but it suffers from four disadvanages: (*a*) it is slow, (*b*) it imposes an

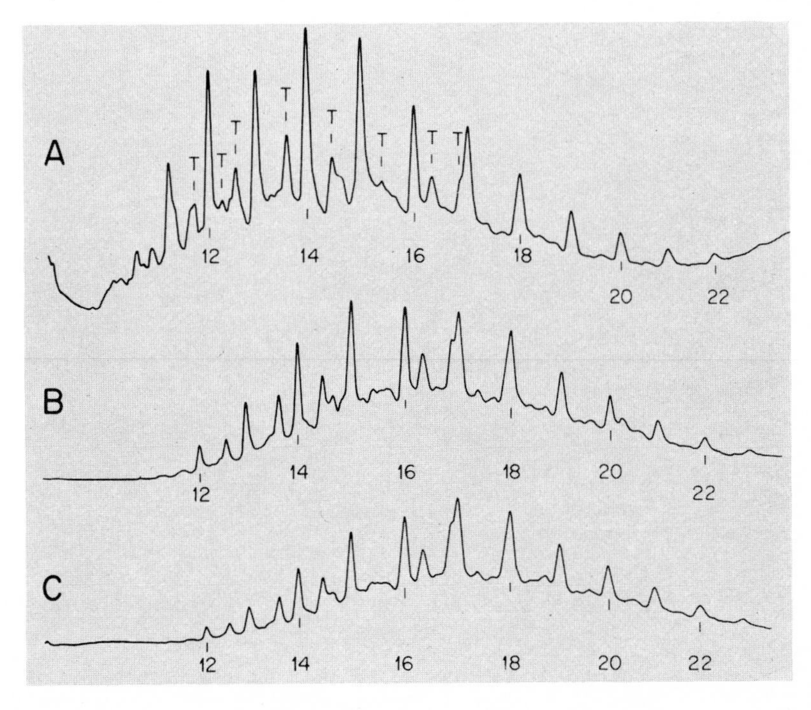

Figure 8.27 Gas chromatograms of the aging West Falmouth (Massachusetts) oil spill: *A*, No. 2 fuel oil; *B*, sediments in Wild Harbor basin 12 days after the spill; *C*, oil recovered from Wild Harbor water 2 months after the spill; *T*, marks the positions of isoprenoid alkanes. (From Blumer et al., with permission of the publisher.)

enormous requirement of precious oxygen (1 gal of crude oil requires all the oxygen dissolved in 320,000 gal of air-saturated seawater, ZoBell, 1969), (*c*) it can be adversely affected by environmental conditions such as low temperature and low oxygen concentration and hindered by the presence of other contaminants such as heavy metals, and (*d*) it is a selective process and tends to be least effective on the most

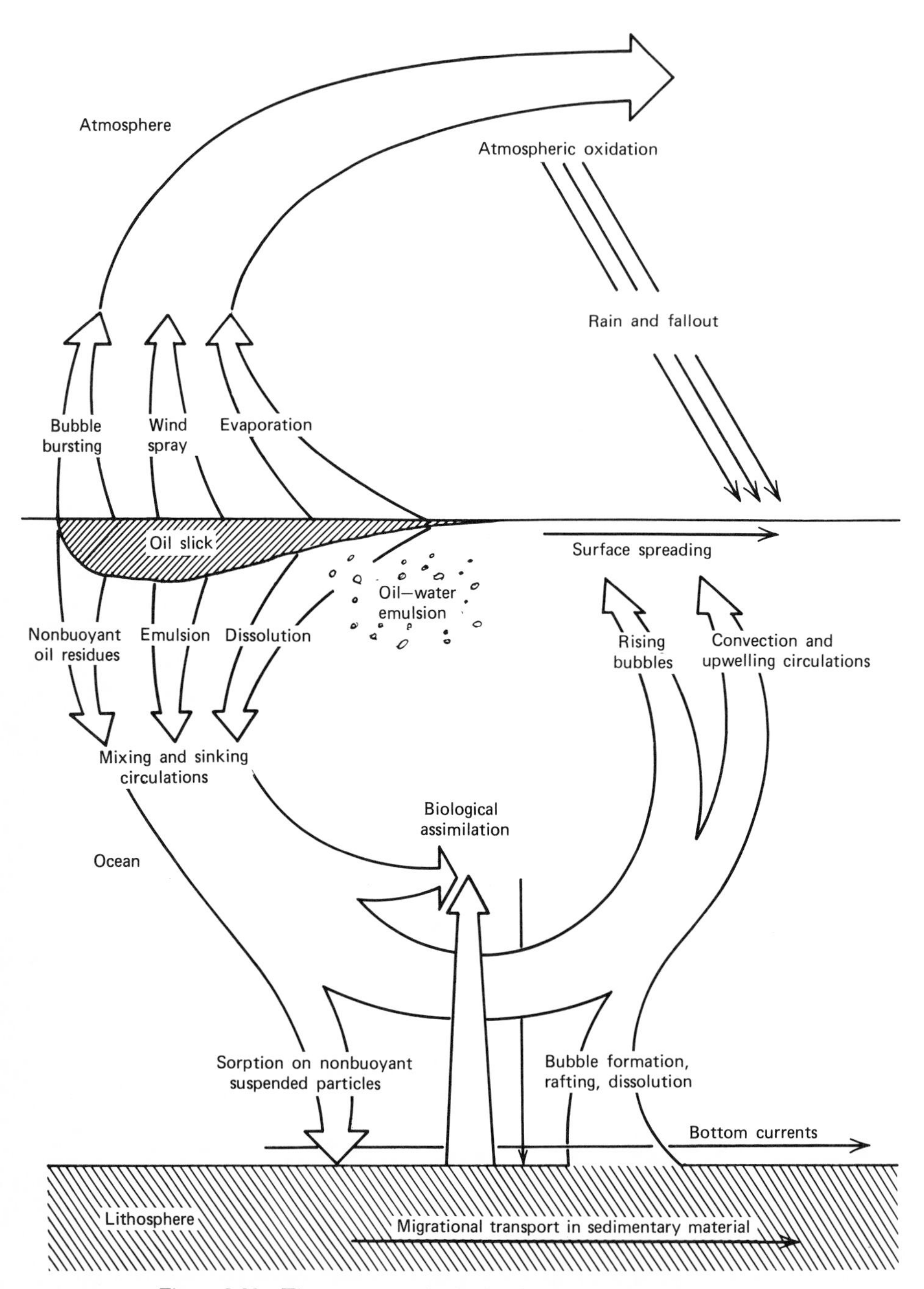

Figure 8.28 The transport of oil spilled in the marine environment.

toxic petroleum constituents. An oil spill will persist and continue to spread for an extended period of time. The peaks of the more volatile and soluble lighter hydrocarbons in an aging spill tend to disappear from the oil's gas chromatogram (Figure 8.27), while the aromatic hydrocarbons that form the base-line curve linger (Blumer, Souza, and Sass, 1970). These are the very compounds that may be responsible for the carcinogenic properties of petroleum products (Cook, Carruthers, and Woodhouse, 1958; Eckardt, 1967), and because of their great toxicity, they are the least likely candidates to be metabollized by microorganisms.

Oil is mixed into the marine environment by a number of processes involving the atmosphere and lithosphere as well as the hydrosphere (Widmark et al., 1972; Figure 8.28). Even when finally taken up by the sediments, the oil is not immobilized but may continue to spread until an extensive patch of the ocean bottom material is persistently contaminated. Oil is also highly mobile within the biosphere where again it rapidly permeates food webs (Blumer et al., 1972, Figure 8.29) (for a bibliography of the biological effects of oil pollution see Radcliffe and Murphy, 1969, and for brief reviews, Mitchell et al., 1970, and Templeton, 1971).

With respect to oil pollution, as with so many other cases of environmental deterioration, it is not always easy to establish who the guilty parties are so that appropriate legislative and enforcement action can be taken. However, this is one area in which technological advancement does hold some promise. The movement of oil films on water can be monitored with infrared spectroradiometry, and techniques

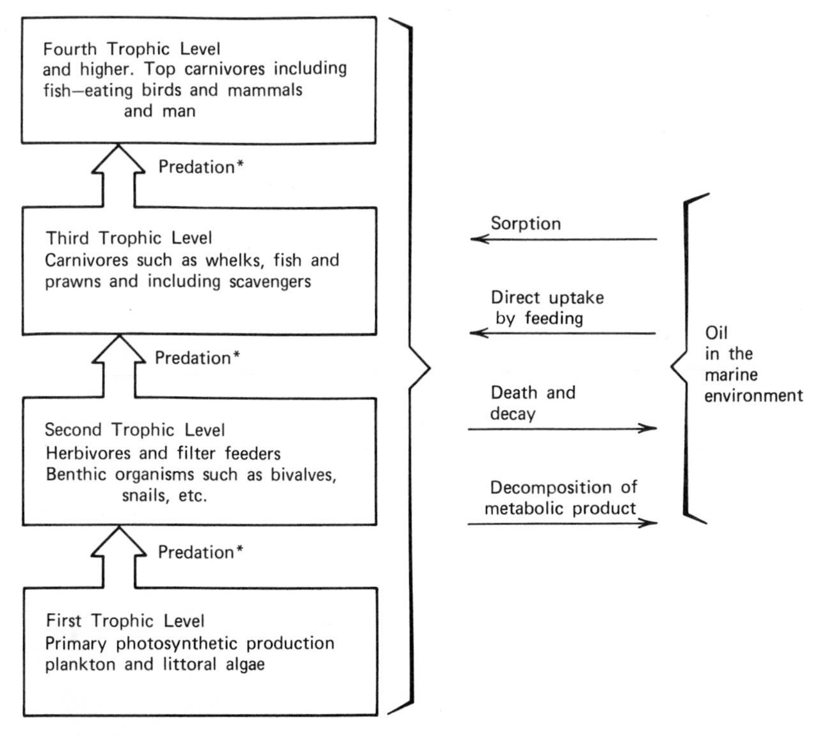

Figure 8.29 Movement and concentration of oil in the marine ecosystem.

Table 8.20 Pollutants of the Marine Environment (From Preston et al., 1972)

A. Radionuclide Concentrations from Global Fallout { pCi/liter seawater; pCi/kg dry sediments; pCi/kg wet biological material }

	Water	Sediments	Biological Materials: Seaweeds	Mollusks	Crustaceans	Fish
^{3}H	$1–10^{2}$					
^{54}Mn	(1)	$10^{3}–10^{4}$	10^{3}			
^{90}Sr	$10^{-1}–1$	$1–10^{2}$	1–10	0.1–10	1–10	1–10
^{95}Zr/Nb	$10^{-1}–10$	$10^{2}–10^{5}$	$10^{3}–10^{4}$			
^{106}Ru	$10^{-2}–1$	$10^{2}–10^{5}$	$10–10^{4}$	$10–10^{2}$	$1–10^{2}$	$1–10^{2}$
^{137}Cs	$10^{-1}–1$	$10–10^{4}$	1–10	1–10	1–10	$1–10^{2}$
^{144}Ce	$10^{-2}–1$	$10^{2}–10^{5}$	$10–10^{4}$	$10–10^{3}$	$1–10^{2}$	$1–10^{2}$
^{239}Pu	$10^{-5}–10^{-3}$	$10^{-1}–1$	1	10^{-1}		10^{-3}

B. Radionuclide Concentrations Resulting from the Controlled Disposal of Radioactive Wastes from Nuclear Power Production { pCi/liter seawater; pCi/Kg dry sediments; pCi/Kg wet biological material }

	Water	Sediments	Seaweeds	Mollusks	Crustaceans	Fish
^{3}H	—					
^{32}P	1–10			$10^{2}–10^{3}$		
^{54}Mn						
^{55}Fe	$10^{-1}–10$		10^{2}	$10^{2}–10^{3}$		
^{60}Co	$10^{-2}–1$	$10^{2}–10^{3}$	$10^{2}–10^{3}$	$10–10^{3}$		
^{65}Zn	$10^{-3}–1$	10^{3}	$10^{2}–10^{3}$	$10^{2}–10^{5}$		
^{90}Sr	$1–10^{2}$		$10–10^{2}$	$10^{2}–10^{3}$	10^{3}	1
^{95}Zr/Nb	$1–10^{3}$	$10^{5}–10^{7}$	$10^{4}–10^{6}$	$10^{4}–10^{6}$	$10^{3}–10^{4}$	
^{106}Ru	$1–10^{3}$	$10^{5}–10^{8}$	$10^{4}–10^{6}$	$10^{4}–10^{6}$	$10^{3}–10^{5}$	10^{3}
^{110}Ag	$10^{-3}–10^{-2}$			$10^{2}–10^{3}$		
^{134}Cs	1–10	10^{3}				10^{3}
^{137}Cs	$1–10^{2}$	$10^{2}–10^{4}$	10^{3}	$10^{2}–10^{3}$		$10^{3}–10^{4}$
^{144}Ce	$1–10^{3}$	$10^{5}–10^{7}$	$10^{4}–10^{6}$			
^{239}Pu		—	—			

(Continued)

such as gas chromatography (Figure 8.27) (Cole, 1971) and flame spectrophotometry (Adlard and Matthews, 1971) now appear to provide a means of "fingerprinting" spills and tracing their sources.

Dredge spoil, sewage sludge, and oil are but a few of the pollutants that man dumps or leaks into the oceans of Earth. Preston et al., 1972; Dryssen et al., 1972; and Widmark, 1972 have tried to compile a list and classify some of the more im-

Table 8.20 (*Continued*)

C. Inorganic Chemicals to Be Considered as Pollutants of the Marine Environment (From Dryssen et al., 1972)

Element[a]	Natural Conc. in Sea Water[b] (μg/liter)	World Production (tons/year) (1967)	Routes of Entry into the Sea[c]	Pollution Categories[d]
H (acids)	pH = 8 (alk. = 0.0024 M)		D	IIIc
Be	0.001		U	IVc ?
Ti	2		A ?	IVb ?
V	2	9,000	A	IVa ?
Cr	0.04	2,000,000	R(U)	IVc ?
Fe	10		D, R	IVc
Cu	1	5,000,000	D, R	IVc
Zn	2	5,000,000	D, R	IIIc
Cd	0.02		A, R	IIc
Hg	0.1	9,000	A, R	Ib
Al	10	8,000,000	D, R	IVc
CN^-			D, R	IIIc
Pb	0.02	3,000,000	A, R	Ia
P			D	IVc
As	2		D	IIc
Sb	0.45	60,000	U	IVc ?
Bi	0.02		U	IVc ?
Se	0.45		U	IIIc ?
F^-	1340		D, R	IVc

[a] Listed in the order of groups in the extended form of the periodic table.

[b] These values are approximate but are representative for low levels in unpolluted seawater.

[c] D, dumping; A, through atmospheric pollution; R, through rivers (runoff) or pipelines; U, unknown.

[d] I–IV order of decreasing menace. a, worldwide; b, regional; c, local (coastal, bays, estuaries, single dumpings).

(*Continued*)

portant of these pollutants (Table 8.20), while Ketchum (1970) has tried to diagram the major processes that determine the distribution and fate of pollutants in the marine environment (Figure 8.30).

But perhaps most frightening of all, we really have relatively little inkling of the dangerous trash that governments and industry seek to hide in the seas, ranging from live ammunition to radioactive wastes. Every once in a while reports of these practices bursts upon the public consciousness like fetid bubbles, and I conclude this chapter with brief accounts of two such incidents. In the United States, a violent public debate was touched off by the discovery of the Army's intent to dump a quantity of nerve gas into Atlantic coastal waters. The aroused conservationists seemed to be

Table 8.20 (*Continued*)

D. Organic Pollutants of Mainly Natural Origin, Partly Changed at Process (Tentative)

Pollutants	Comments
Tannins	Waste from dye industry.
Lignin	Waste from paper and pulp mill industry.
Carbohydrates	Waste from paper and pulp mill industry, breweries, whiskey industry and from sugar production. Oxygen consumption. Of local importance.
Proteins, Included Pepsides Amino Acids, Amines, Fatty Acids, Lipids	Waste from slaughteries, dairies, and fish industry. Oxygen consumption. Of local importance.
Hydroxy Fatty Acids Humic Acids	Waste from the bark chipping of pine.
Pyrenthrines	Insecticides.
Terpenes	Floatation of ore.
Polycyclic Aromatic Hydrocarbons (PAH)	Found in marine organisms and sediments, and in areas with volcanic activity.

E. Organic Pollutants of Synthetic Origin (Tentative) (From Widmark et al., 1972)

Pollutants	Comments
Alkylbenzen Sulfonates (ABS)	Detergent of the "hard" type, toxicity to marine organisms increasing with increasing branching. Not readily biodegradable.
Linear Alkyl Sulfonates (LAS)	Detergent of the "soft" type, less toxic to marine organisms than the former and more rapidly degraded by biological organisms.
Phenols	In waste from industry, coke and gas works, also found in natural seawater (1–3 μg/liter).
Polycyclic Aromatic Hydrocarbons	From oil refineries, heating, and so on.
Aniline and Related Compounds	From dye industries.

more concerned with the principle than with the chemistry of the dumping and the chemistry, that is, the rapid hydrolysis ($t_{1/2}$ = 30 min) of isopropylmethylphosphonofluoridate (Epstein, 1970), appears to indicate that indeed serious long-range environmental damage from marine dumping was unlikely. All of which points to a danger far greater than that from dumping poisons into the oceans, namely of responding in an irrational way to environmental problems with the fire of emotionalism and with little scientific judgement.

A second recent and more grounded furor concerned the dumping of industrial wastes in the North Sea. Protests by Norway, Sweden, and Denmark over a new "poison ship," the *Hudson Stream*, capable of dumping 375,000 ton/yr of wastes, not only delayed its sailing but gave great and badly needed impetus to an international ban on dumping (Anon., 1971*a*). In this instance, the protest was entirely justified: The North Sea is a resource shared by many nations, but because of its shallowness its valuable fisheries are particularly vulnerable to pollution. Furthermore, analyses of some of the chemicals being dumped proved to be a real horror story (Table 8.21). One material, "Mixture D" in Table 8.21, is lethal to fish even when diluted 100,000,000 times (Greve, 1971).

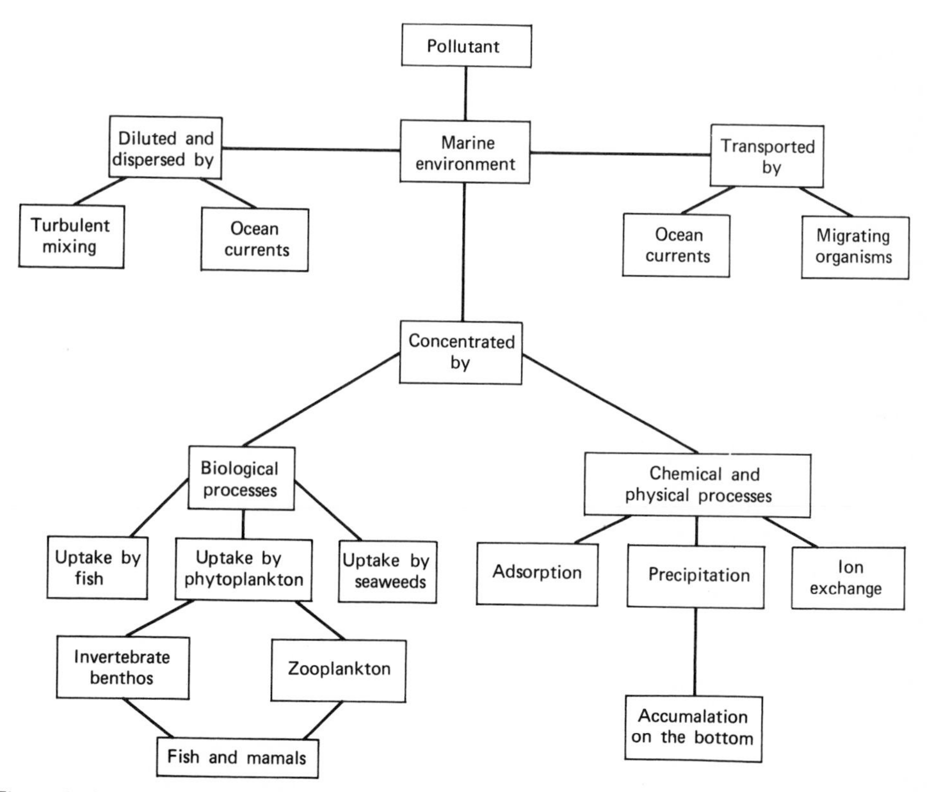

Figure 8.30 Processes which can determine the fate and distribution of a pollutant in the marine environment. Efficiency of energy transfer from one trophic level to the next is approximately 2 to 4%. (Arrows indicate possible movements of oil in the system.) Biodegredation occurs on all return steps and within the oil in the sea.

Table 8.21 Composition and Toxicity of Wastes Found in the North Sea (mixtures A, B, C, D, and E_1) and of Wastes Scheduled for Dumping in the Sea, Ready for Loading at Rotterdam Harbor (mixtures B, D, E_1, E_2, and F)[a] (From Greve, © 1971 by the American Association for the Advancement of Science, with permission of the AAAS.)

Composition					
Mixture[b]	Compound	Approximate Average Content (%)	Toxicity	North Sea Samples Investigated (No.)	Harbor Samples: Estimated Total Weight of Loads (tons)
A	Vinyl acetate and vinyl propionate	90	++	7	
	Phenothiazine	5			
	Isopropenyl acetate	Trace			
	Acetic acid	Trace			
	N-Acetylphenothiazine	Trace			
B	1,2-Dichloropropane	60	+	54	650
	Di(2-chloroisopropyl) ether	25			
	Propylene, butylene, and trimethylene oxides	5			
	Chloropropanoles	5			
	Epichlorohydrin	5			
C	Mineral oils	100	−	4	
D	Endosulfan and derivatives	5	++++	3	350
	Toluene	5			
	Clayish material	55			
	Solution of sodium sulfite, sodium carbonate, and sodium chloride	35	+++	1	
E_1	Trichlorotoluidines	80	+++	1	350*
	Dichlorotoluidines	15			
	Monochlorotoluidines	Trace			
	o-Toluidine	Trace			
E_2	Dichloronitrobenzenes	85	+++		
	Trichlorobenzenes	Trace			
	p-Dichlorobenzene	10			
F	2-Butene-1,4-diol diacetate	95	+++		
	Hexachlorobutadiene	Trace			
	Dichloropropanoles	Trace			
	Epichlorohydrin	Trace			

[a] In each toxicity experiment, three 2-wk-old guppies were used. The toxicity is expressed as follows (ppm, parts per million; ppb, parts per billion): −, 100 ppm gives no kill in 3 days; +, 1000 ppm is lethal within 2 hr but 100 ppm gives no kill in 3 days; ++, 100 ppm is lethal within 2 hr but 10 ppm gives no kill in 3 days; +++, 10 ppm is lethal within 2 hr but 1 ppm gives no kill in 3 days; ++++, 10 ppb is lethal within 2 hr but 1 ppb gives no kill in 3 days.

[b] This value is the total for mixtures E_1, E_2, and F.

BIBLIOGRAPHY

Ad Hoc Committee for the Evaluation of the Influence of Dumping in the New York Bight, U.S. Dept. Interior, Draft Rept., April 13, 1970.

L. H. Adams and R. G. Hall, *J. Phys. Chem.*, *35*, 2145 (1931).

E. R. Adlard and P. H. D. Matthews, *Nature Phys. Sci.*, *233*, 83 (1971).

A. Allen, R. S. Schlueter, and P. G. Mikolaj, *Science*, *170*, 974 (1970).

P. L. Altman and D. S. Dittmer, *Blood and Other Body Fluids*, U.S. Air Force Aeronautical Systems Div. Tech. Rept. 61-199 (June 1961).

Anon., *Nature*, *233*, 160 (1971*a*).

Anon., *Chemical Week*, 24 (July 7, 1971*b*).

F. A. J. Armstrong, in J. P. Riley and G. Skirrow (eds.), *Chemical Oceanography*, Academic, London, 1965, v. I.

H. Barnes, *Apparatus and Methods of Oceanography: I, Chemical*, Allen and Unwin, London, 1969.

W. L. Belser, in M. N. Hills (ed.), *The Sea*, Interscience, New York, 1963.

M. Blumer, in D. P. Hoult (ed.), *Oil on the Sea*, Plenum, New York, 1969.

M. Blumer, "Oil Contamination and the Living Resources of the Sea," paper presented FAO(UN) Conf. Mar. Pollut., Rome (December 1970).

M. Blumer et al., in E. D. Goldberg (ed.), *A Guide to Marine Pollution*, Gordon & Breach, New York, 1972.

M. Blumer, G. Souza, and J. Sass, *Mar. Biol.*, *5*, 195 (1970).

A. Bradshaw and K. E. Schleicher, *Deep-Sea Res.*, *12*, 151 (1965).

W. S. Broecker, in M. N. Hill (ed.), *The Sea*, Interscience, New York, 1963.

W. S. Broecker, *Quarternary Res.*, *1*, 188 (1971).

N. L. Brown and B. V. Hamon, *Deep-Sea Res.*, *8*, 65 (1961).

R. P. Brown and E. H. Shenton, "Evaluating Waste Disposal at Sea," paper presented 7th Ann. Mar. Tech. Soc. Conf., Washington, D.C. (1971).

A. G. Bulavko, *Nature and Resources* (UNESCO), 7(1), 12 (1971).

R. S. Burd, "A Study of Sludge Handling and Disposal," U.S. Dept. Interior, FWPCA Pub. WP-20-4 (May 1968).

J. D. Burton, in J. P. Riley and G. Skirrow (eds.), *Chemical Oceanography*, Academic, London, 1965.

J. D. Burton, *J. Cons., Cons. Int. Explor. Mer.*, *33*, 141. (1970).

D. E. Carritt, in M. N. Hill (ed.), *The Sea*, Interscience, New York, 1963.

L. J. Carter, *Science*, *167*, 1102 (1970).

R. Chester, in J. P. Riley and G. Skirrow (eds.), *Chemical Oceanography*, Academic, London, 1965.

T. J. Chow and A. W. Mantyla, *Nature*, *206*, 383 (1965).

R. D. Cole, *Nature*, *233*, 546 (1971).

J. W. Cook, W. Carruthers, and D. L. Woodhouse, *Brit. Med. Bull.*, *14*, 132 (1958).

R. A. Courant, R. A. Horne, and D. R. Kester, unpublished, 1971.

F. Culkin, in J. P. Riley and G. Skirrow (eds.), *Chemical Oceanography*, Academic, New York, 1965.

D. L. Correll, *Limnol. Oceanogr.*, *10*, 364 (1965).

R. B. Dean, *Tappi*, *52*, 457 (1969).

E. T. Degens (ed.), *Hot Brines and Recent Heavy Metal Deposits in the Red Sea*, Springer-Verlag, New York, 1969.

G. Dietrich, *General Oceanography*, Interscience, New York, 1963.

D. A. Dobbins, P. C. Ragland, and J. D. Johnson, *Environ. Sci. Tech.*, *4*, 743 (1970).

M. A. Dobrzanskaya and A. D. Kovalevsky, *Abstrs. Int. Oceanogr. Congr.*, p. 884, Amer. Assoc. Adv. Sci., Washington, D.C., 1959.

D. Dryssen et al., in E. D. Goldberg (ed.), *A Guide to Marine Pollution*, Gordon & Breach, New York, 1972.

E. K. Duursma, in J. P. Riley and G. Skirrow (eds.), *Chemical Oceanography*, Academic, London, 1965.

R. E. Eckardt, *Internat. J. Cancer*, *2*, 656 (1967).

J. Epstein, *Science*, *170*, 1396 (1970).

B. M. Fabuss and A. Korosi, *J. Chem. Eng. Data*, *11*, 606 (1966).

F. H. Fisher, *J. Phys. Chem.*, *66*, 1607 (1962).

F. H. Fisher, *J. Phys. Chem.*, *69*, 695 (1965).

D. L. Fox, C. M. Oppenheimer, and J. S. Kittredge, *J. Mar. Res.*, *12*, 233 (1953).

H. S. Frank and M. W. Evans, *J. Chem. Phys.*, *13*, 507 (1945).

H. S. Frank and W. Y. Wen, *Disc. Faraday Soc.*, *24*, 133 (1957).

G. M. Friedman and E. Gavish, *J. Sed. Petrol.*, *40*, 930 (1970).

A. B. Gancy and S. B. Brummer, *J. Phys. Chem.*, *73*, 2429 (1969).

J. E. Gannon and A. M. Beeton, *J. Water Poll. Control Fed.*, *43*, 392 (1971).

R. M. Garrels, *Science*, *148*, 69 (1965).

R. M. Garrels and M. E. Thompson, *Amer. J. Sci.*, *260*, 57 (1962).

S. Glasstone, K. L. Laidler, and M. Eyring, *The Theory of Rate Processes*, McGraw-Hill Book Co., New York, 1941.

E. Goldberg, in M. N. Hills (ed.), *The Sea*, Interscience, New York, 1963, v. I.

E. Goldberg, in J. P. Riley and G. Skirrow (eds.), *Chemical Oceanography*, Academic, London, 1965.

R. H. Gould, *Proc. Amer. Soc. Civ. Engr.*, *94*, 1041 (1968).

A. J. Gow, H. T. Ueda, and D. E. Garfield, *Science*, *161*, 1011 (1968).

P. A. Greve, *Science*, *173*, 1021 (1971).

J. J. Groot and M. Ewing, *Science*, *142*, 579 (1963).

M. G. Gross, *Water Resources Res.*, *6*, 927 (1970).

K. J. Hall, W. C. Weimer, and G. F. Lee, *Limnol. Oceanogr.*, *15*, 162 (1970).

R. Hartung and G. W. Klinger, *Environ. Sci. Tech.*, *4*, 407 (1970).

H. W. Harvey, *J. Mar. Biol. Assoc. U.K.*, *14*, 71 (1926).

H. W. Harvey, *The Chemistry and Fertility of Sea Water*, Cambridge Univ. Press, Cambridge (England), 1966.

P. Hepple, *Water Pollution by Oil*, Elsevier, New York, 1969.

J. E. Hobbie, C. C. Crawford, and K. L. Webb, *Science*, *159*, 1463 (1968).

O. Holm-Hansen and C. R. Booth, *Limnol. Oceanogr.*, *11*, 510 (1966).

R. W. Holmes, in D. P. Hoult (ed.), *Oil in the Sea*, Plenum, New York, 1969.

M. H. Horn, J. M. Teal, and R. H. Backus, *Science*, *168*, 245 (1970).

R. A. Horne, *Nature*, *200*, 418 (1963).

R. A. Horne, *Water Resources Res.*, *1*, 263 (1965).

R. A. Horne, *Adv. High Pressure Res.*, *2*, 169 (1969*a*).

R. A. Horne, *Marine Chemistry*, Wiley-Interscience, New York, 1969*b*).

R. A. Horne, *Adv. Hydrosci.*, *6*, 107 (1970*a*).

R. A. Horne, in L. L. Ciaccio (ed.), *Chemical and Microbiological Analysis on Water and Water Pollution*, Dekker, New York, 1971.

R. A. Horne and J. D. Birkett, *Electrochim. Acta*, *12*, 1153 (1967).

R. A. Horne and R. A. Courant, *J. Geophys. Res.*, *69*, 1971 (1964).

R. A. Horne and D. S. Johnson, *J. Phys. Chem.*, *70*, 2182 (1966*a*).

R. A. Horne and D. S. Johnson, *J. Geophys. Res.*, *71*, 5275 (1966*b*).

R. A. Horne and D. S. Johnson, *J. Phys. Chem.*, *71*, 1147 (1967).

R. A. Horne, A. J. Mahler, and R. C. Rossello, "The Marine Disposal of Sewage Sludge and Dredge Spoil in the Waters of the New York Bight," Woods Hole Ocean. Inst. Tech. Memo No. 1-71 (February 1971).

R. A. Horne and R. P. Young, *J. Phys. Chem.*, *72*, 1763 (1968).

R. A. Horne, R. P. Young, and A. F. Day, *J. Phys. Chem.*, *73*, 2782 (1969).

D. P. Hoult (ed.), *Oil on the Sea*, Plenum, New York, 1969.

G. P. Howells, T. J. Kneipe, and M. Eisenbud, *Environ. Sci. Tech.*, *4*, 26 (1970).

J. M. Hunt and M. Blumer, "Oil Pollution in the Marine Environment," paper presented Internat. Symp. Hydrogeochem. Biogeochem., Tokyo (September 1970).

M. B. Jacobs, and M. Ewing, *Science*, *149*, 179 (1965).

H. W. Jannasch, K. Eimhjellen, C. O. Wirsen, and A. Farmanfarmian, *Science*, *171*, 673 (1971).

H. R. Jitts, *Austral. J. Mar. Freshwater Res.*, *10*, 7 (1959).

B. H. Ketchum, "Ecological Effects of Sewer Sludge Disposal at Sea," paper presented Water Pollut. Control Fed. Convention, Boston (October 1970).

B. H. Ketchum and N. Corwin, *Limnol. Oceanogr.*, *10*, R148 (1965).

T. Koyama and T. G. Thompson, *Abstr. Internat. Oceanogr. Congr.*, p. 925, Amer. Assoc. Adv. Sci., Washington, D.C. (1959).

K. B. Krauskopf, *Geochim. Cosmochim. Acta*, *9*, 1 (1956).

J. L. Kulp, *Bull. Geol. Soc. Amer.*, *62*, 326 (1951).

I. Kuziemska, *Rozpr. Hydrotech.* (*Poland*), *26*, 183 (1970).

F. W. Lane et al., "The Effect of Oil Pollution on Marine Wildlife," Rept. to U.S. Commissioner of Fisheries, Appendix 5 (1925).

G. H. Lauff (ed.), *Estuaries*, Amer. Assoc. Adv. Sci., Washington, D.C., 1968.

Y. H. Li, *J. Geophys. Res.*, *72*, 2665 (1967).

A. P. Lisitzin, *Abstr. Int. Ocean. Congr.*, p. 240, Amer. Assoc. Adv. Sci., Washington, D.C., 1959.

D. C. Martin and D. A. Bella, *J. Water Poll. Control Fed.*, *43*, 1865 (1971).

A. R. Miller, *Nature*, *203*, 590 (1964).

S. L. Miller, *Proc. Nat. Acad. Sci. U.S.A.*, *47*, 1515, 1798 (1961).

S. L. Miller, *Proc. 2nd Symp. Underwater Physiol.*, Pub. No. 1181, Nat. Acad. Sci. U.S.A., Washington, D.C., 1963.

S. L. Miller, *Science*, *165*, 489 (1969).

S. L. Miller and W. D. Smythe, *Science*, *170*, 531 (1970).

Y. Miyake, *Bull. Chem. Soc. Japan*, *14*, 58 (1939).

Y. Miyake, in M. N. Hill (ed.), *The Sea*, Interscience, New York, 1963.

A. W. Morris, *Nature*, *233*, 427 (1971).

B. F. Morris, *Science*, *173*, 430 (1971).

C. T. Mitchell, E. K. Anderson, L. G. Jones, and M. J. North, *J. Water Pollut. Control Fed.*, *42*, 812 (1970).

T. A. Murphy, "Environmental Effects of Oil Pollution," paper presented Amer. Soc. Civil Engr. Meeting, Boston (July 1970).

K. V. Natarajan and R. C. Dugdale, *Limnol. Oceanogr.*, *11*, 621 (1966).

Nat. Acad. Sci. U.S.A., *Wastes Management Concepts for the Coastal Zone*, Nat. Acad. Sci.–Nat. Acad. Engr., Washington, D.C., 1970.

B. W. Nelson, *Abstrs. Internat. Oceanogr. Congr.*, p. 640, Amer. Assoc. Adv. Sci., Washington, D.C., 1959.

R. G. Paquette, "Conf. Phys. Chem. Prop. Sea Water," Easton, Md., 1958, Nat. Acad. Sci. U.S.A.–Nat. Res. Council Publ. No. 600, 128 (1959).

K. Park, W. T. Williams, J. M. Prescott, and D. W. Hood, *Science*, *138*, 531 (1962).

T. R. Parsons, *Prog. Oceanogr.*, *1*, 205 (1963).

L. Pauling, *Science*, *134*, 15 (1961).

J. Pearce, *The Effects of Waste Disposal in the New York Bight*, Interim Rept., Sandy Hook Mar. Lab., U.S. Bur. Sport Fish., December 1969.

W. G. Peter, *Bioscience*, *20*, 617, 669 (1970).

N. Pilpel, *Endeavour*, *27*(100), 11 (1968).

H. Postma, *Arch. Neerl. Zool.*, *10*, 1 (1954).

A. Preston et al., in E. D. Goldberg (ed.), *A Guide to Marine Pollution*, Gordon & Breach, New York, 1972.

L. Provasoli, in M. N. Hill (ed.), *The Sea*, Interscience, New York, 1963.

D. R. Radcliffe and T. A. Murphy, *Biological Effects of Oil Pollution*, FWPCA, Washington, D.C., 1969.

M. M. Rhead, G. Eglinton, G. H. Draffan, and P. J. England, *Nature*, *232*, 327 (1971).

H. E. Ries, Jr., and B. L. Meyers, *Science*, *160*, 1449 (1968).

F. A. Richards, in J. P. Riley and G. Skirrow (eds.), *Chemical Oceanography*, Academic, London, 1965*a*.

F. A. Richards, J. D. Cline, W. W. Broenkow, and L. P. Atkinson, *Limnol. Oceanogr.*, *10*, R158 (1965*b*).

G. A. Riley, *Bull. Bingham Oceanogr. Coll.*, *73*, 83 (1959).

G. A. Riley, D. van Hemert, and J. P. Wangersky, *Limnol. Oceanogr.*, *10*, 354 (1965).

J. P. Riley, in J. P. Riley and G. Skirrow (eds.), *Chemical Oceanography*, Vol. II, Academic, New York, 1965.

R. A. Robinson, *J. Mar. Biol. Assoc. U.K.*, *33*, 449 (1954).

J. H. Ryther and W. M. Dunstan, *Science*, *171*, 1008 (1971).

R. G. J. Shelton, *Mar. Pollut. Bull.*, *2*(2), 24 (1971).

R. Siever, *Sedimentology*, *11*, 5 (1958).

L. G. Sillen, *Science*, *156*, 1189 (1967).

E. Sinha, *Coastal/Estuarine Pollution*, Ocean Engr. Inform. Serv., Vol. 3 (1970).

D. D. Smith and R. P. Brown, *Ocean Disposal of Barge-Delivered Liquid and Solid Wastes from U.S. Coastal Cities*, EPA, U.S. Gov. Print. Off., Washington, D.C., 1971.

J. E. Smith (ed.), *Torrey Canyon Pollution and Marine Life*, Cambridge Univ. Press, New York, 1968.

J. N. Spencer and A. F. Voigt, *J. Phys. Chem.*, *72*, 464, 471 (1968).

E. M. Stanley and R. C. Batten, *J. Geophys. Res.*, *74*, 3415 (1969).

R. W. Stoughton and M. H. Lietzke, *J. Chem. Engr. Data*, *12*, 101 (1967).

J. D. H. Strickland, in J. P. Riley and G. Skirrow (eds.), *Chemical Oceanography*, Academic, London, 1965.

H. U. Svedrup, M. W. Johnson, and R. H. Fleming, *The Oceans*, Prentice-Hall, Englewood Cliffs, 1942.

J. C. Swallow and J. Crease, *Nature*, *205*, 165 (1965).

M. Tatsumoto, W. T. Williams, J. M. Prescott, and D. W. Hood, *J. Mar. Res.*, *17*, 247 (1958); *19*, 89 (1961).

W. L. Templeton, *J. Water Pollut. Control Fed.*, *43*, 1081 (1971).

W. H. Thorp, *J. Acoust. Soc. Amer.*, *38*, 648 (1965).

D. K. Todd, *The Water Encyclopedia*, Water Information Center, Port Washington, N.Y., 1970.

W. N. Torpey, *J. Water Pollut. Control Fed.*, *39*, 1797 (1967).

R. E. Train, R. Cahn, and G. J. MacDonald, *Ocean Dumping*, Council Environ. Qual., U.S. Gov. Print. Off., Washington, D.C., 1970.

K. K. Turekian, in J. P. Riley, and G. Skirrow (eds.), *Chemical Oceanography*, Academic, London, 1965.

G. S. Verma, *Rev. Mod. Phys.*, *31*, 1052 (1959).

G. E. Walsh, *Limnol. Oceanogr.*, *10*, 570, 577 (1965).

J. H. Wang, *J. Phys. Chem.*, *69*, 4412 (1965).

M. Whitfield, *Limnol. Oceanogr.*, *14*, 547 (1969).

G. Widmark et al., in E. D. Goldberg (ed.), *A Guide to Marine Pollution*, Gordon & Breach, New York, 1972.

P. M. Williams, *Nature*, *189*, 219 (1961).

P. M. Williams, *J. Fish. Bd. Canad.*, *22*, 1107 (1965).

P. M. Williams and K. S. Chan, *J. Fish. Res. Bd. Canad.*, *23*, 575 (1966).

M. G. Wolman, *Science*, *174*, 905 (1971).

W. A. Zisman, *Phys. Rev.*, *39*, 151 (1932).

C. E. ZoBell, *Int. J. Air Water Pollut.*, *7*, 173 (1963).

C. E. ZoBell, "Microbial Modification of Crude Oil in the Sea," paper presented at API-FWPCA Conf. Prevention Oil Spills (December 1969).

S. Zuckerman et al., *The Torrey Canyon*, H.M. Stationery Office, London, 1967.

ADDITIONAL READING

D. W. Hood (ed.), *Impingement of Man on the Oceans*, Wiley-Interscience, New York, 1971.

9

THE ROLE OF WATER IN OUR TOTAL ENVIRONMENT

9.1 INTRODUCTION

In the previous chapters, we have examined the chemistry of the major divisions of our environment—exosphere, lithosphere, atmosphere, and hydrosphere. The biosphere will be examined in the next section and in the remainder of the book we consider the all important interactions *among* these divisions. We find that water is commonly the mediator of these interactions, that water is, so to speak, the currency of the chemical economy of our environment. This should some as no surprise for we have already repeatedly taken notice of water's central role in our environment. To give further emphasis to its importance here I pause before we begin our study of the interactions and recapitulate and expand the theme of water's significance in our total environment, notably the hydrologic cycle and the various phase changes of water that occur in our environment, including the isotopic fractionation observed in nature as a consequence of these changes and other processes.

9.2 THE HYDROLOGIC CYCLE

There is no sequence of processes in the physical chemistry of our environment more obvious than the hydrologic cycle (Figure 9.1). It is responsible for our weather, and it makes our rivers run. There are not many processes in our world of comparable importance. The hydrologic cycle as a whole is not essential to life on this planet, but the cycle certainly is essential to the life forms that have emerged, and it is very difficult to conceive what Earth would be like, even the physical appearance of the landscape, without the hydrologic cycle.

Table 9.1 (see also Table 8.1) shows the distribution of water on Earth, while Table 9.2 shows water distribution in the United States alone. The largest reservoir of water is locked in the lithosphere. In the United States about 86% of the stored water is ground water, and ground water supplies about 95% of the domestic water in rural areas (Gelhar, 1972). Of the hydrosphere itself (Table 9.3), about 97.2% resides in the oceans, and 2.15% of the remainder is trapped in polar ice and glaciers—reservoirs of little use to man. Man relies on only 0.65% of the hydrosphere, and it is this small yet vital fraction that is renewed by the hydrologic cycle. The hydrologic cycle makes water available to us and replenishes our supplies. Were it not for this solar-powered renewal cycle, man would quickly exhaust all his useful supplies of

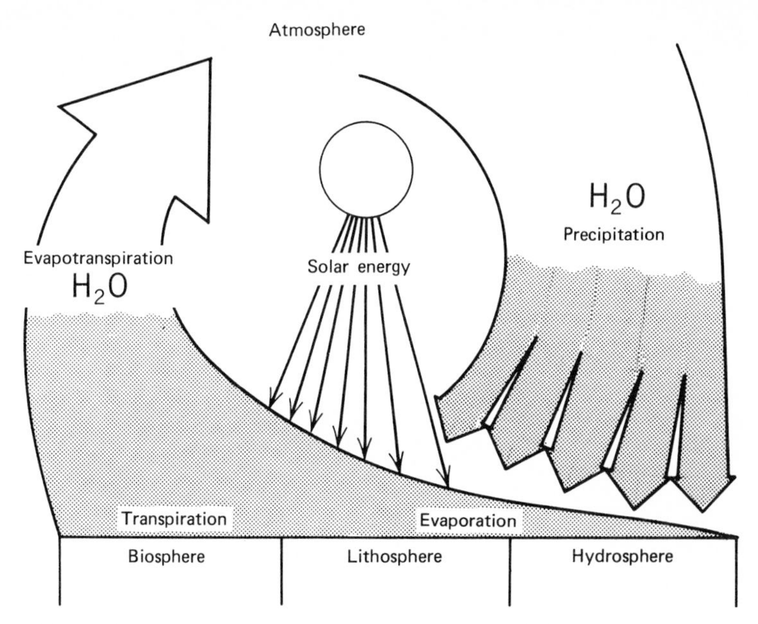

Figure 9.1 The hydrologic cycle. Most atmospheric water originates by evaporation from the marine hydrosphere.

Table 9.1 Water Content of the Various Parts of the Earth (From Hutchinson, 1957)

	Water[a]	
	kg/cm²	Gg
Primary Lithosphere	4900	250,000
Sedimentary Rocks	41	2,100
Ocean	269	13,800
Polar Caps and Other Ice	3.24	167
Inland Waters	0.0049	0.25
Circulating Groundwater	0.049	2.5
Atmospheric Water Vapor	0.0026	0.13

[a] The symbol kg/cm²e means kilograms per square centimeter of Earth's surface; Gg means geogram or 10^{20} g.

water. In the final analysis, the distribution of the human species on this planet and the success of its enclaves is determined, not by historical events, but by the availability of water. To cite an example that has recently become painfully clear; resolution of the United States' energy crisis by strip mining the western coal fields may be limited by the scarcity of water in these arid lands. There is probably insufficient water for the extensive gasification of the coal, and there is certainly insufficient water for the restoration of the strip-mined lands (Gillette, 1973). Again, Winstanley (1973) has noted that "much of the economic development and population explosion

Table 9.2 Distribution of Water in the Conterminous United States (Source: U.S. Geological Survey) (From Todd, 1970, with permission of Water Information Center, Inc.)

	Area (mi^2)	Volume (mi^3)	Annual Circulation (million acre-ft/yr)	Detention period (yr)
Frozen Water				
Glaciers	200	16	1.3	40
Ground Ice		(seasonal only)		
Liquid Water				
Fresh-Water Lakes	61.000	4,500	150	100
Salt Lakes	2,600	14	4.6	10
Average in Stream Channels	—	12	1500	0.03
Ground Water				
Shallow	3,000,000	15,000	250	200
Deep	3,000,000	15,000	5	10,000
Soil Moisture (3-ft root zone)	3,000,000	150	2500	0.2
Gaseous Water				
Atmosphere	3,000,000	45	5000	0.03

[a] United States part of Great Lakes only.

Table 9.3 Hydrospheric Water

Type	Volume (gal.)	Percentage of Total Hydrosphere
World Oceans	$348,700 \times 10^{15}$	97.2
Ice Cover	$7,700 \times 10^{15}$	2.15
Ground Water (to 0.5 mi depth)	$1,000 \times 10^{15}$	0.31
Ground Water (greater than 0.5 mi)	$1,000 \times 10^{15}$	0.31
Atmosphere	34	0.01
Freshwater Lake	33×10^{15}	0.009
Saline Lakes and Inland Seas	28×10^{15}	0.008
Vadose Water (includes soil moisture)	18×10^{15}	0.005
Stream Channels (average)	0.3×10^{15}	0.0001

in the former French [African] colonies, leading to national independence, took place during a period of abnormally good rainfall . . . . The present drought situation and the probable long term trends now seriously threaten the economic and political viability of the Sahelian states of West Africa."

In its most general terms, the hydrologic cycle (Figure 9.1) maintains the water balance of Earth (Table 9.4) by taking water up into the atmosphere as evaporation from the surface of the oceans and terrestrial bodies of water and transpiration from plants and then precipitating it back as snow and rain. The water remains in the atmosphere only a matter of days (Table 9.5). In a particular region, if evaporation is equal to or exceeds precipitation, water courses will diminish and disappear into the desert sands before they reach the sea. But if precipitation over a land area exceeds evaporation (and transpiration), then the difference will be returned to the oceans as run-off. Table 9.6 summarizes the water balance of the continents, while Tables 9.7 and A.30 give the world distribution of run-off. Figure 9.2 shows the average rainfall in the United States, while Figure 9.3 shows the balance between potential evapotranspiration and rainfall precipitation. The former exceeds the latter, it should be noted, in the desert regions of the American Southwest. The fortunate area east of the Mississippi River receives 65% of the country's total rainfall.

Table 9.4 Water Balance of Oceans and Continents (From Hutchinson, 1957)

	Water Balance	
	$g/cm^2/yr$	Gg/yr
Evaporation from Ocean Surfaces	106	3.83
Precipitation on Ocean Surfaces	96	3.47
Evaporation from Land Surfaces	42	0.63
Precipitation on Land Surfaces	67	0.99

Table 9.5 Average Residence Time of Water Vapor in the Atmosphere as a Function of Latitude (From Junge, 1963, with permission of Academic Press, Inc., and the author.)

	Latitude Range (degrees)								
	0–10	10–20	20–30	30–40	40–50	50–60	60–70	70–80	80–90
Average Precipitable water (g/cm_0)	4.1	3.5	2.7	2.1	1.6	1.3	1.0	0.7_8	0.45[a]
Average Precipitation (g/cm_0 yr)	186	114	82	89	91	77	42	19	11
Residence Time (days)	8.1	11.2	12.0	8.7	6.4	6.2	8.7	(13.4)	(15.0)

[a] Values extrapolated.

Table 9.6 Water Balance of the Continents[a] (Source: Budyko, *The Heat Balance of the Earth's Surface*, Leningrad, 1956) (From Todd, 1970, with permission of Water Information Center, Inc.)

	Precipitation	Evaporation	Runoff
Africa	67	51	16
Asia	61	39	22
Australia	47	41	6
Europe	60	36	24
North America	67	40	27
South America	135	86	49

[a] Values in centimeters per year.

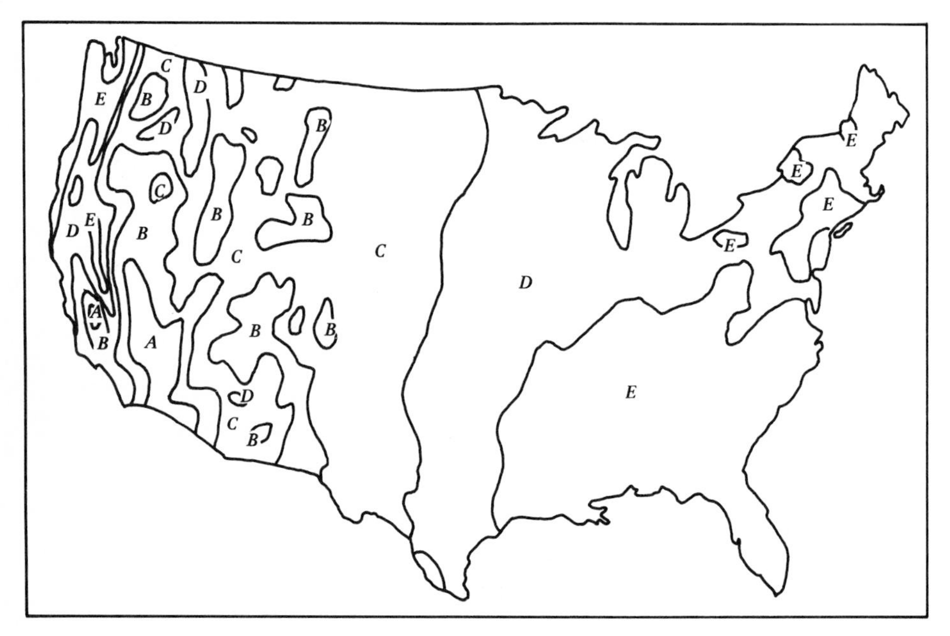

Figure 9.2 Average rainfall on the continental United States. *A*, $<0.25 \times 10^6$ gal/day-mi²; *B*, 0.25×10^6 to 0.5×10^6 gal/day-mi²; c, 0.5×10^6 to 1×10^6 gal/day-mi²; *D*, 1×10^6 to 2×10^6 gal/day-mi²; *E*, $>2 \times 10^6$ gal/day-mi².

Now let us look at the hydrologic cycle in somewhat greater detail (Figure 9.4). Three subcycles can be distinguished: (*a*) the *short cycle* of evaporation and precipitation over ocean areas; (*b*) the *long cycle* of evaporation from ocean areas and evaporation and transpiration from land areas, precipitation over land areas, and eventual partial return to the oceans as river run-off; and (*c*) the *very long cycle* consisting of the outgassing of the lithosphere and the recently discovered return of water to the magma by the slipping under of marine sedimentary deposits at the continental margins (mentioned earlier in Section 3.4). These cycles exchange water among

Table 9.7 World Distribution of Runoff (Source: L'vovich, State Hydrological Institute, Moscow, 1945) (From Todd, 1970, with permission of Water Information Center, Inc.)

	Atlantic Slope		Pacific Slope		Regions of Interior Drainage		Total Land Area	
Continent (or other area)	Area (thousands of mi^2)	Runoff (in.)	Area (thousands of mi^2)	Runoff (in.)	Area (thousands of mi^2)	Runoff (in.)	Area (thousands of mi^2)	Runoff (in.)
Europe (including Iceland)	3,073	11.7	—	—	661	4.3	3,734	10.3
Asia (including Japanese and Philippine Islands)	4,626	6.4	6,422	11.8	5,273	0.66	16,321	6.7
Africa (including Madagascar)	5,110	14.0	2,109	8.6	4,291	0.54	11,510	8.0
Australia (including Tasmania and New Zealand)	—	—	1,634	5.5	1,441	0.24	3,075	3.0
South America	6,041	18.7	519	17.5	381	2.6	6,941	17.7
North America (including West Indies and Central America)	5,657	10.8	1,914	19.1	322	0.43	7,893	12.4
Greenland and Canadian Archipelago	1,499	7.1	—	—	—	—	1,499	7.1
Malayan Archipelago	—	—	1,012	63.0	—	—	1,012	63.0
Total or average	26,006	12.4	13,610	15.5	12,369	0.82	51,985	10.5

seven major reservoirs; the atmosphere, the biosphere, ground water, primary water, the oceans, surface water on land masses, and ice cover (Figure 9.5). Perhaps we should add a fourth subcycle—*human use*. In developed countries, a very significant fraction of long cycle water now passes through human use, which leaves it very much the worse for the experience (Figure 9.6). Man also injects further filth into the hydrologic cycle by his pollution of the atmosphere (Chapter 6) and his dumping and transportation uses of the seas.

Let us turn our attention first to the short cycle. The Sun's energy, some 1.4 kW/m^2 (Figure 9.7), evaporates water from the surface of the oceans and after a period of residence in the atmosphere about 90% of this moisture (Figure 9.4) falls back into the seas as precipitation. Most of the energy input to the coastal zone, as Inman and Busch (1973) point out, comes, however, from the open sea (Figure 9.7). The oceanic evaporation precipitation balance is a major factor in determining the salinity of surface waters. An excess evaporation will result in highly saline waters; an excess of precipitation, in lower salinity. Because of this dependence on the evaporation/precipitation balance, the salinity of surface waters exhibits a diurnal cycle, the salinity being highest in the afternoon (Figure 9.8), as well as a seasonal variation (Figure 9.9). Table 9.8 shows some average values of salinity, evaporation, precipitation, and their difference, while Figure 9.10 shows the surface salinities of Earth's oceans during the northern summer. Of particular interest is the average surface salinity of all oceans plotted as a function of latitude (Figure 9.11). An equatorial minimum in salinity due to higher rainfall and diminished wind speed is bounded by two maxima formed by the high evaporation rates resulting from strong trade winds and high temperatures in the tropical regions. Toward the extreme north and south, the salinity trails off again as rainfall exceeds evaporation.

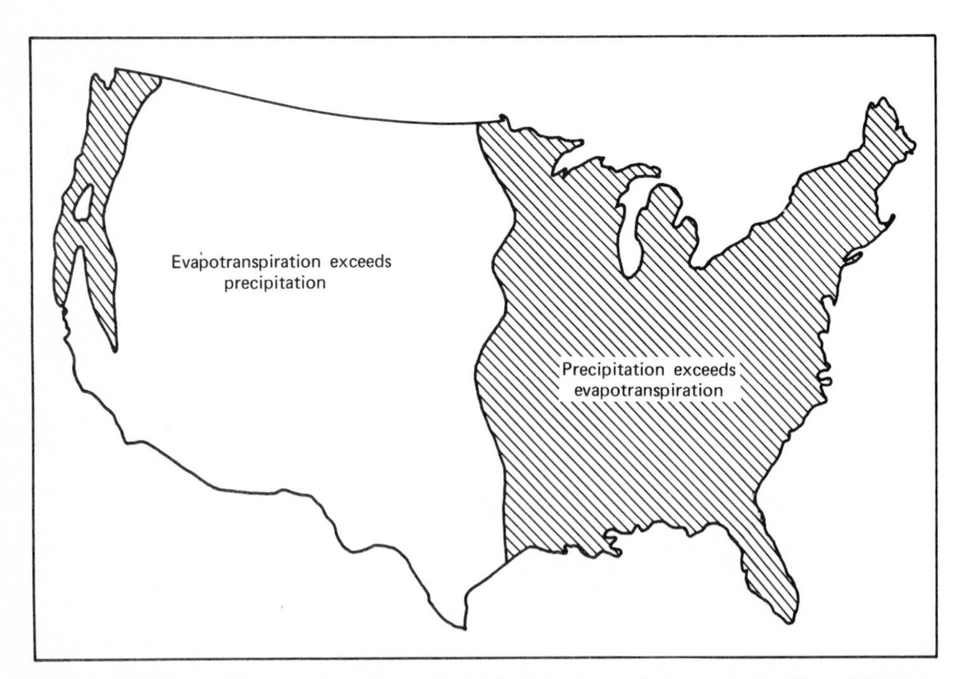

Figure 9.3 Potential evapotranspiration–rainfall precipitation balance in the United States.

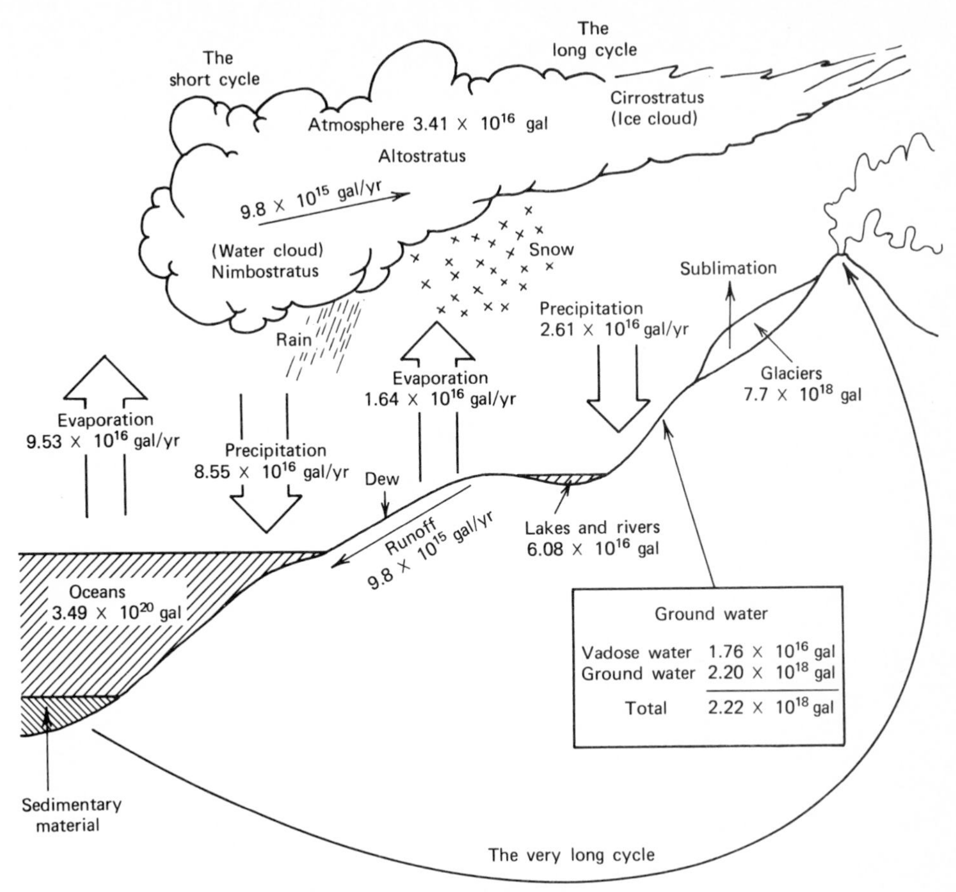

Figure 9.4 Details of the hydrologic cycle.

Table 9.8 Surface Salinity and the Evaporation-Precipitation Balance

	Atlantic Ocean				Pacific Ocean				All Oceans			
Latitude	S	E	P	$E-P$	S	E	P	$E-P$	S	E	P	$E-P$
40°N	35.8	94	76	18	33.6	94	93	1	34.5	94	93	1
30°N	36.8	121	54	67	34.8	116	65	51	35.6	120	65	55
20°N	36.5	149	40	110	34.4	130	62	68	35.4	133	65	68
10°N	35.6	132	101	31	34.3	123	127	−4	34.7	129	127	2
0°N	35.7	116	96	20	34.4	116	98	18	35.1	119	102	17
10°N	36.5	143	22	121	35.4	131	96	35	35.3	130	96	34
20°N	36.5	132	30	102	35.7	121	70	51	35.7	134	70	64
30°N	35.7	116	45	71	35.4	110	64	46	35.6	111	64	47
40°S	35.7	81	72	9	34.6	81	84	−3	34.8	81	84	−3

S = Salinity, ‰
E = Evaporation, cm/yr
P = Precipitation, cm/yr
$E - P$ = Difference, cm/yr

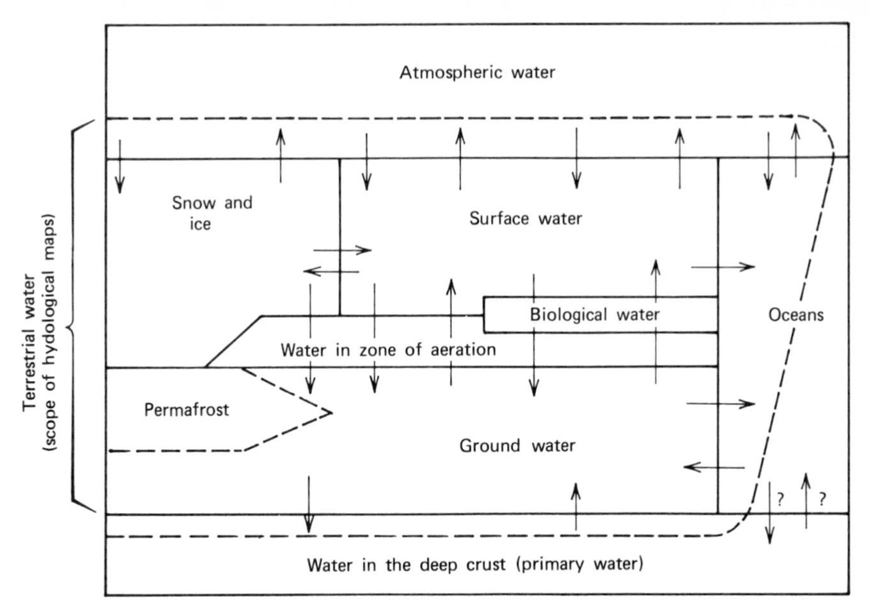

Figure 9.5 Conceptual basis for the classification of hydrological maps (reproduced from Heindl, *Nature and Resources*, *VII*, 1 (1971), by permission of UNESCO, © UNESCO, 1971).

About 10% of the moisture picked up from the oceans, about 9.8×10^{15} gal/yr, is carried by the winds over land areas. Here it is joined in its long cycle travels by some 1.6×10^{16} gal/yr of moisture from land areas (Figure 9.4); some from evaporation losses from open bodies of water such as rivers, lakes, and human impoundments; and some from transpiration from the leaves of plants. The relative effectiveness of evaporation and transpiration in returning rain water to the atmosphere is not known. The contribution of transpiration from vegetation, however, is believed to be very large. Much more water is moved by transpiration than by photosynthesis, the ratio of the former to the latter is about 2000:3 (Penman, 1970). So great are transpiration losses that Rutter (1967) has questioned the practice of planting watersheds with forests that, while they prevent erosion, enormously aggravate water loss. Atmospheric moisture is precipitated as rain, snow, hail, and dew. The excess returns to the seas as run-off by two routes—fast moving surface rivers and streams and very slow migration of subterranean ground water—to complete the long cycle. River run-off, we might note in passing, can account from 7 to 21% of the average annual fluctuations in the sea level along the Atlantic and Gulf of Mexico coasts of the United States (Meade and Emery, 1971). The excess of precipitation over evaporation and transpiration is the stream flow water—the withdrawable or useable fraction. In the United States, this withdrawable water amounts to about 30% of the total rainfall or 4.66×10^{14} gal/yr (Skinner, 1969). In the case of the Colorado River, divertment for human use is so great that only a trickle reaches the sea, illustrating the injustice and the legal problems that river use can create both nationally and internationally. Not only is Arizona fighting California for the precious water, but the United Staes is stealing nearly all of Mexico's fair share.

In emphasizing the importance of the hydrologic cycle in the chemistry of our environment, we must not forget that it also largely determines the energy balance

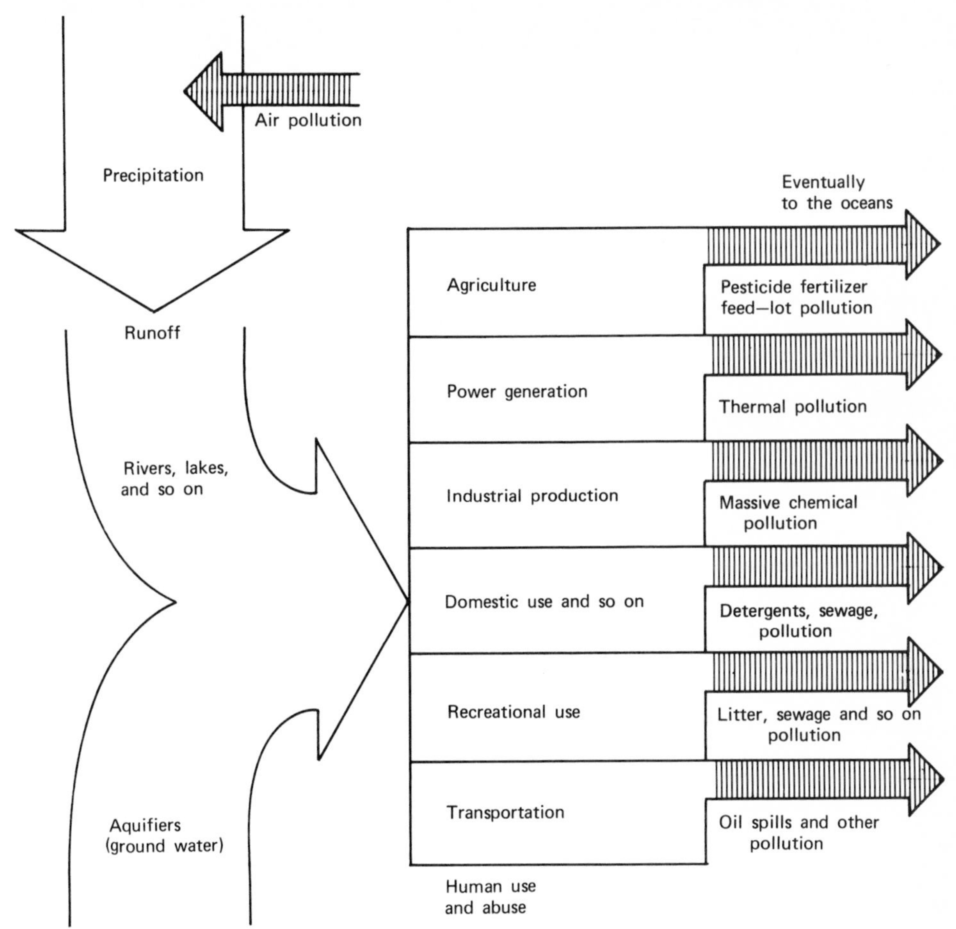

Figure 9.6 Injection of pollution into the long cycle.

of the Earth. The solar energy incident on this planet is distributed by oceanic and atmospheric circulation (Oort, 1970). The high heat capacity of liquid water and the high heat of all its phase transitions (Table 7.3) make it an excellent heat transfer fluid, ideally suited for the purpose. Coastal residents are very much award of the moderating effect of the oceans on their climate. England, remember, is a far north as Labrador. The difference is the Gulf Stream. The thermal energy transported by this ocean current every hour is the equivalent to the burning of 1.75×10^8 tons of coal. All the coal mined in the world in 1 yr could supply warmth at this rate for only 1.5 hr (Franks, 1968)!

9.3 SNOW AND ICE

Water in reservoirs such as the atmosphere, oceans, and even shallow groundwater is relatively mobile, but water trapped in the planet's ice cover is not. This water in the polar ice caps (Figure 9.12) thus forms a record of the history of the chemistry of

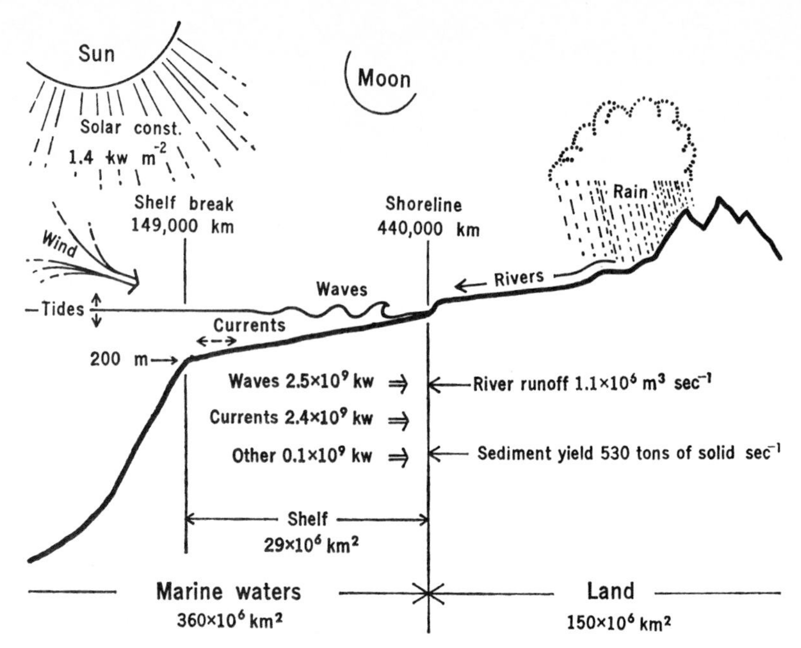

Figure 9.7 Energy budget and water run-off in the coastal zone (from Inman and Busch, © 1973 by the American Association for the Advancement of Science with permission of the AAAS).

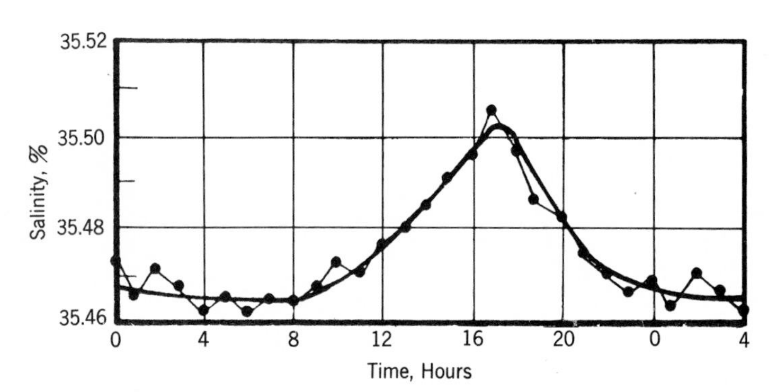

Figure 9.8 Diurnal salinity variations (from Horne, 1969).

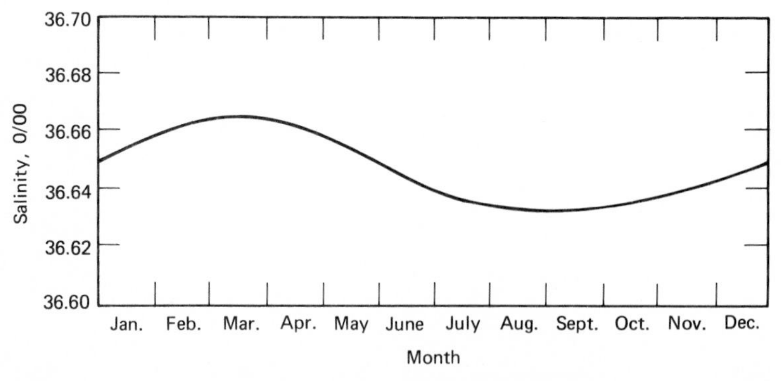

Figure 9.9 Seasonal variations in the salinity of north Atlantic surface waters.

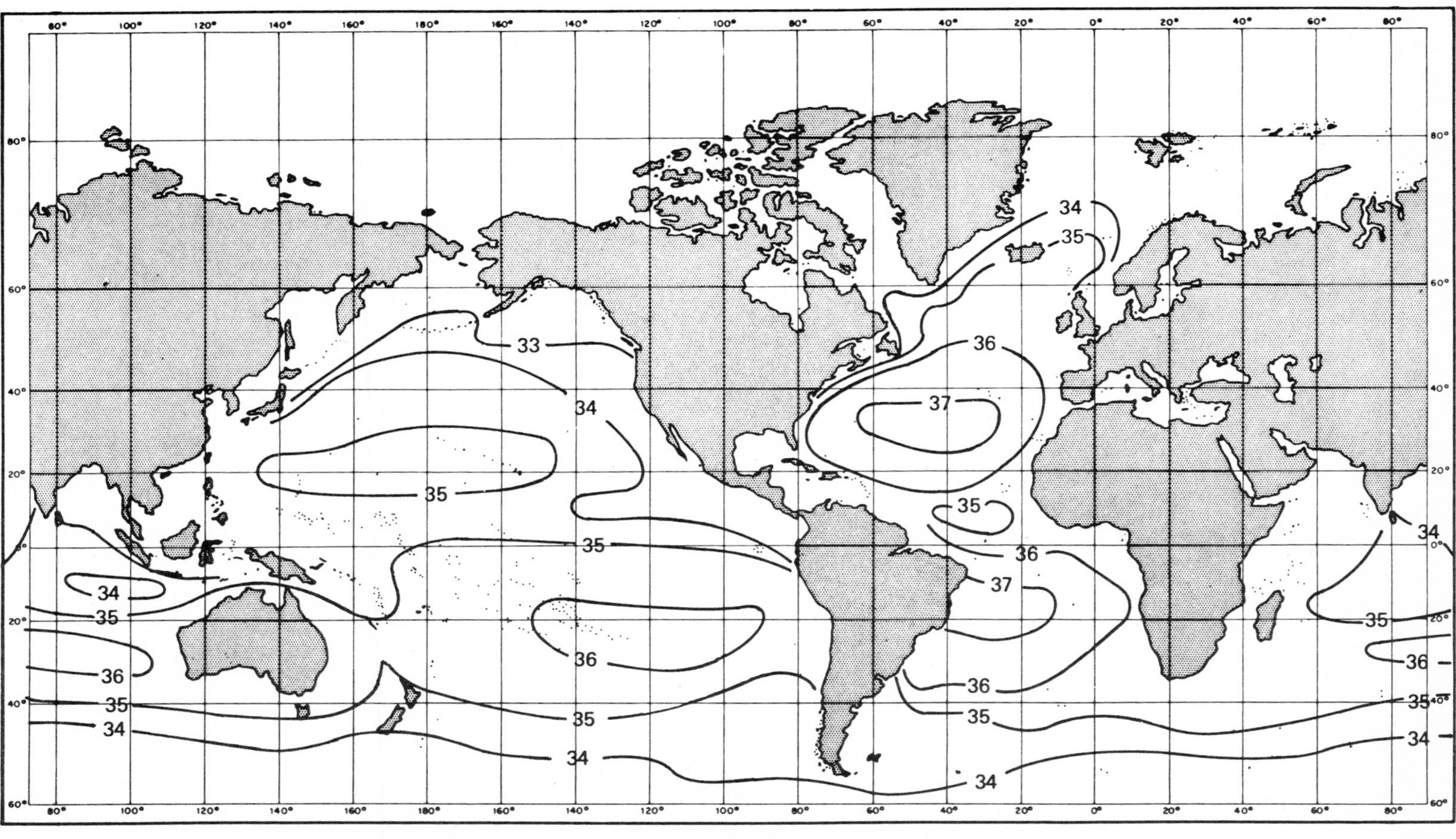

Figure 9.10 Northern summer surface salinities (from Horne, 1969).

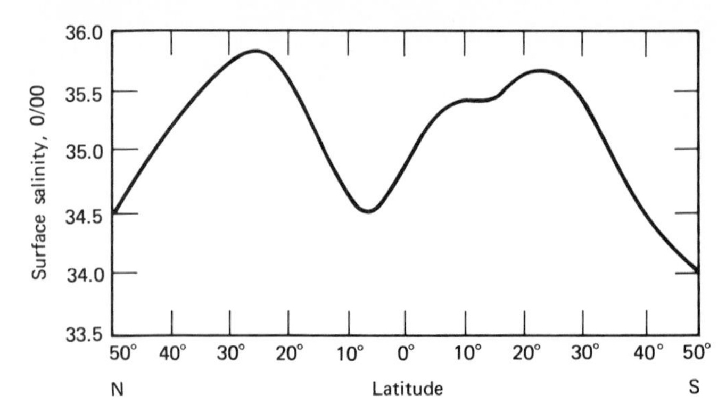

Figure 9.11 Variation with latitude of average surface water salinity values for all oceans.

the Earth's atmosphere just as fascinating as the record of the chemical history of the oceans preserved in marine sediments. Climatic changes (Raynaud and Lorius, 1973), solar activity (Bray, 1970), and, as we have seen in the case of lead, recent atmospheric pollution as well as volcanic activity can all be traced in the chemical deposits in polar ice cores. It is further possible to relate the record of snowfields with the sedimentary record (Windom, 1969). In addition to glaciers and snowfields, great quantities of water are also trapped in the ground as permafrost over extensive regions of the globe (Williams, 1970).

Rivers and lakes, of course, freeze in the winter season, but this fresh water ice has a lot of interesting physics but little interesting chemistry and need not concern us further. The ice found floating in the sea represents a more complicated story. This ice may be either sea ice, that is to say frozen out of seawater, or land ice chipped off from glaciers. The form of an iceberg can give us evidence of its origin (Figure 9.13). Winter ice must be less than a year old, polar ice can be centuries old. Freezing and melting can alter the surface salinity of ocean areas, and the salt and gas content of the ice itself, as well as its crystalline morphology, depends on its history (for a more detailed discussion see Section 13.4 in Horne, 1969). Table 9.9 gives the freezing point

Table 9.9 Freezing (Melting) Point of Seawater (From Horne, 1969)

S	*T*	*S*	*T*	*S*	*T*	*S*	*T*
1	0.055	11	0.587	21	1.129	31	1.683
2	0.108	12	0.640	22	1.184	32	1.740
3	0.161	13	0.694	23	1.239	33	1.797
4	0.214	14	0.748	24	1.294	34	1.853
5	0.267	15	0.802	25	1.349	35	1.910
6	0.320	16	0.856	26	1.405	36	1.967
7	0.373	17	0.910	27	1.460	37	2.024
8	0.427	18	0.965	28	1.516	38	2.081
9	0.480	19	1.019	29	1.572	39	2.138
10	0.534	20	1.074	30	1.627	40	2.196

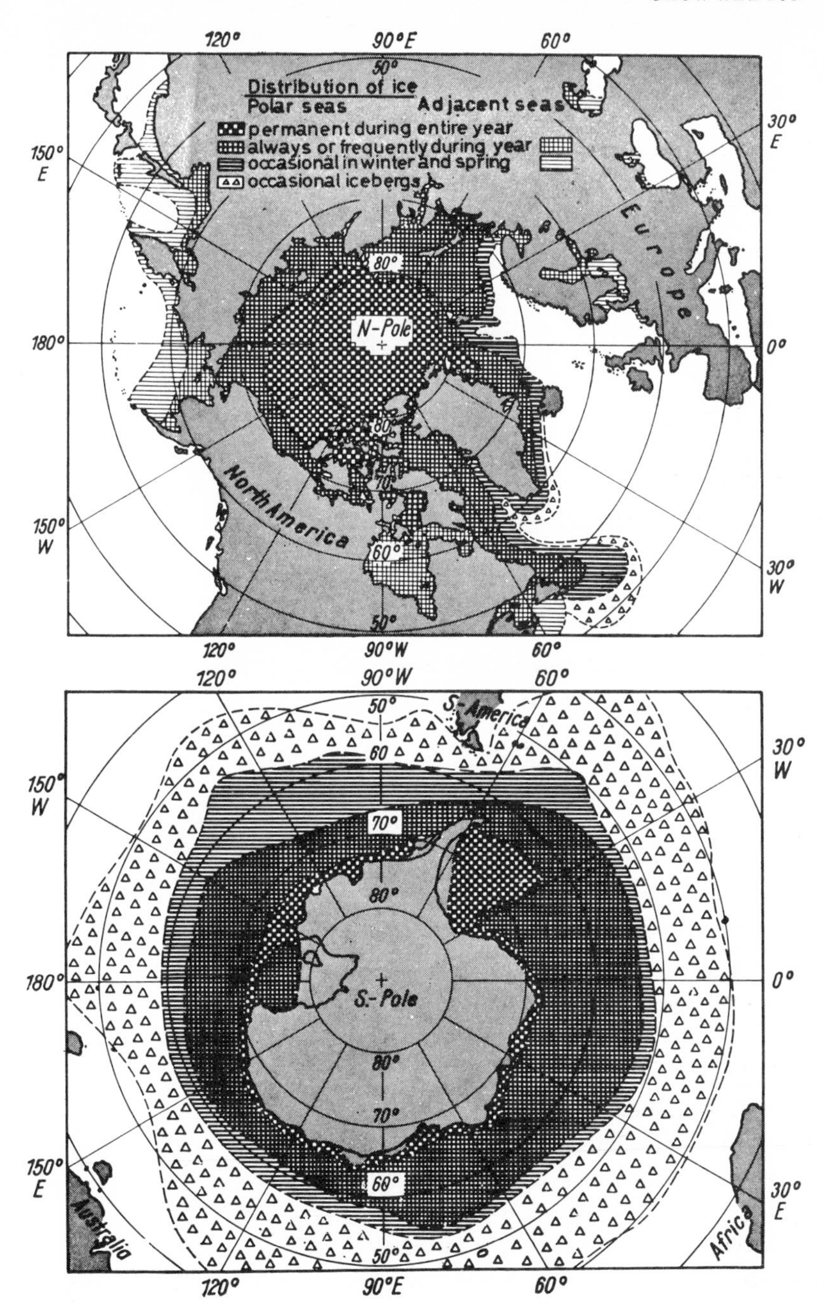

Figure 9.12 Boundaries of the ice cover of the polar regions (from Dietrich, 1963).

of seawater as a function of salinity. The growth and waning of the polar ice caps have caused major fluctuations in sea level, and glaciers have been enormously important in transporting chemical materials and shaping the face of the land, but since these matters are of more interest to geomorphology rather than to the chemistry of our environment, I will say no more about them here.

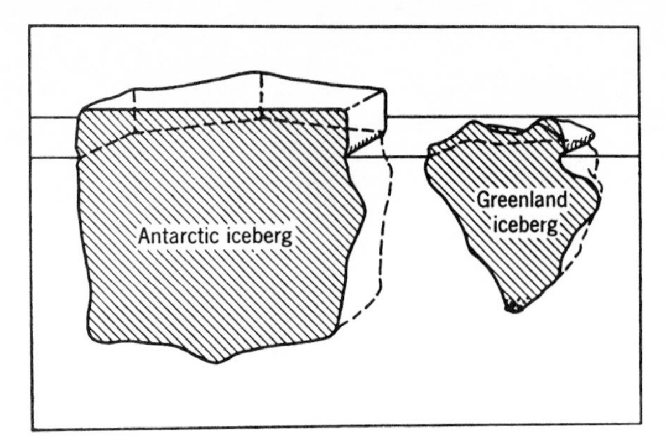

Figure 9.13 Floating icebergs (exaggerated vertically). The Antarctic, composed of barrier-ice, is tabular and about four-fifths submerged. The Arctic, composed of glacier-ice, is tooth-shaped, about six-sevenths submerged and carries moraine debris at the bottom (from Kuenen, 1963).

9.4 NUCLEATION AND PRECIPITATION

Because of the extreme importance of weather and rainfall to human affairs the mechanism of precipitation formation has been intensively studied, both experimentally and theoretically; yet for all the knowledge gained thereby, many details of the process remain unclear.

Atmospheric water vapor can condense upon cooling by two mechanisms—(*a*) homogeneous nucleation and (*b*) nucleation and condensation on foreign nucleating substances. From a theoretical standpoint, homogeneous nucleation appears to be the much better understood (Mason, 1957); however, because it requires very formidable degrees of supersaturation, and because atmospheric impurities are so ubiquitous and provide so much easier a path for nucleation, its role in the real world is evidently a relatively minor one. Here we will concern ourselves solely with ice and droplet formation on nucleating impurities in the atmosphere. But before we turn to this subject I want to inject a doubt concerning our understanding of both types of nucleation. In Chapter 13, it will be proposed that water near an interface, any kind of an interface, is different in its structure and properties from ordinary bulk liquid water. It is further proposed that this layer of altered or "vicinal" water near an interface may be quite thick. If such is the case, remembering that cloud droplets can be quite small, all or most of the water will be close to a liquid–gas interface, that is to say an appreciable fraction of the condensed water in the atmosphere may be vicinal rather than "normal" water. Then too it must be remembered that the nuclei themselves, whether soluble or insoluble, hydrophobic or hydrophilic, will alter water structure (Chapter 8). Although noted by a few atmospheric scientists (see Pruppacher, 1963; Evans, 1970), the full implications of these possibilities are still inadequately appreciated. Even in the case of pure, nuclei-free water, the median freezing temperature of droplets depends on the gas present (−36.4°C for argon and −37.6° for CO_2, Kuhns and Mason, 1968). Pena and De Pena (1970) have theorized that gas clathrate formation inhibits the freezing of supercooled water droplets.

There are a great many different cloud nuclei in the atmosphere (remember Chapter 6). Natural sources include mineral and salt dust swept up from the Earth by wind; smoke materials from forest fires, volcanic smoke, and dust; substances formed by chemical reactions in the atmosphere; meteoric dust; and salt-containing spray from the ocean (Roberts and Hallett, 1968). Tables 6.23 and 6.24 listed particulate nuclei from man-made sources, and even these lists are far from complete. Lead particles from auto exhausts, for example, especially if they react in the atmosphere to form lead iodide, are a potential source of ice nuclei (Schaefer, 1968). Man-made pollution can be the major source of nuclei; in one study in the vicinity of Great Salt Lake, Utah, Reynolds (1970) found that copper smelter was a primary source of ice nuclei in the immediate area, with traffic, refineries, and steel mills also contributing. Iron is an even more common atmospheric pollutant than lead; however, the atmospheric ferromagnetic spherules, possibly of meteoric origin do *not* appear to be effective ice nuclei (Prodi and Wirth, 1972). As an air mass cools, condensation first occurs on the largest hygroscopic nuclei; thus the giant and large particles (Figure 5.12) are the major nuclei with Aitken particles playing a variable and lesser role. The most effective large hygroscopic nuclei are believed to originate from marine salt spray and combustion processes. However because of their relative rarity, there has been a long controversy as to whether or not they are responsible for the largest contribution to nucleation. The relative importance of sea salt's role varies with location, being more significant near and over the oceans than over continental land masses (Woodcock, 1952, 1953; Woodcock and Blanchard, 1955). Even fine salt spray from bubble bursting over vast tracts of the world's oceans (see Chapter 15) is still insufficient. After reviewing the literature Mason (1957) concludes, "The overall evidence suggests that, of the nuclei involved in cloud-droplet formation, perhaps one tenth consist of salt spray, and the rest of mixed nuclei and the products of natural and man-made fires." Cloud and fog droplets have been examined by a variety of microscopic techniques, some of them quite ingenious. These studies have revealed particles and crystals of NaCl, $MgCl_2$, $MgSO_4$, $CaSO_4$, unidentified solids often coated with deliquescent salts, amorphous residues, agglomerates of combustion products, diatoms, carbon granules, tarry particles, hexagonal plates that are possibly silicates, and ill-defined microcrystals of other materials (Table 9.10). It is easy to

Table 9.10 Nature of Nuclei Found in Cloud and Fog Droplets (Kuroiwa, 1953; Yamamoto and Ohtake, 1953) (From Mason, 1957, with permission of the Clarendon Press, Oxford.)

	Fog and Haze Nuclei				Cloud Nuclei			
	Number		Percent		Number		Percent	
Nature of Nuclei	K	Y & O	K	Y & O	K	Y & O	K	Y & O
Sea Salt	16	14	28	23	5	11	13	16
Combustion	28	18	49	30	20	25	51	36
Soil Particles	8	12	14	20	11	16	28	23
Unknown or No Observable Nucleus	5	16	9	27	3	17	8	25
Totals	57	60	100	100	39	69	100	100

Table 9.11 Chemical Composition of Large and Giant Nuclei (From Mason, 1957, with permission of the Clarendon Press, Oxford.)

	Large Nuclei	Giant Nuclei ($\mu g/m^3$)
SO_4^{2-}	4.38	2.30
NH_4^+	1.57	0.31
Cl^-	0.16	0.56

overemphasize the importance of salts since they are so immdeiately obvious from their crystalline forms.

To return to the giant and large particles that are such highly effective nuclei even if present in relatively small numbers, Junge (1963) has found that these two particle types differ in chemical composition (Figure 6.12 and Table 9.11). The giant nuclei appear to have their origin in sea salt spray (note the high Cl^-), whereas the large nuclei, rich in NH_4^+ and SO_4^{2-}, possibly are formed from the coagulation of molecular complexes of ammonium sulfate produced in the gaseous phase from combustion products. We throw some more light on these matters in Section 9.5.

Elsewhere we have mentioned weather modification and deliberate cloud "seeding" by man. This consists simply of introducing artificial nuclei into a supersaturated atmosphere. AgI crystals and solid solutions of Ag—I—Br—Pb are very effective seeding materials (Passarelli et al., 1973). The presence of sulfate appears to enhance ice nucleation by AgI but chloride interferes (Murty and Murty, 1972*a*). Carbon dioxide ("dry ice") and liquid propane also have been used (Vardiman et al., 1971), and more recently Murty and Murty (1972*b*) have demonstrated that even an aqueous suspension of Portland cement is an effective ice nucleation agent. The idea is that the "seed" or nucleus forms a substrate of suitable geometry upon which ice is encouraged to grow.

Over and over again we encounter examples of the critical role played by the unique properties of water, how, even, the eccentricities of this substance are a necessary condition for life. The case at point, water supercools very easily and may be reluctant to freeze even down to temperatures as low as about $-40°C$ (see Mason, 1958). Were it not for this very anomalous (and mysterious) behavior, the Earth's evaporation/precipitation balance and distribution of precipitation would be entirely different from what it is. The residence time of moisture in the atmosphere would be much shorter and the long cycle (Figure 9.4) would be much diminished. Since water would tend to come down closer to its point of evaporation, much of the Earth's land masses would be far more arid than they are today, so arid in fact that the social evolution of man may never have pulled ahead of that of his fellow primates, if indeed under such drastically altered conditions biological evolution had proceeded as far as the primates.

Deserts are interesting environments rather neglected in this book (see McGinnies et al., 1969; McGinnies and Goldman, 1969). Their study as a convincing exercise for the importance of water. As human population on this planet expands unchecked, the lush areas of the world will soon be even more exhausted, and man will be forced to come to more serious grips with arid regions.

Once the nucleus has done its work, a water droplet or ice crystal in the atmosphere will continue to grow by a number of mechanisms—diffusion, collision, coalescence, evaporation and recondensation, and so on. Larger droplets are favored over smaller ones, and droplets with more concentrated solutes will grow at the expense of those with less. When the droplet or ice particle reaches a certain critical size, it will begin to fall. In the process of the movements of water through the atmosphere it may undergo many vapor/liquid/solid phase changes. In the cases of snow and hail, the details of the physical morphology of the particles reaching the ground can give us hints as to some of these transformations.

9.5 THE CHEMICAL COMPOSITION OF RAINWATER

Many scientists find it easier to collect data than to think hard about their findings. Because of the tremendous power and accuracy of the techniques it offers, analytical chemistry seems to have a peculiar fascination for some would-be environmental scientists. I certainly in no sense wish to detract from the enormous importance of analytical chemistry, but I do wish that environmental scientists would show more concern for what is going on in our world and why than in an extra significant figure or a lower "experimental error." A needlessly great amount of time and money have been lavished on determining the concentration of trace metals in seawater (Chapter 8), while leaving the elucidation of substantial chemical processes in the oceans neglected. I have the feeling that the same may be true of the atmospheric sciences, namely that there is more data about the composition of the atmosphere and precipitation than necessary and that this largess contrasts sharply with the paucity of essential, hard-to-obtain information on chemical mechanisms and rates in the atmosphere. Human existence is a rate process; the growth and decline of civilizations are rate processes; our chemical environment is a rate process. The rate of our consumption of metals, water, fossil fuels, and other materals is far more important than their amounts. While the amounts are given, we do have some potential to exercise controls over the rates of their uses. I can say without fear of exaggeration that the *rates* of chemical processes in our environment are far more important than chemical quantities in our environment, not only from the point of view of the continuing dominance of our species, but also from the point of view of the scientific understanding of *what is going on* in our world.

Gases and aerosols, both liquid and particulate, are removed from the atmosphere by rainout and washout. The former refers to removal within the clouds; the latter, to removal by precipitation below the clouds. Such wet removal by precipitation is not the only removal process. Dry removal of aerosols from the troposphere by sedimentation and by impact on obstacles at the Earth's surface (pine trees are particularly effective) also occurs, while other processes for gas removal include absorption and reaction at the Earth's surface, conversion to aerosols or to other gases by chemical reactions in the atmosphere, and escape into the exosphere. Junge (1963) notes that "very little information is available with which to assess the role of dry deposition in comparison with wet removal."

The concentration of chemicals in rain (see Eriksson, 1952, for a review), of course, depends on a great many factors, including human pollution. Turner (1955), for example, found higher chloride concentrations in small rain droplets than in larger

Table 9.12 Average Decrease of Aerosol and Gas Concentrations in Ground Air Due to Rain for Various Constituents in Frankfurt/Main, June 1956–May 1957 (From Junge, 1963, with permission of Academic Press, Inc., and the author.)

	Aerosols			Gases			
	NH_4^+	NO_3^-	SO_4^{2-}	NH_3	NO_2	SO_2	Cl
Air Concentration ($\mu g/m^3$)							
Before Rains	6.7	6.0	16.7	21.6	11.9	328	14.3
After Rains	4.7	1.6	9.7	11.0	9.1	212	5.3
% Decrease	70	27	58	52	76	64	39

ones, an observation he attributed to the more rapid evaporation of the smaller droplets. Similarly, the pH of rain droplets collected near Boston varied from 3 to 5.5 with, again, the smaller droplets containing the greater concentrations of hydrogen ions (for more recent studies of the pH of rainwater see Granat, 1972, and Førland, 1973). Junge (1963) has estimated that 0.04 represents the fraction of aerosols removed per day from the troposphere during the summer over the eastern United States. Table 9.12 shows the decrease in the concentrations of atmospheric aerosols and gases due to rainfall, while Figure 9.14 shows how the chemical content of rain diminishes as precipitation cleanses the air. The chemical composition of droplets in short rainfalls tend to be highly variable because of the effects of washout and evaporation; however during prolonged precipitation, both these effects diminish (the latter because high humidity reduces evaporation) and rainout tends to become the controlling factor. Also the form of precipitation is important: Neumann et al. (1959) found that the organic content of rain over Sweden was nearly twice the value for snow—2.5 mg/liter average compared to 1.3 mg/liter.

Precipitation accounts for 80 to 90% of the deposition of radioactive fallout (Chapter 6) from the atmosphere. In fact, fallout appears to be rather directly proportional to precipitation (Collins and Hallden, 1958; Martell, 1959; Small, 1960). Furthermore, the concentration of radioisotopes in rain water is roughly proportional to their concentrations in the atmosphere: The ratio of the former to the latter is 0.25×10^{-6} to 2.1×10^{-6} (Small, 1960).

The major constituents of rainwater tend to be SO_4^{2-}, Cl^-, Na^+, Ca^{2+}, and K^+ with SO_4^{2-} predominating (there is almost 10 times more SO_4^{2-} in Greenland ice than these other constituents, Junge, 1960). Figures 9.15 to 9.21 show the concentrations of these constituents in rain over the United States, as well as NH_4^+ and NO_3^-. Commonly these constituents are discussed in terms of their ratios and "excesses," but a simple examination of the figures is perhaps just a revealing and corroborates more or less what common sense would lead us to suspect, namely Na^+ and Cl^- are high in coastal areas (Figures 9.16 and 9.17), whereas Ca^{2+} and K^+ present entirely different pictures (Figures 9.18 and 9.19) tending to have highs deep in the conti-

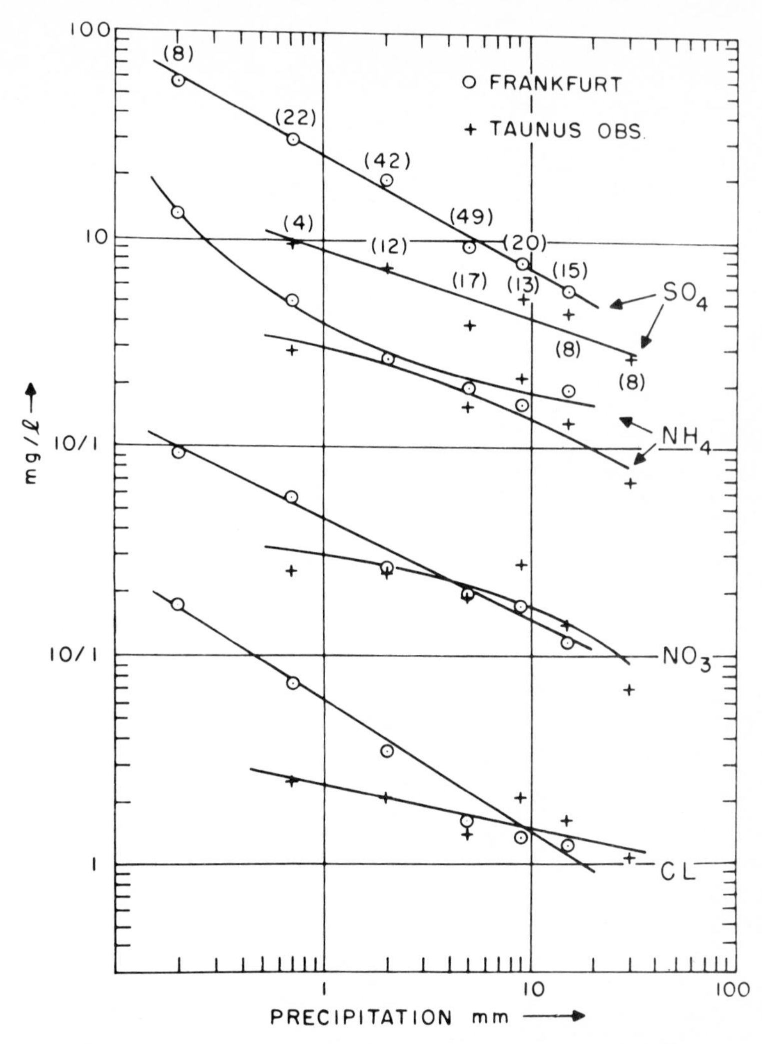

Figure 9.14 Rainwater concentrations of various constituents as a function of rainfall. ○, Polluted atmosphere in Frankfurt, Germany; +, Less polluted atmosphere at Taunus Observatory, Germany. (From Junge, 1963, with permission of Academic Press, Inc., and the author.)

nental land mass—again what one might epxect of these common constituents of crustal mineral dust. However, lest we be too simple minded, note that the Cl^-/Na^+ ratio, even near the ocean is *not* the same as for seawater (1.80; Table 9.13; see also Chapter 15). Sulfate (Figure 9.16) is a curious actor. It appears in high concentrations, sometimes far from coastal areas, and it is not that common an ingredient of soil dust. Junge (1963) notes that "This excess SO_4^{2-} concentration is unpolluted areas seems to be a world-wide phenomenon and apparently keeps within a rather narrow limit. It is likely that it represents that part of the SO_2 which is oxidized photochemically or in rainwater . . ." and that SO_2, as we have seen (Chapter 6), may come in part from the oxidation of biogenic H_2S. Man-made pollution also

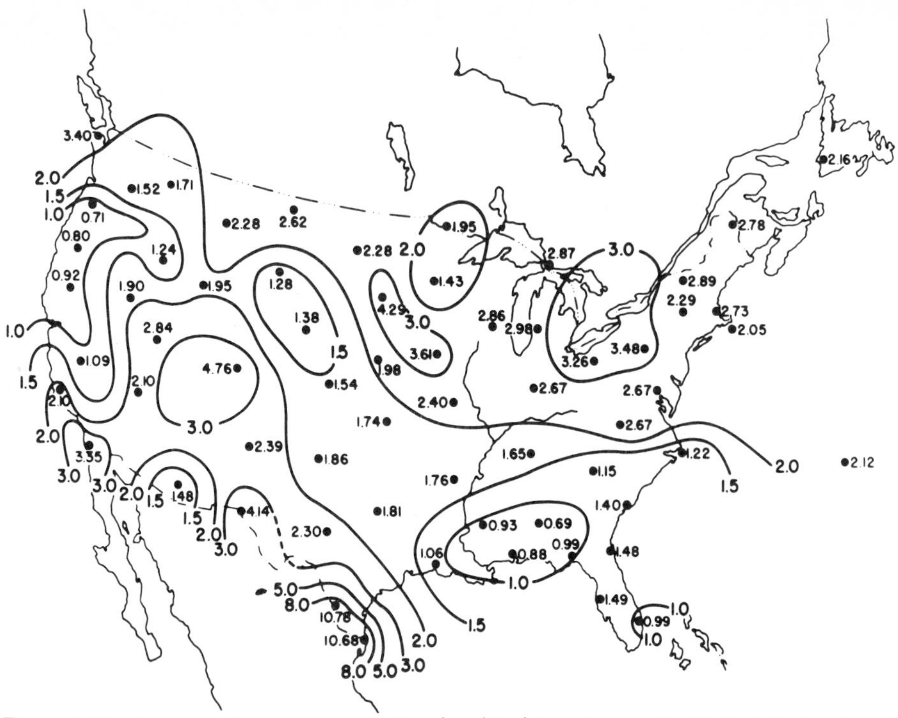

Figure 9.15 Average sulphate concentration (mg/liter) in rain over the United States, July 1955 through June 1956 (from Junge, 1963, with permission of Academic Press, Inc., and the author).

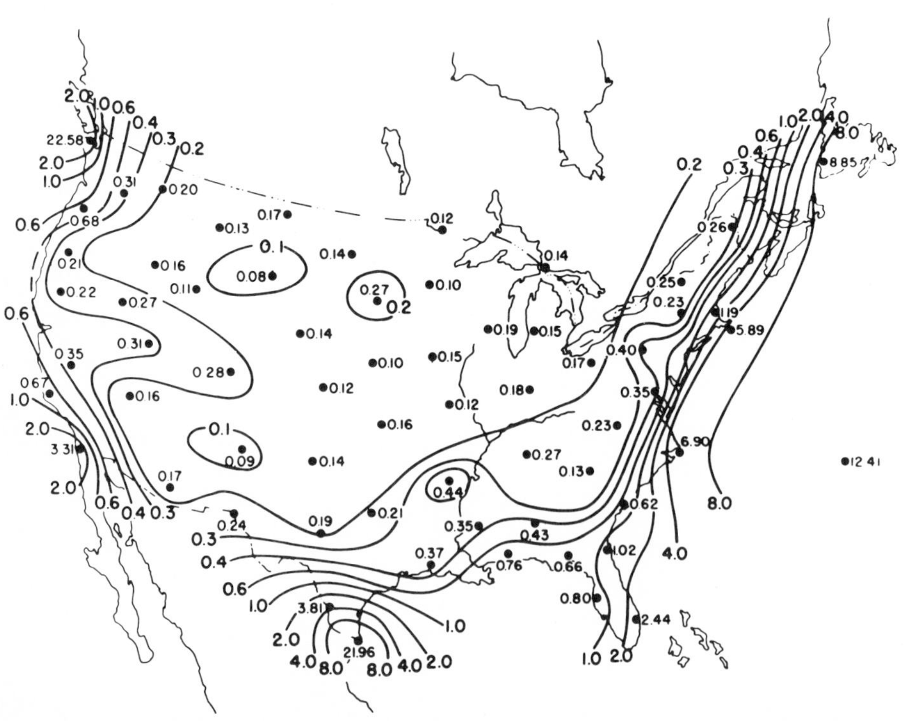

Figure 9.16 Average chloride concentration (mg/liter) in rain over the United States, July 1955 through June 1956 (from Junge, 1963, with permission of Academic Press, Inc., and the author).

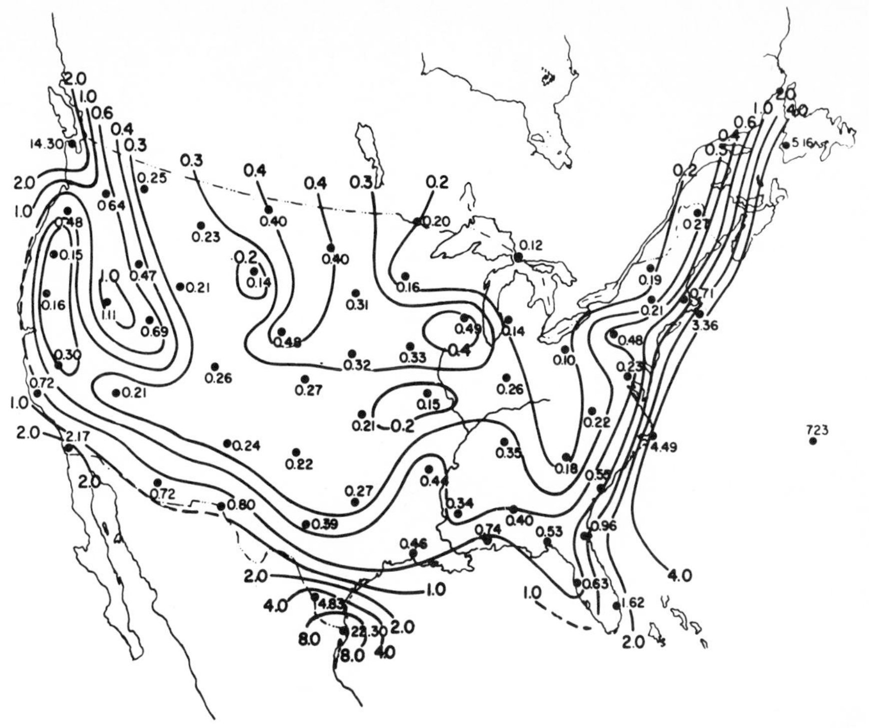

Figure 9.17 Average Na^+ Concentration (mg/liter) in rain over the United States, July 1955 through June 1956 (from Junge, 1963, with permission of Academic Press, Inc., and the author).

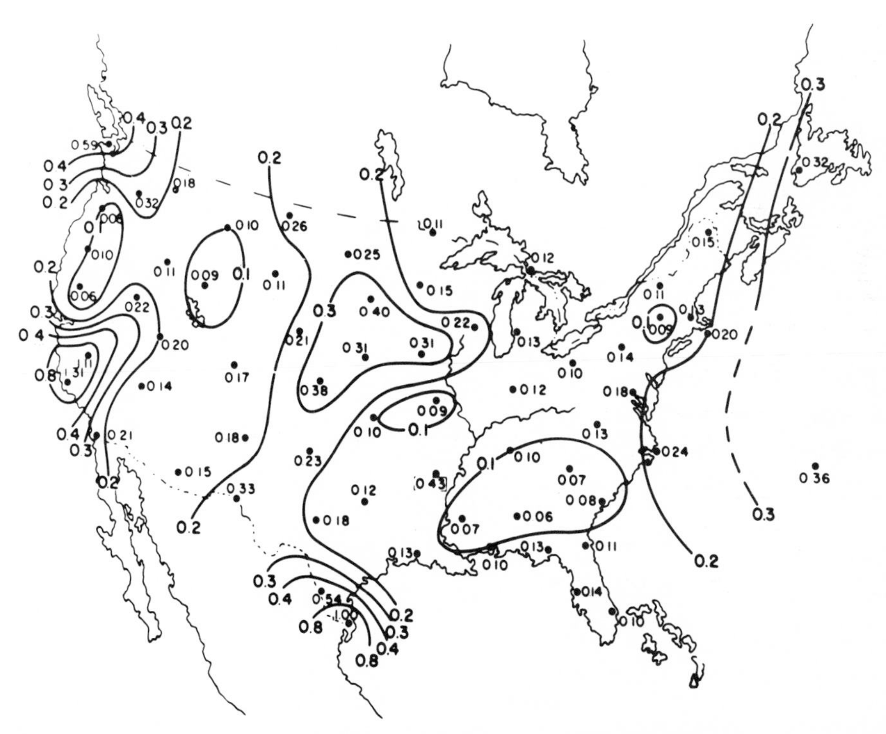

Figure 9.18 Average K^+ Concentration (mg/liter) in rain over the United States, July 1955 through June 1956 (from Junge, 1963, with permission of Academic Press, Inc., and the author).

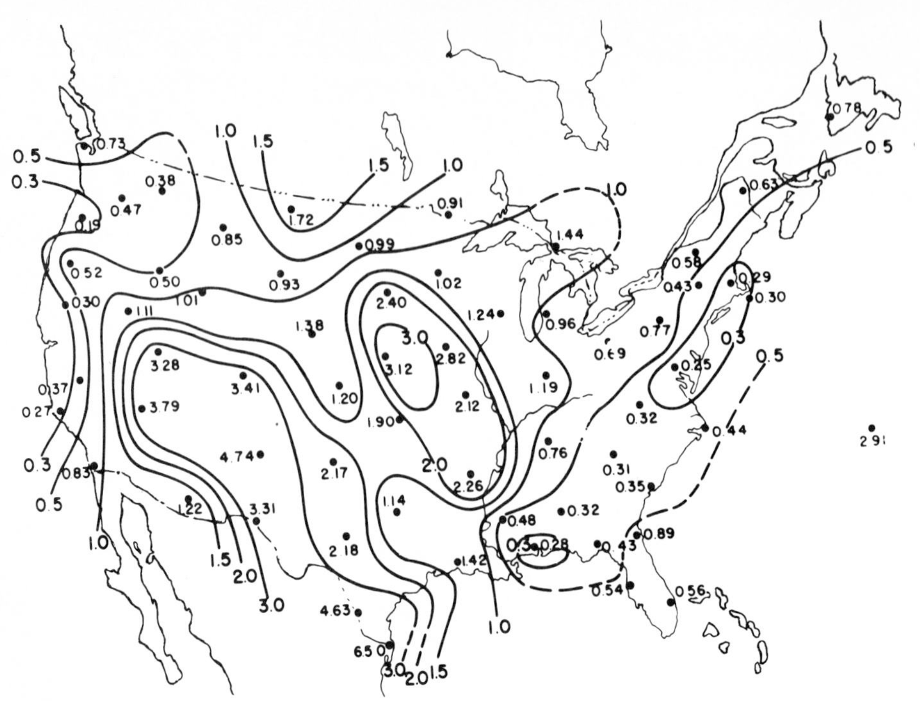

Figure 9.19 Average Ca^{2+} concentration (mg/liter) in rain over the United States, July 1955 through June 1956 (from Junge, 1963, with permission of Academic Press, Inc., and the author).

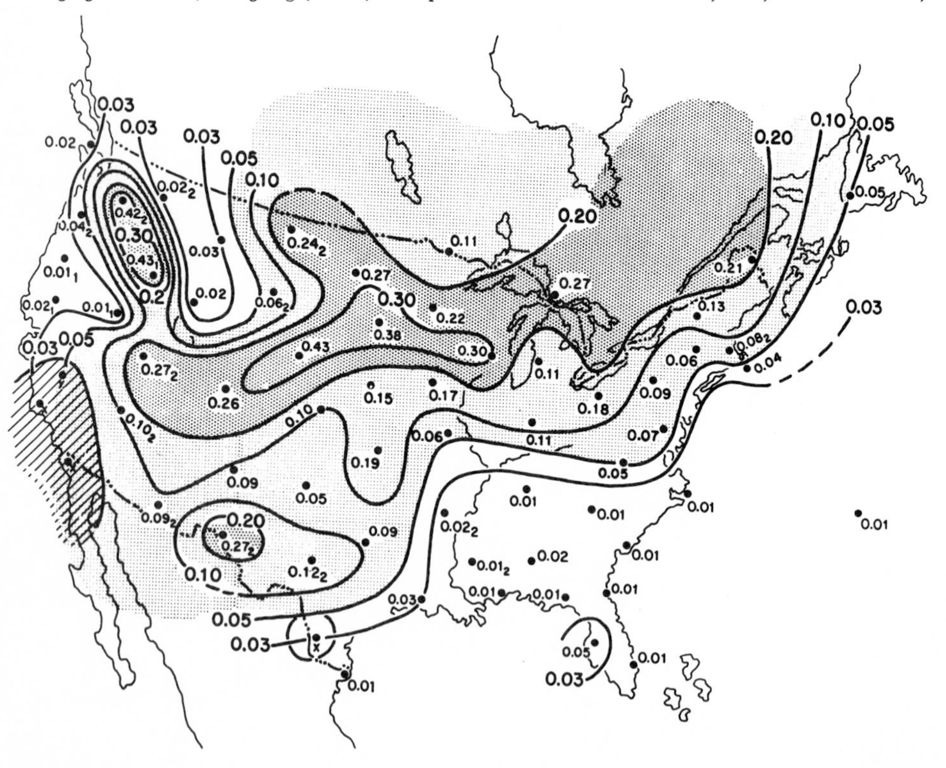

Figure 9.20 Average NH_4^+ concentration (mg/liter) in rain over the United States, July through September 1955 (from Junge, 1963, with permission of Academic Press, Inc., and the author).

Table 9.13 Observed Cl^-/Na^+ Ratio in Rain- and Fresh Water on Islands and in Coastal Areas[a] (From Junge, 1963, with permission of Academic Press, Inc., and the author.)

Location	Ratio
Netherlands—rainwater; average values for several years; variation with distance 0 to 86 km from the coast.	1.90–2.30
Lake District, west coast of England—water from 18 tarns; variation with distance 0 to 50 km from the coast.	1.64–1.96
Lerwick, Shetland Islands—rainwater; variation of monthly averages.	1.17–1.69
Average of 1958 rainwater:	
Lista, southwest coast of Norway	1.71
Rjupnahed, west coast of Iceland	1.78
Lerwick, Shetland Islands	1.62
Den Helder, coast of Netherlands	1.63
Camborne, southwest England	1.79
Bermuda	1.72
Tatoosh Island, Washington	1.58
Falkland Islands—bog waters	1.98
Windward side of Island of Hawaii (near Hilo). Rainwater average for samples from 2 days 0 to 15 km from the coast.	1.90

[a] The ratio in seawater is 1.80.

plays an important part in the SO_2 story, but again, as we have seen, anthropogenic and biogenic SO_2 are readily distinguishable from their seasonal fluctuations—in the temperate and northern latitudes the former is high in the winter, the latter high in the summer. As for NH_4^+ biological processes in the soil appear to be responsible, while for NO_3^- responsibility rests with a combination of such processes and anthropogenic pollution. Still speaking of minor constituents, methane appears to be at equilibrium concentrations in rainwater, but carbon monoxide can be as much as 200-fold supersaturated, suggesting formation by chemical reactions, not in the normal clear atmosphere, but in clouds (Swinnerton et al., 1971). Galbally (1972) has attributed the CO formation to the photodecomposition of aldehydes in rainwater.

Gibbs (1970) divides world surface waters into three types depending on the source of their alkali and alkali metal cations: (*a*) atmospheric precipitation, (*b*) rock dominance, and (*c*) the evaporation–crystallization process. In well-leached, high rainfall tropical regions the Na, K, Mg, and Ca content of rivers comes largely from rainfall rather than from rocks (Table 9.14).

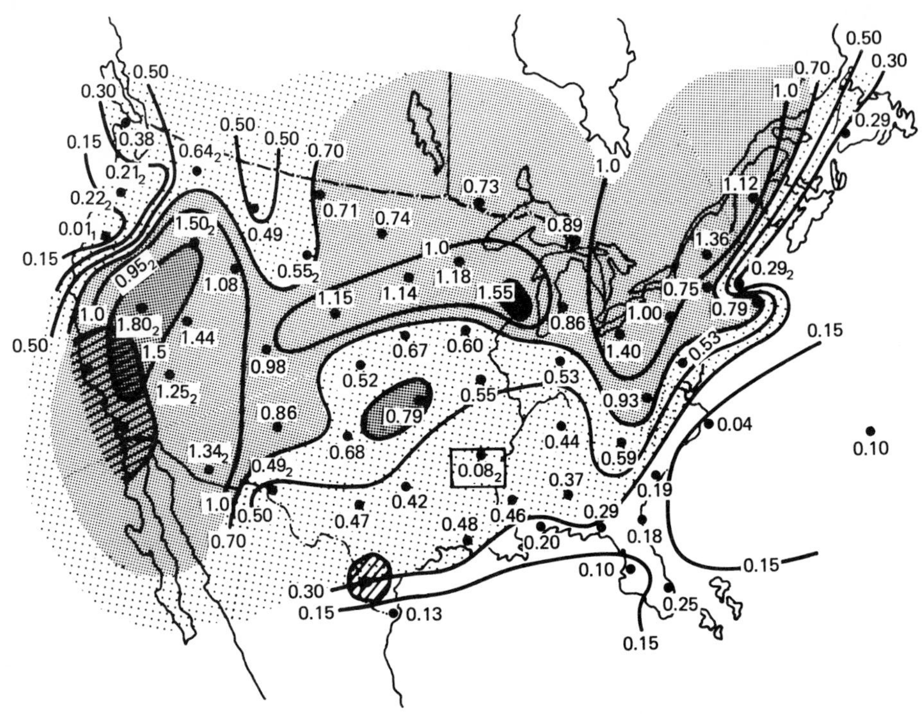

Figure 9.21 Average NO_3^- concentration (mg/liter) in rain over the United States, July through September 1966 (from Junge, 1963, with permission of Academic Press, Inc., and the author).

Table 9.14 Source of Na, K, Mg, and Ca in Various Types of Rivers (From Gibbs, © 1970 by the American Association for the Advancement of Science with permission of the AAAS.)

River	Contribution from Precipitation (%)	Contribution from Rocks (%)
	Rain-Dominated River Type	
Rio Tefe	81	19
	Rock-Dominated River Type	
Ucayali	4.8	95.2
	Evaporation–Crystallization River Type	
Rio Grande[a]	0.1	99.9

[a] Feth (1971) has raised questions concerning this example.

9.6 PHASE CHANGES AND ISOTOPIC FRACTIONATION

In the course of geological time, a given molecule of water is cycled through our environment an enormous number of times. In the course of these endlessly repeated adventures, it is subject to countless phase changes and chemical reactions. Figure 9.22 is a redrawing of the hydrologic cycle (Figures 9.1 and 9.4) intended to emphasize some of the more important of these transformations. Just in the atmosphere along during a single residence, water can be subjected to multiple phase changes between the gas, liquid, and solid states. Then too we must not forget that in addition to being exposed to lithospheric material, water is constantly and repeatedly being passed through the biosphere.

While certainly the most spectacular and important example, water is not, of course, the only chemical being subjected to physical and chemical transformations in our environment. The rates of physical–chemical processes and the stability of chemical bonds depend on the momentum, that is to say, the masses of the atoms involved. Thus the isotopes of a given element will behave ever so slightly differently in our environment, but since the number of transformations is so very great and these effects are cumulative, their consequences may be observable. Such isotope effects (as distinct from the isotopic ratios resulting from radioactive decay—see Chapter 6) should be particularly observable for the lighter elements where the mass differences are relatively greater (Table 9.15). One example of natural isotopic fractionation, possibly arising from ion exchange or diffusion processes has even been found for an element as heavy as calcium (Heumann and Lieser, 1973). Similarly the D isotope fractionation in oil field brines also arises from exchange and diffusional

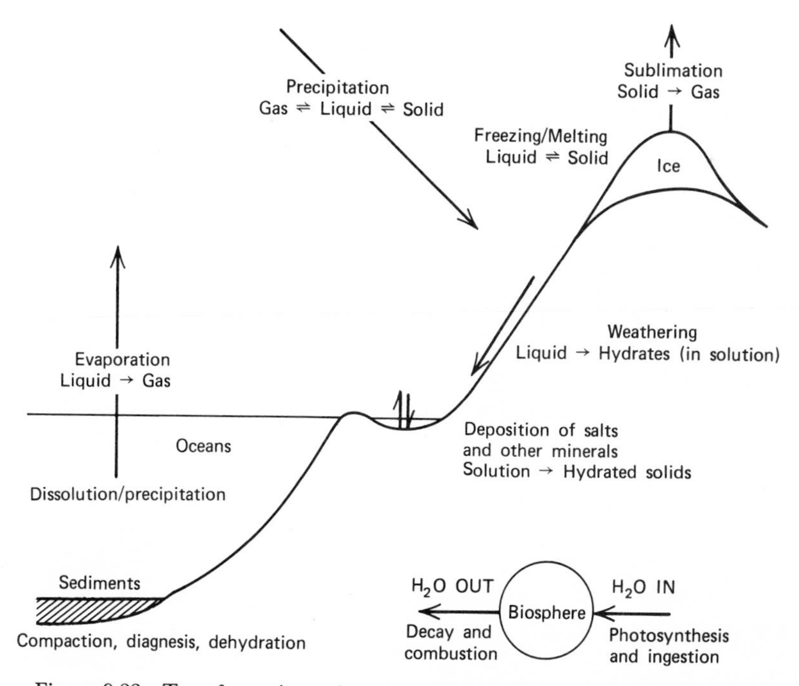

Figure 9.22 Transformations of water in the course of the hydrologic cycle.

Table 9.15 Isotopic Composition of Some Lighter Elements

Element	Atomic Weight	Weight Percent Average Natural Abundance
${}^{1}_{1}H$	1.007825	99.985
${}^{2}_{1}H$	2.01410	0.015
${}^{12}_{6}C$	12.0000	98.89
${}^{13}_{6}C$	13.00335	1.11
${}^{14}_{7}N$	14.00307	99.63
${}^{15}_{7}N$	15.00011	0.37
${}^{16}_{8}O$	15.99491	99.759
${}^{17}_{8}O$	16.99914	0.037
${}^{18}_{8}O$	17.99916	0.204
${}^{28}_{14}Si$	27.97693	92.21
${}^{29}_{14}Si$	28.97649	4.70
${}^{30}_{14}Si$	29.97376	3.09
${}^{32}_{16}S$	31.97207	95.0
${}^{33}_{16}S$	32.97146	0.76
${}^{34}_{16}S$	33.96786	4.22
${}^{36}_{16}S$	35.96709	0.014

processes in the minerals but the ^{18}O fractionation, while increasing with depth, has been attributed to carbonate equilibria (Kharaka et al., 1973). Isotopic ^{18}O and D fractionation upon the transport of aqueous salt solutions under high pressure through montmorillonite has been demonstrated in the laboratory (Coplen and Hanshaw, 1973). We have already seen in Section 6.2 how the relative amounts of ^{14}C and ^{12}C in substances enables us not only to tell whether their carbon is "old" or "new" but even to determine the source of atmospheric pollution. Radiodecay and physical–chemical isotope effects provide the bases of the most powerful means at the disposal of science for unraveling the past history of the chemistry of our environment.

Of the eighteen different types of water molecules in ordinary water, the proportions of $H_2{}^{16}O/HDO/H_2{}^{18}O$ are roughly 1,000,000/320/2000 (Craig and Gordon, 1965). The structure of D_2O is believed to be very similar to that of H_2O, although somewhat more extensive at room temperature (Kirshenbaum, 1951; Bhandari and Sisodia, 1964; Kavanau, 1964; Horne and Johnson, 1967) for the greater mass of deuterium imparts somewhat greater strength to the D-bond than to the H-bond. In natural waters, it should be noted, the ratio of H to D is so great (Table 7.22) that there is essentially no D_2O and each HDO is always surrounded by H_2O neighbors. Because the D-bone is stronger it should be more difficult for HDO to escape from the liquid to the vapor phase, and indeed the boiling point of HDO is higher and its vapor pressure lower (Table 7.23). During evaporation, HDO will tend to be left behind. Such isotopic fractionation on evaporation results in higher D/H ratios in equatorial waters and in surface waters of the ocean compared with deep waters, while during precipitation the reverse occurs and the heavier isotope (D or ^{18}O) is preferentially precipitated (Kirshenbaum, 1951). Due to isotopic fractionation during freezing and melting cycles, Arctic ice contains about 2% more D than the seawater from which it is formed (Weston, 1955; Friedman et al., 1961). Tritium, as well as deuterium, is enriched in rain (Gulliksen, 1970) and can be used to trace the airmass source (McNaughton, 1972). Oxygen as well as hydrogen is also fractionated, but since the relative mass difference between $H_2{}^{18}O$ and $H_2{}^{16}O$ is much less than between HDO and H_2O, the effect is correspondingly smaller (Craig and Gordon, 1965). The physical–chemical isotope effect is temperature dependent, thus enabling the temperature of ancient seas to be estimated on the basis of the ^{18}O content of water and carbonates (Urey, 1947; Epstein et al., 1951; for a recent review of isotopic fractionation in the H_2O–CO_2 system see Horibe et al., 1973). Oxygen and deuterium isotope effects leave a record of past climatic changes in polar ice sheets (Dansgaard, et al., 1969; Epstein et al., 1970; Johnson et al., 1972) and in mountain snow fields (Friedman and Smith, 1972). Even past wind strengths can be estimated (Wilson and Hendy, 1971). Biological processes in our environment are important (Figure 9.22) in giving rise to isotopic fractionation. Plankton fractionate ^{18}O and ^{16}O (Shackleton et al., 1973), and Rees (1973) has described a steady-state model for biological sulfur isotopic fractionation. Cortecci (1973) has attributed the observed $^{18}O/^{16}O$ ratios in Italian lake sulfate by biological processes. Folinsbee et al. (1970) have examined carbon and oxygen isotopic ratios in dinosaur, crocodile, and bird eggshells, but they were not able to conclude whether dinosaurs were warm or cold blooded.

BIBLIOGRAPHY

R. C. Bhandari and M. L. Sisodia, *Indian J. Pure Appl. Phys.*, *2*, 266 (1964).

J. R. Bray, *Science*, *168*, 571 (1970).

W. R. Collins and N. A. Hallden, "A Study of Fallout in Rainfall Collections . . .," U.S. Atomic Energy Comm. Rept. No. HASL-42, 339 (1958).

T. B. Coplen and B. B. Hanshaw, *Geochim. Cosmochim. Acta*, *37*, 2295 (1973).

G. Cortecci, *Geochim. Cosmochim. Acta*, *37*, 1531 (1973).

H. Craig and L. I. Gordon, *Proc. Symp. Mar. Geochem.*, Univ. Rhode Island Publ. No. 3, 277 (1965).

W. Dansgaard et al., *Science*, *166*, 377 (1969).

G. Dietrich, *General Oceanography*, Interscience, New York, 1963.

D. A. Dobbins et al., *Environ. Sci. Tech.*, *4*, 743 (1970).

S. Epstein et al., *Bull. Geol. Soc. Amer.*, *62*, 417 (1951).

S. Epstein et al., *Science*, *168*, 1570 (1970).

E. Eriksson, *Tellus*, *4*, 215, 280 (1952).

L. F. Evans, in J. A. Warburton (ed.), *Conference on Cloud Physics*, Ft. Collins, CO., 1970.

J. H. Feth, *Science*, *172*, 870 (1971).

R. E. Folinsbee et al., *Science*, *168*, 1353 (1970).

E. J. Førland, *Tellus*, *25*, 291 (1973).

F. Franks, *Chem. Ind.*, No. 18, 560 (May 1968).

I. Friedman et al., *J. Geophys. Res.*, *66*, 1861 (1961).
I. Friedman and G. I. Smith, *Science*, *176*, 790 (1972).
I. E. Galbally, *J. Geophys. Res.*, *77*, 7129 (1972).
L. W. Gelhar, *Tech. Rev.* (*M.I.T.*), *74*(5), 45 (March/April 1972).
R. J. Gibbs, *Science*, *170*, 1088 (1970).
R. Gillette, *Science*, *181*, 525 (1973).
L. Granat, *Tellus*, *25*, 550 (1972).
S. Gulliksen, *J. Geophys. Res.*, *75*, 2247 (1970).
L. A. Heindl, *Nature and Resources* (*UNESCO*), *7*, 15 (March 1971).
K. G. Heumann and K. H. Lieser, *Geochim. Cosmochim. Acta*, *37*, 1463 (1973).
Y. Horibe et al., *J. Geophys. Res.*, *78*, 2625 (1973).
R. A. Horne, *Marine Chemistry*, Wiley-Interscience, New York, 1969.
R. A. Horne and D. S. Johnson, *J. Phys. Chem.*, *71*, 1936 (1967).
G. E. Hutchinson, *A Treatise on Limnology*, Wiley, New York, 1957.
D. L. Inman and B. M. Busch, *Science*, *181*, 20 (1973).
S. J. Johnson et al., *Nature*, *235*, 429 (1972).
C. E. Junge, *J. Geophys. Res.*, *65*, 227 (1960).
C. E. Junge, *Air Chemistry and Radioactivity*, Academic, New York, 1963.
I. E. Kuhns and B. J. Mason, *Proc. Roy. Soc. London, Ser. A*, *302*, 437 (1968).
Y. F. Kharaka et al., *Geochim. Cosmochim. Acta*, *37*, 1899 (1973).
P. H. Kuenen, *Realms of Water*, Wiley, New York, 1963.
J. L. Kavanau, *Water and Solute-Water Interactions*, Holden-Day, San Francisco, 1964.
I. Kirshenbaum, *Physical Properties and Analysis of Heavy Water*, McGraw-Hill, New York, 1951.
D. Kuroiwa, in T. Hari (ed.), *Studies on Fogs*, Tanne Trading Co., Sapporo, Hokkaido, Japan, 1953.
E. A. Martell, *Science*, *129*, 1197 (1959).
B. J. Mason, *The Physics of Clouds*, Clarendon, Oxford, 1957.
B. J. Mason, *Adv. Phys.*, *7*, 221 (1958).
R. H. Meade and K. O. Emery, *Science*, *173*, 425 (1971).
W. G. McGinnies et al., *Deserts of the World*, Univ. Arizona Press, Tucson, 1969.
W. G. McGinnies and B. J. Goldman (eds.), *Arid Lands in Perspective*, Univ. Arizona Press, Tucson, 1969.
D. L. McNaughton, *Tellus*, *24*, 255 (1972).
A. S. R. Murty and B. V. R. Murty, *Tellus*, *24*, 581 (1972a).
A. S. R. Murty and B. V. R. Murty, *Tellus*, *24*, 150 (1972b).
G. H. Neumann, S. Fonselius, and L. Wahlman, *Internat. J. Air Poll.*, *2*, 132 (1959).
A. H. Oort, *Sci. Amer.*, *223*(3), 55 (September 1970).
R. E. Passarelli et al., *Science*, *181*, 549 (1973).
J. A. Pena and R. A. De Pena, *J. Geophys. Res.*, *75*, 2831 (1970).
H. L. Penman, *Sci. Amer.*, *223*(3), 100 (September 1970).
F. Prodi and E. Wirth, *Tellus*, *24*, 561 (1972).
H. R. Pruppacher, *J. Chem. Phys.*, *39*, 1586 (1963).
D. Raynaud and C. Lorius, *Nature*, *243*, 283 (1973).
C. E. Rees, *Geochim. Cosmochim. Acta*, *37*, 1141 (1973).
G. W. Reynolds, in J. A. Warburton (ed.) *Conf. Cloud Physics*, Ft. Collins, CO., 1970.

P. Roberts and J. Hallett, *Quart. J. Roy. Meteorol. Soc.*, *94*, 25 (1968).

A. J. Rutter, *Endeavour*, *26*, 39 (1967).

V. J. Schaefer, *J. Appl. Meteorol.*, *7*, 148 (1968).

B. J. Skinner, *Earth Resources*, Prentice-Hall, Englewood Cliffs, 1969.

S. H. Small, *Tellus*, *12*, 308 (1960).

N. J. Shackleton et al., *Nature*, *242*, 177 (1973).

H. U. Sverdrup, M. W. Johnson, and R. H. Fleming, *The Oceans*, Prentice-Hall, Englewood Cliffs, 1942.

J. W. Swinnerton et al., *Science*, *172*, 943 (1971).

D. K. Todd, *The Water Encyclopedia*, Water Information Center, Inc., Port Washington, N.Y., 1970.

J. S. Turner, *Quart. J. Roy. Meteorol. Soc.*, *81*, 418 (1955).

H. C. Urey, *J. Chem. Soc.*, 562 (1947).

E. D. Vardiman et al., *J. Appl. Meteorol.*, *10*, 515 (1971).

R. E. Weston, Jr., *Geochim. Cosmochim. Acta*, *8*, 281 (1955).

J. R. Williams, *Ground Water in the Permafrost Regions of Alaska*, U.S. Geol. Surv. Paper 696, U.S. Gov. Print. Off., Washington, D.C., 1970.

A. T. Wilson and C. H. Hendy, *Nature*, *234*, 344 (1971).

H. L. Windom, *Bull. Geol. Soc. Amer.*, *80*, 761 (1969).

D. Winstanley, *Nature*, *245*, 190 (1973).

A. H. Woodcock, *J. Meteorol.*, *9*, 200 (1952).

A. H. Woodcock, *J. Meteorol.*, *10*, 362 (1953).

A. H. Woodcock and D. C. Blanchard, *Tellus*, *7*, 437 (1955).

G. Yamamoto and T. Ohtake, *Sci. Repts. Tokoku Univ.*, 5th Ser., *5*, 141 (1953).

ADDITIONAL READING

W. H. Amos, *The Infinite River*, Random House, New York, 1970.

W. D. Kingery (ed.), *Ice and Snow*, M.I.T. Press, Cambridge, 1963.

U.S. Atomic Energy Comm., *Precipitation Scavenging*, Nat. Tech. Inform. Serv., Springfield, Va., 1970.

VI

THE BIOSPHERE

To the man who is truly ethical
all life is sacred including that
which from the human point of view
seems lower in the scale.

ALBERT SCHWEITZER
Out of My Life and Thoughts

10

THE COMPOSITION AND STRUCTURE OF THE BIOSPHERE

10.1 INTRODUCTION

What can I say about the biosphere? We are ourselves part of this segment of our environment. How silly, then, it sounds to say that the biosphere thus is of the ultimate importance to us. What can I say about the biosphere without getting hopelessly lost in detail? The biosphere is marvelously complex. Our whole galaxy's structure is a simple-minded production compared to the complexity of the simplest living cell, compared even to a single protein molecule. The rest of our immediate environment, the lithosphere, hydrosphere, and atmosphere, is no more complex than the superficial macroscopic morphology of a diseased onion. And most incomprehensible of all is the biochemistry that permits us, albeit partially, to understand our enviornment.

Since the particulars of the biosphere could easily fill a library of books, I will try to concentrate in this chapter on only the most obvious and essential aspects of the biosphere, especially inasmuch as they relate to the rest of our environment. In this chapter, we are primarily concerned with the substance and the molecular and macroscopic structure of the biomaterial, with form rather than function, leaving the discussion of a few of the more important biochemical processes for the next chapter. But please remember that in the biosphere form and function are inseparable. The enormously complex forms have been elaborated to serve highly specific functions, while the specific functions require highly specialized forms.

10.2 THE BIOSPHERE AS A PERTURBATION OF THE COSMIC ENVIRONMENT

If we compare the relative abundance of the elements in the cosmos with their relative abundance in the biosphere (Tables 1.5 and 10.1), we see that life, while more or less using the materials most at hand rather than exotic elements, has very definitely rearranged their relative amounts. Because of its meager chemistry, life can

COSMOS H > He > O > C > Ne > N > Mg > Si > Fe > S > Al > Ca > Na > P > K

PLANTS O > H > C > K > N > Si = Ca > Mg > P > Na > Fe > S

make little use of He, the second most common element in the universe. On the other hand, elements such as oxygen, potassium, and phosphorus are promoted in im-

Table 10.1 Composition of the Biosphere (Data from Deevey, 1970)

O	104,600 kg/hectare
C	78,500
H	13,500
N	1,000
Ca	754
K	456
Si	241
Mg	196
S	142
Al	111
P	104
Cl	99
Fe	77
Mn	42
Na	38

portance. Oxygen especially is significant. The universe as a whole is a strongly reducing environment; even the atmospheres of our neighboring planets, except Mars, are reducing. A characteristic of life (inseparable from life because it involves water) is the mobilization and concentration of oxygen and oxygenated compounds. Life, so to speak, tends to create oxidative islands in the cosmic reductive sea. The biosphere is responsible for the very peculiar chemistry of the Earthly environment. In textbooks, one finds lists of elements necessary for life, probably necessary, and probably not necessary (Table 10.2). "Probably not necessary" should in all likelihood be changed to "function not yet discovered." Life has a remarkable perseverance, not only adapting to conditions at hand, but utilizing all resourses available. I find it suspect that life has overlooked any terrestrial element, except possibly the noble gases. For necessary elements the peak in Figure 7.33 lies to the right, whereas for the so-called "unnecessary elements" it may be far to the left. Research is steadily increasing the list of necessary elements (see Frieden, 1972, for a brief, popular review), for example vanadium was added to the list for animals a year or two ago (Schwarz

Table 10.2 The Role of the Chemical Elements in the Life Process (From Horne, 1969)

Elements necessary for the life processes

H, B, C, N, O, F, Na, Mg, Si, P, S, Cl, K, Ca, V, Mn, Fe, Co, Ni, Cu, Zn, Br, I

Elements probably necessary for life

Al, Ti, As, Sn, Pb

Elements probably not necessary for life

He, Li, Be, Ne, A, Sc, Cr, Ga, Ge, Se, Kr, Rb, Sr, Y, Zr, Nb, Tc, Ru, Rh, Pd, Ag, Cd, In, Sb, Te, Xe, Cs, Ba, La, Rare earths, Hf, Ta, W, Re, Os, Ir, Pt, Au, Hg, Tl, Bi, Po, At, Rn, Fr, Ac, Th, Pa, U

and Milne, 1971). We notice that life's favorite elements with a few exceptions are all light weight elements from the upper portion of the periodic table (Table A.1). And life has chosen its materials very carefully; consider why carbon rather than its sister element silicon (Firsoff, 1963)? Si is 146 times more abundant than C in the Earth's crust and has many of the same chemical properties, but SiO_2 is insoluble in water, whereas CO_2 is soluble. Carbon can bind to other carbons and form an endless variety of chain and ring compounds, while silicon can form only short chains with other Si atoms. Si can form longer chains if oxygen atoms intervene, but unlike C—C chains these —Si—O—Si— chains are so stable, so inert, that they are functionally useless. From time to time, science fictioneers have postulated noncarbonaceous biosystems, but as I have pointed out elsewhere, "While a life-system based upon silicon cannot be logically eliminated, it is safe to say that its probability is less than

$$\frac{\sum_{i=1}^{N_{Si}} \binom{N_{Si}}{i}(i!)}{\sum_{i=1}^{N_{C}} \binom{N_{C}}{i}(i!)}$$

where N_C is the number of known carbon and N_{Si} the number of known silicon compounds—an exceedingly small number indeed" (Horne, 1971). But the other chemical key, even more important than carbon, is water, of which we have a great deal more to say presently. Life is water chemistry in a carbonaceous flask. This is no accident; this is no consequence of the random concourse of atoms. The physical–chemical properties of aqueous solutions and of the solvation sheaths of solute species and interfaces have dictated the steps that the evolution of the biosphere *had* to take. Life is the direct consequences of these properties. Life is not miraculous; it is inescapable. The chemical progression in which life is an interval is not an exception in this universe; it may very well be the rule. And we can be confidant that, if ever we discover life on other worlds, while its morphological forms may be bizarre, the chemical processes and structures upon which it is based will be those with which we are already familiar.

We might classify the chemical elements that life uses into three broad categories. There are first the basic structural elements—H, C, O, and N—which together with the hydration sheath—H and O—form the matrix in which the functional groups, which are responsible for biochemical processes, are imbedded. These functional groups commonly contain the elements of our first category—C, H, N, and O—and also N, P, and S. P and S can also be structural elements. And then finally in the third category, we lump all the other elements including the trace or catalytic elements. Since several elements fall into more than one category, this classification is a helpful but not entirely satisfactory one. Another not entirely satisfactory breakdown that I have proposed approaches the problem from a gross morphological rather than a chemical point of view: There is first the biomaterial proper, then the body fluids, and finally the largely inorganic skeletal structures (Figure 8.16). This second classification has the advantage of making more clear the importance of such alkali and alkaline earth metals as Na, K, Mg, and Ca and Si. Life has found ingenius uses for these crustally abundant elements.

Turning now to the subject of trace metals, life accomplishes many very difficult chemical tasks, especially certain energy production and utilization processes and very complex organic syntheses, with an ease that makes human technology look monstrously crude by comparison. This is commonly done making use of catalysts (or enzymes), and these catalysts in turn commonly incorporate trace metals in their functional parts. Perhaps the two most familiar examples are the role of Mg in plant photosynthesis and of Fe in animal respiration. Table 10.3 lists some of these metallo-enzymes along with hemoglobin and hemocyanin. To see just how effective these catalysts are, let us look at catalase. Calvin (1959) points out that the rate for the decomposition of hydrogen peroxide by free aqueous ferric ion is 10^{-5}/ml/sec (Figure 10.1) for free heme in which four of the iron's six coordination positions are occupied by organic ligands it is 10^{-2}/ml/sec, but for catalase with all positions occupied, it is 10^{5}/ml/sec. Siegel and Roberts (1968) have reported that catalase retains some of its catalytic activity in nonaqueous solvents, but I feel that their findings, which have been cited to give credence to the possibility of nonaqueous biosystems, are suspect (Horne, 1971). Not uncommonly, the functions of trace elements are intricately linked with one another in a synergistic and even antagonistic manner: A Cu deficient diet produces anemia, since the Cu blood protein ceruloplasm promotes the release of Fe from the liver in order to form the Fe blood protein transferrin necessary for hemoglobin biosynthesis, while Zn and Mo can interfere with the utilization of Cu. In ecosystems, even more elaborate relationships probably exist. Mercury, a powerful poison on everybody's list of "unnecessary elements," is a strong fungicide, and we accidentally noted in our laboratory aquaria that goldfish

Table 10.3 Some Metalloenzymes and Their Biological Function

Ca	Lipase	Lipid digestion, fat hydrolysis
Co	Ribonucleotide reductase	Biosynthesis of DNA
	Glutamate mutase	Amino acid metabolism
Cu	Laccase	Phenol oxidation
	Tyrosinase	Skin pigmentation
	Cytochrome oxidase	Major terminal oxidase
	Ceruloplasmin	Iron utilization
	Ascorbic acid oxidase	Ascorbic acid oxidation
	Plastocyanin	Photosynthesis
Fe	Ferredoxin	Photosynthesis
	Succinate dehydrogenase	Carbohydrate oxidation
	Peroxidase	Oxidant in the presence of hydrogen peroxide
	Catalase	Decomposes hydrogen peroxide (see Fig. 10.1)
	Aldehyde oxidase	Aldehyde oxidation
	Cytochromes	Electron transport
Mg	Hexokinase	Phosphate transfer
Mn	Arginase	Formation of urea
	Pyruvate carboxylase	Pyruvate metabolism
Mo	Xanthine oxidase	Purine metabolism
	Nitrate reductase	Nitrate utilization
Zn	Carbonic anhydrase	Regulation of CO_2 level and pH
	Carboxypeptidase	Digestion of protein
	Alcohol dehydrogenase	Alcohol metabolism

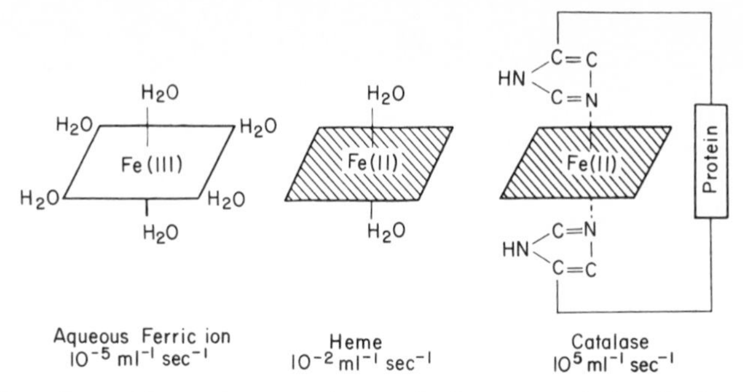

Figure 10.1 The effect of increasing organic complexing on catalytic effectiveness. The shaded areas are the porphyrin ring system (from Horne, 1969).

in the very low Hg tanks developed fatal fungal infections. Might not then low levels of Hg in natural waters be beneficial to fish health? This raises the interesting possibility that if man is too successful at removing certain "pollutants" from the environment he might cause damage (see Horne, 1972*b*).

In addition to being a chemical perturbation of the cosmic environment, the biosphere is also an energetic or thermodynamic perturbation, and we return to this topic subsequently.

10.3 CHEMICAL NATURE OF THE BIOSPHERE

Water is the major constituent of the biosphere, the weight percentage ranging widely in organisms from about 50% for woody trees to over 99% for jellyfish. The rest of the biomaterial proper (Figure 8.16) consists largely of carbohydrates, proteins, and a catch-all of organic solvent soluble substances called lipids. Since land plants alone far outweigh all animals (Figure 10.9), much more of the biosphere is based on cellulose, a carbohydrate, rather than protein. Table 10.4 gives the elemental

Table 10.4 Composition of Marine Biomaterial (% by weight)

	Organic Material						
	O	C	H	P	N	S	Fe
Proteins	22.4	51.3	6.9	0.7	17.8	0.8	0.1
Lipids	17.9	69.1	10.0	2.1	0.6	0.3	
Carbohydrates	49.4	44.4	6.2				

	Skeletal Material							
	Ca	Mg	CO_3^{2-}	SO_4^{2-}	PO_4^{3-}	SiO_2	$(Al, Fe)_2O_3$	Organic
Foraminifera	34.9	3.0	59.7			0.03	0.1	2.3
Coral	38.5	0.1	58.0			0.07	0.05	3.3
Lobster	16.8	1.1	22.4	0.5	5.5			53.4
Siliceous Sponge	0.2		0.2			88.6	0.3	10.7

compositions of these materials. For purposes of chemical stoichiometric calculations the marine biomaterial is often represented by

$$(CH_2O)_{106}(NH_3)_{16}H_3PO_4$$

Carbohydrates, as the name implies, are polyhydroxyl compounds primarily composed of C, H, and O but they may also contain N and S. Table 10.5 gives the carbohydrate composition of algae cells. In each species the major carbohydrate is glucose. D-Glucose is a very important *monosaccharide* found in the blood of animals as well as in the sap of plants, and it can exist in two forms

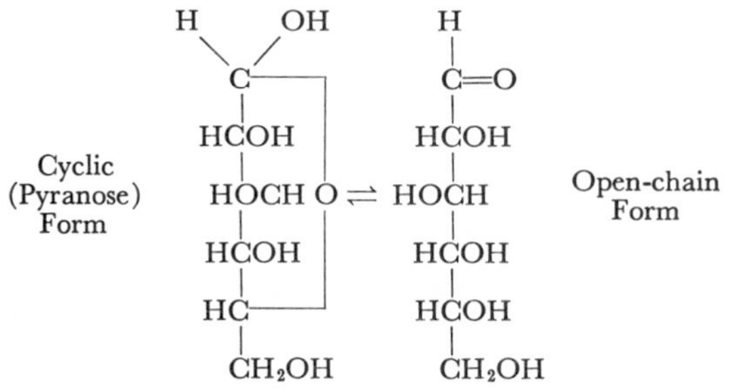

Cyclic (Pyranose) Form ⇌ Open-chain Form

Glucose

Oliosaccharides, such as ordinary sugar (sucrose), are formed by the condensation of the hydroxyl groups of one monosaccharide molecule with the reducing group of another. In organisms, *polysaccharides* containing many units serve two crucial types of function—the structural polysaccharides provide a rigid mechanical support for the organism; the nutrient polysaccharides, such as starch, provide readily utilizable metabolic reserves of monosaccharides in animals and plants. An example of the former is chitin which occurs in large amounts in invertebrate exoskeletons such as the shells of lobsters and crabs.

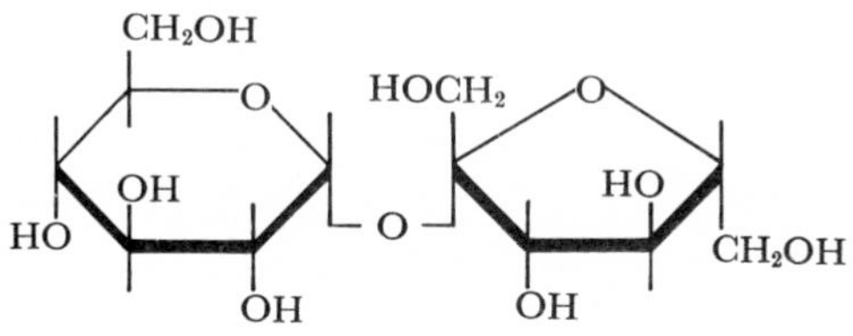

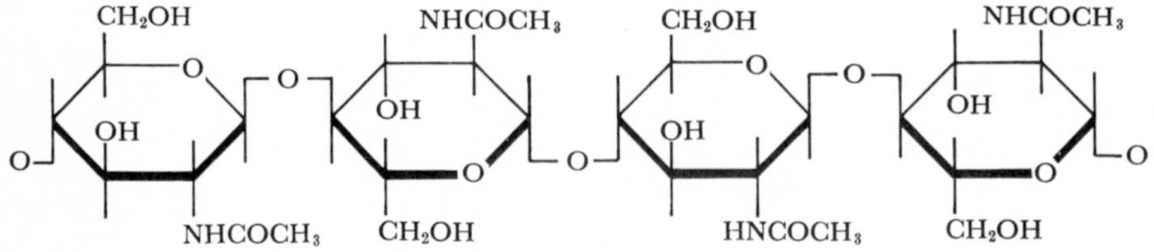

Sucrose

Chitin

Table 10.5 Carbohydrate Composition of Algae Cells as Percent Dry Weight[a] (From Parsons, 1963, with permission of Pergamon Press, Ltd.)

Species	Glucose	Galactose	Mannose	Ribose	Xylose	Arabinose	Rhamnose	Fucose	Fructose	Hexosamine	Hexuronic Acids
Chlorophyceae											
Tetraselmis maculata	11.9	2.3	–	0.95	–	–	–	–	–	–	+
Dunaliella salina	17.2	11.8	–	1.7	–	–	–	–	–	–	+
Chrysophyceae											
Monochrysis lutheri	22.1	4.4	–	1.3	3.5	–	–	–	–	–	+
Syracosphaera carterae	9.2	7.1	–	1.5	0.8	1.9	–	–	–	–	+
Bacillariophyceae											
Chaetoceros sp.	3.3	1.5	0.79	0.71	0.4	–	2.8	+	–	–	+
Skeletonema costatum	16.4	1.8	0.87	1.2	–	–	1.0	0.9	–	–	+
Coscinodiscus sp.	2.1	0.4	0.41	+	–	–	0.7	0.5	–	–	+
Phaeodactylum tricornutum	10.7	2.7	3.7	0.72	0.7	–	1.5	–	–	–	+
Dinophyceae											
Amphidinium carteri	19.0	8.4	–	0.9	–	–	+	–	–	–	–
Exuviella sp.	26.8	8.3	–	+	+	+	+	–	–	–	–
Myxophyceae											
Agmenellum quadruplicatum	17.4	3.2	–	1.5	–	–	–	–	3.5	0.3	+

[a] – Sugars not detected; + sugars detected but not estimated.

Nitrogen, as we can clearly see in Table 10.4 is the characteristic element in proteins. In fact, it is common practice to determine the protein content of plant and animal biomaterial by analyzing for nitrogen and multiplying its weight by a factor of 100/16, the 16% being the average N-content of protein. Proteins are very large macromolecules; egg albumin has a molecular weight around 34,000 and hemoglobin around 64,000. The approximate formula for the milk protein β-lactoglobulin is $C_{1864}H_{3012}O_{506}N_{468}S_{21}$. Proteins can be hydrolyzed, say by warming with acid, into the simpler amino acid units that compose them. Table 10.6 gives the amino acid compositions of some proteins, while Table 10.7 lists the names and formulas of some of the major protein amino acids. It can be seen that the general form of amino acids is

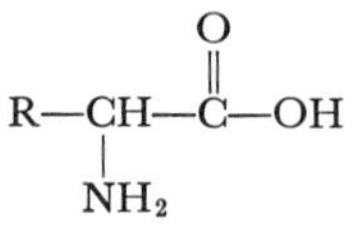

$$\mathrm{R{-}\underset{\underset{\displaystyle NH_2}{|}}{CH}{-}\overset{\overset{\displaystyle O}{\|}}{C}{-}OH}$$

In the protein, the amino acids are linked together by peptide bonds

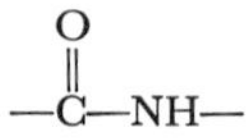

$$\mathrm{{-}\overset{\overset{\displaystyle O}{\|}}{C}{-}NH{-}}$$

to form enormously long polypeptide chains

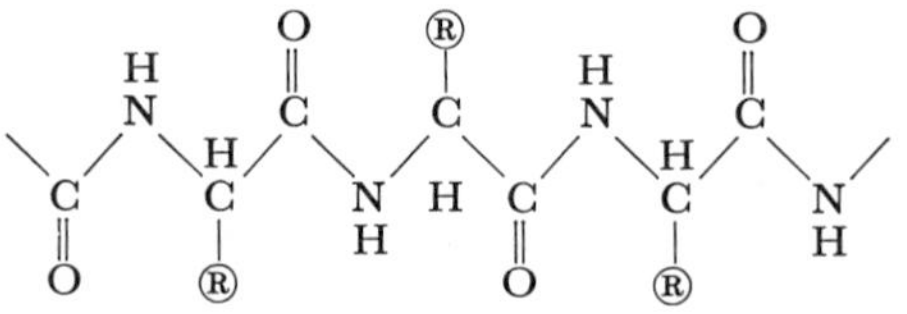

```
            O      (R)         O
    H       ‖       |     H    ‖
    N       C       C     N    C
 \ / \ H / \ / \ / \ H / \ /
  C     C     N  H C     C     N
  ‖     |     H    ‖     |     H
  O    (R)         O    (R)
```

In addition to amino acid linkages, proteins can also contain nucleic acid linkages. Figure 10.2 shows the amino acid sequence in a small segment of a protein. To digress a bit, let me say that such sequences are exceedingly exact, and an irregularity

pGlu -Met-Ser -Tyr -Gly-Tyr-Asp -Glu-Lys -Ser -Ala-Gly -Val -Ser -Val-15
Pro -Gly -Pro-Met-Gly-Pro -Ser -GlysPro -Arg -Gly-Leu -Hyp-Gly-Pro-30
Hyp-Gly -Ala-Hyp-Gly-Pro -Gln -Gly-Phe-Gln -Gly-Pro -Hyp-Gly-Glu-45
Hyp-Gly -Glu-Hyp-Gly-Ala -Ser -Gly-Pro -Met-Gly-Pro -Arg -Gly-Pro-60
Hyp-Gly -Pro-Hyp-Gly-Lys -Asn -Gly-Asp-Asp -Gly-Glu -Ala -Gly-Lys-75
Pro -Gly -Arg-Hyp-Gly-Gln-Arg -Gly-Pro -Hyp-Gly-Pro -Gln -Gly-Ala-90
Arg -Gly -Leu-Hyp-Gly-Thr-Ala -Gly-Leu-Hyp-Gly-Met-Hyl -Gly-His -105
Arg -Gly -Phe-Ser -Gly-Leu-Asp -Gly-Ala -Lys -Gly-Asn -Thr -Gly-Pro-120
Ala -Gly -Pro-Lys -Gly-Glu-Hyp-Gly-Ser -Hyp-Gly-Glx -Asx -Gly-Ala-135
Hyp-Gly -Gln-Met-

Amino acid sequence of 139 residues at the N-terminal end of the α1 chain of rat skin (or tendon) collagen. The order of residues 134–136 has not been determined but can be derived from the assumptions that glycine (Gly) will be in every third position and that hydroxyproline (Hyp) can occur only in a position preceding glycine. Many of the residues shown as Hyp contain proline owing to incomplete hydroxylation. In rat skin collagen, the first four residues are absent (from Traub and Piez, 1971, with permission of Academic Press, Ltd., and the authors).

Figure 10.2 Amino acid sequence of 139 residues at the N-terminal end of the α1 chain of rat skin collagen (from Traub and Piez, 1971, with permission of Academic Press, Ltd., and the authors).

Table 10.6 Weight of Free Amino Acid in 100 g of Some Proteins (From Fruton and Simmonds, 1953, with permission of John Wiley & Sons)

	β-Lacto-globulin	Hemo-globin (Horse)	Insulin	Edestin	Egg Albumin	Chymo-trypsinogen	Silk Fibroin	Gelatin	Salmine
Glycine	1.5	5.6	4.5	—	3.1	5.3	43.6	26.9	2.9
Alanine	6.6	7.4	4.4	4.3	6.7	—	29.7	9.3	1.1
Valine	5.7	8.4	7.5	5.7	7.1	10.1	3.6	3.3	3.1
Leucine	15.5	16.0	12.9	7.4	9.2	10.4	0.9		1.6
Isoleucine	5.9	—	2.8	4.7	7.0	5.7	1.1		—
Serine	4.1	5.8	5.2	6.3	8.2	11.4	16.2	3.2	9.1
Threonine	5.8	4.4	2.1	3.9	4.0	11.4	1.6	2.2	—
Cystine	3.4	1.0	12.5	1.4	1.0	6.6	—	—	—
Cysteine	1.1	0.4	—	—	1.4	1.3	—	—	—
Methionine	3.2	1.5	—	2.3	5.2	1.2	—	0.9	—
Aspartic acid	11.2	10.4	7.5	12.8	9.3	11.3	2.8	5.6	—
Glutamic acid	21.5	8.5	18.6	19.3	16.5	9.0	2.2	11.2	—
Amide NH_3	1.3	1.1	1.7	2.2	1.2	1.9	—	0.1	—
Lysine	11.2	8.6	2.5	2.4	6.3	8.0	0.7	4.6	—
Arginine	2.8	3.6	3.4	16.7	5.7	2.8	1.1	8.6	85.2
Histidine	1.6	7.9	4.9	2.9	2.4	1.2	0.4	0.7	—
Phenylalanine	3.8	7.9	8.0	5.7	7.7	3.6	3.4	2.6	—
Tyrosine	3.8	3.0	12.2	4.5	3.7	3.0	12.8	1.0	—
Tryptophan	1.9	—	—	1.4	1.2	5.6	—	—	—
Proline	5.2	3.9	2.6	4.6	3.6	5.9	0.7	14.8	5.8
Hydroxyproline	—	—	—	—	—	—	—	14.5	—
Total	117.1	105.4	113.3	108.5	110.5	115.7	120.8	114.7	108.8

Table 10.7 Principal Amino Acids Derived from Proteins (From Horne, 1969)

I. Aliphatic Amino Acids

A. Monoaminiomoncarboxylic acids

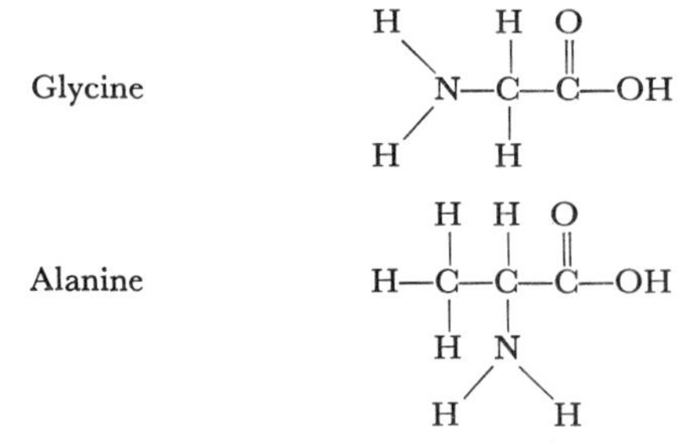

Amino acid	Formula
Valine	$(CH_3)_2CH \cdot CH(NH_2)COOH$
Leucine	$(CH_3)_2CH \cdot CH_2CH(NH_2)COOH$
Isoleucine	$CH_3 \cdot CH_2CH(CH_3) \cdot CH(NH_2)COOH$
Serine	$HOCH_2 \cdot CH(NH_2)COOH$
Threonine	$CH_3 \cdot CH(OH) \cdot CH(NH_2)COOH$
B. Sulfur-containing amino acids	
Cysteine	$HSCH_2 \cdot CH(NH_2)COOH$
Methionine	$CH_3SCH_2 \cdot CH_2 \cdot CH(NH_2)COOH$
C. Monoaminodicarboxylic acids and their amides	
Aspartic acid	$HOOC \cdot CH_2 \cdot CH(NH_2)COOH$
Asparagine	$NH_2CO \cdot CH_2 \cdot CH(NH_2)COOH$
Glutamic acid	$HOOC \cdot CH_2 \cdot CH_2 \cdot CH(NH_2)COOH$
Glutamine	$NH_2CO \cdot CH_2 \cdot CH_2 \cdot CH(NH_2)COOH$
D. Basic amino acids	
Lysine	$NH_2 \cdot CH_2 \cdot CH_2 \cdot CH_2 \cdot CH_2 \cdot CH(NH_2)COOH$
Hydroxylysine	$NH_2 \cdot CH \cdot CH(OH) \cdot CH_2 \cdot CH_2 \cdot CH(OH_2)COOH$
Arginine	$NH_2 \cdot C \cdot NH \cdot CH_2 \cdot CH_2 \cdot CH_2 \cdot CH(NH_2)COOH$
Histidine	H N—C—$CH_2 \cdot CH(NH_2)COOH$ H—C ‖ N—CH

(Continued)

in the sequence can produce a specific disease. In normal hemoglobin (hemoglobin A), the β-polypeptide chains contain 146 amino acid residues with glutamic acid occupying the sixth position. Possibly because it provides some protection from malaria a variant protein molecule, hemoglobin S, evolved with the glutamic acid replaced by valine in both β-chains (Ingram, 1957; Hunt and Ingram, 1959). Blood cells containing this hemoglobin S tend to form crescentic shapes that cannot oxygenate properly. The resulting hereditary disease, sickle-cell anemia, has been aptly characterized by Professor Pauling as a "molecular disease" (see Murayama, 1966). By comparing the structures of glutamic acid and valine (Table 10.7), we can see that substitution by the latter would make that segment of the macromolecule more

Table 10.7 (*Continued*)

II. Aromatic Amino Acids	
Phenylalanine	HC(CH=CH)(CH=CH)C—$CH_2 \cdot CH(NH_2)COOH$
Tyrosine	HO—C_6H_4—$CH_2CH(NH_2)COOH$
Diiodotryrosine	HO—$C_6H_2I_2$—$CH_2 \cdot CH(NH_2)COOH$
Thyroxine	HO—$C_6H_2I_2$—O—$C_6H_2I_2$—$CH_2 \cdot CH(NH_2)COOH$
III. Heterocyclic amino acids	
Tryptophan	indole ring (N—H)—$CH_2 \cdot CH(NH_2)COOH$
Proline	H_2C—CH_2; H_2C $CH \cdot COOH$; N—H
Hydroxyproline	HO—HC—CH_2; H_2C $CH \cdot COOH$; N—H
Histidine (see I.D.)	

hydrophobic. Evidently hydrophobic bonds align the hemoglobin S protein molecules in a special way that gives rise to sickling, and it is fascinating to note that the clinical treatment for sickle-cell anemia crises involves the administration of solutes, such as urea, which are well known for their ability to modify water structure in bulk solution and, presumably, the hydration sheath of macromolecules.

Now to return to the nature of the proteins: the macromolecules can assume several different important categories of spacial configuration both with respect to themselves and their neighbors. The factors determining these spacial configurations and which are thus so crucial to biological form and function include H-bonding, coulombic interactions, sulfhydryl linkages, hydrophobic bonding, and perhaps most important of all, least understood, and related to hydrophobic bonding, constraints imposed by the hydration sheath of the biopolymer. The first category of these spacial configurations is the spiraling of the amino acid chains to form the now well-known α-helix (Figure 10.3). The second category, on a somewhat larger scale, is the cross-linking and folding of these spirals to form a fibrous or globular protein molecule. This is illustrated in Figure 10.4, which shows, not a snapshot of the random writhing of a snake, but, on the contrary, the marvelous way in which the heme functional group in myoglobin (a red respiratory pigment from whale muscle)

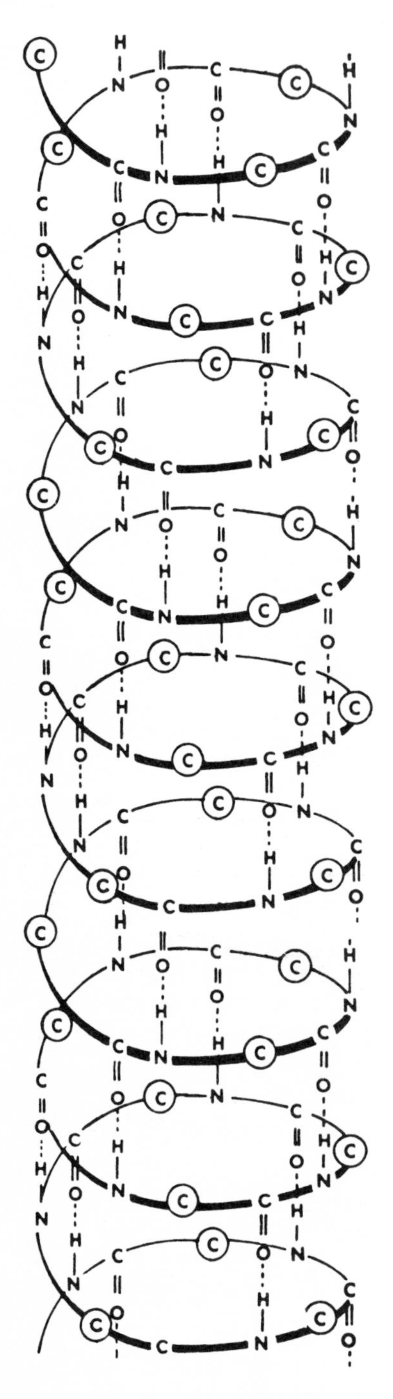

Figure 10.3 The α-helix. Hydrogen bonds that hold the form of the helix are indicated by broken lines (from Horne, 1969).

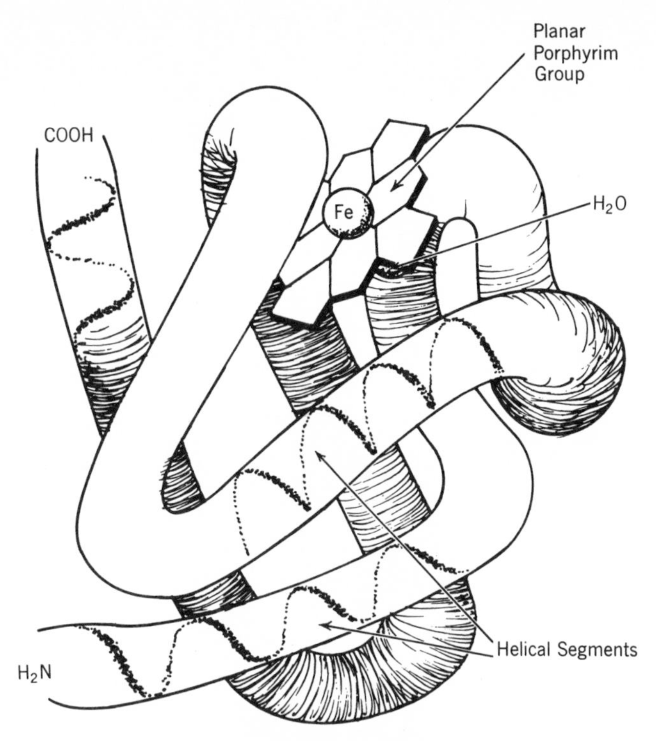

Figure 10.4 A portion of the myoglobin molecule (adapted from Kendrew, 1961, and based on a model constructed from three-dimensional Fourier synthesis by Kendrew and Perutz, see Kendrew, 1963).

is carefully and exactly folded in a highly specific way by the protein chain. The four heme groups in human hemoglobin are similarly enfolded. The third category is the interaction of proteins with other substances such as oxygen in the case of hemoglobin. With respect to this interaction, the four hemes in hemoglobin do not appear to be equivalent, and the interaction with oxygen is now believed to be accompanied by alterations in the configuration of the protein, as well as with one another (of which sickle cell anemia affords an example), to finally the build up to the macroscopic morphology of living cells and tissues.

The second type of biomaterial diagrammed in Figure 8.16 is the body fluids. The obvious can be elusive, and in the past, the role of water in living organisms has been neglected as scientists have concentrated on protein chemistry. In the not too distant past, water was looked upon as simply filling up the empty spaces in cells, but today the biological importance of water is at last being widely if not universally recognized (Dick, 1972; Ling, 1972). In addition, thanks largely to the efforts of G. N. Ling and a handful of other scientists, there is now beginning to be an awareness that the fluid in cells is not simply a saline solution, but that the state of the water and dissolved solutes are modified in some way, still far from clear, by the biological environment and that this modification in turn, in terms of biological form and function, is of the most profound importance. Figure 10.5 schematically represents the relationships among the several fluid compartments in an animal, whereas Figure 10.6 compares the chemical composition of human body fluids with seawater. Except for the cell fluid, whose marked enrichment in K^+ over Na^+, the so-called

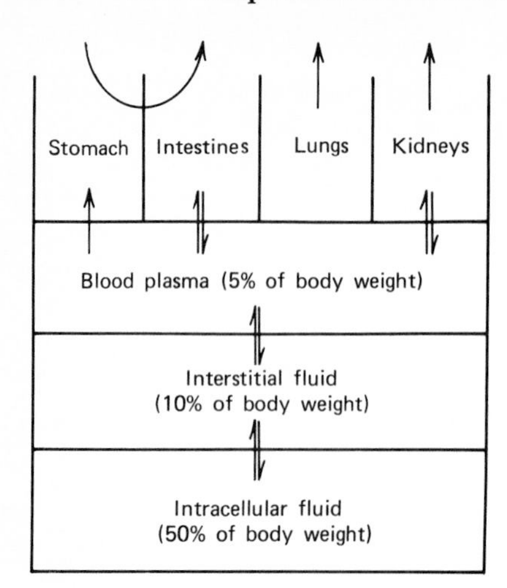

Figure 10.5 Diagram of the relationships between animal fluid compartments.

"sodium-potassium pump" is one of the great unsolved riddles of biology (for a highly speculative hypothesis of my own on this subject see Horne 1972*b*), the composition of the body fluids, not unexpectedly, closely resembles that of seawater (Figure 10.6 and Table 10.8). Human blood and seawater are both buffered NaCl solutions. The pH range of body fluids, including blood, lymph, cerebrospinal fluid, and so on, is also near neutrality but somewhat more narrow than the seawater range, 7.1 to 7.7 compared with 7.5 to 8.4. Departures from seawater composition in body fluids arise through evolutionary changes or as a consequence of the specific biological function of the fluid. This is especially conspicuous in the greater concentrations of some of the minor constituents as shown in Tables 8.14 and 10.8.

The composition of the body fluids of contemporary organisms may give us clues as to the chemical composition of the ancient seas in which their ancestors evolved. This composition may also give us clues as to the evolutionary history of the organism itself. As an example of the latter, the Mg/Ca ratio in the blood serum of land and fresh water animals is quite different from that of primitive marine animals whose values are closer to the present ratio, 311, for seawater (Table 10.9). Yet a very large class of bonefish, including such common species as pollack and cod have Mg/Ca ratios comparable to the former, leading to the conclusion that they were once fresh water forms and that they subsequently returned to the oceans in some more recent geological period.

Our third and final type of biomaterial (Figure 8.16) is what, with gross oversimplification, we categorize as the "inert" and "inorganic" skeletal structural

Table 10.8 Composition of the Body Fluids of Some Marine Organisms (Na = 100)

	Na	Cl	Mg	S(in SO_4^{2-})	Ca	K
Sea urchin	100	182	12	8.5	3.9	3.7
Lobster	100	156	1.5	2.2	5.0	4.7
Seawater	100	180	12.1	8.4	3.8	3.6

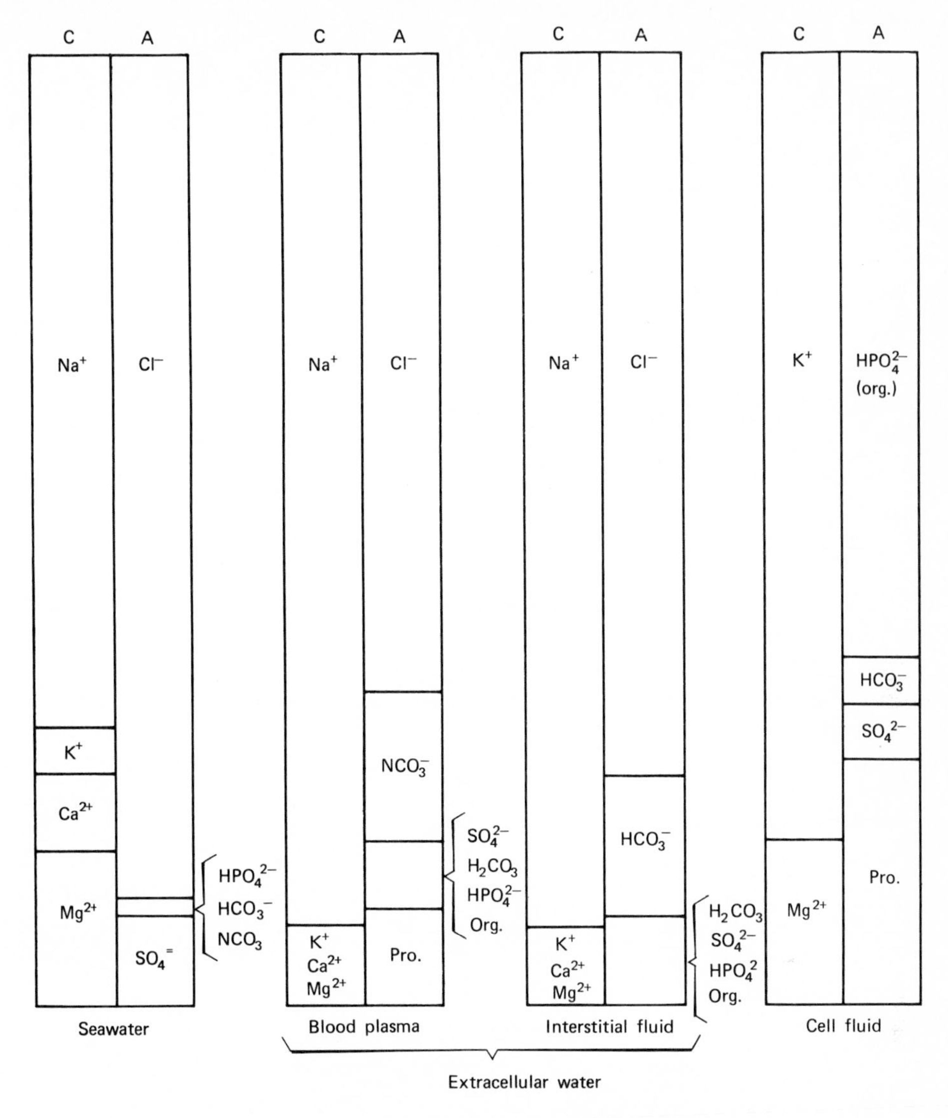

Figure 10.6 Relative compositions of seawater and human body fluids. (Pro., protein).

Table 10.9 Mg/Ca Ratio in the Blood Serum of Selected Animals (From G. Dietrich, 1963, with permission of John Wiley & Sons)

Primitive Marine Animals		Land and Fresh Water Animals		Recent Fish	
Crustacea	130–196	Human	25	*Homaris americanus*	34.7
Echinoderm	240–253	Mammals	18–29	*Gadus collarus* (codfish)	35.9
Mollusk	214–275	Chicken	19	*Pollachius virens* (pollack)	47.0
		Crocodile	33		
		Tench	34		
		Edible snail	11		

material. For the greater part these materials are either $CaCO_3$ or SiO_2 with other species such as magnesium and phosphate playing roles of varying importance. In the oceans, calcareous skeletons tend to be replaced by siliceous ones as temperature decreases and/or pressure increases. Extensive deposits of coral, for example, are only formed in tropical seas. Skeletal structures can be enormously intricate both on the macroscopic and microscopic scale; they can also contain sites, as in the case of the marrow near bone endings, of important biochemical processes. Just how organisms form these supporting deposits is far from obvious. Organisms can, for example, deposit $CaCO_3$ even in seawater undersaturated with respect to that substance (Cloud, 1965). The deposition of these substances represents a major link in the interaction of the biosphere with the lithosphere. We return to this subject later.

Earlier we noted that the earthly biosphere has perturbed the chemistry of the cosmos and has created a tiny island of oxidative environment in the overall reductive environment of the universe. I would like to conclude this section by mentioning an example of perturbation of the chemical environment by organisms on a much smaller but not inconsequential scale. Organisms may build up enormous concentrations of certain trace elements (Tables 10.10, A.24, and A.25). There is clearly some very fascinating chemistry involved in these enrichments, but this topic has been somewhat neglected. Goldberg (1957) has pointed out that the enriched metals complex strongly with ligands and this observation might serve as the basis of a general theory of element enrichment in marine organisms. The question is not devoid of practical significance for the metals concentrated may be radioactive as in the case of the uptake of ^{65}Zn by oysters (Section 7.6).

10.4 LIFE AND ORDER

What is life? Scientists have addressed themselves to this mystery ever since science began in ancient Greece. From time to time the question even surfaces in the popular media, often in a moral or religious context, in connection with social issues such as population control, euthanasia, and "the right to die." Then too the beginnings of our exploration of space, which really boils down to a search of extraterrestrial life,

Table 10.10 Elemental Enrichment in Shellfish (Scallop) Relative to the Marine Environment (Data selected from Brooks and Rumsby, 1965)

Cadmium	2,260,000
Chromium	200,000
Copper	3,000
Iron	291,500
Lead	5,300
Manganese	55,500
Molybdenum	90
Nickel	12,000
Silver	2,300
Vanadium	4,500
Zinc	28,000

has further ramified the question: How shall we look for life and/or its traces? How shall we know it when we find it? The question has been complicated by the discovery of the virus which in one state seem to fit most of the definitions of life while in another dormant state can be crystallized and exhibit other physical and chemical properties and transformations, which seem more characteristic of inanimate matter. Numerous definitive "characteristics" of life have been proposed—movement, replication, metabolism, and so on—but these characteristics are not unique and can be paralleled singly or in combination in clearly nonbiotic and in some instances even inorganic systems.

Perhaps the best answer to the question "What is life?" is contained in a brilliant little monograph of that title by Schrödinger (1944) in which he examines life as a thermodynamic eccentricity. This entropic hypothesis of the nature of life pictures that phenomenon as being a victory, highly localized in space and time, over the universal march into disorder, into randomness—the so-called "heat death" of the universe. With respect to their thermodynamic aspects, I can add nothing here to the arguments of Schrödinger, however I would like to stress one point—life is not only order, it is order of a very special type. The cosmos is strewn with examples of order: Galaxies, stars, and planets are all ordered structures. The interstellar chemical compounds that we examined in Chapter 1 represent a more ordered state than would a random mixture of their elemental constituents. Again, to chose a more familiar example, there is the beautiful systematic order of the repeating patterns of the crystal lattices of mineral substances. But the order represented by life, while it may have taken the simpler level of order of mineral substrates as its point of departure (Matheja and Degens, 1970), is an order that differs so in degree that it seems to differ in kind.

Compare the crystal lattice of Ice-I (Figure 7.7) with the conformation of a small corner of a respiratory pigment protein molecule about the functional heme group (Figure 10.4). These two figures confront us with two distinctly different types of order. To describe these differences, I have followed the examples of artists, musicians, and critics and borrowed a couple of terms from architecture—classical and baroque (Horne, 1971).

> The components of classical order are related to one another by the geometric symmetry of their positions in space, whereas the components of baroque order are related by their . . . function. . . . In contrast to the simple classical order of crystals, the living substance, both intra- and intermolecular, and on every level of complexity up to and including whole living organisms, is characterized by baroque order, that is to say, not a symmetrical, regular, repeating geometric pattern, but an exceedingly complex and superficially irregular composition of inter-related parts. . . . Such baroque order is not merely characteristic of life; it is necessary to realize the complexity and multiplicity of biological function. We tend to define life in terms of its vital functions rather than form, and properly so . . . [for] the complexity of form is a necessary condition of the complexity of function. . . . The complexity of carbon-chemistry is a necessary but not a sufficient condition for baroque order and the life it characterizes . . . [and] such baroque order can come into existence only in aqueous systems by virtue of certain properties [of] water which, if not unique in kind . . . are unquestionably unique in degree (Horne, 1971).

We are not alone in the universe. There is life on other worlds. But the chemistry of life will be similar wherever it is found. There is no noncarbonaceous life. There is no nonaqueous life.

Living organisms are not only highly organized chemical systems, not only do they alter the chemistry of their environment, but many of them in their nest-building activities also *order* the local environment. This is especially true of man who alters and orders his environment on a truly massive scale. Rather than defining him as a naked biped (a definition equally appliable, as some wag has pointed out, to a plucked chicken), we might characterize man as the environment altering and ordering animal. So great is his capability in this direction that in the nineteenth century he became master of his environment and in the twentieth century he has become the destroyer of his environment.

10.5 BIOGENESIS AND THE HISTORY AND FATE OF THE BIOMATERIAL

Life is essentially a product of water.

J. D. Bernal
The Origin of Life

Classical chemical thermodynamics concerns itself only with initial and final states, and while connectable to atomic and molecular processes via statistical mechanics, the connection is so complex that for practical purposes thermodynamics can give us relatively little insight into the mechanisms of atomic and molecular processes. We have said that from an enthropic point of view life is order. So much for "how?", now let us ask that illicit scientific question "why?" What are the chemical forces and properties responsible for the evolution of such a complex and highly ordered chemical system as the biosubstance?

Except in the state of California, life, order gradually evolved over a long period of time from inorganic matter. Pasteur was wrong, and Aristotle was right; there was spontaneous generation of life. But because bioprecoursers are now oxidized and/or consumed by organisms spontaneous generation has probably ceased to occur on this planet. Let me sketch in the remarkable story of biogenesis by quoting at length from my earlier book (Horne, 1969). For a more detailed treatment, consult the many books on the subject of chemical evolution and biogenesis (see the books listed in the Additional Reading at the end of the chapter) I particularly recommend that of Calvin (1969).

A living cell is an enormously complex dynamic structure. Even the most elaborate chemical processing plants built by man are pitifully simple by comparison. Yet, when I first turned my attention to the origin and evolution of life, I was amazed, not because we know so little, but rather because we know so much. In the words of Professor Bernal* (1961), "we now have the skeleton of a story of the origin of life. . . ." While the details remain scarce and while many questions will remain unanswered for many years to come, perhaps forever, the answers to the principal questions now seem to be all at least foreshadowed; the principal conceptual barriers have already been reached.

The first of these barriers was an emotional and metaphysical one—the conviction that "organic" compounds could be formed only through the agency of living cells, that their genesis

* I hope that it does not escape the reader's attention, or seem to be a coincidence, that one of the most prominent scientists concerned with the problem of biogenesis is the father of modern theories of the structure of water.

was in a sense supernatural and necessitated a vital or "life force." This belief was shattered in 1824 when Wöhler demonstrated that the familiar organic compound urea, an end product of protein metabolism, can be prepared simply by the evaporation of a solution of a typical inorganic salt, ammonium cyanate. Just as the Copernican theory of heliocentricity cleansed cosmology of its mystical and anthropormorphic elements and placed the solar system in true perspective in relation to the rest of the natural order, so Wöhler's simple experiment revolutionized biological thought, set organic chemistry in its proper perspective in relation to chemistry in general, and revealed that life is a natural process subject to oxidation by scientific method, thereby establishing the basis of a true science of biochemistry.

The inorganic synthesis of urea opened the possibility that organic compounds, and subsequently life, could have resulted from purely "nonliving" or inorganic substances and processes at some time in the Earth's distant geological past. Yet subsequently the long-flourishing theory of the spontaneous generation of living from nonliving forms was finally laid to rest by the work of Pasteur. Does this mean that life could not have originated on Earth from purely natural causes? No. We now know that it means simply that the process of spontaneous generation requires a span of time far in excess of the total lifetime of the human species and/or that the environmental conditions necessary for spontaneous generation no longer obtain on this planet.

Let me note parenthetically here that theories of the extraterrestrial inoculation of Earth with life (see Shklovskii and Sagan, 1966, for a good review) do not answer the question of the origin of life they simply remove the problem to a different and nonexaminable frame of reference.

More than one hundred years separate Wöhler's demonstration that life *in principle* could have arisen from nonliving substances by natural processes from the first *actual* synthesis of proto-biomaterials from inorganic substances under simulated primitive Earth conditions. At the suggestion of Professor Urey, Miller (1953, 1955) passed an electrical discharge through a gaseous mixture of CH_4, NH_3, H_2O and H_2 gases believed to be constituents of the primitive atmosphere of the earth—for time intervals of roughly a week—and obtained polyhydroxy compounds and amino acids, the fundamental building blocks of the life substance. Not only were yields surprisingly abundant, but also the number of compounds identifiable was considerable [(Table 10.11)].

In recent years this work has been widely repeated, confirmed, and varied. In addition, other processes for the production of simple organic from inorganic materials have been discussed, such as the formation of hydrocarbons from the reaction of certain carbides with water, and Eck, Lippincott, et al. (1966) have calculated the concentration of a number of organic substances such as alcohols, carboxylic acids, hydrocarbons, ketones, amines, ethenes and heterocyclic compounds that might result from simple equilibria with inorganic reactants.

But the stones are not a temple; once in hand, the building blocks must be put together. Now we must confront the third barrier. How are the pieces brought together? The putting together of the pieces was a long, tedious, and delicate sequence and each step in the sequence was highly improbable. Fortunately the time span allotted to the beginnings of life was exceedingly long, perhaps several billion years (Bernal, 1961), so that the improbable was not necessarily the impossible. Biogenesis is pushed further into the realm of possibility if there were mechanisms operative for the concentration of the pieces and, in order to outrace the forces of dissolution, for the stabilization of the pieces and their combinations.

In terms of proto-biological reactants the ancient seas were a very dilute broth, and various mechanisms have been suggested that could have resulted in large enough concentrations of these substances for the improbable life building steps to become probable. Darwin suggested that the proto-biological broth may have become concentrated by the evaporation of shallow pools. However, I think the alternative mechanism suggested by Professor Bernal (1961) is far more more attractive, namely, concentration by interfacial absorption processes, which we [encounter elsewhere] in this book, on bubbles (Wangersky, 1965; Barber, 1966), on the

Table 10.11 Yield of Organic Compounds Resulting from Electrical Discharge in a Simulated Primitive Earth Atmosphere (Reprinted with permission from S. L. Miller, *J. Amer. Chem. Soc.*, *77*, 2351 (1955). © The American Chemical Society.)

A. Yield of Amino Acids

Amino Acid	Moles (× 10⁵)	mg	Mole Ratio (glycine = 1)	Yield (%)
Run 1				
Glycine	63	47	1.00	2.1(4.0)
Alanine	34	30	0.54	1.7(3.2)
Sarcosine	5	4	0.08	0.3(0.5)
β-Alanine	15	13	0.24	0.8(1.5)
B_1	1	1	0.02	0.07(0.13)
o-Aminobutyric Acid	5	5	0.08	0.3(0.6)
B_2	3	3	0.05	0.2(0.4)
Run 2				
Glycine	55	41	1.00	1.8(3.2)
Alanine	36	32	0.65	1.8(3.2)
Sarcosine	2	2	0.04	0.1(0.2)
β-Alanine	18	16	0.33	1.0(1.8)
B_1	0.1	0.1	0.002	0.01(0.02)
o-Aminobutyric Acid	3.0	3.0	0.054	0.2(0.34)
B_1	0.4	0.4	0.007	0.03(0.05)
Run 3				
Glycine	80	60	1.0	0.46(2.1)
Alanine	9	8	0.11	0.08(0.3)
Sarcosine	86	77	1.07	0.74(3.4)
β-Alanine	4	3	0.05	0.03(0.1)
B_1	12.5	12.9	0.16	0.14(0.63)
o-Aminobutyric Acid	1	1	0.01	0.01(0.05)
B_2	14.7	15.2	0.18	0.17(0.76)

B. Yields of Acids

Acid	Moles (× 10⁵)	Mg	Yield (%)
Run 1			
Formic	233	107	3.9(7.4)
Acetic	15.2	9.1	0.5(1.0)
Propionic	12.6	9.1	0.6(1.2)
Glycolic	56	42	1.9(3.5)
Lactic	39	35	1.8(3.4)
Run 3			
Formic	149	69	0.4(2.0)
Acetic	135	81	0.7(3.5)
Propionic	19	14	0.2(.07)
Glycolic	28	21	0.2(0.7)
Lactic	4.3	3.9	0.03(0.2)

sea's surface, or on solid particles. I am taken by his analogy that life, like Aphrodite, the goddess of love, was born of the sea foam. Let us imagine, then, the proto-biological substances being absorbed on bubble surfaces, transported upward to the sea's surface and joined with other material absorbed there, then tossed by the waves and carried by the sea spray up onto the beaches and the estuarine mud where, in the richer, warmer waters the pieces begin to react and the aggregates to grow. Earlier life was described as a struggle against the forces of destruction. These forces were operating long before anything identifiable as life came into being, constantly endangering the existence of not only the aggregates but even the initial organic building blocks as well. In the present world the forces of degradation of organic materials are bacteriological decomposition, oxidation, and the reversal of chemical equilibria.

Bacterial decomposition obviously was not a problem in the prebiotic world. Neither was oxidation. The bulk of evidence indicates that the primitive atmosphere of the Earth was a reducing one, and that the present oxidative atmosphere and hydrosphere are relatively recent and, in fact, the product of biological activity. But, what stayed the hand of dissolution by the reversal of equilibria? The apparent answer is a fascinating one and one which returns us to the central theme of this book—structural stabilization. Once polymerization of the amino acid fragments has occurred the danger of dissolution diminishes, for the polymer may organize itself and its fellows into a stable configuration. Hydrophobic bonding comes into play, and the polymer is surrounded by a hydration sheath which in effect protects it from the solvent powers of the bulk water. These structural configurations may be further stabilized by the electrolytes in seawater. As an example, Wald (1954) calls attention to the astonishing transformation of an aqueous solution of highly random collagen filaments into highly organized collagen filaments on the addition of 1% NaCl.

At this point the aggregates may assume the form of coacervate droplets. Smith and Bellware (1966) have demonstrated that suspended proteinoid microspheres assume a coacervate structure upon repeated dehydration and rehydration such as might have occurred in the rocky crevices along the edge of some ancient sea (Hinton and Blum, 1965). The coacervate droplets are according to Oparin (1961), "the original precursors of primitive organisms. In structure, protoplasm is a complex coacervate. Oparin (1961) also theorizes that it was at this level that the mechanism of evolution, natural selection came into operation. Coacervates more stable and more successful than their fellows in the competition for "nutrient" material survived, while the weaker contemporaries perished and their substance was devoured.

One important structural feature separates the coacervate droplet from the cell; the cell has a wall. Lipids have one hydrophobic and one hydrophitic end on their long-chain molecules; they tend to pack in sheets, and such sheets or membranes made possible the isolation of the interior of the coacervate from its surroundings.

To repeat, Bernal (1961) claims that, with respect to the first beginnings of life, "there are, roughly, only about six [stages of molecular complexity] between simple atoms and something you can see in the microscope." He represents this hierarchy of polymer complexity first by the simple polypeptide chain, then the α-helix, next the folded coil, then linked folded coils as in myoglobin, and finally the linking of different folded coils together with a ribonucleic acid chain as in the tobacco mosaic virus—one of the simplest living things.

Although the beginnings were painfully slow, once certain crucial milestones were passed, the evolutionary life process accelerated at an ever-increasing rate as much more effective measures were realized to accomplish necessary ends, such as the evolution of vastly more effective biocatalysts and the development of superior techniques of supplying the energy necessary to hold at bay the forces of disintegration. The earliest energy source may have been analogous to fermentation, the conversion of organic material to waste products, CO_2, and utilizable energy. But, if no alternative energy source were forthcoming, life would have been doomed as the organic content of seawater was depleted. This catastrophe was averted by the evolution of an improved energy source, the utilization of CO_2, H_2O, and solar energy—photosynthesis. The photosynthetic process in turn released O_2, a powerful oxidant, and thereby laid the basis for a still more effective energy source—respiration. "To use an economic analogy, photosynthesis brought organisms to the subsistence level; respiration provided them with capital. It is mainly this capital that they invested in the great enterprise of organic evolution" (Wald, 1954).

At the same time, the release of O_2 by photosynthesis resulted in the formation of a layer of ozone high in the Earth's atmosphere that shielded the surface from lethal ultraviolet radiation so that life forms could finally give the shelter of the seas and venture upon the dry land. The rest was almost an anticlimax. The stage was set, the actors were in the wings, for the appearance of man.

No creature reigns forever. But, when our kind has gone, possibly as the result of our own folly, if not on the ruined Earth then on other planets in this and distant galaxies, wherever there is adequate water, life will continue to appear, flourish, and evolve.

> Nor can those motions that bring death prevail
> Forever, nor eternally entomb
> The welfare of the world; nor, further, can
> Those motions that give birth to things and growth
> Keep them forever when created there.
> Thus the long war, from everlasting waged,
> With equal strife among the elements
> Goes on and on. Now here, now there, prevail
> The vital forces of the world—or fall.
> Mixed with the funeral is the wildered wail
> Of infants coming to the shores of light.
>
> Lucretius
> *De Rerum Natura*

And now to this account, let me add some further remarks. Dr. Fox, who has prepared polymeric "proteinoids" from inorganic reactants at high temperatures in nonaqueous systems, has pointed out to me quite correctly that the preliminary steps of amino acid formation need not have occurred in an aqueous environment. The first building blocks may have been formed under nonaqueous conditions, but they were soon transferred to the seas so that evolution still remains, except perhaps for a tiny initial segment, an aqueous phenomenon. Nor need these steps have occurred on Earth, for as we saw in Chapter 1, the universe is remarkably rich in simple organic radicals and molecules. Amino acids have been synthesized in the absence of water from gases known to be present in the interstellar material (Wollin and Ericson, 1971). However I feel that the importance of the discovery of astrochemistry lies not in any relevance to Earthly biogenesis, but rather in underscoring the ubiquity of life's raw materials throughout the universe, thereby rendering the widespread distribution of life on other worlds that much more likely.

Textbooks such as this are expected to confine themselves to more or less accepted "facts." I would like to depart from this constraint and interject some of my own speculations at this point, not only because I like to break rules and outrage my colleagues, but even more importantly because I feel that these speculations, for all their lack of precision and confirmation at this time, make clear on the basis of relatively simple chemical principles why the evolution of the biomaterial has, in fact must have, taken the path it has taken. In addition to clarifying how the increased order that is life has been achieved, these speculations also serve to emphasize the major theme running through this book—the importance of water in our environment.

Water has a *unique* ability to form extended, three-dimensional, polymorphic H-bonded structures. We know that many crystalline forms of the solid exist, and there is every reason to believe that there are at least several different types of structural order in the liquid, both in the bulk phase and adjacent to solutes and surfaces (Chapter 7). Two different types of hydration atmosphere are described in Chapter 8—*coulombic hydration* surrounding ions, charge sites, polar solutes, and polar segments of macromolecules and *hydrophobic hydration* enveloping nonpolar solutes and hydrophobic segments of macromolecules and interfaces. While very imperfectly understood, the so-called hydrophobic bond is a formidable one, strong enough in some well-documented instances to dominate the coulombic repulsion of like charges. Just as we describe a chemical bond as having so much ionic and so much covalent character, we can describe the solvation atmosphere of a solute, macromolecule, or surface as having so much hydrophilic (coulombic) and so much hydrophobic character (Figure 8.8). In a previous book I noted that "if careless time has left any clues to the origins of life on this planet and to the course of the long evolutionary process, those clues are probably somewhere in the oceans of Earth" (Horne, 1969). Let us look at Figure 8.1 again: Size, hierarchal order increases as we go down this diagram, while the biomaterial tends to be in the lower right-hand corner of the several categories. Let us now imagine a scale or spectrum of natural chemical substances and aggregates constructed on the basis of the water structure of their hydration envelopes ranging from totally hydrophilic solutes, such as Na^+ ,Cl^-, and Ca^{2+} ions, on the one extreme to totally hydrophobic substances, such as the hydrocarbon constituents of petroleum on the other extreme. As our measure of hydrophilicity–hydrophobicity in the absence of a better parameter, let us be simple minded and take solubility. Furthermore, remembering that life is order and taking molecular size as a rough and ready measure of order, let us superimpose on this spectrum a curve representing the number of atoms in the molecule. Now let us look at what we have (Figure 10.7). Falling on our scale ". . . is a unique range of hydrophobic–hydrophilic character conducive to the formation of exceedingly large, exceedingly complex [and I would like to add here exceedingly highly ordered] carbon molecules and molecular aggregates—the biosubstance. From an architectural point of view life can be described as the proper balance of hydrophilic–hydrophobic character enabling the formation of such a high degree of organization" (Horne, 1968). A second look at the figure reveals that the hydrophilicity–hydrophobicity scale corresponds with the vector of geological time. Figure 10.7 then is a representation of the course of the evolution of life, of prebiological chemical evolution, of biological evolution itself, and even of aging, death, and decay. Analyzing Figure 10.7 and spelling out all of its implications could easily fill a fat volume, but here I will close this discussion by restricting myself only to a few comments, pointing out some of its features in reference to schematic molecular forms (Figure 10.8). Figure 10.8*A* represents the hydrophilically hydrated ionic constituents of the primordial oceans; Figure 10.8*B*, the simple amino acid building blocks with their hybrid hydration envelopes. Figures 10.8*C*, *D*, and *E* represent the build-up of macromolecules: Figures 10.8*F* and, G the formation of films and enclosed structures; Figure 10.8H, the aging process, entailing the destruction of hydrophilic sites and the subsequent crosslinking and dehydration of the biopolymeric material; and Figure 10.8*I*, the end-product of it all, hydrophobic petroleum.

A conspicuous feature of [Figure 10.7] is the gap in the solubility region from about 5×10^{-2} to 5×10^{-4} moles/liter. This gap does not represent a failure of the hypothesis. On the con-

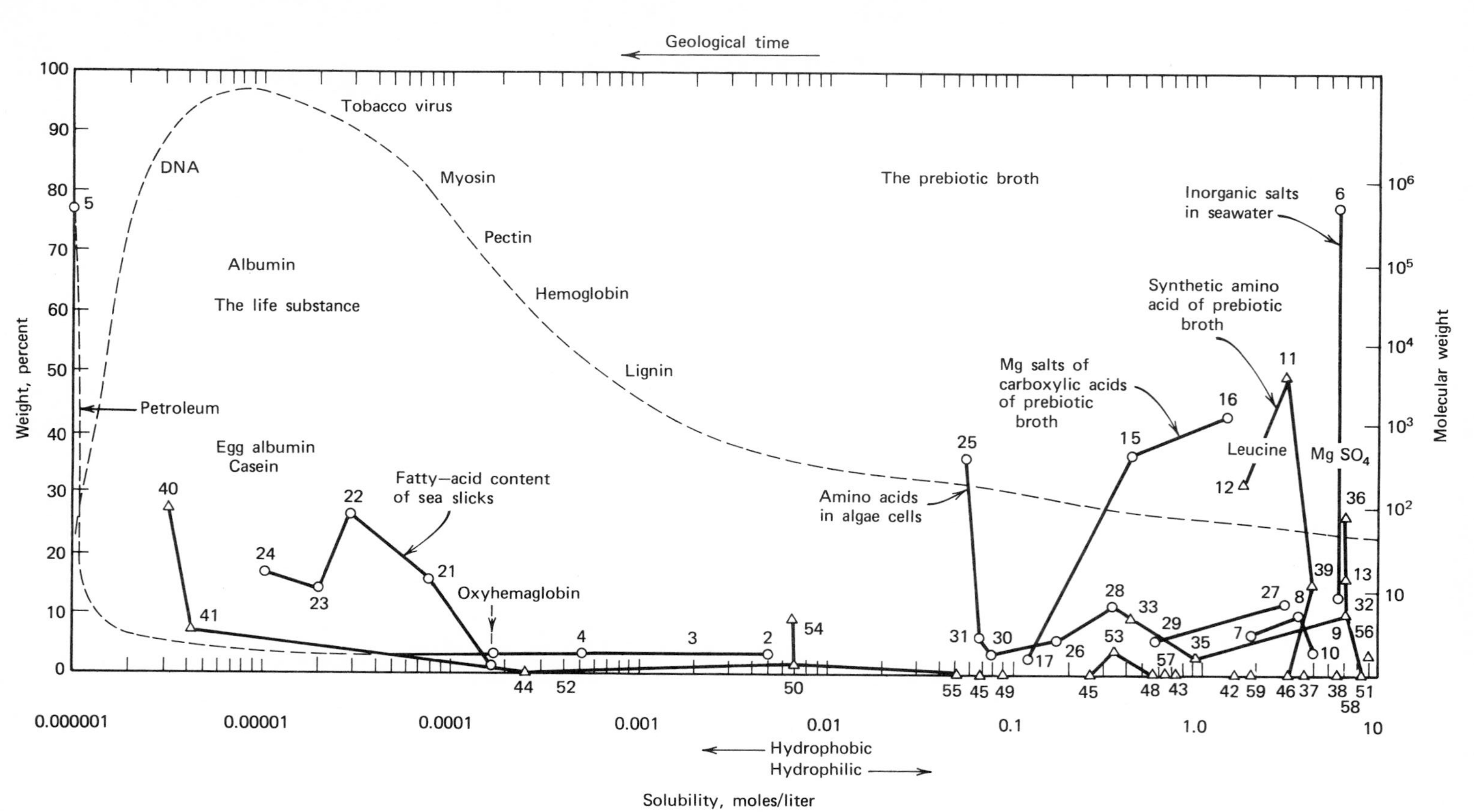

Figure 10.7 Origin, history, and fate of the biomaterial.

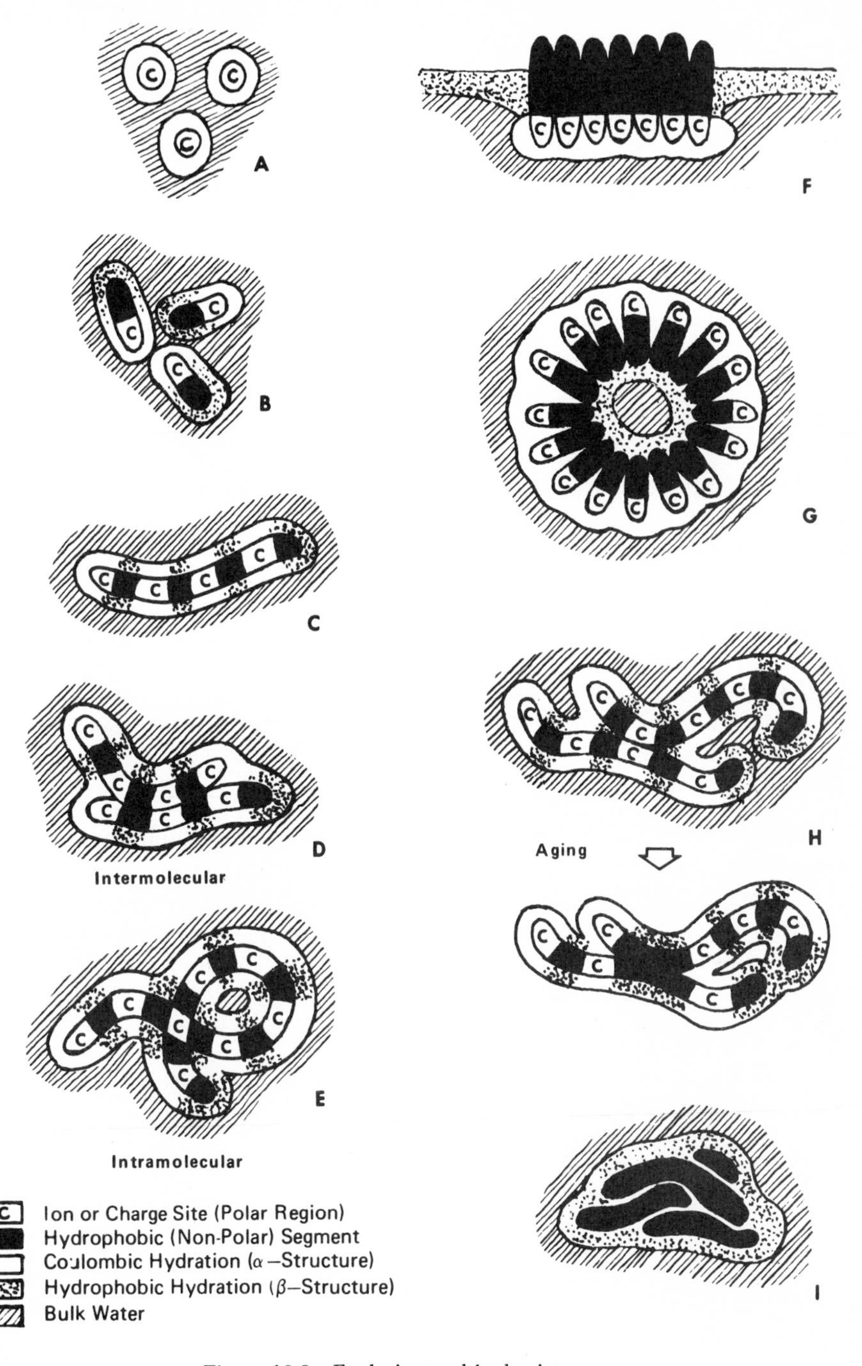

Figure 10.8 Evolution and hydration types.

trary, the intermediate sized life building block molecules that correspond to this region are not found in the . . . [present] biotic marine environment for two very good reasons. The first reason is that they are consumed by oxidation and organisms in the present oceans. The second reason is far more interesting for the gap is testimony to what very well may have been a sequence of the most difficult steps in biogenesis. The nature of the difficulty is twofold—concentration and stability (Horne, 1968).

We have already cited a few possible modes of the concentration of the prebiotic broth; with respect to the latter, in the words of Professor Wald (1954)

spontaneous union, step by step, of amino acid units to form a protein has a certain small probability . . . but the dissolution of the protein or of any intermediate product into its component amino acids is much more probable. . . . The situation we must face is that of the patient Penelope waiting for Odysseus, yet much worse: each night she undid the weaving of the preceding day, but here a night could readily undo the work of a year or a century.

It is perhaps at this point that the various efforts to reconstruct the steps of biogenesis are the weakest, and it is at this very . . . point where the role of water structure may have been most crucial. The smaller intermediates may have been stabilized by absorption, then as the organic molecules grew larger and larger they started to become increasingly stable by [virtue] of their very size. The most frequently offered explanation for such structural or architectural stabilization is inter- and especially intra-molecular H-bonding [Figure 10.8*D* and *E*]. . . . I would like to emphasize the possible stabilizing role of the local water structure in the intermediates' hydration sheathes. The life substance can be looked upon as a delicate balance between hydrophobicity and sufficient isolation from the solvent water to insure stability in the sense of sustained identifiability of the macromolecule and, on the other hand, hydrophilicity and sufficient interaction with the solvent water to enable the vital chemistry of the life processes to occur. . . .

Oxidizing conditions, bacteria, and thermodynamic reversibility in the present ocean have, as we have seen, eaten a gap in our life scale (Figure 10.8). However vestiges of ancient seas still exist, although quite probably in a highly permuted form, in the body fluids of living organisms (Figure 10.7 and Tables 10.8 and 10.9), and, it is significant to note, the substances present in such 'ancient seas' as human blood do indeed fill in and extend across the gap (Horne, 1968).

A word here now about petroleum formation (see also Chapter 14).

The susceptibility of compounds to low temperature chemical attack increases with increasing solubility, that is to say with increasing hydrophilicity or diminishing hydrophobicity. Or in still other words, the functional groups on macrobiopolymers tend to be surrounded by [coloumbic hydration] rather than [hydrophobic hydration] and they are the first to be consumed by chemical attack, either biological or non-biological. Consequently if not eaten *in toto* by some organism, "dead" biomaterial should assume an increasingly hydrophobic character as a consequence of the selective destruction of its hydrophilic constituents. Here, in line with more modern notions (Hanson, 1969), we have a program for the conversion of biomaterial to hydrocarbons without the agency of the high temperatures and pressures assumed by older theories of petroleum-formation. But this leaves the question of how the biomaterial escaped the cycle, escaped consumption by oxygen and organisms. Several possibilities come to mind. The biomaterial might accumulate in an anoxic zone of the oceans. Even in a shallow sea a population explosion could exhaust both nutrients and oxygen to produce a temporary lifeless anoxic zone. With increasing depth in a sediment the oxygen supply and even the population of [anaerobic] organisms decreases very rapidly, thus rapid deposition or slumping of a marine mud could bury biomaterial and remove it from the cycle (Horne, 1968).

10.6 THE STRUCTURE OF THE BIOSPHERE

In general terms, the structure of the Earthly biosphere is simply described—it is a very thin scum (averaging only 580 mg/cm^2) rather spottily covering the surface of this planet, the land as well as the ocean areas, but in more specific and detailed terms it becomes exceedingly intricate. Perhaps the easiest approach to the problem of the structure of the biosphere is to examine the distribution of carbon (Figure 10.9). The bulk of the carbon is in sedimentary deposits, some 20,000,000,000,000,000,000 metric tons of it. Much smaller but next in size is the fossil biosphere reservoir with an estimated 10,000 billion tons. The present dead biosphere represents some 3700 billion tons, 80% of which is in the oceans and 20% in soils. The viable biosphere adds up to only 466 billion tons and only a little more than 2% of this is in the oceans. What then becomes of the starry-eyed proposals to feed the exploding human popu-

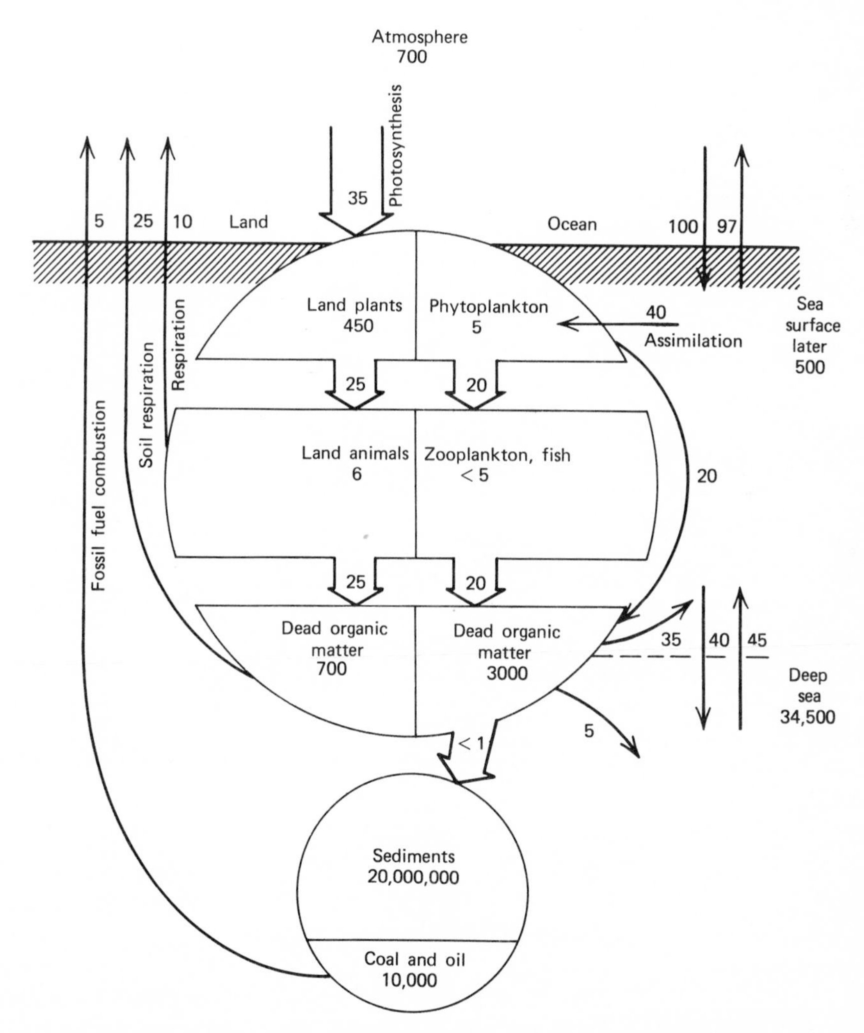

Figure 10.9 Distribution of carbon. Reservoir numbers are in billions of tons (metric) and flux numbers are in billions of tons per year (values, except for land animals, are from Bolin, 1970).

lation from the seas? Even the atmosphere, quantity-wise, is a more important carbon reservoir with 700 billion tons, mostly as CO_2. The carbon bulk of animals and plants in the oceans are roughly equal (animals are a bit less, Figure 10.9), but on land, plant carbon outweighs animal carbon by almost 60-fold. Only about 0.1% of the solar energy incident on Earth is utilized for the production of organic matter by photosynthesis. Nevertheless this corresponds to the annual worldwide production of (dry) organic material of 150,000,000 to 200,000,000 tons (Woodwell, 1970). The horizontal distribution of the biosphere is very spotty, both on land and in the oceans. Land plants depend on the availability of water and on temperature and soil conditions. Rainfall on the land masses is very uneven, thus land productivity of biomaterial is very uneven. And in addition to being uneven it is also highly variable in time, depending on human activity as well as changing climatic conditions. As examples of natural ecosystems, the net bioproduction level of a New York oak and pine forest is about 1200 g/m^2 yr and of a West Indies jungle 6000 g/m^2 yr; as examples of agricultural ecosystems, a Minnesota corn field yields about 1400 g/m^2 yr, while a sugar cane field in Java may yield 9400 g/m^2 yr. Comparing these values now with the hydrosphere, a Danish fresh water pond yields about 1000 g/m^2 yr, while a sewage pond in California yields almost 5000 g/m^2 yr; on the average, the yield of the open ocean is only 100 g/m^2 yr, the coastal zone 200 g/m^2 yr, and rich upwelling areas 600 g/m^2 yr (Woodwell, 1970). Again it is clear that the oceans are not very productive (see also Alverson et al., 1970; Anon., 1973).

The most productive natural areas of the land masses are the tropical jungles and of the oceans the coastal waters and regions of nutrient-rich upwelling. The least productive realms of the oceans are the anoxic waters—the graveyards of the seas, marine deserts characterized by death and decay (Richards, 1965; Horne, 1969), while the least productive land areas may very well be the dry valleys of Antarctica (Cameron, 1969; Horowitz et al., 1972). These valleys are so cold and dessicated that they have been compared with the environmental conditions believed to exist on Mars, and life in them is largely restricted to the microbial with occasional mosses in protected niches.

Man is increasing productivity with fertilizers and creating new arable land with irrigation, nevertheless his destructive practices—urban sprawl, road and other construction, overgrazing, slash-and-burn farming, and so on—are chewing away at the amount of potentially arable land at an even faster rate. In Europe and Asia, 80 to 100% of the potentially arable land is presently in use (Table 10.12). There appears to be further arable lands available in the tropic regions (Table 10.12), but much of this land has been leached of mineral nutrients and can support only so-called "bush-fallow" farming. As population pressures reduce the fallow period, this land is soon rendered useless. Then too a great deal of the land in West Africa, India, Southeast Asia, Australia, and parts of South America consist of high iron content lateritic soils that harden irreversibly with use.

Although the fact is not well known, the bioproductivity of the oceans is just as spotty as the land areas (Figure 10.10). Far from being an endlessly rich and inexhaustable soup, vast stretches of the open ocean, even of surface waters, are nearly lifeless deserts. Then of course as we descent into the deep ocean, since primary production is confined to the very shallow euphotic layer, the quantity of life sharply diminishes. The hope that aquaculture will supply man's growing food needs is

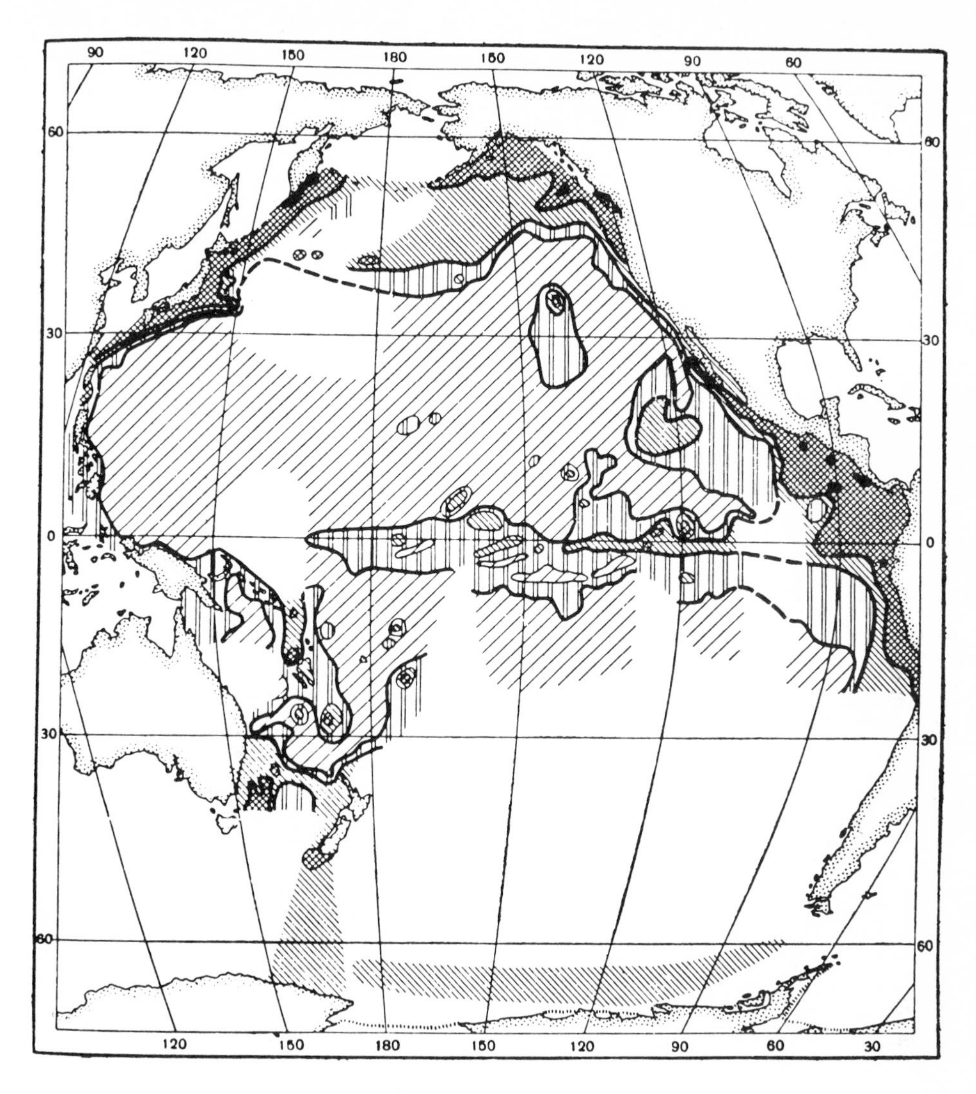

In the total water column

Milligrams of carbon fixed per day per square meter	Grams of carbon fixed per year per square meter
100	36
100 - 150	36 - 55
150 - 250	55 - 91
250 - 650	91 - 237
650+	237+

Figure 10.10 Gross primary production per unit area in the Pacific Ocean. Net production of plant material is about 60% of the values shown (from Roblenz-Mishke, 1965).

Table 10.12 Percentage of Potentially Arable Land Now Cultivated, and Acres Cultivated Per Person, on Different Continents (Hendricks, 1965)

Continent	Percent Cultivated	Acres Cultivated Per Person
Asia	83	0.7
Europe	88	0.9
South America	11[a]	1.0
Africa	22[b]	1.3
North America	51	2.3
U.S.S.R. (Europe–Asia)	64	2.4
Australasia	2[c]	2.9

[a] Tropical limitation.
[b] Desert and tropical limitation.
[c] Desert limitation.

about as realistic as the hope that magnetohydrodynamics will supply his growing energy needs. With optimum cultivation conditions, the biospheric film can utilize photosynthetically as much as a few percent of the incident visible light, but the overall average for the utilization of solar energy for land areas is only 0.1 to 0.3% and the average for water areas falls in the same range (Hutchinson, 1970).

The vertical distribution of the biosphere is every bit as complex and interesting as its horizontal distribution, in fact it often parallels it. In climbing a few miles up a mountain, we can pass through the same succession of ecosystems as we can do by driving hundreds of miles northwards along the coast (1000 vertical feet is the ecological equivalent of about 300 coastal miles). The biosphere is an exceedingly

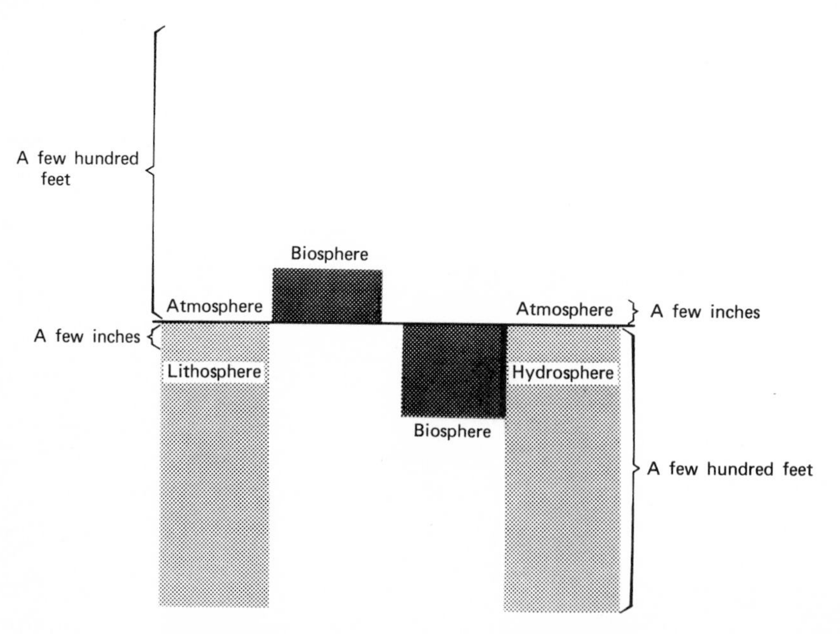

Figure 10.11 Vertical distribution of the biosphere.

thin film indeed and from this fact derives the supreme importance of interfacial chemistry in the chemistry of our environment. Land plants reach only 100 ft into the atmosphere, while above is a very dilute concentration of birds, insects, pollen, spores, bacteria, and so on (Figure 10.11). (Dormant life forms such as spores are sometimes referred to as the "parabiosphere." Modern radar is sufficiently sensitive to detect flying insects. Insect layers are detected as high as 600 m, and the vertical distribution of insects is sometimes correlated with the structure of the lower atmosphere (Richter et al., 1973).) Plant root systems can penetrate many feet in search of water; however, while burrowing macroorganisms may dig deeper, the majority of microorganisms are restricted to the thin upper layer of soil or humus. This is even more evident in the case of the hydrosphere where viable microorganisms penetrate fresh water and marine sediments only for a few inches. Methane from biological processes is found trapped deep in permafrost, but it is not known whether it is produced at such depths. We have already mentioned that in the seas the majority of the biomass is confined to the euphotic zone and shallow coastal waters (Figure 10.12).

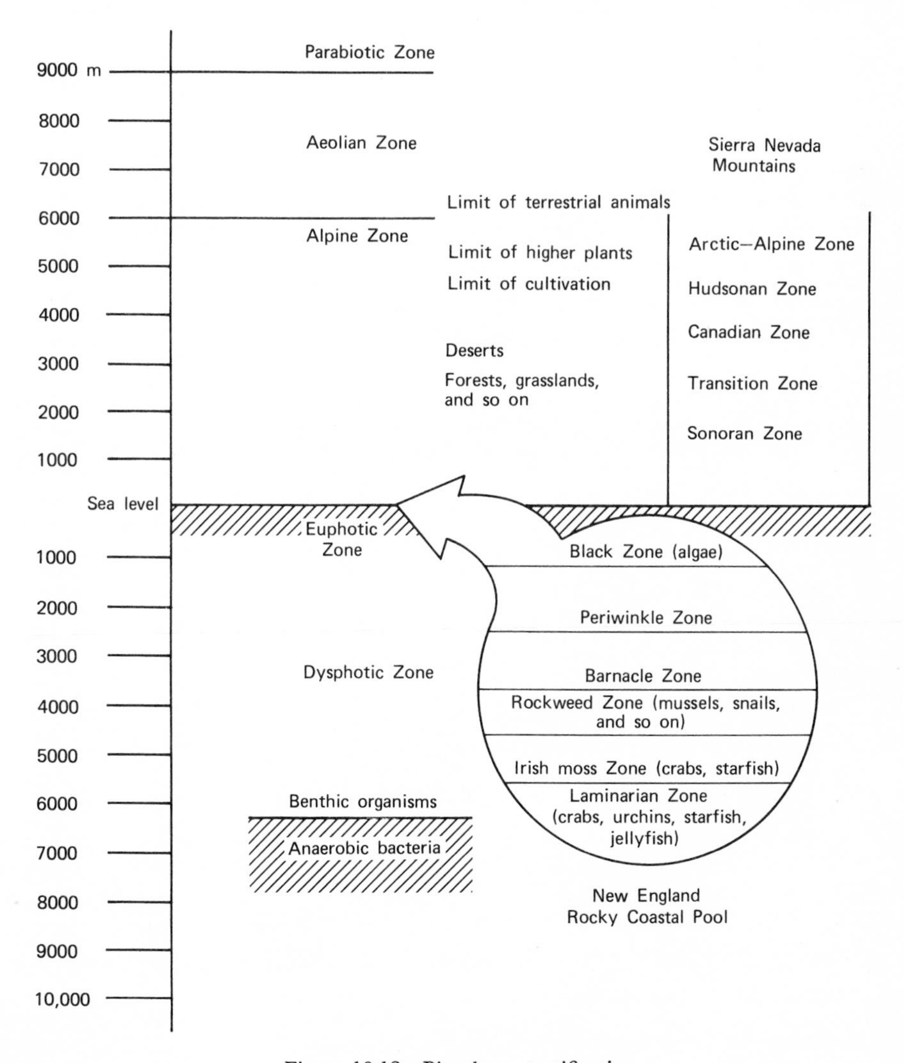

Figure 10.12 Biosphere stratification.

The margin of the dry land and the sea, the tidal zone and the beach, is a showcase of rapid and complex ecosystem stratification (Figure 10.12).

Very conscious of the fact that, although it deals with the largest of subjects, this has been one of the shortest chapters in this book, let us turn now away from the substance and structure of the biosphere to the next chapter, to another far too brief examination of some of the major chemical processes in the biosphere.

BIBLIOGRAPHY

P. L. Altman and D. S. Dittmer, *Blood and Other Body Fluids*, U.S. Air Force Aeronautical Systems Div. Tech. Rept. 61-199 (June 1961).

D. L. Alverson et al., *Science*, *168*, 503 (1970).

Anon., *Atlas of the Living Resources of the Seas*, FAO, UN, UNIPUB, New York, 1973.

R. T. Barber, *Nature*, *211*, 257 (1966).

J. D. Bernal, in M. Sears (ed.), *Oceanography*, Amer. Assoc. Adv. Sci. Pub. No. 67, Washington, D.C., 1961.

B. Bolin, *Sci. Amer.*, *223*(3), 65 (September 1970).

R. R. Brooks and M. G. Rumsby, *Limnol. Oceanogr.*, *10*, 521 (1965).

M. Calvin, *Science*, *130*, 1170 (1959).

M. Calvin, *Chemical Evolution*, Oxford Univ. Press, New York, 1969.

R. E. Cameron, in W. G. McGinnies and B. J. Goldman (eds.), *Arid Lands in Perspective*, Univ. Arizona Press, Tucson, 1969.

P. E. Cloud, Jr., in J. P. Riley and G. Skirrow (eds.), vol. 2, *Chemical Oceanography*, Academic, London, 1965.

E. S. Deevey, Jr., *Sci. Amer.*, *223*(3), 149 (September 1970).

D. A. T. Dick, *Cell Water*, Butterworths, Washington, D.C., 1966.

D. A. T. Dick, in R. A. Horne (ed.), *Water and Aqueous Solutions*, Wiley-Interscience, New York, 1972.

G. Dietrich, *General Oceanography*, Interscience, New York, 1963.

R. V. Eck et al., *Science*, *153*, 628 (1966).

V. A. Firsoff, *Life Beyond the Earth*, Hutchinson, London, 1963.

E. Frieden, *Sci. Amer.*, *227*, 52 (July 1972).

J. S. Fruton and S. Simmonds, *General Biochemistry*, Wiley, New York, 1953.

E. D. Goldberg, *Mem. Geol. Soc. Amer.*, *67*, 345 (1957).

W. E. Hanson, in P. H. Abelson (ed.), *Research in Geochemistry*, Wiley, New York, 1959.

S. B. Hendricks, in Nat. Acad. Sci. U.S.A.–Nat. Res. Council, *Resources and Man*, Freeman, San Francisco, 1969.

H. E. Hinton and M. S. Blum, *New Sci.*, *28*, 270 (1965).

R. A. Horne, *De Generatione et Corruptione: The Origin and Fate of Biomaterial in the Oceans and Its Dependence on Local Water Structure*, Arthur D. Little, Inc., Tech. Rept. No. 34 to Off. Nav. Res., U.S.N. (June 24, 1968). See also in Y. Miyake (ed.), *Proc. Internat. Symp. Hydrogeochem. Biogeochem., Tokyo, 1970*, Clarke, Washington, D.C., 1973.

R. A. Horne, *Marine Chemistry*, Wiley-Interscience, New York, 1969.

R. A. Horne, *Space Life Sci.*, *3*, 34 (1971).

R. A. Horne, *Space Life Sci.*, *3*, 235 (1972*a*).

R. A. Horne, *Science*, *177*, 1152 (1972*b*).

N. H. Horowitz et al., *Science*, *176*, 242 (1972).

J. A. Hunt and V. M. Ingram, *Nature*, *184*, 640 (1959).

G. E. Hutchinson, *Sci. Amer.*, *223*(3), 45 (September 1970).

V. M. Ingram, *Nature*, *180*, 326 (1957).

J. D. Kendrew, *Sci. Amer.*, *205*(6), 96 (1961).

J. D. Kendrew, *Science*, *139*, 1259 (1963).

G. N. Ling, in R. A. Horne (ed.), *Water and Aqueous Solutions*, Wiley-Interscience, New York, 1972.

J. Matheja and E. T. Degens, *Molecular Biology of Phosphates*, Fischer-Verlag, Stuttgart, 1972.

S. L. Miller, *Science*, *117*, 528 (1953).

S. L. Miller, *J. Amer. Chem. Soc.*, *77*, 2351 (1955).

M. Murayama, *Science*, *153*, 145 (1966).

A. I. Oparin, in M. Sears (ed.), *Oceanography*, Amer. Assoc. Adv. Sci. Pub. No. 67, Washington, D.C., 1961.

T. R. Parsons, *Prog. Oceanogr.*, *1*, 205 (1963).

F. A. Richards, in J. P. Riley and G. Skirrow (eds.), *Chemical Oceanography*, Vol. I, Academic, London, 1965.

J. H. Richter et al., *Science*, *180*. 1176 (1973).

W. E. Ricker, in Nat. Acad. Sci. U.S.A.–Nat. Res. Council, *Resources and Man*, Freeman, San Francisco, 1969.

O. I. Roblenz-Mishke, *Okeanologiya*, *5*, 325 (1965).

E. Schrödinger, *What Is Life*, Cambridge Univ. Press, Cambridge (England), 1944.

K. Schwarz and D. B. Milne, *Science*, *174*, 42 (1971).

I. S. Shlovskii and C. Sagan, *Intelligent Life in the Universe*, Holden-Day, San Francisco, 1966.

S. M. Siegel and K. Roberts, *Space Life Sci.*, *1*, 131 (1968).

A. E. Smith and F. T. Bellware, *Science*, *152*, 362 (1966).

H. U. Sverdrup, M. W. Johnson, and R. H. Fleming, *The Oceans*, Prentice-Hall, Englewood Cliffs, 1942.

W. Traub and K. A. Piez, *Adv. Protein Chem.*, *25*, 243 (1971).

G. Wald, *Sci. Amer.*, *191*, 44 (August 1954).

P. J. Wangersky, *Amer. Sci.*, *53*, 358 (1965).

G. M. Woodwell, *Sci. Amer.*, *223*(3), 65 (September 1970).

G. Wollin and D. B. Ericson, *Nature*, *233*, 615 (1971).

ADDITIONAL READING

J. D. Bernal, *The Origin of Life*, World, Cleveland, Ohio, 1967.

J. Brooks and G. Shaw, *Origins and Development of Living Systems*, Academic, London, 1973.

J. D. Costlow, Jr. (ed.), *Fertility of the Sea*, Gordon & Breach, New York, 1971.

S. W. Fox (ed.), *The Origins of Prebiological Systems*, Academic, New York, 1964.

S. W. Fox and K. Dose, *Molecular Evolution and the Origin of Life*, Freeman, San Francisco, 1972.

K, A. Grossenbacker and C. A. Knoght, *Origins of Prebiological Systems and Their Molecular Aggregates*, Academic. New York, 1965.

L. J. Henderson, *The Fitness of the Environment*, Beacon, Boston, 1958.

D. H. Kenyon and G. Steinman, *Biochemical Predestination*, McGraw-Hill, New York, 1969.

L. Margulis (ed.), *Origins of Life*, Gordon & Breach, New York, 1971.

A. I. Oparin, *Origins of Life*, Dover, New York, 1953.

C. Ponnamperuma, *The Origins of Life*, Dutton, New York, 1972.

D. L. Rohlfing and A. I. Oparin, *Molecular Evolution*, Plenum, New York, 1972.

M. G. Rutten, *The Origins of Life*, Elsevier, New York, 1971.

E. Samuel, *Order In Life*, Prentice-Hall, Englewood Cliffs, 1972.

E. Schoffeniels (ed.), *Biochemical Evolution and the Origins of Life*, Elsevier, New York, 1971.

11

THE CHEMISTRY OF LIFE

11.1 INTRODUCTION

In the previous chapter, we examined briefly the origin, composition, and distribution of the biosphere. In this chapter, we turn our attention to a few of the more significant chemical processes occurring in the biosphere, and especially, that result in important chemical mass transport between the biosphere and other segments of our environment. Whole libraries are filled with our accumulated knowledge of the chemistry of life, ridiculously incomplete and fragmentary though that knowledge may be. Consequently in this chapter, as in the previous one, I must be exceedingly Procrustean making no effort to touch upon the countless topics encompassed by biochemistry, but rather confining our attention to only a very few subjects. I have chosen fermentation, photosynthesis, and respiration, the major energy-utilizing metabolic processes because, not only for their biological importance, but also for the enormous impact that these processes have had and continue to have on the chemistry of our environment. Finally, to this chapter I append a short section on the degradation of biological material.

11.2 FERMENTATION AND ANAEROBIC PROCESSES

The biosphere most evident to us is dominated by big organisms, by animals and plants, by elephants and flies, by forests and ferns, and, of course, by man. And the reason why these organisms are so big, are so obvious, are so dominant, is that they run on two remarkably efficient metabolic processes—photosynthesis and respiration. But less conspicuous—in the dirt beneath our feet, in the soil covering the fields (Figure 11.1), in the humus blanketing the forest floor, in the mud of swamps and marshes, and in the sediments in bays, estuaries and even at the bottom of the sea—is a whole and different world of microorganisms. While individually minute these organisms far outnumber us, and their effect on the chemistry of our environment is considerable. We have already seen that they put more carbon monoxide into our atmosphere than do all our engines and quite possibly more sulfur dioxide (initially as H_2S) than do all our furnaces and smelters. These tiny organisms are a major path for keeping carbon, once fixed in the life cycle, cycling and returning it via the soil to plants (Figure 11.2). But in this work, they are not always highly efficient, and some carbon "leaks" out of the biosphere (in addition to CO_2 from respiration) into the lithosphere. Man has reconnected this "lost" material back into the loop by his combustion of fossil fuels and release of CO_2 to the atmosphere. We more flashy

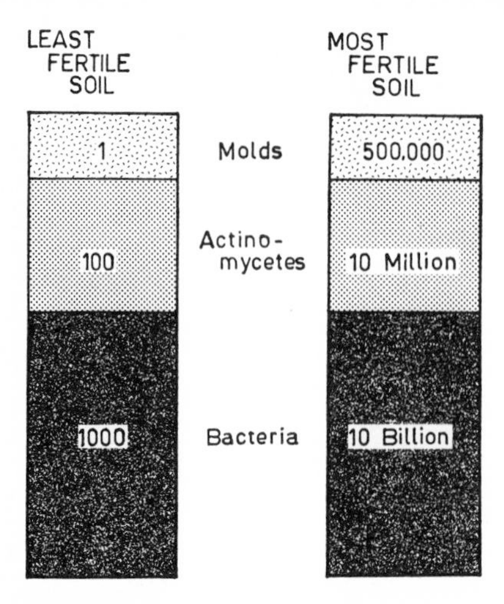

Figure 11.1 Microbial composition of typical soils (from Hahn, © 1968 used with permission of Doubleday & Co., Inc.).

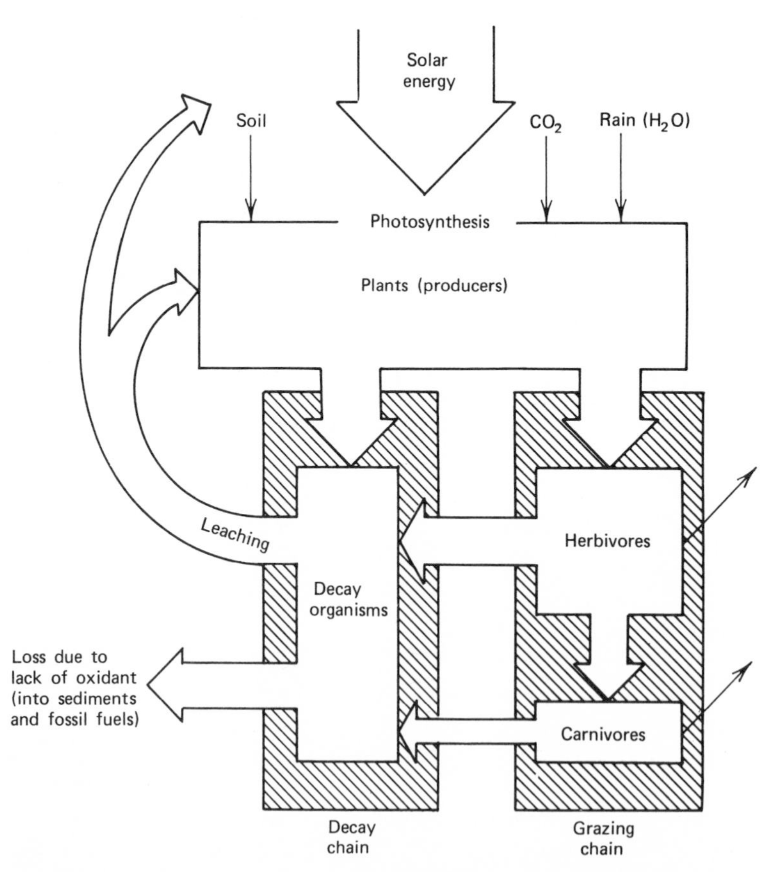

Figure 11.2 Role of the decay chain in completing the loop of the life cycle.

creatures are having our day on this planet, but many of these more modest life forms with their simpler chemistry were at work long before we appeared and with their greater insensitivity to radiation and chemical pollution may very well persist long after we have made our possibly dramatic exit. They are also the life forms most likely to be encountered on planets with more severe environmental conditions than Earth. Pre-Cambrian paleontology has revealed the traces of ancient microorganisms (see Calvin, 1969). Some of the present day organisms may not be too dissimilar from ancient forms, while others are the products of long evolutionary sequences. Some microorganisms are very sensitive to environmental conditions, others have the ability to survive under astonishing conditions of temperature and acidity (Table 11.1). With respect to their chemistry, microorganisms fall into two broad types: *autotrophs* whose energy comes from light (photoautotrophs) or the oxidation of H_2 or inorganic N, S, and Fe compounds (chemoautotrophs) and whose carbon comes mostly from CO_2 and *heterotrophs* whose energy and substance comes from organic matter. Only autotrophs, it should be noted, can create new organic material; heterotrophs depend on existing organic material. Thus prior to the advent of photosynthesis, the total biomass could increase only by extremely slow, tedious, nonbiological synthesis. In the absence of photosynthesis, life would have been slowed to a virtual steady state or may even, since organic material is lost from the life cycle by sedminentation, have declined and vanished.

Pasteur described fermentation as "La vie sans air." Perhaps the most familiar fermentation process is the production of ethanol from carbohydrates by the action of yeast. We have mentioned already the release of methane into the atmosphere from

$$\text{carbohydrate} \xrightarrow{\text{yeast}} CH_3CH_2OH + CO_2 \qquad (11.1)$$

microbial processes. Anaerobic methane-producing microorganisms are important in a host of decay processes including sewage treatment (Pohland, 1971). Microorganisms not only produce methane, but they can also methylate mercury (see Chapter 7) thereby greatly facilitating the mobility of this toxic heavy metal in the environment. We have also mentioned the importance of microorganisms in the

Table 11.1 Dominants in Extreme Environments (From Alexander, 1971)

Environment	Typical Dominant
Hot Springs	*Synechococcus*
Acid Hot Springs	*Cyanidium caldarium*
Heating Dung Piles	*Bacillus*, *Streptomyces*
Snow Surfaces	Various algae
Antarctic Grassland	*Corythion dubium*
Brines, Salty Products	Halophilic bacteria
Great Salt Lake	*Uroleptus*
Solar Salt Pond	*Dunaliella*
Alkaline Lake	*Arthrospira platensis*
Roots of Desert Plants	*Bacillus*
Desert Soil	*Bacillus*
Soil During Drought	*Streptomyces*

world's sulfur budget. We encounter them again in our discussions of soil chemistry (Chapter 4) and of the nitrogen cycle (Chapter 15), while in this chapter we see an example of their chemical ingenuity—in the absence of oxygen they can use nitrate and even sulfate ions as oxidants.

It is interesting to note that anaerobic fermentation processes, like photosynthesis (see Section 11.3), simultaneously produce and environmentally segregate oxidized and reduced material—in Equation 12.1, part of the carbohydrate is oxidized to CO_2, part reduced to an alcohol.

Various clues persist that reveal some of the secrets of prerespiratory biochemistry. The copper-containing enzyme cytochrome oxidase, which activates and binds oxygen, may be a remnant from the first evolutionary experiments in oxidative metabolism, while the low tolerance of obligate anaerobes to oxygen and the presence of oxygen-removing peroxisome enzymes are reminders that life began in the absence of oxygen and that this powerful oxidant was probably a very effective poison for primitive organisms (Cloud and Gibor, 1970).

11.3 PHOTOSYNTHESIS

In Chapter 1 we noted that the incident solar energy arriving at Earth produces two chemical "hotspots" (Figure 1.3); one when this radiation first impinges on the upper reaches of the Earth's atmosphere and a second when the remaining energy impinges on the Earth's surface, especially the biosphere. This energy is utilized to stir the atmosphere and the oceans (Oort, 1970) and to drive the hydrologic cycle on the one hand, and on the other to serve as the primary energy source of all living things (Woodwell, 1970).

The life processes we have just discussed consume organic chemicals, and had evolution stopped short of photosynthesis, life would have soon exhausted the organic material so laboriously synthesized by nonbiological processes. While it may not have dwindled away, the total biomass would have remained fixed at a relatively modest level. A creative process was needed that could synthesize complex organic molecules from simple inorganic ones. Such a process was evolved—photosynthesis.

In the generalized overall photosynthetic process, carbohydrates are synthesized from carbon dioxide (from the atmosphere) and a hydrogen donor, H_2A, such as H_2O, H_2S, or even H_2.

$$nCO_2 + 2nH_2A \rightarrow (CH_2O)_n + nH_2O + 2nA \tag{11.2}$$

Most commonly the hydrogen donor is water, as in the cases of the blue-green algae

$$6CO_2 + 6H_2O \xrightarrow[\text{chlorophyll}]{\text{light}} \underset{\text{glucose}}{C_6H_{12}O_6} + 6CO_2 \tag{11.3}$$

and all the higher plants. However in the case of photosynthetic sulfur bacteria H_2A is H_2S, and in the case of the nonsulfur purple bacteria even organic compounds. In the larger geochemical view, the net result of photosynthesis is twofold: (*a*) the production of complex organic molecules from simple inorganic ones, and (*b*), as Hutchinson (1970) has pointed out, an oxidized realm in our environment (the atmosphere and the hydrosphere) and a reduced realm (the biosphere including the bodies of organisms and the products of their metabolism, death, and decay (Figure 11.3)). This

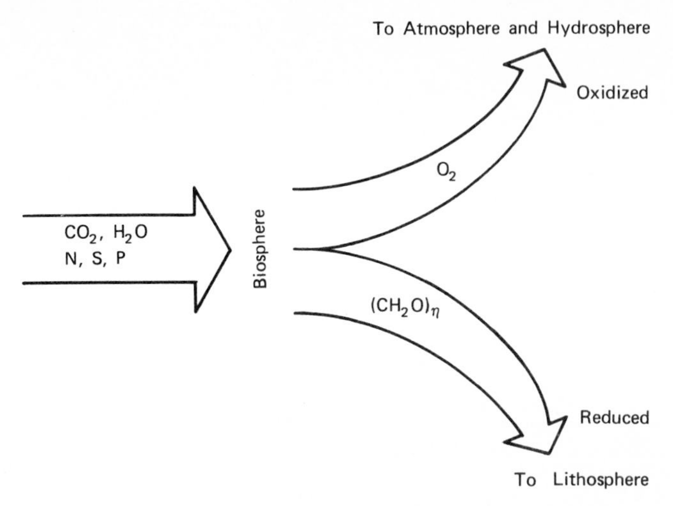

Figure 11.3 Chemical segregation in our environment by photosynthesis.

molecular architectural assault on the cosmic disorder costs dearly, some 690 kcal/mole glucose in the case of Equation 11.3, and this energy is supplied by sunlight. The process is enormously facilitated by the catalyst chlorophyll in which the catalytic function of the metal, magnesium, is enhanced by inclusion in the same sort of porphyrin ring systems that we encountered in the previous chapter in con-

```
             CH2CH3         CH3
              |     H        |
              C     C        C
             // \  // \     / \\        O
            /    C      C      \\     //
   CH3—C         |      |       C—C
         \       |      |      /   |
          C—N           N—C        |   OCH3
         //    \       /   \\      |    |
    H—C          Mg          C—C—C=O
         \     /      \    /    H
 H        C—N           N—C                  CH3              CH3
   \     //                \                  |                |
    C=C—C                   CH(CH2)2—C—O—CH2—C=C—(CH2)3—CH
   /  H      |          |            ||       H                |
 H           C          C            O                         |
           // \       // \                                     |
          C     C          CH         CH3           CH3        |
          |     H          |           |             |         |
         CH3              CH3   CH3—CH—(CH2)3—CH—(CH2)3
```

Chlorophyll

nection with the iron in heme and catalase (Marks, 1969). In 1965, Professor Woodward received the Nobel Prize for the total synthesis of chlorophyll (Woodward et al., 1960). Discovered two centuries ago by Joseph Priestly, the photosynthetic process at last has begun to yield its secrets to modern science (Franck and Loomis, 1949; Rabinowitch, 1951; McElroy and Glass, 1961; Kamen, 1963), and it turns out to be exceedingly complex making Equation 11.3 a very gross oversimplification (Figures 11.4, and 11.5). The primary photosynthetic step is not carbohydrate production from H_2O and CO_2 but the conversion of photoenergy into chemical energy (Figure 11.3), which is then utilized in a number of ways. The electrons of chlorophyll are

Adenosine Diphosphate(ADP)

Adenosine Triphosphate(ATP)

H^+

Reduced Triphosphopyridine nucleotide ($TPNH_2$)

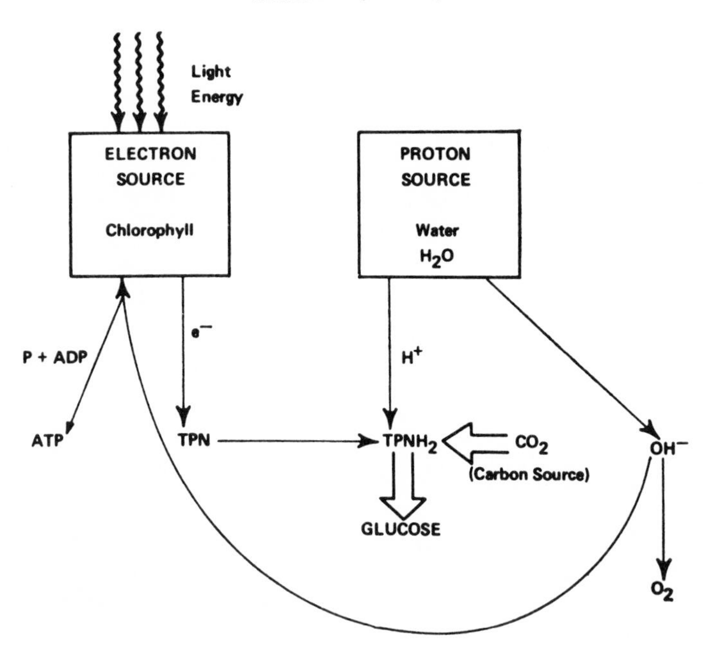

Figure 11.4 The photosynthetic process (from Horne, 1969).

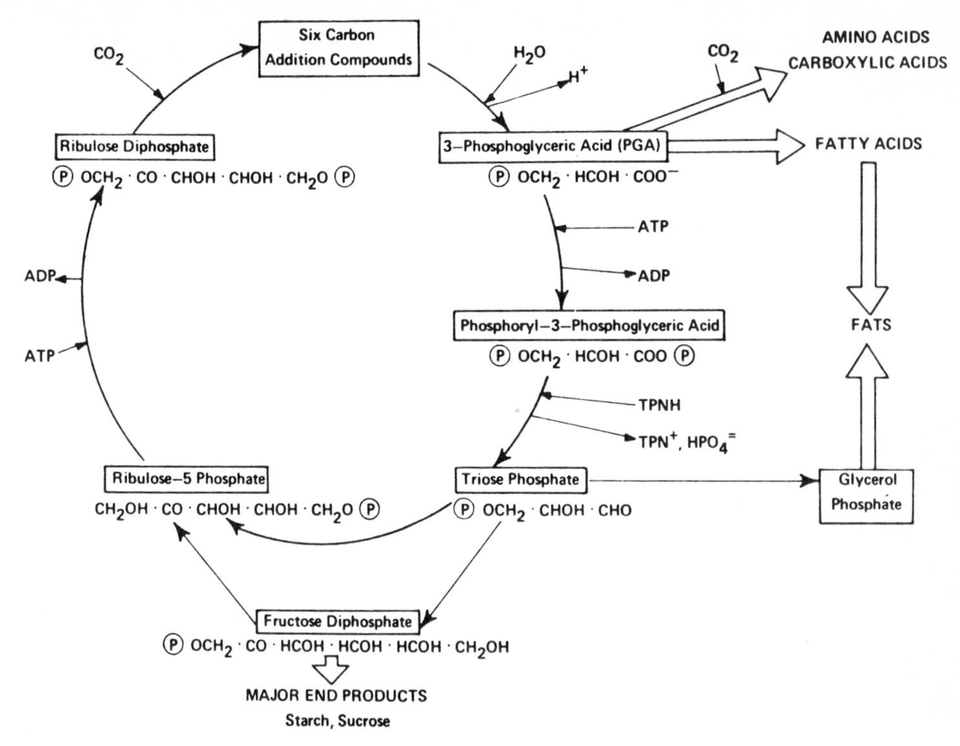

Figure 11.5 Products of the photosynthetic process (from Horne, 1969).

excited by the absorption of a photon of light. This excitation enables the green pigment to act as an electron source for the production of adenosine triphosphate (ATP). With water serving as a proton source the triphosphopyridine nucleotide (TPN) is reduced to form $TPNH_2$ which is turn in responsible for the reduction of CO_2 and the production of glucose. Carbohydrates are far from being the only products of photosynthesis: fatty acids, fats, and carboxylic and amino acids are also formed as shown in Figure 11.5, again highly oversimplified. Table 11.2, which summarizes the results from experiments using ^{14}C tracer, is interesting inasmuch as it shows where the C taken up as CO_2 goes.

The concentration of chlorophyll a in seawater can be measured readily by spectroscopic means [Table 11.3 (see Parsons, 1963)] and interpreted (not so readily) to give an indication of primary production or the bulk of the phytoplankton crop. There is presently concern that anthropogenic contaminants such as pesticides and mercury (see Harriss et al., 1970) might reduce planktonic photosynthesis in the world's oceans. In the case of land plants in an article entitled "The Sun's Work in a Cornfield," Lemon et al. (1971) describe a crop-simulation soil/plant/atmosphere model that takes into account crop, climate, and soil parameters as well as solar energy, thermal energy, and water and CO_2 transport. But remember, the biosphere is at most an exceedingly thin film, the density of solar energy flux is not great, and while chlorophyll may be a remarkable catalyst the overall efficiency to solar energy, utilization is very small, only about 0.1 to 0.3% for land surfaces. Thus "the ma-

Adenosine Diphosphate (ADP)

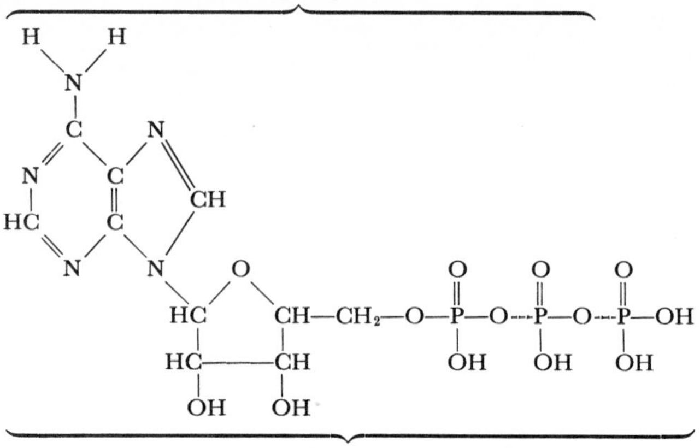

Adenosine Triphosphate (ATP)

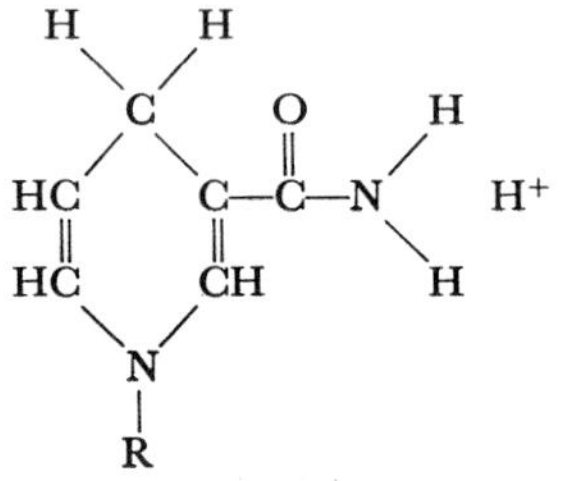

Reduced Triphosphopyridine nucleotide ($TPNH_2$)

chinery by which energy enters the living world is clearly quite tenuous" (Hutchinson, 1970).

Enormous quantities of water pass through the biosphere, but only a tiny fraction, it should be noted, is actually incorporated into organic material by photosynthesis. Twenty tons of fresh weight crop will contain about 15 tons of water in transit and only 3 tons of water fixed chemically in the 5 tons of dry weight organic material, yet in the course of the growing season, the roots will take up 2000 tons of water, nearly all of which is transpired back into the atmosphere (Penman, 1970).

While raising as many questions as it has answered, research on photosynthesis in recent years has come up with some most fascinating findings (Marx, 1973). The efficiency of photosynthetic conversion of solar to chemical energy differs greatly for different plant species. High efficiency plants such as maize and sugar cane have net photosynthetic efficiencies from 42 to 63%, while low efficiency plants, including such major food plants as wheat and rice, have efficiencies of only 12 to 31%. The high efficiency plants are called C_4 plants because their enzymic photosynthetic process involves 4-carbon atom acids such as oxaloacetic, malic, and aspartic acids (Figure 11.6). Low efficiency plants, on the other hand, are called C_3 plants because the first product of their CO_2 assimilation by the Calvin (1957) cycle (Figure 11.7) is the 3-carbon 3-phosphoglyceric acid (PGA). The chemistry of the two types of plant differ markedly. The photosynthetic efficiency of the C_4 plants is at a maximum between 35 and 47°C, whereas the C_3 plants are less temperature dependent and have their maximum around 25°C. Normal concentrations of O_2 inhibit C_3 but not C_4 plants (the Warburg effect), C_4 plants use water more efficiently, and even the

Table 11.2 Gas Exchange and Products of Photosynthesis by *Acetabularia* Chloroplast Preparations in ^{14}C-Bicarbonate (*p*H 8) (From Dodd and Bidwell, 1971)

	μmole CO_2/hr/mg Chlorophyll	
Gas Exchange	Crude Preparation	Purified Preparation
CO_2 Uptake in Light	52	45
CO_2 Production in Dark	12	7
Products of Photosynthesis	^{14}C-content, % of total fixed in 10 min	
Insoluble Products	20.4	25.1
Sucrose	8.9	9.1
Triose Phosphate	1.1	1.2
Hexose Monophosphate	5.6	8.3
Sugar Diphosphate	11.0	7.4
Phosphoglyceric Acid	6.9	5.0
Glycolic Acid	23.3	21.3
Glyceric Acid	0.7	0.5
Malic Acid	0.9	0.3
Citric Acid	0.1	0
Succinic Acid	0.1	0
Alanine	3.9	7.5
Serine + Glycine	12.9	11.4
Aspartic Acid	3.6	2.4
Glutamic Acid	0.6	0.5

morphology of the two plant types differ. Plants also respire, and dark respiration in mitochondria produces needed energy by the oxidation of sugars to water and CO_2 (net photosynthesis is the difference between CO_2 assimilated and CO_2 lost in respiration). Photorespiration, in contrast, does not appear to produce useful energy. Greater photorespiration, therefore, may be at least a partial explanation for the reduced efficiency of C_3 plants. These findings are of far more than academic interest

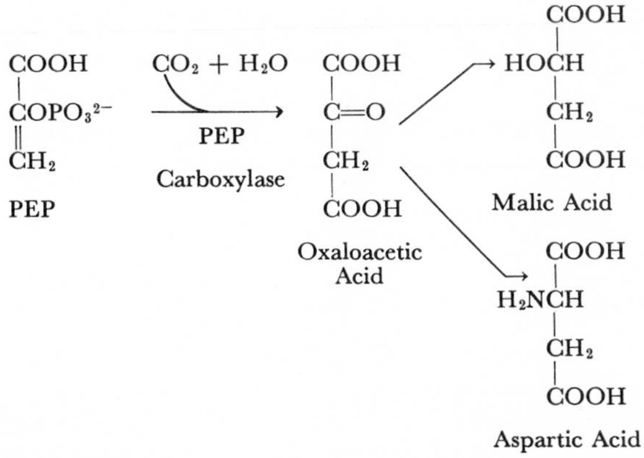

Figure 11.6 Carbon dioxide fixation by the Hatch–Slack pathway (from Marx, © 1973 by the American Association for the AAAS).

Table 11.3 Oceanic Distribution of Chlorophyll a (From Parsons, 1963, with permission of Pergamon Press, Ltd.)

Area		Date	Depth	Chlorophyll a mg/m^3
Atlantic Ocean				
Continental shelf off the coast of New York	<100 m	February	Euphotic av.	1.5–3.5
	<200 m	February	Euphotic av.	0.3–1.0
	<2000 m	February	Euphotic av.	ca. 0.2
	>2000 m	February	Euphotic av.	<0.2
Sargasso Sea		Seasonal av.	Surface	0.3
ca. 40°N		October	Surface	0.1–0.4
ca. 50–60°N		Summer	Surface	0.3–2.0
ca. 60–70°N		July	Surface	0.3–2.8
Pacific Ocean				
Approximate average over 5° square of latitude and longitude regardless of date (East Pacific)	Northern Peru current	—	Surface	0.7
	ca. 10°N	—	Surface	0.1–0.8
	15–40°N	—	Surface	<0.3
	45–55°N	—	Surface	0.2–2.0
Northwest Pacific ca. 160°E	30–40°N	Summer	Euphotic av.	0.2 or less
	40–50°N	Summer	Euphotic av.	0.2–0.6
Southwest Pacific	20–30°S	April	Surface	ca. 0.1
	30–35°S	December	Surface	<0.2
	35–40°S	June	Surface	0.4–0.6
	30–35°S	April	Surface	0.2–0.6
Chukchi Sea		August	Surface to 50 m	ca. 0.2–10

for they might provide a rational scientific basis for increasing the yield of food crops. Another suggestion for crop improvement is to induce plants to switch from sucrose to protein production by means of chemical sprays (Bassham, 1971). Earlier we mentioned that plant yields can be increased by CO_2 enrichment in greenhouses and even in the fields (Allen et al., 1971). These approaches, it should be noted, might help to circumvent the water pollution problems that have been created by the massive application of chemical fertilizers.

11.4 RESPIRATION

In Nature's great scheme of things, photosynthesis in plants and respiration in animals are chemically opposite and complementary processes that maintain the delicate balance of the life cycle. We have already called attention to the similarity of the catalysts involved in these very different reactions. Chemically, respiration is the

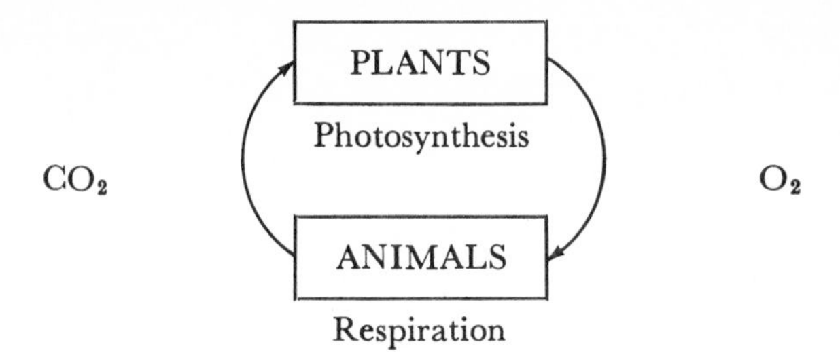

reverse of photosynthesis; the latter builds up complex organic molecules (Eq. 11.3), the former breaks them down; the one utilizes energy, the other makes it available.

$$C_6H_{12}O_6 + 6O_2 \rightarrow 6CO_2 + 6H_2O + \text{Energy} \quad (11.4)$$
$$(686\ \text{kcal/mole glucose})$$

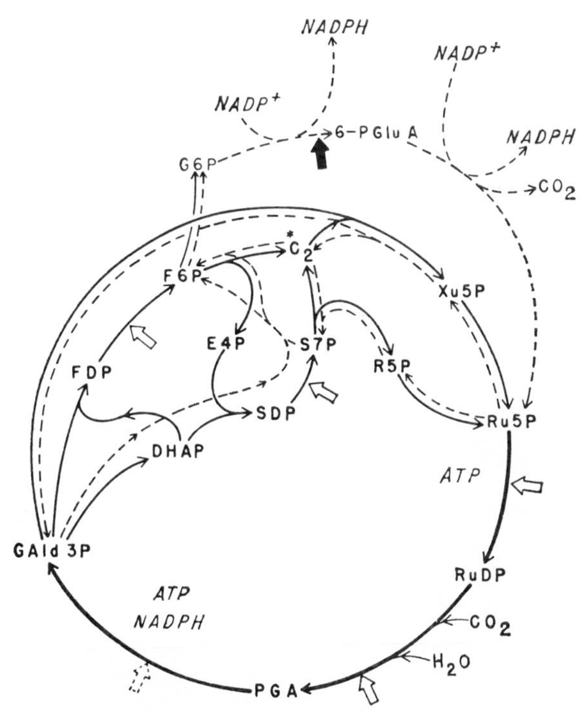

Figure 11.7 Photosynthetic metabolism (solid arrows) and the reductive pentose phosphate cycle (Calvin cycle) (dashed lines). White arrows are at sites of metabolic regulation which are active in the light. Dark arrows show reactions which are activated in the dark. The dashed arrow indicates a reaction for which evidence of regulation is so far limited to studies of the properties of isolated enzymes. The $^*C_2$ indicated in the cycle is actually enzyme-bound thiamine pyrophosphate-glycolaldehyde, an intermediate in the two reactions mediated by transketolase. (PGA, 3-phosphoglycerate; GAld3P, glyceraldehyde-3-phosphate; DHAP, dihydroxyacetone phosphate; FDP, fructose-1,6-diphosphate; F6P, fructose-6-phosphate; G6P, glucose-6-phosphate; E4P, erythrose-4-phosphate; SDP, sedoheptulose-1,7-diphosphate; S7P, sedoheptulose-7-phosphate; R5P, ribose-5-phosphate; Ru5P, ribulose-5-phosphate; Xu5P, xylulose-5-phosphate; RuDP, ribulose-1,5-diphosphate; NADPH and $NADP^+$, nicotinamide adenine dinucleotide phosphate, reduced and oxidized forms, respectively; 6PGluA, 6-phosphogluconate; ATP, adenosine triphosphate.) (From Bassham, © 1971 by the American Association for the Advancement of Science with permission of the AAAS.)

The respiratory oxidation of foodstuffs such as carbohydrates by animal metabolic systems makes enormous quantities of energy available to the organism that the organism can then use in a multitude of ways ranging from maintenance of body temperature in the case of homeothermic animals (thus freeing them from the caprices of the environment) and muscular movement to the synthesis of vastly more intricate organic substances.

The preferred oxidant is, of course, the strongest, oxygen, but if the oxygen supply is depleted, as in anoxic waters and sediments, muds, and soils, anaerobic organisms can utilize nitrate as an oxidant (denitrification). When all the nitrates and nitrites are consumed sulfate bacterial reduction can occur. If we represent the organic matter of the biosphere as having the composition $(CH_2O)_{106}(NH_3)_{16}H_3PO_4$, these three types of oxidative metabolism can be represented by

$$(CH_2O)_{106}(NH_3)_{16}H_3PO_4 + \boxed{138O_2} \rightarrow 106CO_2 + 122H_2O + 16HNO_3 + H_3PO_4 \quad (11.5)$$

$$(CH_2O)_{106}(NH_3)_{16}H_3PO_4 + \boxed{84.8HNO_3} \rightarrow 106CO_2 + 148.8H_2O + 42.4N_2 + 16NH_3 + HP_3O_4 \quad (11.6)$$

and

$$(CH_2O)_{106}(NH_3)_{16}H_3PO_4 + \boxed{53SO_4^{2-}} \rightarrow 106CO_2 + 106H_2O + 16NH_3 + 53S^{2-} + H_3PO_4 \quad (11.7)$$

> In cells of higher organisms the oxidative system of enzymes and electron carriers is located in the special organelles called mitochondra. These organelles can be regarded as efficient low-temperature furnaces where organic molecules are burned with oxygen. Most of the released energy is converted into the high energy bonds of ATP (Cloud and Gibor, 1970).

In other words, the mitochondrial apparatus is "the principal energy transducers in all aerobic organisms" (Green and Hatefi, 1961). This "burning" takes place in two steps. The first step, glycolysis, harks back to anaerobic fermentation processes in prephotosynthetic organisms, and the second step is aerobic respiration, which completes the oxidation of the pyruvic acid resulting from the former all the way to CO_2 and H_2O by means of a series of Fe-containing respiratory enzymes, ubiquinone and cytochromes.

$$CH_3\text{—}\overset{\overset{\displaystyle O}{\|}}{C}\text{—}\overset{\overset{\displaystyle O}{\|}}{C}\text{—}OH \begin{cases} \nearrow CH_3CHOHCOOH,\ CH_3CH_2OH \text{ (anaerobic fermentation)} \\ \quad\ \text{Lactic acid} \qquad\qquad \text{ethanol} \\ \searrow CO_2,\ H_2O \text{ (aerobic respiration)} \end{cases}$$

Pyruvic acid

Respiration occurs not only on land and in the oceans but in sediments as well; however, oxygen-uptake measurements indicate that the respiration of benthic organisms that inhabit the deep ocean bottom (or 76% of the world ocean's total floor) is very much less (0.4 to 0.6 ml/m^2hr) even than in shallower waters on the continental shelves (Pamatmat, 1971; Smith and Teal, 1973).

11.5 DECAY AND BIODEGRADATION

Life is an interval in a sequential history of an exceedingly complex conglomerate of chemical reactions. The nature of these reactions and their kinetics change and these changes represent major milestones in the existence of the organism. Taking ourselves as examples, prior to the severance of the umbilical cord, the fetus' chemistry is linked to that of its mother. Nevertheless prenatal and immediate postnatal chemistry has some unique and fascinating features, for just at the development of the morphology of the fetus parallels the biological evolution of the species, the chemistry of the fetus gives us hints of the ancient forms from which we are descended. A most obvious example is the large water content of the infant, a reminder of our aqueous origin, which falls off to "normal" soon after birth. The new-born infant also has many disease fighting chemical mechanisms that diminish or vanish as he matures. Infants even have a special enzyme for digesting milk. If the person continues to drink cow's milk his body will continue to manufacture this catalyst; however, an adult who stopped drinking milk in infancy may lose his capacity to digest it and can be made ill by milk. Our well-intentioned shipment of powdered milk to undeveloped countries has created problems for this reason.

By the time we have reached 20 yr of age, Nature expects that we have served our biological function (procreation) and she discards us. We begin to die. Since we no longer have the need to be sexually attractive, we get bald and fat. Each day thousands of our brain cells die *and are not replaced*. Even if we do not fall prey to particular chemical malfunction and metabolic diseases, our body chemistry becomes increasingly ineffective. Some of the changes are quite abrupt and traumatic—such as menopause in women. The chemistry of aging (Strehler, 1962; Metzger, 1972, Chapter 2; Rockstein and Baker, 1972) is a fascinating subject, very controversial and, surprisingly enough, rather neglected. Let me mention just one aspect of aging here.

As we age, the protein collagen in our connective tissues loses water. The biopolymer also becomes more highly cross-linked (Sinex, 1968). My own pet hypothesis, and everybody seems to have one, is that polar site annihilation, say by free radicals, not only cross-links but at the same time changes the character of the protein's hydration sheath, edging it along from a more hydrophilic to a more hydrophobic situation (that is to say, pushing the biomaterial more to the left in Figure 10.7). Such a shift to a more hydrophobic nature would render the biosubstance increasingly ineffectual. Chemicals that interfere with free-radical formation, such as butylated hydroxytoluene (BHT) can retard aging, and a number of sulfur compounds, such as

MEA or cysteamine $H_2N—CH_2—CH_2—SH$

WR 2721 $H_2N—(CH_2)_3—NH—(CH_2)_2—S—P(=O)(OH)—OH$

AET $H_2N—(CH_2)_2—S—C(NH_2)=NH$

Cysteine $HOOC—CH(NH_2)—CH_2—SH$

which reduce radiation damage by removing free radicals can lengthen the life expectancy of experimental animals. A quantum mechanical theory of aging emphasizes the role of radiation-induced "proton errors" that confound the transmission of chemical information in the organism. Following spawning in fresh waters, ocean salmon die from accelerated aging. In a couple of weeks, a salmon ages as much as we do in 20 to 40 yr. The fish's controlling pituitary gland swells and runs amuck speeding up the metabolic processes so that the animal literally burns itself away. Another effect is the softening of the fish's bones, a familiar feature of aging in humans, as Ca dissolves out to maintain the Ca level in the bloodstream and can no longer be replaced because of the low Ca content of fresh water.

Death is another chemical experience for an organism. At the "moment of death" some chemical processes stop, some start, some change rate, and some continue unaffected, but the inability to define death, to determine whether it is a process or an event, has raised some serious ethical and legal as well as scientific questions (Morison, 1971; Kass, 1972). In the case of our deaths, one of the most obvious chemical events is the cessation of breathing and the resulting termination of oxygen transport by the respiratory pigments thereby extinguishing the warmth produced by metabolic oxidation.

In the chemistry of our environment, the chemical processes representing death and decay are every bit as important as those representing life and growth. Paradoxically, life is a destructive process, whereas death and decay are processes of regeneration, returning material to the viable biosphere (Figure 11.2). Although even the mythologies of the most primitive peoples universally recognize the significance of death and regeneration (and as I write these lines on Easter Sunday I am reminded that Christianity is no exception), modern technocratic society has lost sight of these principles, and many of our present sociological and environmental problems are the consequence, not only of our failure to control birth, but of the attempt to control death and prolong life beyond its useful bounds.

In Chapter 10, we described the biomaterial as a sort of band on the broad hydrophilicity–hydrophobicity spectrum. The purpose (dirty word in the natural sciences) of the delicate balance between hydrophobic and hydrophilic character is to realize two seemingly contradictory ends—stability and reactivity. The hydrophobic hydrocarbon backbone of the biopolymer enables the species to preserve its identity in the aqueous environment, while the functional sites containing O, N, P, S, and other atoms attached to the backbone enable the macromolecule to perform its biological function. But these reactive sites, not surprisingly, make the species vulnerable, and it is at these sites that the forces of decay focus their attack. Organisms are capable of metabolizing even the most refractory carbon-containing compounds (such as the plastic sheathing on submarine cables), but they much prefer to fasten their digestive "teeth" at points where N, O, P, and S atoms weaken the chemical defenses and the hydrophobic wall of the macromolecule. Since these sites are the points of decompositional attack, we find that as biomaterial decays the relative concentrations of its constituent elements change—N, O, P, and S disappear, and C and H are left (Table 11.4). The younger low grade coals, for example, contain more sulfur, the older, more completely carbonized anthracite less. As we noted earlier in connection with the biodegradation of spilled oil (Chapter 8; see also Blumer and Sass, 1972), the more toxic components also tend to be the more persistent. Some very complex structures are exceedingly stable; the porphyrin ring system, so crucial to biocatalysts such as hemoglobin and chlorophyll, is very persistent and is found in very ancient rocks, a spoor of the existence of the earliest living forms. The other fragment from chlorophyll, the pristane side chain is also found in geological samples even although the parent pigment is not (Maxwell et al., 1971).

Table 11.4 Diagenesis of Life's End Products[a] (Data from Degens and Matheja, 1968)

Bioresidues in Soil after Microbial Degradation	C 55%	H 5%	O 35%	N 3%	P 1%
Bioresidues in Sediments	56	5	22	4	1
Kerogen	80	7	10	2	1
Crude Oil	85	13	1	1	1

[a] Note also the increased reduction reflected in the increase in H.

Our environment, as we have seen, is an unusual one—it is oxidizing rather than reducing, and it is permeated by a biosphere. These two considerations largely determine the fate of chemicals in that environment. Nature long ago devised usages for coping with the elements and chemical structures familiar to her, but she is baffled by many of the exotic organic chemicals, such as halogenated hydrocarbons, that man is dumping into the environment, and she is ill equipped to deal with them. Figure 11.8 lists a few of the types of chemical groups and configurations that are highly resistant to oxidative and/or biological attack. The deposition and great persistence of petroleum in the environment under anoxic conditions testifies to the relative stability of hydrocarbons and the C—C and C—H bonds. Substitution by halogens and branching increases this resistance. On the other hand, unsaturated linkages, partially oxidized groups, and linkages and functional groups containing

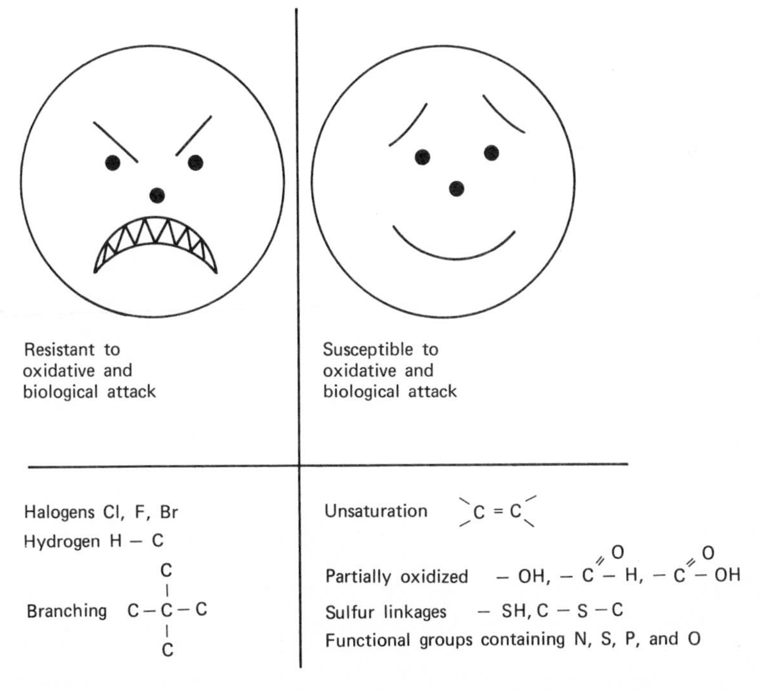

Figure 11.8 Factors determining chemical persistence in our environment.

N, S, P, and O are weak points in a molecule's defenses and provide a ready handle for oxidative and/or biological attack.

Figure 11.9 summarizes some of the more important principles governing the persistence of chemicals in our environment. The atmosphere and the biosphere are reservoirs, but the hydrosphere and the lithosphere (including the fossil biosphere) are sinks. Reactant and/or soluble chemicals will be washed out of the atmosphere by precipitation and will have residence times of only a week or two at the most. Even finely divided particulate material and unreactive gases will remain in the atmosphere for only a matter of years, a century at the most (the atmosphere may be considered a leaky sink for the noble gases). The residence time of chemicals in the viable biosphere is limited by the life expectancy of organisms or the organisms constituting a predation web. Chemicals and their oxidation products that are not precipitated or absorbed tend to be highly persistent in the hydrospheric sink, while insoluble, precipitated, and absorbed species disappear into the lithosphere where, except for some slow leaching and weathering, they remain for very long periods of time, the residence times here being determined by exceedingly slow major geological processes.

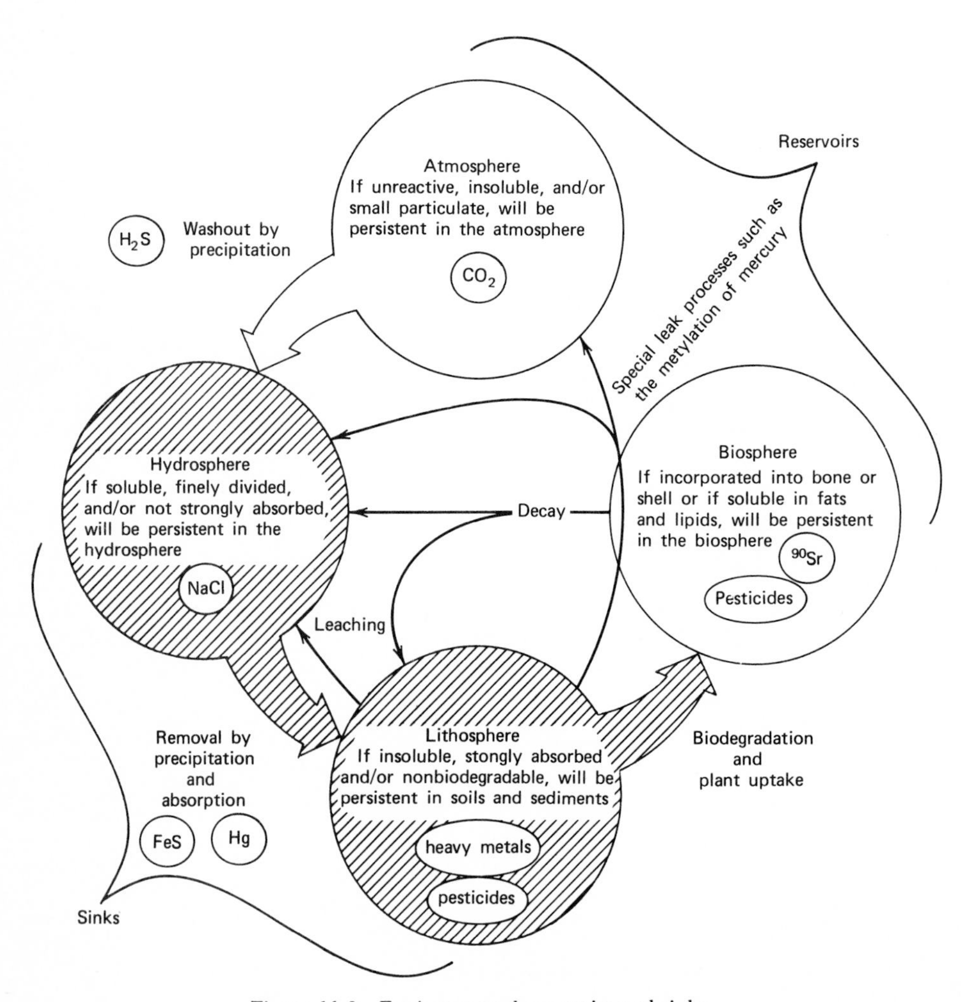

Figure 11.9 Environmental reservoirs and sinks.

The importance of biodecay processes in returning organic material in the life cycle, in consuming the oxygen content of natural waters, and in sanitary engineering practice such as sewage treatment (Chapter 16) has long been recognized. But in recent years, biodegradability has assumed enlarged recognition as the consequence of the accumulation and resulting damage from highly refractory man-made plastics (Table 11.5A) and chemicals such as detergents, pesticides (Table 11.5B) and other halogenated hydrocarbons (Alexander, 1965*a*; Dugan, 1972; see Chapter 14). Inasmuch as their primary source is petroleum, plastics production is a drain on that nonrenewable resource. In 1966, the United States generated some 3,800,000,000 lb of plastic waste (70% of which was packaging waste), and this amount is expected to more than double by 1976 (Srinivasan, 1972). Some of this material is incinerated, but unless the incinerator is specially designed, plastics tend to melt and foul in-

Table 11.5 A. Plastic Products with Long Persistence in the Environment

Polyethylene ("Poly")	H H H H H H H H H H H —C—C—C—C—C—C—C—C—C—C—C— H H H H H H H H H H H
Polyvinyl chloride	H H H H H H H H —C—C—C—C—C—C—C—C— H Cl H Cl H Cl H Cl
Vinylidine chloride ("Saran")	H Cl H Cl H Cl H Cl —C—C—C—C—C—C—C—C— H Cl H Cl H Cl H Cl
Polyisobutylene	H CH_3 H CH_3 H CH_3 H —C—C—C—C—C—C—C— H CH_3 H CH_3 H CH_3 H
Polytetrafluoroethylene ("Teflon")	F F F F F F F F F F F —C—C—C—C—C—C—C—C—C—C—C— F F F F F F F F F F F

B. Approximate Persistence of Chlorinated Hydrocarbon Insecticides in Soil (Data from Alexander, 1965*b*, with permission of the Soil Science Society of America.)

Toxaphene	>6 yr
Heptachlor	>9 yr
Aldrin/Dieldrin	>9 yr
DDT	>10 yr
Hexachlorocyclohexane	>10 yr
Chlordane	>12 yr

cinerator operation. Since these large polymers are alien to Nature, microorganisms do not have the enzymes to attack them. Much of the observed microbial attack on plastics is directed at the plasticizers they contain—much smaller molecules. It is possible, however, to make biodegradable plastics, so what is needed is not so much better technology as stricter laws (for more about the plastic waste problem see Saudinger, 1970; Dugan, 1972; Walters and Hueck-Van Der Plas, 1972).

BIBLIOGRAPHY

M. Alexander, *Microbial Ecology*, Wiley, New York, 1971.

M. Alexander, *Adv. Appl. Microbiol.*, *7*, 35 (1965*a*).

M. Alexander, *Proc. Soil Soc. Amer.*, *29*, 1 (1965*b*).

L. H. Allen, Jr., S. E. Jensen, and E. R. Lemon, *Science*, *173*, 256 (1971).

J. A. Bassham, *Science*, *172*, 526 (1971).

M. Blumer and J. Sass, *Science*, *176*, 1120 (1972).

M. Calvin, *The Path of Carbon in Photosynthesis*, Prentice-Hall, Englewood Cliffs, 1957.

M. Calvin, *Chemical Evolution*, Oxford Univ. Press, New York, 1969.

P. Cloud and A. Gibor, *Sci. Amer.*, *223*(3), 111 (September 1970).

E. T. Degens and J. Matheja, *J. Brit. Interplanet. Sci.*, *21*, 52 (1968).

W. A. Dodd and R. G. S. Bidwell, *Nature*, *234*, 45 (1971).

R. R. Dugan, *Biochemical Ecology of Water Pollution*, Plenum, New York, 1972.

J. Franck and W. E. Loomis, *Photosynthesis in Plants*, Iowa State College Press, Ames, 1949.

D. E. Green and Y. Hatefi, *Science*, *133*, 13 (1961).

P. A. Hahn, *Chemicals from Fermentation*, Doubleday, Garden City, 1968.

R. C. Harriss et al., *Science*, *170*, 736 (1970).

R. A. Horne, *Marine Chemistry*, Wiley-Interscience, New York, 1969.

G. E. Hutchinson, *Sci. Amer.*, *223*(3), 45 (September 1970).

M. D. Kamen, *Primary Processes in Photosynthesis*, Academic, New York, 1963.

L. R. Kass, *Science*, *173*, 698 (1971).

E. Lemon et al., *Science*, *174*, 371 (1971).

G. S. Marks, *Heme and Chlorophyll*, Van Nostrand, London, 1969.

J. L. Marx, *Science*, *179*, 365 (1973).

J. R. Maxwell et al., *Quart. Rev. Chem. Soc.*, *25*, 571 (1971).

W. D. McElroy and B. Glass (eds.), *Light and Life*, Johns Hopkins Press, Baltimore, 1961.

N. Metzger, *Men and Molecules*, Crown, New York, 1972.

R. S. Morison, *Science*, *173*, 694 (1971).

A. H. Oort, *Sci. Amer.*, *223*(3), 55 (September 1970).

M. M. Pamatmat, *Limnol. Oceanogr.*, *16*, 536 (1971).

T. R. Parsons, *Prog. Oceanogr.*, *1*, 205 (1963).

H. L. Penman, *Sci. Amer.*, *223*(3), 100 (September 1970).

F. G. Pohland, *Anaerobic Biological Treatment Processes*, Adv. Chem. Ser. No. 105, Amer. Chem. Soc., Washington, D.C., 1971.

E. Rabinowitch, *Photosynthesis*, Interscience, New York, 1951.

M. Rockstein and G. T. Baker (eds.), *Molecular Genetic Mechanisms in Development and Aging*, Academic, London, 1972.

F. M. Sinex, *Treatise Collagen*, *2*, 409 (1968).

K. L. Smith, Jr., and J. M. Teal, *Science*, *179*, 282 (1973).

V. R. Srinivasan, *Tech. Rev.* (*M.I.T.*), *74*(6), 45 (May 1972).

J. J. Staudinger, *Disposal of Plastics Waste and Litter*, Soc. Chem. Ind. Monogr. No. 35, London, 1970.

B. L. Strehler, *Time, Cells, and Aging*, Academic, New York, 1962.

A. H. Walters and E. H. Hueck-Van Der Plas (eds.), *Biodeterioration of Materials*, Wiley, New York, 1972.

R. B. Woodward et al., *J. Amer. Chem. Soc.*, *82*, 3800 (1960).

G. M. Woodwell, *Sci. Amer.*, *223*(3), 65 (September 1970).

ADDITIONAL READING

E. Baldwin, *Dynamic Concepts of Biochemistry*, Cambridge Univ. Press, Cambridge (England), 1947.

P. D. Boyer, H. Lardy, and K. Myrbäck (eds.), *The Enzymes*, Academic, New York, 1960.

M. Calvin, "The Chemistry of Life," *Chem. Engr. News*, 96 (May 22, 1961).

T. W. Goodwin (ed.), *Biochemistry of Chloroplasts*, Academic, New York, 1967.

D. M. Greenberg (ed.), *Metabolic Pathways*, Academic, New York, 1960.

M. D. Hatch, C. B. Osmond, and R. O. Slayter (eds.), *Photosynthesis and Photorespiration*, Wiley-Interscience, New York, 1971.

D. H. K. Lee and D. Minard (eds.), *Physiology, Environment, and Man*, Academic, New York, 1970.

R. Lemberg and J. W. Legge, *Hematin Compounds and Bile Pigments*, Interscience, New York, 1949.

K. Shibata, A. Takamiya, A. T. Jagendorf, and R. C. Fuller (eds.), *Comparative Biochemistry and Biophysics of Photosynthesis*, Univ. Tokyo Press, Tokyo, 1968.

I. Zelitch, *Photosynthesis, Photorespiration, and Plant Productivity*, Academic, New York, 1971.

12

MAN'S PERTURBATION OF THE BIOSPHERE

12.1 INTRODUCTION

In the two previous chapters, we saw how the biosphere, largely through the process of photosynthesis, has and continues to perturb the chemistry of the environment of this planet. In this chapter, we turn our attention to a few of the countless ways in which one organism—man—has and continues to perturb the biosphere. We examine, not only the impact of his resource and energy demands on the creatures about him, but also the "internal pollution" that he inflicts upon himself by the inhalation and ingestion of alien chemical substances. In an attempt to understand why the biosphere is so sensitive to chemical disruption, we also examine a selected number of the chemical substances that establish and maintain delicate ecological hierarchies.

12.2 MAN AS A CHEMICAL FACTORY: MATERIAL USE AND WASTE

Organisms move chemical materials about in the environment. By far the most important of these transport and redistribution processes is photosynthesis by plants. This process takes highly oxidized carbon in the form of CO_2 from the atmosphere along with water from the hydrosphere (Figure 12.1) and fixes them in the biosphere as partially reduced carbon compounds such as carbohydrates. Here the carbon flux pauses until the organism dies, and upon death, decay follows, amounting to further reduction of the carbon to free carbon (in coal) or hydrocarbons (in petroleum). Through the action of microorganisms some of the carbon is returned from the dead to the living biosphere, some, however, is deposited in the lithosphere where, until the advent of man and his burning of fossil fuels, it remained locked and lost from the biosphere.

While green plants still dominate the chemistry of our environment, the chemical effects produced by man are fast becoming very appreciable. In earlier chapters, we noted that he is removing materials from the lithosphere at rates comparable to the removal by natural weathering processes and that he has succeeded in increasing slightly the CO_2 content of the Earth's atmosphere. At the same time in at least one location (the waters of the New York bight), he is dumping solid wastes into the ocean at a rate exceeding natural sedimentation. And all the while man's activities consume and/or dirty an appreciable fraction of the terrestrial hydrosphere. Even the oceans have not escaped chemical changes that he has brought about. As human population continues to grow, man's chemical impact on our environment will continue to enlarge.

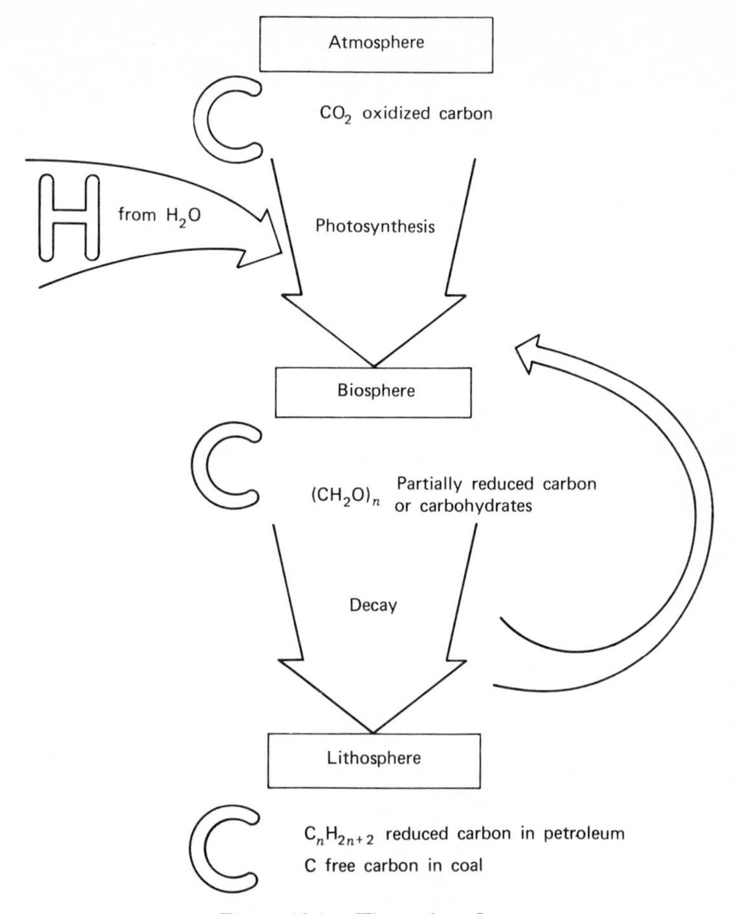

Figure 12.1. The carbon flux.

As we saw in Chapter 4, man takes enormous quantities of materials, metals, minerals, and fossil fuels from the lithosphere (Figure 12.2). The metals and minerals are nonrenewable resources, and many of them will have to be recycled if their supply is not to be exhausted. Fossil fuel formation is so slow a geological process that these materials too for all intent are nonrenewable resources, and they are not recyclable. In these instances, subsitutes must be found. In return for this drain upon the resources of the lithosphere man dumps upon the land a rain of solid wastes (Figure 12.2) ranging from abandoned automobiles (7,000,000/yr in the United States), building rubble and mine tailings to a carpet of litter composed of bottles (28,000,000,000/yr in the United States), cans (48,000,000,000/yr in the United States), and plastics.

Man's attack on the biosphere is a more complicated rape (Figures 12.2 and 12.3). In more enlightened countries, forests are treated like a crop (Figure 12.3*D*) and are carefully replanted or young growth is spared in order to insure that the forests remain a renewable resource. But, alas, such care is often the exception rather than the rule, and even in the United States clear cutting is still permitted—often of the

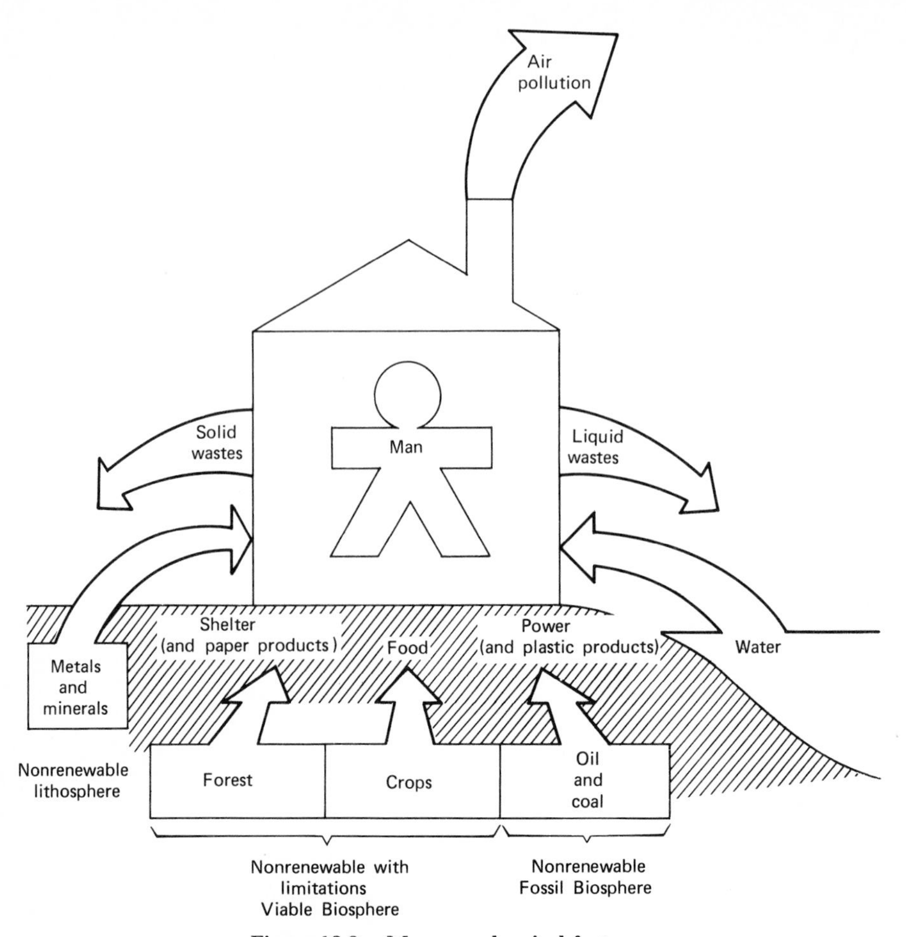

Figure 12.2. Man as a chemical factory.

public's trees for private profit (Wood, 1971). Deforestation as a result of cutting, radiation (Woodwell, 1970), or herbicides (see discussion that follows) gives rise to increased erosion and a dumping of nutrients, especially nitrogen compounds, into waterways (Bormann et al., 1968; Smith et al., 1968; Likens et al., 1969); however, since deforestation can also result from natural causes such as fire and blight, Nature is equipped to respond to such emergencies and revegetation and restoration of the ecosystem is often rapid (Marks and Bormann, 1972).

Forests are still being cleared, especially in tropical areas, for agricultural purposes (Figure 12.3*C*). In principle, this represents the replacement of a lesser by a more necessary green, but again unless precautions are taken, and they usually are not, agricultural malpractice can reduce green to barren, useless desert. Slash and burn tropical farming and the advance of the Sahara have already been mentioned in this connection. Major climatic and geological changes (such as the fall of the water table level) caused by agricultural malpractice are particularly worrisome. These effects are widespread and they may very well be irreversible. At the present time

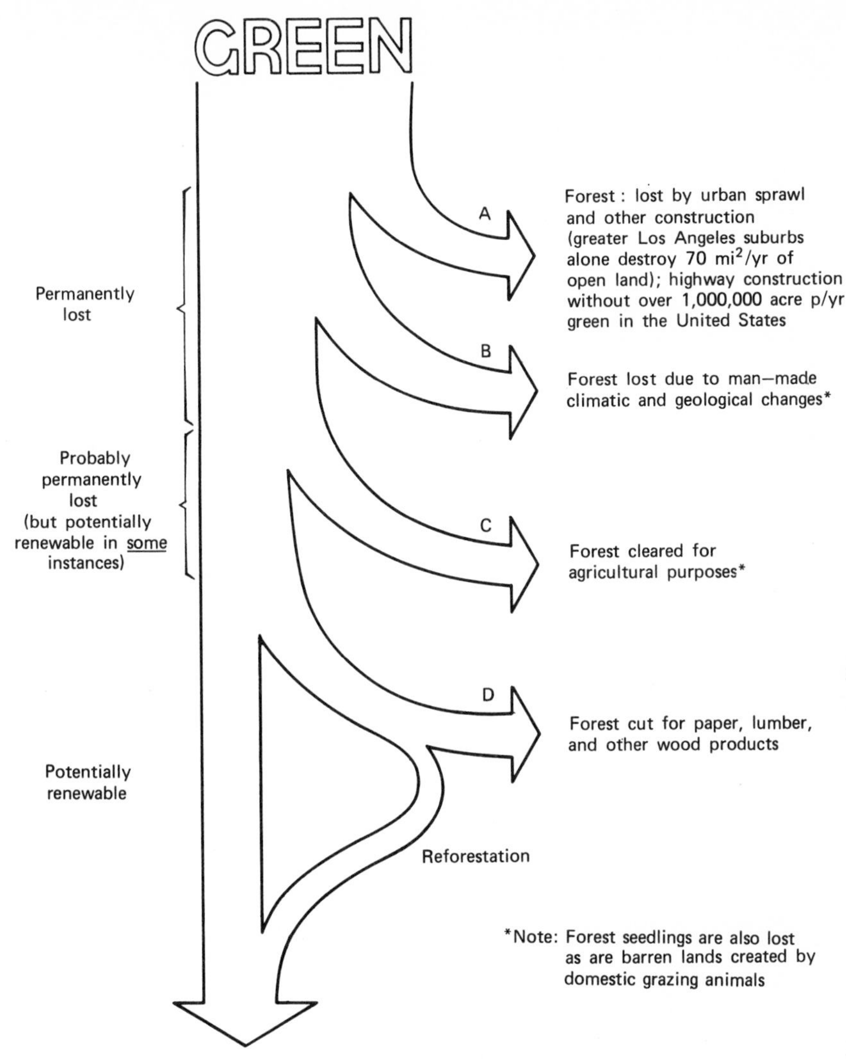

Figure 12.3. Where is the green going?

(1973), the African states of Mali and Upper Volta are in the midst of an extreme 40-month drought. Of Mali's estimated 5,000,000 cattle, at least 1,000,000 have died. The cause of this disaster is, as usual, man. Goat and camel flocks have denuded millions of acres of savanna. Herdsmen have cut the tops off young trees. This stripping of the land has reduced rainfall and the Sahara Desert, as a consequence, is spreading south at a rate of 0.5 mi/yr. We return in more detail to the impact of agriculture on our environment in Chapter 14, and we have more to say about ecological disruption later in this chapter. The important point to remember here is that the biosphere is a renewable resource of foodstuffs and materials *only* if it is exploited with extreme care.

12.3 MAN AS A CHEMICAL FACTORY: ENERGY USE AND THERMAL POLLUTION

Human society runs on materials and energy, and the energy requirements of modern technological civilizations are staggering. Energy consumption is not only directly linked to population growth, worse yet, per capita energy consumption climbs as societies grow and develop (Figure 12.4). The United States with only 6% of the world's population consumes a generous 35% of the world's energy production, and the rate of consumption is growing nearly 5%/yr (Figure 12.5; cf. Chapman et al., 1972). It takes 2000 kcal/day or about 100 thermal W/day simply as food to keep a man alive. The average American consumes 100 times this minimal amount (Singer, 1970). As Table 12.1 shows, industry accounts for the lion's share, and the household and transportation bites are about the same. Table 12.2 gives some additional and more detailed fuel consumption data. *All* energy sources have serious deleterious

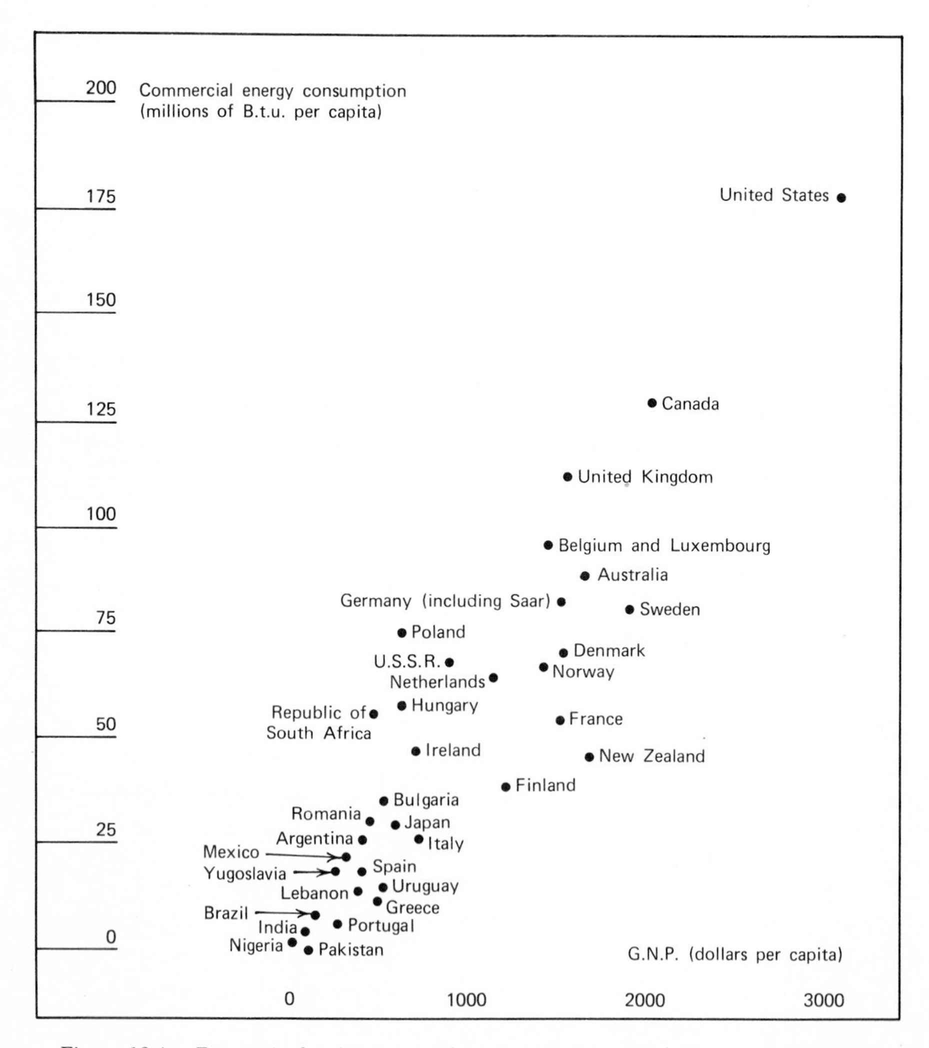

Figure 12.4. Economic development and energy consumption (from Gambel, 1964).

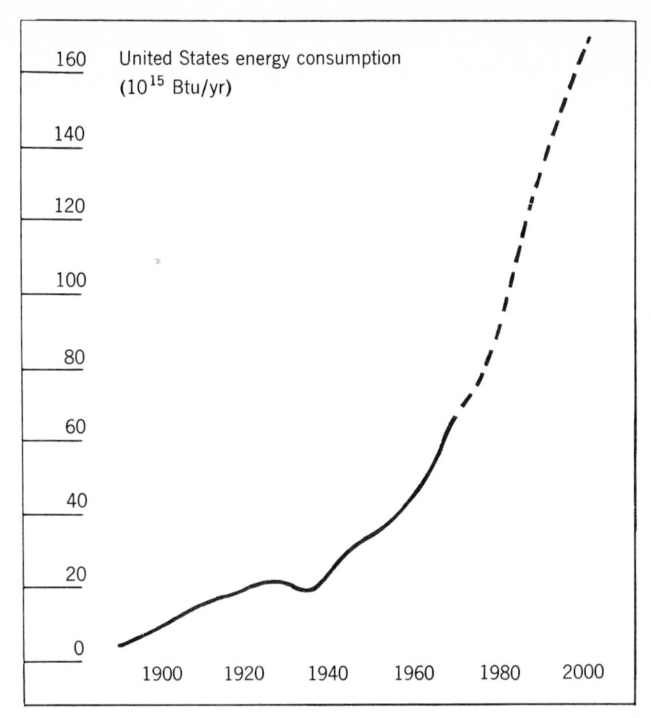

Figure 12.5. The growth of the United States' energy consumption (from White, *Technology Review*, edited at the Massachusetts Institute of Technology, © 1971 by the Alumni Association of M.I.T. with permission).

environmental impacts (Table 12.3). Even hydroelectric power, which now accounts only for a little over 1% of our energy (Figure 12.6), disrupts the natural flow of waterways, floods land area, and alters sediment distribution. As noted earlier, our beaches are disappearing thanks in part to impoundments, and when the lakes behind our great dams become silted in, what do we do then? Solar energy has been called "the largest resource" (Hammond, 1972*b*), and its environmental problems may be fewer than other major energy sources, yet solar energy collectors preempt valuable land from agricultural and other uses. The energy industry consumes enormous quantities of nonrenewable resources, largely fossil fuels, and it also uses prodigious quantities of water (Figure 12.7), most of which is returned to the environment thermally polluted (Figure 12.8). The United States has experienced temporary war time materials shortages, and it now appears to have lapsed into a permanent situation of intensifying energy shortages. If remedies are not forthcoming, our way of life will grind to a halt within the next generation.

In 1971, the United States consumed 6.32×10^{16} Btu of energy and more than half of this energy was wasted (Figure 12.6). In a conventional steam generator, power plant the chemical energy of fossil fuels is first converted to thermal energy (Figure 12.8) by combustion producing as it does so air pollution and ash waste. The thermal energy is next converted to mechanical energy, and finally this mechanical energy is converted to electrical energy. All of these steps are inefficient (Table 12.4), and the energy losses usually appear as heat (Figure 12.8)—"the

Table 12.1 Energy Consumption (Data from Singer, 1970)

	U.S.A., 1967 (kW/day-person)
Household Energy Consumption	
Space heating	1.2
Other heat	0.4
Electricity	0.5
	2.1
Commercial Energy Consumption	
Space heating	0.2
Other heat	0.4
Electricity	0.3
	0.9
Transportation Energy Consumption	
Space heating	0.1
Motive use	2.0
	2.1
Industrial Energy Consumption	
Space heating	0.2
Other heat	2.0
Electricity	1.0
Nonenergy uses	0.4
	3.6
Other Energy Consumption	
Space heating	0.1
Other heat	0.6
Electricity	0.3
Nonenergy uses	0.7
	1.7
Total	10.4

ultimate waste" (Harleman, 1971). It is possible to reduce thermal waste and pollution by controlling combustion temperature and by designing specific combustion chambers for specific purposes (Weinberg, 1971). There are numerous schemes for the production of electrical energy with fewer conversion steps (Table 12.4), but many of these conversion devices are either inefficient or else, despite very considerable research and development, remain uncompetitive with conventional energy conversion techniques. Some years ago, my own work involved the comparative evaluation of energy conversion systems, and it is a disheartening if educational experience to find that again and again, for all the dollars and human ingenuity

Table 12.2 Total Fuel Energy Consumption in the United States by End Use. Electric Utility Consumption Has Been Allocated to Each End Use. (From Berg, © 1973 by the American Association for the Advancement of Science, with permission of the AAAS.)

End Use	Consumption (trillions of Btu) 1960	1968	Annual Rate of Growth (%)	Percent of National Total 1960	1968
Residential					
Space Heating	4,848	6,675	4.1	11.3	11.0
Water Heating	1,159	1,736	5.2	2.7	2.9
Cooking	556	637	1.7	1.3	1.1
Clothes Drying	93	208	10.6	0.2	0.3
Refrigeration	369	692	8.2	0.9	1.1
Air Conditioning	134	427	15.6	0.3	0.7
Other	809	1,241	5.5	1.9	2.1
Total	7,968	11,616	4.8	18.6	19.2
Commercial					
Space Heating	3,111	4,182	3.8	7.2	6.9
Water Heating	544	653	2.3	1.3	1.1
Cooking	93	139	4.5	0.2	0.2
Refrigeration	534	670	2.9	1.2	1.1
Air Conditioning	576	1,113	8.6	1.3	1.8
Feedstock	734	984	3.7	1.7	1.6
Other	145	1,025	28.0	0.3	1.7
Total	5,742	8,766	5.4	13.2	14.4
Industrial					
Process Steam	7,646	10,132	3.6	17.8	16.7
Electric Drive	3,170	4,794	5.3	7.4	7.9
Electrolytic Processes	486	705	4.8	1.1	1.2
Direct Heat	5,550	6,929	2.8	12.9	11.5
Feedstock	1,370	2,202	6.1	3.2	3.6
Other	118	198	6.7	0.3	0.3
Total	18,340	24,960	3.9	42.7	41.2
Transportation					
Fuel	10,873	15,038	4.1	25.2	24.9
Raw Materials	141	146	0.4	0.3	0.3
Total	11,014	15,184	4.1	25.5	25.2
National Total	43,064	60,526	4.3	100.0	100.0

Table 12.3 Energy Sources Have Significant Environmental Impacts (Reprinted with permission from G. A. Mills et al., *Envir. Sci. Tech.* 5, 30 (1971). © The American Chemical Society.)

Energy Source	Impacts on Land Resource			Impacts on Water Resource			Impacts on Air Resource		
	Production	Processing	Utilization	Production	Processing	Utilization	Production	Processing	Utilization
Coal	Disturbed land	Solid wastes	Ash, slag disposal	Acid mine drainage		Increased water temperatures			Sulfur oxides, nitrogen oxides, particulate matter
Uranium	Disturbed land		Disposal of radioactive material		Disposal of radioactive material	Increased water temperatures			
Oil				Oil spills, transfer, brines		Increased water temperatures			Carbon monoxide, nitrogen oxides, hydrocarbons
Natural Gas						Increased water temperatures			
Hydro									

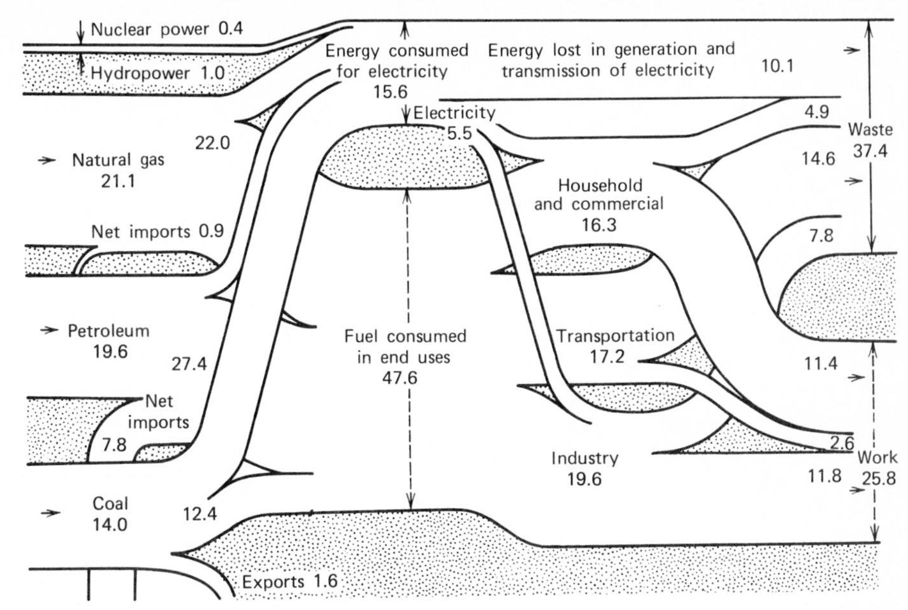

Figure 12.6. Production, consumption, and waste of energy in the United States. Total energy consumption in 1971 was 63.2×10^{13} Btu, excluding nonenergy uses of fossil fuels (from Hammond, © 1972*a* by the American Association for the Advancement of Science, with permission of the AAAS).

spent, the more feasible systems are always those that have been around for some time—the internal combustion engine, the steam generator, and the battery. The only promising newcomer is nuclear power. Any type of heat engine has a theoretical efficiency imposed upon it by the laws of thermodynamics. Some biological processes are highly efficient. Electrochemical batteries are also interesting for if discharged slowly they convert chemical energy directly into electrical energy with very high efficiencies (Table 12.4; Figure 12.9), and for this reason great hopes were placed in the continuous feed electrode electrochemical cell, or "fuel cell," but electrode material problems and the extremely complicated chemistry at a high rate discharge electrode have somewhat dashed these hopes (but not completely, see discussion about the "hydrogen economy" that follows).

At the present time, short of controls on population growth and per capita consumption of energy, nuclear power appears to be our best hope of alleviating the energy crisis, although in 1971 less than 1% of the United States' energy came from this source (Figure 12.6). However nuclear plants, like conventional power plants, thermally pollute waters (Tables 12.3 and 12.5), and there is furthermore the still unsolved problem of the disposal of nuclear wastes (Hambleton, 1972). Conservationists, making the most of a very deep-seated public fear concerning nuclear safety, have managed to impede the more rapid development of nuclear power in the United States (Nelkin, 1971; Gillette, 1972; Lewis, 1972; Wilson, 1973), thereby, since nuclear power is still less destructive to the environment than conventional power, in the longer view damaging the environment. When vast tracts of our western lands (including those of the Indians) have been laid barren by strip mining to supply the coal to meet our energy needs, we shall have largely the conservationists

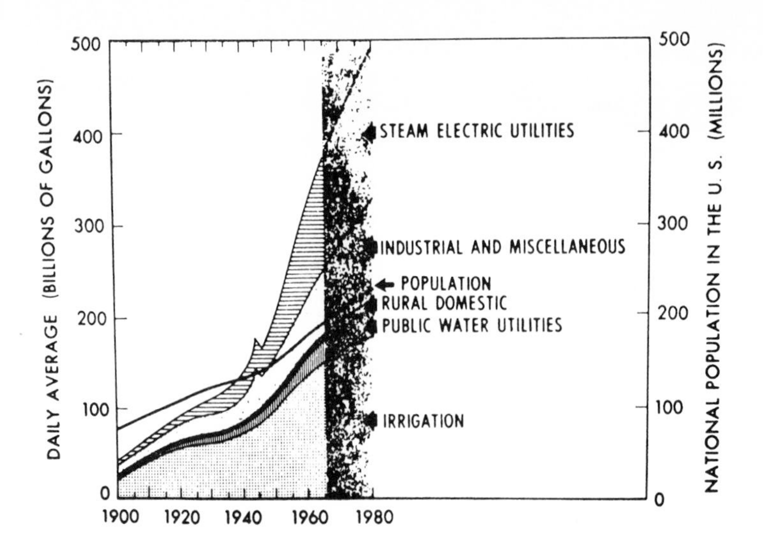

Figure 12.7. Water use in the United States (from Cairns, © 1971 and reprinted with permission of The Water Pollution Control Fecderation).

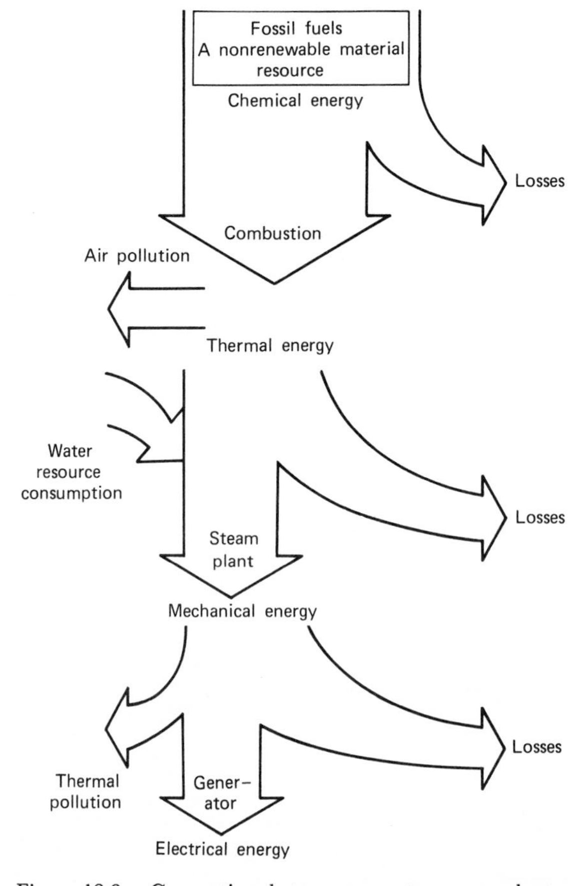

Figure 12.8. Conventional steam generator power plant.

Table 12.4 Energy Conversion Efficiencies

Electrochemical battery (low discharge rate)	Chem → Elect	90%
Fuel cell (low temperature hydrogen cell)	Chem → Elect	40–85%
Fuel cell (high temperature)	Chem → Elect	30–70%
Motor generator	Mech → Elect	70%
Steam turbine	Therm → Mech	40%
Internal combustion engine	Chem → Mech	20%
Solar cell (gallium arsenide)	Solar → Elect	18%
Solar cell (silicon)	Solar → Elect	16%
Thermionic generator	Therm → Elect	15%
Thermoelectric generator	Therm → Elect	10%
Electron-voltaic nuclear battery	Nucl → Elect	2.5%
Direct charge nuclear battery	Nucl → Elect	1%
Contact potential nuclear battery	Nucl → Elect	0.01%
Photogalvanic cell	Solar → Elect	0.01%
Workman–Reynolds effect	Therm → Elect	0.004%
Electrokinetic transducer	Mech → Elect	0.0002%
Thermocell	Therm → Elect	0.00003%
Thermomagnetic generator	Therm → Elect	0.000006%

and our unwise foreign policy in the Near East to thank. Sagan (1972) has published a thoughtful analysis of the "Human Costs of Nuclear Power" in which he points out that no member of the public has yet been injured by or even exposed to excessive radiation as a result of the United States' nuclear power industry. He concludes that "Our society is able to maintain low prices [of conventional power] by evading environmental costs. . . . At the same time, the price of nuclear energy is maintained at an artificially high level by an overprotective government policy. . . ." Weinburg (1972) estimates that electrical power can be produced more cheaply (Table 12.6) from a pressurized water reactor (PWR) with environmental protection (cooling towers to reduce thermal pollution) than from a conventional coal-burning power plant with

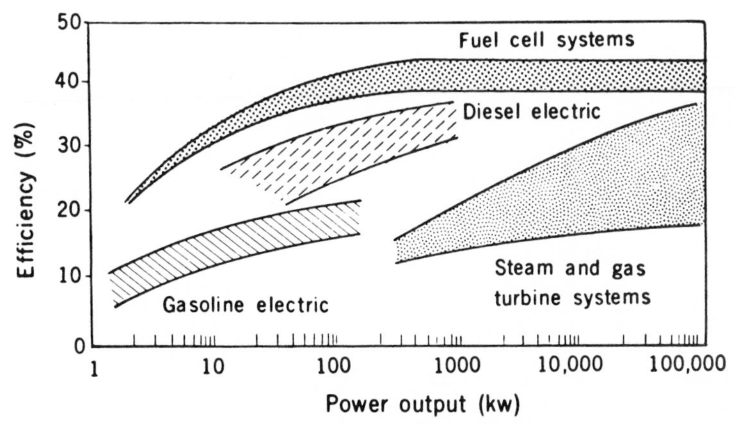

Figure 12.9. Thermal efficiency of electrical production of fuel cells and other types of generators as a function of output (from Maugh, © 1972*b* by the American Association for the Advancement of Science, with permission of the AAAS).

Table 12.5 Summary of the Environmental Impact of Power Generation

Type of Power Generation	Environmental Impact
Conventional Fossil Fuel Burning Power Station	Depletion of nonrenewable resource, oil spills and water pollution, strip mining for coal, acid mine drainage, thermal pollution, SO_2 and other air pollutants, damage to fish (thermal pollution) and plants (air pollution), atmospheric oxygen removal.
Hydroelectric Power	Alteration of water courses, interference with natural silting and sedimentation processes, loss of beaches, earth movement hazards, interference with migrating fish, loss of land area, applicable only to highly restricted regions.
Tidal Power	Alteration of coastal and estuarine hydrology and ecology, could produce significant alterations in the distribution of the planet's mass and momentum, applicable only in highly restricted areas.
Solar Power	Interference with land use over very large areas unless a satellite primary collector is used, applicable only to restricted areas, unsightly.
Nuclear Power, Fission	Depletion of nonrenewable resources, strip-mining (to provide coal burning power station energy for fuel enrichment plants), thermal pollution, radioactive waste disposal (and radioactive mine tailings), potentially catastrophic accident possibility.
Nuclear Power, Fusion	Thermal pollution, potentially catastrophic accident possibility, perturbation of hydrologic cycle.
Geothermal Power	Air and water pollution in some instances (from H_2S, etc.), unknown long-range geological effects, applicable only in very highly restricted areas.
Magnetohydrodynamic, Thermionic, and Thermoelectric Generators and Fuel Cells	If fossil fuels are used, the same as for conventional power stations (see above), if nuclear fuels are used the same as for nuclear power stations (see above), if hydrogen fuel is used, the same as for the hydrogen economy (see below).
The Hydrogen Economy	Same as whatever is used as the primary energy source, perturbation of the hydrologic cycle.

partial environmental protection (cooling towers and SO_2 removal but no reduction of the environmental havoc created by coal mining), and he also calls attention to the real threat of nuclear power—not an accidental escape of radioactivity, but the fact that our political institutions are too unstable to guarantee the necessary very long-range security of deposits of nuclear wastes. The half-lives of some reactor wastes are thousands of years, while the half-lives of many modern governments, in developed as well as undeveloped countries, appear to be much less than 100 yr. Then too, natural disasters such as earthquakes and deliberate human sabotage assume a particularly threatening aspect in connection with thermonuclear power and the safe storage of radioactive wastes.

Table 12.6 Estimated Costs (in 1978) of Power from Pressurized Water Reactors Compared with Conventional Fossil Fuel Power Plants

	Capital Costs ($/kWhe)	Fixed Charges	Fuel Costs	Operational Costs	Total Power Cost (mills/kWhe)
Pressurized Water Reactor					
Into river	365	7.8	1.9	0.6	10.3
With cooling towers	382	8.2	1.9	0.6	10.7
Coal-Burning Power Plant					
No SO_2 control into river	297	6.7	3.9	0.5	10.8
With SO_2 control into river	344	7.4	3.9	0.8	12.1
No SO_2 control cooling towers	311	6.6	3.9	0.5	11.0
With SO_2 control cooling towers	358	7.7	3.9	0.8	12.4

The present liquid water reactors (LWR) are very inefficient, utilizing only 1% of the energy of the uranium fuel, and the thermodynamic efficiency is also rather low—32% (Hammond, 1972*c*). Since uranium resources are much more limited than coal resources in the long view, this type of reactor must be considered as a stop-gap measure. While the purpose of nuclear power is to meet energy needs while at the same time conserving valuable coal reserves for chemical production rather than wasteful burning, perversely enough the energy requirements of the gaseous diffusion plants that enrich the ^{235}U fuel have in fact stimulated strip mining of coal. As an attack on some of these problems, much emphasis has been placed on the development of the necessary technology for breeder reactors which produce more fissionable material than they consume. Japan, the U.S.S.R., the United States, and several European nations are hard at work trying to develop a liquid-metal-cooled fast breeder reactor (LMFBR).

An entirely different approach to nuclear energy and an alternative to fission is fusion—control of the enormous quantities of energy released by the processes occurring in a hydrogen bomb and the sun's fires (see Chapter 1). Not only is the energy potential much greater, but the horrendous waste disposal problems that vex fission power are also absent. Unfortunately, the development of the necessary plasma physics technology has proved to be far more difficult than initially realized, but recent advances in laser heating and magnetic containment of the plasma have again raised hopes that a controlled fusion system might be achieved possibly by about 1980 (Metz, 1972).

As the energy crisis has tightened around us, suggestions of novel power source schemes have proliferated. These have included reexamination of such "nonpolluting" sources as geothermal power, tidal power, and especially solar power. But even though these alternatives may produce little or no chemical pollution, they nevertheless all have deleterious environmental impacts (Table 12.5). Here we discuss only one of these schemes, one that has been receiving increasing attention of late, the so-called "hydrogen economy" (Jones, 1971; Bockris, 1972*a*, *b*; Maugh, 1972*a*, *b*, *c*; Gregory, 1973; Winsche et al., 1973).

In addition to the wastes discussed previously that were produced by the generation of electricity, there are very appreciable losses upon the transmission of electrical energy (Figure 12.6). Power lines strung across the countryside are one of the most unsightly desecrations of our environment (and require massive herbicide applications to keep them free from tree growth). The hydrogen economy is not a new energy source but a novel approach to the distribution of energy (Figure 12.10). The hydrogen economy begins with the generation of electricity by one of the familiar methods, such as hydroelectric or nuclear power (with their attendant environmental impacts), but then this electricity is converted to chemical energy on site by the electrolysis of water, a highly efficient process

$$2H_2O \xrightarrow[\text{energy}]{\text{electrical}} 2H_2 + O_2 \qquad (12.1)$$

Work currently underway at the Euratom center in Ispra, Italy, may also provide a method of dissociating water thermally in a nuclear reactor (Maugh, 1972*a*). The hydrogen would then be distributed to the user by underground pipes—possibly using an extension of the existing pipeline systems for distributing natural gas

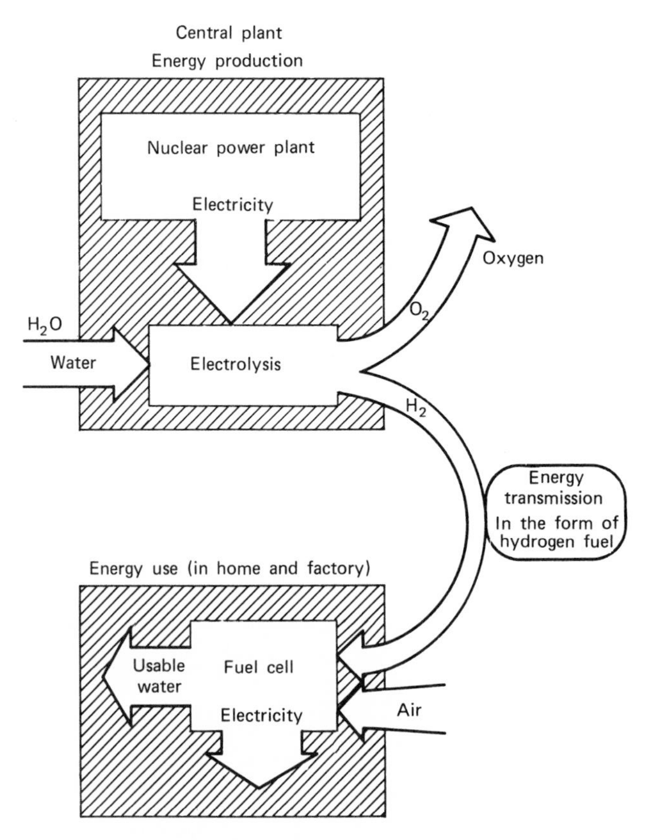

Figure 12.10. The hydrogen economy.

(Germany already has 185 mi of hydrogen pipelines). In the consumer's home or factory, the chemical energy in the form of hydrogen would be converted back into electrical energy in a fuel cell (Maugh, 1972*b*).

$$2H_2O + O_2 \text{ (from air)} \rightarrow 2H_2O + \text{electrical energy} \quad (12.2)$$

also a highly efficient process (Table 12.4). Electrochemical "combination" of H_2 in an air fuel cell does not produce oxides of nitrogen, but conventional burning of H_2 does, although less than for hydrocarbon combustion. Bockris (1972*a*, *b*) has pointed out a number of advantages of the hydrogen economy in addition to its cyclic and nonpolluting features. He estimates that the overall efficiency might be in the range of 36%. The demand for electrical energy is highly variable, and the hydrogen economy, unlike conventional systems, readily provides a storage facility for peak load periods. Average household use of electricity would yield a by-product of 14 liter/day of pure water. In addition to being piped for electricity production, the hydrogen could also be used for other purposes, such as transportation fuel, and chemical uses (about 40% of present H_2 production goes to ammonia synthesis), such as hydrogenation of fats and the reduction of metal ores. As disadvantages, Bockris lists conservatism, the lack of training in electrochemical engineering, and public fear of hydrogen (remember the *Hindenburg*). He overlooks one of the more important environmental aspects of the hydrogen economy: It alters the distribution of oxygen and water in the environment and could become an appreciable new path in the hydrologic cycle. But since water and oxygen are both highly mobile in the atmosphere and hydrosphere, any alteration (unless enormous quantities of hydrogen are stored for long periods of time) should be very transitory. The "least studied of all aspects of the hydrogen economy" (Maugh, 1972*a*) is what to do with the oxygen produced concurrently with the production of hydrogen (Eq. 12.1). Steel production consumes oxygen and even more of it might be used in sewage treatment and for aerating natural waters to restore or improve their dissolved oxygen content.

Jones (1971) has examined the use of liquid hydrogen as a replacement for hydrocarbon fuels for transportation, and he concludes that cost-wise and energy-wise liquid hydrogen fueled fuel cells compare favorably (Figure 12.11 and Table 12.7), that such replacement is technically feasible, that the high energy/weight factor (Table 12.7) for liquid hydrogen gives it a particular advantage for air transportation, and that . . . "as a pollution-free fuel [liquid hydrogen] must be seriously considered

Table 12.7 Comparison of Energy and Costs of Transportation Fuels (From Jones, © 1971 by the American Association for the Advancement of Science, with permission of the AAAS).

Fuel	Energy/mass (cal/g)	Density (g/cm^3)	Energy/volume (cal/cm^3)	Cost ($/cal)
Liquid Hydrogen	29,000	0.07078	2,050	6×10^{-9} at $0.08/lb 8×10^{-9} at $0.11/lb
Gasoline	11,500	0.74	8,500	4.2×10^{-9} at $0.12/lb
Fuel Oil	10,500	0.96	10,000	

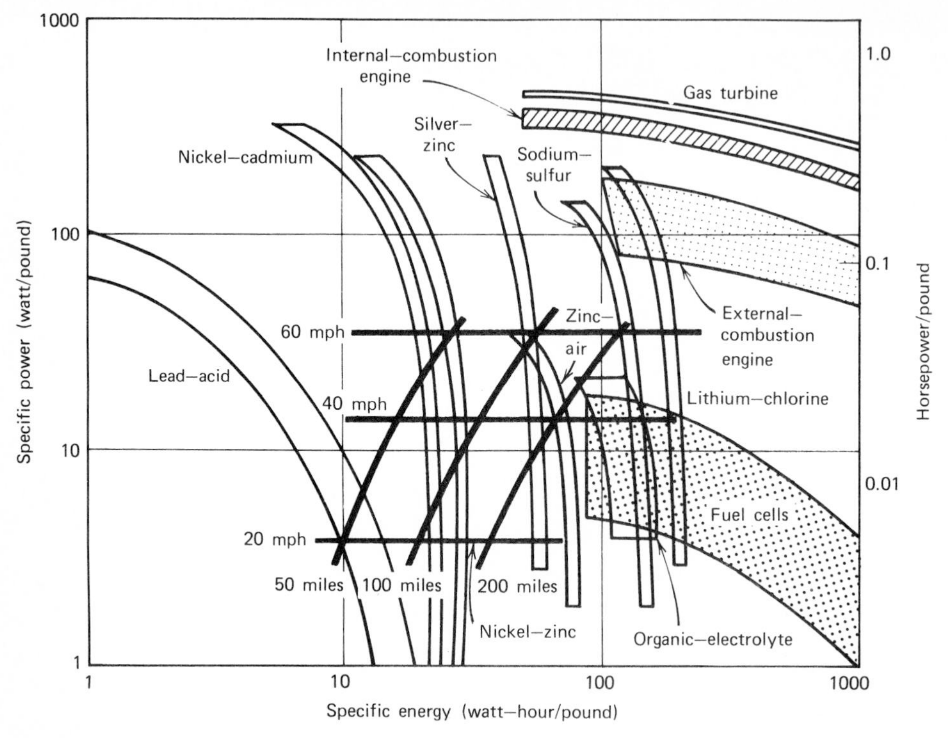

Figure 12.11. Vehicle requirements for a 2000-lb vehicle and the capability of power plant systems. Solid lines indicate the ranges in miles corresponding to different constant speeds (mi/hr) transformed on the specific energy–specific power coordinates (from Jones, © 1971 by the American Association for the Advancement of Science, with permission of the AAAS).

as the logical replacement of hydrocarbons in the 21st century." To which I can only add, if we make it to the twenty-first century.

Professor Cook (1972) has reviewed past and projected future energy sources (Figures 12.12 and 12.13) and he divides man's use of energy into three phases: (*a*) continuation of the present *mining phase*, dependent on fossil fuels, for a few more decades; then (*b*) a *rock-burning phase*, dominated by uranium and thorium breeder nuclear reactors, which will last a few centuries; followed by (*c*) the *solar phase*, which may last until the exit of the human species. It is interesting to note in passing that in fact all our major energy sources at the present time are derived in one way or another from the sun (Figure 12.14). This is well recognized in the case of the fossil fuels produced by ancient photosynthesis, but even nuclear fuels are the legacy of primordial stellar element synthesis (Singer, 1970). Then too it is the sun that supplies the energy for the hydrologic cycle and the movement of air masses so that hydroelectric and even wind power are but further ramifications of solar energy.

To return to losses, conventional power plants derive their energy by heating water. In principle, the colder the water that is returned to the environment, the greater the plant's efficiency. But in practice, water is dumped back into the environment at higher temperatures than it is removed. It has been estimated that by the year 1980 one-sixth and by 2000 one-third of all the fresh water run-off in the United

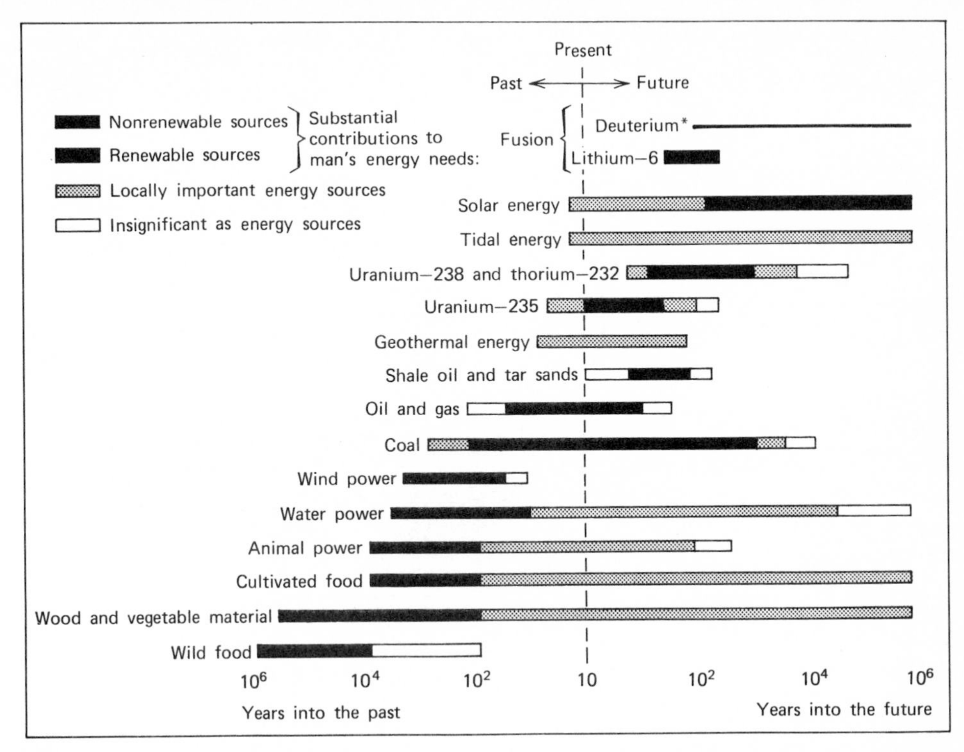

Figure 12.12. Man's energy sources past and future (from Cook, *Technology Review*, edited at the Massachusetts Institute of Technology, © 1972 by the Alumni Association of M.I.T., with permission).

States will be used to cool power plants (Holcomb, 1970). Nuclear power also consumes enormous quantities of cooling water and returns it to the environment thermally polluted. Without cooling towers, the Monticello nuclear power plant would use 65% of the entire flow of the upper Mississippi River, at times raising the river's temperature by as much as 9°C (Cairns, 1971). Thermal pollution directly alters the chemistry of the receiving waters—generally speaking the rates of chemical processes, including biochemical processes such as eutrophication (see Chapter 16) are increased thereby (Krenkel and Parker, 1969; Tarzwell, 1970; Cairns, 1971; Coutant, 1971; Levin et al., 1972) with both deleterious and beneficial environmental consequences. Worse still, gas solubilities, notably of oxygen, markedly decrease as the temperature increases (see Tables A.2, A.3, and A.4). Active game fish such as trout and salmon cannot tolerate warm, oxygen-depeleted water (Figure 12.15). Thermal pollution also disrupts the migrations of anadromous fish, increases their disease rates while at the same time often favoring their less desirable competitors (Snyder and Blahm, 1971; Zaugg et al., 1972). Other organisms such as slimes and algae thrive on thermal pollution (Figure 12.16). Power plant thermal pollution is readily but expensively prevented (Eisenbud and Gleason, 1970; Löf and Ward, 1970). So perhaps the most interesting approach to the problem hinges on the concept of making use of thermal pollution to accelerate the growth of organisms useful to man, especially in estuarine

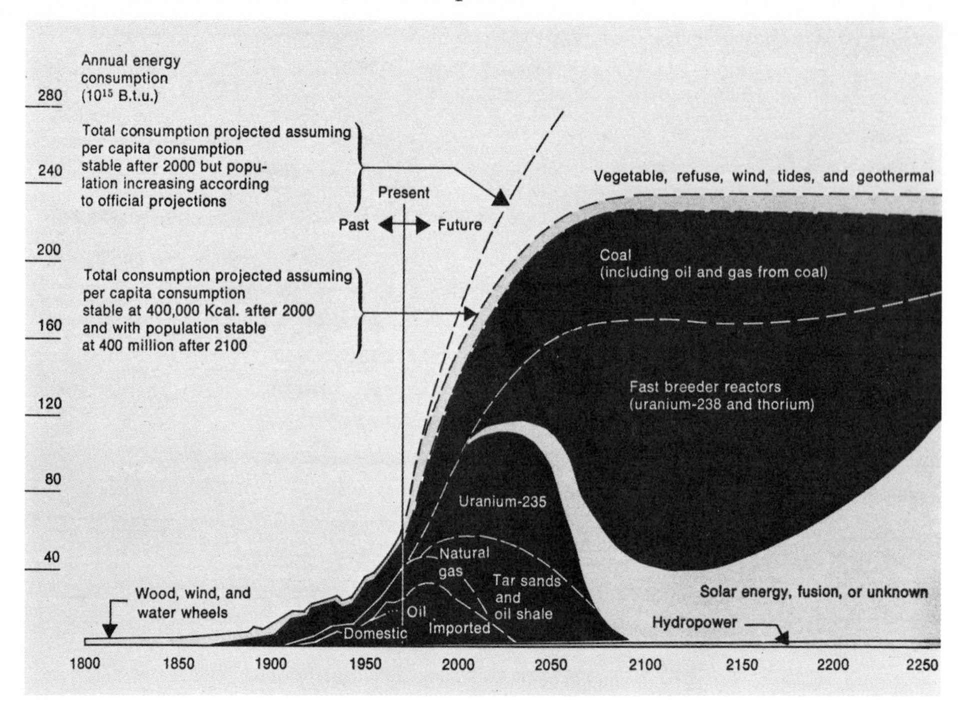

Figure 12.13. Future projections of United States' energy sources (from Cook, *Technology Review*, edited at the Massachusetts Institute of Technology, © 1972 by the Alumni Association of M.I.T., with permission).

and coastal areas where increasing water temperature may be beneficial rather than harmful. Lobsters, for example, grow more rapidly in warm water (Hughes et al., 1972), and thermal aquaculture suggests a host of fascinating possibilities (Yee, 1972; see also Mathur and Stewart, 1971). Terrestrial crop yields might also be increased by warm water irrigation (Anon., 1969*a*) or by nuclear power plant, effluent-warmed greenhouses (Klein, 1970).

In summary then, man the chemical factory consumes enormous quantities of precious nonrenewable resources such as minerals, metals, and fossil fuels (Figure 12.2), enormous quantities of limited renewable resources such as forests and crops, and as he does so he dirties enormous quantities of water and dumps enormous quantities of wastes into the environment on a scale now comparable to natural biogeochemical processes and ever increasing. And what is the product of this factory? Civilization? Greater civilizations were produced in Periclean Athens and Renaissance Florence at much less an environmental price. No, it is not civilization. The terribly expensive product is more polluters, more people.

> *We will not like some solutions we will have to adopt, but unless we can resolve the question of population growth, it not only will aggravate our current problems but may eventually dwarf them. . . .*
>
> THE PRESIDENT'S COMMISSION ON POPULATION GROWTH AND THE AMERICAN FUTURE, 1972

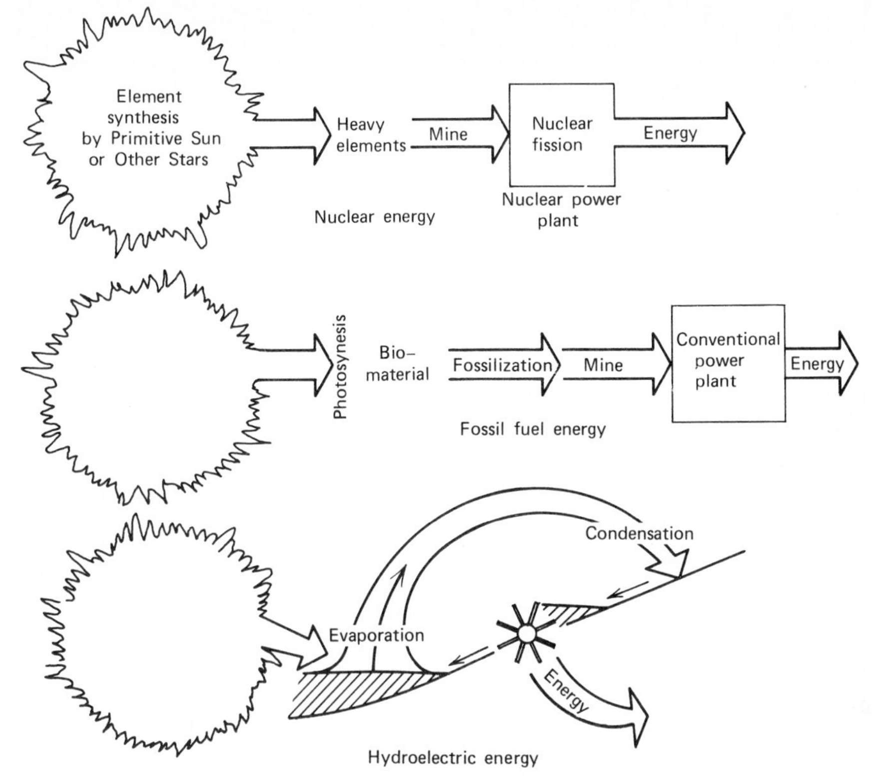

Figure 12.14. The various forms of solar energy as made useful to man.

12.4 ECOLOGICAL DISRUPTION

Man's impact on the ecology of his environment ranges all the way from the extinction of species (the recent extermination of Earth's largest animal, the blue whale, is a particularly mournful example, see Small, 1971) to the most subtle redistributions of animal and plant populations, from the total devastation of strip mining to changes so slight and gradual that man may never become aware of the damage he has done. Man can produce massive ecological changes like the acceleration of the advance of the Sahara (Anon., 1971*b*) or very local changes, such as the alteration of phytoplankton productivity by the chlorination of power plant water (Brook and Baker, 1972). Sometimes while damaging one species, he improves the lot of another: Increasing oil pollution is clotting the underfeathers of the Bermuda longtail and hastening its disappearance, while, on the other hand, the numbers of the nearly extinct Bermuda cahow appear to be increasing, possibly because of the absorption of pesticides by floating oil and tar. Even man's well-intended efforts to safeguard the environment can backfire: His prevention of forest fires has resulted in the accumulation of underbrush and litter that can fuel even more destructive conflagrations (Dodge, 1972), and the lack of the natural cauterizing effects of fire has increased pine beetle damage. Man has proved himself much more effective at destroying ecological balance than preserving it. In the face of exploding population, even his efforts to preserve "unspoiled" nature are probably doomed (Houston, 1971; Platt,

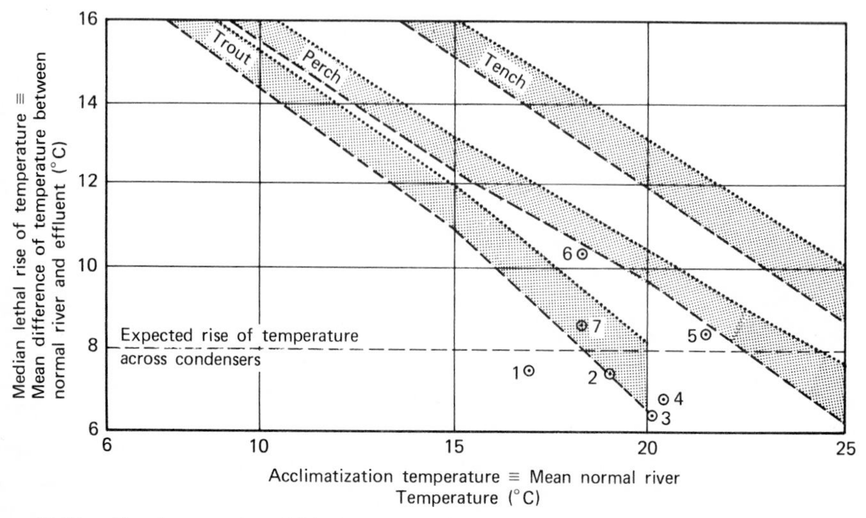

Figure 12.15. For three species of fish the mean lethal rise in temperature decreases as acclimatization increases (from Cairns, © 1971 and reprinted by permission of the Water Pollution Control Federation).

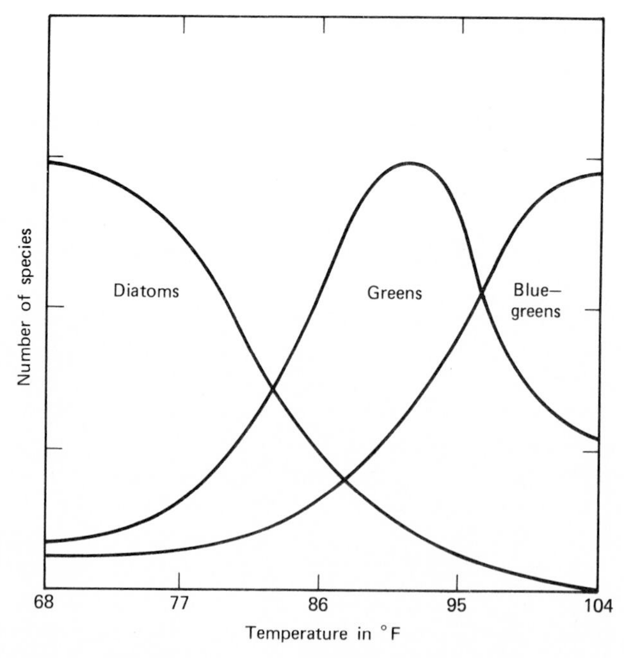

Figure 12.16. How the relationship between three groups of algae shifts as the water temperature rises (from Cairns, © 1971 and reprinted by permission of the Water Pollution Control Federation).

1971; Moir, 1972; for a discussion of African preserves see Myers, 1972). Harsh environments, such as deserts and tundra, are particularly susceptible to ecological disruption and correspondingly slow to heal. In many instances, man's "improvements" have triggered a chain of ecological disasters that, with a certain ironic justice, have made himself a victim. Lake Nasser behind Aswan High Dam is so clogged with weeds that the transpiration losses threaten to reduce the water volume to a level insufficient to drive the power generators. The dam has stopped the natural flow of silt down the Nile and has so increased erosion that the productive land thus lost may exceed the land opened by irrigation. At the same time the curtailment of nutrient-rich silt from reaching the Mediterrean Sea has caused the Egyptian sardine catch to decline from 18,000 tons in 1965 to 500 tons in 1968. And the most sinister retribution of all—increased irrigation has encouraged the schistosomiasis carrying snail to thrive, thereby increasing the incidence of that dread disease. In other parts of Africa, the same dire consequence has been achieved by the wholesale slaughtering of the hippopotami. Serves man right, doesn't it?

Ecological disruption is a recurrent theme throughout this book (note especially Chapter 14 on the impact of agriculture), and here we concern ourselves with a few examples of disruption brought about for chemical purposes or by chemical means, while in Section 12.6 we find some clues as to why chemicals are such potent disruptors of ecosystems.

Rats carry disease, including the dreadful bubonic plague, eat grain (an estimated 33,000,000 ton/yr) and spoil other human foods, and occasionally bite babies. A number of rodent control poisons have been developed (Gutteridge, 1972). Older poisons such as strychnine and zinc phosphide killed friendly animals and were responsible for an occasional human death as well. In recent years, they have been replaced by anticoagulent rodentcides such as warfarin. But the most recent develop-

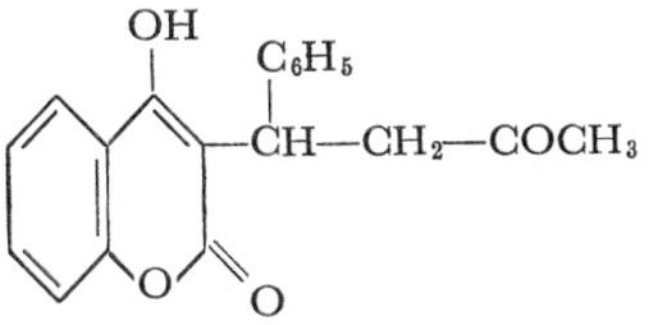

OH C_6H_5 CH—CH_2—$COCH_3$ O O

Warfarin

ment in the ageless war between rats and men is a victory on the side of the rats . . . the appearance of "super rats," a genetic strain resistant to warfarin (Jackson and Kaukeinen, 1972). Rats can also be eradicated by chemical sterilization (androgen treatment), and Gentry (1971) has developed a model for evaluating eradication strategies. Vampire bats, which rarely attack humans but raise havoc with cattle, especially in Latin America (an estimated $250,000,000/yr loss), are very sensitive to anticoagulants such as diphenadione (2-diphenylacetyl-1,3-indandione), and they also have the social habit of preening one another so a bat smeared with a mixture of petroleum jelly and anticoagulant will kill 2 to 3 dozen of his fellows (Anon., 1972*a*; Thompson et al., 1972). Alternatively, the cattle, because of their bulk, can be safely injected with the anticoagulant to kill the bats that feed on them. A Mexico–United States demonstration program has indicated that the bats can be controlled in this manner at a cost of about 0.3% of the annual damage now caused. Any detrimental ecological effects have not made themselves evident. Some plants, it might be noted in passing, such as sweet clover contain anticoagulants, presumably as a defense against the animals that graze upon them.

Because farmers do not like them, 74,000 coyotes/yr are poisoned in the western United States. How many of what are nicely called "nontarget" species are also wiped out by these government sponsored programs, including the fast-vanishing eagle, is a cause of occasional scandals (Anon., 1971*b*).

Man is also poisoning himself (see Section 12.5) and his human enemies, although the peculiar etiquette of his species demands that the latter be done indirectly. Herbicides are used to keep power transmission line rights of way open and in agriculture to impede weed growth. Considerable research effort has been expended in developing herbicides that are more selective in discriminating between crops and weeds (Ashton and Crafts, 1973) and in developing crop strains more resistant to herbicides (Pinthus et al., 1972). Herbicides and pesticides such as arsenic compounds and toxaphene are used to control plant growth and unwanted fish in lakes, ponds, and waterways. Toxaphene soon disappears out of the water column and into the underlying sediments where it is very persistent (Veith and Lee, 1971). The environmental impact of still another use of herbicides has received a great deal of publicity. Air-sprayed herbicides have been used extensively in the Viet-nam war to defoliate jungle areas and remove cover from guerilla forces (Orians and Pfeiffer, 1970; Westing, 1971). The military usefulness of this technique is open to question (Shapley, 1972); however, crop spraying could appreciably reduce the war-making capability especially of primarily agricultural nations. A team of the American Association for the Advancement of Science (AAAS) investigated the ecological impact of the military use of herbicides in South Viet-nam, and they found that some 1400 km² of the area's mangrove forests have been destroyed and that years after the spraying there is no sign of their reforestration, while in another area half of the mature hardwood trees have been destroyed and the region is being invaded by bamboo growth (Boffey, 1971). In Viet-nam, 90% of the spraying activity was in jungle areas and 10% on crops (representing 2000 km² of land or food to feed an estimated 600,000 people for 1 yr). Among the herbicides used were 2,4,5-T, now known to be teratogenic, 2,4-D, picloram, and cacodylic acid—an arsenic compound

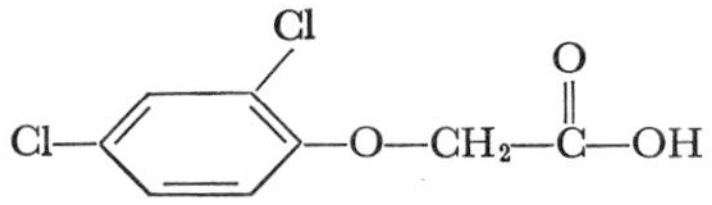

2,4-D
(Also used in the form of esters and amine salts)

that has been the subject of particular concern. The AAAS team found no unequivocal evidence for human injury as a result of the herbicide program; however, refugees have reported minor effects, and rumors of abortions and monstrous births in human and animals have persisted (Rose and Rose, 1972). Except for the arsenicals, the herbicides used in Viet-nam are only moderately toxic to warm blooded animals. Tschirley (1969) earlier concluded that "there is no evidence to suggest that the herbicides used in Viet-nam will cause toxicity problems for man or animals," and he even notes that the fish catch has increased during the period of intensive defoliation. As in the case of the SST, the use of defoliants in the war is an issue over which some scientists have confused science and pacifist politics.

Surprisingly enough, herbicides seem to have few side effects on soil microflora and fauna, possibly because these chemicals are so strongly absorbed on the soil (Anon., 1972*c*). Tschirley (1969) has even noted that the population of some organisms is greater in 2,4-D contaminated soil. The resiliency of the biosphere and its ability not only to resist but even make use of man's abuses never cease to amaze me. Eisner et al. (1971) report the presence of 2,5-dichlorophenol in the odorous defense froth emitted by the grasshopper *Romalea micropetra*, possibly derived from the ingestion of herbicides or herbicide derivatives. A much better known example of Nature's adaptability is the darkening of certain insect species in sooty, heavily polluted areas (Anon., 1971*c*). This melanism is reversible, and in areas where smoke abatement laws are enforced, the darker-colored insect varieties tend to diminish in number.

But even Nature's resilience has its limits, and because of the delicacy of ecological balance, the weapons man aims at the biosphere tend to be very indiscriminate, and nontarget species victims are the rule rather than the exception. When tourists complained about the bugs, massive pesticide spraying was applied to the Rangely Lakes region of Maine. The mosquitos were destroyed but so also was one of the most famous sport fishing regions in the United States. Man's weapons commonly backfire, having exactly the reverse of the desired effect. Insecticides kill the insects, but the birds that normally control insect populations also disappear, and in the absence of their natural enemies, if spraying is not continued indefinitely, the insects may make a massive comeback. Pesticides and their residues tend to concentrate as one goes up a food chain, thus "the pecticides tend to favor the herbivores, the very organisms they were intended to control" (Woodwell, 1970).

12.5 CHEMICAL SENSATION, HORMONAL IMBALANCE, AND MUTAGENS

Organisms interact with their environment by photolytic, thermal, chemical, mechanical, and electrical sensors (Table 12.8), and in the course of evolution, each organism tends to become very highly dependent on one particular type of

Table 12.8 Sensors

Nature of the Stimulus	Examples
Photolytic	*Sight, photosensitivity possible sensitivity to other types of electromagnetic radiation unknown*
Thermal	*Hot and cold sensors, some snakes and insects locate prey by thermal sensors*
Chemical	***Taste and smell***
Mechanical	*Touch and hearing*
Electrical	*Electric eels are believed to locate objects by field perturbation*

sensor: bats upon high-pitched sound, certain vipers on thermal pits, and, perhaps wierdest of all, certain eels on the electrical field they create. Sharks also can detect prey by electrical fields, and even a minor scratch can double a person's voltage gradient in seawater (Bullock, 1973). The voltages arise from multitudenous membrane, concentration gradient, and other voltaic potentials in any living organism. Man is a visual animal. His development of pictures, which is unique among animals, and of art is no irrelevant quirk. Visual litter precipitates far more public irritation than serious but unseen menaces to human health. Some of his senses man has neglected—in some instances deliberately. Puritanical notions of sin have deprived many modern men of the pleasures of the tactile senses (and have exacted a steep price in mental derangement). Especially comic is the desperate panic of Americans to make themselves and their environment odorless. As usual, man has put himself out of step with the rest of the biosphere for chemical sensation is perhaps the most ubiquitous sensory technique in the biosphere. I also suspect that it was the first to become fairly highly developed in the course of biological and even protobiological evolution for no type of organism or protoorganism could have survived long in the ancient seas if it had been unable to avoid chemical hazards. Then, closely behind, I would guess, came thermal sensation, then mechanical sensation, and finally concurrently with the evolution of photosynthesis, photosensitivity. Here we will concern ourselves only with chemical sensation. Because of the extremely great importance of chemical sensation and interaction in the biosphere, chemical pollution possesses a very great potential for ecological disruption. Let me say in passing that for all our neglect of taste and smell, for all the expensive equipment, we have developed to detect chemicals for us, they remain the best pollution detectors we have; if our air, our water, or our food smell and/or taste bad, they are probably contaminated.

Chemical sensation and interactions in the biosphere fall into two categories (Figure 12.17):

1. Interaction of an individual or community of organisms with its or their environment including *interspecies interactions*, and
2. *Intraspecies interactions*, that is to say chemical interaction among the organisms of the same species.

In their excellent short review paper Whittaker and Feeny (1971) call the former *allelochemic effects* and present a much more detailed classification (Table 12.9). Environment identification (Figure 12.17-1) by chemical definition may take two forms. In the first instance, the organism creates a chemical trace—a familiar example is the practice of dogs in establishing a territory by urination. In the second instance, the organism detects a chemical trace preexisting in the environment. This affords us with one of the most extraordinary and mysterious sensing feat in the biosphere—the ability of some fish such as salmon to return to the exact waters of their birth to spawn. Plants in their more passive way are also adept at identifying chemical substances, especially water, nutrients, and other vital chemicals. Plant distribution has been used as a means, not only in locating ground waters, but also in prospecting minerals.

Interspecies chemical interactions (Fig. 12.17-3, -4, and -5) are sometimes benign, as in the case of certain symbiotic relationships of which the most familiar example

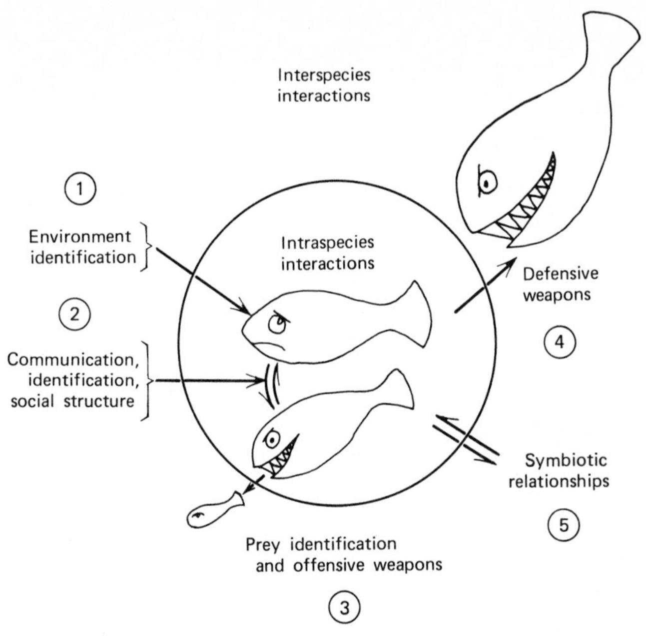

Figure 12.17. Chemical sensation and species interactions.

is the lichen, a fungus incapable of photosynthesis depending upon and at the same time providing an environment for an algae (Kershaw, 1963). The lesson should not be missed in passing that this friendly arrangement is one of the most durable forms of life and if anything is left of the biosphere of the planet after man's depredations it probably will be lichens—they can tolerate very high levels of radioactivity (Woodwell, 1970). More obvious, however, is the endless arsenal of chemical weapons, both defensive and offensive, deployed in the endless struggle for survival and supremacy. We have already mentioned the defensive froth of the wingless grasshopper. Many salamanders exude skin secretions containing tetrodotoxin (tarichatoxin), one of the most poisonous, nonproteinous substances known (Brown and Mosher, 1963; Wakely et al., 1966; Hurlbert, 1970), and the eastern red-spotted newt, both in the aquatic and terrestrial elf stage, is rarely attacked by leeches, thanks to an as yet unidentified repellant (Pough, 1971). The search for alternatives to dangerous artificial pesticides, such as DDT, has focused attention on the chemicals used by plants to protect themselves, for example garlic oil contains the very effective larvicides diallyl disulfide and diallyl trisulfide (Amonkar and Banerjii, 1971). Green and Ryan (1972) have

$$(CH_2{=}CH{-}CH_2)_2S_3 \qquad\qquad (CH_2{=}CH{-}CH_2)_2S_3$$

Diallyl Disulfide   Diallyl Trisulfide

found that when leaves of the potato or tomato plant are wounded by potato beetles or their larvae an animal intestinal proteinase (such as trypsin or chymotrypsin) inhibitor rapidly accumulates in the injured air-exposed plant tissue probably as a defensive mechanism. Some plants use chemical warfare, not only to protect themselves from insects and other herbivores, but also to give them a competitive advantage over other plants competing for the same soil space. The California hard chaparral

Table 12.9 Classes of Interorganismic Chemical Effects.[a] (From Whittaker and Feeny, © 1971 by the American Association for the Advancement of Science, with permission of the AAAS).

I. Allelochemic Effects

  A. Allomones (+ /), which give adaptive advantage to the producing organism

   1. Repellents (+ /), which provide defense against attack or infection (many secondary plant substances, chemical defenses among animals, probably some toxins of other organisms)
   2. Escape substances (+ /) that are not repellents in the usual sense (inks of cephalopods, tension-swimming substances)
   3. Suppressants (+ −), which inhibit competitors (antibiotics, possibly some allelopathics and plankton ectocrines)
   4. Venoms (+ −), which poison prey organisms (venoms of predatory animals and myxobacteria, aggressins of parasites and pathogens)
   5. Inductants (+ /), which modify growth of the second organism (gall, nodule, and mycorrhiza-producing agents)
   6. Counteractants (+ /), which neutralize as a defense the effect of a venom or other agent (antibodies, substances inactivating stinging cells, substances protecting parasites against digestive enzymes)
   7. Attractants (+ /)
      a. Chemical lures (+ −), which attract prey to a predator (attractants of carnivorous plants and fungi)
      b. Pollination attractants, which are without (+ 0) or with (+ +) advantage to the organism attracted (flower scents)

  B. Kairomones (/ +), which give adaptive advantage to the receiving organism

   1. Attractants as food location signals (/ +), which attract the organism to its food source, including (− +) those attracting to a food organism (use of secondary substances as signals by plant consumers, of prey scents by predators or chemical cues by parasites), (+ +) pollination attractants when the attracted organism obtains food, and (0 +) those attracting to nonliving food (response to scent by carrion feeder, chemotactic response by motile bacteria and by fungal hyphae)
   2. Inductants (/ +), which stimulate adaptive development in the receiving organism (hyphal loop factor in nematode-trapping fungi, spine-development factor in rotifers)
   3. Signals (/ +) that warn of danger or toxicity to receiver [repellent signals (A, 1) that have adaptive advantage to the receiver; scents and flavors that indicate unpalatability of nonliving food, predator scents]
   4. Stimulants (/ +), such as hormones, that benefit the second organism by inducing growth

  C. Depressants (0 −), wastes, and so forth, that inhibit or poison the receiver without adaptive advantage to releaser from this effect (some bacterial and parasite toxins, allelopathics that give no competitive advantage, some plankton ectocrines)

II. Intraspecific Chemical Effects

  A. Autotoxins (− /), repellents, wastes, and so forth, that are toxic or inhibitory to individuals of the releasing populations, with or without selective advantage from

---

(Continued)

Table 12.9 (*Continued*)

detriment to some other species (some bacterial toxins, antibiotics, ectocrines, and accumulated wastes of animals in dense culture)

B. Adaptive autoinhibitors (+ /) that limit the population to numbers that do not destroy the host or produce excessive crowding (staling substance of fungi)

C. Pheromones (+ /), chemical messages between members of a species, that are signals for:

1. Reproductive behavior
2. Social regulation and recognition
3. Control of caste differentiation
4. Alarm and defense
5. Territory and trail marking
6. Food location

[a] Adaptive advantage is indicated by +, detriment by −, and adaptive indifference by 0, for the releasing organism first and the receiving organism second. The virgule (/) indicates that adaptive advantage or detriment is not specified for one side of the relationship.

releases a mixture of toxic compounds, many of them phenolic, which rainwater carries into the soil where they inhibit the germination of the seeds of other plant species (McPherson and Muller, 1969). In the same area of California, the soft chaparral emits volatile terpenes that are absorbed out of the air and that again act as herbicides so effective that it is not uncommon for the scrub patches to be surrounded by belts of bare ground 1- to 2-m wide. Whittaker and Feeny (1971) discuss a number of examples of allelopathy and a few of their examples are listed in Table 12.10. The relationship among species can be very complex, for example the beetles that have evolved a mechanism for detoxifying hypericin and now have turned the tables on the plant by using the plant's pesticide as a signal for finding food. The tables have been turned twice in the case of the cabbage aphid, for the mustard oil ingested from the plant by the immune aphid acts as an attractant for a parasite of the aphid. Other tasty species, such as the false monarch butterfly, lacking chemical defense systems of their own, have learned to mock the coloration and even the form of different chemically protected species.

In many instances, allelopathic chemicals are quite simple substances. Phenolic acids and their derivatives are particularly popular, and even a compound as simple as formic acid is used to good effect in bee and ant stings. Popular among plants are phenolic acids, flavonoids, terpenoids, steroids, alkaloids, and organic cyanides (Whittaker and Feeny, 1971). Commonly, the plants protect themselves against their own toxins by tying them up as innoculous glycosides. Whereas animals can rid themselves of poisons by excretion, plants must sequester or otherwise chemically bind harmful chemicals in order to inactivate them. Insects favor lower aliphatic acids, aldehydes, ketones, esters, lactones, hydrogen cyanide, phenols, and quinones, but they also include in their chemical arsenals much more complicated structures as well (Figure 12.18).

The bombardier beetle has a simple but very effective defensive organ (Fig. 12.19). Phenolic precursors from one chamber are mixed with hydrogen peroxide from

Table 12.10 Some Allelopathic Interactions

Source	Target	Chemical Nature
Plant versus Plant		
Hard chaparral	Competing herbs	Phenols
Soft chaparral	Competing herbs	Terpenes (camphor, cineole, etc.)
Walnut tree	Competing plants	Juglone
Grass (*Aristida oligantha*)	N-fixing bacteria and blue-green algae in soil; Lack of N discourages other plants	Phenolic acids
Eucalyptus	An example of self-toxicity	
Plant versus Animal		
Buttercup	Grazing animals	Protoanemonin
Larkspur	Grazing animals	Neurotoxic alkaloids such as delphinine
Tobacco	Aphids	Nicotine
Foxglove	Vertebrates	Steroid cardiac glycosides
Oleander	One leaf can be fatal to man	
Oak tree	Vertebrates, moth larvae and also fungi and viruses	Protein-binding tannins
Balsam fir	Insects	Hormones and hormone-analogue
Hypericum	Herbivores (some beetles (Chrysolina) detoxify hypericin)	Hypericin (causes intense photosensitivity and blindness)
Cruciferae	Cabbage butterflies, moths, weevils, beetles, and aphids	Mustard oils such as allyl isothiocyanate
	(Some herbivores have evolved immunity to plant toxins and may even be attracted by plant chemicals)	
Animal versus Animal		
Skunk	Anything	Butyl mercaptan
Bombardier beetle	Predators	Quinone spray
Monarch caterpillars	Birds	Cardiac glycosides obtained by feeding on milkweed
Grasshopper	Birds	Cardiac glycosides obtained by feeding on milkweed
Animal versus Bacteria		
Man	Bacteria	Lysozyme
Plant versus Fungus		
Orchid	Fungus	Orchinol
Potato	Fungus	Chlorogenic and caffeic acids

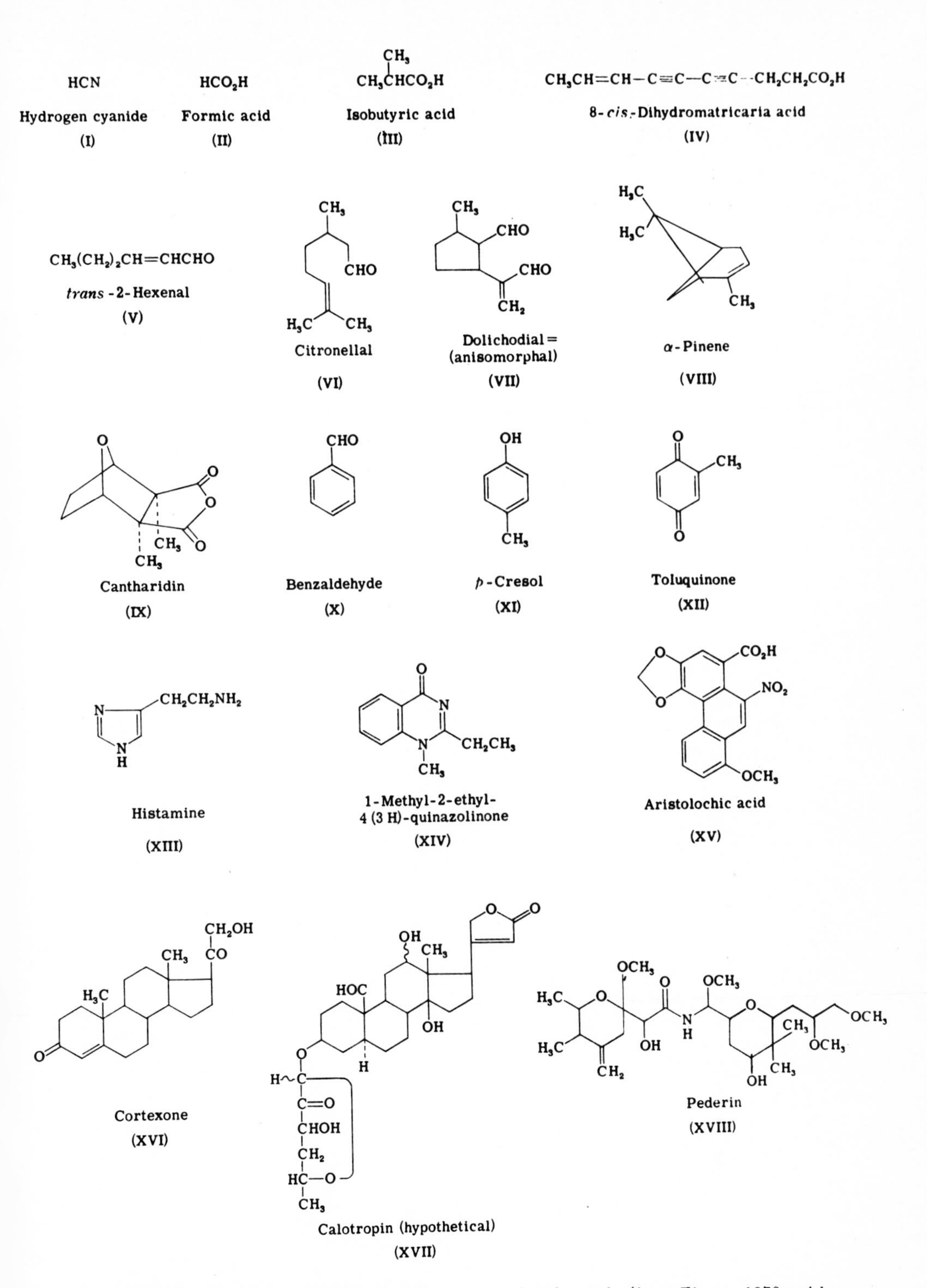

Figure 12.18. Representative defensive substances of arthropods (from Eisner, 1970, with permission of the publisher and the author).

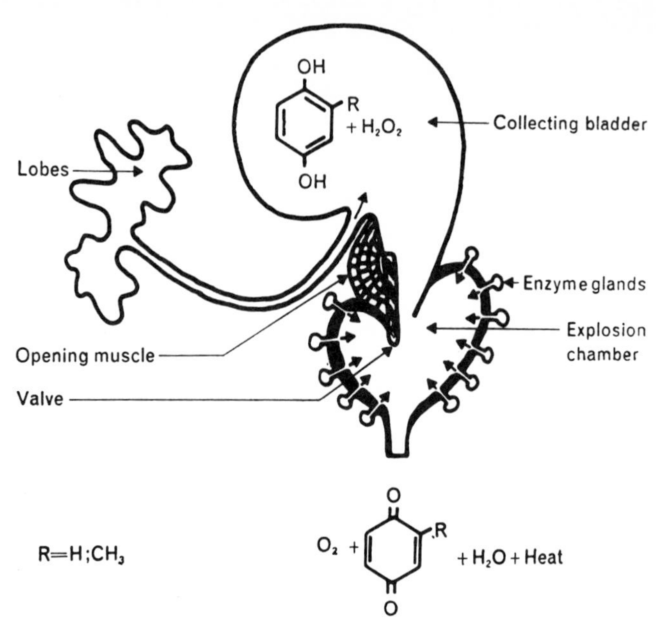

Figure 12.19. The defensive organ of the bombardier beetle (from Schildknecht, 1971, with permission of the author).

another. Not only are the phenols oxidized to quinones but the action of catalase on the hydrogen peroxide produces enough oxygen gas to propel the spray with explosive force (Schildknecht, 1971). The chemical arsenal of an ant is diagrammed in Figure 12.20. In Figure 12.21 Whittaker and Feeny (1971) present a general schema for the major metabolic pathways whereby some of the allomones are formed. The prey-killing venoms of snakes and other animals are commonly proteins or peptides of the toxin or hydrolytic enzyme type. Table 12.11 summarizes some of these substances, while Figure 12.22 shows the protein sequence in melittin (according to Habermann, 1972), the main (50% of dry venom weight) polypeptide toxin in bee venom. These formidable substances attack the nervous system causing heart and respiratory failure, while the enzymes can destroy tissue. Some of these chemicals are exceedingly poisonous; as little as 0.12 μg of cobra venom will kill a mouse. The mucous secretion of the Hawaiian trunkfish can kill a man, and Boylan and Scheuer (1967) have identified the active ingredient as ostracitoxin ("pahutoxin").

$$CH_3{-}(CH_2)_{12}{-}\underset{\displaystyle |\atop O{-}COCH_3}{\overset{\displaystyle H\atop |}{C}}{-}CH_2{-}COO{-}(CH_2)_2{-}N^+(CH_3)_3Cl^-$$

Ostracitoxin

For a very handsome series of volumes on toxic animals, including gruesome clinical photos of their victims (you may never want to go swimming again) see Halstead and Courville (1970). It contains some surprises, for example the liver of the polar bear is poisonous, presumably due to an excess of vitamin A.

Interspecies chemical signals are not always unfriendly. An unknown, heat-labile chemical enables some fish to live commensally among the stinging tenacles of sea

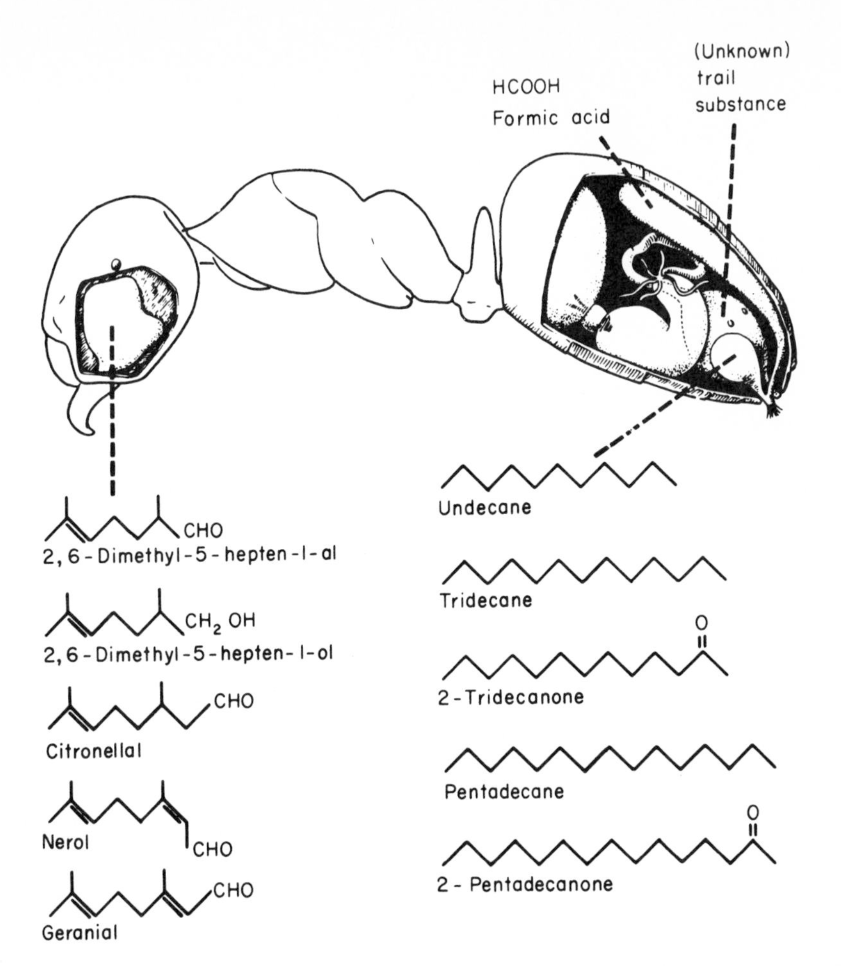

Figure 12.20. The structural formulas of volatile substances found in various of the exocrine glands of *Acanthomyops claviger*. Terpenes are located in the mandibular gland in the head, and alkanes and ketones in the Dufour's gland of the abdomen. All but pentadecane and 2-pentadecanone are efficient alarm substances (from Wilson, 1970, with permission of Academic Press, Inc., and the author).

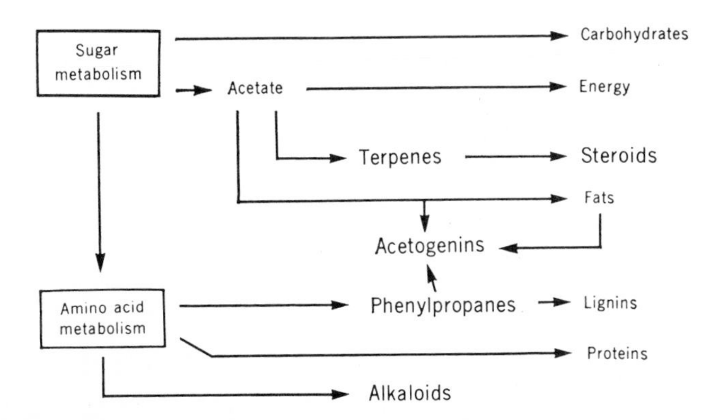

Figure 12.21. Metabolic relationships of the major groups of secondary compounds (shown in large type) to primary metabolism (from Whittaker and Feeny, © 1971 by the American Association for the Advancement of Science, with permission of the AAAS).

Table 12.11 Synopsis of Some Biochemical and Pharmacological Properties of Bee Venom Components and Lysolecithin[a] (From Habermann, 1972)

	Histamine	Melittin	Apamin	MCD-peptide	Hyaluronidase	Phospholipase A	Lysolecithin
Content (percent)[b]	0.1–1	50	2	2	1–3	12	0
Molecular Weight	111	2840	2038	2593	≫20000	14500	524
Sulfur Content	0	0	++	++	+	+	0
Surface Activity	0	++	?	?	0	0	++
General Toxicity[c]	192–445	4	4	>40	0	7.5	150
Local Toxicity; Pain Production	++	++	?	?	0	?	?
Increase of Capillary Permeability	++	++	+	++	Indirectly	+	+
Cellular Damage	0	++	?	+[d]	0	+	++
Neurotoxicity	0	+[e]	++	(+)	0	0	0
"Direct" Hemolysis	0	++	0	0	0	0	++
Circulatory Effects	++	++	0	+	0	+	+
Neuromuscular Effects	0	++	0	?	0	0	?

Smooth Muscular Effects	++	++	0	0	0	+	+
Ganglionic Blockade	0	++	0	?	0	0	++
Histamine Release	0	++	0	++	0	+	++
"Indirect" Hemolysis	0	0	0	0	0	++	0
Thromboplastin Inactivation	0	+	?	?	0	++	+
Interruption of Electron Transport	0	+	?	?	0	++	+
Interruption of Oxidative Phosphorylation	0	+	?	?	0	++	+
Spreading	0	0	0	0	++	0	0
Antigenicity	0	?	?	?	++	++	?

[a] ++, strong; +, significant; ?, not yet tested; 0, not demonstrable.

[b] Approximately, dry venom.

[c] Milligrams per kilogram, mouse, intravenously.

[d] Only known for mast cells.

[e] Only known for local application.

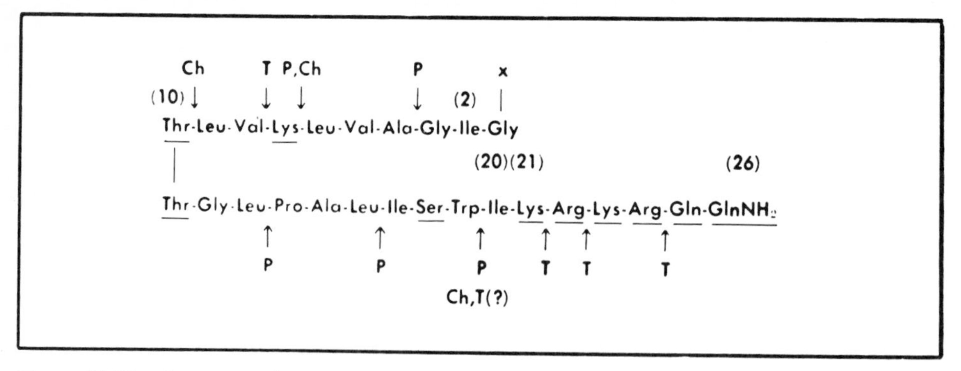

Figure 12.22. Structure of natural and synthetic melittin I. The left-hand side of the diagram indicates the predominantly hydrophobic part (amino acids with a hydrophilic side chain are underlined), and the right-hand side (residues 21 to 26) indicates the hydrophilic, strongly basic part (from Habermann, © 1972 by the American Association for the Advancement of Science, with permission of the AAAS).

anemonies. Another anemone, *Calliaotis parasitica* is commensal with the hermit crab and locates its future companion by responding to the chemistry of the periostracum or coating of the snail shells that the crab occupies.

The location of prey by chemical signals is another important example of interspecies interaction. Lactic and other organic acids in human skin, coupled with respired CO_2, attract the yellow fever mosquito (Acree et al., 1968). Predator mammals follow the chemical scent of their prey, while the prey, just as often, is warned of the approach of its enemy by smell. Starfish are adept at locating the shell fish upon which they feed, probably by tracing to their source minute concentrations of amine and other metabolic and decay products released by the shellfish. Nematodes are popular subjects for chemotaxis experiments. These mindless parasites readily orient themselves in chemical gradients, including amino acids but not sugars (Anon., 1973*a*).

Turning now to intraspecies chemical interactions or pheromones (Table 12.9) perhaps the best known are the sex attractants. Again man's behavior is peculiar; while frantic to disguise his natural scent the female of the species, and now even the male, makes much use of artificial sex pheromones or perfumes. However, man is not alone in trying to disguise his chemical identity: dogs will roll in carrion (dead snakes are particularly popular) in order to disguise their scent for hunting purposes. Aneurotaxis, or attraction of flying insects by airborne pheromones, has recently been demonstrated by Farkas and Shorey (1972) in a very convincing way (Figure 12.23). Insect sex attractants in particular (Jacobson, 1965, 1972) have attracted a great deal of attention as providing alternatives to pesticides for the control of insect pests for both health and agricultural purposes (Beroza and Knipling, 1972). Insects and man, it should be noted, are in direct and intense competition for the vegetable biosphere. Insects may also produce sex attractant inhibitors that might be useful in disrupting the reproduction of insect pests. For example, possibly as a chemical precursor, the sex gland of the female gypsy moth produces 2-methyl-*cis*-7-octadecene which inhibits male attraction to the insect's sex pheromone, *cis*-7,8-epoxy-2-methyloctadecane (Carde et al., 1972). Another potentially very powerful weapon against insect pests is to disrupt the juvenile hormone that controls their life cycle (Williams,

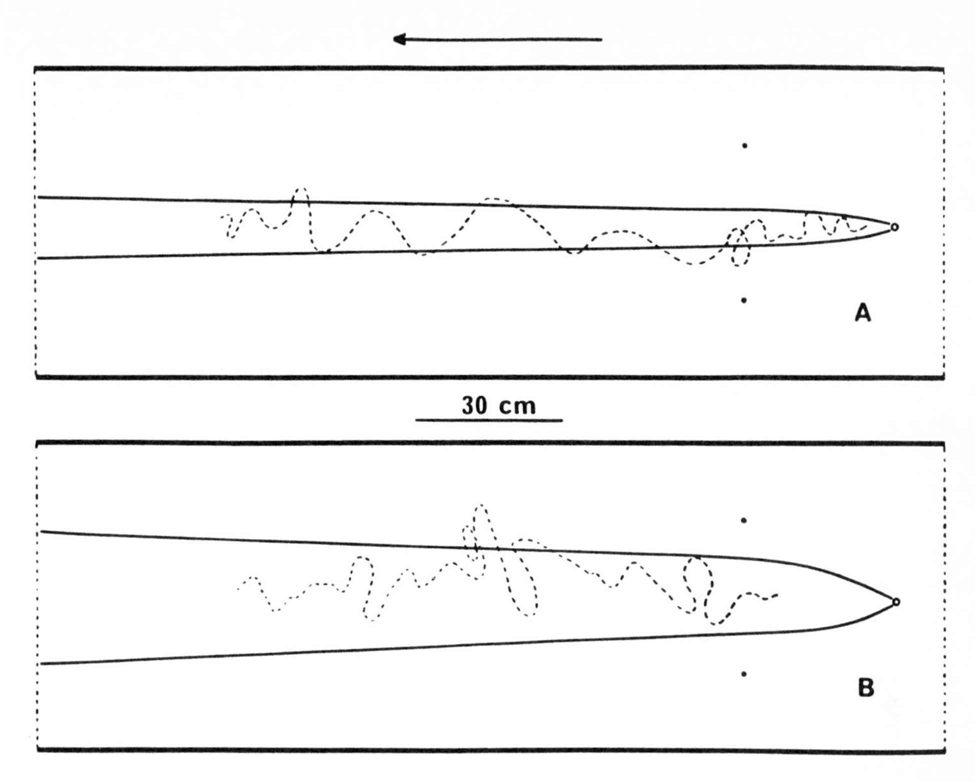

Figure 12.23. Tracings of photographic records of moth flight tracks superimposed on the outline of a time exposure of an odor plume. The arrow indicates the direction of air movement. *A*, track of moth flying through an odor plume in air moving at 25 cm/sec; *B*, track of a moth flying through an odor plume in still air—the plume was produced earlier in air moving at 7 cm/sec. (From Farkas and Shorey, © 1972 by the American Association for the Advancement of Science, with permission of the AAAS).

1970; Menn and Beroza, 1972; Metzger, 1972). Williams (1967) has called these substances the "third generation pesticides" (arsenate of lead is an example of the "first generation," DDT of the second). In the case of plants, some of the analagous

CH_2CH_3 / CH_3 / O / $COOCH_3$ / CH_3CH_2 / CH_3

Cecropia Juvenile Hormone

functions are performed by abscisic acid that regulates growth by promoting senescence and abscission of leaves in the autumn and which can also induce dormancy in

CH_3 / CH_3 / OH / O / CH_3 / COOH

Abscisic Acid

seeds and buds (Wareing and Ryback, 1970). Ants leave trails of odor pheromones to guide themselves about their business. Not only can some ant species follow the odor trail of other ant species (Wilson, 1965, 1968), but the roaches that live in the ants' fungi gardens can also follow their trails (Mosher, 1964). One parasitic wasp leaves a chemical trail over the areas it has searched for hosts to tell its fellows not to duplicate its effort. On the opposite extreme, when it feeds on ponderosa pine, the male bark beetle releases a sort of chemical dinner bell composed of terpene alcohols telling its fellows to come and get it and resulting in mass infestation of the tree. The diners may become dinner for another beetle species uses this same signal to locate high densities of its prey (Wood et al., 1968).

Chemical signals can be remarkably specific. The sex pheromone of the moth *Bryotopha similis* is *cis*-9-tetradecenyl acetate while that of a sibling species is *trans*-9-tetradecenyl acetate, and the sex pheromone of the one actually repels the males

$$\begin{array}{l} \mathrm{\qquad\quad H{-}C{-}CH_2{-}CH_3} \\ \mathrm{\qquad\qquad \|} \\ \mathrm{CH_3{-}CH_2{-}C{-}CH_2{-}CH_2{-}CH_2{-}CH_2{-}CH_2{-}CH_2{-}CH_2{-}CH_2{-}O{-}\overset{\overset{\displaystyle O}{\|}}{C}{-}CH_3} \end{array}$$

cis-9-Tetradecenyl Acetate

$$\begin{array}{l} \mathrm{CH_3{-}CH_2{-}C{-}H} \\ \mathrm{\qquad\qquad \|} \\ \mathrm{CH_3{-}CH_2{-}C{-}CH_2{-}CH_2{-}CH_2{-}CH_2{-}CH_2{-}CH_2{-}CH_2{-}CH_2{-}O{-}\overset{\overset{\displaystyle O}{\|}}{C}{-}CH_3} \end{array}$$

trans-9-Tetradecenyl Acetate

of the other species (Roelofs and Comeau, 1969). Insect olfaction can be exceedingly sensitive. Figure 12.24 shows the threshold sensitivity of an ant's chemical alarm sys-

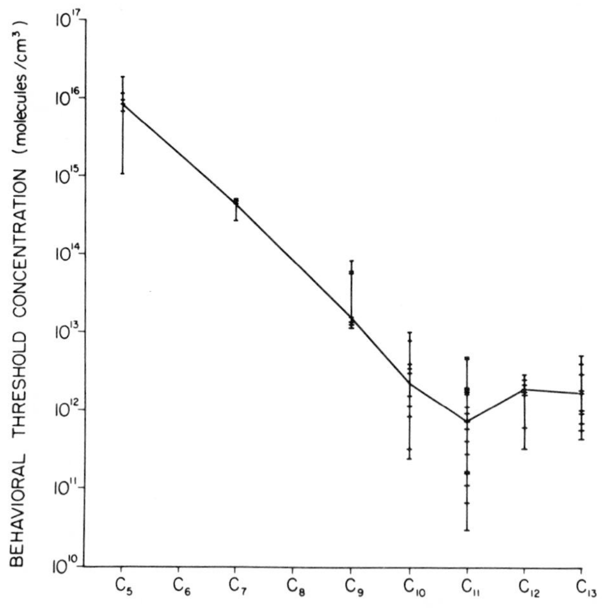

Figure 12.24. Estimates of behavioral threshold concentrations of members of the alkane series from pentane (C_5H_{12}) to tridecane ($C_{13}H_{28}$). Undecane ($C_{11}H_4$) and tridecane are the hydrocarbons that occur naturally in the Dufour's gland of *Acanthomyops claviger* and function as alarm pheromones (from Wilson, 1970, with permission of Academic Press, Inc., and the author).

tem. The antenna of the silkmoth *Bombyx* has a detection threshold for bombykol of only 10^4 molecules (Schneider, 1969)!

Pheromones structure and give order to animal communities, and the bullhead catfish affords us an excellent example. Homeostasis of this fish community is maintained by a strict social order which in turn is dependent upon the identification of each member of the community by smell (Todd, 1971). Chemical pollution is a particularly disruptive stress upon such a community, for if it is strong enough to confuse the social recognition process chaos results; the fish will battle with one another to disability or death and the community is destroyed. In addition to insects and fish, sex pheromones are important behavioral determinates in crustacea such as lobsters (Atema and Engstrom, 1971). In another example of the confusing or masking of essential chemical signals by pollution, it has been demonstrated that low concentrations (0.8 ppb) of the seawater soluble fraction of kerosene (and crude oil) interferes with the ability of the marine snail *Nassarius obsoletus* to locate food extracts (Jacobson and Boylan, 1973). Concentrations of dissolved oil in seawater have been reported as high as 90 ppb 20 mi from the spill site, and the concentration of napthalenes, a major toxic ingredient of the kerosene extract, is around 1 ppb in the polluted Charles River (Boston) basin.

While some species may match him in toxicity, in terms of quantity no species can match man for the release of allelochemicals into the environment, for we must consider human technology as an extension of the human organism. He releases into our environment allelochemicals, which have extinguished some species and endangered and mutated many more—including his own. Whittaker and Feeny (1971) have summarized the situation neatly and grimly: "Civilized man is . . . faced with a phenomenon new in history—the progressive toxification of the biosphere. We are cast, like some pathogens and successional species, in the role of an unstable dominant population that can effect its own demise by autotoxicity and degradation of the environment."

Organisms may release chemicals for reasons other than relating to their own and other species. Certain water skimmers release surfactants that alter the surface tension of water. Fish slimes contain substances, probably polymeric, that can reduce the fluid friction of water 57 to 66%, thus facilitating swimming (Rosen and Cornford, 1971). Taking a clue from Nature, the Navy is interested in additives that can reduce ship drag—these long chain polymeric substances appear to reduce fluid friction by tending to "iron out" micro-eddies and by promoting laminar flow rather than by altering the H-bonded structure of the liquid (Horne et al., 1968).

I do not wish to digress here into the very interesting and complex question of the mechanism of chemical sensation and the nature of the receptors except to say that in the words of Amoore (1971) "although many theories of odor specificity have been proposed, only two have been supported by quantitative measurements with statistical analysis," and these are the stereochemical theory, which correlates odor with the sizes and shapes of molecules, and the vibrational theory, which correlates odor with certain spectral absorption peaks (for a collection of papers on the subject olfaction, see Pfaffmann, 1969).

The mechanism of biological evolution hinges on mutation. Constant mutation provides evolution with its alternatives. Most of the mutant experiments are unsuccessful and disappear, others are unsuccessful yet persistent, giving rise to genetic diseases. Still others are partially successful, offering certain survival advantages, but at the price of imposing certain survival disadvantages—hemoglobin S is an example. The successful mutations give rise eventually to new biological forms better fitted for survival and dominance in the environment. In the past, natural levels of radiation, chemicals, and just plain molecular accidents provided an input of mutations. Boyland (1969) claims that 90% of cancer in man is chemically produced. Solar activity and consequent changes in radiation levels could have made the mutagenic background highly variable in the geological past. One of the most serious aspects of man's polluting activity is that he has greatly increased the inventory of mutagenic radiation and chemicals in the environment (Sutton and Harris, 1972). The relationship between carcinogens and mutagens is not clear. Some carcinogens, such as polycyclic hydrocarbons, are not strongly mutagenic. Current thinking is in the direction of postulating that some kind of metabolic activation, possibly to epoxides, is necessary to produce strong carcinogens or mutagens (Anon., 1972*g*). The metabolite of the parent carcinogen may then bind covalently with cellular macromolecules, such as DNA (Brookes, 1966; Brookes and Duncan, 1971). In the light of the unreactive nature of some of the original hydrocarbons, this suggestion makes some chemical sense.

12.6 INTERNAL POLLUTION

By internal pollution, I mean harmful materials that man ingests, inhales, or applies to his body. In some instances, the contamination derives from unavoidable natural sources. For example, esophageal cancer in the Caspian region is correlated with the occurrence of highly saline littoral soils (Figure 12.25). However, by such a simple expedient as cooking primitive man protected himself from many of the hazards and opened up a great variety and quantity of vegetative food sources (Leopold and Ardrey, 1972). Other animals do not fare as well; the teratogens (i.e., causing congenital malformation of the fetus) cyclopamine, cycloposine, and jervine present

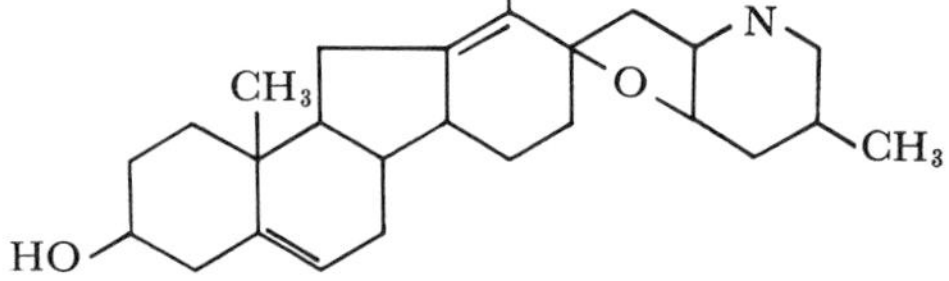

Cyclopamine

in the alpine meadow plant *Veratrum californicum* cause monstrous cyclopean birth defects in the sheep that graze on the plant, while an unknown teratogenic agent in burley tobacco stalks produces arthrogryoptic birth defects in the pigs that feed upon them (Mulvihill, 1972). Today, on the other hand, many of the poisons in our food we have deliberately put there ourselves to either "improve" or preserve the food. The quantities of additives used in the United States have been steadily rising (Figure 12.26), and Americans now consume some 1,060,000,000 lbs/yr of nearly 3000 dif-

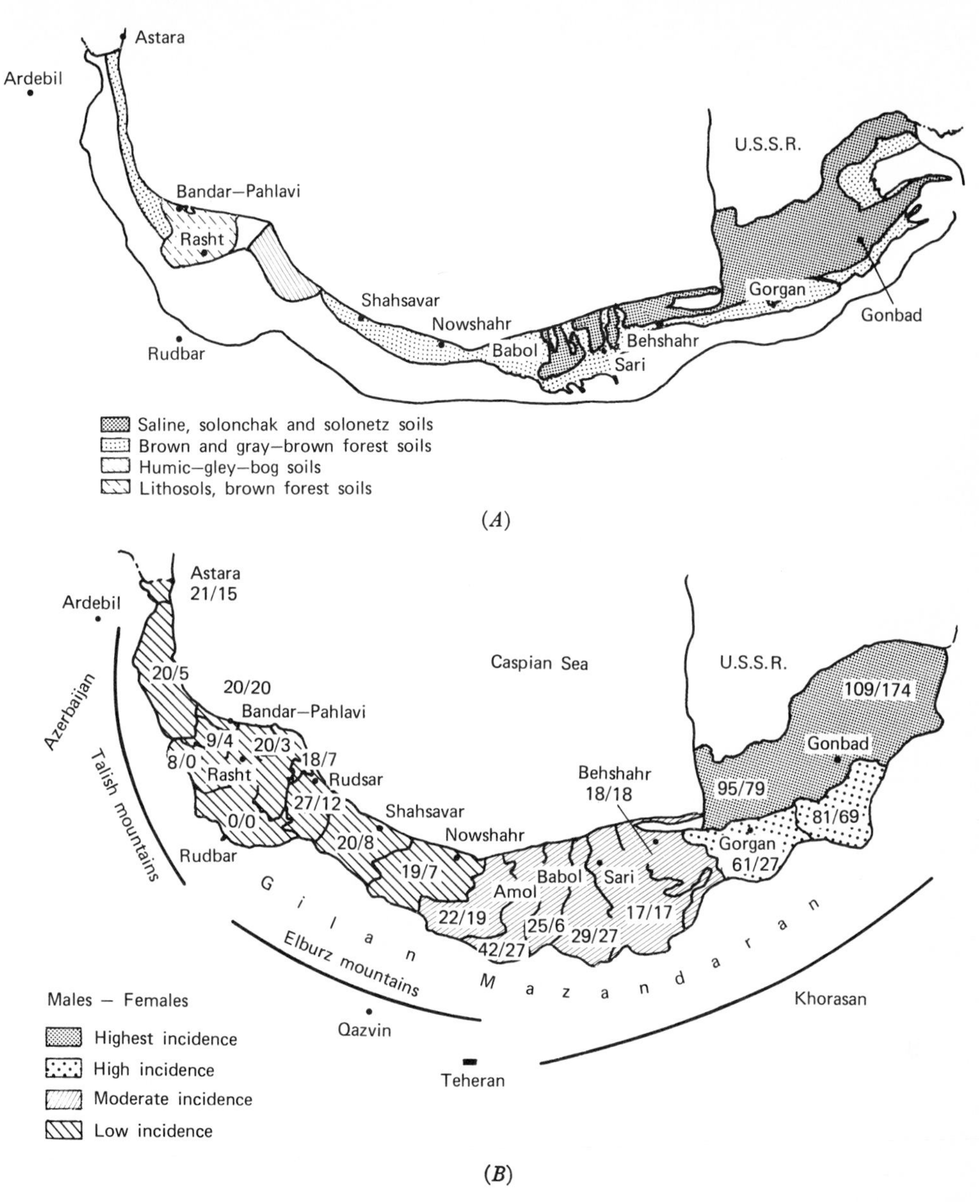

Figure 12.25. Soil type and cancer. *A*, Types of soil in the Caspian littoral of Iran; *B*, age-standardized incidence rates of human esophageal cancer in the Caspian littoral of Iran. (From Kmet and Mahboubi, © 1972 by the American Association for the Advancement of Science, with permission of the AAAS).

ferent food additives or about 3–5 lb/person-yr (Anon., 1972*c*). Quantities and values of some of the major additives are shown in Table 12.12. The additives in a loaf of bread makes a truly impressive list (Table 12.13) (Longgood, 1960; Anon., 1966; Furia, 1968; Linton, 1970).

Let us concentrate on three instances. Perhaps the most highly publicized has seen the ban on cyclamate sweetening agents, and now massive doses of the popular

Table 12.12 (Reprinted with permission from *Anon. Chem. Engr. News* (October 10, 1966), pp. 100ff. © The American Chemical Society.)

	Millions of Pounds of Additives[a] Used in Foods Made in the U.S.						
	1955	1960	1961	1962	1963	1964	1965
Emulsifiers	75.70	106.00	113.10	128.40	132.70	141.20	150.0
Acidulants	92.50	103.40	109.40	113.50	118.30	120.20	120.5
Stabilizers and Thickeners	63.80	79.80	85.40	92.30	96.70	100.80	105.0
Flavoring Agents and Flavor Enhancers	38.10	49.40	56.90	65.90	64.70	70.60	73.0
Leavening Agents	64.50	67.10	66.00	65.30	64.30	64.90	64.0
Preservatives (including antioxidants)	21.85	31.06	32.35	39.15	44.35	47.96	50.0
Artificial Sweeteners	0.25	3.54	4.25	5.00	6.30	8.00	10.0
Colors (FD&C certified)	1.69	2.34	2.02	2.35	2.55	2.60	2.6
Miscellaneous[b]	60.60	68.90	73.30	75.70	77.90	84.90	86.0
Totals	418.99	511.54	542.72	587.60	607.80	641.16	661.1

Millions of Dollars at the Manufacturers' Level of Additives[a] Used in Foods Made in the U.S.

	1955	1960	1961	1962	1963	1964	1965
Emulsifiers	$18.76	$27.28	$29.41	$33.30	$35.18	$38.10	$40.0
Acidulants	17.45	20.97	22.20	22.90	24.76	24.55	24.6
Stabilizers and Thickeners	44.59	54.58	56.98	61.25	63.50	76.16	80.0
Flavoring Agents and Flavor Enhancers	61.63	60.25	66.84	71.16	61.78	60.11	62.5
Leavening Agents	3.52	4.02	4.00	3.96	4.08	4.27	4.2
Preservatives (including antioxidants)	8.20	16.20	18.40	21.40	24.24	26.05	27.5
Artificial Sweeteners	0.50	6.61	7.99	9.40	9.77	7.89	9.0
Colors (FD&C certified)	6.39	9.48	8.46	9.45	10.25	10.40	10.5
Miscellaneous[b]	10.68	14.14	15.52	17.72	20.20	25.77	27.0
Totals	$171.72	$213.53	$229.80	$250.54	$253.76	$273.30	$285.3

[a] Does not include such widely used ingredients as sugar, salt, and starch.

[b] Includes nutritional supplements, enzymes, dough conditioners, anticaking compounds, moisture-retaining agents, yeast foods, clouding agents, sequestrants, and others.

Source: Arthur D. Little, Inc., estimates

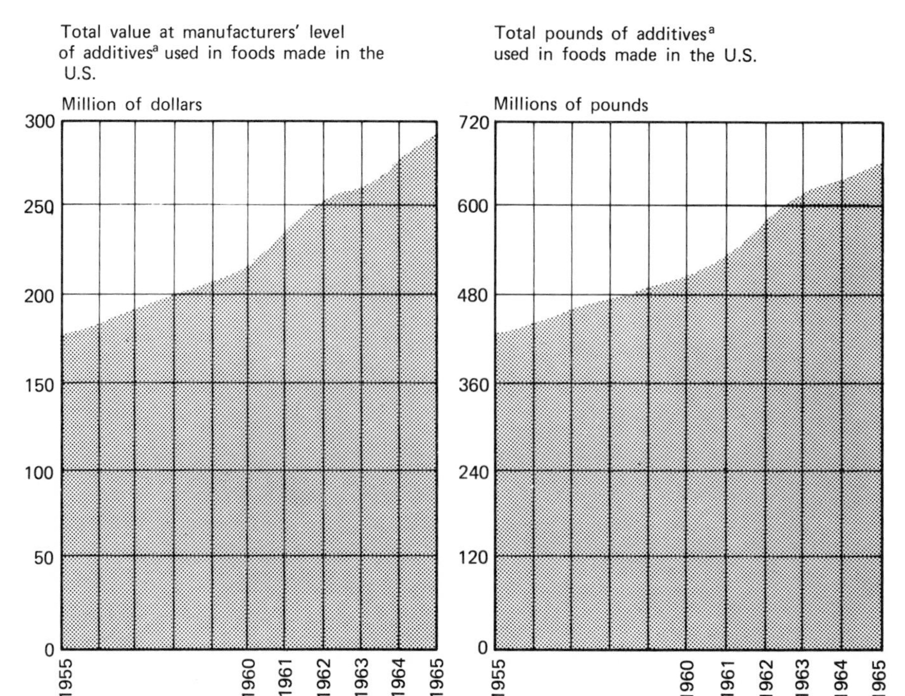

[a] Does not include such widely used ingredients as sugar, salt, and starch.
Source: Arthur D. Little, Inc., estimates

Figure 12.26. The use of food additives has risen more thna 50% in a decade (From Anon., reprinted with permission from *Chem. Engr. News* (October 10, 1966), pp. 100ff. © The American Chemical Society).

sugar substitute saccharine have been shown to be carcinogenic in rats (Anon., 1972*c*) leading one to suspect even sugar (sucrose) itself. Sugar, although linked to

—NH—CO_3Na

Sodium Cyclohexylsulfamate
(Cyclamate)

O, C, NNa, SO_2

Sodium Salt (soluble) of Saccharine

some diseases such as diabetes, appears to be harmless, yet sugar consumption does tend to displace the consumption of more nutritional foods (Yudkin, 1972). Monosodium glutamate, which occurs naturally in some foods, is another very common "taste-improver" about which some reservations have arisen. Our second example,

Table 12.13 Additives in a Loaf of Bread (Based on information in Linton, 1970)

Preservatives—Calcium propionate, sodium diacetate, sodium propionate, acetic acid, lactic acid

Leavening Agents—Potassium acid tartrate, monocalcium phosphate, sodium acid pyrophosphate

Bleaching Agents—Benzoyl peroxide, chlorine dioxide, nitrosyl chloride, oxides of nitrogen

Bread "Improvers"—Potassium bromate, potassium iodate, calcium peroxide

Antioxidants—Butylated hydroxytoluene, butylated hydroxyanisole, propyl gallate, nordihydroquaiaretic acid

Emulsifiers—Lecithin, monoglycerides, diglycerides, sorbitan and polyoxethylene fatty acids

In addition, The Wheat is grown contaminated with fertilizers, herbicides, and pesticides

The Grain on storage is treated with rodenticides, fungicides (some of them mercury compounds) and insecticides

The Water added to the flour may have been purified with alum, chlorine, copper sulfate, and serta ash, and fluoride compounds may have been added (recently epidemics of acute hemolytic anemia in uremic patients have been traced to chlorination of urban drinking water by chloramine bacteriacides—Eaton et al., 1973)

The Sugar (or dextrose) added to the flour may have been refined with lime, sulfur dioxide, phosphates and charcoal

The Salt may contain added iodide and carbonates of calcium and magnesium

The Yeast has been fed on ammonia salts

The Shortening is refined, bleached, and decolorized and has been exposed to traces of nickel; anti-oxidants; citric, ascorbic, and phosphoric acids; and has been glycerinated

nitrite and nitrate compounds, which occur naturally in many foods and are readily reduced to nitrite by bioprocesses, are common preservative food additives. They have been used since Roman times. $NaNO_2$ not only keeps meat red and fresh-looking, but it may also prevent the growth of microorganisms that produce deadly botulism. But on the negative side, large doses of nitrite can reduce the blood's oxygen-carrying capacity by combining with hemoglobin to form methemoglobin, and even small amounts may be dangerous, for evidence is accumulating that nitrites may combine with amines and amides in the stomach to form highly carcinogenic nitrosoamines (Magee and Barnes, 1967; Anon., 1972*d*). Also it has been demonstrated that nitrosopyrrolidine and dimethylnitrosoamine are formed in nitrite preserved bacon when it is cooked (Sen et al., 1973). Because so little is known about the *in vivo* synthesis of nitrosoamines Wolff and Wasserman (1972) caution against banning nitrate preservatives outright and warn that "we should be very sure that we are not foregoing the needed preservative effects of nitrate, which protects us against serious outbreaks of food poisoning." Diethyl pyrocarbonate (DEP or "baycovin") is another popular additive to foods and beverages to prevent microbial

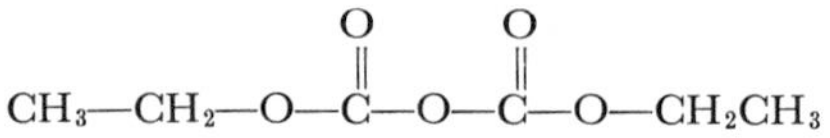

$$CH_3{-}CH_2{-}O{-}\overset{\displaystyle O}{\overset{\|}{C}}{-}O{-}\overset{\displaystyle O}{\overset{\|}{C}}{-}O{-}CH_2CH_3$$

Diethyl Pyrocarbonate

growth, but recently it has been shown in the latter to form the carcinogen urethan (Löfroth and Gejvall, 1971).

Our third example is in many ways the most frightening, and unnecessary. Stilbestrol and diethylstibestrol (DES) are exceedingly powerful and biologically disruptive chemicals. They are potent even although their structural resemblance

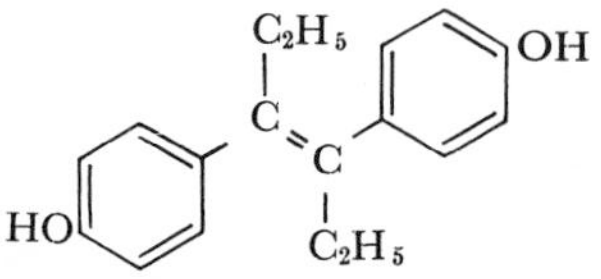

Diethylstibestrol

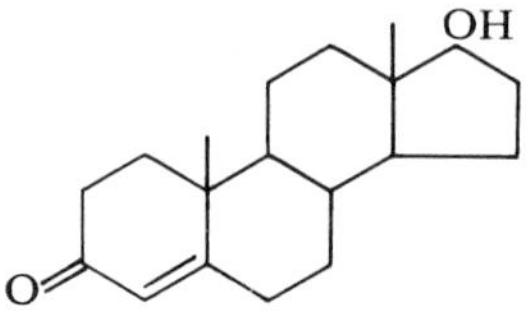

Testosterone (Male)

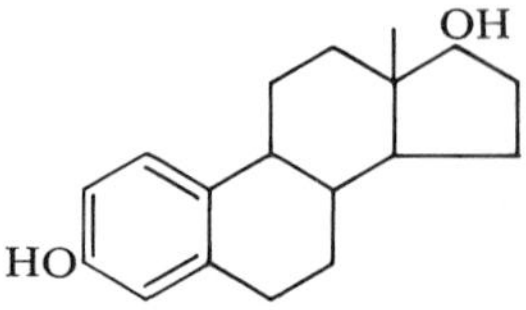

Estrone (Female)

to natural sex hormones such as testosterone and estrone is very superficial, and this becomes even more remarkable when we note the very little chemical difference between the human male and female hormones. Plants, by the way, also have sex hormones: two are sirenin, the first plant sexual hormone ever isolated, and anther-

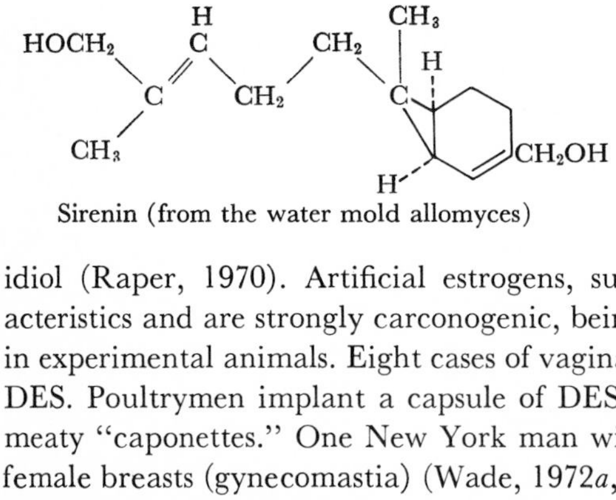

Sirenin (from the water mold allomyces)

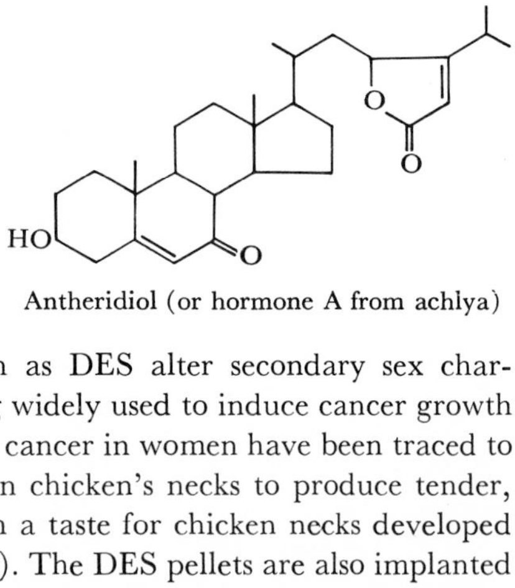

Antheridiol (or hormone A from achlya)

idiol (Raper, 1970). Artificial estrogens, such as DES alter secondary sex characteristics and are strongly carconogenic, being widely used to induce cancer growth in experimental animals. Eight cases of vaginal cancer in women have been traced to DES. Poultrymen implant a capsule of DES in chicken's necks to produce tender, meaty "caponettes." One New York man with a taste for chicken necks developed female breasts (gynecomastia) (Wade, 1972*a*, *b*). The DES pellets are also implanted in the ears or administered in the feed of 75% of the 30,000,000 cattle slaughtered in the United States to fatten them, thus saving cattlemen in the United States an estimated $90,000,000/yr in added feed costs. The pellets are supposed to be absorbed within 48 hr (later extended to 7 days) before the meat of the animal is marketed, but the evidence has shown that such is not always the case and that the carcinogen and its residues can persist. The United States government reluctantly and only after considerable political pressure finally joined the 22 other nations who ban these chemicals from foodstuffs, first in poultry and now at last in beef (Wade, 1972*a*). The whole affair afforded insight into the scandalous degree to which a government "protective" agency is prepared to endanger the public health for political purposes.

We not only eat and drink dangerous chemicals, often unwittingly in our determination to be pheromoneless, some of us wallow in them—perfumes, soaps, lotions, cosmetics, antiperspirants, "hygenic" sprays, preparations, and so on (Figure 12.27). While certainly very much on the increase, this dangerous silliness is not a novelty—the ancients beautified themselves with cosmetics containing arsenic pigments! In passing let me mention one example, hexachlorophene. Hexachlorophene is an

Hexachlorophene

effective antiseptic. It has been used in hospitals around the world for more than 20 yr. In addition to its legitimate uses, "among its most needless uses is in vaginal deodorants, a $53 million-a-year racket founded on high pressure advertising and the ruthless exploitation of modern phobias about body odor" (Wade, 1971). The chemical is capable of causing two diseases—chloasma and burn encephalopathy—and is known to cause serious brain damage in rats. When new-born monkeys were washed daily for 90 days with a 3% hexachlorophene solution (hospitals routinely wash human babies in dilute solutions of the chemical as a precaution against diaper rash and other infections), they developed brain damage similar to that observed in the

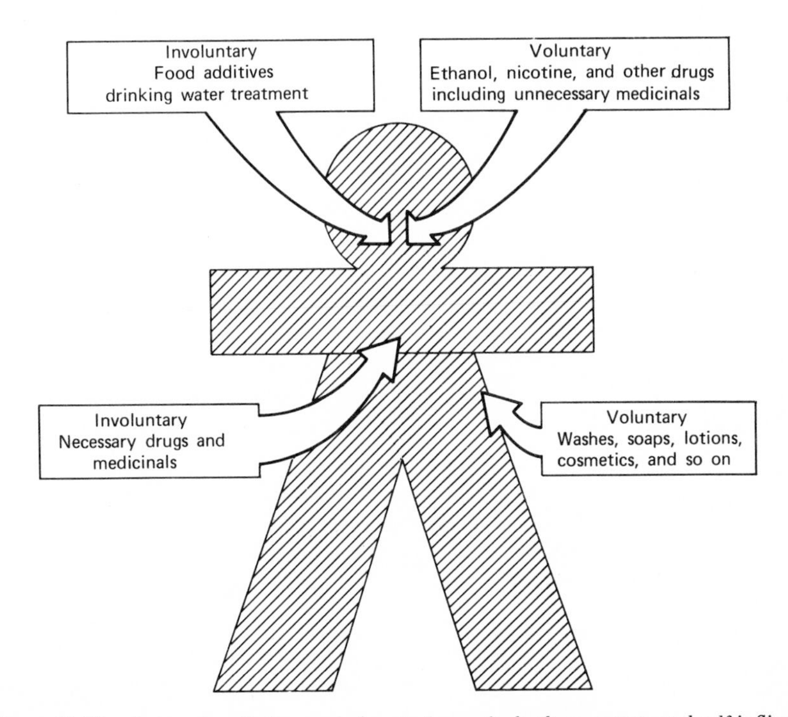

Figure 12.27. Internal pollution and chemical assault, both necessary and self-inflicted.

rats, and the United States Food and Drug Administration issued a "prudence" warning (Wade, 1972*b*). This ban was followed by outbreaks of staphylococcal skin disease in 23 United States hospitals (Anon., 1972*e,f*). Like nitrite food preservatives, hexachlorophene is a highly useful chemical in safeguarding human health, but equally clear, these are both chemicals whose abuse can be dangerous. The failure of government regulatory agencies to distinguish carefully between necessary uses and frivolous additives and applications of chemical substances is not reassuring. The shibboleth "no additives" is as foolish and dangerous as their indiscriminate use. Dinman (1972) has suggested the beginning of an approach to the rational and semiquantitive evaluation of the impact of foreign chemicals on organisms, including man, and experimental definition of Figure 7.31 for suspect chemicals could accomplish this same goal.

To change the subject, despite their fast-growing popularity, there is no scientific evidence that so-called "natural" or organically grown foods are any more nutritious than those grown with artificial fertilizers. Furthermore, since heavy metals, pesticides, and other contaminants rain down so impartially throughout the northern hemisphere, there can be no assurance that "natural" foods are uncontaminated. Far more outrageous to me is the millions of dollars that Americans spend on artificial foods advertised on the basis of their absence of nutritional value, while much of the rest of the world goes hungry. I sometimes hope that natural selection will obtain and that the plastic people will wash, gargle, spray, and noneat themselves into extinction.

While the chemicals we consume, more or less unwittingly, in our food and drinking water are certainly cause for concern, by far the most dangerous forms of internal pollution are the drugs that we ingest knowingly and by our own volition. And of all these deadly chemicals, by far the most dangerous is ethanol. In the United States, *ethanol is responsible for more deaths, incapacity, and social disruption than all other forms of chemical pollution in our environment combined.*

The chemistry of drugs and the mechanism of their modes of action is an exceedingly fascinating and important topic, and I am confidant that research utilizing these chemicals will continue to increase vastly our understanding of the molecular biology of physiological and psychological processes. Sadly enough, I will have to severely restrict our discussion here, and I will confine it to mention of a few of the most familiar drugs that alter human consciousness.

In a corner of the Boston Public Garden is a little noticed monument to the doctor who first used ether during a surgical operation. It is perhaps the world's most deserved monument. The pain and human suffering that has been alleviated by anaesthetics and pain-relieving drugs far exceeds any possible assessment. In addition to ether, a great many chemicals are anaesthetics. Some of them can have very little body chemistry and some of them, the noble gases, have little chemistry at all! How then do they work? Several years ago, Professor Pauling (Pauling, 1961; see also Miller, 1963) ingeniously pointed out that, while the chemical properties of anaesthetics vary enormously, they do all have one thing in common—they all alter the structure of liquid water in their immediate neighborhood. This structuring of water by the dissolution of these substances in the spinocerebral fluid, Pauling argued, interferes with transport processes, specifically the molecular, ionic, and/or charge movements necessary for the transmission of nerve signals. Just as water–clathrate formation physically clogs hydrocarbon pipelines, clathrate formation electrically "clogs" nerve pathways. It is unfortunate that Pauling described this structuring specifically as clathrate formation, for his clever hypothesis has been rather coldly received, partly because the self-clathrate model for pure liquid water was subsequently shown to be untenable (Danford and Levy, 1962). Nevertheless, I suspect there is a great deal of truth in the Pauling hypothesis of anaesthesia: Here we have the abandonment of a very good idea by the scientific establishment because a couple of its details are awry. Sad to say this is an all too common occurrance, and I would like to digress a bit in view of the importance of the lesson to be learned and mention two more examples. For many years a controversy persisted as to whether or not certain "kinks" (abrupt changes in the slope when water properties are plotted versus temperature) were real phenomena or the products of experimental error. The issue was finally settled by Professor Drost-Hansen (1968), one of the proponents of kinks, whose careful measurements showed no kinks in the case of pure, *bulk*, liquid water. So everybody has now forgotten about kinks. But also forgotten is Drost-Hansen's more important finding—the kinks *are* a real property of water near interfaces (vicinal water). Now most of the water in living cells and tissues is vicinal water, and one of the kinks near 37°C is clearly of the most profound biological significance. The kinetics of many biological processes, including growth, exhibit maxima or inflections at this temperature, and the quasiphase change in the hydration atmosphere of biopolymers corresponding to this kink possibly controls temperature in homeothermic animals (Horne et al., 1971). For many years Professor Klotz (see Klotz, 1970) referred to this hydration sheath as "ice-like." Because they could not divorce the idea of a specifically Ice-I structure, most of his colleagues have missed the important point—that the water is structured. A second, more recent, and highly publicized example (even finding its way into the popular press; *Time* December 19, 1969; *Time*, October 19, 1970; *New York Times*, April 2, 1970) is the orthowater (or "polywater") first reported by the Soviet Academician Deryagin (see Deryagin, 1970). Elaborate theories of the stuff were published, so when "polywater" turned out to be dirty water there were some tarnished reputations. The scientific community was relieved to forget about the whole episode—and to forget that the water near interfaces, especially silica surfaces, is quite different from bulk liquid water and that its physical–chemical properties are consequently quite different. And so again the baby is thrown out with the bathwater.

To return to the mechanism of anaesthetic action, the argument runs that relatively weak physical intermolecular forces rather than strong chemical forces must be involved, and pursuing this line of thought, Koski et al. (1973) have reported an interesting correlation between anaesthetic effectiveness and the van der Waals a constant (Figure 12.28; see also Wulf and Featherstone, 1957; Wilson et al., 1969).

Analgesics or pain relievers are also every bit as essential to humane medical practice, and again their mode of chemical action has been the subject of considerable inquiry. The most important analgesic continues to be morphine, the active ingredient of opium (morphine derivatives include heroin and codeine), despite its dangers as an addictive drug. Its physiological function is clearly related to the spacial characteristics of its structure (Figure 12.29). All potent analgesics have certain common structural features: an aromatic nucleus linked to a carbon or nonbasic nitrogen atom not bound to hydrogen and linked through two carbons to a tertiary, or more rarely, secondary basic nitrogen. Morphine's mirror image, (+)morphine, is inactive as are all of its derivatives. Spatial accommodation to a specific receptor site in the central nervous system seems to be involved. The analgesic

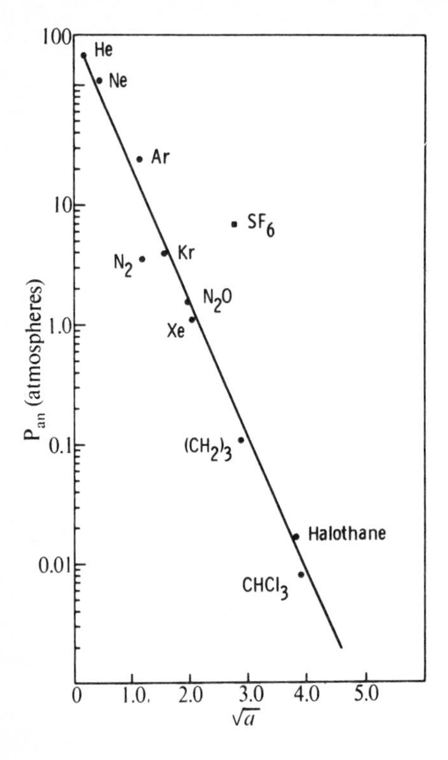

Figure 12.28. Isonarcotic pressure versus the square root of the van der Waals *a* Constant for various anaesthetics (from Koski et al., 1973, with permission of the publisher and authors).

effectiveness of heroin and related substances appears to be inseparably linked to their undesireable drug characteristics; however, the accidental discovery of morphine antagonists such as nalorphine (Figure 12.29) has given new hope that more safe analgesics might be developed (Bentley, 1964; Eddy and May, 1973). In addition narcotic antagonists offer a most promising attack on the addiction problem (Hammond, 1971*a*, *b*; Maugh, 1972*c*).

Ether produces an intense (and very dangerous) high. Another well-known anaesthetic, nitrous oxide (laughing gas), was very popular among the English fashionable set more than 100 yr ago. Powerful natural hallucinogens, notably opium (laudanum or tincture of opium) were also used in the eighteenth and nineteenth centuries in England and France, but the vogue was largely restricted to certain literary and artistic circles. From chewing betel nuts to swilling beer and mainlining heroin, narcotics have always been used universally throughout the world. Schultes (1969) has prepared a fascinating review of plant hallucinogens used by primitive peoples, and Figure 12.30 shows some of the main active ingredients of psychotomimetic plants. These drugs are commonly used in conjunction with religious ceremonies. The role of peyote in the rites of the Indians of the American southwest is well known. And a faint palimpsest of such practices persists even in Christian ritual in the eucharistical wine. Artificial hallucinogens are even more potent; for example LSD (Figure 12.31) is much more powerful than mescaline. The former, the hypothesis goes, occupies completely specific receptor sites, whereas the latter only partially occupies them. Electron donation is now suspected of playing a role in psychoactivity (Anon., 1973*d*; see also Baker et al., 1972). Drugs are remarkably specific. While certain structural configurations such as the indole ring

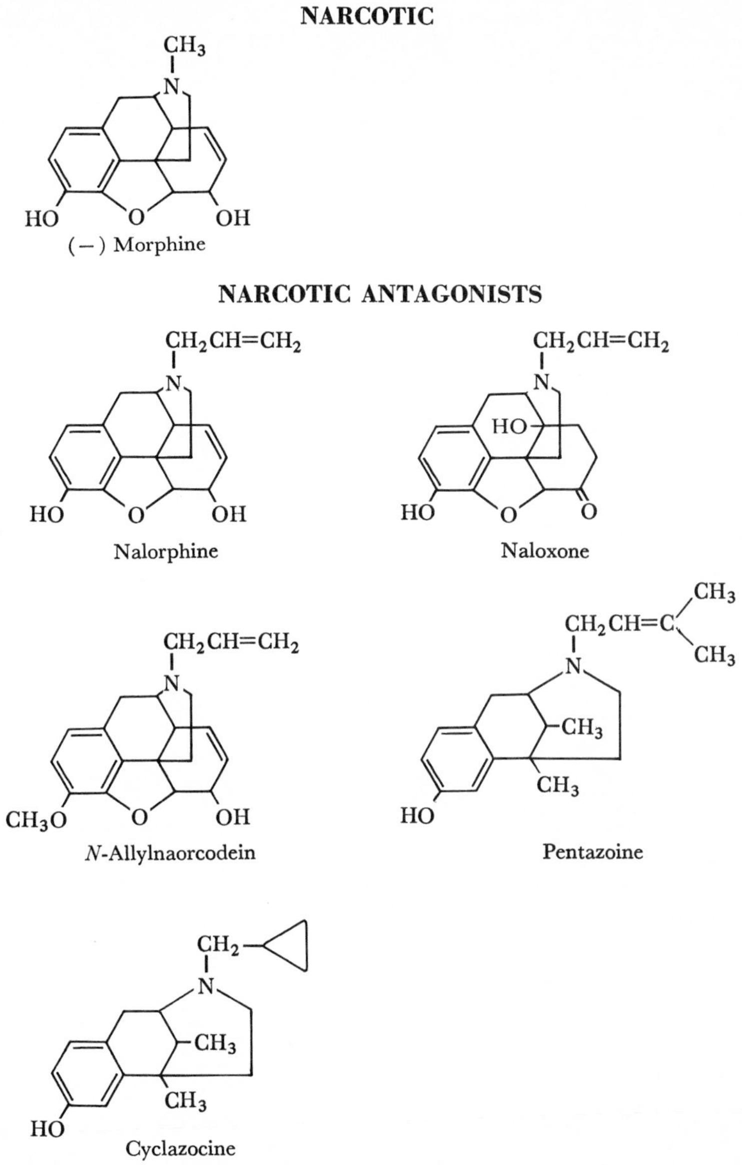

NARCOTIC

(−) Morphine

NARCOTIC ANTAGONISTS

Nalorphine

Naloxone

N-Allylnaorcodein

Pentazoine

Cyclazocine

Figure 12.29. Structural similarity between a narcotic (morphine) and its antagonists (from *Chem. Engr. News*, July 3, 1972, p. 14).

(Figure 12.31) appear to be responsible for the activity, at the same time seemingly minor alterations of the groups attached to these structures can greatly modify the activity (Figure 12.32). Lucky are those people whose opiate is a relatively mild one. Drug addiction (known in Japan as the "American disease") is particularly rampant in the United States, and the social causes and consequences are many and complex, including indifference on the part of the public as long as addiction was largely confined to ethnic minorities, middle-class ennui, and laxness of law enforcement. More relevant to our present concern here with internal pollution is the peculiar

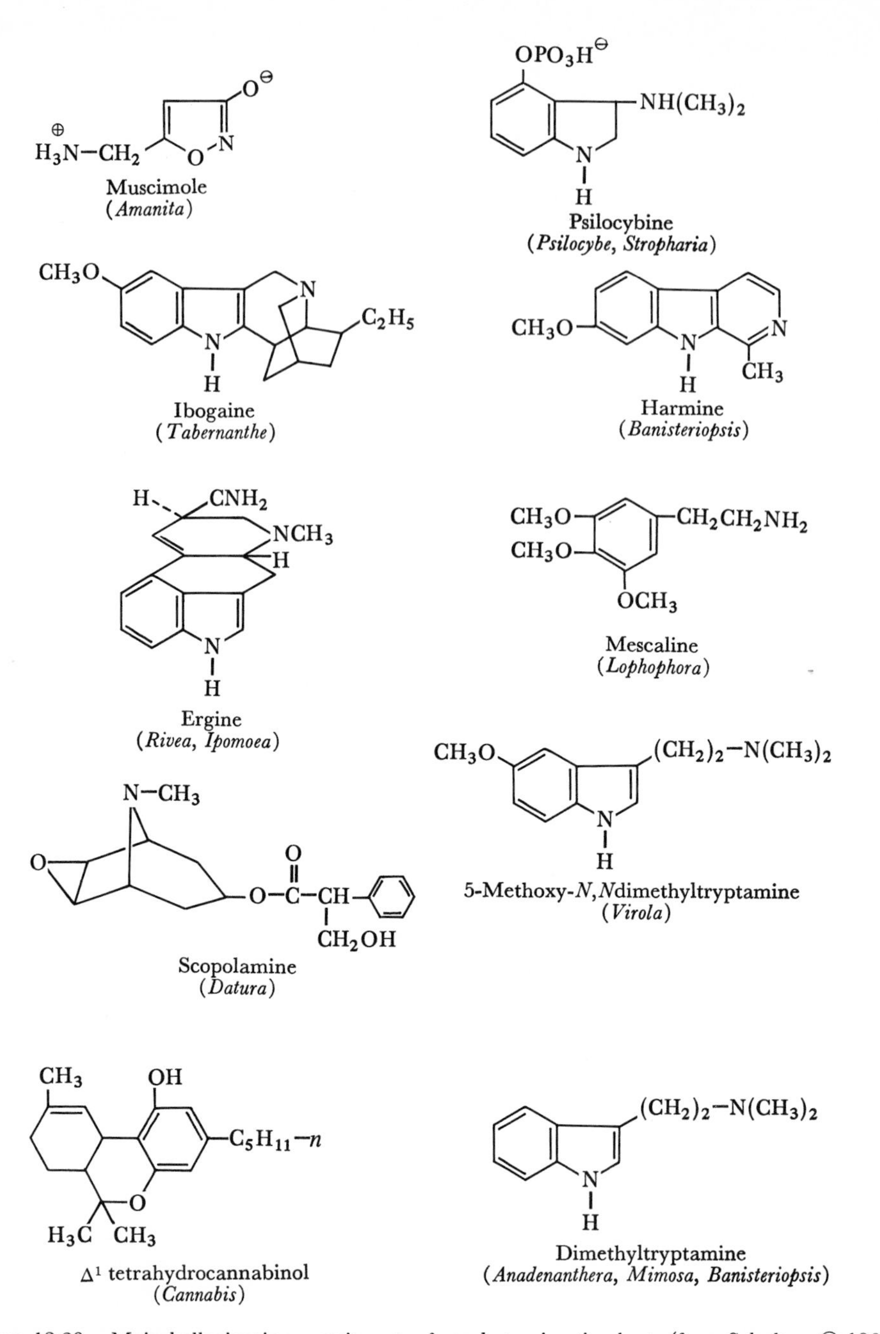

Figure 12.30. Main hallucinating constituents of psychotomimetic plants (from Schultes, © 1969 by the American Association for the Advancement of Science, with permission of the AAAS).

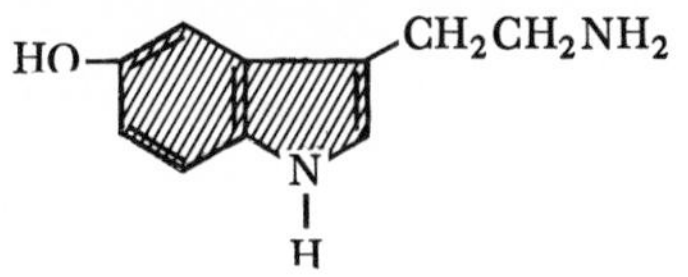

HO— $CH_2CH_2NH_2$ N—H

Serotonin (in the human body imbalance may cause schizophrenia)

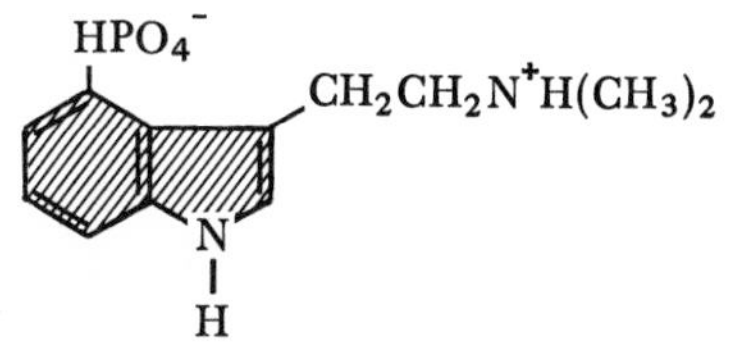

$N(C_2H_5)_2$ O=C N—CH_3 N—H

HPO_4^- $CH_2CH_2N^+H(CH_3)_2$ N—H

Psilocybin (from mushrooms)

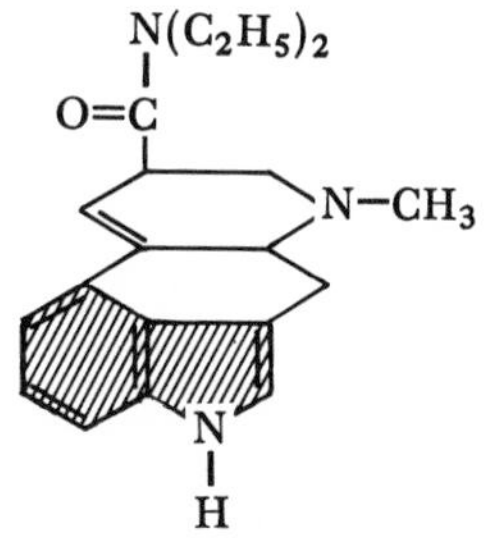

LSD (synthesized from ergot, a creal fungus)

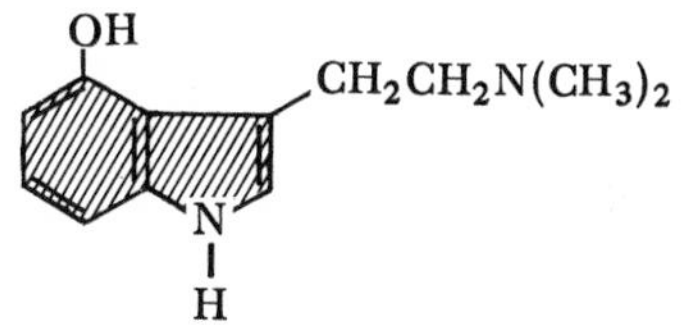

OH $CH_2CH_2N(CH_3)_2$ N—H

CH_3O CH_3O CH_2 CH_2 N H H

Psilocin

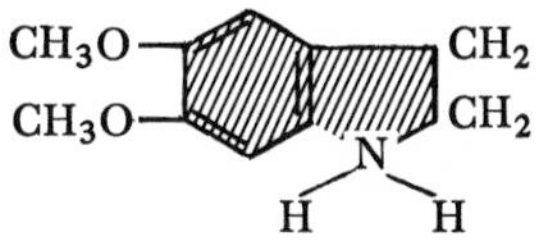

Mescaline (from peyote cactus)

Figure 12.31. Hallucinogens and neurohumors. The shaded areas are indole ring or indol ringlike structures (data from Barron et al., 1964).

American predilection for pills. Excessive use of unnecessary palliatives merges imperceptibly into drug abuse, and it is important to note that several of our most dangerous types of drugs of abuse (Table 12.14) got their initial impetus in this manner, notably the barbiturates and amphetamines. The person who must take a drink or a pill for the slightest discomfort or for "nerves" or simply when feeling frustrated is as much a victim as the heroin addict. In many ways alcoholism provides the broad base of the American drug scene. Although the government has issued numerous pamphlets on alcoholism, the hardest of hard drugs does not appear among its drugs of abuse (Table 12.14). This is in contrast to an older and more objective comparison by Seevers (1962) (Table 12.15) where ethanol appears as comparable to heroin in hazard but devoid of government control. When the never quite sober father tells his children not to smoke grass, little wonder they do not listen. Addiction in America is like an iceberg—only the tiny fraction of "hard" drug addiction tends to be visible in the media. Let me add a word here about Tables 12.14 and 12.15 and the like. They are always out of date. Not only does the list of drugs of abuse grow rapidly year by year, but as any drug is investigated with greater intensity, of course, more and more seriously injurious effects are uncovered.

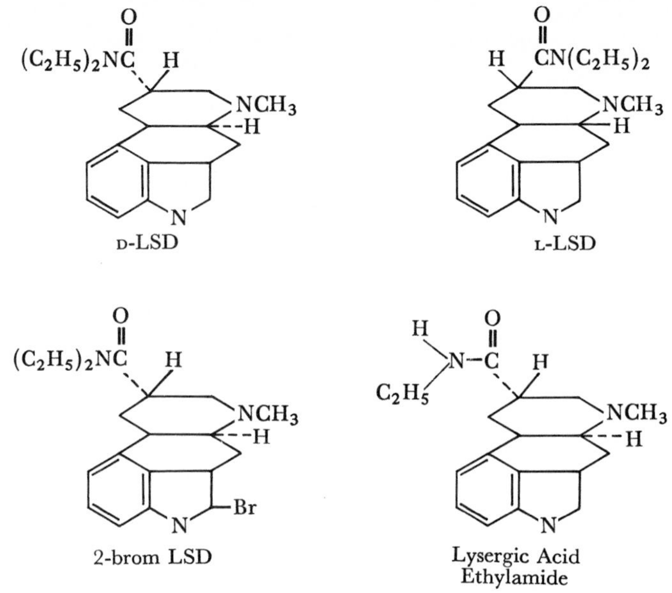

D-LSD

L-LSD

2-brom LSD

Lysergic Acid Ethylamide

Figure 12.32. Relative activity of LSD and related compounds (from Barron et al., 1964).

Table 12.14 Dangers of Drug Abuse (Information abstracted from *Drugs of Abuse*, U.S. Gov. Print. Off., Washington, D.C., 1972).

Drug	Physical Dependence	Psychological Dependence	Convulsions	Unconsciousness	Hepatitis	Psychosis	Death from Withdrawal	Death from Overdose	Possible Chromosome Damage
Morphine	×	×	×	×	×			×	
Heroin	×	×	×	×	×			×	
Codeine	×	×			×			×	
Hydromorphone	×	×		×	×			×	
Meperidine	×	×		×	×			×	
Methadone	×	×		×	×			×	
Cocaine		×	×		×			×	
Marihuana		×							
Amphetamines		×			×	×		×	
Methamphetamine		×			×	×		×	
Barbiturates	×	×	×	×	×		×	×	
LSD		×				×			×
STP		×			×				
Phencyclidine (PCP)		×		×	×				
Peyote		×			×				
Psilocybin		×							
Diethyltryptamine		×			×				

Table 12.15 Medical Characteristics and Regulatory Status of Some Common Drugs

	Medical Characteristics						
Drug	Psychogenetic Dependence	Physical Dependence	Uncontrolled Compulsion to Continued Abuse	Tolerance	Psychotoxic Effects During Administration	Psychotoxic Effects During Withdrawal	Regulatory Status[a]
Opiates and Morphine-like Analgesics	×	×	×	×		×	L, P
Barbiturates and Other Hypnotics	×	×	×	×	×	×	P
Ethanol	×	×	×	×	×	×	C
Bromides	×				×		C
Cocaine	×		×		×		L, P
Amphetamines and Related Stimulants	×		×	×	×		P
Marihuana	×		×	?	×		L
Nicotine	×			×			C
Caffeine	×			×			C

[a] L = Specific control laws
P = Prescription only
C = Available over the counter

Now I would like to turn our attention to three of the most popular drugs in the United States—tobacco, marihuana, and alcohol. As if our air were not dirty enough, what can be more unnatural than deliberately breathing smoke? Smoking is so totally at odds with all the instinctive reactions of mammals that I marvel that primitive man ever invented the practice. I marvel even more that it has taken a mountain of medical evidence to convince *some* people that so obviously dangerous a habit might not be in the best interest of their health? I have even seen people complain about air pollution *while* they are smoking!!! The statistics on death and disease from air pollution quoted in Chapter 6 are for nonsmokers. The comparable figures for smokers are far more grim. Overwhelming evidence demonstrates that "cigarette smoking is not only a major cause of lung cancer and chronic bronchitis, but that it is also associated with illness and death from chronic bronchopulmonary disease, cardiovascular disease and other illness" (Anon., 1973*b*). Cigarette smoke even interferes with mitochondrial energy production (Gariola and Aleem, 1973). Particularly distressing and still more cogent evidence of the potency of the poisons in tobacco smoke—smoking by a pregnant woman affects fetal development and results in lowered birth weight, higher stillborn rates, and higher late fetal and infant mortality. As more and more women smoke they share the risks involved: in 1949–1959, one woman to 12 men (for Westchester County, New York) met sudden death from coronary heart disease, but in 1967–1971 the ratio was one to four (Siegel et al., 1973; Anon., 1973*c*). In addition to a generous bouquet of carcinogens, both cigarette and cigar smoke have high particulate concentrations (Hoffmann and Wynder, 1972). Smoke from cheap cigars even contains inorganic particulate matter, presumably from the diatomaceous earth used in their manufacture (Langer et al., 1971). Another example of a manufacturing procedure introducing additional hazard is the contamination of parenteral drugs by the asbestos used for filtering (Nicholson et al., 1972).

One of the more deadly components of tobacco smoke, nicotine, is now removed from many brands, but a plentiful supply of carcinogens and other toxins remains

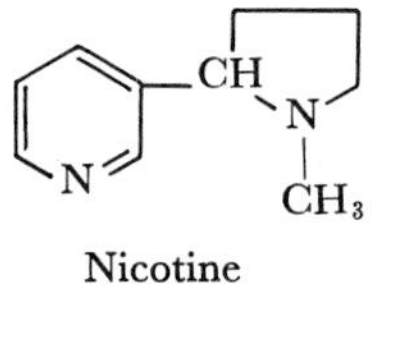

Nicotine

(Table 12.16). The ability of tobacco smoke components to bind transition metals may be related to disease capability (Finelli et al., 1972). In 1970, $5,464,000,000 worth of tobacco products were shipped in the United States. In 1969, the industry spent $174,959,000 on television advertising alone, and in 1970, $64,732,000 on magazine advertising. In view of these sums of money, the fact that the government has moved at all to impose warning requirements is eloquent testimony to the magnitude of the public health menace poised.

In addition to tobacco, marihuana has become a popular smoke in the United States. Although probably no more dangerous than tobacco and certainly less dangerous than ethanol, marihuana is still (1973) illegal in the United States. Narcotics agents sometimes used specially trained dogs to sniff out shipments. As a consequence, the composition of the marihuana aroma ("headspace") has been analyzed, and it differs appreciably from the essential oil (Table 12.17). Marihuana is a mild euphor-

Table 12.16 Selected Compounds in Mainstream Smoke (From Hoffmann and Wynder, © 1972 by the American Association for the Advancement of Science, with permission of the AAAS.)

Smoke Component	Concentration	Nonfilter Cigarette	Filter Cigarette	Little Cigar A	Little Cigar B	Small Cigar C
Carbon Monoxide	Vol. %	4.6	4.5	5.3	11.1	7.7
Carbon Dioxide	Vol. %	9.4	9.6	8.5	13.2	12.7
Hydrogen Cyanide	μg/cig.	536	361	381	697	1029
Acetaldehyde	μg/cig.	770	774	630	1238	1150
Acrolein	μg/cig.	105	71	41	54	66
Pyridine	μg/cig.	24.8	11.0	24.2	35.8	29.5
α-Picoline	μg/cig.	13.1	4.5	9.4	13.6	11.5
β-Picoline	μg/cig.	23.0	5.6	12.6	20.3	19.0
γ-Picoline	μg/cig.	6.8	2.0	3.5	6.4	5.9
Lutidines[a]	μg/cig.	15.1	4.2	8.3	9.2	14.4
Total Pyridines	μg/cig.	82.8	27.3	58.0	85.3	80.3
Phenol	μg/cig.	124.2	33.0	35.1	63.4	94.1
o-Cresol	μg/cig.	24.0	6.8	4.0	10.0	19.5
m- + *p*-Cresol	μg/cig.	75.4	22.2	16.9	37.8	67.1
2,4 + 2,5-Dimethylphenol	μg/cig.	9.4	4.6	1.0	3.7	6.1
m- + *p*-Ethylphenol	μg/cig.	22.1	9.2	6.5	17.6	27.0
Benz[*a*]anthracene	ng/cig.	74	31	34	25	39
Benzo[*a*]pyrene	ng/cig.	47	20	18	22	30

[a] Sum of values for 2,6-, 2,4-, and 3,5-lutidines which were determined individually by gas chromatography.

[b] Value for cigar D, 536.1 μg.

ient and hallucinogen, and not listed in Table 12.17 is its active ingredient, Δ′-tetrahydrocannabinol or THC (see Figure 12.30). Partial pyrolysis components of the smoke may also contribute to the drug's effect. The use of the drug is highly controversial with those under 40 for and those over 40 against its legalization. I had thought that, although socially debilitating, the drug was innoccuous until some brownies convinced me otherwise. Nevertheless, tests have shown that the manual performance of habitual users may even improve while under the influence of the drug. Marihuana is usually smoked with less frequency than tobacco, yet in order to be smoked most effectively the smoke is inhaled and deliberately retained in the lungs. The greater health danger in marihuana use appears to lie in the fact, not that it may be a dangerous drug, but that it is smoked. Marihuana smoke is every bit as deadly as tobacco smoke (Leuchtenberger et al., 1973). The effects of marihuana smoking have recently been the subject of a flurry of investigations, and, of course, numerous ill effects have been uncovered. But one is left wondering whether the many effects are evidence that the stuff is dangerous of simply evidence that it is being studied so intensively. The most serious social impact of marihuana is that the police in the United States seem to be preoccupied with arresting kids for smoking grass rather than trying to reduce serious crime (even my apartment was raided—as

Table 12.17 Comparison of Headspace Composition of Marihuana with Composition of Essential Oil (From Hood et al., 1973, with permission of the publisher and the authors.)

Component	Composition (%)	
	Headspace	Essential Oil
α-Pinene	55.5	3.9
Camphene	0.9	0.7
β-Pinene	16.4	2.2
2-Methyl-2-heptene-6-one[a]	0.4	0.6
Δ^3-Carene	0.6	0.1
Myrcene	8.3	1.0
α-Terpinene	t[b]	t
Limonene } β-Phellandrene	5.4	1.0
cis-Ocimene	1.2	0.2
trans-Ocimene	3.2	0.7
γ-Terpinene	t	t
Terpinolene	0.8	0.6
p-Cymene	—	0.1
Linalool	—	0.5
Fenchyl alcohol[a]	—	0.1
Borneol[a]	—	t
trans-α-Bergamotene	0.7	8.0
β-Caryophyllene	3.4	37.5
β-Farnesene	0.8	9.8
α-Terpinenol	—	1.0
β-Humulene	0.7	13.9
α-Selinene	—	2.2
β-Bisabolene[a]	—	3.2
Curcumene	—	1.4
Caryophyllene Oxide	—	7.4
Total identified components	98.3	96.0

[a] Not previously reported as constituent of marihuana.
[b] t, trace, $<0.1\%$.

usual without benefit of a search warrant). In 1971, California police made 70,000 marihuana arrests but only 100,000 arrests for all crimes of violence, including homicide, robbery, assault, and rape (Kaplan, 1973)! How's that for a perversion of priorities?

We Americans are cursed by the fact that our national opiate is one of the most dangerous drugs of all, ethanol (Anon., 1971*d*). As we noted earlier, ethanol is responsible for more human misery and death than all other chemical pollutants of the environment combined. In the United States, ethanol is responsible for more ruined lives than all other drugs combined. So-called "hard" drug users (such as heroin addicts), if they survive, end up as alcoholics. I lived in Haight-Ashbury

Table 12.18 Nonethanol Addiction in the United States (Data from Lerner, 1972)

	Reported Active Addicts (December 1970)
Heroin	66,040
Morphine	697
Opium	173
Dilaudid	367
Demerol	489
Methadone	167
Codeine	423
Eucodal	87
All other (except ethanol)	821
Total	68,864

during the death of the "hippie" scene there, and the young remnants who hung on discovered that there is more mind-destroying capacity per dollar in cheap California wine than in any other drug on the street. Over two-thirds of all American adults use the drug ethanol. Twelve percent are classified as heavy drinkers. There are 4,000,000 to 5,000,000 ethanol addicts in the United States. That is to say there are about *60 times* more ethanol addicts in the United States than all other types of drug addicts combined (Table 12.18). There are even 20 times more ethanol addicts in the government employ alone (est. 140,000, see *Science*, *179*, 363 (1973)) than the total number of other types of drug addicts in the whole nation. In 1972, the United States government spent an estimated $133,400,000 on ethanol addict rehabilitation—about twice the expenditure for juvenile delinquency or for the fight against organized crime (Lerner, 1972). Sick pay, absenteeism, and decreased efficiency due to alcoholism cost American industry an estimated $32,000,000 *every working day* (Arthur D. Little, Inc., estimate)! The drug is involved in more than 40% of *all* transportation deaths in the United States and is responsible for at least 800,000 accidents per year and some 25,000 deaths per year (Table 12.19; in 1969, for comparison, 15,000 Americans were murdered and about 10,000 were killed in the Viet-nam war, (Nordsiek, 1973). The likelihood of an accident increases steeply with ethanol consumption (Figure 12.33). The crime, suicides, and social costs of ethanol addition are staggering. Industrial losses alone due to ethanol are estimated to be $2,000,000,000/yr (Anon., 1969*b*; for an earlier review see Haddon, 1962).

Ethanol has some extremely interesting chemistry in the body (see Wallgren and Barry, 1970; Kissin and Begleiter, 1972), and with the growth of the attitude that alcoholism is a disease rather than a crime, let us hope that studies of this chemistry will be accelerated. About 20% of the ethanol ingested passes immediately through the stomach walls into the bloodstream, which then carries it directly to the brain's central control areas. The remaining 80% soon follows suit, passing through the gastrointestinal tract. Alcohol does not require "digestion." Moments after it is ingested, ethanol can be found in all of the body's tissues, organs, and secretions. Alcoholism is caused by psychological and sociological factors as well as biochemical factors.

Table 12.19 Accident Involvement of Ethanol in Transportation Accidents

Highway Accidents
17% of the drivers involved in highway accidents are under the influence of ethanol
More than 50% of the drivers killed in highway accidents are under the influence of ethanol
45 to 75% of pedestrian fatalities involve ethanol
51 to 71% of passenger fatalities involve ethanol
Aviation Accidents
24 to 43% of pilot fatalities involve ethanol
Marine Accidents
Less than 5% fatalities due to ethanol in commercial marine transportation accidents
3% fatalities due to ethanol in pleasure boating
13% of the people who "fall overboard" and drown are drunk

You cannot ween a people in a free society from their opiate. Prohibition was tried in the United States and was a colossal failure. But it might be possible to substitute for one opiate another less injurious. Addicts can be maintained on methadone, but some authorities believe that this only substitutes one drug for another and is thus not a satisfactory solution to the problem (Lennard et al., 1972). In a society as corrupt and permissive as ours, the control of anything as lucrative as ethanol

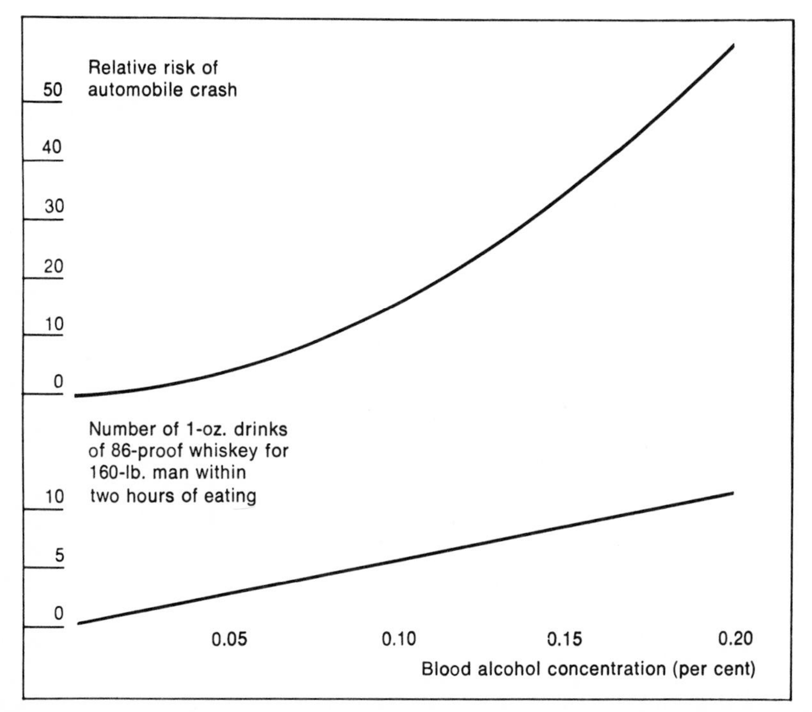

Figure 12.33. Alcohol consumption and accidents (from Nordsiek, *Technology Review*, edited at the Massachusetts Institute of Technology, © 1973 by the Alumni Association of M.I.T., with permission).

($26,000,000,000 worth of drinking alcohol was sold in 1970), tobacco, or other drugs, is next to impossible however injurious to the public health and welfare. From my personal experience with the narcotic police (San Francisco, March 10, 1969), they derive their satisfaction, not from reducing drug traffic, but from the safe sport of terrorizing people too young to know and/or too poor to afford their legal rights. The only solution I can see to the ethanol problem lies in the development of a competing drink, carefully controlled by the government, containing a relatively safe euphorient, possibly THC, and designed to displace ethanol use.

"Hard drugs," that is, somebody else's palliatives, are a highly publicized but miniscule fraction of the overall drug problem in the United States. We have become a nation of pill takers. Some of the mountain of palliatives (in 1970 the United States produced 1,003,000 lb of barbiturates and 1,206,000 lb of tranquilizers, Lerner, 1972) that we consume appear to be relatively innocuous such as that old, highly effective favorite aspirin (1970 United States production—35,170,000 lb) and some have proved to be exceedingly dangerous, and I would like to close this section on

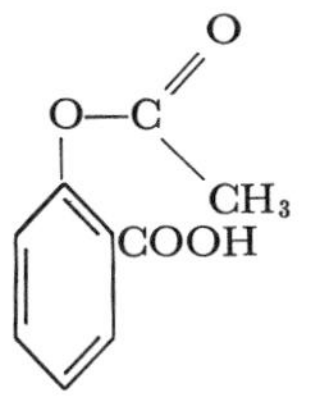

Aspirin

internal pollution by mentioning one of the most tragic examples which this country narrowly escaped, thalidomide.

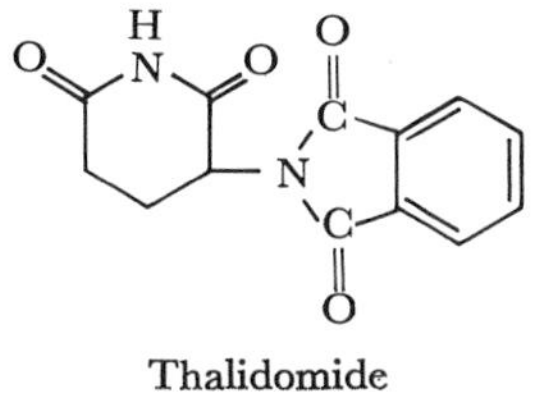

Thalidomide

Carcinogens produce cancer; mutagenic substances result in inheritable changes in cells; and teratogenic substances alter the development of the fetus, but the defect is not passed on to subsequent generations, that is to say, the damage is not genetic. Earlier we noted the effects of a few plant teratogens on the animals that feed upon them. But man has also been busy adding teratogens as well as mutagens and carcinogens to our environment. In about 1960, a rare and dreadful infant deformity (phocomelia or seal limb) suddenly became commonplace in Germany (Taussig, 1962). As many as 4000 to 6000 infants were affected in West Germany alone. The luckier victims, armless and legless and unable to move, soon succumbed in their cribs to pneumonia. The disaster was traced to a "safe, mild" sedative containing thalidomide taken by the mothers between the 30th and 60th day of their pregnancies. Birth defects also occurred in other countries—Canada, England, Australia, Scotland, Sweden, Brazil, Portugal, Switzerland, Lebanon, and Israel—where the drug was marketed. The United States was spared only because of the perspicacity of a FDA employee whose name well deserves to be remembered,

Mrs. F. O. Kelsey. A deformed monster, cursed for all the days of its life, is a disproportionate penalty to pay for the palliation of a mild headache. The thalidomide tragedy is a terribly compelling argument against unnecessary medication, but, alas except for those parents who must watch their child learn to lift a spoon with its feet, it is a lesson already forgotten by the public. A recent Edinburgh study revealed that drugs are prescribed for 82% of women during pregnancy and 65% of the women take nonprescription drugs (Anon., 1973*d*).

BIBLIOGRAPHY

F. Acree, Jr. *et al.*, *Science*, *161*, 1346 (1968).

S. V. Amonkar and A. Banerji, *Science*, *174*, 1343 (1971).

J. E. Amoore, *Nature*, *233*, 270 (1971).

Anon., *Chem. Engr. News*, 100 (October 10, 1966).

Anon., *Chem. Engr. News*, 32 (March 11, 1968).

Anon., *Science*, *165*, 478 (1969*a*).

Anon., *Alcohol Problems and Transportation Safety*, Nat. Trans. Safety Board, Dept. Trans., Washington, D.C., 1969*b*.

Anon., *Nature*, *234*, 326 (1971*a*).

Anon., *Nature*, *234*, 322 (1971*b*).

Anon., *Nature*, *233*, 586 (1971*c*).

Anon., *Alcohol*, U.S. Dept. Health, Education and Welfare Pamphlet (HSM)71-9048, U. S. Gov. Print. Off., Washington, D.C., 1971*d*.

Anon., *Science*, *176*, 495 (1972*a*).

Anon., *Nature*, *235*, 16 (1972*b*).

Anon., *Science*, *177*, 971 (1972*c*).

Anon., *Nature*, *239*, 63 (1972*d*).

Anon., *Nature*, *236*, 3 (1972*e*).

Anon., *Science*, *177*, 1175 (1972*f*).

Anon., *Nature*, *235*, 129 (1972*g*).

Anon., *Nature*, *243*, 186 (1973*a*).

Anon., *Nature*, *241*, 236 (1973*b*).

Anon., *Nature*, *241*, 236 (1973*c*).

Anon., *Nature*, *242*, 367 (1973*d*).

F. M. Ashton and A. S. Crafts, *Mode of Action of Herbicides*, Wiley, New York, 1973.

J. Atema and D. G. Engstrom, *Nature*, *232*, 261 (1971).

R. W. Baker et al., *Science*, *178*, 614 (1972).

K. W. Bentley, *Endeavour*, *23*, 97 (1964).

C. A. Berg, *Science*, *181*, 128 (1973).

M. Beroza and E. F. Knipling, *Science*, *177*, 19 (1972).

J. O'M. Bockris, *Science*, *176*, 1323 (1972*a*).

J. O'M. Bockris (ed.), *Electrochemistry of Cleaner Environments*, Plenum, New York, 1972*b*.

P. M. Boffey, *Science*, *171*, 43 (1971).

F. H. Bormann et al., *Science*, *159*, 882 (1968).

D. B. Boylan and P. J. Scheuer, *Science*, *155*, 52 (1967).
E. Boyland, *Prog. Exp. Tumour Res.*, *11*, 222 (1969).
A. J. Brook and A. L. Baker, *Science*, *176*, 1414 (1972).
P. Brookes, *Cancer Res.*, *26*, 1994 (1966).
P. Brookes and M. E. Duncan, *Nature*, *234*, 40 (1971).
M. S. Brown and H. S. Mosher, *Science*, *140*, 295 (1963).
T. H. Bullock, *Amer. Sci.*, *61*, 316 (1973).
R. T. Carde et al., *Nature*, *241*, 474 (1972).
J. Cairns, Jr., *J. Water Pollut. Control Fed.*, *43*, 55 (1971).
E. Cook, *Tech. Rev.* (*M.I.T.*), *75*, No. 2, 16 (December 1972).
D. Chapman et al., *Science*, *178*, 703 (1972).
C. C. Coutant, *J. Water Pollut. Control Fed.*, *43*, 1292 (1971).
M. D. Danford and H. A. Levy, *J. Amer. Chem. Soc.*, *84*, 3965 (1962).
B. V. Deryagin, *Sci. Amer.*, *223*(5), 52 (1970).
B. D. Dinman, *Science*, *175*, 495 (1972).
M. Dodge, *Science*, *177*, 139 (1972).
W. Drost-Hansen, *Chem. Phys. Lett.*, *2*, 647 (1968).
J. W. Eaton *et al.*, *Science*, *181*, 463 (1973).
N. N. Eddy and E. L. May, *Science*, *181*, 407 (1973).
M. Eisenbud and G. Gleason (eds.), *Electric Power and Thermal Discharges*, Gordon & Breach, New York, 1970.
T. Eisner, in E. Sondheimer and J. B. Simeone (eds.), *Chemical Ecology*, Academic, London, 1970.
T. Eisner et al., *Science*, *172*, 277 (1971).
S. R. Farkas and H. H. Shorey, *Science*, *178*, 67 (1972).
V. N. Finelli, *Environ. Sci. Tech.*, *6*, 740 (1972).
T. E. Furia (ed.), *Handbook of Food Additives*, Chemical Rubber Handbook Co., Cleveland, Ohio, 1968.
C. Gairola and M. I. H. Aleem, *Nature*, *241*, 287 (1973).
A. B. Gambel, *Energy Research and Development and National Progress*, Off. Sci. Tech., Washington, D.C., 1964.
J. W. Gentry, *Environ. Sci. Tech.*, *5*, 704 (1971).
R. Gillette, *Science*, *177*, 771 (1972).
T. R. Green and C. A. Ryan, *Science*, *175*, 776 (1972).
D. P. Gregory, *Sci. Amer.*, *228*(1), 13 (January 1973).
N. J. A. Gutteridge, *Chem. Soc. Rev.*, *1*, 381 (1972).
E. Habermann, *Science*, *177*, 314 (1972).
B. W. Halstead and D. A. Courville, *Poisonous and Venomous Marine Animals of the World*, U.S. Gov. Print. Off., Washington, D.C., 1970.
W. W. Hambleton, *Tech. Rev.* (*M.I.T.*), *74*(5), 15 (March/April 1972).
A. L. Hammond, *Science*, *172*, 660 (1971*a*).
A. L. Hammond, *Science*, *173*, 503 (1971*b*).
A. L. Hammond, *Science*, *178*, 1079 (1972*a*).
A. L. Hammond, *Science*, *177*, 1088 (1972*b*).
A. L. Hammond, *Science*, *178*, 147 (1972*c*).

D. R. F. Harleman, *Tech. Rev. (M.I.T.)*, *74*(2), 44 (December 1971).
D. Hoffmann and E. L. Wynder, *Science*, *178*, 1197 (1972).
R. W. Holcomb, *Science*, *167*, 159 (1970).
L. V. S. Hood et al., *Nature*, *242*, 401 (1973).
R. A. Horne et al., *J. Appl. Polymer Sci.*, *12*, 1484 (1968).
R. A. Horne et al., *J. Colloid Interface Sci.*, *35*, 77 (1971).
R. A. Horne, *Science*, *177*, 1152 (1972).
D. B. Houston, *Science*, *172*, 648 (1971).
J. T. Hughes et al., *Science*, *177*, 1110 (1972).
S. H. Hurlbert, *J. Herpetol.*, *4*, 47 (1970).
W. B. Jackson and D. Kaukeinen, *Science*, *177*, 1343 (1972).
M. Jacobson, *Insect Sex Attractants*, Wiley, New York, 1965.
M. Jacobson, *Insect Sex Pheromones*, Academic, New York, 1972.
S. M. Jacobson and D. B. Boylan, *Nature*, *241*, 213 (1973).
L. W. Jones, *Science*, *174*, 367 (1971).
J. Kaplan, *Science*, *179*, 167 (1973).
K. A. Kershaw, *Endeavour*, *22*, 65 (1963).
B. Kissin and H. Begleiter (eds.), *The Biology of Alcoholism*, Plenum, New York, 1972.
R. M. Klein, *Nat. Hist.*, *79*(2), 10 (February 1970).
I. M. Klotz, *Membranes Ion Trans.*, *1*, 433 (1970).
J. Kmet and E. Mahboubi, *Science*, *175*, 846 (1972).
W. S. Koski et al., *Nature*, *242*, 65 (1973).
P. A. Krenkel and F. L. Parker (eds.), *Biological Aspects of Thermal Pollution*, Vanderbilt Univ. Press, Nashville, 1969.
A. M. Langer et al., *Science*, *174*, 585 (1971).
H. L. Lennard et al., *Science*, *176*, 881 (1972).
A. C. Leopold and R. Ardrey, *Science*, *176*, 512 (1972).
W. Lerner (ed.), *Statistical Abstracts of the United States*, U.S. Dept. Commerce, Washington, D.C., 1972.
C. Leuchtenberger et al., *Nature*, *241*, 137, 403 (1973).
A. A. Levin et al., *Environ. Sci. Tech.*, *6*, 224 (1972).
R. S. Lewis, *The Nuclear Power Rebellion*, Viking, New York, 1972.
G. E. Likens et al., *Science*, *163*, 1205 (1969).
R. M. Linton, *Terricide*, Little, Brown, Boston, 1970.
G. O. G. Löf and J. C. Ward, *J. Water Pollut. Control Fed.*, *32*, 2102 (1970).
G. Löfroth and T. Gejvall, *Science*, *174*, 1248 (1971).
W. Longgood, *The Poisons in Your Food*, Simon & Schuster, New York, 1960.
P. N. Magee and J. M. Barnes, *Adv. Cancer Res.*, *10*, 163 (1967).
P. L. Marks and F. H. Bormann, *Science*, *176*, 914 (1972).
S. P. Mathur and R. Stewart (eds.) *Proc. Conf. Beneficial Uses of Thermal Discharges*, N.Y. State Dept. Environ. Conserv., Albany, N.Y., 1971.
T. H. Maugh, *Science*, *178*, 849 (1972*a*).
T. H. Maugh, *Science*, *178*, 1273 (1972*b*).
T. H. Maugh, *Science*, *177*, 249 (1972*c*).
J. K. McPherson and C. H. Muller, *Ecol. Monogr.*, *39*, 177 (1969).

J. J. Menn and M. Beroza (eds.), *Insect Juvenile Hormones*, Academic, New York, 1972.
W. D. Metz, *Science*, *177*, 180; *178*, 291 (1972).
N. Metzger, *Men and Molecules*, Crown, New York, 1972.
S. L. Miller, in *Proc. 2nd Symp. Underwater Physiol.*, Nat. Acad. Sci. Pub. No. 1181, Washington, D.C., 1963.
G. A. Mills et al., *Environ.*, *Sci. Tech.*, *5*, 30 (1971).
W. H. Moir, *Science*, *177*, 396 (1972).
J. C. Mosher, *Science*, *143*, 1048 (1964).
J. J. Mulvihill, *Science*, *176*, 132 (1972).
N. Myers, *Science*, *178*, 1255 (1972).
D. Nelkin, *Nuclear Power and Its Critics*, Cornell Univ. Press, Ithaca, 1971.
W. J. Nicholson et al., *Science*, *177*, 171 (1972).
F. W. Nordsiek, *Tech. Rev.* (*M.I.T.*), *75*(3), 41 (January 1973).
G. H. Orians and E. W. Pfeiffer, *Science*, *168*, 544 (1970).
L. Pauling, *Science*, *134*, 15 (1961).
C. Pfaffmann (ed.), *Olfaction and Taste*, Rockefeller Univ. Press, New York, 1969.
M. J. Pinthus et al., *Science*, *177*, 715 (1972).
R. Platt, *The Great American Forest*, Prentice-Hall, Englewood Cliffs, 1971.
F. H. Pough, *Science*, *174*, 1144 (1971).
J. R. Raper, in E. Sondheimer and J. B. Simeone (eds.), *Chemical Ecology*, Academic, London, 1970.
W. L. Roelofs and A. Comeau, *Science*, *165*, 398 (1969).
H. A. Rose and S. P. R. Rose, *Science*, *177*, 710 (1972).
M. W. Rosen and N. E. Cornford, *Nature*, *234*, 49 (1971).
L. A. Sagan, *Science*, *177*, 487 (1972).
H. Schildknecht, *Endeavour*, *30*, 136 (1971).
D. Schneider, *Science*, *163*, 1031 (1969).
R. E. Schultes, *Science*, *163*, 245 (1969).
N. P. Sen et al., *Nature*, *241*, 473 (1973).
M. H. Seevers, *J. Amer. Med. Assoc.*, *181*, 92 (1962).
D. Shapley, *Science*, *177*, 776 (1972).
H. Siegel et al., *J. Amer. Med. Assoc.*, *224*, 1005 (1973).
S. F. Singer, *Sci. Amer.*, *223*(3), 175 (September 1970).
G. L. Small, *The Blue Whale*, Columbia Univ. Press, New York, 1971.
W. H. Smith et al., *Soil Sci.*, *106*, 471 (1968).
G. R. Snyder and T. H. Blahm, *J. Water Pollut. Control Fed.*, *43*, 890 (1971).
H. E. Sutton and M. J. Harris (eds.), *Mutagenic Effects of Environmental Contaminants*, Academic, New York, 1972.
C. M. Tarzwell, *J. Water Pollut. Control Fed.*, *42*, 824 (1970).
H. B. Taussig, *Sci. Amer.*, *207*(2), 29 (August 1962).
D. L. Thompson et al., *Science*, *177*, 806 (1972).
J. H. Todd, *Sci. Amer.*, *224*(5), 98 (May 1971).
F. H. Tschirley, *Science*, *163*, 779 (1969).
G. D. Veith and G. F. Lee, *Environ. Sci. Tech.*, *5*, 230 (1971).
N. Wade, *Science*, *174*, 805 (1971).

N. Wade, *Science*, *177*, 335, 503, 588 (1972*a*).
J. F. Wakely et al., *Toxicon*, *3*, 195 (1966).
H. Wallgren and H. Barry, *Actions of Alcohol*, Elsevier, New York, 1970.
P. F. Wareing and G. Ryback, *Endeavour*, *29*, 84 (1970).
A. M. Weinberg, *Science*, *177*, 27 (1972).
F. S. Weinberg, *Nature*, *233*, 239 (1971).
A. H. Westing, *Bioscience*, *21*, 893 (1971).
D. C. White, *Tech. Rev.* (*M.I.T.*), *74*(1), 18 (October/November 1971).
R. H. Whittaker and P. P. Feeny, *Science*, *171*, 757 (1971).
C. M. Williams, in E. Sondheimer and J. B. Simeone (eds.), *Chemical Ecology*, Academic, London, 1970.
C. M. Williams, *Sci. Amer.*, *217*(1), 13 (July 1967).
E. O. Wilson, *Science*, *49*, 1064 (1965).
E. O. Wilson and T. A. Sebeck (eds.), *Animal Communication*, Indiana Univ. Press, Bloomington, 1968.
E. O. Wilson, in E. Sondheimer and J. B. Simeone (eds.), *Chemical Ecology*, Academic, London, 1970.
K. M. Wilson et al., *Physiologist*, *12*, 395 (1969).
R. Wilson, *Nature*, *241*, 317 (1973).
W. E. Winsche et al., *Science*, *180*, 1325 (1973).
I. A. Wolff and A. E. Wasserman, *Science*, *177*, 15 (1972).
D. L. Wood et al., *Science*, *159*, 1373 (1968).
N. Wood, *Clearcut: The Deforestation of America*, Sierra Club, San Francisco, 1971.
G. M. Woodwell, *Science*, *168*, 429 (1970).
R. J. Wulf and R. M. Featherstone, *Anesthesiology*, *18*, 97 (1957).
W. C. Yee, *Environ. Sci. Tech.*, *6*, 232 (1972).
J. Yudkin, *Nature*, *239*, 197 (1972).
W. S. Zaugg et al., *Science*, *176*, 415 (1972).

ADDITIONAL READING

Anon., *Cannabis*, Information Canada, Ottawa, 1972.
Anon., *Marihuana*, U.S. Gov. Print. Off., Washington, D.C., 1972.
E. M. Brecher (ed.), *Licit and Illicit Drugs* (Consumers Union Rept.), Little, Brown, Boston, 1972.
W. Bücherl and E. Buckley, *Venomous Animals and Their Venoms*, Academic, New York, 1971.
K. L. Chambers (ed.), *Biochemical Coevolution*, Oregon State Univ. Press, Corvallis, 1970.
Committee on Resources and Man, *Resources and Man*, Nat. Acad. Sci. U.S.A.–Nat. Res. Council, Freeman, San Francisco, 1969.
J. Darmstadter, P. D. Teitelbaum, and J. G. Polach, *Energy in the World Economy*, Johns Hopkins Press, Baltimore, 1971.
R. A. Deju, R. B. Bhappu, G. C. Evans, and A. B. Baez, *The Environment and Its Resources*, Gordon & Breach, London, 1972.
J. Ebling and K. C. Highnam, *Chemical Communication*, Arnold, London, 1969.
D. W. Ehrenfeld, *Biological Conservation*, Holt, Rinehart, & Winston, New York, 1970.

N. Fabricant and R. M. Hellman, *Toward a Rational Power Policy: Energy, Politics and Pollution*, Braziller, New York, 1971.

L. Fishbein, W. G. Flamm, and H. L. Falk, *Chemical Mutagens*, Academic, New York, 1970.

J. Holdren and P. Herrara, *Energy: A Crisis in Power*, Sierra Club Books, New York, 1972.

E. B. Hook, D. T. Janerich, and I. H. Porter (eds.), *Monitoring Birth Defects and Environment*, Academic, New York, 1971.

M. F. Jacobson, *Eater's Digest*, Doubleday, New York, 1972.

J. W. Johnston et al. (eds.), *Communication by Chemical Signals*, Vol. I, Adv. Chemoreception, Appleton-Century-Crofts, New York, 1970.

J. Levitt, *Response of Plants to Environmental Stress*, Academic, New York, 1972.

I. E. Liener (ed.), *Toxic Constituents of Plant Foodstuffs*, Academic, New York, 1969.

W. H. Matthews and F. E. Smith (eds.), *Man's Impact on Terrestrial and Oceanic Ecosystems*, M.I.T. Press, Cambridge, 1972.

H. T. Odum, *Environment, Power, and Society*, Wiley-Interscience, New York, 1970.

F. J. C. Roe (ed.), *Metabolic Aspects of Food Safety*, Academic, New York, 1971.

J. Rose (ed.), *Technological Injury*, Gordon & Breach, London, 1969.

S. H. Schurr (ed.), *Energy, Economic Growth, and the Environment*, Johns Hopkins Press, Baltimore, 1971.

Secretary of H.E.W., *Alcohol and Health*, Scribners, New York, 1973.

C. H. Southwick, *Ecology and the Quality of Our Environment*, Van Nostrand-Reinhold, New York, 1972.

H. E. Sutton and M. I. Harris (eds.), *Mutagenic Effects of Environmental Contaminants*, Academic, New York, 1972.

A. Turk, J. Turk, and J. T. Wittes, *Ecology, Pollution, Environment*, Saunders, Philadelphia, 1972.

VII

THE CRUCIAL INTERACTIONS

13

HYDROSPHERE–LITHOSPHERE INTERACTION

13.1 INTRODUCTION

In the previous half of this book we examined the chemical characteristics of the major segments of our environment—exosphere, atmosphere, biosphere, lithosphere, and hydrosphere. Now in the second half, we turn our attention to the chemical interactions among these segments, and in many respects this is the more important half. Throughout the first half of this book, I tried to give strong emphasis to the importance of water in the chemistry of our environment, and I continue to press this emphasis. I also tried to point out that what has made the chemistry of this planet so extraordinary—a tiny island of oxidizing environment in an endless sea of reducing environment—is the biosphere. To water chemistry and biochemistry as the predominant features of the chemistry of our environment, I now add a third, surface or interfacial chemistry.

The chemistry of our environment is:

1. Water chemistry.
2. Biochemistry.
3. Interfacial chemistry.

Most of the chemistry going on in our environment right now, as in the geological past, is occurring at interfaces—those crucial zones where atmosphere, hydrosphere, lithosphere, and biosphere meet. There is, for example, except for a few fast simple ionic metathesis equilibria, remarkably little chemistry going on in an element of seawater taken from the deep ocean. The chemistry of the oceans is largely the chemistry of air–sea and sea–sediment interactions. In the most general terms, we can look upon our environment as an exceedingly complex chemical system in State I in process of chemical (and physical) change into a second state, State II (Figure 13.1). This chemical change is occurring by two major routes, each with its own remarkable "catalyst." The first route consists of ultimately reversible equilibria that, on the one hand, by a dissolution mechanism in accord with enthropic considerations tend to distribute chemical substances more randomly in our environment (such as the dissolution of salts in the sea), and on the other hand, tend to concentrate materials by precipitation and other deposition processes out of solution as highly ordered crystalline mineral substances. The "catalyst" for these processes is water, and a very powerful catalyst it is. The very great hydration and solubilization powers of water enable it easily to take ions out of crystal lattices that otherwise can be done only with the expenditure of enormous quantities of thermal energy. Water mobilizes environmental materials.

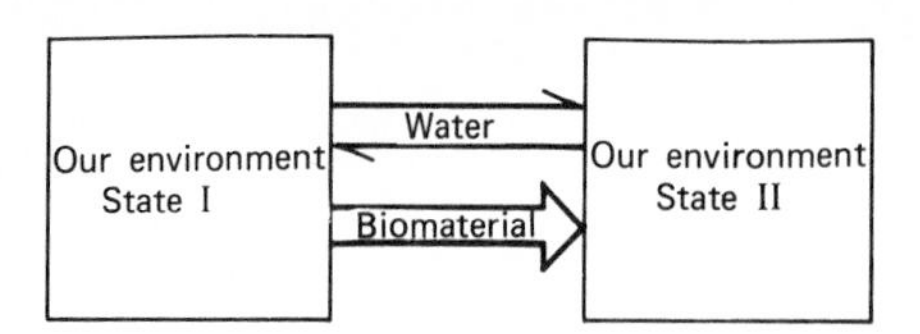

Figure 13.1. Our environment as a changing chemical system.

The second major route segregates oxidizing (O_2 in the atmosphere) and reducing (C in fossil fuels) materials in our environment, an exceedingly difficult chemical chore contrary to the overall thermodynamic inclination of the cosmos. And this is made possible by a catalyst even more powerful than water, a catalyst that in fact utilizes water for nearly all of its functions—and that catalyst is the biomaterial. The biosphere consists of organisms composed of one or more cells. These cells contain membranes, and they are themselves often arranged to form membranes, which are really nothing more than very elaborate surfaces. Within the cells themselves are many interfaces where crucial chemical functions are transacted. In its simplest chemical terms the biosphere is a catalytic interface.

13.2 THE STRUCTURE OF WATER AT AN INTERFACE*

Inasmuch as the three keys to the chemistry of our environment are water chemistry, biochemistry, and interfacial chemistry the properties of interfacial water become of supreme importance. I hope that it is as clear to you as it is to me that in understanding the chemistry of our environment is by far the most important fundamental question that we can ask contemporary science is, What is the structure of liquid (?) water near an interface?

Unfortunately, contemporary science has no satisfactory answer to this all important question. And for a very good reason. In earlier chapters, we saw that the structures of pure liquid water and of aqueous solutions are exceedingly complex and imperfectly understood, that, despite intensive scientific effort, they remain subjects of intense controversy. The structure of interfacial (or vicinal) water appears to be many orders of magnitude more complex. Little wonder then that even less is certain about it, that there is even more controversy surrounding it, that even, since scientists are human and do like to feel that they are making progress, it is an area that has tended to be avoided. Scientists are now in agreement that the presence of an interface, any interface whether charged or neutral, whether hydrophobic or hydrophilic, alters the structure of pure liquid water and of solutions, but they are far from agreement as to the details of the structural changes produced by the presence of the interface and, equally important, as to how deeply the perturbation of the structure extends into the liquid. The latter question is particularly crucial to environmental chemistry for if the perturbation is deep, then most of the active water in our environment—in atmospheric cloud droplets, in recent sedimentary material, and in the biomaterial—will be perturbated water, will be vicinal rather than "normal" water, possessing a structure corresponding to the structure of interfacial water.

* Much of this section is taken from Horne (1969).

Comparison with a "normal" liquid (Table 7.3) illustrates the extraordinary nature of liquid water. The excellent solvent capabilities of water, so important in the chemistry of our environment, are in large part a consequence of its very high dielectric constant, while the high density, specific heat, and viscosity all are evidence for the tendency of water molecules to associate with one another and to resist separation. Earlier, we saw that the boiling point of water is very much higher than those of its sister compounds (Figure 7.2), and we saw in Table 7.3 that the heat of evaporation is also anomalously great. Evidently, it is relatively difficult for a water molecule to escape from the liquid into the gaseous phase across the interface. Notice that I have said *relatively* difficult, for the rapid establishment of vapor–liquid equilibrium in a closed system indicates that many molecules do manage to escape (and return), and the interface is the scene of turmoil with "immense turbulence and disturbance on the molecular scale" (Drost-Hansen, 1965).

Most of our knowledge concerning the gas–aqueous electrolytic solution interface derives from surface potential and surface tension measurements (Randles, 1957, 1963). The surface tension of water is strikingly high compared to that of other liquids (Figure 7.1). If, as in the case of the boiling points (Figure 7.2), we compare water's surface tension and those of a series of similar compounds (Table 13.1) its anomalously high surface tension stands out clearly. As the molecular weight decreases, the surface tension of a series of alcohols is about 23 dynes/cm and decreases slightly, but for water, the smallest "alcohol," the value soars to about 72 dynes/cm. The analogy of a crowded dance hall has been frequently used successfully elsewhere to illustrate the dynamic nature of chemical equilibria, and I think we can put it to good purpose here: In order to "leave," a water molecule must exert itself, not once, but twice. First, it must disengage itself from its partner, then it must make its way through the crowd at the door. The difficulty in the escape from liquid to gaseous phases, as reflected in the high boiling point and heat of evaporation of water, is a consequence both of molecular association or structure in the bulk phase and some kind of barrier or structural enhancement at the interface. Some of our information concerning the gas–solution interface is thermodynamic (Randles, 1963), and as usual, thermodynamics tells us relatively little about what we really would like to know, namely, the spatial orientations of the molecules on a microscopic scale. However, the entropy is a measure of the relative amount of order/disorder obtaining and is thus of some help. On the basis of surface tension measurements, the average molar surface entropy of nonpolar liquids is about 24 joules/degree; whereas, that of water is only about 10 joules/degree. In other words, the water surface is more ordered, and the entropy difference of about 14 joules/degree is attributed to the considerable

Table 13.1 Surface Tension of Water and Related Compounds at 20°C (From Horne, 1969)

Water	HOH	73 dynes/cm
Methanol	CH_3OH	22.6 dynes/cm
Ethanol	C_2H_5OH	22.7 dynes/cm
n-Propanol	C_3H_7OH	23.8 dynes/cm
n-Butanol	C_4H_9OH	24.6 dynes/cm

orientation of the near- and at-surface molecules (Good, 1957). There remains little doubt that a zone or layer of molecules oriented in some particular way exists at the surface of liquids, especially polar liquids such as water (Henniker, 1949; Frenkel, 1955; and for more recent reviews see Drost-Hansen, 1967, 1969), but the all important questions about the nature of these layers in terms of molecular arrangements and their thicknesses are surrounded by controversy and remain largely unanswered.

Guastalla (1947) and more recently Claussen (1967) have theorized that the surface layer of liquid water consists of a plane of puckered hexagons, about 19.5×10^{16} cm^2 each in area similar to the planes in Ice-I((Figure 7.7), and the latter author has had some success in calculating surface energy as a function of temperature, but, perhaps fortunately, the calculation is more dependent on the assumption that the thermal expansion of the surface and bulk water behave similarly than on a particular ice-like model for the surface layer. Liquid water readily supercools; thus any structure existing in the liquid must be below the critical nucleation size and/or they cannot be Ice-I_h-like. Fascinating results from measurements on water in very thin capillaries (Schufle and Venugopalan, 1967, and the references cited therein) indicate that interfacial water is especially loath to freeze. Thus our earlier argument applies with added force to the surface water, and I believe we can safely assert that the structure of interfacial water is *not* ice-like.

Furthermore, I think it probable that we can say that the structure of the interfacial water is not the same as the structure of the Frank–Wen clusters—whatever that may be. The subject of the effect of pressure on surface tension deserves a great deal of further experimental study. The few data available (Slowinski, Gates, and Waring, 1957) appear to indicate that the effects are small and are thus in contrast to the profound effect of pressure on the structure of bulk water [Chapter 7]. The thermodynamic theory of surface tension under pressure (Eriksson, 1962) again tells us virtually nothing about the structural details of the surface layer. The effects of solutes on the surface tension of water are complex and varied (Drost-Hansen, 1965) and, although important relationships with bulk effects undoubtedly exist, it is far from clear at this time what they may be. The most straightforward evidence for the conclusion that the structure of the surface layer and the bulk clusters are different is the difference in their stability with respect to temperature. The relative viscosity of water, dependent on the bulk structure (and, inasmuch as it is the shear viscosity, probably dependent at least in part on surface structure), falls off much more rapidly with increasing temperature than does the relative surface tension (Figure 7.9), which is dependent on the surface structure, and the difference is much greater for water than for a "normal" liquid such as *n*-octane (Horne, et al., 1968).

Wicke (1966) points out that we might expect a breaking-off or loosening of the water surface at the interface. That (quite the reverse) we should observe an apparent strengthening or consolidation of the structure means that any disruption must be masked by a greater increase in short-range order. He further suggests that this order may be analogous to the "hydration of the second kind" such as encountered in the hydration of the tetraalkylammonium cations [Chapter 8].. At the present time my own prejudices on this matter lie very much in the same direction. Perhaps even more germane than the hydration of the tetraalkylammonium cations is the thermodynamics of gas solubilities in water and aqueous solutions [Chapter 8]. There, we found that the large entropy effect indicated extensive water structure enhancement around the dissolved alien molecules, and similar effects are also encountered in the

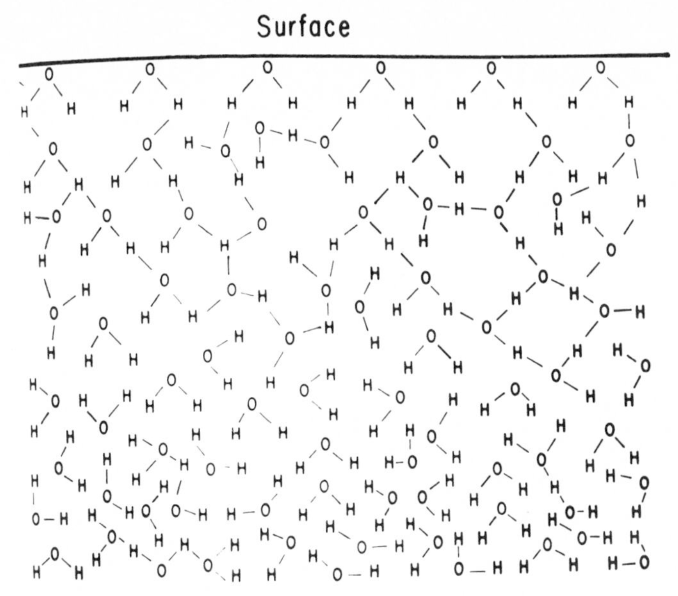

Figure 13.2. Perturbation of the water structure as an interface is approached (from Horne, 1968, with permission of Academic Press, Inc.).

dissolution of hydrocarbons. If a single dissolved gas molecule is capable of profoundly perturbing the structure of liquid water, no great stretch of the imagination is required to imagine a similar perturbation of the water structure as the gas–water interface is approached. Figure 13.2 represents a very crude attempt to represent the alteration of water structure as the interface is approached. This figure shows the H_2O's ranged in a highly random manner and the density of the near-surface water less than that of the more distant water, both of which may or may not be, and probably are not, true. Two features probably correctly represented in Figure 13.2 are the increased H bonding as the interface is approached, and the H_2O's at the interface aligned with their oxygen pointing to the air and their hydrogens to the water phase. With respect to the latter, Fletcher (1962) has calculated that the energy gained as a consequence of the water dipole orienting itself at the surface with the negative (oxygen) vertex outward is about 10^{-12} erg/molecule.

Unfortunately the proponents of the various theories of water structure in the bulk phase [Chapter 7] have been slow to apply their favorite models to the structure near the interface. An exception is Lu et al. (1967), who have attempted to apply the significant structure theory to the phenomenon of surface tension.

As for the very important question of the depth of the surface area, opinion seems to be divided. The older, more established point of view appears to be that it is relatively shallow, comparable in thickness to the electrical double layer, that is, a few, but not much more than, 10 molecules thick. In the paper mentioned above, Fletcher (1962) points out that the preferential orientation of the water dipoles decays exponentially, and he estimates the thickness of the surface zone to be 13 molecular layers or about 26Å. The so-called "Dutch School," on the other hand, believes that the range of surface van der Waals and London forces in liquids is very much greater than for individual molecules and may extend as far as thousands of angstroms (see Henniker's, 1949, old but excellent review). The direction of some recent experimental evidence appears to substantiate the second viewpoint and indicates that the perturbation of the water structure near an interface is quite extensive.

If the liquid at the water–air interface is seawater, in addition to considering the structural changes in the water itself as the boundary is approached, we must also consider the changes in the spatial array of the dissolved solutes, especially ionic species. Certainly, it seems reasonable to expect that the concentration and arrangement of species will not be the same as in bulk solution. Surface potential studies on aqueous solutions (Randles, 1957) indicate that most ions (but not tetra-alkylammonium ions) are repelled from the interface, anions less strongly than cations. The order of repulsion appears to be related with the size and hydration energy of the ions. Three hypotheses have been advanced to account for the diminishment of electrolytes near the interface. From classical electrostatic theory, electrolyte ions should be repulsed from the low dielectric constant air phase into the high dielectric constant water phase and be crowded away from the interface into the interior of the solution (Onsager and Samaras, 1934). McConnell et al. (1964) have proposed a "geometric effect"; they argue that small radius ions can crowd closer to a surface than larger radius ions, thus the concentration of the former can be larger in the boundary zone. I am sure that both of these factors are indeed operative at the interface, but the effects of both will be at most very shallow; they cannot, therefore, account for any deep perturbation of solute concentration. Both fail to take into consideration water structure or even ionic hydration. If vicinal water is structurally different from bulk water, then its properties will be different, including the solubilities of electrolytes and other solutes in it. I propose that vicinal water is structured in a manner similar to the hydrophobic hydration surrounding nonpolar solutes (such as the atmospheric gases). Such being the case, we might expect hydrophobically solvated polar solutes to be more soluble in vicinal than in bulk water and hydrophilically solvated ionic and polar solutes to be less soluble—the greater the hydrophilic character of their solvation envelope the less soluble. Thereby there will arise preferential solubilities of solutes near an interface, with nonpolar solutes being concentrated and ionic solutes preferentially excluded (Figure 13.3).

The surface tension and other interfacial properties of water and aqueous solutions are very sensitive to impurities. The behavior of surface-active material or "surfactants" and of surface films has expanded into a science in its own right and, thanks to practical applications such as the reduction of evaporation from reservoirs by surface films, has even achieved a certain measure of public attention (LaMer and Healy, 1965). Many film-forming molecules are bifunctional; that is, they have hydrophobic and hydrophilic segments, and at the surface they align themselves nicely with the former part directed out of, and the latter part directed into, the water phase.

Extensive areas of the sea's surface have been observed to be covered with organic films or slicks from shipping and other human pollution near shore and from natural biological processes (plankton) in the open sea. A scum sample formed by film collapse collected near the Scripps pier contained about 27% organic material (Ewing, 1950). In addition to measuring film pressures, surface tensions, and surface viscosities of slicks, Jarvis et al. (1967) found that the films were composed of the same type of compounds reported by Garrett (1964), namely, fatty esters, fatty acids

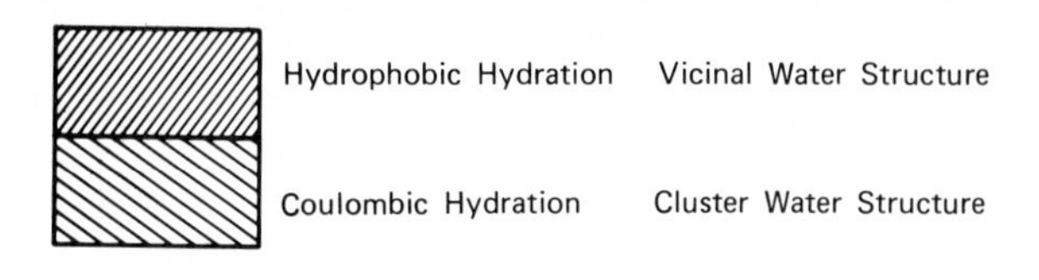

Figure 13.3. Solute fractionation for an ionic (Na^+Cl^-) and a nonpolar (HC being hydrocarbon) solute and water structure at the air–water interface (from Horne, R. A., *J. Geophys. Res.*, 77, 5170 (1972), © American Geophysical Union, with permission of the AGU).

(of 8–12 carbon atoms), fatty alcohols, and both saturated and unsaturated hydrocarbons.

Analyses of the water-insoluble organic components of film materials collected from the Atlantic and Pacific oceans, gulfs of California and Mexico, and the Bay of Panama revealed fatty esters, acids, and alcohols, and hydrocarbons, the relative amounts being dependent on the meteorological and oceanographic conditions prevalent at the location when a sample was taken (Garrett, 1964). These films affect several of the surface properties of the sea, including the surface potential and viscosity and the damping of capillary waves (Jarvis et al., 1967). In addition, by reducing evaporation, these films can also give rise to an excess warming of the surface water (Jarvis, 1962). In the open sea, much of the organic material of the surface film may come from dissolved organic matter in seawater (Parsons, 1964) which is concentrated by adsorption on bubble surfaces; carried upward, and dispersed at the air-sea interface when the bubble bursts. Phosphate, ammonia, and nitrite concentrations also appear to be larger at the sea's surface (Goering and Menzel, 1965).

In addition to the biosphere, which may be looked upon as a highly chemically active film, the two most important interfaces in our environment are between water and the atmosphere and water and mineral substances of the lithosphere. We have just examined the liquid–gas interface, and now we turn our attention to the liquid–solid interface. Only in relatively recent times have problems such as artificial organs focused strong attention on the interaction of water and aqueous solutions with neutral and/or hydrophobic surfaces.

Paradoxically, the solution–solid interfacial situation about which we have the most detailed information is a relatively complex one, namely, the distribution of ions in aqueous solutions near a charged metal surface. This situation constitutes the heart of the science of elecrochemistry and is of enormous fundamental and applied significance. As a consequence it has been the subject of extensive study, both theoretical and experimental, and from this effort has gradually evolved the theory of the electrical double layer, (or, more accurately, the electrical multilayer) (see Delahay, 1965, Conway, 1965; Devanthan and Tilak, 1965). Qualitatively the picture which has emerged (Figure 13.4) is one of two layers of oppositely charged ions, their order depending on the charge of the solid surface. But, for all the effort expended, our understanding of the electrical double layer is still clearly very incomplete. For example, hitherto attention has been largely focused on the spatial configuration of the ions, and few have had the courage to confront the even more difficult problem of how the water molecules arrange themselves in the double layer. An exception has been Bockris et al. (1963) and their effort to say something more specific about what is happening to the water molecules in the double layer (Figure 13.5) represents still, I believe, the deepest penetration of the question.

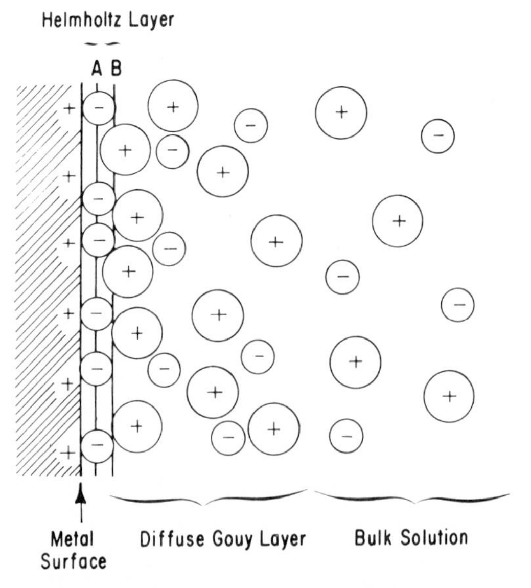

Figure 13.4. The electrical double layer (from Horne, 1968, with permission of Academic Press, Inc.). *A*, Inner Helmholtz or Stern plane of anionic centers; *B*, Outer Helmholtz or limiting Gouy plane of cationic distance of closest approach.

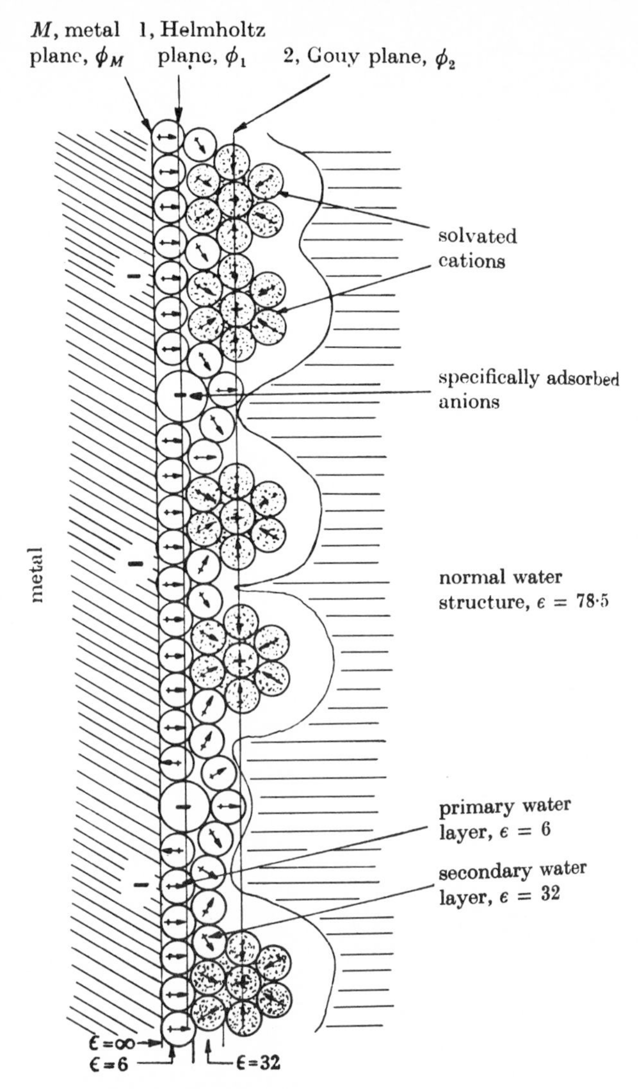

Figure 13.5. Model for the positions and orientations of water molecules in the electrical double layer (from Bockris et al., 1963, with permission of the Royal Society).

In addition to being central to electrochemistry the electrical multilayer is also of great significance in colloid chemistry, specifically the coagulation of suspended material, problems encountered in such disparate areas as estuarine chemistry and water treatment. It also makes its appearance in some rather unexpected places—such as the clotting of blood and, as we have just mentioned, synthetic materials for artificial arteries and organs.

In addition to studies of surface tension and potential and of the electrical double layer, a third source of clues to the structure of water near interfaces has been the investigation of the properties of liquid water and aqueous solutions in porous materials and very thin capillaries—and very remarkable indeed these properties are! Many years ago Bijl (1927) noted that the temperature of maximum density of a mixture of water and finely divided charcoal is lower than for pure water. This finding was interpreted in terms of the theory of liquid water then current, namely, the concentration of water polymers is different in the layer near the surface from the concentration in the bulk water. Even earlier Duff (1905) reported that liquids flowing through capillaries exhibit abnormally high viscosities, and Wolkowa (1934), working with clay and silica gel, found that the phenomenon was particularly marked in the case of water (for a review see Henniker, 1949). But perhaps the most unexpected property of liquid water in porous materials and thin capillaries is its extreme reluctance to solidify. Such water can be cooled to $-40°C$ before it finally solidifies. This and related phenomena have been the subject of several studies (Derjaguin and Fedyakin, 1962; Chahal and Miller, 1965; Schufle, 1965). In particular Schufle and Venugopalan (1967) have measured the specific volume of water in fine capillaries and obtained the extraordinary results shown in Figure 13.6.

Rather than a new form of water, the Ortho-Water (or "polywater") reported by Soviet investigators now appears to be contaminated vicinal water (Derjaguin and Churaev, 1973), and there is a danger that in forgetting about polywater the scientific community will also forget some of the important properties of vicinal water that the Soviet studies have revealed.

Nuclear magnetic resonance studies (Korringa, 1956) indicate that the interfacial water structure layer is as thick as 700 Å, and Drost-Hansen (1959) has estimated that in a 3 μ diameter capillary approximately 25% of the water is abnormal. He gives 0.1–0.4 μ as the thickness of the layer and states that the thickness depends on the ion exchange capacity of the porous material and that it decreases slowly with increasing temperature.

NMR studies have also indicated that the water surface layer in, for example, silica gel has considerable crystallinity (Zimmerman and Brittin, 1957 [for more recent NMR studies on ion-exchange resins see Creekmore and Reilley, 1970]), but just what the structure might be remains unclear. There may even be different structures adjacent to different solids; for example, the water structure on kaolinite appears to be different from that on montmorillonite with the waters more rigidly bound in the former instance (Slonimskaya and Raitburd, 1965). [Fortunately, water transport properties in clay minerals have been investigated in some detail by soil scientists (Low, 1961; Dutt and Low, 1962; Oster and Low, 1963; Olejnik and White, 1972).] Water and silicate minerals have the same crystal lattice parameters; the O—O distance in ice is 4.52 Å and in some silicates 4.51 Å. This has led Macey (1942) to propose that a layer of ice exists on the surface of the solid which extends outward until overcome by thermal agitation and the effect of electrolytes. But earlier we declined to assign an Ice-I_h type of structure to the Frank-Wen clusters in bulk water, and now the great proclivity of water near surfaces to supercool makes our reason even more compelling. Weiss (1966) has pointed out that the dimensions of the water structure on silicate surfaces corrspond well with those of the water structure in gas clathrates and, for want of better evidence, I am inclined to favor this view.

To further complicate matters there also can be definite chemisorption of water on the solid material, and this chemisorption can be very strong—the water monolayer on kaolinite is not removed until the temperature reaches 425°C (Brindley and Millholen, 1966). The most strongly adsorbed H_2O's on silica are chemically bound to solid surface hydroxyl groups (Hambleton et al., 1966), and there may be as many as three chemisorbed layers of immobilized water on the surface (Antoniou, 1964). To

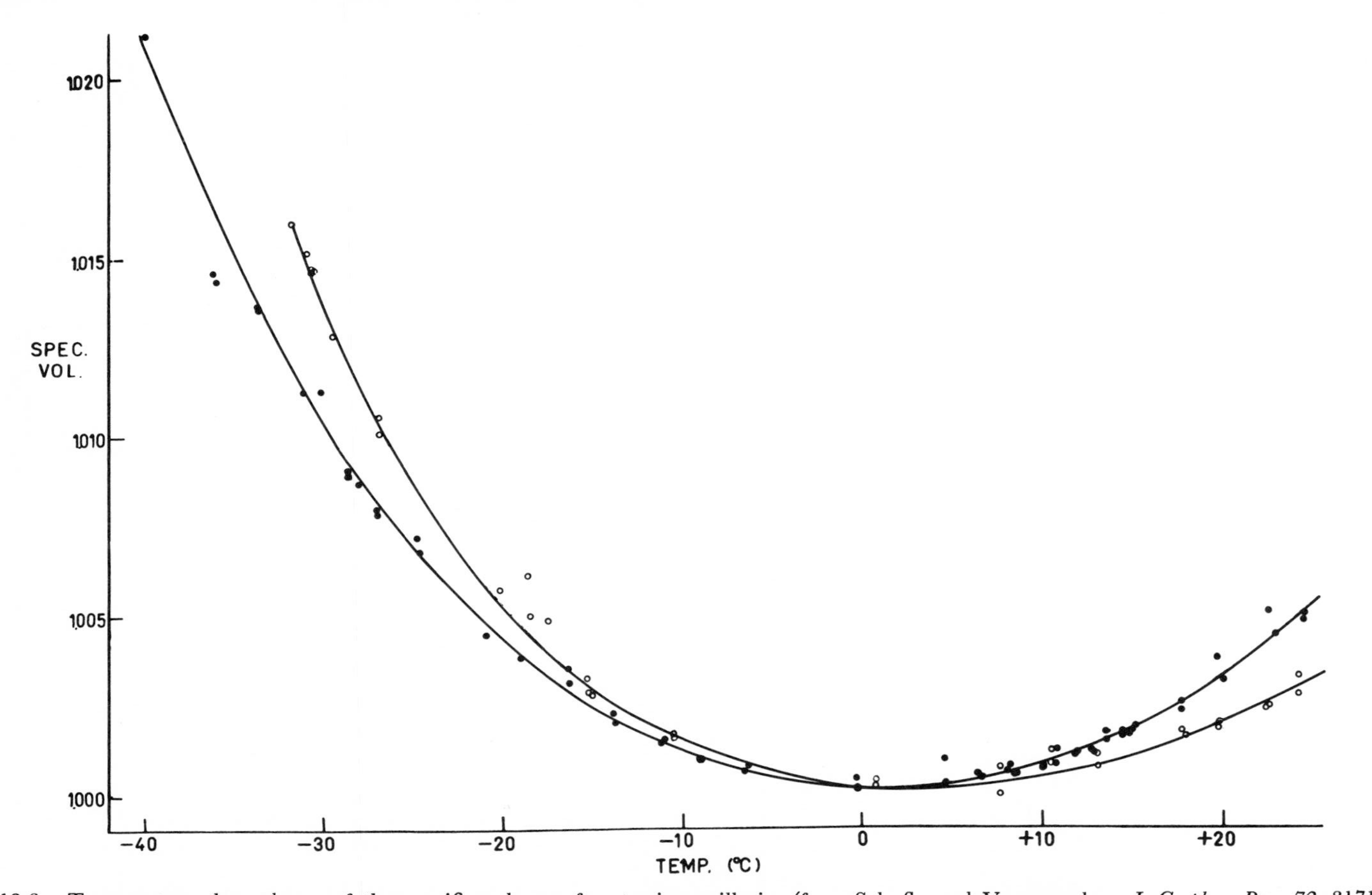

Figure 13.6. Temperature dependence of the specific volume of water in capillaries (from Schufle and Venugopalan, *J. Geophys. Res.*, *72*, 3171 (1967), © American Geophysical Union, with permission of the AGU and the authors).

summarize, then, in a porous material there can be at least three kinds of water: (*a*) strongly and rigidly bound chemisorbed water on the solid surface, (*b*) structured water near the surface, and (*c*) "normal" water at some greater distance from the surface.

[We saw previously that the highly structured water in the interfacial gas/water boundary layer tends to exclude polar solutes] that disrupt water structure; does the water structure near a solid surface behave in a similar manner? The answer appears to be yes. For example, certain cations are selectively excluded from silica gel. Tien (1965) has attributed this exclusion or negative absorption to the increased difficulty experienced in penetrating the smaller pores in the gel. Maatman (1965), however, has rejected this explanation and attributed the exclusion phenomenon to a "geometric effect" mentioned earlier. Inasmuch as ion exclusion by this mechanism can only occur immediately next to the surface, the phenomenon should be observable only when the ratio of surface area to volume of the system is extremely large, that is, gels, which appears to be contrary to observation (see discussion that follows). Nevertheless, despite these difficulties, if one takes the radii of the hydrated ion in the gel as obtained from this theory and treats them simply as a measure of ion exclusion, plotting them against the corresponding viscosity B-coefficients a striking correlation obtains with ions falling mainly into two classes, the structure makers and the struc-

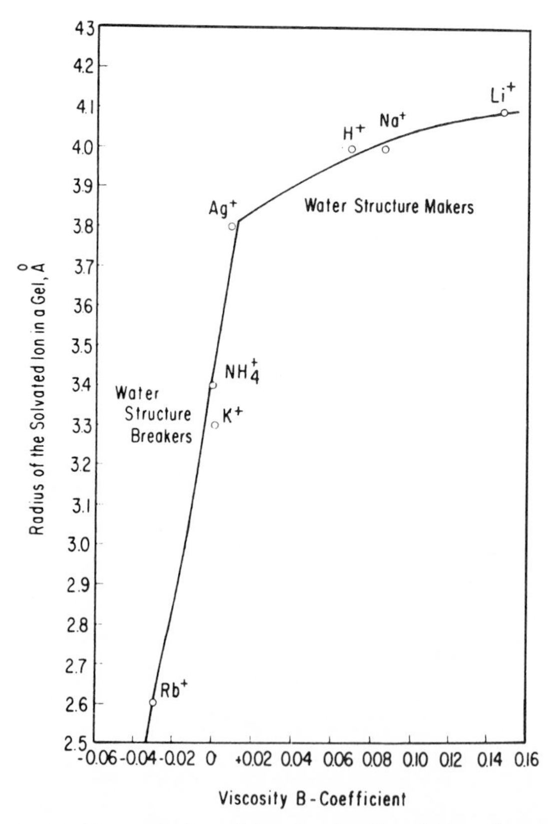

Figure 13.7. Dependence of ion exclusion of the viscosity B-coefficient (reprinted with permission from Horne, *J. Phys. Chem.*, *70*, 1335 (1966), © The American Chemical Society).

ture breakers (Figure 13.7). On the basis of Figure 13.7, the more positive a cation's B-coefficient (or perhaps more accurately the larger the absolute value of an ion's B-coefficient), the more strongly it is excluded.

Elution and high pressure conductivity experiments also yield evidence for the preferential exclusion of ions from the region near the solution/solid interface (Horne et al., 1968). The maximum in the relative electrical conductivity versus pressure curve for an aqueous solution permeated solid powder is 10% less than for the solution alone; in effect the solution mixed with the solid appears to me more concentrated. The fact that any reduction at all is observed at these pressures (about 2000 kg/cm^2) indicates that the structural form of water with which we are dealing, unlike the structure of the Frank-Wen clusters and the hydration atmospheres of ions, is relatively stable with respect to pressure, a conclusion difficult to reconcile with the large specific volume of interfacial water (Figure 13.6).

When a solution is passed through a long column of granular solid material, the first few emergent aliquots are more concentrated than the initial solution—further evidence of electrolyte exclusion. The qualitative picture (Figure [13.8]) is one of enhanced water structure near the solid surfaces with consequent exclusion of electro-

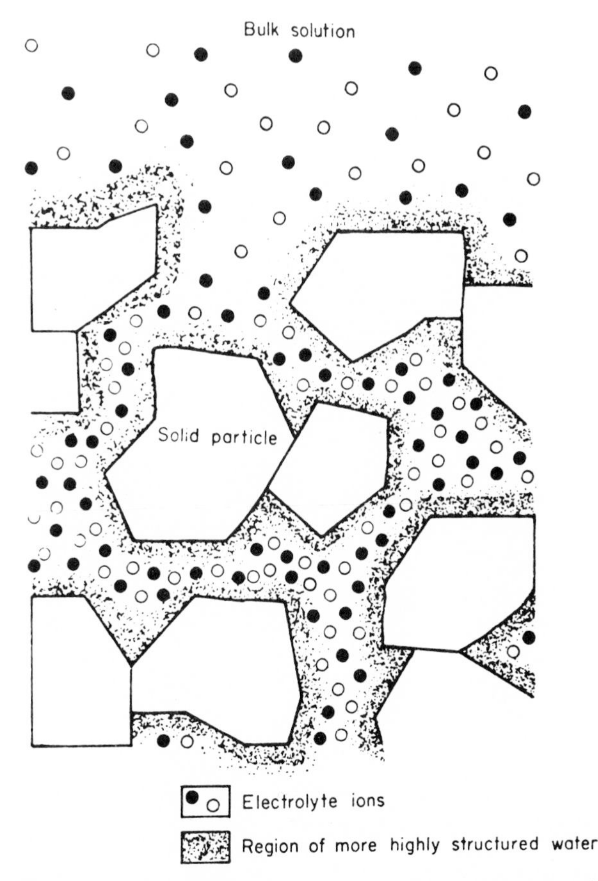

Figure 13.8. Model of particulate solids in an aqueous electrolytic solution (from Horne et al., 1968, with permission of Pergamon Press, Ltd.).

lyte and increased concentration of ions in the interstitial water more removed from the interfaces.

In addition to viscosity, ion transport processes in solution permeated porous materials have also been investigated. Owing to the compensating effect of ionic hydration, the limiting equivalent conductivities of ions in free solution are surprisingly independent of charge—26.5, 31.2, and 34.4 cm^2/ohm-equiv for Na^+ Ca^{2+}, and La^{3+}, respectively but, in an ion exchange resin, a porous medium with fixed charge sites distributed throughout, a strong charge dependence is evident, the corresponding values being 2.82, 0.99, and 0.22 cm^2/ohm-equiv (George and Courant, 1967). The Arrhenius activation energies of electrical conductivity tend to be higher for the membrane than for free solution (George and Courant, 1967), and E_a is also higher for protonic conduction in thin capillaries (Yu, 1966) since that process depends on the rotability of water molecules. Similarly, as one might expect, the self-diffusion coefficient of water in zeolites (1.34×10^{-5} to 1.88×10^{-5} cm^2/sec at 30°C) is less than for pure bulk water (2.5×10^{-5} cm^2/sec), and its activation energy is 0.6 to 1.9 kcal/mole higher (Parravano et al., 1967). Nevertheless, the question of the hinderance of water movement near surfaces in pores and capillaries continues to be controversial with some investigators reporting no or little difference from "normal" bulk water properties (Anderson and Quinn, 1972) and others finding definite evidence of very appreciable retardation (Schufle and Yu, 1968; Beck and Schultz, 1970). Manheim (1970) found ionic diffusion rates in unconsolidated marine sediments reduced from one-half to one-twentieth compared with free solution. Hydrostatic pressure appears to have roughly the same effect on transport process in vicinal water as in bulk water (Horne et al., 1969). These two observations are important in the migration of chemical substances in deep ocean sediments.

Notice that at the solution–solid interface we can get both increases and decreases of solute species population—the concentration of ions by direct sorption on in the electrical multilayer and exclusion owing to the structuring of the solvent. Which of the two effects dominates depends on the nature of the solid surface. For a conductive and/or highly charged surface, such as a metallic electrode, the electrical multilayer is very much in evidence, whereas for a nonconducting solid with few charge sites ion exclusion may be the more significant process.

To summarize our ruminations concerning the structure of aqueous electrolytic solutions, such as seawater, for a final time both in the bulk phase and near interfaces, let us say that in addition to unassociated monomeric water there are *at least* two other structural forms of liquid water with properties summarized in Table 13.2. But above all else we must remember that our conclusions at this time are imperfect in the extreme, that Table 13.2 is both tentative and highly speculative, and that the question of the structure of liquid water, both in the bulk phase and near interfaces, will remain one of the most central and difficult problem area in the chemistry of our environment, especially in marine chemistry, and in molecular biology for the foreseeable future.

13.3 CHEMICAL COMPOSITION OF MINERAL WATERS

Earlier in Chapters 7 and 8, respectively, we described the chemical composition of lakes, ponds, and streams, and of the oceans. But the chemical processes involved in

Table 13.2 Hydration Types (From Horne et al., 1968, with permission of Pergamon Press, Ltd.)

Hydrophilic or Coulombic Hydration	Hydrophobic Hydration
Surrounds ions and charge sites.	Envelops interfaces (liquid–water, gas–water, and solid–water) hydrophobic materials and hydrophobic portions of molecules.
Structure unknown, may be random. Probably is not Ice I.	Structure unknown, possibly Pauling clathrate type. Probably is not Ice I.
Relatively temperature-stable. Gradually destroyed in going from 0 to 100°C. Gives rise to the density maximum at 4°C.	Greater thermal stability. Gives rise to anomalies in water properties near 40°C.
Completely destroyed by 1000 kg/cm^2. 4°C.	
Completely destroyed by 1000 kg/cm^2.	Still evident at 5000 kg/cm^2.

the interactions of hydrosphere and lithosphere are perhaps even more clearly reflected in the composition of mineral waters and pore waters in sediments (see Section 13.7). Mineral waters contain substances dissolved out of the lithosphere; on the other hand, certain mineral substances deposit out of these waters into the lithosphere. Generally speaking, "mineral water" is water containing too much dissolved mineral matter to be potable, that is to say, in excess of 1000 ppm dissolved solids. If the waters issue at temperatures conspicuously greater than usual ground water temperatures, they are called "thermal springs" (Hem, 1963). Mineral waters are usually the result of the contact of circulating meteoric water with soluble rock minerals; however, it may contain varying amounts of juvenile or connate water.

Ever since ancient times therapeutic value has been attributed to mineral waters, and in the eighteenth and nineteenth centuries particularly, it was fashionable to travel from spa to spa "taking the waters." This practice is still popular in Europe, but the great hotels at such once-famous American spas as Saratoga Springs, New York, have fallen into neglect, fire, and ruin. Table 13.3 summarizes the characteristics of some American spa waters (while Table A.28 gives data for mineral waters in the Soviet Union; Table A.29 gives analyses of ground waters). Interest in geothermal waters has suddenly mushroomed in the United States, not for health reasons, but as a possible source of energy.

Hot water under high pressure, and often acidic due to dissolved CO_2, SO_2, and H_2S, is a powerful solvent on many rock minerals. Even the solubility of silica is significantly increased. When discharged, both pressure and temperature drop, dissolution equilibria are reversed, and minerals such as travertine (calcium carbonate) deposit around the mouth of the spring. Mammoth Hot Springs in Yellowstone Park, Wyoming, for example, represents an extensive travertine deposit. Since, as we have noted earlier, the lithosphere is chemically reduced, spring waters frequently contain such reduced materials as ferrous ion and H_2S. On exposure to the atmosphere, these are oxidized, respectively, to the reddish stain of ferric oixde and the white or pale yellow deposits of free sulfur so often evident near hot springs.

Table 13.3 Some American Spa Waters (Data abstracted from Licht, 1963)

Location	Flow Rate (liter/d)	Temperature (°C)	Concentration Dissolved Solids (g/liter)	Major Chemical Species	Indications
Berkeley Springs (West Virginia)	11,000,000	23.3	0.173	Na^+, Ca^{2+}, CO_3^{2-}, SO_4^{2-}	Rheumatism, skin disease, neurologic disease
Buckthorn Mineral Wells (Arizona)		44	1.1	Na^+, Ca^{2+}, Cl^-	Rheumatism, obesity, liver disease, metabolic disease
Dillsboro (Indiana)		0–13	9.1	27 mg H_2S/liter, Na^+, Ca^{2+}, Mg^{2+}, SO_4^{2-}, Cl^-	Rheumatism, gastrointest.
French Lick Springs (Indiana)	980,000	12.8	3.7–4.6	60 mg H_2S/liter, CO_2, Rn, Na^2, Ca^{2+}, Mg^{2+}, Cl^-, SO_4^{2-}	Rheumatism, gall bladder, liver, skin disease, gastro-intest.
Glenwood Springs (Colorado)	19,000,000	52	21.4	H_2S, NaCl	Rheumatism, nervous disease
Hot Springs (Virginia)	6,000,000	36.7–40	0.57	$CaCO_3$, $MgSO_4$	Rheumatism, gastrointest., nerve disease
Hot Springs National Park (Arkansas)	3,000,000	57–63	0.28	$Ca(HCO_3)_2$	Rheumatism, skin disease, convalescence
Marlin (Texas)	1,100,000 (3–1020m wells	55–64	7–10	Na^+, Ca^{2+}, Mg^{2+}, SO_4^{2-}, HCO_3^-, Cl^-	Rheumatism, skin disease, nerve disease, gastrointest.
Saratoga Spa (New York)	1,900,000 (20–30 to 300m wells)	10	5–16	NaCl, Ca^{2+}, Mg^{2+}, CO_2, alkaline	Rheumatism, nerve disease, cardiovasc., metabolic.
Sulphur Springs (Oklahoma)			4–4.8	NaCl, H_2S, Br^-, I^-	Rheumatism, skin disease, gastrointest.
Thermopolis (Wyoming)	70,000,000	57.2	2.2	NaCl, SO_4^{2-}, $CaCO_3$	Rheumatism, skin disease, gastrointest.
White Sulphur Springs (West Virginia)		16.5	2.1	H_2S, Ca^{2+}, Mg^{2+}, SO_4^-	Rheumatism, liver, gall bladder

Barnes (1970) has examined "anomalous" geothermal waters from the tectonic belt of the Pacific coast of the United States, and he attributes the high concentrations of NH_3, B, CO_2, H_2S, and hydrocarbons that he found to low grade metamorphosis of marine sedimentary deposits.

Certainly less spectacular than thermal mineral springs but of greater importance quantitatively in the chemistry of our environment is the more gentle chemistry of ordinary spring waters and of the equilibria of dissolution and deposition that occur in them and in lake waters. Garrels and Mackenzie (1967) have studied the chemical composition of some springs and lake in the Sierra Nevada; however, because the chemical composition of igneous rocks found in this area are reasonably representative of the rocks of the continental crust, their findings and conclusions probably have widespread applicability. The compositions of the spring waters investigated are shown in Table 13.4. The CO_2-rich soil waters rapidly attack the silicate rocks kaolinizing chiefly plagioclase, biotite, and K-spar and picking up much of their silica content in the first few feet of penetration. As they penetrate more deeply, these reactions slow, and both kaolinite and montmorillonite are weathering products. The chemical difference between shallow and deeper penetration are reflected in the composition differences between the waters of ephemeral and perennial springs (Table 13.4), and the different weathering equilibria involved are shown in Table 13.5. The accumulation of these waters and their evaporation in equilibrium with the atmosphere (Figure 13.9) produces highly alkaline Na-HCO_3^-–CO_3^{2-} waters represented by soda lakes with $CaCO_3$, $Mg(OH)_2$, and amorphous silica precipitating out and the more highly soluble univalent cation species becoming concentrated in solution.

Table 13.4 Mean Values for Compositions of Ephemeral and Perennial Springs of the Sierra Nevada (From Feth et al., 1964)

	Ephemeral Springs		Perennial Springs	
	ppm	Molality × 10^4	ppm	Molality × 10^4
SiO_2	16.4	2.73	24.6	4.1
Al	0.03	—	0.018	—
Fe	0.03	—	0.031	—
Ca	3.11	0.78	10.4	2.6
Mg	0.70	0.29	1.70	0.71
Na	3.03	1.34	5.95	2.59
K	1.09	0.28	1.57	0.40
HCO_3	20.0	3.28	54.6	8.95
SO_4	1.00	0.10	2.38	0.25
Cl	0.50	0.14	1.06	0.30
F	0.07	—	0.09	—
NO_3	0.02	—	0.28	—
Dissolved solids	36.0		75.0	
pH	6.2[a]		6.8[a]	

[a] Median Value.

Table 13.5 Shallow and Deep Weathering Equilibria (Selected from Garrels and Mackenzie, 1967)

A Source Minerals for Ephemeral Sierra Nevada Springs (concentrations moles/liter × 10^4)

Reaction (coefficients × 10^4)	Na^+	Ca^{2+}	Mg^{2+}	K^+	HCO_3^-	SO_4^{2-}	Cl^-	SiO_2	Products moles/liter × 10^4
Change kaolinite back into plagioclase Kaolinite				minus plagioclase					
1.23 $Al_2Si_2O_5(OH)_4$ + 1.10 Na^+ + 0.68 Ca^{2+} + 2.44 HCO_3^- + 2.20 SiO_2 = Plagioclase	0.00	0.00	0.22	0.20	0.64	0.00	0.00	0.50	1.77 $Na_{0.62}Ca_{0.38}$ feldspar
1.77 $Na_{0.62}Ca_{0.38}Al_{1.38}Si_{2.62}O_8$ + 2.44 CO_2 + 3.67 H_2O									
Change kaolinite back into biotite kaolinite				minus biotite					
0.037 $Al_2Si_2O_5(OH)_4$ + 0.073 K^+ + 0.22 Mg^{2+} + 0.15 SiO_2 + 0.51 HCO_3 = 0.073 Biotite $KMg_3AlSi_3O_{10}(OH)_2$ + 0.51 CO_2 + 0.26 H_2O	0.00	0.00	0.00	0.13	0.13	0.00	0.00	0.35	0.073 biotite
Change kaolinite back into K-feldspar				minus K-feldspar					
0.065 $Al_2Si_2O_5(OH)_4$ + 0.13 K^+ + 0.13 HCO_3^- + 0.26 SiO_2 = K-feldspar	0.00	0.00	0.00	0.00	0.00	0.00	0.00	0.12	0.13 K-feldspar
0.13 $KAlSi_3O_8$ + 0.13 CO_2 + 0.195 H_2O									

B Source Minerals and Weathering Products of Deeper Circulation (concentrations in moles/liter × 10^4)

Reaction (coefficients × 10^4)	Na^+	Ca^{2+}	Mg^{2+}	K^+	HCO_3^-	SO_4^{2-}	Cl^-	SiO_4	Mineral Altered and Product (moles/liter × 10^4)
Change kaolinite back into biotite									
$0.07\ Al_2Si_2O_5(OH)_4 + 0.42\ Mg^{2+} + 0.14\ K^+ + 0.28\ SiO_2 + 0.98\ HCO_3^- = 0.14\ KMg_3AlSi_3O_{10}(OH)_2 + 0.98\ CO_2 + 0.49\ H_2O$	1.09	1.67	0.00	0.02	4.41	0.00	0.00	1.09	0.14 biotite 0.07 kaolinite
Change kaolinite back into plagioclase									
$0.26\ Al_2Si_2O_5(OH)_4 + 0.235\ Na^+ + 0.144\ Ca^{2+} + 0.47\ SiO_2 + 0.52\ HCO_3^- = 0.38\ Na_{0.62}Ca_{0.38}Al_{1.38}Si_{2.62}O_8 + 0.52\ CO_2 + 0.78\ H_2O$	0.85	1.53	0.00	0.00	3.89	0.00	0.00	0.62	0.38 palgioclase 0.26 kaolinite
Change montmorillonite back into plagioclase									
$0.81\ Ca_{0.17}Al_{2.33}Si_{3.67}O_{10}(OH)_2 + 0.85\ Na^+ + 0.38\ Ca^{2+} + 0.61\ SiO_2 + 1.62\ HCO_3^- = 1.37\ Na_{0.62}Ca_{0.38}Al_{1.38}Si_{2.62}O_8 + 1.62\ CO_2 + 1.62\ H_2O$	0.00	1.15	0.00	0.00	2.27	0.00	0.00	0.01	1.37 plagioclase 0.81 montmorillo-nite
Precipitate $CaCO_3$									
$1.15\ Ca^{2+} + 2.30\ HCO_3^- = 1.15\ CaCO_3 + 1.15\ CO_2 + 1.15\ H_2O$	0.00	0.00	0.00	0.00	0.03	0.00	0.00	0.01	1.15 calcite

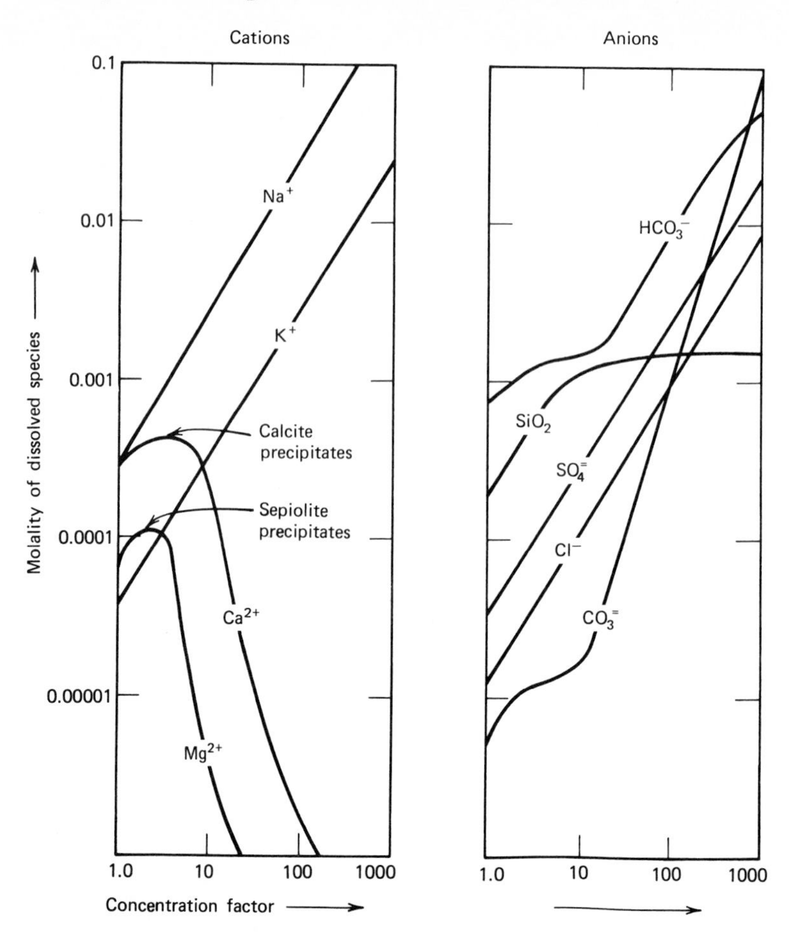

Figure 13.9. Calculated compositional changes upon the evaporation of typical Sierra Nevada spring water at constant temperature and in equilibrium with atmospheric carbon dioxide.

A chemical equilibrium model for the Great Lakes has been postulated by Kramer (1967) in which calcite, dolomite, apatite, kaolinite, gibbsite, and Na- and K-feldspars are in equilibrium with lake waters and the CO_2 of the air ($P_{CO_2} = 3.5 \times 10^{-4}$ atm) at 5°C and 1 atm. Cold lake waters contain excess CO_2 but are undersaturated with respect to calcite, dolomite, and apatite, while warmer waters are nearly all at equilibrium with the atmosphere but somehow supersaturated with respect to these minerals (Figure 13.10).

A great deal of the underground water in the United States is too brackish or saline to be useful without expensive desalination. Table 13.6 presents a comparison of the composition of the waters of a Michigan brine field with that of seawater based on analyses reported by Egleson and Querio (1969), who conclude that the Ca, Mg, and Sr contents are probably determined by dolomitization but that the Li, I, and NH_3 contents are too high to have resulted directly from the evaporation of seawater without some additional concentration process, possibly biological. Subsurface brines, in addition to salts, also may contain hydrocarbons (McAuliffe, 1969). Hydrogen and oxygen isotope effects increasing with depth have been observed in oil field brines and have been attributed to temperature-dependent ion-exchange and carbonate equilibria (Kharaka and Berry, 1973).

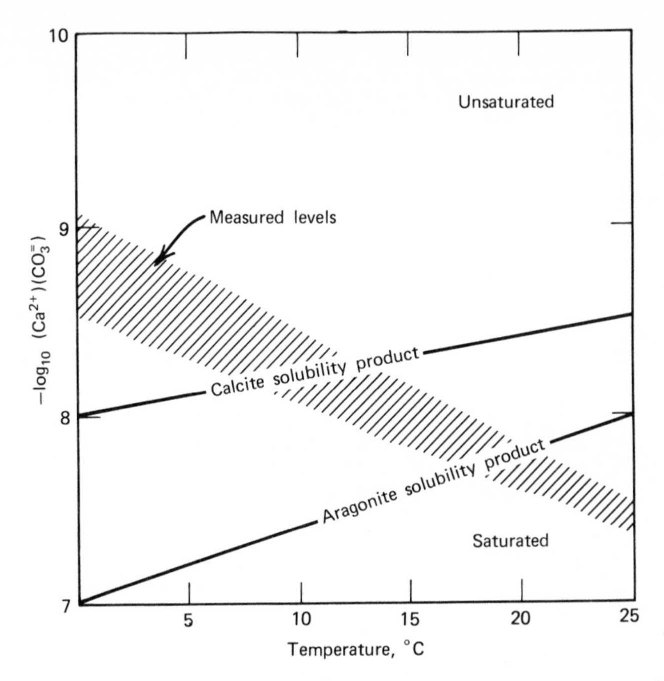

Figure 13.10. Saturation of lake water with respect to calcium carbonate minerals.

Table 13.6 Comparison of the Compositions of Seawater and a Michigan Brine

	Seawater Average (ppm)	Brine Average (ppm)	Concentration Ratio (brine/seawater)
Ca	390	72,200	142
Mg	1300	9,200	7.1
Br	65	2,790	42.9
Sr	8	3,100	388
SO_4	2580	44	0.02
K	370	8,000	21.6
B	4.5	333	74
Rb	0.12	15	125
Li	0.17	58	340
I	0.15	34	680
NH_3	low	398	high

13.4 WEATHERING AND THE CHANGING FACE OF THE LAND

The more important natural chemistry of our environment, which we have already noted is largely aqueous and interfacial chemistry, falls into four major categories:

1. Atmospheric processes,
2. Processes in the biosphere,
3. Precipitation depositions (often biologically mediated) out of the hydrosphere, and
4. Weathering.

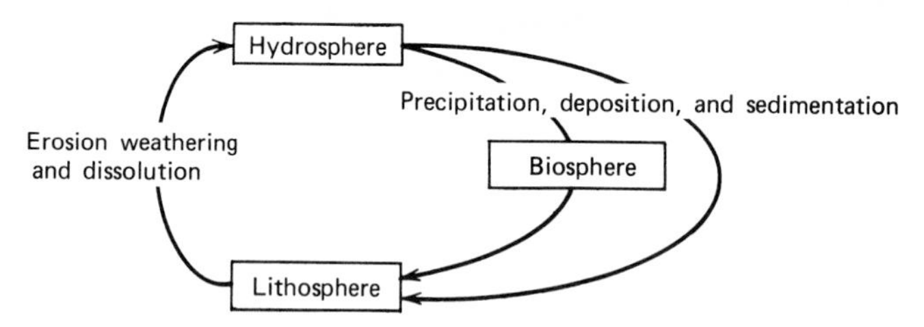

Figure 13.11. Major hydrosphere/lithosphere interactions.

The last, weathering, is the chemical mass transport of substances out of the exposed lithosphere into the hydrosphere and/or atmosphere. It is not uncommon for these materials to be eventually redeposited in the lithosphere (Figure 13.11). Weathering in turn is conveniently divided into two processes (*a*) erosion and (*b*) chemical weathering, which go hand in hand. Erosion applies to physical processes that remove surface lithosphere and the latter to chemical processes. The hydrosphere, it should be noted, acts in weathering both as (*a*) a conveyor of matter and (*b*) a reactant (Stumm and Morgan, 1970). Both atmosphere and hydrosphere attack the exposed lithosphere and erode its surface. Wind carried dust particles sand-blast surface rocks, dislodging particles and carrying them away (Figure 13.12). Spring melt

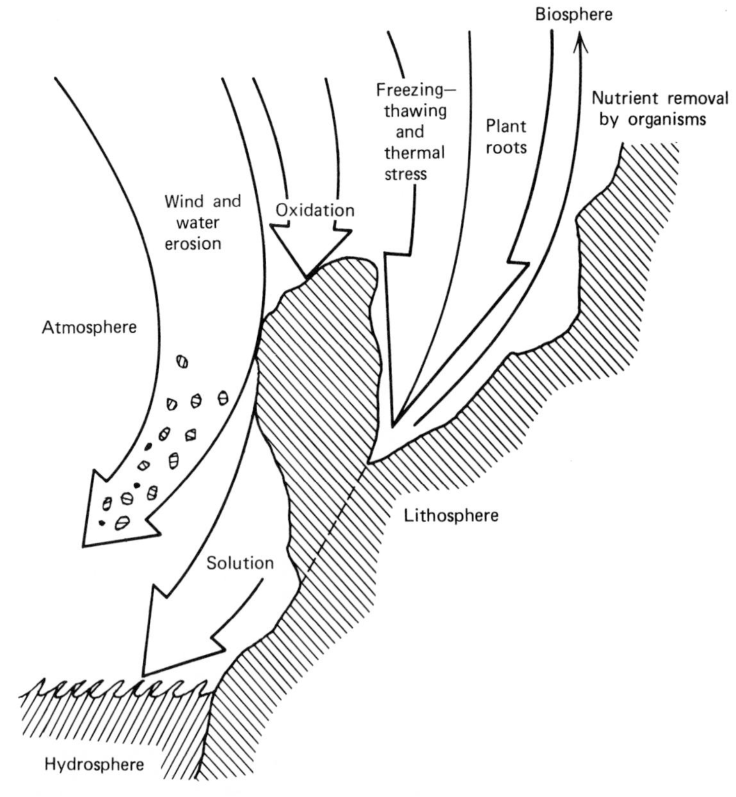

Figure 13.12. Weathering attack.

swollen streams and rivers dash sand, gravel, stones, and even car-size chunks of rock against their banks, each year thus gnawing off and carrying away great quantities of the lithosphere. More subtle processes join the attack on the lithosphere (Figure 13.12). Alternate freezing and thawing drives frost wedges into rocks that can easily split them asunder. Plant roots are also capable of splitting rock and breaking it up to form the soil that encourages further plant growth. The heat from forest fires breaks off large flakes of exposed bedrock (Longwell and Flint, 1955). Even the thermal stresses produced by diurnal warming and cooling sphallate rock surfaces. These physical attacks on the stone are launched in concert with chemical attacks. The physical weathering exposes fresh material for chemical attack, while the volume increases accompanying the formation of clay minerals upon weathering produce mechanical stresses which in turn are a further physical assault on the rock. The chemical material of the unexposed lithosphere has an excess of electrons (i.e., it is in reduced form) and a deficiency of water. Not surprisingly then, when this material is exposed to the highly oxidizing, water-rich atmosphere, it becomes very unstable and susceptible to chemical attack. Then too the crustal materials are basic, some atmospheric gases acidic. Yet rather unexpectedly, cationic denudation of New England watersheds remains low despite increased acidity from SO_2 in the polluted air (Johnson et al., 1972). The organic materials in crustal rock rapidly weathers away as soon as they are exposed (Leythanenser, 1973). Four major types of chemical attack are directed against the face of the Earth:

1. Oxidation,
2. Hydration and dissolution,
3. Neutralization, and
4. Ion-exchange.

(Morgan and Stumm (1970) use a somewhat different classification scheme, and their chemical weathering examples are listed in Table 13.7, which might also be compared to Table 13.5.) As we have said, in unexposed lithosphere mineral substances are often in a reduced state; metals, such as iron for an example, occur in their lower oxidation state. Thus ferrous sulfide will tend to be oxidized up to ferric sulfides, even the sulfur may be oxidized up to sulfite (we touch on these reactions again in our discussion of acid mine drainage). Generally speaking, solubility increases with oxidation hence the oxidation process will accelerate the dissolutive attack on the rock.* Even the strongest ionic crystal lattices are disruptible by water, so great is the energy of hydration. For many minerals, the solubility may be small, yet it is nevertheless finite. Given sufficient time, water not already saturated with these minerals will dissolve them. Next we come to neutralization or the "world titration"

$$\begin{array}{ll} \text{Basic crustal material} + \text{acidic atmospheric gases} & \\ (CaCO_3, MgCO_3, \text{etc.}) \qquad (CO_2, SO_2, \text{etc.}) & \\ \qquad \rightarrow \quad \text{soluble salts} + \text{water} & (13.1) \\ \qquad (Ca^{2+}, Mg^{2+}, HCO_3^-, SO_4^{2-}, \text{etc.}) & \end{array}$$

Remember (Section 9.5) rain water is acidic. To be more specific, let us take the example of the "titration" of the feldspar minerals (recall also Table 13.5), which

* Iron is an exception. Fe(III) species are only move soluble than Fe(II) species in acid water.

Table 13.7 Examples of Typical Weathering Reactions (From Stumm and Morgan, 1970)

I. Congruent Dissolution Reactions

$SiO_2(s) + 2H_2O = H_4SiO_4$
quartz

$CaCO_3(s) + H_2O = Ca^{2+} + HCO_3^- + OH^-$
calcite

$CaCO_3(s) + H_2CO_3 = Ca^{2+} + 2HCO_3^-$

$Al_2O_3 \cdot 3H_2O(s) + 2H_2O = 2Al(OH)_4^- + 2H^+$
gibbsite

$Al_2O_3 \cdot 3H_2O(s) + 6H_2CO_3 = 2Al^{3+} + 6HCO_3^- + 6H_2O$

$Ca_5(PO_4)_3(OH)(s) + 3H_2O = 5Ca^{2+} + 3HPO_4^{2-} + 4OH^-$
apatite

$Ca_5(PO_4)_3(OH)(s) + 4H_2CO_3 \rightleftharpoons 5Ca^{2+} + 3HPO_4^{2-} + 4HCO_3^- + H_2O$

II. Incongruent Dissolution Reactions

$MgCO_3(s) + 2H_2O = HCO_3^- + Mg(OH)_2(s) + H^+$
magnesite; brucite

$Al_2Si_2O_5(OH)_4(s) + 5H_2O = 2H_4SiO_4 + Al_2O_3 \cdot 3H_2O(s)$
kaolinite; gibbsite

$NaAlSi_3O_8(s) + \frac{11}{2}H_2O = Na^+ + OH^- + 2H_4SiO_4 + \frac{1}{2}Al_2Si_2O_5(OH)_4(s)$
albite; kaolinite

$NaAlSi_3O_8(s) + H_2CO_3 + \frac{9}{2}H_2O = Na^+ + HCO_3^- + 2H_4SiO_4 + \frac{1}{2}Al_2Si_2O_5(OH)_4(s)$

$CaAl_2Si_2O_8(s) + 3H_2O = Ca^{2+} + 2OH^- + Al_2Si_2O_5(OH)_4(s)$
anorthite; kaolinite

$CaAl_2Si_2O_8(s) + 2H_2CO_3 + H_2O = Ca^{2+} + 2HCO_3^- + Al_2Si_2O_5(OH)_4(s)$

$4Na_{0.5}Ca_{0.5}Al_{1.5}Si_{2.5}O_8 + 6H_2CO_3 + 11H_2O$
plagioclase (andesine)

$= 2Na^+ + 2Ca^{2+} + 4H_4SiO_4 + 6HCO_3^- + 3Al_2Si_2O_5(OH)_4(s)$
kaolinite

$3KAlSi_3O_8(s) + 2H_2CO_3 + 12H_2O$
K-feldspar (orthoclase)

$= 2K^+ + 2HCO_3^- + 6H_4SiO_4 + KAl_3Si_3O_{10}(OH)_2(s)$
mica

$7NaAlSi_3O_8(s) + 6H^+ + 20H_2O$
albite

$= 6Na^+ + 10H_4SiO_4 + 3Na_{0.33}Al_{2.33}Si_{3.67}O_{10}(OH)_2(s)$
Na⁺-montmorillonite

$KMg_3AlSi_3O_{10}(OH)_2(s) + 7H_2CO_3 + \frac{1}{2}H_2O$
biotite

$= K^+ + 3Mg^{2+} + 7HCO_3^- + 2H_4SiO_4 + \frac{1}{2}Al_2Si_2O_5(OH)_4(s)$
kaolinite

$Ca_5(PO_4)_3F(s) + H_2O = Ca_5(PO_4)_3(OH)(s) + F^- + H^+$
fluoroapatite; hydroxyapatite

$KAlSi_3O_8(s) + Na^+ = K^+ + NaAlSi_3O_8(s)$
orthoclase; albite

$CaMg(CO_3)_2(s) + Ca^{2+} = Mg^{2+} + 2CaCO_3(s)$
dolomite; calcite

III. Redox Reactions

$MnS(s) + 4H_2O = Mn^{2+} + SO_4^{2-} + 8H^+ + 8e$

$3Fe_2O_3(s) + H_2O + 2e = 2Fe_3O_4(s) + 2OH^-$
hematite; magnetite

$FeS_2(s) + 3\frac{3}{4}O_2 + 3\frac{1}{2}H_2O = Fe(OH)_3(s) + 4H^+ + 2SO_4^{2-}$
pyrite

$FbS(s) + 4Mn_3O_4(s) + 12H_2O = Pb^{2+} + SO_4^{2-} + 12Mn^{2+} + 24OH^-$
galena

make up more than half of the bulk of granitic rock; one reaction involved can be represented by

$$2KAlSi_3O_8 + H_2CO_3 + H_2O \rightarrow K_2CO_3 + Al_2Si_2O_5(OH)_4 + 4SiO_2 \quad (13.2)$$

orthoclase carbonic acid kaolinite silica

Of the products, the K_2CO_3 is soluble, the silica partially soluble, and the kaolinite, a major clay mineral (see Weaver and Pollard, 1973, for more about clay chemistry) is insoluble and, unless physically transported, is left behind. Quartz, another component of granite, is also very stable chemically and is left behind as the feldspar rots away. The resulting loosened quartz grains can be physically transported to form the sand of beaches and sedimentary deposits, or being caught up in the rock cycle, they can be consolidated and recemented back together again to form sandstones. We will return to the "world titration" again when we inquire into the origin of the salts in the sea (Section 13.5). We also have something more to say about ion exchange here.

Over 90% of the minerals of the Earth's crust are silicates, thus the weathering of silicate minerals is of particular interest. Under attack by water, especially under the influence of temperature, pressure, and/or dissolved acidic substances, univalent and bivalent cations readily go into solution, but in the cases of aluminum and silicon, the dissolution process is somewhat more complicated. Valence deficiencies on the surface of the mineral crystal and at fractures form the points of attack by water molecules. Hydration and hydrolysis remove strongly basic cations such a K^+, Ca^{2+}, and Mg^{2+}, and the oxygen anions may be partially replaced by hydroxyl ions in the crystal lattice. Aluminum tends to assemble six OH's about itself, Si four, and these elements go initially into solution as free $Al(OH)_6^{-3}$ and $Si(OH)_4^{2-}$. These ions then aggregate to form clusters of colloidal size. Then upon aging, these amorphous clusters begin to form new crystalline lattices characteristic of clay minerals. Some minerals, such as biotite and muscovite, can be transformed directly into clay minerals by ionic substitution without lattice breakdown (Mason, 1966).

Erosion and weathering carve and shape the face of Earth. While the increments may be small, acting over vast stretches of geological time, the alterations they are capable of producing can be enormous. They can level mountains, they can fill valleys and seas, they can reshape coastlines. One of the most spectacular testaments to their work is the Grand Canyon of the Colorado River. For the North American continent alone, the total yearly run-off value listed in Table 13.8 corresponds to about 460,000 kg/yr of material.

13.5 THE ORIGIN OF THE OCEANS*

Where did the water and the chemical constituents of the Earth's oceans come from? How have their quantities varied in the geologic past? Descartes, I believe, remarks that there is no opinion so absurd that it has not been stoutly defended by some philosopher at one time or another, and I think this rather cynical remark nicely applies to theories concerning the chemical origin and evolution of the seas. For example, with respect to the question of the volume of the ocean, the theories ad-

* Most of this section is taken from Horne (1969).

Table 13.8 Discharge and Chemical Denudation of North America (From Livingston, 1963)

Region	Area (thousands square miles)	Runoff (thousands cubic feet per second)	Total Dissolved Solids (ppm)	Chemical Analyses Used
North Atlantic Slope (U.S.)	148	210	116	Hudson River at Hudson
South Atlantic Slope (U.S.)	284	325	155	Tombigbee River near Epes
Mississippi River	1250	620	223	At New Orleans
West Gulf of Mexico	320	55	881	Rio Grande at Laredo
Colorado River	246	23	711	Yuma main canal
Great Basin	215	0	0	
Pacific Basins	117	80	152	Sacramento River at Sacramento
Columbia River	262	345	125	Columbia River at Cascade Locks
St. Lawrence River	498	500	161	At Sorel
Mackenzie River	660	260	214	At Ft. Simpson
Nelson River	450	125	210	At mouth
Fraser River	86	94	82	At New Westminster
Yukon River	360	180	208	At Eagle
Franklin Territory	554	139	91	Mean of arctic Alaska lakes
Keewatin Territory	228	114	214	Mackenzie at Ft. Simpson
Newfoundland	43	43	62	Mean of Moser River, Wallace River, Miramichi River, Andrews lakes and Ellerslie Creek (Newfoundland, Labrador, Maritime Province)
Labrador	112	112	62	
Maritime Province	51	51	62	
Hudson Bay (Quebec and Ontario)	592	592	116	Mean of Abitibi, Mattagami, Rainy, and Kapuskasing Rivers
Alaska south of Yukon	195	214	52	Kenai River
Alaska north of Yukon	195	49	91	Mean of arctic Alaskan lakes
Minor coastal streams, British Columbia and elsewhere	319	351	82	Fraser at New Westminster
Mexico	758	20	881	Rio Grande at Laredo
		150	114	Rio Parana above La Plata
Guatemala	42	65 (Guatemala, British Honduras, Honduras)	114	Rio Parana above La Plata
British Honduras	9			
Honduras	59			
Salvador	13	383 (Salvador, Nicaragua, Costa Rica, Panama)	54	River Orinoco at Puerto Ayacucho
Nicaragua	54			
Costa Rica	23			
Panama	29			
Sum or mean	8172	5100	142	

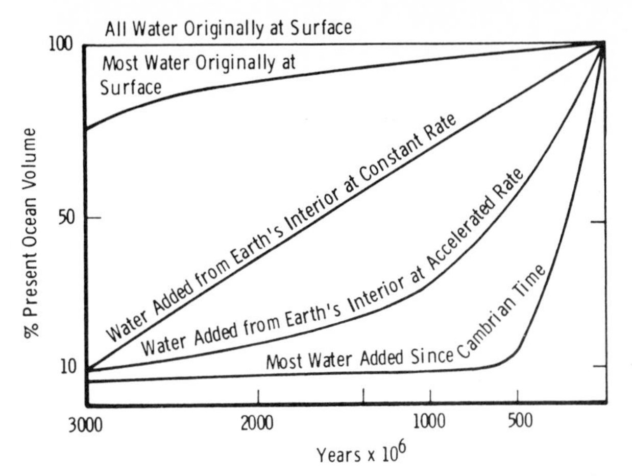

Figure 13.13. Various theories of the volume of the oceans in the geological past (from Revelle, 1955, with permission of the publisher).

vanced cover all possibilities (Figure [13.13]). Some hold that the volume has remained pretty much as it is [a theory supported by a recent consideration of the $^{18}O/^{16}O$ ratios in the marine environment, Chase and Perry, 1972], that the oceans are relatively recent, and the remaining theories run the spectrum from slow to rapid accumulation. But, lest we be too quick to criticize, we must not forget that this diversity of opinion is an immediate consequence of the extreme difficulty and complexity of these problems. Rubey (1951) warns us that the history of the Earth's atmosphere and hydrosphere 'cannot be told until we have solved nearly all other problems of earth hisotry,' and he wryly suggests that his address on the "Geologic History of Sea Water" could be more aptly titled "The Problem of the Source of Sea Water and Its Bearing on Practically Everything Else." In the face of such horrendous complexity and fragmentary data the progress that has been made becomes perhaps much more impressive. Since Conway's (1942, 1943) magnificent pioneering work, Holland (1965) cites three milestones marking this advance: Rubey's (1951) concept of excess volatiles; the appreciation of the geologically brief residence times of cations in the sea (Barth, 1951; Goldberg and Arrhenius, 1958 Goldberg, 1963); and finally Sillen's (1961) recent fresh attack on the problem of the mechanism for pH control in the seas.

Our first thought on addressing the problem of the origin of the water of the world's oceans might be that it was the consequence of the condenstion of a primordial envelope of steam as the hot Earth cooled. This hypothesis appears to be untenable. The gases neon, argon, krypton, and xenon are 10^6 to 10^{10} times less abundant on Earth than in the atmospheres of stars, presumably because these inert gases, not being chemically bound, have escaped from this planet. Indeed, as one would expect, the larger discrepanices are found for the lighter gases. Similarly, gases of comparable molecular weight, H_2O, N_2, O_2, CO, and CO_2, would also have been lost from the atmosphere of the primitive earth. It follows that these substances were not free but were rather chemically bound or occluded in the condensed phase substance of the Earth (Rubey, 1951); that they were subsequently released by volcanic and subterranean hydrothermal activity; and that, after the Earth had sufficiently cooled, they accumulated to form the oceans. The striking differences between Earth and Mars are entirely compatible with this point of view. The atmosphere of Mars is very tenuous as the consequence of the escape of gases from that planet. In contrast with the great oceans of the Earth, undoubtedly the most spectacular visual feature of our world to an out-in-space observer, there is only enough moisture left on the surface of Mars to form an ever-so-light frost. Then, too, while the surfaces of our continents are ravaged with the traces of volcanism and other titanic geological processes, the [relatively] bland face of Mars shows [less] evidence of volcanic or other activity of the type that released the waters of the oceans here (Revelle, 1955).

But the quantity of water contained in the oceans is so immense! Could all this water have conceivably been squeezed out of the interior of the planet? The answer is yes, easily. Rubey (1951) has estimated that, even at their present rate of discharge, which may be considerably less now than earlier in the Earth's geological history, hot springs on the continents and ocean floor release some 6.6×10^{16} g of water per year. Over a period of 3×10^9 years this amounts to 2.0×10^{26} g of water. Yet Rubey estimates the total quantity of water in the atmosphere, hydrosphere, biosphere, and buried in ancient sedimentary rocks to be 1.7×10^{24}, or only about 0.8% of the discharge. Thus, even if only a small fraction of the total quantity was derived from a primary magmatic source, in the long course of geolotic time hot springs can readily account for the total mass of the Earth's oceans.

There now appears to be fairly general agreement that the origin of the oceans was as described above, but there is considerably less agreement about the details of the ocean's development. The fossil record provides arguments that the volumes of the oceans have remained relatively constant for at least the last sixth of geologic time. Yet, on the other hand, the existence of atolls and of flat-topped sea mounts in the Pacific, deeply buried remains of shallow water creatures, and the surprising thinness of certain sedimentary deposits can be interpreted as supporting the idea that the oceans are relatively recent, that as much as one-fourth of all the water in the seas appeared since the Mesozoic Era in the last thirtieth or fortieth of geologic time (Revelle, 1955). Still there seems to be no evidence for the progressive flooding of the continental landmasses which one would expect if the volume of the oceans has been steadily increasing (Nicolls, 1965).

There have been minor (5–10%) variations in the total volume of the oceans as a consequence of the growth and melting of ice sheets.

The answer to our second question—where did the electrolytes in seawater come from?—is fairly obvious: from the condensed phase of the Earth, but the paths by which they found their way into the oceans are much less clearly defined. Careful examination and comparison of the elemental composition of seawater and the Earth's crust is most instructive (Table [13.9]). Again our intuition fails us; it is perfectly obvious from Table [13.9] that there is no direct relationship between marine and lithospheric abundances, that the electrolytes in seawater are not simply leached out of the Earth's crust. But, as we study the last column of this table more closely, the disorder evaporates; and we discover the elements falling into a relatively few remarkably well-defined categories. Let us take sodium as our basis, since it has so little chemistry in the seas. With respect to this element many of the remaining elements fall into three categories: those with percentages in solution comparable to Na; those with very much greater percentages; and those with very much smaller percentages. The very small percentages are associated with elements, such as Si and

Al, whose crustal compounds are highly insoluble [or which readily precipitate out of solution if the pH is raised] and/or with elements, such as Fe, Mn, Ni, V, which are known to be scavenged out of seawater by precipitation, strong sorption process, and/or nodule formation. The concentrations of a subclass of elements are depleted in seawater by the intervention of biological processes. Thus, in Table [13.9] P shows depletion but Ca and C unexpectedly are not conspicuous. However, if we compare the concentration of Ca^{2+} and of CO_3^{2-} in the seas and the amounts added by rivers (Table [13.10]), this depletion by biological precipitation of $CaCO_3$ becomes quite evident. The depletion of the nutrients NO_3^- and SO_4^{2-} is also shown in Table [13.10]; it is about 10 times greater in the case of the NO_3^-, the more important nutrient.

The elements found in seawater in much greater amounts than expected, Cl, B, Br, and S, all occur in anionic form—a first hint that anionic or acidic materials might take a different path from the lithosphere to the hydrosphere than do cationic or basic materials (Figure 13.14). The elements in these anionic electrolytes are the excess volatiles discussed by Rubey (1951) (Table 13.11). If they do not result entirely from the weathering of the crustal material, then what is the additional source of these elements? His answer is that, like the water, they come from within the Earth and are spewed into the seas, directly, via rivers, or via the atmosphere

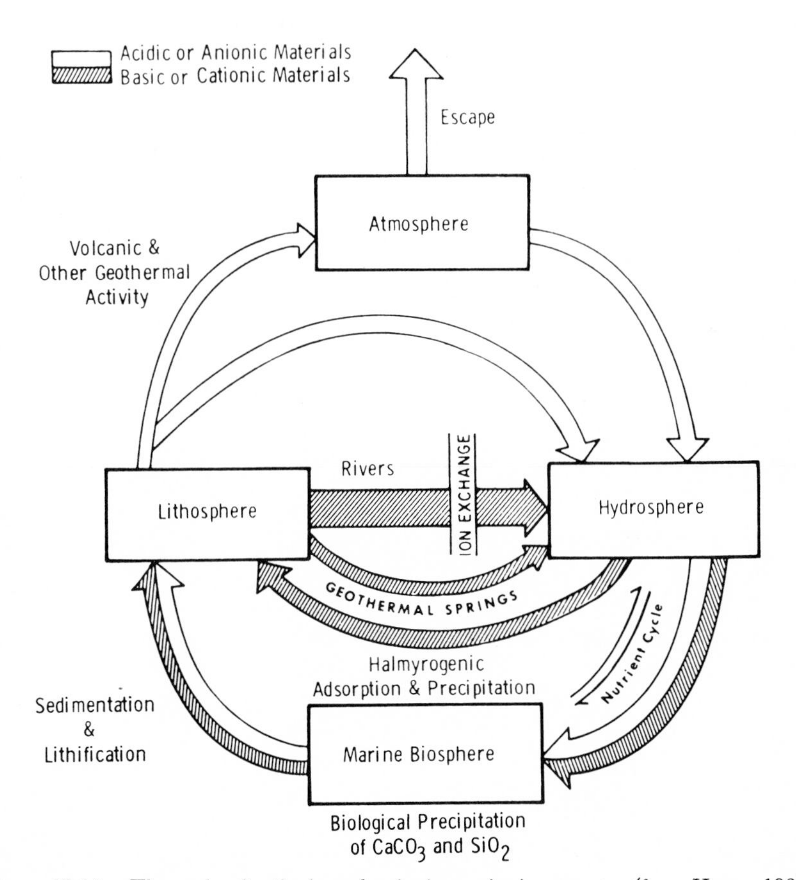

Figure 13.14. The cycle of cationic and anionic species in seawater (from Horne, 1969).

Table 13.9 Comparison of the Elemental Abundances in Seawater and Crustal Rocks

Element	mg/kg in 35‰ *S* Seawater	Potential "Supply" in 600 g of Rock Expressed as mg/kg of Seawater	Percentage in Solution[a]
		Expected Levels	
Ca	408	22,000	2
Na	10,769	17,000	65
K	387	15,000	3
Mg	1,297	13,000	10
C	28	300	9
Sr	13	250	5
F	1.4	160	1
Se	0.004	0.4	1
I	0.05	0.2	25
As	0.02	3	0.7
U	0.015	2	0.8
		Excess Levels	
S	901	300	300
Cl	19,353	290	6,700
Br	66	3	2,000
B	4.7	2	240
		Deficient Levels	
Si	4	165,000	0.002
Al	0.5	53,000	0.001
Fe	0.002	31,000	0.0001
Mn	0.01	560	0.002
P	0.1	470	0.02
Ba	0.05	230	0.02
Rb	0.2	190	0.1
Cu	0.01	60	0.02
Ni	0.0001	60	0.0002
V	0.0003	60	0.0005
Li	0.1	39	0.2
Ce	0.0004	26	0.002
Zn	0.005	24	0.02
Y	0.003	19	0.002
La	0.0003	11	0.003

(Continued)

Table 13.9 (*Continued*)

Element	mg/kg in 35‰ *S* Seawater	Potential "Supply" in 600 g of Rock Expressed as mg/kg of Seawater	Percentage in Solution[a]
Pb	0.004	10	0.04
Mo	0.0005	9	0.005
Cs	0.002	4	0.05
Sc	0.00004	3	0.001
Hg	0.00003	0.3	0.001
Ag	0.0003	0.06	0.5
Au	0.000006	0.003	0.3
Ra	0.0000000003	0.0000006	0.05

[a] Very large excesses and deficiencies are encircled.

Table 13.10 Present Amount of Ions in Sea, and Amount of Dissolved Ions Added by Rivers in 100 Million Years in moles/cm² of Total Earth Surface (From Sillen, 1967a, with permission of the The Chemical Society)

	Na^+	Mg^{2+}	Ca^{2+}	K^+	Cl^-	SO_4^{2-}	CO_3^{2-}	NO_3^-
Present in ocean	129	1	2.8	2.7	150	8	0.3	0.01
Added in 100 million years	196	122	268	42	157	84	342	11

Table 13.11 Estimated Quantities (in units of 10^{20}) of Volatile Materials Now at or Near the Earth's Surface, Compared with Quantities of These Materials that Have Been Supplied by the Weathering of Crystalline Rocks (From Rubey, 1951)

	H_2O	Total C as CO_2	Cl	N	S	H, B, Br, A, F, and so on
In present atmosphere, hydrosphere, and biosphere	14,600	1.5	276	39	13	1.7
Buried in ancient sedimentary rocks	2,100	920	30	4.0	15	15
Total	16,700	921	306	43	28	16.7
Supplied by weathering of crystalline rocks	130	11	5	0.6	6	3.5
"Excess" volatiles unaccounted for by rock weathering	16,600	910	300	42	22	13

from volcanic and other geothermal activity—an explanation supported by their presence in present day emanations (Table 13.12). Recently, Gunter (1973) has reported on the argon, and nitrogen and Mazor and Fournier (1973), on the noble gas concentrations in Yellowstone hydrothermal waters; Rosenberg (1973) has looked for fluorine compounds in volcanic gases, and Muenow (1973) found mostly H_2O, CO_2, and CO, but also H_2, O_2, N_2, S_2, H_2S, SO_2, SO_3, COS, HCl, HF, NH_3 and minor concentrations of organic constituents released from bubbles in Hawaiian volcanic glass. On rare occasions, true flames are observed on Kilauea Volcano, probably due to the burning of hydrogen (Naughton, 1973). The Cl concentration of seawater is particularly large. Most of the elements have definite finite residence times in the oceans (Table 8.10) indicating that they are circulated in the dynamics of the atmosphere–lithosphere–hydrosphere system but chlorine does not. Of the chlorine added from the interior of the Earth, 75% is still present in the oceans; that is, the hydrosphere is a sink for Cl as well as H_2O (Holland, 1965). Rubey (1951) compares two alternative hypotheses concerning the rates at which the excess volatiles were added to seawater; the "quick-soak" (columns 1, 2, and 3 in Table [13.13]) and the "slow-soak" (columns 4 and 5) hypotheses. According to the quick-soak view, all the excess volatiles were originally dissolved into a primitive ocean, according to the slow-soak view, these materials have been dissolving steadily throughout geologic time. The dissolution of the excess volatiles produces acid substances; therefore the primitive quick-soak oceans should have been very acid with a pH of only about 0.3 (Table [13.13]) and the present ocean would have a pH of about 7.3 and a salinity roughly twice its present value, and more than half of the original CO_2 would still be in the atmosphere. Then, too, the neutralization of the highly acid primitive ocean would have required the weathering of a much more igneous rock and the subsequent deposition of much greater quantities or carbonate rocks than indicated by the existing geological record. Rubey (1951) in the light of these insurmountable difficulties is obliged to reject the quick-soak hypothesis and to conclude that both the quantity of water and the excess volatiles have increased gradually, so that the composition of both the hydrosphere and the atmosphere "has varied surprisingly little, at least since early in geologic time" and that "these variations have probably been within relatively narrow limits."

Whereas the anion content of seawater is largely plutonic in origin, the cation content is largely a product of the weathering of the Earth's crust (Figure 13.14). Notice in Tables 13.9 and 13.10 that K, although like Na it has little chemistry, is noticeably depleted in seawater. Conway (1942, 1943) believes that the cation contents in seawater are intimately interdependent and controlled primarily by the removal of K, and that furthermore the rate of this removal has not deviated widely from its present value. While the total amount of Na + K has remained relatively constant (Table 13.14), the Na/K ratio has undergone some rather drastic changes, especially around 270×10^6 yr ago. To weathering and vulcanism, we must now add a third source of seawater solutes—continental drift, the spreading of the ocean floor exposes fresh lithosphere to the forces of dissolution in the area of the midocean rifts (MacIntyre, 1970), and ocean floor basalts may contribute to the geochemistry of the oceans (Hart, 1973).

Another important point that must be borne in mind is that the sedimentary material weathered from the lithosphere and carried to the sea by rivers probably undergoes profound chemical changes at rates rapid on an oceanographic or geo-

Table 13.12 Volume Percentages of Gases from Volcanoes, Rocks, and Hot Springs (From Rubey, 1951)

	Volcano Gases from Kilauca and Mauna Loa			Gases from Rocks: Basaltic Lava and Diabase			Gases from Rocks: Obsidian, Andesitic Lava, and Granite			Gases from Fumaroles of the Katma Region and from Steam Wells and Geysers of California and Wyoming			
	Min.	Max.	Median	Min.	Max.	Median	Min.	Max.	Median		Min.	Max.	Median
CO_2	0.87	47.68	11.8	0.89	15.30	8.1	0.08	20.26	2.0	CO_2	0.03	1.24	0.2
CO	0.00	3.92	0.5	0.02	8.28	0.2	0.01	2.22	0.5	CO	—	0.01	tr
H_2	0.00	4.22	0.4	0.38	6.18	1.2	0.08	11.60	0.4	O_2	0.00	0.08	tr
N_2	0.68	37.84	4.7	0.27	7.21	2.0	0.03	3.90	1.2	CH_4	0.00	0.30	0.11
A	0.00	0.66	0.2	0.00	0.04	tr	0.00	0.02	tr	H_2	0.00	0.29	0.15
SO_2	0.00	29.83	6.4	—	—	—	—	—	—	$N_2 + A$	0.00	0.31	0.02
S_2	0.00	8.61	0.2	0.08	1.96	1.1	0.00	2.89	0.2	NH_3	—	0.02	0.01
SO_3	0.00	8.12	2.3	—	—	—	—	—	—	H_2S	0.00	0.10	0.02
Cl_2	0.00	4.08	0.05	0.06	1.33	0.5	0.01	10.59	0.5	HCl	0.01	0.57	0.06
F_2	—	—	—	0.00	14.12	3.8	0.25	7.80	2.3	HF	0.00	0.10	0.03
H_2O	17.97	97.09	73.5	71.32	92.40	83.1	69.44	98.55	92.9	H_2O	98.04	99.99	99.58
			100.0			100.0			100.0				100.00

Table 13.13 Composition of Atmosphere and Seawater under Alternative Hypotheses of Origin, Compared with Present-Day Conditions (From Rubey, 1951)

	All "Excess" Volatiles in Primitive Atmosphere and Ocean (original P_{CO_2} very high)			Only a fraction of Total Volatiles in Primitive Atmosphere and Ocean (original $P_{CO_2} \leq 1.0$; life begins early)		
	Initial Stage (before rock weathering)	Intermediate Stage ($CaCO_2$ begins to precipitate)	Late Stage (life begins at $P_{CO_2} = 1.0$)	Initial Stage (before rock weathering)	Intermediate Stage ($CaCO_2$ begins to precipitate)	Present Day Conditions
Atmosphere (kg/cm^2)	14.2	13.8	2.1	<1.1	<1.1	1.0
N_2 (%) by volume	9	9	50	7	7	78
CO_2 (%) by volume	89	89	47	90	90	0.03
H_2S (%) by volume	2	2	3	3	3	—
O_2, others (%) by volume	—	—	—	—	tr	22
Ocean ($\times 10^{20}$ g)	16,600	16,600	16,600	<990	<990	14,250
Cl, F, Br (g/kg)	18.3	18.3	18.3	18.3	18.3	19.4
ΣS, *Z*B, others (g/kg)	0.8	0.8	1.3	0.1	0.1	2.5
ΣCO_2 (g/kg)	14.3	15.8	25.2	<1.1	<1.7	0.1
Ca (g/kg)	—	5.9	tr	—	<5.5	0.4
Mg (g/kg)	—	1.3	5.2	—	<1.2	1.3
Na (g/kg)	—	3.1	12.5	—	<2.9	10.3
K (g/kg)	—	1.2	4.7	—	<1.1	0.4
H (g/kg)	0.5	tr	—	0.5	tr	—
Salinity (‰)	33.9	46.4	67.2	<20.0	<30.8	35.2
pH	0.3	5.1	7.3	0.3	5.7	8.2
$CaCO_3$ pptd ($\times 10^{20}$ g)	None	None	980	None	None	1,500±
Igneous rock eroded ($\times 10^{20}$ g)	None	4,200±	17,000±	None	<240	11,000±

Table 13.14 Average Contents of Sodium and Potassium and Average Values of Na/K Ratio in Argillaceous Rocks of Various Ages (From Nicolls, 1965, with permission of Academic Press, Inc.)

Time Scale (in 10^6 yr)	Period	No. of Analyses	Na (%)	K (%)	(Na + K) (%)	Na/K
1	Quaternary	5	0.55	2.6	3.15	0.21
70	Tertiary	12	0.88	1.9	2.78	0.46
	Cretaceous, upper	4	0.85	1.6	2.45	0.53
		17	0.81	2.07	2.88	0.39
	Cretaceous, lower	8	0.99	2.2	3.19	0.45
135	Jurassic	16	1.04	2.6	3.64	0.40
180	Triassic	4	0.76	1.9	2.66	0.40
225						
	Permian	28	0.96	2.2	3.16	0.44
270						
	Carboniferous, upper	9	0.36	2.51	2.87	0.14
	Carboniferous, middle	19	0.54	4.2	4.74	0.13
	Carboniferous, lower	17	0.39	2.2	2.59	0.18
350						
	Devonian	18	0.86	3.07	3.93	0.28
	Devonian, top upper	9	0.47	3.4	3.87	0.14
	Devonian, bottom upper	41	0.43	2.9	3.33	0.15
	Devonian, middle	41	0.45	2.9	3.35	0.16
400						
	Silurian	6	0.51	3.7	4.21	0.14
440						
	Ordovician	8	0.59	3.3	3.89	0.18
500						
	Cambrian	14	0.36	4.0	4.36	0.09
600		7	0.84	2.86	3.70	0.29
	Keweenawen	4	1.68	2.38	4.06	0.71
?900						
	Upper Huronian	10	1.10	3.38	4.48	0.32
	Lower Huronian	15	0.26	3.54	3.80	0.07
	Pre-Huronian	4	2.07	2.45	4.52	0.88

logical scale when entering the marine environment as a result of ion exchange. In particular Mg^{2+}, Na^+, and K^+ are exchanged for Ca^{2+}.

The electrolyte concentration of the Earth's oceans is, as we have seen, largely the result of a massive neutralization of plutonic acids by crustal bases, and we are brought back again to the question what are the chemical processes which have managed to attain and maintain the pH of the oceans. The traditional answer—the carbonate system (both fast control by the $CO_2/HCO_3^-/CaCO_3$ system and slow control by the HCO_3^-/H^+-clay system (MacIntyre, 1970))— has by no means been invalidated, but Sillen's (1961) suggestion that silicate minerals and equilibria of the sort

$$1.5Al_2Si_2O_5(OH)_4(s) + K^+ \rightleftharpoons KAl_3Si_3O_{10}(OH)_2(s) + 1.5H_2O + H^+ \quad (13.3)$$

may also be playing an important role in gaining ground (Holland, 1965; Garrels, 1965; Sillen, 1965, 1966, 1967*a*, *b*).

13.6 SEDIMENTATION AND THE DEPOSITION OF MATERIALS FROM THE HYDROSPHERE

Much of the material removed from the lithosphere by physical erosion and chemical weathering eventually ends up, either settled out or precipitated out of the hydrosphere, back in the lithosphere as sedimentary deposits (Figure 13.11) in lake, river, and ocean bottoms. Erosion, weathering, and sedimentation are but steps in the greater geochemical cycle (Figure 3.12) or, more specifically, the sedimentary rock cycle (Figure 3.15). Man has interposed himself into this weathering/sedimentation cycle with the usual unfortunate consequences (Figure 13.15). His mining, construction, and farming operations expose crustal materials to the atmosphere and hydrosphere, thus accelerating weathering. His impoundments of streams and rivers prevent suspended material from finding its way to the seas to replenish the shifting beaches, while at the same time his use of the oceans as a dump for his usually contaminated solid wastes (see Chapter 4) substitutes a new form of sedimentation.

Quantitatively marine sediments enormously outweigh their fresh water counterparts, so in our discussion (largely taken from Horne, 1969) we give them greater emphasis.

Only rarely does the solid interface at the bottom of the sea consist of the firm rock of the Earth's crust. Almost the entire ocean floor is covered with a slowly accumulating deposit of sedimentary material. Pelagic sediments alone (see classifi-

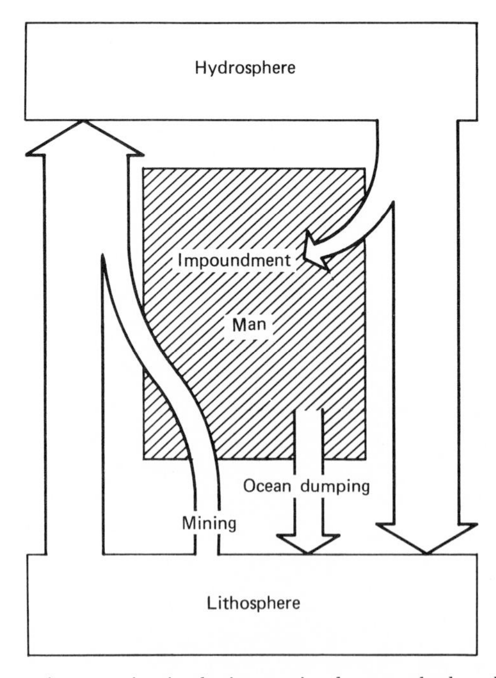

Figure 13.15. Human intervention in the interaction between hydrosphere and lithosphere.

Table 13.15 Ocean Areas (Percentage) Covered by Different Pelagic Sediment Types

	Atlantic Ocean	Pacific Ocean	Indian Ocean	All Oceans
Calcareous Oozes	67	36	54	48
Siliceous Oozes	7	15	20	14
Red Clay	25	49	25	38

cation below) cover some 2.68×10^8 km^2 of the Earth's surface or some 74% of the sea bottom (Table [13.15]). The geologically recent marine sediments have been classified in a number of ways. They can, for example, be classified chemically as calcareous ($CaCO_3$–$MgCO_3$), siliceous (SiO_2), etc., but more useful classifications are based on their origin or their distribution, and Table [13.16] represents an effort to combine in a simplified form both of these classifications. Detail is sometimes added to this classification; for example, the calcareous oozes can be divided into globigerina, pteropod, and cocolith oozes, and the siliceous oozes into diatom and radiolarian oozes on the basis of the most prevalent organism remains, while terrigenous deposits can be classified on the basis of the coarseness of their particles. The origins and the means of transportation of sedimentary material in the sea has been reviewed by Kuenen (1965). The finer the particles, the more easily they are carried great distances from their source (Slabaugh and Stump, 1964), since their settling rates are slower. Consequently, there is a sorting of terrigenous material, and the littoral deposits tend to be coarser and more highly varied in the distribution of their particle sizes and their chemistry, the latter reflecting the complex mineralogical chemistry of their continental sources. Sedimentary material from the land masses, we must remember, is carried to the sea by wind (Bonatti and Arrhenius, 1965) as well as by river run-off. We also should note in passing (Table 13.18) that some sedimentary material comes from the exosphere, ranging from meteoric spherules of Fe, Ni, and Co (Castaing and Fredriksson, 1958; Pettersson and Fredriksson, 1958; Yamakoshi and Tazawa, 1971) to glassy tektites (O'Keefe, 1963; Glass and Heezen, 1967), the latter may or may not be of extraterrestrial origin.

The rock exposed on the ocean floor represents a small potentially active surface area. Thus the chemistry of the sea–ocean bottom interface is largely the surface chemistry of the marine sediments, and the particle sizes, degree of compaction, and water content of the sediment are important, not only because they determine the physical characteristics of the material but also because of the role they play in its chemistry. Coprecipitation and, especially, absorption take trace elements out of seawater (Krauskopf, 1956); the smaller the particles are, the greater the surface area, the more absorption, and the greater the concentration of absorbing trace elements (Table [13.17]). On the basis of this finding we might expect the finer deep sea sediments to be richer in trace elements than coarser near shore material, and such does indeed seem to be the case (Table 13.18 and Figure 13.16). Turning now for a moment to fresh waters, Oliver (1973) found 26 to 42 ppm Pb, 0.20 to 0.28 ppm Hg, 84 to 86 ppm Zn, 24 to 28 ppm Cu, 22 to 23 ppm Ni, 11 to 13 ppm Co, 9200 to 12,700 ppm Fe, 118 to 241 ppm Mn, and 21 to 22 ppm Cr in the sediments of the Ottawa and Rideau (Canada) Rivers and an excellent correlation between metal

Table 13.16 Classification of Marine Sediments (From Horne, 1969)

Source	Origin	Distribution	Product
	Volcanic		
TERRIGENOUS From the Lithosphere (the continental landmasses)	*Clastic* (Mechanical and/or chemical destruction of rocks larger particles →	LITTORAL DEPOSITS Shallow, coastal waters Highly variable distribution Large quantities	Sands, Silts, Muds, Clays
	HEMIPELAGIC (Transitional Littoral–Pelagic found on a continental slope)	Airborne dust Seaborne turbidity	
	Planktonic (Insol. residua from near-surface life forms)	PELAGIC DEPOSITS Deep sea Wide distribution Largest quantities	<30% organic origin RED CLAY >30% organic origin OOZE
BIOGENOUS From the Biosphere (organic origin)	*Benthonic* (Bottom-living animals and plants)	More abundant in shallow, coastal water Relatively small quantities	
HALMYROGENOUS From the Hydrosphere (flocculation and chemical precipitation)		Wide and irregular distribution Relatively small quantities	CONCRETIONS NODULES
COSMOGENOUS Extraterrestrial origin		Wide Distribution Very small quantities	METEORITES

Table 13.17 The Dependence of Trace Element Concentration on Particle Size in a $CaCO_3$ Ooze (From Turekian, 1965, with permission of Academic Press, Inc. and the author.)

Particle Size Range (μm)	Parts per Million						% $CaCO_2$
	Pb	Sn	Ni	Mn	Cu	Sr	
700–1000	15	9	13	320	54	1400	~100
500–700	15	7	15	400	47	1200	~100
250–500	15	4	15	290	23	1400	~100
180–250	9	10	10	210	28	1200	~100
140–180	15	20	15	190	28	1700	~100
125–140	18	10	15	130	27	1300	~100
100–125	27	<4	13	64	17	1400	~100
89–100	18	<4	5	50	13	1300	~100
63–89	18	9	8	96	31	1400	~100
45–63	12	7	13	170	70	1300	~100
32–45	33	28	40	170	310	1400	~100
22–32	40	27	72	270	130	1200	97.3
16–22	70	44	105	600	410	1400	96.1
11–16	54	44	120	1000	330	1600	94.2
8–11	42	66	220	>1000	730	2280	91.8
5.6–8	230	67	900	>1000	1650	2700	90.3
4.0–5.6	110	42	550	≫1000	1650	2940	86.0
2.8–4.0	94	46	510	≫1000	1650	>3000	86.9
2.0–2.8	82	54	490	≫1000	1100	2820	88.2
1.4–2.0	82	42	390	≫1000	1300	1950	87.9
1.0–1.4	88	58	230	>1000	620	2150	88.5
0.7–1.0	150	62	170	>1000	1300	2100	88.0
0.5–0.7	15	10	15	210	43	1500	91.1

content, except for Hg, and the surface area (m^2/g) of the sediment. Organisms are capable of building up very large relative concentrations of trace elements. But, inasmuch as these materials are concentrated largely in the soft tissues my own suspicion is that the processes of dissolution and decay return most of these elements to the seawater long before the skeletal remains finally settle to the bottom, and that as a consequence Table 13.18 and Figure 13.16 represent trace element concentration for the most part by absorption processes rather than by biological activity. The concentrations are particularly high in manganese nodules, objects formed by slow absorptive accretion.

The water content of a freshly settled sediment increases with decreasing particle size, but chemical processes and the pressure of the accumulation of overyling material tend to compress and consolidate the sediment so that the water content decreases with deeper penetration of the sedimentary layer. For example, the water content of a fine-grained (1.7–2.7 μm) sediment decreased from 81% at the top to 75% at a depth of 2 m, whereas a coarser sample (16–20 μm) decreased from 65 to 56%.

Table 13.18 The Distribution of Trace Elements in Marine Environments and Igneous Rocks (Concentrations in ppm) (From Chester, 1965, with permission of Academic Press, Inc. and the author)

Trace Element	Igneous Rocks	Near-Shore Sediments	Deep-Sea Agrillaceous Clays (Pacific)	Manganese Nodules	Ratio, Near-Shore Sediments/ Igneous Rocks	Ratio, Deep Sea Clays/ Near-Shore Sediments	Ratio, Manganese Nodules/Deep Sea Clays
Cr	65	100	77	<10	1.5	0.77	<0.13
V	127	130	330	590	1.0	2.5	1.8
Cu	42	48	570	3300	1.1	11.9	5.8
Pb	13	20	162	1500	1.5	8.1	9.2
Ni	50	55	293	5700	1.1	5.3	19.4
Co	19	13	116	3400	0.69	8.9	29.2
Sn	2	21	20	300	10.5	0.95	15.0
Ba	530	750	2237	3100	1.4	3.0	1.4
Sr	335	<250	587	1000	0.75	2.3	1.7
Zr	152	160	145	340	1.0	0.91	2.3
Ga	17	19	20	17	1.1	1.1	0.85

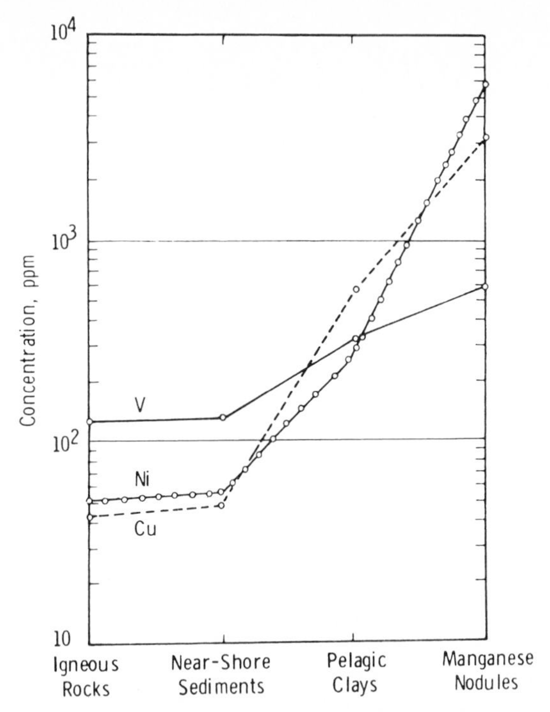

Figure 13.16. Concentration of some trace elements in rocks, sediments, and nodules (from Chester, 1965, with permission of Academic Press, Inc., and the author).

Hamilton (1959, 1964) has examined the gravitational consolidation and lithification of deep sea sediments in some detail, both experimentally and theoretically. The void ratio and porosity decrease with increasing depth. Several studies of the surface properties and pore structure of marine sediments have been made (Kulp and Carr, 1955; Slabaugh and Stump, 1964), and the results of Weiler and Mills (1965) have examined pore and surface properties (Table 13.19). These authors have observed that the chemically available surface area in the sediments is reduced by the blockage of pores by organic material and that surface sorption is adequate for accounting for the measured quantities of organic matter in the sediment. This should remind us once again of the ubiquity of biogenic organic material in the seas. In the ocean, at the surface or bottom interfaces, or in the bulk phase, inorganic chemistry is never seperable from biological processes. Absorbed organic coatings may reduce the rate of solution of $CaCO_3$ and siliceous material in seawater. Inorganic scales of Fe, Mg, and Ca aluminosilicates may also reduce the solution rates of sedimentary components such as biogenic opal (Hurd, 1973). Little wonder that the oceans are in disequilibrium with respect to many important minerals. The dissolution rate of opal, it is interesting to note, also depends upon the amount of water with which it was originally deposited (Huang and Vogler, 1972).

As indicated in Table 13.15, the pelagic sediments are the predominate bottom material over vast reaches of the world's oceans, and now we turn specifically to the question of their chemical composition. The oceanic average composition of pelagic sediments is given in Table 13.20, which is an abstract of a much more detailed

Table 13.19 Surface Properties and Pore Structure of Marine Sediments (From Weiler and Mills, 1965, with permission of Pergamon Press, Ltd.)

Sample No.	Sediment Type	Pretreatment of Sample	Carbonate Content	In situ Density	Bulk Density	True Density	Calculated Geometric Surface Area	True Surface Area	Average Particle Diameter	Total Pore Volume	Average Pore Radius
			%	g/cm³	g/cm³	g/cm³	m³/g	m²/g	Å	cm³/g	Å
1	Red Clay	None		1.42	0.92	2.57		28.2	830		
		Washed with distilled water						28.5			
		Cleaned with H_2O_2	33.9			2.56	0.68	32.4	720		
2	Red Clay	None	27.2		0.66	2.64	0.79	86.7	270		
3	Red Clay	None						54.7	86.7		
		Cleaned with H_2O_2	5.4		1.00	2.51		69.2	350		
4	Globigerina ooze	None		1.42	1.48	3.08	<0.01	4.19	4700	0.017	81
		Cleaned with H_2O_2	94.2		0.91	2.28		6.14	4300	0.022	72
5	Globigerina ooze	None	46.9		0.75	2.65		13.0	1700		
6	Foraminiferal ooze	None	78.9		0.70	2.26	0.94	28.5	930	0.154	108
7	Foraminiferal ooze	None	47.7		0.96	2.07	0.60	10.0	2900		
8	Pteropod ooze	None			0.64	2.94	<0.01	2.81	7200		
		Cleaned with H_2O_2	95.6			2.59		4.77	4900	0.046	193
9	Pteropod ooze	None	96.5		0.79	2.43		5.60	4400		
10	Radiolarian ooze	None			0.66	2.62		71.6	320	0.363	101
		Washed with distilled water						96.0	260		
		Cleaned with H_2O_2	7.0		0.68	2.42		99.2	250	0.358	72
		Cleaned with HCl						149	170	0.767	103
11[8]	Radiolarian ooze	None			0.32	1.95		12.8	2400		
		Washed with distilled water	3.7					24.9	1200		
12	Diatom ooze	None			0.41	1.96		17.9	1700	0.168	188
		Cleaned with H_2O_2	15.8		0.41	2.05		23.9	1200	0.116	197
13	Diatom ooze	None	31.7		0.42	1.95		31.7			
14	Shelf (terrigenous) sediment	None		1.33	0.91	2.52		12.4 (110°, 150°C)	1900		
								14.6 (215°C)	1600	0.136	186
		None, but air-dried						10.9			
								(−14 + 32 mesh)		0.106	195
						2.36		11.7			
								(−60 mesh)	2300	0.106	181
		Washed with distilled water						16.6			
		Cleaned with H_2O_2	2.9		1.12	2.55		17.9			
								(110°, 150°C)	1300		
								16.7 (210°C)	1400	0.123	147

Table 13.20 Chemical Composition of Pelagic Sediments (From Horne, 1969)

A. Oceanic Average of Major Elements as Weight Percent Oxide

SiO_2	44	CaO	0.6
Al_2O_3	12	MgO	2.3
Fe_2O_3	6.5	Na_2O	1.8
TiO_2	0.7	K_2O	2.4
MnO_2	0.9	P_2O_5	0.3

B. Oceanic Average of Trace Elements Given as Parts per Million

Ba	2852	Ni	236
B	3	Pb	109
Co	3	Sn	11
Cr	69	Sr	755
Cu	345	V	193
Ga	18	Zr	180
Ge	1.6		

table given in the review paper of Chester (1965). The values given, it should be emphasized, are averages, and inspection of his original table shows a wide variation in any one of the major or trace elements, for example, a range of 5.3 to 63.9 wt % SiO_2 or 120 to 1110 ppm Sr. There are marked variations depending on the type of pelagic sediment (Table 13.21), and there are other variations even in a given type or, for that matter, with sediment depth in a given location. Although based on data now quite old, two other tables (Table 13.22 and 13.13) are instructive in giving some hint of the variety and relative amounts of biogenic and terrestrial mineral substances in pelagic sediments. On the basis of the similarity of the chemical composition of igneous rocks and Pacific pelagic sediments, the major elements Na, K, Mg, Ca, Al, Ti, and Fe appear to be neither enriched nor depleted in the sediments as a whole (Goldberg and Arrhenius, 1958; Goldberg, 1961) in contrast to the minor elements which are frequently enriched in the sediment (Table 13.18).

The rare earth content of marine sediments (Goldberg et al., 1963) provides an interesting footnote; there appears to be a definite systematic dependence on the ionic radius with members of the series with ionic radii less than that for Sm being increasingly excluded from the sediment as ionic radius decreases.

As a participant in a complicated chemical system (Chapter 15), the $CaCO_3$ content of deep sea sediments shows some interesting behavior. A puzzling feature of the $CaCO_3$ distribution is the abrupt decrease in the $CaCO_3$ content of sediments below 4000 m, sometimes called the "compensation depth." The phenomenon is presumably due to the increased solubility of $CaCO_3$ with increasing hydrostatic pressure (Pytkowicz and Connors, 1964). Seawater is undersaturated with respect to $CaCO_3$ at depths of only 500 m; why, then, the sudden increased solubility at much greater depths? Again the rate of solution, in particular the retardation of the dissolu-

Table 13.21 Chemical Analyses of the Principal Types of Pelagic Sediments (weight % oxides) (From Wakeel and Riley, 1961, with permission of Pergamon Press, Ltd.)

	Red Clay (%)	Calcareous Ooze (%)	Siliceous Ooze (%)
SiO_2	53.93	24.23	67.36
TiO_2	0.96	0.25	0.59
Al_2O_3	17.46	6.60	11.33
Fe_2O_3	8.53	2.43	3.40
FeO	0.45	0.64	1.42
MnO	0.78	0.31	0.19
CaO	1.34	0.20	0.89
MgO	4.35	1.07	1.71
Na_2O	1.27	0.75	1.64
K_2O	3.65	1.40	2.15
P_2O_5	0.09	0.10	0.10
H_2O	6.30	3.31	6.33
$CaCO_2$	0.39	56.73	1.52
$MgCO_3$	0.44	1.78	1.21
Available O	0.11	0.050	N.D.
Organic C	0.13	0.30	0.26
Organic N	0.016	0.017	—
Total	100.20	100.17	100.10
Total Fe_2O_2	9.02	3.14	4.98

tion process by biogenetic organic materials, may provide an answer. Turekian (1965) notes that in areas of the Atlantic Ocean such as the Cape Basin area, where $CaCO_3$ production is very high, high $CaCO_3$ contents are found in the sediments at all depths, whereas in areas such as the Argentine Basin, where $CaCO_3$ solution is rapid, low $CaCO_3$ contents are found.

A second interesting distribution "anomaly" is the high concentration of Ba relative to Al along the East Pacific Rise. The high concentration of barite ($BaSO_4$)

Table 13.22 Chemical and Physical Composition and Particle Sizes of Pelagic Sediments

	$CaCO_3$	Plaktonic Foraminifera (%)	Benthic Foraminifera (%)	Siliceous Remains (%)	Particles less than 0.05 mm diameter (%)
Radiolarian Ooze	4	3	0.1	54	40
Diatom Ooze	23	3	1.6	41	20
Globigerina Ooze	65	53	2.1	2	31
Pteropod Ooze	74	35	3.6	2	20
Red Clay	7	5	1	2	85

Table 13.23 Minerals in Pelagic Sediments (% samples less 0.05mm diameter particles analyzed where present)

	Radiolarian Ooze (%)	Diatom Ooze (%)	Globigerina Ooze (%)	Pteropod Ooze (%)	Red Clay (%)
Allogenic Minerals					
Amphibole	19	100	50	60	44
Feldspar	90	60	73	50	76
Magnetite	—	100	80	40	89
Mica	10	40	26	70	27
Augite	30	40	70	40	61
Quartz	—	80	42	70	30
Authigenic Minerals					
Glauconite	—	20	11	—	9
Manganese Grains	70	—	31	—	79
Palagonite	40	20	8	40	31
Phillipsite	40	—	—	—	14
Other					
Volcanic Glass	70	80	—	65	64
Pumice	20	40	—	30	49

in the sediments of this region has been attributed to volcanic injection (Arrhenius, 1966).

The atmospherically produced radionuclides and the fission products have half-lives short compared with the oceanic residence times of the particular elements. Hence, except under very special circumstances in which the radionuclides chance to be in particulate form (including organisms) or in estuaries of rivers carrying great burdens of activity (see, for example, Gross, 1966), they rarely reach the ocean floor to be incorporated into the sediments. The elements important in the radiochemistry of marine sediments are the members of the long-lived natural radioactive series (Picciotto, 1961).

Koczy (1956) has made quantitative estimates of the partitioning of radionuclides in the marine environment, including sediments based on a steady state model, while Burton (1965) has drawn the following qualitative conclusions concerning the origin and fate of the more important radionuclides in marine sediments:

U and Th^{232} from continental sources, carried into the sea by rivers; small volcanic contribution; lost from hydrosphere by sedimentation rather than decay

Th^{230} from decay of U; deposited chiefly in pelagic sediments

Ra^{223} largely from decay of Th^{230} in sediments; small river contribution lost from hydrosphere largely by radioactive decay.

Earlier I theorized that the larger part of the organic content of deep-sea sediments is absorbed on solid surfaces and that the amount of biomaterial surviving the long settling process is relatively small. We might expect this adsorbed nutrient material to support considerable animal life. But we must remember that the water in the marine sediment is stagnant, that the rapid processes of macroscopic oceanic circulation are no longer operative, and that the renewal of oxygen on which animal life depends becomes in turn directly dependent on the relatively slow microscopic process of molecular diffusion. In the first 5–10 cm of sediment, aerobic bacterial forms may still flourish, but the supply of available oxygen soon becomes depleted and anaerobic bacteria are much in evidence at depths of 40–60 cm, below which even they begin to die off (ZoBell, 1939). The important point to note is that anaerobic conditions commonly obtain in marine sediments with the characteristic chemical processes of sulfur metabolism [Gardner, 1973], H_2S and hydrocarbon production, and oxidation potential and metal valency changes mentioned earlier. Thus the chemistry within a marine sediment may be and often is entirely different from the chemistry of the free ocean water above the sediment. The sea-bottom interface therefore can mark the boundary between two drastically different chemical environments, and this can have consequences of practical significance; for example, metal structures erected in the sea often exhibit extensive corrosion at the mud line.

If we are to single out a few chemical features that distinguish marine from fresh water sediments, they are probably the greater importance of organic material in the latter and the generally far greater quantities of biogenic calcareous and siliceous material in the former. The chemical composition of fresh water sediments and their properties are highly dependent on the quantity and nature of the organic components of the mud. For example, about 98% of the nitrogen in Wisconsin lake sediments is in organic form, the small remainder being clay-fixed and exchangeable NH_4^+ (Keeney et al., 1970). We have already mentioned how the toxic heavy metal mercury rapidly disappears out of the water column into the underlying sediments where it is strongly held by combination with —S— and —SH groups in organic material. Highly organic sediments retain far more Hg than do sandy ones.

Dead leaves can accumulate rapidly on lake bottoms, especially if the body of water is highly eutrophic. Land plants also contribute everything ranging from pollen (Davis, 1968) to leaves to the bottom accumulation; the latter may be a significant source of phosophorus (Cowen and Lee, 1973). The diversity of pigments in lake sediments has been described by Sanger and Gorham (1970).

As in the case of marine environments, the organic content of fresh water sediments and the availability of oxygen (see discussion that follows) determine whether oxidizing or reducing conditions obtain, and this in turn affects the chemical species present (Gorham and Swaine, 1965). Curiously enough, the concentration of some species in lake sediments, such as P, Fe, and Mn, increase with increasing depth of the overlying water, possibly due to some kind of grading process involving the particles upon which these elements readily absorb (Delfino et al., 1969).

Inasmuch as copper sulfate is added in massive doses to drinking water supplies to control algal growth, one might well inquire what this practice does to the level of this toxic metal in lake sediments. While an untreated lake may have 7 to 85 mg Cu/kg dry sediment, treated lakes may have 100 to 1093 mg Cu/kg dry sediment, probably in the form of sulfides or basic carbonates. The accumulation of copper "is clearly having a generally deleterious effect on the whole biocoenosis of the lake in which it is enriched" (Hutchinson, 1957).

The chemical nature, whether oxidizing or reducing, of the sedimentary environment, I repeat, is determined by the availability of dissolved oxygen and its rate of uptake. Carey (1967) has studied oxygen uptake by marine sediments in shallow waters, Pamatmat and Banse (1969) in deeper waters, Edwards and Rolley (1965) have examined uptake by river muds, and Gardner and Lee (1965) by lake sediments. Not surprisingly, the most important variable appears to be physical mixing of the overlying waters.

Concern with lake and pond eutrophication (see Chapter 16) has focused a great deal of attention on soluble phosphorus, the concentration of phosphate in the water column and in the sediments, and on the rate of exchange between water and sediment (Livingston and Boykin, 1962; Harter, 1968; Frink, 1969; Wentz and Lee, 1969; Williams et al., 1970). Lake sediments can represent enormous reservoirs of phosphate; for example, interstitial water may contain 50 times more apparently soluble orthophosphate than overlying water. This sedimentary reservoir contrasts with the atmospheric reservoirs in the environmental cycles for other nutrient elements, such as C and N. It should be noted that this reservoir can be so great that a lake can continue to be eutorphic even if all man-made phosophorus inputs are cut off (Frink, 1967). As in the case of other species, the rate of P-exchange between water and sediment depends on such factors as the depth of the lake, the ratio of water volume to sediment surface area, thermal stratification, and physical mixing. Fortunately, as we will see, P-release from sediments is retarded by the oxidized microzone at the water/sediment boundary. Iron appears to control P-release, and P-mobilization is increased if iron is removed either by reduction of Fe(III) to Fe(II) and precipitation as FeS, as a consequence of anaerobic bacterial sulfate reduction or by oxidation of Fe(II) to Fe(III) by MnO_2 and precipitation of Fe(III) hydroxides (Lee, 1970). With respect to the former mechanism, let me say that it illustrates the importance of the sulfur cycle in fresh water, estuarine, and marine sedimentary chemistry, a subject reviewed in recent papers by Hallberg (1972, 1973).

Finally, we might note that fresh water sediments, unlike deep sea sediments, can frequently and rapidly change the morphology of the body of water. Lakes can become completely silted in or filled with organic debris, while river sediments build up extensive flood plains (Ritter et al., 1973).

Hitherto, we have confined our attention to terrigenous and biogenous marine sediments. Now we turn to halmyrogenous deposits (Table 13.16). These materials are composed of species that flocculated or precipitated out of the hydrosphere into the solid phase once in solution.

Although most of the $CaCO_3$ in marine sediments is biogenous, there is some inorganic precipitation of this mineral. But, inasmuch as most of the seas, except for surface waters, are undersaturated with respect to $CaCO_3$, its inorganic precipitation is rare and confined to special localities. If CO_2 is removed from seawater, for example by warming, aragonite and calcite may precipitate. In certain Bahamian sediments, the $CaCO_3$ may be as much as 65 to 75% the results of chemical precipitation (Cloud, 1965). Aragonite and calcite can also precipitate at grain boundaries in sands and on beach rock; and the calcite and dolomite in some Pacific pelagic red clays are believed to have precipitated from hydrothermal solutions of volcanic origin (Bonatti, 1966).

The existence of manganese-rich concretions on the floors of the world's three major oceans has been known since the *Challenger* expedition of 1873–1876, and in recent years, perhaps as an outgrowth of the talk of the economic feasibility of mining these deposits (Tooms, 1972), there has been a great deal of interest in the form of both research and speculation on these deposits.

The forms of ferromanganese deposits are highly varied. They range from thin stains and coatings on just about everything imaginable, including naval shell fragments, which can accumulate coatings "several millimeters thick in the span of a few tens of years" (Mero, 1965), to granules, to nodules the size of potatoes or cannonballs. The nodules can be so closely spaced as to form a mosiac. One spectacular formation is a continuous manganese pavement on the Blake Plateau some 5000 km^2 in area and grading into round Mn nodules on the south and east and into phosphorite nodules in the westerly direction (Pratt and McFarlin, 1966). In this area, the current of the Gulf Stream assures a fresh supply of seawater, while at the same time preventing burial of the concretions by sediment accumulation. Nodules are relatively scarce in the Atlantic Ocean compared with the Pacific. They are especially common in red clay regions. They are formed in the deep oceans, in shallow waters, in bays and seas, and even in fresh waters such as bogs, swamps, and lakes (Manheim, 1965). Table 13.24 gives some inkling of the enormous total quantities involved.

Table 13.24 Statistics on Total Manganese Nodule Quantities in the Pacific Ocean (Reproduced from Mero, *The Mineral Resources of the Sea*, Elsevier, 1965, with permission of the publisher and the author.)

Statistics	Eastern Region	Central Region	Western Region	Pacific Ocean
Number of Photographs	11	13	5	29
Concentration Estimates				
Maximum, g/cm^2	1.2	2.5	1.5	2.5
Minimum, g/cm^2	0.36	0.9	0.46	0.36
Average, g/cm^2	0.86	1.60	0.90	0.97
Number of Grab Samples	5	5	0	10
Concentration Estimates				
Maximum, g/cm^2	0.23	1.00	—	1.00
Minimum, g/cm^2	0.05	0.17	—	0.05
Average, g/cm^2	0.14	0.56	—	0.35
Number of Cores	24	33	5	62
Concentration Estimates				
Maximum, g/cm^2	2.3	3.8	1.2	3.8
Minimum, g/cm^2	0.1	0.5	0.4	0.1
Average, g/cm^2	0.89	1.71	0.82	1.32
Total of All Measurements	40	51	10	101
Average Concentrations of All Methods (g/cm^2)	0.78	1.45	0.86	1.12
Area in Region (km^2 $\times$ 10^6)	44.9	62.1	47.2	154.2
Tonnage of Nodules (billions of metric tons)	350	900	406	1,656

The details of the structure of the nodules are curious. At the center is commonly some foreign material such as a shark's tooth, and then the nodule is formed around this nucleus in a series of concentric rings, reminiscent of the growth rings of trees. The alternating light and dark layers correspond to accretion zones of high and low geothite (FeO·OH) content (Arrhenius, 1963).

Extensive analytic results have been obtained on the chemical composition of ferromanganese concretions. The nodules consist largely of hydrated oxides of manganese and iron (Table13.25). Mineral substances common in the environment are also incorporated, and the deep sea nodules have a pronounced ability to concentrate trace elements, such as Co, Ni, Zn, and Pb, from seawater. This is apparent in Table 13.26, which compares concretion concentrations with those in seawater, sediments,

Table 13.25 Chemical Composition of Ferromanganese Nodules

	Weight Percent			
Element	Minimum[a]	Maximum[a]	Mean[a]	Weight Percent Average[b]
Fe	2.4	26.6	14.0	14.58
Mn	8.2	52.2	24.2	22.06
Al	0.8	6.9	2.9	2.36
Ba	0.08	0.64	0.18	0.31
B	0.007	0.06	0.029	0.0295
Cd				0.0010
Ca	0.8	4.4	1.9	1.78
Cr	0.001	0.007	0.001	<0.001
Co	0.014	2.3	0.35	0.34
Cu	0.028	1.6	0.53	0.33
Ga	0.0002	0.003	0.001	0.0017
La	0.009	0.024	0.016	0.023
Pb	0.02	0.36	0.09	0.15
Mg	1.0	2.4	1.7	1.47
Mo	0.01	0.15	0.052	0.068
Ni	0.16	2.0	0.99	0.57
P				0.28
K	0.3	3.1	0.8	1.00
Sc	0.001	0.003	0.001	0.0013
Ag		0.0006	0.0003	0.0011
Na	1.5	4.7	2.6	2.55
Si	1.3	20.1	9.4	8.35
Sr	0.024	0.16	0.081	0.10
Tl				0.007
Ti	0.11	1.7	0.67	0.61
V	0.021	0.11	0.094	0.059
Y	0.016	0.045	0.033	0.012
Zn	0.04	0.08	0.047	0.35
Zr	0.009	0.12	0.063	0.034

[a] From Skornyakova et al. (1964).
[b] Averaged by Chester (1965) from several sources including Skornyakova et al. (1964).

Table 13.26 Comparison of the Mean Chemical Composition of Pacific Ocean Concretions with the Clarke Content of Elements in Seawater and the Crust Contents (wt. %) (From Skornyakova, Andruschchenko, and Fumina, 1964, with permission of Pergamon Press, Ltd.)

							Recent Sediments	
	Element[a]	Concretions	Seawater[b]	Crust	Basalts	Granites	Carbonate	Red Clay
B	—	0.029	4.5×10^{-4}	5×10^{-8}	0.0005	0.008	0.0055	0.023
Na	—	2.6	1.05	2.40	1.8	2.71	2.0	4.0
Mg	1.47	1.7	0.13	2.35	4.6	5.5	0.4	2.1
Al	2.87	2.9	1×10^{-6}	7.45	7.8	7.7	2.0	8.4
Si	8.05	9.4	—	—	23.0	33.0	3.2	25.0
P	0.18	—	varies	0.12	0.11	0.076	0.035	0.15
K	—	0.8	3.8×10^{-2}	2.35	0.83	3.6	0.29	2.5
Ca	1.81	1.9	4×10^{-2}	3.25	7.6	1.29	31.24	2.9
Sc	—	0.001	4×10^{-7}	6×10^{-4}	0.003	0.001	0.0002	0.0019
Ti	0.60	0.67	1×10^{-7}	0.61	1.3	0.23	0.077	0.46
V	—	0.054	3×10^{-7}	0.02	0.0230	0.0066	0.002	0.012

Cr	—	0.001	1×10^{-2}	0.03	0.017	0.0013	0.0011	0.009
Mn	19.18	24.2	1×10^{-7}	0.10	0.15	0.0046	0.1	0.67
Fe	12.16	14.0	1×10^{-7}	4.20	8.65	2.19	0.9	6.5
Co	0.36	0.35	4×10^{-8}	2×10^{-3}	0.0048	0.0004	0.0007	0.0074
Ni	0.47	0.99	2×10^{-7}	0.02	0.013	0.001	0.003	0.022
Cu	—	0.55	2×10^{-7}	0.01	0.0087	0.002	0.003	0.025
Zn	—	0.047	5×10^{-7}	0.02	0.01	0.005	0.0035	0.016
Ga	—	0.001	3×10^{9}	1×10^{-4}	0.0017	0.0017	0.0013	0.002
Sr	—	0.081	8×10^{-4}	0.035	0.046	0.027	0.2	0.018
Y	—	0.018	—	—	0.0021	0.0038	0.0042	0.009
Zr	—	0.063	—	0.025	0.014	0.016	0.002	0.015
Mo	—	0.052	1.2×10^{-2}	1×10^{-3}	1.1×10^{-4}	1.1×10^{-4}	0.0003	0.0027
Ag	—	0.0003	1.5×10^{-8}	1×10^{-5}	1.1×10^{-5}	4×10^{-5}	1×10^{-5}	1.1×10^{-5}
Ba	—	0.18	5×10^{-8}	0.05	0.033	0.063	0.019	0.25
La	—	0.016	3×10^{-8}	—	0.0015	0.005	0.001	0.011
Yb	—	0.0031	—	—	2×10^{-4}	4×10^{-4}	0.0001	0.0015
Pb	—	0.09	5×10^{-7}	1.6×10^{-3}	6×10^{-4}	0.0017	9×10^{-4}	0.008

[a] Mean according to the data of chemical analyses at the I.O.A.N. and the I.G.Ye.M.
[b] Mean according to Mero's X-ray spectral analyses.

and crustal rock. The sorption of metal ions appears to involve replacement of Mn in the MnO_2 lattice in the case of ions such as Co^{2+} and Zn^{2+} and replacement of surface protons in the case of ions such as Na^+ and Ca^{2+} (Loganathan and Burau, 1973). The composition of nodules is highly variable from place to place as the maximum, minimum, and mean value in Table 13.25 show. These variations are sometimes systematic and may be the result of different origins of the nodules. For example, mid-Pacific nodules tend to have high cobalt contents (Mn/Co < 300), while samples taken near Japan or off the coasts of North and South America tend to be Co-deficient with Mn/Co ratios greater than 300. The latter concretions may have formed slowly from dilute Mn solutions of continental origin, whereas the former may have been precipitated rapidly from volcanogenic waters (Arrhenius et al., 1964).

This brings us to the question of the origin of the nodules. The idea that organisms are involved has been persistent. Indeed the nodules do contain an appreciable amount of organic carbon, but so does just about every other solid surface in the sea; in fact, the C/N ratios are similar to those for fine-grained sediments (Manheim, 1965). So ubiquitous are organic materials, and even living organisms on solid surfaces in the sea, that their presence is certainly no evidence that they are playing a functional role. Until evidence to the contrary is forthcoming, the intermediacy of biological processes in nodule formation remains an unnecessary hypothesis. To a chemist the mystery which has been confected concerning nodule formation is itself something of a puzzle. Elements capable of existing in many valence states commonly have one in which they appear to be especially happy, and this is particularly true of manganese, whose valence ranges from +2 to +7 but which under ordinary circumstances has a very strong preference for $Mn(IV)O_2$. Manganese solutions all tend to oxidize slowly or reduce to this preferred state—the brown stains which form on flasks of old potassium permanganate solutions are an example familiar to every chemist. Barnes (1967) has given the reactions for nodule formation as

$$Mn^{2+} + \tfrac{1}{2}O_2 + 2OH^- \rightleftharpoons MnO_2 + H_2O \qquad [13.4]$$

$$5Mn^{2+} + 2O_2 + 10OH^- \rightleftharpoons 4MnO_2 \cdot Mn(OH)_2 \cdot 2H_2O + 2H_2O \qquad [13.5]$$

and

$$4MnO_2 \cdot Mn(OH)_2 \cdot 2H_2O + \tfrac{1}{2}O_2 \rightleftharpoons 5MnO_2 + 3H_2O \qquad [13.6]$$

He points out that the partial pressure of dissolved oxygen is a crucial parameter and proposes that the hydrostatic pressure may be controlling in the equilibria and kinetic processes, albeit in ways not yet thoroughly understood. The problem, then, becomes one, not so much of the chemistry of the precipitation process as of the source of the Mn^{2+}. There are, in all likelihood, two sources, both important: the free seawater above the sea–sediment interface, and the interstitial water in the sediment below (Manheim, 1965). Volcanic activity contributes to the manganese content of seawater, and Goldberg and Arrhenius (1958) have argued that the contribution of streams and rivers alone is adequate to account for the Mn and Fe contents of seawater and marine sediments. The probable sequence of events is as follows:

1. Slightly alkaline seawater (pH 8) rich in dissolved O_2 is saturated with Mn and Fe in solution.

2. Hydroxides of these elements precipitate and form colloidal particles, which as they aggregate and settle act as effective scavengers for trace elements.
3. Once on the sea floor, the particles are swept along until they intercept some surface, in particular conducting surfaces which can remove their charge, to which they can adhere.
4. The growing nodule can concentrate further trace elements by adsoprtion.
5. One hypothesis proposes that the growing nodules, by means unknown, appear to be rolled along the ocean floor until they reach a certain critical size, too heavy to be moved further; they are then covered over by sediment and preserved.

Growth rates as slow as 1 mm/100,000 years and as fast as 10 cm/100 years (on a conductive naval shell) have been determined (Manheim, 1965). However, Bender et al. (1966), utilizing the results of Th^{230} dating, have concluded that, generally speaking, the rate of Mn deposition is nearly constant over the world's oceans in both nodules and sediments, [and] they have developed a simple model which gives the rate of growth of the nodule.

In addition to its occurrence in the ferromanganese nodules, iron is also found in marine sedimentary ores deposited in bogs, marshes, and near-shore waters. (See the review by Borchert, 1965). The particular mineral obtained, whether limonite (60–70% Fe_2O_3), siderite ($FeCO_3$), or pyrite (FeS_2), depends on the redox potential (E_h) and the acidity (pH) of the environment (Figure [13.17]). Then, too, we must

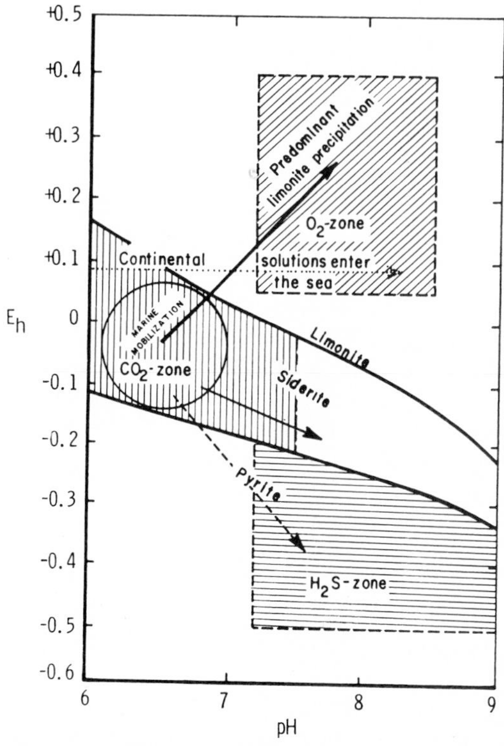

Figure 13.17. Precipitation of iron minerals in the marine environment (from Borchert, 1965, with pernission of Academic Press, Inc., and the author).

Table 13.27 Phosphorite Compositions (Reproduced from Mero, *The Mineral Resources of the Sea*, Elsevier, 1965, with permission of the publisher and author.)

A Chemical Composition of Phosphorite Nodules from the California Borderland Area[a]

Constituent	Geographic Location					
	Forty Mile Bank	Santa Monica Canyon	Redondo Canyon	Outer Banks	Thirty Mile Bank	Patton Escarpment
CaO	47.35	45.43	45.52	46.58	37.19	47.41
R_2O_3[b]	0.43	0.30	2.03	0.70	3.93	1.40
P_2O_5	29.56	29.19	28.96	29.09	22.43	29.66
CO_2	3.91	4.01	4.30	4.54	4.63	4.87
F	3.31	3.12	3.07	3.15	2.47	3.36
Organic	0.10	1.90	2.25	0.44	0.35	1.50
Insoluble in HCl	2.59	3.57	4.45	3.57	20.99	2.12
Totals	87.25	87.52	90.58	88.07	9.199	90.32

[a] Compositions in weight percentages. Remaining portions largely $MgCO_3$, H_2O, and soluble SiO_2.
[b] R denotes metals.

B Chemical Composition of Phosphorite from Various Locations of the World[a]

	Sea Floor		Land					
Constituent	Forty Mile Bank off California	Agulbas Bank off South Africa	Idaho	Florida	Russia	Curaco Islands	Tunisia	Morocco
CaO	47.4	37.3	48.0	36.4	27.9	50.0	44.3	51.6
R_2O_3	0.43	9.4	1.2	12.7	3.5	—	—	—
P_2O_5	29.6	22.7	32.3	31.2	17.9	37.9	29.9	32.1
CO_2	3.9	7.1	3.1	2.2	3.7	3.9	5.8	5.5
F	3.3	—	0.5	2.0	2.0	0.7	3.6	4.2
Organic	0.1	—	—	6.2	3.2	—	—	—
Totals	84.7	76.5	85.1	90.8	58.2	92.5	83.6	93.4

[a] Compositions in weight percentages.

not overlook the fact that iron oxides are responsible for the color of the red clay which represents such a large fraction of pelagic sediments.

Also we must not forget phosphorite deposits and, although they are biogenous in origin, I should like to mention them briefly here. Phosphorite or phosphate rock is a common continental shelf material in certain areas, notably off the coast of California. Phosphorite deposition is still occurring off the coast of Peru (Veeh et al., 1973). This mineral, like MnO_2, is found as nodules; Table 13.27 summarizes the results of some analyses, and Baturin (1971) has described the chemistry of their formation.

Let me say a few words here about the great deposits of salts laid down by the evaporation of shallow, isolated, ancient seas (and some not so ancient ones such as Great Salt Lake, Utah) (for reviews see Borchert, 1965; Kirkland and Evans, 1973). As the seawater evaporates various salts precipitate out from the brine (Table 13.28), the more insoluble first, the more soluble last, in a regular sequence beginning with limestone ($CaCO_3$) and dolomite ($CaMg(CO_3)_2$) and ending with salts such as the chlorides of Na and K (Figure 4.6).

Finally, recall that earlier we touched briefly on the deposition of mineral substances out of the hydrosphere and on ore enrichment (in the cases of Fe and Al) by water leaching. For all the very great importance of these subjects, we do not touch on them again. As metal-bearing minerals become more scarce, prospecting for them will be intensified, and one of the best guides for this activity will be scientific knowledge of the geochemistry of their formation.

Table 13.28 Salts Precipitated During the Concentration of Seawater (From Borchert, 1965, with permission of Academic Press, Inc.)

Density	Volume (liter)	$CaCO_3$ (g/kg)	$CaSO_4$ $2H_2O$ (g/kg)	NaCl (g/kg)	$MgCl_2$ (g/kg)	$MgSO_4$ (g/kg)	NaBr (g/kg)	KCl (g/kg)
1.026	1.000							
1.050	0.533	0.0642						
1.126	0.190	0.0530	0.5600					
1.202	0.112		0.9070					
1.214	0.095		0.0508	3.2614	0.0040	0.0078		
1.221	0.064		0.1476	9.6500	0.0130	0.0356		
1.236	0.039		0.0700	7.8960	0.0262	0.0434	0.0728	
1.257	0.0302		0.0144	2.6240	0.0174	0.0150	0.0358	
1.278	0.023			2.2720	0.0254	0.0240	0.0518	
1.307	0.0162			1.4040	0.5382	0.0274	0.0620	
Total Deposit		0.1172	1.7488	27.1074	0.6242	0.1532	0.2224	
Salts in last bittern (g)				2.5885	1.8545	3.1640	0.3300	0.5339
Sum (g)		0.1172	1.7488	29.6959	2.4787	3.3172	0.5524	0.5339

13.7 CHEMICAL EXCHANGE BETWEEN SEDIMENTS AND THE WATER COLUMN

In short range geological time, chemical mass transfer between the hydrosphere and lithosphere appears to be largely a one way street, for while sedimentation and silting proceed apace, chemical exchange from undisturbed sediments to the water column is generally slight and slow. Furthermore, exchange with undisturbed sediments appears to be restricted to only the upper 5 to 10 cm (Lee, 1970). Water circulation rather than chemical release usually controls water/sediment exchange (Lee, 1970). Sedimentary material is constantly dissolving but Lerman and Brunskill (1971) have found that in lakes this flux (0.02–0.04 cm/yr) is less for the major cations, Ca^{2+}, Mg^{2+}, Na^{+}, and K^{+} than typical deposition rates and accounts for less than 15 to 30% of these ions in lake water. A number of substances have been examined in detail, especially in fresh waters, including oxygen, phosphate, toxaphene, and mercury (for an old but fine review of chemical exchange between mud and waters in lakes see Mortimer (1942), and for a more recent review Lee (1970)). Many dangerous pollutants, including radionuclides, pesticides, and heavy metals rapidly disappear from the water column into the sediments where they are highly persistent (Howells et al., 1970). Toxaphene penetrates 0.6 to 1.1 cm/day of sediment and is very persistent (Veith and Lee, 1971), but reported copper penetrations as great as 8 ft in 20 yr have been questioned (Lee, 1970). Lead concentrations as high as 500 ppm have been observed in the first few centimeters of lake sediments, but below about 50 cm, the Pb levels are low and consant (Lee, 1970). In addition to copper and lead, other heavy metals recorded in lake sediment cores include Zn,Cd, Cr, and Ni (Iskandar and Keeney, 1974).

Three factors may account for the slowness of release of a particular chemical species:

1. Movement through the liquid boundary layer and through the interstitial water in the sediment is limited by relatively slow diffusional processes,
2. The chemical species may be strongly bound in the sediment, and
3. There may be a chemical barrier at the water–sediment boundary.

In many instances factors 1 and 3 may be more significant than 2, for in the case of lake sediments, the mineralogical components are probably usually *not* in equilibrium either with other sediment constituents or the water column (Lee, 1970). Many solutes have comparable diffusion coefficients in bulk liquid water, about 1×10^{-5} to 3×10^{-5} cm^3/sec (Horne, 1969), but the coefficient in the vicinal water in sediment pores may be considerably slower (see foregoing discussion), and, in addition, we must also consider the tortuosity of the diffusional paths in sediments (these two considerations are commonly lumped together).

The mechanisms whereby chemical species might be held in sediments are several:

1. Ion-exchange,
2. Surface sorption,
3. Strong chemical bonding with sedimentary material, and
4. Chemical reaction to form new insoluble compounds.

Earlier we mentioned chemical barriers; exchange from sediment to water column can be further hindered by the formation of an oxidized microzone (Gorham, 1958; Hayes, 1964) or layer of hydrous ferric oxides as ferrous ions diffusing through the anoxic sediment meet the oxygenated waters. At the same time that the layer of hydrous ferric oxides controls the rate of iron exchange between water and sediment, the formation of highly insoluble ferrous sulfide in anoxic muds may control the equilibrium quantities of iron in the aquatic system (Hutchinson, 1957). Hydrous ferric oxide is an excellent scavenger, and this layer is thus highly impervious. If you have a home aquarium, you may have noticed such a layer appearing on the glass walls as a brown stain at the surface of the bottom material. These chemical processes usually occur at the water–mud interface, but if the boundary between reducing and oxidizing conditions is in the water column, then iron precipitation may originate there. Thus Horne and Woernle (1972) found maxima in the Fe and Mn profiles of a coastal pond with a seasonally fluctuating anoxic zone. In addition to ferric oxides, the water–mud boundary layer may also contain ferric phosphates, and there is good experimental reason for believing that the phosphate flux from mud to the water column is controlled by this layer. If the oxidized microzone of the mud surface disappears, there is a concurrent increase in the flux of phosphate *and* ferrous ions into the water (Hutchinson, 1957). Upchurch et al. (1974) have observed a correlation between Fe and P in estuarine muds, and they conclude that P is held in suspended sediments by some type of Fe-inorganic P-complex of limited stability.

I have specified that exchange is slow between the *undisturbed* sediment and the water column, and before we discuss exchange and binding in further detail, I would like to mention ways in which exchange can be facilitated by disturbing the sediment or by certain "leaks." Bottom sediments can be disturbed by many means including burrowing organisms, fish movement, floods, slippage and earth movements, bubble formation in the muds (of CO, CO_2, CH_4; N_2 can also be produced from nitrate or nitrite in anoxic lake sediments, Chen et al., 1972), rafting of mats of benthic organisms by photosynthetically produced oxygen (Lee, 1970), bottom currents, dredging, and boat propellers. Clouds of sediment resuspended by organisms and bottom currents in the sea have been observed just above the water–sediment interface. It has been conjectured that some deep sediments are in a state of semisuspension, and Emery (1973) has found that the sediments in deep Canadian lakes behave like a "weak jelly" that becomes flocculent only when violently disturbed. In addition to burrowing organisms, bottom-feeding organisms can accelerate the escape of some pollutants, such as heavy metals, from bottom sediments into the water column and into food webs. Mercury, and possibly arsenic, have special mechanism by whch they can leak back into the environmental cycle, namely, bacteriological methylation. Some leaks are astonishing: more iron (1.9 metric tons Fe/yr) is transported out of Lake Mendota sediments by hatching flying insect larva than leaves the lake by its 100 ft^3/sec natural outlet discharge (Lee, 1970)!

Now let us return to the mechanisms that hold chemical species in sedimentary material.* We are inclined, I think, to look upon mineral substances as insoluble and highly inert chemically. From this opinion it is a short step to the conclusion that the terrigenous material in marine sediments should be chemically identical with the

* The following section is largely taken from Horne (1969).

parent mineral as it is found on land. But such need not be the case, for we have overlooked one very important type of mineral chemical reaction—ion exchange.

Remember that in aqueous solution the ionic species are in constant *dynamic* equilibrium. For example, we can write the simple metathesis

$$Na^+Cl^- + K^+ \rightleftharpoons Na^+ + K^+Cl^- \qquad [13.7]$$

but this equilibrium is so rapid and the formation constants, $K_{Na^+Cl^-}$ and $K_{K^+Cl^-}$, are so small that for all practical purposes it has no chemical consequences and we can just as well imagine that it does not exist. All the species in system [13.7] are highly mobile. If, however, we can somehow immobilize one of the participants in a simple metathesis such as equilibrium [13.7], then the situation becomes one of great theoretical and practical importance:

$$Na^+R + K^+ \rightleftharpoons Na^+ + K^+R, \qquad K = \frac{(Na^+)}{(K^+)} \qquad [13.8]$$

where R is now a solid material, an enormous polymeric substance either inorganic or organic in nature and, in our particular example, interspersed throughout with negative charge sites capable of attracting cations. Substances such as R, consisting of a solution-penetratable insoluble matrix studded with charge sites, are called ion exchangers.

Although ion exchange was known in antiquity and duly mentioned by Aristotle, the modern history of ion exchange technology began in the middle of the nineteenth century with the rediscovery of the cation-exchanging capability of certain soils. These natural materials, notably the zeolites, found application as water softeners:

$$2Na^+R + Ca^{2+} \text{ (responsible for water hardness)} \rightleftharpoons Ca^{2+}R_2 + 2Na^+ \qquad [13.9]$$

[We also must not forget that selective ion-permeable membranes play roles of enormous importance in the life processes.] Between World War I and World War II, synthetic, organic ion exchange resins came into being. These substances were found to be very useful in separating fission product elements and, as a by-product of the development of the bomb, a whole new area in solution physical chemistry was opened up, an area which was exploited with very great vigor in the postwar years.

But, in spite of this enormous effort, and the tremendous amount of information available about ion exchange (for a good review see Helfferich, 1962), the details of the microscopic processes which go in these materials are far from being adeqately understood. Hence here we content ourselves simply with a listing of the factors, first those of the exchanger and second those of the exchanging ions, which are almost certainly determining in the ion exchange process:

ACCESSIBILITY OF THE CHARGE SITES

If the interior of the material is not accessible, exchange is limited to surface sites. Accessibility depends on many factors: [the particle sites] the pore sizes, the tortuosity, the degree of cross-linking in synthetic resins, the extent of fractures in mineral substances. The pores may be large enough to admit small ions but block large ones. Many exchangers swell when moistened. Accessibility can affect the exchange equilibria, and it is of the greatest importance in determining the rates of ion exchange. [Fine river clays with large surface-to-volume ratios exchange 75% of their exchangeable cations in 3 sec, whereas coarse river gravels can take as long a 12 hr (Malcolm and Kennedy, 1970).]

NUMBER, DISTRIBUTION, AND NATURE OF THE CHARGE SITES

The ion exchange capacity of the material depends on the number of available charge sites. Particularly when dealing with polyvalent ions, the spacing of the charge sites and whatever limited mobility they might have are significant. The nature of the charge sites is half of the story in determining the strength of the mobile ion-charge site interaction and its ionic and/or covalent character. Cation exchangers have negative sites, anion exchangers positive ones.

DIELECTRIC CONSTANT AND WATER STRUCTURE

The strength of ion–ion interactions depends on the dielectric constant, yet there has never been agreement as to what is the value of the dielectric constant within the exchanger, if indeed the microscopic dielectric constant has any meaning at all in such a system. The water structure in the exchanger is certainly very important, and virtually nothing is known about it. If the pores are small and/or the concentration of charge sites and internal mobile electrolytes is large, most of the water will be "abnormal." I venture, on the basis of our earlier discussion, that the water near the uncharged portion of the matrix is in the same form as the hydrophobic hydration associated with interfaces, and that that near charge sites is of the coulombic hydration form.

The Mobile, Exchangeable Ion

CHARGE DENSITY

The most important characteristic of the exchangeable ion appears to be its charge density, that is, its charge divided by its *effective* surface area. From this the important conclusion can be drawn that the ion–ion exchanger–charge site interaction is, in the majority of instances, largely coulombic in nature.

CHARGE

Generally speaking, the greater its charge, the more strongly is an ion associated with the exchanger, and polyvalent ions tend to displace monovalent ions on the exchanger. Thus Na^+ is readily replaced by Ca^{2+}, so that sodium is rarely found in clays that have been in contact with calcium-containing waters. For a given series of comparable ions such as the alkali metal cations, the affinity of the exchange sites sometimes increases with decreasing crystal radii, but more commonly the affinity increases with decreasing radii of the hydrated ion (which is just the reverse of the order of crystal radii) giving the following series of affinities: $Cs > Rb > K > Na > Li$. Exceptions to these tendencies are not rare; the order for basic sodalite, for example, is $Na > Li > K$. Li^+ sometimes seems to be held more strongly than it ought to be, and the suggestion has been made that its very small crystal radius enables it to fit very tightly into mineral lattices. In addition to its direct effect on affinities, the charge also strongly influences the mobility of ions in the solution-filled matrix. Cations diffuse 100 to 10,000 times slower in synthetic organic ion exchange resin than in free solutions, and the retardation of their movement increases precipitously with increasing charge, the self-diffusion coefficients for Na^+, Zn^{2+}, and

La^{3+} being 2.4×10^{-7}, 11.2×10^{-8}, and 5.1×10^{-10} cm^2 sec^{-1}, respectively (Boyd and Soldano, 1953). But of course we must not forget that the higher-charged ions tend to be more heavily hydrated and thus bigger. The isolation of properties due specifically to ionic charge from those due to hydration is not an easy task. Typically the affinities follow the well-known Hofmeister or lyotropic series: $Sr^{2+} > Ca^{2+} > Mg^{2+} > H^+ > Cs^+ > Rb^+ > NH_4^+ > K^+ > Na^+ > Li^+$ for cations, and Cl^- $Br^- > NO_3^- > I^- > SCN^-$ for anions. In general, the charge effects are much more pronounced for cations than for anions.

Some progress has been made in calculating the absorption constants for monovalent cations on bentonite from a consideration of the differences in hydration and polarization energies (Shainberg and Kemper, 1967).

Montmorillonite, a familiar mineral constituent of marine sediments, consists ideally of silicate sheets [Figure 13.18] with 1–4 layers of water molecules sandwiched between them. If a higher-valence element is isomorphically replaced in the crystal lattice by a lower-valence one, such as Al(III) by Mg(II) or Si(IV) by Al(III), then the sheets are left with a negative charge. This negative charge is then neutralized by exchangeable cations (K^+ in Figure [13.18]). If one Si(IV) out of 400 is replaced by Al(III), an exchange capacity as large as 2 meq/100 g can result (Schofield and Samson, 1954). Kaolinite and a number of other minerals behave similarly. [Typical clay minerals have cation exchange capacities in the 10–75 meq/100 g range, but some such as montimorillonite and vermiculite have capacities around 100 meq/100 g (Lee, 1970).] The surface of the mineral substance can also acquire a charge through

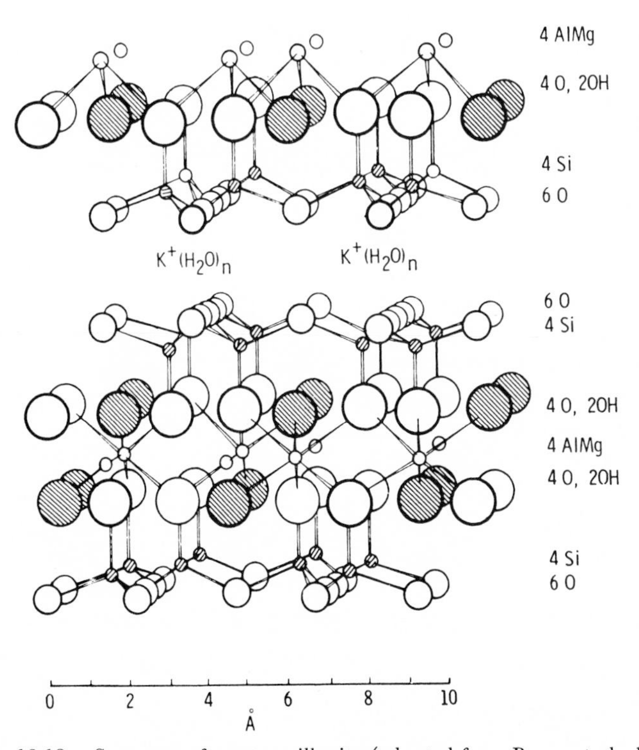

Figure 13.18. Structure of montmorillonite (adapted from Bragg et al., 1965).

lattic defects or by gross mechanical rupture of the crystal. In the latter case, depending on which bonds are broken, the charges can be either positive or negative and hydrolysis may occur:

$$—Si—O—Si—O—Si— \xrightarrow{\text{break}} —Si—O—Si—O— \underset{}{\overset{H_2O}{\rightleftharpoons}} —Si—O—Si—O—H \quad [13.10]$$

$$\rightleftharpoons$$

$$—Si—O—Si—O^-H^+$$

and, as a final complication, the charged surface probably builds up an electrical double layer (Wayman, 1967).

Ionic strength, temperature, and pressure all affect ion exchange processes. In the case of a synthetic organic ion exchange resin, pressure exerted no discernible influence on the K^+–H^+ exchange, but for the exchange Sr^{3+}–H^+ the resin's preference for the more highly charged ion increased by about 3% at 1000 atm. This increase was attributed to the greater importance of pressure-induced dehydration of the initially more heavily hydrated ion (Horne et al., 1964). [In addition to exchanging ions, clay minerals also sorb silica (Siever and Woodford, 1973) and organic matter (Lee, 1970), including organic nutrients (Button, 1969) and possibly even organic free radicals (Wauchope and Haque, 1971). Clay minerals tend to sorb some anions such as phosphate, but not others such as nitrate—herein lies the reason why groundwaters tend to transport the latter but not the former (Lee, 1970).]

Ion exchange is, I repeat, the most important chemical exchange across the sea-sea bottom interface, especially if we include ionic lattice substitution as well as surface effects, and it has only begun to receive the attention it deserves. Once it enters the sea, a terrigenous material suffers chemical change at a rate proportional to its permeability by seawater as a result of adsorption and ion exchange. In the case of the more prevalent and exchangeable cations the changes can be quite marked in very short times, geologically speaking (Table 13.29). Even after it has settled to the ocean floor, the sedimentary particle can continue to interact with its environment, in particular to absorb trace elements from seawater. Mg and Na are abundant in marine sediments, the former because of its strong affinity for exchangers, the latter by virtue of its high concentration in seawater. Cs appears to be enriched in marine sediments (Welby, 1958).

Although Berner (1965) has reported dolomitization of Pacific atoll material by the reaction

$$Mg^{2+} + 2CaCO_3 \rightleftharpoons CaMg(CO_3)_2 + Ca^{2+} \quad [13.11]$$

he found no evidence for any adsorption or exchange of Mg^{2+} or Sr^{2+} with fine-grained carbonate sediments, and he attributed the absence of any exchange to the blockage of ion transport across grain boundaries by adsorbed Mg^{2+} or coatings of organic matter (Berner, 1966). Yet, although he found their exchange capacity (7–30 meq/100 g) less noticeable than for the corresponding materials in soils, Zaitseva (1959) states that the organic content of Pacific sediment clays is too small to affect their exchange capacity.

This contrasts with river sediments where the organic content is so important in determining their chemistry that it has been used as the basis of a classification scheme (Ballinger and McKee, 1971). In the lower reaches of the Hudson River the bottom material is about 5 to 6% humic material, 50% silt, and 20 to 45% clay. Because of the predominately reducing conditions in this mud, most of the ion exchange sites

Table 13.29 Changes in the Composition of Clay Minerals on Exposure to Seawater (wt % oxides) (From Chester, 1965, with permission of Academic Press, Inc. and the author)

	Illitic Clay		Kaolinitic Clay		Montmorillonitic Clay[a]		
	Original Composition	After 54–60 Months in Seawater	Original Composition	After 54–60 Months in Seawater	Original Composition	After 54–60 Months in Seawater	
SiO_2	52.75	52.55	45.38	45.47	58.56	50.8	59.3
Al_2O_2	24.83	24.75	38.24	38.38	18.25	16.3	18.5
Fe_2O_3	4.12	4.10	0.19	0.20	2.82	2.5	2.8
FeO	0.26	0.25	0.00	0.00	0.00	0.0	0.0
MgO	2.29	3.16	0.16	0.58	2.30	12.8	3.3
CaO	0.32	0.10	0.28	0.12	0.41	0.7	0.3
Na_2O	0.35	0.21	0.60	0.27	2.60	0.6	1.3
K_2O	5.71	6.48	0.42	0.27	0.55	3.9	0.6
TiO_2	0.62	0.61	0.76	0.77	0.08	0.08	0.09
H_2O+	7.94	7.85	13.18	13.34	7.35	8.8	7.3

[a] The montmorillonitic clay was diagenetically modified to a chlorite type (A) and an illitic type (B).

are occupied by hydrogen; thus the cation exchange capacity of the sediment is high even in the brackish reaches of the river where there are ample cations in the overlaying water. The organic matter in the mud appears to be responsible for more than 65% of its cation exchange capacity. This mud is rich in ferrous iron and in manganese (0.72 μg/g wet mud compared to 0.01 μg/ml in the ambient river water) (Howells et al., 1970).

In view of water structuring at interfaces, ion exclusion, anaerobic bacterial action, and ion exchange and absorption processes, there is little reason for expecting that the composition of the interstitial water in marine sediments will bear any resemblance to the composition of the seawater through which the material settled or to that of the free seawater above the sediment at the time of sampling. Berner's (1966) observation that the Mg^{2+}/Cl^- and Sr^{2+}/Cl^- ratios for the interstitial water in a fine-grained carbonate sediment did not differ appreciably from the corresponding ratios in seawater may very well be the exception rather than the rule. [Recent studies (Somagajulu and Church, 1973) have also revealed radionuclide concentrations in interstitial water much higher than for seawater.]

Recent marine muds taken off Cape Cod showed higher salinities and lower pH's than the free ocean water near the bottom in the same area, and the salt content decreased with increasing penetration of the sediment (Figure [13.19]) (Siever et al., 1961). The investigators attributed the lower pH values to the production of CO_2 from the bacterial decomposition of organic matter, but the high salinities are more problematical. A decrease in electrolyte concentration could be due to ion exclusion, but the fact that the salinities are higher than that of the free seawater immediately above the sediment is not readily reconciled with such an explanation. The investigators advance two hypotheses: (*a*) the clay acts as a semipermeable system which on

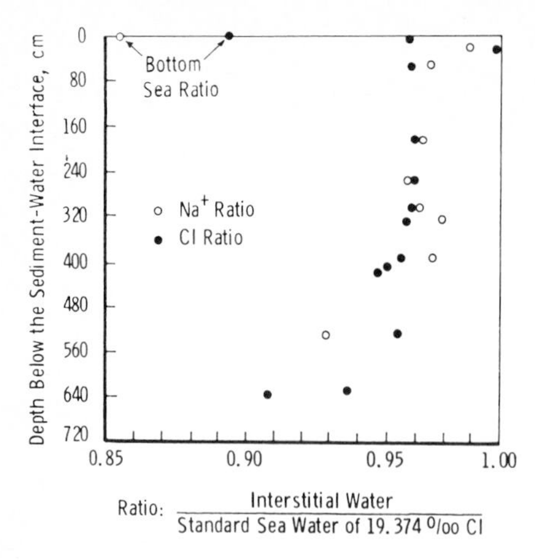

Figure 13.19. Composition of interstitial water in a marine mud taken off Cape Cod, Massachusettts (from Siever et al., © 1961 by the American Association for the Advancement of Science, with permission of the AAAS).

compression expresses the water upwards but retains the hydrated ions; or (*b*) when the seawater was trapped in the accumulating sediment, there was less fresh water dilution from the Gulf of Maine than at the present time. In the latter case, the interstitial water represents a paleosalinity, that is, water trapped sometime in the geological past from ancient seas whose composition was different from that of the modern water [see also Spears, 1973]. As I implied earlier, my own view is that structural and chemical processes are much more important and should completely erase all hint of the composition of the original seawater, except possibly in special circumstances where the compositional changes in the free seawater in a given area have been drastic and very rapid. Another puzzling feature of Figure [13.19] is the failure of the compositional ratios in the interstitial water to approach the free water values as the surface of the sedimentary deposit is approached. As the surface is approached, compaction is less and the forces of ionic diffusion should tend to reduce any concentration gradients. A much more detailed examination of a greater number of cores of different lithology from a variety of sites (Siever et al., 1965) gave similar results: significant salinity differences which again are attributed to compaction phenomena (although the authors describe an experiment showing that squeezing a clay does not affect water composition!) and a lowering of pH. The changes in NaCl content with penetration in deep-sea cores have also been attributed to compaction (Wangersky, 1967). I might mention in passing that H_2S production from anaerobic decay also would contribute to the increased acidity. Mg^{2+} tends to be depleted in the interstitial water, possibly because of uptake by chlorides and incipient dolomitization, and K^+ is [depleted and Ca^{2+} enriched (Sayles et al., 1973). Earlier studies had shown K^+ enrichment in recent marine (Mangelsdorf et al., 1969) and estuarine sediments (Dobbins et al., 1970)].

Dissolved SiO_2 is also enriched, in this case by the solution of diatoms. Some typical composition and profiles of cores taken from the Gulf of California are shown in Figure [13.20]. Further data on the composition of pore solution can be found in Goldberg and Arrhenius (1958) [and data for Gulf of Mexico cores can be found in Manheim and Sayles (1970)]. Berner (1964) has developed an idealized model for the dissolved SO_4^{2-} distribution in recent sediments. Ammonia and phosphorus are also enriched in the interstitial water, sometimes by a factor or 10 or more, as one might

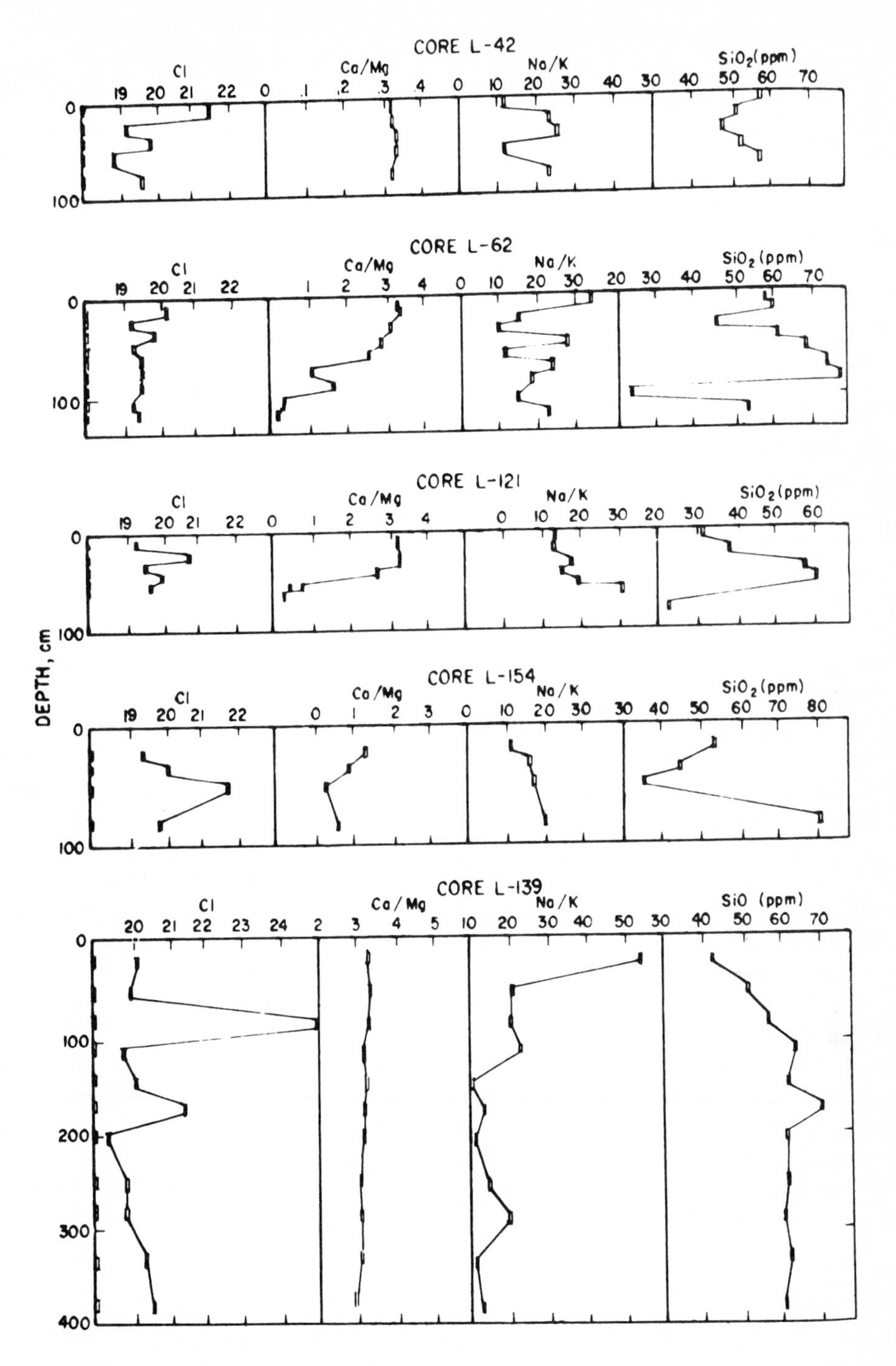

Figure 13.20. Compositional profiles of Gulf of California cores (reprinted from Siever et al. in the *Journal of Geology* (1965) by permission of The University of Chicago Press).

expect, from the decay of organic material (Brujewicz and Zaitseva, 1959). On the other hand, anaerobic utilization of SO_4^{2-} depletes its concentration from about 55 mg equiv/kg in seawater to 2–3 mg equiv/kg in the interstitial water (Shishkina, 1959). [In anoxic marine sediments Mg replaces Fe in clay minerals (Drever, 1971) Berner et al. (1970) has examined carbonate alkalinity in anoxic pore waters.] Particularly extreme conditions are found in the Black and Azov seas, where the interstitial water is characterized by a high content of organic carbon and a very low redox potential. Here the ammonia concentration is more than 4 times greater, the phosphorus concentration is less, and the silicon concentration is about the same as in the free water (Brujewicz and Zaitseva, 1959). The total salt and, especially, the carbonate CO_2 concentrations are higher (Marchenko, 1966). Interstitial waters from the Azov Sea showed a depletion of SO_4^{2-}, an increase in Br^-, considerable accumulation of ammonia, and an increase in the Mg^{2+} with increasing penetration of the sediment (Shishkina, 1961). The way in which the salts are removed from the core can give misleading results, thus casting futher uncertainty about the question of the composition of the interstitial water. For example, Wangersky (1967) has noted that the suspension of air-dried sedimentary material in distilled water removes the Ca^{2+} but not the Mg^{2+} originally in the water. [More recently the whole question of pore water composition was plummeted into something of an uproar by the observation that the temperature effects upon squeezing the water out of the core sample can easily cause significant shift in exchange equilibria (Bishoff et al., 1970).]

Perturbation of the chemical content of seawater by solid material, as we have mentioned, can be very rapid, Keuhl and Mann (1966) studied changes in the interstitial water on a beach during a tide and found that the concentration of sulfur, calcium, and magnesium decreased and that of nitrate and phosphate increased.

Friedman (1965) has found that the interstitial water from a core taken as a part of Project Mohole ". . . is depleted in deuterium by an amount that is very large in comparison with the variations in deuterium in the modern oceans," and no entirely satisfactory explanation for this very fascinating phenomenon is evident. My own suspicion is that the isotopic fractionation arises from the water structuring near interfaces, but one thing is clear, namely, here, as so often elsewhere in the chemistry of the sea–sediment interface, further experimental and theoretical research is badly needed. If undertaken, it holds promise of interesting results. [Among the more esoteric chemical processes in sediments is the racemization of biogenic amino acids (Kvenvolden et al., 1973).]

Our knowledge of the chemistry of the sea–sediment interaction is obviously in its infancy, and speculation about the processes responsible for the chemical differences in interstitial water remains for the greater part exactly that—speculation. Yet in order to end this section on a note of hope I should like to mention a quantitative model which has been applied with some measure of success, for it illustrates that, complicated though these systems may be, they are nevertheless amenable to attack. The mathematical model of Anikouchine (1967), which, unlike earlier models, takes into consideration the movement of interstitial water caused by compaction, yields a relatively simple expression for the distribution of dissolved species in interstitial waters. Although it incorporates a number of simplifying assumptions such as constant rate of sedimentation, homogeneity of the sediment, attainment of steady state, constant [diffusivity], negligible temperature and pressure effects, [the model] yields a very impressive fit of the data of Siever et al. (1965) for the concentration of dissolved

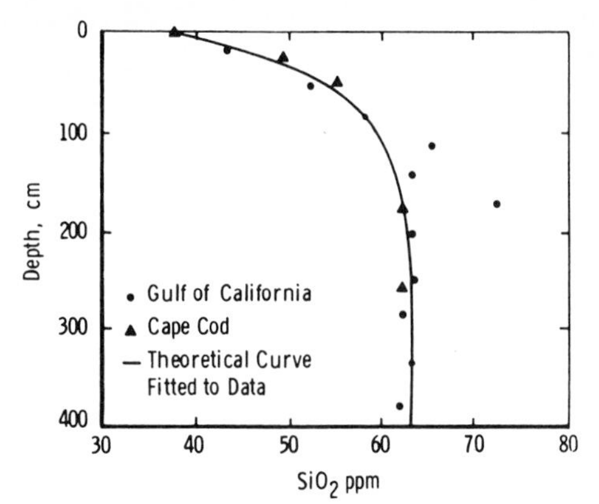

Figure 13.21. Comparison of experimental and theoretical distribution of dissolved silicate in the interstitial water in marine sediments (from Anikouchine, *J. Geophys. Res.*, *72*, 505 (1967), © the American Geophysical Union, with permission of the AGU and the author).

silica in interstitial water [Figure 13.21]. Application of this model also leads to the suggestion that the source of manganese nodules is the interstitial water of the underlying sediment.

BIBLIOGRAPHY

J. L. Anderson and J. A. Quinn, *J. Chem. Soc., Faraday Trans.*, *1*, *68*, 744 (1972).

W. A. Anikouchine, *J. Geophys. Res.*, *72*, 505 (1967).

A. A. Antoniou, *J. Phys. Chem.*, *68*, 2754 (1964).

G. O. S. Arrhenius, in M. N. Hill (ed.), *The Sea*, Interscience, New York, 1963.

G. O. S. Arrhenius et al., *Science*, *144*, 172 (1964).

G. O. S. Arrhenius, in P. M. Hurley (ed.), *Advances in Earth Science*, M.I.T. Press, Cambridge, 1966.

D. G. Ballinger and G. D. McKee, *J. Water Pollut. Control Fed.*, *43*, 216 (1971).

I. Barnes, *Science*, *168*, 973 (1970).

S. S. Barnes, *Science*, *157*, 63 (1967).

T. F. W. Barth, *Theoretical Petrology*, Wiley, New York, 1952.

G. N. Baturin, *Nature*, *232*, 61 (1971).

R. E. Beck and J. S. Schultz, *Science*, *170*, 1302 (1970).

M. L. Bender, *Science*, *151*, 325 (1966).

R. A. Berner, *Geochim. Cosmochim. Acta*, *28*, 1497 (1964).

R. A. Berner, *Science*, *147*, 1297 (1965).

R. A. Berner, *Amer. J. Sci.*, *264*, 1 (1966).

R. A. Berner et al., *Limnol. Oceanogr.*, *15*, 544 (1970).

A. J. Bijl, *Rec. Trav. Chim.*, *46*, 767 (1927).

J. L. Bischoff et al., *Science*, *167*, 1245 (1970).

J. O'M. Bockris et al., *Proc. Roy. Soc. London Ser. A*, *274*, 55 (1963).

E. Bonatti, *Science*, *153*, 534 (1966).

E. Bonatti and G. O. S. Arrhenius, *Mar. Geol.*, *3*, 337 (1965).

H. Borchert, in J. P. Riley and G. Skirrow (eds.), *Chemical Oceanography*, Academic, London, 1965.

G. E. Boyd and B. A. Soldano, *J. Amer. Chem. Soc.*, *75*, 6105 (1953).

L. Bragg, G. F. Claringbull, and W. H. Taylor, *Crystal Structure of Minerals*, Cornell Univ. Press, Ithaca, N.Y., 1965.

G. W. Brindley and G. L. Millhollen, *Science*, *152*, 1385 (1966).

S. W. Brujewicz and E. D. Zaitseva, in *Preprint Internat. Oceanogr. Congr.*, Amer. Assoc. Adv. Sci., Washington, D.C., 1959.

J. D. Burton, in J. P. Riley and G. Skirrow (eds.), *Chemical Oceanography*, Academic, London, 1965.

D. K. Button, *Limnol. Oceanogr.*, *14*, 95 (1969).

A. G. Carey, *Bull. Bingham Oceanogr. Collect.*, *19*, 136 (1967).

R. Castaing and K. Fredriksson, *Geochim. Cosmochim. Acta*, *14*, 114 (1958).

R. S. Chahal and R. D. Miller, *Brit. J. Appl. Phys.*, *16*, 231 (1965).

R. L. Chen et al,, *J. Environ. Qual.*, *1*, 166 (1972).

C. G. Chase and E. C. Perry, Jr., *Science*, *177*, 992 (1972).

R. Chester, in J. P. Riley and G. Skirrow (eds.), *Chemical Oceanography*, Academic, London, 1965.

W. F. Claussen, *Science*, *156*, 1226 (1967).

P. E. Cloud, Jr., in J. P. Riley and G. Skirrow (eds.), *Chemical Oceanography*, Academic, London, 1965.

B. E. Conway, *Theory and Principles of Electrode Processes*, Ronald, New York, 1965.

E. J. Conway, *Proc. Roy. Irish Acad.*, *B48*, 119 (1942); *B48*, 161 (1943).

W. F. Cowen and G. F. Lee, *Environ. Sci. Tech.*, *7*, 853 (1973).

R. W. Creekmore and C. N. Reilley, *Anal. Chem.*, *42*, 725 (1970).

M. B. Davis, *Science*, *162*, 796 (1958).

P. Delahay, *Double Layer and Electrode Kinetics*, Interscience, New York, 1965.

J. J. Delfino et al., *Environ. Sci. Tech.*, *3*, 1189 (1969).

B. V. Derjaguin and N. V. Churaev, *Nature*, *244*, 430 (1973).

B. V. Derjaguin and N. N. Fedyakin, *Dokl. Akad. Nauk. SSSR*, *147*, 403 (1962).

M. A. V. Devanthan and B. Tilak, *Chem. Revs.*, *65*, 635 (1965).

D. A. Dobbins et al., *Environ. Sci. Tech.*, *4*, 743 (1970).

J. E. Drever, *Science*, *172*, 1334 (1971).

W. Drost-Hansen, "The Resistivity of Brines in Capillaries, "Pan-Amer. Petrol. Corp., Res. Dept. Rept. No. F59-G-2 (July 10, 1959).

W. Drost-Hansen, *Ind. Eng. Chem.*, *57*(4), 18 (1965).

W. Drost-Hansen, *J. Colloid Interface Sci.*, *25*, 131 (1967).

W. Drost-Hansen, *Ind. Eng. Chem.*, *61*, 10 (1969).

A. W. Duff, *Phil. Mag.*, *9*, 685 (1905).

G. R. Dutt and P. F. Low, *Soil Sci.*, *93*, 195 (1962).

R. W. Edwards and C. J. Rolley, *J. Ecol.*, *53*, 1 (1965).

G. C. Egleson and C. W. Querio, *Environ. Sci. Tech.*, *3*, 367 (1969).

A. R. Emery, *Science*, *181*, 655 (1973).

J. C. Eriksson, *Acta Chem. Scand.*, *16*, 2199 (1962).

G. Ewing, *J. Mar. Res.*, *9*, 161 (1950).

J. H. Feth et al., U.S. Geol. Surv. Water Supply Paper 1535-I, 170 (1964).

N. H. Fletcher, *Phil. Mag.*, *7*, 255 (1962).

J. Frenkel, *Kinetic Theory of Liquids*, Chapter 6, Dover, New York, 1955.

I. Friedman, *J. Geophys. Res.*, *70*, 4066 (1965).

C. R. Frink, *Environ. Sci. Tech.*, *1*, 425 (1967).

C. R. Frink, *Soil Sci. Soc. Amer. Proc.*, *33*, 326 (1969).

L. R. Gardner, *Geochim. Cosmochim. Acta*, *37*, 53 (1973).

W. Gardner and G. F. Lee, *J. Air Water Pollut.*, *9*, 553 (1965).

R. M. Garrels, *Science*, *148*, 69 (1965).

R. M. Garrels and F. T. Mackenzie, in W. Stumm (ed.), *Equilibrium Concepts in Natural Water Systems*, Adv. Chem. Ser. No. 67, Amer. Chem. Soc., Washington, D.C., 1967.

W. D. Garrett, "The Organic Chemical Composition of the Ocean Surface," U. S. Nav. Res. Lab. Rept. No. 6201 (1964).

J. H. B. George and R. A. Courant, *J. Phys. Chem.*, *71*, 246 (1967).

W. Glass and B. C. Heezen, Paper Presented 48th Ann. Meet. Amer. Geophys. Union, Washington, D.C., 1967.

J. J. Goering and D. W. Menzel, *Deep-Sea Res.*, *12*, 839 (1965).

E. D. Goldberg, in L. H. Ahrens et al. (eds.), *Physics and Chemistry of the Earth*, Chapter 4, Pergamon, London, 1961.

E. D. Goldberg, in M. N. Hill (ed.), *The Sea*, Interscience, New York, 1963.

E. D. Goldberg and G. O. S. Arrhenius, *Geochim. Cosmochim. Acta*, *13*, 153 (1958).

E. D. Goldberg et al., *J. Geophys. Res.*, *67*, 4209 (1963).

R. J. Good, *J. Phys. Chem.*, *61*, 810 (1957).

E. Gorham, *Limnol. Oceanogr.*, *3*, 291 (1958).

E. Gorham and D. J. Swaine, *Limnol. Oceanogr.*, *10*, 268 (1965).

M. G. Gross, *J. Geophys. Res.*, *71*, 2017 (1966).

J. Guastalla, *J. Chim. Phys.*, *44*, 306 (1947).

B. D. Gunter, *Geochim. Cosmochim. Acta*, *37*, 495 (1973).

R. O. Hallberg, *Mineral Deposits*, 7, 189 (1972).

R. O. Hallberg, *Oikos*, Suppl. 15 (1973).

F. H. Hambleton et al., *Trans. Faraday Soc.*, *62*, 795 (1966).

E. L. Hamilton, *Bull. Geol. Soc. Amer.*, *70*, 1399 (1959).

E. L. Hamilton, *J. Geophys. Res.*, *69*, 4257 (1964).

R. A. Hart, *Nature*, *243*, 76 (1973).

F. R. Hayes, *Ocean. Mar. Biol. Ann. Rev.*, *2*, 121 (1964).

R. D. Harter, *Soil Sci. Soc. Amer. Proc.*, *32*, 514 (1968).

F. Helfferich, *Ion Exchange*, McGraw-Hill, New York, 1962.

J. D. Hem, in S. Licht (ed.), *Medical Hydrology*, Licht, New Haven, 1963.

J. C. Henniker, *Rev. Mod. Phys.*, *21*, 322 (1949).

H. D. Holland, *Proc. Nat. Acad. Sci. U.S.A.*, *53*, 1173 (1965).

R. A. Horne, *J. Phys. Chem.*, *70*, 1335 (1966).

R. A. Horne, *Marine Chemistry*, Wiley-Interscience, New York, 1969.

R. A. Horne, *Surv. Prog. Chem.*, *4*, 1 (1968).

R. A. Horne, *J. Geophys. Res.*, *77*, 5170 (1972).

R. A. Horne et al., *J. Phys. Chem.*, *68*, 2578 (1964).

R. A. Horne et al., *Electrochim. Acta*, *13*, 397 (1968).
R. A. Horne et al., *J. Phys. Chem.*, *73*, 2782 (1969).
R. A. Horne and C. H. Woernle, *Chem. Geol.*, *9*, 299 (1972).
G. P. Howells et al., *Environ. Sci. Tech.*, *4*, 26 (1970).
W. H. Huang and D. L. Vogler, *Nature Phys. Sci.*, *235*, 157 (1972).
D. C. Hurd, *Geochim. Cosmochim. Acta*, *37*, 2257 (1973).
G. E. Hutchinson, *Treatise on Limnology*, Wiley, New York, 1957.
I. K. Iskandar and D. R. Keeney, *Environ. Sci. Tech.*, *8*, 165 (1974).
N. L. Jarvis, *J. Colloid Sci.*, *17*, 512 (1962).
N. L. Jarvis et al., *Limnol. Oceanogr.*, *12*, 88 (1967).
N. M. Johnson et al., *Science*, *177*, 514 (1972).
D. R. Keeney et al., *J. Water Pollut. Control Fed.*, *42*, 411 (1970).
Y. K. Kharaka and F. A. F. Berry, *Geochim. Cosmochim. Acta*, *37*, 1899 (1973).
D. W. Kirkland and R. Evans (eds.), *Marine Evaporites*, Wiley, London, 1973.
F. F. Koczy, *Deep-Sea Res.*, *3*, 93 (1956).
J. Korringa, *Bull. Amer. Phys. Soc.*, *1*, Ser. II, 216 (1956).
J. R. Kramer, in W. Stumm (ed.), *Equilibrium Concepts in Natural Water Systems*, Adv. Chem. Ser. No. 67, Amer. Chem. Soc., Washington, D.C., 1967.
K. B. Krauskopf, *Geochim. Cosmochim. Acta*, *9*, 1 (1956).
H. Kuehl and H. Mann, *Heglolaender Wiss. Meeresuntersuch.*, *13*, 238 (1966).
P. H. Kuenen, in J. P. Riley and G. Skirrow (eds.), *Chemical Oceanography*, Academic, London, 1965.
J. L. Kulp and D. R. Carr, *J. Geol.*, *60*, 148 (1952).
K. A. Kvenvolden, *Geochim. Cosmochim. Acta*, *37*, 2215 (1973).
V. K. LaMer and T. W. Healy, *Science*, *148*, 658 (1965).
G. F. Lee, "Factors Affecting the Transfer of Materials Between Water and Sediments," Univ. Wisconsin Water Resources Center Lit. Rev. No. 1 (July 1970).
A. Lerman and G. J. Brunskill, *Limnol. Oceanogr.*, *16*, 880 (1971).
D. Leythanenser, *Geochim. Cosmochim. Acta*, *37*, 113 (1973).
D. A. Livingston, in M. Fleischer (ed.), *Data of Geochemistry*, U.S. Geol. Surv. Prof. Paper No. 440-G, U.S. Gov. Print. Off., Washington, D.C., 1963.
D. A. Livingston and J. C. Boykin, *Limnol. Oceanogr.*, *7*, 57 (1962).
S. Licht (ed.), *Medical Hydrology*, Licht, New Haven, 1963.
P. Loganathan and R. G. Burau, *Geochim. Cosmochim. Acta*, *37*, 1277 (1973).
C. R. Longwell and R. F. Flint, *Introduction to Physical Geology*, Wiley, New York, 1955.
P. F. Low, *Adv. Agronomy*, *13*, 269 (1961).
W. C. Lu et al., *J. Chem. Phys.*, *46*, 1075 (1967).
R. W. Maatman, *J. Phys. Chem.*, *69*, 3196 (1965).
H. H. Macey, *Trans. Brit. Ceram. Soc.*, *41*, 73 (1942).
F. MacIntyre, *Sci. Amer.*, *223*, 104 (November 1970).
R. L. Malcolm and V. C. Kennedy, *J. Water Pollut. Control Fed.*, *42*, R153 (1970).
P. C. Mangelsdorf et al., *Science*, *165*, 171 (1969).
F. T. Manheim, in D. K. Schink and J. T. Corless (eds.), *Symposium on Marine Geology*, Univ. Rhode Island Occas. Pub. No. 3, Kingston, 1965.

F. T. Manheim, *Earth Planet, Sci. Lett.*, *9*, 307 (1970).
F. T. Manheim and F. L. Sayles, *Science*, *170*, 57 (1970).
A. S. Marchenko, *Sov. Geol.*, *9*, 155 (1966).
B. Mason, *Principles of Geochemistry*, Wiley, New York, 1966.
E. Mazor and R. O. Fournier, *Geochim. Cosmochim. Acta*, *37*, 515 (1973).
C. M. McAuliffe, *Chem. Geol.*, *4*, 225 (1969).
B. L. McConnell et al., *J. Phys. Chem.*, *68*, 2941 (1964).
J. L. Mero, *The Mineral Resources of the Sea*, Elsevier, Amsterdam, 1965.
C. H. Mortimer, *J. Ecol.*, *29*, 280 (1941); *30*, 147 (1942).
D. W. Muenow, *Geochim. Cosmochim. Acta*, *37*, 1551 (1973).
J. J. Naughton, *Geochim. Cosmochim. Acta*, *37*, 1163 (1973).
G. D. Nicolls, in J. P. Riley and G. Skirrow (eds.), *Chemical Oceanography*, Academic, London, 1965.
J. A. O'Keefe (ed.), *Tektites*, Univ. Chicago Press, Chicago, 1963.
S. Olejnik and J. W. White, *Nature Phys. Sci.*, *236*, 15 (1972).
B. G. Oliver, *Environ. Sci. Tech.*, *7*, 135 (1973).
L. Onsager and N. Samaras, *J. Chem. Phys.*, *2*, 528 (1934).
J. D. Oster and P. F. Low, *Proc. Soil Sci. Soc. Amer.*, *27*, 369 (1963).
M. M. Pamatmat and K. Banse, *Limnol. Oceanogr.*, *14*, 250 (1969).
C. Parravano et al., *Science*, *155*, 1535 (1967).
T. R. Parsons, *Prog. Oceanogr.*, *1*, 203 (1964).
H. Pettersson and K. Fredriksson, *Pacif. Sci.*, *12*, 71 (1958).
E. E. Picciotto, in M. Sears (ed.), *Oceanography*, Amer. Assoc. Adv. Sci., Pub. No. 67, Washington, D.C., 1961.
R. M. Pratt and P. F. McFarlin, *Science*, *151*, 1080 (1966).
R. M. Pytkowicz and D. N. Connors, *Science*, *144*, 840 (1964).
J. E. B. Randles, *Disc. Faraday Soc.*, *24*, 194 (1957).
J. E. B. Randles, in P. Delahay (ed.), *Electrochemistry*, Wiley, New York, 1963.
R. Revelle, *J. Mar. Res.*, *14*, 446 (1955).
D. F. Ritter et al., *Science*, *179*, 374 (1973).
P. E. Rosenberg, *Geochim. Cosmochim. Acta*, *37*, 109 (1973).
W. W. Rubey, *Bull. Geol. Soc. Amer.*, *62*, 1111 (1951).
F. L. Sayles et al., *Science*, *181*, 154 (1973).
J. E. Sanger and E. Gorham, *Limnol. Oceanogr.*, *15*, 59 (1970).
R. K. Schofield and H. R. Samson, *Disc. Faraday Soc.*, *18*, 135 (1954).
J. A. Schufle, *Chem. Ind.*, 690 (1965).
J. A. Schufle and M. Venugopalan, *J. Geophys. Res.*, *72*, 3271 (1967).
J. A. Schufle and N-T. Yu, *J. Colloid Interface Sci.*, *26*, 395 (1968).
I. Shainberg and W. D. Kemper, *Soil Sci.*, 1034 (1967).
O. V. Shishkina, "On the Salt Composition of the Marine Interstitial Waters," Preprint Internat. Oceanogr. Congr., Amer. Assoc. Adv. Sci., Washington, D.C., 1959.
O. V. Shishkina, *Okeanol.*, *1*, 646 (1961).
R. Siever et al., *Science*, *134*, 1071 (1961).
R. Siever et al., *J. Geol.*, *73*, 39 (1965).

R. Siever and N. Woodford, *Geochim. Cosmochim. Acta*, *37*, 1851 (1973).

L. G. Sillen, in M. Sears (ed.), *Oceanography*, Amer. Assoc. Adv. Sci., Pub. No. 67, Washington, D.C., 1961.

L. G. Sillen, *Arkiv Kemi*, *24*, 431 (1965); *25*, 159 (1966).

L. G. Sillen, *Chem. Britain*, *3*, 291 (1967a).

L. G. Sillen, *Science*, *156*, 1189 (1976b).

N. S. Skornyakova et al., *Deep-Sea Res.*, *11*, 93 (1964).

W. H. Slabaugh and A. D. Stump, *J. Phys. Chem.*, *68*, 1251 (1964); *J. Geophys. Res.*, *69*, 4773 (1964).

M. V. Slonimskaya and T. M. Raitburd, *Dokl. Akad. Nauk SSSR*, *162*, 176 (1965).

E. J. Slowinski, Jr. et al., *J. Phys. Chem.*, *61*, 808 (1957).

B. L. K. Somayajulu and T. M. Church, *J. Geophys. Res.*, *78*, 4529 (1973).

D. A. Spears, *Geochim. Cosmochim. Acta*, *37*, 77 (1923).

W. Stumm and J. J. Morgan, *Aquatic Chemistry*, Wiley-Interscience, New York, 1970.

H. U. Sverdrup et al., *The Oceans*, Prentice-Hall, Englewood Cliffs, 1942.

H. T. Tien, *J. Phys. Chem.*, *69*, 350 (1965).

J. S. Tooms, *Endeavour*, *31*, 113 (1972).

K. K. Turekian, in J. P. Riley and G. Skirrow (eds.), *Chemical Oceanography*, Academic, New York, 1965.

J. B. Upchurch et al., *Environ. Sci. Tech.*, *8*, 56 (1974).

H. H. Veeh et al., *Science*, *181*, 844 (1973).

G. D. Veith and G. F. Lee, *Environ. Sci. Tech.*, *5*, 230 (1971).

S. K. Wakeel and J. P. Riley, *Geochim. Cosmochim. Acta*, *25*, 110 (1961).

P. J. Wangersky, *J. Geol.*, *75*, 733 (1967).

R. D. Wauchope and R. Haque, *Nature Phys. Sci.*, *233*, 141 (1971).

C. H. Wayman, in S. D. Faust and J. V. Hunter (eds.), *Principles and Application of Water Chemistry*, Wiley, New York, 1967.

C. E. Weaver and L. D. Pollard, *The Chemistry of Clay Minerals*, Elsevier, Amsterdam, 1973.

R. R. Weiler and A. A. Mills, *Deep-Sea Res.*, *12*, 511 (1965).

A. Weiss, *Kolloid-Z.*, *Z. Polymer.*, *211*, 94 (1966).

G. W. Welby, *J. Sediment. Petrol.*, *28*, 431 (1958).

D. A. Wentz and G. F. Lee, *Environ. Sci. Tech.*, *3*, 754 (1969).

E. Wicke, *Angew. Chem.*, *5*, 106 (1966).

J. D. H. Williams et al., *Environ. Sci. Tech.*, *4*, 517 (1970).

Z. W. Wolkowa, *Kolloid-Z.*, *67*, 280 (1934).

K. Yamakoshi and Y. Tazawa, *Nature*, *233*, 542 (1971).

N-T. Yu, M.S. Thesis, New Mexico Highlands Univ., 1966.

E. D. Zaitseva, "Exchange Capacity and Exchangeable Cations in Sea Sediments," Preprint Int. Oceanogr. Congr., Amer. Assoc. Adv. Sci., Washington, D.C., 1959.

J. R. Zimmerman and W. G. Brittin, *J. Phys. Chem.*, *61*, 1328 (1957).

C. E. ZoBell, in P. D. Trask (ed.), *Recent Marine Sediments*, Amer. Assoc. Petroleum Geologists, Tulsa, Oklahoma, 1939.

ADDITIONAL READING

R. A. Berner, *Principles of Chemical Sedimentology*, McGraw-Hill, New York, 1971.

P. W. Birkeland, *Pedology, Weathering and Geomorphological Research*, Oxford Univ. Press, New York, 1974.

H. Blatt, G. Middleton, and R. Murray, *Origin of Sedimentary Rocks*, Prentice-Hall, Englewood Cliffs, 1972.

O. Braitsch, *Salt Deposits, Their Origin and Composition*, Springer-Verlag, New York, 1971.

P. J. Brancazio and A. G. W. Cameron (eds.), *The Origin and Evolution of Atmospheres and Oceans*, Wiley, New York, 1964.

W. S. Broecker, *Chemical Equilibria in the Earth*, McGraw-Hill, New York, 1970.

D. Carroll, *Rock Weathering*, Plenum, New York, 1970.

E. T. Degens, *Geochemistry of Sediments*, Prentice-Hall, Englewood Cliffs, 1965.

R. M. Garrels, *Mineral Equilibria*, Harper & Brothers, New York, 1960.

R. M. Garrels and C. L. Christ, *Solutions, Minerals, and Equilibria*, Harper & Row, New York, 1965.

R. M. Garrels and F. T. Mackenzie, *Evolution of Sedimentary Rocks*, Norton, New York, 1971.

H. H. H. Jellinek (ed.), *Water Structure at the Water-Polymer Interface*, Plenum, New York, 1972.

Z. Kukal, *Geology of Recent Sediments*, Academic, New York, 1971.

G. Larsen and G. V. Chilingar (eds.), *Diagenesis in Sediments*, Elsevier, Amsterdam, 1967.

F. J. Pettijohn, *Sedimentary Rocks*, Harper & Row, New York, 1957.

W. Stumm (ed.), *Equilibrium Concepts in Natural Water Systems*, Adv. Chem. Ser. No. 67, Amer. Chem. Soc., Washington, D.C., 1967.

14

LITHOSPHERE–BIOSPHERE INTERACTIONS

14.1 INTRODUCTION

Any worthwhile book should have surprises, not only for the reader, but for the writer. Writing as well as reading should be an educational process. Preparing this book, I am happy to report, has had some surprises for me, and two of these surprises are discussed in this chapter. The first of these surprises for me was the realization of the importance of soil chemistry in the chemistry of our environment. Of course I had some awareness of the significance of soil quality in the production of food, but initially I had little inkling of the importance of the role played by the soil in atmospheric chemistry, specifically its significance as both a source and a sink of atmospheric gases (in addition to Chapter 6 see also Abeles et al., 1971; Schütz et al., 1970; Bates and Hayes, 1967; Inman et al., 1971). Then, too, like most people, I associated pollution with industrial activity. It thus came as a second shock to discover that agriculture is so serious an offender. The abatement of industrial pollution may be costly, but the means are at our disposal for its accomplishment. Agricultural pollution, in part because it is so widespread rather than concentrated in urban areas, is not so easily stopped. Some idealists like to imagine that the problem of the pollution of our environment can somehow be solved simply by turning back the technological clock. What can be more antitechnological and "natural" than growing grains and vegetables? But the production of food has become just as contingent on new technology as the production of automobiles. And this technology is necessary to support our swollen populations.

The biosphere may be looked upon as a sort of transducer (Figure 14.1) taking carbon from the atmosphere, fixing it into organic material by photosynthesis, and finally adding it to the lithosphere. In the hydrosphere, the biosphere also adds a great deal of material to the lithosphere by fixing ions in solution into relatively insoluble skeletal materials such as $CaCO_3$ and SiO_2, which, when the organisms die, sink to the bottom and accumulate in the lithosphere. In many respects, the interaction between biosphere and lithosphere is a one way street with the patient lithosphere getting the better part of the bargain. Nevertheless, the biosphere could not exist without the essential trace elements and nutrients supplied by the lithosphere, so neither party in this great exchange has any grounds for complaint.

14.2 SOIL CHEMISTRY

Soil is the product of physical–chemical erosion and weathering combined with biological action. Soil must contain organic matter; thus in desert regions, erosion

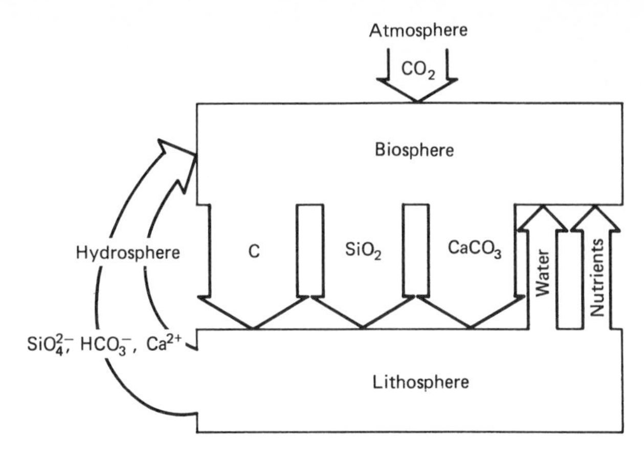

Figure 14.1. Biosphere/lithosphere interactions.

and weathering may fail to produce a true soil. Of the substances and mixtures in our environment, even if we subtract its content of living organisms, only the viable biosphere exceeds soil in complexity. In keeping with this complexity, soils are exceedingly variable, and the classification of soils, a chore pioneered by the Soviets, is a science in itself. Soil has been aptly described as "the great bridge between the inanimate and the living" (Longwell and Flint, 1955). Soil, that is to say that thin layer of the lithospheric mantle that is capable of supporting rooted plants, is, like water, a basic economic resource, and as in the case of water, "man's culture and civilization can be closely corrleated with the pattern of soil fertility" (Mason, 1966).

Once the hard crustal material, or regolith, broken up by erosion and weathering, begins to accumulate organic matter, it is on its way to becoming a soil, and gradually as the soil "matures" a "soil profile" (Figure 14.2) develops. In tropical areas where there is plenty of organic matter, plenty of rain, and high temperatures to hurry matters along, this development may be rapid, but in colder climates, the process may take thounsands of years. As we have noted, under some conditions a soil may never develop or a truncated profile may form with some of its feature zones or "horizons" ill formed or missing entirely. The soil profile is developed into three of these horizons (Figure 14.2). The deepest of these horizons, the C-horizon, is the zone of slightly weathered parent material; it may be bed rock, alluvial or glacial material, or even the soil of a preceding cycle. The B-horizon is more extensively weathered and may be rich in materials such as clays and iron oxides that have been leached from the A-horizon above and transported downward by the percolation of rain water. Trace elements are also commonly enriched in the B-horizon (Swaine and Mitchell, 1960; Mitchell, 1964). Figure 14.3 shows the range of trace element compositions of normal soils. In many places such as near highways and smelters, very high trace metal concentrations in the soil have been observed (Lagerwerff, 1967; John et al., 1972; Klein, 1972; Buchauer, 1973; Gish and Christensen, 1973). Some trace elements (Fe, Mn, B, Cu, Zn) in soil are crucial to the growth and health of plants and the animals that feed on them. Much of the soil in the United States is deficient in boron. Grass will grow on Co-deficient soil, but animals that feed on this grass will sicken. As one proceeds from rock to soil (Figure 14.4), there is generally

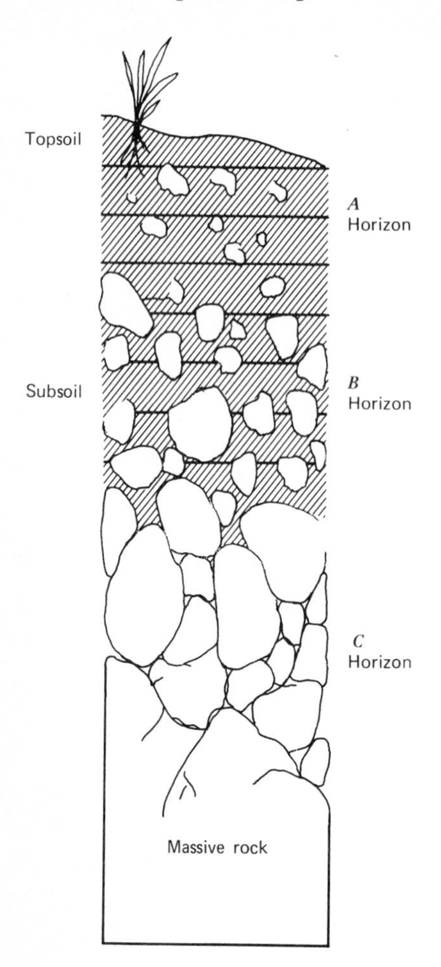

Figure 14.2. Typical soil profile.

a decrease of the more soluble elements such as Ca, Mg, Na, and K, a somewhat lesser decrease of Al and Fe, and a relative increase in Si and, of course, organic matter.

Figure 14.5 summarizes the general nature of the major chemical substances in soil. The dark-colored organic matter or "humus" formed from the decomposition of plant, and to a far lesser degree, animal material (Figure 14.6) for all its enormous importance is very imperfectly understood and little is known about its exact chemical composition (Goring and Hamaker, 1972; Schnitzer and Khan, 1972; Allison, 1973). Humus hastens mineral degradation and increases the mobilization of metals (Baker, 1973). The organic content of soils ranges from about 2 to 95% (Miyake, 1965) and is commonly at the lower end of this scale, that is, less than 5%. N, P, and S are vital to plant growth and 30 to 60% of the total P in surface soil, most of the S, and almost all of the N are contained in the organic matter.

The gases in soil are those of the atmosphere, but with considerably more CO_2 and slightly less O_2. Good soil ventilation is necessary to prevent biological and chemical processes from depleting the O_2, and aeration depends on such factors as water drainage, soil texture, and tillage.

Primary minerals, such as silica, limestone, and dolomites commonly account for 70% of the soil mass. Present in lesser amounts, but of greater importance in deter-

10,000
1000
100
10
1
0.1
Metal content (ppm)
Ti
Mn
Ba
Zr
Sr
C
Rb
V
Ni
Zn
Li
Pb
Cu
B
Ga
As
Co
Sn
Be
Mo
Se
Ag

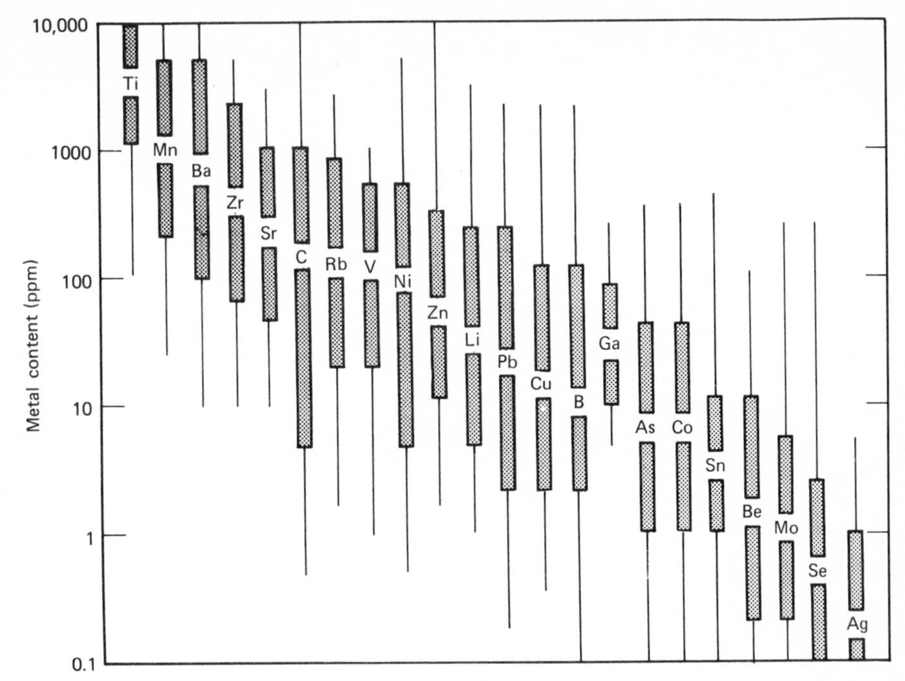

Figure 14.3 Range of trace element levels in normal soils. The thin lines indicate more unusual values (from Mason, 1966).

Organic material
Rock
Si
Soil
Marked loss of Ca, Mg, Na, and K
Smaller loss of Al and Fe

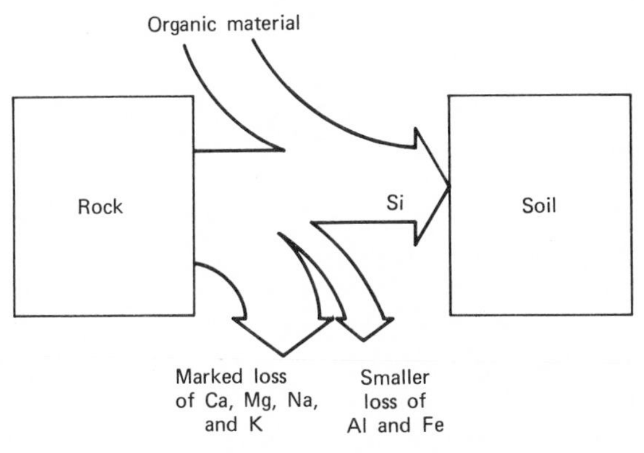

Figure 14.4 The transition from rock to soil.

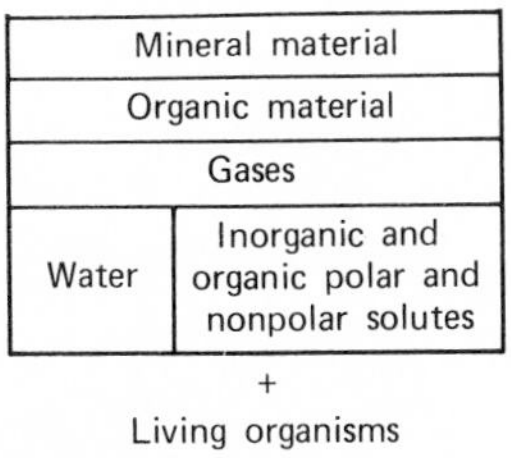

Mineral material
Organic material
Gases
Water
Inorganic and organic polar and nonpolar solutes
+
Living organisms

Figure 14.5. Types of chemical substances present in soils.

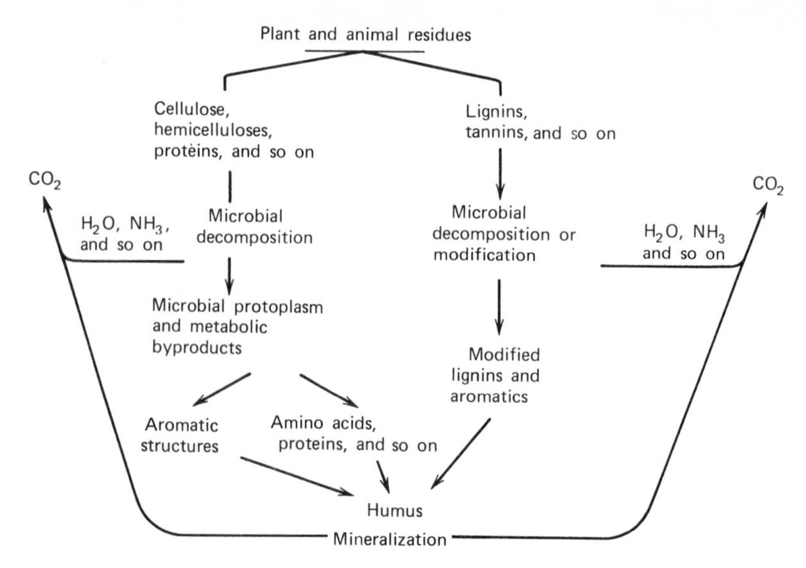

Figure 14.6. Organic matter decomposition and the formation of humic substances in soil (from Stevenson, 1964, with permission of the editor).

mining soil properties, are the secondary or clay minerals. Their particle size range includes the colloidal (Toth, 1964). Since they result from the weathering away of bases these colloidal clay particles are negatively charged. They are thus responsible for the cation exchange properties of soils (Wiklander, 1964), which ranges from about 2 to 40 meq/100-g sample, including the exchange of protons. The acidity of soils ranges from pH 4.8 for virgin, humid, temperate regions to pH 8.3 and even greater for high alkalinity soils. Highly alkaline and/or saline soils create agricultural problems in many parts of the world (Seatz and Peterson, 1964), and irrigation unbalanced by proper drainage can accelerate the latter to form "hardpan." Soils also exchange anions, but anion exchange is less well understood than cation exchange, and mineral fracture surfaces appear to be involved.

The humus (Figure 14.7) is also partially present in colloidal form. The mineral and other colloids have very high surface-to-volume ratios, and they are highly hygroscopic and hydrated, which brings us to one of the most important components of soil—water. The several forms of water in soils are summarized in Figure 14.8. Biological activity is strongly dependent on the form of soil water; For example, it is greater in clay soils than in clay gels (Low et al., 1968).

Finally, we come to the organisms so important in soil chemistry (Alexander, 1961; McLaren, 1967, 1971; Bonner, 1970; Raper, 1970; Hattori, 1973). These include both autotropic (that get their C from CO_2) and heterotropic (that get their C from organic matter) bacteria (Doetsch and Cook, 1973), fungi (Domsch and Gams, 1973), actinomycetes, algae, and protazoa (See Figure 11.1). Generally, highly productive soils will have an abundance of microorganisms, whereas less productive soils support smaller populations. In addition to the decomposition of organic material, these organisms can play an important role in mineral breakdown and soil formation. One fungus, for example, produces citric acid capable of solubilizing the Si, Al, Fe, and Mg in basalt, granite, and quartzite (Silverman and Munoz, 1970). And it is the

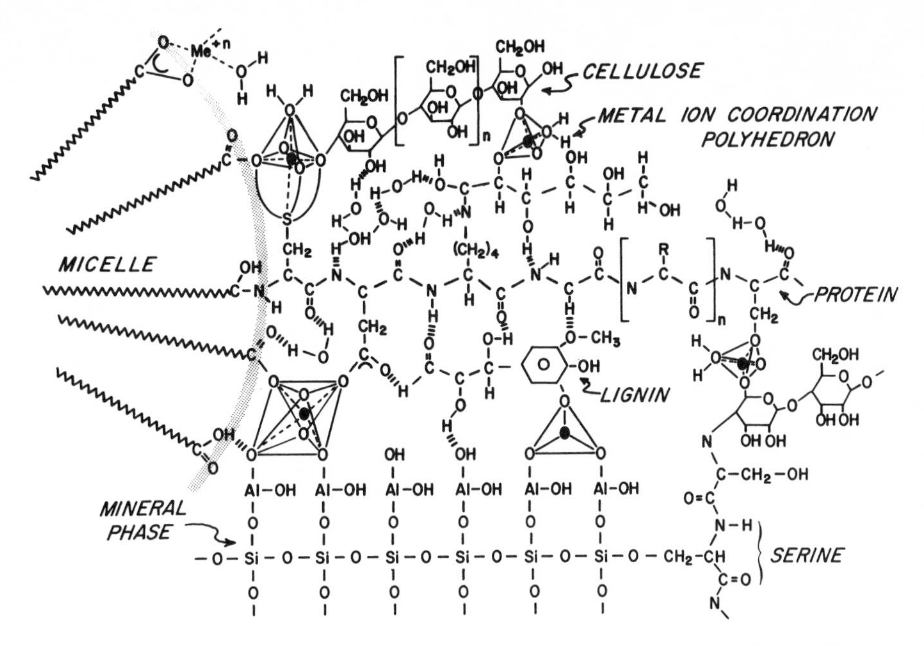

Figure 14.7. Schematic representation of a sedimentary humic substance (from Mopper and Degens, 1972).

Hygroscopic moisture	Thin film on soil surfaces	Unavailable to plants
Capillary moisture	Interstitial water held between particle by surface tension	Available to plants
Gravitational water (free, excess, or surface water)	Difference between precipitation and a soil's capacity to hold water	Can "drown" a soil and be inimical to plants

Figure 14.8. The types of soil water.

handiwork of these organisms that is responsible for nitrogen fixation by ammonification and nitrification and which places soil so squarely into the great cycles of the crucial elements N, S, and P (Figure 14.9). Organic-N must be converted to inorganic-N before it can be utilized by growing crops. Leguminous plant crops have nodules of bacteria (rhizobia) that form on their roots and that are capable of fixing atmospheric N_2 into useable forms. Under favorable conditions, such crops can fix 50 to 150 lb atmos-N/acre. Organisms, we might note, are not an unmixed blessing; soil in the humid tropics tend to be much less fertile, not only as a consequence of the rapid leaching action of warm rain to form lateritic soils (discussed in an earlier chapter, see also UNESCO, 1971; Gomez-Pompa et al., 1972; Mohr et al., 1973), but also as a consequence of the very rapid removal of organic matter by microorganisms. We must not forget the role of larger organisms such as earthworms (and in the tropics, termites, Lee and Wood, 1971) as processers of litter and turners of soil.

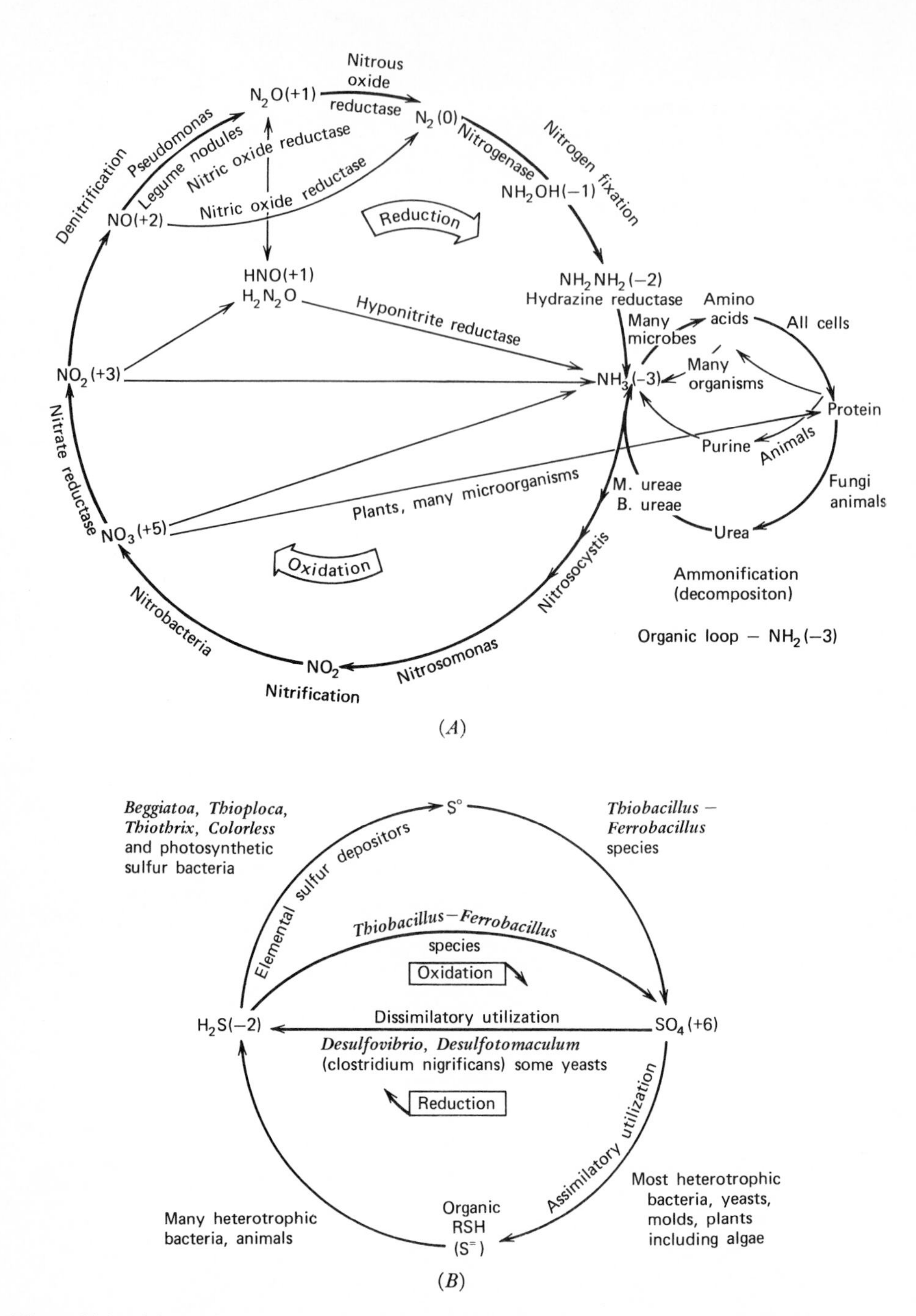

Figure 14.9. Major element cycles. *A*, The nitrogen cycle; *B*, the sulfur cycle; and *C*, the phosphorus cycle (from Odum, E. P., *Fundamentals of Ecology*, 2nd ed., Saunders, Philadelphia, 1959. With permission of the publisher and author).

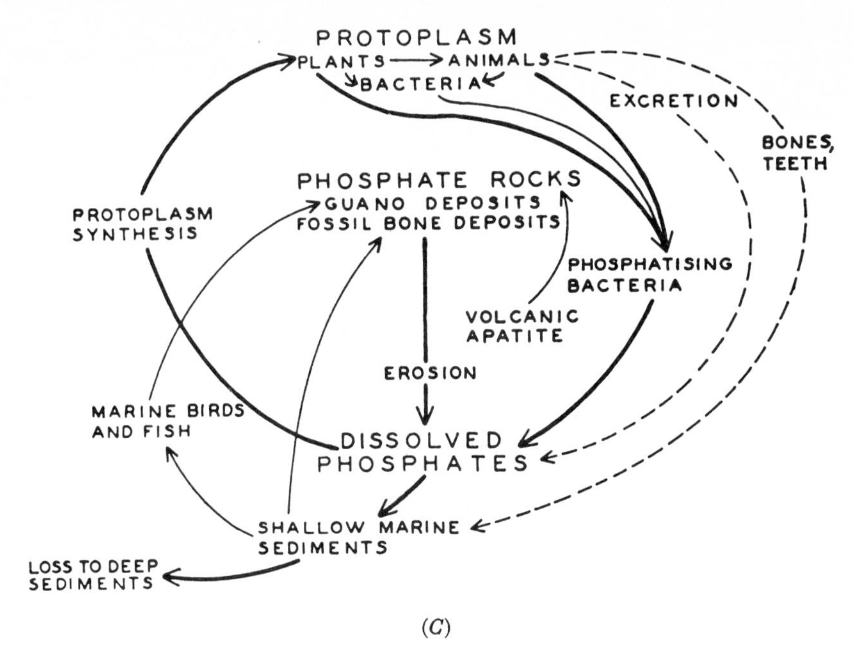

(*C*)

Figure 14.9. (*continued*).

14.3 THE PROSPECTS OF AGRICULTURE

In earlier chapters, we touched upon the problem of the exhaustion of the non-renewable resources of the lithosphere. Now we must move into an even more critical area, the exhaustion of the supposedly renewable resources of the biosphere. Ironically, although it is perhaps the most crucial problem of our environment, we treat it only briefly here, since it lies largely outside of the specific title of this book.

Ten years ago, Dr. R. Ewell (1964) warned

> The world is on the threshold of the biggest famine in history . . . the production of food in [Asia, Africa, and Latin America] is lagging the population growth. . . . This is the greatest . . . problem in the history of the world. And it is almost here. . . . Such a famine will be of massive proportions affecting hundreds of millions, possibly even billions of persons. If this happens, as appears probable, it will be the most colossal catastrophe in history.

His voice is only one of a chorus of experts, and now 10 yr later their dreadful prophecies are coming to pass. Famine is on the march in North Africa and in the Indian subcontinent. Most of the people of the world are poorly fed, millions are suffering from malnutrition at this very moment throughout the world. And although the problems of the developed countries are trivial in comparison, shortages and an inflationary increase in food costs are being experienced in the United States. For the moment in America, we are preoccupied with an energy crisis (perhaps real, perhaps partially imaginary), but hot on its heels is a very real food crisis. The energy crisis threatens our luxuries and comforts; the food crisis threatens our necessities and our survival.

What are the prospects that agricultural technology can hold back the hand of famine? Improved technology can double or triple grain production (Brown, 1970*a*, *b*)

Table 14.1 Major Grain Yields in Developing and Developed Nations in 1961–1962

	Wheat Yield (lb/acre)	Rice Yield (lb/acre)
Brazil	475	1020
India	760	870
Australia	1010	3840
United States	1440	2210
Japan	2450	2720

(in Asia about 70% of the total food supply comes from grain), thus the yield of the soil is very much greater in the developed than the undeveloped nations (Table 14.1). Over the past 20 yr, crops and livestock production in the United States has climbed steadily (Figure 14.10), even outpacing population growth, so that, as in the cases of other developed countries, total agricultural production per capita has increased (Figure 14.11). In the critical less-developed countries, however, per capita food production has remained low. More ominous still, in 1971 agricultural production in these countries dipped downward and per capita production declined (Figure 14.11).

The food supply of the biosphere can be increased by three technological advances—the so-called "green revolution"—massive application of artificial fertilizers, massive application of pesticides, and the substitution of higher yield crop strains. Increased irrigation has long been an important factor, and if large-scale desalination becomes more economical, could create considerable tracts of new farmland (Young,

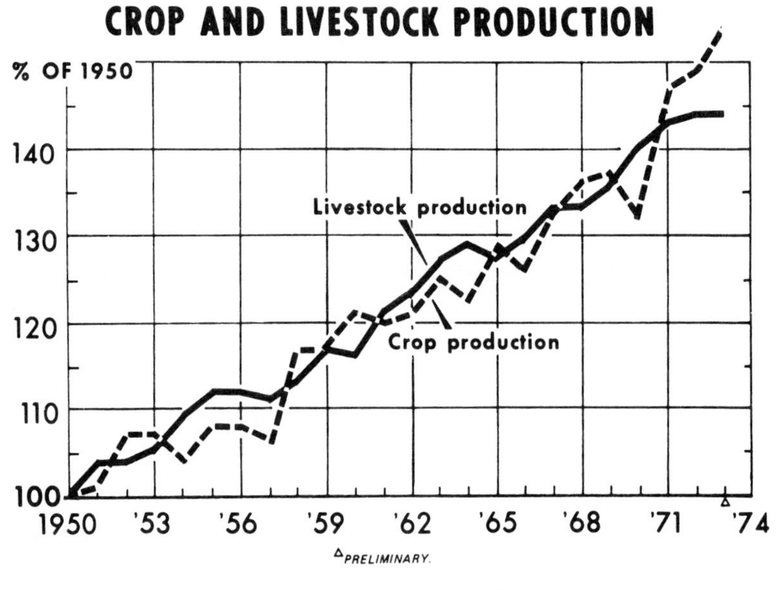

Figure 14.10. Crop and livestock production (from Anon., 1973*a*).

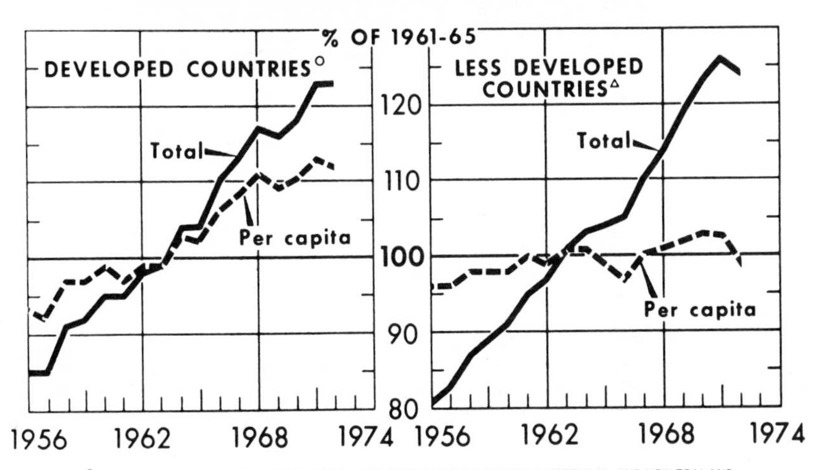

Figure 14.11. World agricultural production (from Anon., 1973*a*).

1970). Ewell (1964) analyzed the fertilizer needs of undeveloped areas for the 1960–1980 period (Table 14.2), but he noted that these are the very areas lagging hopelessly behind in fertilizer use (Table 14.3). The presently rapidly climbing price of fuels in the world market is expected to be a direct and very serious obstacle in meeting fertilizer needs. For example, NH_3 production capacity in the United States leveled off in 1969, and unless we can find supplies of cheap gas the United States will go from an exporter to a progressively larger importer of nitrogen products (Anon., 1974*b*).

Table 14.2 Grain and Fertilizer Needs of Undeveloped Areas

Needed Grain Production (in million metric tons) to Provide a Minimum Diet of 16 oz/Person-Day			
	1960	1970	1980
Asia[a]	140[b]	191	245
Africa	40[b]	52	67
Latin America	42[b]	54	69
Needed Fertilizer (in million metric tons)			
Asia[a]	1.4[c]	9	20
Africa	0.7[c]	3	5
Latin America	0.9[c]	3	5

[a] Excluding China and Japan.
[b] Actual consumption.
[c] Actual production.

Table 14.3 Per Capita Use of Fertilizer, 1962–1963

	Oceania	United States/ Canada	Europe	Japan	U.S.S.R.	Latin America	Africa	Asia[a]
Nitrogen (N)	50	3554	4920	669	1070	356	400	1181
Phosphorus (P_2O_5)	890	2937	4830	465	853	323	310	405
Potash (K_2O)	90	2291	4880	506	826	219	100	194
Total	1030	8782	14630	1640	2749	898	810	1780
1962 Population	17.2	206	434	95	221	224	269	952
(millions)	132	94	74	38	27	8.9	6.6	4.1

[a] Excluding China and Japan.

As we see in the following sections, the massive applications of fertilizers and pesticides together with agricultural solid waste disposal are having a profoundly deleterious impact on the chemistry of our environment. The reduction of crop strains is a highly risky invitation to ecological disaster as we noted earlier (see also Wade, 1972). Agriculture is making increasing demands on our already overtaxed water resources and contributes to the increasing degradation of soil and water by forming excessively high salinities (Thorne and Peterson, 1967). And fertilizer production and the use of farm machinery will be highly constrained by the rapidly worsening shortages of materials, fuels, and energy (Table 14.4).

Table 14.4 Average Energy Inputs in Corn Production During Different Years (all figures per acre) (From Pimental et al. © 1973 by the American Association for the Advancement of Science with permission of the AAAS.)

	1945	1950	1954	1959	1964	1970
Labor[a]	23	18	17	14	11	9
Machinery (kcal $\times$ 10^3)	180	250	300	350	420	420
Gasoline (gal)	15	17	19	20	21	22
Nitrogen (lb)	7	15	27	41	58	112
Phosphorus (lb)	7	10	12	16	18	31
Potassium (lb)	5	10	18	30	29	60
Seeds for Planting (bushels)	0.17	0.20	0.25	0.30	0.33	0.33
Irrigation (kcal $\times$ 10^3)	19	23	27	31	34	34
Insecticides (lb)	0	0.10	0.30	0.70	1.00	1.00
Herbicides (lb)	0	0.05	0.10	0.25	0.38	1.00
Drying (kcal $\times$ 10^3)	10	30	60	100	120	120
Electricity (kcal $\times$ 10^3)	32	54	100	140	203	310
Transportation (kcal $\times$ 10^3)	20	30	45	60	70	70
Corn yields (bushel)	34	38	41	54	68	81

[a] Mean hours of labor per crop acre in United States.

In order to meet the fast growing world food crisis, the farming of new land areas and increasing the crop yield of soils is receiving much attention. In addition, many alternative food sources have been suggested, notably aquaculture (Bardach et al., 1972; Yee, 1972) and the more intense exploitation of marine food resources, but the oceans are as finite a potential food source as are the land areas. In any event, the question of the capacity of the oceans for primary photosynthetic production of organic material looms as a very crucial one (Ryther, 1969). Protein material is in particularly short supply (Walsh, 1973). The use of fish flour for human consumption (Holden, 1971) remains controversial, and questions of health have slowed efforts to produce protein from petroleum (Anon., 1973*c*). Sorghum strains with higher protein levels are another possible way of aleviating the situation (Shapley, 1973). Some scientists are looking at cellulose as a possible starting material for the production of protein, and it has even been suggested that scrap automobile tires might be used as a substrate for growing a yeast-based food (Anon., 1972*a*). More serious are attempts to increase food supply by increasing symbiotic N-fixation (Phillips et al., 1971). While these alternative food sources if realized may give the human species a temporary stay of execution, in the final analysis, they cannot be a panacea and add up to little more than further exploitation of the already over-exploited resources of the biosphere. The prospects of agriculture, therefore, are not good.

14.4 AGRICULTURAL POLLUTION

Human food production is a significant perturbation of the biosphere of Earth, and in addition agricultural activity directly impacts the chemistry of our environment in the following major ways:

1. The disposal of animal wastes,
2. The disposal of plant wastes,
3. The disposal of food processing wastes,
4. Contamination from pesticide application,
5. Contamination from fertilizer application,
6. Depletion of natural soil nutrients and fertilizers, and
7. Salt and mineral accumulation from irrigation.

Furthermore, agriculture affects land and water use and plays a role in erosion and sedimentation (Small, 1971). Few people appear to realize the relative magnitude of the agricultural problem. A survey (Golueke, 1971) of the San Francisco Bay area found that in 1966 livestock and livestock products solid wastes amounted to 1,386,460 tons; other agricultural products, 942,014 tons; and food processing, 678,712 tons. Chemicals and applied products accounted for only 115,312 tons; petroleum refining, 151,692 tons; primary metals, 95,141 tons; and paper and allied products, 131,280 tons. In the United States a few years ago, 49,000,000 cattle, 57,000,000 hogs, 21,000,000 sheep, and 3,000,000,000 poultry produced 2,000,000,000 ton/yr waste—the equivalent of a human population of 2,000,000,000 people (Anon., 1970; see also Taiganides, 1967; Wadleigh, 1968), and as we can clearly see from Figure 14.10, the problem is worsening every year. Wastes from swine, chicken, and cattle alone in the

United States are ten times greater than the waste from the human population (Taiganides, 1963). Nonessential animals add to the mountain of waste—pet dogs in the United States produce 3500 ton/day of feces (Anon., 1973*b*), a presence all too familiar to city dwellers.

In order to get some grasp of the enormous dimensions of the problem of agricultural and food production wastes, let us examine a few examples. Americans consume more dairy products than any other type of foodstuff—286 lb/person-yr in 1967 compared to 176 lb/person-yr for the next largest item, red meats—59,000,000,000 lb of milk alone (Mercer and Ralls, 1970). The wastes of the dairy farm are mostly the same as those for feedlots (see discussion that follows) namely, manure. Dairy processing plants in the United States produced about 600,000,000 lb/yr of whey from cheese making, the major solid waste (Ben-Gera and Kramer, 1969). Dairy liquid effluents are various dilutions of milk. Although their solid content is low (4500 ppm, Nemerow, 1971) their biological oxygen demand (BOD) can be large, and they are commonly dumped into the closest water course. Table 14.5 gives the estimated pollution loadings for the dairy and other agriculturally based processing industries.

Cotton agriculture produces two types of solid waste—field waste and gin waste. A gin processing 15 bales of cotton/hr produces 22,500 lb/hr of waste (Bush, 1967). In the United States, the problem is being acerbated by the increasing use of dirtier machine-picked in place of hand-picked cotton (Moore and McCaskill, 1967; Pendelton, 1967). The disposal of gin waste is complicated by the fact that it is usually contaminated with pesticides and defoliants and is often infested with weevils (Elliot, 1967). This waste has little value as livestock feed. It cannot be spread on the soil if it contains bollworms. It must be buried or burned, and the latter creates an air pollution problem.

The production of vegetable foods produces immense quantities of waste since the edible part of the plant represents only a fraction of its total weight. Producing 1 lb of food for the consumer leaves 5 to 10 lb of solid waste in the field or factory (Mercer, 1970), again a problem made worse by the increasing use of highly indiscriminate mechanical harvesting (Hart, 1968). In addition to solid wastes, the food industry is the fifth biggest water polluter with some 700,000,000,000 gal/yr of polluted effluent (Downing, 1971), and the problem is particularly severe because of the seasonal nature of harvesting and processing. To take tomatoes, for example, the state of California accounts for 70% of the total United States production of 6,000,000,000 lb/yr, which corresponds to 400,000,000 lb/yr of net solid wastes and 5,000,000,000 gal/yr of liquid waste (Schultz et al., 1971). As in the case of cotton (see previous discussion), field waste has only limited value as livestock feed. Most of it is now left to rot in the fields where eventually it is plowed under. (For a recent review of some new techniques for reducing food processing wastes, see Hoover, 1974.)

Even citrus groves are potential polluters of the environment. In 1967–1968, Florida produced 9,000,000 tons of citrus fruit; 75% of the oranges and 80% of the grapefruit production of the United States (Coca-Cola Co., 1970). The groves themselves produce waste (and are doused with pesticides and fertilizers, Sheffield et al, 1968), but the processing of the fruit is the worst offender. About 45 to 60% of the total weight of the fruit is waste, and at the height of the season, Florida's citrus processing plants must dispose of 200,000,000 to 250,000,000 lb/wk of waste material (Cross and Ross, 1959). Part of this material is returned to the groves where, if it is not covered, it creates a fly nuisance. Another solution to the problem may lie in converting the waste into cattle feed and citrus molasses; the production of these in 1965–1966 was 350,000 and 48,000 tons, respectively (Coca-Cola Co., 1970; see also Kirk and Koger, 1970).

Some poultry farms contain a million or more birds—the waste equivalent of a city of 68,000 people (Taiganides, 1967). In 1951–1960, the poultry population of the United States averaged 394,900,000 birds (Taiganides, 1963) corresponding to almost 100,000,000 ton/yr of manure. An average egg farm of 100,000 birds produces 25,000 lb/day of manure. Some progress has been made in processing chicken shit and feeding it back to the birds (Biely et al., 1972; see also Sheppard, 1970). Is such recycling in store for us if we do not control our numbers? While manure is the major poultry solid waste disposal problem, what to do with dead birds is another. In a laying hen production unit, this amounts to about 1%/month of the birds (Taiganides, 1963; Hart and Fairbank, 1964).

The meat processing industry likes to boast that it makes use of all of the animal except the squeal. Nevertheless in 1967, the slaughter of 3,200,000 head of cattle in California alone produced 100,000 tons of solid waste (Cornelius, 1969), and on the national scale, it is estimated that the packing industry will produce 2,400,000,000 lb/yr of waste in 1975 (Koenig and Barker, 1969). Three different areas are waste sources—the stockyards, the slaughterhouse, and the packinghouse, with the last producing the least waste. The prime stockyard waste, like the feedlots (see discussion that follows), is manure, while blood and paunch manure come from the killing floor of the slaughterhouse. More than 95% of the blood is recovered and sold (Nemerow, 1971). Viscera and bones are recovered, as are also hides, skin, and hair. The meat processing industry also consumes and pollutes great quantities of water; in 1964 an estimated 84,000,000,000 gal/yr for slaughtering and an additional 4,000,000,000 gal/yr for processing (Siegelman and Ullman, 1969).

Finally, we come to the most notorious and polluting innovations of modern agricultural technology, the feedlot. The practice of fattening cattle prior to slaughter (and as we have seen, injecting them with hormones) crowded together in very high density feedlots is rapidly increasing (Miner et al., 1970; Townshend et al., 1970; Miner, 1971). Feedlots of 20,000 head capacity are not uncommon, and 100,000 head feedlots have been constructed (Loehr, 1971). Feedlot waste is particularly offensive: it stinks, it attracts clouds of flies and makes clouds of dust (Loomis, 1973), it contaminates adjacent waters (Gilbertson et al., 1971; Madden and Dornbush, 1971) that may already be contaminated with nitogenous matter from fertilizer run-off, and it can pose a health hazard to humans (Loehr, 1969). In addition to contaminating soil and water with ammonia and other nitrogen compounds, feedlots also contaminate the air with ammonia and other volatiles and highly odorous aliphatic amines (Mosier et al., 1973). Numerous attacks on the problem of feedlot waste disposal are being made. Processing and using the manure as fertilizer has often proved to be unattractive economically in the United States. Already severe, the problem will continue to worsen.

Pesticide contamination will be treated in the next section, so now we turn our attention to water pollution by fertilizers. Expert opinion appears to be somewhat unclear. The text of one article (Anon., 1970) states that "thus far, there is no positive evidence that the use of chemical fertilizers causes water pollution" while a figure caption in this same article reads "Increasing fertilizer use is linked with nearby

Table 14.5 Estimated Pollution Loadings of Selected Agricultural Processing Industries[a] (From Hoover and Jasewicz, in N. C. Brady (ed.), *Agriculture and the Quality of Our Environment*, Pub. No. 85. © 1967 by the American Association for the Advancement of Science with permission of the AAAS and the authors.)

Processing Industry	Annual Production (million pounds)	5-Day BOD: Data in Literature	5-Day BOD: Pounds BOD/1000 lb Processed	5-Day BOD: Potential Daily Load (1000 lb)	Potential Daily Population Equivalent (millions)
Canneries					
Apples	1,218	32 gal 3600 ppm BOD/case of 24 no. 2½ cans	13.3	44	0.26
Peaches	2,970	50 gal 2000 ppm BOD/case of 24 no. 2½ cans	20.8	169	1.02
Corn	2,364	19.5 lb BOD/ton corn processed	9.8	63	0.38
Tomatoes	9,790	8.4 lb BOD/ton tomatoes processed	4.2	113	0.68
Canning, Total	—	—	—	1,370	8
Corn Wet Milling	10,800	1 bu = 1–2 PE	4.5	133	0.80
Cotton, Processed Through Basic Dyeing Step	4,600	PE (per 1000 lb goods): Sizing 2; Desizing 96; Kiering 108; Bleaching 17; Scouring 12; Mercerizing 83; Basic dyeing 90	68	857	5.14

Dairy		Pounds BOD/10,000 lb milk equivalent			
Fluid Milk	59,000	10	1.0	162	1.0
Evaporated Milk	1,888	10.5	2.25	11.6	0.07
Nonfat Dry Milk	2,176	25.0	26.4	157	0.95
Cheddar Cheese	1,157	24.5	23.6	77.6	0.47
Cheddar Whey, Dried	20% of total	17.0	25.0	9.7	0.06
Cheddar Whey	50% of total	350	—	500	3.0
Cottage Cheese	1,424	350	16.5	64.5	0.38
Cottage Cheese Whey	7,500	350	—	1,000	6.0
Hides and Leather	1,300	650 gal 1500 ppm per 100 lb hides	81	300	0.18
Meat					
Slaughtering and Packing	59,400	14 lb BOD/1000 lb live weight	14.0	2,300	13.0
Paper and Pulp					
Wood Pulp	66,000	300 lb BOD/ton of pulp	150.0	27,000	162
Paper and Paper Board	96,600	68 lb BOD/ton of paper	34.0	9,000	54
Potatoes					
Chips	7.1	29.3 lb BOD/ton raw potatoes	14.6	106	0.64
Dehydrated	2.7	71.1 lb BOD/ton raw potatoes	35.6	93	0.58
Flour and Starch	3.2	57.0 lb BOD/ton raw potatoes	28.5	91	0.55
Frozen French Fries	5.4	22.0 lb BOD/ton raw potatoes	11.0	57	0.34
Poultry	8,200	33 lb BOD/1000 birds	10.0	225	1.3
Soybean	300	1.7 lb BOD/1000 bu	0.19	0.16	0.085
Sugar Refining					
Cane	48,000	5.31 lb BOD/ton	Av. 3.0	800	4.8
Beet	47,000	6.64 lb BOD/ton			
Wool Scouring	130	8 gal 4000 ppm/lb wool	267	100	0.6

[a] Hoover (private communication, 1975) warns that more recent data may be significantly different, for example, more recent BOD levels for the dairy industry are much higher.

stream pollution," and in a paragraph later in the text is the admission that "The use of fertilizers may result in excess nitrogen and phosphorus in the soil. The build up of elements in the soil may result in injury to vegetation, but more likely, it becomes a contaminant of waterways." The problem is obscured by the difficulty of tracing nitrate sources unambiguously. In addition to fertilizers (Stout and Burau, 1967), other sources include barnyard and feedlot run-off (Table 14.6), septic tanks (Table 14.7) and other untreated sewage, and the nitrate occurring naturally in soils. A study by the well-known environmentalist Barry Commoner (Kohl et al., 1971) using radioactively tagged nitrogen found that 55 to 60% of the nitrate in a typical Illinois corn belt watershed originated from fertilizer nitrogen, but even these results have been questioned (Hauck et al., 1972). In any event, whatever its specific source, one fact remains clear, nitrate pollution "is inherent in high yield farming" (Cowen, 1971). Since ordinary sewage treatment does not remove nitrate, even treated waters can contribute this pollutant and other soluble salts. The comparative

Table 14.6 Nitrate Nitrogen Content of Two Soils at Varying Distances from Old Feedlots (From Smith, in N. C. Brady (ed.), *Agriculture and the Quality of Our Environment*, Pub. No. 85. © 1967 by the American Association for the Advancement of Science, with permission of the AAAS and the author.)

Distance from Contaminated Area (ft)	Farm A[a] Pounds N/acre 0–18 ft	Farm A[a] NO_3 N in groundwater (ppm)	Farm B[b] Pounds N/acre 0–13 ft
0	2425	73	1375
150	1475	48	357
300	1014	13	317
600	780	tr	275
5280	958	—	227

[a] Loess soil, level topography, approximately 80% silt.
[b] Silt loam, but with 25 to 40% clay below 24 in.

Table 14.7 Nitrate Nitrogen Content of Soil and Groundwater at Varying Distances from Septic-Tank Drainage Field, Saline County, Missouri (From Smith, in N. C. Brady (ed.), *Agriculture and the Quality of Our Environment*, Pub. No. 85. © 1967 by the American Association for the Advancement of Science, with permission of the AAAS and the author.)

Distance from septic tank (ft)	Pounds N/acre 0–13 ft soil	N groundwater (ppm)
60	474	22.0
86	375	8.8
112	308	7.4

ease with which nitrate ion moves through soils, noted earlier, also makes source location difficult (Smith, 1967). Nitrate is a dangerous pollutant. Both human infants and ruminate livestock have sickened and even died after drinking nitrate polluted waters. Nitrate is reduced to nitrite in the stomach, which in turn causes methemoglobinemia ("blue babies"). From 1947 to 1950, there were 139 cases of this disease, including 14 deaths, from nitrate in farm well water in the state of Minnesota alone (Amer. Chem. Soc., 1969). Table 14.8 compares fish kills resulting from fertilizers with those resulting from pesticides.

Phosphate is also a widespread pollutant from agriculture (and from detergents, see Chapter 16), and high levels have built up in many of our waters (Figure 14.12). Phosphate does not travel as rapidly through soil as does nitrate, and it is not as dangerous, although as a major contributor, along with nitrate and other nutrient to eutrophication (see Chapter 16), it creates serious nuisances. It may have some rather unexpected side effects; for example, the increased uranium run-off into the Gulf of Mexico has been attributed to the increased use of phosphate fertilizers (Spaulding and Sackett, 1972).

Finally, we should note that the fertilizer industry itself is a major depleter of resources and polluter. Phosphate plant settling lagoons are a nuisance (Cox, 1968; Timerlake, 1970), and ammonia compound effluents from nitrogenous fertilizer

Table 14.8 A Fish Kills Resulting from Pesticides (From Dawson et al., 1970)

Year	Number of Reports	Total Kill Reported	Average Kill of Incidents Reporting Kill Totals
1963	60	401,415	10,849
1964	93	191,167	2,583
1965	74	770,557	12,039
1966	51	217,406	4,941
1967	43	329,130	7,654
1968	51	325,194	7,742

B Fish Kills Resulting from Fertilizer Compounds

Year	Number of Reports	Total Kill Reported	Average Kill of Incidents Reporting Kill Totals
1963	3	1,400	700
1964	5	67,040	16,760
1965	4	2,697	674
1966	1	1,200	1,200
1967	2	10,000	5,000
1968	5	15,116	3,023

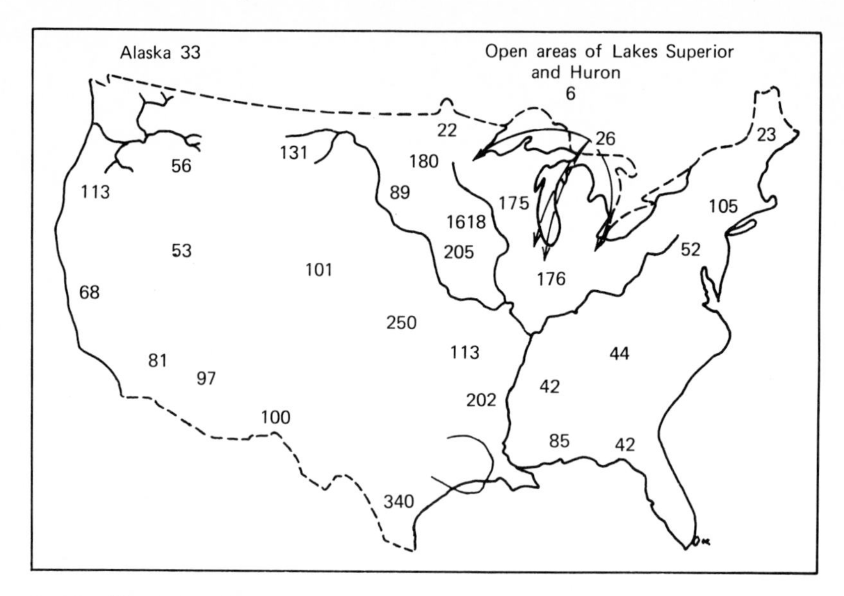

Figure 14.12. Total phosphorus concentrations in drinking water sources over the United States, 1965–1966 (from Verduin, © 1967 by the American Association for the Advancement of Science, with permission of the AAAS and the author).

production have been shown to be toxic to marine phytoplankton and to interfere with photosynthesis if the level exceeds 1.1 mg/liter (Natarajan, 1970).

Nitrogen and phosphorus nicely illustrate the more far reaching chemical consequences of man's agricultural activity. Agriculture speeds up the carbon cycle. It also consumes monstrous quantities of water, but most of this water is eventually restored to the hydrologic cycle. Nevertheless the short range local impact of this water reallocation can be severe. With respect to nitrogen, phosphorus, and other vital

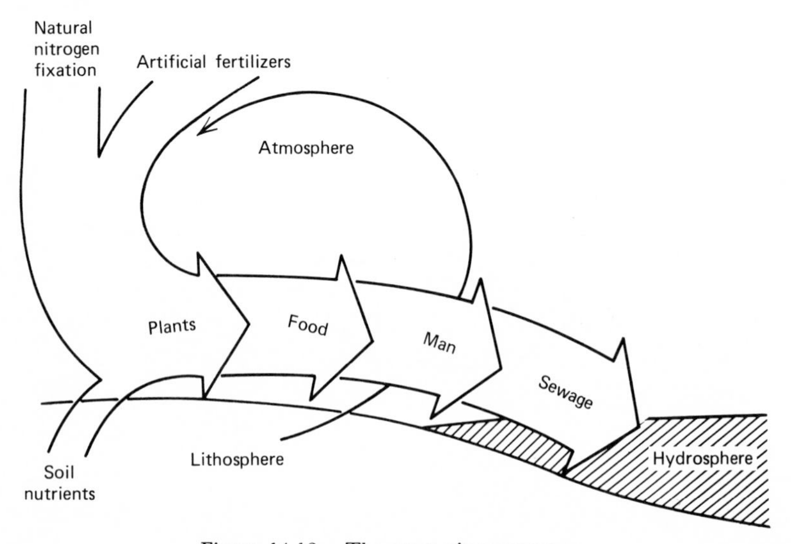

Figure 14.13. The great nitrogen pump.

plant nutrients and essential trace elements, agriculture acts like an enormous pump (Figure 14.13), removing these materials from the lithosphere and dumping them into the oceans, which, for all practical purposes, is an infinite and *permanent* sink for them. In the case of nitrogen the natural return paths from hydrosphere to lithosphere are, relatively speaking the merest of trickles. However man accelerates this return by the preparation of artificial N-fertilizers. Nitrogen (in addition to Figure 14.9A, see also the next chapter for more about the N-cycle) is unique in that in returning this element to the lithosphere man can draw upon an exceedingly large reservoir—the atmosphere. Such is not the case with the other nutrient elements; here he must mine his artificial fertilizers, such as phosphate, from the lithosphere, just as plants must "mine" them from the soil. For these nutrients agriculture represents a permanent and irreversible depletion of the resources of the lithosphere. Taken as a whole, therefore, agriculture represents an acceleration of the biosphere's exhaustion of the lithosphere.

14.5 PESTICIDES AND OTHER PERSISTENT POLLUTANTS

For several years now one of the most highly publicized class of environmental pollutants has been pesticides. Pesticides include, not only the now familiar insecticides, but herbicides, fungicides, rodenticides, and so on. In scale, the contamination of our environment by persistent pesticides and related substances has been global; but it is also a problem that sharply divides the interests of the more from the less-developed nations. Table 14.9 gives the common and chemical names of some pesticides, while Table A.30 gives their chemical structures as well. Table A.15 gives United States production of pesticides in 1970. It is clear from these tables that many of the pesticides currently in use are chlorinated hydrocarbons and their derivatives. While the domains of some pollutants tend to be restricted, pesticide contamination has found its way to every corner of our environment—soils (Bailey and White, 1964; Sheets, 1967; Nasim et al., 1971) where they are sorbed on minerals and other substances (Haque and Coshow, 1971); sediments (Albone et al., 1972; Leland et al., 1973) where pesticides can be further concentrated by solvent extraction by petrochemical pollution (Hartung and Klingler, 1970); aquifers (Lee, 1970; Boucher and Lee, 1972); water (Table 14.10, Green et al., 1967) including drinking water (Schafer et al., 1969), lakes, ponds, rivers (Eichelberger and Lichtenberg, 1971), streams, and ground waters as well as the oceans (Eisler, 1968); Antarctic snow (Peterle, 1969); even the air (at 0.1 to 2520 ng/m^3 levels, Stanley et al., 1971, see also Antommaria et al., 1965; Abbott et al., 1966; Tabor, 1966; Risebrough et al, 1968) is dirtied with them. Chlorinated hydrocarbon pesticides (and polychlorinated biphenyls (PCB's), Bidleman and Olney, 1974) are heavily concentrated in sea slicks with concentration factors up to 105 (Seba and Corcoran, 1969). Indeed, they tend to concentrate in the worst possible place, the biosphere, thanks to their relatively high solubility in the lipids of fatty tissues (Risebrough et al., 1967, 1969; Wurster and Wingate, 1968). Many of the paths that these dangerous chemicals follow in our environment lead back to us (Figure 14.14)—to our food (Gunther and Gunther, 1971; Table 14.11) and to our own bodies (Hayes et al., 1965; Curley et al., 1973; Table 14.12). There is good reason to believe that these poisons are cummulative in the body and that there is no "physiological equilibrium" that limits their levels in the body (Deichmann

Table 14.9 Common and Chemical Names of Some Pesticides (From Sheets, in N. C. Brady (ed.), *Agriculture and the Quality of Our Environment*, Pub. No. 85. © 1967 by the American Association for the Advancement of Science, with permission of the AAAS and the author.)

Common Name	Chemical Name
Aldrin	1,2,3,4,10,10-Hexachloro-1,4,4*a*,5,8,8*a*-hexahydro-1,4-*endo-exo*-5,8-dimethanonaphthalene
Atrazine	2-Chloro-4-ethylamino-6-isopropylamino-*s*-triazine
BHC	1,2,3,4,5,6-Hexachlorocyclohexane (several isomers)
Calcium arsenate	Mixtures of tricalcium arsenate $Ca_3(AsO\text{-})_2$ and acid calcium arsenate $CaHAsO_4$
Carbaryl	1-Naphthyl methylcarbamate
Chlordane	1,2,4,5,6,7,8,8-Octachloro-2,3,3*a*,4,7,7*a*-hexahydro-4,7-methanoindene (not over 40% of related compounds)
2,4-D	2,4-Dichlorophenoxyacetic acid
DDE	1,1-Dichloro-2,2-bis(*p*-chlorophenyl)ethylene
DDT	1,1,1-Trichloro-2,2-bis(*p*-chlorophenyl)ethane
Diazinon	O,O-Diethyl O-(2-isopropyl-4-methyl-6-pyrimidinyl)phosphorothioate
Dichlobenil	2,6-Dichlorobenzonitrile
Dieldrin	1,2,3,4,10,10-Hexachloro-6,7-epoxy1,4,4*a*,5,6,7,8,8*a*-octahydro-1,4-*endo-exo*-5,8-dimethanonaphthalene
Diphenamid	*N,N*-Dimethyl-2,2-diphenylacetamide
Diuron	3-(3,4-Dichlorophenyl)-1,1-dimethylurea
Endrin	1,2,3,4,10,10-Hexachloro-6,7-epoxy-1,4,4*a*,5,6,7,8,8*a*-octahydro-1,4-*endo-endo*-5,8-dimethanonaphthalene
Fenac	2,3,6-Trichlorophenylacetic acid
Guthion	O,O-Dimethyl-S-4-oxo-1,2,3-benzotriazin-3(4H)-ylmethyl phosphorodithioate
Heptachlor	1,4,5,6,7,8,8-Heptachloro-3*a*,4,7,7*a*-tetrahydro-4,7-methanoindene
Heptachlor Epoxide	1,4,5,6,7,8,8-Heptachloro-2,3-3poxy-2,3,3*a*,4,7,7*a*-hexahydro-4,7-methanoindene
Lead Arsenate	Acid orthoarsenate PbHAsO- and basic orthoarsenate Pb-(PbOH) $(AsO\text{-})_3$
Lindane	1,2,3,4,5,6-Hexachlorocyclohexane (99% or more gamma BHC)
Malathion	S-[1,2-bis(Ethoxycarbonyl)ethyl]O,O-dimethyl phosphorodithioate
Monuron	3-(*p*-Chlorophenyl)-1,1-dimethylurea
Neburon	1-*n*-Butyl-3-(3,4-dichlorophenyl)-1-methylurea
Parathion	O,O-Diethyl-O-*p*-nitrophenylphosphorothioate
Phorate	O,O-Diethyl S-(ethylthio)methylphosphorodithioate
Picloram	4-Amino-3,5,6-trichloropicolinic acid
Propazine	2-Chloro-4,6-bis(isopropylamino)-*s*-triazine
Simazine	2-Chloro-4,6-bis(ethylamino)-*s*-triazine
2,4,5-T	2,4,5-Trichlorophenoxyacetic acid
2,3,6-TBA	2,3,6-Trichlorobenzoic acid
TDE, DDD	1,1-Dichloro-2,2-bis(*p*-chlorophenyl)ethane
Toxaphene	Chlorinated camphene containing 67–69% chlorine

Table 14.10 Levels of Chlorinated Hydrocarbon by Order of Decreasing Concentrations in Micrograms per Liter at Top Ten Stations at Which Highest Levels Were Observed (From Green et al., in N. C. Brady (ed.), *Agriculture and the Quality of Our Environment*, Pub. No. 85, © 1967 by the American Association for the Advancement of Science, with permission of the AAAS and the authors.)

CAM Samples 1958–1964		Synoptic Surveys (Grab Samples) 1964		1965		1966	
Dieldrin							
Mississippi: W. Memphis, Ark.	0.122	Savannah: N. Augusta, S.C.	>0.118	Tombigbee: Columbus, Miss.	0.100	Merrimack: Lowell, Mass.	0.167
Savannah: N. Augusta, S.C.	0.056	Merrimack: Lowell, Mass.	>0.071	Merrimack: Lowell, Mass.	0.068	Savannah: N. Augusta, S.C.	0.110
Ohio: Cincinnati, Ohio	0.055	Potomac: Great Falls, Md.	>0.040	Savannah: N. Augusta, S.C.	0.051	Savannah: Pt. Wentworth, Ga.	0.048
Schuylkill: Philadelphia, Pa.	0.035	Schuylkill: Philadelphia, Pa.	>0.032	Kanawha: Enfield Dam, Conn.	0.045	Susquehanna: Conowingo, Md.	0.031
Mississippi: New Orleans, La.	0.034	Rio Grande: El Paso, Tex.	0.032	Rio Grande: Alamosa, Colo.	0.029	Delaware Bay	0.025
Delaware: Philadelphia, Pa.	0.033	Platte: Plattsmouth, Nebr.	0.023	Tennessee: Lenoir City, Tenn.	0.028	Connecticut: Northfield, Mass.	0.017
Apalachicola: Chattahoochee, Fla.	0.024	Connecticut: Northfield, Mass.	>0.022	Ohio: Cairo, Ill.	0.028	Connecticut: Enfield Dam, Conn.	0.016
Mississippi: Vicksburg, Miss.	0.023	Savannah: Port Wentworth, Ga.	0.020	Mississippi: Dubuque, Iowa	0.024	Schuylkill: Philadelphia, Pa.	0.015
Mississippi: Delta, La.	0.022	Mississippi: Vicksburg, Miss.	0.017	Missouri: Kansas City, Kans.	0.023	Chattahoochee: Lanett, Ala.	0.015
Savannah: Pt. Wentworth, Ga.[a]	0.016	Mississippi: New Roads, La.[b]	0.016	Savannah: Pt. Wentworth, Ga.	0.022	Kanawha: Winfield Dam, W. Va.	0.015
Endrin							
Mississippi: W. Memphis, Ark.	0.214	Potomac: Great Falls, Md.	>0.094	Mississippi: W. Memphis, Ark.	0.116	Hudson: Narrows, N.Y.	0.069
New Orleans, La.	0.160	Rio Grande: El Paso, Tex.	0.067	Atchafalaya: Morgan City, La.	0.019	S. Platte: Julesburg, Colo.	0.063
Vicksburg, Miss.	0.072	Big Horn: Hardin, Mont.	0.026	Delaware: Trenton, N.J.	0.018	Savannah: Pt. Wentworth, Ga.	0.031
Delta, La.	0.044	Mississippi: Vicksburg, Miss.	0.025	Tombigbee: Columbus, Miss.	0.015	St. Joseph: Benton Harbor, Mich.	0.029
Connecticut: Enfield Dam, Conn.	0.023	Connecticut: Northfield, Mass.	0.025	Clinch: Kingston, Tenn.	0.015	L. Superior: Duluth, Minn.	0.022
Atchafalaya: Morgan City, La.	0.015	Red (N): Grand Forks, N.D.	0.023	Rio Grande: Alamosa, Colo.	0.014	Savannah: N. Augusta, S.C.	0.022
Mississippi: Cape Girardeau, Mo.	0.013	Mississippi: New Roads, La.	0.023	Monongahela: Pittsburgh, Pa.	0.014	Bear: Preston, Idaho	0.015
Allegheny: Pittsburgh, Pa.	0.012	Yellowstone: Sidney, Mont.	0.021	Tennessee: Lenoir City, Tenn.	0.009	Clearwater: Lewiston, Idaho	0.019
Rio Grande: Brownsville, Tex.	0.011	Columbia: Clatskanie, Oreg.	0.019	Red (N): Grand Forks, N.D.	0.009	Connecticut: Northfield, Mass.	0.014
Mississippi: New Roads, La.	0.010	Atchafalaya: Morgan City, La.	0.018	Mississippi: Delta, La.	0.008	Mississippi: Delta, La.	0.014
DDT							
Rio Grande: Brownsville, Tex.	0.144	Maumee: Toledo, Ohio	0.087	Rio Grande: Alamosa, Colo.	0.149	Brazos: Arcola, Tex.	0.123
Laredo, Tex.	0.052	Red (N): Grand Forks, N.D.	0.072	San Juan: Shiprock, N. Mex.	0.125	Rio Grande: El Paso, Tex.	0.046
El Paso, Tex.	0.032	San Joaquin: Vernalis, Calif.	0.066	Colorado: Page, Ariz.	0.058	Mississippi: Vicksburg, Miss.	0.044
Ohio: Cairo, Ill.	0.023	Atchafalaya: Morgan City, La.	0.047	Platte: Plattsmouth, Nebr.	0.039	Arkansas: Ft. Smith, Ark.	0.042
Mississippi: New Orleans, La.	0.019	Mississippi: Vicksburg, Miss.	0.041	Spokane: Post Falls Dam, Idaho	0.037	Potomac: Great Falls, Md.	0.038
Delaware: Philadelphia, Pa.	0.015	Bear: Preston, Idaho	0.034	Red (N): Grand Forks, N.D.	0.034	Mississippi: Delta, La.	0.031

(Continued)

Table 14.10 (*Continued*)

CAM Samples 1958–1964		Synoptic Surveys (Grab Samples) 1964		1965		1966	
Chattahoochee: Lanett, Ala.	0.011	Columbia: Clatskanic, Oreg.	0.034	Ohio: Cairo, Ill.	0.023	Missouri: Kansas City, Kans.	0.029
Tennessee: Pickwick, Ldg., Tenn.	0.011	Red (S): Alexandria, La.	0.031	S. Platte: Julesburg, Colo.	0.023	Delaware: Trenton, N.J.	0.028
Mississippi: Vicksburg, Miss.	0.010	Willamette: Portland, Oreg.	0.029	Mississippi: Delta, La.	0.019	L. Superior: Duluth, Minn.	0.026
Sacramento: Green's Landing, Calif.[c]	0.009	Appalachicola: Chattahoochee, Fla.	0.027	Mississippi: Vicksburg, Miss.[d]	0.017	Snake: American Falls, Idaho	0.025
			DDE				
Delaware: Philadelphia, Pa.	0.012	Maumee: Toledo, Ohio	0.015	San Juan: Shiprock, N. Mex.	0.009	Brazos: Arcola, Tex.	0.004
Mississippi: Vicksburg, Miss.	0.011	Bear: Preston, Idaho	0.011	Detroit: Detroit, Mich.	0.008	San Joaquin: Vernalis, Calif.	0.003
Hudson: Poughkeepsie, N.Y.	0.006	Mississippi: St. Paul, Minn.	0.011	Yellowstone: Sidney, Mont.	0.002	St. Lawrence: Massena, N.Y.	0.002
South Platte: Julesburg, Colo.	0.005	S. Platte: Julesburg, Colo.	0.009	Platte: Plattsmouth, Nebr.	*P*	Columbia: Clatskanie, Oreg.	0.001
Mississippi: New Orleans, La.	0.004	Delaware: Martins Creek, Pa.	0.008	Rainy: Baudette, Minn.	*P*	Arkansas: Pendleton Ferry, Ark.	*P*
Rio Grande: Brownsville, Tex.	0.004	Mississippi: W. Memphis, Ark.	0.007			Red (S): Alexandria, La.	*P*
Laredo, Tex.	0.004	Columbia: Clatskanie, Oreg.	0.005			Rio Grande: El Paso, Tex.	*P*
Lake Superior, Duluth, Minn.	0.004	San Joaquin: Vernalis, Calif.	0.005			Lake Superior: Duluth, Minn.	*P*
12 stations in various river basins	0.002	Snake: Payette, Idaho	0.005			Hudson: Poughkeepsie, N.Y.	*P*
		7 different sampling points	0.004			Hudson: Narrows, N.Y.[e]	*P*
			DDD				
Delaware: Philadelphia, Pa.	0.080	Shenandoah: Berryville, Va.	0.083	Rio Grande: Brownsville, Tex.	0.026	Connecticut: Enfield Dam, Conn.	0.013
Savannah: N. Augusta, S.C.	0.031	All other stations	<0.075	Delaware: Trenton, N.J.	0.018	Rio Grande: Brownsville, Tex.	0.013
Rio Grande: Brownsville, Tex.	0.019			Willamette: Portland, Oreg.	0.013	St. Joseph: Benton Harbor, Mich.	0.013
Rio Grande: El Paso, Tex.	0.012			Missouri: Kansas City, Kans.	0.011	Raritan: Perth Amboy, N.J.	0.012
Mississippi: New Roads, La.	0.012			St. Lawrence: Massena, N.Y.	0.010	Detroit: Grosse Isle, Mich.	0.012
Red (S): Alexandria, La.	0.011			Platte: Plattsmouth, Nebr.	0.010	Potomac: Great Falls, Md.	0.012
San Joaquin: Vernalis, Calif.	0.010			Waikele Stream: Oahu, Hawaii	0.008	Arkansas: Pendleton Ferry, Ark.	0.012
Rio Grande: Laredo, Tex.	0.009			Red (S): Alexandria, La.	0.008	Chattahoochee: Lanett, Ala.	0.011
Apalachicola: Chattahoochee, Fla.	0.008			Merrimack: Lowell, Mass.	0.007	Atchafalaya: Morgan City, La.	0.010
Sacramento: Green's Landing, Calif.[f]	0.006			Potomac: Washington, D.C.	0.007	Missouri: Kansas City, Kans.	0.010
			Aldrin				
Red (S): Alexandria, La.	0.006	Colorado: Page, Ariz.	0.085	ND		Rio Grande: El Paso, Tex.	*P*
Snake: Wawawai, Wash.	0.003	10 stations	*P*				
Chattahoochee: Lanett, Ala.	0.002						
Savannah: N. Augusta, S.C.	<0.001						
Merrimack: Lowell, Mass.	<0.001						
Yakima: Richland, Wash.	<0.001						
Yellowstone: Sidney, Mont.	<0.001						
19 stations in various river basins	*P*						

			Heptachlor				
Atchafalaya: Morgan City, La.	0.002	16 stations	*P*	Red (N): Grand Forks, N.D.	0.155	Missouri: Kansas City, Kans.	0.004
Mississippi: W. Memphis, Ark.	*P*			Mississippi: Dubuque, Iowa	0.048		
Potomac: Great Falls, Md.	*P*			Rio Grande: Brownsville, Tex.	0.035		
Detroit: Detroit, Mich.	*P*			St. Lawrence: Massena, N.Y.	0.031		
None detected at other stations				Delaware: Martins Creek, Pa.	0.025		
				Ohio: Cincinnati, Ohio	0.024		
				Missouri: St. Louis, Mo.			
				Tennessee: Lenoir City, Tenn.	0.020		
				Sacramento: Green's Ldg., Calif.	0.020		
				Detroit: Detroit, Mich.	0.015		
			Heptachlor Epoxide				
Mississippi: W. Memphis, Ark.	0.020	ND-all	<0.075	Mississippi: Dubuque, Iowa	0.067	South Platte: Julesburg, Colo.	0.019
Missouri: St. Louis, Mo.	0.002			Red (N): Grand Forks, N.D.	0.020	Lake Superior: Duluth, Minn.	0.010
Mississippi: New Orleans, La.	0.001			Ohio: Addison, Ohio	0.020	Neuse: Raleigh, N.C.	0.008
St. Lawrence: Massena, N.Y.	0.001			Mississippi: W. Memphis, Ark.	0.020	Hudson: Narrows, N.Y.	0.007
Potomac: Great Falls, Md.	<0.001			Sacramento: Green's Ldg., Calif.	0.019	Mississippi: Delta, La.	0.007
6 stations in various river basins	*P*			St. Lawrence: Massena, N.Y.	0.017	Savannah: Pt. Wentworth, Ga.	0.006
				Missouri: Kansas City, Kans.	0.014	Bear: Preston, Idaho	0.005
				Wabash: New Harmony, Ind.	0.012	Chattahoochee: Lanett, Ala.	0.004
				Missouri: St. Louis, Mo.	0.007	Mississippi: St. Paul, Minn.	0.004
				Potomac: Washington, D.C.	0.003	North Platte: Henry, Nebr.	0.004
			Benzene Hexachloride				
Apalachicola: Chattahoochee, Fla.	0.022	Delaware: Martins Creek, Pa.	*P*	Red (N): Grand Forks, N.D.	0.004	Ohio: Cincinnati, Ohio	0.056
Sacramento: Green's Ldg., Calif.	0.011	Mississippi: W. Memphis, Ark.	*P*	Ohio: Cairo, Ill.	0.002	Hudson: Narrows, N.Y.	0.034
Red (N): Grand Forks, N.D.	0.004	All other stations	<0.025	Verdigris: Nowata, Okla.	*P*	Ohio: Addison, Ohio	0.026
St. Lawrence: Massena, N.Y.	0.003			Connecticut: Enfield Dam, Conn.	*P*	Rio Grande: El Paso, Tex.	0.023
Missouri: Kansas City, Kans.	0.003			Monongahela: Pittsburgh, Pa.	*P*	South Platte: Julesburg, Colo.	0.022
Ohio: Cairo, Ill.	0.002					Trinity: Livingston, Tex.	0.013
Savannah: N. Augusta, S.C.	<0.001					Allegheny: Pittsburgh, Pa.	0.013
15 stations in various river basins	*P*					Mississippi: St. Paul, Minn.	0.012
						Mississippi: Vicksburg, Miss.	0.011
						San Joaquin: Vernalis, Calif.[g]	0.008

[a] Same concentration in Merrimack River at Lowell, Massachusetts.
[b] Same concentration in Apalachicola River at Chattahoochee, Florida.
[c] Same concentration in Tombigbee River at Columbus, Mississippi.
[d] Same concentration in Chattahoochee River at Lanett, Alabama.
[e] Same concentration in Tennessee River at Bridgeport, Alabama.
[f] Same concentration in Tennessee River at Pickwick Landing Dam, Tennessee.
[g] Same concentration in Chattahoochee River at Lanett, Alabama; Arkansas River at Ponca City, Oklahoma.

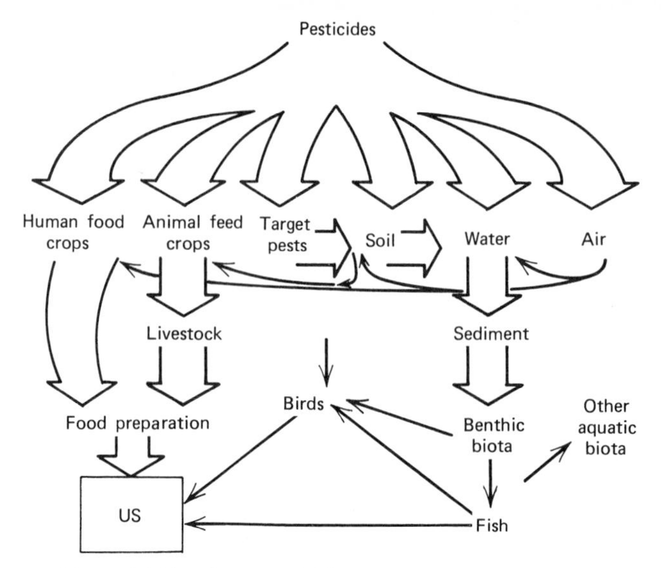

Figure 14.14. Some pesticide paths in our environment.

et al., 1971). Inasmuch as man is now top predator on this planet, any chemical that concentrates in a food chain represents a particular hazard to him. Nature too, administers justice, and it is sobering to pause a moment and reflect on what species is the "pest" in pesticides. Figure 14.15, which traces the concentrations of DDT in an aquatic environment, illustrates this concentration of the chemical as we go up a food chain (from top to bottom in the figure). This same sinister tendency is shown again in Figure 14.16.

Although they are highly persistent chemicals in our environment (see Figure 11.7 and Table 11.5B) pesticides nevertheless undergo metabolic and other degradation processes (Alexander, 1967), including photolytic decomposition. "Photodieldrin," for example, is even more dangerous than the parent pesticide and can be formed by soil microorganisms as well (Matsumura et al., 1970), and the possibility has been raised that PCB's (see the discussion that follows) can be formed by solar irradiation of DDT (Maugh, 1973). It is not exceptional for the residues formed from pesticide breakdown to be just as dangerous as the parent compound. Among the major types of general reactions involved in pesticide breakdown in the environment are:

1. Oxidation: (*a*) hydroxylation of aromatic rings; (*b*) oxidation of side chains to alcohols, ketones, or carboxyl groups; (*c*) dealkylation from oxygen or sulfur (ether cleavage); (*d*) Sulfoxide formation; (*e*) *N*-oxide formation.
2. Dehydrogenation and dehydrohalogenation;
3. Reduction;
4. Conjugation: (*a*) amide formation, (*b*) metal complex, (*c*) glucoside or glucuraonic acid, (*d*) sulfate;
5. Hydrolytic reactions: (*a*) cleavage of esters, (*b*) cleavage of amides;
6. Exchange reactions;
7. Isomerization.

Table 14.11 Estimated Daily Content of DDT and DDE in Complete Meals in the United States (From Hayes, 1966, with permission of the National Academy of Sciences and the author.)

Year Location	Source	Number	DDT[a]	DDE[a] as DDT	Total[a] as DDT	DDE as DDT (% of total)
1953–1954						
Wenatchee, Washington	Restaurant	18	0.178	0.102	0.280	37
Tacoma, Washington	Prison	7	0.116	0.063	0.179	35
1954–1955						
Tallahassee, Florida	Prison	12	0.202	0.056	0.258	21
1956–1957						
Walla Walla, Washington	College dining room for meat abstainers	11	0.041	0.027	0.068	39
1959–1960						
Anchorage, Alaska	Hospital	3	0.184	0.029	0.213	14
1961–1962						
Washington, D. C. Baltimore, Maryland Atlanta, Georgia Minneapolis, Minnesota San Francisco, California	Market basket survey	36[b]	0.026[c]	0.017[c]	0.043[c]	40[c]
1962–1964						
Wenatchee, Washington	Restaurant	12	0.038	0.049	0.087	56
Wenatchee, Washington	Household	17	0.314	0.193	0.507	40
1962–1964						
Atlanta, Georgia Baltimore, Maryland Minneapolis, Minnesota St. Louis, Missouri San Francisco, California	Market basket survey	23[b]	0.023[c]	0.013[c]	0.036[c]	36[c]
1964						
Baltimore, Maryland	Market basket survey	1[b]	0.023[c]	0.017[c]	0.040[c]	43[c]

[a] Total daily content in milligrams.

[b] This figure refers to the number of diet samples, each consisting of the total normal 14-day food intake for males 16 to 19 years old, which were tested. In some instances, additional diet samples were taken and aliquots were analyzed for pesticide content of various classes of foodstuffs, but no composite value was given.

[c] The author did not calculate the daily DDT or DDE intake. However, using the author's mean dietary concentrations of DDT and DDE and the mean daily food intake of 3.78 kg from the Market Basket survey, the reviewer has calculated the values shown.

Table 14.12 Pesticides in Humans (From Hayes, 1966)

A Concentration of DDT-Derived Material in Body Fat

Country	Year	No. of Samples	Analysis Method	DDT[a] (ppm)	DDE as DDT (ppm)	Total[b] as DDT (ppm)	DDE as DDT (% of total)
United States	1961–1962	28	Colorimetric	3.7	6.9	10.6	65
			GLC[c]	2.4	4.3	6.7	64
France	1962–1963	5	Colorimetric	3.1	6.5	9.6	68
			GLC[c]	3.5	5.3	8.8	60
India (Delhi area, civilian)	1964	24	Colorimetric	18	12	30	40
			GLC[c]	14.3	12.9	27.2	47
India (other cities, military)	1964	11	Colorimetric	7	4	11	36
			GLC[c]	4.7	7.1	11.8	60

[a] Includes *p,p'*- and *o,p'*-DDT.
[b] Includes all detected isomers of DDT and DDE, but not DDD.
[c] Gas–liquid chromatography.

B Average Concentration of Chlorinated Hydrocarbon Pesticides (other than those derived from DDT) in Body Fat of the General Population of Various Countries[a]

Country	Year	No. of Samples	Storage Level in Body Fat (ppm): BHC Isomers	Dieldrin	Endrin	Heptachlor Epoxide
United States	1961–1962	28	0.20	0.15	[b]	[b]
United States	1962–1963	282	0.57	0.11[c]	nd	[b]
United States	1964	64	[b]	0.31	<0.02	0.10
United States	1964	25	0.60	0.29	<0.03	0.24
England	1961–1962	131	[b]	0.21	[b]	[b]
England	1963–1964	65	0.42	0.26	[b]	<0.1
England	1964	100	0.02[d]	0.21	<0.02	<0.01
France	1961	10	1.19	[b]	[b]	[b]
India	1964	35	1.43	0.04	nd	nd

[a] All analyses were carried out by gas–liquid chromatography.
[b] The specific pesticide was not tested for, or at least such testing was not mentioned in the published paper. If the pesticide was looked for but not detected, the result is shown as less than the sensitivity of the method used when this value is known; or, if the limit of sensitivity is not known, the result is simply shown as not detectable (nd).
[c] Only 64 samples were tested for dieldrin content.
[d] Only 20 samples were tested for BHC content.

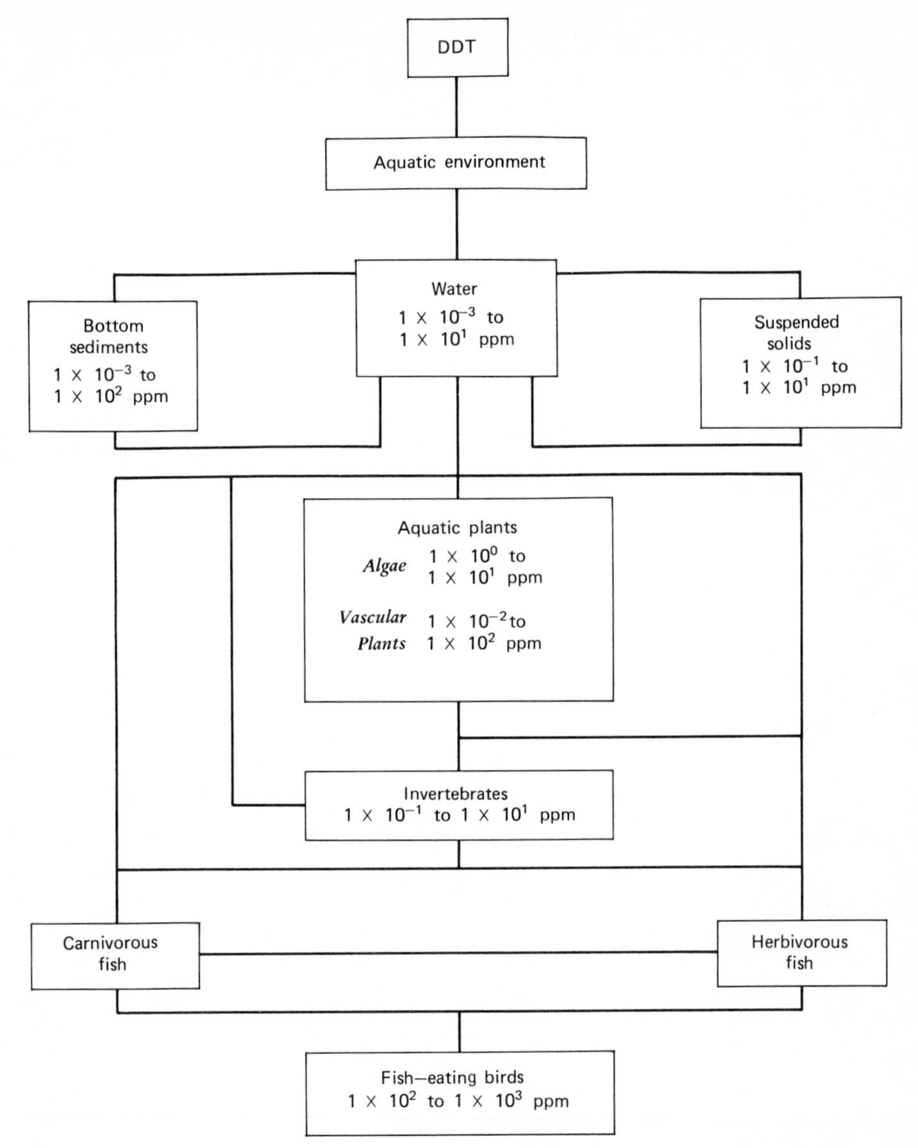

Figure 14.15. Model for the distribution of DDT in the aquatic ecosystem (from Dawson et al., 1970).

Table 14.13 lists examples of these reactions (for a further recent review of pesticide metabolism and its environmental impact, see Conney and Burns, 1972). Some of the reactions of pesticides are fascinating, for example, on the basis of their studies of the conversion of DDT to DDD under anaerobic conditions (including sewage sludge) Zoro et al. (1974) suggest that iron porphyrins—ubiquitous in both living and dead organic matter—are involved.

Some of these dangerous chemicals and their residues are carcinogenic (but not at levels likely to be found in man; Anon., 1972*b*), mutagenic (Epstein and Legator, 1971), their effects are cummulative and synergistic (Lichtenstein et al., 1973), and

Table 14.13 Some Metabolic Reactions Common to Pesticides (From Dugan, 1972)

Pathway	Schematic	Pesticide That Undergoes This Reaction
1. Oxidation		
a. Hydroxylation	$Cl—C_6H_3(Cl)—O—CH_2—C(=O)—OH \rightarrow Cl—C_6H_3(OH)—O—CH_2—C(=O)—OH$	2,4-D
b. Side-chain oxidation	$Cl—C_6H_4—C(H)(CCl_3)—C_6H_4—Cl \rightarrow Cl—C_6H_4—C(OH)(CCl_3)—C_6H_4—Cl$	DDT
c. Ether cleavage	$Cl—C_6H_3(Cl)—O—CH_2—C(=O)—OH \rightarrow Cl—C_6H_3(Cl)—OH$	2,4-D
d. Sulfoxide formation	$R—S—CH_3 \rightarrow R—S(O)—CH_3$	Phorate
e. *N*-oxide formation	$R—O—P(N—(CH_3)_2)(N—(CH_3)_2) \rightarrow R—O—P(N(O)—(CH_3)_2)(N—(CH_3)_2)$	Schraden
2. Dehydrogenation and dehydrohalogenation	$Cl—C_6H_4—CH(C—Cl_3)—C_6H_4—Cl \rightarrow Cl—C_6H_4—C(=CCl_2)—C_6H_4—Cl$	DDT
3. Reduction	$R—NO_2 \rightarrow RNH_2$	DNOC
4. Conjugation		
a. Amide formation	$R—NH_2 + R'—COOH \rightarrow R—N(H)—C(=O)—R$	Amitrole
b. Metal complex	$R_2—N + MeX \rightarrow (R_2N)Me)X$	Amitrole
c. Glucoside and glucuronic acid	R—OH + glucuronic acid → R—OCH(O—CH(COOH)—CH—OH)—CH(OH)—CH(OH)	Barthrin

(*Continued*)

Table 14.13 (*Continued*)

Pathway	Schematic	Pesticide That Undergoes This Reaction
d. Sulfate	$TOH + SO_4^{=} \rightarrow R{-}O{-}S(=O)_2{-}O^{+}$	Biphenyl
5. Hydrolytic		
a. Cleavage of esters	$R{-}C(=O){-}OR' \rightarrow R{-}C(=O){-}OH + HOR'$	Malathion
b. Cleavage of amides	$R{-}C(=O){-}N(H){-}R' \rightarrow R{-}C(=O){-}OH + R'{-}NH_2$	Dimethoate
6. Exchange reactions	$R{-}O{-}P(=S){-}(OR)_2 \rightarrow R{-}O{-}P(=O){-}(OR)_2$	Parathion
7. Isomerization	$R{-}O{-}P(=S){-}(OR)_2 \rightarrow R{-}S{-}P(=O){-}(OR)_2$	Most organophosphates

there is evidence that their toxicity is greatly increased in protein deficient animals (Boyd, 1973). Pesticides are exceedingly persistent in soils and sediments (Table 14.14). Although pesticide use in the United States (NOT production) declined from about 70,000,000 ton/yr in 1962 to 30,000,000 ton/yr in 1969, there was no parallel decrease in their levels in soils or even waters (orchard soils may contain as much as 52 ppm DDT, and DDT levels in rivers and lakes are still close to their

Table 14.14 Persistence of Some Common Pesticides in Soil (Reprinted from *Cleaning Our Environment—The Chemical Basis for Action*, a report by the Subcommittee on Environmental Improvement, Committee on Chemistry and Public Affairs, American Chemical Society, 1969, p. 205. Reprinted by permission of the copyright owner.)

Insecticide	Type	Time for 50% of Applied Dose to Disappear	Time to Reach Residue Level of 0.1 ppm (3% of applied dose)
Aldrin	Chlorinated Hydrocarbon	2 months	—
Carbaryl (Sevin)	Carbamate	1 month	—
Phorate (Thimet)	Organophosphorus	1 month	—
Azinphosmethyl (Guthion)	Organophosphorus	20 days	—
Parathion	Organophosphorus	20 days	90 days
Methyl Parathion	Organophosphorus	—	30 days
Malathion	Organophosphorus	—	8 days

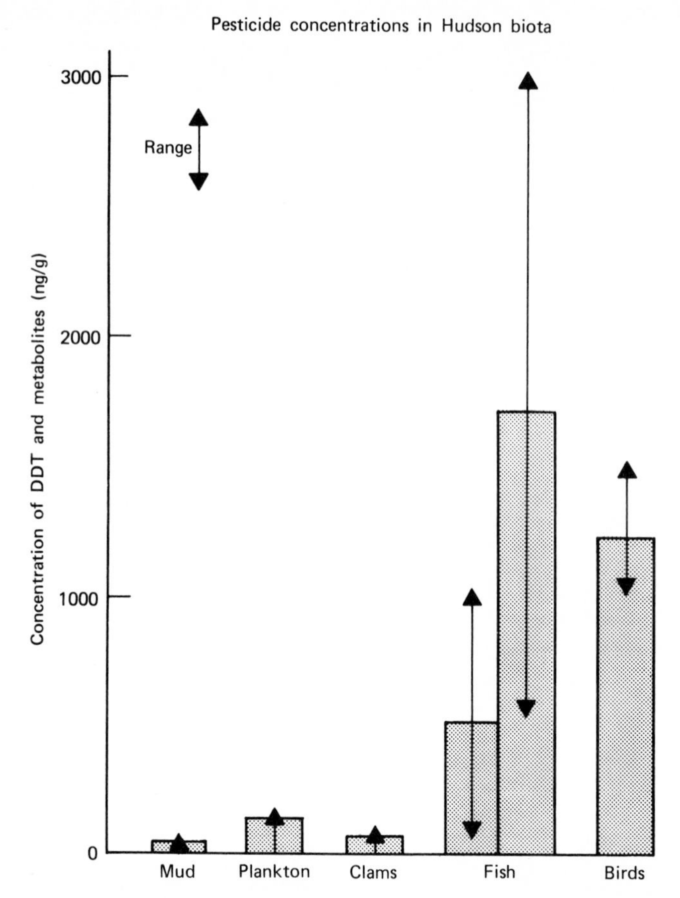

Figure 14.16. Pesticide concentrations in Hudson River area biota (Reprinted with permission from G. P. Howells et al., *Environ. Sci. Tech.*, *4*, 26 (Z1970), © The American Chemical Society).

solubility limits, Anon., 1971*a*). Studies of the microbial degradation of the herbicide atraxine found the process much slower in submerged sediments than in aerated soils (Goswami and Green, 1971) or in laboratory samples (Albone et al., 1972, see also Graetz et al., 1970).

For years now every month in the scientific literature, more and more reports of the damage of pesticides and their residues to the living biosphere appear. Pesticides and their metabolic products have been found in just about every organism in which they have been sought including harbor porpoises (Gaskin et al., 1971), seals (Hensen et al., 1969), whales (Robinson et al., 1969), lake bacteria (Leshniowsky et al., 1970), lake trout (Youngs et al., 1972) and salmon (Anderson and Everhart, 1966), eels (Janicki and Kinter, 1971), crabs (Burnett, 1971), California sea lions (LeBoeuf and Bonnell, 1971), where they appear to increase the frequency of abortion and premature birth (DeLong, 1973), and, of course, man (in addition to Table 14.12 see Curley et al., 1973). Pesticides even damage the plants they are supposed to protect (Anon., 1974*a*). Some orchards have been so heavily treated with arsenic pesticides that the soil has become unproductive (Sheets, 1967). Inasmuch as pesticides tend to be highly indiscriminate, they may increase the population of a pest by destroying

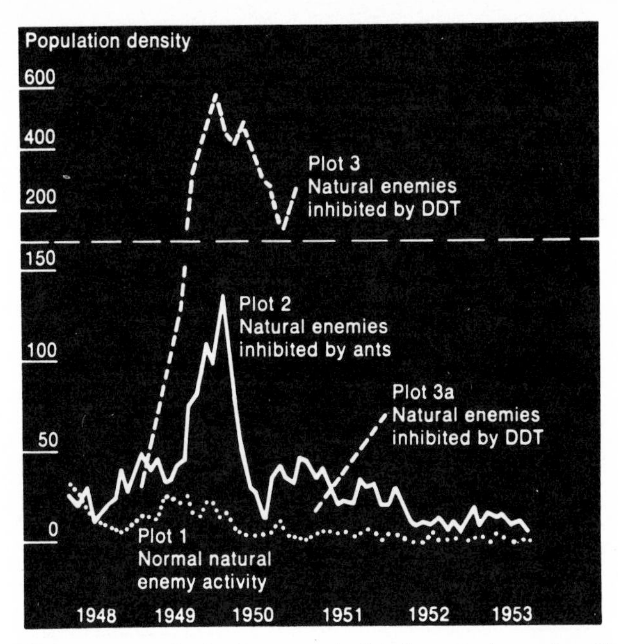

Figure 14.17. DDT is an enemy both of pests and of their natural enemies. Plot 1 describes the density of red scale on California lemon trees in experimental patches under conditions of excellent control by natural enemies. Plot 2 shows the density increase as the scale's natural enemies are inhibited by ants. Plot 3 shows the scale density as its natural enemies are inhibited by applications of DDT, which affected the growth of the scale only indirectly, by reducing the numbers of its enemies (from Huffake, *Technology Review*, edited at the Massachusetts Institute of Technology, © 1971 by the Alumni Association of M.I.T. with permission).

instead its natural enemies (Figure 14.17). There is evidence that pesticides contribute to fish diseases (Mawdesley-Thomas, 1972). Exposure to dieldrin appears to lower the resistance of fresh water fish to thermal stress (Silbergeld, 1973). The insect *Daphnia magna*, the first animal link in an important aquatic food chain concentrates DDT by a factor of 16,000 to 23,000 (Crosby and Tucker, 1971). Pesticide contamination may even be endangering the primary productivity of the oceans. DDT residues have been found in marine phytoplankton (Cox, 1970, who found a threefold increase in Monterey Bay, California, from 1955 to 1969) and is capable of reducing photosynthesis (Wurster, 1968). DDT reduces the salt tolerance of blue-green algae (Batterton et al., 1972), and its presence alters species population composition in mixed cultures of algae (Mosser et al., 1972*b*). Menzel et al. (1970) found that the sensitivity of response of marine phytoplankton to chlorinated hydrocarbons covers a wide range.

DDT alters fertility in experimental mammals (Heinrichs et al., 1971). Birds, particularly, have been the victims of pesticides because of their relatively high positions in predation chains (Figures 14.15 and 14.16). Not only have there been massive outright kills—for example, 50,000 birds were killed in the fall of 1973 in the Cote de Donna (Spain) bird refuge as a result of unauthorized pesticide use by nearby rice farmers—but pesticides are believed to increase drastically egg mortality (Cooke, 1973; Peakall, 1974). Eggshells consist largely of calcite ($CaCO_3$), and the Ca comes from the bird's bone reservoir. A hen can mobilize as much as 10% of her

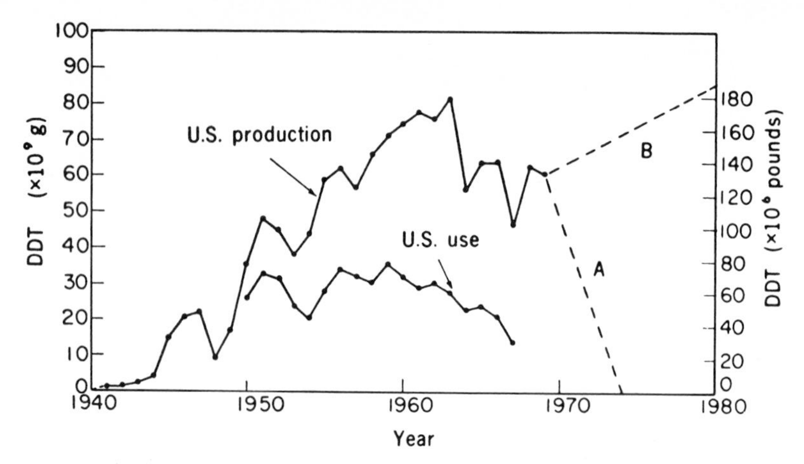

Figure 14.18. DDT production and use in the United States. Curve *A*, based on the assumption of declining use through 1974; Curve *B*, based on the assumption of increasing use through 1980 (from Woodwell et al., © 1971 by the American Association for the Advancement of Science, with permission of the AAAS).

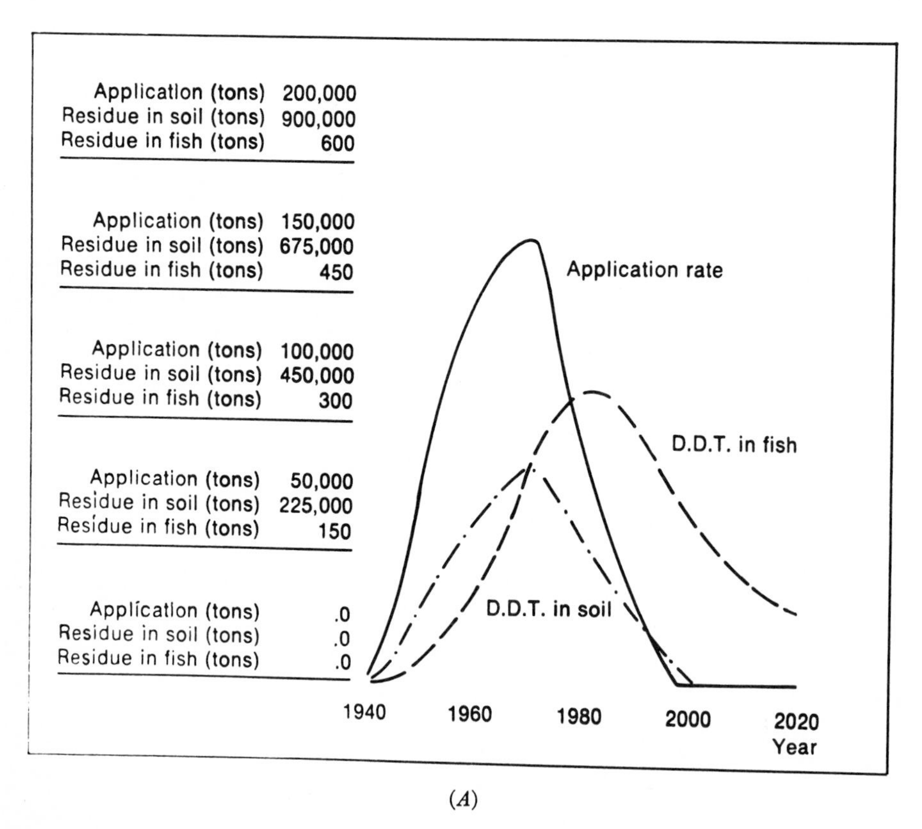

(*A*)

Figure 14.19. DDT in the environment. *A*, if the application of DDT is gradually stopped over the next 30 yr, the levels in soils and biota will decline; *B*, if DDT continues to be applied at its present rate while the levels in the soil would stabilize, those in fish would continue to climb. (From Boehm, *Technology Review* edited at the Massachusetts Institute of Technology, © 1972 by the Alumni Association of M.I.T. with permission.)

total bone substance in a day. This is made possible by the enzyme carbonic anhydrase, and it has been theorized that pesticides such as DDT inhibit the action of this enzyme (although some experiments fail to confirm this widely held notion; Dvorchik et al., 1971; Pocker et al., 1971). As a result, the eggshell becomes thin and is readily crushed by the brooding parent (Bitman et al., 1969; Heath et al., 1969; Lehner and Egbert, 1969; Porter and Wiemeyer, 1969; Weimeyer and Porter, 1970; Cade et al., 1971; Blus et al., 1972; Lakhani, 1973; Nisbet, 1973). If the hens are administered supplemental Ca their shells do not thin (Whitehead et al., 1972).

Pesticides pose a particularly acute problem for mankind.

> Agriculture is expected to maintain and increase food productivity in order to feed the increasing millions. . . . We can no longer afford to give up so large a share of the potential world food supply to pests and disease. At the same time, with more people crowded closer together the need for protecting the environment from pollution is more acute (Wilson, 1970).

And we must remember that pesticides are also essential in the battle against insect-carried diseases such as malaria. Pesticide manufacturers are strongly opposed to reductions in the use of DDT and other pesticides, and, even though pesticide use has been restricted in the United States, we continue to produce pesticides for overseas sale (Figure 14.18), despite the global nature of the contamination threat (Carter, 1973). Even if pesticide application were to continue at its present level or even cease contamination of soil and biosphere would continue (Figure 14.19).

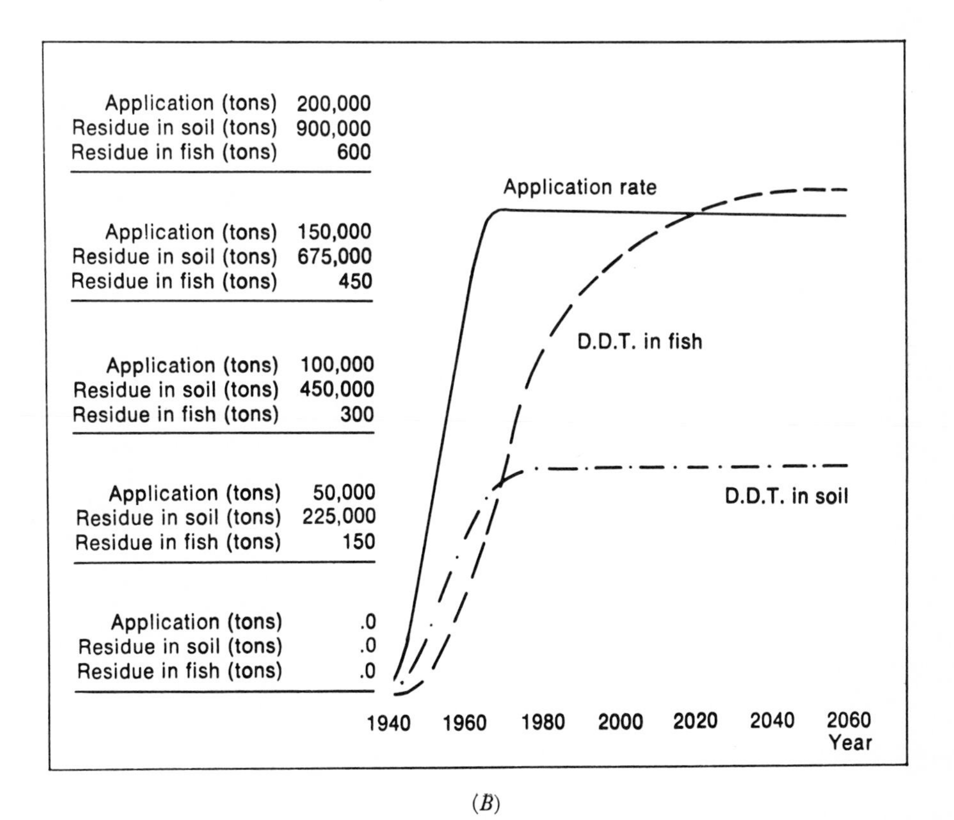

(*B*)

Figure 14.19. (*Continued*).

Cramer (1972) estimates that it will take the environment 25 to 110 yr to return to equilibrium after termination of DDT usage. Then too other voices have been raised in defense of the stuff, notably by Dr. Borlaug, "father of the Green Revolution" (Anon., 1971*b*, *d*). The role of pesticides in controlling disease carrying pests is often cited. It is interesting to note, however, that one African state, Sierra Leone, has officially expressed gratitude to the malarial mosquito for effectively preventing European exploitation. One fact remains clear, the worst pest on this planet has become man himself; "the major problem in past management lies in stabilizing the human population at a level appropriate to the needs and resources of the biosphere" (Brussard, 1971).

A large part of the problem with pesticides is that their application is so indiscriminate, falling on soil, water, air, and nontarget species as well as the pests. One way to circumvent this difficulty might be controlled release of pesticides from bio- or water-degradable polymers (Allan et al., 1971). Alternatives to pesticides in the war for the world's plant stocks, some of them mentioned elsewhere in this book, are more resistent crop varieties, controlled release of parasites (Huffake, 1971; Boehm, 1972), predators, and insect diseases including viruses (Marx, 1973*a*; Wade, 1973; for an account of a successful application of a rust to control a weed see Cullen et al., 1973); chemical and radioactive induction of insect sterility; natural insecticides such as pyrethrum (Anon., 1971*a*; Casida, 1973); chromosome rearrangement (Foster et al., 1972) and lethal mutation (Smith and von Borstel, 1972); and attractants and other phermones (Marx, 1973*b*) and hormones (Williams, 1967).

Let me return to the subject of pesticide paths in our environment. Only a small fraction of the pesticide applied to the field hits the target pest (Figure 14.14).Much of it ends upon the leaves of crops or in the soil from whence it is washed into waterways (Ritter et al., 1974). Another appreciable fraction ends up in the atmosphere, not only from spraying application, but by volatilization from soil and leaves under the hot sun (Swoboda et al., 1971). Bidleman and Olney (1974) have recently found that most of the DDT and other chlorinated hydrocarbons in the air over the Sargasso Sea is in gaseous form. Pesticide degradation products tend to have higher vapor pressures than the parent compounds and thus volatilize more readily (Cliath and Spencer, 1972). In the case of an arsenical herbicide under aerobic conditions, 35% of the amount applied to soil was converted to volatile compounds within 24 wk, and under anaerobic conditions, 71% (Woolson and Kearney, 1973).

Woodwell et al. (1971) and more recently Cramer (1973) have published models of DDT circulation in our environment, both considering atmospheric transport and the former investigators concentrating on the distribution in the biosphere. The paths taken by DDT are many (Figures 14.14 and 14.20). DDT is commonly applied as a liquid suspension spray. In the case of aerial forest spraying, only about 50% of the chemical hits the forest with much or the rest being dispersed in the air(Woodwell et al., 1971). Chlorinated hydrocarbons are also transported in the atmosphere and widely distributed throughout the world on airborne dusts (Risebrough et al., 1968) Table 14.15 summarizes DDT levels in various types of soil where it is retained with a mean lifetime of 4 to 5 yr. Woodwell et al. (1971) estimate that in the 1960's agricultural soils in the United States were retaining some 1.42×10^{10} g of DDT, and they list four mechanisms accounting for most of the DDT loss from soils:

1. Volatilization (including wind erosion of small soil particles),
2. Removal by the harvesting of organic material,

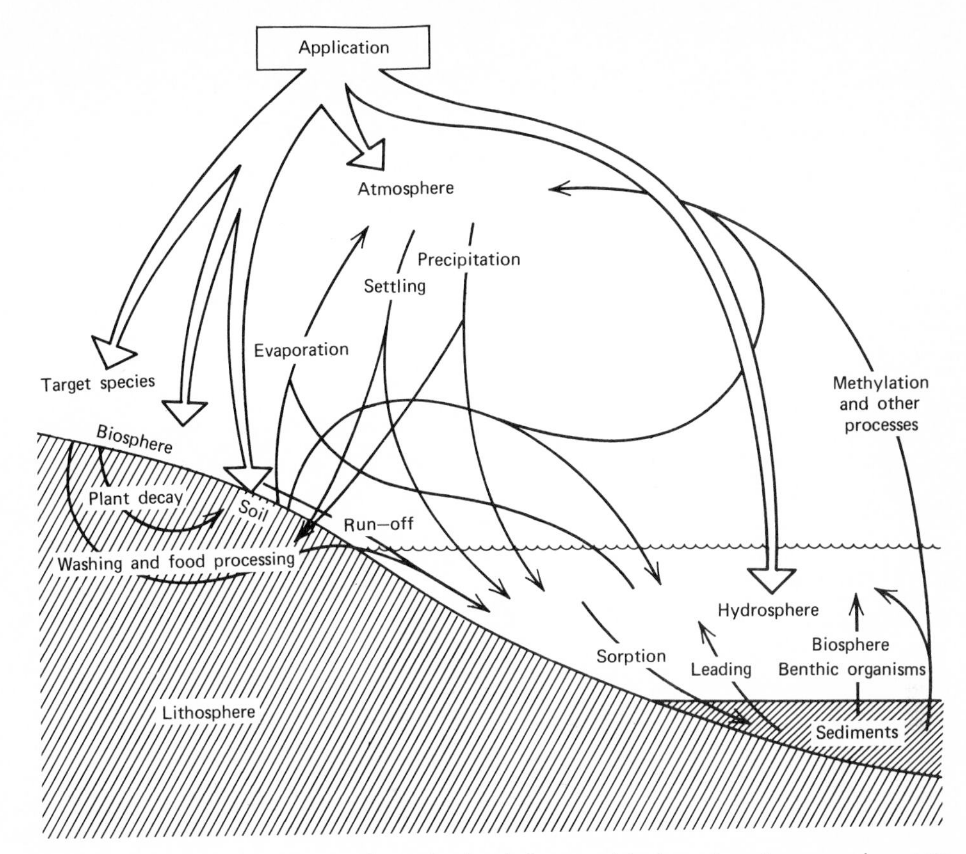

Figure 14.20. The paths of pesticides and other halogenated hydrocarbons in our environment. (1) volatilization (including wind erosion of small soil particles); (2) removal by the harvesting of organic material; (3) water run-off; and (4) chemical (including biotic) degradation.

3. Water run-off, and
4. Chemical (including biotic) degradation.

At 20°C, DDT has an appreciable vapor pressure (1.5×10^{-7} mm Hg), thus it readily gets into the atmosphere, as we have said, by volatilization as well as stray aerosols from spray application. It is removed from the atmosphere by four mechanisms:

1. Rainfall,
2. Settling of aerosols and contaminated dust particles,
3. Diffusion across the air/sea interface, and
4. Chemical degradation (including photodegradation)

of which rainfall is probably the most important. Rainfall removes an estimated 3×10^{10} g of DDT/yr from the atmosphere, mostly into the oceans, and this removal time is about 3 yr (Woodwell et al., 1971). More recently, Bidleman and Olney (1974) have estimated much shorter times of 40 to 50 days for the residence of PCB's and DDT in the marine atmosphere. In the oceans, DDT concentrates in surface slick

Table 14.15 DDT Residues in Soils of Agricultural and Nonagricultural Land in the United States[a] (From Woodwell et al., © 1971 by the American Association for the Advancement of Science with permission of the AAAS.)

Soil Sites	Sites Sampled (No.)	DDT Residues (g/m^2) Range	Mean
	Agricultural		
Orchards	14	0.34–22.1	6.0
Crops	24	0–0.87	0.24
Root Crops	48	0.045–5.73	1.25
Vineyards	2	2.13–3.18	2.69
Orchards	2	8.18–14.60	11.4
Vegetable Crops	10	0.07–9.52	2.62
Randomly Selected	41	0.002–1.30	0.148
Alfalfa Crops	12	0.06–0.98	0.336
Soybean Crops	43	0.004–4.03	0.986
	Nonagricultural		
Boreal Forests (sprayed)	3	0.179–0.258	0.213
Boreal Forests (unsprayed)			0.0045
Forest in Pennsylvania (unsprayed)		0.0003–0.0006	0.0004

[a] Data were selected because of large sample size or because they are the only data available.

lipids (Garrett, 1967; Seba and Corcoran, 1969), and circulates in the upper mixed layer. Transfer to deeper waters is slow, and in view of the high solubilities in fats and the low solubility in water, probably is proportional to the rate or general sedimentation of carbon into the abyss. Another report (Panel on Monitoring Persistent Pesticides in the Marine Environment, 1971) has concluded that of the nearly 20,000,000 tons of DDT manufactured since 1947 as much as 25% has found its way to the oceans where 0.1% of that production or 20,000 tons resides in marine biota. Finally, Woodwell et al. (1971), adding up the various biological reservoirs (Table 14.16), conclude that in the late 1960's there was about 5.4×10^9 g of DDT in the world's viable biosphere. The major paths of DDT envisioned by these authors are shown in Figure 14.21. They conclude that DDT levels in the atmosphere and in the mixed layer of the oceans lags behind application of the pesticide by only a few years. They also conclude that "despite the abundance, persistence, and worldwide distribution of DDT residues, they are not as freely available to the biota as might be assumed." The reasons for the biota's good luck in being so spared are unclear.

Pesticides are only one class of the many man-made organic chemicals, some of them very persistent and dangerous, that are being indiscriminately distributed throughout our environment. Goldberg (1971) has estimated that world production of chlorinated aliphatic hydrocarbons alone probably exceeds 3×10^6 ton/yr. These additional pollutants include detergents (discussed in Chapter 16); poly-

Table 14.16 DDT Residues in the Biota in the Late 1960s[a] (From Woodwell et al., © 1971 by the American Association for the Advancement of Science, with permission of the AAAS.)

Location	Dry Biomass (× 10^9 metric tons)	DDT Content (ppm)	Total DDT (× 10^8 g)
	Plant Biomass		
On Land			
Lakes and streams	0.04	0.010	0.004
Swamps and marshes	24	0.001	0.240
Terrestrial vegetation (forests, desert, savanna, grassland, tundra)	1814	0.0001	1.814
Agriculture	14	0.1	14.000
Total land	1852		15.058
In Oceans			
Open ocean algae	1.0	0.1	1.0
Continental shelf algae	0.3	1.0	3.0
Attached algae	2.0	1.0	20.0
Total ocean	3.3		24.0
Total plants	1855		39.06
	Animal Biomass		
On Land			
Feral mammals	0.009	1.0	0.09
Domestic mammals	0.17	1.0	1.7
Man	0.30	1.0	3.0
Birds	0.00024	1.0	0.002
In Oceans			
Fish	0.65	1.0	6.5
Mammals	0.055	1.0	0.55
Others (protozoa, coelenterates, annelids, nematodes, mollusks, echinoderms, arthropods)–	3.02	0.1	3.02
Total animals	4.20		14.86
Total DDT in the biota		5.4×10^9 g	

[a] Concentrations are expressed to the nearest order of magnitude only; ppm, parts per million.

chlorinated biphenyls (PCB's); chlorophenols (Higginbotham et al., 1968); heptachlor (Andrews et al., 1966); degreasing and cleaning solvents; "benzene hexachloride" (BHC) (Benezet and Matsumura, 1973); hexachlorobenzene (HCB), which caused 5000 cases of porphyria cutanea in Turkey (Schmid, 1960) and has been found in cow and human milk (Brady and Siyali, 1972); polyvinyl chlorides (PVC); various plastics, which now cover the surface of the oceans (Carpenter and Smith, 1972; Carpenter et al., 1972); chlorofluorocarbon (Su and Goldberg, 1973) aerosol pro-

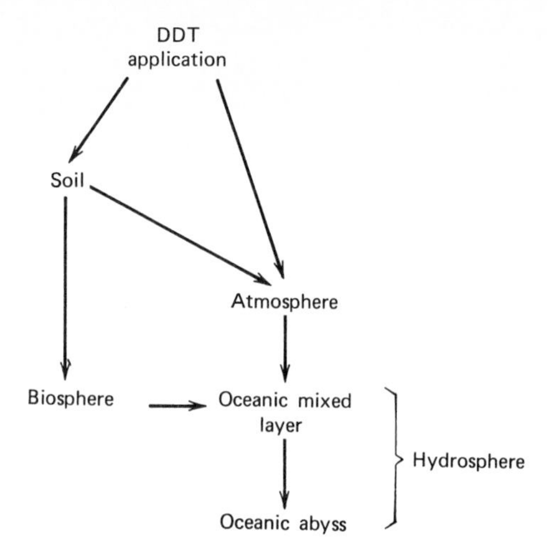

Figure 14.21. Major paths of DDT in our environment as discussed by Woodwell et al. (1971).

pellants that have been found in samples of Atlantic air and seawater (Lovelock et al., 1973; Murray and Riley, 1973) and fluoro-olefins (Cook and Pierce, 1973); bis-(chloromethyl)ether (Collier, 1972; Van Duuren et al., 1973); and phthalic acid esters (Marx, 1972). Like the pesticides, PCB's, for example, appear to have insinuated their way into every corner of our environment including the oceans (Harvey et al., 1973). In 1968, more than 1000 persons in Japan were poisoned by the chronic ingestion of rice oil contaminated with PCB's (Katsuki, 1969; Higuchi, 1971), and in 1971, another controversy arose when a number of Americans ate eggs containing PCB's from contaminated fish meal chicken feed (Pichirallo, 1971; see also Westoo et al., 1970, and Biros et al., 1970). In addition to eggs, PCB's have also been detected in milk and cheese (Maugh, 1972). PCB's are used primarily as a heat transfer fluid and as an additive to such things as sealants, rubber, paints (a Swedish study (Jensen et al., 1972) traced plankton contamination to the PCB's from paints on boat bottoms), inks, carbonless copy paper (Masuda et al., 1972), adhesives, plastics, and insecticides. Like pesticides, PCB's tend to concentrate in the organic segments of our environment, ranging from oil-polluted marine sediments to the fatty tissues of animals; however, they can be solubilized and thus mobilized by surfactants (Zitko, 1970). Dissolved salts tend to decrease pesticide (and presumably PCB) solubility as expected ("salting out") (Masterton and Lee, 1972), but, perhaps surprisingly, salt concentration and pH and temperature as well appear to have little effect on pesticide uptake and release by aquatic sediments (Huang, 1971). Some commercial PCB mixtures may contain exceedingly toxic impurities such as dibenzofurans. One estimate places the leakage of PCB's from plasticized materials into the atmosphere at 1000 to 2000 ton/yr and from lubricants, hydraulic fluids, and heat transfer fluids into waterways at 4000 ton/yr (Hammond, 1972). PCB residues are commonly found in human adipose tissue, possibly the result of eating contaminated fish. The PCB level in lake trout, it is significant to note, increases with their age (Figure 14.22). PCB's have serious biological effects that appear to be synergistic with pesticides; the presence of PCB's renders DDT and some organophosphate insecticides as much as 100% more toxic (Hammond, 1972). PCB's can reduce the growth rate of marine

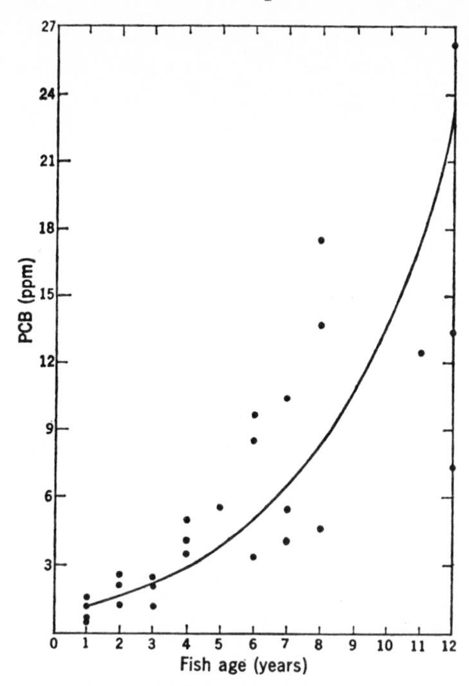

Figure 14.22. The concentration of PCB's in Cayuga Lake trout as a function of age (from Bache et al., © 1972 by the American Association for the Advancement of Science, with permission of the AAAS).

phytoplankton but do not appear to inhibit either fresh water or marine algae (Mosser et al., 1972*b*; see also Moore and Harriss, 1972, and Fisher et al., 1973). PCB's have been found in sea birds (Bogan and Bounne, 1972). Commercial PCB's are a mixture and experiments with birds have indicated that the ease of metabolism decreases with increasing chlorination (Bailey and Bunyan, 1972). The metabolism of PCB's in pigeons, rats, and trout has been investigated by Hutzinger et al. (1972).

14.6 THE DEPOSITION OF COAL AND PETROLEUM

The biosphere deposits many materials in the lithosphere ranging from limestone ($CaCO_3$) and chert (SiO_2) and coral reefs, through phosphate and nitrate deposits, sulfur and possibly iron formations, to the enrichment and deposition of some metal ores. But in this section, we confine our attention to only two such examples of biosphere–lithosphere interaction—the deposition of coal and petroleum. As we saw earlier, only a tiny fraction of the carbon inventory of our environment is in the viable biosphere (Figure 10.9), most is stored or dead material of biogenic origin. Of these reservoirs, next only to the 20,000,000 million metric tons of C stored in sediments, is the estimated 10,000 million metric tons stored as coal and oil. The world distribution of oil deposits was shown in Figure 4.15. There are many similarities and a few very important differences between the chemistries of coal and oil formation (Figure 14.23). Available evidence seems to point in the direction that coal had as

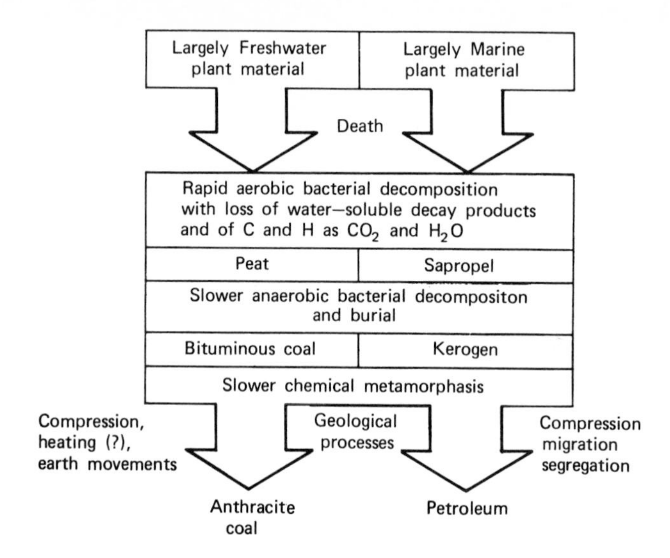

Figure 14.23. Comparison of the formation processes for coal and petroleum.

its original material plants in swamps, bogs, and other shallow fresh waters, whereas petroleum, while also derived from plant material, traces its origins to the marine environment. In both cases, the dead plant material underwent a long sequence of chemical transformations. As we saw in Chapter 11, biological and chemical attack is focused on the more susceptible portions of the large biomacromolecules, so there is a progressive loss of elements associated with reactive functional groups and a persistence of the carbon and hydrogen of the more refractory hydrocarbon backbone. This is shown in Tables 11.4 and 14.17. In the case of coal, every intermediate stage along the way has long been known to man from the origial woody tissue to the final anthracite. As the relative carbon content increases (Table 14.17), the "rank" or thermal content of the material as a fuel increases (Table 4.6) (Van Krevelen, 1961). In the case of petroleum, however, it has taken modern science to identify some of the more elusive intermediate steps. Organisms play a major role in the initial decomposition of plant material in moist soils (Dickinson and Pugh, 1974), fresh waters, and the marine environment. The chief constituents of plant material, as we saw earlier, are cellulose and lignin. Coal tar and other breakdown products of coal are largely aromatic, and since lignin is aromatic but cellulose is aliphatic, this leads to the conclusion that it is the lignin that is the primary material from which

Table 14.17 The Average Composition of Fuels (From Mason, 1966)

	C(%)	H(%)	N(%)	O(%)
Wood	49.65	6.23	0.92	43.20
Peat	55.44	6.28	1.72	36.56
Lignite	72.95	5.24	1.31	20.50
Bituminous Coal	84.24	5.55	1.52	8.69
Anthracite	93.50	2.81	0.93	2.72

coal is formed. This conclusion also nicely fits the fact that lignin is the more stable of the two, while cellulose is readily decomposed by microbial attack into H_2O, CO_2, CH_4, and aliphatic acids (Mason, 1966). Next to the increasing percentage of carbon, the other most obvious change is the loss of oxygen (Table 14.17). As rapid aerobic bacterial decomposition depletes the oxygen both from the dead plant material itself and from the surrounding environment, anoxic conditions will obtain, a condition that was enhanced by physical burial by more decomposing plant material and/or silt and other mineral depositions. The increase in the relative amount of nitrogen (Table 14.17) is revealing, since neither cellulose or lignin contain this element. This increase has been attributed to the death and decomposition of the decomposing bacteria, which may contain as much as 13% nitrogen. Finally, with deeper burial, as the supply of all oxidants becomes exhausted, biological processes cease and slow chemical metamorphoses alone continue. Inasmuch as fixed carbon tends to increase and volative substances to decrease with depth in coal seams, pressure and possibly temperature were probably important in accelerating these changes. If pressure and temperature are great, the coal may undergo a final phase transition to graphite. Geologically solid coal in contrast to liquid petroleum tends to remain where it is formed and is moved only by processes that displace the rocks of the Earth's crust.

Turning now from coal to petroleum, oil also is of plant origin, presumably marine plankton, although some petroleum is nonmarine in origin (Didyk and McCarthy, 1971). Hypotheses postulating a nonbiological origin of petroleum are now out of fashion, and only recently has evidence, in the form of steroid carboxylic acids, come to light of a possible animal contribution to the genesis of oil (Anon., 1973*d*). Oil is a much less oxidized form than is coal, that is to say it still retains a great deal of its hydrogen unoxidized, yet oil is not an intermediate in the formation of coal. Clearly there must be some very important differences between the circumstances of petroleum and coal formation. Oil is a more delicate substance than coal, thus principle among these differences must be the possibility that petroleum is formed under more gentle conditions than is coal. Crude oil contains porphyrins and complex N-containing hydrocarbons that are unstable above 200°C; therefore, high temperature cannot be a factor in oil formation for oil would be readily destroyed by thermal conditions that would only cause an increase in the rank of coal. Second, it would seem reasonable to postulate that the transition from aerobic to anaerobic conditions is more rapid in the case of oil, thereby sparing the H's of the hydrocarbons from oxidative removal and slowing the rapid destruction of cellulose material. This makes sense inasmuch as deeper marine waters are less penetrable by oxygen than the shallow waters of swamps and bogs, and sedimentation, especially in delta regions, and burial might be more rapid in the former case. In fact, we can envision very rapid burial by the slumping of marine sediment deposits. Just as peat is the precursor of coal, the black mud known as sapropel (Figure 14.23), formed by putrefaction in anoxic waters, is believed to be the precursor of petroleum. Black Sea sapropel contains as much as 35% organic matter, while the average marine sediments contain only about one-tenth this amount. Other factors that may contribute are the small size of planktonic organisms in contrast to land plants such as mosses, trees and ferns, and the ample supply of water for a longer stretch of time during the metamorphic processes in the marine as compared to the terrestrial environment. The high salinity of seawater may also be still another factor by "salting out" or decreasing the solubility of hydrocarbons. The aliphatic acids, lost in the case of coal (see previous discussion), in the case of petroleum are retained and reduced further to hydrocarbons. Just as coal is still being formed in areas such as the Great Dismal Swamp of Virginia, so oil is still probably being formed in places such as the Gulf of Mexico. The hydrocarbons in recent sediments show a strong preponderance of molecules with an even number of carbon atoms and low molecular weights are absent, whereas in oil there is no such preference for even numbers, and the low molecular weight fraction is considerable (Mason, 1966). The organic material in some Gulf of Mexico cores, however, reveals a preference for odd-numbered carbon molecules (Aizenshtat et al., 1973).

Chlorophyll itself is unstable and is not found in geological samples, yet its pristane and porphyrin fragments are highly persistent and frequently encountered (Dilcher et al., 1970; Maxwell et al., 1971). Calvin (1969) has pointed out that nonphotosynthetic organisms tend to be composed of higher hydrocarbon derivatives than

Table 14.18 Geological Time-scale for Chemical and Biological Evolution (from Calvin, © Oxford University Press, 1969, with permission of the publisher).

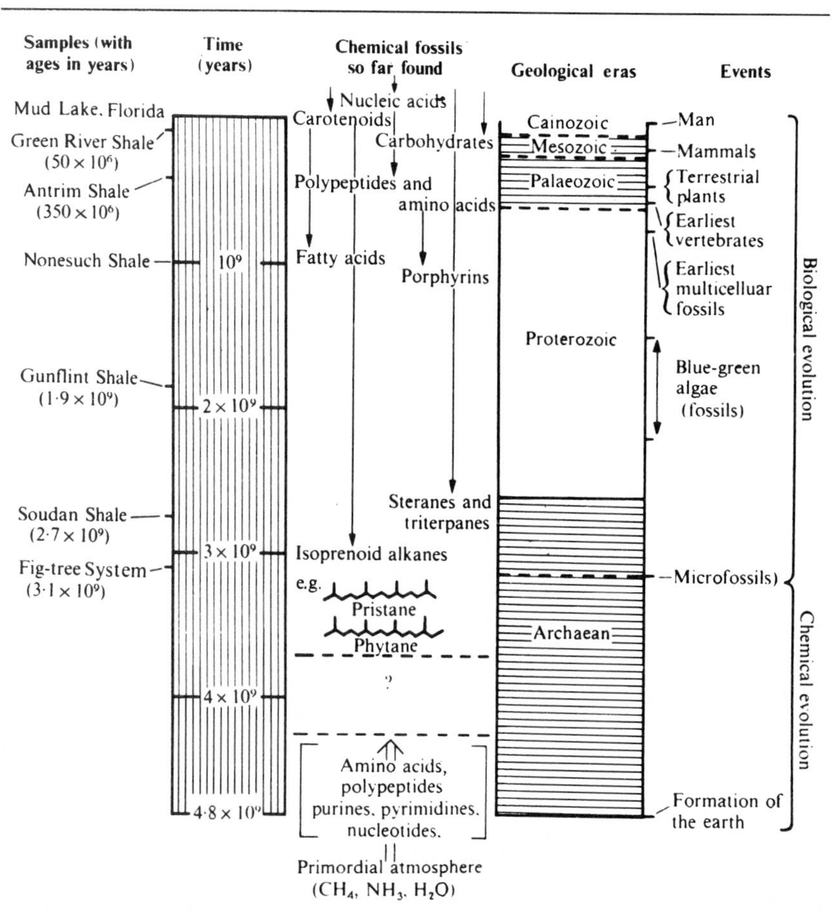

photosynthesizing organisms, from which he concludes that the 5000-yr-old organic-rich sediments from Mud Lake, Florida, result, not from the original biomaterial, but rather from a secondary product of the anaerobic bioprocesses that digests the original rain of carbohydrate material. Calvin's book (1969), by the way, is a fascinating account of the origin and evolution of the biosphere and how it changed the chemical composition of the lithosphere and of the clues that it has left in rocks. Table 14.18, taken from this book, parallels biological, geological, and chemical time tables (for a further table of geological time periods see Table A.19). Among the clues that have proved to be most helpful to scientists in tracing the changes that have and are occurring in biogenic lithospheric material are certain trace elements such as V and Ni, carbon isotope data (Oehler et al., 1972), optical isomers and racematization of optically active molecules (Bada et al., 1972; Kvenvolden et al., 1973), carbon chain length, chain branching, odd versus even numbers of carbon (Van Hoeven et al., 1969), and certain persistent key compounds such as purines, pristane, steranes (Balogh et al., 1971), perylene (Aizenshtat, 1973) porphyrins, amino acids (Hodgson et al., 1969), and fatty acids (Eglinton, 1969; Shimoyama and Johns, 1971; especially useful in recent sediments—see Farrington and Quinn, 1973, and Johnson and Calder, 1973).

Oil, unlike coal, segregates itself from saline solution and can migrate both horizontally and vertically through sediments, sandstones, and other porous rocks (see, for example, Al Shahristani and Al Atyia, 1972). The small temperature increases sometimes associated with continental drift at land mass margins may accelerate this migration (Tarling, 1973). The lighter the fraction, the more mobile it will be, thus the lightest fraction—natural gas—in addition to forming a capping over an oil pool may wander off to form a separate large accumulation. This ability to stray far from its point of formation has further helped to obscure the chemistry of petroleum's origins. Hydrocarbons are ubiquitous in the lithosphere. In addition to deposits of oil and gas, oil tars, oil sands, and oil shales, aliphatic hydrocarbons are even found in weathered limestone (Noonter et al., 1972).

BIBLIOGRAPHY

D. C. Abbott et al., *Nature*, *211*, 259 (1966).

F. B. Abeles et al., *Science*, *173*, 914 (1971).

Z. Aizenshtat, *Geochim. Cosmochim. Acta*, *37*, 559 (1973).

Z. Aizenshtat et al., *Geochim. Cosmochim. Acta*, *37*, 1881 (1973).

H. Al Shahristani and M. J. Al Atyia, *Geochim. Cosmochim. Acta*, *36*, 929 (1972).

E. S. Albone et al., *Environ. Sci. Tech.*, *6*, 914 (1972).

M. Alexander, in N. C. Brady (ed.), *Agriculture and the Quality of Our Environment*, Amer. Assoc. Adv. Sci., Pub. No. 85, Washington, D.C., 1967.

M. Alexander, *Introduction to Soil Microbiology*, Wiley, New York, 1961.

G. G. Allan et al., *Nature*, *234*, 349 (1971).

F. E. Allison, *Soil Organic Matter and Its Role in Crop Production*, Elsevier, Amsterdam, 1973.

Amer. Chem. Soc., *Cleaning Our Environment*, Amer. Chem. Soc., Washington, D.C., 1969.

R. B. Anderson and W. H. Everhart, *Trans. Amer. Fish. Soc.*, *95*, 160 (1966).

A. K. Andrews et al., *Trans. Amer. Fish. Soc.*, *95*, 297 (1966).

Anon., *Environ. Sci. Tech.*, *4*, 1098 (1970).

Anon., *Nature*, *233*, 299 (1971*a*).

Anon., *Nature*, *233*, 444 (1971*b*).

Anon., *Nature*, *233*, 441 (1971*c*).

Anon., *Nature*, *233*, 437 (i971*c*).

Anon., *Chem. Engr. News*, 19 (December 4, 1972*a*).

Anon., *Nature*, *234*, 420 (1972*b*).

Anon., *Handbook of Agricultural Charts*, U.S. Dept. Agriculture Handbook No. 455, U.S. Gov. Print. Off., Washington, D.C., 1973*a*.

Anon., *Chem. Engr. News*, 35 (May 21, 1973*b*).

Anon., *Chem. Engr. News*, 9 (March 5, 1973*c*).

Anon., *Chem. Engr. News*, 13 (June 25, 1973*d*).

Anon., *Nature*, *247*, 337 (1974*a*).

Anon., *Chem. Engr. News*, 10 (February 18, 1974*b*).

P. Antommaria et al., *Science*, *150*, 1476 (1965).

C. A. Bache et al., *Science*, *177*, 1191 (1972).

J. L. Bada et al., *Science*, *170*, 730 (1970).

S. Bailey and P. J. Bunyan, *Nature*, *236*, 34 (1972).

G. W. Bailey and J. L. White, *Agric. Food Chem.*, *12*, 324 (1964).

W. E. Baker, *Geochim. Cosmochim. Acta*, *37*, 269 (1973).

B. Balogh et al., *Nature*, *233*, 261 (1971).

J. E. Bardach, J. H. Ryther, and W. O. McLarney, *Aquaculture*, Wiley-Interscience, New York, 1972.

D. R. Bates and P. B. Hayes, *Planet. Space Sci.*, *15*, 189 (1967).

J. C. Batterton et al., *Science*, *176*, 1141 (1972).

I. Ben-Gera and A. Kramer, *Adv. Food Res.*, *17*, 132 (1969).

H. J. Benezet and F. Matsumura, *Nature*, *243*, 480 (1973).

T. F. Bidleman and C. E. Olney, *Science*, *183*, 516 (1974).

J. Biely et al., *Poultry Sci.*, 1502 (September 1972).

F. J. Biros et al., *Bull. Environ. Cont. Tox.*, *5*, 317 (1970).

J. Bitman et al., *Nature*, *224*, 44 (1969).

L. J. Blus et al., *Nature*, *235*, 376 (1972).

G. A. Boehm, *Tech. Rev.* (*M.I.T.*), *74*(8), 26 (July/August 1972).

J. A. Bogan and W. R. P. Bourne, *Nature*, *240*, 358 (1972).

J. T. Bonner, in E. Sondheimer and J. B. Simeone (eds.), *Chemical Ecology*, Academic, London, 1970.

F. R. Boucher and G. F. Lee, *Environ. Sci. Tech.*, *6*, 538 (1972).

E. M. Boyd, *Protein Deficiency and Pesticide Toxicity*, Thomas, Springfield, Ill., 1973.

M. N. Brady and D. S. Siyali, *Med. J. Austral.*, *1*, 158 (1972).

L. R. Brown, *Seeds of Change*, Praeger, London, 1970*a*.

L. R. Brown, *Sci. Amer.*, *223*(3), 162 (September 1970*b*).

P. F. Brussard, *Science*, *174*, 397 (1971).

M. J. Buchauer, *Environ. Sci. Tech.*, *7*, 131 (1973).

R. Burnett, *Science*, *174*, 606 (1971).

E. H. Bush, in Symp. Control Dispos. Cotton-Ginning Wastes, Dallas, Texas, 1966, U.S. Dept. HEW, Durham, N.C., 1967.

T. J. Cade et al., *Science*, *172*, 955 (1971).

M. Calvin, *Chemical Evolution*, Oxford Univ. Press, New York, 1969.

E. J. Carpenter and K. L. Smith, Jr., *Science*, *175*, 1240 (1972).

E. J. Carpenter et al., *Science*, *178*, 749 (1972).

L. J. Carter, *Science*, *181*, 143 (1973).

J. E. Casida (ed.), *Pyrethrum*, Academic, New York, 1973.

Coca-Cola Co., Foods Div., "Treatment of Citrus Processing Wastes," Rept. to Environ. Protect. Agency, Orlando, Fla., October 1970.

M. M. Cliath and W. F. Spencer, *Environ. Sci. Tech.*, *6*, 910 (1972).

L. Collier, *Environ. Sci. Tech.*, *6*, 930 (1972).

A. H. Conney and J. J. Burns, *Science*, *178*, 576 (1972).

E. W. Cook and J. S. Pierce, *Nature*, *242*, 337 (1973).

A. S. Cooke, *Environ. Pollut.*, *4*, 85 (1973).

J. Cornelius, *Calif. Vector News*, *16*, 1 (May 1969).

R. C. Cowen, *Tech. Rev.* (*M.I.T.*), *73*(6), 6 (April 1971).

J. L. Cox, in M. A. Schwartz (ed.), *Proc. 1st Mineral Waste Utilization Symp.*, Chicago, 1968.

J. L. Cox, *Science*, *170*, 71 (1970).

J. Cramer, *Atm. Environ.*, *7*, 241 (1973).

D. G. Crosby and R. K. Tucker, *Environ. Sci. Tech.*, *5*, 714 (1971).

F. L. Cross, Jr., and R. W. Ross, *Food Engineering*, *41*, 83 (1969).

J. M. Cullen et al., *Nature*, *244*, 462 (1973).

A. Curley et al., *Nature*, *242*, 338 (1973).

G. W. Dawson et al., "Control of Spillage of Hazardous Polluting Substances," Water Pollut. Control Res. Ser. 15090 Foz 10/70, U.S. Gov. Print. Off., Washington, D.C., 1970.

W. B. Deichmann et al., *Science*, *172*, 275 (1971).

R. L. DeLong, *Science*, *181*, 1168 (1973).

C. H. Dickinson and G. J. F. Pugh (ed.), *Biology of Plant Litter Decomposition*, Academic, New York, 1974.

B. M. Didyk and E. D. McCarthy, *Nature Phys. Sci.*, *232*, 103 (1971).

D. L. Dilcher et al., *Science*, *168*, 1447 (1970).

R. N. Doetsch and T. M. Cook, *Introduction to Bacteria and Their Ecology*, University Park Press, Baltimore, 1973.

K. H. Domsch and W. Gams, *Fungi in Agricultural Soils*, Halsted, New York, 1973.

D. L. Downing, in *Agricultural Wastes*, Cornell Univ. Conf. Agric. Waste Manag., Syracuse, N.Y., 1971.

P. R. Dugan, *Biochemical Ecology of Water Pollution*, Plenum, New York, 1972.

B. H. Dvorchik et al., *Science*, *172*, 728 (1971).

G. Eglinton, *Adv. Organ. Geochem.*, *31*, 1 (1969).

J. W. Eichelberger and J. J. Lichtenberg, *Environ. Sci. Tech.*, *5*, 541 (1971).

R. Eisler, *Underwater Naturalist*, *5*, 11 (1968).

F. C. Elliot, in *Symposium on the Control and Deposit of Cotton-Ginning Wastes, Dallas, Texas, 1966*, U.S. Dept. HEW, Durham, N.C., 1967.

S. S. Epstein and M. S. Legator, *The Mutagenicity of Pesticides*, M.I.T. Press, Cambridge, 1971.

R. Ewell, *Chem. Engr. News*, 106 (December 14, 1964).

J. W. Farrington and J. G. Quinn, *Geochim. Cosmochim. Acta*, *37*, 259 (1973).

N. S. Fisher et al., *Nature*, *241*, 548 (1973).

G. G. Foster et al., *Science*, *176*, 875 (1972).

W. D. Garrett, *Deep-Sea Res.*, *14*, 221 (1967).

E. E. Gaskin et al., *Nature*, *233*, 499 (1971).

C. B. Gilbertson et al., *J. Water Pollut. Control Fed.*, *43*, 483 (1971).

C. D. Gish and G. E. Christensen, *Environ. Sci. Tech.*, *7*, 1060 (1973).

E. D. Goldberg, in D. W. Hood (ed.), *Impingement of Man on the Oceans*, Wiley-Interscience, New York, 1971.

C. G. Golueke, "Comprehensive Studies of Solid Waste Management," 3rd Ann. Rept., EPA, U.S. Gov. Print. Off., Washington, D.C., 1971.

A. Gomez-Pompa et al., *Science*, *177*, 762 (1972).

K. Goswani and R. E. Green, *Envir. Sci. Tech.*, *5*, 426 (1971).

C. A. I. Goring and J. W. Hamaker (eds.), *Organic Chemicals in the Soil Environment*, Dekker, New York, 1972.

D. A. Graetz et al., *J. Water Pollut. Control Fed.*, *42*, R76 (1970).

R. S. Green et al., in N. C. Brady (ed.), *Agriculture and the Quality of Our Environment*, Amer. Assoc. Adv. Sci., Pub. No. 85, Washington, D.C., 1967.

F. A. Gunther and J. D. Gunther (eds.), *Residues of Pesticides and Other Foreign Chemicals in Foods and Feeds*, *Residue Review*, Vol. 35, Springer-Verlag, New York, 1971.

A. L. Hammond, *Science*, *175*, 155 (1972).

R. Haque and W. R. Coshow, *Environ. Sci. Tech.*, *5*, 139 (1971).

S. A. Hart, *Agric. Engr.*, 72A (December 1968).

S. A. Hart and W. C. Fairbank, in *2nd Nat. Symp. Poultry Ind. Waste Manag.*, 1964.

R. Hartung and G. W. Klinger, *Environ. Sci. Tech.*, *4*, 407 (1970).

G. R. Harvey et al., *Science*, *180*, 643 (1973).

T. Hattori, *Microbial Life in the Soil*, Dekker, New York, 1973.

R. D. Hauck et al., *Science*, *177*, 453 (1972).

W. J. Hayes et al., *Life Sci.*, *4*, 1611 (1965).

W. J. Hayes, in *Scientific Aspects of Pest Control*, Nat. Acad. Sci. U.S.A.–Nat. Res. Council, Pub. No. 1402, Washington, D.C., 1966.

R. G. Heath et al., *Nature*, *224*, 47 (1969).

W. L. Heinrichs et al., *Science*, *173*, 642 (1971).

G. R. Higginbotham et al., *Nature*, *220*, 703 (1968).

C. Holden, *Science*, *173*, 410 (1971).

G. W. Hodgson et al., *Geochim. Cosmochim. Acta*, *33*, 532 (1969).

S. R. Hoover, *Science*, *183*, 824 (1974).

S. R. Hoover and L. B. Jasewicz, in N. C. Brady (ed.), *Agriculture and the Quality of Our Environment*, Amer. Assoc. Adv. Sci., Pub. No. 85, Washington, D.C., 1967.

G. P. Howells et al., *Environ. Sci. Tech.*, *4*, 26 (1970).

O. Hutzinger et al., *Science*, 178, 312 (1972).

J. C. Huang, *J. Water Pollut. Control Fed.*, *43*, 1739 (1971).

C. B. Huffake, *Tech. Rev.* (*M.I.T.*), *73*, 31 (June 1971).

P. E. Inman et al., *Science*, *172*, 1229 (1971).

R. H. Janicki and W. B. Kinter, *Science*, *173*, 1146 (1971).

S. A. Jensen et al., *Nature, 224*, 247 (1969).

S. Jensen et al., *Nature, 240*, 358 (1972).

R. W. Johnson and J. A. Calder, *Geochim. Cosmochim. Acta, 37*, 1943 (1973).

M. K. John et al., *Environ. Sci. Tech., 6*, 555 (1972).

S. Katsuki, *Fukuoka Igaku Zasshi, 60*, 409 (1969).

W. G. Kirk and M. Koger, Inst. Food Agric. Sci. Bull. No. 739, Univ. Florida, Gainesville (July 1970).

D. H. Klein, *Environ. Sci. Tech., 6*, 560 (1972).

L. Koenig and W. Barker, in *Tech.-Econ. Study Solid Waste Disposal Needs Practices*, U.S. Dept. HEW, Rockville, Md., 1969.

D. H. Kohl et al., *Science, 174*, 1331 (1971).

K. A. Kvenvolden et al., *Geochim. Cosmochim. Acta, 37*, 2215 (1973).

J. V. Lagerwerff, in N. C. Brady (ed.), *Agriculture and the Quality of Our Environment*, Amer. Assoc. Adv. Sci. Pub. No. 85, Washington, D.C., 1967.

K. H. Lakhani, *Nature, 242*, 340 (1973).

B. J. LeBoeuf and M. L. Bonnell, *Nature, 234*, 108 (1971).

G. F. Lee, *Factors Affecting the Transfer of Materials Between Water and Sediments*, Univ. Wisconsin Water Resources Center Lit. Res. No. 1 (July, 1970).

K. E. Lee and T. G. Wood, *Termites and Soil*, Academic, New York, 1971.

P. N. Lehner and A. Egbert, *Nature, 224*, 1218 (1969).

H. V. Leland et al., *Environ. Sci. Tech., 7*, 833 (1973).

W. O. Leshniowsky et al., *Science, 169*, 993 (1970).

E. P. Lichtenstein et al., *Science, 181*, 847 (1973).

R. C. Loehr, *J. Sanit. Engr. Div. Proc. Amer. Soc. Civil Engrs., 95*, 189 (1969).

R. C. Lehr, *J. Water Pollut. Control Fed., 43*, 668 (1971).

E. C. Loomis, *J. Milk Food Tech., 36*, 57 (1973).

C. R. Longwell and R. F. Flint, *Introduction to Physical Geology*, Wiley, New York, 1955.

J. E. Lovelock et al., *Nature, 241*, 194 (1973).

P. F. Low et al., *Science, 161*, 897 (1968).

J. M. Madden and J. N. Dornbush, in *Proc. Internat. Symp. Livestock Wastes*, Amer. Soc. Agric. Engrs., St. Joseph, Mich., 1971.

J. L. Marx, *Science, 178*, 46 (1972).

J. L. Marx, *Science, 181*, 833 (1973*a*).

J. L. Marx, *Science, 181*, 736 (1973*b*).

B. Mason, *Principles of Geochemistry*, Wiley, New York, 1966.

Y. Masuda et al., *Nature, 237*, 41 (1972).

W. L. Masterton and T. P. Lee, *Environ. Sci. Tech., 6*, 919 (1972).

F. Matsumura et al., *Science, 170*, 1206 (1970).

T. H. Maugh, *Science, 180*, 578 (1973).

T. H. Maugh, *Science, 178*, 388 (1972).

L. E. Mawdesley-Thomas, *Nature, 235*, 17 (1972).

J. R. Maxwell et al., *Quart. Rev. Chem. Soc., 25*, 571 (1971).

D. McLaren, *Soil Biochemistry*, Dekker, New York, 1967, 1971.

D. W. Menzel et al., *Science, 167*, 1724 (1970).

W. A. Mercer, *U.S. Agric. Res. Serv.*, ARS-72, 80, 4 (March 1970).

W. A. Mercer and J. W. Ralls, *Proc. Nat. Ind. Solid Waste Manag. Conf.*, Houston, Texas, 1970.

J. R. Miner, *J. Water Pollut. Control Fed.*, *43*, 991 (1971).

J. R. Miner et al., *J. Water Pollut. Control Fed.*, *42*, 391 (1970).

R. L. Mitchell, in F. E. Bear (ed.), *Chemistry of the Soil*, Reinhold, New York, 1964.

Y. Miyake, *Elements of Geochemistry*, Maruzen, Tokyo, 1965.

E. C. J. Mohr et al., *Tropical Soils*, Wiley, London, 1973.

S. A. Moore, Jr., and R. C. Harriss, *Nature*, *240*, 356 (1972).

V. P. Moore and D. L. McCaskill, in *Symp. Control Dispos. Cotton-Ginning Wastes*, Dallas, Texas, 1966, U.S. Dept. HEW, Durham, N.C., 1967.

K. Mopper and E. T. Degens, "Aspects of the Biochemistry of Carbohydrates and Proteins in Aquatic Environments," Woods Hole Oceanogr. Inst. Tech. Rept., No. 68 (September 1972).

A. R. Mosier et al., *Environ. Sci. Tech.*, *7*, 642 (1973).

J. L. Mosser et al., *Science*, *175*, 191 (1972b).

J. L. Mosser et al., *Science*, *176*, 533 (1972a).

A. J. Murray and J. P. Riley, *Nature*, *242*, 37 (1973).

A. I. Nasim et al., *Environ. Pollut.*, *2*, 1 (1971).

K. V. Natarajan, *J. Water Pollut. Control Fed.*, *42*, R184 (1970).

N. L. Nemerow, *Liquid Waste of Industry*, Addison-Wesley, Reading, Mass., 1971.

I. C. T. Nisbet, *Nature*, *242*, 341 (1973).

H. W. Noonter et al., *Geochim. Cosmochim. Acta*, *36*, 953 (1972).

E. P. Odum, *Fundamentals of Ecology*, Saunders, Philadelphia, 1959.

D. Z. Oehler et al., *Science*, *175*, 1246 (1972).

Panel on Monitoring Persistent Pesticides in the Marine Environment, *Chlorinated Hydrocarbons in the Marine Environment*, Nat. Res. Council, Washington, D.C., 1971.

D. B. Peakall, *Science*, *183*, 673 (1974).

A. M. Pendleton, in *Symp. Control Dispos. Cotton-Ginning Wastes*, Dallas, Texas, 1966, U.S. Dept. HEW, Durham, N.C., 1967.

T. J. Peterle, *Nature*, *244*, 620 (1969).

D. A. Phillips et al., *Science*, *174*, 169 (1971).

J. Pichirallo, *Science*, *173*, 899 (1971).

D. Pimental et al., *Science*, *182*, 443 (1973).

Y. Pocker et al., *Science*, *174*, 1336 (1971).

R. D. Porter and S. N. Wiemeyer, *Science*, *165*, 199 (1969).

J. R. Raper, in E. Sondheimer and J. B. Simeone (eds.), *Chemical Ecology*, Academic, London, 1970.

R. W. Risebrough et al., *Nature*, *216*, 589 (1967).

R. W. Risebrough et al., *Science*, *159*, 1233 (1968).

R. W. Risebrough, in M. W. Miller and G. G. Berg (eds.), *Chemical Fallout*, Thomas, Springfield, Ill., 1969.

W. F. Ritter et al., *Environ. Sci. Tech.*, *8*, 38 (1974).

J. Robinson et al., *Nature*, *214*, 1307 (1967).

J. H. Ryther, *Science*, *166*, 72 (1969).

M. L. Schafer et al., *Envir. Sci. Tech.*, *3*, 1261 (1969).

R. Schmid, *New England J. Med.*, *263*, 397 (1960).

M. Schnitzer and S. U. Khan, *Humic Substances in the Environment*, Dekker, New York, 1972.

J. Schultz et al., *J. Food Sci.*, *36*, 397 (1971).

K. Schutz et al., *J. Geophys. Res.*, *75*, 2230 (1970).

L. F. Seatz and H. B. Peterson, in F. E. Bear (ed.), *Chemistry of the Soil*, Reinhold, New York, 1964.

D. B. Seba and E. F. Corcoran, *Pestic. Monit. J.*, *3*, 190 (1969).

D. Shapley, *Science*, *182*, 147 (1973).

T. J. Sheets, in N. C. Brady (ed.), *Agriculture and the Quality of Our Environment*, Amer. Assoc. Adv. Sci., Pub. No. 85, Washington, D.C., 1967.

C. C. Sheppard (ed.), *Poultry Pollution*, Mich. State Univ. Res. Rept. No. 117 (July 1970).

C. W. Sheffield et al., in *Proc. 23rd Indust. Waste Conf.*, Purdue Univ. (May 1968).

A. Shimoyama and W. D. Johns, *Nature Phys. Sci.*, *232*, 140 (1971).

H. J. Siegelman and J. E. Ullman (eds.), *Waste Disposal in Selected Industries*, Hofstra Univ., Hempstead, N.Y., 1969.

E. K. Silbergeld, *Environ. Sci. Tech.*, *7*, 846 (1973).

M. P. Silverman and E. F. Munoz, *Science*, *169*, 985 (1970).

W. E. Small, *Tech. Rev. (M.I.T.)*, *73*, 49 (April 1971).

G. E. Smith, in N. C. Brady (ed.), *Agriculture and the Quality of Our Environment*, Amer. Assoc. Adv. Sci., Pub. No. 85, Washington, D.C., 1967.

R. H. Smith and R. C. von Borstel, *Science*, *178*, 1164 (1972).

R. F. Spalding and W. M. Sackett, *Science*, *175*, 629 (1972).

C. W. Stanley et al., *Environ. Sci. Tech.*, *5*, 430 (1971).

P. R. Stout and R. G. Burau, in N. C. Brady (ed.), *Agriculture and the Quality of Our Environment*, Amer. Assoc. Adv. Sci., Pub. No. 85, Washington, D.C., 1967.

I. L. Stevenson, in F. E. Bear (ed.), *Chemistry of the Soil*, Reinhold, New York, 1964.

C-W. Su and E. D. Goldberg, *Nature*, *245*, 27 (1973).

D. J. Swaine and R. L. Mitchell, *J. Soil Sci.*, *11*, 347 (1960).

A. R. Swoboda et al., *Environ. Sci. Tech.*, *5*, 141 (1971).

E. C. Tabor, *Trans. N.Y. Acad. Sci.*, *28*, 659 (1966).

E. P. Taiganides, "Agricultural Solid Wastes," in *Proc. Nat. Conf. Solid Waste Res.*, Chicago, 1963.

E. P. Taiganides, in N. C. Brady (ed.), *Agriculture and the Quality of Our Environment*, Amer. Assoc. Adv. Sci., Pub. No. 85, Washington, D.C., 1967.

D. H. Tarling, *Nature*, *243*, 277 (1973).

W. Thorne and H. B. Peterson, in N. C. Brady (ed.), *Agriculture and the Quality of Our Environment*, Amer. Assoc. Adv. Sci. Pub. No. 85, Washington, D.C., 1967.

R. C. Timberlake, in M. A. Schwartz (ed.), *Proc. 2nd Mineral Waste Utilization Symposium*, Chicago, 1970.

S. J. Toth, in F. E. Bear (ed.), *Chemistry of the Soil*, Reinhold, New York, 1964.

A. R. Townshend et al., *J. Water Pollut. Control Fed.*, *42*, 195 (1970).

UNESCO, *Soils and Tropical Weathering*, Paris, 1971.

B. L. Van Duuren et al., *Environ. Sci. Tech.*, *7*, 744 (1973).

W. Van Hoeven et al., *Geochim. Cosmochim. Acta*, *33*, 877 (1969).

D. W. Van Krevelen, *Coal*, Elsevier, New York, 1961.

J. Verduin, in N. C. Brady (ed.), *Agriculture and the Quality of Our Environment*, Amer. Assoc. Adv. Sci. Pub. No. 85, Washington, D.C., 1967.

N. Wade, *Science*, *177*, 678 (1972).

N. Wade, *Science*, *181*, 925 (1973).

J. Walsh, *Science*, *181*, 634 (1973).

C. H. Wadleigh, *Wastes in Relation to Agriculture and Forestry*, U.S. Dept. Agric. Misc. Pub. No. 1065, Washington, D.C., 1968.

G. Westoo et al., *Var Foda*, *22*, 9 (1970).

C. C. Whitehead et al., *Nature*, *239*, 411 (1972).

S. N. Wiemeyer and R. D. Porter, *Nature*, *227*, 737 (1970).

L. Wiklander, in F. E. Bear (ed.), *Chemistry of the Soil*, Reinhold, New York, 1964.

C. M. Williams, *Sci. Amer.*, *217*, 13 (July 1967).

G. W. Wilson, Jr., *Science*, *168*, 1419 (1970).

G. M. Woodwell et al., *Science*, *174*, 1101 (1971).

E. A. Woolson and P. C. Kearney, *Environ. Sci. Tech.*, *7*, 47 (1973).

C. F. Wurster, Jr., *Science*, *159*, 1474 (1968).

C. F. Wurster, Jr., and D. B. Wingate, *Science*, *159*, 979 (1968).

W. C. Yee, *Environ. Sci. Tech.*, *6*, 232 (1972).

G. Young, *Science*, *167*, 339 (1970).

W. D. Youngs et al., *Environ. Sci. Tech.*, *6*, 451 (1972).

V. Zitko, *Bull. Environ. Contam. Toxicol.*, *5*, 279 (1970).

J. A. Zoro et al., *Nature*, *247*, 235 (1974).

ADDITIONAL READING

Anon., *Fate of Organic Pesticides in the Aquatic Environment*, Adv. Chem. Ser. No. 111, Amer. Chem. Soc., Washington, D.C., 1971.

Anon., *Degradation of Synthetic Organic Molecules in the Biosphere*, Nat. Acad. Sci., Washington, D.C., 1972.

Anon., *Effects of DDT on Man and Other Mammals*, MSS Information Corp., New York, 1973.

Anon., *Evaluation of Some Pesticide Residues in Food*, World Health Organization, Geneva, 1973.

Anon., *Safe Use of Pesticides*, World Health Organization, Geneva, 1973.

F. M. Ashton and A. S. Crafts, *Mode of Action of Herbicides*, Wiley-Interscience, New York, 1973.

R. G. C. Bathurst, *Carbonate Sediments and Their Diagenesis*, Elsevier, Amsterdam, 1971.

A. Berg (ed.), *Nutrition, National Development and Planning*, M.I.T. Press, Cambridge, 1973.

P. W. Birkeland, *Pedology, Weathering, and Geomorphological Research*, Oxford Univ. Press, New York, 1974.

L. R. Brown and G. W. Finsterbusch, *Man and His Environment. Food*, Harper & Row, New York, 1972.

C. Clark, *The Economics of Irrigation*, Pergamon, Oxford, 1970.

J. C. Cruockshank, *Soil Geography*, Wiley, New York, 1972.

J. Faust and J. V. Hunter, *Organic Compounds in Aquatic Environments*, Dekker, New York, 1971.

C. Fest and K-J. Schmidt, *The Chemistry of Organophosphorous Pesticides*, Springer-Verlag, New York, 1973.

C. B. Heiser, Jr., *Seed to Civilization: The Story of Man's Food*, Freeman, San Francisco, 1973.

D. Hillel, *Soil and Water*, Academic, New York, 1971.

D. Hillel (ed.), *Optimizing the Soil Physical Environment Toward Greater Crop Yields*, Academic, New York, 1972.

C. B. Hunt, *Geology of Soils*, Freeman, San Francisco, 1972.

D. E. G. Irvine and B. Knights (eds.), *Pollution and the Use of Chemicals in Agriculture*, Butterworths, London, 1973.

P. J. Kramer, *Plant and Soil Water Relationships*, McGraw-Hill, New York, 1969.

H. F. Kraybill (ed.), *Biological Effects of Pesticides on Mammalian Systems*, N.Y. Acad. Sci., New York, 1969.

A. I. Levorsen and F. A. F. Berry, *Geology of Petroleum*, Freeman, San Francisco, 1967.

T. L. Lyon, H. O. Buckman, and N. C. Brady, *The Nature and Properties of Soils*, Macmillan, New York, 1952.

F. Matsumura, G. M. Boush, and T. Misato (eds.), *Environmental Toxicology of Pesticides*, Academic, New York, 1972.

N. N. Melnikov, *Chemistry of Pesticides*, Springer-Verlag, New York, 1971.

M. W. Miller and G. G. Berg (eds.), *Chemical Fallout*, Thomas, Springfield, Ill., 1969.

R. D. O'Brien, *Insecticides: Action and Metabolism*, Academic, New York, 1967.

B. Phillips (ed.), *Chemurgy for Better Environment and Profits*, The Chemurgic Council, New York, 1970.

R. L. Rabb and F. E. Guthrie (eds.), *Concepts of Pest Management*, North Carolina State Univ., Raleigh, 1970.

H. E. Sutton and M. I. Harris (eds.), *Mutagenic Effects of Environmental Contaminants*, Academic, New York, 1972.

F. M. Swain, *Non-Marine Organic Geochemistry*, Cambridge Univ. Press, New York, 1970.

A. S. Tahori (ed.), *Pesticide Chemistry*, Gordon & Breach, New York, 1972.

J. A. Taylor (ed.), *The Role of Water in Agriculture*, Pergamon, Oxford, 1970.

J. A. Wallwork, *Ecology of Soil Animals*, McGraw-Hill, New York, 1970.

W. C. White and D. N. Collins (eds.), *The Fertilizer Handbook*, The Fertilizer Institute, Washington, D.C., 1972.

R. White-Stevens (ed.), *Pesticides in the Environment*, Dekker, New York, 1971.

T. L. Willrich and G. E. Smith (eds.), *Agricultural Practices and Water Quality*, Iowa State Univ. Press, Ames, 1971.

E. J. Winter, *Water, Soil, and the Plant*, Macmillan, London, 1974.

15

ATMOSPHERE–BIOSPHERE AND ATMOSPHERE–HYDROSPHERE INTERACTIONS

15.1 INTRODUCTION

Earlier in this book (Chapter 2), we saw how the atmosphere of Earth differs chemically not only from the other planets of our solar system but also from the cosmos as a whole. For emphasis, its peculiarities are worth repeating and this is done so in Tables 15.1 and 15.2. Although alternative hypotheses have been advanced, the consensus of learned opinion is that the biosphere is responsible for the oxidizing nature of our atmosphere. The biosphere is responsible, not only for the creation of this exceptional chemical mixture, but also for the maintenance of its composition as well. In the Earth's distant geological past, the transition from a typical-reducing to an atypical-oxidizing environment was easily the most traumatic chemical event in the life of our planet. In addition to a history of the Earth's atmosphere, largely in terms of its oxygen content, this chapter explores the chemical cycles of the other two major constituents of our atmosphere—nitrogen and carbon dioxide. The contributions of the biosphere to the levels of such minor atmospheric constituents as CH_4, CO, and H_2S were touched on in Chapter 6. Finally, in the last part of this chapter, we turn our attention to chemical mass transport between atmosphere and oceans, the sea as a *source* of the marine aerosol as well as the settling out and washout of materials into the sea and the dissolution of atmospheric gases.

15.2 THE HISTORY OF THE EARTH'S ATMOSPHERE

There now appears to be universal consent that the present gaseous envelope of our planet is a secondary, not the primitive original atmosphere. Originally, the atmospheres of the planets possibly reflected cosmic abundances (Table 15.1), that is to say they were composed largely of hydrogen with lesser amounts of helium and perhaps methane. The massive major planets (Table 15.2) seem to have retained much of their original atmospheres intact, but it is clear that the atmospheres of the less massive terrestrial planets, including our Earth, have been the subjects of extensive changes with respect to their chemical compositions. The major processes responsible for these changes fall into three main categories:

1. Escape of lighter constituents,
2. Continuing outgassing of the planet's lithosphere, and
3. Specific chemical changes.

Table 15.1 Comparison of the Chemical Composition of the Earth's Atmosphere–Hydrosphere with the Cosmos as a Whole

Element	Cosmos	Earth Atmosphere–Hydrosphere	Crust
Hydrogen	1000	2.0	0.03
Helium	140		
Oxygen	0.68	9.9	0.62
Carbon	0.30	0.0001	0.0005

Table 15.2 Comparison of the Chemical Composition of the Earth's Atmosphere with the Other Planets in Our Solar System

Chondrites (Stony meteorites)		35% Oxygen	
Igneous Rocks		47% Oxygen	
Venus	terrestrial planets	CO_2	
Earth	terrestrial planets	N_2,	O_2, CO_2, H_2O
Mars	terrestrial planets	CO_2,	H_2O, CO, H, O
Jupiter	major planets	H_2,	He, CH_4, NH_3
Saturn	major planets	H_2,	CH_4
Uranus	major planets	H_2,	CH_4

Mason (1966) has given a more detailed classification broken down into two major classes, atmospheric additions and losses. For Earth:

Atmospheric Additions During Geological Time

1. Gases released by the crystallization of magmas (lithospheric outgassing),
2. Oxygen produced by photochemical dissociation of water vapor,
3. Oxygen produced by photosynthesis, and
4. Helium and argon from the radioactive decay of crustal radionuclides.

Atmospheric Losses During Geological Time

5. Loss of O_2 by the oxidation of H_2 to H_2O, Fe(II) to Fe(III), S-compounds to sulfate, Mn-compounds to MnO_2, and similar reactions,
6. Loss of CO_2 by the formation (via the biosphere) of fossil fuel deposits,
7. Loss of CO_2 by the formation of Ca and Mg carbonates,
8. Loss of N_2 by the formation of N-oxides in the air and by the action of nitrifying soil bacteria, and
9. Escape of H_2 and He from the Earth's gravitational field

Table 15.3 Production and Use of Oxygen (From Mason, 1966)

Total Estimated Production	
By photosynthesis in excess of decay	181×10^{20} g
By photodissociation of water vapor followed by hydrogen escape	1×10^{20} g
Total	182×10^{20} g
Total Estimated Use	
Oxidation of ferrous iron to ferric iron during weathering	14×10^{20} g
Oxidation of S^{2-} to SO_4^{2-} during weathering	12×10^{20} g
Oxidation of volcanic gases	
CO to CO_2	10×10^{20} g
SO_2 to SO_3	11×10^{20} g
H_2 to H_2O	$<150 \times 10^{20}$ g
Total	$<197 \times 10^{20}$ g
Total in Atmosphere	$\sim 12 \times 10^{20}$ g

and in the case of oxygen he gives a balance sheet based on the calculations of Holland (1962) (Table 15.3). Hydrogen disappears by escape, but also, at least in the case of Earth (and possibly Mars) by oxidation. As Rubey (1951) has suggested (recall our discussion of excess volatiles in the earlier section on the origin of the solutes in seawater), continued outgassing of the Earth's crust would have added those gases still associated with the emissions of volcanoes and hot springs (Table 15.4), notably H_2O, CO_2, CO, N_2, SO_2, and HCl. Some of these gases are oxidizing and would have tended to react with and remove highly reduced constituents of the primitive atmosphere, such as H_2, NH_3, and CH_4. An interesting hypothesis has been advanced to account, at least in part, for the disappearance of methane from the Earth's atmosphere. Lasaga et al. (1971) have proposed that the methane polymerized under the influence of intense ultraviolet solar radiation and that an "oil slick" as thick as 1 to 10 m may have been deposited on the planet's surface.

Holland (1962) has divided the evolution of the Earth's atmosphere into three stages (Table 15.5). Notice that his first-stage gases, in contrast to what we suggested previously, do not reflect cosmic abundances and that the chemical composition of the outgassing was different, while there still was appreciable free iron in the Earth's mantle before the segregation of the core.

Here we restrict out attention to the third stage, since this is the period (still continuing) during which our planet's atmosphere acquired its most striking chemical peculiarity—a high concentration of highly oxidizing molecular oxygen. And since it is the biosphere that is responsible for the separation of oxidized and reduced material (Figure 11.4), the evolution of the Earth's atmosphere is inseparable from the origin and evolution of life. The earliest organisms were certainly anaerobic and heterotropic (Chapter 10). If their metabolism produced a gas that eventually reached the atmosphere, that gas was probably the same as that produced by present day fermentationlike processes, namely CO_2 (Figure 15.1) with possibly some CH_4, H_2S, N_2, or NH_3 (see Ycas, 1972). The appearance of gaseous oxygen in the Earth's atmosphere was not simultaneous with the advent of photosynthesis (for a discussion

Table 15.4 Volume Percentages of Gases from Volcanoes, Rocks, and Hot Springs (From Rubey, 1964)

	Volcano Gases from Kilauea and Mauna Loa (26 samples)			Gases from Rocks: Basaltic Lava and Diabase (13 samples)			Gases from Rocks: Obsidian, Andesitic Lava, and Granite (17 samples)				Gases from Fumaroles of the Katmai Region and from Steam Wells and Geysers of California and Wyoming (23 samples)		
	Minimum	Maximum	Median	Minimum	Maximum	Median	Minimum	Maximum	Median		Minimum	Maximum	Median
CO_2	0.87	47.68	11.8	0.89	15.30	8.1	0.08	20.26	2.0	CO_2	0.03	1.24	0.02
CO	0.00	3.92	0.5	0.02	8.28	0.2	0.01	2.22	0.5	CO	—	0.01	tr
H_2	0.00	4.22	0.4	0.38	6.18	1.2	0.08	11.60	0.4	O_2	0.00	0.08	tr
N_2	0.68	37.84	4.7	0.27	7.21	2.0	0.03	3.90	1.2	CH_4	0.00	0.30	0.11
A	0.00	0.66	0.2	0.00	0.04	tr	0.00	0.02	tr	H_2	0.00	0.29	0.15
SO_2	0.00	29.83	6.4	—	—	—	—	—	—	$N_2 + A$	0.00	0.31	0.02
S_2	0.00	8.61	0.2	0.08	1.96	1.1	0.00	2.89	0.2	NH_3	—	0.02	0.01
SO_3	0.00	8.12	2.3	—	—	—	—	—	—	H_2S	0.00	0.10	0.02
Cl_2	0.00	4.08	0.05	0.06	1.33	0.5	0.01	10.59	0.5	HCl	0.01	0.57	0.06
F_2	—	—	—	0.00	14.12	3.8	0.25	7.80	2.3	HF	0.00	0.10	0.03
H_2O	17.97	97.09	73.5	71.32	92.40	83.1	69.44	98.55	92.9	H_2O	98.04	99.99	99.58
			100.0			100.0			100.0				100.00

Table 15.5 The Three Stages of the Chemical Evolution of the Earth's Atmosphere

	First Phase Original Exothermic Planetesimal Accretion and Outgassing of the Lithosphere Prior to Core Formation	Second Phase Outgassing after Core Formation (similar to contemporary volcanic emissions)	Third Phase Oxygen Production Exceeds Oxygen Losses—Photosynthesis
Major Components ($P > 10^{-2}$ atm)	CH_4, H_2(?)	N_2	N_2, O_2
Minor Components ($10^{-2} < P < 10^{-4}$ atm)	H_2(?), H_2O, N_2, H_2S, NH_3, Ar	H_2O, CO_2, Ar	H_2O, CO_2
Trace Components ($10^{-4} < P < 10^{-6}$ atm)	He	Ne, He, CH_4, NH_3(?), SO_2(?), H_2S(?)	Ne, He, CH_4, Kr

of the evolution of photosynthesis see Olson, 1970; Raff and Mahler, 1972), not only did the oceans have to become saturated and the reduced solutes, such as Fe(II), in the oceans oxidized before the atmospheric oxygen level began to climb, but, as Cloud (1968) has pointed out, "only one of the three known types of photosynthesis . . . releases free oxygen, and it is unlikely that it was primitive. Nevertheless, oxygen-releasing photosynthesis did [eventually] arise." At first, evidenced by the sulfur isotope ratios in ancient sedimentary barites (Perry et al., 1971), the oxygen pressure of the atmosphere remained low, and oxidation was restricted to the locale of the marine photosynthetic zone. As oxygen began to invade the atmosphere, prior to the formation of the protective ozone shield (Chapter 5), very powerful and deadly oxidants, such as atomic oxygen and ozone, probably were formed in the lower atmosphere and were responsible for intensive oxidation of surface lithospheric and hydrospheric materials (such as the deposition of red beds—Table 15.6) and in fact probably temporarily restricted the habitable domains of life. Before the delicate balance between O_2 from photosynthesis and CO_2 from decay and animal respiration was finally established, there may have been some fluctuations in the levels of these gases (Figure 15.2). As Table 15.6, based largely on material in Cloud (1968), and Figures 15.1 and 15.2 are designed to show these atmospheric changes were paralleled by changes of equally enormous importance in lithosphere, hydrosphere, and biosphere. There are few events more splendid to the intellect than the metamorphasis of a scientific hypothesis into a theory, of the falling together of countless circumstantial pieces of a great puzzle so that the picture finally emerges in all its novelty and grandeur. In the nineteenth century, it was the Darwinian theory of biological evolution, and in our own times, the theory of continental drift. Although some shreds of evidence remain, many of the signs of the chemical history of Earth have been erased by time so that perhaps there can never be a great marshaling and coherence of evidence. Nevertheless, dim and partial though our insight must remain, I cannot help a certain feeling of awe when confronted with the two most stupendous chemical events in the history of our planet—the phase separation of the material of Earth into core, mantle, hydrosphere, and atmosphere, and the transition of the Earthly environment from a reducing to an oxidizing one. Just

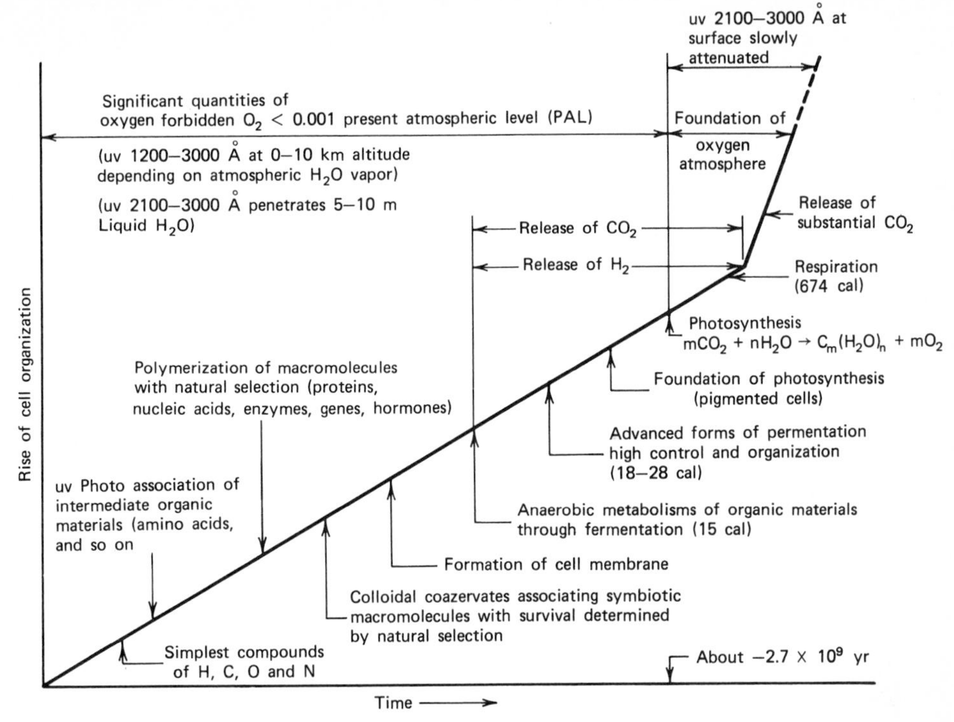

Figure 15.1. Diagrammatic visualization of the evolution of the simple living cell (from Berkner and Marshall, 1964).

imagine the oceans of the world undergoing this second transition, all the dissolved Fe(II) in the seas being oxidized up to Fe(III) and precipitated out as the hydrous ferric oxides whose spectacular trace still remains in the numerous, widespread, and thick-banded red iron deposits throughout the world!

There is at present 39×10^{20} g of oxygen in the atmosphere, SO_4^{2-} in seawater, and other sources that, combined with the 551×10^{20} g of oxygen in carbonate rocks, sedimentary sulfate, organic matter in sedimentary rocks, and the oxidation of FeO to Fe_2O_3, gives a total of 590×10^{20} g oxygen. This total is in the expected 32/12 ratio with the 252×10^{20} g carbon total in the atmosphere, biosphere, hydrosphere, and carbonate rocks, confirming the suggestion that this mobilized oxygen derived from CO_2 through photosynthesis (Cloud and Gibor, 1970). The present oxygen cycle is represented in Figure 15.3.

The emergence of theory from hypothesis does not imply the final resolutions of all questions. Quite the opposite, contrary to the misconception of laymen (and some scientists), the vitality of the scientific quest derives directly from the fact that our understanding of the world about us can never be complete, that our answers can never be final. Biological evolution and continental drift have both left and raised questions. Similarly, the majority opinion that the oxidizing nature of the Earth's atmosphere is the consequence of photosynthesis (with a lesser contribution from the photodissociation of water vapor) has its dissenters. Earlier, Brinkmann (1969) had argued that the contribution of photodissociation might have been underestimated, and more recently Van Valen (1971) has raised this and some further questions.

Table 15.6 Parallel Chemical Events in the Earth's Geohistory

Time Past	Atmosphere	Lithosphere	Hydrosphere	Biosphere
1.0×10^9 yr	3% O_2 decreasing CO_2	Sedimentary limestone and dolomite Sedimentary $CaSO_4$, glaciation	Increasing pH	METAZOA EUCARYOTA 0.7×10^9 yr Ozone ultraviolet shield
	Ozone increase in O_2 and O_1 1% O_2	Oxidation by O_3 and deposition of red beds (1.8×10^9 to 2×10^9 yr)		PROCARYOTA
	O_2 invasion of the atmosphere (1.8×10^9 to 2×10^9 yr)	Stromatolite rocks (indicate lunar tides)	Excess O_2 lunar tides	Blue-green algae
2.0×10^9 yr	Biological production of O_2 (but little getting into atmosphere)	Deposition of banded iron formations	Oxidation of Fe(II) by biologically produced O_2	PROCARYOTA No oxygen- and peroxide-mediating enzymes
3.0×10^9 yr	H_2O, CO, N_2, SO_2, HCl	Oldest sedimentary rocks (2.5×10^9 to $3.0 \times 10_6$ yr)	Hydrosphere exists by this time	Photoantotrophic life
	(little CH_4 or NH_3, no O_2) (degassing from partial melting following lunar capture, loss of any previous atmosphere	The oldest rocks (3.3×10^9 to 3.6×10^9 yr) THERMAL EVENT (possibly capture of moon)	H_2O from outgassing Loss of any previous hydrosphere	Anaerobic, heterotropic life BIOGENESIS Chemical evolution High energy ultraviolet at sea surface
4.0×10^9 yr		Oldest meteorites and terrestrial Pb		
4.6×10^9 yr	SOME KIND OF HOMOGENIZATION EVENT			

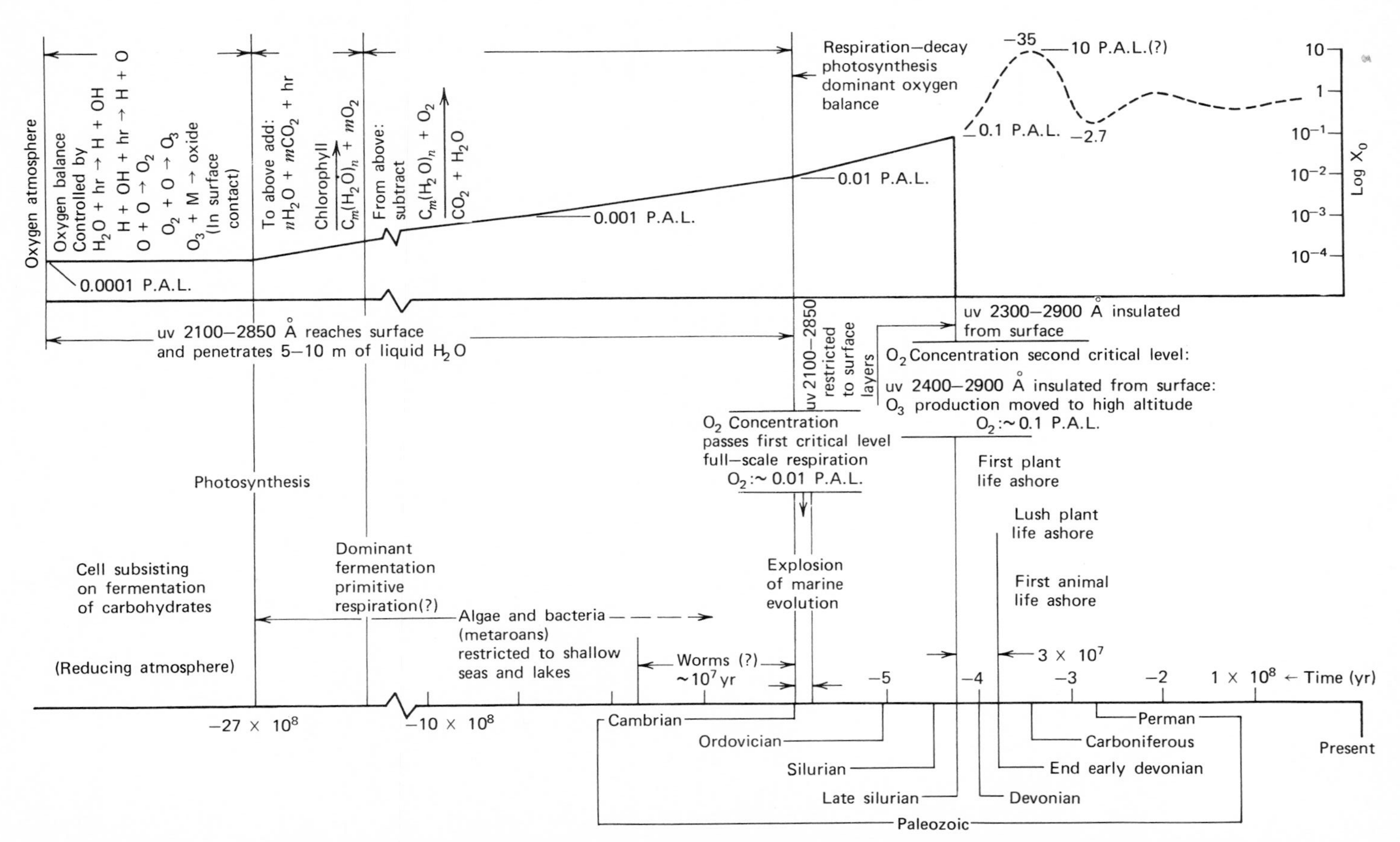

Figure 15.2. Tentative model of growth of oxygen in the atmosphere (x_0 is the penetration of atmosphere and liquid water) (from Berkner and Marshall, 1964).

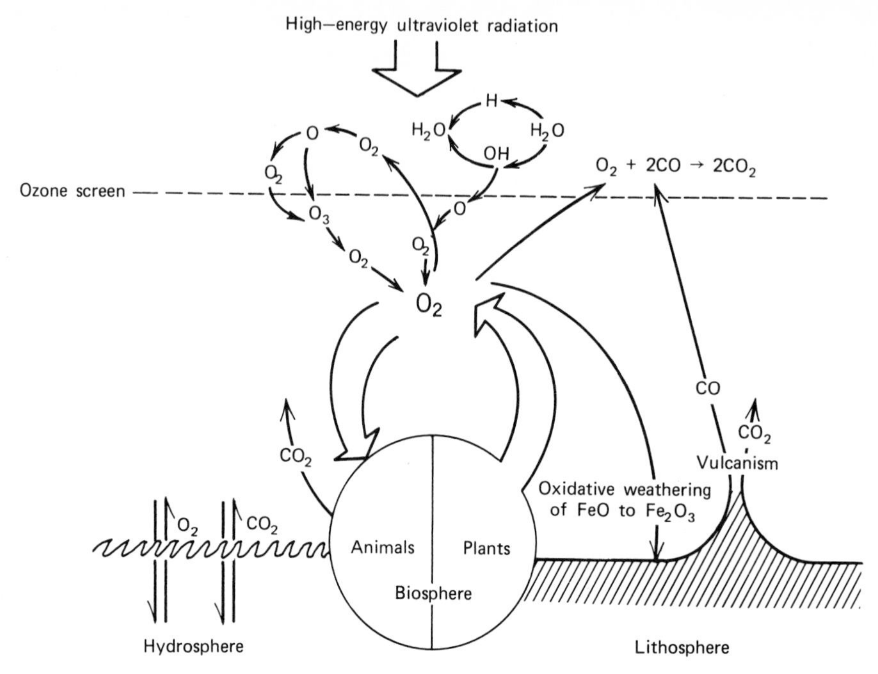

Figure 15.3. The present oxygen cycle on Earth.

Among factors he considers are oxygen inhibition of net photosynthesis, oxygen loss by the oxidation of hydrogen passing through the atmosphere, and the massive continuing burial of reduced carbon. He feels that the causes both of the initial oxygen increase and of its regulation at its present level remain unclear. For a discussion of a possible future scenario for the Earth's atmosphere, see the end of Section 15.4.

15.3 THE NITROGEN CYCLE

In contrast to the highly reactive oxygen, the other major constituent of our atmosphere, molecular nitrogen, N_2, is highly inert chemically. Nevertheless, it is still involved in some very complex and highly important chemistry in our environment so, although we have mentioned it earlier (see for example Figure 14.9A, here we dwell on the N-cycle in somewhat greater detail. There are a number of different ways of representing the N-cycle. Figure 14.9A emphasized the chemical reactions involved; Figure 15.4, the nitrogen reservoirs and the exchanges among them; Figure 15.5A, the major chemical forms of nitrogen –N_2, NH_3, NO_2^-, NO_3^-, and $-\overset{\displaystyle O}{\overset{\|}{C}}-N-$; and Figure 15.5B, the chemical energy levels. Figure 15.6 gives the values of the nitrogen reservoirs and fluxes. These diagrams do not include a significant path for the return of nitrogen minerals from plants to the soil—the burning of vegetation (Christensen, 1973), another of the benefits that we are beginning to realize accrue from even such destructive natural events as forest fires. Nitrogen is removed from its atmospheric reservoir by both inorganic and biological processes

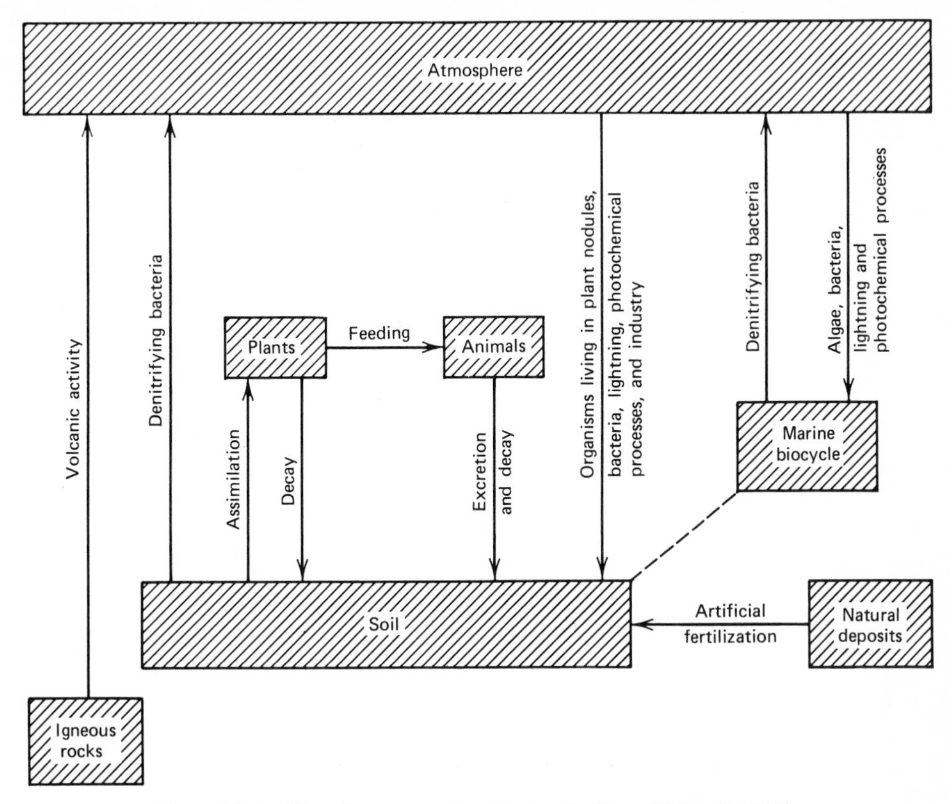

Figure 15.4. The nitrogen geochemical cycle (from Holland, 1964).

(Figures 15.4 and 15.5). The latter includes microbial nitrogen fixation by a host of organisms including those living in the root nodules of plants that we mentioned in Chapter 14. These symbiotic organisms are more important in terms of the quantities of nitrogen fixed than are free-living soil bacteria that are significant only in relatively barren soils (Stewart, 1969). Blue-green lake algae and bacteria in sediments also fix nitrogen (Howard, 1970), as do blue-green algae in the open ocean (Carpenter, 1972). Bacteria in termites appear to be capable of nitrogen fixation (Bennemann, 1973; Breznak et al., 1973), and even some primitive human tribes may have gut bacteria capable of fixing nitrogen and thus supplementing their hosts' N-deficient diet (Metzger, 1972). The chemistry of biological nitrogen fixation for all its great importance to man and his future is very complex and imperfectly understood. The trace metal molybdenum is essential for bacterial nitrogen fixation (Chatt and Leigh, 1972), and the enzyme responsible for N-fixation, nitrogenase, also contains iron, just as do those other great biocatalysts, the respiratory pigments.

Biological fixation is greater than fixation from such nonbiological natural processes as electrical discharge and photochemical reactions in the atmosphere—an estimated 0.008 to 0.07 mg N_2/cm^2 yr for the land surfaces of Earth compared to 0.0035 mg N_2/cm^2 yr (Hutchinson, 1954). The decay of organic matter returns much of the N to the atmosphere, but some remains in soils and sediments, in some cases in sufficient concentrations to form nitrogenous deposits as in Chile. We have noted previously that man in his production of NH_3 and other chemicals including artificial fertilizers

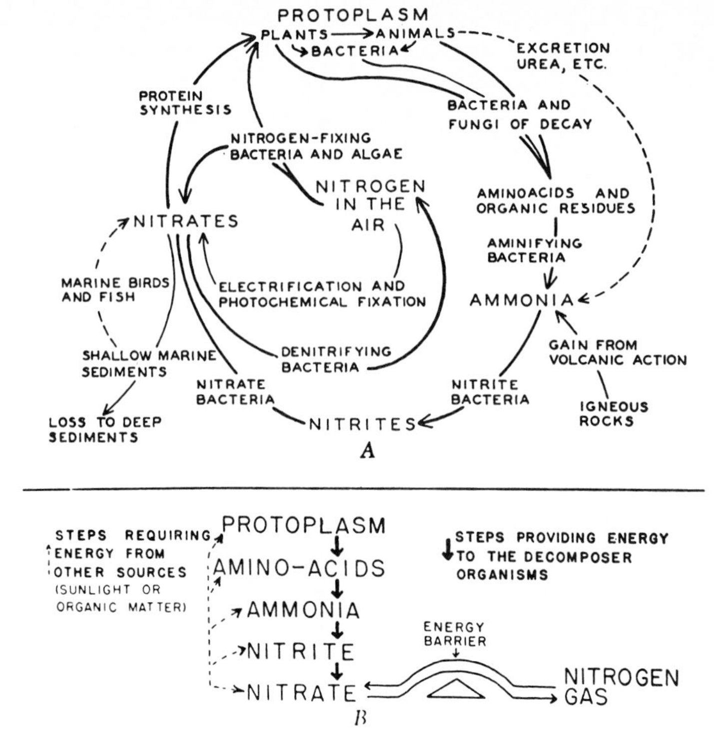

Figure 15.5. The nitrogen biogeochemical cycle (from E. P. Odum, *Fundamentals of Ecology*, 2nd ed., Saunders, Philadelphia, 1959. With permission of the publisher and the author).

now fixes significant quantities of atmospheric nitrogen (Table 15.7). Its large energy requirements may impose restraints on man's fixation of nitrogen. In view of this limitation on his capacity to produce artificial nitrogenous fertilizers, research in methods of extending "natural" N-fixation becomes even more essential (Dixon and Postgate, 1971; Phillips et al., 1971; Metzger, 1972). Also we must not forget that the N-cycle is not confined to the land masses; the oceans are also a big participant (Figure 15.6, see also Vaccaro, 1965 and Chapter 9 in Horne, 1969). There is, of course, an exchange of N_2 gas in the atmosphere with that dissolved in seawater. In addition, rainfall adds an estimated 28 mg NO_3^-/m^2 of ocean surface and 56 to 240 mg NH_4^+/m^2 (Harvey, 1966). The potential productivity of the sea hinges upon its content of nutrient nitrogen, and Ryther (1963) has made the interesting if discouraging observation that the ocean water richest in this nutrient still has concentrations four orders of magnitude less than fertile soil. Earlier (Chapter 8), we noted that nitrate, like its fellow nutrient phosphate, is one of the variable constituents of seawater and that its relative amount can be used as a sort of biological clock to identify and date water masses. Because of their high solubilities, NH_4^+, NO_2^-, and NO_3^- do not escape readily from the marine environment, and starting with the NH_4^+ from decay and metabolic processes, there is a progressive oxidation in the oceans up to nitrate (Figure 15.7). Later in this chapter, we examine the marine aerosol and chemical mass transport *from* the oceans. This may provide one avenue of escape for nitrogen. Bloch and Luecke (1970) note that the subsequent

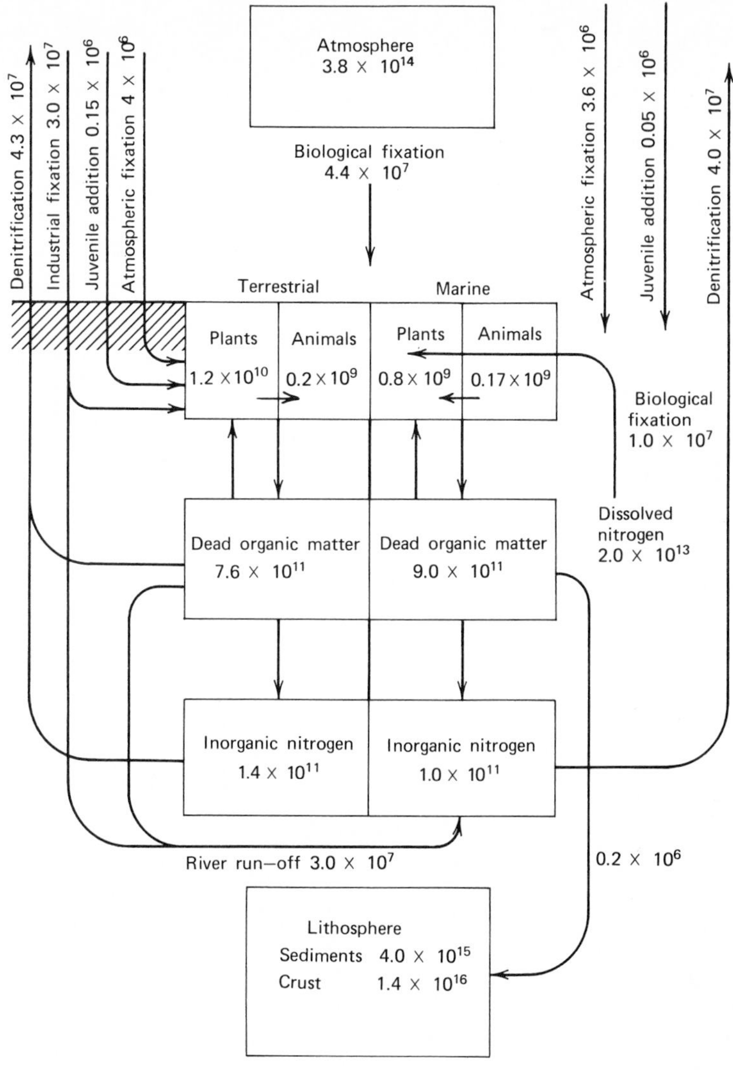

Figure 15.6. Nitrogen reservoirs and fluxes (in metric tons, based on Delwiche, 1970).

oxidation of the relatively high ammonia content of ocean bubble spray may contribute to plant life's supply of nitrogen. Conversion of NH_3 to NO_3^- need not be essential for plant utilization; Porter et al. (1970) report that the leaves of growing plants absorb ammonia, and they suggest therefrom that this may represent a significant sink for atmospheric ammonia (see also Hutchinson et al., 1972).

Before we leave nitrogen, we mention a fascinating mystery; at the present pH of the oceans and partial pressure of O_2 in the atmosphere, if thermodynamic equilibrium obtained, the N in the oceans should be present as NO_3^- rather than N_2. Sillen (1966) has suggested one solution to the "nitrogen problem": There may be some organism in the upper layers of the ocean that destroys nitrate ions.

Table 15.7 The Nitrogen Inventory (Based on values of Delwiche, 1970)

Nitrogen Fixation	
Biological fixation—Terrestrial	
(Historic)	30×10^6 metric tons
Legume crops	14×10^6 metric tons
Marine	10×10^6 metric tons
Industrial fixation	30×10^6 metric tons
Atmospheric fixation	7.6×10^6 metric tons
Juvenile addition	0.2×10^6 metric tons
Total	92×10^6 metric tons
Denitrification	
Terrestrial	43×10^6 metric tons
Marine	40×10^6 metric tons
To sediments	0.2×10^6 metric tons
Total	83×10^6 metric tons

$\Delta N - 9 \times 10^6$ metric tons (nitrogen build-up in biosphere, soils, and the hydrosphere).

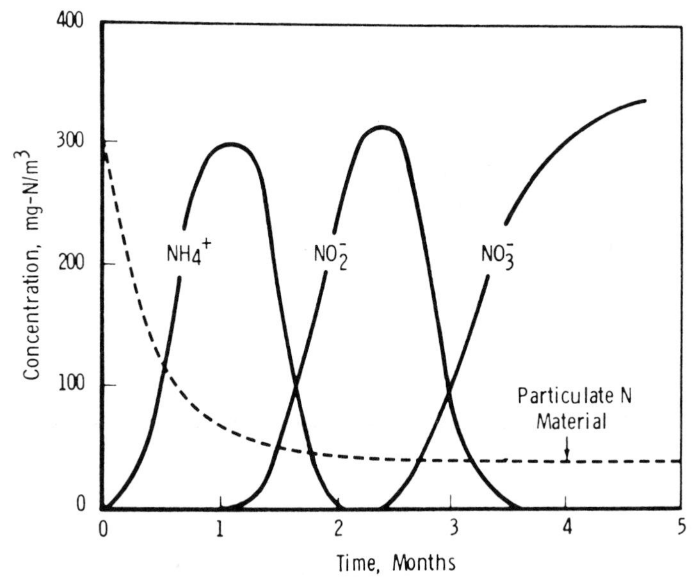

Figure 15.7. Production of nitrogenous material from the decomposition of phytoplankton in aerated seawater stored in the dark (from Horne, 1969).

15.4 THE CARBON DIOXIDE CYCLE

Plant photosynthesis and animal respiration closely link the oxygen and carbon dioxide cycles (Chapter 11 and Figure 15.8). Earlier we examined the Earth's carbon inventory (Figure 10.9 and summarized again Table 15.8). Figure 15.9 presents this inventory in another way in terms of the weight of carbon per square centimeter of Earth's surface and shows its relationship to the CO_2–O_2 cycle. Annually

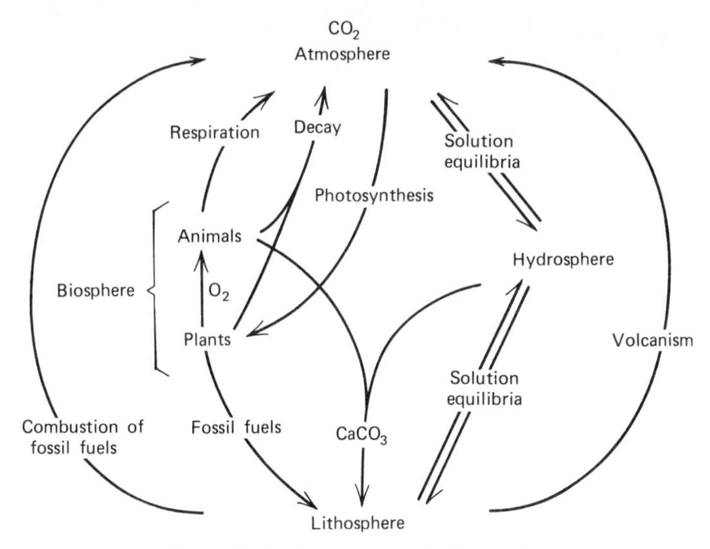

Figure 15.8. The carbon dioxide cycle.

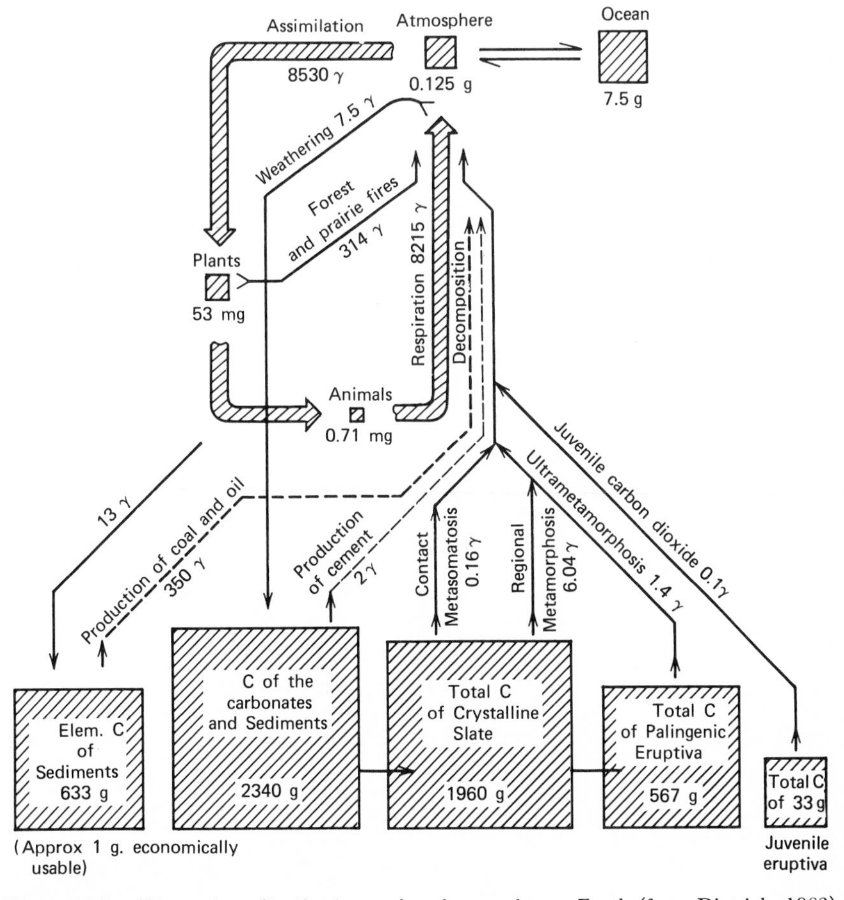

Figure 15.9. The carbon distribution and carbon cycles on Earth (from Dietrich, 1963).

Table 15.8 Inventory of Total Carbon (as CO,) in Atmosphere, Hydrosphere, Biosphere, and Sedimentary Rocks (From Rubey, 1964)

	$\times 10^{1\text{-}}$ g	
Atmosphere	2.33	
Ocean and fresh water	130	147
Living organisms and undecayed organic matter	14.5	
Sedimentary rocks (including interstitial water)		
Carbonates	67,000	
Organic C	25,000	92,000
Coal, oil, and so on	27	

the biosphere uses about 3% of the total atmospheric CO_2 (Junge, 1963). During the day, growing plants assimilate CO_2 from the atmosphere, whereas after darkness falls, this process ceases and the release of CO_2 into the atmosphere from decaying organic material in the soil predominates (Figure 15.10) giving rise to a daily fluctuation in the near ground level of atmospheric CO_2 (Figure 15.11). The minimum occurs just after sunset, and the maximum, just before sunrise. Soil, as we noted earlier, may contain as much as 100 times more CO_2 than the air in the atmosphere. Even frozen tundra soil releases CO_2 (Kelley et al., 1968; Coyne and Kelley, 1971). In contrast to this relatively rapid "breathing" of the biosphere, the oceans represent a more sluggish but far vaster reservoir of CO_2 (Table 15.8). Therefore, that part of the carbon cycle which has the greater chemical impact on our environment is the exchange of CO_2 between atmosphere and hydrosphere (Plass, 1972), and, even though this chapter is entitled biosphere–atmosphere interactions, we concentrate here on that aspect of the cycle. In a sense, the carbon "cycle" is more of a gradual flux (Figure 15.12) taking CO_2 from the atmosphere, transforming it in the biosphere, and finally depositing it in the lithosphere (Bolin, 1970). Since the paths of

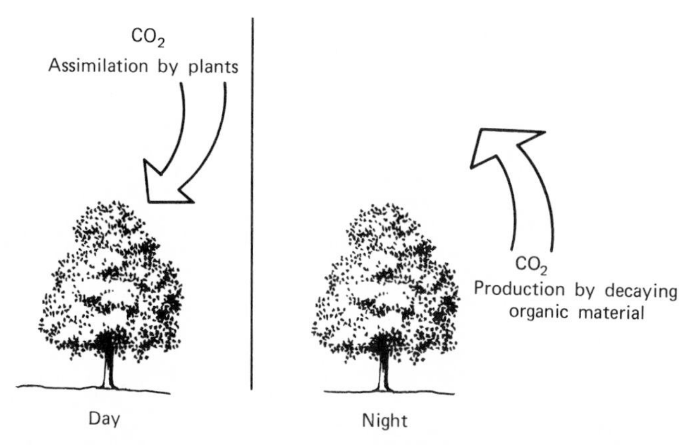

Figure 15.10. Daily fluctuations in carbon dioxide exchange between biosphere and atmosphere.

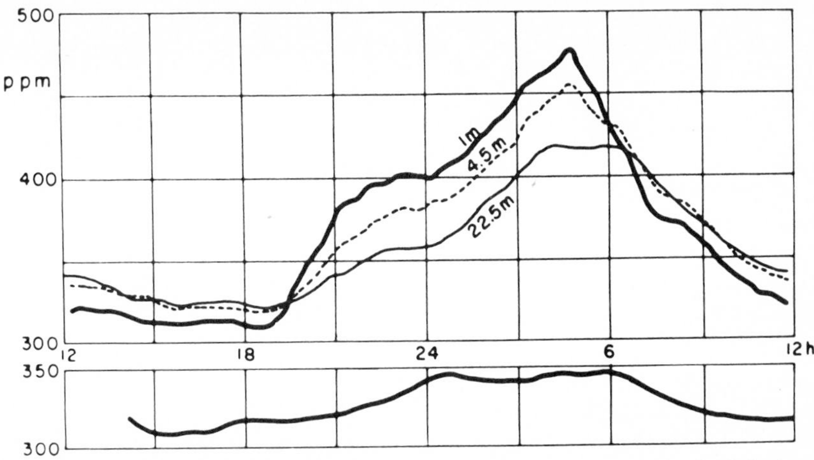

Figure 15.11. Daily variations of CO_2 at 1, 4.5, and 22.5 m above a wheat field during sunny weather (upper plots) and overcast weather (lower plot, the differences between the three heights are too small to be plotted) (from Junge, 1963, with permission of Academic Press, Inc., and the author).

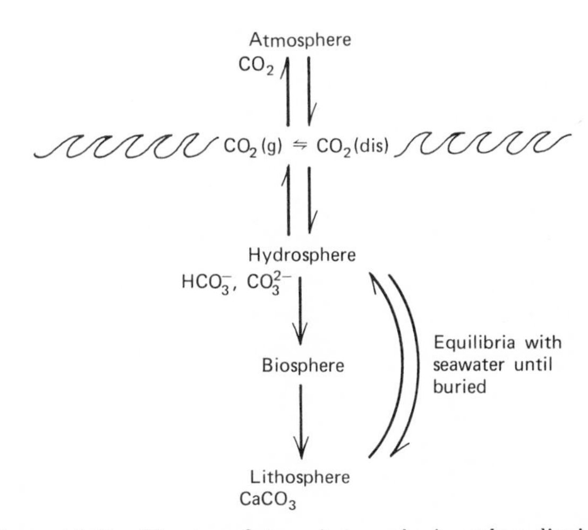

Figure 15.12. The transfixing of atmospheric carbon dioxide.

CO_2 back into the atmosphere are relative trickles, in the course of geological time, the CO_2 level of the atmosphere probably has decreased (in contrast to short-range increases due to human activity). Since the pH of the oceans depends on the CO_2 partial pressure of the atmosphere, such a decrease should be accompanied by a decline in seawater acidity. Unfortunately, in examining the history of the Earth's atmosphere, scientists have been so preoccupied with oxygen that the question how the CO_2 level has been changing in geological time, except since the Industrial Revolution, has been somewhat neglected.

The exchange of CO_2 between atmosphere and the oceans involves a number of equilibria (Figure 7.16) of which the dissolution of the gas into the liquid phase

$$CO_2\ (g) \rightleftharpoons CO_2\ (\text{dissolved}) \tag{15.1}$$

may be the all-important rate-controlling step. Berger and Libby (1969) have suggested that an enzyme such as carbonic anhydrase may be involved. Next the carbonic acid formed

$$CO_2 \text{ (dissolved)} + H_2O \rightleftharpoons H_2CO_3 \tag{15.2}$$

dissociates stepwise

$$H_2CO_3 \rightleftharpoons H^+ + HCO_3^- \qquad K_1 = \frac{(H^+)(HCO_3^-)}{H_2CO_3} \tag{15.3}$$

$$HCO_3^- \rightleftharpoons H^+ + CO^{2-} \qquad K_2 = \frac{(H^+)(CO^{2-})}{HCO_3^-} \tag{15.4}$$

Then, finally, there are the equilibria for the precipitation and deposition of mineral carbonates, usually mediated by organisms

$$Mg^{2+} + CO_3^{2-} \rightleftharpoons MgCO_3(s) \tag{15.5}$$

and

$$Ca^{2+} + CO^{2-} \rightleftharpoons CaCO_3(s) \tag{15.6}$$

The thermodynamic equilibrium constants for the stepwise ionization

$$K^\circ_1 = \frac{a_H + a_{HCO_3^-}}{a_{CO_2}} \tag{15.7}$$

and

$$K^\circ_2 = \frac{a_H + a_{CO_3^{2-}}}{a_{HCO_3^-}} \tag{15.8}$$

where the a's are activities, are given in Table 15.9 as a function of temperature. It is more instructive, however, to examine a diagram such as Figure 7.17 from which it is clear that, both in seawater and fresh waters in the near neutral pH range that predominates in the environment, the major species is bicarbonate ion, HCO_3^-, with little CO_2 and even less CO_3^{2-}. It is this bicarbonate ion that plays a central role in maintaining the pH of the oceans, both in fast equilibria (Equations 15.1, 15.2, and 15.3) with CO_2 in the atmosphere and in its release of protons (Equations 15.2 and 15.4) and their subsequent slow ion-exchange equilibria with clay minersls (MacIntyre, 1970). For a more detailed treatment of the $CO_2/HCO_3^-/CO_3^{2-}$

Table 15.9 Thermodynamic Dissociation Constants of Carbonic Acid at 1 atm (From Horne, 1969)

Temperature (°C)	K°_1	K°_2
0	2.64×10^7	2.36×10^{-11}
5	3.04×10^7	2.77×10^{-11}
10	3.44×10^7	3.27×10^{-11}
15	3.81×10^7	3.71×10^{-11}
20	4.16×10^7	4.20×10^{-11}
25	4.44×10^7	4.69×10^{-11}
30	4.71×10^7	5.13×10^{-11}

equilibria in the oceans see Cloud (1965), Chapter 7 in Horne (1969), and the references cited therein.

How do organisms deposit calcium and magnesium carbonates in their skeletons and shells? This is an interesting question, since they can do so even in seawater which is undersaturated with respect to the equilibrium solubilities of the minerals formed. At the present time, most of the oceans are supersaturated with respect to calcite and aragonite (see Li et al., 1969). However, should the oceans become widely undersaturated as the result of the dissolution of excess CO_2 from the combustion of fossil fuels, coral reefs would tend to dissolve, and organisms might encounter difficulties in depositing their protective shells. As Fairhall (1973) has cautioned, "The consequences [of the dissolution of excess CO_2] could be very serious for life in the sea." And why are the minerals formed in the marine environment thermodynamically the wrong ones? For example, dolomite, $CaMg(CO_3)_2$, should be formed readily in the marine environment (Garrels et al., 1960; Cloud, 1965), but it is rare, and calcite, $CaCO_3$, common. Furthermore, organisms precipitate aragonite, a different crystalline form of $CaCO_3$, rather than calcite. The dissolution of these minerals, as well as their formation, does not behave as expected (Peterson, 1966; Berger, 1967), and Chave (1965) has proposed that organic surface coatings may impede the dissolution of the mineral particles. There is a lesson contained in these difficulties, namely in constructing chemical equilibrium models of atmosphere and hydrosphere; it is easy to forget that the chemistry of our environment is highly complex and dynamic, that much of it occurs at interfaces, and that disequilibrium rather than equilibrium is the rule. Biochemistry, as well as surface chemistry, further works against our efforts to apply equilibrium conditions. To get back to the question of how organisms secrete their nacreous shells, two hypotheses have been advanced; the "compartment" hypothesis, which envisions crystal nucleation as occurring in preexisting hollows in the organic matrix, and the "template" hypothesis, which envisions crystal growth as nucleating on certain active sites on the open surface of biomembranes. In a recent paper, Erben and Watube (1974) report experimental findings supporting the second hypothesis.

The length of this chapter is not commensurate with the importance of its subject matter because some of the topics that might have been treated here are dealt with elsewhere, notably in the sections on atmosphere and biosphere. For this reason, Table 15.10 summarizes the major biosphere–atmosphere interactions in terms of biological sources and sinks of the major and some minor atmospheric constituents.

I would like also to insert some further material here on the effect of human activity on the composition of the atmosphere, for man is, after all, a part of the biosphere. Analyses of gases trapped in Greenland and Antarctic ice samples 100 to 2500 yr old indicate that the Industrial Revolution, as noted earlier in Chapter 6, has not produced a large increase in the carbon monoxide content of the atmosphere (Robbins et al., 1973). In the case of carbon dioxide, however, man's perturbation of the carbon cycle by the combustion of fossil fuels and other means may be significant. Keeling (1973) has estimated that man's industrial, domestic, and transportation activities up to the year 1970 have cumulatively increased the CO_2 in the short-term carbon cycle by 4×10^{17} g or about 18% of the atmospheric CO_2. As for the future, Cramer and Myers (1972) project an atmospheric CO_2 level of 380 ppm by the year 2000 and 495 ppm by the year 2100. The climatic effects of these increases in the atmospheric level of CO_2 have been discussed by Moller (1963) and Pytkowicz

Table 15.10 Biospheric Sources and Sinks of Atmospheric Constituents

	Biospheric Sources	Biospheric Sinks
N_2	Denitrifying microorganisms in soils, sediments, and waters	Nitrogen fixing bacteria and algae in soils and waters
O_2	Photosynthesis	Animal respiration and oxidation of decaying organic matter
CO_2	Animal respiration, decay and oxidation of biogenic organic material, fermentation, and related processes	Photosynthesis, dissolution into the oceans and subsequent biological fixation as carbonate minerals
CO, CH_4, NH_3, $H_2S(SO_2)$	Decay and bacterial metabolic processes in anaerobic soils, muds, sediments, and waters. Leakage from biogenic deposits. The SO_2 comes from the oxidation of biogenic H_2S	Soils are a sink for CO, NH_3, and $H_2S(SO_2)$. Plant foliage may be a sink for CO

(1972), while Peterson (1969) has summarized these and many other possible environmental aftermaths including redistribution of plant species, changes in marine life, and rise of sea level. While this anthropogenic production of CO_2 has certainly consumed oxygen, photosynthesis has apparently been able to replenish the supply for, as we noted in Chapter 6, at least since 1910 there appears to be no detectable decrease (or increase) in the oxygen content of our atmosphere (Broecker, 1970; Machta and Hughes, 1970). Any man-caused increases in the "greenhouse effect" from CO_2 may be far outweighed by more long-range events on a far vaster scale. Sagan and Mullen (1972) speculate that with continuing solar evolution the temperature of the Earth's surface will increase, with increasing temperature the water vapor in the atmosphere will increase, and this in turn will increase the absorption of solar energy resulting in a runaway greenhouse effect such as may have occurred in the case of Venus (Ingersoll, 1969; Rasool and de Berg, 1970; Pollack, 1971). The Earth will develop an atmosphere of 300 bars of steam, but the authors conclude, "at the same epoch the global temperature of Mars will become very similar to that of present-day Earth. If there are any organisms left on our planet in that remote epoch, they may wish to take advantage of this coincidence."

15.5 AIR–SEA INTERACTIONS

Finally, we come to the matter of the interaction between the oceans and the atmosphere. The paths of transport of matter and energy across the air/sea interface are many and their mechanisms complex (Figure 15.13). Of the utmost importance to these processes is the nature of the liquid boundary layer through which energy and matter must pass and which undoubtedly in the majority of cases is the region that determines the rate of these processes. In Chapter 13, we speculated about the nature of this boundary layer of vicinal water, and we suggested that its structure and properties are different from those of "ordinary" bulk liquid water and that among

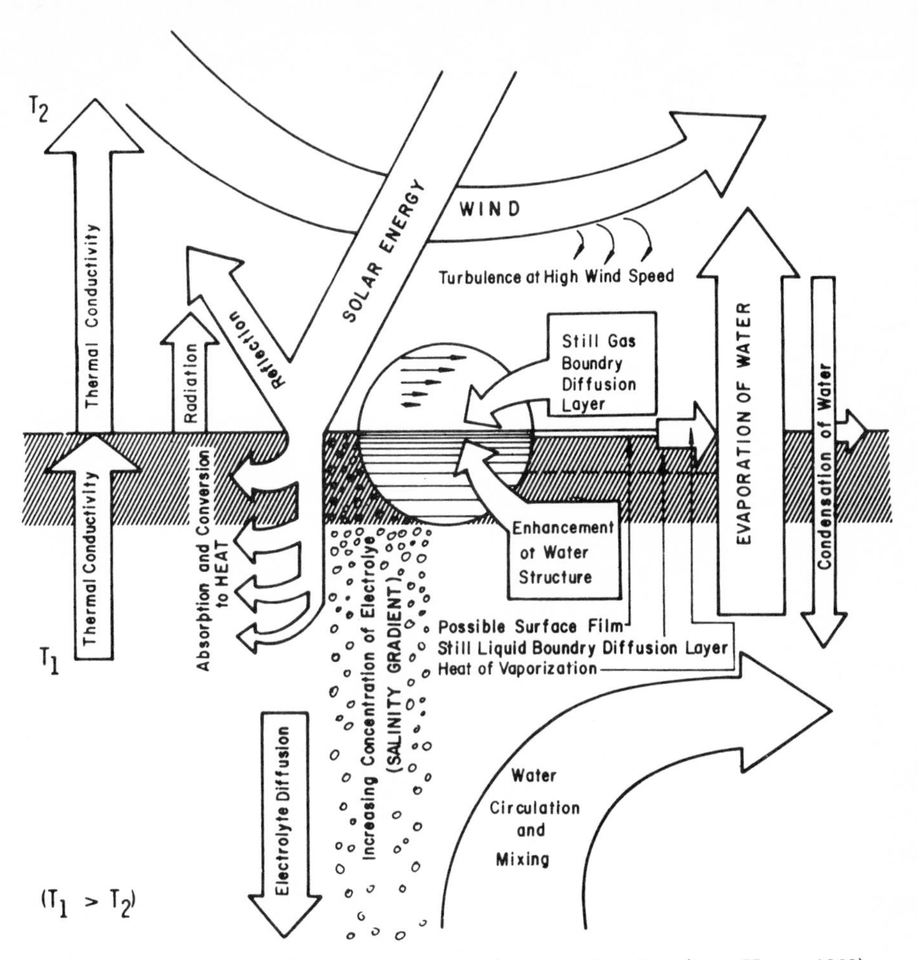

Figure 15.13. Matter and energy transport at the air–sea interface (from Horne, 1969).

the different properties are solubilities. We further guessed that if the structure of the boundary layer has the character of the hydration envelope of nonpolar solutes, then these solutes might be more soluble in the hydrophobically structured zone, whereas ionic solutes with their coulombic or hydrophilic hydration envelopes will be preferentially excluded (Figure 13.3; see also Horne, 1972). As we see presently, these earlier conclusions have a crucial bearing on the chemical composition of the marine atmosphere.

The minor chemical species in the marine atmosphere come from three principle sources (Figure 15.14; *vide* Folger, 1970). There is, of course, the man-made filth that pollutes the atmosphere on a global scale over the oceans as well as over the land masses where it largely originates. As Goldberg (1972) has pointed out, in the case of elements these pollutants are identifiable by their enrichment with respect to the iron in the atmosphere as compared to average crustal material, and they include Zn, Sn, Pb, Ag, Cu, As, and Sb (Chester and Stoner, 1973; Imboden and Stumm, 1973). Then there are the natural sources that we have examined previously such as continental dust (Bonatti and Arrhenius, 1965; Griffin, 1967; Windom et al., 1967;

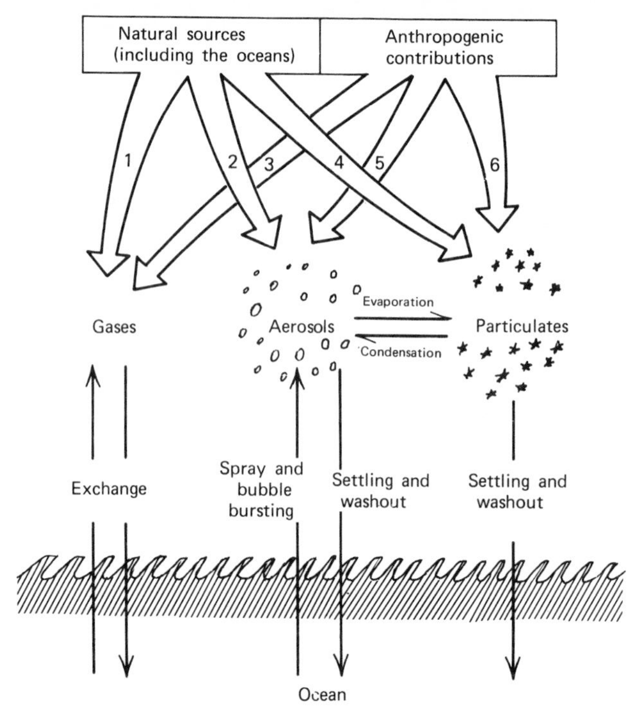

Figure 15.14. Sources of "pollutants" in the marine atmosphere. 1—CO_2, O_2, SO_2, CO, CH_4, and so on; 2—H_2O, NaCl, $MgSO_4$, and so on; 3—SO_2, CO_2, CO, NO_x, and so on; 4—SiO_2, Fe, Sr, minerals, and so on; 5—Pb, Hg, SO_4^{2-}, and so on; and 6—Zn, Sn, Pb, and so on.

Chester et al., 1971; Prospero and Carlson, 1972), volcanic material, and the products of biological processes. But in the marine atmosphere and spreading over coastal regions, there is still a third major source, and this is the oceans themselves (Martens and Harriss, 1973). Delany (1973) conclude that the aerosol of the middle and upper troposphere contains 90 to 95% continental material and 5 to 10% marine material. Matter passes not only from the atmosphere into the ocean but also in the reverse direction from the oceans into the atmosphere (Table 15.11). We have already mentioned the most obvious of these exchanges—that of water—the evaporation–precipitation balance (Chapter 9). Settling and precipitation washout carry atmospheric material into the oceans. Gases also exchange between the atmosphere and seawater (see Chapter 11 in Horne, 1969, and the references cited therein), including gases such as CO and CH_4 (Lamontagne et al., 1973; Linnenbom et al., 1973), often thought of as pollutants. Gas exchange is highly dependent on conditions such as temperature, salinity, and especially wind velocity and sea state. Table 15.12 summarizes some measurements of the rate or gas transfer. Figure 15.15 shows the solubilities of oxygen and nitrogen in 16‰ and 20‰ Cl seawater, while Table 15.13 gives the solubilities of the noble gases. Since CO_2 is not simply soluble in seawater but reacts with it (Equations 15.1 to 15.4), its solubility is correspondingly higher (Table 15.14). As noted earlier, the phases of our environment are seldom in equilibrium with one another and disequilibrium is the rule rather than the exception.

Table 15.11 Chemical Mass Transport Between Air and Oceans

From the Atmosphere *Into the Oceans*		Gas solubility equilibria Settling of aerosols and particulates Washout by precipitation
Out of the Oceans Into the Atmosphere		Gas solubility equilibria Evaporation Salt spray Marine aerosol Special chemical mechanisms
And for Water	In by	Precipitation Gas–liquid equilibrium Sorbed and bound water on settled and washed-out material
	Out by	Evaporation Spray and aerosol formation

Table 15.12 Gas Transfer across the Gas–Liquid Interface (From Horne, 1969)

Gas	Transport Rate across the Air–Sea Interface	Conditions	Reference
O_2, N_2	—	Solution of bubbles	Wyman et al. (1952)
O_2	30×10^4 ml O_2/m²/yr	Left surface of Gulf of Maine, October–March	Redfield (1948)
CO_2	24 moles/m² yr	Ocean average	Craig (1957)
O_2	1 to 60 cm/hr exit coefficient	Dependent on stirring wind, velocity, wave height film, and so on	Downing and Truesdale (1955)
Air	2 to 60 cm/hr exit coefficient	Strongly dependent on wind velocity	Kanwisher (1963)
CO_2	0.32 mg/cm²/min/atm	—	Suguira et al. (1963)
CO_2	0.020 mg/cm²/min/atm	—	Miyake and Hamanda (1960)
CO_2	3 to 5 moles/m²/yr	Lakes	Broecker and Walton (1959)
O_2	1.23×10^4 ml O_2/m²/month	In situ, Oregon coast	Pytkowicz (1964)
CO_2	18 moles/cm²/atm/yr	Pacific Ocean	Keeling (1965)
CO_2	8×10^{-3} cm/sec	Gulf of Mexico	Park and Hood (1963)

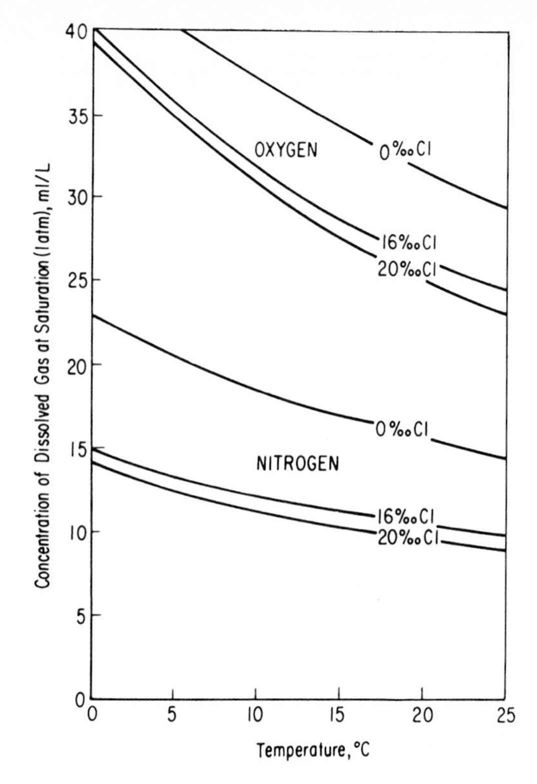

Figure 15.15. Solubility of the major atmospheric gases in seawater (from Horne, 1969).

Seawater at any given location is usually either undersaturated or supersaturated with respect to any particular gaseous component of the atmosphere. Figure 15.16 shows the degree of saturation with respect to oxygen of some South Atlantic surface waters measured in the month of February. Just as they have characteristic nutrient ratios (Chapter 8), the major oceans of the world tend to have characteristic oxygen profiles (Figure 15.17) with the Atlantic being peculiar because of the refreshment of its deep waters with oxygen-rich water from the polar regions (Figure 15.18).

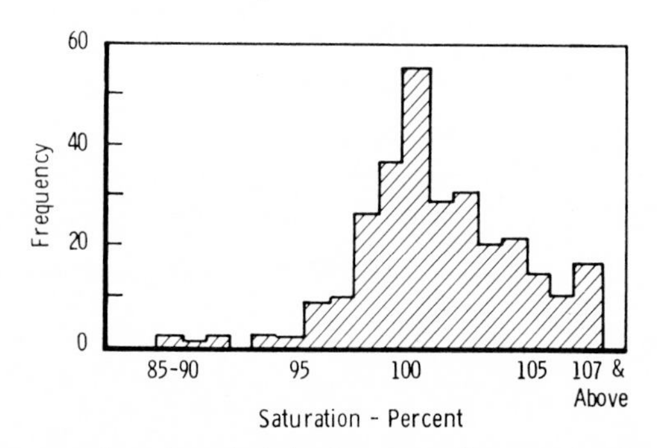

Figure 15.16. Oxygen saturation of south Atlantic waters (from Richards, 1965, with permission of Academic Press, Inc., and the author).

Table 15.13 Solubility of Carbon Dioxide in Seawater (moles/liter) × 10^4, STP) (From Riley and Skirrow, 1965, with permission of Academic Press, Inc.)

	T(°C)							
Cl‰	0	2	4	6	8	10	12	14
0	770	712	662	619	576	536	502	472 × 10^{-4}
15	674	623	578	538	504	472	442	416
16	667	617	573	533	499	468	438	413
17	660	611	567	528	495	464	434	410
18	653	605	562	524	490	460	431	406
19	646	599	557	519	486	456	428	403
20	640	593	551	514	482	452	424	400
21	633	587	546	509	477	448	421	396

	T(°C)							
Cl‰	16	18	20	22	24	26	28	30
0	442	417	394	372	351	332	314	299 × 10^{-4}
15	393	371	351	331	314	299	284	270
16	390	368	348	329	312	297	281	268
17	387	365	346	327	310	294	279	266
18	384	362	343	324	307	292	277	264
19	381	359	340	321	304	289	275	262
20	377	356	337	319	302	287	273	260
21	374	354	335	317	300	285	271	258

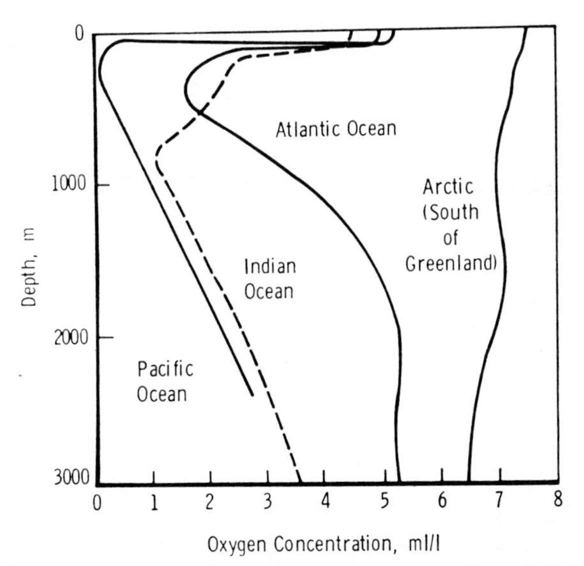

Figure 15.17. Characteristic oxygen profiles of the Atlantic, Pacific, and Indian Oceans and of the area of formation of the deep Atlantic waters (from Dietrich, 1963).

Table 15.14 Solubilities of the Noble Gases in Seawater (ml of gas at STP/kg seawater) (From von König, 1963, with permission of the publisher)

Temperature (°C)	He		Ne		Ar		Kr		Xe	
	Pure	Air	Pure	Air	Pure	Air	Pure	Air	Pure	Air
0	—	—	9.37	171×10^{-6}	—	—	71.5	81.5×10^{-6}	136	11.70×10^{-6}
1	7.91	41.4×10^{-6}	—	—	38.5	0.359	—	—	—	—
5	—	—	9.02	164×10^{-6}	35.3	0.329	63.9	72.8×10^{-6}	115	9.89×10^{-6}
10	7.40	38.8×10^{-6}	8.67	158×10^{-6}	32.8	0.306	58.2	66.3×10^{-6}	103	8.86×10^{-6}
15	6.95	36.4×10^{-6}	8.35	152×10^{-6}	30.2	0.282	51.6	58.5×10^{-6}	90.0	7.74×10^{-6}
17.5	—	—	—	—	—	—	50.2	57.2×10^{-6}	—	—
20	7.00	36.7×10^{-6}	8.20	149×10^{-6}	26.3	0.245	44.8	51.1×10^{-6}	80.0	6.88×10^{-6}
22.8	—	—	—	—	—	—	44.2	50.4×10^{-6}	—	—
24	—	—	—	—	—	—	42.9	48.9×10^{-6}	—	—
25	—	—	8.07	147×10^{-6}	—	—	—	—	70.2	6.04×10^{-6}

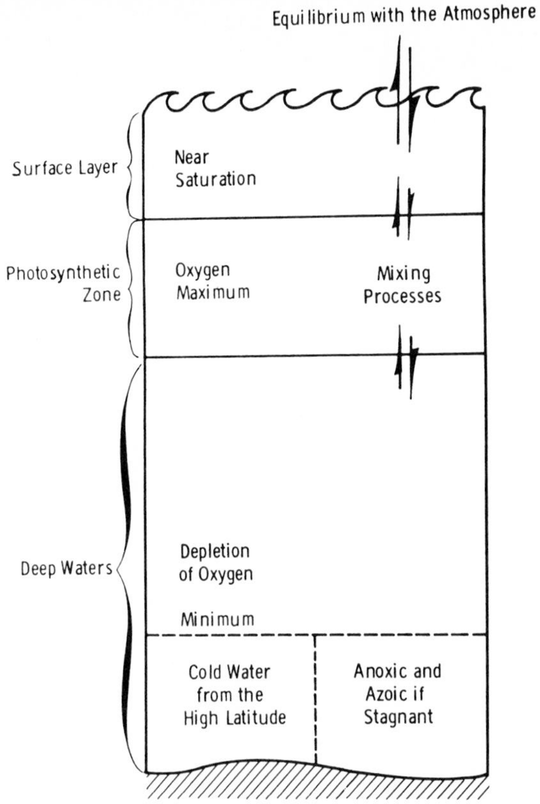

Figure 15.18. Diagrammatic representation of the several conditions of dissolved oxygen (from Horne, 1969).

Material is thrown up into the atmosphere from the oceans by wave spray and bubble bursting (Figure 15.14), yet the composition of the marine aerosol, as we can plainly see in Table 15.15, is appreciably different from the seawater that is presumably its source. While the question of whether the sea is a source or sink for a great many atmospheric materials remains controversial (see Duce et al., 1972; in the case of some constituents, such as light hydrocarbons, the sea can be both a source and a sink, Brooks and Sackett, 1973), still it is clear that the salts in the marine aerosol (excluding ammonium sulfate) must come from the sea for the salt (or chloride) content of the atmosphere falls off with increasing distance from the coast (Figure 15.19) and with altitude (Figure 15.20). Why then the ionic fractionation evident in Table 20.15 as material is transferred from the ocean's surface to the atmosphere? Rather than from wave spray and surf, the bulk of the salts in the marine aerosol are believed to be ejected from the ocean by the continuous bursting of minute bubbles (or largely unknown origin) over vast stretches of the ocean's surface. When these bubbles burst (Figure 15.21), the larger jet drops fall back into the sea, but the minute droplets formed by the collapse of the upper wall of the bubble remain suspended. This process acts as a sort of microtome, sampling only the water of the sea's surface boundary layer (MacIntyre, 1968, 1970). Thus the marine aerosol reflects the solute composition, not of bulk seawater, but of the oceans's surface

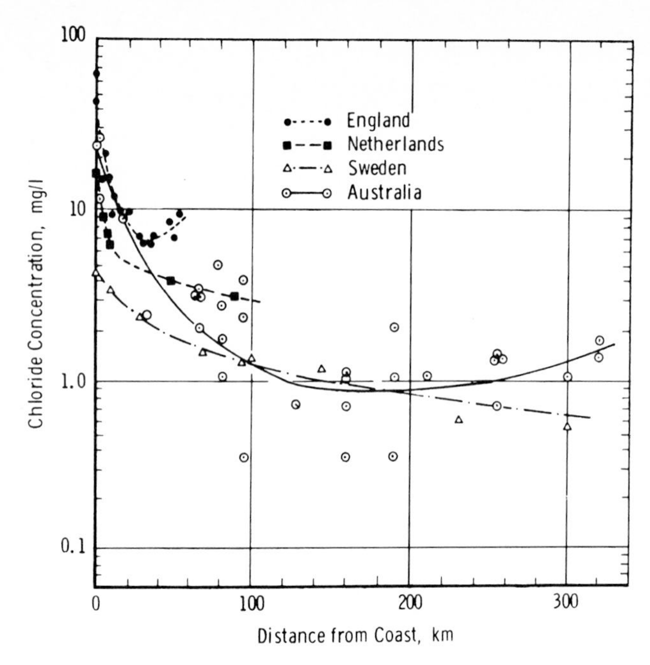

Figure 15.19. Decrease in the chloride content of rain water with increasing distance from the coast (from Junge, 1963 with permission of Academic Press, Inc., and the author).

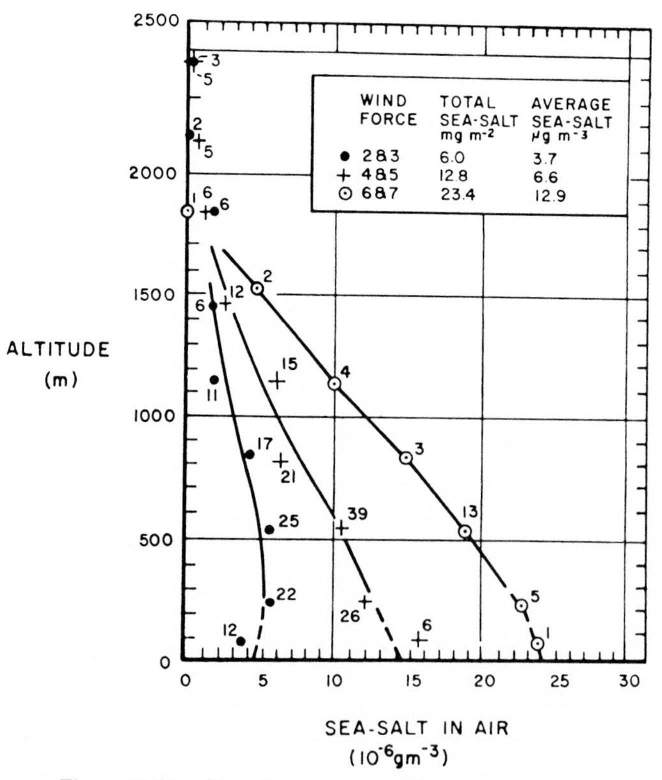

Figure 15.20. Sea salt content as a function of altitude.

Table 15.15 Comparison of Ionic Ratios in Seawater and Precipitation (Abstracted from Table 11.14 in Horne, 1969)

Na^+/K^+	Seawater	27
	Rain, Southwestern Iceland	18
	Rain, Western Ireland	6
	Rain, Jerusalem	7
	Snow, Rasman Glacier	1
Cl^-/Br^-	Seawater	302
	Rain, Jerusalem	45
	Rain, Göttingen	86
	Snow, Bolschewo	38
Cl^-/Na^+	Seawater	1.8
	Rain, Netherlands	1.6
	Rain, Norway	1.7
	Rain, Hawaii	1.9
	Rain, Boston	1.6
Mg^{++}/Na^+	Seawater	0.12
	Rain, Iceland	0.14
	Rain, Norway	0.17
	Rain, Ireland	0.16
Ca^{++}/Na^+	Seawater	0.038
	Rain, Bermuda	0.40
	Rain, Newfoundland	0.15
	Rain, Boston	0.41
	Rain, Iceland	0.11
	Rain, Norway	0.14
	Rain, Netherlands	0.085

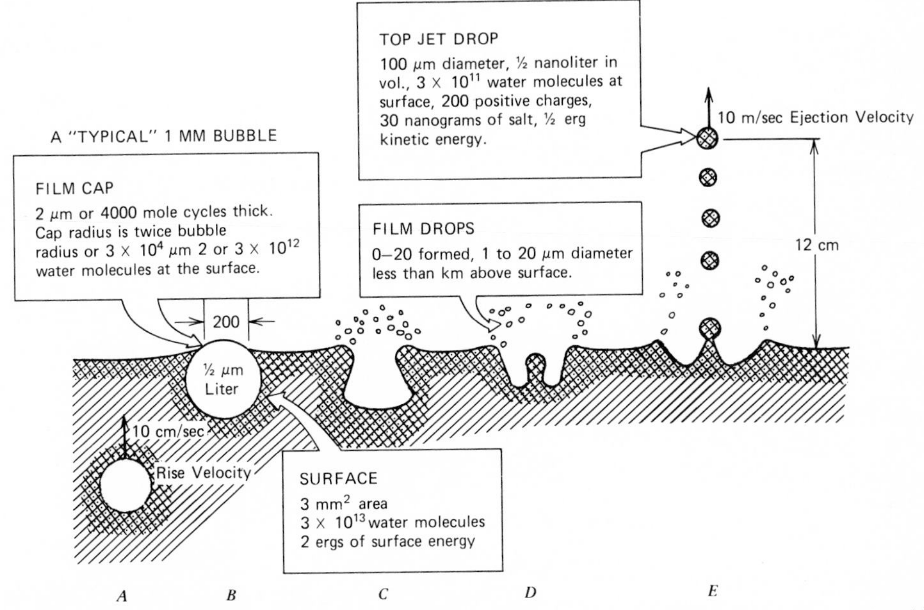

Figure 15.21. Diagrammatic representation of bubble collapse (from MacIntyre, 1965, with permission of the author).

layer, which as we noted in Chapter 13 and earlier in this section can be quite different. Sea salts are not the only materials fractionated in the surface layer by bubble bursting; trace metals (Piotrowicz et al., 1972; Szekielda et al., 1972), organic substances (Riley, 1963; Barber, 1966; Garrett, 1967; Williams, 1967), mineral materials (Szekielda et al., 1972), and even bacteria (Barber, 1966; Blanchard and Syzdek, 1970, 1972) can be transported in this manner. Bubbles also play a central role in the air–sea exchange of gases (Atkinson, 1973). Isotope fractionation as well occurs at the air–sea interface (Weiss, 1970; Kroopnick and Craig, 1972). The salt content of the marine aerosol increases with the breaking-up of sea ice. Inasmuch as the contribution from bubble bursting appears to be small, some other as yet unknown mechanism is evidently at work (Scott and Levin, 1972). Finally, we must not forget that some elements such as fluorine and boron are fractionated from the sea into the atmosphere by special chemical processes that need not involve bubble bursting.

BIBLIOGRAPHY

L. P. Atkinson, *J. Geophys. Res.*, *78*, 962 (1973).

R. Barber, *Nature*, *211*, 257 (1966).

R. Berger and W. F. Libby, *Science*, *164*, 1395 (1969).

W. K. Berger, *Science*, *156*, 383 (1967).

J. R. Benemann, *Science*, *181*, 164 (1973).

L. V. Berkner and L. C. Marshall, in P. J. Brancazio and A. G. W. Cameron (eds.), *The Origin and Evolution of the Atmosphere and Oceans*, Wiley, New York, 1964.

D. C. Blanchard and L. Syzdek, *Science*, *170*, 626 (1970).

D. C. Blanchard and L. Syzdek, *J. Geophys. Res.*, *77*, 5087 (1972).

M. R. Bloch and W. Luecke, *Israel J. Earth Sci.*, *19*, 41 (1970).

B. Bolin, *Sci. Amer.*, *223*, 125 (September 1970).

E. Bonatti and G. Arrhenius, *Mar. Geol.*, *3*, 337 (1965).

J. A. Breznak et al., *Nature*, *244*, 577 (1973).

R. T. Brinkmann, *J. Geophys. Res.*, *74*, 5355 (1969).

W. S. Broecker, *Science*, *168*, 1537 (1970).

W. S. Broecker and A. Walton, Abstr. Intern. Oceanog. Congr. 1959, Amer. Assoc. Adv. Sci., Washington, D.C., 1959, p. 856.

J. M. Brooks and W. M. Sackett, *J. Geophys. Res.*, *78*, 5248 (1973).

E. J. Carpenter, *Science*, *178*, 1207 (1972).

J. Chatt and G. J. Leigh, *Chem. Soc. Rev.*, *1*, 121 (1972).

K. E. Chave, *Science*, *148*, 1723 (1965).

R. Chester et al., *Nature*, *233*, 474 (1971).

R. Chester and J. H. Stoner, *Nature*, *246*, 138 (1973).

N. L. Christensen, *Science*, *181*, 66 (1973).

P. E. Cloud, Jr., in J. P. Riley and G. Skirrow (eds.), *Chemical Oceanography*, Academic, London, 1965.

P. E. Cloud, Jr., *Science*, *160*, 729 (1968).

P. Cloud and A. Gibor, *Sci. Amer.*, *223*, 111 (September 1970).

P. I. Coyne and J. J. Kelley, *Nature*, *234*, 407 (1971).

H. Craig, *Tellus*, *9*, 1 (1957).

J. Cramer and A. L. Myers, *Atmos. Environ. 6*, 563 (1972).

A. C. Delany, *J. Geophys. Res., 78*, 6249 (1973).

C. C. Delwiche, *Sci. Amer., 223*, 137 (September 1970).

G. Dietrich, *General Oceanography*, Interscience, New York, 1963.

R. A. Dixon and J. R. Postgate, *Nature, 234*, 47 (1971).

A. L. Downing and G. A. Truesdale, *J. Appl. Chem., 5*, 570 (1955).

R. A. Duce et al., "Working Symposium on Sea–Air Chemistry," *J. Geophys. Res., 77*(27) (1972).

H. K. Erben and N. Watube, *Nature, 248*, 128 (1974).

A. W. Fairhall, *Nature, 245*, 20 (1973).

D. W. Folger, *Deep-Sea Res., 17*, 337 (1970).

R. M. Garrels et al., *Amer. J. Sci., 258*, 402 (1960).

W. D. Garrett, *Deep-Sea Res., 14*, 221 (1967).

E. D. Goldberg (ed.), *Baseline Studies of Pollutants in the Marine Environment and Research Recommendations*, IDOE Baseline Conf., New York, 1972.

J. J. Griffin, *Geochim. Cosmochim. Acta, 31*, 885 (1967).

H. W. Harvey, *The Chemistry and Fertility of Sea Water*, Cambridge Univ. Press, Cambridge (England), 1966.

H. D. Holland, *Geol. Soc. Amer., Buddington Vol.*, 447 (1962).

H. D. Holland, in P. J. Brancazio and A. G. W. Cameron (eds.), *The Origin and Evolution of the Atmosphere and Oceans*, Wiley, New York, 1964.

R. A. Horne, *Marine Chemistry*, Wiley-Interscience, New York, 1969.

R. A. Horne, *J. Geophys. Res., 77*, 5170 (1972).

D. L. Howard, *Science, 169*, 61 (1970).

G. L. Hutchinson et al., *Science, 175*, 771 (1972).

G. E. Hutchinson, in G. P. Kuiper (ed.), *The Earth as a Planet*, Univ. Chicago Press, Chicago, 1954.

D. M. Imboden and W. Stumm, *Chimica, 27*, 155 (1973).

A. P. Ingersoll, *J. Atm. Sci., 26*, 1191 (1969).

C. E. Junge, *Air Chemistry and Radioactivity*, Academic, London, 1963.

J. Kanwisher, *Deep–Sea Res., 10*, 195 (1967).

C. D. Keeling, *J. Geophys. Res., 70*, 6099 (1965).

C. D. Keeling, *Tellus, 25*, 174 (1973).

H. von König, *Z. Naturforsch., 18a*, 363 (1963).

P. Kroopnick and H. Craig, *Science, 175*, 54 (1972).

J. J. Kelley et al., *Ecology, 49*, 358 (1968).

R. A. Lamontagne et al., *J. Geophys. Res., 78*, 5317 (1973).

A. C. Lasaga et al., *Science, 174*, 53 (1971).

Y-H. Li et al., *J. Geophys. Res., 74*, 5507 (1969).

V. J. Linnenbom et al., *J. Geophys. Res., 78*, 5333 (1973).

L. Machta and E. Hughes, *Science, 168*, 1582 (1970).

F. MacIntyre, *Ph.D. Thesis*, M.I.T., 1965.

F. MacIntyre, *J. Phys. Chem., 72*, 589 (1968).

F. MacIntyre, *Sci. Amer., 223*, 104 (November 1970).

F. MacIntyre, *Tellus, 22*, 451 (1970).

C. S. Martens and R. C. Harriss, *J. Geophys. Res.*, *78*, 949 (1973).
B. Mason, *Principles of Geochemistry*, Wiley, New York, 1966.
N. Metzger, *Men and Molecules*, Crown, New York, 1972.
Y. Miyake and A. Hamada, *Oceanog. Congr.*, Itelsinki Comm., SK13 (1960).
F. Moller, *J. Geophys. Res.*, *68*, 3877 (1963).
E. P. Odum, *Fundamentals of Ecology*, Saunders, Philadelphia, 1959.
J. M. Olson, *Science*, *168*, 438 (1970).
K. Park and D. W. Hood, *Limnol. Oceanog. 8*, 287 (1969).
E. C. Perry, Jr. et al., *Science*, *171*, 1015 (1971).
E. K. Peterson, *Environ. Sci. Tech.*, *3*, 1162 (1969).
M. N. A. Peterson, *Science*, *154*, 1542 (1966).
D. A. Phillips et al., *Science*, *174*, 169 (1971).
S. R. Piotrowicz et al., *J. Geophys. Res.*, *77*, 5243 (1972).
G. N. Plass, *Environ. Sci. Tech.*, *6*, 736 (1972).
J. B. Pollack, *Icarus*, *14*, 295 (1971).
L. K. Porter et al., *Science*, *175*, 759 (1970).
J. M. Prospero and T. N. Carlson, *J. Geophys. Res.*, *77*, 5255 (1972).
R. M. Pytkowicz, *Deep-Sea Res.*, *11*, 381 (1969).
R. M. Pytkowicz, *Geophysics*, *3*, 15 (1972).
R. A. Raff and H. R. Mahler, *Science*, *177*, 575 (1972).
S. I. Rasool and C. de Berg, *Nature*, *226*, 1037 (1970).
A. C. Redfield, *J. Mar. Res.*, *7*, 347 (1948).
G. A. Riley, *Limnol. Oceanogr. 8*, 372 (1963).
J. P. Riley and G. Skirrow (eds.), *Chemical Oceanography*, Academic, London, 1965.
F. A. Richards, in J. P. Riley and G. Skirrow (eds.), *Chemical Oceanography*, Academic, London, 1965.
R. C. Robbins et al., *J. Geophys. Res.*, *78*, 5341 (1973).
W. W. Rubey, *Bull. Geol. Soc. Amer.*, *62*, 111 (1951).
W. W. Rubey, in P. J. Brancazio and A. G. W. Cameron (eds.), *The Origin and Evolution of the Atmosphere and Oceans*, Wiley, New York, 1964.
J. H. Ryther, in M. N. Hill (ed.), *The Sea*, Interscience, New York, 1963.
C. Sagan and G. Mullen, *Science*, *177*, 52 (1972).
W. D. Scott and Z. Levin, *Science*, *177*, 425 (1972).
L. G. Sillen, *Arkiv. Kemi*, *25*, 159 (1966).
W. D. P. Stewart, *Proc. Roy. Soc. London Ser. B*, *172*, 367 (1969).
Y. Suguira et al., *J. Mar. Res.*, *21*, 11 (1963).
K-H. Szekielda et al., *J. Geophys. Res.*, *77*, 5278 (1972).
R. F. Vaccaro, in J. P. Riley and G. Skirrow (eds.), *Chemical Oceanography*, Academic, London, 1965.
L. Van Valen, *Science*, *171*, 439 (1971).
R. Weiss, *Science*, *168*, 247 (1970).
P. M. Williams, *Deep-Sea Res.*, *14*, 9791 (1967).
H. Windom et al., *Environ. Sci. Tech.*, *1*, 923 (1967).
J. Wyman, *J. Mar. Res.*, *11*, 47 (1952).
M. Ycas, *Nature*, *238*, 163 (1972).

ADDITIONAL READING

E. B. Kraus, *Atmosphere-Ocean Interaction*, Oxford Univ. Press, New York, 1972.

G. P. Kuiper (ed.), *The Atmospheres of the Earth and Planets*, Univ. Chicago Press, Chicago, 1952.

G. P. Kuiper (ed.), *The Earth as a Planet*, Chapter 8, Univ. Chicago Press, Chicago, 1954.

E. N. Mishustin and V. K. Shil'nikova, *Biological Fixation of Atmospheric Nitrogen*, Macmillan, London, 1971.

J. R. Postgate, *The Chemistry and Biochemistry of Nitrogen Fixation*, Plenum, London, 1971.

A. San Pietro (ed.), *Photosynthesis and Nitrogen Fixation*, Academic, New York, 1972.

H. C. Urey, *The Planets*, Yale Univ. Press, New Haven, Conn., 1952.

16

BIOSPHERE–HYDROSPHERE INTERACTIONS

. . . The aborigenes of the northwest provinces of South America . . . go to the rivers for gold only in sufficient amount to buy what they need. Any surplus they returned to the stream. They say that if they borrow more than they really need the river-god will not lend them any more.

W. G. Sumner
Folkways

16.1 INTRODUCTION

How appropriate it is that in this final chapter of this book we should return to that substance which is the key to the chemistry of our environment—water. Again and again throughout this book we have emphasized the importance of water. Water (and molecular oxygen) has given our environment its chemical character.

This chapter is devoted to biosphere–hydrosphere interactions. Now man is a part of the biosphere, a renegade perhaps, but undeniably a part. Man's agricultural, industrial, and other polluting activities are a part of his biological activity. They are if you like, extensions of his organism. Thus this chapter is the appropriate place to discuss anthropogenic water pollution. The enormity and rapidity of the chemical changes that man creates in the environment should not mislead us into thinking that these changes are any different in type from those made by any other organism. He consumes materials, and he generates wastes.

Organisms can only exist in close association with water. In the course of their evolution, some creatures have strayed further from the aqueous environment than others. Certain eggs, larvae, and dormant forms remain viable even when extraordinarily dessicated. On the one extreme, there is the desert kangaroo rat that neither imbibes nor excretes liquid water, whereas on the other extreme are, of course, the biota of the seas still living in the primeval aqueous habitat of their ancestors. As a sort of reminder of the aqueous origins of life, the feeding of organisms refers back in one way or another through food webs to water. Similarly, the waste products are eliminated in aqueous form. Even plant expired O_2 and animal expired CO_2 do not represent exceptions to this rule—the solution–gas phase interface is simply within the organism.

Water is the usual recipient of human pollution (Figure 16.1) in our environment. It is commonly the vehicle of pollution, and all too often the hydrosphere is the final repository or sink of pollution. There are some good reasons why this is so. To put a pollutant into the atmosphere usually requires a great deal of energy—of gasification,

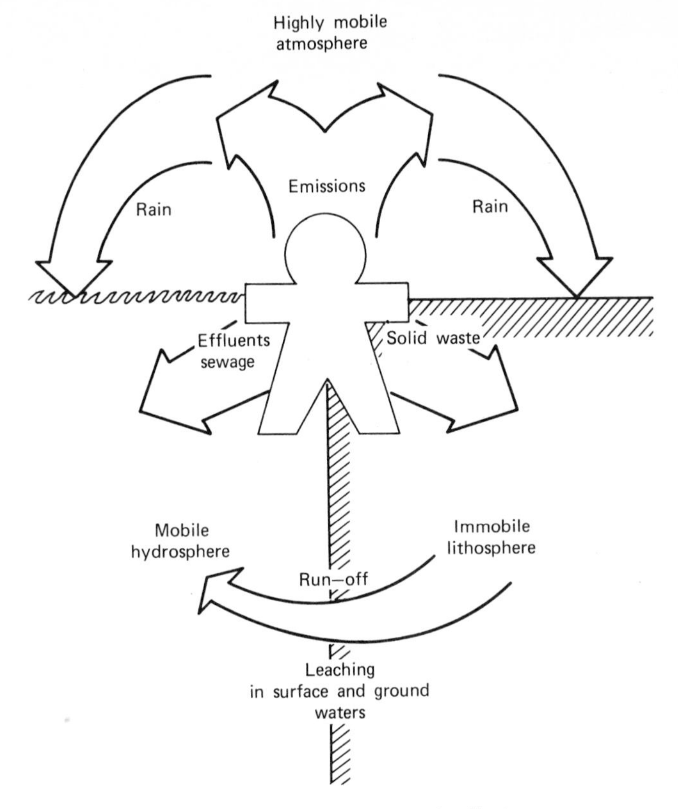

Figure 16.1. The pathways of pollution.

combustion, vaporization, or of very fine pulverization. However to put a pollutant into the hydrosphere or lithosphere requires very little energy. A pollutant is toxic or at least inimical to the organism producing it, thus pollution must be transported away from its point of origin. Pollutants are highly mobile in the atmosphere (Figure 16.1) yet immobile in the lithosphere. Pollutants, such as solid wastes, deposited in the lithosphere, tend not to disperse and to accumulate, eventually (thanks to our old friend, The Principle of Le Chatelier) simply exhausting space in which to put them. However pollutants dumped into the hydrosphere are also mobile and are readily transported away from the point of origin if they are soluble or dispersable, often to create a problem elsewhere. There is a pollution disposal dilemma here. We want to get rid of the pollutant, to mobilize it, to disperse it. Yet the greater the mobility, the greater the dispersal, the greater the area contaminated. Confinement or dispersal? Except in the cases of very dangerous chemicals and radioactive wastes, man usually opts for the latter alternative. He abrogates responsibility and *spreads* the filth to others. The greatest pollution threat on a global scale is the most mobile vehicle—air pollution. The biota of Earth, including the human species, have already suffered damage of undetermined extent from radioactive fallout. Now it appears that we may have entered into a new grim period in which global pollution of the hydrosphere, of the world's oceans, has become a real threat. Lithospheric pollution alone tends to be local and contained, although through leaching by ground waters, stream flooding, and so on, lithospheric pollution all too often can become hydrospheric pollution. Generally, the natural processes that remove pollutants from the hydrosphere, unlike the atmosphere, are much slower than the rates of input, especially cultural stress. Thus in addition to being the major vehicle for the transport of pollutants in our environment, the hydrosphere, notably the oceans, tends to become the sink or final repository where pollution accumulates (Figure 16.1).

16.2 THE CHEMISTRY OF WATER POLLUTION

Water pollutants, both natural and anthropogenic, can be divided into four more or less overlapping categories:

1. Nutrients	Examples: nitrate, phosphate
2. Organic material	sewage
3. Fouling substances	oil spills
4. Poisons	mercury, phenols, pesticides

Nutrients encourage organism growth and can accelerate eutrophication (see discussion that follows); the decay of organic material can deplete the water's vital dissolved oxygen supply (chemical oxygen demand, COD; biological oxygen demand, BOD). By fouling substances, I mean materials that discolor the water, increase its turbidity, or coat its surface, sediments, and shores. Finally, there are chemicals specifically harmful to aquatic life and other organisms, including man, that might come in contact with or ingest the waters. Pollutants can also alter the pH of waters or impart to them noxious tastes and/or smells. Because they cannot quantize it, scientists habitually neglect the aesthetics of water quality and other forms of environmental damage, yet this is the very aspect of environmental deterioration most obvious to the general public. The public is more readily riled by visual damage, such as litter and billboards, than by threats to health, such as lead poisoning, alcoholism, and the burning of high sulfur fuels, and I am not certain that I wish to disagree strongly with the wisdom of this priority. In addition to being chemically polluted, as we noted earlier, water can also be thermally polluted, and this form of pollution can have disastrous chemical consequences such as the reduction of the level of dissolved oxygen.

Some human activities are notoriously serious water polluters, and we examine a few of them here briefly—industrial wastes (Nemerow, 1971), pulp and paper production, acid mine drainage, detergents, and finally sewage.

Industrial Wastes

Industry not only uses enormous quantities of water, 3.4% of annual precipitation—a fraction equaled only by irrigation use and nearly six times greater than municipal use (Table 16.1), but some industries have been long recognized as very serious polluters (Tables 16.2 and 16.3). These include the agriculture and food processing industry (remember Chapter 14; see also Litchfield, 1971; Richter and Soderquist, 1971); the fermentation industry (Gandy, 1971); the metals production, fabrication, and finishing industries (Smith, 1971); textiles (Alspaugh, 1971; Porter et al., 1972; Jones, 1973*a*); and tanning (Eye, 1971; Eye and Liu, 1971); pulp and paper (Allan

Table 16.1 The Fate of the Annual Average (30 in.) Rainfall on the Continental United States (From *Environmental Protection* by E. T. Chanlett. © 1973 by McGraw-Hill. Used with permission of McGraw Hill Book Co.)

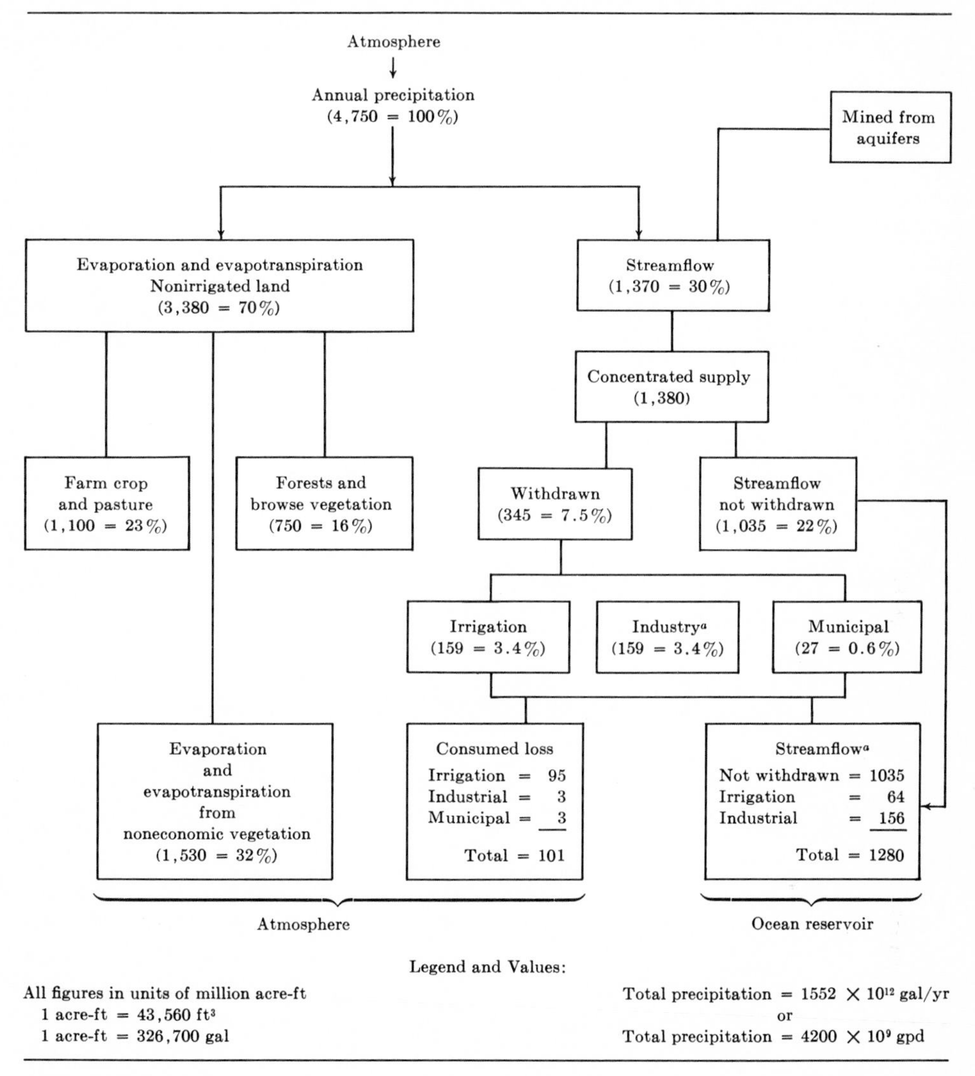

[a] The same water may be reused at points spaced along a single stream.

et al., 1972; Jones, 1973*b*); and petrochemical-based industries (Hall, 1971). Table 16.2 gives the amounts of some industrial effluents, while Table 16.4 indicates the sources of heavy metals in major industries. A great deal of work has been done on treating industrial effluents (Lund, 1971), so much in fact that we must exclude that very large and important topic from further mention in this book. A very great deal more remains to be done, and the cost is large (Table 16.5). Industry consumes on an average 65% of public water supplies in the United States (Table 16.6). Table 16.7 lists some fish kills from chemical plant discharges, and it is interesting to compare these disasters with pesticide fish kills (Table 14.8).

Table 16.2 Estimated Volumes of Industrial and Domestic Wastes Before Treatment, 1963 (Reprinted from *Cleaning Our Environment—The Chemical Basis for Action*, a report by the Subcommittee on Environmental Improvement, Committee on Chemistry and Public Affairs, American Chemical Society, 1969. Reprinted by permission of the copyright owner.)

Industry	Waste Water (billion gallons)	Standard Biochemical Oxygen Demand (million pounds)	Settleable and Suspended Solids (million pounds)
Food and kindred products	690	4,300	6,600
Textile mill products	140	890	n.a.
Paper and allied products	1,900	5,900	3,000
Chemical and allied products	3,700	9,700	1,900
Petroleum and coal	1,300	500	460
Rubber and plastics	160	40	50
Primary metals	4,300	480	4,700
Machinery	150	60	50
Electrical machinery	91	70	20
Transportation equipment	240	120	n.a.
All other manufacturing	450	390	930
All manufacturing	13,100	22,000	18,000
Domestic			
Served by sewers (120 million people)	5,300[a]	7,300[b]	8,800[c]

[a] Number of persons × 120 gal/person/day × 365 days.
[b] Number of persons × 1/6 lb/person/day × 365 days.
[c] Number of persons × 1/5 lb/person/day × 365 days.

Wastes from the natural fibers textile industry can impose a heavy biological oxygen demand on receiving waters, while synthetic fabrics production can add organic solvents and alter the pH as well (Poon, 1970), and dyeing and the related industry of dye production can be sources of very toxic pollutants. Benzidine is such a pollutant associated with synthetic dyes. Other synthetic dyes have intermediates such as aniline and its derivatives (Table 8.20), which are highly poisonous. Dye synthesis may also involve heavy metal catalysts such as mercury. Even natural dye production may release excessive tannin into the receiving waters (Table 8.20).

Pulp and Paper

Conservationists have long looked upon the pulp and paper industry as a villain, and not without good cause for the waters polluted by this industry, because of their location in wooded regions, are often among the scenic waters in our nation. The tragedy of this pollution is acerbated by the fact that these once handsome waters often receive their deadly, foul-smelling slug far up in their headwaters, thus (as in the case of the Merrimac, see Table 17.21) being ruined for their entire length. Forests are despoiled and air as well as water is fouled. Paper mills and the rivers

Table 16.3 Objectionable Components of Industrial Wastewaters, Their Effects, and Typical Sources (From *Environmental Protection* by E. T. Chanlett, © 1973 by McGraw-Hill. Used with permission of McGraw-Hill Book Co.)

Component Group	Effects	Typical Sources
1 Bio-oxidizables expressed as BOD	Deoxygenation, anaerobic conditions, fish kills, stinks	Large amounts of soluble carbohydrates: sugar refining, canning, distilleries, breweries, milk processing, pulping, and paper making
2 Primary toxicants: As, CN, Cr, Cd, Cu, F, Hg, Pb, Zn	Fish kills, cattle poisoning, plankton kills, accumulations in flesh of fish and mollusks	Metal cleaning, plating, pickling; phosphate and bauxite refining; chlorine generation; battery making; tanning
3 Acids and alkalines	Disruption of pH buffer systems disordering previous ecological system	Coal mine drainage, steel pickling, textiles, chemical manufacture, wool scouring, laundries
4 Disinfectants: Cl_2, H_2O_2, formalin, phenol	Selective kills of microorganisms, taste, and odors	Bleaching of paper and textiles; rocketry; resin synthesis; penicillin preparation; gas, coke, and coal tar making; dye and chemical manufacture
5 Ionic forms: Fe, Ca, Mg, Mn, Cl, SO_4	Changed water characteristics: staining, hardness, salinity, encrustations	Metallurgy, cement making, ceramics, oil well pumpage
6 Oxidizing and reducing agents: NH_3, NO_2^-, NO_3^-, S^{2-}, SO_3^{2-}	Altered chemical balances ranging from rapid oxygen depletion to over nutrition, odors, selective microbial growths	Gas and coke making, fertilizer plants, explosive manufacture, dyeing and synthetic fiber making, wood pulping, bleaching
7 Evident to sight and smell	Foaming, floating, and settleable solids; stinks; anaerobic bottom deposits; oils, fats, and grease; waterfowl and fish injuries	Detergent wastes, tanning, food and meat processing, beet sugar mills, woolen mills, poultry dressing, petroleum refining
8 Pathogenic organisms: *B. anthracis*, *Leptospira*, toxic fungi, viruses	Infections in man, reinfection of livestock, plant diseases from fungi-contaminated irrigation water; risks to man slight	Abattoir wastes, wool processing, fungi growths in waste treatment works; poultry processing wastewaters

Table 16.4

Metals (and Fluorine) Found in Major Industries

	Al	Ag	As	Cd	Cr	Cu	F	Fe	Hg	Mn	Pb	Ni	Sb	Sn	Zn
Alkalis, chlorine, inorganics	●		●	●	●		●	●	●		●			●	●
Basic nonferrous metals	●	●	●	●	●	●	●		●		●		●		●
Basic steel works			●	●	●	●	●	●	●		●	●	●	●	●
Fertilizers	●		●	●	●	●	●	●	●	●	●	●			●
Flat glass, cement, asbestos products					●										
Leather tanning					●										
Motor vehicles, aircraft	●	●		●	●	●			●			●			
Organic chemicals and petrochemicals	●		●	●	●		●	●	●		●			●	●
Petroleum refining	●		●	●	●	●	●	●			●	●			●
Pulp and paper					●	●			●		●	●			●
Steam power plants					●										●
Textiles					●										

below them for miles are incredibly putrid. Table 16.8 breaks down and summarizes the effluent volumes for various types of pulp and paper production activities, while Figure 16.2 diagrams the mechanical and chemical pulping operations and gives more detailed information on the effluent volumes of each of the major processing steps. The waste liquor from a pulp mill contains about 115 g/liter total solids, 7.8 g/liter of sulfur, ash, and sulfur dioxide; in other terms, the production of 1 ton of dry paper pulp leaves 600 kg of lignin, 200 kg SO_2 combined with lignin, 90 kg of CaO combined with lignin sulfonic acid, 325 kg of carbohydrates, 15 kg of proteins, and 30 kg of rosin and fat (Wilbur, 1969). The oxygen demand of all this organic material is huge, the sulfur compounds (including methyl mercaptan as well as SO_2 and sulfonic acid and its derivatives) are strong direct poisons, and the excess solid matter blankets the bottom of the receiving waters resulting in anaerobic decompo-

Table 16.5 Current and Projected Waste Water Treatment Needs of Major Industrial Establishments[a] (Reprinted from *Cleaning Our Environment—The Chemical Basis for Action*, a report by the Subcommittee on Environmental Improvement Committee on Chemistry and Public Afiairs, American Chemical Society, 1969. Reprinted by permission of the copyright owner.)

Industry	Required Investment, Constant Dollars (millions)	
	By Expert[b] Estimate	By Census[b] Projection
Current Need		
Food and kindred products	740	670
Textile mill products	170	170
Paper and allied products	320	920
Chemical and allied products	380	1000
Petroleum and coal	380	270
Rubber and plastics	41	59
Primary metals	1500	1400
Machinery	39	56
Electrical machinery	36	51
Transportation equipment	220	160
All other manufacturing	200	290
Total Current Need	4000	5000
Plant now provided		
By industry	2200	1800
Through municipal facilities	730	640
Total Current Backlog	1100	2600
New Facilities, Fiscal Years 1969–1973	700	1000
Replacing Obsolete Equipment	800	1000
Total Capital Requirement, Fiscal Years 1969–1973	2600	4600

[a] Assumes at least 85% removal of BOD and settleable and suspended solids.
[b] All values rounded to two significant figures.

Table 16.6 Public Water-Supply Use in the United States (From *Environmental Protection* by E. T. Chanlett, © 1973 by McGraw-Hill. Used with permission of McGraw-Hill Book Co.)

Use	Gallons per capita per day	
	Range	Average
In Home	50 to 70	50
Commercial and Industrial	10 to 100	65
Public Services	5 to 20	10
Unaccounted (leakage)	10 to 40	25
Total	75 to 230	150

Table 16.7 Fish Kills Resulting from Chemical Plant Releases (From Dawson et al., 1970)

Year	Number of Reports	Total Kill Reported	Average Kill of Incidents Reporting Kill Totals
1963	34	224,441	7,739
1964	26	525,739	20,220
1965	37	218,661	7,053
1966	36	708,815	19,689
1967	24	43,732	1,987
1968	39	731,881	18,766

sition that produces toxic sulfides (Ziebell et al., 1970). Other toxic materials include pentachlorophenol and its sodium salt, which is sometimes added as a pulp preservative. Although the practice has now been largely discontinued in the United States, the industry also used a liberal application of mercurial slimicides (Marton and Marton, 1972), and this practice has left a legacy of pollution of the aqueous environment in some places. The effects of all these chemicals on biota has been reviewed by Wilbur (1969). Returning to the solid material, even such a simple matter as blockage of water passages in bottom gravel by decaying bark debris can reduce the survival of fish eggs to fry (Servizi, 1971). Harvesting the raw material also creates pollution problems—solid wastes from lumber and pulp are now largely utilized to form pressboard and other products. The logs themselves are commonly transported by floating downstream to the mills. This puts a lot of bark and organic debris into the receiving waters. But in cases where the receiving waters are fast flowing and well oxygenated, oxygen depletion does not become serious. Insects from the floating logs provide a feast for game fish so this may even be considered a beneficial impact. Not much can be said for a lake or river bank strewn with logs, and, as we noted in an earlier chapter, tree cutting, particularly such drastic harm as clear cutting, results in a release of nutrients from the soil into the drainage system (recall Section 12.2). Because of the seriousness of the pollution problem that the pulp, paper, and lumbering industries have created, a great deal of attention has been focused on ways of combating the problem ranging all the way to using less water (the so-called "waterless web" method of paper production) to a wide variety of waste treatment processes (Bolker, 1971; Gellman and Blosser, 1971; Jones, 1973*b*).

Metals Industry

Another very serious industrial offender responsible for the pollution of our water is the metals industry. Every step of this industry appears to endanger our environment from mining and ore dressing through metal finishing and electroplating (we have already seen how its products—automobiles and construction—are laying waste to our environment). Pickling and plating waste effluents are especially troublesome, for they are not only highly acidic (0.5 to 10% acid; HCl is now replacing H_2SO_4 in popularity—Anon., 1970*g*), but are also rich in iron (12%) and

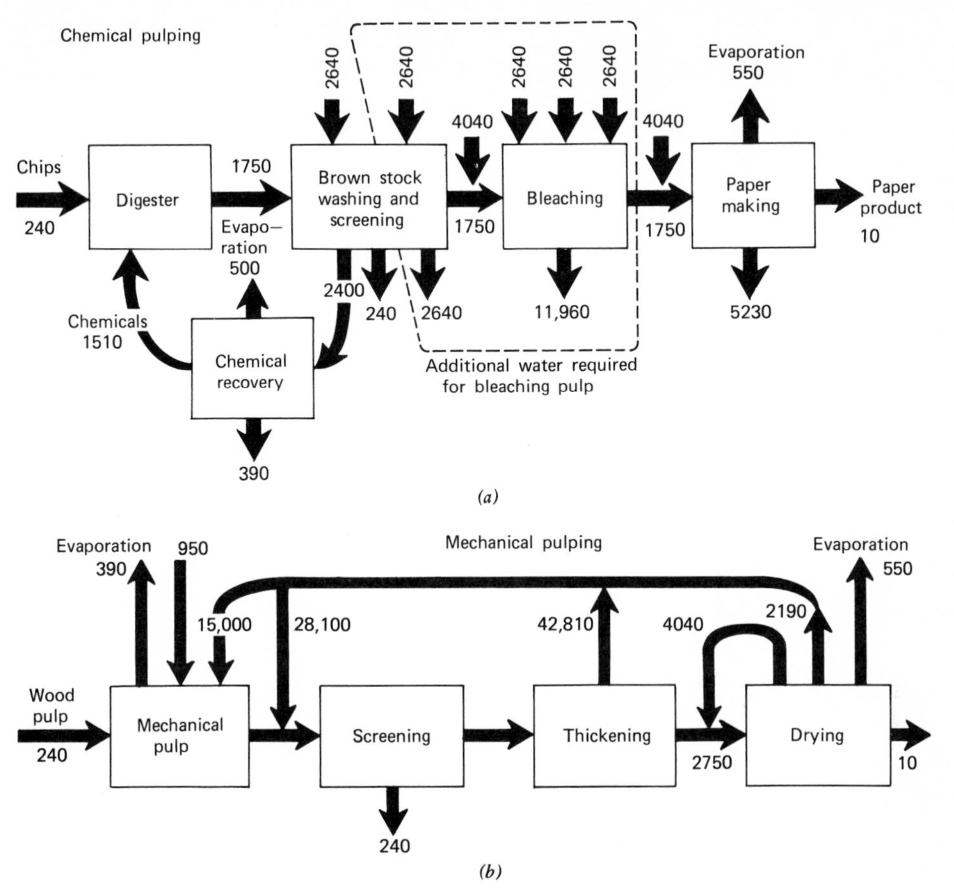

Figure 16.2. Flow diagrams for chemical and mechanical pulping in gallons of water per ton of product (from Federal Water Pollut. Control Adm., 1968).

Table 16.8 Average Waste Discharges per Ton of Paper Product (From C. G. Wilbur, *The Biological Aspects of Water Pollution*, 1969. Courtesy of Charles C Thomas, Springfield, Ill., and C. G. Wilbur)

Product	Gallons of Waste Per Ton of Product Daily
Pulp Mills	
Groundwood	5,000
Soda	85,000
Sulfate (kraft)	64,000
Sulfite	60,000
Misc. Paper: No Bleaching	39,000
Misc. Paper: With Bleaching	47,000
Paperboard	14,000
Strawboard	26,000
Deinking: Old Paper Stock	83,000

other heavy metals (Tables 16.3 and 16.4). Plating effluents may contain high levels of cyanide (Grune, 1971). Fortunately, this exceedingly deadly poison is not very persistent in the environment. Since cyanide kills the bacteria necessary for activated sludge sewage treatment (see discussion that follows) many municipalities have placed stringent restrictions on its discharge. A number of methods are available for detoxifying cyanide wastes including oxidation to cyanates by alkaline chlorination (the cyanate can then be converted to CO_2 and NH_3 by acid hydrolysis), electrolytic oxidation, oxidation with H_2O_2 and precipitation of metals, ozonation, ion exchange, and even radiation (Anon., 1971*e*). Other polluting ingredients of metal-finishing wastes include phosphates, phenols, and oil. Sometimes it is possible to reduce the cost of metal-finishing wastes treatment by recovering materials (Zievers and Novotny, 1973). Of the nearly 1,000,000,000,000 gal of process water used by the primary metals industry the thirstiest metal production is iron and steel; iron and steel production uses over 800,000,000,000 gal, foundaries, another 12,000,000,000 gal, while the copper and aluminum industries use only somewhat more than 30,000,000,000 and 20,000,000,000 gal, respectively (Federal Water Pollut. Control Adm., 1968). Figure 16.3 identifies the major sources of air and solid waste as well as water pollution associated with the steps in the steelmaking process, while Table 16.9 gives the percentage water use, types of pollution, and abatement treatments.

Table 16.9 Water Pollution in an Integrated Steel Mill

Process	Percentage Water Use	Type of Pollution	Abatement Treatment
Coke Production	12.5%	Ammonia still wastes, light oil decanter wastes (containing ammonia, phenols, cyanides, chlorides, and sulfur compounds	Reuse for coke quenching biological treatment chemical oxidation carbon absorption
Blast Furnace	25	Gas cleaning by wet washing (containing ammonia, phenols, cyanides, and suspended solids	Recycling
Open Hearth Hot Mills	12.5 2.5	Gas cleaning by wet washing (containing suspended solids)	Recycling chemical coagulation magnetic agglomeration
Rolling Mills		Suspended scale particles	Settling chambers (scale pits)
		lubricating oils	settling and/or skimming
		spent pickle liquor pickling rinse water	neutralization
Finishing Mills	20	Heavy metal salts from plating and galvanizing oil emulsions	Recycling chemical treatment
Sanitary, Boiler, and Other Uses	5		

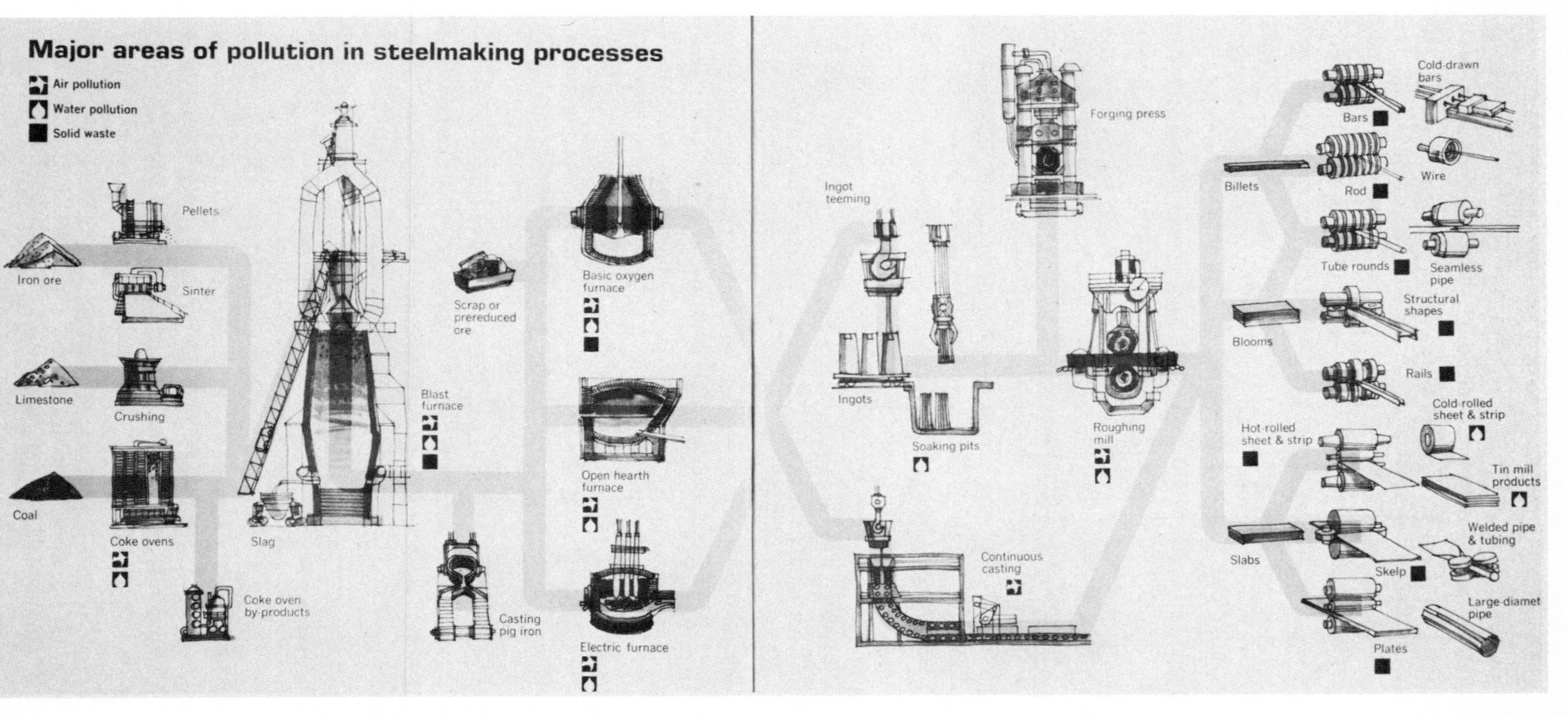

Figure 16.3. Major sources of pollution in steelmaking (reprinted from H. C. Bramer, *Environ. Sci. Tech.*, *5*, 1004 (1971), © The American Chemical Society).

Refineries and the numerous different types of chemical industries are also among the worst offenders. Users of petroleum products are also polluters; in the United States, 50% of the 1.1 billion gal of automobile lubricants and 30% of the 1.5 million gal of industrial lubricating oils are not consumed and are disposed of in some way—in the former case as much as 23% may be dumped on the ground near the service station (Anon., 1972*b*). Metals production and finishing, refining, and chemical production, it should be noted, commonly have effluents with very high levels of pollutants. Worse still, these pollutants are substances very alien to the environment, and as a consequence, while their impact will tend to be lessened by dilution and in some instances by precipitation and/or sorption, the natural cleaning processes in the environment are ill-equipped to cope with these pollutants. Natural waters, up to a point, can handle nontoxic organic material; they can even to some degree respond to and recover from excessive injections of nutrients, but environmentally exotic materials, such as phenols, halogenated hydrocarbons, cyanide, high acidity, and high concentrations of heavy metals, pose a more serious problem. Not only may there be no adequate natural mechanism for rendering them innocuous, but they may sterilize or otherwise damage the aquatic environment thereby destroying much of whatever assimilative capacity the waters may have had originally.

Dean (1971) has distinguished two types of industrial wastes: those containing one or more useable or potentially useable substances that are uneconomic to salvage, and those containing unwanted by-products inherent in the purification of a raw material. What is waste is in large part determined by economics; one man's waste may be another's resource. For example, in the early days of the chlor-alkali industry, soda was the product and chlorine the waste; today chlorine is usually the major product, and sodium and its compounds are the waste. Economics and demand also determine "waste" utilization. Waste lignin from paper making can be oxidized to vanillin, but one small sulfite paper mill could produce vanillin enough for all the ice cream in the country; flyash can be used to make building blocks, but ash production is far greater than block production; pickle liquor could be a cheap source of iron salts to remove phosphorus from sewage, but today there are not a sufficient number of sewage treatment plants to keep pace with the available supplies of pickle liquor.

As our water resources are diminished and water becomes a more expensive commodity, industry will cease to waste it in such enormous amounts. Recycling will become much more common and this in turn will help to alleviate the problem of industrial water pollution (Gloyna et al., 1970; Rey et al., 1971).

Acid Mine Drainage

Acid mine drainage is another example of severe industrial pollution that has finally raised some public outcry. "Few water pollution problems have effects as insidious as mine drainage" (Anon., 1969*b*). It is one of the primary causes of fish kills in the United States, in 1967 killing more than 1,000,000 fish. All types of mining activity that expose sulfur-bearing minerals to the Earth's atmosphere, thereby accelerating the exposure and oxidation of crustal material, pose an acid mine drainage threat, but by far the worst offender is coal mining because of the greater quantities of crustal materials involved. And if the so-called "energy crisis" stampedes us into returning to the more extensive use of that fuel without environmental safeguards, the problem will worsen. In the United States, 75% of the mine drainage problem, originating from both operating and abandoned mines and from both surface and deep shaft mines, afflicts the Appalachia area (Anon., 1969*d*). In this once beautiful land alone more than 10,000 miles of surface streams are degraded (Anon., 1969*b*).

Pyrite, FeS_2, is often found in close association with coal deposits. Mine wastes consisting of mountainous heaps of pyritic low grade coal are called "gob piles." Upon exposure to moisture and atmospheric oxygen the pyrite becomes oxidized (Stumm and Lee, 1961) losing an electron

$$Fe^{2+} \rightarrow Fe^{3+} + e^- \qquad (16.2)$$

and this oxidation is the rate controlling step of the whole sequence of reactions (Singer and Stumm, 1970). Bacteria such as the acidophilic Thiobacillus–Ferrobacillus group can then utilize the energy represented by this released electron to partially satisfy their nutritional demands for the ultimate reduction of CO_2 into new cell material (Dugan, 1972). Although Fe(II) oxidation can take place in their absence, the process is accelerated several hundredfold by the presence of the bacteria. Aeration and the exposed surface area of the pyrite are further important considerations in determining the rate of the process (Baker and Wilshire, 1970). The ferric ion produced can then react nonbiologically with sulfide

$$8Fe^{3+} + S^{2-} + 4H_2O \rightarrow 8Fe^{2+} + SO_4^{2-} + 8H^+ \qquad (16.3)$$

or water

$$Fe^{3+} + 3H_2O \rightarrow Fe(OH)_3 + 3H^+ \qquad (16.4)$$

Both of these reactions produce acid, and further acidity can result from the direct microbiotic oxidation of reduced sulfur compounds (Trudinger, 1967). The Fe(II) produced in Equation 16.3 can then be used again as an energy source for bacteria, while the ferric hydroxide precipitated in reaction 16.4 is responsible for the all too familiar brown stain ("yellow boy") of contaminated streams. The overall reaction for acid mine drainage from pyritic material can be represented by

$$2FeS_2 + 7O_2 + 2H_2O \rightarrow 4SO_4^{2-} + 2Fe^{2+} + 4H^+ \qquad (16.1)$$

Excessive sedimentation is also associated with acid mine drainage because the highly acid water aggravates erosion.

Several treatment methods are available to combat the destructive consequences of acid mine drainage (Anon., 1969*b*, 1970*a*, *b*); these include the use of lime or carbonate treatment (Anon., 1970*c*) to precipitate out Fe(III), heterotrophic anaerobic bacteria to reduce sulfate (Tuttle, 1969), and the application of certain chemicals, such as α-keto and carboxylic acids, which inhibit the iron- and sulfur-oxidizing bacteria and which are not dangerous to most other organisms. Still other treatment methods examined include silicate (Anon., 1971*b*), partial freezing (Anon., 1971*c*), and foam fractionation (Anon., 1970*d*). Fly ash (pH 8 to 12) has been used to neutralize acidic strip mine wastes (pH 2.5 to 3.5) and make a viable soil (Anon., 1971*d*). Leaching can be a serious problem in mine disposal areas, and burial may be necessary to prevent oxidation (Grube et al., 1972). Work is also being done on the prevention of acid mine drainage by sealing abandoned mines (Anon., 1970*e*).

Table 16.10 Levels of Pollutants in the Rhine River

	Concentrations	
	1965 (mg/liter)	1972 (mg/liter)
Chloride ions	130	280
Ammonia (as N_2)	1	3
Total N_2 (Kjeldahl, except nitrate)	n.a.	5
Orthophosphate (as PO_4)	0.6	1.2
Total Phosphates	n.a.	2 to 4
Detergents	0.2	0.4
Phenols	25 μg/liter	40 μg/liter

The Rhine River has been called Europe's longest sewer, and there are fears that it may become a "dead" river because the quantities of industrial and municipal pollutants have increased so steeply in recent years (Anon., 1972*a*; Table 16.10). The Dutch at the mouth of the river are particularly distressed. The river is their main source of drinking water, and they use it to dilute the salt from seawater seepage in much of their below sea level agricultural areas. But the French potash-mining operations in Alsace are dumping salt into the river, as much as 200 to 250 kg/sec, near Mulhouse. An international commission has been asked to tackle the problem, and it will be interesting to follow what measure of success, or lack of it, can be achieved.

Mineral extraction and production plant effluents are by no means the only sources of industrial chemical pollution of our environment. Chemicals must be transported from place to place, and they can be accidently spilled during such transfer and handling operations. Large oil spills have received the most public attention, but chemical spills from accidents and simply from sloppy handling procedures are an everyday occurrence. In Table 16.11, some hazardous chemical transported are listed by the United States production rank. Figure 16.4 shows the locations of fish kills in the United States that resulted from the transportation of chemicals, while Figure 16.5 shows that the number of accidents in the United States involving hazardous materials is climbing steadily.

Detergents

Agriculture and industry certainly are not alone in dumping chemical pollutants into the aquatic environment. You are probably guilty too. One of the most disruptive types of water pollutants and one traceable in enormous amounts to home use as well as to commercial laundries is synthetic detergents (Banerji, 1971), especially phosphate detergents. Sodium tripolyphosphate makes up about 40% of most detergents. Phosphate, as we see presently, is a major cause of "cultural" (i.e., nonnatural) eutrophication of surface waters. Grundy (1971) estimates that 10^9 kg/yr of phosphorus enters United States waters from phosphate detergents, and this amount represents 30 to 40% of the total phosphorus entering the aquatic environment. Phosphate detergent pollution can be successfully controlled by legislation. Following

Table 16.11 Hazardous Transported Chemicals Ranked by United States Production (From Dawson et al., 1970)

Name	Rank
Sulfuric Acid	1
Ammonia	2
Oxygen	3
Ammonia CPDS	4
Sodium Hydroxide	5
Sodium Carbonate	6
Nitric Acid	7
Ammonium Nitrate	8
Phosphoric Acid	9
Benzene	10
Urea	11
Sodium CPDS	12
Ethylene Dichloride	13
Toluene	14
Halogenated Hydrocarbons	15
Acids Acylhalides and Anhydrides	16
Ammonia Sulfate	17
Styrene	18
Potassium CPDS	19
Ethyl Benzene	20
Methyl Alcohol	21
Formaldehyde	22
Vinyl Chloride	23
Hydrochloric Acid	24
Propane	25
Xylenes	26
Aldehydes and Ketones	27
Insoluble with Density LE 1	
Cyclohexane	1
Cumene	2
Heptane Mixed	3
Nonene	4
Hexane	5
Decyl Alcohol	6
Ethyl Acrylate	7
Butyl Acrylate	8
Divinyl Benzene	9
Insoluble with Density GT 1	
Ethylene Dichloride	1
Sulfur	2
Perchloroethylene	3
Trichloroethane	4
Dioctyl Phthalates	5
Trichlorofluoromethane	6
Dinitro Aniline	7
Epichlorohydrin	8
Benzyl Chloride	9
Dibutyl Phthalate	10
Methyl Bromide	11
2,4,5-Trichlorophenol and Salts	12
Lauroyl Peroxide	13
Beryllium Dust	14
Mercury	15
Soluble Substances	
Phenol	1
Methyl Alcohol	2
Insecticides Rodenticides Cyclic	3
Acrylonitrile	4
Chlorosulfonic Acid	5
Benzene	6
Ammonia	7
Miscellaneous Cyclic Insecticides	8
Phosphorus Pentasulfide	9
Styrene	10
Acetone Cyanohydrin	11
Chlorine	12
Nonyl Phenol	13
DDT	14
Isoprene	15
Hazardous Chemicals Which Because of Volatility or Other Reasons Present Only a Temporary Spill Threat	
Vinyl Chloride	1
Calcium Hydroxide R	2
Propylene	3
Butane	4
Propane	5
Oxygen	6
Hydrogen Sulfide R	7
Dichlorodifluoromethane	8
Isobutane	9
Butenes	10
Butadilne Inhibited	11
Isopentane	12

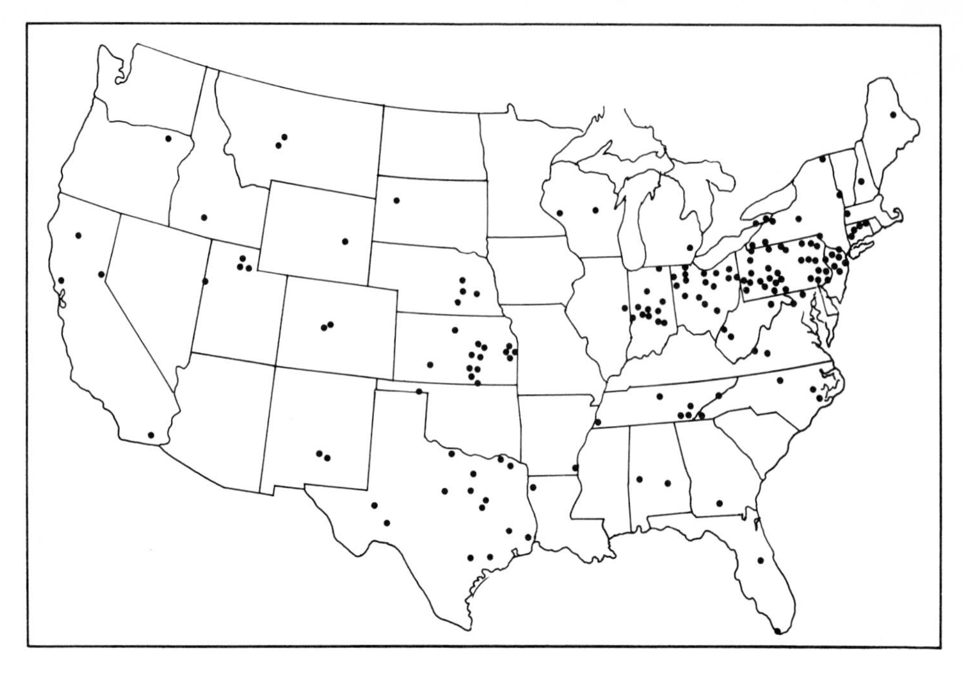

Figure 16.4. Geographical distribution of fish kills originating from transportation activities, 1960–1968 (from Dawson et al., 1970).

implementation of city and state control laws, the average inorganic phosphate level in the epilimnion and hypolimnion of Onandago Lake, New York, decreased 85 and 76%, respectively, and of orthophosphate 47 and 15% (Murphy, 1973). The issue of the role of phosphate detergents, however, remains highly controversial; Mitchell (1971), for example, has reported little difference in eutrophication potential of activated sludge sewage treatment effluents for phosphate detergents, phosphate-free detergents, and even detergent-free water. Phosphate is not the only problem associated with detergents. Foaming and persistence are others. Earlier (around 1965) legislation had proved effective in prodding the detergent manufacturers to substitute biodegradable linear alkylate sulfonates (LAS) for the alkyl benzene sulfonates (ABS) that were clogging our sewer systems and covering our streams with wet foam and that were also accumulating in soils (Fink et al., 1970) and sediments (Ambe, 1973). High levels of LAS, however, can increase fish fry mortality (Pickering and Thatcher, 1970).

$$\text{—C—C—C(C)—C—C(C)(C)—C—C(C)—C—OSO}_3^-$$

Alkylbenzene Sulfonate (ABS)

$$\text{—C—C—C—C—C—C—C—C—C—C—C—C—OSO}_3^-$$

Linear Alkylbenzene Sulfonate (LAS)

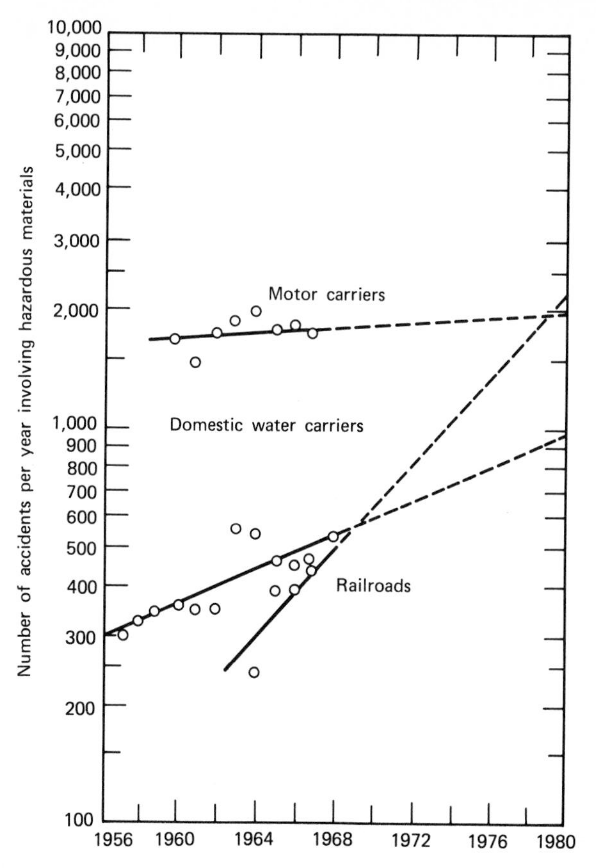

Figure 16.5. Hazardous materials accidents predictions (from Dawson et al., 1970).

The addition of enzyme from *Bacillus subtilis* to "pre-soak" detergents to remove difficult dirt was phased out after reports began to accumulate that they caused allergic reactions in some persons, particularly workers involved in the production and packaging of these materials (Gruchow, 1970; Lichtenstein et al., 1971). In experimental animals, preparations of these additives produced losses in body weight, increased susceptibility to some infections, and cytoxic effects, and the activity was not destroyed by boiling (Dubas, 1971). The elimination of phosphate is even more problematical. The possible use of nitrilotriacetic acid, NTA, $N(CH_2CCOH)_3$, a powerful chelating agent that complexes in hard water with Ca^{2+} and Mg^{2+} ions and that is readily biodegradable in natural waters (Warren and Malec, 1972) and in conventional activated sludge water treatment installations (Shumate et al., 1970) was dropped after it was realized that this chemical might be a health hazard. Also, the nitrogen contained in this material could, like phosphorus, contribute to eutrofication. NTA, it is interesting to note in passing, can inhibit cadmium toxicity (Scharpf et al., 1972), probably because of its strong complexing affinity for bivalent cations. A number of phosphate-free detergents have appeared on the market based on sodium carbonate, but, not only are they strongly alkaline enough to hurt the eyes and mucous membranes if improperly used, they simply do not do the cleaning job as well as phosphates. Borate has also been considered as a phosphate substitute, but boron can be toxic at relatively low concentrations. Then too there has been a controversy over the presence of arsenic in phosphate detergents (Angino et al., 1970), but it has been argued that this element's presence does not represent a health hazard because it is in the pentavalent state (as arsenate) and not in the highly toxic trivalent state (Pattison, 1970; Sollins, 1970). Polyelectrolytes containing neither N nor P have been tested as detergents (Anon., 1969*c*). Clearly, getting rid of the phosphate is proving to be a much more difficult task than the change-over from ABS to LAS. In view of the greater hazards presented by phosphate substitutes, for all the sound and fury it looks like we are right back where we started—with phosphate detergents (see Anon., 1971*a*).

16.3 SEWAGE

All organisms produce metabolic wastes. In the case of animals, these wastes often contain materials inimical to the health of the generating organism and must therefore be physically separated from it. Man is no exception. As long as primitive man was nomadic, following the migrations of the prey he hunted, he left his wastes behind. But when he became a prisoner of the land, a grower of grains, a keeper of animals, a harvester of crops, immobilized in one place, a builder and dweller in permanent towns and cities, then a serious problem of waste disposal came into being. Solid wastes were thrown into middens, usually outside the walls. It is a wry truth that we shall be known to future archaeologists, if there are any, by our dumps. Flowing water proved to be a serviceable media for the removal of excreta and garbage. The growth of the larger cities on rivers is no accident of history. The waters of the river provided water for transportation; for drinking, cooking, and washing; for agriculture; and, equally important, for sewage. The cities of the ancients are often remarkable for their feats of sanitary engineering. The Romans, particularly, excelled in this field. Some of their magnificent aquaducts are still in service. But the Dark Ages were also the Filthy Ages, and very little thought was given to sanitation in Europe until the nineteenth century. The Versailles of Louis XIV reeked with the smell of urine. And it has only been within the past few decades that serious attention has been given to waste disposal other than dumping it into the nearest waterway. This is not as unpardonable as it may appear. Offensive though it might be to our sensibilities, raw sewage, uncontaminated with synthetic chemicals, is a *relatively* innocuous pollutant. However several factors now militate to make more sophisticated forms of disposal and water treatment mandatory. These are:

1. Increased quantities of sewage resulting from increased population growth,
2. Increased clean water demand resulting from increased population growth, and
3. Increased chemical pollution.

The last is especially important for, not only may no natural mechanism exist for the removal of dangerous chemical pollutants, but their presence may reduce or even destroy the natural assimilative capacity of the receiving waters for sewage and other biogenic wastes.

Table 16.12 summarizes the composition of typical domestic sewage, exclusive of some components that may be present in the original domestic water supply ("carriage

Table 16.12 Typical Composition of Domestic Sewage[a] (From *Wastewater Engineering* by Metcalf & Eddy, Inc., © 1972. Used with permission of the McGraw-Hill Book Co.)

Constituent	Concentration Strong	Medium	Weak
Solids, Total	1200	700	350
Dissolved, Total	850	500	250
Fixed	525	300	145
Volatile	325	200	105
Suspended, Total	350	200	100
Fixed	75	50	30
Volatile	275	150	70
Settleable Solids (ml/liter)	20	10	5
Biochemical Oxygen Demand, 5-day, 20°C (BOD_5-20°)	300	200	100
Total Organic Carbon (TOC)	300	200	100
Chemical Oxygen Demand (COD)	1000	500	250
Nitrogen (total as N)	85	40	20
Organic	35	15	8
Free Ammonia	50	25	12
Nitrites	0	0	0
Nitrates	0	0	0
Phosphorus (total as P)	20	10	6
Organic	5	3	2
Inorganic	15	7	4
Chlorides[b]	100	50	30
Alkalinity (as $CaCO_3$)[b]	200	100	50
Grease	150	100	50

[a] All values except settleable solids are expressed in milligrams per liter.
[b] Values should be increased by amount in carriage water.

water"); Table 16.13 breaks down the solid components of waste water with respect to type, while Table 16.14 lists the effects and critical levels of waste water pollutants.

Sewage, it must be remembered, can contain storm drainage (Hayes et al., 1970; Economics Systems Corp., 1970) as well as the water that has been used by man, and some idea of the relative importance of this contribution is given in Table 16.15. Storms and flooding can create very severe pollution problems. If the volume of the rain-swollen waters exceeds the capacity of a community's treatment facility, overflow raw sewage may find its way into the water course. Under flood conditions when the land is inundated, septic tank material may be washed into streams, fertilizers and pesticides may be washed from flooded fields, oil may be flooded out of home and industrial storage tanks (a major cause of damage to monuments and works of art during the monstrous Florence flood), and settling lagoon dikes may be broken and their chemical contents carried away. Another ancillary problem that we might mention in passing is that the formation of H_2S can create accelerated corrosion in sewer systems (Hawthorn, 1970).

Table 16.13 Estimate of the Components of Total Solids in Wastewater (From *Wastewater Engineering* by Metcalf & Eddy, Inc., © 1972. Used with permission of the McGraw-Hill Book Co.)

Component	Dry Weight (gpcd)
Water Supplies and Ground Water (assumed to have little hardness)	12.7
Feces (solids, 23%)	20.5
Urine (solids, 3.7%)	43.3
Toilet (including paper)	20.0
Sinks, Baths, Laundries, and Other Sources of	86.5
Ground Garbage	30.0
Water Softeners	[a]
Total for Domestic Sewage from Separate Sewerage Systems (excluding contribution from water softeners)	213.0
Industrial Wastes	200.0[c]
Total for Industrial and Domestic Wastes from Separate Sewerage System	413.0
Storm Water	25.0[b]
Total for Industrial and Domestic Wastes from Combined Sewerage System	438.0

[a] Variable.
[b] Will vary with the season.
[c] Will vary with the type and size of industries.

Waste water treatment can be classified as primary (Figure 16.6) ("straining out the dead cats"); secondary, designed to remove the biogenic organic material responsible for large biological oxygen demand (BOD); and advanced, if it is necessary to remove the dissolved materials that escape primary and secondary treatment. After screening out coarse solid material and letting the finer suspended solids settle out (primary treatment), in secondary treatment the degradation processes that occur naturally are simulated and accelerated by some kind of biological oxidation process such as trickling-filter or activated sludge type treatment (Sawyer, 1965; secondary treatment, see Figure 16.7). Anaerobic bacteriological treatment of wastes has also been examined (Gould, 1971). Table 16.16 lists the removal efficiencies and costs of primary and secondary treatment of municipal waste water, while Table 16.17 summarizes the impurities present in secondary treatment plant effluents. These components that are not removed, it should be noted, are soluble salts, and, as the last column in the table shows, they are still present after treatment in amounts equal to or greater than the amounts added by water use. The organic materials include humic substances (mainly fulvic acids), detergents, and proteins (Rebhun and Manka, 1971). The forms of nitrogen include free NH_3 (approximately 60%, urea–nitrogen (8%), and amino–nitrogen (13%) (Hanson and Lee, 1971). Methods

Table 16.14 Typical Chemical Constituents That May Be Found in Waste Water and Their Effects (From *Wastewater Engineering* by Metcalf & Eddy, Inc., © 1972. Used with permission of the McGraw-Hill Book Co.)

Constituent	Effect	Critical Concentration (mg/liter)
	Inorganic	
Ammonia	Increases chlorine demand	Any amount
	Toxic to fish	2.5
	Can be converted to nitrates	Any amount
Calcium and Magnesium	Increase hardness	Over 100
Chloride	Imparts salty taste	250
	Interferes with industrial processes	75–200
Mercury	Toxic to humans and aquatic life	0.005
Nitrate	Stimulates algal and aquatic growth	0.3[a]
	Can cause methemoglobinemia in infants (blue babies)	10[b]
Phosphate	Stimulates algal and aquatic growth	0.015[a]
	Interferes with coagulation	0.2–0.4
	Interferes with lime soda softening	0.3
Sulfate	Cathartic action	600–1000
	Organic	
DDT	Toxic to fish and other aquatic life	0.001
Hexachloride	May cause taste and odor problems in water	0.02
Petrochemicals	May cause taste and odor problems in water	0.005–0.1
Phenolic Compounds	May cause taste and odor problems in water	0.0005–0.001
Surfactants	Cause foaming and may interfere with coagulation	1.0–3.0

[a] For quiescent lakes.
[b] USPHS Recommended Drinking Water Standards, 1962.

of removing nitrogen from waste waters, including secondary treatment effluents, are being studied (Wild et al., 1971; Smith, 1972).

Treatment methods applicable to both industrial and municipal wastes include screening, flocculation, chemical coagulation, flotation, sedimentation, centrifuging, filtration, stripping, neutralization, chemical oxidation, chemical reduction, wet oxidation, fermentation, emulsion breaking, evaporation, distillation, incineration, biological filtration, activated sludge, anaerobic digestion, stabilization lagoons, spray irrigation, and disinfection, and this long list by no means exhausts the possibilities. Advanced treatment includes the application of techniques such as chemical

Table 16.15 Amounts of Pollution from Urban Drainage (1000 lb/mi²/yr) (Reprinted from *Cleaning Our Environment—The Chemical Basis for Action*, a report by the Subcommittee on Environmental Improvement, Committee on Chemistry and Public Affairs, American Chemical Society, 1969. Reprinted by permission of the copyright owner.)

Constituent	Rainfall[a]	Separate Sewer Run-off[b]	Combined Sewer Overflow	Community Sewage (mostly domestic) 10 persons/acre[c]			
				Raw	3%	5%	10%
Suspended Solids	57	366	162	390	12	20	39
BOD	—	27	—	390	12	20	39
Total Nitrogen	5.5	5.0	3.9	78	2.3	3.9	7.8
Inorganic Nitrogen	3.0	1.6	3.5	58	1.7	2.9	5.8
Total Phosphate (as phosphorus)	0.4	0.6	3.2	20	0.6	1.0	2.0

[a] 30 in./yr.

[b] 30 in./yr and 0.37 run-off coefficient.

[c] Population density in Cincinnati. Percentages show magnitude of loss of pollutant to combined sewer overflows at 3, 5, and 10% loss.

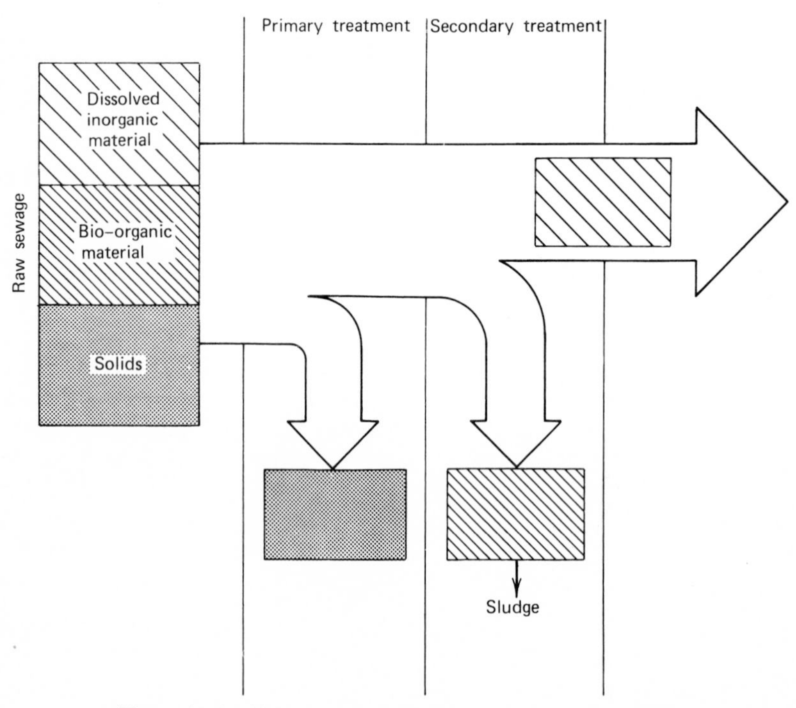

Figure 16.6. Primary and secondary wastewater treatment.

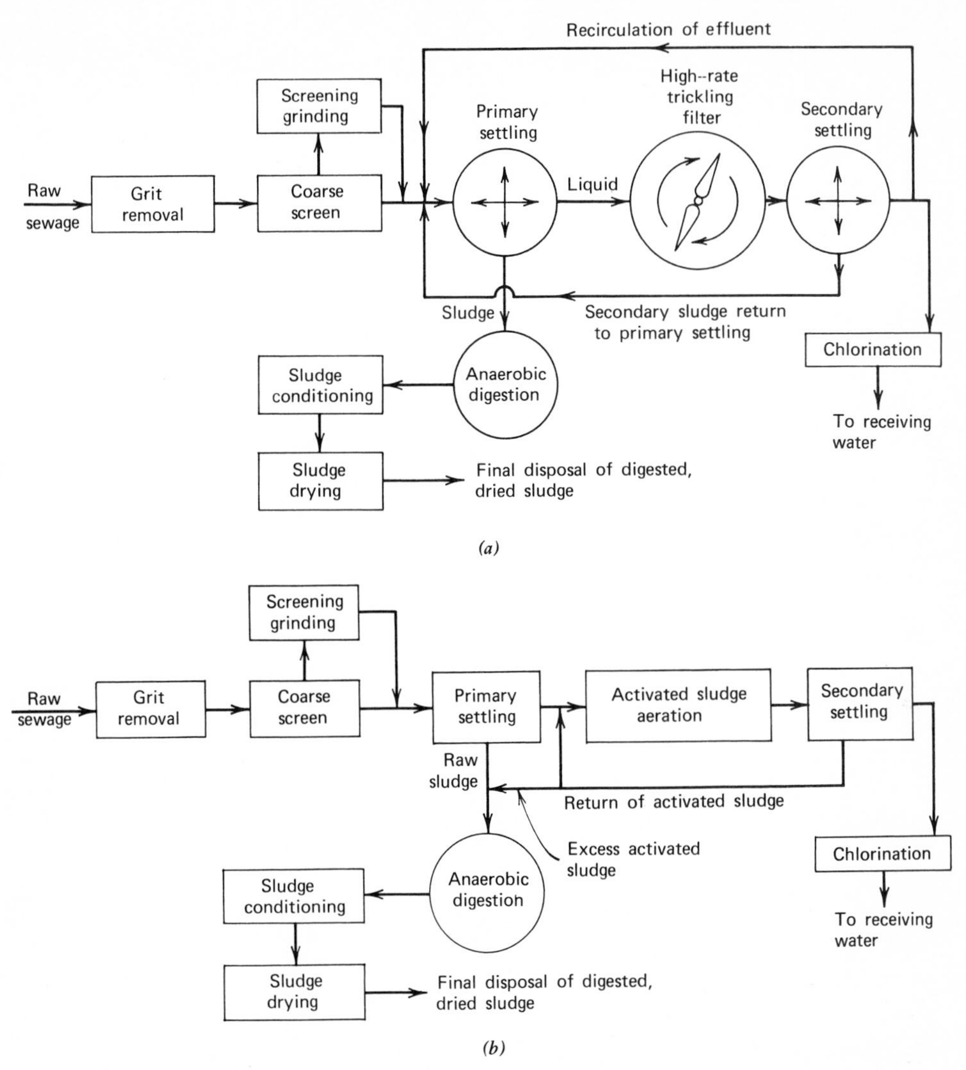

Figure 16.7. Schematic flow diagrams of sewage treatment plants: *A*, high-rate, trickling-filter plant; *B*, Activated sludge-type plant. (From *Environmental Protection* by E. T. Chanlett, © 1973 by McGraw-Hill. Used with permission of McGraw-Hill Book Co.)

precipitation, adsorption, disinfection, and a group of processes such as ion exchange, electrodialysis, and so on, aimed at the most difficult to remove soluble contaminants (see Stenberg et al., 1968). With carefully controlled chemical precipitation, it is possible to remove both suspended and colloidal matter—80 to 90% of the total suspended matter, 50 to 55% of the total organic matter, and 80 to 90% of the bacteria, thus improving on plain sedimentation (primary treatment) by 10 to 30% (Metcalf & Eddy, 1972). Recently, interest in chemical precipitation has enjoyed a revival because it can be used to remove phosphorus (Levin et al., 1972), and, combined with activated carbon absorption, it can provide a complete waste water treatment process by-passing the need for biological treatment. Commonly used waste water precipitants include alum ($Al_2(SO_4)_3 \cdot 18H_2O$), ferrous sulfate (copperas,

Table 16.16 Approximate Performance and Cost of Conventional Treatment of Municipal Wastes[a] (Based on raw waste concentrations) (Reprinted from *Cleaning Our Environment—The Chemical Basis for Action*, a report by the Subcommittee on Environmental Improvement, Committee on Chemistry and Public Affairs, American Chemical Society, 1969. Reprinted by permission of the copyright owner.)

	Removal Efficiency of Treatment	
	Primary	Primary Plus Secondary
Biochemical Oxygen Demand	35%	90%
Chemical Oxygen Demand	30%	80%
Refractory Organics	20%	60%
Suspended Solids	60%	90%
Total Nitrogen	20%	50%
Total Phosphorus	10%	30%
Dissolved Minerals	—	5%
Cost/1000 gal[b]	3–4 cents	5–10 cents

[a] Figures may differ significantly in specific instances.

[b] For industrial wastes, the costs cover a wider range: 2 to 5 cents for primary treatment, 2 to 20 cents for secondary treatment.

$FeSO_4 \cdot 7H_2O$), lime ($Ca(OH)_2$), sulfuric acid (H_2SO_4), sulfur dioxide (SO_2), ferric chloride ($FeCl_3$), and ferric sulfate ($Fe_2(SO_4)_3$).

The addition of alum to sewage containing Ca (or Mg) bicarbonate alkalinity forms a gelatinous floc of insoluble $Al(OH)_3$

$$Al_2(SO_4)_3 \cdot 18H_2O + 3Ca(HCO_3)_2 \rightleftharpoons 3CaSO_4 + \underline{2Al(OH)_3} + 6CO_2 + 18H_2O \tag{16.5}$$

which sweeps suspended material out of the sewage as it slowly settles out. The gelatinous floc $Fe(OH)_3$ behaves similarly to $Al(OH)_3$ floc, and it is formed first by adding ferrous sulfate

$$FeSO_4 \cdot 7H_2O + \underset{\text{or } Mg(HCO_3)_2}{Ca(HCO_3)_2} \rightleftharpoons Fe(HCO_3)_2 + CaSO_3 + 7H_2O \tag{16.6}$$

then lime

$$Fe(HCO_3)_2 + 2Ca(OH)_2 \rightleftharpoons Fe(OH)_2 + 2CaCO_3 + 2H_2O \tag{16.7}$$

and finally the ferrous hydroxide is oxidized up to ferric hydroxide by oxygen dissolved in the sewage

$$4Fe(OH)_2 + O_2 + 2H_2O \rightleftharpoons \underline{4Fe(OH)_3} \tag{16.8}$$

Table 16.17 Average Composition of Effluent from Secondary Treatment of Municipal Waste Water[a] (Reprinted from *Cleaning Our Environment—The Chemical Basis for Action*, a report by the Subcommittee on Environmental Improvement, Committee on Chemistry and Public Affairs, American Chemical Society, 1969. Reprinted by permission of the copyright owner.)

Component	Concentration (mg/liter)	Average Increment Added During Water Use (mg/liter)
Gross Organics	55	52
Biodegradable Organics (as biochemical oxygen demand)	25	25
Sodium	135	70
Potassium	15	10
Ammonium	20	20
Calcium	60	15
Magnesium	25	7
Chloride	130	75
Nitrate	15	10
Nitrite	1	1
Bicarbonate	300	100
Sulfate	100	30
Silica	50	15
Phosphate	25	25
Hardness (as calcium carbonate)	270	70
Alkalinity (as calcium carbonate)	250	85
Total Dissolved Solids	730	320

[a] Figures may vary significantly in specific instances.

[b] Concentration increase from tap water to secondary effluent.

Lime alone can be used as a precipitant

$$Ca(OH)_2 + H_2CO_3 \rightleftharpoons \underline{CaCO_3} + 2H_2O \tag{16.9}$$

$$Ca(OH)_2 + Ca(HCO_3)_2 \rightleftharpoons \underline{2CaCO_3} + 2H_2O \tag{16.10}$$

with the $CaCO_3$ formed acting as the coagulant. But the amount of lime added is critical: too little and the effluent will not be clear, too much and some suspended organic material will dissolve in the caustic solution to make an effluent worst than the original sewage. Flyash has been used in conjunction with lime coagulation to reduce chemical oxygen demand, COD (Eye and Basu, 1970).

Neither sulfuric acid nor sulfur dioxide are floc-forming chemicals. In sewage, they react with calcium (or magnesium) bicarbonates to form soluble sulfates. These electrolytes in turn encourage repulsive charge removal (decrease the zeta-potential) from colloidal material and thus aggragation. Polyelectrolytes, both natural and synthetic, accomplish this same purpose, not only by charge removal but by an interparticle bridging mechanism as well (Gutche, 1972).

The ferric salts, $FeCl_3$ and $Fe_2(SO_4)_3$, used in conjunction with lime, like alum, form a $Fe(OH)_3$ floc. Multivalent cations also possess the important capability of removing phosphate (Stumm and Morgan, 1970)

$$10Ca^{2+} + 6PO^{3-} + 20H^- \rightleftharpoons Ca_{10}(PO_4)_6(OH)_2 \tag{16.11}$$

$$Al^{3+} + H_nPO_4{}^{3-n} \rightleftharpoons AlPO_4 + nH^+ \tag{16.12}$$

$$Fe^{3+} + H_nPO^3{}_4{}^{-n} \rightleftharpoons FePO_4 + nH^+ \tag{16.13}$$

Activated carbon absorption can remove trace metals as well as organic material, for example 86, 93, 94, and 34% Ag, Cd, Cr, and Se are removed, respectively, compared to 92, 99.9, 96, and 99.7% by ion exchange columns (Linstedt et al., 1971). Heat, sunlight (or ultraviolet lamps), filtration, radiation, and chemicals can all be used to disinfect waste waters. Their various efficiencies for removing bacteria are compared in Table 16.18. Chemical disinfectants include phenols and phenolic compounds, ozone, alcohols, iodine and bromine, chlorine and chlorine-releasing compounds, heavy metals, dyes, soaps, detergents, quaternary ammonia compounds, chloroamines, hydrogen peroxide, and various acids and bases. Of these, chlorine is by far the most commonly used, although ozone is gaining in popularity for, unlike chlorine, it leaves no toxic residue (Robert A. Taft Water Res. Center, 1969). Chlorination has been a cause of environmental damage, changing fish community composition and migration habits (Anon., 1970*f*) and, in the case of power plants, depressing the rate of photosynthesis (Brook and Baker, 1972).

Water treatment creates its own pollution and waste disposal problems (Dean, 1969). Sewage sludge from treatment plants is a semiliquid waste. Because of their solubility, nutrients tend to end up in the plant effluent rather than in the sludge (Table 16.19). The organic content of the sludge is high; the total loss on ignition ranging from 46 to 80% of the dry weight of the material (Gross, 1970; see also Horne, Mahler, and Rossello, 1971), and some of its components that have been identified are listed in Table 16.20. The lipid polymer cutin has recently been identified as an important component of the organic fraction of sludge (Kolattukudy and Purdy, 1973). Lipids from both treated and raw sewage often end up in the

Table 16.18 Removal or Destruction of Bacteria by Different Treatment Processes (*Wastewater Engineering* by Metcalf & Eddy, Inc., © 1972. Used with permission of McGraw-Hill Book Co.)

Process	Percent Removal
Coarse Screens	0–5
Fine Screens	10–20
Grit Chambers	10–25
Plain Sedimentation	25–75
Chemical Precipitation	40–80
Trickling Filters	90–95
Activated Sludge	90–98
Chlorination of Treated Sewage	98–99

Table 16.19 Summary of Nutrients in the Alki Point Plant (Seattle, Washington) Effluent and Digested Sludge (From Horne, Mahler, and Rossello, 1971)

Source	Minimum	Maximum	Mean	lb/day
Effluent (mg/liter)				
Soluble PO_4	0.17	6.72	1.75	124
Total PO_4	2.3	10.5	5.4	382
Ammonia N	1.1	11.3	6.1	432
Nitrate N	0.22	5.8	1.9	134
Kjeldahl N	0.55	4.8	2.1	148
Digested Sludge (mg/g dry solids)				
Total PO_4	1.8	11.2	5.1	9
Nitrate N	0.3	12	4.9	9
Kjeldahl N	2.8	6.1	4.3	8
Seawater (mg/liter)				
Soluble PO_4	0.076	0.315	0.226	—

Table 16.20 Percent Total Carbon Composition of the Suspended Solid Material in Sewage (From Horne, Mahler, and Rossello, 1971)

Unidentified	63%
Protein	20%
Carbohydrates	10%
Amino Sugars	3%
Soluble Acids	1%
Fats—Ester	1%
Fats—Acid	1%
Muramic Acid	0.4%
Anionic Detergents	0.4%
Amide	0.2%

sedimentary pollution accummulating on harbor bottoms (Shaw, 1973). Some of the materials in the sludge might be valuable and their recovery used to help underwrite the costs of disposal (Wallen and Davis, 1972). The sewage sludge resulting from secondary treatment is a serious problem for in some ways it is a far more dangerous and concentrated pollutant than the original sewage. Toxic heavy metals, for example, can be concentrated in the sludge (Table 16.21) (Train et al., 1970). There are many strong organic complexers in the sludge that tightly hold metals. Bender et al (1970), for example, have found that Fe and Cu are complexed and held in two distinct molecular weight organic fractions. A numer of sludge treatment and disposal processes are in use (James, 1972, Table 16.22). As we have noted,

Table 16.21 Heavy Metals Concentrations in Sewage Sludge (ppm) (From Train et al., 1970)

Metal	Concentrations in Sewage Sludge Minimum	Average	Maximum	Natural Concentrations in Seawater	Concentrations Toxic to Marine Life
Copper	315	643	1980	0.003	0.1
Zinc	1350	2459	3700	0.01	10.0
Manganese	30	262	790	0.002	—

Table 16.22 Sewage Sludge Handling and Disposal Methods in Common Use (Reprinted from *Cleaning Our Environment—The Chemical Basis for Action*, a report by the Subcommittee on Environmental Improvement, Committee on Chemistry and Public Affairs, American Chemical Society, 1969. Reprinted by permission of the copyright owner.)

1. Concentration
    - Clarifier thickening
    - Separate concentration
        - Gravity thickening
        - Flotation
2. Digestion
    - Aerobic bacteria (which utilize free oxygen)
    - Anaerobic bacteria (which utilize oxygen contained in chemical compounds)
3. Dewatering
    - Drying beds
    - Lagoons
    - Vacuum filtration
    - Centrifugation
4. Heat drying and combustion
    - Heat drying
    - Incineration
        - Multiple hearth
        - Fluidized solids
    - Wet oxidation
5. Final sludge disposal
    - Landfill
    - Soil conditioning
    - Discharge to sea

because of their toxicity, the ultimate disposal of the sludge and of advanced waste water treatment residues (Table 16.23) can present a problem. Improper incineration produces air pollution. The fertilizer value of these wastes is relatively low, making that course of reuse economically unattractive in the United States at the present time. However this disposal alternative is so attractive from an environmental point of view that it continues to be studied intensively. Landfill sites are rapidly disappear-

Table 16.23 Methods for the Ultimate Disposal of Concentrated Contaminants Resulting from Advanced Wastewater Treatment (From *Wastewater Engineering* by Metcalf & Eddy, Inc., © 1972. Used with permission of the McGraw-Hill Book Co.)

Disposal Method	Remarks
	Liquid
Evaporation Ponds	Provisions must be made to prevent ground water contamination
Spreading on Soil	Provisions must be made to prevent ground water contamination
Shallow-Well Injection	Provisions must be made to prevent ground water contamination
Deep-well Injection	Porous strata, natural or artificial cavities should be available
Landfill	Liquid used as a wetting agent to increase compaction
Controlled Evaporation	Depends on liquid volume, power costs, and local conditions
Ocean Discharge	Truck, rail hauling, or pipeline needed for transportation
	Sludge
Spreading on Soil	Sludge may be pretreated to aid dewatering or to remove objectionable components
Lagooning	Provisions must be made to prevent ground water contamination
Landfill	Sludge used as a wetting agent to increase compaction
Recovery of Products	Depends on sludge characteristics, recovery technology, and costs
Wet Combustion	Heat value may be recovered for use. Disposal of ash required.
Incineration	Concentration of sludge is needed. Disposal of ash required.
Ocean Discharge	May not be allowed in the future
	Ash
Landfill	Mixed with refuse to increase compacted density of landfill
Soil Conditioner	Depends on waste characteristics
Ocean Discharge	May not be allowed in the future.

ing, and there is always the danger of leaching even from the most careful landfill disposal. Ocean disposal of sewer sludge has already done very severe damage to the marine environment as we saw in Chapter 8. Reducing the volume and mass of the sludge is highly desireable since the transportation costs are so high. In addition to the methods for dewatering the sludge listed in Table 16.22, more exotic techniques have been tried including dewatering by freezing (Farrell, 1971).

Table 16.24 summarizes the costs of primary, secondary, and various advanced treatments (see also Culp and Culp, 1971), and it makes clear why treatment beyond secondary is rare even in affluent nations such as the United States. The cost of clean water is high, but the costs of a despoiled environment are much higher.

Let me add a note of caution—the removal of substances from the natural environment can be as injurious to biota as their addition (recall the biphastic nature of the dose-response curve, Figure 17.13), and I have suggested elsewhere (Horne, 1972) that over-zealous clean-up of waters might result in environmental damage. While this may not be a real danger at the present time (Dinman, 1972), it does appear to

Table 16.24 Uses and Costs of Water from Sample Renovation System[a] **(Reprinted from *Cleaning Our Environment—The Chemical Basis for Action*, a report by the Subcommittee on Environmental Improvement, Committee on Chemistry and Public Affairs, American Chemical Society, 1969. Reprinted by permission of the copyright owner.)**

Treatment Sequence	Estimated Cumulative Capital Cost (million dollars) 15 mgd[b]	Estimated Cumulative Capital Cost (million dollars) 100 mgd[b]	Estimated Cumulative Operating Cost (cents/1000 gal) 15 mgd[b]	Estimated Cumulative Operating Cost (cents/1000 gal) 100 mgd[b]	Uses of Treated Water
Raw Waste Water	0	0	0	0	None. Highly polluting
Primary Treatment	2.2	9.5	5.2	3.5	Partial pollution control. No direct reuse possible
Secondary Treatment (activated sludge)	4.5	20	11	8.3	Conventional pollution control. Nonfood crop irrigation
Coagulation–Sedimentation	5.1	24	15	13	Improved pollution control. General irrigation supply; low quality industrial supply; recreational water supply; short-term water recharge
Carbon Adsorption	7.3	30	23	17	Complete organic pollution control. High quality irrigation supply; good quality industrial supply; body contact recreational supply; long-term ground water recharge
Electrodialysis	11	47	37	26	Complete organic-inorganic pollution control
Brine Disposal	25	77	53	33	High quality industrial supply; indefinite ground water recharge
Disinfection	25	77	54	34	Absolute pollution control; potable water supply

[a] A 75,000 gal/day system of this configuration is installed at FWPCA's field station at Lebanon, Ohio.

[b] Million gallons per day.

be within the potential capability of modern waste water treatment technology. Also, earlier we noted the plateau (Figure 8.24) in progressive waste loading and the deleterious ecological effects of disinfection. Considerations such as these are cause for caution and suggest that the *total* environmental impact of waste water treatment, especially of sewage, is in need of some careful rethinking.

16.4 EUTROPHICATION

Because their lives are so much longer than ours, we look upon geological and even ecological features as being permanent, but they too submit to change, changes ranging from as fast as the alteration of a river channel in a few years or even during the course of a single flood to as slow as weathering and erosion and the drift of the continents. Ponds and lakes are temporary features of the landscape, and, like us, they grow old and die. Sedimentation fills them in, shore plants encroach, they become bogs and disappear. Deforestation and other human activity lowers the ground water level and lakes and ponds shrink. If there are ample nutrients, aquatic plant growth flourishes, and when the organisms die they leave decaying organic matter that consumes the vital oxygen of the waters in addition to re-releasing on the average about 50 to 73% of their nutrient material (Jewell, 1971). The oxygen may be completely depleted, the waters become anoxic, fish die. Excess organic debris accumulates on the bottom, filling the pond. This "sickness" of bodies of water is called eutrophication. It is a *natural* process, but the course of the malady can be greatly accelerated by anthropogenic pollution (Fruh, 1967).

A great deal of controversy continues to surround the causes of eutrophication (Anon., 1970*h*; Likens, 1972). To me, this controversy seems somewhat foolish. An excessive amount of any nutrient element—nitrogen, phosphorus, even carbon (King, 1970)—can in principle give rise to eutrophication. Table 16.25 summarizes differences in opinion between those who point the accusing finger at phosphorus and those who claim that carbon is the guilty element. A recently published controlled experiment (Schindler, 1974) on several small Canadian lakes appears to support strongly those who accuse phosphorus: Eutrophication induced by the addition of N, P, and C was arrested when the addition of P *only* was discontinued. Schindler and his co-workers (1972) had earlier demonstrated that in the presence of adequate P and N uptake of atmospheric CO_2 might control eutrophication. Furthermore, there is evidence that the CO_2 level is important in determining the species composition of the algal community (Shapiro, 1973*a*). Algal blooms also appear to depend on pH (Anon., 1973*b*), which in turn is controlled by CO_2 and other factors (see Goldman, 1973, and Shapiro, 1973*b*, for further discussion). In a recent interesting paper, Morton and Lee (1974) report that the iron level, but not the manganese level, while not altering the total algal biomass can also induce shifts in the composition of the algal population. We must remember that some 15 to 20 nutrients play an essential role in eutrophication including Fe, K, and other trace metals as well as C, N, P, and Si (Grundy, 1971). It depends on which nutrient in any given body of water, fresh, estuarine, or marine, is rate controlling in bioproductivity. As we noted in Chapter 7, in most surface waters in the United States, phosphate is the limiting nutrient. Therefore in these waters the addition of excessive amounts of phosphate, often from human sources such as detergents, sewage, and fertilizer

Table 16.25 Two Schools of Thought Clash on Many Points (Reprinted with permission from Anon., *Envir. Sci. Tech.*, 4, 725 (1970), © The American Chemical Society.)

Carbon-Is-Key School Believes	Phosphorus-Is-Key School Believes
Carbon controls algal growth	Phosphorus controls algal growth
Phosphorus is recycled again and again during and after each bloom	Recycling is inefficient: some of the phosphorus is lost in bottom sediment
Phosphorus in sediment is a vast reservoir always available to stimulate growth	Sediments are sinks for phosphorus, not sources
Massive blooms can occur even when dissolved phosphorus concentration is low	Phosphorus concentrations are low during massive blooms because phosphorus is in algal cells, not water
When large supplies of CO_2 and bicarbonate are present, very small amounts of phosphorus cause growth	No matter how much CO_2 is present, a certain minimum amount of phosphorus is needed for growth
CO_2 supplied by the bacterial decomposition of organic matter is the key source of carbon for algal growth	CO_2 produced by bacteria may be used in algal growth, but main supply is from dissociation of bicarbonates
By and large, severe reduction in phosphorus discharges will not result in reduced algal growth	Reduction in phosphorus discharges will materially curtail algal growth

run-off (Verduin, 1967) is responsible for eutrophication (Weiss, 1969). Phosphorus, however, does not appear to be the limiting nutrient in many estuarine waters, and Ryther and Dunstan (1971) have found that N, rather than P, is limiting in coastal marine waters. Nitrogen fixation also appears to be the important factor in controlling the high productivity of a tropical lake examined by Horne and Viner (1971). Bacterial N-fixation can occur in bottom sediments as well as in the water column (Howard et al., 1970). Table 16.26 compares N and P inputs into our environment. Also it should be noted that phosphorus is the most controllable of the major nu-

Table 16.26 Phosphorus and Nitrogen Inputs in Our Environment (United States) (Reprinted with permission from R. D. Grundy, *Envir. Sci. Tech.*, 5, 1184 (1971), © The American Chemical Society.)

Source	P 10^6 lb/yr	%	N 10^6 lb/yr	%
Natural	245–711	64	1035–4210	22
Anthropogenic	686–1015	74–57	3990	49–46
Domestic sewage	(387–446)		(1330)	
Urban run-off	(19)		(200)	
Farmland run-off	(110–380)		(2040)	
Livestock run-off	(170)		(420)	
Total	931–1726	100	5025–8200	100

trients, while nitrogen is difficult to control because of the large atmospheric reservoir, and carbon is impossible to control (Likens et al., 1971). P-Pollution tends to deplete silica, and even though silica does not become limiting, the composition of the algal population can be drastically altered thereby (Schelske and Stoermer, 1971).

In carefully controlled laboratory experiments, Azad and Borchardt (1970) have found three different types of phosphate uptake by algae: (*a*) metabolic uptake for maximum growth, (*b*) storage or "luxury" uptake, and (*c*) starvation uptake by P-deprived algae. More than 90% of the soluble phosphorus in a lake can be in complex organic compounds with about 20% of the recoverable organic P in materials with molecular weights in excess of 50,000 (Minear, 1972). The organophosphate excreted by lake plankton appears to be in colloidal form (Lean, 1973).

In the United States, the death of Lake Erie has attracted considerable public attention. The use of this once magnificent body of water as an industrial, agricultural, and municipal cesspool has drastically altered its chemistry over a period of about 10 yr (Table 16.27) as the population, particularly on the United States shore, has mushroomed. And the biology of the lake has simultaneously suffered drastic degradation as reflected in commercial fish catches (Figure 16.8; see also Regier and Hartman, 1972). But Lake Erie is not the only sick Great Lake. Anthropogenic nutrients are also invading Lake Ontario (Johnson and Owen, 1971), which receives an annual phosphate discharge of 13,700 tons compared to 30,100

Table 16.27 A. Chemical Changes Observed in Western Lake Erie During 1948–1962 (From Verduin, © 1967 by the American Association for the Advancement of Science, with permission of the AAAS and the author.)

Chemical	Values Before 1950	Values in 1960–1961
Hydrogen ion (pH)	Maxima of 8.7	Maxima of 9.2
Oxygen near bottom	Minima of 80% of saturation	Minima below 40% of saturation, near zero at times
Gross photosynthesis CO_2 fixed per day	32 μmol/liter	70 μmol/liter
Nitrate nitrogen in Maumee Bay	520 μg/liter	810 μg/liter
Soluble phosphorus in Maumee Bay	35 μg/liter	148 μg/liter

B. Comparison of Nitrogen and Phosphorus Data in Western Lake Erie for 1942 with Data for 1965–1966

Year	Available Nitrogen, NH_3N plus NO_3N (μg/liter)	Soluble Phosphorus (μg P/liter)	N/P Ratio
1942	261	7.5	35
1965–1966	330	36	9.2

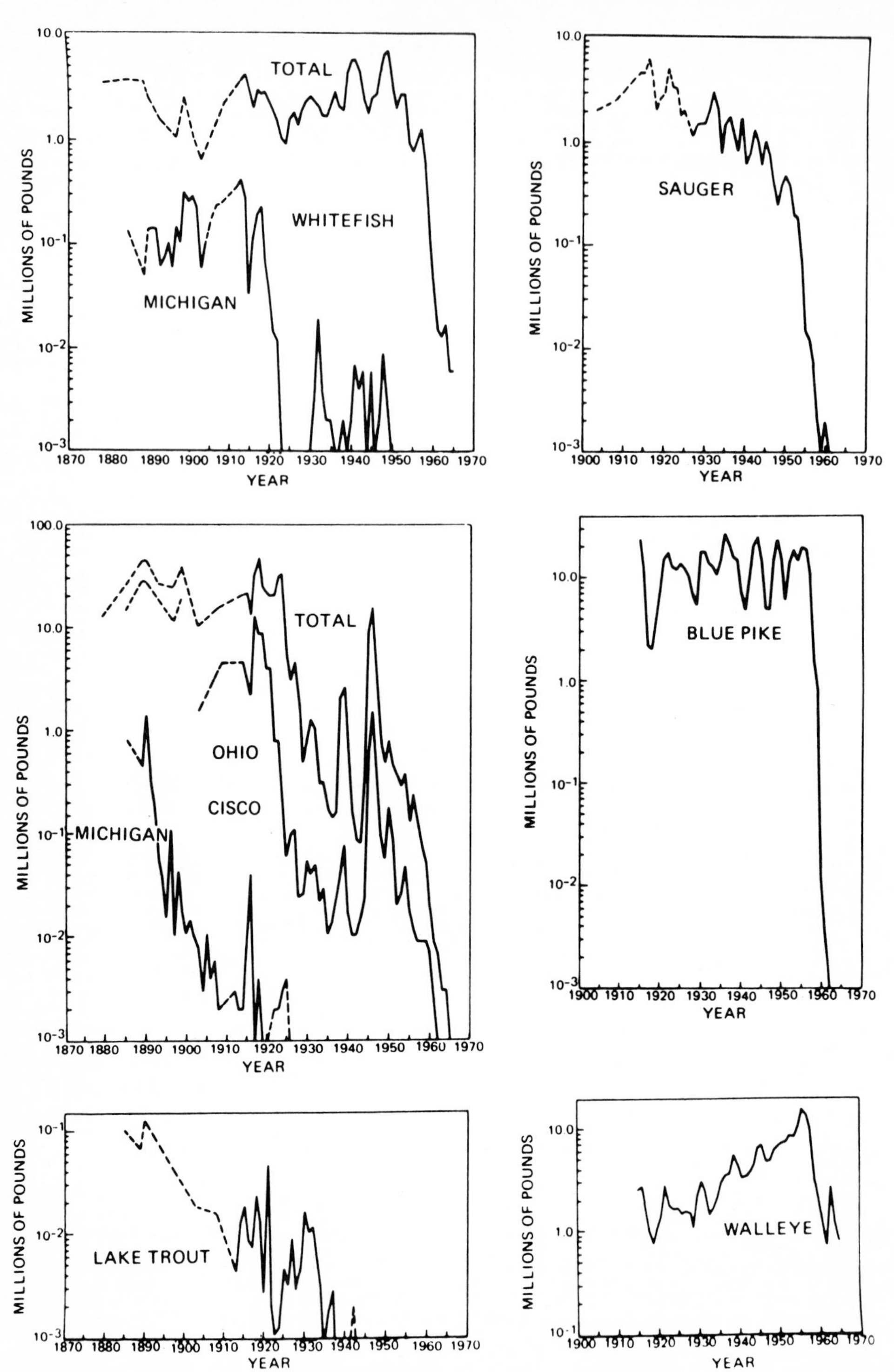

Figure 16.8. Commercial production of blue pike, cisco (lake herring), lake trout, sauger, walleye, and whitefish in Lake Erie (broken lines represent production during periods when annual data were not available) (from Beeton in *Eutrophication: Causes, Consequences, and Correctives*, ISBN No. 0-309-01700-9, Div. Biol. Agric. Nat. Acad. Sci.–Nat. Res. Council, Washington, D.C., 1969, with permission of the NAS and the aurhor).

tons for Lake Erie (Anon., 1969*e*), and the chemistry and biology even of the upper Great Lakes is also changing as the consequence of the slower but still relentless population growth along their shores (Figure 16.9; Beeton, 1969). Can the lower Great Lakes be salvaged? Lake Ontario, with stringent phosphate control (detergent reformulation and tertiary treatment) could be returned to an oligotrophic condition, but Lake Erie apparently was a mesotrophic lake even before the recent anthropogenic phosphate enrichment, and control probably could not reduce it to below mesotrophy (Figure 16.10). Because of the great number of other P-sources, such as urban and rural (fertilizer) run-off and natural silting removal of phosphates from detergents or even 88% removal of phosphates from waste water treatment plant effluents (and there are a number of techniques for accomplishing this end, see Daniels and Parker, 1973; methods cf N-removal have been reviewed by Adams, 1973) still might fail to improve the water quality of Lake Erie substantially (Hammond, 1971). One proposal would simply build a dam across the lake so that the western basin could be used for what would be in effect a sewage lagoon for the Detroit area, while a somewhat more acceptable scheme would utilize the eutrophication and harvest the biomass resulting from the lake's excessive productivity (Hubschman, 1971). Rosenblum and Hollocher (1971), however, question whether a sufficient fraction of the nutrient content could be removed by biomass harvesting. They note that the P-input to the lake is 2.3×10^{10} g/yr (24% from run-off, 76% from municipal and industrial wastes; see Table 16.28 and also Table 16.29, which show how the relative importance of the different P- and N-sources differs markedly for different bodies of water), while present commercial fishing now removes only 1.4×10^{8} g P/yr or only 0.6% of the P-input.

Another body of water that has received a particularly great amount of public concern in Lake Tahoe (Smith and Ludwig, 1968; Middlebrooks et al., 1971*a*, *b*). The aesthetic and recreational value of this lake has been jeopardized by overdevelopment along its shores. How ironic it is that our most beautiful natural splendors are destroyed by their own attractiveness!

In his study of an "examplary," unpolluted, eutrophic Connecticut lake, Frink (1967) found more than sufficient nutrient input from the largely forested watershed to cause the eutrophied condition. In addition, the bottom sediments represented a vast nutrient reservoir (the upper 1 cm along contained ten times the estimated annual N- and P-input), which could still keep the lake eutrophic even if all other nutrient sources could be excluded. Nitrogen and carbon levels in lake sediments have been examined (Keeney et al., 1970; Konrad et al., 1970) as well as the phosphorus content and partitioning (Frink, 1969; Harter, 1968; MacPherson et al.,

Table 16.28 Regional Variation of Phosphate Sources (Based on data in Grundy, 1971)

	Potomac River (%)	Alafia River (%)	Lake Erie (%)
Land Run-off	13		24
Industrial Wastes	9	96	7
Municipal Wastes	78	4	69

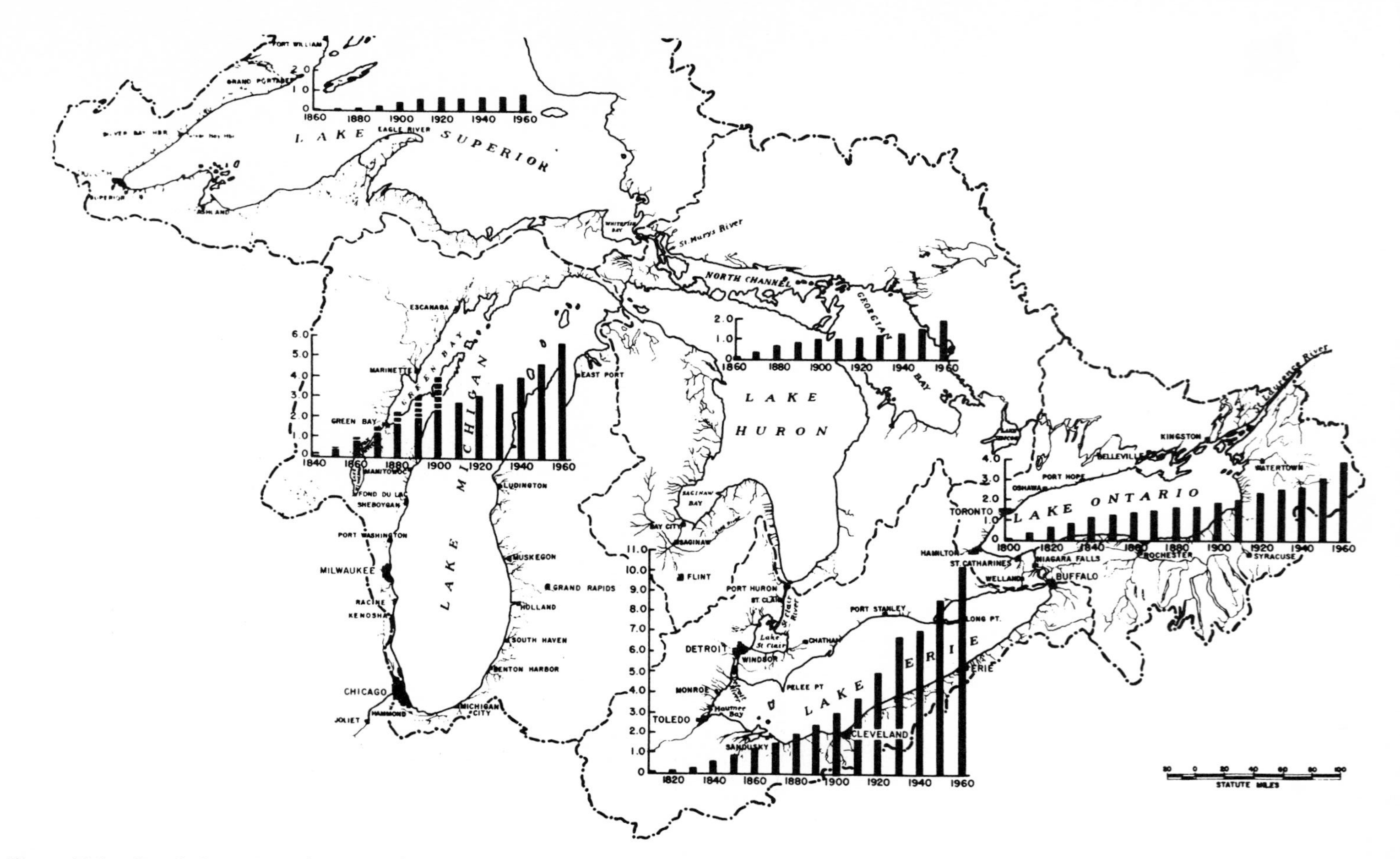

Figure 16.9. Population growth (in millions) in the basins of the Great Lakes (from Beeton in *Eutrophication: Causes, Consequences, and Correctives*, ISBN No. 0-309-01700-9, Div. Biol. Agric., Nat. Acad. Sci.–Nat. Res. Council, Washington, D.C., 1969, with permission of the NSA and the author).

Table 16.29 Percentage Contributions from Various Cultural and Natural Sources for Selected Lakes (Reprinted with permission from E. E. Shannon and P. L. Brezonik, *Envir. Sci. Tech.*, *6*, 719 (1972), © The American Chemical Society.)

Lake and Type[a]	Nutrient	Sewage	Urban Run-off	Fertilized Area	Pasture Area	Unproductive Cleared Area	Forest Area	Septic Tanks	Rainfall on Lake Surface	% Cultural
Santa Rosa (U)	N	0	0	0	0	0	47	13	40	13
	P	0	0	0	0	0	30	11	59	11
Santa Fe (O)	N	0	7	5	7	1	41	2	37	22
	P	0	15	1	3	N.S.	25	2	54	21
Orange (M)	N	0	1	10	11	4	57	N.S.	17	21
	P	0	5	2	6	3	49	N.S.	35	13
Newman's (E)	N	0	8	2	14	4	56	1	15	25
	P	0	22	N.S.[b]	7	3	41	1	26	30
Hawthorne (E)	N	0	36	N.S.	N.S.	5	14	32	13	68
	P	0	57	N.S.	N.S.	2	6	23	12	80
Dora (H)	N	13	4	74	N.S.	N.S.	1	2	6	93
	P	60	12	14	N.S.	N.S.	1	1	12	87

[a] Ultraoligotrophic; O, oligotrophic; M, mesotrophic; E, eutrophic; H, hypereutrophic.
[b] Not significant (less than 1%).

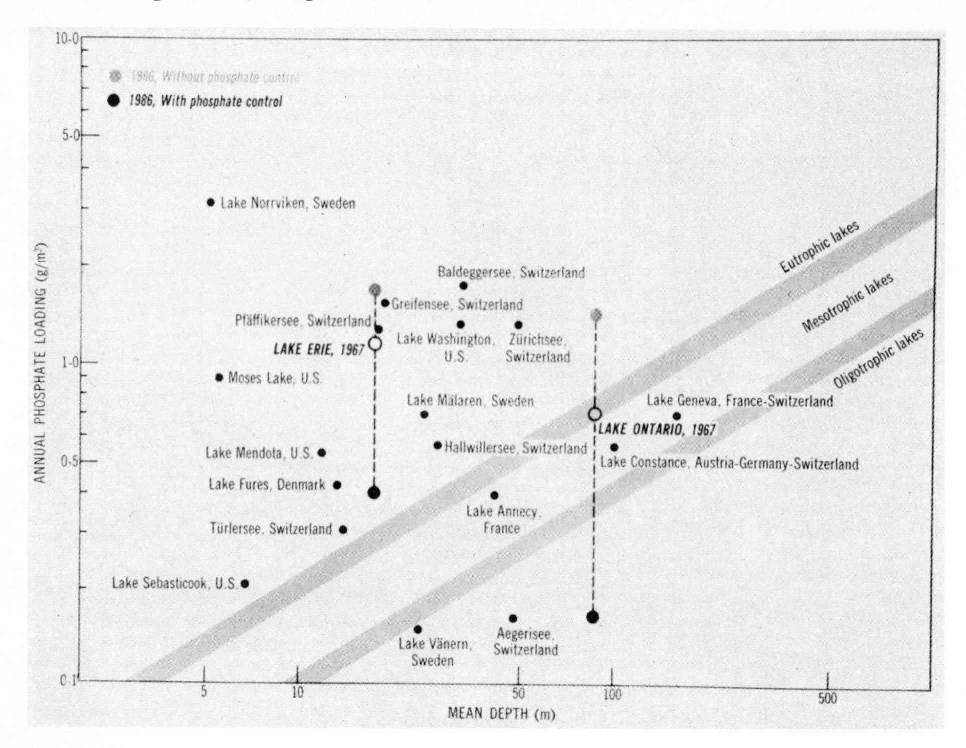

Figure 16.10. Comparison of the phosphate loadings of some lakes (reprinted with permission from Anon., *Environ. Sci. Tech.*, 1243 (1969*e*), © The American Chemical Society).

1958, Hayes and Phillips, 1958). However due to the slowness of nutrient exchange between bottom sediments and the water column (recall Chapter 13), there is some controversy as to the significance of this reservoir in determining the eutrophic state of a lake (see Lee, 1970). However in their study of 55 Florida lakes, Shannon and Brezonik (1972) found that a number of the eutrophied bodies of water had very evident anthropogenic sources of N and P (Table 16.29). They conclude that trophic state depends largely on the gross supply of N and P and that the P-loading is the more significant (or limiting) factor. In the case of rivers and streams, flow rate can be important for the level of P in the waters tends to decrease with flow rate but N increases (Wang and Evans, 1970). Lake organisms are capable of fixing molecular nitrogen, thus adding to the level of that nutrient (Granhall and Lundren, 1971). Some of the products of heterotrophic nitrification may represent hazards to human health even at low concentrations (Verstraete and Alexander, 1973).

Some success has been achieved in reviving *small* dead lakes with direct aeration with pure oxygen (rather than air to avoid fish kills caused by nitrogen supersaturation, Anon., 1973*a*) in order to restore the oxygen consumed by excessive eutrophic growth. Also, as was suggested previously, the nutrients levels can be reduced by growing aquatic plants, such as water hyacinths, harvesting them, and removing the N, P, and C they contain from the system (Yount and Grossman, 1970). Plants such as rye grass might also be grown at a secondary sewage treatment plant to remove nutrients (Law, 1969), although the nutrient P is commonly removed by chemical precipitation (Ferguson and McCarty, 1971). A number of proposals have been advanced for coagulating, precipitating, and removing the algal cells that so heavily concentrate phosphorus (Tenney et al., 1969; Golueke and Oswald, 1970; McGarry, 1970). Another corrective step might be the removal of orthophosphate on activated alumina (Ames and Dean, 1970; Winkler and Thodos, 1971; a bibliography on P-removal was released in 1973 by the U.S. Dept. Interior, WR SIC 73-208). Improving the efficiency of nutrient removal by treatment plants would seem to be a more rational approach than trying, after the fact, to clean up the receiving waters.

16.5 WATER RESOURCES AND MANAGEMENT

Recently I visited Israel. It is evident even to the most casual visitor that the quality of life that has been achieved there is entirely dependent on the careful marshalling of the nation's scant water resources (see Shamir, 1972). Through very hard work the Israelis have literally "made the desert bloom." Israel utilizes 90% of its total water resources, and one scientist jokingly remarked to me that he thought that 110% might be a more accurate figure. This small country already has had to face up to its water problems, but much larger nations more rich in water resources like the United States have not. But they will.

Earlier in Chapter 9, we examined the forms and travels of water in our environment. These are, of course, closely related to the many ways in which man collects, stores, treats, and uses water (Figure 16.11). In regions where the supply is ample,

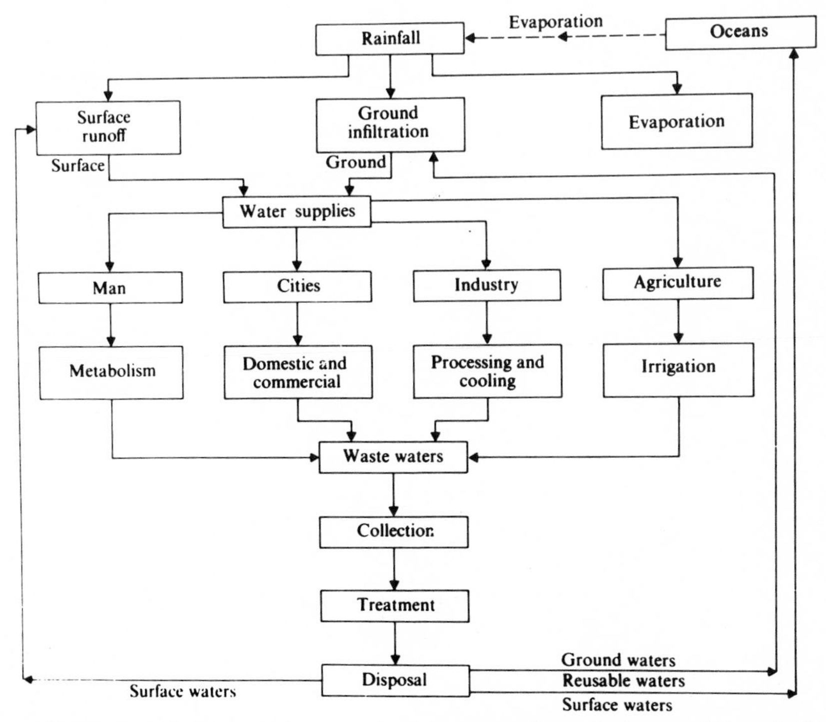

Figure 16.11. The environmental system and use of water (from *Environmental Protection* by E. T. Chanlett, © 1973 by McGraw-Hill. Used with permission of McGraw-Hill Book Co.).

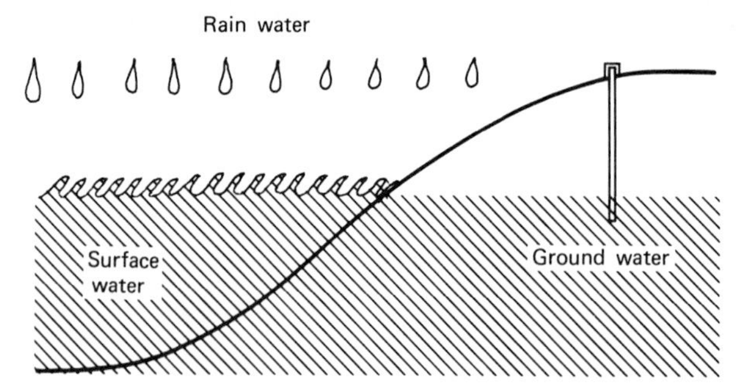

Figure 16.12. The three major sweet water resources.

sweet surface waters (Figure 16.12) meet man's needs with a minimum expenditure of his effort. If his demands are excessive, surface water will be exhausted and lakes and rivers will simply disappear, as has been the case with the lower Colorado River. The next most accessible reservoir upon which man can draw is ground water (Figure 16.13), but he must spend money and effort to drill wells to tap this source. But here again excessive withdrawal can create very serious consequences, such as saline water invasion, sinking of the water table level, and even subsidence. A large part of the responsibility for the plight of sinking Venice can be assigned to water withdrawal by the high volume wells in the Mestre industrial area. In the absence of adequate sweet surface and ground waters, but where there is rainfall, as in the instance of many island and coastal regions throughout the world, that resource can be collected on the roofs of homes and other collection areas. Until recent years, any region that did not have adequate sweet surface or ground waters or rainfall was simply uninhabitable. But now thanks to science and technology, an enormous

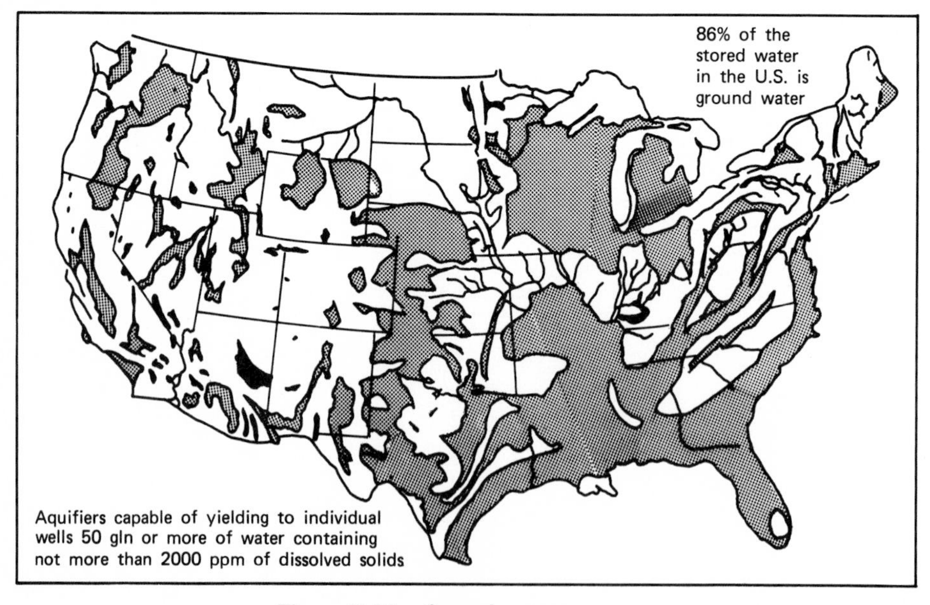

Figure 16.13. Ground water resources.

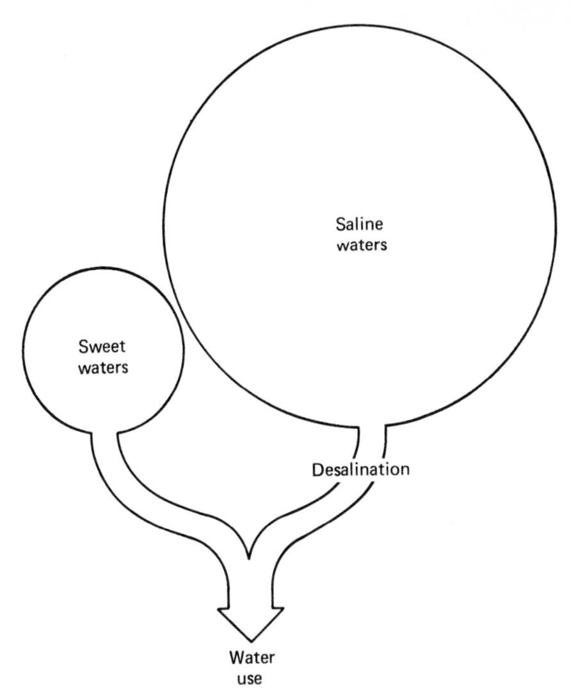

Figure 16.14. Enlargement of water resources by desalination.

water resource has been made potentially available. Brakish and saline waters, hitherto useless, can be desalinated and made fresh (Figure 16.14). I have summarized the various methods of desalination elsewhere (Horne, 1969, Chapter 13) and a number of good reviews and monographs describing the technology, including nuclear desalination (Internat. Atomic Energy Agency, 1969) and especially reverse osmosis (Hauck and Sourirajan, 1969; McDermott, 1970; Sourirajan, 1970; Lonsdale and Podall, 1972; Luttinger and Hoche, 1974), are available (Helfferich, 1962; Spiegler, 1966; see also the series of volumes containing the proceedings of the International Symposia on Fresh Water from the Sea, Athens, 1962, 1967, 1970). But desalinated water is expensive. However, as water becomes more scarce, more precious, then more costly methods of water conservation, utilization, and treatment will steadily become economically feasible. The era of cheap energy and cheap materials including water is being hastened to an end by the demands of exploding world population. Also passing, I am glad to report, is the era in which the cost of environmental damage was ignored (Solow, 1971). As the cost of water increases, recourse will be taken to more expensive means of marshaling water resources. Fortunately, modern science and water resource management technology have placed a considerable number of means at our disposal at every strategic point in the total water utilization sequence (Figure 16.15). First, as we mentioned in an earlier chapter, the precipitation source of all surface waters can be controlled by cloud seeding, but such weather modification for the most part in effect simply changes the distribution of water resources, taking vital moisture from some and giving it to others, thereby raising very complex and far-reaching social, economic, and legal issues. In many instances, the water supply is very uneven with spring floods and

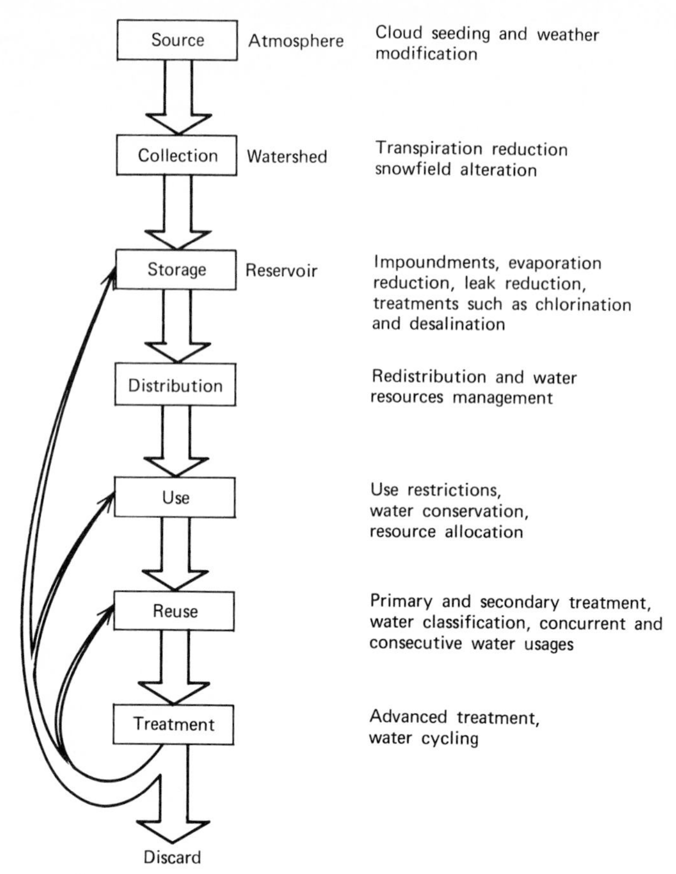

Figure 16.15. The steps of water resource management.

summer drought. Such fluctuations can be "dampened," notably by impoundments, but also some thought has been given to exercising some kind of moderating influence over the snow field sources (Anderson, 1966). Transpiration losses from forested watersheds (and from aquatic plants) can be considerable, yet it is relatively difficult to reduce these losses and conserve water without causing aesthetic and other types of environmental damage. Economically, the reduction of evaporative losses from reservoirs by surfactant films has been disappointing. Wind and other factors tend to disperse the film so that the quantities of surfactant that must be added continuously to maintain a coherent surface film can be so large that their value exceeds that of the water conserved. Plastic sheets have been used to stop leakage losses through the bottom of water supplies into the ground, but again this is expensive and feasible only for small reservoirs. In classical and Byzantine times, underground water reservoirs were constructed, but clearly this approach is economically impractical in the case of the staggering quantities of water used by modern industrialized urban centers.

Who will get the water and who will be deprived? These are the Solomonic judgments involved in the allocation and distribution of water resources, and they are inseparatably enmeshed in social, legal, economic, and political difficulties (Sewall et al., 1969), problems that are already critical in arid regions (Dregne, 1970; Kelso et al., 1973). Yet, it is abundantly clear that the usefulness of water resources can be stretched and maximum benefit derived therefrom by carefully planned distribution by a responsible and powerful centralized authority.

There is equally little doubt that a large fraction of our water resources are wasted, that there are many water uses that can be lessened and even eliminated, if not voluntarily then by legal compulsion. Luxury uses must not continue to be allowed to preempt necessary uses. Should some ghetto child go to bed a little more hungry so that some affluent sourthwestern suburbanite can enjoy a green lawn and a swimming pool? Should crops be sacrificed for car washes? The ways to conserve water fortunately are many. Some few summers ago, New York City felt the impact of drought, and waiters stopped setting tables with glasses of water unless specifically requested by the customer. Again, someone proposed saving water by putting a brick in toilet reservoirs and research and development is currently being pursued on a nonaqueous toilet. Are these savings trivial? More than 40% of the domestic water used in the United States is used for flushing toilets (Chanlett, 1973). Even more important, these precautions that touch every person and home are invaluable in educating the public of the preciousness of our water resources. A summer restriction on lawn watering is far more immediate and impressive than any lengthy government report on the necessity of water conservation and management.

While restricting and banning certain water uses serve an invaluable public educational function, water, luckily, admits of many simultaneous and sequential uses. We must abandon the luxury of single use and discard. We are entering a period when water increasingly, like other precious materials, must be recycled and used again and again.

In the United States, the federal government has required the states to classify their waters with respect to use. Typically, these classification schemes distinguish four levels of use and water quality (Figure 16.16): (*a*) water fit to drink, (*b*) water fit to swim in, (*c*) water fit for boating and fishing, and (*d*) navigational waters. In all too many places, the waters do not even meet the criteria for the last and most degraded category so a fifth, and hopefully temporary, category has been added (Figure 16.16). Table 16.30 gives the water quality characteristics of each category as defined by the Commonwealth of Massachusetts whose classification is fairly typical.

The goal of clean waters can be realized by preventive and corrective action (Figure 16.17). In the United States and elsewhere, widespread pollution has resulted, at least in part, from an overly optimistic estimation of the assimilative capacity of streams (Busch, 1971) as well as from indifference to the destruction of water resources. In our permissive, capitalistic society, where the voice of industry and special interest groups is often much louder and more strongly organized than the voice of the people, the corrective rather than the preventative approach has been emphasized. While allocating heavy funding for water treatment facilities, the government has been most delinquent in cracking down on polluters. Only recently has there been a serious legal attempt to put a stop to the contamination of our environment. And even this attempt has been rather timid and now appears to be moving in two questionable directions: (*a*) moving enforcement responsibility away

Table 16.30 Water Quality Standards of the Commonwealth of Massachusetts

	Class A	Class B	Class C	Class D
Dissolved Oxygen	≥75% saturation for at least 16 out of 24 hr. Never less than 5 mg/liter	≥75% saturation for at least 16 out of 24 hr. Never less than 5 mg/liter	≥5 mg/liter for at least 16 out of 24 hr. Never less than 3 mg/liter	Not less than 2 mg/liter at any time
Sludges and Scums	None	None	Those resulting from waste treatment facilities only	Those resulting from waste treatment facilities only
Color and Turbidity	Natural origin only	None that impairs usages assigned	None that impairs usages assigned	None that impairs usages assigned
Coliform	≤50/100 ml	≤1000/100 ml	None that impairs usages assigned	None that impairs usages assigned
Taste and Odor	Natural origin only	None that impairs usages assigned or causes taste and odor in edible fish	None that impairs usages assigned or causes taste and odor in edible fish	None that impairs usages assigned
Radioactivity	Natural origin only	None in concentrations harmful to human, animal, or aquatic life		
pH	As naturally occurs	6.5–8.0	6.0–8.5	6.0–9.0
Total Phosphate		≤0.05 mg P/liter	≤0.05 mg P/liter	
Temperature Increase		≤4°F	≤4°F	Water temperature not to exceed 90°F
Chemicals	None in concentrations harmful or offensive to human, animal, or aquatic life	As in Class A, none in concentrations harmful to usages assigned	As in Class B	None in concentrations harmful to human, animal, or aquatic life as to usages assigned

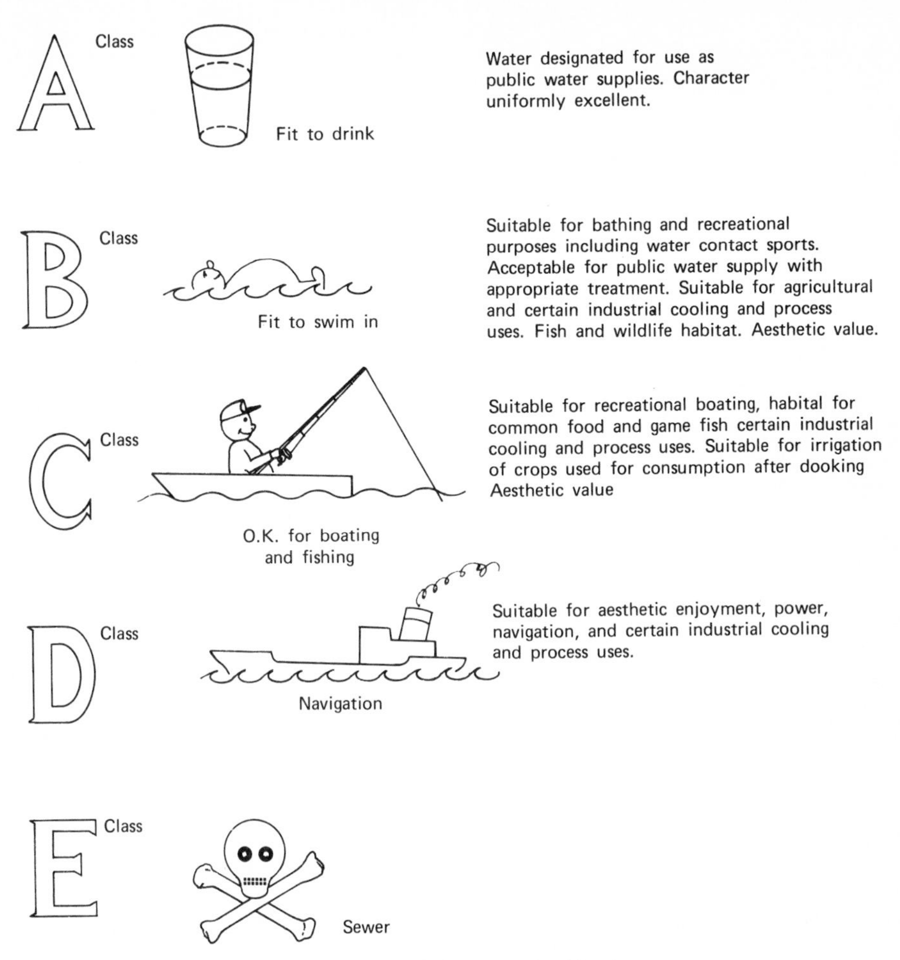

Figure 16.16. Water use classification.

from the federal to the local level where it will be more susceptible to pressures by industry and to popular outcry against the price of a clean environment and (*b*) imposing monetary penalties (Freeman and Haveman, 1972) rather than stopping the discharge of pollutants, in other words, selling licenses to pollute. It has also been argued convincingly (Cicchetti et al., 1973) that in the United States we may, paradoxically, be swinging from a condition of too little resource management to one of too much, that is to say, to excessively large land and water resource development programs.

Technology has given us a considerable measure of potential control over our future. The future of our species will be largely determined on if and how we exercise this technology, on how we control or fail to control our fertility, and how we carefully manage or fail to manage our water resources. Careful husbanding of water resources should enable us not only to maintain but to improve the quality of life, but will we choose to do so and in time?

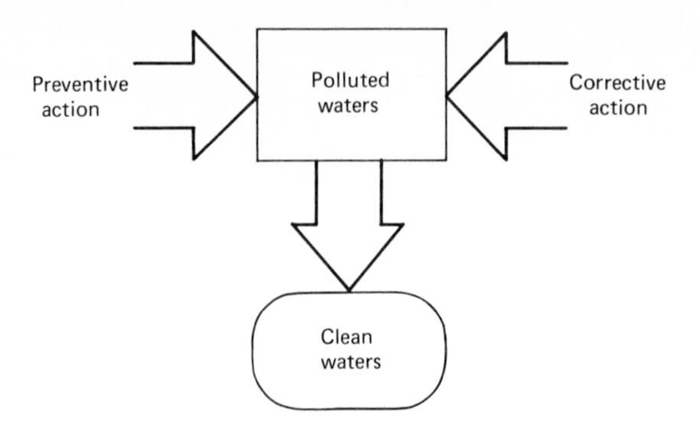

Figure 16.17. Achieving clean waters.

16.6 WILL EARTH SURVIVE?

> This is the way the world ends
> This is the way the world ends
> This is the way the world ends
> Not with a bang but a whimper.
>
> T. S. ELIOT
> *The Hollow Men*

Now that I am nearly done writing this book I am filled with a sense of futility. Hundreds of pages, thousands of words about the chemistry of our environment and about the pollution and degradation of that environment, but only an occasional remark on the causes of that devastation—the population explosion, the failure of man to control his fertility, the people-plague (Stanford, 1972; Frejka, 1973; Piotrow, 1973; Pohlman, 1973). The human population on this small planet is exploding (Figure 16.18). It took centuries to reach a population of 1 billion, but only 80 yr to reach 2 billion, and only 41 years to reach 3.7 billion (in 1971). "There will be 7 billion standing in line for their rains in the year 2000. By 2050, perhaps 30 billion will be fighting like animals for a share of the once-green earth" (Friedrich, 1971). This population explosion is not only the cause of most of our environmental and social problems, it is without the slightest doubt the most grave problem confronting the human race. As Table 16.31 shows, the problem is particularly critical in undeveloped nations, the very areas least able to cope (and least willing to protect the environment as illustrated at the Stockholm Conference, Hawkes, 1974) with it.

Some have argued that the problem arises from too much technology rather than from too many people. This argument is specious. While it is true that people in developed nations produce more pollution per capita than persons in countries with lower levels of technology, technology is just a multiplication factor and the basic figure multiplied is the population (Ehrlich and Holdren, 1971). Any advancement that technology might be able to make in pollution abatement will be negated by continued population growth. In the words of an editorial in *Environmental Science and Technology*, clean-up without population control is "merely playing a game" (Bowen, 1969). If technology must take its share of guilt for our danger, it is because of its reduction of infant-death rates and extension of life expectancy. Certainly

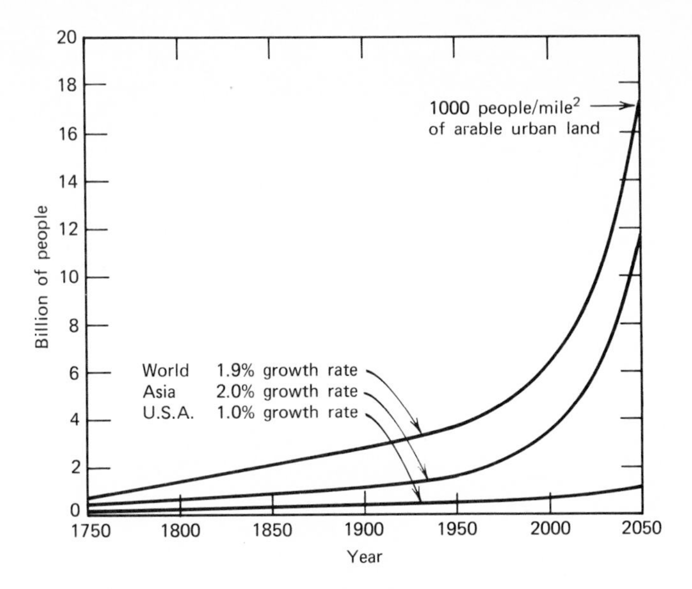

Figure 16.18. The exploding world population.

warning has been sounded by many shrill voices. Books are pouring out of the presses spelling out in lurid detail the calamitous consequences of unchecked population growth (Benarde, 1970; Dorst, 1970; Johnson, 1970; Linton, 1970; Anon., 1970*i*; Day, Fost, and Rose, 1971; Commoner, 1971; Falk, 1971; Helfrich, 1971; Revelle, Khosla, and Vinovskis, 1971; Brown and Hutchings, 1972; Brubaker, 1972; Cox and Peel, 1972; Ehrlich and Ehrlich, 1972; Polunin, 1972; Fraser, 1973; Smith

Table 16.31 Estimated Populations of the World and Its Major Regions, in Millions, in 1900, 1950, 1960, and 2000, with Comparisons of Increases from 1900 to 1950 and from 1950 to 2000 Numerically and as a Ratio (From *Environmental Protection* by E. T. Chanlett, © 1973 by McGraw-Hill. Used with permission of McGraw-Hill Book Co.)

Area	Estimated Population (millions)			Projected Population (millions)	Increase (millions)		Ratio of Increase, 1950–2000 to 1900–1950
	1900	1950	1960	year 2000	1900–1950	1950–2000	
World	1550	2518	2995	6907	968	4389	4.6
Africa	120	209	254	663	89	454	5.1
North America	81	168	199	326	87	158	1.7
Latin America	63	163	206	651	100	488	4.9
Asia	857	1389	1679	4250	532	2861	5.4
Europe (including U.S.S.R.)	423	576	641	987	153	411	2.7
Oceania	6	13	17	30	7	17	2.4

et al., 1974). Of these predictions of doom, one of the most controversial has been a computer prognostication of inevitable future disaster entitled *The Limits to Growth* (Meadows et al., 1972; see also Forrester, 1971, and Meadows and Meadows, 1973). These analyses have been widely criticized (Burke, 1973; Cole et al., 1973). They are highly sensitive to some of their input assumptions with respect to capital investment and pollution (Salerno, 1973), and they do contain some errors (Boyle, 1973) that do not seem to have seriously altered their dire conclusions (Meadows and Meadows, 1974). These difficulties have provided an excuse for those who do not want to believe the report's conclusions to do so. No one listens to prophecies of doom whether by Cassandra or computer. There is always faith that there will be a miracle or that somehow natural checks and balances will intervene to avert disaster. I feet that the latter view is well founded, while remembering that the intervention can be exceedingly cruel. I do not believe that we are all doomed, that the world will end, that we shall all be poisoned, or that atmospheric pollution will precipitate a premature ice age. But I do believe that life for man on this planet is fast becoming increasingly unpleasant as a consequence of population pressures, and that it probably will become very unpleasant indeed—recalling Hobbes' insight Professor Adelman (1973) observed, "man's life will soon again become nasty, brutish, and short."

I have called Table 16.32 a "Doomsday Timetable," and it lists in chronological order what I believe to be the likely sequence of future events. Various nations have traveled different distances down this dismal road. India* and the Sahara States of Northern Africa already find themselves in possibly insoluble difficulties, and even developed countries such as Japan, Italy, and the United States find themselves on the critical threshold. In my opinion, the situation in the United States is worse than in a number of western European countries. Overcrowding is a new problem for America, and we are less able to cope with it or even admit it, while older nations who have experienced crowding for much longer periods of time have developed suitable attitudes and responses to the problem that are at least partial and temporary solutions.

In terms of disruption, the effects of overpopulation in producing chemical pollution are trivial compared to the social upheavals arising from crowding (Nat. Acad. Sci., 1971; Parsons, 1971; Campbell and Wade, 1972; Glass and Revelle, 1972; Hines, 1973). Future generations will not be poisoned; they will be killed. Man will become more brutish. Population pressure will make the altruistic formulations of the world's great religions increasingly untenable. Mutual respect among persons is one of the first victims of overcrowding. Yes, Earth will survive, but it will not be a pleasant place for man to live. He will live in constant danger from his fellows, in a constant miasma of inconvenience, discomfort, and nastiness; and the richness and diversity of the biosphere that we now enjoy will be very much diminished.

What hope I have in survival is based on our old friend, the Principle of LeChatelier, and the multiplicity of alternatives still at our disposal, if only we shall use them, for checking population growth (Table 16.33). These range from insuring fundamental human rights, which should have been done long ago quite apart from any environmental considerations, to more drastic measures that are probably compatible with our moral standards, legal codes, and democratic institutions, to far

* Recently, India seems to have made some headway in controlling population, but at the expense of her democratic institutions.

Table 16.32 Doomsday Timetable[a]

	Disappearance of the frontier	
	Crowding	
	Diminishment of social and environmental amenities	
	Local environmental pollution	Western Europe
	Shortages of energy and materials reflected in increasing prices	United States, Japan, Italy
	The unpleasant life becomes the norm	
Critical		
	Food shortages	
	Growth stoppage	
	Abandonment of environmental protection	
	Rationing	
	Malnutrition	
	Class struggles	
	Inability of democratic institutions to cope with population-pressure created problems	India
	Global environmental pollution	
	Significant reduction in life expectancy due to pollution	
	Famine	
	Pestilence	
	Tyranny	Sahara States
Point of No Return		
	International wars	
	Global nuclear war	
	Reduction in habitable areas of the Earth	
	Civil wars	
The End		
	Anarchy	
	Barbarism	

[a] The Principle of Le Chatelier operates as a conservative force tending at all times to return the stressed system to an earlier less stressed state.

more drastic measures about which certainly we should have very serious reservations. If we fail to take mild steps now, we will be *forced* to take harsher steps later. The unacceptable will become necessary. "The desperate and repressive measures for population control which might have to be contemplated then are reason in themselves to proceed with foresight, alacrity, and compassion today" (Ehrlich and Holdren, 1971). Within the first grouping of alternatives in Table 16.33, it is important to notice women's rights and social security. The women's liberation movement has probably given more real hope to the future of our environment than have all the environmental protection laws together. Assuring the economic security of the elderly is also crucial, especially in the less developed nations, for in the absence of such security there is a very strong incentive for large families so that the children

Table 16.33 Population Control Measures

Fundamental Human Rights	1.	Women's rights
	2.	Free birth control information and treatment (including abortion and voluntary sterilization) for all who need them
	3.	End of tax discrimination against single persons
	4.	End of family tax advantages
	5.	Social security (and medical assistance)
	6.	End of discrimination against nonprocreative sexual relationships
Probably Compatible with Democratic Institutions	7.	Euthanasia
	8.	Gender selection
	9.	Tax disincentives for families
	10.	Sterilization as a qualification for welfare
	11.	Anti-family propaganda
	12.	Sterilization of defectives, criminals, and indigents
	13.	Treatment of foodstuffs and water supplies with fertility-reducing substances
Drastic and Morally Unacceptable	14.	Random compulsory sterilization
	15.	Genocide
	16.	Decimation

can support their aged parents. Male children are also needed to provide a farm labor base. If gender selection becomes a real possiility, then redundant female births can be reduced (Sheldrake, 1974) to provide some relief from population pressures. However a largely male population in itself would create social dislocations that it might take polyandry or homosexuality to resolve.

I am proud to say that chemistry is contributing mightily to population control thanks to "the pill." Birth control pills contain progesterone or other sex hormones

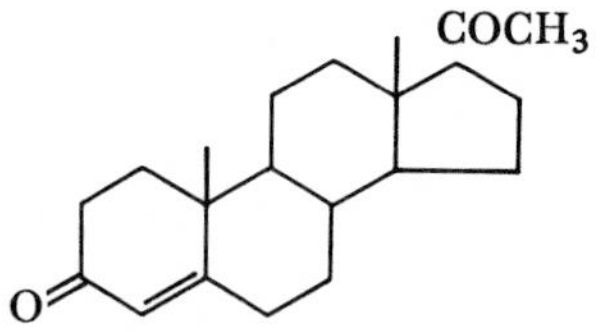

$COCH_3$

Progesterone

and related substances (recall Section 12.6). Chemistry holds even further hope, for it appears to be technically feasible to control fertility, notably of large urban centers, by the addition of powerful chemicals to water supplies. Is such a measure compatible with our democratic institutions? There is no "right" to make babies, just as there is no right to commit any other form of harmful antisocial behavior.

The Principle of LeChatelier is on our side, but it can be a most stern corrective vector. Experiments with some mammals have shown that excessive crowding can induce not only drastic behavioral changes but a sharp decline in fertility as well. In the human population, long before such severe compensation becomes operative, let us hope that voluntary reduction of fertility can achieve some relief. Happily this seems to be happening in the United States at present, where we seem to be moving in the direction of zero growth. The future has become so uncertain, the consequences of overpopulation such as soaring crime rates have become so apparent, and the quality of life has so declined that now more responsible and thoughtful couples are disinclined to invest in the future and bring more children into the world. The rapidity of change in such profoundly fundamental attitudes has astonished me. Not too many years ago, birth-control and abortion were forbidden topics in the United States, but now, despite continuing hostility from certain segments or our society, it is clear that they are ideas whose time, as they say, has come (Blake, 1971; Friedrich, 1971; Westoff and Westoff, 1971).

Food, fuel, energy, and materials shortages; inflation; rationing and further deterioration of the quality of life will be additional and stronger incentives for family planning.

Everybody says that they want a clean environment, but who is ready to pay for it? The bitter fact is that the "Environmental Movement" in the United States, despite some more widespread popular lip-service, is a vocal minority group. When it comes right down to basic issues such as increasing prices to absorb pollution abatement costs, such as sacrificing jobs to protect fish, such as even a little inconvenience to keep the air cleaner, a clean environment frankly does not rank very high in the list of real priorities of the overwhelming majority of Americans. In the past, our society has been predicated on growth. Progress has been equated with the "American Way," and most of the nations of the world now seem determined to follow our bad example. Environmental deterioration is not unique to capitalistic democracies. Other governments appear to be equally impotent in controlling the problems that unregulated growth creates (Goldman, 1970, 1972). The choice between poverty and starvation and the environment is particularly critical in developing countries (de Almeida et al., 1972), which are almost forced to adopt short-sighted and ultimately very destructive "solutions." At least the concept of unlimited growth has now finally been called into question (Ridker, 1972). There may even be an effort to draw a step back from unlimited growth. But the concepts of the "dominion of man" (Black, 1970) and unlimited growth are so interwoven into the fabric of our way of life, into our basic outlook and underlying social philosophy, that any retreat from growth is going to cause most severe economic and social disruption, not to mention psychological re-orientation. But these disruptions of our thinking and of our institutions are insignificant compared to the disruptions, to the destruction and chaos, toward which unlimited growth is propelling us.

How has environmental protection fared in the United States? Not very well I am sorry to report. About 5 yr ago, Congress passed the National Environmental Protection Act (NEPA). It sounded like an action bound to secure the approbation of the electrorate. Congress was for the flag, for apple pie, and for a clean environment. But, much to the dismay of many people in high places, including congressmen, the courts began to take the law (and the Clean Air Act) seriously and to interpret the letter and spirit of the law literally. Consequently, there has been a great deal of propaganda on the need to relax NEPA, talk, which became a loud outcry during the recent so-called energy crisis. This crisis has demonstrated the alacrity with which politicians and the people are willing to set the environmental protection laws aside (Anon., 1973*c*). NEPA and the Clean Air Act are doomed. We are going back to extensive strip mining, we are going back to high sulfur fuels, we are going back to dirty air. While NEPA has worked too well for the comfort of politicians and industrialists, it has not worked at all for the very people whose environment most desperately needs amelioration. Why weaken NEPA? It is already fatally weak. It makes no provision for enforcement; no government agency has the *real* responsibility and authority to make it work. The impacted victims have to sue. This is all very well for the affluent, they have the money to sue to protect their game preserves, their green space, and their recreational areas. But what can the poor do? Does anybody give a damn about their environment? Does anybody think that cities are as important as marshes?

The pressures of population growth produce stresses that are irresistable. Environmental deterioration and pollution, as I have said, will be among the least of our future problems. Can a pluralistic, egalitarian society that has failed in its primary responsibility of protecting the security of the individual's person and property secure such amenities for its citizens as clean air and water? What is the good in cleaning the air so we can see the stars in the night sky again if we do not dare to venture outside from behind the locked and barricaded doors of our homes to look at them?

And even if as a nation we are able to control our environmental problems, what about our neighbors? On spacecraft *Earth*, if anyone is dirty, the rest will have to live in that filth. We have entered into a curel and ugly world, albeit one of our own making. For whatever sloganistic trappings and ideological camouflage they have been given, most wars since the beginning of history and long before have been for turf (a point nicely illustrated in V. G. Childe's little book *What Happened in History*. Penguin Books, Harmondsworth, 1954), for livingspace, for crop and grazing land, for mineral and water resources. When segments of our species, in distant lands or near at home, through ignorance, apathy or obstinacy, fail to control their numbers, rather than trying to rescue them from self-inflicted disaster, we must learn to cast aside the outdated and now even dangerous altruistic notions that we have inherited from a less crowded age and be prepared to stand aside and let the Principle of LeChatelier perform its corrective function.

Our environment is going to continue to deteriorate. We who are concerned with the quality of our environment are going to lose. But what does this mean? What should we do? Should we acknowledge defeat and quit? No! When the women of ancient Sparta sent their husbands off to war, they admonished them to return with their shields or on them, that is to say, in victory or death. We must never give up the fight for a decent environment. We shall lose the war, but we may win important skirmishes. We cannot alter the direction of the inevitable, but perhaps we can slow its pace. If we work hard, if we do not despair, perhaps we can keep the environment a little bit better, a little bit longer in our own times.

Yes, Earth will survive. But it will be a different and harsher place. Like Mr. Thoreau, I am relieved that I am not going to have to live in too much of the future.

BIBLIOGRAPHY

L. E. Adams, Jr., *Environ. Sci. Tech.*, 7, 696 –(973).

M. A. Adelman, *Science*, *181*, 151 (1973).

L. Allan et al., *Paper Profits: Pollution in the Pulp and Paper Industry*, M.I.T. Press, Cambridge, 1972.

M. O. de Almeida et al., *Environment and Development*, Carneigie Endow, Internat. Peace, New York, 1972.

T. A. Alspaugh, *J. Water Pollut. Control Fed,. 43*, 1001 (1971).

Y. Ambe, *Environ. Sci. Tech.*, *7*, 542 (1973).

H. W. Anderson, in A. V. Kneese and S. C. Smith (eds.), *Water Research*, Johns Hopkins Press, Baltimore, 1966.

E. E. Angino et al., *Science*, *168*, 489 (1970).

L. L. Ames and R. B. Dean, *J. Water Pollut. Control Fed.*, *42*, R161 (1970).

Anon., *Cleaning Our Environment*, Amer. Chem. Soc., Washington, D.C., 1969*a*.

Anon., *Environ. Sci. Tech.*, *3*, 1237 (1969*b*).

Anon., *Chem. Engr. News*. 7 (Dec. 22, 1969*c*).

Anon., *Stream Pollution by Coal Mine Draining in Appalacia*, U.S. Dept. Interior, FWPCA, Cincinnati, Ohio 1969*d*.

Anon., *Environ. Sci. Tech.*, *3*, 1243 (1969*e*).

Anon., *Oxygenation of Ferrous Iron*, U.S. Dept. Interior, U.S. Gov. Print. Off., Washington, D.C., 1970*a*.

Anon., *Feasibility Study Manual and Mine Water Pollution Control Demonstrations*, U.S. Dept. Interior, U.S. Gov. Print. Off., Washington, D.C., 1970*b*.

Anon., *Studies on Limestone Treatment of Acid Mine Drainage*, Bituminous Coal Res. Inst., U.S. Gov. Print. Off., Washington, D.C., 1970*c*.

Anon., "Treatment of Acid Mine Draingae" (Horizons, Inc.), U.S. Dept. Interior, U.S. Gov. Print. Off., Washington, D.C., 1970*d*.

Anon., "New Mine Sealing Techniques for Water Pollution Abadement," FWQA, U.S. Dept. Interior, Washington, D.C., 1970*a*.

Anon., *Mar. Pollut. Bull.*, *1*, 99 (1970*f*).

Anon., *Environ. Sci. Tech.*, *4*, 380 (1970*g*).

Anon., *Environ. Sci. Tech.*, *4*, 725 (1970*h*).

Anon., *Man's Impact on the Global Environment*, M.I.T. Press, Cambridge, 1970*i*.

Anon., *Nature*, *233*, 295 (1971*a*).

Anon., "Silicate Treatment for Acid Mine Drainage," (Tyco Labs., Inc) EPA, U.S. Gov. Print. Off., Washington, D.C., 1971*b*.

Anon., "Purification of Mine Water by Freezing," (Appl. Sci. Lab. Inc.) EPA, U.S. Gov. Print. Off., Washington, D.C., 1971*c*.

Anon., *Tech. Rev.* (*M.I.T.*), 74, 64 (December 1971*d*).

Anon., *Environ. Sci, Tech.. 5*, 496 (1971*e*).

Anon., *Chem. Engr. News*, (November 20, 1972*a*).

Anon., *Environ. Sci. Tech.*, *6*, 25 (1972*b*).

Anon., *Chem. Engr. News*, 20 (December 10, 1973*a*).

Anon., *Nature*, *241*, 166 (1973*b*).

Anon., *Science*, *181*, 641 (1973*c*).

H. S. Azad and J. A. Borchardt, *Environ. Sci. Tech.*, *4*, 737 (1970).

R. A. Baker and A. G. Wilshire, *Environ. Sci. Tech.*, *4*, 401 (1970).

S. K. Banerji, *J. Water Pollut. Control Fed.*, *43*, 1123 (1971).

A. M. Beeton in *Eutrophication Causes, Consequences, and Correcteves*, Nat. Acad. Sci. U.S.A., Washington, D.C., 1969.

M. A. Bernarde, *Our Precarious Habitat*, Norton, New York, 1970.

M. E. Bender et al., *Environ. Sci. Tech.*, *4*, 520 (1970).

J. Black, *The Dominion of Man*, Aldine, Chicago, 1970.

J. Blake, *Science*, *171*, 540 (1971).

H. I. Bolker, *Tech. Rev.* (*M.I.T.*), *73*, 23 (April 1971).

D. H. M. Bowen, *Environ. Sci. Tech.*, *3*, 1225 (1969).

T. J. Boyle, *Nature*, *245*, 127 (1973).

H. C. Bramer, *Environ. Sci. Tech.*, *5*, 1004 (1971).

A. J. Brook and A. L. Baker, *Science*, *176*, 1414 (1972).

H. Brown and E. Hutchings (eds.), *Are Our Descendents Doomed?*, Viking, New York, 1972.

S. Brubaker, *To Live on Earth*, Johns Hopkins Press, Baltimore, 1972.

F. E. Burke, *Nature*, *246*, 226 (1973).

A. W. Busch, *J. Water Pollut. Control Fed.*, *43*, 1480 (1971).

R. R. Campbell and J. L. Wade (eds.), *Society and Environment*, Allyn & Bacon, Boston, 1972.

E. T. Chanlett, *Environmental Protection*, McGraw-Hill, New York, 1973.

C. J. Cicchetti et al., *Science*, *181*, 723 (1973).

H. S. D. Cole et al. (eds.), *Models of Doom*, Universe Books, New York, 1973.

P. R. Cox and J. Peel (eds.), *Population and Pollution*, Academic, New York, 1972.

B. Commoner, *The Closing Circle*. Knopf, New York, 1971.

R. L. Culp and G. L. Culp, *Advanced Wastewater Treatment*, Van Nostrand-Reinhold, New York, 1971.

S. L. Daniels and D. G. Parker, *Environ. Sci. Tech.*, *7*, 690 (1973).

G. W. Dawson et al., "Control of Spillage of Hazardous Polluting Substances," Water Pollut. Control Serv. 15090 Foz 10/17, U.S. Gov. Print. Off., Washington, D.C., 1970.

J. A. Day, F. F. Fost, and P. Rose, *Dimensions of The Environmental Crises*, Wiley, New York, 1971.

R. B. Dean, *Tappi*, *52*, 457 (1969).

R. B. Dean, *Tech. Rev.* (*M.I.T.*), *73*, (March 1971).

B. D. Dinman, *Science*, *177*, 1154 (1972).

J. Dorst, *Before Nature Dies*, Penguin, Baltimore, 1970.

H. E. Dregne (ed.), *Arid Lands in Transition*, Univ. Arizona Press, Tucson, 1970.

R. Dubas, *Science*, *173*, 259 (1971).

P. R. Dugan, *Biochemical Ecology of Water Pollution*, Plenum, New York, 1972.

Economics Systems Corp., "Storm Water Pollution from Urban Land Activity," U.S. Gov. Print. Off., Washington, D.C., 1970.

P. R. Erhlich and A. H. Ehrlich, *Population, Resources, and Environment*, Freeman, San Francisco, 1972.

P. R. Ehrlich and J. P. Holdren, *Science*, *171*, 1212 (1971).

J. D. Eye, *J. Water Pollut. Control Fed.*, *43*, 998 (1971).

J. D. Eye and T. K. Basu, *J. Water Pollut. Control. Fed.* *42*, R125 (1970).

J. D. Eye, *J. Water Pollut. Control Feed.*, *43*, 998 (1971).

J. D. Eye and L. Liu, *J. Water Pollut. Control Fed.*, *43*, 2291 (1971).

R. A. Falk, *This Endangered Planet*, Random House, New York, 1971.

J. B. Farrell, *Environ. Sci. Tech.*, *5*, 716 (1971).

Federal Water Pollut. Control Adm., *Water Quality Criteria*, U.S. Gov. Print. Off., Washington, D.C., 1986.

J. F. Ferguson and P. L. McCarty, *Environ. Sci. Tech.*, *5*, 534 (1971).

D. H. Fink et al., *J. Water Pollut. Control Red.*, *42*, 265 (1970).

J. W. Forrester, *World dynamics*, Wright-Allen Press, Cambridge, 1971.

D. Fraser, *The People Problem*, Indiana Univ. Press, Bloomington, 1973.

A. M. Freeman and R. H. Haveman, *Science*, *177*, 322 (1972).

O. Freidrich, *Time*, 58 (September 13, 1971).

T. Frejka, *The Future of Population Growth*, Wiley-Interscience, New York, 1973.

C. R. Frink, *Environ. Sci. Tech.*, *1*, 425 (1967).

C. R. Frink, *Soil Sci. Soc. Amer. Proc.*, *33*, 326, 369 (1969).

E. G. Fruh, *J. Water Pollut. Control Fed.*, *39*, 1449 (1967).

A. F. Gandy Jr., *J. Water Pollut. Control Fed.*, *43*, 952 (1971).

I. Gellman and R. O. Blosser, *J. Water Pollut. Control Fed.*, *43*, 1546 (1971).

D. V. Glass and R. Revelle (eds.), *Population and Social Change*, Arnold, London, 1972.

J. C. Goldman, *Science*, *182*, 306 (1973).

M. I. Goldman, *Science*, *170*, 37 (1970).

M. I. Goldman, *The Spoils of Progress: Environmental Pollution in the Soviet Union*, M.I.T. Press, Cambridge, 1972.

C. G. Golueke, Jr., and W. J. Oswald, *J. Water Pollut. Control Fed.*, *42*, R304 (1970).

R. F. Gould (ed.), *Anaerobic Biological Treatment Processes*, Amer. Chem. Soc., Washington, D.C., 1972.

E. F. Gloyna et al., *J. Water Pollut. Control Fed.*, *42*, 237 (1970).

M. G. Gross, Mar. Sci. Res. Center, State Univ. New York, Stony Brook, 1970 (Specia. Rept. Nos. 5 and 7).

U. Granhall and A. Lundren, *Limnol. Oceanogr.*, *16*, 711 (1971).

R. D. Grundy, *Environ. Sci. Tech.*, *5*, 1184 (1971).

W. E. Grube, Jr. et al., *Nature*, *236*, 70 (1972).

W. N. Grune, *J. Water Pollut. Control Fed.*, *43*, 1024 (1971).

N. Grunchow, *Science*, *167*, 151 (1970).

S. Gutche, *Waste Treatment with Polyelectrolytes*, Noyes Data Corp., Park Ridge, N.J., 1972.

M. W. Hall, *J. Water Pollut. Control Fed.*, *43*, 1020 (1971).

A. L. Hammond, *Science*, *172*, 361 (1971).

A. M. Hanson and G. F. Lee, *J. Water Pollut. Control Fed.*, *43*, 2271 (1971).

A. R. Hauck and S. Sourirajan, *Environ. Sci. Tech.*, *3*, 1269 (1969).

R. D. Harter, *Soil. Sci. Soc. Amer. Proc.*, *32*, 514 (1968).

J. E. Hawthorn, *J. Water Pollut. Control Fed.*, *42*, 425 (1970).

N. Hawkes, *Science*, *176*, 1308 (1972).

F. R. Hayes and J. E. Phillips, *Linmol. Oceanogr.*, *3*, 459 (1958).

S. Hayes, Mattern, and Mattern, "Engineering Investigation of Sewer Overflow Problems," FWQA, U.S. Gov. Printing Office, Washington, D.C., 1970.

F. Helfferich, *Ion Exchange*, McGraw-Hill, New York, 1962.

H. W. Helfrich, Jr., *Agenda for Survival*, Yale Univ. Press, New Haven, Conn., 1971.

L. G. Hines, *Environmental Issues*, Norton, New York, 1973.

A. J. Horne and A. B. Viner, *Nature*, *232*, 417 (1971).

R. A. Horne, *Marine Chemistry*, Wiley-Interscience, New York, 1969.

R. A. Horne, A. J. Mahler, and R. C. Rosello, "The Marine Disposal of Sewage Sludge and Dredge Spoil in the Waters of the New York Bight," Woods Hole Oceanogr. Inst., Rept. to U.S. Corps Engrs. (January 29, 1971).

R. A. Horne, *Science*, *177*, 1153 (1972).

D. L. Howard et al., *Science*, *169*, 61 (1970).

J. H. Hubschman, *Science*, *171*, 536 (1971).

Internat. Atomic Energy Agency, *Nuclear Desalination*, Elsevier, New York, 1969.

R. W. James, *Sewage Sludge Treatment*, Noyes Data Corp., Park Ridge, N. J., 1972.

W. J. Jewell, *J. Water Pollut. Control. Fed*, *43*, 1457 (1971).

C. E. Johnson (ed.), *Eco-Crisis*, Wiley, New York, 1970.

M. G. Johnson and G. E. Owen, *J. Water Pollut. Control Fed.*, *43*, 836 (1971).

H. R. Jones, *Pollution Control in the Textile Industry*, Noyes Data Corp.. Park Ridge, N.J., 1973*a*.

H. R. Jones, *Pollution Control and Chemical Recovery in the Pulp and Paper Industry*, Noyes Data Corp., Park Ridge, N. J.,1973*b*.

D. R. Keeney et al., *J. Water Pollut. Control Fed.*, *42*, 318 (1970).

M. M. Kelso et al., *Water Supplies and Economic Growth in an Arid Environment*, Univ. Ariz. Press, Tucson, 1973.

D. L. King, *J. Water Pollut. Control Fed.*, *42*, 2035 (1970).

R. E. Kolattukudy and R. E. Purdy, *Environ. Sci. Tech.*, *7*, 619 (1973).

J. G. Konrad et al., *J. Water Pollut. Control Fed.*, *42*, 2094 (1079).

J. P. Law, Jr., "Nutrient Removal...," U.S. Gov. Print. Off., Washington, D.C., 1969.

D. R. S. Lean, *Science*, *179*, 678 (1973).

G. F. Lee, "Factors Affecting the Transfer of Materials Between Water and Sediment," Univ. Wisconsin Water Resources Center Lit. Rev. No. 1 (July 1970).

G. V. Levin et al., *Environ. Sci. Tech.*, *6*, 280 (1973).

L. M. Lichtenstein et al., *J. Allergy*, *47*, 53 (1971).

G. E. Likens et al., *Science*, *172*, 873 (1971).

G. E. Likens (ed.), *Nutrients and Eutrophication*, Amer. Soc. Limnol. Oceanogr., Lawrence, Kans., 1972.

K. D. Linstedt et al., *J. Water Pollut. Control Fed.*, *43*, 1507 (1971).

R. M. Linton, *Terricade*, Little, Brown, Boston, 1970.

J. H. Litchfield, *J. Water Pollut. Control Fed.*, *43*, 948 (1971).

H. K. Lonsdale and H. E. Podall (eds.), *Reverse Osmosis Membrane Research*, Plenum, New York, 1972.

H. F. Lund (ed.), *Industrial Pollution Control Handbook*, McGraw-Hill, New York, 1971.

L. B. Luttinger and G. Hoche, *Environ. Sci. Tech.*, *8*, 614 (1974).

L. B. MacPherson et al., *Limnol.*, *Oceanogr.*, *3*, 318 (1958).

J. Marton and T. Marton, *Tappi*, *55*, 1614 (1972).

J. McDermott, *Desalination by Reverse Osmosis*, Noyes Data Corp., Park Ridge, N.J., 1970.

M. G. McGarry, *J. Water Pollut. Control Fed.*, *42*, R191 (1970).

D. H. Meadows et al., *The Limits to Growth*, Universe Books, New York, 1972.

D. L. Meadows and D. H. Meadows (ed.), *Toward Global Equilibrium*, Wright-Allen, Cambridge, 1973.

D. H. Meadows and D. L. Meadows, *Nature*, *247*, 97 (1974).

Metcalf & Eddy, Inc,. *Wastewater Engineering*, McGraw-Hill, New York, 1972.

E. J. Middlebrooks et al., *J. Water Pollut. Control Fed.*, *43*, 454 (1971*a*).

E. J. Middlebrooks et al., *J. Water Pollut. Control Fed.*, *43*, 242 (1971*b*).

R. A. Minear, *Environ. Sci.*, *Tech. 6*, 431 (1972).

D. Mitchell, *Science*, *174*, 827 (1971).

S. D. Morton and T. H. Lee, *Envir. Sci. Tech.*, *8*, 673 (1974).

C. B. Murphy, Jr., *Science*, *182*, 379 (1973).

Nat. Acad. Sci., *Rapid Population Growth*, Johns Hopkins Press, Baltimore, 1971.

N. L. Nemerow, *Liquid Waste of Industry*, Addison-Wesley, Reading Mass., 1971.

J. Parsons, *Population Versus Liberty*, Pemberton, London, 1971.

E. S. Pattison, *Science*, *170*, 870 (1970).

Q. H. Pickering and T. O. Thatcher, *J. Water Pollut. Control Fed.*, *42*, 243 (1970).

P. T. Piotrow, *World Population Crisis*, Praeger, New York, 1973.

E. Pohlman (ed.), *Population*, Mentor Books, New York, 1973.

N. Polunin (ed.), *The Environmental Future*, Barnes & Noble, New York, 1972.

C. P. C. Poon, *J. Water Pollut. Control Fed.*, *42*, 100 (1970).

J. J. Porter et al., *Environ. Sci. Tech.*, *6*, 37 (1972).

M. Rebhun and J. Manka, *Environ. Sci. Tech.*, *5*, 606 (1971).

H. A. Regier and W. L. Hartman, *Science*, *180*, 1248 (1972).

R. Revelle, A. Khosla, and M. Vinovskis (eds.), *The Survival Equation*, Houghton Mifflin, Boston, 1971.

G. Rey et al., *Environ. Sci. Tech.*, *5*, 760 (1971).

G. A. Richter and M. R. Soderquist, *J. Water Pollut. Control Fed.*, *43*, 983 (1971).

R. G. Ridker (ed.), *Population, Resources, and the Environment*, U.S. Gov. Print. Off., Washington, D.C., 1972.

Robert A. Taft Water Res. Center, "Oxone Treatment of Secondary Effluents from Wastewater Treatment Plants," U.S. Dept. Interior, FWPCA, Cincinnati, Ohio, 1969.

S. M. Rosenblum and T. C. Hollocher, *Science*, *172*, 1294 (1971).

J. H. Ryther and W. M. Dunstan, *Science*, *171*, 1008 (1971).

J. Salerno, *Nature*, *244*, 488 (1973).

C. N. Sawyer, *J. Water Pollut. Control Fed.*, *37*, 151 (1965).

L. G. Scharpf et al., *Nature*, *239*, 231 (1972).

C. L. Schelske and E. F. Stoermer, *Science*, *173*, 423 (1971).

D. W. Schlindler et al., *Science*, *177*, 1192 (1972).

J. A. Servizi et al., *J. Water Pollut. Control Fed.*, *43*, 278 (1971).

W. R. D. Sewell et al., *Water Management Research: Social Science Priorities*, Dept. Energy, Mines, and Resources, Canada, 1969.

U. Shamir, *Tech. Rev.* (M.I.T.), *74*, 41 (June 1972).

E. E. Shannon and P. L. Brezonik, *Environ. Sci. Tech.*, *6*, 719 (1972).

J. Shapiro, *Sicence*, *179*, 382 (1973*a*).

J. Shapiro, *Science*, *182*, 307 (1937*b*).

A. R. Sheldrake, *Nature*, *250*, 180 (1974).

K. S. Shumate et al., *J. Water Pollut. Control Fed.*, *42*, 631 (1970).

D. G. Shaw, *Environ. Sci. Tech.*, *7*, 740 (1973).

P. C. Singer and W. Stumm, *Science*, *167*, 1121 (1970).

G. J. C. Smith et al. (eds.), *Our Environmental Crisis*, Macmillan, New York, 1974.

J. M. Smith, *Environ. Sci. Tech.*, *6*, 260 (1972).

R. F. Smith and H. F. Ludwig, *Water Res.*, *2*, 615 (1968).

S. E. Smith, *J. Water Pollut. Control Fed.*, *43*, 1014 (1971).

I. V. Sollins, *Science*, *170*, 871 (1970).

R. M. Solow, *Science*, *173*, 498 (1971).

S. Sourirajan, *Reverse Osmosis*, Academic, New York, 1970.

K. S. Spiegler (ed.), *Principles of Desalination*, Academic, London, 1966.

Q. H. Stanford, *The World's Population*, Oxford Univ. Press, New York, 1972.

R. L. Stenberg, *J. Sanit. Eng. Div., Proc., Amer. Soc. Civil Engrs.*, *94*, 1121 (1968).

W. Stumm and G. F. Lee, *Ind. Engr. Chem.*, *53*, 143 (1961).

W. Stumm and J. J. Morgan, *Aquatic Chemistry*, Wiley-Interscience, New York, 1970.

M. W. Tenney et al., *Appl. Microbiol.*, *18*, 965 (1969).

R. E. Train et al., *Ocean Dumping*, Council Environ. Qual., U.S. Gov. Print. Off., Washington, D.C., 1970.

P. A. Trudinger, *Rev. Pure Appl. Chem.*, *17*, 1 (1967).

J. H. Tuttle, *J. Bacteriol.*, *97*, 594 (1969).

J. Verduin in N. C. Brady (ed.), *Agriculture and the Quality of Our Environment*, Amer. Assoc. Sci. Pub. No. 85, Washington, D.C., 1967.

W. Verstraete and M. Alexander, *Environ. Sci. Tech.*, *7*, 39 (1973).

L. L. Wallen and E. N. Davis, *Environ. Sci. Tech.*, *6*, 161 (1972).

L. Walter, *Water and Sewage Works*, 478 (December 1961).

W-C. Wang and R. L. Evans, *J. Water Pollut. Control Fed.*, *42*, 2117 (1970).

C. B. Warren, and E. J. Malec, *Science*, *176*, 277 (1972).

L. A. Westoff and C. F. Westoff, *From Now to Zero*, Little, Brown, Boston, 1971.

C. G. Wilbur, *The Biological Aspects of Water Pollution*, Thomas, Springfield, Ill., 1969.

H. E. Wild, Jr. et al., *J. Water Pollut. Control Fed.*, *43*, 1845 (1971).

C. M. Weiss, *J. Amer. Water Works Assoc.*, *51*, 387 (1969).

B. F. Winkler and Thodos, *J. Water Pollut. Control Fed.*, *43*, 474 (1971).

J. L. Yount and R. A. Grossman, Jr., *J. Water Pollut. Control Fed.*, *42*, R173 (1970).

C. D. Ziebell et al., *J. Water Pollut. Control Fed.*, *42*, 229 (1970).

J. F. Zievers and C. J. Novotny, *Environ. Sci. Tech.*, *7*, 209 (1973).

ADDITIONAL READING

H. E. Allen and J. R. Kramer, *Nutrients in Natural Waters*, Wiley-Interscience, New York, 1973.

Amer. Water Works Assoc., *Water Quality and Treatment*, McGraw-Hill, New York, 1971.

R. H. Boyle, J. Graves, and T. H. Watkins, *The Water Hustlers*, Sierra Club, San Francisco, 1971.

N. Buras, *Scientific Allocation of Water Resources*, Elsevier, New York, 1972.

D. E. Carr, *Death of Sweet Waters*, Norton, New York, 1971.

L. L. Ciccacio (ed.), *Water and Water Pollution Handbook*, Dekker, New York, 1972.

J. W. Clark, W. Viessman, Jr., and M. J. Hammer, *Water Supply and Pollution Control*, International Textbook Co., Scranton, Pa., 1971.

Council Environ. Qual., *Environmental Quality*, U.S. Gov. Print. Off., Washington, D.C., 1970.

E. S. Deevey, Jr,. "Mineral Cycles," *Sci. Amer.*, *223*, (3), 149 (September 1970).

P. R. Erlich and A. H. Erlich, *Human Ecology: Problems and Solutions*, Freeman, San Francisco, 1973.

G. M. Fair, J. C. Geyer, and D. A. Okum, *Elements of Water Supply and Wastewater Disposal*, Wiley, New York, 1971.

E. F. Gloyna and W. W. Eckenfelder, Jr., *Water Quality Improvement by Physical and Chemical Processes*, Univ. Texas Press, Austin, 1970.

C. R. Goldman, M. J. McEvoy, and P. J. Richerson (eds.), *Environmental Quality and Water Development*, Freemen, San Francisco, 1973.

R. F. Gould (ed.), *Equilibrium Concepts in Natural Water Systems*, Amer. Chem. Soc., Washington, D.C., 1967.

C. W. Howe and K. W. Easter, *Interbasin Transfers of Water: Economic Issues and Impacts*, Johns Hopkins Press, Baltimore, 1971.

K. Imhoff, W. J. Muller, and D. K. E. Thistlethwayte, *Disposal of Sewage and Other Waterborne Wastes*, Ann Arbor Sci. Pub., Ann Arbor, Mich., 1971.

L. D. James and R. R. Lee, *Economics of Water Resource Planning*, McGraw-Hill, New York, 1970.

S. H. Jenkins, (ed.), *Water Quality*, *Pergamon*, New York, 1973.

G. F. Lee, "Eutrophication," Univ. Wisconsin Water Res. Center Occas. Paper No. 2 (September 1970).

R. K. Linsley and J. B. Franzini, *Water Resources Engineering*, McGraw-Hill, New York, 1972.

R. Mitchell (ed.), *Water Pollution Microbiology*, Wiley-Interscience, New York, 1972.

Nat. Acad. Sci., *Eutrophication: Causes, Consequences, Correctives*, Nat. Acad. Sci., U.S.A., Washington, D.C., 1969.

R. T. Oglesby, C. A. Carlson, and J. A. McCann (eds.), *River Ecology and Man*. Academic, New York, 1972.

H. C. Periera, *Land Use and Water Resources in Temperate and Tropical Climates*, Cambridge Univ. Press, New York, 1973.

A. J. Rubin, *Chemistry of Water Supply, Treatment, and Distribution*, Ann Arbor Sci. Pub., Ann Arbor, Mich., 1974.

J. A. Salvato, Jr., *Environmental Engineering and Sanitation*, Wiley-Interscience, New York, 1972.

D. Seckler (ed.), *California Water*, Univ. Calif. Press, Berkeley, 1971.

R. D. Swisher, *Surfactant Biodegradation*, Dekker, New York, 1970.

G. Sykes and F. A. Skinner (eds.), *Microbial Aspects of Water Pollution*, Academic, New York, 1971.

T. H. Y. Tebutt, *Principles of Water Quality Control.*, Pergamon, New York, 1971.

C. J. Velz, *Applied Stream Sanitation*, Wiley-Interscience, New York, 1970.

C. E. Warren, *Biology and Water Pollution Control*, Saunders, Philadelphia, 1971.

G. F. White, *Strategies of American Water Management*, Univ. Mich. Press, Ann Arbor, 1969.

J. E. Zajic, *Water Pollution, Disposal, and Reuse* Dekker, New York, 1971.

D. Zwick and M. Benstock, *Water Wasteland*, Grossman, New York, 1971.

APPENDICES

Table A.1 Periodic Table of the Elements (From T. R. Dickson, *Introduction to Chemistry*, 2nd ed., Wiley, New York, 1971)

Periodic Table of the Elements

Key: Atomic No. / Symbol / Atomic Wt. (common oxidation numbers shown before the atomic number)

Representative Elements s block (IA, IIA); Transition Elements d block (IIIB–IIB); Representative Elements p block (IIIA–VIIA); Noble gases (O)

Periods	IA	IIA	IIIB	IVB	VB	VIB	VIIB	VIII	VIII	VIII	IB	IIB	IIIA	IVA	VA	VIA	VIIA	O
1	+1 1 H 1.0079																	2 He 4.003
2	+1 3 Li 6.941	+2 4 Be 9.012											+3 5 B 10.81	+4 +2 6 C 12.011	+5 +3 −3 7 N 14.007	−2 8 O 15.999	−1 9 F 18.998	10 Ne 20.18
3	+1 11 Na 22.99	+2 12 Mg 24.30											+3 13 Al 26.98	+4 +2 14 Si 28.08	+5 +3 −3 15 P 30.97	+6 +4 −2 16 S 32.06	+7 +5 +3 +1 −1 17 Cl 35.45	18 Ar 39.95
4	+1 19 K 39.10	+2 20 Ca 40.08	21 Sc 44.96	22 Ti 47.90	23 V 50.94	+6 +3 24 Cr 52.00	+5 +4 +2 25 Mn 54.94	+3 +2 26 Fe 55.85	+3 +2 27 Co 58.93	+3 +2 28 Ni 58.71	+2 +1 29 Cu 63.55	+2 30 Zn 65.38	+3 31 Ga 69.72	+4 +2 32 Ge 72.59	+5 +3 33 As 74.92	+6 +4 −2 34 Se 78.96	+7 +5 +3 +1 −1 35 Br 79.90	36 Kr 83.80
5	+1 37 Rb 85.47	+2 38 Sr 87.62	39 Y 88.91	40 Zr 91.22	41 Nb 92.91	42 Mo 95.94	43 Tc 98.91	44 Ru 101.07	45 Rh 102.91	46 Pd 106.4	+1 47 Ag 107.87	+2 48 Cd 112.40	+3 49 In 114.82	+4 +2 50 Sn 118.69	+5 +3 51 Sb 121.75	+6 +4 −2 52 Te 127.60	+7 +5 +3 +1 −1 53 I 126.90	54 Xe 131.30
6	+1 55 Cs 132.91	+2 56 Ba 137.34	57 La 138.91	72 Hf 178.49	73 Ta 180.95	74 W 183.85	75 Re 186.2	76 Os 190.2	77 Ir 192.22	78 Pt 195.09	79 Au 196.97	+2 +1 80 Hg 200.6	+3 +1 81 Tl 204.4	+4 +2 82 Pb 207.2	+5 +3 83 Bi 209.0	+6 +4 84 Po (210)	85 At (210)	86 Rn (222)
7	+1 87 Fr (223)	+2 88 Ra 226.0	89 Ac (227)	104 Ku*	105 Ha*													

Inner Transition Elements f block

Lanthanum Series	58 Ce 140.12	59 Pr 140.1	60 Nd 144.24	61 Pm (147)	62 Sm 150.4	63 Eu 151.96	64 Gd 157.2	65 Tb 158.93	66 Dy 162.50	67 Ho 164.93	68 Er 167.26	69 Tm 168.93	70 Yb 173.04	71 Lu 174.97
Actinium Series	90 Th 232.0	91 Pa 231.0	92 U 238.0	93 Np 237.0	94 Pu (242)	95 Am (243)	96 Cm (247)	97 Bk (247)	98 Cf (247)	99 Es (254)	100 Fm (253)	101 Md (256)	102 No (254)	103 Lr (257)

Mass numbers of the most stable or most abundant isotopes are shown in parentheses
The elements to the right of the bold lines are called the nonmetals and the elements to the left of the bold line are called the metals.
Common oxidation numbers are given for the representative elements and some transition elements

Table A.2A Solubility of Oxygen from a Wet Atmosphere at a pressure of 760 mm Hg (mg/liter) at Temperatures from 0 to 35°C (From G. E. Hutchinson, *A Treatise on Limnology*, Wiley, New York, 1957)

Temp.	0.0	0.1	0.2	0.3	0.4	0.5	0.6	0.7	0.8	0.9
0	14.16	14.12	14.08	14.04	14.00	13.97	13.93	13.89	13.85	13.81
1	13.77	13.74	13.70	13.66	13.63	13.59	13.55	13.51	13.48	13.44
2	13.40	13.37	13.33	13.30	13.26	13.22	13.19	13.15	13.12	13.08
3	13.05	13.01	12.98	12.94	12.91	12.87	12.84	12.81	12.77	12.74
4	12.70	12.67	12.64	12.60	12.57	12.54	12.51	12.47	12.44	12.41
5	12.37	12.34	12.31	12.28	12.25	12.22	12.18	12.15	12.12	12.09
6	12.06	12.03	12.00	11.97	11.94	11.91	11.88	11.85	11.82	11.79
7	11.76	11.73	11.70	11.67	11.64	11.61	11.58	11.55	11.52	11.50
8	11.47	11.44	11.41	11.38	11.36	11.33	11.30	11.27	11.25	11.22
9	11.19	11.16	11.14	11.11	11.08	11.06	11.03	11.00	10.98	10.95
10	10.92	10.90	10.87	10.85	10.82	10.80	10.77	10.75	10.72	10.70
11	10.67	10.65	10.62	10.60	10.57	10.55	10.53	10.50	10.48	10.45
12	10.43	10.40	10.38	10.36	10.34	10.31	10.29	10.27	10.24	10.22
13	10.20	10.17	10.15	10.13	10.11	10.09	10.06	10.04	10.02	10.00
14	9.98	9.95	9.93	9.91	9.89	9.87	9.85	9.83	9.81	9.78
15	9.76	9.74	9.72	9.70	9.68	9.66	9.64	9.62	9.60	9.58
16	9.56	9.54	9.52	9.50	9.48	9.46	9.45	9.43	9.41	9.39
17	9.37	9.35	9.33	9.31	9.30	9.28	9.26	9.24	9.22	9.20
18	9.18	9.17	9.15	9.13	9.12	9.10	9.08	9.06	9.04	9.03
19	9.01	8.99	8.98	8.96	8.94	8.93	8.91	8.89	8.88	8.86
20	8.84	8.83	8.81	8.79	8.78	8.76	8.75	8.73	8.71	8.70
21	8.68	8.67	8.65	8.64	8.62	8.61	8.59	8.58	8.56	8.55
22	8.53	8.52	8.50	8.49	8.47	8.46	8.44	8.43	8.41	8.40
23	8.38	8.37	8.36	8.34	8.33	8.32	8.30	8.29	8.27	8.26
24	8.25	8.23	8.22	8.21	8.19	8.18	8.17	8.15	8.14	8.13
25	8.11	8.10	8.09	8.07	8.06	8.05	8.04	8.02	8.01	8.00
26	7.99	7.97	7.96	7.95	7.94	7.92	7.91	7.90	7.89	7.88
27	7.86	7.85	7.84	7.83	7.82	7.81	7.79	7.78	7.77	7.76
28	7.75	7.74	7.72	7.71	7.70	7.69	7.68	7.67	7.66	7.65
29	7.64	7.62	7.61	7.60	7.59	7.58	7.57	7.56	7.55	7.54
30	7.53	7.52	7.51	7.50	7.48	7.47	7.46	7.45	7.44	7.43
31	7.42	7.41	7.40	7.39	7.38	7.37	7.36	7.35	7.34	7.33
32	7.32	7.31	7.30	7.29	7.28	7.27	7.26	7.25	7.24	7.23
33	7.22	7.21	7.20	7.20	7.19	7.18	7.17	7.16	7.15	7.14
34	7.13	7.12	7.11	7.10	7.09	7.08	7.07	7.06	7.05	7.05
35	7.04	7.03	7.02	7.01	7.00	6.99	6.98	6.97	6.96	6.95

* From Truesdale, Downing, and Lowden (1955).

Table A.2B Solubility of Oxygen in Seawater (ml/liter seawater from a dry atmosphere 0.2094 mole fraction O_2) (From R. A. Horne *Marine Chemistry*, Wiley-Interscience, New York, 1969)

Temp., °C	Chlorinity, ‰						
0	0	5	10	15	20	25	30
5	10.35	9.72	9.11	8.56	8.03	7.52	7.07
10	9.08	8.54	8.04	7.57	7.13	6.73	6.34
15	7.20	6.83	6.47	6.11	5.80	5.48	5.20
20	6.53	6.18	5.88	5.56	5.29	5.02	4.76
25	5.95	5.65	5.38	5.11	4.84	4.61	4.38
30	5.49	5.23	4.97	4.71	4.49	4.25	4.04
35	5.12	4.85	4.61	4.38	4.18	3.97	3.76

Table A.3 The Solubility of Carbon Dioxide in Seawater (in moles/liter × 10^4, STP) (Data selected from J. P. Riley and G. Skirrow, eds., *Chemical Oceanography*, Academic, London, 1965)

Chlorinity (Cl ‰)	0°C	4°C	10°C	16°C	20°C	26°C
0	770	662	536	442	394	332
15	674	578	472	393	351	299
16	667	573	468	390	348	297
17	660	567	464	387	346	294
18	653	562	460	384	343	292
19	646	557	456	381	340	289
20	640	551	452	377	337	287
21	633	546	448	374	335	285

Table A.4 The Solubility of Nitrogen in Seawater (ml N_2/ml H_2O, STP) (Data selected from E. Douglas, *J. Phys. Chem.*, *69*, 2608 (1965))

Chlorinity (Cl ‰)	0°C	4°C	10°C	16°C	20°C	26°C
15	0.0193	0.0177	0.0155	0.0140	0.0131	0.0120
16	0.0190	0.0174	0.0154	0.0138	0.0129	0.0119
17	0.0188	0.0172	0.0152	0.0136	0.0128	0.0117
18	0.0185	0.0170	0.0150	0.0135	0.0126	0.0116
19	0.0182	0.0167	0.0148	0.0133	0.0125	0.0115
20	0.0180	0.0165	0.0146	0.0131	0.0133	0.0114
21	0.0177	0.0163	0.0145	0.0130	0.0122	0.0113

Table A.5 Self-Dissociation Constant of Water

$$H_2O \rightleftharpoons H^+ + OH^-, K = (H^+)(OH^-)$$

Temperature (°C)	$K_W \times 10^{14}$
16	0.63
17	0.68
18	0.74
19	0.79
20	0.86
21	0.93
22	1.01
23	1.10
24	1.19
25	1.27
30	1.89
40	3.80
50	5.95
60	12.6
70	21.2
80	35
90	53
99	72

Table A.6 Thermal Dissociation of Water

Temperature (°K)	Percent Dissociation
1000	0.00003
1200	0.00081
1400	0.00861
1600	0.051
1800	0.199
2000	0.588
2200	1.42
2400	2.92

Table A.7 Phase Diagram for Water—Water System at High Pressures

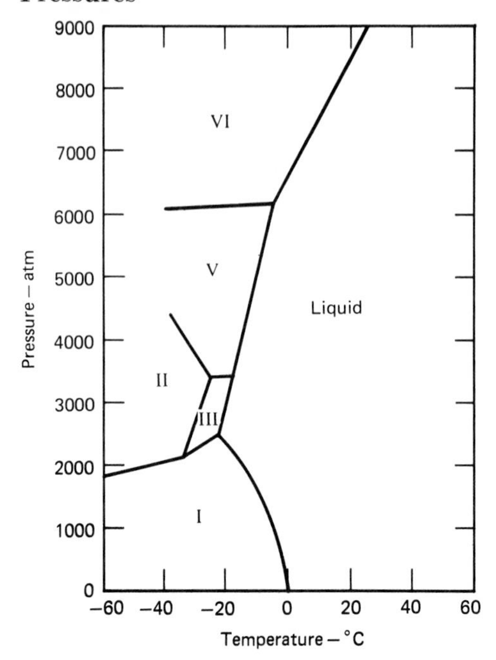

Table A.8 Difiusion Coefficients of Some Seawater Electrolytes at 25°C (From R. A. Horne, *Marine Chemistry*, Wiley-Interscience, New York, 1969)

Concentration, *M*	NaCl	KCl	CsCl	$CaCl_2$	$SrCl_2$	Na_2SO_4	$MgSO_4$
				$cm^2/sec \times 10^5$			
0.005	1.560	1.934	1.978	1.179	1.219	1.123	0.710
0.01	1.545	1.917	1.958	—	—	—	—
0.05	1.507	1.864	—	—	—	—	—
0.10	1.483	1.844	1.971	—	—	—	—
0.50	1.474	1.850	1.860	—	—	—	—
1.00	1.404	1.892	1.902	—	—	—	—
1.50	1.473	1.943	—	—	—	—	—

Table A.9 Gas Molecular Diffusion Coefficients in Pure Water at 1 atm (From R. A. Horne, *Marine Chemistry*, Wiley-Interscience, New York, 1969)

Gas	T	*D*
CO_2[a]	25°C	1.92×10^{-5} cm²/sec
N_2[a]	10	1.29
	25	2.01
	40	2.83
	55	3.80
O_2[a]	10	1.54
	25	2.20
	40	3.33
	55	4.50
O_2[b]	30	3.49
N_2[b]	30	3.47
H_2[b]	30	5.42

[a] From R. T. Ferrell and D. M. Himmelbau, *J. Chem. Eng. Data*, **12**, 111 (1967).
[b] From I. M. Kreiger, G. W. Mulholland, and C. S. Dickey, *J. Phys. Chem.*, **71**, 1123 (1967).
[c] These coefficients were determined from measurements in the rate of decrease in size of small gas bubbles.

Table A.10 Comparison of Transport Phenomena in Pure Water and in Seawater at 1 atm (From R. B. Montgomery in *American Institute of Physics Handbook*, D. E. Gray (ed.), 1957. Used with permission of McGraw-Hill Book Co.)

	Pure Water		Seawater Salinity 35/mille ‰	
Name, Symbol, Units	0°C	20°C	0°C	20°C
Dynamic viscosity, η, g/cm/sec = poise	0.01787	0.01022	0.01877	0.01075
Thermal conductivity, k, W/cm/°C	0.00566	0.00599	0.00563	0.00596
Kinematic viscosity, $\nu = \eta/\rho$, cm²/sec	0.01787	0.01004	0.01826	.01049
Thermal diffusivity, $\kappa = k/c_P\rho$ cm²/sec	0.00134	0.00143	0.00138	0.00149
Diffusivity, D, cm²/sec				
NaCl	0.0000074	0.0000141	0.0000068	0.0000129
N_2	0.0000106	0.0000169	—	—
O_2		0.000021	—	—
Prandtl number, $N_P = \nu/\kappa$	13.3	7.0	13.1	7.0

Table A.11 Osmotic Pressure P, Freezing Point ϑ, Boiling-Point Elevation Δt_s, and Vapor-Pressure Lowering b as a Function of Salinity (From G. Dietrich, *General Oceanography*, Wiley, New York, 1963)

S‰	P_0, atm.	P_{20}, atm.	ϑ, °C.	Δt_s, °C.	b, mm. Hg
4	2.59	2.78	−0.214	0.06	1.69
8	5.16	5.54	−0.427	0.12	3.39
12	7.73	8.30	−0.640	0.19	5.08
16	10.34	11.10	−0.856	0.25	6.82
20	12.97	13.92	−1.074	0.31	8.55
24	15.63	16.78	−1.294	0.38	10.30
28	18.31	19.65	−1.516	0.44	12.13
32	21.02	22.56	−1.740	0.51	13.97
36	23.76	25.50	−1.967	0.57	15.79
40	26.53	28.48	−2.196	0.64	17.73

Table A.12 The Sizes of the World's Oceans and Seas

Body of Water		Area (10^6 km^2)	Volume (10^6 km^3)	Mean Depth (m)
Pacific Ocean	including	174.68	723.70	4028
Atlantic Ocean	adjacent	106.46	354.68	3332
Indian Ocean	seas	74.92	291.95	3897
Total		361.06	1370.32	3795
Pacific Ocean	excluding	165.25	707.56	4282
Atlantic Ocean	adjacent	82.44	323.61	3926
Indian Ocean	seas	73.44	291.03	3963
Total		321.13	1322.20	4117
Baltic Sea		0.42	0.02	55
Red Sea		0.44	0.22	491
North Sea		0.58	0.05	94
Irish Sea		0.10	0.006	60
Bering Sea		2.27	3.26	1437
Mediterranean and Black Seas		2.97	4.24	1429

Table A.13 Temperature of Maximum Density of Seawater

S (‰)	T_m (°C)	S (‰)	T_m (°C)
0	+3.947	25	−1.398
5	+2.926	30	−2.473
10	+1.860	35	−3.524
15	+0.772	40	−4.451
20	−0.310		

Table A.14 Specific Volume of Seawater (cm^3/gm) (From R. A. Horne, *Marine Chemistry*, Wiley-Interscience, New York, 1969)

		0°C	10°C	20°C
Pure Water	1 bar	1.0000	1.0003	1.0017
	500 bar	0.9767	0.9781	0.9804
	1000 bar	0.9567	0.9590	0.9618
10‰ Cl	1 bar	0.9921	0.9926	0.9942
	500 bar	0.9695	0.9711	0.9735
	1000 bar	0.9501	0.9526	0.9554
20‰ Cl	1 bar	0.9842	0.9850	0.9868
	500 bar	0.9623	0.9642	0.9666
	1000 bar	0.9435	0.9461	0.9491
30‰ Cl	1 bar	0.9764	0.9775	0.9794
	500 bar	0.9552	0.9572	0.9598
	1000 bar	0.9370	0.9397	0.9428
40‰ Cl	1 bar	0.9687	0.9770	0.9720
	500 bar	0.9481	0.9504	0.9530
	1000 bar	0.9305	0.9334	0.9365

Table A.15 Velocity of Sound in Seawater[a] (From W. D. Wilson, *J. Acoust. Soc. Amer.*, *32*, 641 (1960) with permission of the American Institute of Physics and the author)

$T = -3°C$	0°C	5°C	10°C	20°C
$\Delta V_T = 14.37$	0.00	+21.79	+41.05	+72.81
$P = 1$ bar	200 bar	400 bar	600 bar	800 bar
$\Delta V_P = +0.16$	+32.64	+65.98	+100.30	+135.33
$S = 33‰$ S	34‰ S	35‰ S	36‰ S	37‰ S
$\Delta V_S = -3.09$	−1.47	−0.00	+1.31	+2.47
$P = 1$ bar	200 bar	400 bar	600 bar	800 bar
$\Delta V_{STP} = 0.00$[b]	+0.31	+0.73	+1.30	+2.05

[a] The velocity of sound V in seawater is given (m/sec) by the empirical equation:

$$V = 1449.22 + \Delta V_T + \Delta V_P + \Delta V_S + \Delta V_{STP}$$

[b] At 35‰ S.

Table A.16 Abundance of the Elements (From V. M. Goldschmidt, *Geochemistry*, Clarendon Press, Oxford, 1954 with permission of the publishers)

		Lithosphere		Meteorites			Stellar
Z	Element	ppm	Atoms per 100 Sl	ppm	Atoms per 100 Sl	Solar Atmosphere Atoms per 100 Sl	Atmosphere Atoms per 100 Sl
1	H	—	—	—	—	1,550,000	—
2	He	—	—	—	—	—	—
3	Li	65	0.1	4	0.01	0.0003	—
4	Be	6 (S)[a]	0.0067	1	0.0020	0.0002	—
5	B	10	0.0095	1.5	0.0024	0.32 :	—
6	C	320	0.27	300	0.33	1,000	—
7	N	—	—	—	—	2,100	—
8	O	466,000	200	323,000	347	2,800	—
9	F	800	0.43	3.3	0.03	0.32 :	—
10	Ne	—	—	—	—	—	—
11	Na	28,300	12.4	5,950	4.42	98	100
12	Mg	20,900	8.76	123,000	87.24	170	71
13	Al	81,300	30.5	13,800	8.79	11	65
14	Si	277,200	100	163,000	100	100	—
15	P	1,200	0.391	1,050	0.58	0.03 :	—
16	S	520	0.16	21,200	11.4	43	—
17	Cl	480	0.14	1,000–1,500 ?	0.4–0.6 ?	—	—
18	A	—	—	—	—	—	—
19	K	25,900	6.70	1,540	0.69	0.81	2.2
20	Ca	36,300	9.17	13,300	5.71	8.7	50
21	Sc	5 (S)	0.0011	4	0.0015	0.011	—
22	Ti	4,400	0.92	1,320	0.47	0.47	1.3
23	V	150	0.030	39	0.013	0.058	0.3
24	Cr	200	0.039	3,430	1.13	1.95	0.8
25	Mn	1,000	0.18	2,080	0.66	1.48	2.1
26	Fe	50,000	9.13	288,000	89.1	270	12
27	Co	40	0.0069	1,200	0.35	0.55	—

28	Ni	100	0.0175	15,680	4.60	4.6	—
29	Cu	70	0.010	170	0.046	0.090	—
30	Zn	80	0.0124	138	0.0362	0.31	0.69
31	Ga	15	0.0022	35.6	0.0088	0.0003 :	—
32	Ge	7	0.00097	55	0.0130	0.003	—
33	As	5	0.00067	detected	—	0.000019	—
34	Se	0.09	0.000012	7	0.0015	—	—
35	Br	2.5	0.00032	20	0.0043	—	—
36	Kr	—	—	—	—	—	—
37	Rb	280	0.033	0.45	0.0007	0.0002	—
38	Sr	150	0.017	20	0.0004	0.011	0.035
39	Y	28.1 (S)	0.00307	4.72	0.000974	0.000974	—
40	Zr	220	0.026	73	0.0139	0.0012	—
41	Nb	29	0.002	—	—	0.00003	—
42	Mo	2.3	0.00024	5.3	0.00095	0.0003	—
43	Ma	—	—	—	—	—	—
44	Ru	—	—	2.23	0.00036	0.00016	—
45	Rh	0.001	0.0000001	0.80	0.00013	0.00001	—
46	Pd	0.010	0.0000009	1.54	0.00025	0.00004	—
47	Ag	0.02	0.0000018	2	0.00032	0.00003	—
48	Cd	0.18	0.000016	2.4	—	0.0005 :	—
49	In	0.1	0.0000071	0.15	0.000023	0.000003 ?	—
50	Sn	40	0.00343	20	0.0029	0.00002	—
51	Sb	(1)	0.000083	detected	—	0.00002 :	—
52	Te	(0.0018) ?	—	(0.1) ?	—	—	—
53	I	0.3	0.000024	1	0.000136	—	—
54	Xe	—	—	—	—	—	—
55	Cs	3.2	0.0006	0.01	0.00001	?	—
56	Ba	430	0.0312	0.9	0.00083	0.00046	0.0082
57	La	18.3 (S)	0.00128	1.58	0.000208	0.002	—
58	Ce	41.6 (S)	0.00321	1.77 (?)	0.000232 (?)	0.0008	—
59	Pr	5.53 (S)	0.000389	0.75	0.000096	0.00001 :	—
60	Nd	23.9 (S)	0.00162	2.59	0.000331	—	—
61	—	—	—	—	—	—	—

(*Continued*)

Table A.16 (*Continued*)

Z	Element	Lithosphere		Meteorites		Solar Atmosphere[b]	Stellar Atmosphere-
		ppm	Atoms per 100 Sl	ppm	Atoms per 100 Sl	Atoms per 100 Sl	Atoms per 100 Sl
62	Sm	0.47 (S)	0.000410	0.95	0.000115	0.0001	—
63	Eu	1.06 (S)	0.000068	0.25	0.000028	0.00008 :	—
64	Gd	0.36 (S)	0.000394	1.42	0.000165	0.00004 :	—
65	Tb	0.91 (S)	0.000056	0.45	0.000052	—	—
66	Dy	4.47 (S)	0.000269	1.80	0.000203	0.001 :	—
67	Ho	1.15 (S)	0.000068	0.51	0.000057	—	—
68	Er	2.47 (S)	0.000144	1.48	0.000163	0.000004 :	—
69	Tm	0.20 (S)	0.0000115	0.26	0.000029	0.00001 :	—
70	Yb	2.66 (S)	0.000149	1.42	0.000150	0.00003 :	—
71	Lu	0.75 (S)	0.000037	0.46	0.000048	0.00003 :	—
72	Hf	4.5	0.00030	1.6	0.00015	0.000008 :	—
73	Ta	2.1	0.000117	—	—	0.000003 :	—
74	W	1	0.000055	15	0.00145	0.000005 :	—
75	Re	0.001	0.000000054	0.0020	0.00000018	—	—
76	Os	—	—	1.92	0.000174	0.00001 :	—
77	Ir	0.001	0.00000005	0.65	0.000058	0.000002 ?	—
78	Pt	0.005	0.00000027	3.25	0.000287	0.001	—
79	Au	0.001	0.00000005	0.7	0.000057	—	—
80	Hg	0.5 (S)	0.000025	—	—	—	—
81	Tl	0.3	0.000015	detected	0.000015	—	—
82	Pb	16	0.00080	11	0.00091	0.0018	—
83	Bi	0.2	0.000009	detected	—	—	—
90	Th	11.5 (S)	0.00050	0.8	0.00059	—	—
92	U	4	0.00016	0.36	0.000023	—	—

[a] (S) indicates sediments, colon (:) signifies doubtful value, (?) very doubtful.

Table A.17 Glossary of Some Common Mineral Terms Used in the Text

Amphibole	$Na_3(Fe, Mg)_5Si_8O_{22}(OH, F)$
Amphibolite	Metamorphic rock derived from argillaceous limestones
Analcite	$NaAlH_2Si_2O_7$
Andalusite	Al_2SiO_5, an aluminum ore
Andesine	$NaCaAl_3Si_5O_{16}$
Andesite	Andesine, hornblende, and mica
Anorthite	$CaAl_2Si_2O_8$, an aluminum ore
Anorthosite	Igneous gabbro rock composed mostly of feldspar
Anthracite	C, hard coal
Apatite	$Ca_5(PO_4)_3OH$
Argonite	$CaCO_3$
Armalcolite	$(Fe, Mg)Ti_2O_5$
Asbestos	Fibrous form of minerals such as chrysolite
Augite	$CaMg_2Al_2Si_3O_{10}$
Baddelyite	ZrO_2
Barite	$BaSO_4$
Basalt	Compact igneous rock composed mostly of particles of feldspar, augite, and iron minerals
Bauxite	$Al_2O_3 \cdot 2H_2O$, a major aluminum ore
Bentonite	Sodium montmorillonite clay
Biotite	A ferrous mica
Boehmite	$HALO_2$, an aluminum ore
Bronzite	$MgSiO_3$, rhombic pyroxene with 10% $FeSiO_3$
Calcite	$CaCO_3$
Carnallite	$KCl \cdot MgCl_2 \cdot 6H_2O$
Chalcopyrite	$CuFeS_2$
Chalcocite	Cu_2S
Chamosite	$Fe_2Al_2SiO_5(OH)_4$
Chert	SiO_2
Chlorite	Clinochlore, penninite, prochlorite
Chrysolite	Olivine
Chrysotile	$Mg_3Si_2O_5(OH)_4$
Clay	Hydrous silicate of alumina formed by the decomposition of feldspar and other aluminum minerals
Clinopyroxene	$Ca(Mg, Fe)Si_2O_6$
Cohenite	Fe_3C
Cristobalite	SiO_2
Cryolite	Na_3AlF_6
Diamond	C
Diaspore	$HAlO_2$, an aluminum ore
Diopside	$CaMgSi_2O_6$ containing Cr
Diorite	Igneous rock containing quartz, plagioclase, and ferric minerals
Dolomite	$CaMg(CO_3)_2$
Dunite	Peridotite rock consisting of chrysolite and olivine
Dysanalyte	$(Ca, Na, Fe, Ce)(Nb, Ti)O_3$
Eclogite	Augite and hornblende
Enstatite	$MgSiO_3$
Epidote	$HCa_2(Al, Fe)_3Si_3O_{16}$
Evaporite	Salts crystallized upon aqueous evaporation

(Continued)

Table A.17 (*Continued*)

Feldspar	K_2O, Al_2O_3, $6SiO_2$
Ferropseudobrookite	A lunar chromium–titanium spinel
Fluoropatite	$Ca_5(PO_4)_3F$
Gabbro	Group of igneous rocks consisting of plagioclase and pyroxenes
Galena	PbS
Garnet	$(Ca, Mg, Fe, Mn)_3(Al, Fe, Cr)_2Si_3O_{12}$
Gibbsite	H_3AlO_3, an aluminum ore
Glauconite	Fe, K, Al, Mg, Ca silicate formed from ocean sediments
Gneiss	Group of crystalline metamorphic rocks consisting typically of quartz or feldspar
Goethite	$HFeO_2$, an iron ore
Granite	Crystalline igneous rock consisting of quartz, orthoclase, muscovite and biotite formed by cooling under great pressure.
Graphite	C
Gypsum	$CaSO_4$ hydrate
Hematite	Fe_2O_3, an iron ore
Hornblende	$Ca(MgF_3)(SiO_3)_4$
Hydrotroilite	FeS
Hydroxyapatite	$Ca_5(PO_4)_3(OH)$
Hypersthene	(Fe, Mg)O, SiO_2
Ilmenite	$FeTiO_3$
Jadeite	$3MgO \cdot CaO \cdot 2SiO_2$, jade
Kamacite	Fe
Kaolinite	$Al_2Si_5O_5(OH)_4$, an aluminum ore
Kyanite	Al_2SiO_5, an aluminum ore
Lignite	Soft, coal-like material
Limestone	$CaCO_3$
Limonite	$HFeO_2$, an iron ore
Magnetite	Fe_3O_4, an iron ore
Marble	Recrystallized $CaCO_3$
Mica	$K_2(Mg, Fe)_5(Al, Ti)(Si, Al)_8O_{20}(OH)_4$
Montmorillonite	$(Na\ or\ K)_{0.33}Al_{2.33}Si_{3.67}O_{10}(OH)_2$
Nepheline	An aluminum ore consisting of nephelite and pyroxene
Nephelite	20% $KAlSiO_4$; 75% $NaAlSiO_4$; 5% $NaAlSi_3O_8$
Olivine	$(Mg, Fe)_2SiO_4$
Opal	$SiO_2 \cdot xH_2O$, amorphous
Orthoclase	$KAlSi_3O_8$
Peridotite	Igneous rock composed mostly of olivine
Perovskite	$CaTiO_3$
Phillipsite	Complex clay containing Na, K, and Ca
Pitchblende	A uranium ore
Plagioclase	$CaAl_2Si_2O_8$–$NaAlSi_3O_8$
Polyhalite	K_2SO_4–$MgSO_4 \cdot 2CaSO_4 \cdot 2H_2O$
Porhyry	An igneous rock
Pumice	Porous volcanic stone consisting mostly of silicates
Pyrite	FeS_2
Pyromanganite	A lunar triclinic pyroxene-like mineral
Pyroxene	CaO, MgO, $2SiO_2$
Pyroxferroite	$(Fe, Ca)_2Si_2O_6$

(*Continued*)

Table A.17 (*Continued*)

Quartz	SiO_2
Rutile	TiO_2
Sanidine	Glassy form of orthoclase
Schist	Crystalline rock which splits into flakes
Schreibersite	$(Fe, Ni)_3P$
Serpentine	$Mg_3Si_2O_7 \cdot 2H_2O$
Shale	Fine-grained sedimentary rock
Siderite	$FeCO_3$, an iron ore
Siderolite	Spongy iron meteorite with silicate mineral grains
Sillimanite	Al_2SiO_5, an aluminum ore
Sphalerite	ZnS
Spinel	$MgAl_2O_4$
Syenite	Granular igneous rock
Sylvite	KCl
Taconite	Low-grade iron ore
Taenite	Fe, Ni
Talc	$3MgO \cdot 4SiO_2 \cdot H_2O$
Tektite	Glassy spherules of controversial origin
Tetrahedrite	$Cu_{12}Sb_4S_3$
Tourmaline	Aluminum silicates containing B, Li, and so on, and often colored
Tridymite	SiO_2
Troilite	FeS
Ulvöspinel	Fe_2TiO_4
Uraninite	Pitchblende
Vermiculite	$3MgO(Fe, Al)_2O_3$
Whitlockite	$Ca_3(PO_4)_2$
Wollastonite	$CaSiO_3$
Zircon	$ZrSiO_4$

Table A.18 The Average Amounts of the Elements in Crustal Rocks in Grams per Ton or Parts per Million (omitting the rare gases and the short-lived radioactive elements) (From B. Mason, *Principles of Geochemistry*, Wiley, New York, 1966 (3rd ed.)

Atomic Number	Element	Crustal Average	Granite (G-1)	Diabase (W-1)
1	H	1,400	400	600
3	Li	20	24	12
4	Be	2.8	3	0.8
5	B	10	2	17
6	C	200	200	100
7	N	20	8	14
8	O	466,000	485,000	449,000
9	F	625	700	250
11	Na	28,300	24,600	15,400
12	Mg	20,900	2,400	39,900
13	Al	81,300	74,300	78,600
14	Si	277,200	339,600	246,100
15	P	1,050	390	650
16	S	260	175	135
17	Cl	130	50	
19	K	25,900	45,100	5,300
20	Ca	36,300	9,900	78,300
21	Sc	22	3	34
22	Ti	4,400	1,500	6,400
23	V	135	16	240
24	Cr	100	22	120
25	Mn	950	230	1,320
26	Fe	50,000	13,700	77,600
27	Co	25	2.4	50
28	Ni	75	2	78
29	Cu	55	13	110
30	Zn	70	45	82
31	Ga	15	18	16
32	Ge	1.5	1.0	1.6
33	As	1.8	0.8	2.2
34	Se	0.05		
35	Br	2.5	0.5	0.5
37	Rb	90	220	22
38	Sr	375	250	180
39	Y	33	13	25
40	Zr	165	210	100
41	Nb	20	20	10
42	Mo	1.5	7	0.05
44	Ru	0.01		

(Continued)

Table A.18 (cont.) The Average Amounts of the Elements in Crustal Rocks in Grams per Ton or Parts per Million

Atomic Number	Element	Crustal Average	Granite (G-1)	Diabase (W-1)
45	Rh	0.005		
46	Pd	0.01	0.01	0.02
47	Ag	0.07	0.04	0.06
48	Cd	0.2	0.06	0.3
49	In	0.1	0.03	0.08
50	Sn	2	4	3
51	Sb	0.2	0.4	1.1
52	Te	0.01		
53	I	0.5		
55	Cs	3	1.5	1.1
56	Ba	425	1,220	180
57	La	30	120	30
58	Ce	60	230	30
59	Pr	8.2	20	2
60	Nd	28	55	15
62	Sm	6.0	11	5
63	Eu	1.2	1.0	1.1
64	Gd	5.4	5	4
65	Tb	0.9	1.1	0.6
66	Dy	3.0	2	4
67	Ho	1.2	0.5	1.3
68	Er	2.8	2	3
69	Tm	0.5	0.2	0.3
70	Yb	3.4	1	3
71	Lu	0.5	0.1	0.3
72	Hf	3	5.2	1.5
73	Ta	2	1.6	0.7
74	W	1.5	0.4	0.45
75	Re	0.001	0.0006	0.0004
76	Os	0.005	0.0001	0.0004
77	Ir	0.001	0.006	
78	Pt	0.01	0.008	0.009
79	Au	0.004	0.002	0.005
80	Hg	0.08	0.2	0.2
81	Tl	0.5	1.3	0.13
82	Pb	13	49	8
83	Bi	0.2	0.1	0.2
90	Th	7.2	52	2.4
92	U	1.8	3.7	0.52

Table A.19 Geological Time Scale

Era Period Epoch	Began (yr ago × 10^6)	Duration (yr) 10^6	Events
Cenzoic			
Quaternary			
Recent	0.011		Modern man
Pleistocene	0.6	2	Early man
Tertiary			
Pliocene	12	5	Large carnivores
Miocene	20	19	Whales, apes, grazing forms
Oligocene	35	11	Large browsing animals
Eocene	55	16	Flowering plants
Paleocene	65	12	First placental animals
Mesozoic			Extinction of dinosaurs, flora with modern aspects
Cretaceous			
Upper	90		
(Base Santonian)			
Middle	120		
(Base Albian)			
Lower	140		
Jurassic		45	Dinosaurs zenith, primitive birds, and first small mammals
Upper	155		
(Base Callovian)			
Middle	170		
(Base Bajocian)			
Lower	185		
Triassic		45	Appearance of dinosaurs
Upper	200		
Middle	215		
Lower	230		
Paleozoic			
Permian		45	Conifers abundant and reptiles developed
Upper	245		
(Base Ochoan)			
Middle	260		
(Base Guadalupian)			
Lower	275		

(Continued)

Table A.19 (*Continued*)

Era Period Epoch		Began (yr ago × 10^6)	Euration (yr) 10^6	Events
Pennsylvanian	Carboniferous		35	First reptiles, great forests from which coal was later formed
Upper				
Middle				
Lower				
Mississippian		310	40	Sharks abundant
Upper				
Middle				
Lower		350		
Devonian			50	Fish abundant, amphibians appear
Upper		365		
Middle		385		
Lower		400		
Silurian			20	Earliest land plants and animals
Upper				
Middle				
Lower		420		
Ordovician			70	First primitive fish
Upper		440		
Middle		460		
Lower		490		
Cambrian			50+	Large fauna of marine invertebrates
Upper		510		
(Base Croixian)				
Middle		540		
Lower				
Pre-Cambrian				
(Proterozoic)		about 3500		Plants and animals with soft tissues, few fossils
(Archaean)		about 4500		Chemical evolution
		about 5000		Origin of Earth

Table A.20 Estimated Annual World Consumption of the Elements (in tons unless otherwise stated)

Element	Principal Sources	Amount of Element	Price of Element (U.S. $ per ton)
H	water, methane	2×10^6	150
He	natural gas	3500	7000
Li	petalite, lepidolite, spodumene, lake brines	1500	20/kg (Li) 1000 ($LiCO_3$)
Be	beryl	300	120/kg
B	Na, Ca borates, brines	400,000 (B_2O_3)	330 (B_2O_3)
C	diamond	6 (industrial)	
	graphite	300,000	
	coal	1.8×10^9	
	petroleum	0.9×10^9	
	natural gas	0.4×10^9	
N	air	1.7×10^7	44 ($NaNO_3$)
	soda niter	1×10^6 ($NaNO_3$)	
O	air	2×10^7	30
F	fluorite	1.1×10^6	
Ne	air		
Na	halite	1.0×10^8 (NaCl)	400
Mg	sea water, magnesite	150,000 (Mg) 9×10^6 ($MgCO_3$)	750
Al	bauxite	6.1×10^6	560
Si	quartz	700,000	350
P	apatite (phosphorite)	7×10^6	1000
S	sulfur, pyrite, natural gas	2.0×10^7	27
Cl	halite	5×10^6	60
Ar	air	20,000	400
K	sylvite, carnallite	1.0×10^7	20/kg (K) 25 (KCl)
Ca	calcite	6.0×10^7 (CaO)	14 (CaO)
Sc	thortveitite	50kg (Sc_2O_3)	2000/kg (Sc_2O_3)
Ti	ilmenite, rutile	10,000 (Ti) 1×10^6 (TiO_2)	2500 (Ti) 500 (TiO_2)
V	U, Pb vanadates	7000	7000
Cr	chromite	1.4×10^6	2500
Mn	pyrolusite, psilomelane	6×10^6	700
Fe	hematite, magnetite	3.1×10^8	50
Co	Co sulfides, arsenides	13,000	3000
Ni	pentlandite, garnierite	400,000	1800
Cu	chalcopyrite, chalcocite	5.4×10^6	800
Zn	sphalerite	3.8×10^6	300
Ga	bauxite	10	1000/kg
Ge	germanite	100	250/kg
As	arsenopyrite, enargite	40,000	1000 (As) 80 (As_2O_3)
Se	byproduct of Cu smelting	1000	9000
Br	sea water, brines	110,000	500

(Continued)

Table A.20 (cont.) Estimated Annual World Consumption of the Elements

Kr	air		
Rb	pollucite, salt deposits		1000/kg
Sr	celestite	8000	250 ($Sr(NO_3)_2$)
Y	monazite, euxenite	5 (Y_2O_3)	60/kg (Y_2O_3)
Zr	zircon	1000 (Zr) 200,000 (zircon)	10/kg (Zr)
Nb	columbite, pyrochlore	1300	90/kg
Mo	molybdenite	45,000	7000
Ru	platinum ores	120 kg	1500/kg
Rh	platinum ores	3000 kg	4000/kg
Pd	platinum ores	24	900/kg
Ag	silver sulfides	8000	40/kg
Cd	sphalerite	13,000	6000
In	byproduct of Zn and Pb smelting	10	50/kg
Sn	cassiterite	190,000	5000
Sb	stibnite	60,000	1100
Te	byproduct of Zn and Pb smelting	200	12/kg
I	brines, byproduct of Chilean nitrate	4000	2000
Cs	pollucite	1000 kg	200/kg
Ba	barite	3.2×10^6 (barite)	100 ($BaCO_3$)
La-Lu	monazite, bastnasite	2000 (oxides)	8000
Hf	zircon	50	150/kg
Ta	tantalite	300	60/kg
W	scheelite, wolframite	30,000	6000
Re	molybdenite	1000 kg	1300/kg
Os	platinum ores	60 kg	1600/kg
Ir	platinum ores	3000 kg	2500/kg
Pt	platinum ores	30	4000/kg
Au	gold, gold tellurides	1600	1150/kg
Hg	cinnabar	9000	15/kg
Tl	byproduct of Zn and Pb smelting	30	17/kg
Pb	galena	2.8×10^6	440
Bi	byproduct of Pb smelting	3000	4500
Th	monazite, byproduct of U extraction	50	50/kg
U	uraninite	30,000 (U_3O_8)	16/kg (U_3O_8)

(Data mainly from 1964 Minerals Yearbook)

Table A.21 Production of Most Major Chemical Products Is Rising Strongly (Data selected from *Chem. Engr. News*, December 18, 1972)

	1970	1973
Inorganic Chemicals		
(thousands of short tons unless otherwise noted)		
Ammonia, Synthetic Anhydrous	13,824	15,450
Ammonium Nitrate, Original Solution	6,456	7,450
Carbon Dioxide, Solid, Liquid, and Gas	1,135	1,450
Chlorine Gas	9,764	10,450
Hydrochloric Acid	2,014	2,350
Nitric Acid	6,679	7,420
Nitrogen (billions of cubic feet)	151.2	190
Oxygen (billions of cubic feet)	283.9	360
Phosphoric Acid, 100% P_2O_5	5,683	6,700
Sodium Carbonate, Synthetic	4,393	4,400
Sodium Hydroxide	10,141	10,800
Sulfuric Acid	29,525	32,700
Titanium Dioxide	655	710
Organic Chemicals		
(millions of pounds unless otherwise noted)		
Acetic Anhydride	1,589	1,580
Acetone	1,615	2,100
Acrylonitrile	1,039	1,200
Benzene (millions of gallons)	1,134	1,500
Butadiene	3,101	4,300
Carbon Tetrachloride	1,011	1,050
Cyclohexane	1,841	2,500
Ethylene	18,089	22,500
Ethylene Oxide	3,865	4,200
Formaldehyde, 37%	4,427	5,500
Methanol, Synthetic	4,932	6,100
Perchloroethylene	707	810
Phenol, Synthetic	1,708	2,080
Phthalic Anhydride	734	960
Propylene	6,641	8,500
Propylene Oxide	1,179	1,650
Styrene	4,335	6,050
Toluene (millions of gallons)	830	960
Urea	6,500	7,500
Vinyl Acetate	803	1,350
o-Xylene	799	850
p-Xylene	1,590	2,350

(Continued)

Table A.21 (*Continued*)

	1970	1973
Plastics (millions of pounds)		
Phenolics	1,186	1,600
Polyethylene, Low-Density	4,240	5,650
Polyethylene, High-Density	1,604	2,700
Polypropylene	1,031	2,100
Polystyrene and Copolymers	3,550	5,200
Polyvinyl Chloride and Copolymers	3,115	4,750
Man-Made Fibers (millions of pounds)		
Acrylic and Modacrylic	492	670
Cellulose Acetate	498	425
Nylon	1,355	2,150
Polyester, Staple and Tow	1,022	1,450
Rayon	875	950
Olefin and Vinyon	262	470

Table A.22 Relative Elemental Composition of Some Marine Animals (Na = 100) and Enrichment Factors (in Parentheses) Compared to Seawater

	Copepod	Fish	Nudibranch
Ca	7.4 (2)	52 (14)	262 (69)
C	1113 (4,300)	4100 (15,800)	480 (1850)
Cl	194 (1)	—	180 (1)
Cu	—	0.008 (80)	0.4 (4300)
Fe	1.3 (6000)	1.3 (6000)	0.2
I	0.04 (80)	—	—
K	54 (15)	383 (109)	20 (6)
Mg	5.6 (0.5)	36 (3)	156 (13)
N	280 (280,000)	1276 (1,276,000)	107 (107,000)
P	24 (241,000)	256 (2,560,000)	6 (60,000)
Si	1.3 (13,000)	—	—
S	26 (3)	259 (31)	7 (0.9)

Table A.23 Concentrations of Trace Elements in Individual Organs of Shellfish[a] (Data selected from M. G. Rumsby, *Limnol. Oceanogr.*, *10*, 521 (1965))

Sample	Percent of Whole Animal	Mn	Pb	Cr	Mo	V	Cu	Ag	Cd	Zn	Ni	Fe	Sb
Scallop													
Mantle	13	45	<5	<3	1.8	7	15	0.2	<20	<100	<2	1,540	<30
Gills	10	353	52	145	3.1	3	36	1.0	<20	<100	68	21,600	<30
Muscle	24	2	<5	<3	<0.1	<2	1	<0.1	<20	108	<2	34	<30
Visceral Mass[b]	17	24	8	8	2.0	30	24	1.8	2000	400	2	2,200	<30
Intestine[b]	1	435	28	24	3.6	16	131	2.9	<20	392	52	6,090	<30
Kidney	1	2660	137	17	3.4	4	78	4.8	<20	2630	106	2,470	<30
Foot	1	27	14	8	0.4	4	17	1.1	<20	210	22	2,380	<30
Gonads	20	5	78	<3	<0.1	32	9	0.2	<20	256	<2	228	<30
Shell	—	1	<5	<3	<0.1	130	2	<0.1	<20	<100	<2	2,000	<30
Sediment	—	693	<5	307	1.5	84	102	<0.1	<20	<100	219	73,000	<30
Seawater (ppb)	—	2	3	0.05	10	2	3	0.3	0.11	10	0.5	10	0.5

[a] All concentrations in ppm on material dried at 110°C.
[b] Includes the gut content.

Table A.24 1971–1972 United States Production of Major Industrial Chemicals and Synthetic Materials[a] (Selected from *Chem. Eng. News*, May 7, 1973)

Rank 1972	Rank 1971[a]		Production (billions of pounds) 1972	Production (billions of pounds) 1971
1	1	Sulfuric Acid, total	60.09	58.84
2	3	Oxygen, high and low purity	29.40	28.00
3	2	Ammonia, synthetic anhydrous	28.60	28.05
4	4	Sodium Hydroxide, 100% liquid	20.53	19.33
4	6	Ethylene	20.53	18.30
6	5	Chlorine, gas	19.74	18.70
7	7	Sodium Carbonate, synthetic and natural	14.86	14.31
8	8	Nitric Acid, total	14.04	13.48
9	9	Ammonium Nitrate, original solution	13.74	13.21
10	11	Nitrogen, high and low purity	13.40	11.98
11	10	Phosphoric Acid, total	12.53	12.48
12	12	Benzene, all grades	8.95	10.32
13	13	Ethylene Dichloride	8.84	7.15
14	17	Propylene	7.96	5.56
15	14	Polyethylene, high- and low-density	7.63	6.40
16	16	Urea, primary solution	7.02	6.14
17	19	Ethylbenzene	6.65	4.98
18	15	Toluene, all grades	5.96	6.21
18	18	Methanol, synthetic	5.96	5.01
20	21	Styrene	5.87	4.41
21	23	Xylene, all grades	5.81	4.35
22	22	Formaldehyde, 37% by weight	5.50	4.37
23	25	Vinyl Chloride	5.18	4.19
24	26	Polystyrene and Copolymers	4.60	3.75
25	24	Hydrochloric Acid, total	4.40	4.20
26	28	Polyvinyl Chloride and Copolymers	4.29	3.44
27	27	Ethylene Oxide	4.08	3.61
28	29	Butadiene (1,3-), rubber grade	3.81	3.06
29	20	Ammonium Sulfate	3.73	4.64
30	30	Ethylene Glycol	3.30	3.03
31	30	Carbon Black	3.21	3.03
32	36	Cumene	2.75	2.14
33	32	Sodium Sulfate, high and low purity	2.73	2.71
34	33	Carbon Dioxide, all forms	2.69	2.54
35	42	Cyclohexane	2.29	1.73
36	35	Aluminum Sulfate, commercial	2.25	2.39
37	38	Acetic Acid	2.15	2.05
38	37	Sodium Tripolyphosphate	2.06	2.08

(*Continued*)

Table A.24 (*Continued*)

Rank 1972	Rank 1971[a]		Production (billions of pounds) 1972	Production (billions of pounds) 1971
39	40	Dimethyl Terephthalate	2.00	1.74
40	40	Phenol, total	1.93	1.74
41	34	Calcium Chloride, solid and liquid	1.90	2.43
42	44	Ethanol, synthetic	1.87	1.64
43	39	Isopropanol	1.85	1.82
44	43	Acetone	1.76	1.65
45	49	Polypropylene	1.73	1.26
46	46	Acetic Anhydride	1.56	1.55
47	—	Propylene Oxide	1.51	1.17
48	47	Titanium Dioxide	1.38	1.36
49	50	Adipic Acid	1.35	1.25
50	49	Sodium silicate (water glass)	1.33	1.26

	Production 1972 (billions of pounds)	Production 1971 (millions of pounds)
Synthetic Rubber, total	5.45	2,241[b]
Styrene-Butadiene Rubber	3.35	1,416[b]
Synthetic Fibers, total	7.32	6,153
Cellulosics, total	1.39	1,392
Rayon	0.97	915
Noncellulosics, total	5.36	4,293
Nylon	1.98	1,595
Polyester	2.34	1,831
Plastics, total	24.20	21,100
Polyethylene, total	7.63	6,356
High Density	2.34	1,905
Low Density	5.29	4,451
Polypropylene	1.73	1,271
Polyvinyl Chloride[b]	4.29	3,440
Polystyrene[b]	4.60	3,840
Phenolics	1.45	1,453
Urea and Melamine Resins	0.91	795
Polyester Resins[c]	0.93	724

[a] Sources: Bureau of Mines, Bureau of the Census, Tariff Commission, Society of the Plastics Industry, industry and C&EN estimates.

[b] Includes copolymers

[c] Unsaturated Sources: Rubber Manufacturers Association, Textile Economics Bureau, Tariff Commission, and Society of the Plastics Industry

Table A.25 1970 United States Production and Sales of Natural and Synthetic Chemicals (Abstracted from U.S. Tariff Commission, *Synthetic Organic Chemicals*, T.C. Pub. No. 479)

Chemical	1970 Production (lb)	1970 Sales (lb)
Tar	7,609,000,000	3,712,000,000
Tar, crudes	9,300,000,000	6,533,000,000
Crude petroleum and natural gas products	77,879,000,000	43,439,000,000
Synthetic Organic Chemicals, total	138,322,000,000	74,974,000,000
Dyes	235,000,000	223,000,000
Organic Pigments	57,000,000	47,000,000
Medicinal Chemicals	214,000,000	155,000,000
Flavor and Perfume Materials	100,000,000	92,000,000
Plastic and Resin Materials	19,210,000,000	17,074,000,000
Rubber Processing Chemicals	298,000,000	228,000,000
Elastomers	4,438,000,000	3,820,000,000
Plasticizers	1,336,000,000	1,239,000,000
Surface Active Agents	3,886,000,000	2,061,000,000
Pesticides and Related Products	1,034,000,000	881,000,000
Miscellaneous	79,257,000,000	35,998,000,000
Total	233,110,000,000	128,478,000,000

Table A.26 World River Run-off (Based on data in M. I. Lvovitch, *US IHD Bull., 28* (January 1973))

	Stable River Runoff				Total (km^3)
	Underground Origin (km^3)	Regulated by Lakes (km^3)	Regulated by Reservoirs (km^3)	Total (km^3)	
Europe	1,065	60	200	1,325	3,120
Asia	3,410	35	560	4,005	13,190
Africa	1,465	40	400	1,905	4,225
North America	1,440	150	490	2,380	5,960
South America	3,740	—	160	3,900	10,380
Australia[a]	465	—	30	495	1,965
Total	11,885[b]	285[b]	1,840[b]	14,010[b]	41,730

[a] Includes Tasmania, New Guinea, and New Zealand.
[b] Excludes polar zones.

Table A.27 Water Balance and Water Resources of the Continents of the World and of the Land as a Whole (From M. D. Lvovitch, *U.S. IHD Bull.*, *28* (January 1973))

Indices	Europe[a]	Asia	Africa	North America	South America	Australia[c]	The Whole of the Land[d]
1	2	3	4	5	6	7	8
Area (millions of km^2)	9.8	45.0	30.3	20.7	17.8	8.7	132.3
			in millimeters				
Precipitation, P	734	726	686	670	1,648	736	834
River Run-off							
Total, R	319	293	139	287	583	226	294
Underground, U	109	76	48	84	210	54	90
Surface, S	210	217	91	203	373	172	204
Total Moistening of an area, W	524	509	595	467	1,275	564	630
Evaporation, R	415	433	547	383	1,065	510	540

			in cubic kilometers				
Precipitation, P	7165	32,690	20,780	13,910	29,355	6405	110,303
River Run-off							
Total, R	3110	13,190	4,225	5,960	10,380	1965	38,830
Underground, N	1065	3,410	1,465	1,740	3,740	465	11,885
Surface, S	2045	9,780	2,760	4,220	6,640	1500	26,945
Total Moistening of an area, W	5120	22,910	18,020	9,690	22,715	4905	83,360
Evaporation, E	4055	19,500	16,555	7,950	18,975	4440	71,475
			Relative Values				
Underground Run-off in p.c. of the Total One	34	26	35	32	36	24	31
Coefficient of Ground Water Discharge into Rivers, K_u	0.21	0.15	0.08	0.18	0.16	0.10	0.14
Coefficient of Run-off, K_R	0.43	0.40	0.23	0.31	0.35	0.31	0.36

[a] Including Iceland.

[b] Excluding the Canadian archipelago and including Central America.

[c] Including Tasmania, New Guinea, and New Zealand, only within the limits of the continent: P—440 mm, R—47 mm, U—7 mm, S—40 mm, W—400 mm, E—393 mm.

[d] Excluding Greenland, Canadian archipelago, and Antarctica.

Table A.28 Composition of Some Soviet Spas Waters (From S. Licht (ed.), *Medical Hydrology*, Licht Pub., New York, 1963, with permission of the publisher)

Name of Spa	Province or Republic	Tempera-ture (°C)	pH	Minerals (g/liter)	Principal Ions	Gases	Peloids
Abalakh	Yakutiya		8.2	1.6	HCO_3, Na, Cl	H_2S	Mineral Sulfur
	(Spring Lake)		9.5	40–130			
Ak-su	Kirgizstan	56–59	8.3	0.4	SO_4, Cl, Na, Mg		
Alma-Arasan	Kazakhstan	37	7.0	0.26	HCO_3, SO_4, Na, Si		
Annenskie	Khabarovsk	51.5	9.3	0.3	HCO_3, SO_4, Na, Si		
Arasan-Kopal	Kazakhstan	37.5	9.3	0.4	SO_4, Cl, Na, Si		
Archman	Turkistan	28		1.4	SO_4, Cl, HCO_3, Na, Ca, Mg	H_2S	
Arshan	Buryatia	18	6.1	4.2	HCO_3, SO_4, Ca, Mg, Si	CO_2	
Badamly	Azerbaidzhan	17	6.4	5.7	HCO_3, Cl, Na,	CO_2	
Baldone	Latvia	7.1	7.2	2.4	SO_4, Ca	H_2S	Peat
Bauntovsk	Buryatia	54	7.3	0.5	HCO_3, SO_4, Na, Ca, Si		
Beriosovskic	Ukraine			0.8	HCO_3, Ca, Na, Mg		
Birshtomas	Lithuania	8.4	7.3	6.9	Cl, Na		Peat
Bobruisk	Byelorussia	11.0	6.7	14.3	Cl, Na		
		9.0	7.7	6.6	Cl, SO_4, Na, Ca		
Bolshie Soli	Yaroslav	7.4	7.1	16	Cl, SO_4, Na		
Chapaevskye	Saratov	11.0	7.2	18	Cl, Na	H_2S	
Chartak	Uzbekistan	49	6.9	59	Cl, Na, Ca		
Cherche	Ukraine			0.8	HCO_3, Ca, Na, Mg		
Chernovsky	Ukraine	12.8	7.6	5.4	SO_4, Na		
Chimion	Uzbekistan	27	7.2	6.1	Cl, Na, Ca	H_2S	
Dorokhovo	Moscow	9	6.9	2.9	SO_4, Ca		
			7.6	115	Cl, Na		
Dzhalal-Abad	Kirgizstan	28–43	6.9	0.5–2.6	SO_4, HCO_3, Na, Ca		
		25		2.2	SO_4, Cl, Na		
Dzhety-Oguz	Kirgizstan	25–42	6.9–7.2	3.2–11.5	Cl, Na, Ca	Rn	
Garm-Chashma	Tadzhikistan	62	6.6	3.2	HCO_3, Cl, Na	CO_2, H_2S	
Gormaya Tissa	Ukraine	10–11	7.4	33.2	Cl, HCO_3, Na, As	CO_2	
Goryachinsk	Buryatia	54.5	9.3	0.6	SO_4, Na, Si		
Irkutsk Ongara	Irkutsk	24	7.6	57	Cl, Na	H_2S	Mineral Sulfur
Isti-Su	Azerbaidzhan	55	6.6	4.9	HCO_3, Cl, Na	CO_2	
Ivanova	Ivanova	7.0	7.6	2.6	SO_4, Na, Ca, Mg		
		9.7	7.2	119	Cl, Na		
Izhevskie M.S.	Tatarstan	7.1	7.3	4.9	SO_4, Cl, Na, Ca, Mg		
Karmadon	North Osetia	56.5	6.5	8.2	Cl, HCO_4, Na		Peat
		7	7.7	2–34	SO_4, Ca, Mg, Na		
		7–7.8	7–7.7	10–28	Cl, SO_4, Na		
Khadyzhensk	Krasuodar	70		10.3	Cl, HCO_3, Br, I, Na	CH_4	
Khilovo	Pskov	6.8	7.6	1.9	SO_4, HCO_3, Ca	H_2S	Peat
Khmelnik	Ukraine	10	7.3	5.0	HCO_3, Cl, Na, Ca	CO_2, Rn	
Kirillovka	Ukraine	17		7.0	Cl, Na	H_2S	Mineral Sulfur
Kisly Kluch	Kunshir Tusel	55.6	2.4	3.9	Cl, SO_4, Al, Na, Si		
Klyuchi	Perm	6.4	6.8	3.5	SO_4, Cl, Na, Ca, Mg	H_2S	Mineral Sulfur

(*Continued*)

Table A.28 (*Continued*)

Name of Spa	Province or Republic	Temperature (°C)	pH	Minerals (g/liter)	Principal Ions	Gases	Peloids
Krasnousolsk	Bashkiria	12	7.3	58–69	Cl, Na	H_2S	
		11.5	7.5	9.8	Cl, Na	Rn	
		13.6	7.4	3	Cl, Na	H_2S	
Kuibyshev	Kuibyshev		7.4	26.8	Cl, HCO_3, Na	H_2S	
Kuka	Chita	0.2	6.0	3.2	HCO_3, Mg, Ca	CO_2	
Kumagorsk	Stavropol	32.5	8.2	2.0	HCO_3, Cl, Na	H_2S	
Likenai	Lithuania	7.0	7.0	2.4	SO_4, Ca	H_2S	Peat
Lipetsk	Lipetsk	12	7.8	4.0	SO_4, Cl, Na		Mineral
Lugansk	Ukraine	20.6		30.2	Cl, Na		
Lyuben-Vyeliki	Ukraine	11	6.8	2.1	SO_4, Ca	H_2S	
Maikop	Krasnodar	70	7.6	23.1	Cl, Br, I, Na	CH_4	
Makhinjauri	Adzharia	21	8.4	0.3	Cl, HCO_3, Na		
Mazisyalnye S.	Karelia		6.5	0.2–0.7	SO_4, HCO_3, Ca		
Medvezhie L.	Kurgan	6.6	7.8	2.0	Cl, HCO_3, SO_4, Na, Mg		Mineral Sulfur
Mendzhi	Georgia	22	7.4	12.8	Cl, Na	H_2S	
Minsk	Byclorussia	8.4–0.0	7–6.6	4.8–17	Cl, Na		Peat
Mirgorod	Ukraine	21.5	7.5	2.9	Cl, Na		
Mironovka	Ukraine	12		2.5	HCO_3, Ca, Mg	Rn	
Molokovka	Chita	1.0	5.5	0.9	HCO_3, Ca, Mg	CO_2, Rn	
Moscow	Moscow	28.8	7.8	269	Cl, Na		
Nachik	Kamchatka	80.6	8.4	1.2	SO_4, Cl, Na, Si		
Nizhnie Sergi	Sverdlovsk	7.0	7.2	6.7	Cl, Na		
Obi Garm	Tadzhikistan	46	8.4	1.0	SO_4, Cl, Na, Ca, Si		
Olentui	Chita	1.5	6.5	1.3	HCO_3, Ca, Mg	CO_2	
Paratunka	Kamchatka	61	8.2	1.5	SO_4, Cl, Na, Ca, Si		
Polyana	Ukraine	8	6.6	9.4	HCO_3, Na	CO_2	
Sairmye	Georgia	10.5	6.6	6.6	HCO_3, Na, Ca	CO_2	
Saratov	Saratov	18	7.4	10	Cl, Na	H_2S	
Seregovo	Komi	9.4	7.1	77	Cl, Na	H_2S	
Sernovodsk	Checheno	65.8	7.9	3.4	Cl, HCO_3, Na	H_2S	
Shayan	Ukraine	10	6.7	6.3	HCO_3, Na	CO_2	
		11	6.5	4.3	HCO_3, Cl, Na	CO_2	
Sherbintsy	Ukraine	11.8	6.8	3.3	SO_4, HCO_3, Ca, Na	H_2S	
Shikhov	Azerbaidzhan	68	8.3	16	Cl, HCO_3, Na	H_2S	
Shivanda	Chita	—0.1	5.9	1.1	HCO_3, Ca, Mg, Na	CO_2	
Sinegorsk	Sakhalin	7.0	6.7	24.8	Cl, HCO_3, Na, As	CO_2	
Sinyak	Ukraine	9.5	7.2	1.1	SO_4, Ca	H_2S	
Soimy	Ukraine	12	6.9	7.3	Cl, HCO_3, Na, Ca	CO_2, H_2S	
Solonikha	Archangelsk	4	7.6	13	Cl, SO_4, Na		Mineral Sulfur
Solyvichegodsk	Archangelsk	6	7.5–8.4	8–17.2	Cl, SO_4, Na		Mineral Sulfur
Starobyelsk	Ukraine			15	Cl, Na		
Surakhany	Azerbaidzhan	16.6	8.2	23	Cl, Na	H_2S	
Talgi	Dagestan	38	6.6	5.9	Cl, Na, Ca	H_2S	
Talitsa	Sverdlovsk	26.6		9.5	Cl, Na, Br, I	CH_4	
Tanisek	North Ossetia	16		2.6	SO_4, Ca		
Tashkent	Uzbekistan	53	8.2	0.8	HCO_3, Cl, Na		
Tbilisi	Georgia	43.5	9.2	0.3	HCO_3, Cl, Na, Ca	H_2S	
Tkvarcheli	Abhaia	34–39		0.5	SO_4, Cl, Na, Ca		
Totma	Vologda	7.2	7.8	52.6	Cl, SO_4		
Tsaishi	Georgia	76	7.3	1.75	SO_4, Cl, Ca, Na, Si		
Turinsk	Sverdlovsk	38	6.6	14.3	Cl, Na	CH_4	

(*Continued*)

Table A.28 (*Continued*)

Name of Spa	Province or Republic	Temperature (°C)	pH	Minerals (g/liter)	Principal Ions	Gases	Peloids
Tyumen (Yar)	Tyumen	40	7.1	11.4	Cl, Na	CH_4	
Uchum	Krasnoyarsk	25	7.9	4.1	SO_4, Cl, Na		Mineral Sulfur
Ungenyi	Moldavia	20	8	3.2	HCO_3, Cl, Na	H_2S	
Urguchan	Chitz	0.5	5.3	1.2	HCO_3, Cl, Na	CO_2, Rn	
Ust-Kut	Irkutsk	7.5	8.7	128	Cl, Na		Mineral Sulfur
Uvildy	Chelyabinsk	2.1	6.6	2.4	HCO_3, Ca, Mg	CO_2	
Valmiera	Latvia	12	7.8	6	Cl, Na, Ca		
Vangoi	Primorskii	32	9.4	0.14	HCO_3, Na		
Varzi-Yatchi	Udmurtia	6.5	6.6	1.5	SO_4, Ca, Mg	H_2S	Mineral Peat
Velikye Luki	Pskov	9.0	7.3	0.5	HCO_3, Mg, Ca	H_2S	
Vologda	Vologda		8.5	185	Cl, Na, Br		
Yamarovka	Chita	2.1	6.6	2.4	HCO_3, Ca, Mg	Rn	Sapropel
Yamkun	Chitz	18.4	6.6	1.8	HCO_3, Ca, Mg	CO_2, Rn	
Yaroslavl	Yaroslavl	12	8.0	11.6	SO_4, Cl, Na		
Yeisk	Krasnodar	12	7.5	4.3	Cl, Na	H_2S	Mineral Sulfur
Zavodouskovskain	Tyumen	40	7.0	17.7	Cl, Na	CH_4	
Zvenigorod	Moscow	11.5	6.9	2.5	SO_4, Ca		

Table A.29 Chemical Analyses of Ground Waters (ppm)

Sample[a]	A	B	C	D	E	F	G	H	I	J
Bicarbonate (HCO_3)	146	143	213	4090	445	1,490	77	—	111	30
Carbonate (CO_3)	—	—	—	—	—	—	—	—	—	—
Sulfate (SO_4)	4.0	1570	4.9	6.0	303	882	15	1570	30	5.9
Chloride (Cl)	3.5	24	2.0	390	80	15,960	10	3.5	10	1.8
Fluoride (F)	—	—	—	—	1.2	—	1.6	1.1	22	0.1
Nitrate (NO_3)	7.3	18	4.8	—	17	—	0.4	—	0.5	0.4
Silica (SiO_2)	8.4	29	8.4	80	13	45	103	216	99	7.9
Aluminum (Al)	1.0	—	1.4	—	—	—	—	56	—	0.6
Iron (Fe)	0.04	—	0.24	5.6	—	—	—	33	0.04	11
Manganese (Mn)	—	—	—	—	—	—	—	3.3	—	0.32
Calcium (Ca)	46	636	40	26	30	496	0.65	185	2.4	8.4
Magnesium (Mg)	4.2	43	22	614	31	286	1.1	52	1.4	1.5
Sodium (Na)	1.5	17	0.4	624	279	9,910	40	6.7	100	1.5
Potassium (K)	0.8		1.2			167		24	2.9	3.6
Dissolved Solids	139	2410	180	3760	973	28,400	—	—	348	44
Temperature, °F	61	—	58	64	—	70	101	150	122	63
pH	7.0	—	7.4	—	—	7.1	6.7	1.9	9.2	6.3

[a] Sample: A—Big Spring, Huntsville, Alabama (limestone); B—Jumping Springs, Eddy County, New Mexico (gypsum); C—Jefferson City, Tennessee (dolomite); D—Cooks Springs, Colusa, California (serpentine); E—Well, San Miguel County, New Mexico (shale); F—Salt Banks, Chrysotile, Arizona; G—Rio San Antonio, Sandoval County, New Mexico (rhyolite); H—Lemonade Spring, Sandoval County, New Mexico; I—Well, Owybye County, Idaho; J—Well, Fulton, Mississippi (sand).

Table A.30 Pesticide Nomenclature and Structure (From Anon., *Cleaning Our Environment*, Amer. Chem. Soc., Washington, D.C., 1969)

Common Name	Chemical Name	Structural Formula
Chlorinated Hydrocarbon Insecticides		
Aldrin	1,2,3,4,10-hexachloro-1,4-4a,5,8,8a-hexadhydro-endoexo-5,8-dimethanonaphthalene	H Cl C H C HC C CCl CH_2 CCl_2 HC C CCl C H C H Cl
Dieldrin	1,2,3,4,10,10-hexachloro-6,7-epoxy-1,4,4a,5,6,7,8,8a-octahydro-1,4-endo-exo-5,8-dimethanonaphthalene	H Cl C H C CH C CCl O CH_2 CCl_2 CH C CCl C H C H Cl
Endrin	1,2,3,4,10,10-hexachloro-6,7-epoxy-1,4,4a,5,6,7,8,8a-octahydro-1,4-endo-endo-5,8-dimethanonaphthalene	CH CCl CH CH CCl O CH_2 CCl_2 CH CH CCl CH CCl
Heptachlor	1,4,5,6,7,8,8-heptachloro-3a,4,7,7a-tetrahydro-4,7-endomethanoindene	Cl C HC—CH CCl CCl_2 HC CH CCl C C H Cl Cl
DDT	2,2-bis[*p*-chlorophenyl]-1,1,1-trichloroethane	Cl–⟨ring⟩–CH–⟨ring⟩–Cl CCl_3
DDD, TDE	2,2-bis [*p*-chlorophenyl]-1-1-dichlorocthane	Cl–⟨ring⟩–CH–⟨ring⟩–Cl $CHCl_2$

(Continued)

Table A.30 (*Continued*)

Common Name	Chemical Name	Structural Formula
DDE	Dichlorodiphenyl dichloroethylene	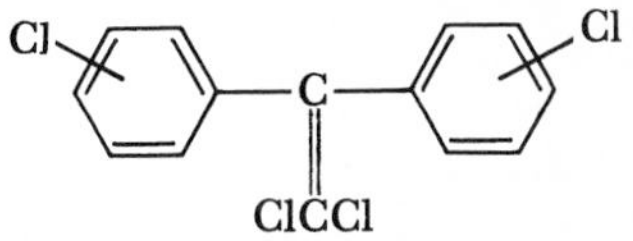
Methoxychlor	2,2-bis [*p*-methoxyphenyl]-1,1,1-trichloroethane	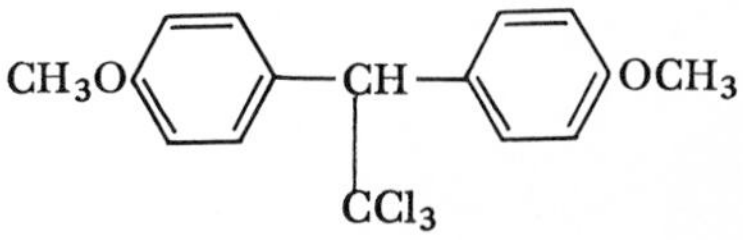
BHC lindane (gamma isomer)	1,2,3,4,5,6-Hexachlorocyclohexane	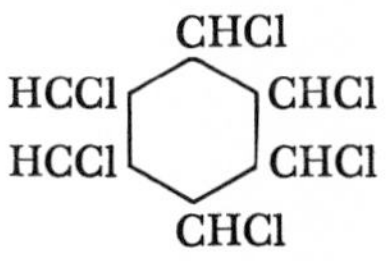
Toxaphene	Synthesized by chlorination of camphene to chloride content of 67 to 69→	
Tetradifon	2,3,5,4′-tetrachlorodiphenyl sulfone	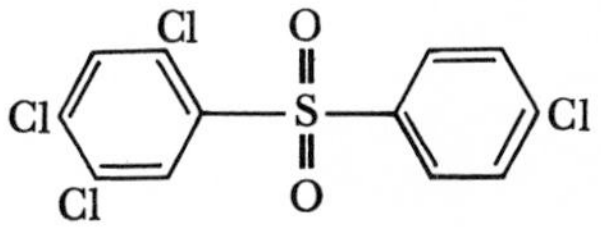
Mirex	Dodecachlorooctahydro-1,3,4-metheno-2H-cyclobuta [cd] pentalene	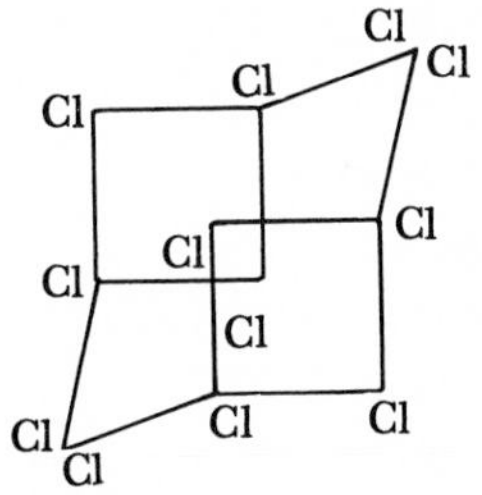
Organophosphorus Insecticides		
Fenthion	*O*,*O*-Dimethyl *O*-[4-methylthio[-*m*-tolyl]	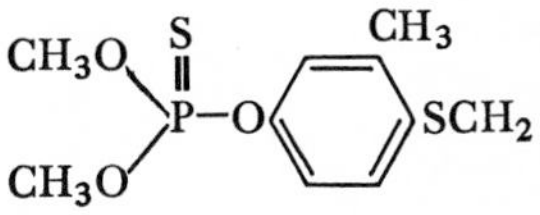
Diazinon	*O*,*O*-Diethyl *O*-(2-isopropyl-4-methyl-6-pyrimidyl) phosphorothioate	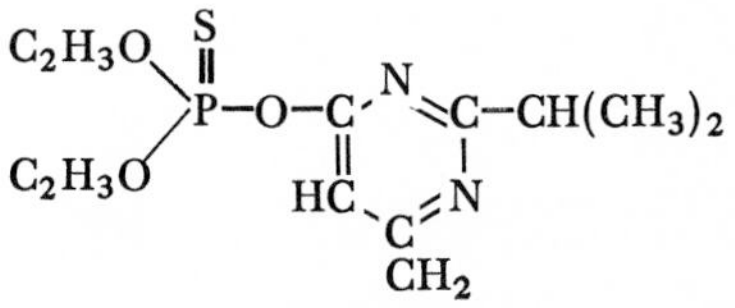

(*Continued*)

Table A.30 (*Continued*)

Common Name	Chemical Name	Structural Formula
Azinphosmethyl	*O,O*-Dimethyl *S*-[4-oxo-1m2m3-benzotriazin-3(4H)-ylmethyl] phosphorodithioate	$(CH_3O)_2P(=S)-S-CH_2-N$ (benzotriazinone ring: $N-C(=O)-C_6H_4-N=N$)
Malathion	*O,O*-dimethyl S-(1,2-dicarbethoxyethyl) phosphorodithioate	$(CH_3O)_2P(=S)-S-CH(COOC_2H_5)CH_2COOC_2H_5$
Methyl parathion	*O,O*-Cimethyl-*O*-*p*-nitrophenyl phosphorothioate	$(CH_3O)_2P(=S)-O-C_6H_4-NO_2$
Parathion	*O,O*-Diethyl-*O*-*p*-nitrophenyl phosphorothioate	$(C_2H_5O)_2P(=S)-O-C_6H_4-NO_2$
Phorate	*O,O*-Diethyl *S*-(ethylthio) methyl phosphorodithioate	$(C_2H_5O)_2P(=S)-SCH_2SC_2H_5$
Mevinphos	2-carbomethoxy-1-propen-2yl dimethyl phosphate	$(CH_3O)_2P(=O)-OC(CH_3)=CHC(=O)-OCH_3$
Carbamate Insecticides		
Carbaryl	*N*-Methyl-1-naphthylcarbamate	$CH_3NH-C(=O)-O-C_{10}H_7$
	4-dimethylamino-3,5-xylyl *N*-methylcarbamate	$CH_3-NH-C(=O)-O-C_6H_2(CH_3)_2-N(CH_3)_2$

(*Continued*)

Table A.30 (*Continued*)

Chemical Name	Structural Formula
Herbicides	
2,4-D 2,4-dichlorophenoxyacetic acid	Cl, Cl, $-OCH_2COOH$
2,4,5-T 2,3,5-trichlorophenoxyacetic acid	Cl, Cl, Cl, $-OCH_2COOH$
Fenoprop 2-(2,4,5-Trichlorophenoxy) propionic acid	Cl, Cl, Cl, $-OCH(CH_3)COOH$
2,3,6-TBA 2,3,6-trichlorobenzoic acid	Cl, Cl, Cl, $-COOH$
Fenac 2,3,6-trichlorophenylacetate	Cl, Cl, Cl, $-CH_2COOH$

Table A.31 Composition of Natural Gas, Petroleum, and Tar

A Natural Gas

Up to 95% methane (plus varying amounts of CO_2, N_2, H_2S, He, and H_2O)
"Sour" gas contains appreciable amounts of H_2S
"Wet" gas contains appreciable amounts of liquifiable hydrocarbons

B Petroleum Oil

Nearly all crude oils have similar analyses
C 84 to 87% by weight
H 11 to 14% by weight
S 0.06 to 2.0% by weight
N 0.1 to 2.0% by weight
O 0.1 to 2.0% by weight

C Tar and Asphalt

(Bitumens, waxes, resins, and pitch)
Very complex, poorly understood. Formed by the loss of the more volatile constituents and the polymerization and partial oxidation of oil.

D Hydrocarbon Content of Midwestern United States Petroleum

		Amount by Volume (*n*-octane = 1.0)
Paraffinic		
C_7H_{16}	*n*-Heptane	1.1
C_8H_{18}	*n*-Octane	1.0
C_9H_{20}	*n*-Nonane	1.0
$C_{10}H_{22}$	*n*-Decane	0.8
C_6H_{14}	*n*-Hexane	0.7
C_7H_{16}	2-Methylhexane	0.3
C_6H_{14}	3-Methylpentane	0.2
C_6H_{16}	3-Methylhexane	0.2
C_9H_{20}	2-Methyloctane	0.2
C_6H_{14}	2-Methylpentane	0.1
C_9H_{20}	2,6-Dimethylheptane	0.1
Naphthenic		
C_7H_{14}	Methylcyclohexane	0.3
C_6H_{12}	Cyclohexane	0.3
C_6H_{12}	Methylcyclopentane	0.2
C_8H_{16}	1,3-Dimethylcyclohexane	0.2
C_8H_{16}	Ethylcyclohexane	0.1
C_9H_{18}	Nonanaphthene	0.1
Aromatic		
C_7H_8	Toluene	0.3
C_9H_{12}	1,2,4-Trimethylbenzene	0.2
C_8H_{10}	*m*-Xylene	0.1
C_8H_{10}	*o*-Xylene	0.1
C_6H_6	Benzene	0.08
C_9H_{12}	1,2,3-Trimethylbenzene	0.06
C_8H_{10}	*p*-Xylene	0.04
C_8H_{10}	Ethylbenzene	0.03
C_9H_{12}	Isopropylbenzene	0.03

(*Continued*)

Table A.31 (*Continued*)

E Heavy Metal Content (in ppm) of Some Crude Oils

Source	Fe	Ni	V	Cu
Texas	3–5	2–5	1–8	0.4
Kansas	5.8	5.8	21	0.4
California	28–31	33–46	41–49	0.6–1.1
Kuwait	0.7	6.0	23	0.1
Morocco	—	0.8	0.6	0.1

F Some Sulfur Compounds in Texas Crude Oil

Compound	Weight Percent
Ethanethiol	0.0053
2-Butanethiol	0.0039
2-Methylthiacyclohexane	0.0029
2-Hexanethiol	0.0028
Trans-2,5-dimethylthiacyclopentane	0.0025
Methanethiol	0.0024
Cis-2,5-Dimethylthiacyclopentane	0.0024
2-Methylthiacyclopentane	0.0023

Author Index

For references by more than two co-authors only the first-named author may be listed.

content of atmosphere, 142, 303
hydrosphere, and lithosphere, 303